Advanced Applications of Supercritical Fluids in Energy Systems

Lin Chen
Tohoku University, Japan & Japan Society for the Promotion of Science (JSPS), Japan

Yuhiro Iwamoto
Nagoya Institute of Technology, Japan

A volume in the Advances in Chemical and Materials Engineering (ACME) Book Series

IGI Global
DISSEMINATOR OF KNOWLEDGE
www.igi-global.com

Published in the United States of America by
IGI Global
Engineering Science Reference (an imprint of IGI Global)
701 E. Chocolate Avenue
Hershey PA, USA 17033
Tel: 717-533-8845
Fax: 717-533-8661
E-mail: cust@igi-global.com
Web site: http://www.igi-global.com

Library of Congress Cataloging-in-Publication Data

Names: Chen, Lin, 1987- editor. | Iwamoto, Yuhiro, 1984-
Title: Advanced applications of supercritical fluids in energy systems / Lin Chen, Yuhiro Iwamoto, editors.
Description: Hershey, PA : Engineering Science Reference, [2017] | Includes bibliographical references.
Identifiers: LCCN 2016054199| ISBN 9781522520474 (hardcover) | ISBN 9781522520481 (eISBN)
Subjects: LCSH: Electric power-plants--Materials. | Supercritical fluids--Industrial applications. | Fluids--Thermal properties. | Heat-transfer media.
Classification: LCC TK1041 .A39 2017 | DDC 621.31028/4--dc23 LC record available at https://lccn.loc.gov/2016054199

This book is published in the IGI Global book series Advances in Chemical and Materials Engineering (ACME) (ISSN: 2327-5448; eISSN: 2327-5456)

British Cataloguing in Publication Data
A Cataloguing in Publication record for this book is available from the British Library.

For electronic access to this publication, please contact: eresources@igi-global.com.

Advances in Chemical and Materials Engineering (ACME) Book Series

J. Paulo Davim
University of Aveiro, Portugal

ISSN:2327-5448
EISSN:2327-5456

Mission

The cross disciplinary approach of chemical and materials engineering is rapidly growing as it applies to the study of educational, scientific and industrial research activities by solving complex chemical problems using computational techniques and statistical methods.

The **Advances in Chemical and Materials Engineering (ACME) Book Series** provides research on the recent advances throughout computational and statistical methods of analysis and modeling. This series brings together collaboration between chemists, engineers, statisticians, and computer scientists and offers a wealth of knowledge and useful tools to academics, practitioners, and professionals through high quality publications.

Coverage

- Ductility and Crack-Resistance
- Numerical Techniques
- Artificial Intelligence Methods
- Statistical methods
- Chemical Engineering
- Metallic Alloys
- Fatigue and Creep
- Nanomaterials
- Biomaterials
- Sustainable Materials

IGI Global is currently accepting manuscripts for publication within this series. To submit a proposal for a volume in this series, please contact our Acquisition Editors at Acquisitions@igi-global.com or visit: http://www.igi-global.com/publish/.

The Advances in Chemical and Materials Engineering (ACME) Book Series (ISSN 2327-5448) is published by IGI Global, 701 E. Chocolate Avenue, Hershey, PA 17033-1240, USA, www.igi-global.com. This series is composed of titles available for purchase individually; each title is edited to be contextually exclusive from any other title within the series. For pricing and ordering information please visit http://www.igi-global.com/book-series/advances-chemical-materials-engineering/73687. Postmaster: Send all address changes to above address.

Titles in this Series

For a list of additional titles in this series, please visit: www.igi-global.com/book-series

Innovative Applications of Mo(W)-Based Catalysts in the Petroleum and Chemical Industry Emerging Research and Opportunities
Hui Ge (Chinese Academy of Sciences, China) Xingchen Liu (Chinese Academy of Sciences, China) Shanmin Wang (Oak Ridge National Laboratory, USA) Tao Yang (China University of Petroleum, China) and Xiaodong Wen (Synfuels China, China)
Engineering Science Reference • copyright 2017 • 162pp • H/C (ISBN: 9781522522744) • US $135.00 (our price)

Sustainable Nanosystems Development, Properties, and Applications
Mihai V. Putz (West University of Timişoara, Romania & Research and Development National Institute for Electrochemistry and Condensed Matter (INCEMC) Timişoara, Romania) and Marius Constantin Mirica (Research and Development National Institute for Electrochemistry and Condensed Matter (INCEMC) Timişoara, Romania)
Engineering Science Reference • copyright 2017 • 794pp • H/C (ISBN: 9781522504924) • US $245.00 (our price)

Computational Approaches to Materials Design Theoretical and Practical Aspects
Shubhabrata Datta (Calcutta Institute of Engineering and Management, India) and J. Paulo Davim (University of Aveiro, Portugal)
Engineering Science Reference • copyright 2016 • 475pp • H/C (ISBN: 9781522502906) • US $215.00 (our price)

Green Approaches to Biocomposite Materials Science and Engineering
Deepak Verma (Graphic Era Hill University, Dehradun, India) Siddharth Jain (College of Engineering Roorkee, India and University of Alberta, Canada) Xiaolei Zhang (Queens University, Belfast, UK) and Prakash Chandra Gope (College of Technology, G.B.Pant University of Agriculure and Technology, Pantnagar, India)
Engineering Science Reference • copyright 2016 • 322pp • H/C (ISBN: 9781522504245) • US $165.00 (our price)

Position-Sensitive Gaseous Photomultipliers Research and Applications
Tom Francke (Myon, Sweden) and Vladimir Peskov (CERN, Switzerland)
Engineering Science Reference • copyright 2016 • 562pp • H/C (ISBN: 9781522502425) • US $240.00 (our price)

Research Perspectives on Functional Micro- and Nanoscale Coatings
Ana Zuzuarregui (CIC nanoGUNE Consolider, Spain) and Maria Carmen Morant-Miñana (CIC nanoGUNE Consolider, Spain)
Information Science Reference • copyright 2016 • 511pp • H/C (ISBN: 9781522500667) • US $215.00 (our price)

IGI GLOBAL
DISSEMINATOR OF KNOWLEDGE
www.igi-global.com

701 East Chocolate Avenue, Hershey, PA 17033, USA
Tel: 717-533-8845 x100 • Fax: 717-533-8661
E-Mail: cust@igi-global.com • www.igi-global.com

Editorial Advisory Board and List of Reviewers

Table of Contents

Foreword xviii

Preface xx

Acknowledgment xxviii

Section 1
Fundamentals of Supercritical Fluids Flow and Heat Transfer

Chapter 1
Thermal Effects in Near-Critical Fluids: Piston Effect and Related Phenomena 1
Daniel A. Beysens, Ecole Supérieure de Physique et Chimie Industrielles ParisTech, France
Yves Garrabos, Institut de Chimie de la Matière Condensée de Bordeaux, France
Bernard Zappoli, Centre National d'Etudes Spatiales, France

Chapter 2
Numerical Modelling of Hydrodynamic Instabilities in Supercritical Fluids 33
Sakir Amiroudine, University of Bordeaux, France

Chapter 3
Thermal-Mechanical Effects and Near-Critical Fluid Dynamic Behaviors in Micro-Scale 55
Lin Chen, Tohoku University, Japan & Japan Society for the Promotion of Science (JSPS), Japan

Chapter 4
X-Ray Scattering Studies of Expanded Fluid Metals 102
Kazuhiro Matsuda, Kyoto University, Japan
Masanori Inui, Hiroshima University, Japan

Section 2
Supercritical Natural Circulation Flow and Its Applications

Chapter 5
Principles, Experiments, and Numerical Studies of Supercritical Fluid Natural Circulation System 136
Lin Chen, Tohoku University, Japan & Japan Society for the Promotion of Science (JSPS), Japan

Chapter 6
Supercritical Natural Circulation Loop: A Technology for Future Reactors 188
Dipankar Narayan Basu, Indian Institute of Technology Guwahati, India
Milan Krishna Singha Sarkar, Indian Institute of Technology Guwahati, India

Chapter 7
Application of Supercritical Carbon Dioxide for Solar Water Heater 215
Yuhiro Iwamoto, Nagoya Institute of Technology, Japan
Hiroshi Yamaguchi, Doshisha University, Japan

Chapter 8
Heat Transfer and Fluid Flow of Supercritical Fluids in Advanced Energy Systems 235
Hongzhi Li, Xi'an Thermal Power Research Institute, China
Yifan Zhang, Xi'an Thermal Power Research Institute, China

Section 3
Supercritical Thermal Conversion of Different Scales

Chapter 9
Heat Transfer in Supercritical Fluids: Going to Microscale Times and Sizes 271
Sergey B. Rutin, Institute of Thermal Physics, Ural Branch, Russian Academy of Sciences, Russia
Aleksandr D. Yampol'skiy, Institute of Thermal Physics, Ural Branch, Russian Academy of Sciences, Russia
Pavel V. Skripov, Institute of Thermal Physics, Ural Branch, Russian Academy of Sciences, Russia

Chapter 10
Study on Cooling Heat Transfer of Supercritical Carbon Dioxide Applied to Transcritical Carbon Dioxide Heat Pump 292
Chaobin Dang, University of Tokyo, Japan
Eiji Hihara, University of Tokyo, Japan

Chapter 11
Supercritical Carbon Dioxide and Its Application to Rankine Cycle 335
Hiroshi Yamaguchi, Doshisha University, Japan

Chapter 12
Supercritical Fluids and Their Applications in Power Generation 369
Huijuan Chen, GE Global Research Center, USA
Ricardo Vasquez Padilla, Southern Cross University, Australia
Saeb Besarati, SolarReserve, USA

Section 4
Supercritical Power Generation Systems and Green Energy Harvesting

Chapter 13
Application of Supercritical Pressures in Power Engineering 404
Igor Pioro, University of Ontario Institute of Technology, Canada
Mohammed Mahdi, University of Ontario Institute of Technology, Canada
Roman Popov, University of Ontario Institute of Technology, Canada

Chapter 14
Energy Conversion Using the Supercritical Steam Cycle 458
Thomas Schulenberg, Karlsruhe Institute of Technology, Germany

Chapter 15
The Basic Thermal Hydraulic Issues of Applying Supercritical Fluid to Nuclear Reactors 481
Xiao Yan, Nuclear Power Institute of China, China
Jinguang Zang, Nuclear Power Institute of China, China
Ting Xiong, China Nuclear Energy Industry Corp., China
Xi Sui, Nuclear Power Institute of China, China
Yanping Huang, Nuclear Power Institute of China, China
Zejun Xiao, Nuclear Power Institute of China, China

Chapter 16
Supercritical Fluids as a Tool for Green Energy and Chemicals 554
Maša Knez Hrnčič, University of Maribor, Slovenia
Darija Cör, University of Maribor, Slovenia
Željko Knez, University of Maribor, Slovenia

Chapter 17
Application of Supercritical Technologies in Clean Energy Production: A Review 588
Vladimir Anikeev, Institute of Catalysis, Siberian Branch, Russian Academy of Sciences, Russia
Khaled A. M. Gasem, University of Wyoming – Laramie, USA
Maohong Fan, University of Wyoming – Laramie, USA

Compilation of References 617

About the Contributors 671

Index 678

Detailed Table of Contents

Foreword xviii

Preface xx

Acknowledgment xxviii

Section 1
Fundamentals of Supercritical Fluids Flow and Heat Transfer

In this section, fundamentals of supercritical fluids and near-critical phenomena are reported and discussed, including the very basic thermophysics of critical transition under thermal and flow conditions. Thermal-mechanical effects of critical fluids and the related heat transfer and flow stability problems are discussed by major research groups in this field. Recent developments on the critical transition and a special liquid metal critical study are also included.

Chapter 1
Thermal Effects in Near-Critical Fluids: Piston Effect and Related Phenomena 1

Daniel A. Beysens, Ecole Supérieure de Physique et Chimie Industrielles ParisTech, France
Yves Garrabos, Institut de Chimie de la Matière Condensée de Bordeaux, France
Bernard Zappoli, Centre National d'Etudes Spatiales, France

In this chapter is addressed the very particular thermal behavior that supercritical fluids exhibit when nearing their critical point. In this region, supercritical fluids exhibit strong anomalies in their thermodynamic and transport properties. Pressure change associated to a temperature variation leads to a nearly isentropic thermalization of the fluid, the "piston effect", which leads to a paradoxical "critical speeding-up". Bulk fluid temperature is uniform, temperature gradients are confined in thermal boundary layers, making the bulk fluid a thermal short-circuit. It follows very particular behavior, as dynamic heat pipes or heat going seemingly backward, in apparent contradiction with the 2nd principle of thermodynamics. Under an acceleration field, thermal convection occurs only in the boundary layers, which paradoxically can enhance the fluid stability or even cool the fluid after a heat pulse. These effects can deeply modify the supercritical fluids thermal behavior in space and energy activities, giving to these effects socio-economic relevance.

Chapter 2
Numerical Modelling of Hydrodynamic Instabilities in Supercritical Fluids 33
Sakir Amiroudine, University of Bordeaux, France

The case of a supercritical fluid heated from below (Rayleigh-Bénard) in a rectangular cavity is first presented. The stability of the two boundary layers (hot and cold) is analyzed by numerically solving the Navier-Stokes equations with a van der Waals gas and stability diagrams are derived. The very large compressibility and the very low heat diffusivity of near critical pure fluids induce very large density gradients which lead to a Rayleigh–Taylor-like gravitational instability of the heat diffusion layer and results in terms of growth rates and wave numbers are presented. Depending on the relative direction of the interface or the boundary layer with respect to vibration, vibrational forces can destabilize a thermal boundary layer, resulting in parametric / Rayleigh vibrational instabilities. This has recently been achieved by using a numerical model which does not require any equation of state and directly calculates properties from NIST data base for instance.

Chapter 3
Thermal-Mechanical Effects and Near-Critical Fluid Dynamic Behaviors in Micro-Scale 55
Lin Chen, Tohoku University, Japan & Japan Society for the Promotion of Science (JSPS), Japan

Supercritical CO_2 fluid has been widely used in chemical extraction, chemical synthesis, micro-manufacturing and heat transfer apparatus, and so forth. The current chapter deals with near-critical CO_2 micro-scale thermal convective flow and the effects of thermal-mechanical process. When the scale becomes smaller, new and detailed figures of near-critical thermal effects emerges. To explore this new area, theoretical developments and numerical investigations discussed and explained in this chapter. From a theoretical point of view, the thermal-mechanical nature of near-critical fluid would play a leading role in small time and spatial scales. This effect is found to dominants the thermal dynamic responses and convective structures of micro-scale fluid behaviors. The scaling effects, boundary thermal-mechanical process, instability evolutions, mixing flows and characteristics, possible extensions and applications are also discussed in this chapter.

Chapter 4
X-Ray Scattering Studies of Expanded Fluid Metals 102
Kazuhiro Matsuda, Kyoto University, Japan
Masanori Inui, Hiroshima University, Japan

Fluid metals exhibit significant properties of thermodynamic-state dependence, since the inter-particle interaction among the constituents (electrons and ions) considerably changes depending on their thermodynamic conditions. The authors have thus far carried out X-ray scattering experiments of fluid metals in the expanded state, which have enabled us to gain insight into microscopic understanding of the structural and electronic properties of fluid metals. The purpose of this chapter is to provide intriguing aspects of fluid metals originated from the existence of conduction electrons, which distinguishes fluid metals from non-conducting fluids, through the results of fluid rubidium and mercury.

Section 2
Supercritical Natural Circulation Flow and Its Applications

Chapters in this section are selected to discuss the convective behaviors and circulation flow heat transfer characteristics. Supercritical loop systems are very basic concept for various applications of effective energy conversion and thermal utilization processes. This section is then focused on the basic principles, mechanisms and applications in this aspect. The results and proposals in this section may contribute to related natural circulation system designs of supercritical fluids.

Chapter 5
Principles, Experiments, and Numerical Studies of Supercritical Fluid Natural Circulation System 136
Lin Chen, Tohoku University, Japan & Japan Society for the Promotion of Science (JSPS), Japan

Due to the unique thermal and transport properties, Supercritical natural circulation loop (NCL, or thermosyphon) has been proposed in many energy conversion systems. This chapter presents the principals of supercritical natural circulation loop system and its application challenges. A specially designed experimental prototype system is introduced and compared with numerical findings. The system is operated in wide range of high pressures in the critical region. It is found that in a supercritical NCL system, very high Reynolds number natural convection flow can be achieved only by simple heating and cooling. Thermal performance analysis and parameter effects are carried out along with the experimental development. The heat transfer dependency on operation and its mechanisms are also explained and summarized in this chapter. The comparison of experimental and numerical results contributes to better understanding of NCL stability phenomena and applications in energy systems.

Chapter 6
Supercritical Natural Circulation Loop: A Technology for Future Reactors 188
Dipankar Narayan Basu, Indian Institute of Technology Guwahati, India
Milan Krishna Singha Sarkar, Indian Institute of Technology Guwahati, India

Supercritical natural circulation loop is a compelling technology for cooling of modern nuclear reactors, which promises enhanced thermalhydraulic performance in a simple design. Being a new concept, related knowledge base is relatively thin and involves several conflicting theories and controversies. Present chapter summarizes the observation till date, starting from the very fundamentals. The phenomenon of natural circulation and suitability of supercritical medium as working fluid are discussed in details. Different methods of analyses, including analytical, simple 1-D numerical and multidimensional computational codes, as well as experimental, are elucidated. A comprehensive discussion is presented about the effect of various geometric and operating parameters on the system behavior, from both thermalhydraulic and stability point of view. Finally, a few recommendations are included about the operation of such loops and future direction of research.

Chapter 7
Application of Supercritical Carbon Dioxide for Solar Water Heater 215
Yuhiro Iwamoto, Nagoya Institute of Technology, Japan
Hiroshi Yamaguchi, Doshisha University, Japan

For supercritical CO_2, a small change in temperature or pressure can result in large change in density, especially in the state close to the critical point. The large change in density can easily induce the natural convective flow. In this chapter, a solar water heater using supercritical CO_2 which is originally designed and constructed will be introduced. The solar water heater is a closed loop system with main components of an evacuated solar collector and a heat exchanger. The working fluid of CO_2 is naturally driven by the large change in density with absorbing and transporting heat in the solar collector. And the heat energy (hot water) is produced by exchanging the transferred heat with water in the heat exchanger. This chapter will describe the typical system operation and performance at different season and climates.

Chapter 8
Heat Transfer and Fluid Flow of Supercritical Fluids in Advanced Energy Systems......................... 235
Hongzhi Li, Xi'an Thermal Power Research Institute, China
Yifan Zhang, Xi'an Thermal Power Research Institute, China

This chapter aims to clarify the supercritical fluids thermal hydraulics characteristics including heat transfer, pressure drops and flow instabilities for the purpose of accurate design and efficient safe operation of advanced energy systems. At first, the convection heat transfer models considering the effect of nonlinear properties and the effect of buoyancy and acceleration have been provided and discussed. Secondly, the hydraulic resistance models for supercritical fluids have been selected and suggested for different conditions. Thirdly, the published models for supercritical flow instabilities based on four different regional partitions are summarized and clarified. At last, two typical case studies have been provided to further intuitively elaborate the thermal hydraulics of supercritical fluids within the advanced energy systems.

Section 3
Supercritical Thermal Conversion of Different Scales

In this section, multi-scale analysis is discussed for supercritical and near-critical fluid systems: from simple channels to complex heat transfer system designs. The confinement and scaling effect of supercritical fluid heat transfer flow is focused in this section. The knowledge obtained by major researchers in this field will contribute to various applications of supercritical fluid based heating, cooling, microscale transportation, reaction, etc.

Chapter 9
Heat Transfer in Supercritical Fluids: Going to Microscale Times and Sizes.................................... 271
Sergey B. Rutin, Institute of Thermal Physics, Ural Branch, Russian Academy of Sciences, Russia
Aleksandr D. Yampol'skiy, Institute of Thermal Physics, Ural Branch, Russian Academy of Sciences, Russia
Pavel V. Skripov, Institute of Thermal Physics, Ural Branch, Russian Academy of Sciences, Russia

Results of experimental study of non-stationary heat transfer in supercritical fluids, which were obtained using the method of controlled pulse heating of low-inertia wire probe, are discussed. The aim of this study was to clarify the peculiarities of heat conduction mode at significant heat loads. A threshold decrease in the "instant" heat transfer coefficient, the more pronounced the closer the pressure value to critical pressure, has been found, as well as the absence of impact of the isobaric heat capacity peak known from stationary measurements on the experimental results. These results give new insights into selection of the operating pressure of supercritical heat transfer agent. Small time and spatial scale in the experiments (units of millisecond and units of micrometer) in combination with high-power heat release (up to 20 MW/m2) makes it possible to associate the results with the behavior of boundary layer region of heat transfer agent.

Chapter 10
Study on Cooling Heat Transfer of Supercritical Carbon Dioxide Applied to Transcritical Carbon Dioxide Heat Pump 292
Chaobin Dang, University of Tokyo, Japan
Eiji Hihara, University of Tokyo, Japan

Understanding the heat transfer characteristics of supercritical fluids is of fundamental importance in many industrial processes such as transcritical heat pump system, supercritical water-cooled reactor, supercritical separation and supercritical extraction processes. This paper addresses recent experimental, theoretical and numerical studies on cooling heat transfer of supercritical CO_2. A systematic study on heat transfer coefficient and pressure drop of supercritical CO_2 was carried out at wide ranges of tube diameter, mass flux, heat flux, temperature and pressure. Based on the understanding of temperature and velocity distributions at cross-sectional direction provided by the numerical simulation, a new prediction model was proposed, which agreed well with the experimental results. In addition, the effect of lubricating oil was also discussed with the focus on the change in flow pattern and heat transfer performance of oil and supercritical CO_2.

Chapter 11
Supercritical Carbon Dioxide and Its Application to Rankine Cycle 335
Hiroshi Yamaguchi, Doshisha University, Japan

Supercritical CO_2 has been given much attention to be a working fluid in a power cycle due to its unique properties. The supercritical CO_2 solar Rankine cycle system was designed and developed by using the benefit of supercritical state of CO_2 to generate electric power and supply heat energy in environmentally friendly manner. The development of main components in the system are introduced and discussed particularly by focusing on the properties of CO_2 for obtaining higher performance. The properties of CO_2 in near critical region are also discussed in this chapter. Operating the power cycle in the supercritical region of CO_2 enhances the heat transfer in energy exchanging process and improves the cycle efficiency.

Chapter 12
Supercritical Fluids and Their Applications in Power Generation 369
Huijuan Chen, GE Global Research Center, USA
Ricardo Vasquez Padilla, Southern Cross University, Australia
Saeb Besarati, SolarReserve, USA

Supercritical fluids have been studied and used as the working fluids in power generation system for both high- and low-grade heat conversions. Low-grade heat sources, typically defined as below 300 °C, are abundantly available as industrial waste heat, solar thermal, and geothermal, to name a few. However, they are under-exploited for power conversion because of the low conversion efficiency. Technologies that allow the efficient conversion of low-grade heat into mechanical or electrical power are very important to develop. First part of this chapter investigates the potential of supercritical Rankine cycles in the conversion of low-grade heat to power, while the second part discusses supercritical fluids used in higher grade heat conversion system. The selection of supercritical working fluids for a supercritical Rankine cycle is of key importance. This chapter discusses supercritical fluids fundamentals, selection of supercritical working fluids for different heat sources, and the current research, development, and commercial status of supercritical power generation systems.

Section 4
Supercritical Power Generation Systems and Green Energy Harvesting

Chapters in this section are mainly focused on power generation systems assisted by supercritical fluids and supercritical cycles. Based on the current understanding of supercritical transport mechanism (heat, mass, chemical reaction, etc.), system scale analysis is focused for the discussion of power efficiency and generation system designs. The compact and high efficiency in power engineering, chemical engineering, thermal engineering as well as nuclear engineering are found attractive for scientists and engineers in this field. Still, the contribution of green energy harvesting and other related supercritical energy systems are also included.

Chapter 13
Application of Supercritical Pressures in Power Engineering............ 404
Igor Pioro, University of Ontario Institute of Technology, Canada
Mohammed Mahdi, University of Ontario Institute of Technology, Canada
Roman Popov, University of Ontario Institute of Technology, Canada

SuperCritical Fluids (SCFs) have unique thermophysical properties and heat-transfer characteristics, which make them very attractive for use in power industry. In this chapter, specifics of thermophysical properties and heat transfer of SCFs such as water, carbon dioxide and helium are considered and discussed. Also, particularities of heat transfer at SuperCritical Pressures (SCPs) are presented, and the most accurate heat-transfer correlations are listed. SuperCritical Water (SCW) is widely used as the working fluid in the SCP Rankine "steam"-turbine cycle in fossil-fuel thermal power plants. This increase in thermal efficiency is possible by application of high-temperature reactors and power cycles. Currently, six concepts of Generation-IV reactors are being developed, with coolant outlet temperatures of 500°C~1000°C. SCFs will be used as coolants (helium in GFRs and VHTRs; and SCW in SCWRs) and/or working fluids in power cycles (helium; mixture of nitrogen (80%) and helium [20%]; nitrogen, and carbon dioxide in Brayton gas-turbine cycles; and SCW "steam" in Rankine cycle).

Chapter 14
Energy Conversion Using the Supercritical Steam Cycle............ 458
Thomas Schulenberg, Karlsruhe Institute of Technology, Germany

A supercritical steam (or Rankine) cycle is used today for more most of the new coal-fired power plants. More recently, it has been proposed as well for future water-cooled nuclear reactors to enhance their efficiency and to reduce their costs. This chapter provides the technical background explaining this technology. Some criteria for boiler design and operation, like drum or once-through boiler design, fixed or sliding pressure operation and coolant mixing, are discussed in general to explain the particular challenges of supercritical steam cycles. Examples of technical solutions are given for two large-scale applications: a coal-fired power plant and a supercritical water-cooled reactor, both producing around 1000 MW electric power.

Chapter 15
The Basic Thermal Hydraulic Issues of Applying Supercritical Fluid to Nuclear Reactors 481
Xiao Yan, Nuclear Power Institute of China, China
Jinguang Zang, Nuclear Power Institute of China, China
Ting Xiong, China Nuclear Energy Industry Corp., China
Xi Sui, Nuclear Power Institute of China, China
Yanping Huang, Nuclear Power Institute of China, China
Zejun Xiao, Nuclear Power Institute of China, China

This chapter is mainly focused on illustrating some introductory progress on thermal hydraulic issues of supercritical water, including heat transfer characteristics, pressure loss characteristics, flow stability issues and numerical method. These works are mainly performed in Nuclear Power Institute of China (NPIC) these years, to give a basic idea of elementary but important topics in this area. An analytical method was proposed up to predict the heat transfer coefficient and friction coefficient based on the two-layer wall function. Flow instability experiments have been carried out in a two-parallel-channel system with supercritical water, aiming to provide an up-to-date knowledge of supercritical flow instability phenomena and initial validation data for numerical analysis. An in-house code has been developed in NPIC in order to better utilize and further expand the experimental results on supercritical flow instability. At last, some future research directions are suggested for reference.

Chapter 16
Supercritical Fluids as a Tool for Green Energy and Chemicals 554
Maša Knez Hrnčič, University of Maribor, Slovenia
Darija Cör, University of Maribor, Slovenia
Željko Knez, University of Maribor, Slovenia

Hydrothermal conversion of biomass is a promising technology for the conversion of biomass into biofuels and biobased chemicals. This chapter is focused on the waste biomass conversion for production of biofuels and chemicals by applying sub- and supercritical fluids. One of the biggest disadvantages in biomass conversion by SCF is the extremely high energy requirement for heating the media above the water critical point (374 °C, 221 bar). The idea behind the recent research is to reduce the operating temperature and energy requirements by processing biomass with water at much higher pressures. The importance of knowledge on behavior of multicomponent systems at elevated pressures and temperatures is underlined. Methods, developed by the authors of this chapter for determination of thermodynamic and transport properties for multicomponent systems of different solid compounds and supercritical fluid under extreme conditions are described. Future perspective of hydrothermal technology as a tool to obtain advanced materials and the possible scope for future research is also discussed.

Chapter 17
Application of Supercritical Technologies in Clean Energy Production: A Review 588
Vladimir Anikeev, Institute of Catalysis, Siberian Branch, Russian Academy of Sciences, Russia
Khaled A. M. Gasem, University of Wyoming – Laramie, USA
Maohong Fan, University of Wyoming – Laramie, USA

In this chapter the authors seek to acquaint the reader with unusual properties of supercritical fluids, and how these properties are used for various applications in the synthesis of biofuels and the intensification of energy processes and thermal cycles. The review includes the studies of leading scientists from around the world related to effective and environmentally friendly methods and technologies of energy acquisition and conversion involving the synthesis of motor fuels, materials and chemicals, solar energy conversion and thermal cycle efficiencies using supercritical fluids. A special place in the chapter is occupied by studies of the transformation of vegetable oils, algae and wood in supercritical alcohols, esters and water to produce biodiesel fuels and hydrogen.

Compilation of References 617

About the Contributors 671

Index 678

Foreword

This book treats fundamentals and applications of supercritical fluids. To my understanding, the clear explanation of supercritical states and the critical processes are very new and challenging. However, applications of supercritical fluids have already been proposed more than half a century ago. There are a lot many publications of supercritical fluids in chemical extraction using their unique thermal and chemical properties. However, the wide applications of such fluids in energy related fields are truly new and important. My personal research experience is mainly focused on the basics of fluid phase and thermodynamics. I'm very glad to see the publication of this book as it covers from the fundamentals to practical designs and applications, which should be welcomed by both scientists and engineers who are working related topics.

We have recognized rather recently that the thermodynamic states of supercritical fluids can be changed almost instantaneously by a heat input through a boundary. These changes are adiabatic in the bulk (at constant entropy per unit mass), while heat exchange between the solid and the fluid produces thermal diffusion layers. These layers are very thin (~10^{-3} cm, typically) as compared to a macroscopic cell size (~1 cm), but they can play the role of an efficient thermal-mechanical "piston" upon their thermal expansion or shrinkage. Particularly near the gas-liquid critical point, where the isobaric thermal expansion is largely enhanced, this effect gives rise to rapid heat transport through the cell in the form of sounds. In the literature of physics and hydrodynamics, however, not enough attention has yet been paid to the underlying fast acoustic processes taking place on a timescale of $L=c$, where L is the cell length and c is the sound velocity. These dynamical processes are in fact ubiquitous in confined compressible fluids and are thus crucial in heat-exchange applications, as discussed in this book. We stress that they are even fundamental in understanding traditional thermodynamics, where adiabatic and isothermal changes of thermodynamic states are just assumed without analysis of the actual physical processes. I noticed that several chapters of this book are doing such topics and the mechanisms of the very critical process will be helpful for the understanding of practical system performances.

Recently, a number of simulations have been performed on convection motions in supercritical fluids in two and three dimensions, not only in small scale models but also in large scale energy conversion cycles, which have been revealing unique features of supercritical hydrodynamics. First of all, in such fluids, convective fluid motions can easily be induced even by small thermal disturbances owing to the enhanced thermal expansion and a relatively small shear viscosity with and without earth gravity. For example, jet-like fluid ow has been observed around a heated boundary even in gravity-free conditions. As another unique effect, overall temperature changes in the cell take place due to the piston effect, for example, upon arrivals of thermal plumes at the boundary walls, which is unfamiliar in the usual

Rayleigh-Benard convection. Namely, local and global dynamical processes are inseparably coupled in supercritical fluids, leading to unique hydrodynamics different from that of incompressible fluids.

Supercritical fluids can be used in various technological applications because of their high ability of heat transport. In one phase states, the piston effect yields energy transport in the form of sounds. In two-phase states, latent heat is adsorbed or released at the gas-liquid interfaces upon evaporation and condensation, which is indeed the well-known mechanism of heat pipes used in refrigerators. Hydrodynamics of highly compressible fluids in the supercritical and two-phase regions is not easy to understand, but it should be new and crucial in science and technology. Also, the large density gradients may lead to a preferable potential for natural circulation flow systems, such as solar and mechanical heating/cooling systems. Then the focus of this book is on energy conversion, which includes the heat transfer and flow dynamics details, natural convection systems, power generation, solar process, nuclear thermo-hydraulic problems, various energy cycles, as well as related thermodynamic topics, fluid dynamics, multi-scale experiments and analysis.

I believe that this book provides timely and adequate accounts of the above-mentioned aspects of supercritical fluids and very new and important energy conversion applications. Many of the contributors are my friends and in this book they summarized the very best research works in fluid system and energy conversion fields. I would like to thank Dr. Lin Chen for inviting me to be the first reader of the raw chapters and write a foreword. Sincerely I recommend this book to readers who have interest in this field. I'm sure this book will be one useful reference for researchers, engineers and even graduate students who have interest in flow dynamics studies and energy topics.

Akira Onuki
Kyoto University, Japan

Akira Onuki *obtained his B.S (1971) and PhD (1976) from the University of Tokyo. Then he worked as a research associate in Kyushu University (1977), associate professor (1983) and full professor in Kyoto University (1991-2012). Now he is a Professor Emeritus of Kyoto University. His research is focused on soft matters and physical understanding of phase transitions, equilibriums, as well as thermodynamics and some applications in real systems. In recent years, he has published many highly cited scientific papers on those topics in PRL, PRB, PRE, JCP, EPL, etc. He has authored many professional books/chapters, including the most famous one "Phase Transition Dynamics" (Cambridge, 2004).*

Preface

Supercritical fluids are now increasingly utilized in various energy related systems. The applications of supercritical fluids contribute to both the improvement of energy efficiency as well as reducing greenhouse gas emissions. Applications of supercritical fluids in recent years could be found in solar heater, geothermal process, extraction, refrigeration, cooling of the turbine blades and electronic components, heat pumps, power generations, nuclear systems and many other occasions.

A supercritical fluid is one fluid state that at a temperature and pressure above the critical point, where no distinct liquid or gas phases exist. This fluid status is specially controlled or maintained to take advantage of the preferable thermal and transport properties. Near to the critical point, small changes in pressure or temperature will result in large changes in thermos-dynamic properties. It is only for those thirty-years or so that scientists and engineers began to treat the in-depth exploration and utilization of such fluids. As it is well known that supercritical fluids are suitable as a substitute solvent, energy conversion working fluid, more and more attentions have been attracted in this topic in recent decades.

Flow dynamics and fundamental heat transfer of supercritical fluids are of critical importance in both practical supercritical devices design and fluid sciences. For example, the most often studied/used supercritical fluid CO_2, as a natural working fluid, is a nonflammable, nontoxic alternative that could be safely used as substitute of traditional CFCs, HCFCs and hydrocarbons, ammonia, etc. Such fluid has potential for future energy systems and transportation mechanisms. CO_2 is also proved to be environmentally benign due to the fact that it's ODP (Ozone Depletion Potential) = 0 and GWP (Global Warming Potential) = 1. Now both in the laboratory and industrial fields, supercritical CO_2 is widely employed. Supercritical water is also widely used for high temperature applications (mainly power cycles) for its suitable critical temperature and pressure operation ranges. The critical point data is shown in Table 1 for some representative fluids, which can also be easily referenced in engineering handbooks. Such utilization of supercritical fluids indeed has provided new choices of future green working fluids with both high efficiency and benign environmental impacts.

The basic flow and thermal physical behaviors of supercritical fluids have been discussed by many researchers since early 20s century. The basic property changes and heat transfer characteristics, in a unique range of temperature and pressure for specific applications, of supercritical working fluids have shown new challenges to scientists and engineers of that time. The main question is: what is the nature of supercritical states and how to improve heat transfer? It is only in recent years that such mechanisms are revealed to the scientific community. System behavior and optimal design of supercritical flow have also been widely investigated after 1980s. And later one very important Space Mission of Europe (D-1 mission in 1984), quick and homogeneous heating up of critical fluids are experimentally identified under microgravity. This opens a new world of critical phenomena and it attracted scientist to explore in detail what is happening when a fluid goes approaching/across the critical point.

Table 1. Critical parameters of some fluids

Fluid	Critical T_c (K)	Critical p_c (MPa)	Critical ρ_c (kg/m^3)
H_2O	647.1	22.06	322.2
CO_2	304.13	7.378	467.8
N_2	126.24	3.398	313.9
O_2	154.58	5.043	436.2
Xe	289.74	5.840	1113
H_2	32.976	3.761	31.43
^3He	3.317	1.2928	41.45
SF_6	318.717	3.7545	742
R13	302.0	3.879	582.88
R23	299.29	4.832	526.5
R134a	374.21	4.0593	511.9
R410a (50%R32+50%R125)	344.49	4.9012	459.03

Data from: NIST Standard Reference Database-REFPROP, Version 8.0, 2006.

Supercritical fluid and the "abnormal" behaviors are welcomed by engineers so as to utilize the large properties' changes in the critical region, in a belief that such phenomena can be understood and controlled. For such large and drastic changes of properties, especially the density variations, viscosity fluctuations, thermal conductivity changes, and thermal capacity peaks are all dependent on how the status is close to the critical point, namely the critical distance, which can be defined by a reduced temperature parameter by the critical temperature. Figure 1 shows the schematic picture of the developments in near-critical phenomena, which is of critical importance for the understanding of the choices in supercritical application systems. Some detailed explanations of such basics as well as the heat mass transfer characteristics will be covered by chapters of Section 1 in this book.

Among the many applications, natural circulation flow and heat transfer attracted a lot of attention. In the absence of pumps and their associated mechanical devices, a natural circulation system can work well and thus it has advantage over a forced circulation system in terms of cost, system reliability and maintenance. Natural circulation systems using supercritical fluids give good performances with the large density changes inside a loop system when it is near the critical region. Such system yields both high thermal efficiency and good fluid mobility. However, as for near critical fluid the physical properties and transport coefficients would experience dramatic changes, leading to chaotic behaviors and instabilities. These instabilities are harmful in most application systems. Nevertheless, the mechanisms of these behaviors, especially system instabilities are due to the complexities of both system geometries and various boundary conditions. Section 2 of this book deals with such topics and the applications in solar and nuclear systems is also covered. The factors that might affect system stability are also identified through experiments and numerical simulations on this kind of fluid model.

The detailed mechanisms of the abnormal behaviors of supercritical/trans-critical CO_2 thermal cycles are still less studied and systematic reports just can be found in very recent years. Some recent studies have focused on the heat and mass transport details in supercritical/trans-critical systems. In those supercritical energy conversion systems, typical generation feature was identified and unsteady cases are

Figure 1. Piston effect and thermal-mechanical effect related research history and major topic flow

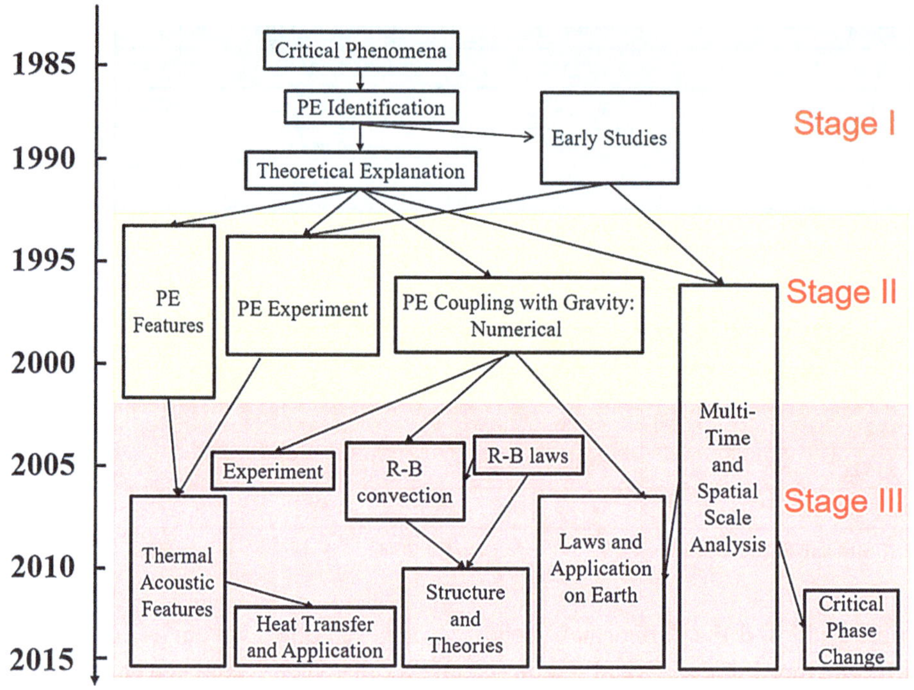

more often seen than normal water cases due to the unique thermo-physical properties of near-critical fluid. The peculiar changes of properties, especially for conditions in the vicinity of critical point, are typical for supercritical fluid and it brings additional difficulties in related scaling and modeling process. Many recent studies have also set up transient multi-scale models and systematically analyzed the basic heat transfer, hydraulic behaviors and strategies for stability control. Such topics are covered in Section 3 of this book.

After the nuclear accident in Japan, though new generations of nuclear power plant are still under development, more and more people are becoming skeptical on the future of nuclear power. Now the major developments of new generation nuclear power plant focus on the working fluid and respective safety design. And the thermal and hydrodynamic behaviors of fluid are one of the core problems to be investigated, especially for non-linear fluids like supercritical water or supercritical CO_2, which are considered for utilization in new generation nuclear cooling cycles. Supercritical fluids have become one substitute for future nuclear power plant is partly due to their preferable loading performance with high heat capacity and high circulation rate even under pure natural convection condition. Chapters in Section 4 of this book mainly focus on the basic thermal and hydrodynamic behaviors of trans-critical/ supercritical fluid in nuclear systems. Indeed the application of supercritical CO_2 should not be limited in nuclear engineering or related, in fact it has been proved to have equal or higher performance in heat pump or solar conversion system, and it has also been proposed and widely used in many other engineering systems.

ORGANIZATION OF THE BOOK

This book has gathered altogether 17 chapter contributions from representative groups of supercritical fluids research from Australia, China, Canada, Germany, France, India, Japan, Russia, Slovenia, United States, etc. Those chapters are organized into four major sections. The contents of the book chapters cover first the basic thermo-dynamic behaviors and 'abnormal' properties from a thermophysical aspect; then basic heat transfer and flow characteristics, recent findings in physical aspect, engineering, micro-/nano-scale phenomena, multi-scale analysis and transient behaviors in energy systems are discussed; after that, most recent and challenging problems and outlook for the applications and innovations of supercritical fluids in energy systems are discussed.

Section covers the fundamentals of supercritical fluids flow and heat transfer studies. Four chapters (Chapter 1-4) are included in this section. In those chapters, fundamentals of supercritical fluids and near-critical phenomena are reported and discussed, including the very basic thermophysics of critical transition under thermal and flow conditions. Thermal-mechanical effects of critical fluids and the related heat transfer and flow stability problems are discussed by major research groups in this field. Recent developments on the critical transition and a special liquid metal critical study are also included.

Chapter 1 is "Thermal Effects in Near-Critical Fluids: Piston Effect and Related Phenomena". This chapter deals with the very particular thermal behavior that supercritical fluids exhibit when nearing their critical point. In this region, supercritical fluids exhibit strong anomalies in their thermodynamic and transport properties. Pressure change associated to a temperature variation leads to a nearly isentropic thermalization of the fluid, the "piston effect", which leads to a paradoxical "critical speeding-up" and very interesting convection phenomena. These effects can deeply modify the supercritical fluids thermal behavior in space and energy activities, giving to these effects socio-economic relevance.

Chapter 2 is "Numerical Modelling of Hydrodynamic Instabilities in Supercritical Fluids". The case of a supercritical fluid heated from below (Rayleigh-Bénard) in a rectangular cavity is first presented. The stability of the two boundary layers (hot and cold) is analyzed by numerically solving the Navier-Stokes equations with a van der Waals gas and stability diagrams are derived. The very large compressibility and the very low heat diffusivity of near critical pure fluids induce very large density gradients which lead to a Rayleigh–Taylor-like gravitational instability of the heat diffusion layer. Depending on the relative direction of the interface or the boundary layer with respect to vibration, vibrational forces can destabilize a thermal boundary layer.

Chapter 3 is "Thermal-Mechanical Effects and Near-Critical Fluid Dynamic Behaviors in Micro-Scale". The chapter deals with near-critical CO_2 micro-scale thermal convective flow and the effects of thermal-mechanical process. When the scale becomes smaller, new and detailed figures of near-critical thermal effects emerges. To explore this new area, theoretical developments and numerical investigations discussed and explained in this chapter. The thermal-mechanical nature of near-critical fluid would play a leading role in small time and spatial scales. That effect dominants the thermal dynamic responses and convective structures in micro-scale. The scaling effects, boundary process, instability evolutions, mixing flows and characteristics, and applications are also discussed in this chapter.

Chapter 4 is "X-Ray Scattering Studies of Expanded Fluid Metals". Fluid metals exhibit significant properties of thermodynamic-state dependence, since the inter-particle interaction among the constituents (electrons and ions) considerably changes depending on their thermodynamic conditions. X-ray scattering experiments of fluid metals in the expanded state gives the microscopic understanding of the structural and electronic properties of fluid metals. The purpose of this chapter is to provide intriguing aspects of fluid metals originated from the existence of conduction electrons, which distinguishes fluid metals from non-conducting fluids, through the results of fluid rubidium and mercury.

Section 2 deals with the convection flow and heat transfer characteristics of supercritical fluids. Four chapters (Chapter 5-8) are included in this section. Chapters in this section are selected to discuss the convective behaviors and circulation flow heat transfer characteristics. Supercritical loop systems are very basic concept for various applications of effective energy conversion and thermal utilization processes. This section is then focused on the basic principles, mechanisms and applications in this aspect. The results and proposals in this section may contribute to related natural circulation system designs of supercritical fluids.

Chapter 5 is "Principles, Experiments, and Numerical Studies of Supercritical Fluid Natural Circulation System". The chapter presents the principals of supercritical natural circulation loop system and its application challenges. In this chapter, a specially designed experimental prototype system is introduced and compared with numerical findings. The system can be operated in wide range of high pressure in the critical region. Very high Reynolds number natural convection flow can be achieved only by simple heating and cooling. Thermal performance analysis and parameter effects are carried out along with the experimental development. The heat transfer dependency on operation and its mechanisms are also explained and summarized.

Chapter 6 is "Supercritical Natural Circulation Loop: A Technology for Future Reactors". Supercritical natural circulation loop is a compelling technology for cooling of modern nuclear reactors, which promises enhanced thermalhydraulic performance in a simple design. This chapter summarizes the observation till date, starting from the very fundamentals. Different methods of analyses, including analytical, simple 1-D numerical and multi-dimensional codes, as well as experimental are elucidated. A comprehensive discussion is made on various geometric and operating parameters from both thermalhydraulic and stability point of view.

Chapter 7 is "Application of Supercritical Carbon Dioxide for Solar Water Heater". The large variation of thermophysical properties of supercritical fluids allows a strong natural convection flow with absorbing as well as transporting a large amount of heat energy. Taking the low critical temperature, this study proposed a solar water heater using supercritical carbon dioxide. The new solar water heater has a great potential to more effectively supply heat energy from solar energy, comparing with the conventional solar water heaters of water-in-glass tube type. The measured collector and heat recovery efficiencies are over 60%. In this chapter, the overview of the new solar water heater using supercritical carbon dioxide is also included.

Chapter 8 is "Heat Transfer and Fluid Flow of Supercritical Fluids in Advanced Energy Systems". This chapter aims to clarify the supercritical fluids thermal hydraulics characteristics including heat transfer, pressure drops and flow instabilities for advanced energy systems. The convection heat transfer models considering the effect of nonlinear properties and buoyancy acceleration are discussed. Then, the hydraulic resistance models are suggested for different conditions. The published models for supercritical flow instabilities are also summarized and clarified. At last, two typical case studies have been provided to further intuitively elaborate the thermal hydraulics of supercritical fluids within the advanced energy systems.

Section 3 is the supercritical thermal conversion of different scales. Four chapters (Chapter 9-12) are included in this section. In those chapters, multi-scale analysis is discussed for supercritical and near-critical fluid systems: from simple channels to complex heat transfer system designs. The confinement and scaling effect of supercritical fluid heat transfer flow is focused in this section. The knowledge obtained by major researchers in this field will contribute to various applications of supercritical fluid based heating, cooling, microscale transportation, reaction, etc.

Chapter 9 is "Heat Transfer in Supercritical Fluids: Going to Microscale Times and Sizes". Results of experimental study of non-stationary heat transfer in supercritical fluids, which were obtained using the method of controlled pulse heating of low-inertia wire probe, are discussed. The aim of this study was to clarify the peculiarities of heat conduction mode at significant heat loads. A threshold decrease in the "instant" heat transfer coefficient, the more pronounced the closer the pressure value to critical pressure, has been found. Small time and spatial scale in the experiments (units of millisecond and units of micrometer) in combination with high-power heat release (up to 20 MW/m^2) makes it possible to associate the results with the behavior of boundary layer of heat transfer agent.

Chapter 10 is "Study on Cooling Heat Transfer of Supercritical Carbon Dioxide Applied to Trans-Critical Carbon Dioxide Heat Pump". This chapter addresses to recent experimental, theoretical and numerical studies on cooling heat transfer of supercritical CO_2 under the effect of lubricating oil. A systematic study on heat transfer coefficient and pressure drop of supercritical CO_2 was carried out at wide ranges of tube diameter, mass flux, heat flux, temperature and pressure. Effects of parameters such as mass flux, pressure, heat flux, and cooling-tube diameter on the heat transfer coefficient and pressure drop were experimentally analyzed. It was found that the varied flow pattern of two-phase fluid caused by the difference in miscibility of oil with supercritical CO_2 lead to different heat transfers.

Chapter 11 is "Supercritical CO_2 and Its Application to Rankine Cycle". Supercritical CO_2 has been given much attention to be a working fluid in a power cycle due to its unique properties. The supercritical CO_2 solar Rankine cycle system was designed and developed by using the benefit of supercritical state of CO_2 to generate electric power. Operating the power cycle in the supercritical region of CO_2 enhances the heat transfer in energy exchanging process and improves the cycle efficiency. In this chapter, the development of main components in the system are introduced and discussed particularly by focusing on the properties of CO_2 for higher performance.

Chapter 12 is "Supercritical Fluids and Their Applications in Heat Conversion". The chapter presents the importance of the selection of supercritical fluids for both high- and low-grade heat sources for power generation. Low-grade heat sources, typically defined as below 300 °C, are abundantly available as industrial waste heat, solar thermal, and geothermal, to name a few. This chapter investigates the potential of supercritical Rankine cycles in the conversion of low-grade heat to power, while the second part discusses supercritical fluids used in higher grade heat conversion system. This chapter discusses supercritical fluids fundamentals, selection of supercritical working fluids for different heat sources, and the current research, development, and commercial status of supercritical power generation systems.

Section 4 is supercritical power generation systems and green energy harvestation. Five chapters (Chapter 13-17) are included in this section. Chapters in this section are mainly focused on power generation systems assisted by supercritical fluids and supercritical cycles. Based on the current understanding of supercritical transport mechanism (heat, mass, chemical reaction, etc.), system scale analysis is focused for chapters in the discussion of power efficiency and generation system designs. The compact and high efficiency in power engineering, chemical engineering, thermal engineering as well as nuclear engineering would be attractive for scientists and engineers in this field. Still, the contribution of green energy harvestation and other related supercritical energy systems are also included.

Chapter 13 is "Application of Supercritical Pressures in Power Engineering". This chapter discusses the use of SuperCritical Fluids (SCF) based thermal power plants and their various applications including nuclear power plants for Generation IV reactors. The historic development (since 1930s) and recent challenges of SuperCritical Fluids (SCF) in nuclear reactors, power generation systems and others are introduced. The primary objectives for using SuperCritical Water (SCW) as a coolant in nuclear reactors are introduced in this chapter. Generation IV nuclear reactors with SCF are introduced and discussed into detail on the basic system designs, features, and future challenges.

Chapter 14 is "Energy Conversion Using the Supercritical Steam Cycle". A supercritical steam (or Rankine) cycle is used for new coal-fired power plants, and it has also been proposed for future water-cooled nuclear reactors to enhance their efficiency and to reduce their costs. This chapter provides the technical background explaining this technology. Some criteria for boiler design and operation, like drum or once-through boiler design, fixed or sliding pressure operation and coolant mixing, are discussed in general to explain the particular challenges of supercritical steam cycles.

Chapter 15 is "The Basic Thermal Hydraulic Issues of Applying Supercritical Fluid to Nuclear Reactors". This chapter is mainly focused on illustrating some major progress on thermal hydraulic issues of supercritical water performed in Nuclear Power Institute of China (NPIC). An analytical method and in-house code are proposed to predict the heat transfer coefficient and friction coefficient based on the two layer wall function. Flow instability experiments in a two-parallel-channel system with supercritical water are made to provide an up-to-date knowledge of supercritical flow instability phenomena and initial validation data for numerical analysis.

Chapter 16 is "Supercritical Fluids as a Tool for Green Energy and Chemicals". Hydrothermal conversion of biomass is a promising technology for the conversion of biomass into biofuels and bio-based chemicals. This chapter is focused on the waste biomass conversion for production of biofuels and chemicals by applying sub- and supercritical fluids. The target is to reduce the operating temperature and energy requirements by processing biomass with water. Methods developed in this chapter for determination of thermodynamic and transport properties for multicomponent systems of different solid compounds under extreme conditions are described.

Chapter 17 is "Application of Supercritical Technologies in Clean Energy Production: A Review". In this chapter the authors seek to acquaint the reader with unusual properties of supercritical fluids, and how these properties are used for various applications in the synthesis of biofuels and the intensification of energy processes and thermal cycles. The review includes the studies of leading scientists from around the world related to effective and environmentally friendly methods and technologies of energy acquisition and conversion involving the synthesis of motor fuels, materials and chemicals, solar energy conversion and thermal cycle efficiencies using supercritical fluids.

TARGET OF THE BOOK

This book is focused on gathering and organizing the current developments in supercritical fluid related energy conversion studies and engineering developments. There are several books published in supercritical fluids, however, nearly all of them are focused in chemical extraction fields or fundamental physical aspects of supercritical fluids. Such demand in the current status-of-art in this field triggered the thinking of gather the main researchers together for publication of a book that summarizing the aspect of increasingly important field in heat transfer and energy conversion. Therefore, this book tries to cover from the very basics to the current applications of supercritical fluids. The understanding of the fundamentals will surely contribute to the development in related system designs.

It is hoped that this book will not only contributions to scientific world but also to the engineers. Indeed, such new applications of the supercritical field are also related to the energy efficiency enhancement and greenhouse gas emission reduction, which will be of great importance to the general wellbeing of humankind. For the above reason, the editors tried their best to invite the major research groups of supercritical fluids around the world to discuss and contribute to the current book. The research and publication in supercritical fluid dynamics and engineering should be of great importance in both scientific advancement and industrial innovations. The following research topics are well covered in this book:

- Supercritical fluid based heat transfer mechanisms,
- Numerical development of supercritical/near-critical fluids,
- Experiments on supercritical/near-critical fluid flow and heat transfer,
- Natural circulation flows and heat mass transfer and applications,
- Supercritical fluid based chemical reaction and manufacturing processes,
- Supercritical fluid based energy conversion and harvestation,
- Supercritical fluid utilization and performances in nuclear engineering,
- Efficiency analysis and apparatus design in supercritical fluid based systems,
- Design and optimizations of supercritical fluid/binary fluid based apparatus.

The target audience of this book will be composed of graduate students, young and professional researchers, engineers and other public policy makers interested in the field of supercritical fluids, energy conversion and environmentally friendly systems. In addition, the book could be used as handbook or reference for supercritical fluid and energy conversion related researches and engineering.

Lin Chen
Tohoku University, Japan & Japan Society for the Promotion of Science (JSPS), Japan

Yuhiro Iwamoto
Nagoya Institute of Technology, Japan

Acknowledgment

The idea of organizing a book on supercritical fluids related energy conversion systems comes from the increasing demand from scientific world and a lack of state-of-art analysis. The editors and contributors of this book have all explored in this field for a couple of years and the application of supercritical fluids indeed diverges from muti- time and spatial scales in real world. Though the applications in chemical engineering are already a big industry, the thermal and energy aspect is still under fast development. The innovations in energy conversion fields utilizing the preferable thermal and transport properties of supercritical fluids indeed give new possibilities for higher efficiency and compact designs.

The editors would like to acknowledge the help of all the people involved in this project and, more specifically, to the authors and reviewers that took part in the review process. Without their support, this book would not have become a reality. Our sincere gratitude goes to the chapter's authors who contributed their time and expertise to this book. Also, the editors wish to acknowledge the valuable contributions of the reviewers regarding the improvement of quality, coherence, and content presentation of chapters. It is the contributors' work that makes it possible to cover from the very basics to various energy applications of supercritical fluids in this book. Most of the authors also served as referees; we highly appreciate their double task.

The support from JST-CREST, and JSPS Overseas Researcher Grant in Japan (No. 16F16068) are gratefully acknowledged by Dr. Lin Chen. The support and encouragement from Prof. Shigenao Maruyama, Prof. Atsuki Komiya and Prof. Junnosuke Okajima in Tohoku University, and the kind help and suggestions from Prof. Hiroshi Yamaguchi in Doshisha University and Prof. Xin-Rong Zhang in Peking University indeed helped a lot in the editing process of this book. Prof. Benard Zappoli (CNES, France), Prof. Daniel Beysens, Prof. Yves Garrabos, Prof. Sakir Amiroudine (University of Bordeaux, France) and many friends provided valuable comments about the publication of this book. And it is a great honor to have Prof. Akira Onuki in Kyoto University to write the foreword and recommend this book, who was also the first reader of the raw chapters.

The editors would also like to thank the managing editors/assistants/secretaries: Ms. Janine Haughton, Ms. Kayla Wolfe, Ms. Lindsay Johnston, Ms. Mary Imgrund, Mr. Joshua Herring (from IGI global) and Ms. Hitomi Kobayashi, Ms. Shizu Sato and Ms. Chieko Nakaya (Heat Transfer Control Lab, Tohoku University), and many others who were involved in this project. Their kind help and hard work during the book development made it possible for this book to meet readers on time.

Lin Chen
Tohoku University, Japan & Japan Society for the Promotion of Science (JSPS), Japan

Yuhiro Iwamoto
Nagoya Institute of Technology, Japan

Section 1

Fundamentals of Supercritical Fluids Flow and Heat Transfer

In this section, fundamentals of supercritical fluids and near-critical phenomena are reported and discussed, including the very basic thermophysics of critical transition under thermal and flow conditions. Thermal-mechanical effects of critical fluids and the related heat transfer and flow stability problems are discussed by major research groups in this field. Recent developments on the critical transition and a special liquid metal critical study are also included.

Chapter 1
Thermal Effects in Near-Critical Fluids:
Piston Effect and Related Phenomena

Daniel A. Beysens
Ecole Supérieure de Physique et Chimie Industrielles ParisTech, France

Yves Garrabos
Institut de Chimie de la Matière Condensée de Bordeaux, France

Bernard Zappoli
Centre National d'Etudes Spatiales, France

ABSTRACT

In this chapter is addressed the very particular thermal behavior that supercritical fluids exhibit when nearing their critical point. In this region, supercritical fluids exhibit strong anomalies in their thermodynamic and transport properties. Pressure change associated to a temperature variation leads to a nearly isentropic thermalization of the fluid, the "piston effect", which leads to a paradoxical "critical speeding-up". Bulk fluid temperature is uniform, temperature gradients are confined in thermal boundary layers, making the bulk fluid a thermal short-circuit. It follows very particular behavior, as dynamic heat pipes or heat going seemingly backward, in apparent contradiction with the 2nd principle of thermodynamics. Under an acceleration field, thermal convection occurs only in the boundary layers, which paradoxically can enhance the fluid stability or even cool the fluid after a heat pulse. These effects can deeply modify the supercritical fluids thermal behavior in space and energy activities, giving to these effects socio-economic relevance.

INTRODUCTION

Classically, one considers that fluid can be thermalized by three different modes: radiation (where heat is exchanged by emission and absorption of electro-magnetic waves), diffusion (as in a solid), and convection by flow motion. In the following, we focus on diffusion and convection and see that in compressible

DOI: 10.4018/978-1-5225-2047-4.ch001

fluids like supercritical fluids nearing their critical point (near supercritical fluids), another thermalization effect, the « Piston effect» (PE), where a thermal boundary layer expands and near adiabatically heats the fluid, can be of high importance. This is especially true in space weightlessness where convection no more occurs and where it has been observed the first time. On earth, this PE, although in competition with convection to thermalize near supercritical fluids, eventually reveals to be the main process responsible of their thermalization. In addition, this thermo-compressible effect leads to very peculiar and paradoxical phenomena in heat transport and thermalization, seemingly in disagreement with the 2nd thermodynamics principle.

These effects can modify in a significant manner the thermo-mechanical response of supercritical fluids when they reach the (immediate or not) vicinity of their critical point. This is particularly the case for supercritical fluid used in the space industry and propulsion under weightlessness. For instance, the supercritical water oxidation (SCWO) process was one among the selected processes to treat the waste materials involved in the life support for the space exploration (see Major, 2014; Phillips, 2014). Similarly, near-critical fluids can be used as relevant models (Nikolayev, 2015) to study the boiling crisis dynamics in high power thermal plants including nuclear plants and the analysis of safety risks (Juhaszl et al., 2009).

GENERAL BACKGROUND

Supercritical fluids, that is, fluids at pressure and temperature above their critical point (CP) coordinates, exhibit particular properties (large density, low viscosity, large mass diffusivity), which make them intermediate between liquids and gases. It was indeed well-recognized from the end of the 90's that the fine pressure/temperature control of the supercritical conditions is very appealing to the industry as an easy mean to tune their non-polluting solvatation power and host the chemical reaction rates with high yield efficiency (see for example Noyori, R., 1999 and related papers in the same special issue). Fluids in such supercritical conditions are now mainly used to open new routes in green synthesis of innovative materials (Adschiri et al., 2015; Dumas et al., 2016) or in hydrothermal biomass conversion processes (Kruse & Dahmen, 2015).

In addition, the fact that temperature is large increases the yield of thermo-mechanical processes in boiling water, steam or molten salt reactors. Under reduced gravity, the storage of cryogenic propellants is sometimes made in their supercritical conditions to avoid a non-controlled two-phase distribution.

However, when temperature and pressure approach the CP values, such fluids show strong anomalies in a number of static and dynamical properties. In particular, isothermal compressibility, thermal expansion and specific heats at constant pressure and volume can become extremely large. Dynamical properties can also be much affected. The so-called "critical slowing down" corresponds to a strong decrease of thermal diffusivity while, in contrast, heat conductivity and specific heat diverge. In the vicinity of the critical point temperature and pressure, such fluids are called « near critical" or "near supercritical ». We address below the main properties of such near supercritical fluids.

Thermodynamic Scaling Laws

An important aspect of the critical region (see e.g. Stanley (1971) is that most of the anomalies of the thermodynamic and transport properties can be set in the form of scaled, universal (power law) functions with respect to the CP coordinates (see Table 1). Then any results obtained with one single fluid can be immediately generalized to a whole class of systems, the "class of fluids" (to which liquid mixtures, including polymer melts and solutions, micro-emulsions, molten salts, monotectic liquid metals, etc. also belongs). The standard representative of this class is a magnet, the 3D uniaxial Ising model where magnetization corresponds to the reduced fluid density difference $\frac{\varrho - \rho_c}{\rho_c}$. ρ is the density and ρ_c is the critical density. This scaling is of fundamental nature and stems from the universal behavior that asymptotically the free energy F must obey to fulfill the conditions of the CP 2nd order phase transition.

Excited by the thermal fluctuations in the system, the local density fluctuates more and more strongly as the CP is approached due to the divergence of the thermal expansion coefficient. The vicinity of the CP is then characterized by the presence of large fluctuations in density, whose amplitude and extent (the fluctuation correlation length) diverge. Fluctuations give rise to an unusual strong scattering of light, the so-called "critical opalescence".

Because relationships exist between the thermodynamics coefficients that undergo critical anomalies (e.g. compressibility and specific heat) a number of relations can be found between the exponents and the diverging amplitudes. It can be shown that only two exponents are independent. For the same

Table 1. Asymptotic power law (critical behavior) of some fluid properties. τ is the distance to the critical point defined by $\tau = \frac{T - T_c}{T_c}$. $D = 3$.

Property	Power Law	Thermodynamic Path	Critical Exponent	
			Ising $D = 3$; $n = 1$	Mean Field
Compressibility	$\kappa_T \sim \lvert\tau\rvert^{-\gamma}$	Critical isochore	$\gamma \approx 1.239$	1
Specific heat at constant pressure	$c_p \sim \lvert\tau\rvert^{-\gamma}$	Critical isochore	$\gamma \approx 1.239$	1
Specific heat at constant volume	$c_V \sim \lvert\tau\rvert^{-\alpha}$	Critical isochore	$\alpha \approx 0.110$	0 (jump)
Density of two phases	$(\rho_\ell - \rho_\Delta) \sim \lvert\tau\rvert^{\beta}$	Critical isochore	$\beta \approx 0.326$	1/2
Surface tension	$\sigma \sim \lvert\tau\rvert^{\mu}$	Critical isochore	$\mu = 2\nu \approx 1.26$	1
pressure	$\lvert\Delta p\rvert \sim \lvert\Delta\rho\rvert^{\delta}$	Critical isotherm	$\delta \approx 4.80$	3
Correlation function	$H(r) \sim r^{2-D-\eta}$ $(\xi^{-1} = 0)$	Critical point Large r	$\eta \approx 0.034$	0
Correlation length	$\xi \sim \lvert\tau\rvert^{-\nu}$	Critical isochore	$\nu \approx 0.630$	1/2

reason, only two amplitudes are independent. As an important consequence, the behavior of all systems can be expressed as scaled, universal functions within only two system dependent amplitudes. This is the so-called "two-scale factor" universality (Stauffer et al., 1972).

Thermal Critical Slowing Down

As above for the statics, it is possible to define universal dynamics scaling laws for the transport coefficients, in particular thermal phenomena. The universality classes are not so large as for the case of statics. The dynamics of the fluctuations of the OP appears to be the natural timescale, as the correlation length for space, and is ruled by the decay time of a fluctuation of size ξ. The fluctuation vanishes by a diffusion process, with a diffusion coefficient D_T which can be estimated from the Brownian diffusion of a cluster of size ξ (Kawasaki, 1970):

$$D_T = \frac{k_B T}{6\pi\eta\xi}. \tag{1}$$

Here k_B is the Boltzmann constant and η is the shear viscosity. A typical time $t=t_\xi$ is naturally defined as the diffusion time of a fluctuation of size ξ on the typical lengthscale ξ, i.e.

$$t_\xi = \frac{6\pi\eta\xi^3}{k_B T}. \tag{2}$$

From the temperature dependence of ξ, it is easy to determine the temperature dependence of t_ξ, i.e.

$$t_\xi \sim \tau^{-3\nu}, \tag{3}$$

which shows that the fluctuations of the OP relax more and more slowly as the system approaches its CP. This is the well-known "critical slowing-down". Although pressure fluctuations equilibrate nearly instantaneously (at the sound velocity), density and temperature fluctuations are slowed down.

The diffusion coefficient can be written also as the isothermal heat diffusion coefficient

$$D_T = \frac{\Lambda}{\rho c_p} \tag{4}$$

with Λ the thermal conductivity and c the specific heat at constant pressure. Although Λ and c diverge near the CP, D_T goes to zero. This is due to the power law exponents which are different: $c \sim \tau^{-\gamma}$, and $\Lambda \sim \tau^{-\nu}$. The fact that D_T goes to zero means that any heat diffusion process is dramatically slowed down. To give an idea of this critical slowing down, the thermal equilibration at 1 mK from T_c of a CO_2 sample with thickness $2L$=1cm would need a time $t_D \approx L^2/D_T \approx$ 1 month to reach thermal equilibrium. Note that this equilibrium process is at constant pressure, that is, in a semi-infinite medium. For real, finite volume samples, thermalization is accompanied by a pressure change, which can adiabatically change the fluid temperature. This is the foundation of the PE as detailed in a next section. Under a gravity field, thermal instabilities can in addition induce convective flows that can also speed up thermalization.

Weightlessness

At the beginning of the space experiments under weightlessness, it thus seemed hopeless to the scientists to try to homogenize in temperature and density a sample of fluid close to the CP in a reasonable experiment time. Weightlessness is indeed mandatory to go very close the CP as the anomalies in the thermodynamic and transport properties make the fluids under gravity field very sensitive to the effects of gravity. Three effects are impairing the experimentation on earth. In the supercritical region, the divergence of the isothermal compressibility makes the liquid compressed under its own weight. Then, in a closed cell, the nearest the temperature to the CP, the more the fluid moves away from it by the effect of the hydrostatic pressure, which makes density deviates from its critical value. In the sub-critical region, where liquid and vapor coexist, the Bond number that measures the effect of the gravity forces compared to the capillary forces, tends towards infinity while approaching the CP. In addition, in the two regions sub- and super-critical, the critical slowing down of thermal diffusion results in strong convections in the fluid. The latter can become very unstable with even negligible temperature variations. The Rayleigh and Grashoff numbers, which measure the effect of buoyancy convection, indeed tend to infinite.

Thanks to those convection-free weightless experiments, it has been discovered that the fast-thermal equilibration observed on earth was due to a process which was misleadingly attributed to buoyancy-driven convections. This process, the "piston effect", is detailed below.

THE PISTON EFFECT MECHANISM

A supercritical fluid can be considered in a first approximation as a Newtonian, compressible van der Waals fluid. Although the van der Wall equation of state comes from a mean field theory with classical values of the critical exponents (see last column of Table 1), it describes the main hydrodynamic features of critical fluids: isothermal compressibility and heat capacity at constant pressure are diverging on the critical isochore. The corresponding governing equations are the compressible Navier-Stokes equations with heat conduction, completed by the van der Waals equation of state.

The following problem is considered. Heat is deposited at the left hand side wall of a slab-like container of width L (=12 mm), filled with a near supercritical fluid at critical density, initially at rest and at the thermodynamics equilibrium. Typical heating is a heat flux of 1 Wm^{-2} during 0.5 s, such as the temperature increase at the boundary is a small perturbation (from 1 mK to some mK). The right hand side wall is maintained at constant temperature. The time scale of the heat addition is equal to the heat diffusion time scale in a gas that would have the background value of the heat diffusion coefficient. Thermal conductivity of the walls is considered as infinite and a given heat flux is imposed to the wall. The equations are solved by the matched asymptotic expansions technique; details are given in Zappoli & Carlès (1995). In the following, the time variable is reported to the acoustic time of the slab, $t_a = \frac{L}{C_{s0}}$, where C_{s0} is the (constant) sound velocity in the ideal gas that would be set in the same temperature condition (i.e., $T \approx T_c$). The values $C_{s0} = 284\,ms^{-1}$ and then $t_a = 35.2\mu s$ are the typical values used in the present 1D modelling of a van der Waals-like CO_2 fluid.

Several periods of time can be distinguished during the fluid heating process. They are detailed in the next sections.

The Acoustic Period

On the initial time scale a very thin thermal boundary layer forms, with a thickness δ_1 that scales as $\tau^{0.25}$, with τ the reduced temperature distance to the CP. Such a behavior of δ_1 is consistent with the thermal diffusivity coefficient D_T that scales as $\tau^{0.5}$ as the CP is neared along the critical isochore in the supercritical region.

The fluid contained in this layer expands and its maximum velocity is reached at the edge of the thermal boundary layer. The expanding fluid acts as a piston that generates a compression wave in the bulk region. This compression wave heats the bulk fluid, reflects on the opposite, thermostated wall as a compression wave, heats up in his turn the bulk fluid, and so on. The piston velocity is increasingly high as the CP is approached; it is $\tau^{-3/2}$ times larger than calculated in an ideal gas (Kassoy, 1979) and is directly proportional to the heat flux at the wall. The temperature plot during the acoustic period is given in Figure 1. A second, cooling boundary layer of thickness δ_2 forms at the thermostated wall.

Figure 1. The thermal field plotted every 25 μs after the beginning of the heating period shows the compression wave emitted at the edge of the boundary layer that flashes back and forth, thus provoking a homogeneous increase of the bulk temperature. (x is the space variable reported to the fluid sample size L)

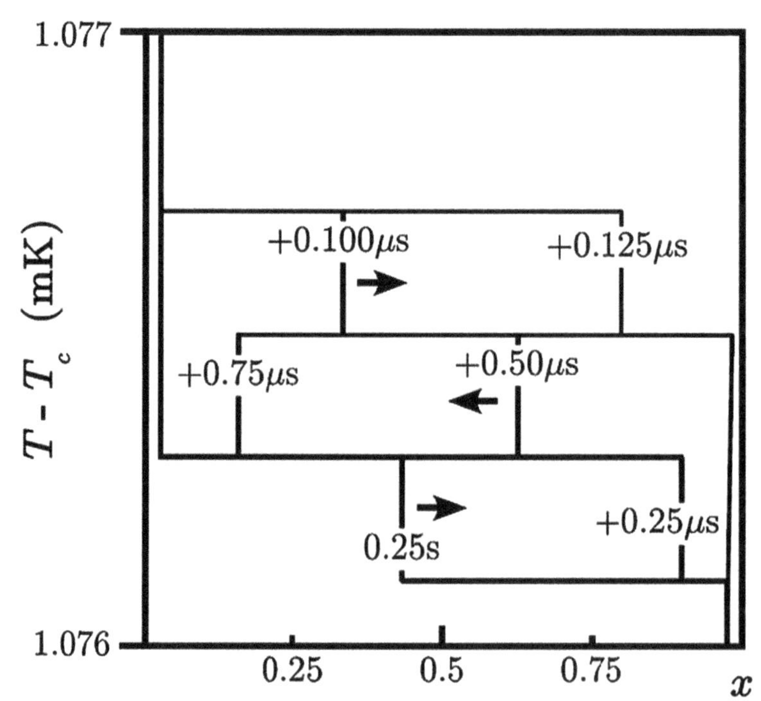

Real time is used in Figure 1 to illustrate the back and forth heating process, which occurs at the sound velocity (leading to $t_a = 35.2\mu s$ as noted above). The density change associated to this fast heating process was indeed observed by Miura et al. (2006) using an ultrasensitive interferometer.

The Heat Equilibration Period

Piston velocity appears to be much larger than in an ideal gas, thanks to CP scaling. Efficiency of the thermoacoustic heating is much increased. The characteristic time scale of the heat equilibration as given by an asymptotic analysis scales as $\tau^{3/2}$. This means that heat equilibrates the faster the closer to the CP. This is counterintuitive since the heat diffusion time is divertingly large (it scales as $\tau^{-1/2}$ as the CP is neared). A "fourth" heat equilibration mechanism then does exist, which is due to the cumulative effects of the acoustic waves continuously emitted by the heating boundary layer, waves which are much stronger than in an ideal gas. The boundary layer transforms heat into kinetic energy, the latter being transformed in its turn into heat by adiabatic compression of the bulk phase. This is the reason why we have coined this effect the "piston effect" (Zappoli et al., 1990) in reference to the work of Kassoy (1979). The PE characteristic time can be written as follows:

$$t_{PE} = \frac{t_D}{\gamma_0^2 - 1} \tag{5}$$

where t_D is the diffusion time scale in the fluid approaching its CP and $\gamma_0 = \frac{c_p}{c_V}$ is the ratio of the specific heat at constant pressure and constant volume. The PE time scale obeys universal power law $t_{PE} \sim \tau^{3/2}$ since $\gamma_0 \sim \tau^{-1}$ diverges near the CP. Therefore, instead of critical slowing down, "critical speeding up" is rather observed (Boukari et al., 1990). For example, in a van der Waals-like CO_2 sample of 1cm size in the temperature range $T - T_c < 1K$, typical thermalization would take weeks by diffusion and less than one second for equilibration by the PE. Note that a simple energy balance estimation (Onuki & Ferrell, 1990; Onuki et al., 1990) can give the time t_c needed to equilibrate the mean temperature of the boundary layer with the bulk temperature, leading to

$$t_c = \frac{t_{PE}}{\gamma_0^2}. \tag{6}$$

The time close to the critical temperature where $\gamma_0 >> 1$. The temperature evolution of the fluid sample on the PE time scale is given in Figure 2a. The density evolution is displayed on Figure 2b, which clearly shows a strong density decrease in the heating boundary layer (x=0) and a similar density increase in the cooling boundary layer (x=1).

Experimental validation of such a PE behavior is illustrated in Figure 3 obtained from interferometer observation of a near supercritical SF_6 cell using the Critical Point Facility of the European Space Agency

Figure 2. Temperature (a) and density (b) plot on the PE time scale. Time is expressed in seconds and x is normalized to the fluid sample length L.

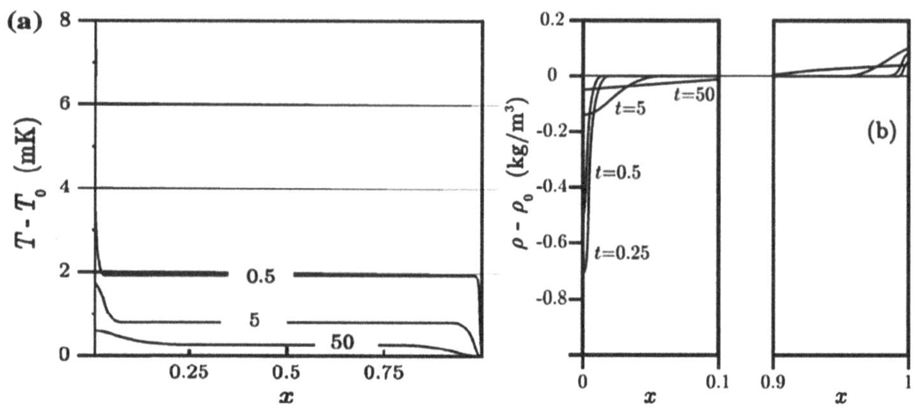

Figure 3. Optical interferometer cell filled with SF_6 confined between two conductive walls at distance L. Temperature of the "hot" wall is maintained constant (HW, $T=T_c+100mK$, $T_c=45.564K$) while that of the "cold" wall has just been decreased by 10 mK (CW, $T=T_c+90$ mK). The fluid is under weightless conditions; the black arrow indicates the heating boundary layer (thickness δ_1) and the white arrow the cooling boundary layer (thickness δ_2).

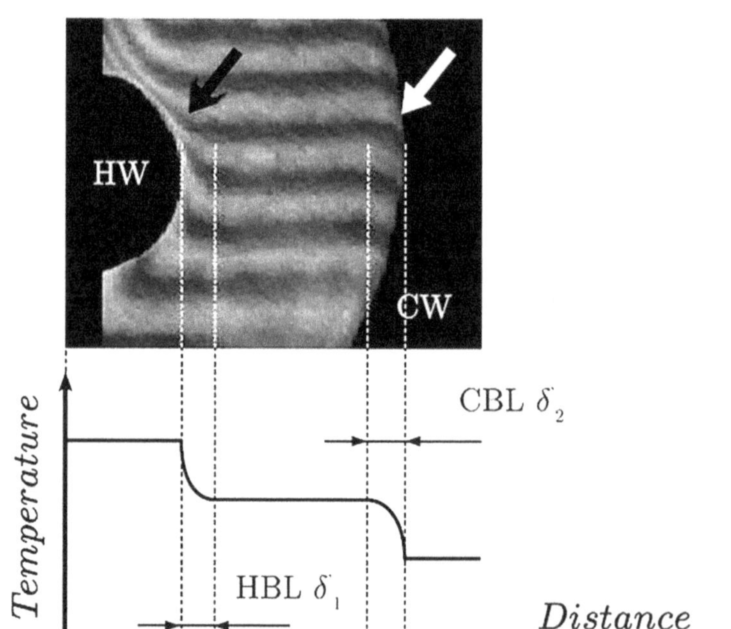

during the IML1 mission (2002). A detailed validation of this 1D modelling of the PE mechanism accounting for true Ising-like fluid singularities (see column 4 of Table 1) are given in Garrabos et al. (1998) for the case of supercritical CO_2.

In the following are summarized the basic properties of the piston effect:

- PE is an adiabatic thermo-compressible effect with faster and faster dynamics when nearing the CP.
- PE is a homogeneous process. The temperature of the bulk phase is homogeneous since it is the result of the superposition of linear compression and expansions waves, as observed in the experiment of Miura et al. (2006).
- PE is an adiabatic temperature equilibration mechanism when the non-heated wall is insulated: compression waves bring energy in the heated wall, no expansion wave takes energy out from the insulated boundary. In such a configuration, the piston effect can be also called "adiabatic effect" (Boukari et al., 1990; Onuki & Ferrell, 1990; Onuki et al., 1990) since the bulk phase only undergoes an adiabatic compression.
- PE is an adiabatic heat transfer process in the case when the non-heated wall is thermostated: compression waves bring energy in at the heated wall, expansion waves take energy out from the thermostated boundary. Such a configuration is typically encountered in the experimental studies of the PE mechanisms (see for example Wilkinson, 1998; Wilkinson et al., 1998; Garrabos et al., 1998; de Briujn et al., 1997); Straub et al., 1995).

It is worthy to note that the PE thermalization process is isentropic, except in the boundary layers where heat conduction at constant pressure takes place. Such a PE process can lead to some particular and remarkable behaviors, as illustrated below in the three last sections.

Density Relaxation

The PE homogenizes temperature on a much shorter time scale than heat diffusion does. During this time period where time scales as $\tau^{-3/2}$, the temperature perturbation decays from its initial order of magnitude, say α, to $\alpha\tau$. During the same period of time, the density perturbation, which is τ^{-1} larger than the temperature perturbation, owing to the fluid large compressibility, decays from its initial order of magnitude $\alpha\tau^{-1}$ to α. During the next period of time (the heat diffusion time, which scales as $\sqrt{\tau}$), the temperature perturbation decays from $\alpha\tau$ to $\alpha\tau^2$, while the density perturbation decays from α to $\alpha\tau$. (Details are given in Bailly et al. (2000)). The process can be summarized in formulating that at first order PE homogenizes temperature while heat diffusion homogenizes density. Experimental validations of the diffusive nature of the density relaxation are given in Garrabos et al. (1998), Wilkinson (1998) and Wilkinson et al. (1998). (The latter reference reports measurements of the thermal diffusion coefficient until 1 mK distance from the critical temperature of SF_6).

In the presence of gravity field, strong convection can occur on the long-lasting heat diffusion period where strong density gradients are still present even though the medium is quasi thermally homogenized. 2-D numerical simulations confirm the presence of long lasting strong convective motion.

The Close Vicinity of the Critical Point

The behavior of the piston effect when going closer and closer to the CP can be summarized as follows. The amplitude of the acoustic waves emitted by the heated boundary layer is larger and larger. Ultimately, after a critical distance is reached, the amplitude of the acoustic wave becomes the same as the temperature increase at the wall (Zappoli et al., 2015). This means that heat propagates at the speed of sound. Then, this compression wave reflects on the thermostated wall. If the reflection process was the same as in an ideal gas, the reflected wave would have been a compression wave and the temperature in the bulk would have been larger than the temperature at the boundary. Instead, the reflection process in the supercritical fluid is reversed and the reflected wave is an expansion wave that takes back the temperature in the bulk to its initial value. The mean value of the temperature increase in the bulk is zero: the fluid behaves as a thermal short-circuit. The reason for this consequence is the following. When interacting with the thermostated wall, the initial, compression wave creates a cooling adaptation layer (thermal boundary layer that cools down the fluid). The fluid in this layer contracts very much since it is very expandable, and creates a flow towards the wall (a flow that does not exist in an ideal gas); the wave behaves as if it interacted with an opened end and reflects as an expansion wave that restores the initial temperature of the bulk fluid following the present theory and hypothesis. The value of the distance to the CP where occurs inversion of the reflection rules and temperature propagation at the speed of sound, is a few mK for CO_2. Note that the presence of non-negligible adsorption layers (see e.g. Zhao et al., 1995) very close to the CP, could also somewhat modify the reflection of the sound waves.

HEAT TRANSPORT PHENOMENA

The PE is thus characterized by the fact that the fluid temperature outside the thermal boundary layers is homogeneous. In a sample heated at one side and maintained at another side at constant temperature, where two boundary layers are present, hot at one end, cold at the other end, the bulk fluid acts as a thermal short circuit, as illustrated above in Figure 2 and Figure 3. A question arises whether it is possible to use this PE as a kind of heat pipe to carry heat on long distances under weightlessness.

Transient Heat-Pipes Under Weightlessness

Experiments (Figure 4) and simulations have been performed with magnetically levitated H_2 and extrapolated to CO_2 (Beysens et al., 2010). One obtains a very fast conduction at short times, its dynamics being only limited by the thermal properties of the materials forming the pipe (Figure 5). The output in terms of ratio of transported/injected powers, does not exceed however 10-30% and is limited by the heat capacity of the pipe. These results remain valid in a large range of temperature around the critical temperature. Note that, as hot and cold boundary layers are diffusing with time, the final stage will be always conduction. This time can be, nonetheless, very long near the CP (see above in section General background): thermal equilibration at 1 mK from T_c of a CO_2 sample with thickness $2L$=1cm would need a time $t_D \approx L^2/D_T \approx$ 1 month. Although these properties look promising to speed up transient thermal exchange, they have not yet given industry-related applications.

Figure 4. Piston effect heat transport in H_2 under magnetic compensation of gravity. Cell head and base are connected to a helium bath via heat conductor's thin wires. Heat is sent from the head and flows to the base. On the right is schematized the spatial variation in the cell of the fluid temperature, with cold and hot boundary layers.

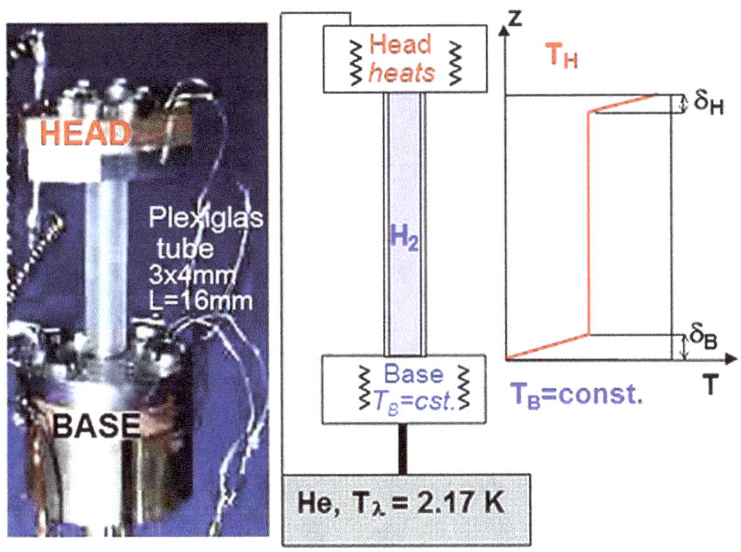

HEAT CAN FLOW BACKWARDS

The PE can be at the origin of paradoxical non-equilibrium behaviors in a heated two-phase fluid tank in weightlessness. During heating from the sample walls a striking overheating of the vapor phase can be observed. The latter becomes hotter than the heating wall in a seemingly violation of the second principle of thermodynamics as detailed in the next sections.

Two-Phase Fluids Experiments and Measurements

In the absence of Earth's gravity, the phase distribution of a closed cell filled at critical density ($\rho = \rho_c$) consists of a vapor bubble of volume fraction close to 1/2, surrounded by the liquid which wets the cell walls. Both pure SF_6 and CO_2 were used in cylindrical samples. They show under weightlessness similar configurations (Figure 6). Part (a) of this Figure is an interferometer image using the Twyman–Green interferometer of the ALICE 2 facility (Marcout et al., 1995) onboard the Mir space station. Part (b) is a schematic cross-section showing the gas and liquid phases in the internal cylindrical volume of a body cell. The latter is made of copper alloy and closed by two sapphire windows, one with a layered reflective mirror. The vapor bubble is slightly constrained by the two windows. Since liquid always wets the walls near the CP, the liquid–gas meniscus between the two parallel windows form a semi-circular interface in the plane perpendicular to the windows. This interface appears dark in the image Figure 6a because the liquid–gas meniscus refracts the incident light away from the axis of the cell. The detailed descriptions of this experimental arrangement can be found in Garrabos et al (1999).

Figure 5. Heat transport (solid line) measured in the H_2 experimental configuration of Figure 4. The fluid is at near critical density at T_c + 0.1 K (T_c = 33 K). The heat diffusion time is 2.2×10^5 s and the piston-effect time is 3 s. A constant power (Q_0 = 7.5mW = 1 kWm^{-2}) is initially supplied to the head (dashed squared curve). Q is the power received by the base. Heat conduction by the sample walls when the sample is empty (Q_v, broken line) is also shown for comparison. The diffusion time in Plexiglas is 450 s. The difference $Q-Q_v$, which qualitatively represents the heat transfer by H_2 alone, is also plotted. The arrows represent the behavior at short times for a filled cell (solid arrow) and an empty) cells (dotted arrow), highlighting very fast heat transport at short time

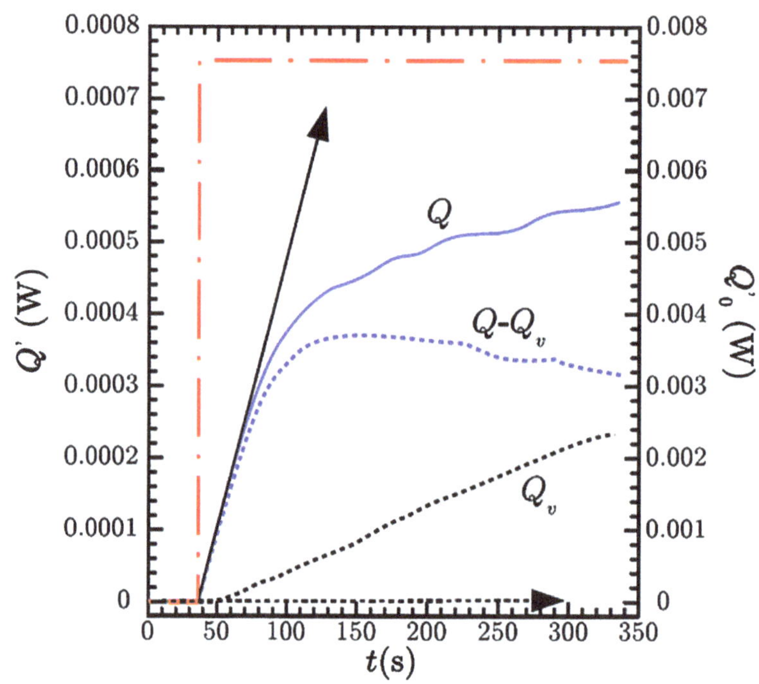

Figure 6. Interferometer image of the cell (thickness 6.7 mm) used to perform in situ temperature measurements White circles show the positions of the thermistors Th1, Th2 and Th3, whose measurements are labeled T_L^1, T_L^2 and T_V, respectively. (b) Schematic cross-section of the cell showing the constrained gas and wetting liquid phases.

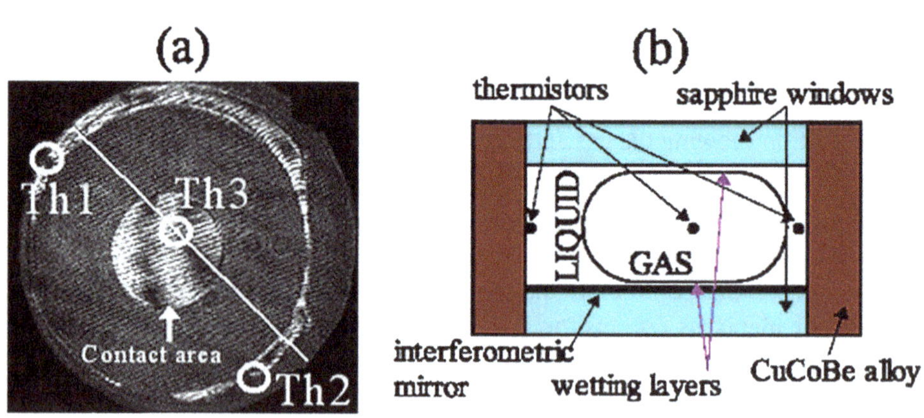

The vicinity of the CP enables the thermo-physical properties of the fluid to be varied continuously by simply changing temperature, thus taking advantage of the well-known universal scaling laws of critical phenomena. The experiments consisted in performing *in situ* temperature measurements during temperature steps of the sample cell to reach the final temperature (labeled T_f) (Wunenburger et al., 2000). A series of rapid, positive temperature quenches of amplitude $\Delta T = T_f - T_i = 100\,\text{mK}$ and 50 mK were performed for initial temperatures T_i ranging from amplitude $T_c - 10.1\text{K}$ to $T_c - 0.1\text{K}$. A quench of amplitude ΔT consisted of a sharp linear increase of the temperature of the cell walls (labeled T_W) up to 80% of ΔT in less than 10 s, followed by a smooth evolution up to T_f.

Temperature is measured in the cell wall (T_W) and in the fluid at three locations (Figure 6). Th1 and Th2 are located close to the cell wall (at a distance of about 1 mm), and are always in the liquid phase. The temperatures measured by Th1 and Th2 are labeled T_L^1 and T_L^2. The third thermistor (Th3) is mounted in the center of the cell, always in the vapor phase. Temperature measured by Th3 is labeled T_V. The evolutions of the temperature rises at the cell wall $\delta T_W = T_W - T_i$, in the two liquid parts $\delta T_L^1 = T_L^1 - T_i$ and $\delta T_L^2 = T_L^2 - T_i$, and in the vapor $\delta T_V = T_V - T_i$, during a temperature quench of $\Delta T = +100\,\text{mK}$ from $T_i = T_c - 10.1\text{K}$, are shown in Figure 7.

At the beginning of the quench, δT_W reaches 90% of ΔT in less than 10 s, and its evolution is linear for at least 5 s (Figure 7). During this initial time period (*t* = 0-25 s), the evolutions of δT_L^1, δT_L^2, and δT_V are also found to increase almost linearly over time for at least 5 s (Figure 7), with T_L^1 and T_L^2 increasing less rapidly than T_V and remaining smaller than T_W. It was pointed out by Onuki & Ferrell

Figure 7. Temperature rises at the cell wall $\delta T_W = T_W - T_i$, in the liquid $\delta T_L^1 = T_L^1 - T_i$ and $\delta T_L^2 = T_L^2 - T_i$, and in the vapor $\delta T_V = T_V - T_i$ (see text), during a quench of amplitude $\Delta T = +100\,mK$ from the initial temperature $T_i = T_c - 10.1K$

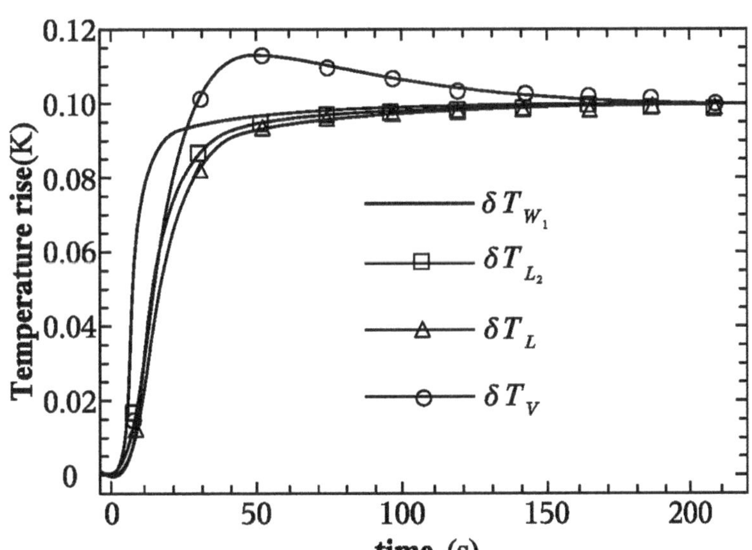

(1990) that, during the short period when the PE is effective, neither heat nor mass can be transferred between a vapor and a liquid that are initially coexisting, because the thermal diffusivity is very weak near the CP. Consequently, liquid and vapor phases should behave as if they were independent and the temperature rise in the bulk liquid (δT_L) and the bulk vapor (δT_V) should be homogeneous and isentropic. Here "bulk" is the fluid region beyond the thermal diffusion layers, which is heated only by adiabatic compression. Therefore, in each bulk (liquid or vapor) phase of homogeneous temperature labeled $T_{L,V}^{\text{bulk}}$, the ratio $\frac{\delta T_V}{\delta T_L}$ is expected to be

$$\frac{\delta T_V}{\delta T_L} = \frac{\left(\frac{\partial T}{\partial p}\right)_S^V}{\left(\frac{\partial T}{\partial p}\right)_S^L}. \tag{7}$$

Here pressure is labeled *p*, the subscript *S* means a derivative at constant entropy and subscripts *V* and *L* denote the adiabatic derivative of *T* with respect to *p* in one-phase vapor and one-phase liquid. Since $\left(\frac{\partial T}{\partial p}\right)_S^V > \left(\frac{\partial T}{\partial p}\right)_S^{VL}$, vapor should then be more rapidly heated than the liquid - which is actually observed. However, the comparison of the experimental data with the theoretical prediction Equation 7 shows that the data are smaller than the calculated values. This disagreement is presumably due to a weak heat and mass exchange between liquid and vapor during the adiabatic compression. According to Onuki (1991), the rate of vaporization of the liquid is negative at the start of adiabatic heating. Hence, the adiabatic heating process causes vapor to condense at the interface at the beginning of the temperature rise. This condensation was numerically observed by Zhong & Meyer (1999). Due to condensation, vapor is thus less heated than the liquid, leading to smaller temperature rise than deduced from Equation 7.

Vapor Overheating

Before the end of the quench, vapor temperature becomes much higher than that of the heating wall. Then T_V decreases slowly toward its final value, T_f . Similar observations are made during quenches of different amplitudes. This striking vapor-overheating phenomenon, in apparent contradiction with the second principle of thermodynamics where heat has to flow from hot to cold, can be explained as follows. The vapor temperature rise, δT_V , is firstly scaled with the cell wall temperature rise, δT_W , for more convenient comparison between quench amplitudes, and the behavior of the maximum $\left(\frac{\delta T_V}{\delta T_W}\right)_{\text{MAX}}$ is plotted as a function of $T_c - T_i$ (Figure 8). A maximum of 123% is observed around $T_c - T_i \approx 6$ K. The vanishing differences in thermophysical properties between vapor and liquid at the CP imply that - asymptotically near the CP - vapor and liquid exhibit the same thermal response. On the other hand, far from the CP, adiabatic heating efficiency is reduced and heat transfer is mainly diffusive, i.e., a situation

Figure 8. Vapor overheating measured by $\left(\frac{\delta T_V}{\delta T_W}\right)_{MAX}$ *as a function of* $T_c - T_i$. *Full squares +100 mK quenches. Full triangles +50 mK quenches. The occurrence of such a high level of overheating is due to the geometry of the liquid and vapor phases.*

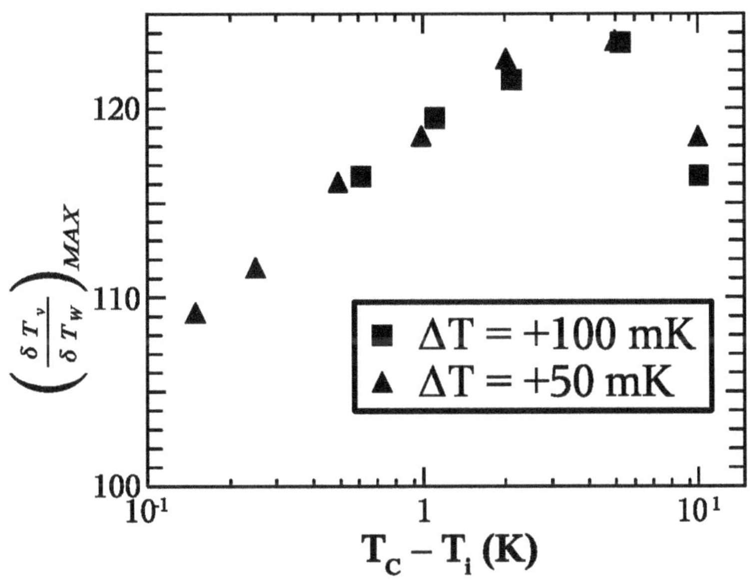

that prevents vapor from being overheated. The existence of the overheating maximum at $T < T_c$ is thus due to the competition between these two limiting behaviors.

In the above cell configuration, the vapor bubble is almost completely isolated from the thermostated cell walls by the liquid (the area of contact between vapor and windows is less than 6% of the overall heating area). During the quench, the adiabatic heating process stops when T_L^{bulk} has reached T_W (there is no longer any temperature gradient at the boundary). Before equilibration of T_L^{bulk} and T_W, the bulk vapor is heated more than the bulk liquid by the homogeneous pressure increase. Since the vapor bubble is not in contact with the heating wall, its temperature has no influence on the temperature gradient that drives the expansion of the liquid hot boundary layer (HBL); T_V^{bulk} can thus exhibit a large overshoot.

The situation is different on Earth, where gravity causes the vapor to be in contact with the upper part of the cell. The HBL developing in the vapor stops expanding when the bulk vapor temperature reaches T_W. When $T_{L,V}^{\text{bulk}}$ exceeds T_W, the vapor HBL may even contract (cooling PE) if the liquid HBL continues to expand in order to compensate for vapor overheating. Numerical simulations (Zhong & Meyer, 1996; Wunenburger et al., 2002) confirm this influence of these two-phase distributions on the temperature evolution with or without Earth's acceleration.

Figure 9. Temperature dependence of the characteristic time t_R of the exponential relaxation of gas temperature (log-log plot). Full squares +100 mK quenches. Full triangles +50 mK quenches. Continuous line is a fit to Equation 9

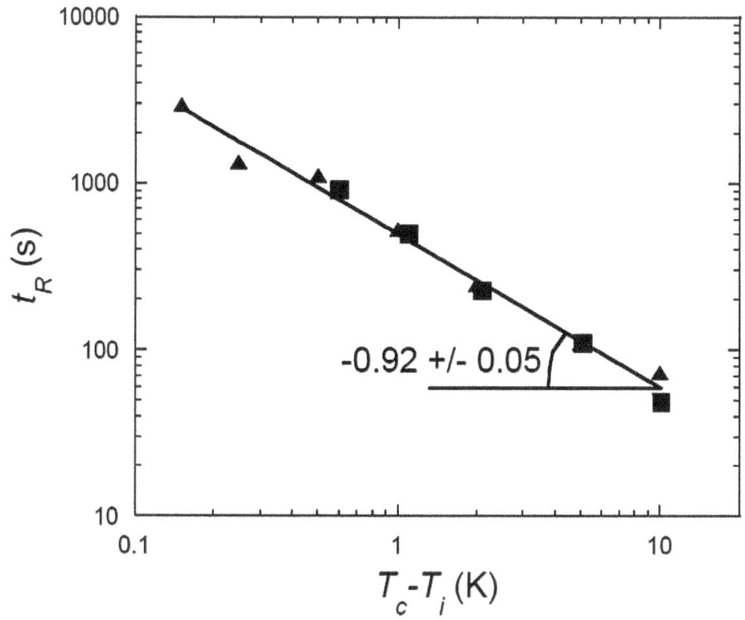

Vapor Temperature Relaxation

The long-time thermal relaxation of T_V towards T_f after the maximum overheating follows an exponential law

$$T_V - T_f = \theta e^{-\frac{t}{t_R}}, \tag{8}$$

with amplitude of θ and relaxation time t_R (Wunenburger et al., 2000). This analytical form is justified by the fact that, at long times, only the (exponential) relaxation of the slowest thermal mode can be observed (Behringer et al., 1990).

The variation of t_R as a function of $T_c - T_i$ (see Figure 9) follows the equation

$$t_R = t_0 \left(T_c - T_i\right)^x. \tag{9}$$

Here, t_0 and x as free fitting parameters. The best fit value of the exponent is $x = -0.92 \pm 0.05$, which appears to be comparable to the measured value (-0.89) of the effective critical exponent of the thermal diffusivity for both vapor and liquid (Jany & Straub, 1987). It can be thus concluded that the relaxation of the temperature inhomogeneities generated by the adiabatic heat transfer is diffusive, although the density inhomogeneities are large. This is due to the thermal isolation of the gas bubble from the cell walls by the liquid layer. The overheating of the gas is not seen by the walls, so fast relaxation by the cold PE is prevented.

EFFECT OF AN ACCELERATION FIELD

Under acceleration or under earth gravity fields, convection phenomena occur in one phase, supercritical fluids. These convection phenomena couple with the PE. The thermal processes can be much affected and lead to quite peculiar behaviors, as described in the following.

Buoyant Convection in a Squared, 2-D, Side Heated Cavity

The simplest configuration to study the basic properties of buoyant convection is a 2-D side heated cavity. A 1 cm high 2-D cavity containing a critical fluid set at 1 K above its CP on the critical isochore is heated from its left-hand side boundary (the temperature is raised by 10 mK in 1 s), the others being thermostated. The following properties are outlined (see Zappoli et al., 1996).

The Piston Effect Period

During the PE period, a strong density gradient forms at the heated boundary. A single roll convective pattern is displayed in Figure 10, whose evolution is much faster than in an ideal gas. The convective motion is generated by the very large density gradient even though the temperature shows little differences. Note that the fluid velocity (on the order of 1 mm/s) is much larger than usually found in an ideal gas (some μm/s). The topology of the convective motion is also very different: a strong uprising flow located close to the heated wall in the near supercritical fluid, a circular roll that takes the whole cell in the ideal gas.

Figure 10. The velocity field is plotted for the same heating conditions in (a) a ideal gas and (b) in a near supercritical fluid. (Note that the velocity plot is 6.5 μm/s in (a) whereas it is 2.2 mm/s in (b))

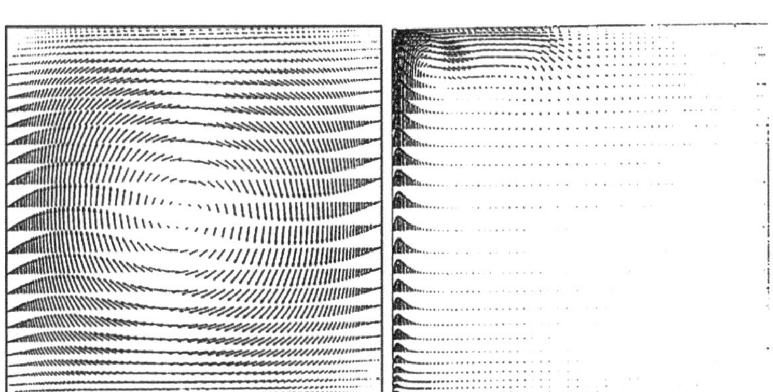

(a) (b)

The Heat Diffusion Period

During the heat diffusion period the convective pattern is quite different than in an ideal gas in the same conditions. Its duration is longer the closer to the CP, several tenths of minutes at 1K from the CP. As a matter of fact, although the temperature is homogenized by the PE, the remaining minute temperature gradients relax by heat diffusion, which is divergently long. The associated density gradients are still of first order because of the high fluid compressibility. They generate the observed strong and long lasting convective motion. However, whereas the convection topology is unchanged compared to a ideal gas, a double counter-rotative roll appears. The reason why this happens is the following. As the strong above mentioned uprising single roll impinges the top wall in the vicinity of the upper left corner, the temperature rises at the stagnation point on the top, insulated wall. The fluid is hotter than the heated wall because the kinetic energy is transformed into internal energy. In the region where the fluid temperature overpasses the wall temperature, the heated wall cools down the fluid that moves down, forming the second roll that becomes larger and larger as heat diffusion plays, to form the above mentioned long lasting double roll pattern displayed on Figure 11.

Rayleigh-Bénard Convection

Rayleigh-Bénard configuration is the most studied configuration for buoyancy-induced convection. We study below the specific topics induced by the PE in this configuration.

Specific Configuration

A squared shaped cavity filled with a supercritical fluid is heated from below and its top wall is maintained at the constant, initial temperature (see Zappoli et al., 2015). The side walls are insulated (Figure 12).

Figure 11. Velocity field (a) and temperature field (b) about 700 s after heating has stopped: the nearly isothermal convection field is still on the order of 0.1 mms^{-1}, which has to be compared to the maximum velocity, 6.5 mms^{-1} in an ideal gas in the same conditions, just after heating has stopped.

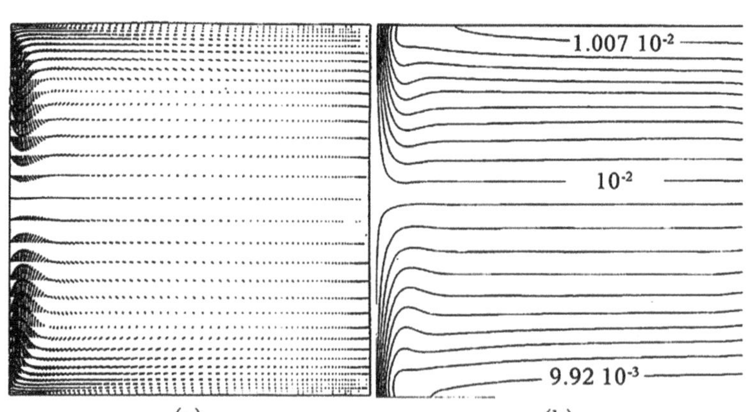

Figure 12. Rayleigh-Bénard configuration. The height of the cavity is L=10 mm. Gravity is collinear with the y-axis. The fluid is set above its critical temperature. Characteristic zones are visualized: an isothermal bulk in the core region and two, hot and cold, boundary layers along the bottom and top plates, respectively. The evolution of the layers is shown by arrows. Thickness and temperature difference in the hot (bottom) boundary layer are h and δT, respectively.

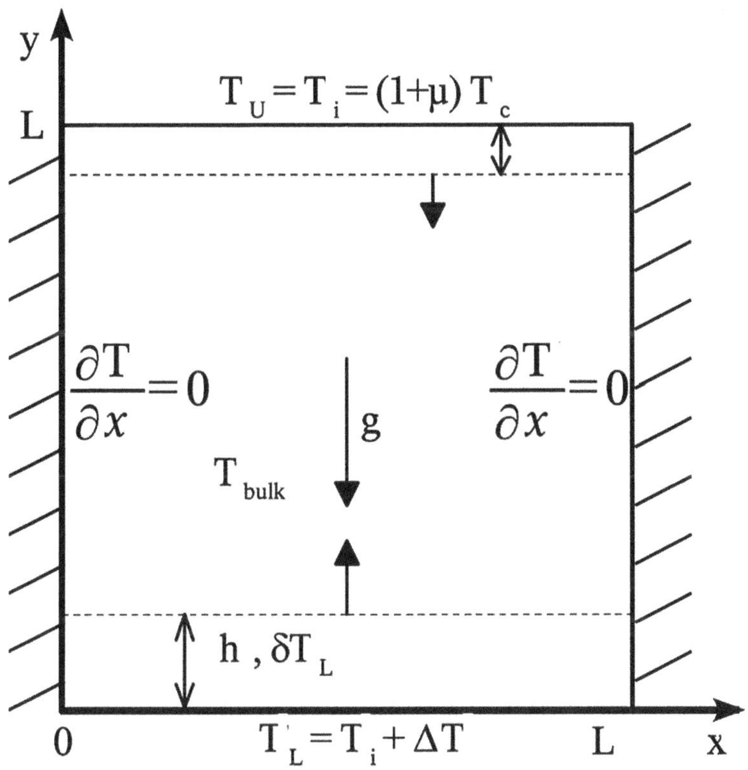

Convective stability of the cavity does not depend on its height because the PE homogenizes the bulk temperature independently of height. Instead, the stability of the system is driven by the stability of the bottom layer since the temperature difference is larger than in the top layer, both having the same thickness equal to the diffusion length $\sqrt{D_T t}$. It follows from Gitterman & Steinberg (1970) that the threshold temperature difference for the instability onset can be written as:

$$\delta T_{onset} = Ra_c \frac{\lambda\mu}{g_0 h^3 \beta_\rho \rho^2 c_p} + \frac{g_0 T \beta_p h}{c_p} \tag{10}$$

Ra_c is the critical Rayleigh number equal to 1100 for a fluid layer in contact with a solid surface at the bottom and stress-free at the top. The latter case is considered in the following. β_p and β_ρ are the isobaric and the isochoric expansion coefficients, respectively. Far from the critical temperature (further than 1 K) the stability is driven by the Rayleigh criterion, the first term on the right hand side of the above equation. The second term corresponding to the adiabatic temperature gradient, the

Figure 13. Temperature onset δT_{onset} (K) as a function of height h (m) of the hot, bottom layer as given by the stability criterion expression. The decreasing branch of the stability curve is the Rayleigh line, the increasing part is the Schwarzschild line. The curves with arrows represent the evolution of the temperature difference between the layer and the bulk for different values of the bottom plate temperature rise ΔT. As the PE heats up the bulk fluid, the temperature onset decreases and the bottom layer height (heat diffusion) increases. The crossing of the arrow curves with the stability curve gives a good estimate of the temperature difference and the corresponding height at the onset in this bottom layer (the minimum is 65 mK)

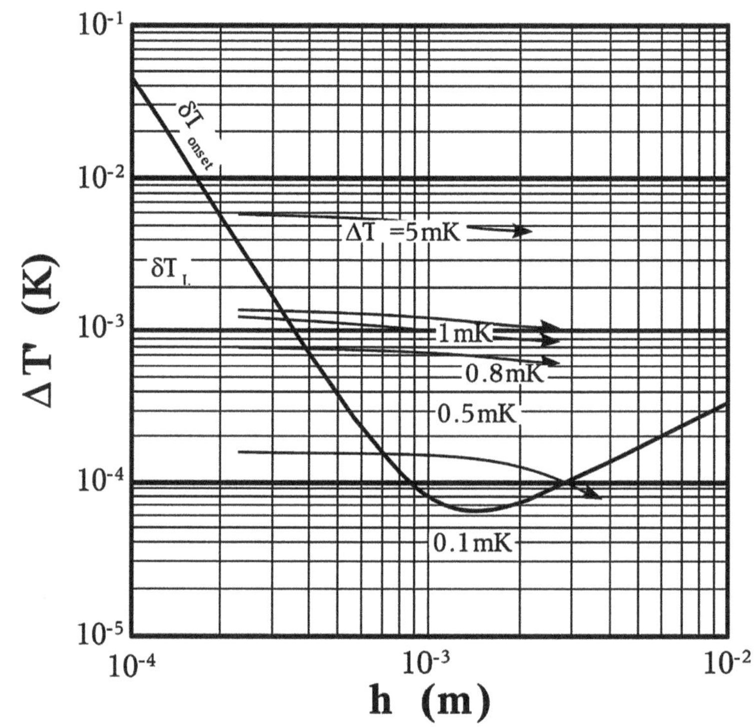

decrease in temperature due to the adiabatic cooling of a fluid particle rising in a hydrostatic pressure gradient. Close to T_c this term shows small variations since the isobaric expansion coefficient and the heat capacity at constant pressure diverge at the same rate as the CP is approached. When nearing the CP, the first term becomes asymptotically small and the stability criterion is given by the second term (adiabatic gradient).

When the bottom wall temperature is raised from values ranging from 0.01 to 10 mK over time periods of 0.1 to 1 s the heated layer thickens according to the very slow heat diffusion process. Using the above equation the stability curve is plotted giving the temperature difference between the top, stress-free boundary and the heated bottom plate, as a function of the height of the layer. The representing point follows the curves with arrows in Figure 13. For small values of the temperature increase of the bottom wall (0.1 mK for example) the convection can start by crossing the Rayleigh line and fades away when crossing the Schwarzschild line.

Figure 14. Temperature pattern and relevant velocity fields at different times during the heating process for an increase of temperature of 0.5 mK of the lower boundary. Note the onset of six thermal plumes. The maximum velocity at 39.2 s is 0.61 ms^{-1}, and the corresponding Reynolds number calculated with respect to the height of the layer is Re=8.3.

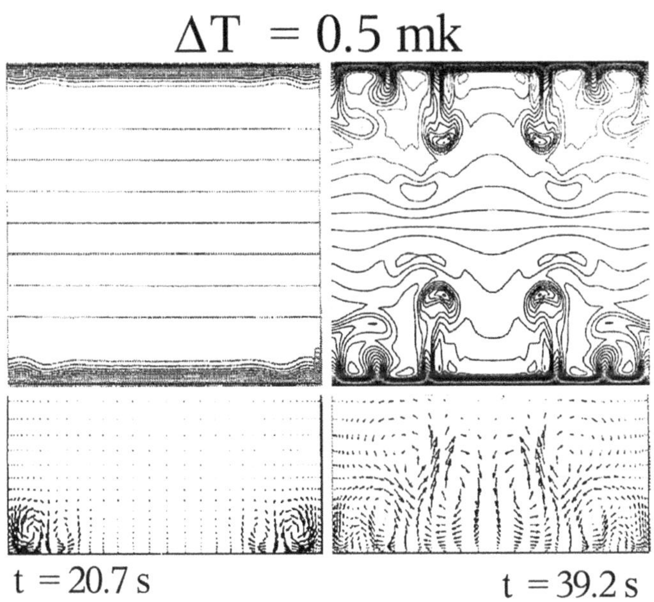

A typical convective pattern is shown in Figure 14. In addition to the upward-moving plumes are found downward cold plumes from the upper, cold layer in the bulk cavity. The stability of this layer has not been studied analytically. No model involving a rigid upper wall and a stress free boundary below are currently available. The numerical studies have shown that it is more stable than the lower layer.

Specific Behaviors

As shown in Figure 15, both Rayleigh and Schwarzschild criteria are relevant for comparable length scales. Striking non-Bousinesq effects can be observed. First, the reverse transitions to stability through the Schwarzschild line (Accary et al., 2005); second the destabilization of a layer of a few mm thick by crossing the Schwarzschild line, as observed in atmospheric sciences (Accary et al., 2007).

Reverse Transition to Stability Through the Schwarzschild Line

A 15 mm high cavity with periodic boundary conditions is used to mimic an infinite width. A critical isochoric sample of CO_2 is set at 1 K above its critical temperature. The bottom plate temperature is increased by a few mK in 1 s, the top plate is maintained at the initial temperature. For such a weak heating, the convective transfer caused by the vortices does not deform the isotherms very much. The boundary layer keeps growing and stabilizes again after crossing the Schwarzschild line. The scenario is shown on Figure 15 as the reverse transition to stability (RTS) path. Snapshots of the corresponding temperature field are given in Figure 16a, b, c, and d.

Figure 15. Evolution of the temperature difference δT in the boundary layer as a function of its thickness h, according to the Rayleigh criterion (Rayleigh line) and the Schwarzschild criterion (COS, Schwarzschild line). The return to stability corresponds to the RTS path (see text), and the convective onset according to the Schwarzschild criterion corresponds to the COS path. Each open circle corresponds to the time when the instability of the bottom plate vortices increase exponentially. Points A and D represent the time when the boundary layer convectively collapses. The filled circles A, B, C and D correspond to the snapshots of the temperature field given in Figure 16.

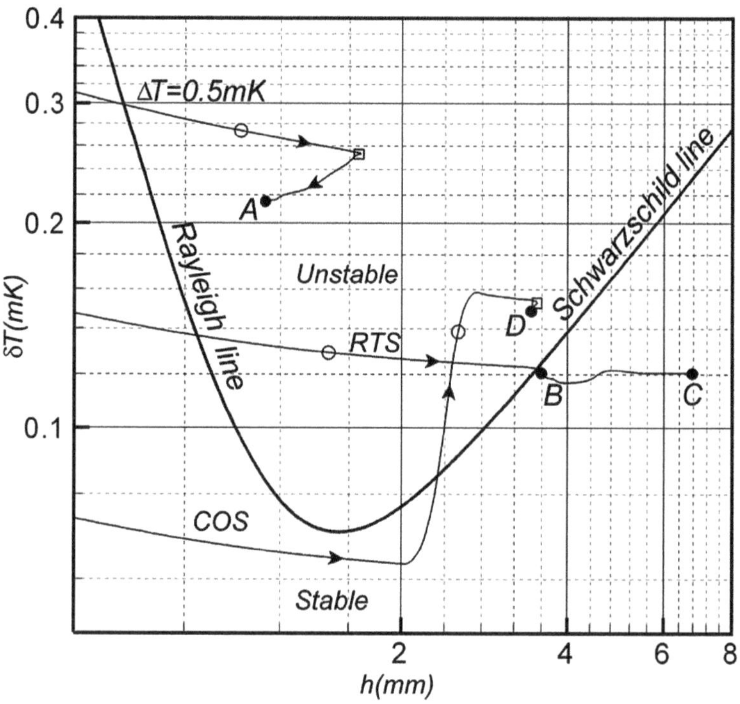

Destabilization Through the Schwarzschild Line

The onset of convection according to the Schwarzschild criterion (COS) requires a two-times phase heating process that corresponds to the COS line in Figure 16 (Accary et al., 2007). In the first phase, a constant increase of 0.12 mK in the lower plate temperature is imposed. For this weak heating the evolution of the curve $\delta T(h)$ remains in the stable zone. This first phase is maintained as long as necessary to bring $\delta T(h)$ under the Schwarzschild line ($h \approx 2mm$, Figure 15). This phase duration is 58 s. The second heating phase consists of increasing the temperature to 0.3 mK in 87 s (according to a cosine law). The new heating phase induces a new homogeneous temperature increase by the PE. The second heating phase is slow enough to prevent convection inside the new boundary layer according to the Rayleigh criterion. However, this second heating phase is fast enough for the temperature gradient inside the modified first boundary layer to exceed the adiabatic gradient. Once the curve $\delta T(h)$ has crossed the Schwarzschild line, convective cells appear in the first boundary layer, not in the second. At the point noted with a circle in Figure 15, the intensity of the vortices increases exponentially with time, producing the collapse of the boundary layer (Figure 16d).

Figure 16. Snapshots of the temperature field, from left to right, from top to bottom, (a, b, c, d). (a) Convection at t= 69.2 s for $\Delta T = 0.5 mK$ (point A in Figure 15. (b) Beginning of the restabilization process at t=196.9 s for $\Delta T = 0.24 mK$ (point B on the RTS line in Figure 15. (c) Restabilized pattern (point C in Figure 5). (d) Convection at t=238.6 s for the second heating phase (point D on the COS line)

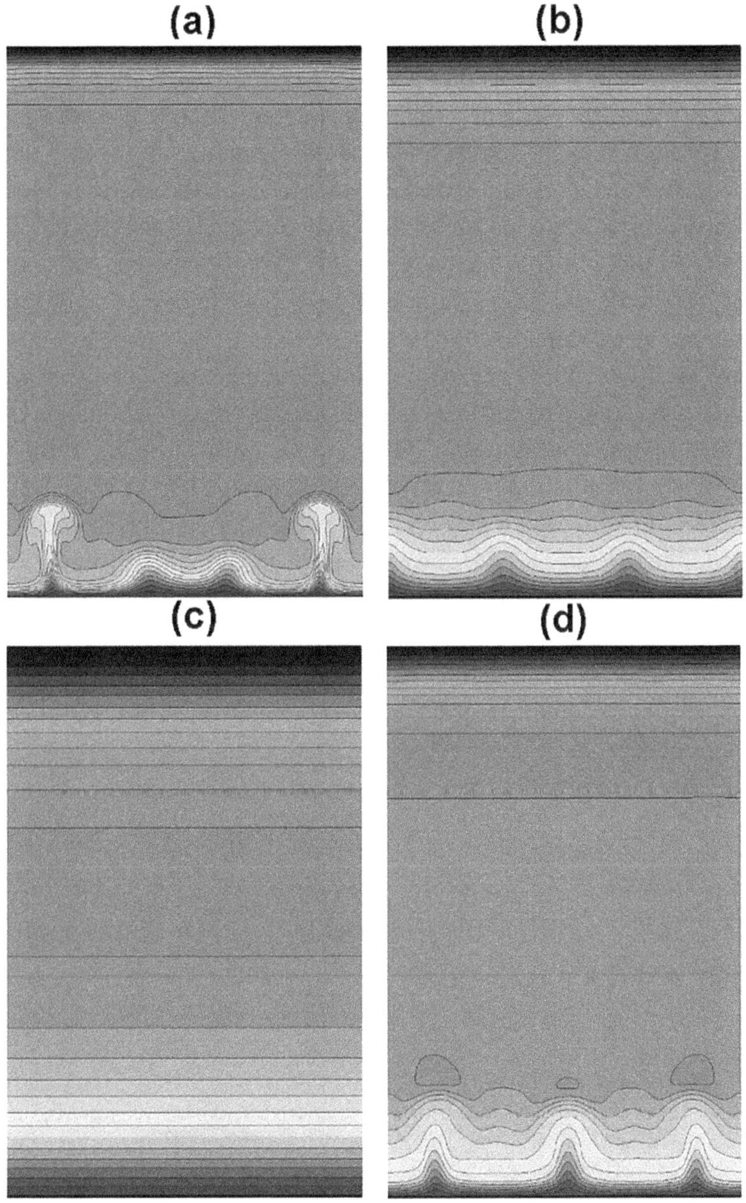

Heating-Induced Cooling Behavior Under Acceleration Field

This last section describes and analyzes thermalization experiments (Beysens et al., 2008; Beysens et al., 2011) performed under weak acceleration field or Earth acceleration. A near supercritical fluid sample is used in a one-phase configuration closer to most situations currently encountered in the industrial context of the heat exchangers or tankers. When heat pulses are applied to the fluid under acceleration or in the Earth's gravity field, the hot boundary layer around the heat source rapidly disappears by convection. However, the cold boundary layer can persists longer and makes pressure decrease. It results a long thermal behavior at the end of the heat pulse where the bulk fluid temperature reaches significantly *below* the initial temperature. This unconventional cooling, created by the fast decompression of the fluid, is observed in a large range of temperature, density, and under various thermodynamic conditions. Cooling can persist for an appreciable period of time corresponding to the diffusive destruction of the cold boundary layer and is also more pronounced when the free fall acceleration is large and/or the fluid is close to T_c. The process can be analyzed by using a simple one-dimensional model with *ad hoc* convective heat losses as explained below.

Experiments were carried out with fluids (SF_6 at liquid-like density or CO_2 at gas-like density) confined in a temperature-controlled (cylindrical) optical interferometer sample, as already described in Figure 16 above for a two-phase configuration. Two or three thermistors, depending on the setup, were used to measure the temperatures in the bulk fluid or close to the HBL or the CBL. One (labeled Th1) was used to send a heat pulse between 10 and 100 mW in the near-central region of the fluid. The pulse duration can be varied. The density variations were visualized using a Twyman-Green interferometer.

In such configurations, the thermal behavior is governed by the balance of the heat flux between (i) the heat source and the diffuse HBL, which compresses the bulk fluid, and (ii) the thermostated walls and the diffuse CBL, which depresses the bulk fluid, as schematically illustrated in Figure 17a in zero gravity. In the presence of an acceleration field, the thickness of the HBL is not or is only weakly affected by convection, as expected in Figure 17b under an acceleration γ and shown in the experiments described below on Earth or under maneuvers, rotations and positioning corrections of a Space Shuttle.

Figure 17. Schematic of thermal behavior under an acceleration γ for a fluid confined in a sample at constant temperature T_i. (a) $\gamma = 0$. The temperature in the bulk is the result of the balance between a hot PE due to the expansion of the HBL and a cold PE arising from the contraction of the CBL. (b) $\gamma \neq 0$. The HBL is convected to the wall and the CBL thins out at the zone of contact [$(CBL)_c$], increasing the cooling effect

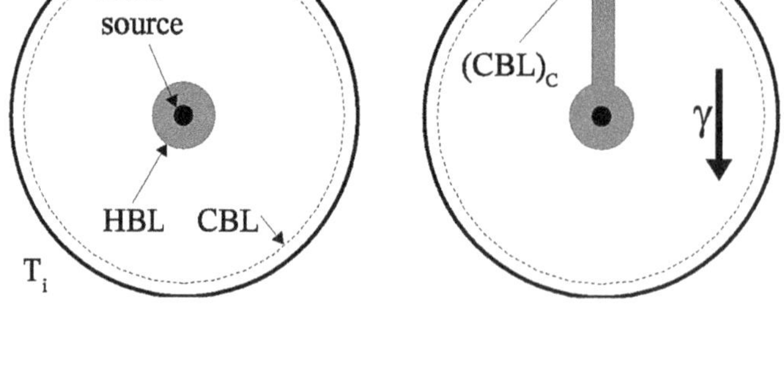

In contrast, the CBL becomes considerably thinner under convection, which appreciably increases the heat losses and lowers the amplitude of the PE.

When heating stops, the compression caused by the HBL also stops. However, the decompression caused by the CBL is still present and vanishes only through a diffusion process. The pressure therefore drops, which can lead, as we will see through Figures 19 & 20 below, to a bulk fluid temperature *below* the thermostat temperature. This cooling phenomenon due to a coupling between the buoyancy convection of a hot thermal boundary layer and the PE, is observed for a significant period of time, corresponding to the diffusive vanishing of the CBL. In this sense, heating the fluid results in cooling, which is also in apparent contradiction with the second law of thermodynamics. This is another spectacular illustration of the thermo-mechanical nature of the heat exchange by the PE which complements the above behavior (section: Heat can flow backwards above) observed in a two-phase, gas–liquid sample where the gas becomes hotter than the heating wall (Wunenburger et al.,2000a; Wunenburger et al., 2002; Beysens et al., 2002).

Figure 18. (a) $t = 0$, homogeneous near supercritical state for the off-critical SF_6 ($\rho = 1.27\rho_c$; $T = T_{coex}$ +0.95K). Black circles: highlighted position of the thermistors. Th1: Heating and measuring thermistor. Th2, Th3: Measuring thermistors. (b-c) 100 mW, 15 s heat pulse sent into the fluid sample under Earth's gravity (1 g). (b) t = 15 s, end of the heat pulse. (c) t ≈ 20s, relaxation towards equilibrium. (d-f) 100 mW, 15 s heat pulse is sent into the fluid sample under a vector acceleration (indicated by the arrow) of modulus $\gamma = 3\times10^{-2}$ g as provided by the Space Shuttle Columbia (STS-55, IML-2, July 1994). (d) t=5 s, the HBL develops symmetrically around Th1. (e) t=20 s, i.e., 5 s after the end of the heat pulse. The HBL is convected towards the cell wall, in the direction perpendicular to the Th2, Th1, Th3 thermistor alignment. (f) t=42 s, during the relaxation period after the end of the heat pulse.

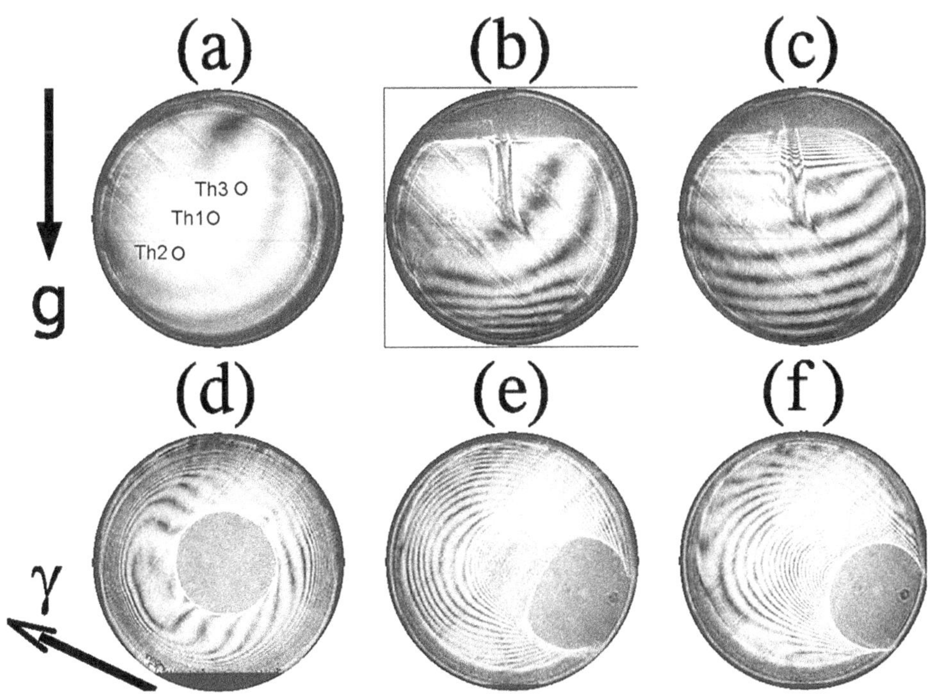

A series of images given in Figure 18 illustrate experimental observations where a near-central heat source is used in a thermostated cylindrical one-phase supercritical sample typically 12 mm in diameter and 6.8 mm thick, similar to the two-phase near supercritical sample shown in previous section: Heat can flow backwards. The fluid is SF_6 at off-critical density $\rho = 1.27\rho_c$ and the initial equilibrium temperature $T = T_{coex} + 0.95\,K$, where T_{coex} is the liquid-vapor coexisting temperature (see Figure 18a). The hot source is a thermistor (Th1), 0.25 mm in diameter, which is in the central position of the fluid sample. The heating power is 100 mW, supplied during 15 s. Th1 is running as a thermometer once the heating has stopped. During the experiment, the bulk fluid temperature is measured by two similar thermistors (Th2 and Th3), 0.25 mm in diameter, which are in the central plane and within ~2.5 mm in the symmetrical opposite direction with respect to Th1.

On Earth, the evolution of the thermal boundary layers during a 15 s heat pulse is shown in Figures 18b-c. During the early stages, a thin HBL develops at the thermistor surface and becomes unstable (Figure 18b). The layer then remains at a constant thickness of 0.2-0.3 mm and hot, expanded fluid is convected (Figure 18c), leading to an accumulation of hot fluid in the upper part of the sample cell. In the part of the cell which is not perturbed by convections, a uniform shift in the fringes shows that density and temperature homogeneously increase in the bulk, demonstrating the presence of the PE.

This behavior is qualitatively the same as the behavior seen during much weaker acceleration ($\gamma = 3 \times 10^{-2} g$), as provided by the rotation of the Space Shuttle (Figure 18d-f). During convection, the HBL remains nearly spherical.

Temperature measurements after the heat pulse, as taken by Th2 and Th3, are shown in Figure 19. A clear temperature decrease, about 2 mK below the cell wall temperature, is observed 60 s after the end of the pulse. At the beginning of the relaxation period (Figure 18e), the HBL is still convected, as it remains hotter than the bulk, explaining the cooling effect.

Figure 19. Time responses of thermistors Th1 and Th2 during relaxation after the 100 mW, 15 s heat pulse of Figure 13d-e with 3×10^{-2} g acceleration. A temperature undershoot below the thermostat temperature $T_i = T_c + 0.95$ K is observed.

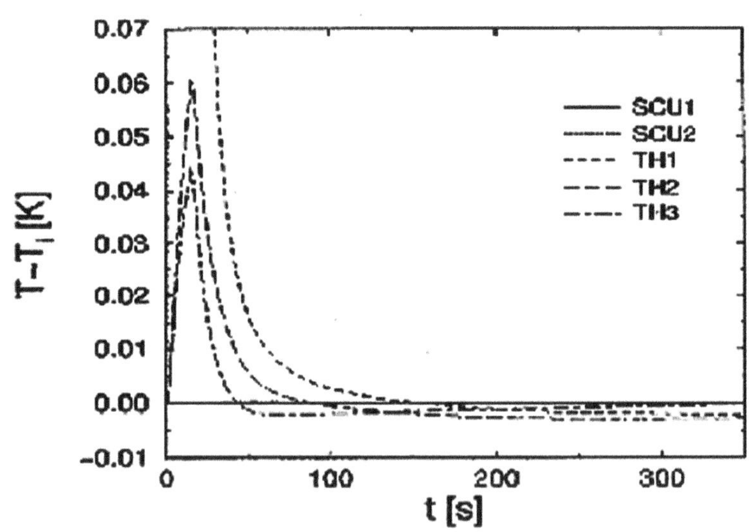

The whole evolution can be well reproduced by a simplified, phenomenological 1-D model (Beysens et al., 2008; Beysens et al., 2011) that only retains the energy balance equation already used by Boukari et al. (1990). As the pressure drop due to the thin CBL persists longer than the pressure rise due to the HBL, which stops with heating, the influence of the thin CBL was accounted for with a "convective" heat loss due to an *ad hoc* additional cold source, i.e., by including the following negative heat source in the simplified energy balance equation:

$$q_{HBL}\left(r,t\right) = C_c\left[T\left(r,t\right) - T_b\right] \tag{11}$$

Such a cold source acts only in the HBL region and its convective contribution is proportional to the difference between the local temperature $T\left(r,t\right)$ and the homogeneous bulk temperature T_b. The pref-actor parameter C_c is phenomenological and accounts for the amplitude of the cooling effect. It must be determined from measurements to adapt the model in order to preserve the ratio of heating (thermistor bead) to cooling (cylindrical body plus plane windows) surfaces and the mass conservation from the spherical to the cylindrical transformation of the sample geometry.

This model well describes the data obtained both on earth (using $C_c = 1\,\text{W}K^{-1}$ to account for a strong 16 mK cooling, 12 s after the end of the pulse, see Figure 20) and in the presence of small accelerations (using $C_c = 0.006\,\text{W}K^{-1}$ to account for a weak cooling effect of 2 mK, 60 s after the end

Figure 20. Time response on earth of the bulk temperature thermistor (dots) during relaxation after having sent a 100 mW, 4 s heat pulse in the middle of a CO_2 sample. Fluid density is off by 18% from the critical density; cell walls are maintained at 16.5 K from T_c. Temperature clearly undershoots below the thermostat temperature. The line is a 1D simulation using Equation 11.

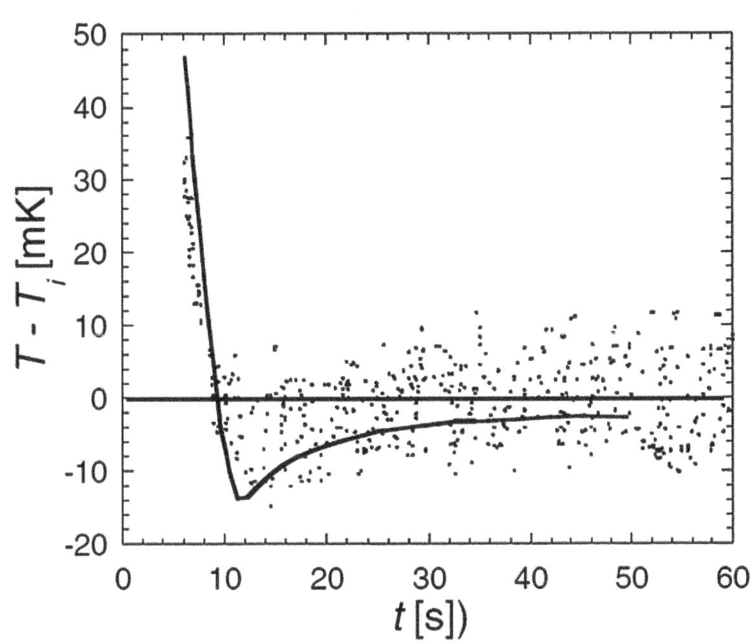

of the pulse, see Figure 19). In the case of Space Shuttle maneuvers, small accelerations, between 10^{-4} and 10^{-2} times the earth acceleration amplitude, act as "gravity" pulses with accurately determined start and end times. The coupling between the buoyancy convection of a hot thermal boundary layer and the PE can then be analyzed in detail, which contrasts with the investigation on Earth, where convection dominates the thermalization process, even at very short times.

This effect can have applications to describe transient behavior of near supercritical fluids stored in space during maneuvers. The effect is all the more important than acceleration is large and could then be used in transient regimes in thermal plants.

CONCLUSION

The vicinity of the critical point of fluids, by drastically enhancing or reducing some of thermodynamic or transport properties, make ostensible very particular thermal phenomena whose piston effect – a near isentropic thermalization process – is the most representative. Very paradoxical behavior can follow, as thermalization critical speeding up instead of the well-known thermalization critical slowing down or the apparent violation of the 2nd principle of thermodynamics when heat can flow backward. Although the applications of such phenomena as e.g. heat pipes remain till now of restricted interest, the above-mentioned effects open the way to a new thermo-compressible approach of fluids. It is worthy to note that, although such aforesaid effects have been detected firstly under weightlessness conditions to isolate them from earth-bound buoyancy-induced convections, it appears that they are also present on Earth. Thermal instabilities indeed appear there much lowered thanks to the piston effect thermal short circuit, making instabilities form only in the thermal boundary layers. More details and more phenomena can be found in the book by Zappoli et al. (2015) untitled: *Heat Transfer and Related Effects in Near Supercritical Fluids*. These effects can deeply modify the supercritical fluids thermal behavior in space and more generally of high yield energy activities. In particular, they can improve thermo-mechanics yields, giving them potential important socio-economic relevance.

REFERENCES

Accary, G., Bontoux, P., & Zappoli, B. (2007). Convection in a supercritical fluid: A reduced model for geophysical flows. *Physics of Fluids*, *19*(1), 014104. doi:10.1063/1.2432159

Accary, G., Raspo, I., Bontoux, P., & Zappoli, B. (2005). Reverse transition to hydrodynamic stability through the Schwarzschild line in a supercritical fluid layer. *Physical Review E: Statistical, Nonlinear, and Soft Matter Physics*, *72*(3), 035301. doi:10.1103/PhysRevE.72.035301 PMID:16241504

Adschiri, T., Takami, S., Suzuki, A., Hojo, D., Aoki, N., Aida, T., & Gimyeong, S. (2015). Recent advances in science and technology of supercritical hydrothermal synthesis—towards super hybrid materials and green nanotechnology (Review). *Review of High Pressure Science and Technology/Koatsuryoku No Kagaku To Gijutsu, 25*, 225-233.

Bailly, D., & Zappoli, B. (2000). Hydrodynamic theory of density relaxation in supercritical fluids. *Physical Review E: Statistical Physics, Plasmas, Fluids, and Related Interdisciplinary Topics*, *62*(2), 2353–2368. doi:10.1103/PhysRevE.62.2353 PMID:11088715

Behringer, R. P., Onuki, A., & Meyer, H. (1990). Thermal equilibration of fluids near the liquidvapor critical point: ^{3}He and ^{3}He-^{4}He mixtures. *Journal of Low Temperature Physics*, *81*(1-2), 71–102. doi:10.1007/BF00683153

Beysens, D., Chatain, D., Nikolayev, V. S., Ouazzani, J., & Garrabos, Y. (2010). Possibility of long-distance heat transport in weightlessness using supercritical fluids. *Phys. Rev. E*, *82* 061126 (11p).

Beysens, D., Fröhlich, T., & Garrabos, Y. (2011). Heat can cool near-critical fluids. *Physical Review E: Statistical, Nonlinear, and Soft Matter Physics*, 84, 051201 (8p).

Beysens, D., Garrabos, Y., & Fröhlich, T. (2008). Heat can cool near critical fluids. *Proceedings of 59th IAC Congress.*

Beysens, D., Garrabos, Y., Wunenburger, R., & Lecoutre-Chabot, C. (2002). Can heat flow backward? Unusual thermal phenomena observed in near-critical fluids. *Physica A*, *314*, 427.

Boukari, H., Shaumeyer, J. N., Briggs, M. E., & Gammon, R. W. (1990). Critical speeding up in pure fluids. *Physical Review A.*, *41*(4), 2260–2263. doi:10.1103/PhysRevA.41.2260 PMID:9903350

de Bruijn, R., van Diest, R. J. J., Karapantsios, T. D., Michels, A. C., Wakeham, W. A., & Trusler, J. P. M. (1997). Heat transfer in pure critical fluids surrounded by finitely conducting boundaries in micro-gravity. *Physica A*, *242*(1-2), 119–140. doi:10.1016/S0378-4371(97)00197-0

Dumas, A., Claverie, M., Slostowski, C., Aubert, G., Carême, C., Le Roux, C., & Aymonier, C. et al. (2016). *Fast-geomimicking using chemistry in supercritical water* (International Edition). Angewandte Chemie; doi:10.1002/anie.201604096

Garrabos, Y., Bonetti, M., Beysens, D., Perrot, F., Fröhlich, T., Carlès, P., & Zappoli, B. (1998). Relaxation of a supercritical fluid after a heat pulse in the absence of gravity effects: Theory and experiments. *Physical Review E: Statistical Physics, Plasmas, Fluids, and Related Interdisciplinary Topics*, *57*(5), 5665 5681. doi:10.1103/PhysRevE.57.5665

Gitterman, M., & Steinberg, V. A. (1970). Criteria for the commencement of convection in a liquid close to its critical point. *High Temperature (USSR)*, *8*(4), 754.

Jany, P., & Straub, J. (1987). Thermal diffusivity of fluids in a broad region around the critical point. *International Journal of Thermophysics*, *8*(2), 165–180. doi:10.1007/BF00515200

Juhaszl, A. J., Rarick, R. A., & Rajmohan, R. (2009). High Efficiency Nuclear Power Plants Using Liquid Fluoride Thorium Reactor Technology. In *Proceedings of the 7th International Energy Conversion Engineering Conference*, 1-12. doi:10.2514/6.2009-4565

Kassoy, D. R. (1979). The response of a confined gas to a thermal disturbance). I. Slow transient. *SIAM Journal on Applied Mathematics*, *36*(3), 3. doi:10.1137/0136044

Kawasaki, K. (1970). Kinetic equations and time correlation functions of critical fluctuations. *Annals of Physics*, *61*(1), 1–56. doi:10.1016/0003-4916(70)90375-1

Kruse, A., & Dahmen, N. (2015). Water – A magic solvent for biomass conversion. *The Journal of Supercritical Fluids*, *96*, 36–45. doi:10.1016/j.supflu.2014.09.038

Major, J. (2014). *Astronauts Work to Make Water That Burns*. Retrieved November 11, from http://news.discovery.com/space/astronauts-work-to-make-water-that-burns-140103.htm

Marcout, R., Zwilling, J. F., Laherrere, J. M., Garrabos, Y., & Beysens, D. (1995). ALICE 2, an advanced facility for the analysis of fluids close to their critical point in microgravity. *Microgravity Quarterly*, *5*, 162–170.

Miura, Y., Yoshihara, S., Ohnishi, M., Honda, K., Matsumoto, M., Kawai, J., & Onuki, A. et al. (2006). High-speed observation of the piston effect near the gasliquid critical point. [R]. *Physical Review E: Statistical, Nonlinear, and Soft Matter Physics*, *74*(1), 010101. doi:10.1103/PhysRevE.74.010101 PMID:16907038

Nikolayev, V., Garrabos, Y., Lecoutre, C., Charignon, T., Hitz, D., Chatain, D., & Beysens, D. et al. (2015). *Boiling crisis dynamics: Low gravity experiments at high pressure*. Micogravity Sci. Technol; doi:10.1007/s12217-015-9447-8

Noyori, R. (1999). Supercritical Fluids. *Chemical Reviews, 99,* 353-634.

Onuki, A. (1991). Sound propagation in phase-separating fluids. *Physical Review A.*, *43*(12), 6740–6755. doi:10.1103/PhysRevA.43.6740 PMID:9905025

Onuki, A., & Ferrell, R. A. (1990). Adiabatic heating effect near the gasliquid critical point. *Physica A*, *164*(2), 245–264. doi:10.1016/0378-4371(90)90198-2

Onuki, A., Hao, H., & Ferrel, R. A. (1990a). Fast adiabatic equilibration near the liquid-vapor critical point. *Physical Review A.*, *41*(4), 2256–2259. doi:10.1103/PhysRevA.41.2256 PMID:9903349

Phillips, T. (2014). *Starting Fire With Water.* Retrieved November 11, from https://science.nasa.gov/science-news/science-at-nasa/2014/10jan_firewater

Stanley, H. E. (1971). *Introduction to Phase Transitions and Critical Phenomena*. New York: Oxford University Press.

Stauffer, D., Ferer, M., & Wortis, M. (1972). Universality of second-order phase transitions: The scale factor for the correlation length. *Physical Review Letters*, *29*(6), 345–349. doi:10.1103/PhysRevLett.29.345

Straub, J., Eicher, L., & Haupt, A. (1995). Dynamic temperature propagation in a pure fluid near its critical point observed under microgravity during the German Spacelab mission D-2. *Physical Review E: Statistical Physics, Plasmas, Fluids, and Related Interdisciplinary Topics*, *51*(6), 5556–5563. doi:10.1103/PhysRevE.51.5556 PMID:9963290

Wilkinson, R. A. (1998). Density relaxation of liquid-vapor critical fluids in Earths gravity. *International Journal of Thermophysics*, *19*(4), 1175–1183. doi:10.1023/A:1022650011721

Wilkinson, R. A., Zimmerli, G. A., Hao, H., Moldover, M. R., Berg, R. F., Johnson, W. L., & Gammon, R. W. et al. (1998). Equilibration near the liquidvapor critical point in microgravity. *Physical Review E: Statistical Physics, Plasmas, Fluids, and Related Interdisciplinary Topics*, *57*(1), 436–448. doi:10.1103/PhysRevE.57.436

Wunenburger, R., Garrabos, Y., Lecoutre, C., Beysens, D., Hegseth, J., Zhong, F., & Barmatz, M. (2002). Thermal response of a two-phase near-critical fluid in low gravity: Strong gas overheating as due to a particular phase distribution. *International Journal of Thermophysics*, *23*(1), 103–115. doi:10.1023/A:1013948925906

Wunenburger, R., Garrabos, Y., Lecoutre-Chabot, C., Beysens, D., & Hegseth, J. (2000). Thermalization of a two-phase fluid in low gravity: Heat transferred from cold to hot. *Phys. Rev. Lett. American Physical Society*, *84*, 4100–4103.

Zappoli, B., Amiroudine, S., Carles, P., & Ouazani, J. (1996). Thermoacoustic and buoyancy-driven transport in a square side-heated cavityfilled with a near-critical pure fluid. *Journal of Fluid Mechanics*, *316*(-1), 53–72. doi:10.1017/S0022112096000444

Zappoli, B., Bailly, D., Garrabos, Y., Le Neindre, B., Guenoun, P., & Beysens, D. (1990). Anomalous heat transport by the piston effect in supercritical fluids under zero gravity. *Physical Review A.*, *41*(4), 2264–2267. doi:10.1103/PhysRevA.41.2264 PMID:9903351

Zappoli, B., Beysens, D., & Garrabos, Y. (2015). *Heat Transfer and related effects in supercritical fluids*. Dordrecht: Springer. doi:10.1007/978-94-017-9187-8

Zappoli, B., & Carles, P. (1995). The thermoacoustic nature of the critical speeding up. *Europ. J. Mech. B*, *14*, 41–65.

Zhao, H., Pennincks-Sans, A., Lee, L.-T., Beysens, D., & Jannink, G. (1995). Probing the universal critical-adsorption profile by neutron reflectometry. *Physical Review Letters*, *75*(10), 1977–1980. doi:10.1103/PhysRevLett.75.1977 PMID:10059177

Zhong, F., & Meyer, H. (1996). Density equilibration near the liquid-vapor critical point of a pure fluid. II. Coexisting phases for $T<T_c$. *Physical Review E: Statistical Physics, Plasmas, Fluids, and Related Interdisciplinary Topics*, *53*(6), 5935–5948. doi:10.1103/PhysRevE.53.5935 PMID:9964952

Zhong, F., & Meyer, H. (1999). Heat transport in a pure fluid near the critical point: Steady state and relaxation dynamics. *Journal of Low Temperature Physics*, *114*(1/2), 231–255. doi:10.1023/A:1021862122617

KEY TERMS AND DEFINITIONS

Buoyancy Convection: Under an acceleration field, hot, lighter fluid directs to the top and cold, denser fluid to the bottom.

Critical Point: Ending point of the saturation curve.

Critical Slowing Down: Thermal diffusion slows down near the critical point of fluids at constant pressure.

Critical Speeding Up: Thermal properties slow down near the critical point of fluids at constant volume due to the piston effect.

Heat Transfer: Transfer of heat, classically by conduction, convection and radiation.

Near Supercritical Fluids: Fluids in the vicinity of their critical point.

Piston Effect: Near adiabatic heat transfer through the volume change of thermal boundary layers.

Rayleigh-Bénard Convection: Buoyancy convection can organize as a roll pattern (instability) above a given threshold depending on fluid properties, sample dimension and acceleration amplitude.

Thermalization: Process of temperature equilibration of a sample after a change of container temperature.

Weightlessness: Absence of gravity field.

Chapter 2
Numerical Modelling of Hydrodynamic Instabilities in Supercritical Fluids

Sakir Amiroudine
University of Bordeaux, France

ABSTRACT

The case of a supercritical fluid heated from below (Rayleigh-Bénard) in a rectangular cavity is first presented. The stability of the two boundary layers (hot and cold) is analyzed by numerically solving the Navier-Stokes equations with a van der Waals gas and stability diagrams are derived. The very large compressibility and the very low heat diffusivity of near critical pure fluids induce very large density gradients which lead to a Rayleigh–Taylor-like gravitational instability of the heat diffusion layer and results in terms of growth rates and wave numbers are presented. Depending on the relative direction of the interface or the boundary layer with respect to vibration, vibrational forces can destabilize a thermal boundary layer, resulting in parametric / Rayleigh vibrational instabilities. This has recently been achieved by using a numerical model which does not require any equation of state and directly calculates properties from NIST data base (NIST, 2000) for instance.

INTRODUCTION

Hydrodynamics of near-critical fluids have gained considerable interest since the identification of the thermo-acoustic effect which is responsible for the fast thermal equilibration of a cell heated on one boundary (Zappoli *et al.*, 2015). Transport coefficients exhibit strong deviations near the critical point (Zappoli *et al.*, 2015; Stanley, 1971). Near-critical fluids are characterised by a large density like a liquid and a low viscosity and a high compressibility like a gas. Their thermal diffusivity goes to zero whereas the isothermal compressibility (and on the same way the heat capacity at constant pressure and thermal expansion) diverge.

Over the past several decades, there has been a growing demand for supercritical fluids in industrial applications as varied as alternative eco-friendly refrigerants, cold energy storage of Liquefied Natu-

DOI: 10.4018/978-1-5225-2047-4.ch002

ral Gas (LNG) (Nakano & Shiraishi, 2004), superconducting magnet cooling (Dobashi *et al.*, 1998), chemical extraction/separation processes (Peter *et al.*, 2007), supercritical chromatography, drying and catalysis. In space, in the absence of gravity effects, the behaviour of fluids is markedly different than on earth. The management of fluids in such conditions (flow control, heat exchange, etc.) is often a challenge and "artificial" gravity can be looked for. It happens that fluids submitted to vibrations of "high" frequency, e.g. frequency larger than the inverse hydrodynamics times (typically thermal diffusion and viscous dissipation times) and "low" amplitude (e.g. amplitudes smaller than the sample size) exhibit convective flows that are similar to buoyancy flows under earth gravity. The interest of studying fluids in such conditions is manifold. Firstly, supercritical oxygen, hydrogen, and helium are already used by the space industry. Secondly, their high compressibility and slow dynamics (critical slowing down) emphasize the behavior encountered in regular fluids. Thirdly, fluids in such conditions obey universal, scaled power laws, valid for all fluids.

We present here an overview of Rayleigh-Bénard (Amiroudine *et al.*, 2001; Chiwata & Onuki, 2001; Meyer & Kogan, 2002), Rayleigh-Taylor (Amiroudine *et al.*, 2005; Boutrouft *et al.*, 2006) and thermo-vibrational instabilities in near-critical fluids (parametric, Rayleigh-vibrational) (Amiroudine & Beysens, 2008; Gandikota *et al.*, 2013). The Rayleigh-Bénard and Rayleigh-Taylor instabilities will be briefly presented and more emphasis will be considered on Rayleigh-vibrational instabilities.

The case of a supercritical fluid heated from below (Rayleigh-Bénard (RB) configuration) in a rectangular cavity is first presented. Owing to the homogeneous thermo-acoustic heating (piston effect), the thermal field exhibits a very specific structure in the vertical direction. A very thin hot thermal boundary layer is formed at the bottom, then a homogeneously heated bulk settles in the core at a lower temperature; at the top, a cooler boundary layer forms in order to continuously match the bulk temperature with the colder temperature of the upper wall. We analyze the stability of the two boundary layers (hot and cold) by numerically solving the Navier-Stokes equations with a van der Waals gas and slightly above its critical point. A Finite-Volume method is used together with an acoustic filtering procedure. The onset of the instabilities in the two different layers is discussed with respect to the results of the theoretical stability analyses available in the literature and stability diagrams are derived.

The stability analysis of a two-layer fluid system (Rayleigh-Taylor like (RTL) instabilities) is then developed. The very large compressibility and the very low heat diffusivity of near critical pure fluids induce very large density gradients which lead to a RTL gravitational instability of the heat diffusion layer when the top layer temperature is some milli-Kelvin cooler than the bottom one. This instability in a one-phase fluid seems to be similar to that which occurs in between two miscible liquids where the species diffusion is replaced by the heat diffusion coefficient.

Finally, thermovibrational instabilities will be presented at the end with a new approach for the calculation of the density. It was indeed surprisingly observed in several terrestrial and weightless experiments that a destabilization of thermal boundary layers occurred when the fluid was vibrated. The interaction of a thermal boundary layer (TBL) with vibration is a stimulating problem of fluid physics. Depending on the relative direction of the interface or the boundary layer with respect to vibration, vibrational forces can destabilize a TBL, resulting in parametric / Rayleigh vibrational instabilities. Until now, the mathematical model has assumed a van der Waals state equation or a linear equation of state being valid near the critical point. This poses limitations on the quench (or heating) conditions and a more sophisticated numerical and mathematical model is further detailed. The evaluation of the effect of longitudinal vibrations is thus not limited by temperature conditions. This has been achieved using a numerical model which does not require any equation of state. The density is directly calculated by mass conservation equation.

THERMAL INSTABILITIES IN SUPERCRITICAL FLUIDS

The thermal diffusivity of supercritical fluids goes to zero as $\varepsilon_T^{\,0.67}$ (here $\varepsilon_T = \dfrac{T - T_c}{T_c}$), whereas the isothermal compressibility, the heat capacity at constant pressure and the coefficient of thermal expansion diverge as $\varepsilon_T^{\,-1.24}$ (Stanley, 1971). The vanishing thermal diffusivity slows down the heat transfer processes ("critical slowing down"). These singular properties imply that classical dimensionless parameters like Rayleigh or vibrational Rayleigh numbers diverge at the critical point, making the system more and more unstable and ultimately turbulent.

Onset of Rayleigh-Bénard Instabilities

The problem that is addressed here is the stability of a horizontal supercritical fluid layer, infinite in lateral extent, heated from below and in the presence of gravity. According to Rayleigh, above a critical temperature difference, the quiescent system becomes unstable and a cellular flow develops. If the variation of density with pressure is negligible compared to the variation induced by temperature, the onset of convection is determined by the Rayleigh criterion. For the incompressible case, it writes as,

$$\mathrm{Ra} = \frac{\rho' C_p' g'}{\lambda' \nu'} \left(\frac{\partial \rho'}{\partial T'} \right)_{P'} L'^3 \left(T_L' - T_U' \right) \geq \mathrm{Ra}_c \tag{1}$$

where T_L' and T_U' are temperatures of the lower and upper walls, respectively, C_p', λ', ρ' and ν' are specific heat at constant pressure, thermal conductivity, density and kinematic viscosity, respectively; g' is the Earth acceleration constant; Ra_c is the critical Rayleigh number for the onset of convection. It is equal to 657.5, 1707.8 and 1100.6, resp., depending on whether the upper and bottom boundaries are both stress-free, both solids, or whether one is solid and the other a free boundary (Chandrasekar, 1961).

Another limiting case which can be relevant occurs when viscosity and thermal conductivity are neglected and compressibility is taken into account; in this case, the stability criterion is the criterion due to Schwarzchild (Landau & Lifschitz, 1959),

$$\frac{\partial T'}{\partial y'} > -\frac{\rho' g'}{\beta'} \left(\chi_T' - \chi_S' \right) = -\frac{g' T' \beta'}{C_p'} \tag{2}$$

where y' is the spatial ordinate; χ_T' and χ_S' are the isothermal and isentropic compressibility and β' is the thermal expansion coefficient, respectively. The definition of the local Rayleigh number for accounting these two effects in compressible flows is (Chandrasekar, 1961; Spiegel, 1965; Gitterman & Steinberg, 1970),

$$Ra = \frac{g'L'^4\rho'\beta'C_p'\left[\frac{\left(T_L' - T_U'\right)}{L'} - \frac{g'T'\beta'}{C_p'}\right]}{\lambda'\nu'} \quad (3)$$

It is worthy to note that the second term of this expression (Schwarzchild term) is constant when approaching the critical point as β' and C_p' diverge with the same critical exponent. The critical temperature gradient at which convection arises is then,

$$\left(\delta T'\right)_{onset} = Ra_c \frac{\lambda'\nu'}{g'L'^3\rho'\beta'C_p'} + \frac{g'T'\beta'L'}{C_p'} \quad (4)$$

The second term on the right-hand side (corresponding to the Schwarzchild term or adiabatic gradient) shows small variations regardless of the proximity to the critical point. The first term becomes asymptotically small close to the critical point, whereas it is dominant far from it. The critical Rayleigh number for compressible fluids is given in (Gitterman & Steinberg, 1970).

Figure 1 is directly derived from relation in Equation 4 and it gives the dependence of critical $\delta T'_{onset}$ with *h'* over a range of *h'* which are relevant with the size of the cavity, *h'*<10mm and corresponds to the thermal boundary layer thickness as time evolves. The variation of the critical $\delta T'_{onset}$ with reduced temperature is significant. Also significant is the dependence on the thermal boundary layer thickness *h'*. One sees a gradual transition from adiabatic gradient condition (the Schwarzchild criterion) for inviscid, compressible fluids, near the critical point, to the Rayleigh condition corresponding to the onset of viscous convection flow far from the critical point. The Schwarzchild criterion (or adiabatic temperature gradient) is important and well known in atmospheric science when orders of magnitude of the dimensions are kilometers and the second term in Equation 4 thus becomes important. In our case, an identical physical phenomenon can be studied in millimeters cells as demonstrated in Figure 1.

The characteristic length scale is the local height of the thermal boundary layer h', $0 < h' < L'$ where L' represents the total height of the cavity. The respective contributions of the Rayleigh and Schwarzchild criteria are identified and zoomed at the left top-hand side of the figure for $h' = 6$ mm.

Numerical Modelling and Results of the Rayleigh-Bénard Instabilities

The governing Navier-Stokes equations are those of a Newtonian, viscous, hypercompressible, heat-conducting van der Waals gas (for more details please see references Amiroudine *et al.*, 1997; Zappoli *et al.*, 1996) and the fluid is chosen to be CO_2. The equations are made dimensionless. The density and the temperature are referred with respect to their critical values ρ'_c and T'_c respectively (T'_c=304.13 K, ρ'_c=467.8 kg.m^{-3}), and the pressure with respect to its perfect gas value, i.e. $\rho'_c R'T'_c$ (R'=188.8 J.kg^{-1}.K^{-1} is the perfect gas constant for CO_2). All quantities with superscript prime stand for dimensional quantities.

The following non-dimensional piston effect velocity and time scales ($\mathbf{V}_{\ddot{A}}$ and τ) have then been defined (Zappoli, 1992; Zappoli *et al.*, 1996):

Figure 1. Critical $\delta T'_{onset}$ (K) at the onset of convection as derived from Rayleigh and Schwarzchild criteria, vs. reduced temperature

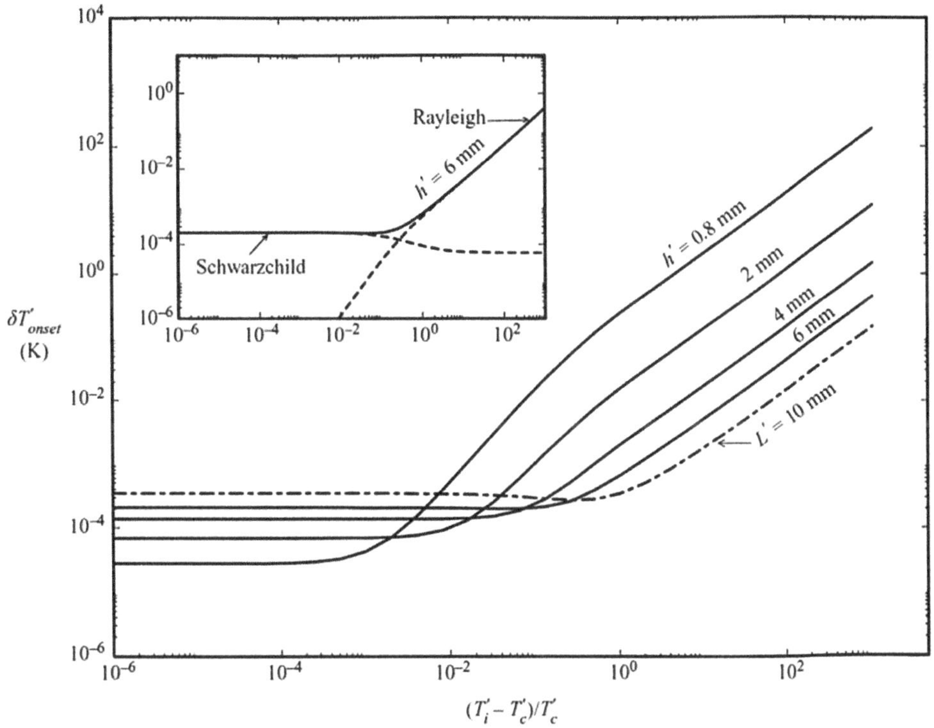

$$V_{\tau} = \frac{\mu^2}{\varepsilon f\left(\varepsilon_T, \Lambda\right)} V, \tau = \frac{\varepsilon f\left(\varepsilon_T, \Lambda\right)}{\varepsilon_T^2} t \text{ with } f\left(\varepsilon_T, \Lambda\right) = \mu\left(\frac{1}{\Lambda} + \frac{1}{\varepsilon_T^{1/2}}\right)$$

where **V** and t are the dimensionless velocity and time referred with respect to the perfect gas acoustic scale: $\mathbf{V} = \mathbf{V}'/ c_0'$ and $t = t'c_0' / L'$. Here, $c_0' = \left(\gamma R'T_c'\right)^{1/2}$ represents the sound velocity for perfect gas and γ is the ratio of specific heats.

The small parameter ε is defined by $\varepsilon = \Pr t_a' / t_d'$ (ε=2.6 10^{-8}) where Pr is the perfect gas Prandtl number ($\Pr = \frac{\nu_0'}{\kappa_0'} \cong 2.27$, ν_0' and κ_0' represent the kinematic viscosity and thermal diffusivity respectively) with $t_a' = L' / c_0'$ and $t_d' = L'^2 / \kappa_0'$. It must be noted that t_d' is not the characteristic time for diffusion in a supercritical fluid. Taking into account the vanishing thermal diffusivity of near-critical fluids, this characteristic time would be of order $t_d' / \varepsilon_T^{1/2}$ which is even longer.

The dimensionless equations for continuity, momentum, energy and state (van der Waals) are then written at the τ timescale,

$$\frac{\partial \rho}{\partial \tau} + \nabla.\left(\rho \boldsymbol{V}_\tau\right) = 0$$

$$\frac{\partial\left(\rho \boldsymbol{V}_\tau\right)}{\partial \tau} + \nabla.\left(\rho \boldsymbol{V}_\tau \boldsymbol{V}_\tau\right) = -\frac{1}{\gamma c_\tau^2}\nabla P + \frac{\varepsilon}{c_\tau}\left[\Delta \boldsymbol{V}_\tau + \frac{1}{3}\nabla\left(\nabla.\boldsymbol{V}_\tau\right)\right] + \frac{1}{F_r c_\tau^2}\rho \boldsymbol{g} \tag{5}$$

$$\frac{\partial\left(\rho T\right)}{\partial \tau} + \nabla.\left(\rho \boldsymbol{V}_\tau T\right) = -\left(\gamma - 1\right)\left(P + a\rho^2\right)\left(\nabla.\boldsymbol{V}_\tau\right)$$
$$+\frac{\varepsilon\gamma}{P_r c_\tau}\nabla.\left[\left\{1 + \Lambda\left(T-1\right)^{-0.5}\right\}\nabla T\right] + \varepsilon\gamma\left(\gamma-1\right)c_\tau \varphi$$

$$P = \frac{\rho T}{1 - b\rho} - a\rho^2$$

where $F_r = \frac{c_0'}{\sqrt{L' g_0'}}$ is the acoustic Froude number with $g_0' = 9.8\ m.s^{-2}$ representing the earth gravity and

$$\varphi = \left(\boldsymbol{V}_\tau\right)_{i,j}\left(\boldsymbol{V}_\tau\right)_{j,i} + \left(\boldsymbol{V}_\tau\right)_{i,j}\left(\boldsymbol{V}_\tau\right)_{i,j} - 2/3\left(\boldsymbol{V}_\tau\right)_{i,i}\left(\boldsymbol{V}_\tau\right)_{j,j}$$

is the heat dissipation. Here a=9/8 and b=1/3 correspond to the dimensionless coefficients of a' and b' which are calculated from the van der Waals equation, i.e. $T'_c = 8a'/27b'$, $\rho'_c = 1/3b'$, $P'_c = a'/27b'^2$.

These equations are solved numerically by the finite volume method with the SIMPLER algorithm which is the revised-SIMPLE (Semi-Implicit for Pressure Linked Equations) (Patankar, 1980, Amiroudine *et al.*, 1997). The time discretization is of Euler's type and first order accurate, and the discretization in space uses the power-law scheme (Patankar, 1980). When approaching the critical point, the boundary layer becomes thinner and one needs a small grid size in order to ensure a sufficiently small grid Peclet number. A non-uniform staggered mesh has then been used and numerical stability has been carefully tested in terms of time step and grid size. In order to reduce computational time, the acoustic filtering procedure has been considered (Paolucci, 1982, Amiroudine *et al.*, 2001) when the acoustic wave description is not needed.

We consider a two-dimensional square cavity of height L' (10 mm) and heated from below, the vertical walls being insulated and the whole cell is subjected to the gravity field. The configuration is displayed with respect to horizontal (x') and vertical (y') axes in Figure 2.

Their expansion in the core is visualised by arrows and characterised by the height h' and the temperature differences $\delta T'_L$ of the hot layer.

The convection inside the cavity is made visible by the temperature pattern (see Figure 3). There is an upward and downward driving of hot plumes of gas from the hot layer below and of cold drops from the cold layer above into the bulk in the middle of the cavity. The convection cells also bring cold mass

Figure 2. Geometry of the cavity, boundary conditions and characteristic zones: isothermal bulk in the core, hot and cold layers near the horizontal walls

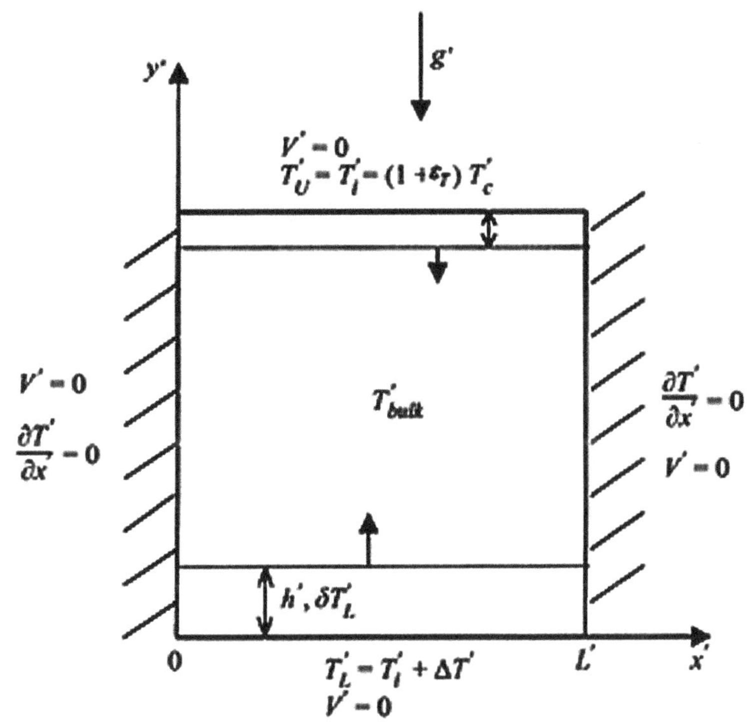

of gas near the hot wall and constrain the thermal layer to restrict its height by zone. This is seen by comparing the temperature patterns before and after the rise on convection. Several modes of instability are seen on both walls. The disturbances generate a similar number of cells in the cold layer but that is relatively weakened and delayed. The confinement and the thermal conditions give rise to the initial perturbations that occur at the threshold. Successive cells progressively grow and fill the entire layer. They grow in magnitude and in size and move upward in the bulk region. We note that on that range there is no or very weak interaction between the vortices originating from the hot and from the cold layers in the bulk. The interesting features of this RB configuration are: (i) the existence of the Schwarzchild criterion (important in atmospheric layers) which is dominant near the critical point even in these millimeters cells, (ii) the appearance of the instability on the top layer maintained initially at the same temperature as the bulk (due to the cold piston effect in comparison with the hot piston effect holding on the bottom layer). This instability is absent or weak in Boussinesq or normal compressible fluids.

Rayleigh-Taylor Like Instabilities: Mathematical Model

The very large compressibility and the very low heat diffusivity of near-critical fluids lead to a Rayleigh-Taylor like gravitational instability of the heat diffusion layer. This instability is similar to the one of two miscible fluids where molecular species diffusion coefficient is replaced by the heat diffusion coefficient.

We consider a rectangular cavity (see Figure 4) of height H' and length L' enclosing a supercritical fluid subjected with gravity $\vec{g}\,' = (0, -g')$ (all quantities with superscript prime stand for dimensional

Figure 3. Temperature patterns for ΔT'=10mK at several stages during the heating t'=3.6s, 6.4s, 6.9s, 7.3s, 7.8s, 8.3s and 8.8s: onset of nine thermal plumes after the threshold of convection

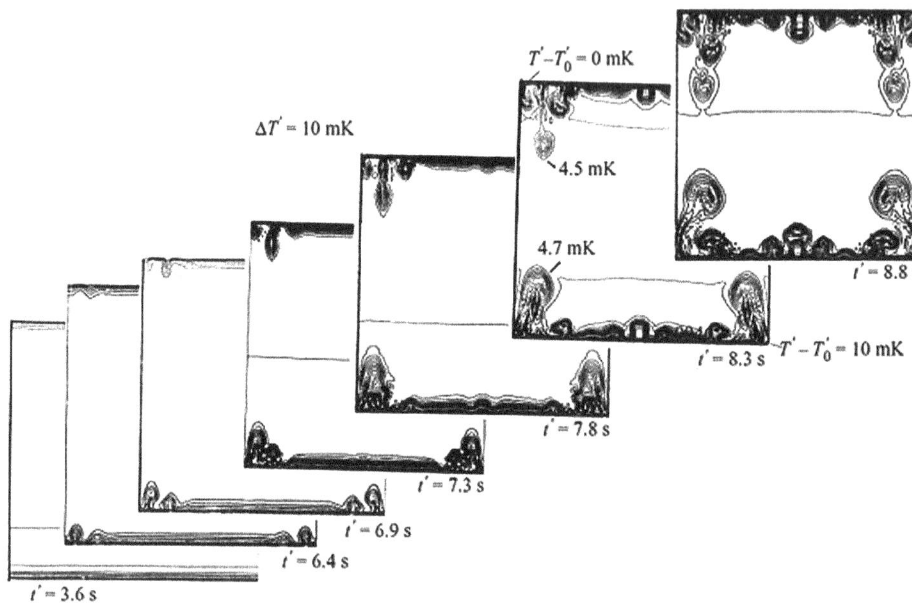

quantities). The conditions without the effect of initial stratification are: at t'=0 and in the upper part of the cavity of height H'-h', the fluid is maintained at a temperature $T'_2=(1+\varepsilon_T)T'_c$ ($\varepsilon_T<<1$) and a density $\rho'_2=\rho'_c$ whereas in the lower part of height h' a density $\rho'_1=(1-\varepsilon_\rho)\rho'_2$ is imposed ($\varepsilon_\rho=\frac{\rho'_2-\rho'_1}{\rho'_2}$ is a given quantity and $\rho'_1<\rho'_2$). In these conditions the temperature T'_1 is determined by the state equation. The pressure field is supposed to be constant and is equal to the critical pressure in the whole cell.

In the proposed model, the following dimensionless quantities are used: $\rho=\frac{\rho'}{\rho'_c}$, $P=\frac{P'}{\rho'_c C'^2}$, $T=\frac{T'}{T'_c}$, $\vec{x}=\frac{\vec{x}'}{H'}$ $\vec{V}=\frac{\vec{V}'}{C'}$, $t=\frac{t'}{t'_a}$ where $t'_a=\frac{H'}{C'}$ is the acoustic timescale and C' is the sound velocity derived from the generalised Mayer equation:

$$C'=\sqrt{\frac{P'_c}{\rho'_c}\left(\frac{P'_c\alpha^2_{\rho_2}}{\rho'_c C'_v T'_c}+\frac{1}{\chi_{T_2}}\right)}$$

with $\alpha_{\rho_2}=\frac{T'_c}{P'_c}\left(\frac{\partial P'}{\partial T'}\right)_{\rho'_2}$, $\chi_{T_2}=\chi_{T'_2}\rho'_c C'^2$ is the non-dimensional isothermal compressibility and C'_v is the heat capacity at constant volume.

Figure 4. The geometry of the problem in the Rayleigh-Taylor type configuration

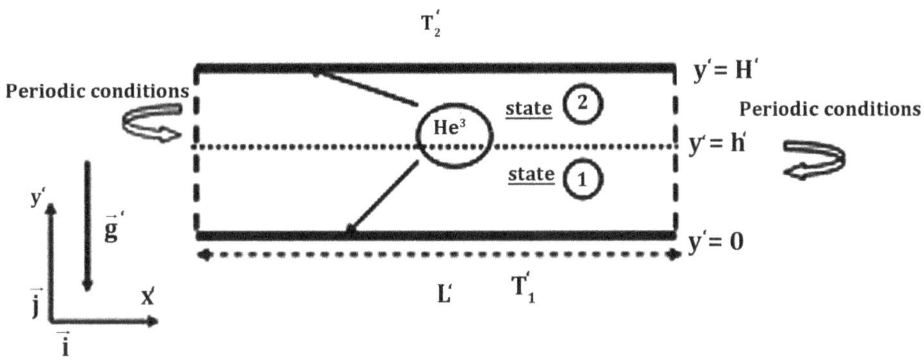

The thermal expansion coefficient $\beta'_{P'}$ and the isothermal compressibility $\chi'_{T'}$ are supposed to be constant and are equal to the values at state 2

$$(P_2^{'}, T_2^{'}) : \chi'_{T'} \cong \chi_{T_2^{'}} \cong \frac{1}{\rho_c^{'}} \left(\frac{\partial \rho^{'}}{\partial P^{'}} \right)_{T^{'}=T_2^{'}}$$

and

$$\beta'_{p^{'}} \cong \beta_{p_2^{'}} \cong -\frac{1}{\rho_c^{'}} \left(\frac{\partial \rho^{'}}{\partial T^{'}} \right)_{p^{'}=p_2^{'}} .$$

The state equation can thus be linearized and be written as follows:

$$\rho^{'} = \rho_2^{'} + \rho_c^{'} \chi'_{T_2^{'}} \left(P^{'} - P_2^{'} \right) - \rho_c^{'} \beta'_{P_2^{'}} \left(T^{'} - T_2^{'} \right).$$

The dimensionless equations are deduced from above,

$$\begin{cases} \dfrac{\partial \rho}{\partial t} + \nabla \cdot (\rho \vec{V}) = 0 \\ \rho[\dfrac{\partial \vec{V}}{\partial t} + (\vec{V}.\vec{\nabla})\vec{V}] = -\vec{\nabla} P + \dfrac{1}{\text{Re}}[\Delta \vec{V} + \dfrac{1}{3}\vec{\nabla}(\vec{\nabla}.\vec{V})] - \dfrac{1}{Fr^2}\rho \vec{j} \\ \rho(\dfrac{\partial T}{\partial t} + \vec{V} \cdot \vec{\nabla} T) = \beta_{p_2} \, Ec \, T \, \dfrac{dP}{dt} + \dfrac{1}{\text{Re}\,\text{Pr}} \vec{\nabla}.[\vec{\nabla} T] + \dfrac{Ec}{\text{Re}} \Phi \\ \rho = \rho_2 + \chi_{T_2}(P - P_2) - \beta_{P_2}(T - T_2) \end{cases} \tag{6}$$

The different dimensionless coefficients appearing in the above equations are the following:

$$Fr = \frac{C'}{\sqrt{H'g'}} \text{ (Froude number),}$$

$$\Pr = \frac{\nu'}{\kappa'} \text{ (Prandtl number)}$$

$$\mathrm{Re} = \frac{H'C'}{\nu'} \text{ (Reynolds number),}$$

$$Ec = \frac{C'^2}{C'_p T'_c} \text{ (Eckert number)}$$

$$\beta_{P_2} = \beta'_{P'_2} T'_c \text{ (thermal expansion coefficient)}$$

Unlike the previous model (Rayleigh-Bénard), all the above dimensionless parameters are defined with respect to supercritical physical properties and depend on only one control parameter ε_T which corresponds to the proximity to critical point.

A compressible fluid breaks down under its own weight and forms density gradients which can be very large as the critical point is approached. As the compressibility diverges at the critical point, for a small variation of pressure, large variations of density gradients may appear in the cell. These stratification effects which exist usually in atmospheric layers due to the large dimension exist also in millimetre cells due to the large value of the compressibility (see also concluding remarks in Section I.1.b) and have been taken into account in this model.

The numerical model is the same as defined in Section I.1.b. The 2D rectangular cavity (Figure 4) is filled with supercritical He^3 and holds two characteristic lengths: $L' = \tilde{\lambda}'$ and H', where $\tilde{\lambda}'$ corresponds to the wavelength of the perturbation. The two horizontal limits of the cavity are no-slip walls which are maintained at the respective temperatures of the two layers. These layers (top and bottom) are initially maintained at densities ρ_2 and ρ_1. Periodic conditions are set on the lateral boundaries which permit to focus on the stability of this two-layer system between the two infinite walls with an imposed wavelength.

Transition to the Rayleigh-Bénard Case

The stability analysis has been established in (Zappoli *et al.*, 1997, Boutrouft *et al.*, 2006) and the method of determining the dispersion relation is reminded here. At t=0, a perturbation of the vertical velocity is performed in the following form:

$$V'(0 \le x' \le L', y' = y'_{interface}) = A' \cos\left(\frac{2\pi}{\tilde{\lambda}'} x'\right)$$

where A' is the given amplitude of the perturbation and $\tilde{\lambda}'$ is the corresponding wavelength. The dynamical growth of the fluctuations (in dimensionless form), i.e. the calculation of the growth rate σ as a function of the wave number $k = \frac{2\pi}{\tilde{\lambda}}$ is numerically evaluated in the linear regime ($\frac{A'\sigma'}{g'} \leq 1$) by considering the plateau profile in the temporal evolution of this growth rate σ which is: $\sigma = \frac{\ln(V_1 / V_2)}{t_1 - t_2}$, where V_1 and V_2 are the vertical components of the velocity at times $t=t_1$ and $t=t_2$ respectively.

Figure 5 shows an example of the evolution of the temperature field during 0.78s for $\frac{\delta\rho}{\rho_1} = \frac{\rho_1 - \rho_2}{\rho_1} = 10^{-2}$, H'=10mm, $\tilde{\lambda}' = 0.5\ mm$, A'=$10^{-4}$ m/s and gives the signature of RTL instability in this situation where the cavity is filled with a pure supercritical fluid and where stability is governed by heat diffusion and compressibility.

Figure 6 shows the evolution of the growth rate of instability σ as a function of the wave number k' for the case: $\varepsilon_T = 0.08$, H'=1mm and $\frac{\delta\rho}{\rho_1} = 10^{-4}$ (which corresponds to $\delta T' = \frac{\delta\rho'}{\rho'_1 \beta'_P} \approx 22\ \mu K$). This value is almost identical to the one of $(\delta T')_{onset}$ in the Rayleigh-Bénard configuration determined by Equation 4 and which is of the order of 15 μK. The different curves in this Figure 6 represent the evolution of $\sigma = f(k)$ for different values of the heights of the two layers and $\frac{\delta\rho}{\rho_1}$. When the thickness of the bottom layer diminishes, the Rayleigh-Taylor like configuration becomes stable $(\sigma < 0)$ and tends

Figure 5. An example of the Rayleigh-Taylor like instability for the case of $\varepsilon_T = 8\ 10^{-2}$, $\frac{\delta\rho}{\rho_1} = 10^{-2}$, wavelength of 0.5mm and A'=10-4 m/s

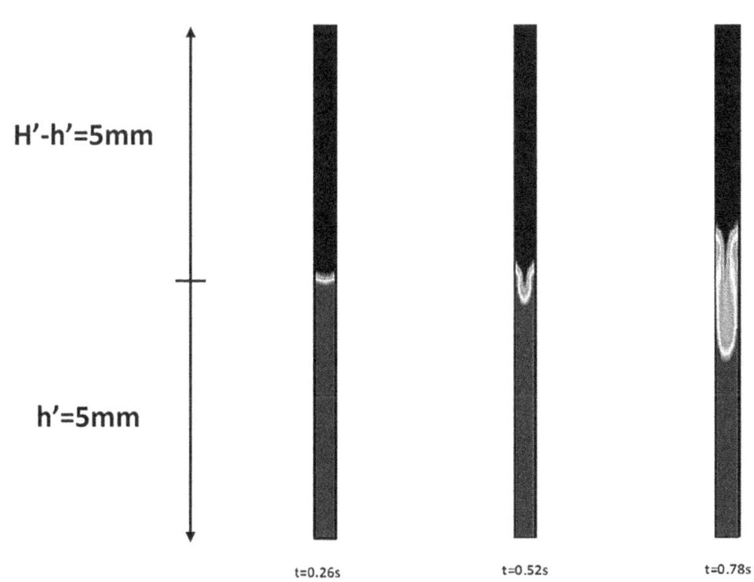

Figure 6. Growth rate of instability as a function of wave-number for two values of $\frac{\delta\rho}{\rho_1} = 10^{-4}$

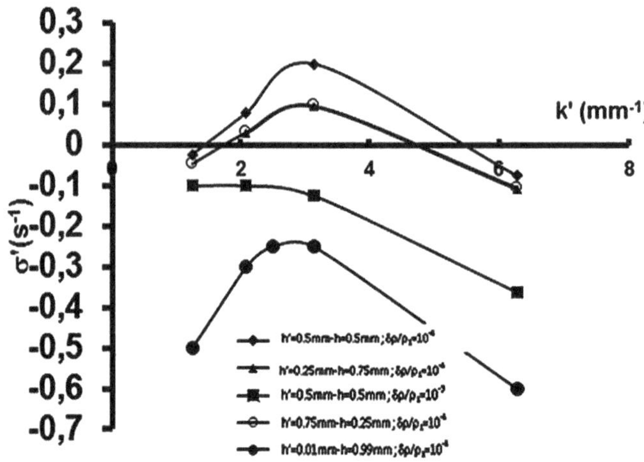

to the Rayleigh-Bénard one which is almost stable for the values of ε_T and $\frac{\delta\rho}{\rho_1}$ considered here. For values of $\frac{\delta\rho}{\rho_1}$ as low as 10^{-5} corresponding to $\delta T' \cong 2.2\ \mu K$ which is almost 7 times lower than $\left(\delta T'\right)_{onset} \cong 15\ \mu K$, the calculations show that, in this situation where the initial discontinuity of the temperature and density fields, is at the mid-high of the cavity, the Rayleigh-Taylor configuration is stable.

THERMAL VIBRATIONS NEAR THE CRITICAL POINT

The experiments pertaining to supercritical fluid subjected to vibrations under weightlessness have been performed using CO_2 on-board a sounding rocket (see Beysens *et al.*, 2009 for details). These experiments have reported the appearance of fingers which shows destabilization of thermal boundary layer under the aforementioned conditions. Amiroudine & Beysens (2008) were the first to perform a numerical study for supercritical fluids subjected to vibrations and observed the fingering structure in the thermal boundary layers with different proximities to the critical point. It was found that the fingering wavelength and the vibrational Rayleigh number ($Ra_v = \frac{\left(A'\omega'\beta_p'\delta T'\delta_{TBL}'\right)^2}{2\nu' D_T'}$, where δ_{TBL}' represents the thermal boundary layer thickness and $\delta T'$ is defined if Figure 7) decrease as a power law with the distance in temperature to the critical point. Gandikota *et al.* (2013) then performed numerical simulation with supercritical hydrogen subjected to longitudinal vibrations. They evaluated the effect of vibrations using three different combinations of boundary conditions thus providing insight into the mechanisms of corner, parametric and Rayleigh vibrational instability. The first model which will be first presented uses a very simple numerical modelling in order to characterize the different instabilities. A further more sophisticated model with the first results will then be detailed.

Figure 7. Three different computational configurations: (a) isothermal conditions on all 4 walls, (b) isothermal conditions on vertical walls and adiabatic conditions on horizontal walls, (c) isothermal conditions on horizontal walls and adiabatic conditions on vertical walls

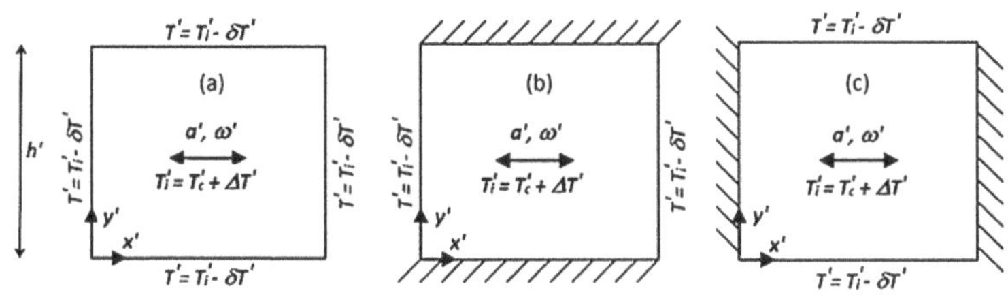

Model, Mesh, Timescales

The problem under study is mathematically modeled using the same model as defined in Section I.2.a with the same numerical method defined above. It is understood to be sufficiently accurate given that the relative thermal quench considered ($\delta T'/\Delta T'$) is kept quite small which is typically fixed as $\Delta T'/10$. In the supercritical region, 2D numerical simulations have been carried out for a 7 mm square cavity filled with supercritical H_2 subjected to vibration with simultaneous thermal quench (very fast temperature change $\Delta T'$) at the boundaries (see Figure 7).

The initial conditions are set as: $\rho_0 = 1$, $p_0 = \dfrac{p'_c}{\rho'_c c'^2}$, $T_0 = 1+\varepsilon_T$ (where $\varepsilon_T = \dfrac{T'_0 - T'_c}{T'_0}$). No slip velocity boundary conditions are applied on the walls. Different boundary conditions on the temperature field are defined in Figure 7 in order to model the different types of instabilities as described in the subsequent Section (parametric, corner, Rayleigh-vibrational). As stated above, the problem under study is a multi-scale hyper compressible problem involving various time scales viz., the thermal diffusion time scale $t'_D = \dfrac{h'^2}{D'_T}$, viscous diffusion time scale $t'_\nu = \dfrac{h'^2}{\nu'}$, piston effect time scale $t'_{PE} \cong \dfrac{h'^2}{\gamma^2 D'_T}$, vibration time scale $t'_{vib} = \dfrac{1}{f'}$ and the acoustic time scale $t'_a = \dfrac{h'}{c'}$, where γ is the ratio of specific heats. The simulation time step should be smaller than all the characteristic time scales including the piston effect time scale t'_{PE}. This is the reason why the computational times can be very high for small values of ε. The hydrodynamic times t'_D and t'_ν are large compared to the vibration time t'_{vib}. When the physical phenomenon at the acoustic timescale is not the primary interest, low-Mach number based calculations can be used (Paolucci, 1982). They filter (or average out) the acoustic waves at time steps larger than the acoustic timescales. A linearized equation of state (as defined in Amiroudine *et al.*, 2005, Amiroudine *et al.*, 2008) and the finite volume method, which integrates (and thus filters) the partial differential equations are used in the present study and did not need any acoustic filtering process. Fixed time steps (10^{-3} s) much larger than the acoustic timescales (of the order of 20 μs) were used without making the calculations to diverge (physical phenomena at the acoustic timescale are not of primary interest). The residue at iteration "k" can be expressed as: $RES = \underset{(i,j)}{Max}\left|\dfrac{\phi_{ij}^k - \phi_{ij}^{k-1}}{\phi_{ij}^k}\right| < \varepsilon$, where ¦ is the field variable (components of the velocity field, pressure or temperature) and (i, j) are the nodes in the x and y directions. The largest residue obtained by the pressure correction equation was as small as $\varepsilon = 10^{-10}$ which ensures the convergence of the present method with time steps as large as 10^{-3}s.

The thickness of the viscous boundary layer is calculated as $\delta'_\nu = (2\nu'/\omega')^{-1/2}$ which ranges from 49 μm (for $f' = 25$ Hz) to 150 μm (for $f' = 2.77$ Hz). A spatial discretization of 80x80 cells using a geometrical power factor of 0.2 (with the first point of the mesh at a distance of about 10 μm from the origin) is sufficient to ensure that at least 5 to 10 cells are present inside the boundary layer. The field parameters (pressure, temperature, density and velocity) are recorded once every half vibrational time period.

Stability Diagram

In the simulations with isothermal conditions on all the 4 walls, three types of instabilities are observed as shown in Figure 8.

- Corner instability forming fingers in the thermal field at the corners of the 2D cell. They appear at the beginning of the process with a frequency equal to the frequency of vibration of the cell.
- Parametric instability from the vertical walls with half the frequency of vibration.
- Rayleigh-vibrational fingers appearing on the horizontal walls.

It was observed in all our numerical simulations that for low vibrational accelerations the Rayleigh vibrational instability is favored, while the parametric instability is seen only for high vibrational accelerations. For a fixed frequency it was observed that as the amplitude of vibration is increased (from 1 mm to 30 mm), Rayleigh vibrational instability appears at lower amplitudes, a coexistence of the Rayleigh-vibrational and parametric instabilities happens for intermediate amplitudes and at higher amplitudes parametric instability predominates.

A strong interaction of these three types of instabilities is observed from the simulations. For higher amplitudes of vibration the parametric instability tends to suppress the apparition of the Rayleigh vibrational instability whereas for moderate amplitudes the Rayleigh-vibrational instability wins over the parametric instability. The corner instability that forms due to the interaction of the fluid with the corners of the domain also has a significant influence on the parametric instability.

In order to study each one of these instabilities, one needs to isolate the influence of one over another. For isolating the effect of the corner instability on the parametric instability, adiabatic conditions are imposed on the horizontal walls and simulations are carried out. The configuration is the one shown

Figure 8. Evolution of the thermal field for the configuration of Fig. 7(a) (4 isothermal walls) with $f' = 5.56$ Hz, $a' = 20$ mm, $\Delta T' = 1$ K and $\delta T' = 100$ mK. Different types of instabilities are evidenced (1): corner instability; (2)-(6): Parametric instability on the vertical walls; (4)-(6): Rayleigh vibrational instability on the horizontal walls

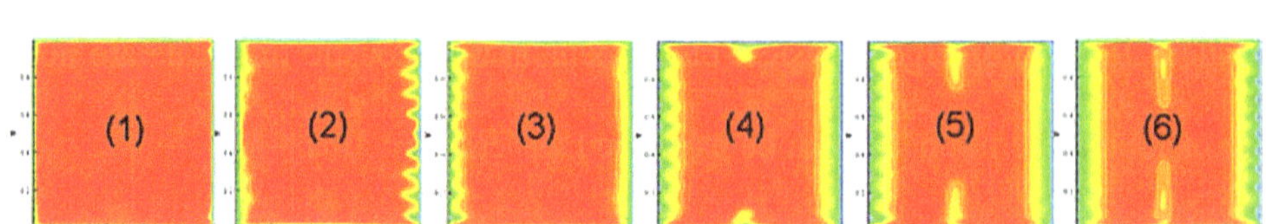

in Figure 7b. Similarly, simulations are carried out with adiabatic vertical walls (Figure 7c) to isolate the effect of parametric instability on the corner instability. Figure 9 shows for example the results of a typical simulation case showing the parametric instability in the thermal field of the fluid oscillating at a frequency equal to half the frequency of vibration. The wavelength versus acceleration and the stability curves for two typical critical point proximities and two typical quench amplitudes demonstrate the equivalence between parametric instability in the supercritical zone and the Faraday-type instability in immiscible fluids (Kumar & Tuckerman, 1994). This suggests the existence of a vanishing surface tension associated to the three types of instabilities mentioned above. The existence of such weak surface tension has been evidenced in miscible liquids (Zoueshtiagh *et al.*, 2009).

Figures 10a and 10b show the stability phase diagram of the three types of instabilities for $\Delta T'$ = 0.1K and 0.01K. It can be seen from the plots that there are zones in the stability diagram where all the instabilities are simultaneously observable. For example, the region shown by point '1' in Fig.10a indicates the region in which corner, Rayleigh vibrational and parametric instabilities are simultaneously observable while the region shown by point '2' indicates simultaneous presence of corner instability as well as Rayleigh vibrational instability. The regions of points '3' and '4' indicate the presence of only Rayleigh vibrational instability or only corner instability respectively. A cross-over between the Rayleigh-vibrational instabilities and the corner instabilities can be observed. The behavior of the three stability curves does not seem to change much with the change in the critical point proximity as can be observed in Figure 10b.

Comparing these three kinds of instabilities, it is seen that for a given frequency of vibration, the parametric instability needs substantially higher amplitudes to appear than the two other types of instabilities. Simulations with the four isothermal walls subjected to a thermal quench showed that the parametric instability has a stabilizing effect on the Rayleigh vibrational instability. For the corner instability the stability domain reduces as the critical point is approached. Similar kind of behavior is observed for the Rayleigh-vibrational and parametric instabilities.

Figure 9. Parametric instability at 4 consecutive time periods at 14.4 s (a), 14.76 s (b), 15.12 s (c) and 15.48 s (d) for $f = 2.78$ Hz, $a = 20$ mm, $\Delta T = 100$ mK and $\delta T = 10$ mK

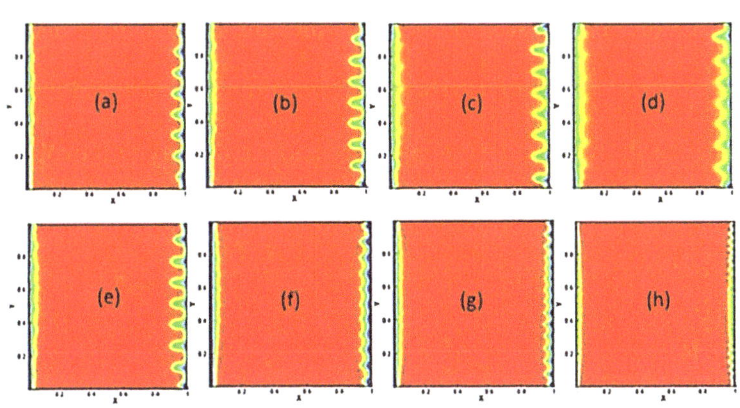

Figure 10. Stability domain for the three types of instabilities: corner (solid line with dots), parametric instability (solid line with boxes) and Rayleigh vibrational instability (solid line with circles) for (a) $\Delta T'$ = -0.1K and (b) $\Delta T'$ = -0.01K

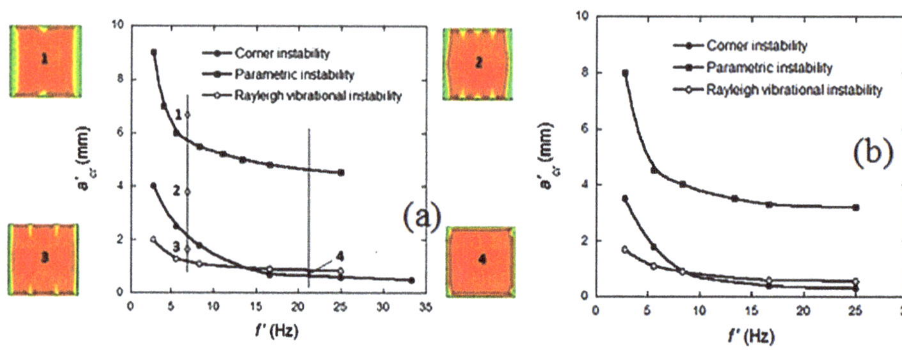

FUTURE RESEARCH DIRECTIONS: A LAGRANGE-EULER MODELLING OF THERMOVIBRATIONAL INSTABILITIES IN SUPERCRITICAL FLUIDS

The work mentioned so far has assumed a linear equation of state being valid near the critical point. This poses limitations on the quench (or heating) conditions that may be considered in the numerical study (assumed to be 10% of initial temperature distance from critical point). The purpose of this study is to evaluate the effect of longitudinal vibrations not limited by temperature conditions. This has been achieved using a numerical model which does not require any equation of state and directly calculates properties from NIST data base or renormalization group theory (Ising model, see Guenoun *et al.*, 1993).

Lagrange-Euler Mathematical Model

The following model solves the full set of governing equations (momentum and conservation of mass) in a Lagrangian-Eulerian formulation (Amiroudine *et al.*, 2014) and is presented hereafter.

Let us consider a Newtonian fluid contained in a domain Ω limited by a surface Γ. This fluid is characterized by its main thermo-physical properties which are the density ρ and the isothermal compressibility $\chi_T = \frac{1}{\rho}\left(\frac{\partial \rho}{\partial P}\right)_T$. The state of a system varies as a function of the velocity (V). The thermodynamic state quantities of the system, i.e. the pressure and the density will evolve as a function of the velocity (V). The infinitesimal variations of the pressure and the density can be written as follows,

$$\begin{cases} dP = \left(\frac{\partial P}{\partial \rho}\right)_T d\rho + \left(\frac{\partial P}{\partial T}\right)_\rho dT \\ d\rho = -dt\rho\nabla \cdot V \end{cases} \tag{7}$$

Here the differential time constant dt is a characteristic time corresponding to a very important physical parameter which will be further detailed below. The last equation of the system corresponds to mass conservation. The state variables (ρ, P) are explicitly calculated from Equation 7 and momentum equations (given below). A constant value of isothermal compressibility (χ_T) is used in the present study while this can also be obtained directly from experimental data or the NIST database (NIST, 2000) for any fluid. Thus, the model can be described without any state equation. Furthermore, owing to the nature of the problem under study (low compressibility of the fluid) it is assumed that any variation of pressure with temperature are negligible. Thus, the variation of pressure in Equation 7 can be simplified to $dP = \left(\frac{\partial P}{\partial \rho}\right)_T d\rho$.

The momentum conservation equation in a non-conservative form and for a Newtonian fluid in the absence of any external force can be written as follows,

$$\rho \frac{d\boldsymbol{V}}{dt} = -\nabla\left[\mathrm{P} - \left(\lambda + \frac{2}{3}\mu\right)\nabla\cdot\boldsymbol{V}\right] + \nabla\cdot\left[\mu\left(\nabla\boldsymbol{V} + \nabla^t\boldsymbol{V} - \frac{2}{3}\nabla\cdot\boldsymbol{V}\underline{\underline{I}}\right)\right] \tag{8}$$

where $\frac{d}{dt}$ is the material derivative, μ and λ correspond to the classical shear viscosity and compression viscosity (or second coefficient of viscosity), and P is the thermodynamic pressure. From Equation 7, the temporal variations of the pressure and density can be re-written as,

$$\begin{cases} \dfrac{dP}{dt} = -\left(\dfrac{1}{\chi_T}\right)\nabla\cdot\boldsymbol{V} \\ \dfrac{d\rho}{dt} = -\rho\nabla\cdot\boldsymbol{V} \end{cases} \tag{9}$$

The scalar quantities (P and ρ) can be finally written, in a Lagrangian form, as,

$$\begin{cases} P = P^0 - \delta t\left(\dfrac{1}{\chi_T}\right)\nabla\cdot\boldsymbol{V} \\ \rho = \rho^0\, \mathrm{e}^{-\delta t\,\nabla\cdot\boldsymbol{V}} \end{cases} \tag{10}$$

The pressure P^0 in Equation 10 is the equilibrium pressure at each time-step δt and the successive term corresponds to the non-equilibrium variations due to the divergence of velocity. Note that divergence of the velocity ($\nabla\cdot\boldsymbol{V}$) is assumed to be independent of time in the time interval δt. The set of Equation 10 thus represents the Lagrangian form of the density and the pressure fields which have to be further advected in the Eulerian form (see below).

The final system includes the conservation of momentum ($\rho\boldsymbol{V}$). The new equilibrium state is fixed with the expressions of the thermodynamic pressure P and the thermodynamic coefficient (isothermal compressibility). The shear viscosity μ is classically given by measurements whereas the second coefficient of viscosity λ verifies the Stoke's law ($\lambda + \frac{2}{3}\mu = 0$). The implicit resolution of the velocity vector equation thus gives,

$$\rho\left(\frac{\partial \boldsymbol{V}}{\partial t}+\boldsymbol{V}.\nabla \boldsymbol{V}\right)=-\nabla\left[P^{0}-\delta t\left(\frac{1}{\chi_{T}}\right)\nabla\cdot\boldsymbol{V}\right]+\nabla\cdot\left[\mu\left(\nabla\boldsymbol{V}+\nabla^{t}\boldsymbol{V}-\frac{2}{3}\nabla\cdot\boldsymbol{V}\underline{\underline{I}}\right)\right] \tag{11}$$

The unknown potentials (P , ρ) are explicitly determined from ($\nabla\cdot\boldsymbol{V}$) (Equation 10). The above set of equations (10 and 11 includes all the important physics and some interesting features can be extracted as follows: (i) the momentum equation in Equation 8 is completely autonomous as it does not contain any unknown pressures unlike most of the usual cases where specific pressure-velocity coupling algorithms have to be considered, (ii) the density is directly calculated from the continuity equation which ensures mass conservation to within machine precision and the model written above does not need any local thermodynamic equilibrium, (iii) the desirable thermophysical properties, isothermal compressibility and the viscosity, required to solve the numerical model are assumed to be constant in the present study though in more complex situations with non-uniform properties, these can be evaluated from the NIST database (NIST, 2000) o renormalization group theory (Ising model).-

Equation 8 can be discretized and solved independently. The density and the pressure are explicitly given by Equation 10 in a Lagrangian form and are then advected from the total derivatives as,

$$\begin{cases}\dfrac{\partial \rho}{\partial t}=\dfrac{D\rho}{Dt}-\boldsymbol{V}\cdot\nabla\rho\\[2mm] \dfrac{\partial P}{\partial t}=\dfrac{DP}{Dt}-\boldsymbol{V}\cdot\nabla P\end{cases} \tag{12}$$

The complete set of the above-mentioned equations gives the solution in the Lagrange-Euler formulation where the thermodynamic effects are solved in the Lagrangian phase and the advection effects in the Eulerian phase.

Validation with Experiments

The first results are briefly presented in this Section. The problem to be investigated consists of supercritical Hydrogen (as in experiments of Gurunath *et al.*, 2014a, 2014b) at its critical density filled in a 2-D square cavity of length $h = 7mm$ subjected to longitudinal vibrations $A\sin\omega t$ (see also Figure 7, the primes have been omitted).

Figure 11 shows one single example of our numerical results in comparison with experimental data (Gandikota *et al.*, private communication, 2016) in zero-g (based on the magnetic field levitation, see details about experimental set-up details in Gurunath *et al.*, 2014a, 2014b) when the cell is subjected to vibrations with an amplitude of 0.875 mm and frequency of 20 Hz (T_i=33.739 K). The wall boundary condition used in the numerical simulation corresponds to a cubic polynomial fit to experimental values. Owing to the vertical symmetry only upper half of the cell has been shown. One can observe that the numerical results match closely with the experimental data. Further, if one tries to superimpose the two figures, the height of the fingers matches within a reasonable accuracy. The appearance of fingers is attributed to Rayleigh vibrational instability (see details in Section II.2.) which appears due to Bernoulli like pressure difference $\left(\Delta p \approx \rho_c \Delta V^2 = \rho\left(\frac{\Delta\rho}{\rho_c}\right)^2 A^2\omega^2\right)$ in thermal boundary layers. The results

Figure 11. Evolution of the temperature field when subjected to longitudinal vibrations (a'=0.875mm, ω=20hz); (a) numerical, (b) experiment at time 1.6s

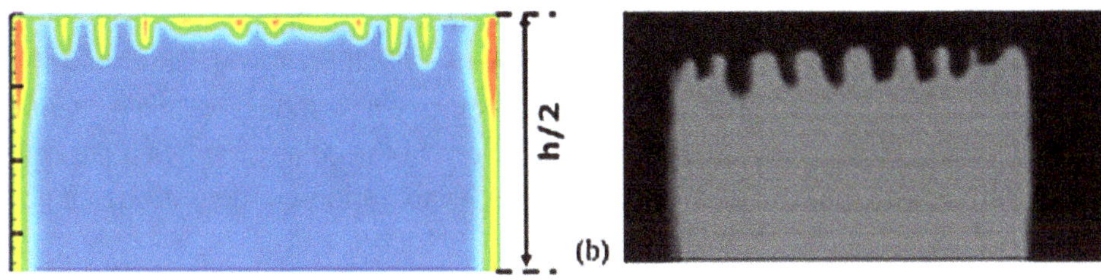

show that the present model is thus able to predict instabilities without any constraint on the heating rate. With an increase in the distance from the critical point, the number of fingers decreases while their amplitude grows faster. The latter is attributed to the fact that diffusion time increases significantly as we move closer to the critical point thereby causing a delay in the growth of instability in the thermal boundary layer. A stability analysis curve is under way for different proximities to the critical point, different values of frequency and amplitude and different heating rates (long time runs).

The next goal is to recover the different instabilities observed in experiments (Faraday and Frozen-wave instabilities, see Gurunath *et al.*, 2014a, 2014b) in two phase-flows (below the critical point). The discussions on the right model to be used are under study.

CONCLUSION

An overview of the numerical modelling of hydrodynamic instabilities in supercritical fluids has been addressed in this Chapter. The numerical model is based on Finite Volume method with the SIMPLER algorithm for solving the velocity-pressure coupling. The Navier-Stokes coupled with energy and state equations were solved numerically. The first model considered the van der Waals state equation with the acoustic filtering procedure in order not to consider the physics of acoustic timescales but to focus on the piston time scale. The linearized state equation has been considered thereafter which did not necessitate any acoustic filtering procedure. The whole model has been recently been reconsidered with the Lagrange-Euler methodology which estimates the density from the mass conservation equation. The state equation is not necessary as the thermophysical properties could be given by experiments or renormalization group theory (Ising model) or NIST database. The density is calculated with machine precision.

The Rayleigh-Bénard case has first been presented which shows the importance of Schwarzchild criterion in this hypercompressible system of supercritical fluids. The Schwarzchild criterion is very important in atmospheric layers. In the case considered in this study (millimeter cells), this criterion is as important as in kilometers layers (atmospheric layers) and could be used to study flow dynamics in lab cells. The Rayleigh-Taylor like instability (large density difference for the monophasic state) showed also unstable zones in the growth rate-wavenumber diagram similar to classical Rayleigh-Taylor configuration of two fluids. Moreover, it tends to the Rayleigh-Bénard case when the depth of the bottom layer decreases.

The thermo-vibrational instability in supercritical fluids confirmed the finger-type instability which appeared on the thermal boundary layers. When the vibration acceleration is parallel to the isotherms, the Rayleigh-vibrational instability showed fingering structures on the boundaries whereas it showed parametric instability in the case of the acceleration perpendicular to isotherms. A stability diagram has been established for two proximities to the critical point. A new model is then presented which calculates the density from the conservation of mass and does not need any explicit state equation. These investigations together with the case of subcritical states (evaporation-condensation) are now under way of investigation.

ACKNOWLEDGMENT

All my co-authors in the reference list are acknowledged without whom this work could not have been addressed in this paper. The author wishes to acknowledge also the French Space Agency (CNES) for the financial support throughout the many years of research.

REFERENCES

Amiroudine, S., & Beysens, D. (2008). Thermovibrational instability in supercritical fluids under weightlessness. *Physical Review E: Statistical, Nonlinear, and Soft Matter Physics*, *78*(3), 036323. doi:10.1103/PhysRevE.78.036325 PMID:18851161

Amiroudine, S., Bontoux, P., Larroudé, P., Gilly, B., & Zappoli, B. (2001). Direct numerical simulation of unsteady instabilities inside a near-critical fluid layer heated from below. *Journal of Fluid Mechanics*, *442*, 119. doi:10.1017/S0022112001004967

Amiroudine, S., Boutrouft, K., & Zappoli, B. (2005). The stability analysis of two layers in a supercritical pure fluid: Application to Rayleigh-Taylor like instabilities. *Physics of Fluids*, *17*(5), 054102. doi:10.1063/1.1868033

Amiroudine, S., Caltagirone, J.-P., & Erriguible, A. (2014). A Lagrangian–Eulerian compressible model for the trans-critical path of near-critical fluids. *International Journal of Multiphase Flow*, *59*, 15–23. doi:10.1016/j.ijmultiphaseflow.2013.10.008

Amiroudine, S., Ouazzani, J., Carles, P., & Zappoli, B. (1997). Numerical solutions of 1-D unsteady near-critical fluid flows using finite volume methods. *European Journal of Mechanics - B/Fluids*, *16*(5), 665–680.

Beysens, D., Garrabos, Y., Chatain, D., & Evesque, P. (2009). Phase transition under forced vibrations in critical CO_2. *Europhysics Letters*, *86*(1), 16003. doi:10.1209/0295-5075/86/16003

Boutrouft, K., Amiroudine, S., & Ambari, A. (2006). Stability diagram and effect of initial density stratification for a two-layer system in a supercritical fluid. *Physics of Fluids*, *18*(12), 124106. doi:10.1063/1.2424497

Chandrasekar, S. (1961). *Hydrodynamic and Hydromagnetic Stability*. Oxford, UK: Clarendon Press.

Chiwata, Y., & Onuki, A. (2001). Thermal Plumes and Convection in Highly Compressible Fluids. *Physical Review Letters*, *87*(14), 144301. doi:10.1103/PhysRevLett.87.144301 PMID:11580652

Dobashi, K., Kimura, A., Oka, Y., & Koshizuka, S. (1998). Conceptual design of a high temperature power reactor cooled and moderated by supercritical light water. *Annals of Nuclear Energy*, *25*(8), 487–505. doi:10.1016/S0306-4549(97)00079-0

Gandikota, G., Amiroudine, S., Chatain, D., Lyubimova, T., & Beysens, D. (2013). Rayleigh and parametric thermo-vibrational instabilities in supercritical fluids under weightlessness. *Physics of Fluids*, *25*(6), 064103. doi:10.1063/1.4811400

Gandikota, G., Chatain, D., Amiroudine, S., Lyubimova, T., & Beysens, D. (2014a). Frozen-wave instability in near-critical hydrogen subjected to horizontal vibration under various gravity fields. *Physical Review E: Statistical, Nonlinear, and Soft Matter Physics*, *89*(1), 012309. doi:10.1103/PhysRevE.89.012309 PMID:24580229

Gandikota, G., Chatain, D., Amiroudine, S., Lyubimova, T., & Beysens, D. (2014b). Faraday instability in a near-critical fluid under weightlessness. *Physical Review E: Statistical, Nonlinear, and Soft Matter Physics*, *89*(1), 013022. doi:10.1103/PhysRevE.89.013022 PMID:24580335

Gitterman, M., & Steinberg, V. A. (1970). Criteria for the commencement of convection in a liquid close to the critical point. *High Temperature, USSR*, *8*(4), 754.

Guenoun, P., Beysens, D., Khalil, B., Garrabos, Y., Kammoun, B., Le Neindre, B., & Zappoli, B. (1993). A thermo cycle around the critical point of CO_2 under reduced gravity. *Physical Review E: Statistical Physics, Plasmas, Fluids, and Related Interdisciplinary Topics*, *47*(3), 1531–1540. doi:10.1103/PhysRevE.47.1531

Kumar, K., & Tuckerman, L. S. (1994). Parametric instability of the interface between two fluids. *Journal of Fluid Mechanics*, *279*(-1), 49. doi:10.1017/S0022112094003812

Landau, L. D., & Lifschitz, E. M. (1959). *Fluid Mechanics* (Vol. 6). Pergamon Press.

Meyer, H., & Kogan, A. B. (2002). Onset of convection in a very compressible fluid: The transient toward steady state. *Physical Review E: Statistical, Nonlinear, and Soft Matter Physics*, *63*(5), 056310. doi:10.1103/PhysRevE.66.056310 PMID:12513599

Nakano, A., & Shiraishi, M. (2004). Numerical simulation for the Piston effect and thermal diffusion observed in supercritical nitrogen. *Cryogenics*, *44*(12), 867–873. doi:10.1016/j.cryogenics.2004.06.002

NIST. (2000). *Thermophysical properties of pure fluids database*. NIST.

Paolucci S. (1982). *On the filtering of sound from the Navier-Stokes equations*. SAND 82-8257.

Patankar, S. V. (1980). *Numerical Heat Transfer and Fluid flow*. New York: Hemisphere.

Peter, L., David, W., Michael, F., Fincke, E., Leroy, C., & William, M. (2007). Microgravity Phase Separation Near the Critical Point in Attractive Colloids. *45th AIAA Aerospace Sciences Meeting and Exhibit*.

Spiegel, E. A. (1965). Convective instability in a compressible atmosphere. *I. The Astrophysical Journal*, *141*, 1068. doi:10.1086/148197

Stanley, H. E. (1971). *Introduction to Phase Transitions and Critical Point Phenomena*. Oxford, UK: Oxford University Press.

Zappoli, B. (1992). The response of a nearly supercritical pure fluid to a thermal disturbance. *Physics of Fluids. A, Fluid Dynamics*, *4*(5), 1040–1048. doi:10.1063/1.858255

Zappoli, B., Amiroudine, S., Carlès, P., & Ouazzani, J. (1996). Thermoacoustic and buoyancy-driven transport in a square side-heated cavity filled with a near-critical fluid. *Journal of Fluid Mechanics*, *316*(-1), 53. doi:10.1017/S0022112096000444

Zappoli B., Amiroudine S., Gauthier S. (1997). Instabilité gravitationnelle dans un fluide supercritique pur. *C.R. Acad. Sci. Paris*, *324*, Série II b.

Zappoli, B., Beysens, D., & Garrabos, Y. (2015). *Heat Transfer and Related Phenomena in Supercritical Fluids*. Berlin: Springer-Verlag.

Zoueshtiagh, F., Amiroudine, S., & Narayanan, R. (2009). Experimental and numerical study of miscible Faraday instability. *Journal of Fluid Mechanics*, *628*, 43. doi:10.1017/S0022112009006156

KEY TERMS AND DEFINITIONS

Faraday Instability: When the vibration frequency exceeds a critical value, the flat hydrostatic surface between two liquids enclosed in a vibrating receptacle becomes unstable. This is known as the Faraday instability.

Hydrodynamic Instability: The field which analyses the stability and the onset of instability of fluid flows. The study of hydrodynamic stability aims to find out if a given flow is stable or unstable, and if so, how these instabilities will cause the development of turbulence.

Rayleigh-Bénard Instability: The Rayleigh-Bénard instability, or natural convection, is a thermo-convective instability developed in a fluid medium subjected to a destabilizing temperature gradient. It results in the formation of convective structures called Bénard cells.

Rayleigh-Taylor Instability: The Rayleigh-Taylor instability is an instability of the interface separating two fluids of different densities, which results from the thrust of the heaviest fluid on the lightest fluid (acceleration in the case of a dynamic system or gravity for an initially static system directed to the light phase).

Rayleigh-Vibrational Instability: The Rayleigh vibrational instability is formed when the vibration direction is tangential to the thermal boundary layer. This instability becomes important when a boundary is subject to cooling or heating.

Schwarzchild Criterion: If a gas is unstable against convection then if an element is displaced upwards its buoyancy will cause it to keep rising or, if it is displaced downwards, it is denser than its surroundings and will continue to sink. This is the essence of Schwarzchild criterion.

Supercritical Fluid: When a fluid is heated beyond its critical temperature and when it is compressed above its critical pressure, the fluid is said supercritical. This state of matter was discovered in 1822 by Charles Cagniard de Latour. The physical properties of a supercritical fluid (density, viscosity, diffusivity) are intermediate between those of liquids and those of gases.

Chapter 3
Thermal-Mechanical Effects and Near-Critical Fluid Dynamic Behaviors in Micro-Scale

Lin Chen
Tohoku University, Japan & Japan Society for the Promotion of Science (JSPS), Japan

ABSTRACT

Supercritical CO_2 fluid has been widely used in chemical extraction, chemical synthesis, micro-manufacturing and heat transfer apparatus, and so forth. The current chapter deals with near-critical CO_2 micro-scale thermal convective flow and the effects of thermal-mechanical process. When the scale becomes smaller, new and detailed figures of near-critical thermal effects emerges. To explore this new area, theoretical developments and numerical investigations discussed and explained in this chapter. From a theoretical point of view, the thermal-mechanical nature of near-critical fluid would play a leading role in small time and spatial scales. This effect is found to dominants the thermal dynamic responses and convective structures of micro-scale fluid behaviors. The scaling effects, boundary thermal-mechanical process, instability evolutions, mixing flows and characteristics, possible extensions and applications are also discussed in this chapter.

BASICS OF CRITICAL FLUID FLOW AND HEAT TRANSFER: FROM SUPERCRITICAL TO NEAR-CRITICAL

Supercritical Flow and Heat Transfer in Microchannels

Near-Critical Background and General Applications

In recent years, supercritical fluid and near-critical fluid have been utilized in micro-/nano-chemical synthesis, microscale fluid flow and heat transfer devices, chemical extraction and separation, biological synthesis and many other fields. The applications of critical fluids have been welcomed in many advanced strategic technologies of 21st century, such as novel chemicals, energy technology, biology and

DOI: 10.4018/978-1-5225-2047-4.ch003

other fields (Kuang et al., 2004; Kumar et al., 2011; Zhao et al., 2011). For example, the utilization of supercritical H_2O and CO_2, by using their special density and diffusivity characteristics, has been found very effective for chemical extraction and preparation processes, which have become matured 'green' technology nowadays (Wang et al., 2011). Also, based on the preferable thermal and transport properties of supercritical/near-critical fluids, new heat transportation designs have been proposed and tested, which have shown potential in new generation of renewable energy technologies and energy conversion systems (Zhang et al., 2007; Zhang and Yamaguchi, 2008; Chen and Zhang, 2011). For example, for new generation of nuclear power plant, supercritical fluids based system has become one of the promising choices in recent framework for the global development (Dimmic et al., 2002). The application fields of critical fluids also include solar conversion, heat pump design, cooling and drying, micro-scale cooling, micro-devices and heat transfer designs. Such system design and performance analysis have drawn the attention of both engineers and scientists around the world and have become one major direction of future development in fluid and energy technology fields.

At the same time, the fluid flow and heat transfer engineering field has been more focused on micro-scales in recent years, such as for molecular level understanding of human life, chemical engineering, medicine, energy and resources engineering (Ameel et al., 1997). Several representative projects of the supercritical/near-critical fluids in microscale engineering can be found in Figure 1, where the micro-/nano-scale chemical synthesis, chemical mixing and reaction, novel energy conversion and heat transfer heating/cooling applications are shown. The scientific analysis and design of those processes have to consider the microscale effects for critical fluid flow and heat transfer under designed conditions. There are a lot of question to be asked before one can try to see into the new world of supercritical/near-critical world! For instance, if the nanoparticle is to be synthesized in a microscale chamber, which is one typical situation of nanoparticle synthesis, the dynamics of fluid confinement and local heat transfer (equilibrium or irreversible procedures) of critical fluids should also be considered, besides the generally known microscale temperature and pressure control (Kuang et al., 2004; Wang et al., 2011; Zhao et al., 2011).

For supercritical/near-critical fluid, the challenges for system control and stability formation are much different from traditional kinds of operation systems. This is due to the microscale effect. The near-critical synthesis of nanoparticles in micro-chamber can provide not only much better products for customers but it also yields much critical conditions of formation (Beckman, 2004; Sierra-Pallares, et al., 2011). The near-critical fluids have very large changes of properties when they go across (or near) the critical point. The specific heat tends to infinity while the thermal diffusivity goes to zero, which effect will lead to special thermal convective structures and operation time scales. Therefore, the applications and their optimizations need more deep understanding into the fundamentals of supercritical/near-critical fluid dynamics and transport characteristics. In recent years, many new findings and breakthroughs have been made by scientists (Zappoli, 2003; Carles, 2010).

Characteristics of Near-Critical Microchannel Flows

In real designs, near-critical fluid flow and heat transfer would largely be affected by its thermal status of existence and operation parameters. For example, when Wang and his colleagues (2011) tried to use a channel of several hundred micrometers for the preparation of micro/nano-particles (sized from 0.5 μm to 5.0 μm), they utilized a precise control of the operation temperate and pressure in each procedure, thus to give good synthesis route, rate and product quality. The detailed changes and evolutions of such

Figure 1. Applications of near-critical mini-/micro-channel flow and heat transfer: (a) nanoparticle synthesis inside near-critical CO_2 fluid; (b) Microscale T-channel mixing and reaction inside near-critical H_2O fluid; (c) solar conversion using trans-critical CO_2 fluid natural circulation; (d) Electronic cooling tests with supercritical CO_2 jet flow
Wang et al., 2011; Blood et al., 2004; Chen and Zhang, 2014; Ducoulombier et al., 2011.

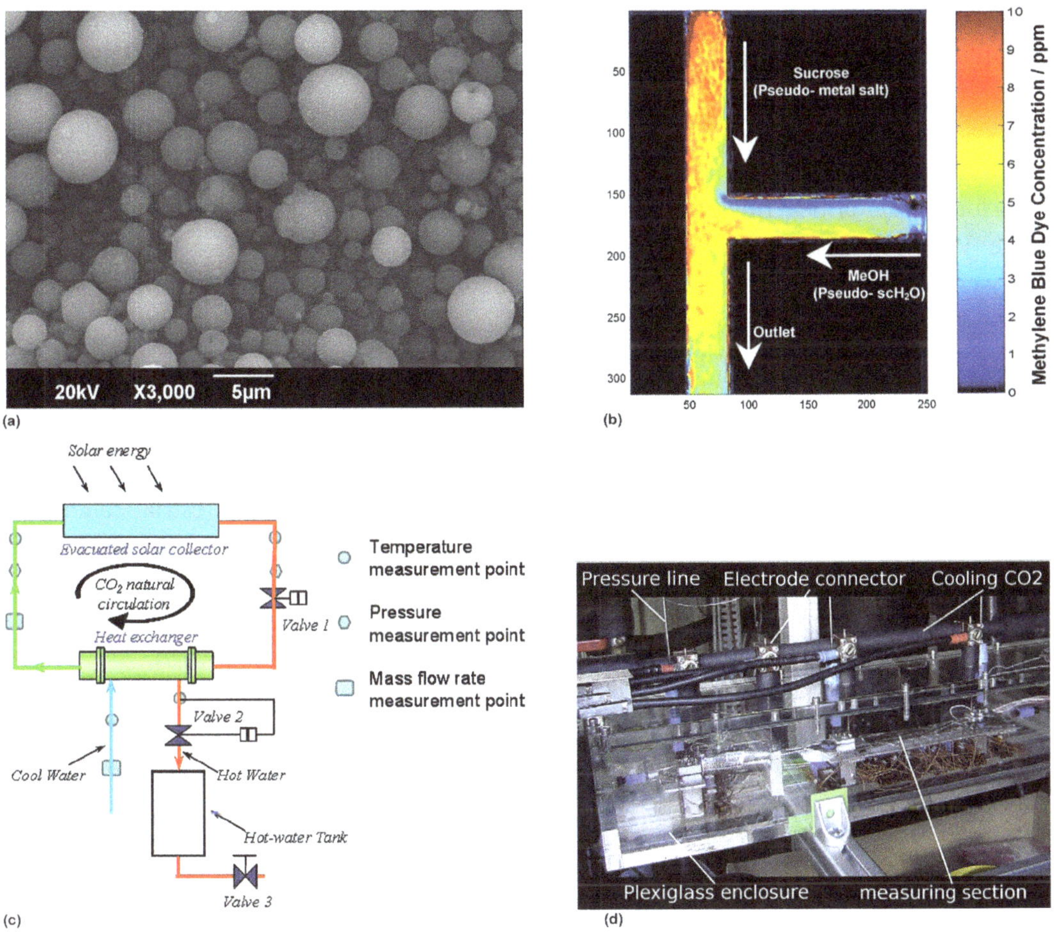

processes inside microscale for near-critical fluids are quite complicated and sensitive as for the control of production or fluid manipulations, which brings new challenges in the design of heat and mass transportation processes in both theoretical analysis and real applications.

Near-critical fluid microchannel flows should also take the above mentioned effects into consideration. The entrance effects, geometry and surface effects are similar with general sized channel. However, for supercritical/near-critical fluids, there are special points to be clarified: (1) rarefied gas effect is not important as the density of critical fluids are still comparable large (similar to the density of liquid), and a calculation of Knudson number would easily be helpful; (2) the surface tension of near-critical fluids would be zero when the condition goes above the critical point; therefore the supercritical fluid flow would have high diffusivity and solubility as a host fluid and also have low flow friction and pressure loss. Thus, it affects the energy consumption and also stability (Chen et al., 2010; Zhang et al., 2010);

(3) the nonlinear changes of near-critical fluid properties is the major difference from normal fluids; as the properties would change dramatically when it goes near the critical point, some new and different behaviors may become possible in microscale.

Near-Critical Difficulties and New Thermodynamic Challenges

Challenges of Near-Critical Microchannel Flows

In summary, many advanced technologies for modern industry, the society and the environment are based on the fundamental development in microscale fluid flow of normal kinds and near-critical kinds. Especially for the near-critical fluids, which would be advantageous for their high efficiency and compact design, the flow dynamics and heat transfer would be critical for the enhancement of system upgrading and energy efficiency. At the same time, the microscale heat and mass transfer mechanism should also welcome new findings and new laws that are much different from traditional transport process for near-critical fluids. The following section will cover the importance of such findings.

New Findings in Thermodynamics of Near-Critical Fluids

Traditional near-critical heat transfer studies are generally conducted under terrestrial conditions (1g), and the major controlling factor of convection is buoyancy effect. However, under microgravity condition, the effect of buoyancy convection is largely limited, which then makes it possible to see clearly the pure heat transfer process of near-critical fluid. According to classical thermal diffusion theory, the characteristic time scale for thermal diffusion can be expressed as (Kawasaki, 1970):

$$t_{\xi} = D^{-1}\xi^{2} \tag{1}$$

$$t_{\xi} \sim \tau^{-3\left(3\nu + Y_{\eta}\right)} = \tau^{-1.93} \tag{2}$$

where t_{ξ} is the characteristic time scale of fluid thermal diffusion process inside a space with characteristic length scale ξ. It can be found that when the fluid approaches the critical point, t_{ξ} goes to infinity, which means that when the fluid go near the critical point, the thermal diffusion would be slowed down (the so-called "critical slowing down") (Carles, 2010; Zappoli et al., 2015).

In mid-1980s, an astonishing finding was reported on the abnormal thermal equilibrium process of near-critical CO_2 fluid inside close chamber by space experiment of European Spacelab mission. In this experiment on a space mission, scientists found that when a fluid is paced in a closed chamber just 1 mK above the critical point is subjected to boundary heating, heat transfer speed can be much faster than previous theoretical estimation. The thermal equilibrium in the 1 cm size chamber is predicted to be around tens of seconds, but the real process in space mission is smaller than 0.1 s (Zappoli, 2003; Carles, 2010).

After the experiments in the Europe space mission, a series of experimental studies have been conducted both on earth surface and in space mission. Those tests exposed to the public a new world of thermal transportation and convection process of near-critical fluid, and meanwhile raised new questions/ challenges to the scientific world.

Development and Overview of This Chapter

Near-Critical Thermal Equilibrium and Microchannel Flow

The 'abnormal' near-critical fluid thermal behaviors under microgravity soon attracted many physicists in the analysis of new features of the unique fluid status. The basic timeline of the development of this field can be found in Figure 2. A series of experimental and theoretical investigations have revealed that the near-critical process was much more complicated than previous explanations (Zappoli, 2003; Carles, 2010; Zappoli, et al., 2015). In early 1990s, three groups reported their findings: near-critical fluids have highly nonlinear thermal fluctuation diverges and compressibility changes, which would lead to unusual thermal boundary compression and reflection in the form of pressure waves in a critical fluid chamber (Zappoli et al., 1990; Boukari et al., 1990; Onuki et al., 1990). Such process was called the "Piston Effect" (PE) as the fluid chamber can be heated up by the work of pressure waves like a piston, which was also recognized as the fourth heat transfer mechanism besides conduction, convection and radiation. Onuki et al. (1990) firstly developed a Heuristic model to analyze the thermal boundary equilibrium process in near-critical fluid, where the boundary thickness (or characteristic length scale of it), thermal diffusivity is used to estimate the characteristic time scale of the Piston Effect:

$$t_{PE} = \frac{L^2}{4\left(\gamma - 1\right)^2 D_T} \sim \tau^{1.6} \tag{3}$$

where L is the characteristic length, γ is the specific heat ratio, D_T is the thermal diffusivity. This model explained the boundary thermal expansion process with a time scale much larger than diffusion time scale.

Based on that mechanism, several new topics and challenges emerged: (1) from the near-critical microscale heat and mass transfer theory, the highly expandable fluid interaction with confinement of geometry would have new behaviors in microscale (Zappoli, 2003); (2) microscale convection structure and evolution of near-critical fluid would be challenging for both microgravity and terrestrial conditions (Zappoli et al, 1990); (3) near-critical region would have two-phase or single phase, which would be challenging for the consideration of surface tension and fluid flow behaviors; (4) the complex coupling effects with gravity, which is much more difficult than traditional situations (Jounet et al., 2000; Chiwata and Onuki, 2001). The above mentioned main studies raised a lot many new problems and challenges to the scientific world, which is of both importance and original interest.

In addition, the microchannel studies of supercritical CO_2 fluid flow have limited parameter ranges. The physics behind the behaviors are not discussed, due to the complex processes in near-critical regions, which is difficult to describe and analyze. The variation of properties in the near-critical region largely affects the transient heat transfer and flow dynamics in microscale. When the fluid goes to critical, the isothermal compressibility diverges and the fluid diffusivity goes to zero. However, how such kind of

Figure 2. Piston effect and thermal-mechanical effect related research history and major topic flow

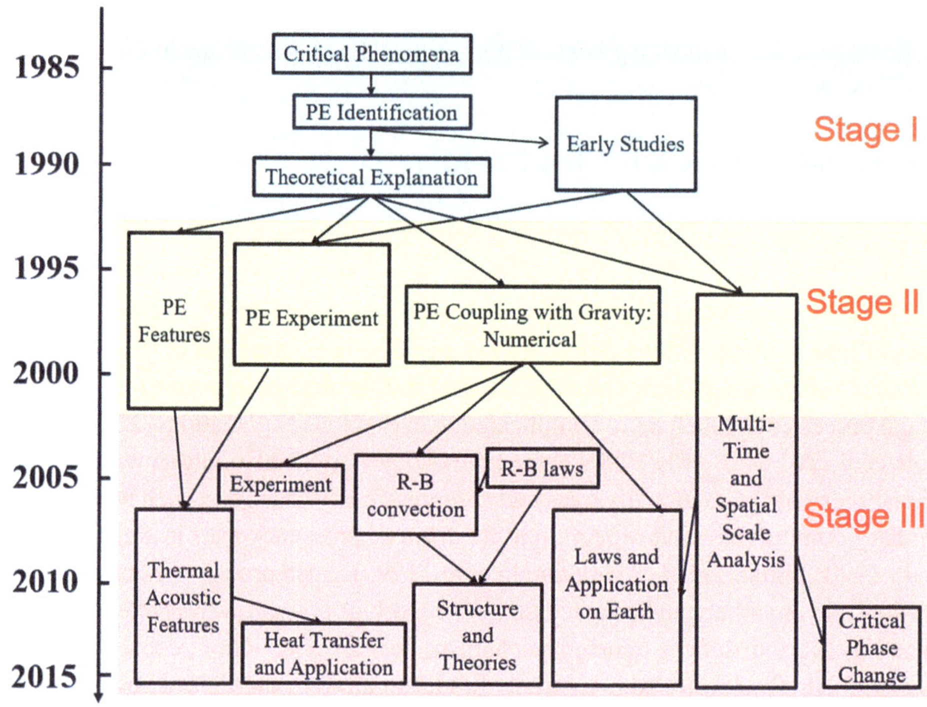

variations in the microchannel will affect the fluid transient behavior and heat transfer? How is the result under different flow rate and intensity of heat input? What is the response of the near-critical fluid? And how the interactions affect the fluid flow itself? Those questions remain un-answered. It has been proved that the fluid property changes in near-critical region may become the dominant the convection process in microscale, which is much different from normal scale process (Jounet et al., 2000; Chiwata and Onuki, 2001). New convective structures and fluid dynamic behaviors have been found for near-critical CO_2 (Jounet et al., 2000; Chiwata and Onuki, 2001; Zappoli, 2003; Frohlich et al., 2006; Carles, 2010; Chen and Zhang, 2014). The investigation of the basics behaviors for near-critical fluid in microscale has become an urgent and important task in recent years.

Overview of This Chapter

This chapter covers the theoretical and numerical analysis into the near-critical microchannels. The overall system parameter studies and theoretical analysis of near-critical fluid through flows are carefully conducted. The numerical results then extend the analysis to multi-scale and it provides a general view about near-critical perturbation, hot boundary evolution as well as the coupling of gravity effect. The mechanisms and governing laws for microchannel near-critical flows are developed in this chapter. This chapter firstly presents the numerical formulation of near-critical microchannel flow and solving method, where the near-critical vortex flow and novel stability variations are discussed. The detailed analysis of near-critical stability evolution, characteristics of the microchannel vortex formation, flow behaviors, parameter dependency is also conducted. Based on the multi-scale analysis model, the Thermal-Mechanical nature of near-critical fluid is applied to microchannel flows. The Kelvin-Helmholtz stability evolution is extended to microscale and the basic perturbation sources are identified. Finally, the generalization and summary are made.

NUMEICAL FORULATION OF NEAR-CRITICAL MICROCHANNEL FLOW PROBLEMS

To further explore the critical thermal convective flow, numerical simulations of basic microchannel models with CO_2 flow near its critical point are reported in this section.

Physical Model and Simplifications

For the single phase microchannel flow, previous studies have concluded that conventional Stokes and Poiseuille theories can provide reliable predictions (Steinke and Kandlikar, 2006), while other studies found them to be unsuitable under certain specific conditions (Koo and Kleinstreuer, 2005). The main reviews have shown that macrochannel correlations can still give reliable predictions as long as the entrance or exit losses are appropriately accounted for (Rostami et al., 2000; Rostami et al., 2002; Morini, 2004; Steinke and Kandlikar, 2006) and the Reynolds number is low (generally below 600) in microchannels (with height larger than 50 μm) (Flockhart and Dhariwal, 1998). For high Reynolds flow, large deviations can be found, and the condition should be dependent on the specific fluid (Cui and Li, 2004). Further, special care should be taken with fluids that feature nonlinearly varying properties, such as supercritical fluids (Rosa et al., 2009).

System Design and Realization

In the current study, a simplified two-dimensional microchannel model with length (*L*) and height (*D*) is constructed, which is representative for microfluidic devices that are generally studied. As shown in Figure 3, near-critical fluid flows into the microchannel from the left side and exits through the outlet. Heat fluxes can be applied at the walls on either side of the microchannel. Initially, an isothermal wall condition is assumed to obtain an isothermal steady flow field. In previous studies, the microchannel flow correlations are shown to be greatly affected by the 'thermal entrance length' for thermal development at the inlet (Ducoulombier et al., 2011). However, the present long, thin (with large *L/D* values) microchannels can generate stable near-critical fluid flow (Chen et al., 2013) and are convenient for real operations such as the nano-particle synthesis process in microchannels (Zhao et al., 2011). Compressibility effect should be considered as in previous studies the thermal perturbation (Zappoli and Carles, 1995; Bailly and Zappoli, 2000; Jounet et al., 2000) and "thermal jet" phenomena (Frohlich et al., 1996, 2006; Garrabos et al., 1998) are often found in the stability analysis. Gravity is also considered in the minus-y direction, as shown in Figure 3. Indeed, the physical behavior of the microchannel flows will be dependent on the height (*D*), the gravity effect in microchannels can be limited, and transitions can be found due to the variations of Froude numbers for microchannels (Chen et al., 2013a, 2013b, 2016). The basic transition of gravity effects is also shown for the current near-critical CO_2 convective flow.

Figure 3. Physical model of the microchannel studied

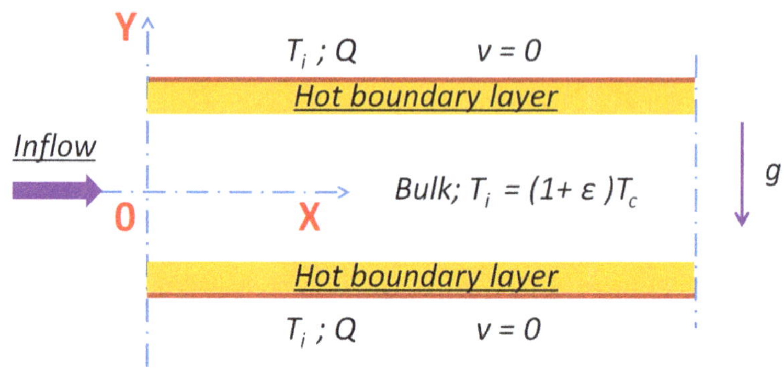

The first goal of the current study is to investigate the basic hydraulic behaviors of near-critical fluid flow in microchannels. However, special care should be taken when treating the micro-scaling effects (Rosa et al., 2009) in microchannels, as discussed in the Introduction, such as rarefied gas effect, non-linear fluid properties, fluid compressibility, boundary slip effect, effects of wall roughness and wall structure, surface tension and wettability (Flockhart and Dhariwal, 1998; Cui and Li, 2004; Kumar et al., 2011). For the current case, the inlet and outlet conditions are carefully addressed with the use of long, thin channels with a supercritical flow condition such that no capillary effect occurs. Additionally, for low Reynolds (Re) number flow (generally less than 600) in microchannels (height larger than 50 μm), the Navier-Stokes law still applies (Flockhart and Dhariwal, 1998; Cui and Li, 2004). Therefore, in the current study, the Navier-Stokes law is used and the nonlinearly varying properties are manipulated to give reasonable predictions of near-critical CO_2 convective flow.

The basic thermal-physical properties of near-critical CO_2 fluid are plotted in Figure 4. It can be seen that from sub-critical pressure (6.0 MPa and 7.0 MPa) to supercritical pressure conditions (8.0 MPa and 9.0 MPa), the properties experience large changes as to the value and trends. In the near-critical region, utilization of the high peaks (sometimes also diverges) is often made in real applications. However, the real process and understanding of the detailed evolutions are still far from enough.

Additionally, near-critical fluid can be unique and local thermal relaxation processes can be very different in microchannels (Zappoli et al., 1990; Frohlich et al., 1996; Onuki et al., 1990; Zhong and Meyer, 1995; Zappoli, 2003; Miura et al., 2006). Thermal vibrations and local perturbations from the HBL (Hot Boundary Layer) and 'thermal jets', respectively, are found for near-critical fluids (Zappoli et al., 1990; Jounet et al., 2000; Chiwata and Onuki, 2001; Frohlich et al., 2006), which is different from general expected stable fluid flow under confined microchannel conditions. The current study attempts to take advantage of the nonlinear (but preferable) thermal-transport properties for micro-fluid manipulation and hydraulic control. As the first step of the simulation, isothermally developed flow in the microchannel is conducted for each configuration. The microchannel walls are initially maintained to be isothermal until the near-critical flow reaches steady state. Then, the fluid is subjected to a two-sided heat flux Q to investigate the fluid responses from isothermal flow to a sudden heat flux input. This study utilizes the compressible Navier-Stokes equation to solve the basic flow field in a near-critical microchannel to give a basic picture of the convection flow behaviors.

Figure 4. Variation of thermal properties with temperature and pressure in the critical region. (a) specific heat; (b) density; (c) thermal conductivity; (d) viscosity

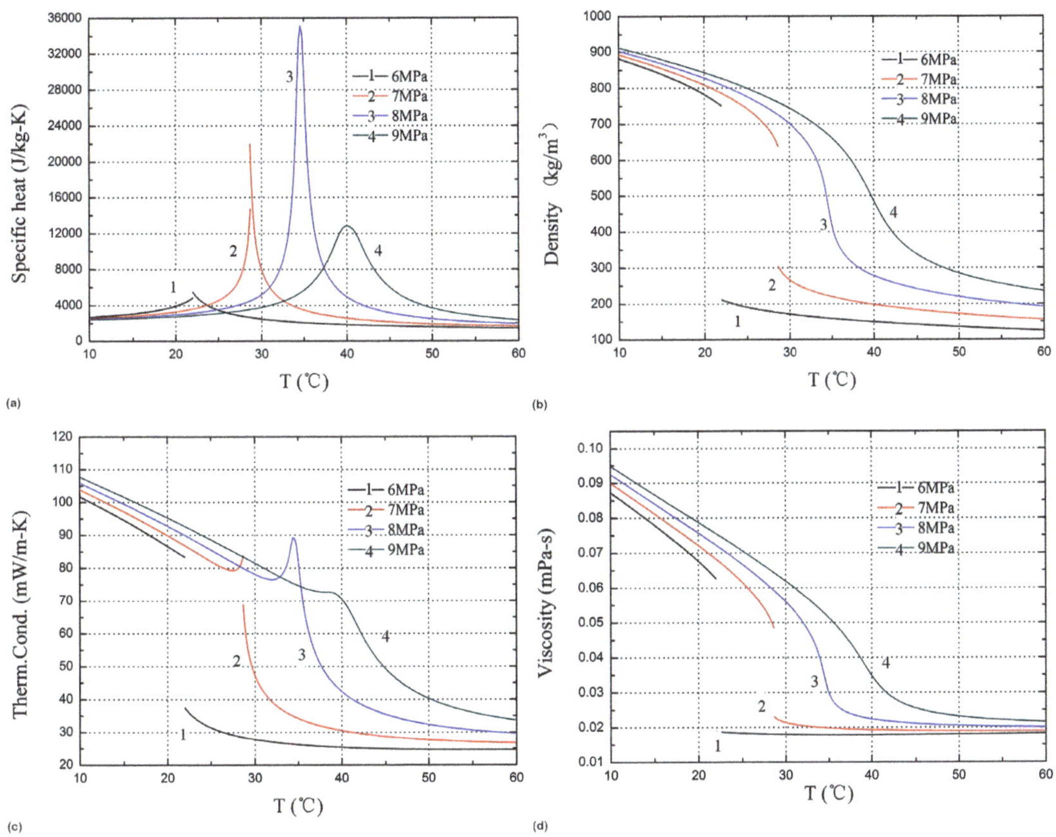

Governing Equations

In the current simulation, continuity assumption is still used as discussed in former sections. For the numerical model, in addition to the compressible Navier-Stokes equations, the current model equations are solved together with the energy equation and the equation of state written for compressible fluid:

Mass equation:

$$\frac{\partial \rho}{\partial t} + \nabla\left(\rho \boldsymbol{V}\right) = 0 \tag{4}$$

Navier-Stokes equation:

$$\frac{\partial \rho \boldsymbol{V}}{\partial t} + \nabla \cdot (\rho \boldsymbol{V}\boldsymbol{V}) = -\nabla P + \nabla\left[\mu(\nabla \cdot \boldsymbol{V} + \nabla \boldsymbol{V}^{T})\right] + \rho \vec{g} \tag{5}$$

Energy equation:

$$\frac{\partial\left(\rho C_p T\right)}{\partial t}+\nabla\cdot\left(\rho V C_p T\right)=\nabla\left(\lambda\nabla T\right)+T\beta_P\frac{\partial P}{\partial t}+\Phi \tag{6}$$

Equation of state:

$$\rho=\rho_0+\rho_c\chi_T\left(P-P_0\right)-\rho_c\beta_P\left(T-T_0\right) \tag{7}$$

where β_P is the isobaric thermal expansion coefficient, χ_T is the isothermal compressibility, both of them are non-linear parameters as the properties diverges near to the critical point. λ is the thermal conductivity, C_P is the specific heat, $\Phi=\sigma_{ij}\left(\partial u_i/\partial x_i\right)$ is the dissipation function with σ_{ij} the viscous stress tensor and u_i the velocity components, P_0 and T_0 are initial pressure and temperature. It should be noted that non-Boussineq assumption for the equation of state is used for the description of pseudo critical region and the critical scaling is used for close-critical region.

Numerical Methods and Conditions

Non-uniform grids are used in the calculation domain with mesh grids ranging from 3×10^5 to 2×10^6 for different case configurations and are refined specifically for the boundary and internal areas during the initial studies to ensure that mesh-independent results are obtained. The current model refers to the NIST-REFPROP standard database (2006) for the nonlinear thermal-transport coefficients and properties, including β_P and X_T development, thus providing relatively accurate qualitative predictions (Chen et al., 2013a, 2013b; Zappoli et al., 2015; Chen et al., 2016). For the near-critical region, the critical exponent (power law) is used, which obeys the universality law of critical fluids (Jounet et al., 2000; Zappoli et al., 2015).

For the transient stability evolution analysis, a finite volume solution method was used, and a numerical code was developed and validated in previous studies (Amiroudine and Zappoli, 2003; Chen et al., 2013). Indeed, special difficulties have been found in solving the coupled convection and thermal propagation process in near-critical fluids. For the numerical algorithm, the pressure-implicit with splitting of operators (PISO) algorithm is used to solve the coupling between the velocity and pressure. The PISO algorithm has been previously validated for the analysis of both acoustic time scale and normal time scale phenomena (Amiroudine and Zappoli, 2003; Chen et al., 2013). The momentum terms and energy terms in the governing equations are iterated with a second-order upwind scheme, which uses the upstream values and gradients to compute the control volume face values. For the near-wall simulation, a non-slip boundary condition is used.

In addition, the comparison between the mesh size and the small scale diverges (or fluctuations) of near-critical fluid would be useful for the study of such near-critical phenomena. The ratio of mesh size to correlation length ($\sim\left|\varepsilon_T\right|^{\nu}$) across the current near-critical boundary layers ranges from 0.2 to 1.0 and in the bulk the effect is generalized as the critical fluctuations will decay with the increase of 'distance' from critical point during heating (Zappoli et al., 2015). Respective correlation time for this diverges is also very much smaller (less than 0.01%) than that of the current pressure relaxation or thermal-mechanical relaxation time scales.

Boundary and Initial Conditions

The focus of the current study is the near-critical convection flow thermal and hydraulic behaviors in microchannels. Steady state theories for the transient convection onset exist for incompressible and moderately compressible fluids; however, interesting phenomena occur when the fluid approaches near-critical region (Zappoli et al., 1990; Jounet et al., 2000; Chiwata and Onuki, 2001). Currently, we define CO_2 inlet condition with inlet temperature T_{in}, initial temperature T_0, and inlet pressure P_{in}, initial pressure P_0, and refer to T_c and P_c as respectively the critical temperature and pressure. Generally, the dimensionless proximity to the critical point can be written by factor ε_T and ε_p (or referred to as 'critical distance' in literature) and respective parameters can be written as bellow:

$$T_{in} = T_0 = \left(1+\varepsilon_T\right)T_c \tag{8}$$

$$P_{in} = P_0 = \left(1+\varepsilon_P\right)P_c \tag{9}$$

In the current numerical study, the iteration time step of the numerical iteration ranges between 10^{-6} and 10^{-5} to capture the details of the boundary thermal-mechanical process (Amiroudine and Zappoli, 2003; Shen and Zhang, 2011; Chen et al., 2013). The convergence limitation is 10^{-3} for the residuals of continuity and velocity and 10^{-6} for the energy term in each step of iteration. Other numerical details can be found in previous studies (Amiroudine and Zappoli, 2003; Shen and Zhang, 2011; Chen et al., 2013). For the first step, the microchannel length is set as $L = 5.0$ cm with channel height D ranging from 1.0×10^{-4} m to 5.0×10^{-4} m ($L = 2.0$ cm and $L = 8.0$ cm cases were also set for testing and comparison).

Basic Near-Critical Microchannel Convective Flows and New Findings

System Stability Trends

Pressure Drop and Friction Analysis

Before the discussion of convective flow and heat transfer characteristics, the analysis of friction-pressure drop would be useful for the initial understanding of the near-critical flow. The Re number of the current discussion is around 3-300, which is the major work range of microfluidics (Kumar et al., 2011). In the discussion of basic hydrodynamic behaviors of CO_2 flow, the friction factor plays a key role in microchannels, major review studies (Cheng and Thome, 2009; Ducoulombier et al., 2011) all recommend the classical Churchill's Law defined as below (Ducoulombier et al. 2011):

$$f_{Churchill} = 2\left[\left(\frac{8}{\mathrm{Re}}\right)^{12} + \left(\frac{1}{\left(A+B\right)^{3/2}}\right)\right]^{1/12} \tag{10}$$

with Re the Reynolds number of CO_2 flow. The A and B quantities are calculated from the following equations:

$$A = \left[2.457 \ln \frac{1}{\left(\frac{7}{\text{Re}}\right)^{0.9} + \frac{0.27k}{D}} \right]^{16} \quad (11)$$

$$B = \left(\frac{37530}{\text{Re}} \right)^{16} \quad (12)$$

where k is the roughness and D is the inner height of microchannel.

For the isothermal flow conditions, basic flow friction factors are calculated and compared with the classical Churchill's correlation. As shown in Figure 5, the general trends for the current study fit relatively well with Churchill's correlation, with the maximum deviations below 25%. In the current low Reynolds region, the friction factors generally decrease with an increasing Reynolds number.

Churchill's law has been identified to be capable of accurately predicting the friction factor of supercritical CO_2 flow. However, deviations have also been found in literature (Ducoulombier et al., 2011), in which the main experimental data points fall slightly below Churchill's curve and acceptable points show deviations within ±30%. Figure 5 shows that the friction factor in the current microchannel study also falls below the theoretical prediction, and friction reduction has been typically found for supercritical CO_2 pipe and channel flows in previous studies (Dimmic et al., 2002; Zhang et al., 2010; Chen et al., 2010, 2012; Chen and Zhang, 2011).

In addition, the comparison of the friction factors calculated from experiments and from theoretical analyses (usually defined as $C^* = (f{\cdot}Re)_{exp}/(f{\cdot}Re)_{th}$) generally falls between 0.5 and 1.50 for microchannels; several very large deviations have also been reported (Kumar et al., 2011). In the current study, the calculated values were about 0.80. The near-critical deviations are also partly due to the dramatic change of properties near the critical region, where very low viscosity is seen for slight changes of temperature, as shown in Figure 5. The Churchill's Law can still be considered capable of predicting trends for near-critical flows in microchannels from the current results.

Unstable Flow and New Findings

The representative microchannel flow stability evolutions are shown in Figure 6. It can be seen that after heat application from the two-sided walls, the near-critical fluid will experience several instable stages toward vortex flow region. Similar to the previous microchannel investigations, in Figure 6, the microchannel flow for $D = 100$ μm condition is plotted. At early stage after heat application, the flow is conduction dominated (see $t = 20$ ms). During this stage, very thin hot boundary layers are formed near the two-sided walls. This stage is due to the very low thermal diffusivity found for near-critical fluids. Then it is found that from $t = 40$ ms, as shown in Figure 6, boundary vortex is formed first near the boundary layer and then it expands toward the mainstream flows. The nonlinear perturbations and thermal-acoustic process dominants the basic expanding process (Boukari et al. 1990; Onuki et al., 1990; Zappoli et al., 1990; Zappoli, 2003; Carles, 2010). This stage is the major characteristic evolution of near-critical fluid convective flow inside confined channels. The unstable source is firstly located near

Figure 5. Isothermal friction factor for near-critical microchannel flows and comparison with classical Churchill's correlation

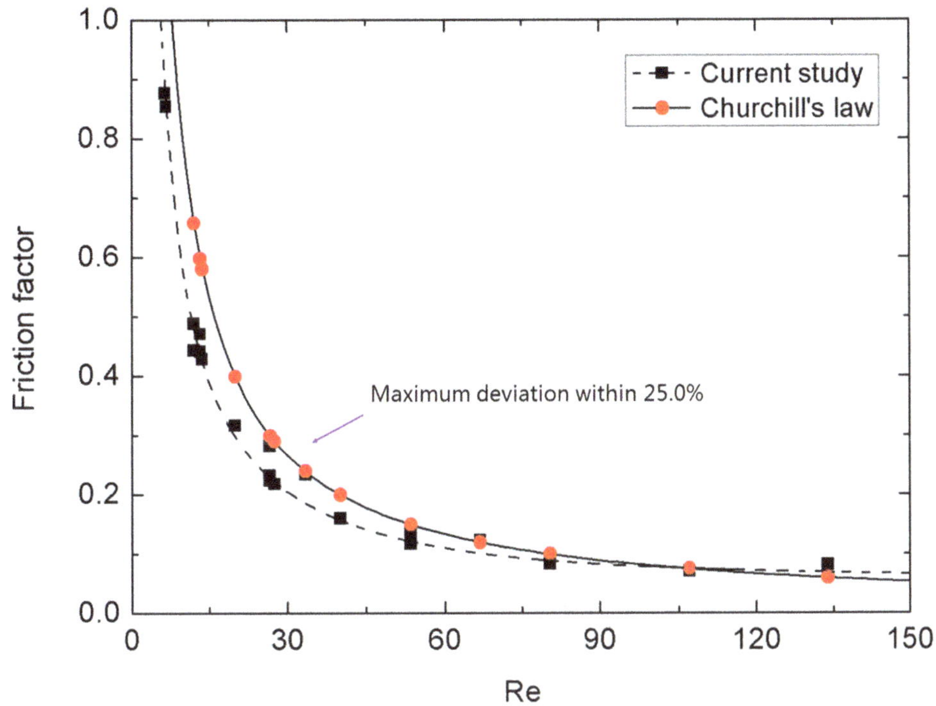

Figure 6. Vortex mixing profiles (near x/L = 0.90) for microchannel cases under gravity with microchannel D = 100 μm, Q = 8000 W/m², V_{in} = 0.005 m/s, Re_{in} = 3.35)

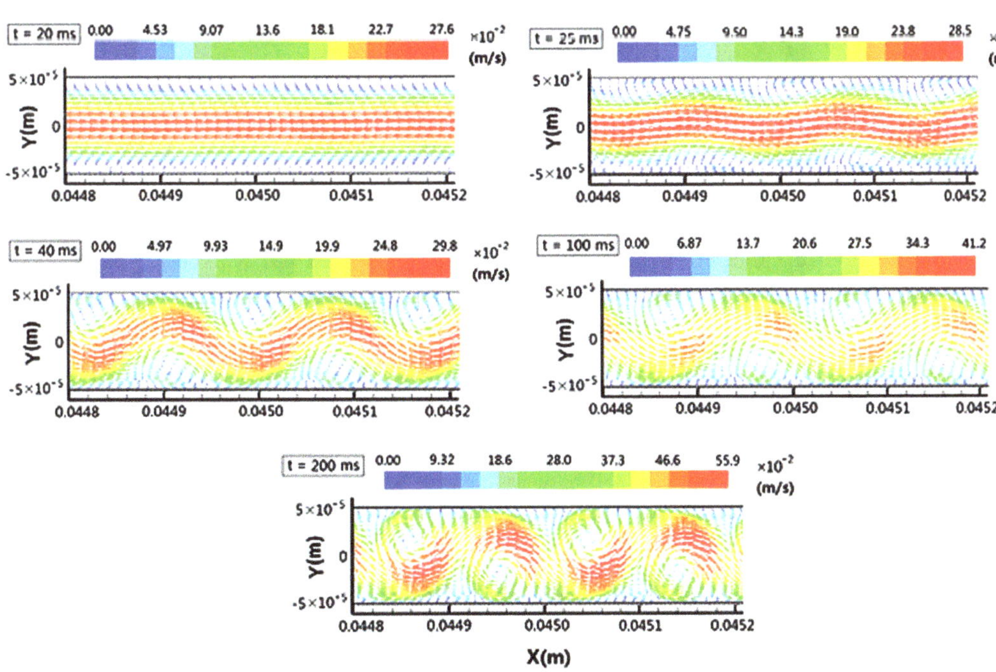

the thin hot boundary and such thin hot boundary expands and breaks up the mainstream flow stability. Finally, the flow will become vortex dominated condition, as shown in Figure 6 (t = 200 ms). Indeed, the microscale vortex flow and mixing process is very challenging in chemical engineering field (Kuang et al., 2004; Kumar et al., 2011; Zhao et al., 2011; Wang et al., 2011), thus the current flow situation may be one possible choice for future discussion in this field.

Such vortex flow happens in very short time scale after the heat flux application, which may be characterized similar with that of thermal-mechanical effect time scale for near-critical fluids (Boukari et al. 1990; Onuki et al., 1990; Zappoli et al., 1990; Jounet et al., 2000; Zappoli, 2003; Carles, 2010; Chen et al., 2013a). After that, the vortex velocity gradually increase and the flow will become strong mixing dominated. At the same time the temperature evolution also shows abnormal behavior. Instead of gradual increase of near-boundary layer fluid temperature, sudden drops of fluid temperatures are found in the microchannels where the vortex happens.

In one former study of near-critical fluid microchannel convection, thermal collapse phenomena have also been reported (Chen et al., 2013a). For boundary flow, such collapse indicates that fast thermal mixing and equilibrium process can happen under the self-organized mixing process. Similarly, for the bulk temperature, it has been reported (Boukari et al. 1990; Onuki et al., 1990; Zappoli et al., 1990; Zhong and Meyer, 1995; Jounet et al., 2000; Zappoli, 2003; Carles, 2010; Chen et al., 2013a) that fast homogeneous heating up can be seen in closed systems, however in the current open channel, such heating still exists due to the same hot boundary layer thermal-mechanical effect (which is also very much expanding under near-critical conditions), which will also be discussed later in this chapter. Such kind of microchanel vortex evolution in response to sudden boundary heating is special for near-critical fluid and it is rarely found in normal fluids based ones.

The Perturbation Flow and Dynamic Characteristics of Microchannel Mixing Flow

Thermal and Density Relaxation on Near-Critical Flows

The transient evolution of wall temperatures is plotted in Figure 7. In the current microchannel model, gravity is neglected as the calculated Froude number is much smaller than unit (Chiwata and Onuki, 2001; Frohlich et al. 2006; Zappoli et al., 2015), therefore it is possible to ignore the effect of gravity. Indeed, the upper wall (Figure 7) temperature evolution is almost the same as lower wall (not shown here). It is seen from Figure 7 that after heat flux was applied, and thermal convection commences, the temperature collapses are found even in small time scales as 10^{-3} s. It is shown that $(T\text{-}T_c)$ drops by 50% after the heat flux is applied for 10^{-2} s, which happens for both 100 μm and 200 μm channels. Also the temperature drops and collapse grow from channel outlet side through the flow while the main wall temperature increases with time. Such collapse indicates much faster thermal equilibrium process in the microchannel than steady thermal convective (or steady thermal diffusion) channel flow condition in normal fluid.

Carles et al. (1990) has summarized the characteristic time scale of near-critical Thermal-Mechanical process that resembles the current evolution, but it still shows much difference from case to case. The current evolution in Figure 7 shows that the temperature drop happens in microchannel will grow and expand toward the main flow. And together with the vortex flow situation (shown in Figure 6), it become possible to reveal the details in near-critical microchannel process.

Figure 7. General temperature relaxation and saturation process in the microchannel. (T_{in} = 304.2 K; P_{in} = 7.5 MPa; V_{in} = 0.01 m/s,). (a) D = 100 μm; q = 10000 W/m²; Re_{in} = 13.40; (b) D = 200 μm; q = 10000W/m²; Re_{in} = 26.80

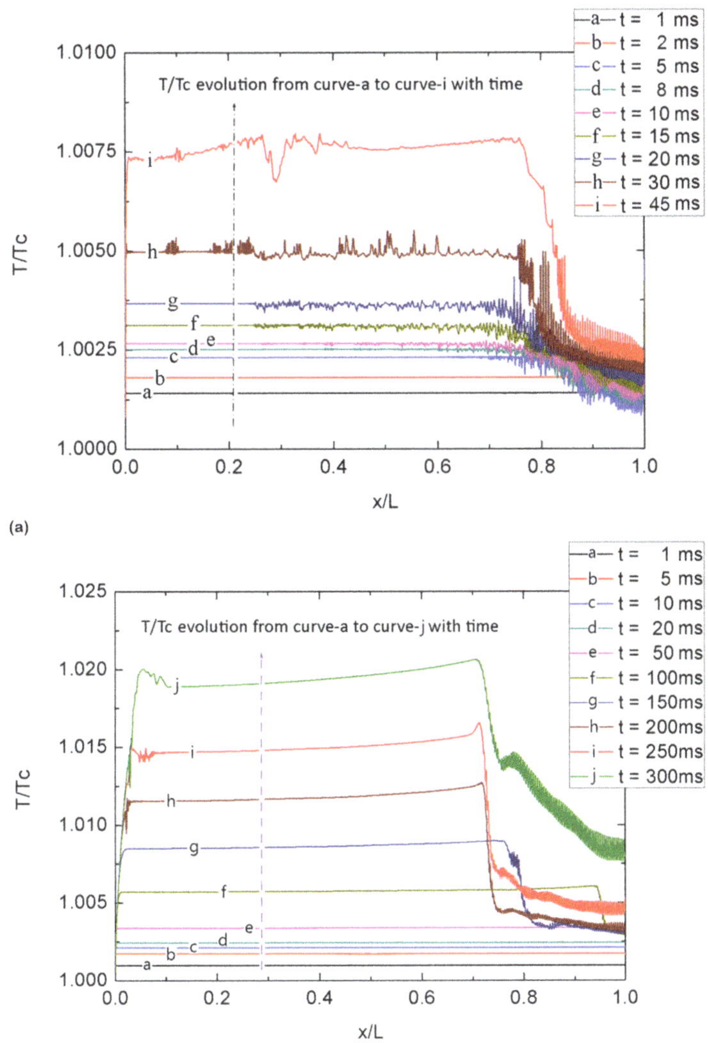

Before the discussion of what happened during the convection onset, the propagation and fluctuation process of near-critical flow should be introduced. The local evolution of fluid temperature including bulk profiles are plotted in Figure 8 for $D = 100$ μm microchannel. As shown in Figure 8, the temperature profiles are generally symmetric and its boundary gradient changes with time due to the peculiar divergences of near critical specific heat. Also the temperature profiles in Figure 8 are all nearly flat in the bulk, which indicates the homogeneous increase of bulk temperature. That homogeneous characteristic of near-critical fluid temperature relaxation is typically seen and ascribed to the fast thermal mechanical effect of fluid (Onuki et al., 1990; Boukari et al. 1990; Zappoli et al., 1990; Zhong and Meyer, 1995; Jounet et al., 2000; Zappoli, 2003; Carles, 2010; Chen et al., 2010; Chen et al., 2013a; Zappoli et al., 2015), and it is fundamentally different from the temperature evolutions shown in Figure 7.

Figure 8. Local bulk-temperature saturation with time (T_{in} = 304.2 K; P_{in} = 7.5 MPa. x/L = 0.85, D = 100 µm; Re_{in} = 13.40, q = 10000W/m²). Thermal equilibrium speeding-up is found for near critical CO_2 (ε_T = 0.00023 and ε_P = 0.01626), the grow of perturbation and temperature collapse indicate critical convection onset instability

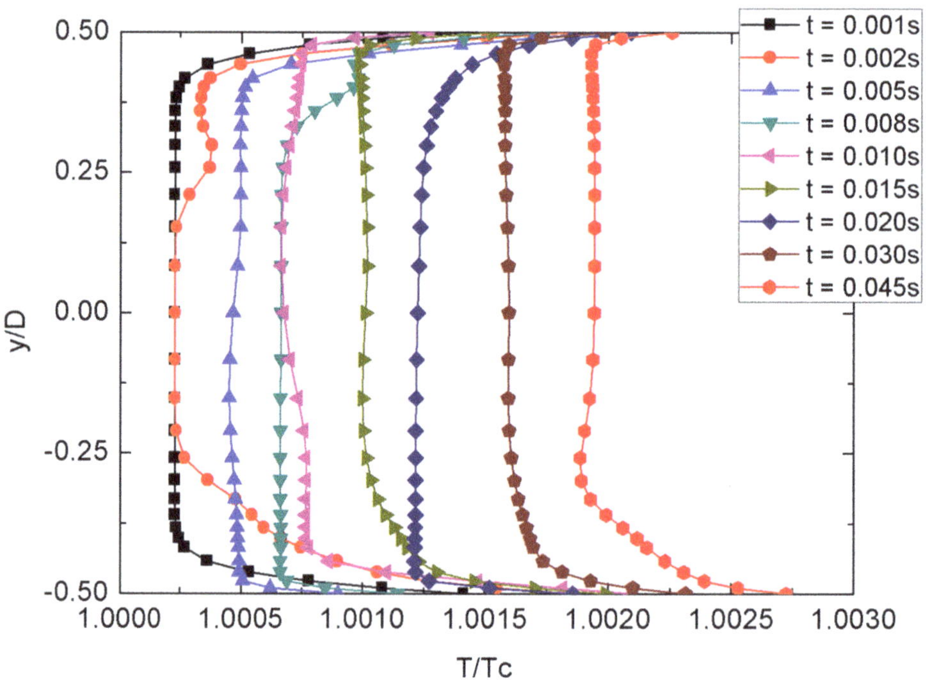

The bulk temperature increase from T/T_c = 1.00023 to T/T_c = 1.002 happens only within 45 ms as shown in Figure 8. Calculation on the characteristic time scales of pure thermal diffusion/convection and the current process can help clarify the 'critical speeding up' instead of 'critical slowing down' due to the very small thermal diffusivity for near-critical fluids (Onuki et al., 1990; Boukari et al. 1990; Zappoli et al., 1990; Zhong and Meyer, 1995; Jounet et al., 2000; Zappoli, 2003; Carles, 2010; Chen et al., 2013a; Zappoli et al., 2015).

In addition, it should be noted that the vortex evolution is triggered by the boundary heat flux. Such process is found to happen in both gravity and micro-gravity conditions in the current study. The vortex and mixing process is indeed difficult for microscale channels or chambers in real applications with traditional fluids. This process involves the fundamental theories of stability flow in microscale and it will be discussed in later chapters. The nonlinear near-critical fluid in this microchannel shows "abnormal" process, which is also differed from previous Piston Effect in closed chambers as the current case is open through flow. Basically, the interactions of thin hot thermal boundary would be responsible for the bulk stability evolutions.

Characteristics of Near-Critical Vortex Flows

Critical Disturbance and Vortex Formation

As discussed, the basic features of near-critical flow may include: (1) the happening of vortex flow is in microscale, which may provide new choice for engineering mixing process; (2) the time scale is another important point as it indicates a transient mixing in micro-to-mini seconds, which is of high efficiency; (3) the parameter ranges of the investigation shows that from practical viewpoint the application should be wide and of high potential. Later chapters would also discuss the scales and application from a more theoretically in-depth analysis.

Figure 9. Vortex region velocity profile for x- and y- components against time. (a) x-velocity component; (b) y-velocity component. Local vortex evolves periodically with flow time but symmetrically in D = 200 μm microchannel (plotted for x/L = 0.85; $q = 8000W/m^2$, $Re_{in} = 26.80$)

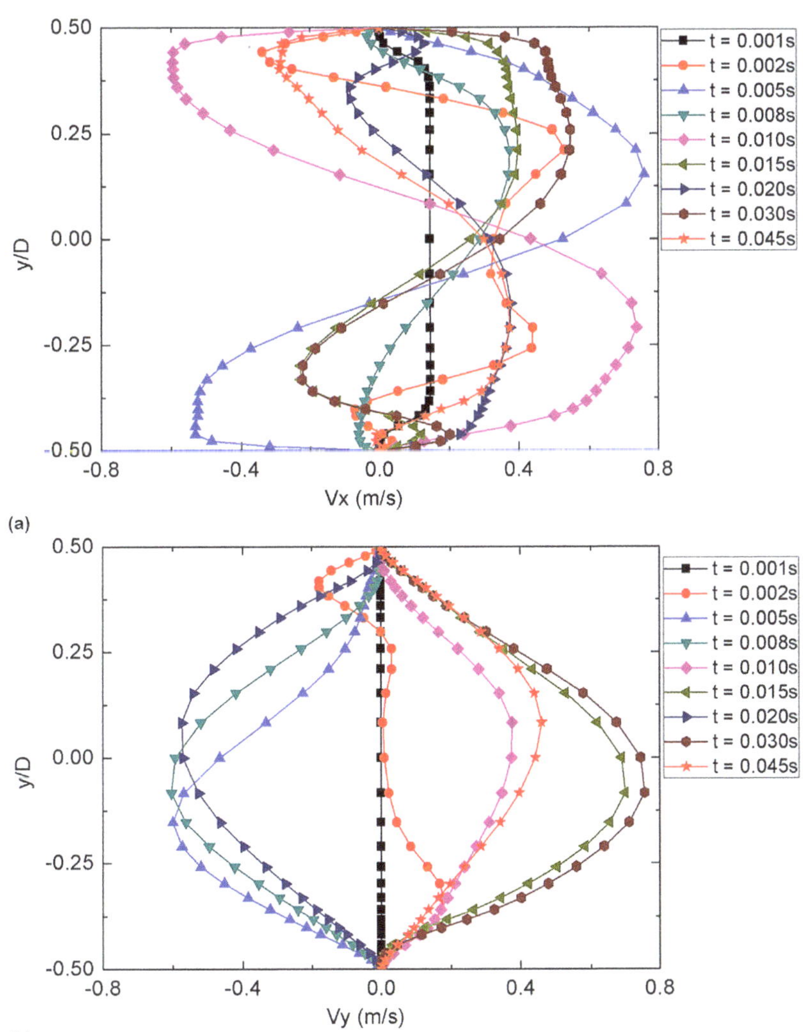

But before that, several interesting questions should be asked: why the boundary temperature collapse and bulk homogeneous heating up happen at the same time for the convection onset of near-critical fluids? What the basic structures and time evolution are of cross vortex parameters? The unsteady evolution of vortex x- and y- velocity against time is plotted in Figure 9. It is seen that vortex happens very quickly as identified negative velocity component in Figure 9 ((a) is x-velocity; (b) is y-velocity). In Figure 9, the vortex velocity component can be as high as 0.4 m/s, which is due to the coupled fluid expanding and thermal acceleration. From the beginning, the x-velocity is less regulated and later it turns to be nearly symmetric, which indicates the vortex happens in the boundary and respective 'hot spot' growth may be dependent on the local condition. Later, while the vortex is formed across the microchannel, both x- and y-velocity show regulated profiles. As in real micro engineering field the vortex is not easily realizable (Kumar et al., 2011), the current vortex growth by thermal perturbation from boundary can be both useful and interesting.

Further Parameter Analysis

For the application consideration, the information on the transient characteristic mixing time and mixing characteristic length scales can be useful. In microchannel mixers, the transient mixing time is generally very small due to the micro-size of geometries (Kumar et al., 2011). In the current study, steady transient vortex mixing can last for approximately 0.1 s for 7.5 MPa tests, or about 0.2 s to 0.3 s for 8.0 MPa and 9.0 MPa tests.

However, in microchannel mixers, in general, there are characteristic mixing times classified to be either laminar type or turbulent type, and the laminar type of characteristic mixing time will be longer than the turbulent type. The characteristic mixing time falls in the range from several milliseconds to several hundred milliseconds in microchannel mixers, and it will decrease with increasing energy dissipation, while the dissipation rates range from 1.0 W/kg to 1.0×10^5 W/kg, as summarized by several groups (Falk and Commenge, 2010; Kumar et al., 2011). In their review studies, the mixing performances are analyzed by Mixing Effectiveness (ME) numbers, and the best performance is found for an asymmetric T-mixer due to the strong flow vibrations of T-passages. Another study (Johnson and Prud'home, 2003) reported a typical 5 ms to 320 ms characteristic microchannel mixing (impinging jets) time for representative chemical reaction process. Special attention is also paid to the energy dissipation across the micro-mixing process, as it partly represents the mixing strength in the microchannel/chamber (impinging jets can provide rapid mixing and energy dissipation rates up to 10^5 W/kg and single turbulent jets with 10^4 W/kg, and rotor stator mixers can provide energy dissipation rates of about 10^2 W/kg, as summarized in the above study).

During micro-mixing process, turbulent-like motion with frequent collision and redirection of very small 'fluid segments' can be important for mixing. That process is also related with energy dissipation and pressure losses inside the microchannel. In the current discussion, the micro-mixing characteristic numbers are also calculated. Indeed, mixing process includes macromixing at the scale of vessel, mesoscale mixing at the scale of turbulent eddies and micro-mixing at the scale of molecular diffusion. The respective time scales can be

$$\tau_{m,t} \propto \tau_s \propto \varepsilon^{-1/3},\ \tau_m \propto \left(\frac{\nu}{\varepsilon}\right)^{1/2},\ \tau_{m,d} \propto \left(\frac{\nu}{\varepsilon}\right)^{1/2} Sc \tag{13}$$

where τ is characteristic time, ε is energy dissipation rate, ν is kinematic viscosity (m^2/s) and Sc ($=\nu/D$) is the Schmidt number. Due to the thermally induced flow expansion and acceleration in the channels, the flow will become turbulent. Following classical definitions, the energy dissipation rate ε (W/kg) taking pressure loss ΔP (Pa) as representation of energy dissipation is proposed as:

$$\varepsilon = \frac{Q_V \Delta P}{\rho V_m} = \frac{\Delta P}{\rho \tau_r} \tag{14}$$

where Q_V (m^3/s) is the volume flow rate, ρ is the density (kg/m^3), V_m (m^3) is the volume of focused space and the residence time can be defined as

$$\tau_r = \frac{V_m}{Q_V} \tag{15}$$

for the chosen volume, thus to give the dissipation rate ε with the unit of (W/kg). In order to obtain large energy dissipation rates, large pressure drop and large variation in pressure per time are desired (Kumar et al., 2011). The residence time in microchannels is usually of the order of milliseconds.

Then the characteristic mixing time can be defined according to previous researches (Johnson and Prud'homme, 2003) as the following:

$$\tau_m = \frac{1}{4}\left(\frac{\nu}{\varepsilon}\right)^{1/2} \tag{16}$$

The current time scale discussions show the effectiveness of mixing very small fluid volumes in a microchannel, which is very important for the design of real systems. This time scale differs from the characteristic time scale of the thermal-mechanical effect in the discussion of the mixing origin and vortex evolution in the current study. For this calculation, the microchannel section from $x/L = 0.80$ to $x/L = 1.00$ is chosen to calculate the energy dissipation rate and other related micro-mixing parameters, as the main steady mixing occurs in this region. Respective parameters are monitored and calculated according to the definitions in Eq. 15-Eq. 16 and plotted in Figure 10.

It is seen that the near-critical CO_2 microchannel mixing characteristic numbers are well correlated with the function of:

$$y = ax^b \ (a = 0.15002,\ b = 4.1167\times10^{-5}) \tag{17}$$

The R-squared value for the correlation is 0.9999. In Figure 10, the calculated characteristic mixing time is plotted against energy dissipation rates for different cases.

The residence time of the near-critical CO_2 fluid flow is plotted in Figure 10 for comparison. As shown in Figure 10, the mixing energy dissipation rate is shown to be quite small in comparison to other types of micro-mixing methods (Kumar et al., 2011) due to the very low near-critical viscosity and fast expanding property. Therefore, a lower energy loss is found for the current high-speed vortex mixing.

Figure 10. Micro-mixing characteristic parameters mixing time vs energy dissipation rate

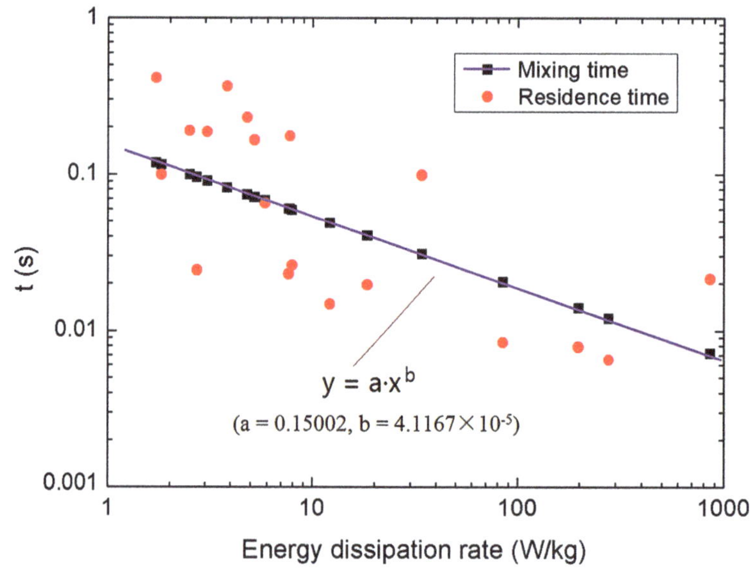

The general trend is similar to previous microchannel mixing studies, where a shorter characteristic mixing time is found for larger energy dissipation rates. However, the residence time plotted in Figure 10 falls randomly at about that of the characteristic mixing time due to the strong velocity differences across the vortex in different locations. In the current discussion, the major mixing time falls between 0 and 0.2 s, which cover the general time scale of traditional micro-mixers (Kumar et al., 2011; Luong et al., 2011; Zhang and Wang, 2011). Further studies are still needed in this problem to obtain optimal control and mixing strategies.

Near Critical Dissipation

The length scale of this microchannel mixing of small fluid volumes is also an important parameter to answer the question of how small a vortex the mixing can create and to provide comparisons with macroscale mixing and molecular diffusion. Indeed, strong and effective micro-mixing will rely on a smaller mixing length, and the characteristic mixing length will asymptotically decrease with the increase of the energy dissipation rate. Moreover, it is very critical in applications such as micro-chemical synthesis and related product formation processes for the microchannel flows to change from parabolic velocity profiles to a segmented vortex mixing flow pattern (Zhao et al., 2011).

The characteristic length (or fluid mixing segment size) is called Kolmogorov length defined as (Aoki et al., 2011):

$$\lambda = \left(\frac{\nu^3}{\varepsilon}\right)^{1/4} \tag{18}$$

To note the micro mixing characteristic of small fluid volume/segment, another parameter called split-number is defined (Aoki et al., 2011) for comparison the strength of mixing effect:

$$n = \frac{D/2}{W} = \frac{D}{2W} \tag{19}$$

where D is the microchannel height and W represents fluid segmental size. Here W equals to the value of Kolmogrov length as defined in the above equations.

The characteristic Kolmogrov length and split numbers are also calculated according to Eq. 18 - Eq. 19 and plotted against energy dissipation rate of cased in Figure 11(a) and Figure 11(b). It should be noted that the current parameters are developed for judging how small the mixing can realize and in the current vortex flow directly generated in microchannels by sudden heating of near-critical fluids. The mixing length can be much smaller than traditional kinds due to very low viscosity and high solubility of supercritical CO_2 fluid. The evolution trends in Figure 11(a) and Figure 11(b) are similar with previous studies (Aoki et al., 2011). Also it is found that the current mixing process follows strict exponential law instead of linear function found for traditional fluid (Aoki et al., 2011; Kumar et al., 2011). The R-squared value for the fitting is 0.9957 for Kolmogrov length-energy dissipation relation (Figure 11(a)):

$$y = ax^b \ (a = 4.5831, b = 0.03917) \tag{20}$$

and 0.9828 for split number-energy dissipation relation (Figure 11(b):

$$y = ax^b \ (a = 18.3197, b = 0.6915) \tag{21}$$

which indicates highly correlated of the characteristic numbers in near-critical micro-mixing. It can be found in Figure 11(a) that the Kolmogorov length is several micro-meter, indicating that very small fluid volumes are formed and the high speed micro-mixing can be effectively operated. The split numbers in the current study are two or three times that of previous studies (Aoki et al., 2011) with channel confluence and bend.

PRESSURE EFFECTS AND HEAT TRANFER CONSIDERATIONS

In near-critical fluid, the unique feature is the coupling of convection dynamics and the heat transfer process (Amiroudine and Zappoli, 2003; Carles, 2010; Chen et al., 2013a; Zappoli et al., 2015). It is very special for the findings of convective structure and heat transfer enhancement in novel near-critical systems. However, how will the vortex flow affect wall heat transfer as a result of the Thermal-Mechanical process? The parameter dependency of microchannel flow behavior will be discussed in this chapter. The extension of near-critical heat transfer to possible wide parameter ranges and the development of heat transfer applications and strategies are also summarized in this chapter.

Figure 11. Micro-mixing characteristic parameters and energy dissipation. (a) effective Kolmogrov length vs energy dissipation rate; (b) split number vs energy dissipation rate

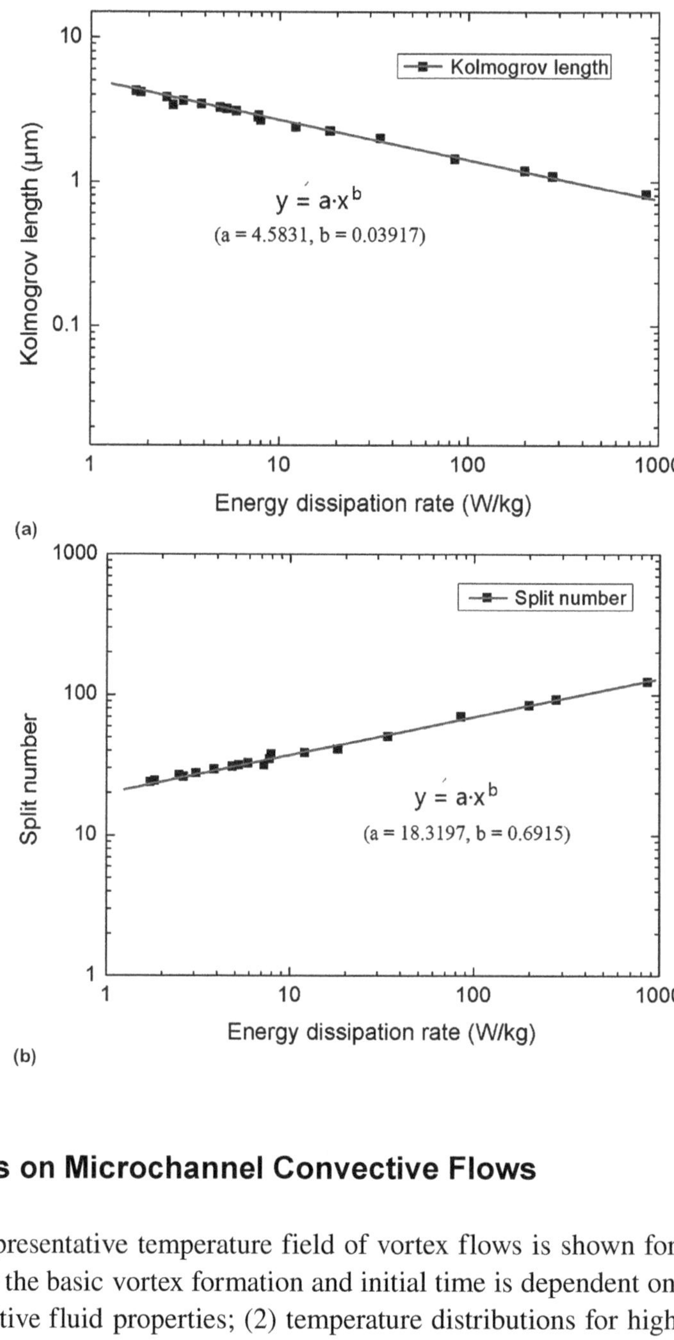

Pressure Effects on Microchannel Convective Flows

In Figure 12, the representative temperature field of vortex flows is shown for different pressures. It can be seen that: (1) the basic vortex formation and initial time is dependent on the pressure, which is due to the less sensitive fluid properties; (2) temperature distributions for higher pressure conditions are more homogeneous, which is due to the relative longer equilibrium time; (3) temperature gradient and profiles follows the basic picture of vortex structure (see previous section), which also indicates the different strength of vortex flow inside microchannels. Indeed, the thermal collapse or thermal equilibrium process is found to be faster for relative more near-critical conditions, which follows similar basic time scale trends of temperature profiles discussed above (and for more discussions on thermal collapse details, see refs (Amiroudine and Zappoli, 2003; Carles and Dadzie, 2005; Accary et al., 2005, 2007).

Figure 12. Detailed microchannel temperature profiles under different operation pressures. (T_{in} = 304.2 K, P_{in} = 7.5 MPa for D = 100 μm; q = 10000 W/m²; V_{in} = 0.01 m/s). (a) P_{in} = 7.5 MPa; Re_{in} = 13.4; t = 0.010 s; (b) P_{in} = 8.0 MPa; Re_{in} = 12.76; t = 0.075 s; (c) P_{in} = 9.0 MPa; Re_{in} = 12.18; t = 0.200 s

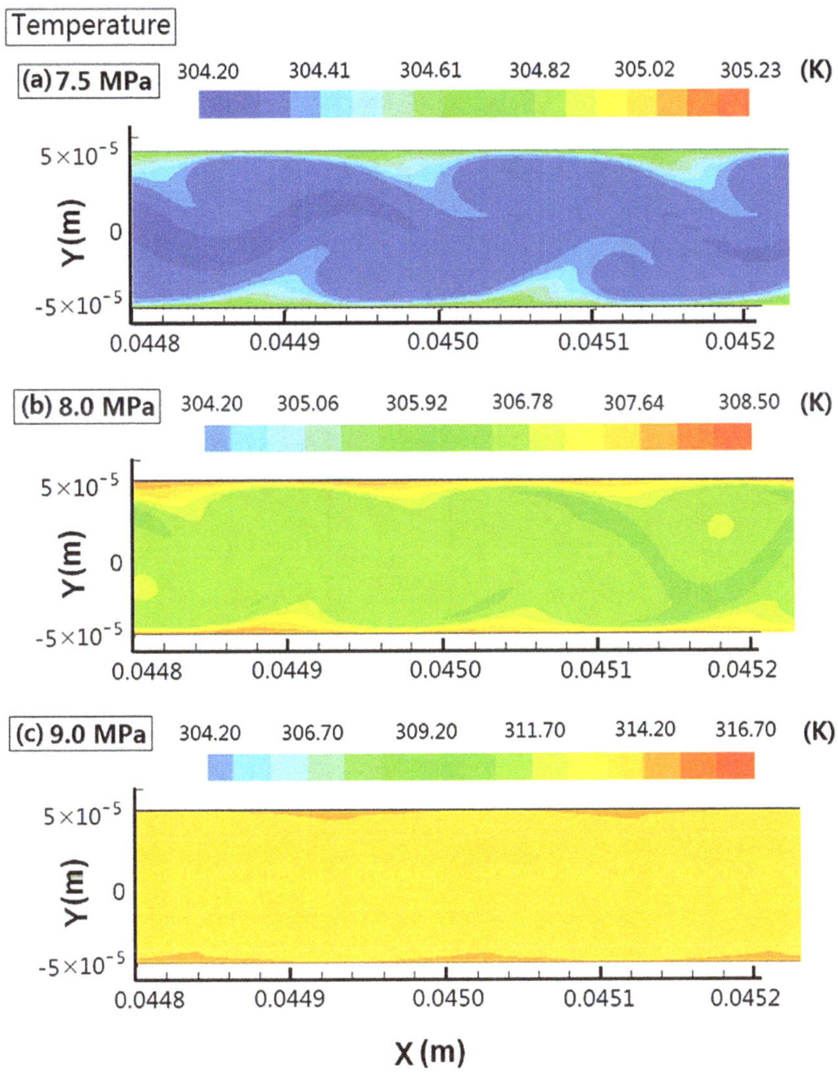

Acceleration and Adverse Pressure Field

The discussion of pressure loss and flow acceleration in near-critical fluid flow would be useful for flow characteristic analysis. However, for near-critical fluids, if the details are not provided, one cannot touch the complex transitions in the microscale flows. Following this basic understanding, the basic velocity and pressure behaviors in representative sections of the microchannels are plotted and compared in Figure 13. The basic low heat flux (stable) flow condition and high heat flux (vortex) flow cases are plotted and compared in Figure 13.

Figure 13. Evolution of velocity and pressure along the horizontal direction of the microchannel. (D = 100 μm case, V_{in} = 0.005 m/s and Re_{in} = 6.70) (a) velocity evolution for Q = 4000 W/m² (stable flow); (b) velocity evolution for Q = 8000 W/m²(vortex flow); (c) pressure evolution for Q = 4000 W/m²(stable flow); (d) pressure evolution for Q = 8000 W/m²(vortex flow)

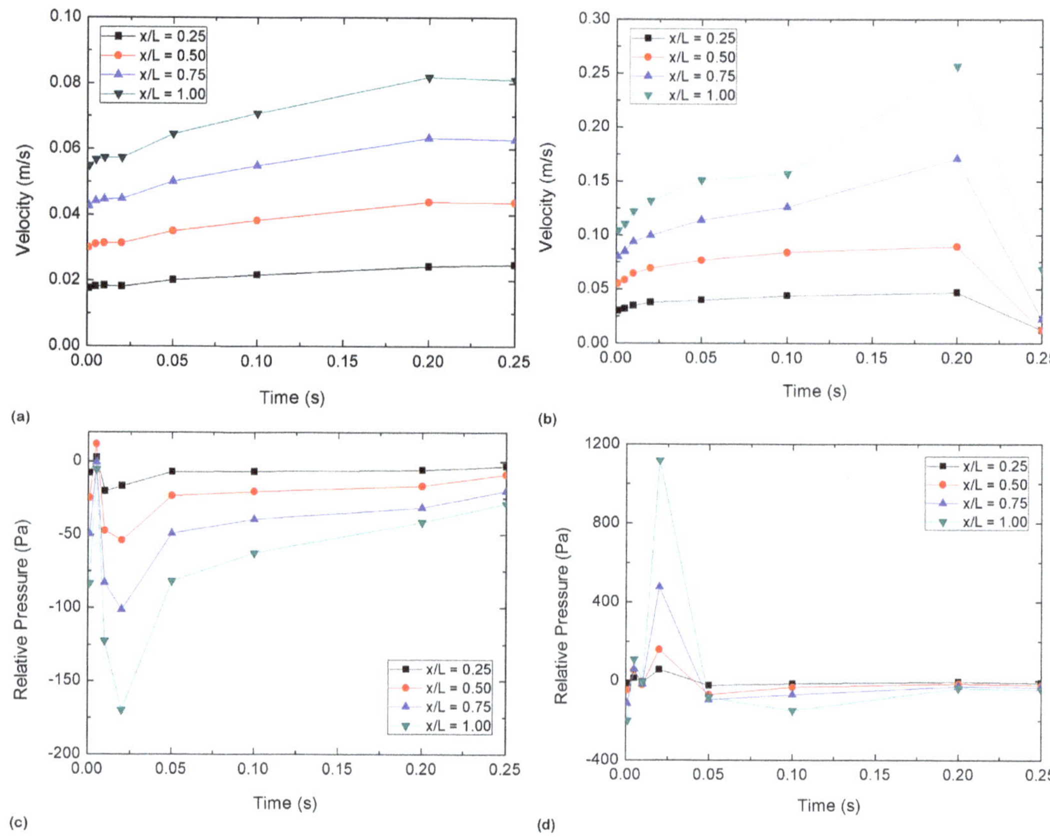

As shown in Figure 13(a) and Figure 13(b), the stable flow case velocity shows linear increase with both time and position, while the vortex flow case show much higher horizontal velocity and with velocity drop back at around 0.25s due to final stabilization process. The stabilization process means that the transient vortex mixing flow will gradually fade out and become stable flow finally. It can also be seen that for both stable flow and vortex flow cases the acceleration process happen. However, for vortex case the acceleration along with horizontal direction is much stronger than stable flow case, which indicates that much stronger expanding of fluid can be seen for downstream flow (as shown in Figure 13(a) and Figure 13(b); for .x/L = 0.75 and x/L = 1.00).

The horizontal relative pressure evolutions are shown in Figure 13(c) and Figure 13(d). It can be seen that the fundamental differences exist between stable flow and vortex flow cases. For stable flow case, as shown in Figure 13(c), basic pressure drop trend from upstream flow towards downstream flow is noticed. And for the early stage, larger horizontal pressure drops are found, which means that self-acceleration and stabilization processes happen for stable flow case. As also discussed in former studies, with large horizontal pressure drop, flow acceleration will enhance the boundary heat transfer condition and this serve as one stabilization factor for the flow (when the heat influx is relatively low) (Chen et al.,

2012, 2013a, 2013b, 2013c, 2014). However, for the vortex flow case, as shown in Figure 13(d), due to the strong local perturbation and expanding/compressing process under relative higher boundary heat influx, transient pressure increases along horizontal direction can be found, which phenomena happens within the characteristic time scale as discussed in former parts.

The adverse pressure gradient phenomena in the vortex case pressure curves are also a very important characteristic of near-critical flows, which indicates the pressure loss and energy dissipation in vortex formation process and the strong expanding effect in near-critical flow field. After the vortex formation and growth period, normal pressure field in horizontal direction can then be formed again inside the microchannel, which also contributes to the stabilization of near-critical flow.

Heat Transfer Analysis

Boundary Heat Transfer Parameters

During the same time of boundary expanding and thermal perturbation process, the boundary heat transfer condition also changes dramatically for the convection onset period. The evolutions of boundary heat transfer coefficient (h) and Nusselt number (Nu) number are plotted in Figure 14 and Figure 15. In transient heat transfer process, the thermal collapse process as discussed in previous sections, the microchannel heat transfer parameters also have large changes. Generally, enhancement can be found in the vortex region.

As shown in Figure 14, due to boundary expanding and vortex mixing, the heat transfer coefficient increase hundreds of times within 40 ms. That is also comparable to that of thermal-mechanical (or called Piston Effect for closed systems) heating found in previous studies (Amiroudine and Zappoli, 2003; Carles, 2010; Zappoli et al., 2015). Then it is clear that together with the temperature collapse, enhanced heat transfer is found. In Figure 14, the conditions of $D = 100$ μm and $D = 200$ μm cases are also compared. It is shown that $D = 100$ μm case will have higher heat transfer rate. Though the heat flux is slightly different, the main reason is the much higher temperature gradient generated in the former one.

Also, it can be seen that for both microchannel heights, the heat transfer rate changes in a decrease direction with the increase of time for upward steady conduction area. That is due to the decrease of thermal diffusivity when the fluid is heated to near-critical condition, where large properties diverge can be seen. However, during this evolution process, heat is conducted to the boundary fluid and only a very small part is transported to the main flow due to very low thermal diffusivity (Zappoli et al., 2015). Thus the thermal boundary evolutions and equilibrium process becomes very critical for the thermal relaxation process, which will be discussed in more detail in next sections.

Critical Paradox: Heating Up or Cooling Down?

The heat transfer process in horizontal direction is also discussed in this study. Basic density and temperature evolution during thermal convection onset period of near-critical CO_2 microchannel flow are plotted in Figure 16. The density and temperature evolution features along the horizontal flow direction (with representative position parameters plotted) and time-dependent trends can be seen in this figure. As shown in Figure 16, both the density and temperature curves show differences between stable flow condition and vortex flow case.

Figure 14. Evolution of near-critical microchannel heat transfer parameters. (V_{in} = 0.01 m/s; Re_{in} = 6.70). (a) D = 100 μm, q = 10000 W/m²; (b) D = 200 μm, q = 8000 W/m²

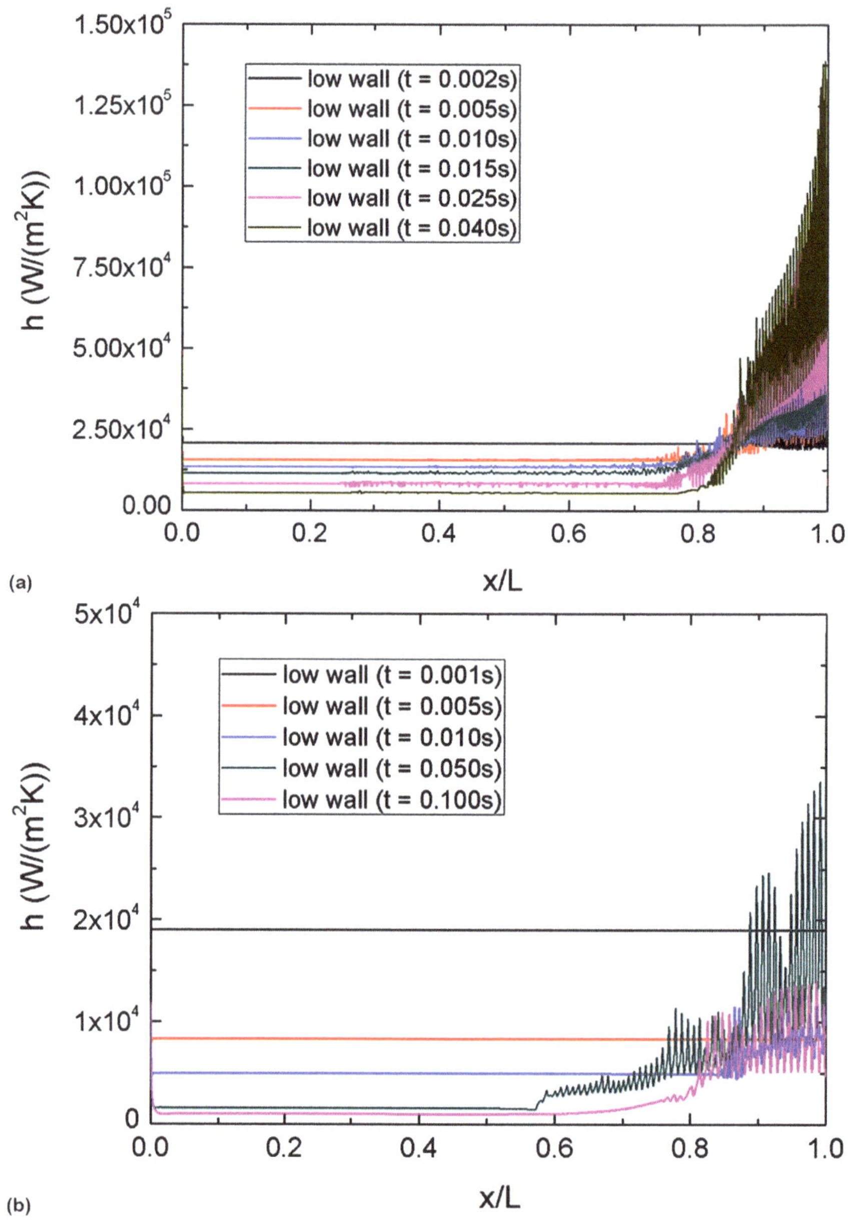

It can be seen from Figure 16 (a) and Figure 16(c) that for stable flow case, both the density and temperature show very small changes along with horizontal direction. For vortex flow case, as shown in Figure 16(b) and Figure 16(d), it can be seen that for the very early stage (~0.02s) the density and temperature evolutions are nearly the same for both stable case and vortex case, and when the unstable boundary layer perturbation begins at characteristic time (PE-time) scale the parameters for vortex case show relative large deviations along flow direction.

Figure 15. Evolution of near-critical microchannel Nu number. (V_{in} = 0.01 m/s; Re_{in} = 6.70). (a) D = 100 μm, q = 10000 W/m²; (b) D = 200 μm, q = 8000 W/m²

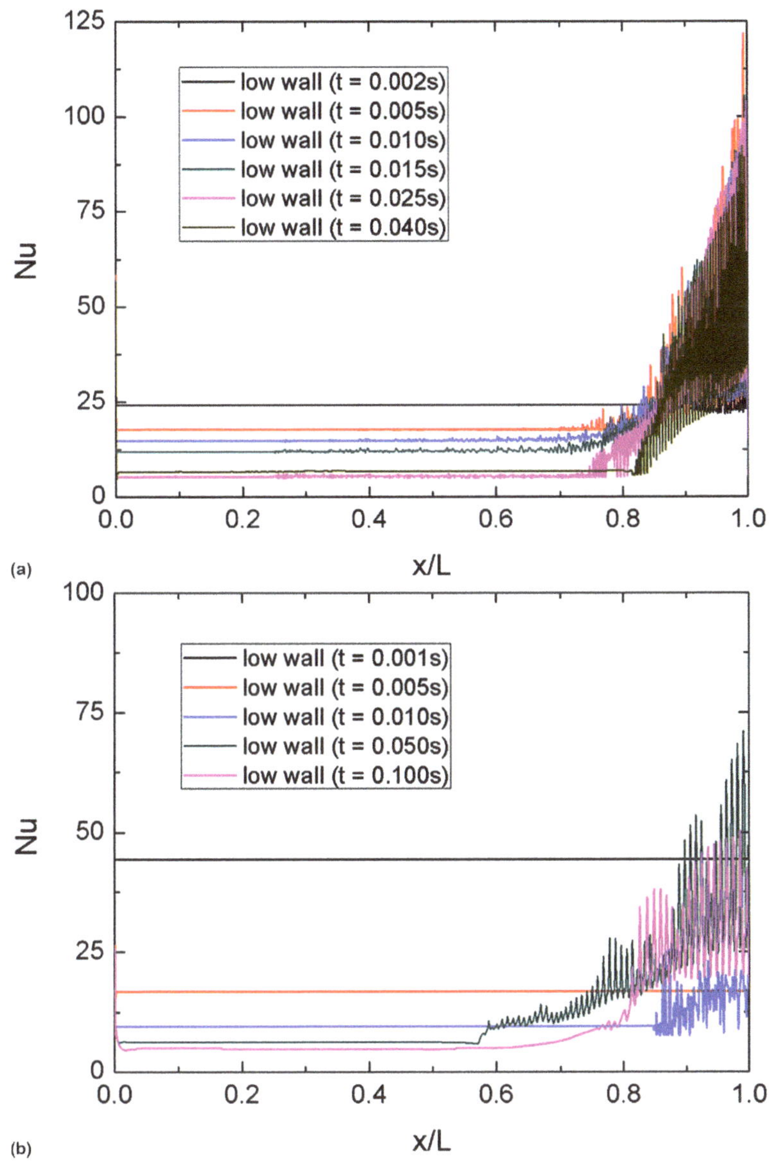

Compared with stable flow case, the vortex flow case shows larger density decrease and temperature increase, as shown in Figure 16. The reason is that higher heat flux input is set for vortex flow condition. And detailed analysis shows that for stable flow case, the density will gradually decrease with horizontal direction (and also with time). However, for vortex flow case, the downstream flow shows denser and colder parameter behaviors along with horizontal direction and the curves forms two groups. For x/L = 0.25 and x/L = 0.50 positions, the basic trend is the same as stable cases, where the density slightly decrease and the temperature slightly increase in the horizontal flow direction. However, for x/L = 0.75

Figure 16. Evolution of density and temperature along the horizontal direction of the microchannel. (D = 100 μm case, V_{in} = 0.005 s and Re_{in} = 6.70). (a) density evolution for Q = 4000 W/m² (stable flow); (b) density evolution for Q = 8000 W/m² (vortex flow); (c) temperature evolution for Q = 4000 W/m² (stable flow); (d) temperature evolution for Q = 8000 W/m² (vortex flow)

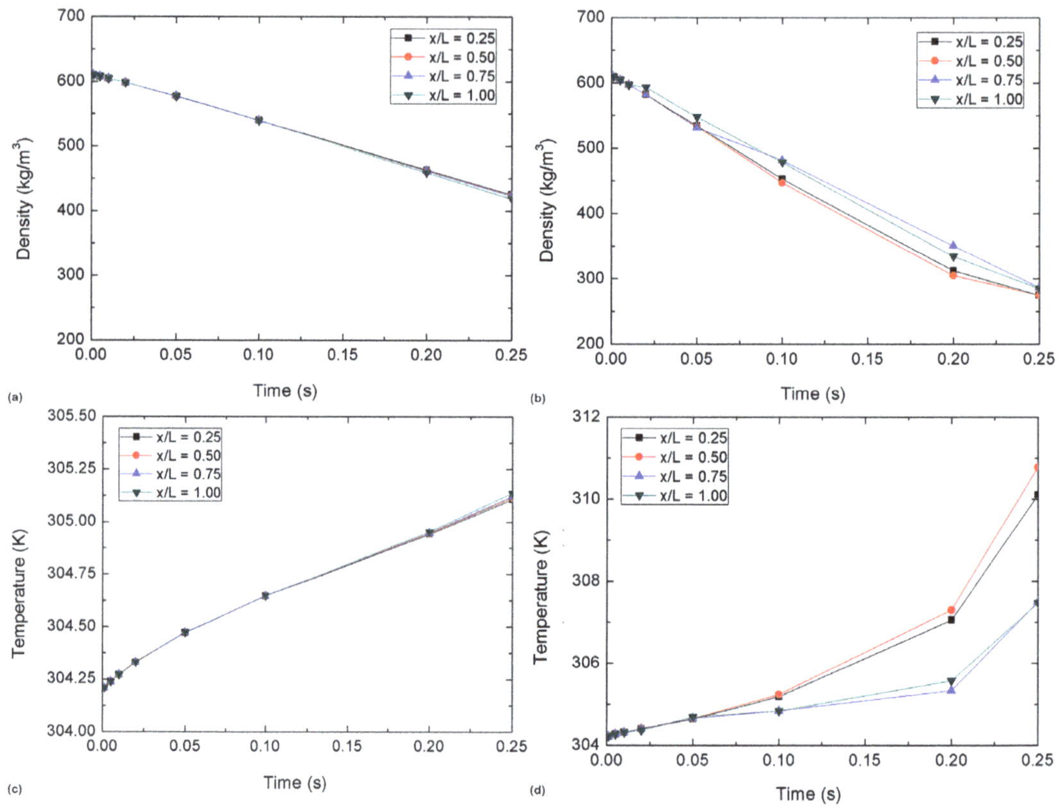

and x/L = 1.00 positions, the above mentioned new behaviors are seen. That is to say that, with heat flux input from the boundary, the downstream flow will become colder than upstream flow as specific time after vortex flow begins (in the region as shown in Figure 16), though the general time evolution of each position follows the general cases (where the downward flow yields relative higher temperature due to heating up), as shown in Figure 16.

Further Comments on Critical Heat Transfer

Here a short summary on the horizontal heat transfer and the expanding nature for near-critical micro-channel flow can be made. It is well known that near-critical fluids can be very much expandable due to the large density changes near the critical region. However, previous researchers have not reported this aspect when examining the heat transfer performances of supercritical flow inside micro-scale channels. In the present study, even in microchannels, besides the span-wise parameter deviations, horizontal parameter evolutions also show different behaviors when compared with traditional fluid based cases (Frohlich et al., 2006).

As discussed in the stability evolution of cases (Amiroudine and Zappoli, 2003; Chen et al., 2013a), basic stability evolution is greatly related with the horizontal parameter behaviors as well as the span-wise perturbation flows in the hot boundary layers. As there has many studies that treat the fast transient process of supercritical chemical synthesis, chemical extraction or micro-fluidics, few studies have discussed the self-acceleration and expanding nature of the flows (Amiroudine and Zappoli, 2003, Jounet et al., 2000; Carles and Dadzie, 2005; Carles, 2010, Chen et al., 2014; Zappoli et al., 2015). Indeed, as discussed in former studies, such expanding nature and transient adverse pressure field will lead to both 'abnormal cooling down' and unstable channel/pipe flows (Amiroudine and Zappoli, 2003, Jounet et al., 2000; Chiwata and Onuki, 2001; Carles, 2010, Shen and Zhang, 2011; Zappoli et al., 2015), where the differences are compressible fluids and sensitivity of fluid properties. In previous limited experimental studies, the flow acceleration and heat transfer conditions are also investigated in circular pipes under stable conditions (Asinari, 2005; Jiang et al., 2008). For the current case, sudden applied boundary heat flux may lead to new thermal convection behaviors and the fluid local expanding nature and horizontal acceleration/expansion can also affect the local heat transfer behaviors of near-critical flows, which will be discussed in the next section.

THEORETICAL ASPECT AND NEAR-CRITICAL STABILITY ANALYSIS

This chapter is focusing on the theoretical analysis of near-critical stability behaviors under the current model settings. The extension from phenomenological study to theoretical development is very important for in-depth understanding of the near-critical physics. The stability evolution of the current through flow microchannels is different from traditional turbulent transition or stability channels, where the major transition is based on Reynolds number description. However, the stability condition for the current microchannel flow is still different from the near-critical closed chamber system, where the closed chamber Rayleigh-Benard stability evolution and chamber reflective pressure waves and disturbances are the major factors. Also, the cases of gravity controlled system and near-critical effects controlled system are different in nature. In this section, the basic Kelvin-Helmholtz instability with new perturbation source of near-critical Thermal-Mechanical process is introduced.

New Kelvin-Helmholtz Stability Region: Near-Critical Findings

Local "Hot Spot" and Stratification of Critical Fluid

As it has been discussed in the previous section that the current microchannel cases with two-sided heat flux are different from the classical PE chamber (Zappoli, 2003; Carles, 2010; Zappoli et al., 2015), where the 'hot spot' and vortex will originate from two boundaries. The concept of "hot spot" indicates the high temperature centers in the boundary layer of near-critical fluid. It is seen that the pressure and perturbation velocity (in y-direction) both fluctuates with time, where the evolution of acoustic expanding process happens. Similar oscillations have also been found in refs (Zappoli et al., 1990; Beckman, 2004), when near critical fluid confined between rigid planes (with relative larger channel size, so the perturbations velocity is 3~4 times larger than in the current study. However, the occurrence of "hot spot" in near-critical fluid is firstly found in the current study.

Origin and Evolution of Local "Hot Spot"

The basic flow and density fields of near-critical microchannel convection are plotted in Figure 17(a). It can be seen from Figure 17(a) that in the first stage very thin boundary layers are formed after heat is applied at two walls, where the fluid density drops quickly. In Figure 17(a), the density soon drops about 8% at $t = 0.002$s. The thermal physical properties also have large fluctuations (Zappoli 2003). Consequently, local hot spots are found inside the thin hot boundaries (with low density centers). Indeed, local pressure gradients are found in both horizontal and vertical directions as discussed in former sections, which then forms a local intrusion (with higher temperature, "hot spot" expansion) form boundary layer into the main flow. Then the vortex will continue growing up till it reaches the walls (Figure 17(a), $t = 0.01$s).

In fact, the hot spot here is the result of vortex grow, where the heterogeneities in temperature, density and pressure also contribute to the formation of vortex. It can be examined that the hot spot center is also low pressure center and vortex center at the same time. The related microchannel vorticity field has also been discussed in another application study (Chen et al., 2013a), it is reported that the two opposite rows are of opposite sign. And in the current case the rows show opposite sign and the order of magnitude of the circulation around each vortex can be up to 10^4, which shows very strong mixing potential inside the microchannel. Similar hot spot formation and disturbance have also been reported (Zhang et al., 2010; Chen et al., 2010) for the steady near-critical diffusion process in a Rayleigh-Benard configuration under gravity.

In Figure 17(b) and (c), the detailed flow velocity and density distributions are shown. It can be seen that both $D = 100\ \mu m$ and $D = 200\ \mu m$ cases, the large density differences are seen, while the flow vortex both expanding to the full width of the microscale channels. However, the homogeneous increase of bulk temperature is one typical temperature evolution mode due to thermal acoustic characteristics of near critical fluids (Zappoli, 2003; Zappoli et al., 2015). The symmetric development of unstable hot spots and vortices is rarely seen in the literature and it represents a new mechanism of unstable relaxation induced by near-critical fluid thermal-mechanical effect.

Local "Hot Spot" Parameter Analysis

In Figure 18, the representative local pressure and density profiles are shown for early stage of vortex formation in the microchannel. It is seen that when vortex is formed for the downstream flow area in microchannel, relative large pressure differences are generated in the span-wise direction (as discussed in (Section 4.4.2)). Also, as discussed in the former section, the characteristic time for $D = 100$ μm channel case is 2.02 ms, the current time shown in Figure 18 is around 2 ms, the vortex size is around 40% of the microchannel height. At that stage, the local pressure profile also shown periodic 'low pressure center' and 'high pressure center' in mainstream direction. The periodic local pressure centers, thus, help in sustaining the periodic (or self-similar) evolution of vortex growth in the microchannels.

In one former study of Chen et al. (2010, 2013d) the cross vortex pressure values are reported in microchannels, which also show very large span-wise differences. The density profiles are also shown in Figure 18. It can be noticed that characteristic thin hot boundary layers for near-critical fluids are also formed for such small channels (Jounet et al., 2000; Chiwata and Onuki, 2001; Amiroudine and Zappoli, 2003; Zappoli et al., 2015). As shown in Figure 18, the density differences inside the microchanel section can be higher than 80 kg/m^3 in the channel span-wise direction. Such large local deviations in

Figure 17(a). Near-critical fluid microchannel convective flow status in 100 μm and 200μm scales (near x/L = 0.90; $Q = 10000 W/m^2$, $Re_{in} = 13.4$ $T_{in} = 304.2$ K, $P_{in} = 7.5$ MPa). (a) evolution of the microchannel flow field and density field with time (D = 100 μm); (b) local velocity and density for D = 100 μm, t = 0.08 s ($q = 8000 W/m^2$, $V_{in} = 0.005$ m/s, $Re_{in} = 6.70$); (c) local velocity and density for D = 200 μm, t = 0.10 s ($q = 5000 W/m^2$, $V_{in} = 0.005$ m/s, $Re_{in} = 13.40$).

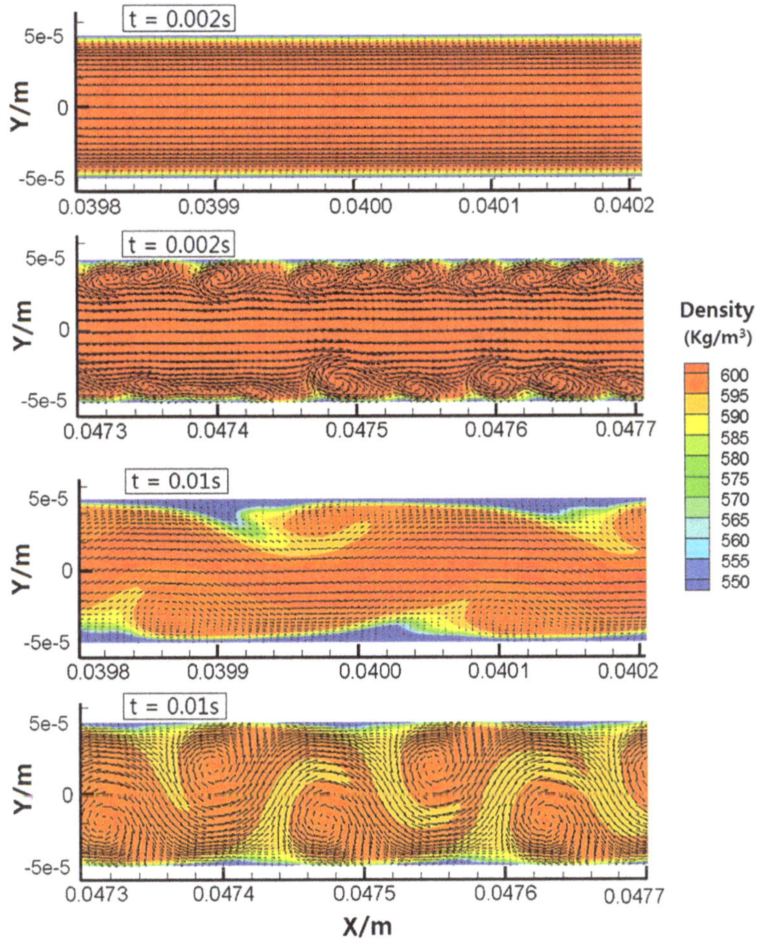

pressure and density contribute to the basic self-similar thermal convection structure and K-H instability evolution in micro-scales. Also, the heat transfer behaviors of the microchannels also show differences from traditional microchannel fluid flows, which will be discussed in later part of this chapter.

New Source of Unstable Disturbance

Basic Explanation on Microchannel Kelvin-Helmholtz Instability

Vortex mixing flow is of engineering interest due to the fast development in supercritical technology and MEMs related fields. Theoretical explanations of the current micro-vortex instability (where gravity is negligible in microchannels) can be identified under the fact that density and velocity stratifications

Figure 18. Representative local relative pressure and density profiles (near x/L = 0.95) at early stage of vortex formation. (x/L = 0.95; D = 100 μm, t = 0.002 s; q = 10000 W/m², V_{in} = 0.01 m/s, Re_{in} = 13.40)

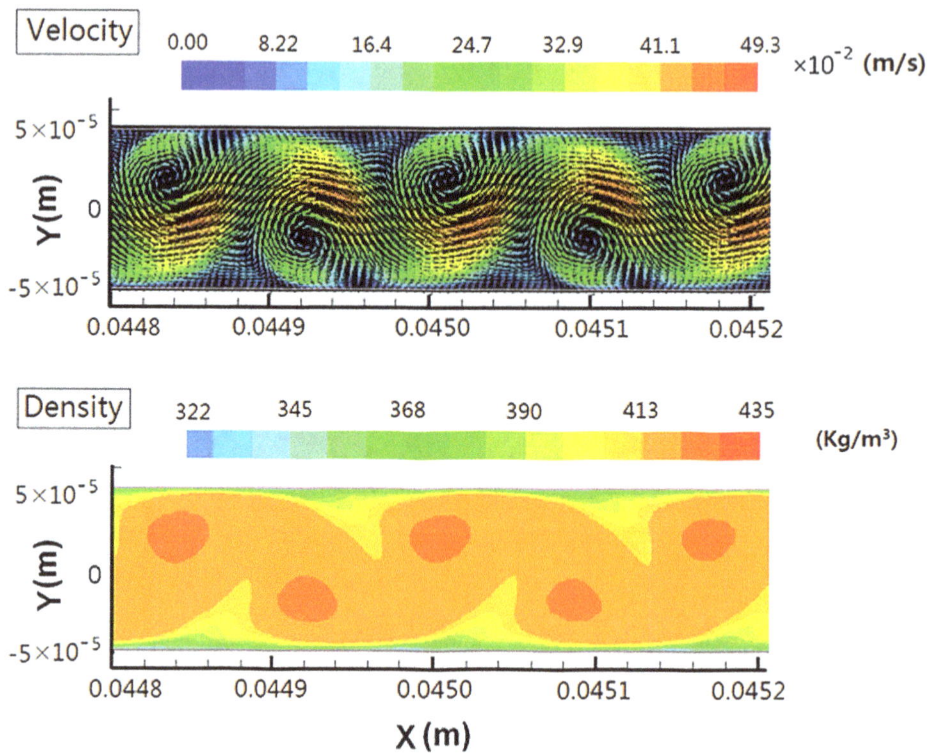

are important factors. Indeed, the current microchannel instability can be categorized to be a type of Kelvin-Helmholtz (K-H) instability as with the characteristic of largely stratified layers of shear fluid flow under thermally induced density and pressure heterogeneities.

The classical K-H instability, due to velocity shear was first introduced by Lord Kelvin (1871) and Helmholtz (1868), where gravity is one usual instability source. The current instability in microchannel still falls within the general K-H instability range with its perturbation source from hot boundary layer (HBL) thermal-mechanical expansion. In the current model, it is new that in confined microchannels the gravity effect is replaced by thermal mechanical perturbation waves across the microchannels.

From classical K-H instability models, any small stratification will lead to fluid 'tear off' however small the perturbation is (with perturbation source provided) (Chen et al., 2012). An important difference of the current study is that usually, Kelvin-Helmholtz is studied in open, un-stationary situation where two streams are meeting tangentially with a difference of velocity, but in the current study the K-H instability is formed in confined geometries (van der Waals, 1873; Hirschfelder et al., 1954). The wavelength is initially small and increases progressively by merging of vortices in the downstream for open cases, while in the current cases it in fact dictating the final wavelength (that saturates). There are also studies that show similar mixing layer trend in normal channels, where in confined geometries the saturated wavelength is in fact determined (Justin, 2002; Onuki, 2004). However, for the current case with expanding layers show unstable behaviors and the evolution process are also greatly affected by the input conditions and heat transfer inside the fluid flow.

K-H Stability Evolution and Its New Perturbation Source

Due to its thermal-mechanical origin, such 'adiabatic heating' or 'Piston Effect' has attracted a lot of research interests, especially in micro-gravity heat transport and thermal convection anomalies under terrestrial conditions. The physical mechanisms of the fast thermal homogeneity by that expansion and compression process in closed cell (like a piston) have been put forward (Zappoli et al., 1990; Onuki et al., 1990; Boukari et al., 1990). Analytically, Onuki et al. (1990) first established the typical time scale of Piston Effect as introduced in the introduction part of this chapter. For example, if D = 100μm, the characteristic diffusion time scale is $t_d = D^2/D_T = 1.15$ s, while the vortex Thermal-Mechanical characteristic time is $t_{PE} = t_d/(\gamma-1)^2 = 8.0$ ms. This result comparison shows the large difference and enhancement of the near-critical Thermal-Mechanical effect. Indeed, the effect of compressibility and ensuing thermal acoustic perturbations, respective thermal relaxation processes happen much more quickly than pure thermal diffusion process, and it may bring new phenomena when coupled with the effect of natural convection under gravity (Jounet et al., 2000; Chiwata and Onuki, 2001).

The near-critical boundary pressure and velocity perturbations with time are plotted in Figure 19. Fluctuations of perturbation curves can be found in Figure 19 with time. The parameters are found to fluctuate forth and back and the peaks are found near the thermal boundary region. The fluctuations indeed contribute to the boundary "hot spot" unstable evolution. Also the fluctuations in pressure show increase at $t = 2.5\times10^{-4}$ s and decrease after $t = 4.5\times10^{-4}$ s, which is an example of how the local parameters changes. This result means that the "hot spot" and local centers are not stationary. Such transient perturbations from boundary layers show the complex process inside the near-critical boundary. The Thermal-Mechanical boundary process dominants the thermal boundary evolution stage.

Indeed, such fluctuation process in near-critical fluid (or compressible fluids) has been discussed as a boundary heating process; however, the mechanisms are still not clearly explained (Zappoli, 2003; Carles, 2010; Zappoli et al., 2015). In the discussion of thermal-acoustic waves near the critical point, Carles (2010) also reported the instability of density/pressure evolution in the heat applied initial state (with several initial density stages measured). Such boundary compressing waves and the propagations form the basic stability source in the Kelvin-Helmholtz instability in the current microscale channels (Straub et al., 1995; Frohlich et al., 1996; Garrabos et al., 1998; Wilkinson, 1998). It should be noted again that the mechanisms of the current situation are different from the gravity induced buoyancy conduction process, though the coupling with gravity is still an interesting topic.

Time Scale and Near-Critical Thermal-Mechanical Effects

Time Scale Analysis

The representative vortex size growth inside microchannels is plotted against time in Figure 20. In Figure 20, the micro-sized vortex size is reduced by the microchannel width as finally it will grow to that size (expanding to the full channel width). The x-axis in Figure 20 is the reduced time by Thermal-Mechanical effect time scale, which is the characteristic time scale of the current microchannel near-critical process. It can be seen from Figure 20 that the growth of vortex mainly happens within 0.8-1.2 times of characteristic time scale.

Figure 19. Boundary Thermal-Mechanical effects the perturbation towards the bulk flow. (a) pressure perturbation; (b) velocity perturbation (near x/L = 0.98; D = 100 μm case, Q = 10000 W/m², V_{in} = 0.01 m/s, Re_{in} = 13.4). x-axis takes the mid-line of horizontal channel direction. P/P_{lmax} is pressure divided by local maximum values

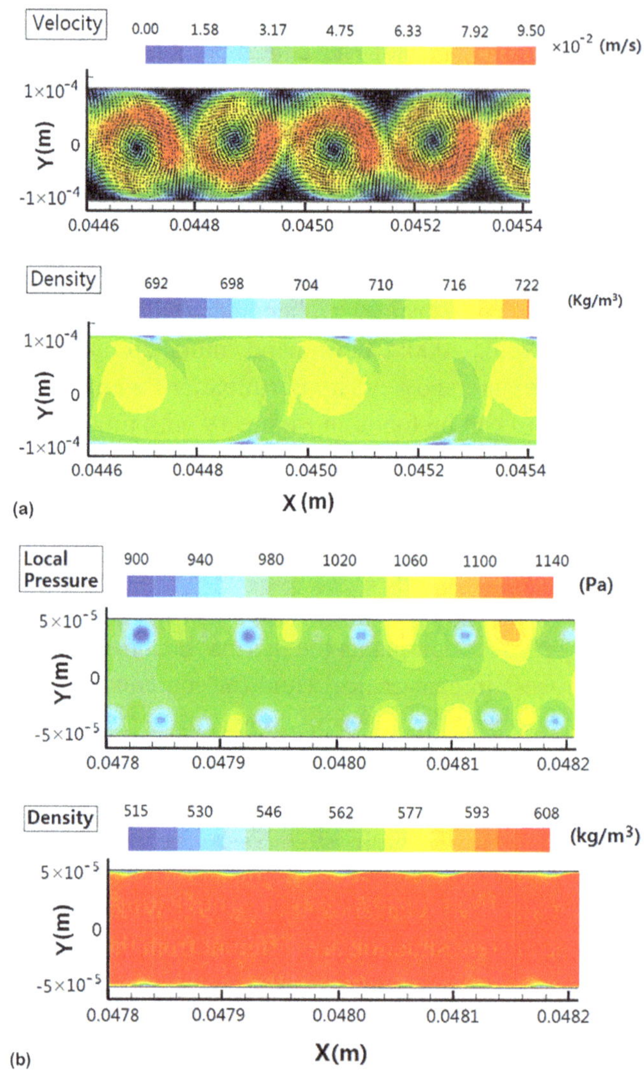

During the growth process, the mixing and thermal equilibrium happens from the boundary to the bulk flow. Both the vortex growth and energy transport happen within this typical time period, which indicates the fast equilibrium nature of near-critical process. In this discussion, the characteristic time is defined as the *t-PE*, which is from the analytical results of Onuki et al. (1990). In the case conditions of Figure 20, the characteristic time scales for *D* = 100 μm and *D* = 200μm are 2.02 ms and 8.06 ms, respectively.

Figure 20. Vortex growth in the near-critical microchannels (near x/L = 0.90; T_{in} = 304.2 K, P_{in} = 7.5 MPa, V_{in} = 0.01 m/s)

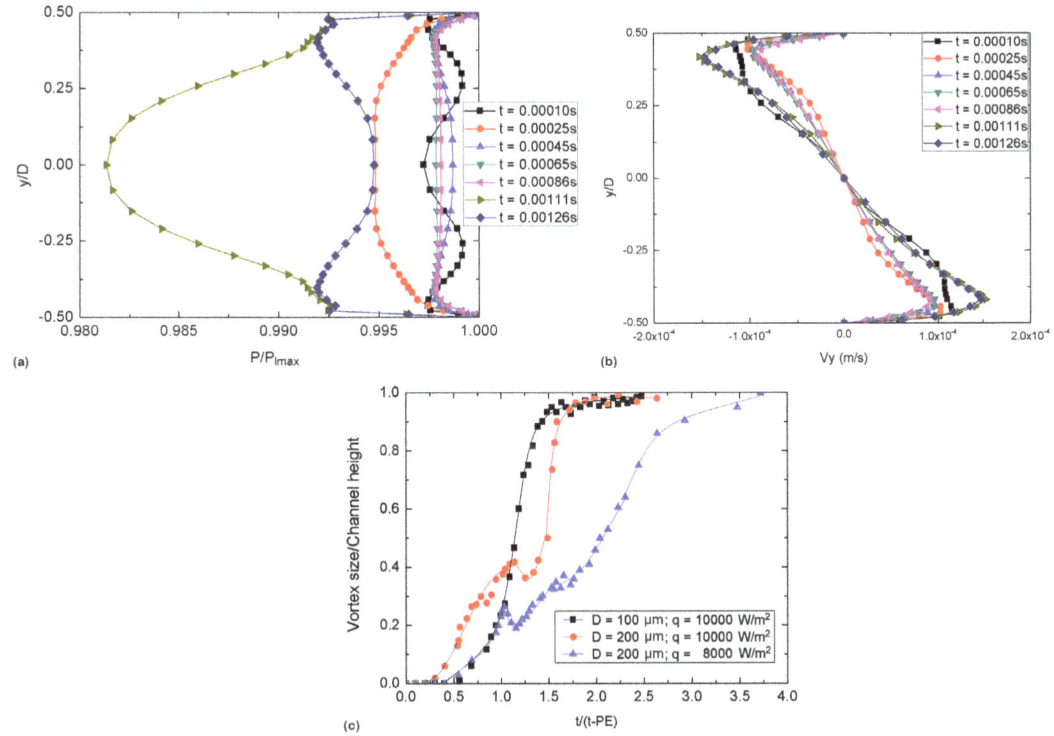

Piston Effect, Thermal-Mechanical Effect, and K-H Instability

It should be noted that in the microchannel vortex flow development, the flow will finally tend to stable laminar flow. The overall thermal equilibrium time is dependent on the geometric and critical distance parameters. Carles (2010) has summarized the near-critical time scale problems in a wide range of critical parameters. In the current discussion, the open channel Thermal-Mechanical process is fundamentally different from previous Piston Effect, which happens inside closed chambers. However, the Thermal-Mechanical nature from the near-critical compressible fluid flow under very low thermal diffusivity conditions is the same (Klein and Feuerbacher, 1987; Onuki et al., 1990; Zappoli, 2003; Carles, 2010; Zappoli et al., 2015).

Thermal-Mechanical effect is a concept that treats the near-critical fluid thermal equilibrium process with thin hot boundary compression wave evolutions. It is a general description of the near-critical thermal-acoustic nature of near-critical fluids. When the near-critical fluid is placed inside closed chambers, the compression wave will propagate toward the bulk fluid region and travel through the fluid chamber to heat up the fluid. The thermal-acoustic wave will be reflected by the chamber wall and heat up the fluid in a speed of sound (theoretically it is the case, but dependent on real system designs). This closed system process is called Piston Effect as the heating up process is like a "piston". In the current discussion with an open microchannel, a representative case of open system, the condition is different from the Piston Effect controlled system. The thermal equilibrium process in the microchannel through flow system is not the compression wave.

The Thermal-Mechanical process happens in the form of boundary perturbation process and local vortex formation. Then fast equilibrium process is achieved in the microchannel. The fundamental nature of the thermal equilibrium process for the microchannel evolutions and the Piston Effect is the same Thermal-Mechanical effect. However, different geometries of closed and open system lead to totally different flow process and convective structures. More importantly, the design of system would affect not only the detailed fluid flow behaviors, it will also lead to much more complex convective process and reveals a new scientific aspect of near-critical convective heat transfer field with respect to the development of different size-effect, convective heat transfer efficiency and thermal-acoustic procedures.

Further Analysis on Gravity Effect

The discussion of size effect and gravity coupling is very important for the explanation of transition phenomena in near-critical flows. It has been explained that the coupling of gravity effect is related to the time scale analysis. The coupling is previously found in relative open systems, where the tendency of thermal plume can be found (Chiwata and Onuki, 2001; Accary et al., 2005, 2007; Carles, 2010). However, the coupling in near-critical confined channels is much different.

The gravity time scale is relatively larger than the Thermal-Mechanical time scale. The first stage tests are conducted under gravity, where the effect of natural convection should be taken into consideration. Also in the highly expandable near-critical thermal boundary, for normal channels, or so called Poiseuille Rayleigh-Bernard configurations show deviations of convection structure and temperature field from traditional fluids (Amiroudine and Zappoli, 2003; Beysens et al., 2010), due to similar coupled effect of thermal-mechanical effect of near-critical fluid and the effect of buoyancy convection (Chiwata and Onuki, 2001; Amiroudine and Zappoli, 2003). The characteristic time scales of thermal-mechanical effects and natural convection are different (Amiroudine and Zappoli, 2003; Carles, 2010; Zappoli et al., 2015), but the initial stage of the later will be affected by the former kind, therefore oscillations and discrepancies are seen in the temperature and density evolutions during the convection onset (Amiroudine and Zappoli, 2003).

For the current microchannels, however, the effect of gravity will be dependent on the height and the effect of gravity waves will be limited. One former study has tested a range of channel height conditions and reported that the effect of natural convection should be considered when the channel height D exceeds 300 μm (Chen et al., 2013a, 2013b). For the current cases, the temperature evolutions of D = 500 μm channel are shown in Figure 21. Here the temperature is rescaled by the critical temperature (Tc = 304.13K). Time is rescaled by characteristic time of t_{PE} = 218 ms (calculated according to Eq. 3). As discussed, temperature discrepancies are seen near the convection onset in Figure 21. The coupled effect of similar convection plume structure and temperature variation have already been formerly studied and validated both theoretically and experimentally in former studies of near-critical channel thermal convection (Chiwata and Onuki, 2001; Amiroudine and Zappoli, 2003). The next section will focus on the more interesting cases of microchannel near-critical fluid convection in smaller channels (D smaller than 300μm).

Figure 21. Evolutions of fluid temperature with time. (near x/L = 0.25, 0.50, 0.75 and 1.00; D = 500 μm; Re_{in} = 67.0; q = 10000 W/m²). The initial evolution indicates the interaction between thermal-mechanical effects and natural convection induced by gravity and density gradient in the boundary, which is typical for moderate channel sizes with near critical fluid under gravity

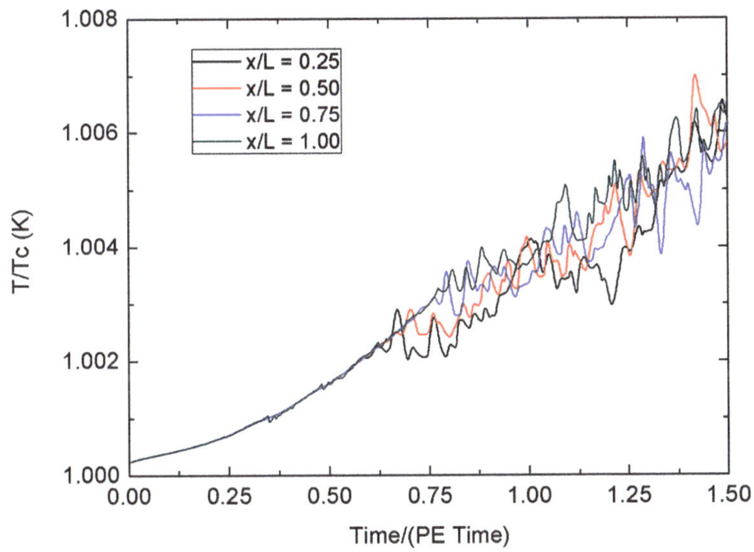

FUTURE RESEARCH DIRECTIONS

- **Near-Critical Thermal-Physical Properties and Critical Phenomena:** The near-critical thermodynamics theory is still insufficient to describe the various new phenomena found in recent experiments. Major theories are in one dimensional and asymptotic analysis. The understanding and explorations in near-critical thermal-physical property induced process is still needed. The coupling of gravity and Thermal-Mechanical effects is dependent on the property changes in the near-critical region.
- **Near-Critical Fluid Convective Structure and Heat Transfer Process:** Near-critical Thermal-Mechanical process and its nature are still not well explained. The thermal convection and heat transfer are dominated by the process but not well understood. New points on heat transfer mechanisms (or the so-called "Fourth Heat Transfer Mechanism" need more deep studies and clarifications.
- **Near-Critical Channel Adverse Pressure Gradient:** Theoretical and three-dimensional analysis into the adverse pressure gradient which happened in multi-scales is a new point for near-critical fluids. Different from conventional adverse pressure gradient, the local behaviors and detailed procedures of the fluid flow transitions in near-critical fluids are still not clear.
- **Multi-Phase Equilibrium Process in the Near-Critical Region:** The thermal-acoustic waves or compression process in the bulk near-critical fluid will become dominant in small scales. However, the possibility of two-phase region inside a critical chamber give rise to the more complex phenomena of pressure wave induced phase separation and phase merging process. The equilibrium process and the transitions should be explained by both microscale and macroscale happenings.

- **Near-Critical Phase Equilibrium and Transitions:** The experimental systems of near-critical studies should be established for both visualization and other fluid parameter tests. Especially, the near-critical fluid control and its interactions with the fluid itself should be considered. The boundary effect and phase process are very critical for near-critical systems. The detailed discussion of critical passages would be interesting and useful for real flow system designs.
- **Critical Effects and Its Applications:** The critical dynamics and the mechanisms in applications of chemical extraction, chemical and biological synthesis, and energy systems are insufficient to sustain the current fast development in related fields. The near-critical control of fluid parameters and the control of behaviors are another key point for the optimization of reliable, robust system design.

CONCLUSION

In summary, a new possibility of near-critical fluid mixing mechanism and heat transfer process is identified in this study. The near-critical dynamics can be extended not only in the extension of parameters, but also in the aspect of near-critical physics. In microscale, the evolutions and transitions in near-critical fluid flow would be useful for real system designs and provide basic information for future understanding of the Thermal-Mechanical coupling with gravity conditions. It is also hoped that the study could be useful for related system designs. The following main points are summarized for the current study:

1. The near-critical fluid has become a hot topic in recent thirty years. Though many scientists are doing supercritical/trans-critical fluid studies, but currently only several research groups are trying to analyze the thermal-physical details and there are a lot many unknowns in the fundamental aspects.
2. Numerical models are established to test the dynamic responses and thermal equilibrium process inside near-critical microchannels. The friction and pressure loss correlations are validated with classic Churchill's predictions. In small time scales (10^{-2}s) the thermal collapse and transient vortex flow behaviors are found for microscale. The boundary expanding and compression to the bulk flow process are identified for the micro-channel unstable evolution process. The so-called "critical speeding-up" process involves the boundary heat accumulation and thermal equilibrium process in a very small time scale. The heat transfer is increased dramatically under the boundary of Thermal-Mechanical effects of near-critical fluids.
3. The dynamic features of near-critical thermal equilibrium process in microchannels are found much different from traditional fluids or in closed chambers. This type of micro-mixing process also shows new behaviors as the energy dissipation is magnitude smaller than traditional kinds. The basic mixing flow follows the classical trends as the lower mixing time would have higher energy dissipation rate.
4. In the heated horizontal microchannel, self-accelerating flows are found. The acceleration process is associated with the vortex flow evolution. The near-critical expanding-accelerating happens under the background of near-critical heating up, at which time the density drops dramatically. Relatively large disturbance is found under vortex flow, and together with the energy dissipation an adverse pressure gradient (APG) is identified to be responsible for the disturbance in the vortex region. The downward flow shows a lower temperature than the upward flow. This "near-critical cooling

down" is an adverse process of the so-called "critical heating up", which is very interesting and merits further study in the future.

5. The mechanism of such boundary-bulk interaction and typical near-critical process is ascribed to a new kind of Kelvin-Helmholtz instability. The time scales of critical disturbance, near-critical Thermal-Mechanical effect, and buoyancy induced convection are different but possibly coupled inside the near-critical microchannels.

ACKNOWLEDGMENT

The support of National Science Foundation of China and JSPS Overseas Researcher Grant (No. 16F16068) are gratefully acknowledged. The author is very thankful for the kind cooperation of Prof. X.R. Zhang in College of Engineering, Peking University. This chapter is also a revised version of the numerical section developed from Dr. Lin Chen's PhD thesis. Dr. Lin Chen is also very grateful for the discussion/ suggestions from Prof. Hiroshi Yamaguchi and Dr. Yuhiro Iwamoto in Energy Conversion Research Center of Doshisha University (Japan), and from Prof. Shigenao Maruyama, Prof. Atsuki Komiya and Prof. Junnosuke Okajima in Institute of Fluid Science, Tohoku University (Japan).

REFERENCES

Accary, G., Bontoux, P., & Zappoli, B. (2007). Convection in a supercritical fluid: A reduced model for geophysical flows. *Physics of Fluids*, *19*(1), 014104. doi:10.1063/1.2432159

Accary, G., Raspo, I., Bontoux, P., & Zappoli, B. (2005). Rayleith-Benard and Schwarzschild instability in a supercritical fluid. *Advances in Space Research*, *36*(1), 11–16. doi:10.1016/j.asr.2005.02.082

Ameel, T. A., Warrington, R. O., Wegeng, R. S., & Drost, M. K. (1997). Miniaturization technologies applied to energy systems. *Energy Conversion and Management*, *38*(10-13), 969–982. doi:10.1016/S0196-8904(96)00127-6

Amiroudine, S., & Zappoli, B. (2003). Piston Effect induced thermal oscillations at the Rayleigh-Benard threshold in supercritical ^{3}He. *Physical Review Letters*, *90*(10), 105303. doi:10.1103/PhysRevLett.90.105303 PMID:12689007

Aoki, N., Umei, R., Yoshida, A., & Mae, K. (2011). Design method for micromixers considering influence of channel confluence and bend on diffusion length. *Chemical Engineering Journal*, *167*(2-3), 643–650. doi:10.1016/j.cej.2010.08.084

Asinari, P. (2005). Numerical prediction of turbulent convective heat transfer in mini/micro channels for carbon dioxide at supercritical pressure. *International Journal of Heat and Mass Transfer*, *48*(18), 3864–3879. doi:10.1016/j.ijheatmasstransfer.2005.03.028

Bailly, D., & Zappoli, B. (2000). Hydrodynamic theory of density relaxation in near-critical fluids. *Physical Review E: Statistical Physics, Plasmas, Fluids, and Related Interdisciplinary Topics*, *62*(2), 2353–2368. doi:10.1103/PhysRevE.62.2353 PMID:11088715

Beckman, E. J. (2004). Supercritical and near-critical CO_2 in green chemical synthesis and processing. *The Journal of Supercritical Fluids*, *28*(2-3), 121–191. doi:10.1016/S0896-8446(03)00029-9

Beysens, D., Chatain, D., Nikolayev, V. S., Ouazzani, J., & Garrabos, Y. (2010). Possibility of long-distance heat transport in weightlessness using supercritical fluids. *Physical Review E: Statistical, Nonlinear, and Soft Matter Physics*, *82*(6), 061126. doi:10.1103/PhysRevE.82.061126 PMID:21230663

Blood, P. J., Denyer, J. P., Azzopardi, B. J., Poliakoff, M., & Lester, E. (2004). A versatile flow visualization technique for quantifying mixing in a binary system: Application to continuous supercritical water hydrothermal synthesis (SWHS). *Chemical Engineering Science*, *59*(14), 2853–2861. doi:10.1016/j.ces.2004.04.021

Boukari, H., Shaumeyer, J. N., Briggs, M. E., & Gammon, R. W. (1990). Critical speeding up in pure fluids. *Physical Review A.*, *41*(4), 2260–2264. doi:10.1103/PhysRevA.41.2260 PMID:9903350

Carles, P. (2010). A brief review of the thermophysical properties of supercritical fluids. *The Journal of Supercritical Fluids*, *53*(1-3), 2–11. doi:10.1016/j.supflu.2010.02.017

Carles, P., & Dadzie, K. (2005). Two typical time scales of the piston effect. *Physical Review E: Statistical, Nonlinear, and Soft Matter Physics*, *71*(6), 066310. doi:10.1103/PhysRevE.71.066310 PMID:16089870

Chandrasekhar, S. (1961). *Hydrodynamic and Hydromagnetic Stability*. Oxford, UK: Clarendon Press.

Chen, L., Deng, B. L., & Zhang, X. R. (2013). Experimental Study of Trans-critical and Supercritical CO_2 Natural Circulation Flow in a Closed Loop. *Applied Thermal Engineering*, *59*(1-2), 1–13. doi:10.1016/j.applthermaleng.2013.05.017

Chen, L., Deng, B. L., & Zhang, X. R. (2013). Experimental Investigation of CO_2 thermosyphon Flow and Heat Transfer in the Supercritical Region. *International Journal of Heat and Mass Transfer*, *64*, 202–211. doi:10.1016/j.ijheatmasstransfer.2013.03.077

Chen, L., & Zhang, X. R. (2011). Simulation of heat transfer and system behavior in a supercritical CO2 based thermosyphon: effect of pipe diameter. ASME Journal of Heat Transfer, 133(12).

Chen, L., & Zhang, X. R. (2014). Experimental Analysis on a Novel Solar Collector System achieved by Supercritical CO_2 Natural Convection. *Energy Conversion and Management*, *77*, 173–182. doi:10.1016/j.enconman.2013.08.059

Chen, L., Zhang, X. R., Cao, S. M., & Bai, H. (2012). Study of trans-critical CO_2 natural convection flow with unsteady heat input and its implications on system control. *International Journal of Heat and Mass Transfer*, *55*(23-24), 7119–7132. doi:10.1016/j.ijheatmasstransfer.2012.07.027

Chen, L., Zhang, X. R., & Jiang, B. (2014). Effects of Heater Orientations on the Natural Circulation and Heat Transfer in a Supercritical CO2 Rectangular Loop. ASME Journal of Heat Transfer, 136(5).

Chen, L., Zhang, X. R., Okajima, J., Komiya, A., & Maruyama, S. (2016). Numerical simulation of stability behaviors and heat transfer characteristics for near-critical fluid microchannel flows. *Energy Conversion and Management*, *110*, 407–418. doi:10.1016/j.enconman.2015.12.031

Chen, L., Zhang, X. R., Okajima, J., & Maruyama, S. (2013). Numerical simulation of near-critical fluid convective flow mixing in microchannels. *Chemical Engineering Science*, *97*, 67–80. doi:10.1016/j.ces.2013.04.010

Chen, L., Zhang, X. R., Okajima, J., & Maruyama, S. (2013). Thermal relaxation and critical instability of near-critical fluid microchannel flow. *Physical Review E: Statistical, Nonlinear, and Soft Matter Physics*, *87*(4), 043016. doi:10.1103/PhysRevE.87.043016 PMID:23679522

Chen, L., Zhang, X.R., Okajima, J., & Maruyama, S. (2013d). Abnormal microchannel convective fluid flow near the gas–liquid critical point. *Physica A, Statistical Mechanics and its Applications*, *398*, 10-24.

Chen, L., Zhang, X. R., Yamaguchi, H., & Liu, Z. S. (2010). Effect of heat transfer on the instabilities and transitions of supercritical CO_2 flow in a natural circulation loop. *International Journal of Heat and Mass Transfer*, *53*(19-20), 4101–4111. doi:10.1016/j.ijheatmasstransfer.2010.05.030

Cheng, L., & Thome, J. R. (2009). Cooling of microprocessors using flow boiling CO_2 in a micro-evaporator: Preliminary analysis and performance comparison. *Applied Thermal Engineering*, *29*(11-12), 2426–2432. doi:10.1016/j.applthermaleng.2008.12.019

Chiwata, Y., & Onuki, A. (2001). Thermal Plumes and Convection in Highly Compressible Fluids. *Physical Review Letters*, *87*(14), 114301. doi:10.1103/PhysRevLett.87.144301 PMID:11580652

Cui, H. H., & Li, Z. H. (2004). Flow characteristics of liquids in micro-tubes driven by a high pressure. *Physics of Fluids*, *16*(5), 1803–1810. doi:10.1063/1.1691457

Dimmic, G. R., Chatoorgoon, V. V., Khartabil, H. F., & Duffey, R. B. (2002). Natural-convection studies for advanced CANDU reactor concepts. *Nuclear Engineering and Design*, *215*(1-2), 27–38. doi:10.1016/S0029-5493(02)00039-0

Ducoulombier, M., Colasson, S., Haberschill, P., & Tingaud, F. (2011). Charge reduction experimental investigation of CO_2 single-phase flow in a horizontal micro-channel with constant heat flux conditions. *International Journal of Refrigeration*, *34*(4), 827–833. doi:10.1016/j.ijrefrig.2011.01.013

Falk, F. L., & Commenge, J. M. (2010). Performance comparison of micromixers. *Chemical Engineering Science*, *65*(1), 405–411. doi:10.1016/j.ces.2009.05.045

Flockhart, S. M., & Dhariwal, R. S. (1998). Experimental and numerical investigation into the flow characteristics of channels etched in <100> silicon. *ASME Journal of Fluid Engineering*, *120*(2), 291–295. doi:10.1115/1.2820647

Frohlich, T., Beysens, D., & Garrabos, Y. (2006). Piston effect induced thermal jets in near-critical fluids. *Physical Review E: Statistical, Nonlinear, and Soft Matter Physics*, *74*(4), 046307. doi:10.1103/PhysRevE.74.046307 PMID:17155173

Frohlich, T., Guenoun, P., Bonetti, M., Perrot, F., Beysens, D., Garrabos, Y., & Bravais, P. et al. (1996). Adiabatic versus conductive heat transfer in off-critical SF_6 in the absence of convection. *Physical Review E: Statistical Physics, Plasmas, Fluids, and Related Interdisciplinary Topics*, *54*(2), 1544–1549. doi:10.1103/PhysRevE.54.1544 PMID:9965226

Garrabos, Y., Beysens, D., Lecountre, C., Dejoan, A., Polezhaev, V., & Emelianov, V. (2007). Thermo-convectional phenomena induced by vibrations in supercritical SF_6 under weightlessness. *Physical Review E: Statistical, Nonlinear, and Soft Matter Physics*, *75*(5), 056317. doi:10.1103/PhysRevE.75.056317 PMID:17677174

Garrabos, Y., Bonetti, M., Beysens, D., Perrot, F., Frohlich, T., Carles, P., & Zappoli, B. (1998). Relaxation of a supercritical fluid after a heat pulse in the absence of gravity effects: Theory and experiments. *Physical Review E: Statistical Physics, Plasmas, Fluids, and Related Interdisciplinary Topics*, *57*(5), 5665–5681. doi:10.1103/PhysRevE.57.5665

Garrabos, Y., Bonetti, M., Beysens, D., Perrot, F., Frohlich, T., Carles, P., & Zappoli, B. (1998). Relaxation of a supercritical fluid after a heat pulse in the absence of gravity effects: Theory and experiments. *Physical Review E: Statistical Physics, Plasmas, Fluids, and Related Interdisciplinary Topics*, *57*(5), 5665–5681. doi:10.1103/PhysRevE.57.5665

Helmholtz, H. (1868). Über discontinuierliche Flüssigkeits-Bewegungen[On the discontinuous movements of fluids]. *Monatsberichte der Königlichen Preussische Akademie der Wissenschaften zu Berlin*, *23*, 215–228.

Hirschfelder, J. O., Curtiss, C. F., & Bird, R. B. (1954). *Molecular Theory of Gases and Liquids*. New York: Wiley & Sons Inc.

Jiang, P. X., Zhang, Y., & Shi, R. F. (2008). Experimental and numerical investigation of convection heat transfer of CO_2 at supercritical pressures in a vertical tube at low Reynolds numbers. *International Journal of Thermal Sciences*, *47*(8), 998–1011. doi:10.1016/j.ijthermalsci.2007.08.003

Johnson, B. K., & Prudhomme, P. K. (2003). Chemical processing and micromixing in confined impinging jets. *AIChE Journal. American Institute of Chemical Engineers*, *49*(9), 2264–2282. doi:10.1002/aic.690490905

Jounet, A., Mojtabi, A., Ouazzani, J., & Zappoli, B. (2000). Low-frequency vibrations in a near critical fluid. *Physics of Fluids*, *12*(1), 197–205. doi:10.1063/1.870295

Jounet, A., Zappoli, B., & Mojtabi, A. (2000). Rapid thermal relaxation in near-critical fluids and critical speeding up: Discrepancies caused by boundary effects. *Physical Review Letters*, *84*(15), 3224–3228. doi:10.1103/PhysRevLett.84.3224 PMID:11019056

Justin, J. Z. (2002). *Quantum field theory and critical phenomena* (4th ed.). Oxford, UK: Oxford University Press. doi:10.1093/acprof:oso/9780198509233.001.0001

Kawasaki, K. (1970). Kinetic equations and time correlation functions of critical fluctuations. *Annals of Physics*, *61*(1), 1–56. doi:10.1016/0003-4916(70)90375-1

Klein, H., & Feuerbacher, B. (1987). Gravity influence on thermal relaxation near the critical point. *Physics Letters. [Part A]*, *123*(4), 183–187. doi:10.1016/0375-9601(87)90701-8

Koo, J., & Kleinstreuer, C. (2005). Analysis of surface roughness effects on heat transfer in micro-conduits. *International Journal of Heat and Mass Transfer*, *48*(13), 2625–2634. doi:10.1016/j.ijheatmasstransfer.2005.01.024

Kuang, G., Ohadi, M. M., & Zhao, Y. (2004). Experimental study on gas cooling heat transfer for supercritical CO2 in microchannels. *Proc.: The 2nd Int Conf on Microchannels and Minichannels*, 325-332.

Kumar, V., Paraschivoiu, M., & Nigam, K. D. P. (2011). Single-phase fluid flow and mixing in microchannels. *Chemical Engineering Science*, *66*(7), 1329–1373. doi:10.1016/j.ces.2010.08.016

Lord Kelvin (William Thomson). (1871). Hydrokinetic solutions and observations. *Philosophical Magazine*, *42*, 362–377.

Luong, T. D., Phan, V. N., & Nguyen, N. T. (2011). High-throughput micromixers based on acoustic streaming induced by surface acoustic wave. *Microfluidics and Nanofluidics*, *10*(3), 619–625. doi:10.1007/s10404-010-0694-0

Miura, Y., Yoshihara, S., Ohnishi, M., Honda, K., Matsumoto, M., Kawai, J., & Onuki, A. et al. (2006). High-speed observation of the piston effect near the gas-liquid critical point. *Physical Review E: Statistical, Nonlinear, and Soft Matter Physics*, *74*(1), 010101. doi:10.1103/PhysRevE.74.010101 PMID:16907038

Morini, G. L. (2004). Single-phase convective heat transfer in microchannels: A review of experimental results. *International Journal of Thermal Sciences*, *43*(7), 631–651. doi:10.1016/j.ijthermalsci.2004.01.003

NIST Standard Reference Database-REFPROP, Version 8.0, 2006.

Onuki, A. (2004). *Phase Transition Dynamics*. Cambridge University Press.

Onuki, A., Hao, H., & Ferrell, R. A. (1990). Fast adiabatic equilibrium in a single-component fluid near the liquid-vapor critical point. *Physical Review A.*, *41*(4), 2256–2260. doi:10.1103/PhysRevA.41.2256 PMID:9903349

Rosa, P., Karayiannis, T. G., & Collins, M. W. (2009). Single-phase heat transfer in microchannels: The importance of scaling effects. *Applied Thermal Engineering*, *29*(17-18), 3447–3468. doi:10.1016/j.applthermaleng.2009.05.015

Rostami, A. A., Mujumdar, A. S., & Saniei, N. (2002). Flow and heat transfer for gas flowing in microchannels: A review. *Heat and Mass Transfer*, *38*(4-5), 359–367. doi:10.1007/s002310100247

Rostami, A. A., Saniei, N., & Mujumdar, A. S. (2000). Liquid flow and heat transfer in microchannels: A review. *Health Technology*, *18*, 59–68.

Shen, B., & Zhang, P. (2011). Thermoacoustic waves along the critical isochore. *Physical Review E: Statistical, Nonlinear, and Soft Matter Physics*, *83*(1), 011115. doi:10.1103/PhysRevE.83.011115 PMID:21405669

Sierra-Pallares, J., Marchisio, D. L., Alonso, E., Parra-Santos, M. T., Castro, F., & Cocero, M. J. (2011). Quantification of mixing efficiency in turbulent supercritical water hydrothermal reactors. *Chemical Engineering Science*, *66*(8), 1576–1589. doi:10.1016/j.ces.2010.12.039

Steinke, M. E., & Kandlikar, S. G. (2006). Single-phase liquid friction factors in microchannels. *International Journal of Thermal Sciences*, *45*(11), 1073–1083. doi:10.1016/j.ijthermalsci.2006.01.016

Straub, J., Eicher, L., & Haupt, A. (1995). Dynamic temperature propagation in a pure fluid near its critical point observed under microgravity during the German Spacelab Mission D-2. *Physical Review E: Statistical Physics, Plasmas, Fluids, and Related Interdisciplinary Topics*, *51*(6), 5556–5560. doi:10.1103/PhysRevE.51.5556 PMID:9963290

van der Waals. (1873). *Over de Continuiteit van den Gas- en Voleistoftoestand*. Leiden, The Netherlands: Ed. Sijthoff.

Wang, Q., Guan, Y. X., Yao, S. J., & Zhu, Z. Q. (2011). Controllable preparation and formation mechanism of BSA microparticles using supercritical assisted atomization with an enhanced mixer. *The Journal of Supercritical Fluids*, *56*(1), 97–104. doi:10.1016/j.supflu.2010.12.002

Wilkinson, R. A. (1998). Density relaxation of liquid-vapor critical fluids in earths gravity. *International Journal of Thermophysics*, *19*(4), 1175–1183. doi:10.1023/A:1022650011721

Zappoli, B. (2003). Near-critical fluid hydrodynamics. *Comptes Rendus. Mécanique*, *331*(10), 713–726. doi:10.1016/j.crme.2003.05.001

Zappoli, B., Bailly, D., Garrabos, Y., Neindre, B. L., Guenoun, P., & Beysens, D. (1990). Anomalous heat transport by the piston effect in supercritical fluids under zero gravity. *Physical Review A.*, *41*(4), 2264–2268. doi:10.1103/PhysRevA.41.2264 PMID:9903351

Zappoli, B., Beysens, D., & Garrabos, Y. (2015). *Heat transfer and related effects in supercritical fluids*. New York: Springer. doi:10.1007/978-94-017-9187-8

Zappoli, B., & Carles, P. (1995). Thermoacoustic nature of the critical speeding-up. *European Journal of Mechanics - B/Fluids*, *14*, 41–65.

Zhang, X. R., Chen, L., & Yamaguchi, H. (2010). Natural convective flow and heat transfer of supercritical CO_2 in a rectangular circulation loop. *International Journal of Heat and Mass Transfer*, *53*(19-20), 4112–4122. doi:10.1016/j.ijheatmasstransfer.2010.05.031

Zhang, X. R., & Yamaguchi, H. (2008). An experimental study on evacuated tube solar collector using supercritical CO_2. *Applied Thermal Engineering*, *28*(10), 1225–1233. doi:10.1016/j.applthermaleng.2007.07.013

Zhang, X. R., Yamaguchi, H., & Uneno, D. (2007). Experimental study on the performance of solar Rankine system using supercritical CO_2. *Renewable Energy*, *32*(15), 2617–2628. doi:10.1016/j.renene.2007.01.003

Zhang, Y., & Wang, T. H. (2012). Micro magnetic gyromixer for speeding up reactions in droplets. *Microfluidics and Nanofluidics*, *12*(5), 787–794. doi:10.1007/s10404-011-0922-2 PMID:22707926

Zhao, C. X., He, L. Z., Qiao, S. Z., & Middelberg, A. P. J. (2011). Nanoparticle synthesis in microreactors. *Chemical Engineering Science*, *66*(7), 1463–1479. doi:10.1016/j.ces.2010.08.039

Zhong, F., & Meyer, H. (1995). Density equilibration near the liquid-vapor critical point of a pure fluid: Single phase T>Tc. *Physical Review E: Statistical Physics, Plasmas, Fluids, and Related Interdisciplinary Topics*, *51*(4), 3223–3241. doi:10.1103/PhysRevE.51.3223 PMID:9963001

KEY TERMS AND DEFINITIONS

Critical Diverges: The diverging phenomena of fluid properties, fluid dynamic responses and convection behaviors of near-critical fluids when the temperature and pressure goes near the critical point.

Critical Distance: The "distance" of fluid states (judge by temperature and pressure condition), which is usually notified as a reduced parameter by temperature distance from the critical point.

Critical Point: The point where boundary between gas and liquid phase merges, and above which point (temperature and pressure condition) there is no phase change happen between the two phases.

Hot Spot: The local spot with high temperature and low density, which represents the centers of unstable evolution in a fluid.

Kelvin-Helmholtz Instability: The widely seen instability mechanism which is due to stratification of fluid layers (with shear flow) and the gravity induced perturbations. The growth and stabilize of the stability process introduces various kinds of convection structures in nature.

Near-Critical Fluid: The fluid status in the very close region to the critical point, which is a region with obvious critical diverges phenomena, with fluctuation of properties and break up of general order fluid behaviors. The phase boundary merges at this region.

Piston Effect: The periodic heating due to thermos-acoustic waves for near-critical fluid inside closed chambers, which is typical for fast heating and quick energy conversion process.

Thermal-Mechanical Effect: The fundamental evolution process from thermal effect to mechanical wave traveling through the fluid cell or apparatus in the critical thermal boundary, which may include both hot or cool boundary conditions, having strong stratification process and the flow will be expansive and much compressible.

Thin Hot Boundary: The near-critical fluid thermal boundary which is very thin but hot with high temperature due to the very low thermal diffusivity and large compressibility.

APPENDIX: NOMENCLATURE

A: Area (m^2)
C_p: Specific heat at constant pressure ($J \cdot kg^{-1} \cdot K^{-1}$)
C_p: Specific heat at constant volume ($J \cdot kg^{-1} \cdot K^{-1}$)
d or D: Width (m, μm)
F: Viscous dissipation (Pa)
g: Gravity: ($m \cdot s^2$)
h: Averaged heat transfer coefficient ($W \cdot m^{-2} \cdot K^{-1}$)
h: Enthalpy ($J \cdot kg^{-1}$)
k: Roughness (μm)
L: Length (m)
L_m: Critical fluctuation length (m)
m: Mass (kg)
$\bar{m}$: Molar mass ($g \cdot mol^{-1}$, $g \cdot kmol^{-1}$)
$\dot{m}$: Mass flow rate ($kg \cdot h^{-1}$)
n: Reflectance
$\bar{m}$: Molar number density (mol^{-1})
N: Molecular number per unit volume (m^{-3})
Nu: Nusselt number
p: Pressure (MPa)
q or Q: Heat flux ($W \cdot m^{-2}$; $W \cdot m^{-3}$)
r: Coordinate parameter
R: Gas constant ($J \cdot mol^{-1} \cdot K^{-1}$)
Re: Reynolds number
Ri: Richardson number
Sc: Schmidt number
T: Temperature (K, °C)
v: Velocity ($m \cdot s^{-1}$)
V: Volume (m^3)

Greek Symbols

α : Scalar for critical specific heat
β : Scalar for critical equilibrium process at controlled volume ::
β_p : Isobaric expansion coefficient (K^{-1})
γ : Ratio of specific heat capacities (-, K^{-1})
ε : Critical distance, energy dissipation rate (-, $W \cdot kg^{-1}$)
θ : Non-dimensional temperature
λ : Thermal conductivity ($W \cdot m^{-1} \cdot K^{-1}$)
μ : Dynamic viscosity ($kg \cdot m^{-1} \cdot s^{-1}$)
ν : Viscosity ($m^2 \cdot s^{-1}$)
ξ : Characteristic length (m)

ρ: Density ($kg \cdot m^{-3}$)
σ : Thickness (m)
σ_{ij} : Viscous stress (Pa)
τ : Non-dimensional temperature, characteristic time (-, s)
ν : Scalar of critical process at controlled volume
Φ : Parameter in the dissipation equation
X_T or K_T: Isothermal compressibility (Pa^{-1})

Super and Sub Scripts

0: Initial state
ambient: Ambient condition
b: Bulk flow
c: Critical point
exp: Experimental value
in: Inlet
l: Liquid phase
out: Outlet
pc: Pseudo-critical point
PE: Piston Effect/Thermal-Mechanical Effect related process
Sup: Supercritical
th: Theoretical value
v or *g*: Gas pahse
vdW: van der Waals
w: Wall value

Chapter 4
X-Ray Scattering Studies of Expanded Fluid Metals

Kazuhiro Matsuda
Kyoto University, Japan

Masanori Inui
Hiroshima University, Japan

ABSTRACT

Fluid metals exhibit significant properties of thermodynamic-state dependence, since the inter-particle interaction among the constituents (electrons and ions) considerably changes depending on their thermodynamic conditions. The authors have thus far carried out X-ray scattering experiments of fluid metals in the expanded state, which have enabled us to gain insight into microscopic understanding of the structural and electronic properties of fluid metals. The purpose of this chapter is to provide intriguing aspects of fluid metals originated from the existence of conduction electrons, which distinguishes fluid metals from non-conducting fluids, through the results of fluid rubidium and mercury.

INTRODUCTION

Liquid Metals

Liquid metals can be regarded as a two component system which consists of highly mobile conduction electrons and much heavier ions. In particular, the existence of conduction electrons is inherent characteristics of liquid metals, which clearly distinguishes liquid metals from a typical insulating liquid such as inert gas liquids, where the constituents at ambient conditions are basically neutral atoms which interact with each other via van der Waals forces. Liquid metals near the melting point exhibit a typical metallic character: the electrical conductivities of simple liquid metals such as alkali metals actually take metallic values (10 -40 μΩ cm in resistivity) near the melting point (Faber, 1972; Shimoji, 1977). These values are within the same order of those in the solid state and also explain high thermal conductivity of liquid metals. With the success of the pseudopotential approach (Harrison, 1966), it was demonstrated

DOI: 10.4018/978-1-5225-2047-4.ch004

that the electron-ion interaction near the melting point can be treated as a weak perturbation. As a result, conduction electrons are well described with the nearly free electron model even though the atomic arrangements become disordered upon melting. The Ziman formula (Ziman, 1961) obtained within the framework of the nearly free electron approximation has reasonably explained most of the conductivities of liquid metals (Faber, 1972).

Although a liquid metal is a mixture of ions and conduction electrons, a liquid metal near the melting point behaves as a simple monatomic liquid. This monatomic picture is based on the fact that ions in a liquid metal can be regarded as neutral pseudo atoms (Ziman, 1979), i.e., the Coulomb forces between ions are well-screened by conduction electrons.

Fluid Metals: Liquid Metals in the Expanded State

When a liquid is heated across the liquid-vapor coexistence curve, the density sharply decreases accompanying the liquid-vapor first order phase transition. Vaporization of the metallic elements across the liquid-vapor coexistence curve accompanies the transformation from a metal to an insulator of the elements. Thus the liquid-vapor phase transition involves the electronic transition. The electrical conductivity abruptly decreases on evaporation across the liquid-vapor coexistence curve, which indicates its discrete nature of this electronic transition, where the itinerant nature of valence electrons vanishes and valence electrons are bound by atoms or molecules, i.e., the constituents of a vapor phase of metals.

However, a substantial and continuous variation in the density becomes possible on a pressure-temperature phase diagram, and the fluid phase provides a great opportunity to investigate variation of the physical properties of materials as a function of density over a wide range of thermodynamic conditions from a liquid phase near the triple point to a vapor one.

A more unique and interesting situation is brought about when a liquid metal circumnavigates the liquid-vapor critical point without crossing the saturated-vapor-pressure curve by controlling temperature and pressure, and "intermediate" density between a liquid and a vapor can be realized as shown in the arrow in Figure 1. Correspondence relation between the liquid-vapor and the metal-nonmetal transitions is no longer straightforward (Landau & Zeldovich, 1943). Earlier investigation indicated that the electrical conductivity gradually decreases with volume expansion and more sharply decreases when the fluid reaches around the critical point (Franz et al., 1980). In this continuous expansion of the fluid, a dilute metallic state is realized where the conduction electrons and ions forced to rarefaction to a large extent. The physical properties of fluid metals are considerably dependent on their thermodynamic state (Hensel & Warren, 1999). The situation makes fluid metals intriguing in terms of basic scientific interest, stimulating the investigation for the nature of an electronic transition from a metal to an insulator, its relation with the liquid-vapor phase transition, and the phase behavior particular to the electronic system.

Alkali Metals

Among the metallic elements, alkali metals are of special concern since they are classified as simple metals. Valence electrons in alkali metals originate from s-state and alkali metals have an almost spherical Fermi surface, thus the electronic states in alkali metals are well described with the nearly free electron approximation. This physical picture of alkali metals is basically preserved in the melts. Volume change of alkali metals upon melting is only about 2.5% increase, which means reduction in the valence electron density is only a small amount. The melting points of alkali metals are also at most a few hundred Celsius

Figure 1. Pressure and temperature phase diagram of a typical fluid metal

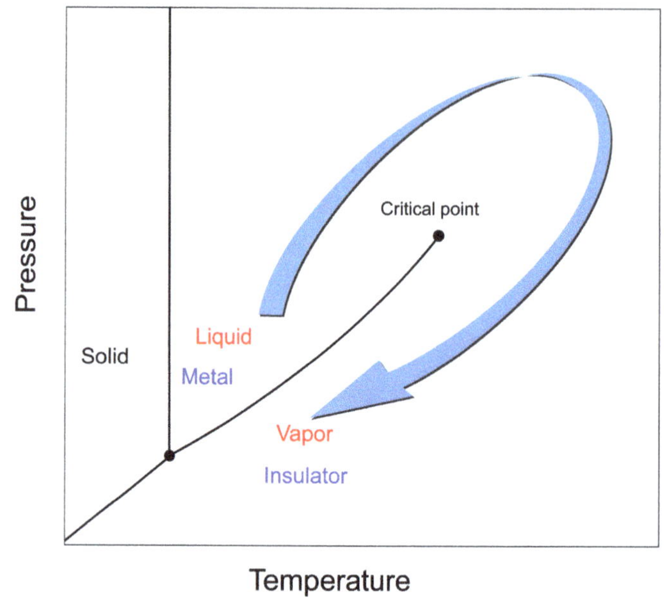

and very low compared with their Fermi temperatures, thus the valence electrons in liquid alkali metals are still in condition of Fermi degenerate gas. Furthermore, as mentioned in the previous section, the electron-ion interaction can be regarded as a weak perturbation even though the atomic arrangements of alkali metals become disordered on melting. These facts validate the nearly free electron description of the valence electron of liquid alkali metals near the melting point. In fact, the experimental investigations such as, the electrical conductivity (Cusack & Enderby, 1960), thermoelectric power (Faber, 1972), Hall coefficients (Greenfield, 1964) support such a nearly free electron picture of liquid alkali metals near the melting point.

According to these facts, extending a liquid phase of alkali metals towards a “vapor” phase provides quite an interesting subject. Since the expansion along the liquid-vapor coexistence curve leads to a reduction in the density of valence electrons and ions in fluid alkali metals, the interaction among valence electrons and also the electron-ion interaction are significantly affected, which should make the system depart from the nearly free electron picture and bring to correlated electronic states.

Several Physical Properties of Fluid Alkali Metals

A considerable amount of experimental investigations on the physical properties of expanded fluid alkali metals have been carried out. As summarized by Hensel and Warren (Hensel & Warren, 1999), an equation of state (Pfeifer et al., 1979; Jüngst et al., 1985), electrical conductivity (Franz et al., 1980; Freyland, 1981), magnetic susceptibility (Freyland, 1979; Freyland, 1980) and nuclear magnetic resonance (Warren et al., 1989) were reported. These results revealed significant thermodynamic-state dependence of the physical properties of fluid alkali metals. The comparison between the experimental conductivity and that estimated from the Ziman formula using experimentally measured structure factor indicated that fluid alkali metals start to deviate from the nearly free electron character although the volume expansion is still small (Gonzalez et al., 1990; Winter et al., 1987). At the density region where such departure from the nearly free electron picture occurs, the magnetic susceptibility (Freyland, 1980) also showed a noticeable enhancement, and the results of nuclear magnetic resonance measurements (Warren et al., 1989) suggest the existence of antiferromagnetic spin fluctuations in the fluid. These experimental results indicate that the electrons in fluid alkali metals become highly correlated as the density decreases.

Electron Gas

Another significance to investigate metallic nature of fluid alkali metals is a reference to understand the electron gas model deeply. The electron gas model is a system where many electrons interact via Coulomb force with each other under the uniform (compensating) positive background (so called jellium model), which has thus far been a basic model in the description for behaviors of the conduction electrons in metals. Referring to a basic model helps to gain an insight into the underlying physics of expanded fluid alkali metals.

Since a pioneering work of Wigner (Wigner, 1934) that predicts the electron gas crystallizes in a dilute limit (Wigner crystal), due to Coulomb repulsion among electrons, theoretical investigations of properties for the dilute electron gas have intensively been carried out. When the electron density is reduced, both of the Coulomb interaction energy and the kinetic one are reduced. When the former dominates over the latter, there appear various intriguing properties in the dilute electron gas. Among them, a significant one is a compressibility anomaly that has been theoretically predicted. The compressibility of the electron gas diverges and turns to negative as the electron density decreases, which means that the uniform electron gas is not stable in its dilute state. Such behavior of the compressibility has been theoretically established, by evaluating the ground state energy of the electron gas (Giuliani & Vignale, 2005). However, experimental information to support the theoretical works on dilute electron systems has been limited so far. Expanded fluid alkali metals are indeed an ideal system in reality which can provide an experimentally accessible system where a dilute electron gas can be realized.

Nevertheless, the readers should notice real fluid alkali metals, where the positive charge is not uniformly distributed as a jellium but discretely distributed as ions. Moreover, ions in the fluids are mobile. Thus, it is also scientifically interesting to investigate the properties of fluid alkali metals if such electron gas problem can be viewed as how the coupling between electrons and ions occurs under the situation where the electron gas suffers the compressional instability.

Mercury

Among sp valent metals, a question "why divalent metals are not semiconductors with a filled valence s-band" may be an origin of the investigation for the metal-nonmetal (M-NM) transition in fluid Hg. The question was answered in 1930's by a theory of Bloch-Wilson's band crossing transition as divalent atoms are condensed (Wilson, 1931). The transition had inevitably predicted that divalent metals should undergo the M-NM transition with volume expansion. As Hg has the lowest critical temperature among divalent metals, the electronic, thermodynamic and optical properties have intensively investigated since the middle of 20th century. The electrical conductivity in metallic fluid Hg was classified into a weak scattering regime to strong diffusive one with decreasing density. At that time, validity between the inhomogeneous structure model proposed by Cohen and Jortner (Cohen & Jortner, 1973) and the pseudo-gap model by Mott (Mott, 1966) had been debated.

A more severe problem on the M-NM transition in expanding fluid Hg in 1970's was that the transition density experimentally determined could not be explained by any theory at that time. The Knight shift obtained by the nuclear magnetic resonance suggests the M-NM transition in fluid Hg at 9 g cm^{-3} (El-Hanany & Warren, Jr., 1975). However theoretical studies needed much lower density to obtain the insulating band structure with a gap at the Fermi level. This discrepancy was finally solved by a first-principles molecular dynamics simulation by Kresse and Hafner (Kresse & Hafner, 1997) in 1990's. Despite that the simulation for a very small system of 50 atoms, they could predict the band crossing in fluid Hg at 9 g cm^{-3} and the local structure consistent with that obtained by X-ray diffraction measurements by Tamura and Hosokawa (Tamura & Hosokawa, 1998). Thus, although the essential problem in the M-NM transition seemed to be resolved, the relationship between the M-NM transition and various theories such as an inhomogeneous structure and the Anderson localization in conduction electrons was left for future investigations.

SAMPLE ENVIRONMENTAL TECHNIQUE

High Pressure Gas Apparatus

On the experimental side, the treatment of fluid metals requires severe experimental conditions. In general, the critical temperatures of metallic elements are very high compared to insulating molecular fluids. Among the elemental metals, heavy alkali metals (rubidium (Rb) and cesium (Cs)) and mercury (Hg) have relatively moderate critical temperatures (T_c (Rb)=1744 °C, T_c (Cs)=1651 °C, T_c (Hg)=1478 °C) (Hensel & Warren, 1999). These are experimentally accessible in a static way of sample heating. However, even for these metals, the critical temperatures are beyond 1400 °C. Moreover, while the critical pressures of Rb and Cs are 124.5 bar and 92.5 bar, respectively, the critical pressure of Hg is 1673 bar, much higher than those of alkali metals (Hensel & Warren, 1999). Thus, in order to realize high temperature and pressure conditions of fluid metals, the authors have developed specially designed high pressure vessels which allow X-ray scattering experiments up to the supercritical conditions of these fluids (as shown in Figure 2). The vessel was originally developed by Tamura and co-workers (Tamura et al., 1999) for X-ray diffraction measurements for fluid Hg, and has thus far been developed and extended to various types of X-ray experiments.

Sample Cell

Our experiments require severe conditions of high temperatures and high pressures. The thermodynamic conditions are located over a wide range from those in liquid and vapor phases. In order to keep a hot fluid sample stable during measurements, the sample should be contained in a cell made of materials which are resistant to corrosive nature of hot fluid metals and also transparent to X-rays. In the scattering experiments, the following two kinds of cells were fabricated. One is made of sapphire and another is made of molybdenum.

Figure 2. The cross-section surface of a high-pressure vessel for large-angle X-ray diffraction measurements from (a) a top-view and (b) a side-view

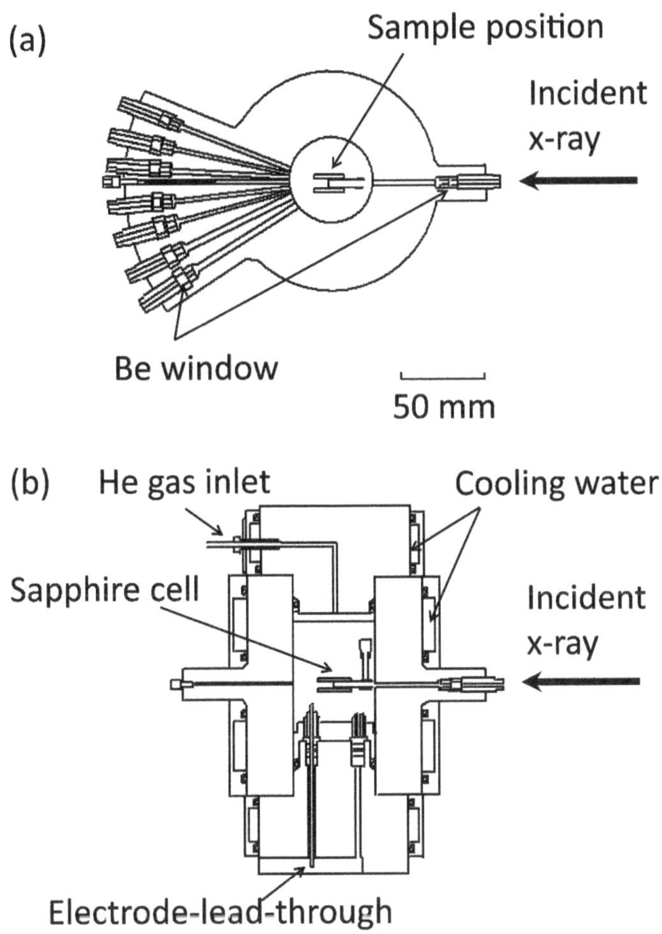

Sapphire Cell for Fluid Hg

For fluid Hg, a single crystalline sapphire was selected for the X-ray diffraction experiments (Tamura et al., 1999). The construction of the cell is illustrated in Figure 3. The cell is constituted from a thin sample space where the sample is locally heated up and a sample reservoir where the liquid sample is kept near the melting point. The liquid sample in the reservoir is loaded to the thin sample space through a narrow channel. When the vessel is filled with a compressed He gas, the pressure is transmitted from the liquid surface in the reservoir to the hot sample in the thin space.

Molybdenum Cell for Fluid Alkali Metals

A sapphire cell can also be applicable to fluid alkali metals at relatively moderate temperature conditions up to approximately 1000 °C. However, high chemical reactivity of hot alkali metals prevents its usage from the experiments at higher temperatures. Instead, refractory metal molybdenum that is resistant to

Figure 3. (a) The cross-section surface of a sample cell made of sapphire and heating accessories. (b) A close-up of the cross-section surface near the thin sample space

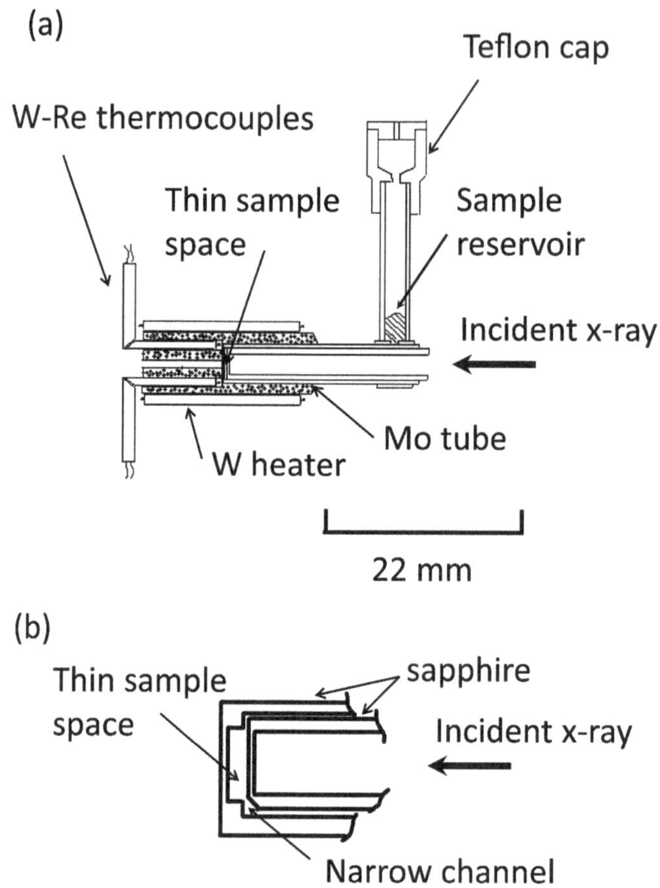

hot alkali metals can be used for the experiments up to the critical regions. In the previous studies on hot alkali metals, molybdenum was used as a suitable cell material for various experiments performed around the critical temperatures, such as the measurement of the equation of state (Pfeifer et al., 1979), electrical conductivity (Pfeifer et al., 1979), and neutron diffraction (Franz et al., 1980). However, the use of molybdenum for X-ray scattering experiments involves several technical problems to be overcome. In order to realize high signal-to-noise ratio measurements, the thickness of X-ray windows made of molybdenum should be much thinner than a sapphire thickness because of a relatively high absorption coefficient of molybdenum for X-rays. The second merit of using thin X-ray windows is that the background noises by scattered X-rays from the windows can be reduced.

Figure 4(a) shows the construction of a molybdenum cell used in X-ray scattering experiments (Matsuda et al., 2004). The concept of the cell design is essentially the same as that of sapphire cell. As in the case of the sapphire cell, it was designed for the transmission geometry of X-rays. All parts of the cell are made of molybdenum. The cell has following several components, i.e., an outer pipe, an inner pipe, a reservoir, a sample spacer, and two thinned disks for X-ray windows. While the first four components two disks are made of polycrystalline molybdenum, which were fabricated by conventional machining techniques, the last two disks for X-ray windows are made of single crystalline molybdenum prepared by mechanical and electrolytic polishing techniques as described in the next paragraph. The use of the single crystalline molybdenum as the material for X-ray windows is of crucial importance especially for X-ray diffraction measurements in the energy-dispersive mode (which will be described later), because a significant background including Debye-Scherrer lines is produced when the polycrystalline molybdenum is used, which generates background noises much larger than the signals from a fluid sample. Although a single crystal also produces larger diffraction peaks, these peaks can be avoided from the detector angles, by adjusting the orientation of the single crystal normal to the surface of disks in the cell preparation (Matsuda et al., 2006).

Figure 4(b) shows an enlarged illustration of the sample space area. A spacer for the determination of the sample thickness is inserted between two thinned single crystalline molybdenum disks. Thickness of the disks is about 30 μm. First, the disks with thickness of 250 μm were cut by electro-discharge machining from a bulk single crystalline molybdenum rod and then mechanically polished to 100 μm and finally electrolytically polished to 30 μm. After all the components of the cell were prepared, they were connected with each other by a high temperature brazing technique using Ru-Mo powder. Prior to X-ray scattering experiments, a liquid sample was introduced into the sample space from the reservoir through the narrow path between outer and inner pipes as shown in Figure 4(a).

X-RAY SCATTERING EXPERIMENTS

X-Ray Diffraction, Small-Angle X-Ray Scattering, Inelastic X-Ray Scattering

The advent of third generation synchrotron radiation facilities and the utilizing high brilliant and high flux X-rays provide a great opportunity to perform X-ray experiments of very accurate spectroscopies and scatterings. In particular, utilizing synchrotron radiation is quite effective for experiments of fluid metals which require extreme conditions of high temperatures and high pressures where the signal from a fluid sample becomes quite weak. Thus, synchrotron radiations were used as a probe for investigating the structural and electronic properties of fluid metals, in order to obtain a microscopic insight into expanded fluid metals. X-ray experiments using a synchrotron radiation have been carrying out at SPring-8 (The third generation synchrotron radiation facility, Hyogo prefecture in Japan). In this section, experimental

Figure 4. (a) Construction of a sample cell made of molybdenum and (b) the enlargement of the area enclosed with the broken square in (a)

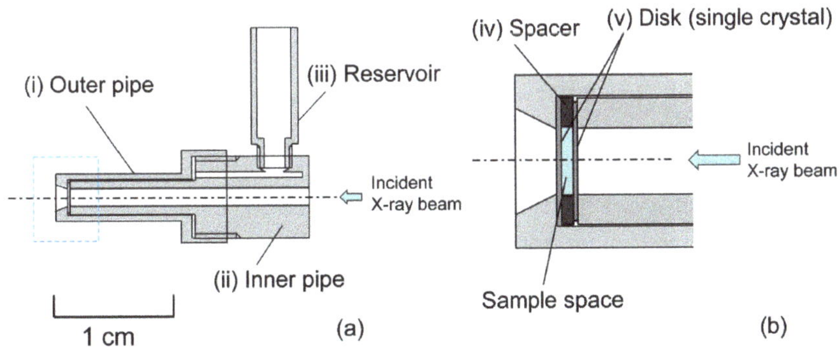

details of X-ray diffraction, small-angle X-ray scattering, and inelastic X-ray scattering measurements, for expanded fluid metals are described.

X-ray diffraction experiments of fluid Hg were carried out at BL04B2/SPring-8 and those of fluid alkali metals at BL28B2/SPring-8 with the energy dispersive mode. White X-rays were incident on the sample and the scattered X-rays were detected up to 150 keV by pure Ge solid state detector. The scattering angles 2θ were fixed to seven values of 4, 11.5, 15, 20, 25 and 33 degs. at the positions of the Be windows of the vessel. To deduce the total structure factors $S(Q)$, where Q is the modulus of the scattering vector ($Q=4\pi E\sin\theta/hc$, where h is Planck's constant, c is the velocity of light, and E is the energy of the X-ray), the numerical data of the absorption coefficients, atomic form factors and Compton factors theoretically calculated for the elements were used. $S(Q)$ has the relationship with the pair distribution function $g(r)$ through the inverse Fourier transform of $g(r)$ as the following formula,

$$S(Q) = 1 + \rho \int_0^\infty (g(r) - 1)\frac{\sin Qr}{Qr} 4\pi r^2 dr \tag{1}$$

where ρ denotes the average number density of a fluid. Hence, by the Fourier transform of $S(Q)$, $g(r)$ can be obtained as,

$$g(r) = 1 + \frac{1}{8\pi^3\rho} \int_0^\infty (S(Q) - 1)\frac{\sin Qr}{Qr} 4\pi Q^2 dQ \tag{2}$$

$g(r)$ describes spatial correlations between atoms in the fluid, which is proportional to the probability finding atoms at the distance r apart from a given atom. When $g(r)$ is obtained, the local structural information can be evaluated, such as the nearest neighbor distance, which corresponds to the first peak position of $g(r)$, and the coordination number N between the distances of r_1 and r_2, which is calculated by the integration,

$$N = 4\pi\rho \int_{r_1}^{r_2} r^2 g(r) dr \tag{3}$$

Small-angle X-ray scattering (SAXS) experiments were performed at BL04B2/SPring-8. The design of the beamline is described in the reference by Isshiki et al. (Isshiki et al., 2001). For the measurements at high temperatures and high pressures, 38 keV X-rays monochromatized using Si(1 1 1) surface were incident on the sample and the scattered X-rays were detected with an imaging plate of 300 times 300 mm^2 located at approximately 2.9 m from the sample position. The accessible Q is from 0.04 to 0.4 $Å^{-1}$. A vacuum path was mounted between the sample stage and the imaging plate to reduce the background from the air. The imaging plate was exposed for 20 minutes per pattern. To make absorption corrections, the intensity of the transmitted X-ray was monitored using an ionization chamber before and after each SAXS measurement. The density of the fluid sample was estimated using the transmission data.

Inelastic X-ray scattering experiments for observing atomic dynamics have been performed at BL35XU /SPring-8. The design of the beamline is described in the reference by Baron et al. (Baron et al., 2000). In our experimental set-up, backscattering at the Si (11 11 11) reflection was used to provide an incident beam in a 0.8 meV bandwidth onto the sample. The energy of the incident beam and the Bragg angle of the backscattering were approximately 21.747 keV and 89.98 deg., respectively. The flux of the incident beam on the sample was 3×10^9 photons/sec in the experiments before 2004, and several improvements at the beamline realize flux several times higher now. Three spherical analyzer crystals among twelve ones at the end of the 10 m horizontal arm were used due to the restriction of the aperture of the high-pressure windows. The total spectrometer resolution of the full width at half maximum (FWHM) was 1.5-1.8 meV depending on the analyzer crystal.

Inelastic X-ray Compton scattering experiments for observing momentum densities have been performed at BL08W/SPring-8 under the standard setup installed at the beamline for high-resolution Compton scattering experiments which consists of a Cauchois-type bent-crystal analyzer and a position sensitive detector. The energy of the incident X-rays is 115.6 keV and the energy of the scattered X-rays is in the range from 70 keV to 90 keV and the scattering angle is 165 degrees. An X-ray Image Intensifier camera was used as a position sensitive detector (Itou et al., 2001). The experimental momentum resolution was 0.13 a.u. (atomic units) in this experiment.

Inelastic X-ray scattering experiments for observing electronic excitation (plasmon) have been performed at BL12XU (National Synchrotron Radiation Research Center, Taiwan)/SPring-8, which was designed for the studies on various types of electronic excitations. The incident X-ray beam from the undulator was first monochromatized by a Si(111) double crystal monochromator and secondary by a Si(400) four-crystal high-resolution monochromator. Inelastically scattered X-rays were collected with a diced, spherical, crystal analyzer and a Si detector in a near backscattering geometry. The energy of the scattered X-rays was fixed at 13835.3 eV ($=E_0$), while the energy of incident X-rays was scanned by 6 eV near E_0. The energy resolution, which is determined by the full width at half maximum (FWHM) of the quasielastic peak, was 0.22eV. The Q resolution was determined by the opening of a mask (85 mm diameter) in front of the analyzer and was evaluated to be 0.25 Å^{-1}.

ALKALI METALS

Structural Studies of Fluid Alkali Metals (Static Structure)

The energy dispersive X-ray diffraction experiments for fluid alkali metals were carried out at BL28B2. Here the experiments for fluid Rb are described (Matsuda et al., 2007). Before the X-ray irradiation, Rb sample was filled into the molybdenum cell in a glove box in which highly purified helium gas was circulated. Then, the cell containing a liquid sample was set in the high pressure vessel also inside the glove box. After the setup was finished, the inlet of helium gas of the vessel was closed and the vessel was took out of the glove box, and carried to the beamline.

In the diffraction experiments, scattered X-rays were detected at 2θ of 4°, 7°, and 11.5°, which provide a sufficient range of the modulus of the scattering wave number Q for fluid Rb. Thermodynamic conditions at which the measurements were performed were set over a wide range of temperatures and pressures from near the melting point of Rb (100 °C and 10 bar) up to a supercritical condition (1850 °C and 160 bar). Owing to the adjustment of the crystal orientation of the single crystalline Mo windows of the cell, no parasitic diffraction peaks (Laue spots) originated from the cell were observed in raw spectra (Matsuda et al., 2004; Matsuda et al., 2006) except for the peak due to fluorescent X-rays. This result indicates the importance of the adjustment of the crystal orientation for diffraction measurements.

Figure 5 shows the structure factors $S(Q)$ and the pair distribution functions $g(r)$ of expanded fluid Rb. At 100 °C (near the melting point of Rb), the first and second maxima in $S(Q)$ are located at 1.5 Å^{-1} and 2.8 Å^{-1}, respectively. With increasing temperature (decreasing density), a clear thermodynamic dependence is observed in the $S(Q)$ profile. As seen in the figure, the magnitudes of the first and the second peaks in $S(Q)$ decrease as the fluid density decreases. The position of the first peak slightly shifts to a lower Q region with decreasing density, whereas that of the broad second peak shifts to a higher Q region. It should be noted that the oscillatory behavior of $S(Q)$ remains even in a much lower density region at 1850 °C. The $S(Q)$ data at the low-Q region were obtained by small-angle X-ray scattering (SAXS) measurements that were separately carried out at BL04B2. Note that the $S(Q)$ at Q less than 0.3 Å^{-1} starts to increase with decreasing Q, which indicates the appearance of a density fluctuation.

As shown in Figure 5, the height of the first maximum of $g(r)$ progressively decreases and the second maximum is almost damped in the density region of 0.60 g cm^{-3}. An important feature to be noticed is that the position of the first maximum r_1 starts to shift to a smaller r with decreasing density. As shown in the figure, a noticeable shift of r_1 occurs below the density of approximately 1.1 g cm^{-3}. Hereafter r_1 is defined as the nearest neighbor distance.

Figure 5. Structure factor S(Q) (left panel) and pair distribution function g(r) (right panel) of expanded fluid Rb. Full curves in S(Q) denote Fourier transforms of g(r)

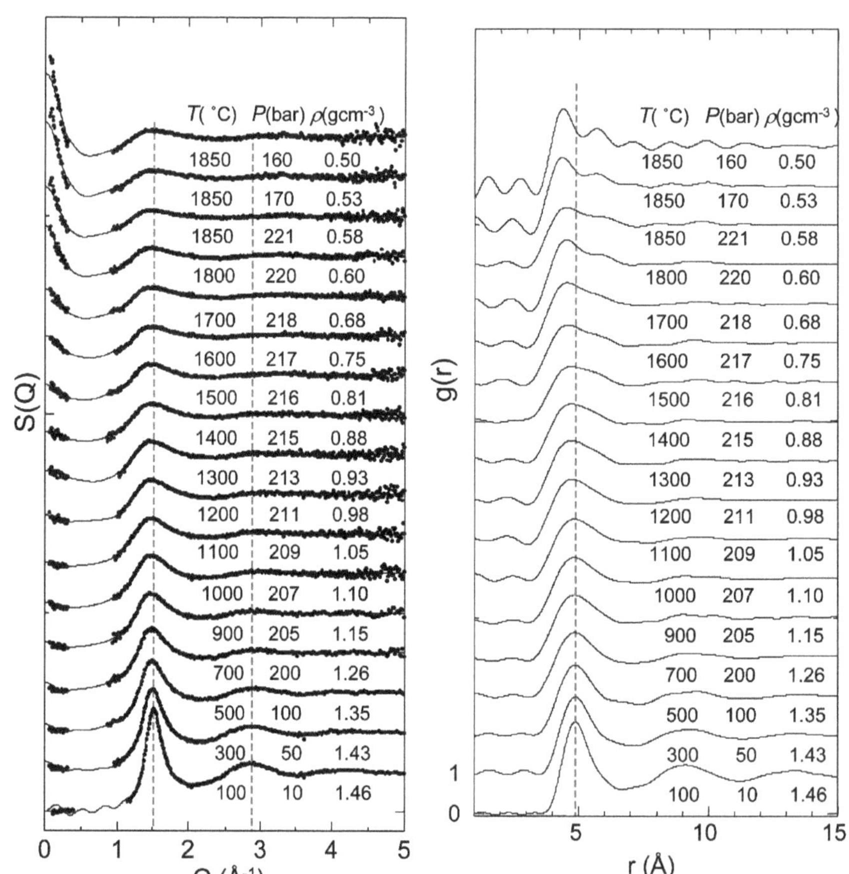

Figure 6(a) shows density dependence of r_1. It exhibits approximately 4.85 Å at 100 °C. This value of r_1 is almost the same with that reported for the solid state (Kittel, 1986). With decreasing density, r_1 slightly decreases and then progressively decreases below approximately 1.1 g cm^{-3}. In the density region lower than 0.5 g cm^{-3}, the value of r_1 seems to approach approximately 4.2 Å. This value is close to the equilibrium interatomic distance of an isolated Rb dimer (Urban & Sadlej, 1995). The contraction of r_1 is consistent with the results obtained by the previous magnetic susceptibility (Freyland, 1980) and inelastic neutron scattering (Pilgrim et al., 1997) measurements where the existence of a considerable amount of dimers in the fluid was suggested.

Figure 6(b) shows density dependence of the density fluctuation $S(0)$ for fluid Rb. The $S(0)$ was deduced by fitting $S(Q)$ at low Q using the Ornstein-Zernike formula $S(Q)=S(0)/(1+\xi^2Q^2)$ where ξ is the correlation length of a density fluctuation. It gradually increases with decreasing density. The increase in $S(0)$ becomes remarkable below the density of approximately 1.1 g cm^{-3}.

A notable feature of these structural changes is that the decrease in the value of r_1 and the increase in $S(0)$ already start to appear at the density region far apart from the critical one (0.29 g cm^{-3}), that is, structural changes are pronounced at an early stage of the volume expansion. The result indicates that a structural inhomogeneity appears in the density range where the fluid is still metallic.

Figure 6. (a) The nearest neighbor distance r_1 and (b) density fluctuation $S(0)$ of expanded fluid Rb

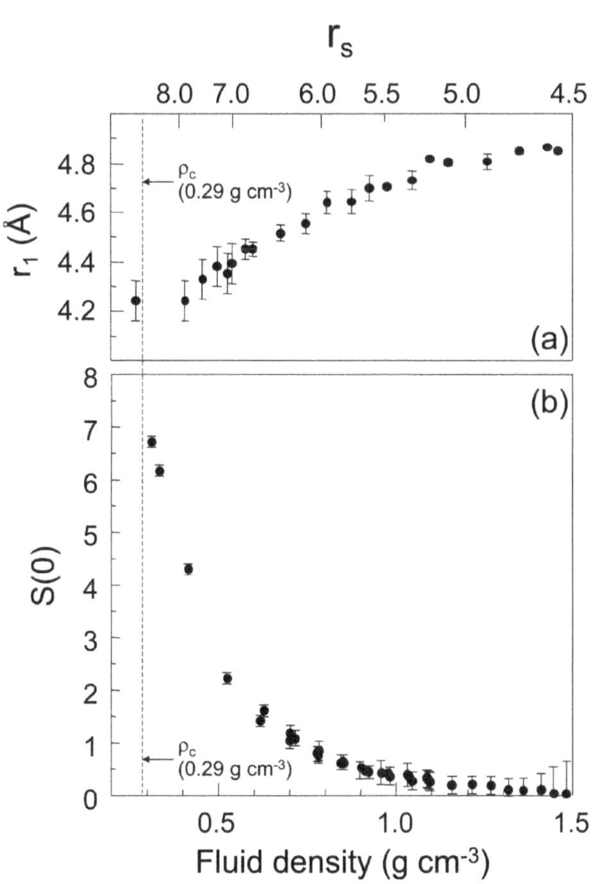

For interpreting these structural changes with volume expansion, Matsuda et al. (Matsuda et al., 2007) focused on the fact that valence electrons in alkali metals can be treated as an ideal realization of the electron gas. The properties of the electron gas are characterized with the electron density parameter r_s (the Wigner-Seitz radius in units of Bohr radius), and the value of r_s is proportional to the ratio of the interaction energy to the kinetic energy per electron. Thus, it is of crucial importance to evaluate the observed structural features on this r_s scale. It was pointed out that the effect of core polarization by ions reduces the effective interaction strength among electrons in real metals. According to Kukkonen and Wilkins (Kukkonen & Wilkins, 1979), the ionic background can be approximated as a uniform background with constant core polarizability. In this picture, r_s reflecting the effective interaction among electrons in metals should be reduced by $r^*_s = r_s/\varepsilon_b$ rather than with r_s, where ε_b is the dielectric constant of the polarizable background defined by $\varepsilon_b = 1+4\pi n\alpha$ (α: ionic polarizability, n: the number density of ions). This corrected r_s ($= r^*_s$) is displayed on the upper abscissa in Figure 6(a).

As can be seen in Figure 6, the density region where the structural changes are pronounced corresponds to the r_s range of approximately 5.0-5.5. According to many-body theories on the electron gas, the compressibility of the electron gas is predicted to diverge and turn to negative (Mahan, 2000), suggesting that the electron gas becomes unstable below the critical r_s (=5.25 (Giuliani et al., 2005)) if the uniform positive background is deformable. The density range where a noticeable structural change starts to appear in fluid Rb is located near this critical r_s, and a close correlation can be envisaged between the stability condition of the electron gas and the observed structural inhomogeneity. The experimental results suggest that the emergence of a structural inhomogeneity occurs in the density range where the uniform electron gas picture for conduction electrons becomes invalid.

The observed decrease of r_1 also suggests that the effective interionic interaction becomes attractive with decreasing density. As Coulomb interaction among ions in liquid metals is screened by conduction electrons, the response of conduction electrons in a dilute regime is of crucial importance for variation in the effective interionic interaction. A theoretical prediction was given that the decrease of r_1 is essentially caused by the interionic attractive force being developed with an enhanced attraction between electrons in the low-density electron gas, combined with the exclusion effect of the electrons from the ionic cores (Maebashi and Takada, 2009). These experimental and theoretical results indicate growing importance of knowing charge fluctuations in conduction electrons to understand the interionic interaction in a dilute metallic system.

Since it was indicated that there are an appreciable number of diatomic molecules in the fluid near the liquid-vapor critical point (Freyland, 1979; Freyland, 1980), the great concern thus far has been whether such molecular structural units persist even in the dense metallic fluid. Pilgrim et al. (Pilgrim et al., 1997) observed a vibration mode in the dynamic structure factor of fluid Rb at the density approximately 2 times higher than the critical density. The result suggests the existence of molecular structural units in the fluid that stays at a metallic regime. Redmer and Warren (Redmer & Warren, 1993) carried out quantum statistical calculations for fluid Rb and Cs, and reported that the presence of ionized molecular species such as Rb_2^+ or Cs_2^+ can explain the enhancement in magnetic susceptibility reported (Freyland, 1979; Freyland, 1980) in the metallic density range about 2 to 3 times higher than the critical density. The emergence of molecular species in the metallic fluid can be closely associated with the observed structural features by X-ray diffraction measurements, such that structural change appears in the early stage of expansion where the fluid is still metallic. In this sense, the experimental investigation for the electronic states in fluid alkali metals is quite important.

Electronic States of Fluid Alkali Metals

X-Ray Compton Scattering

X-ray Compton scattering using a synchrotron radiation has been a unique technique to investigate the electronic states. From Compton scattering experiments, the information on the momentum distribution of electrons can be obtained. The technique is applicable for samples under extreme conditions at high temperatures and high pressures, and its availability over a wide variety of experimental conditions has been demonstrated (Hämäläinen et al., 2000; Reiter et al., 2013; Matsuda et al., 2013). Motivated by the structural features mentioned in the previous section, Matsuda et al. (Matsuda et al., 2011) paid attention to the possibility of observing the variation in the electronic states, and its thermodynamic state dependence in fluid alkali metals using this technique.

In a Compton scattering experiment, the double differential scattering cross section is connected with the electron momentum density, $\rho(\mathbf{p})$, of the scattering system under the condition of the impulse approximation as follows.

$$\frac{d^2\sigma}{d\Omega d\omega} \propto J(p_z) = \iint \rho(\mathbf{p}) dp_x dp_y \tag{4}$$

Here, $J(p_z)$ is so called Compton profiles (CP): the projection of $\rho(\mathbf{p})$ along the photon scattering vector. The experiments have been focusing on single particle behaviors of valence electrons in metals. Variation in itinerant or localized nature of valence electrons is sensitively reflected in Compton profiles.

The Compton scattering experiments for fluid Rb were carried out at BL08W. Thermodynamic conditions ranged from 60 °C and 5 bar up to 1700 °C and 136 bar close to the critical point of Rb. The fluid sample was contained in the sample cell made of molybdenum, the design of which is customized for Compton scattering measurements (Matsuda et al., 2015). A measurement for fluid Rb took approximately 10 h to obtain a spectrum with reasonable statistics, in which total counts at the peak position reached ~1 $\times 10^5$.

Figure 7 shows the raw spectrum of Compton scattering from the cell containing liquid Rb at 60 °C. A shoulder-like feature appears near the peak of the spectrum. As indicated by the arrows in the inset, two dips can be seen near the peak position. As shown in Figure 8, they represent the cross over area beyond which the contribution of core electrons to the spectrum becomes dominant whereas that of valence electrons becomes quite small. The scattering spectrum from an empty cell is also shown in Figure 7. The overall shape of the spectrum of the empty cell is broad, therefore the observed shoulder-like features in the total (sample and cell) spectra persist even after subtracting the empty cell spectrum from the total spectrum in their data analysis of CP derivation described in the next paragraph.

From these raw spectra, CPs were derived as the following procedures. First, the energy calibration was carried out using fluorescence X-ray lines of Bi and Tl, and the channel number was converted to the energy. The contribution of the cell was estimated from the absorption-corrected raw spectrum of the empty cell, which was subtracted from the total (sample and cell) spectrum. Next, the correction of the efficiency of the detector and the Compton scattering cross-section were carried out. After these procedures, the energy was converted to the momentum and then the positive and negative side of the momentum spectrum was averaged. Finally the spectrum was normalized to the number of electrons

Figure 7. Raw X-ray Compton scattering spectrum of liquid Rb at 60 °C in comparison with that of the empty cell. The inset shows the enlargement of the energy channel from 340 to 410 channel

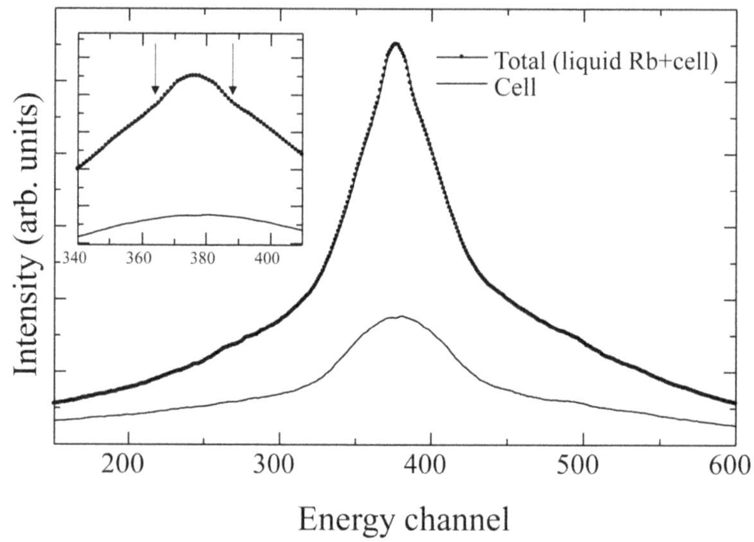

per atom. The double scattering contribution was estimated to be about 3.5% of the total profile (-5 a.u. to +5 a.u.) by a Monte Carlo simulation technique (Sakai, 1987). In this analysis, the subtraction of the simulated double scattering component from the observed spectrum was not carried out because it requires further refinement in the data analysis.

In the normalization of a CP, it is usually assumed that the obtained CP has only the core contribution in the region outside $|p| \leq p_c$ (p_c is the cutoff momentum) and the core contribution is well described with a theoretical free atomic profile. The cutoff momentum often depends on the experimental conditions. The normalization procedure was carried out using following formula

$$\int_{-p_c}^{p_c} J(p)dp = Z_{\mathrm{val}} + \int_{-p_c}^{p_c} J_{\mathrm{core}}(p)dp \tag{5}$$

where Z_{val} is the number of valence electrons per atom which was set to 1 in this case, and J_{core}(p) is the theoretical Hartree-Fock core profile (Biggs. et al., 1975) broadened by the experimental resolution. Here a simple relation $J(p)=C \times I(p)$ was assumed (C is the normalization constant and $I(p)$ is the momentum spectrum before the normalization.) and p_c was set to ±5.0 a.u. This value was determined by referring to the behavior of the theoretical Hartree-Fock total CP of Rb, beyond which valence contribution becomes negligibly small beyond ±5.0 a.u. (Matsuda et al., 2015)

Figure 8(a) shows CP of liquid Rb at 60 °C, which is normalized to number of electrons per atom as mentioned in the above paragraph. As already pointed out, the valence-core crossing is clearly observed in the CP. The total CP approaches to the theoretical Hartree-Fock core electron profile as the absolute value of the momentum increases. The dip position of the CP is located at approximately 0.3-0.4 a.u. (at the negative side, from -0.4 to -0.3 a.u.), which is reasonable for considering the fact that the Fermi momentum of liquid Rb at this density is estimated to be 0.36 a.u. based on the electron gas model. Next, the valence CP was derived from the total CP by subtracting the core profile (Biggs et al., 1975).

Figure 8. (a) Total Compton profile of liquid Rb at 60 °C. Also shown is the profile of the core electrons in atomic Rb by Hartree-Fock calculation (Biggs et al., 1975) that is broadened by the experimental resolution. (b) Valence CP of liquid Rb at 60 °C - 5 bar and 1000 °C - 128 bar obtained by subtracting the core CP from total CPs

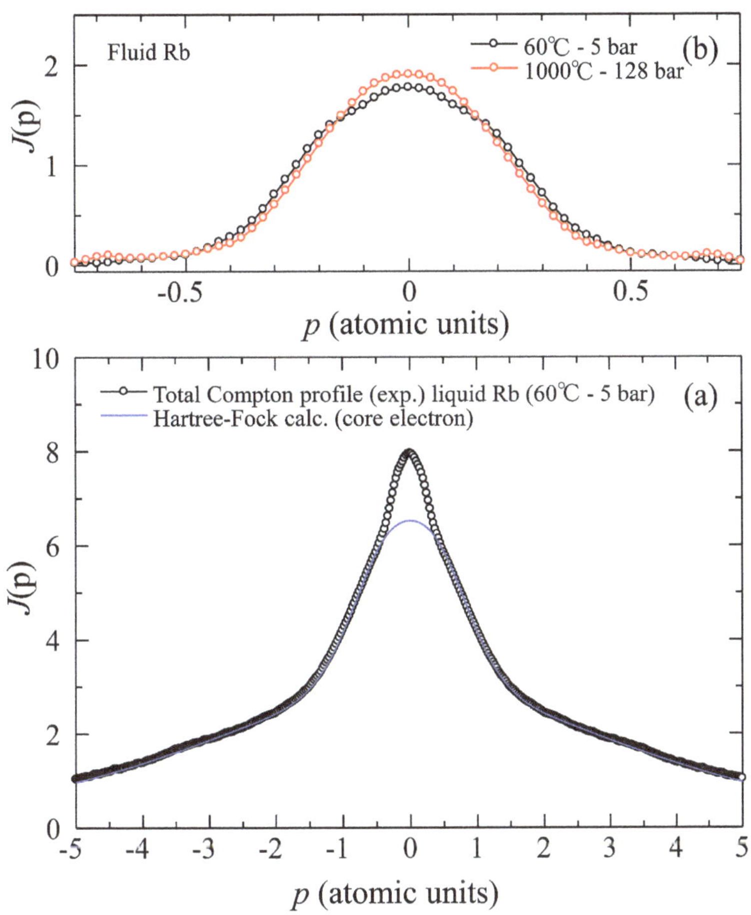

In Figure 8(a), a reasonable agreement can be seen between the experimental CP and the theoretical core CP in the momentum region beyond the valence-core crossing. However, a subtle deviation due to an additional nonlinear background appears. Therefore, to derive the valence CP by subtracting the core profile from the total CP, the iterative procedures proposed by Schülke (Schülke, 2007) were carried out. In this analysis, the normalization constant and a nonlinear background were determined iteratively, assuming that the valence contribution to the total CP is negligible beyond the cutoff $p_{c,}$ and p_c of 0.75 a.u. was chosen.

Figure 8(b) shows CPs of valence electrons in fluid Rb at two different thermodynamic conditions (60 °C - 5 bar and 1000 °C - 128 bar). Thermodynamic state dependence of Compton profiles is clearly visible. As can be seen in the figure, the width of the valence CP at 1000 °C becomes smaller than that at 60 °C. The decrease in the width of valence CP suggests that the Fermi momentum shifts to lower momentum as the density of valence electron in the fluid decreases. From these valence CPs, the Fermi momentum was evaluated by the method of the Fourier-transformed CPs (Pattison and Williams, 1976). It becomes

0.36 a.u. for 60 °C and 0.33 a.u. for 1000 °C, respectively, which indicates that the narrowing of valence CPs is caused by the reduction in the valence electron density. Preliminary data analysis indicates that this tendency of narrowing of the width of valence CPs becomes saturated at higher temperatures (the rate of narrowing with decreasing density becomes slow below the fluid density of approximately 1.1 gcm^{-3}). From the coincidence of the density, such saturation in narrowing of the width of valence CPs seems to be correlated with the structural inhomogeneity observed by X-ray diffraction measurements.

X-Ray Raman Scattering for the Observation of Plasmons

Plasmon is an elementary excitation that originates from the collective motion of valence electrons. The excitation spectrum of plasmon in solid (crystalline) alkali metals was previously investigated by the electron energy loss spectroscopy (EELS) experiments (vom Felde et al., 1989). However, the application of EELS to a liquid sample is difficult. X-ray scattering may be easier than the EELS for observations of plasmon excitations in liquid metals. Nevertheless, experimental studies on plasmon excitations in liquid alkali metals using X-ray scattering were thus far limited only to light elements (Hill et al., 1996).

Using X-ray Raman scattering, plasmon excitations at low momentum transfers can be observable. The dispersion relation of plasmon (the excitation energy vs. momentum transfer relation) and the plasmon line width contain information on the electron-electron and also electron-ion interaction in liquid metals. Although the thermodynamic conditions were limited to a moderate temperature region near melting point, X-ray Raman scattering experiments for liquid and solid Rb were carried out by Kimura et al. (Kimura et al., 2014) and the dispersion relation of plasmon in liquid Rb near the melting point was compared with that in solid Rb.

The X-ray Raman scattering experiments for solid and liquid Rb at 60 °C were carried out at BL12XU/SPring-8. In this experiment, a sapphire cell was used owing to a smaller X-ray absorption of the cell compared to a molybdenum one. The thickness of sample space was 300 μm. The energy transfer ranged from 0.5 to 6.5 eV while the momentum transfer ranged from 0.18 to 0.90 $Å^{-1}$ for the solid state and 0.18 - 0.80 $Å^{-1}$ for the liquid state, respectively. The measurement took three hours for each momentum transfer. The scattering from the empty cell was also measured and it was found that the inelastic contribution from the empty cell was negligibly small compared with that from the sample.

Figure 9 shows the inelastic X-ray scattering spectra of solid and liquid Rb as a function of the energy transfer. At approximately 3.5 eV, a plasmon peak is clearly observed for both in solid and liquid Rb. To obtain the excitation energy and the line width of plasmon, the spectra were fitted by the sum of two Lorentzians, one representing the inelastic component (plasmon), and the other the quasielastic one. The fitting functions were convolved with the Gaussian peak whose line width was 0.22 eV FWHM corresponding to the energy resolution of the spectrometer.

Figure 10(a) shows the plasmon energy of solid and liquid Rb as a function of the momentum transfer q. As can be seen in the figure, the dispersion curve is the cusp-like shape for solid Rb, that is, with increasing q the plasmon energy gradually increases at $0.18\ Å^{-1} \leq q \leq 0.50\ Å^{-1}$, and then decreases at $0.60\ Å^{-1} \leq q \leq 0.90\ Å^{-1}$. This behavior is consistent with the previous results of EELS (vom Felde et al., 1989). In contrast, the plasmon energy of liquid Rb monotonically increases with q. Note that the plasmon energy of liquid Rb becomes larger than that of solid Rb at $q \geq 0.60\ Å^{-1}$. In Figure 10(a), the boundary of the particle-hole (p-h) continuum that was calculated based on a non-interacting electron gas model is displayed. The plasmon dispersion curve intersects with this boundary at approximately 0.5 $Å^{-1}$. Thus the experimental plasmon cut-off wave vector q_c was evaluated to be 0.5 $Å^{-1}$, beyond which the

Figure 9. Inelastic X-ray scattering spectra of (a) solid Rb and (b) liquid Rb normalized by the incident-beam intensity. Solid curves indicate the results using the fitting function. The momentum transfers (denoted as q) are indicated at the right side of each figure. The vertical line denotes the peak position of the spectrum at $q = 0.18$ Å^{-1}

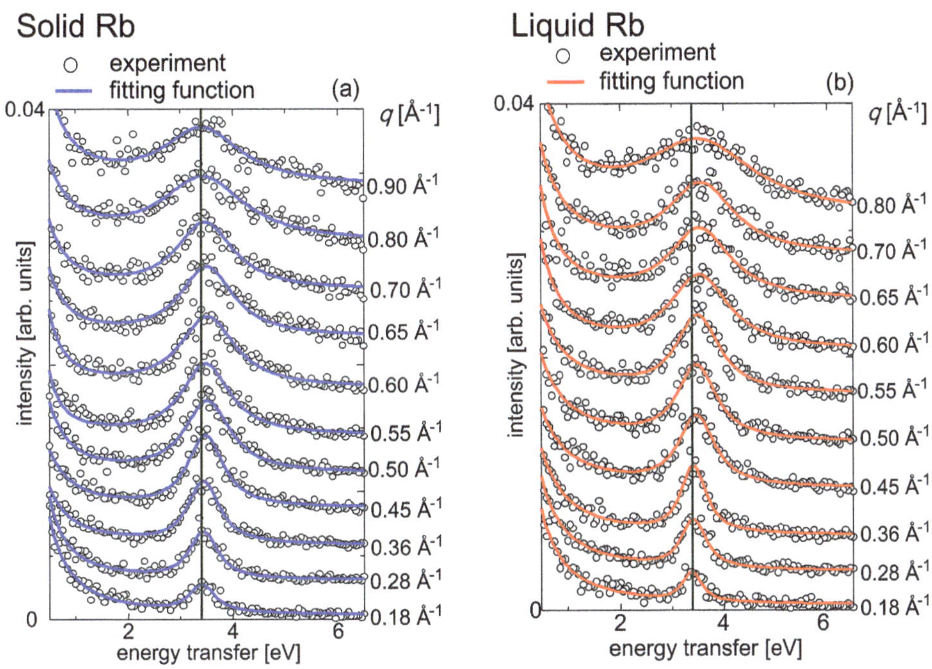

Figure 10. (a) Plasmon dispersions of solid and liquid Rb. The boundary of the electron-hole (e-h) continuum is indicated by the line. (b) q dependence of the plasmon line width $\Delta E_{1/2}(q)$ *of solid and liquid Rb. The modulus of plasmon cut-off wave vector* (q_c) *is indicated by a line*

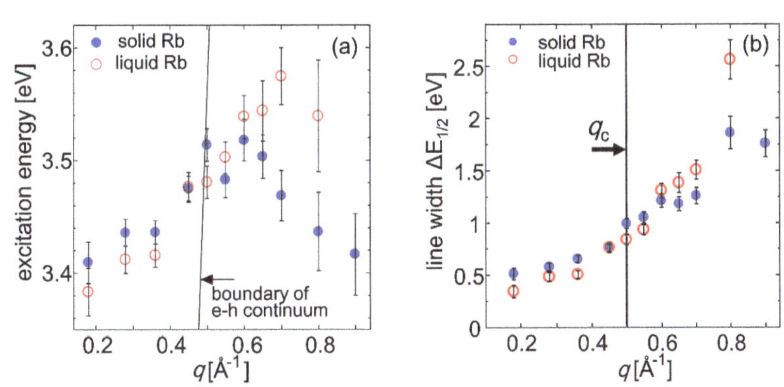

decay of a plasmon into a p-h pair occurs. Figure 10(b) shows the plasmon linewidth $\Delta E_{1/2}$ (q) plotted as a function of q. The values of $\Delta E_{1/2}(q)$ for liquid Rb tend to be smaller in $q < q_c$ and larger in $q > q_c$ than those of solid Rb.

It was indicated that the plasmon energy in solid Rb is considerably modified by the interband transitions to unoccupied d states (Aryasetiawan et al., 1994). The proximity of unoccupied d states to the plasmon energy (approximately 3.5 eV) in the density of states (DOS) affects the observed plasmon energy. In fact, the DOS of solid Rb has several peaks of unoccupied d state near the plasmon peak position (Söderlind & Ross, 2000), whereas peaks are not clearly seen in the DOS of liquid Rb (Mirzoev et al., 2008) near the plasmon peak positions. Although the interpretation was still qualitative, Kimura et al. (Kimura et al., 2014) considered the effect of interband transition, which has been suggested by Aryasetiawan et al. (Aryasetiawan et al., 1994), and explained that the reduction of the interband transition effect upon melting leads to such monotonous increase of the plasmon energy with increasing q as observed in the dispersion curve.

Alternatively, note that the line width $\Delta E_{1/2}(q)$ of liquid Rb tends to become smaller than that of solid Rb in the range $q < q_c$ (=0.5 Å^{-1}) as shown in Figure 10(b). This tendency is quite intriguing because it suggests that the lifetime of the plasmon is longer in the liquid state than in the solid state. To understand this behavior, Kimura et al. (Kimura et al., 2013) calculated the linewidth at $q = 0$ for both solid and liquid Rb under the assumption that the line width includes two effects, i.e., interband transition induced by the ionic potential and intraband (or interband) transitions assisted by phonons (Hasegawa, 1971; Sturm & Oliveira, 1981). Further assumption was made that $\Delta E_{1/2}$ (0) consists of the sum of each contribution as follows, $\Delta E_{1/2}(0) = \Delta E^{ion}_{1/2}(0) + \Delta E^{ph}_{1/2}(0)$. Details of the formulations and calculations are reported in the paper (Kimura et al., 2013). The calculated $\Delta E_{1/2}$ (0) becomes 0.44 eV for solid Rb, and 0.31 eV for liquid Rb. The calculation of the line width was carried out for q=0, whereas the experimental line widths are obtained at finite q (the minimum q experimentally obtained is 0.18 Å^{-1}). However, the results quantitatively reproduce the experimentally observed narrowing of the line width at the low q region upon melting. Moreover, it should be noted that the contribution arising from the interband transition induced by the ionic potential $\Delta E^{ion}_{1/2}$ (0) is found to be dominant over $\Delta E^{ph}_{1/2}$ (0). Thus the variation of $\Delta E^{ion}_{1/2}$ (0) determines calculated $\Delta E_{1/2}(0)$, and the observed $\Delta E_{1/2}(q)$ smaller in the liquid state than in the solid state can be attributed to the reduction in $\Delta E^{ion}_{1/2}(q)$ upon melting. The narrowing in liquid Rb suggests that the vanishing of the Bragg reflections makes the effect of interband transition reduced.

These observed features of plasmon energy and line width indicate that plasmon in the liquid state behaves more closely to plasmon in the electron gas than plasmon in the solid state. A cusp like dispersion in the solid state is turned into a monotonous one in the liquid state. This latter q dependence is also similar to that of the electron gas. Moreover, narrowing of the line width upon melting is considered to be due to the reduction in the interband transition effects once the periodic structure in a solid vanishes. The results indicate that plasmon behaves quite sensitively by variation in the ionic configuration.

MERCURY

Static Structure

To understand the M-NM transition in expanding fluid Hg, the structural information is important. However, the very high critical pressure of fluid Hg had made diffraction experiments very difficult. When Tamura and Hosokawa (Tamura & Hosokawa, 1991) developed the high pressure apparatus combined with an energy-dispersive X-ray diffraction method, the difficulties were overcome. Later, using a synchrotron radiation at SPring-8, Inui et al. (Inui et al., 2003) could obtain the structural data of fluid Hg in the wide density region from a liquid to a dense vapor. The experiments were carried out at BL04B2/

SPring-8. Figure 11 shows *S(Q)* and *g(r)* of fluid Hg. Even in the dense vapor phase at 1.9 g cm^{-3}, *S(Q)* exhibits small oscillations suggesting the correlation between neighboring atoms.

The structural change in the M-NM transition is clearly observed in *g(r)* from 13.6 to 9.0 g cm^{-3}. The first peak position r_1 in *g(r)* does not change much in this density region, as more clearly shown by r_1 plotted in Figure 12, where r_1 and the first nearest coordination numbers N_A and N_B are depicted as a function of density ρ. N_A is defined as twice the integration of $4\pi nr^2g(r)$, where n is the number density, from the cutoff distance (corresponding to an atomic diameter) to r_1 while N_B is defined as the integration of $4\pi nr^2g(r)$ from the cutoff distance to the first minimum position, as shown in the inset of Figure 12. N_A and N_B decrease with decreasing density ρ from 13.6 to 9.0 g cm^{-3}. In contrast, r_1 is approximately constant in this density range. The result suggests that fluid Hg expands with inhomogeneous manner in the metallic region as illustrated in Figure 13. The result was first reported by Tamura and Hosokawa (Tamura & Hosokawa, 1991).

With further expanding toward the critical density of 5.8 g cm^{-3}, r_1 slightly increases and the decreasing rate of N_A becomes small as shown in Figure 12. In the dense vapor phase at $\rho < 5.8$ g cm^{-3}, r_1 of approximately 3.2 Å exhibits little r dependence.

Since the M-NM transition in expanding fluid Hg was observed in the liquid phase near the liquid-vapor critical point, it has been interested in the correlation between the electronic transition and large density fluctuations at the critical point. To investigate this problem, Inui et al. (Inui et al., 2007) carried out the small angle X-ray scattering (SAXS) measurements using the synchrotron radiation source at BL04B2/SPring-8.

Figure 11. The structure factor S(Q) and the pair distribution function g(r) of expanded fluid Hg

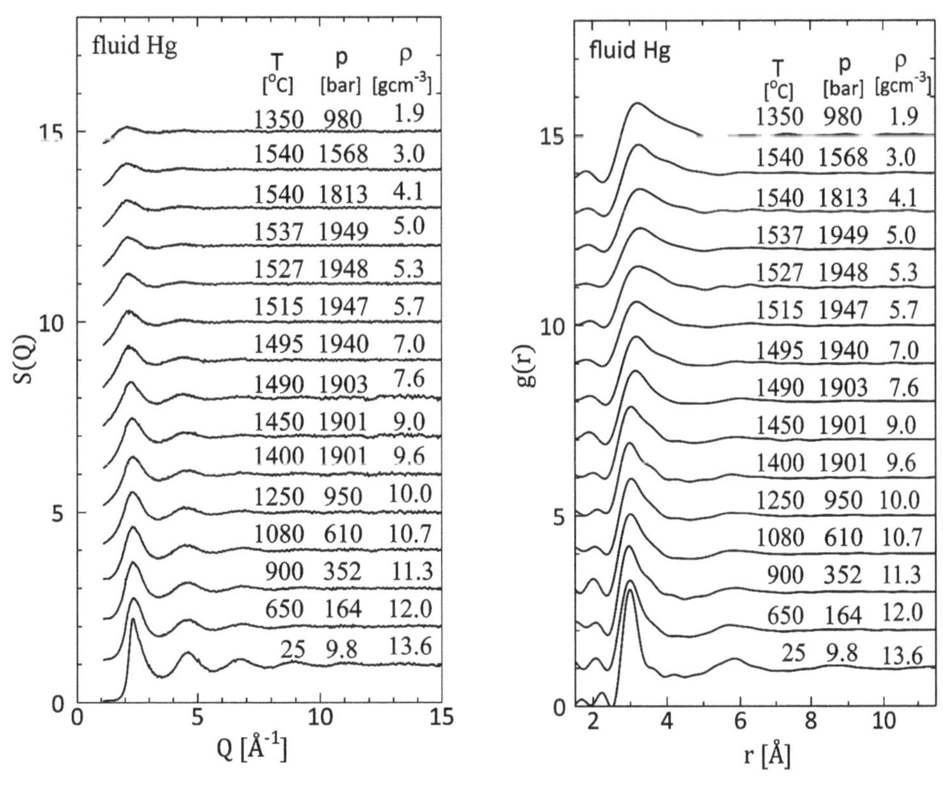

Figure 12. The nearest neighbor coordination number and the nearest neighbor distance r_1 of expanded fluid Hg as a function of density, ρ. The definition of N_A and N_B is described in the text

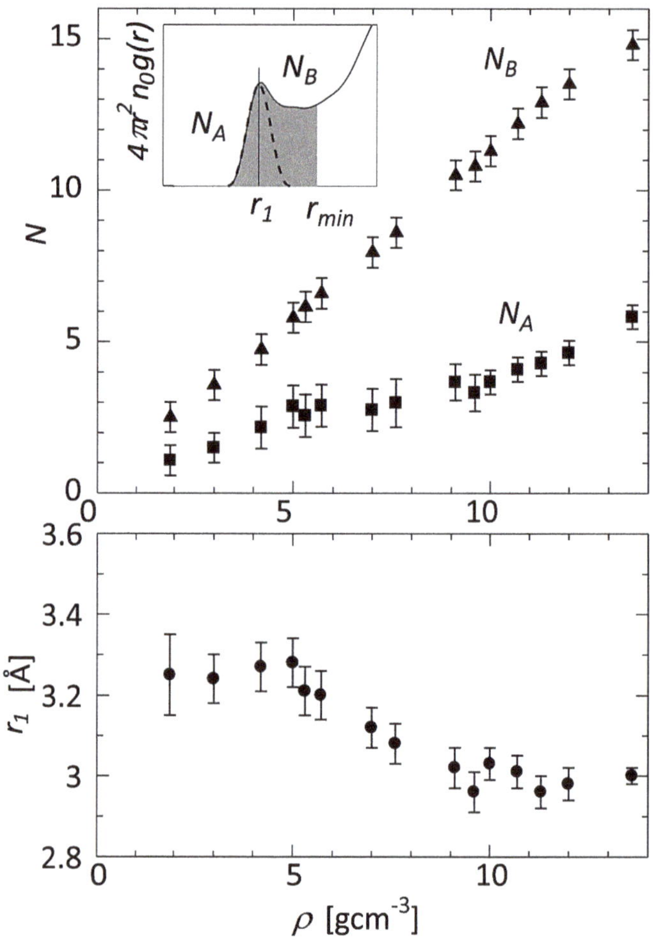

Figure 13. A schematic illustration of homogeneous and inhomogeneous volume expansion

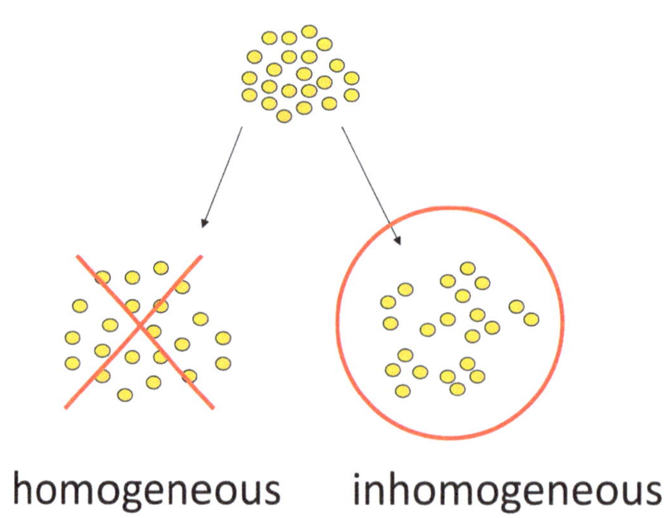

With decreasing ρ, SAXS intensity exhibits clear increase near the M-NM transition. Inui et al. (Inui et al., 2007) deduced $S(Q)$ from the SAXS spectra and obtained $S(0)$ and the correlation length ξ of a density fluctuation by fitting the observed $S(Q)$ using the Ornstein-Zernike formula, $S(Q)=S(0)/(1+\xi^2Q^2)$. Figure 14 shows $S(0)$, ξ and the short range correlation length R ($= \xi / \sqrt{S(0)}$) of fluid Hg as a function of ρ. $S(0)$ and ξ become very large near the critical density and they exhibit strong pressure dependence as the critical point is approached . On the other hand, $S(0)$ and ξ exhibit little pressure dependence near the M-NM transition and these values at the M-NM transition are not small. The results suggest that density fluctuations inherent to the M-NM transition appear at 9 g cm^{-3} in fluid Hg. This interpretation may be supported by the cusp profile in R at 9 g cm^{-3}.

Figure 14. Density fluctuation S(0), the correlation length ξ and short range correlation length R in expanding fluid Hg. The red belt in the figure denotes the density where the M-NM transition occurs. At the M-NM transition, S(0) and ξ are approximately 10 and 10 Å, respectively, while R exhibits a large value of 6 Å

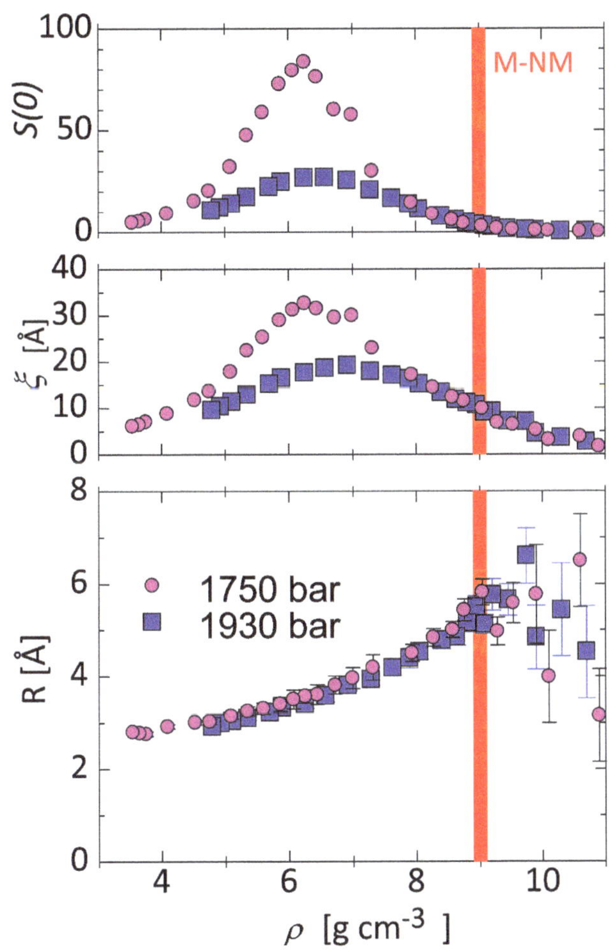

Dynamical Structure

Atomic dynamics in expanding fluid Hg has been of great interest but inelastic neutron scattering measurements were never carried out at such high-pressure conditions. Ishikawa et al. (Ishikawa et al., 2004, Inui et al., 2005) applied the experimental technique for X-ray scattering at high temperatures and high pressures to high-resolution inelastic X-ray scattering, and obtained the dynamic structure factor, $S(Q,E)/S(Q)$ as shown in Figure 15. The experiments were carried out at BL35XU/SPring-8. The $S(Q,E)/S(Q)$ at ambient conditions (13.6 g cm^{-3}) is consistent with that reported by Hosokawa et al (Hosokawa et al., 2002) and the inelastic excitation disperses with increasing Q. At the M-NM transition, however, the inelastic excitation exhibits curious Q dependence. At $Q > 0.8$ Å^{-1}, the inelastic excitation seems to be merged into the central quasi-elastic peak while that at lower Q seems to remain at both sides of the central peak. The least-square fitting for $S(Q,E)/S(Q)$ was carried out using the model function consisting of the Lorentzian and the damped harmonic oscillator and determined the energies of inelastic excitations in the spectra. Figure 16 shows the dispersion relation of the excitation energy for liquid Hg at ambient conditions, fluid Hg at the M-NM transition and dense Hg vapor. The dispersion curve at ambient conditions shows a sinusoidal curve as reported for many liquid metals and the profile is consistent with that of the longitudinal acoustic mode in the liquid. At the M-NM transition, the energy at Q < 0.5 Å^{-1} disperses as fast as that at ambient conditions. With further expanding, the energy at Q < 0.5 Å^{-1} disperses as slow as several hundred m/s. Figure 17 shows the dynamical sound speed determined the dispersion relation at Q < 0.5 Å^{-1} in fluid Hg as a function of ρ. The dynamical sound speed at the M-NM transition is as fast as that at ambient conditions while, as the broken line shows, the ultrasonic sound speed in fluid Hg decreases rapidly from 13.6 to 9 g cm^{-3}. The result suggests that the dynamical sound speed is three times faster than the ultrasonic sound speed at the M-NM transition. The discrepancy between these speeds is recovered with further volume expansion to dense Hg vapor. Inui et al. (Inui et al., 2007) discussed a possibility that density fluctuations inherent to the M-NM transition are correlated with anomalous behavior of the dynamical sound speed at the transition in expanding fluid Hg.

Critical Phenomena

A theory of critical phenomena is now refined based on the scaling hypothesis and renormalization group analysis. To investigate universality of the critical phenomena in fluids, studies on metallic fluids should be needed. In practice, the coexistence curve in fluid Hg is known to deviate largely from the 'law of rectilinear diameters in Van der Waals fluids' as shown in Figure 18. However, the critical indices α and β have been found to be close to the 3D Ising values of 0.110 and 0.325, respectively (Hensel & Warren, 1999). Kohno and Yao (Kohno & Yao, 2002) investigated dynamical critical phenomena in fluid Hg with an ultrasonic technique, and their analysis assuming a 3D Ising universality class provided reasonable results. These results suggest that insulating fluid Hg has an interatomic interaction like the van der Waals one.

The critical indices reported (Hensel & Warren, 1999) were obtained from the data below the critical temperature. Now it is interesting to study the critical indices above the critical temperature. Figure 19 shows $\xi(T, \rho)$ on a temperature-density plane using a bird-eyes view. The data obtained by SAXS measurements were corrected and compiled for the plot. Then Inui et al. (Inui et al., 2016) evaluated ξ at the critical density and plot it as a function of the reduced temperature as shown in the right panel

Figure 15. The dynamic structure factor of liquid and expanded fluid Hg

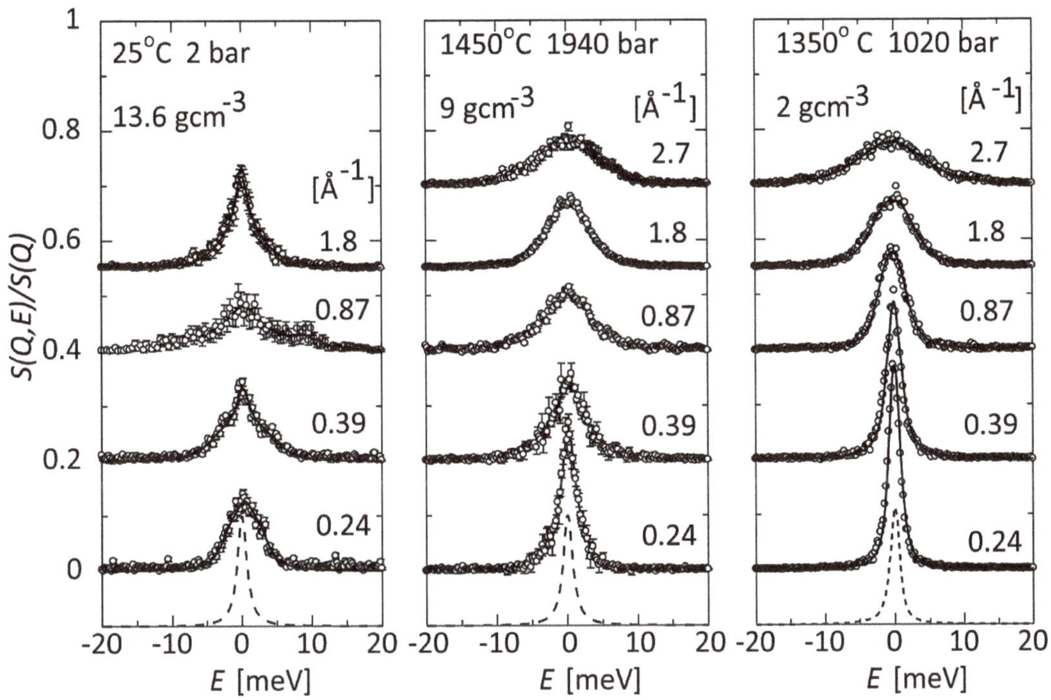

Figure 16. The dispersion relation of the excitation energy for the longitudinal acoustic mode in fluid Hg

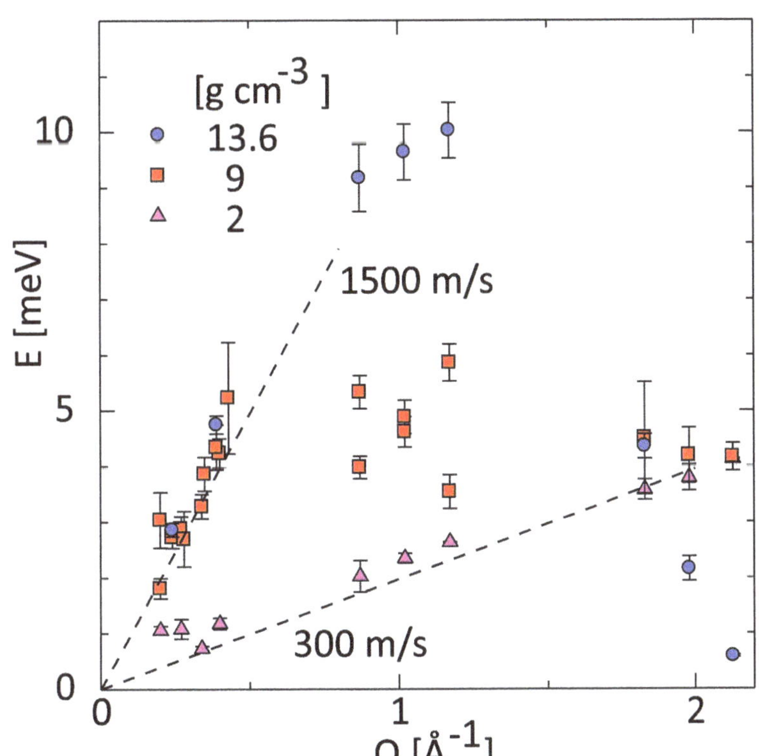

Figure 17. The dynamical and ultrasonic sound speeds in fluid Hg as a function of density

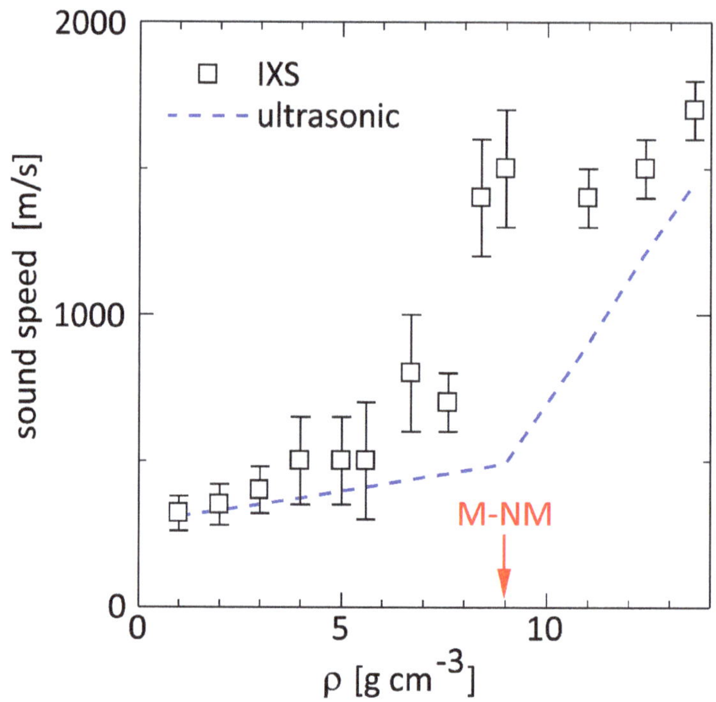

Figure 18. The Co-existence curve and diameter in fluid Hg (solid line) and fluid Xe (broken line) obtained by the density measurements
The data were taken from the book (Hensel & Warren, 1999).

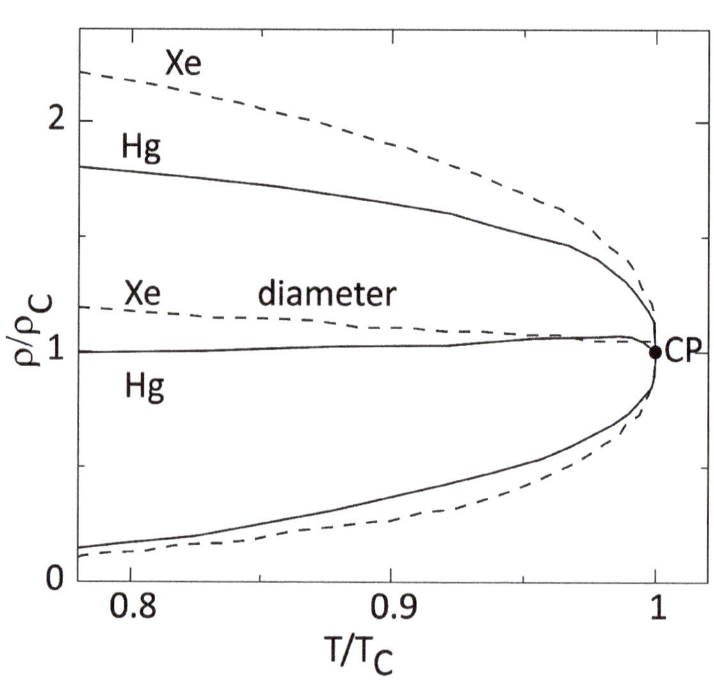

in Figure 20. A solid line denotes the best fits of $\nu = 0.51 \pm 0.02$ while the broken line indicates the fits using the 3D Ising value of $\nu = 0.63$. The left panel in Figure 20 depicts *S(0)* similarly obtained. Solid and broken lines denote the best fits of $\gamma = 1.01 \pm 0.01$ and the fits using the 3D Ising value of $\gamma = 1.2402$, respectively. The results indicate a possibility that fluid Hg belongs to 3D Ising universality class when the accuracy of experimental data is considered. This result suggests that insulating fluid Hg near the critical density behaves like a van der Waals fluid.

Figure 20. S(0) and ξ in fluid Hg as a function of T/Tc - 1 on log–log scales. The solid line denotes the best fit while the broken line indicates the fit using the 3D Ising values

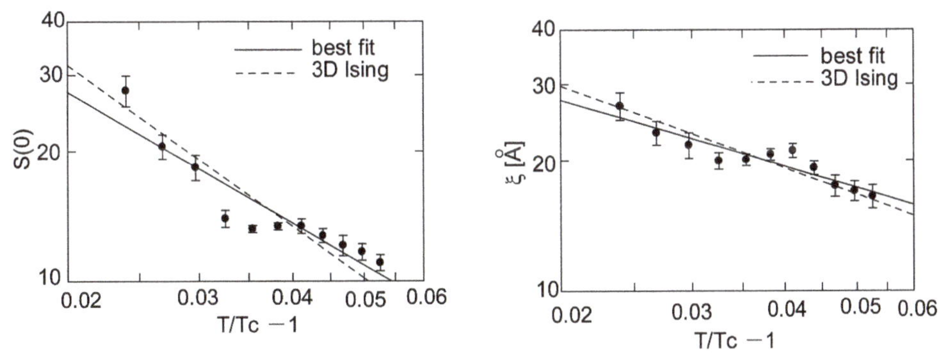

Figure 19. ξ(T,ρ) on a temperature–density plane

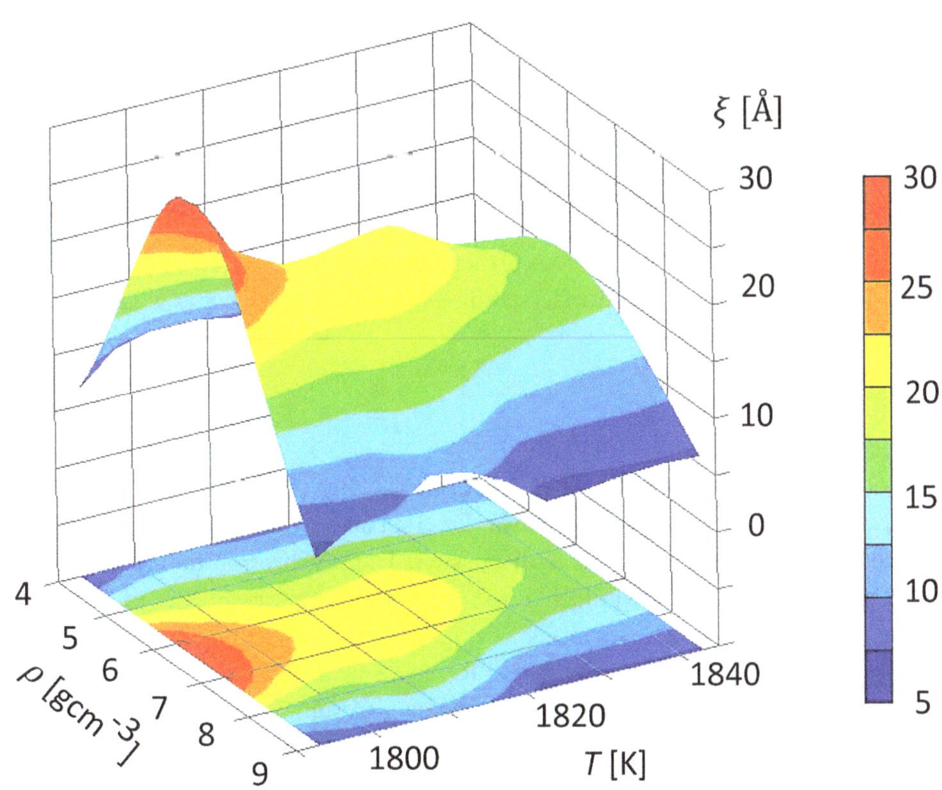

FUTURE RESEARCH DIRECTIONS

The observation of structural properties (static and dynamical) using the X-ray scattering techniques reveals intriguing features of expanded fluid metals. The most important one, commonly observed in both of fluid Rb and fluid Hg, is that structural inhomogeneity appears with the volume expansion. The experimental results suggest that such a structural inhomogeneity cannot simply arise from fluctuations near the ordinary liquid-vapor critical point. Instead, the stability of metallic nature in the fluids is highly correlated with such a inhomogeneity. In fluid systems, it is essential to take into account the flexibility of the atomic arrangements responding to variations in the electronic state, and the observed structural anomalies might be a manifestation of underlying variation in the valence electrons. Hence, a direct observation of the valence electron is of great significance to advance the understandings of expanded fluid metals. An important point is that the charge distribution of valence electrons also becomes inhomogeneous through the electron-ion interactions. For fluid alkali metals, Matsuda and Inui are planning to perform inelastic X-ray Raman scattering experiments at high temperatures and high pressures to investigate plasmon excitations. In fact, a high pressure vessel has already been developed for this purpose, and the experimental techniques are gradually improved to measure X-ray Raman spectra with a high signal-to-noise ratio.

CONCLUSION

In this chapter, X-ray scattering studies for structural and electronic properties of fluid Rb and fluid Hg are reported. In order to realize scattering experiments of hot metallic fluids in highly stable conditions, the techniques of high pressure vessels and sample cells specially designed for fluid metals have been developed. The results of the structural properties of both fluid Rb and fluid Hg revealed that expanded fluid metals have different characteristics from ordinary non-conducting fluids. Furthermore, results suggest that there exists a particular density range where a structural inhomogeneity associated with the electronic transition (metal-nonmetal transition or the electron gas collapse) appears in the fluids. The valuable information experimentally obtained for expanded fluid metals may be useful to understand hot dense matter realized in the deep earth or gaseous planets, and the advanced knowledges of inhomogeneous electronic systems must contribute to deep understandings of the electronic states in many exotic materials.

ACKNOWLEDGMENT

The authors would greatly like to thank Prof. Tamura, Prof. Yao, Prof. Hosokawa, Dr. Kimura, Dr. Sakurai, Dr. Baron, Dr. Nagaya, Dr. Kajihara, Dr. Ishikawa, Dr. Tsutsui, Dr. Itou, Dr. Hiraoka, Dr. Kohara, Dr. Ohishi, Dr. Isshiki and Dr. Ohara for valuable discussion. The synchrotron radiation experiments were performed at SPring-8. The authors would like to acknowledge the organizers of SPring-8 and Prof. Tamura because X-ray scattering experiments at SPring-8 were first promoted by Prof. Tamura who was supported by Grant in Aid for Specially Promoted Research from the Ministry of Education, Culture, Sports, Science and Technology of Japan. This work was also supported by JSPS KAKENHI and the Toray Science Foundation. The technical supports were provided by Kobe Steel Co., Ltd., A.L.M.T. Corp., Mitsubishi Elec. Co., Ltd., and Koyo Assetsu Co., Ltd.

REFERENCES

Aryasetiawan, F., & Karlsson, K. (1994). Energy Loss Spectra and Plasmon Dispersions in Alkali Metals: Negative Plasmon Dispersion in Cs. *Physical Review Letters*, *73*(12), 1679–1682. doi:10.1103/PhysRevLett.73.1679 PMID:10056856

Baron, A. Q. R., Tanaka, Y., Goto, S., Takeshita, K., Matsushita, T., & Ishikawa, T. (2000). An X-ray scattering beamline for studying dynamics. *Journal of Physics and Chemistry of Solids*, *61*(3), 461–465. doi:10.1016/S0022-3697(99)00337-6

Biggs, F., Mendelsohn, L. B., & Mann, J. B. (1975). Hartree-Fock Compton profiles for the elements. *Atomic Data and Nuclear Data Tables*, *16*(3), 201–309. doi:10.1016/0092-640X(75)90030-3

Cohen, M. H., & Jortner, J. (1973). Inhomogeneous Transport Regime in Disordered Materials. *Physical Review Letters*, *30*(15), 699–702. doi:10.1103/PhysRevLett.30.699

Cusack, N., & Enderby, J. E. (1960). A Note on the Resistivity of Liquid Alkali and Noble Metals. *Proceedings of the Physical Society*, *75*(3), 395–401. doi:10.1088/0370-1328/75/3/310

El-Hanany, U., & Warren, W. W. Jr. (1975). Knight Shift in Expanded Liquid Mercury. *Physical Review Letters*, *34*(20), 1276–1279. doi:10.1103/PhysRevLett.34.1276

Faber, T. E. (1972). *Introduction to the theory of liquid metals*. Cambridge, UK: Cambridge University Press.

Franz, G., Freyland, W., Gläser, W., Hensel, F., & Schneider, E. (1980). Structure of expanded liquid rubidium by neutron diffraction. *Journal de Physique. Colloque*, *C8*, 194–198.

Franz, G., Freyland, W., & Hensel, F. (1980). Thermodynamic and electric transport properties of fluid cesium and rubidium in the M-NM transition region. *Journal de Physique. Colloque*, *C8*, 70–73.

Freyland, W. (1979). Magnetic susceptibility of metallic and nonmetallic expanded fluid cesium. *Physical Review B: Condensed Matter and Materials Physics*, *20*(12), 5104–5110. doi:10.1103/PhysRevB.20.5104

Freyland, W. (1980). Electronic correlation and the M-NM transition in fluid alkali metals. *Journal de Physique. Colloque*, *C8*, 74–76.

Freyland, W. (1981). Metal-nonmetal transition in expanded fluid alkali metals. *Comments Solid State Physics*, *10*, 1–10.

Giuliani, G., & Vignale, G. (2005). *Quantum Theory of the Electron Liquid*. Cambridge, UK: Cambridge University Press. doi:10.1017/CBO9780511619915

Gonzalez, D. J., Ng, D. A., & Silbert, M. (1990). Variational MHNC calculations of the structure of expanded liquid rubidium. *Journal of Non-Crystalline Solids*, *117/118*, 469–472. doi:10.1016/0022-3093(90)90976-S

Greenfield, A. J. (1964). Hall coefficients of liquid metals. *Physical Review*, *135*(6A), A1589–A1595. doi:10.1103/PhysRev.135.A1589

Hämäläinen, K., Huotari, S., Laukkanen, J., Soininen, A., Manninen, S., Kao, C.-C., & Mezouar, M. et al. (2000). Free electron gas under high pressure. *Physical Review B: Condensed Matter and Materials Physics*, *62*(2), R735–R738. doi:10.1103/PhysRevB.62.R735

Harrison, W. A. (1966). *Pseudopotential in the Theory of Metals*. New York: Benjamin.

Hasegawa, M. (1971). Theory of Plasmon Damping in Metals. II. Effects of Electron-Ion Interaction. *Journal of the Physical Society of Japan*, *31*(3), 649–667. doi:10.1143/JPSJ.31.649

Hensel, F., & Warren, W. W. Jr. (1999). *Fluid Metals: Liquid–Vapor Transitions of Metals*. Princeton, NJ: Princeton University Press. doi:10.1515/9781400865000

Hill, J. P., Kao, C.-C., Caliebe, W. A. C., Gibbs, D., & Hastings, J. B. (1996). Inelastic X-Ray Scattering Study of Solid and Liquid Li and Na. *Physical Review Letters*, *77*(17), 3665–3668. doi:10.1103/PhysRevLett.77.3665 PMID:10062277

Hosokawa, S., Sinn, H., Hensel, F., Alatas, A., Alp, E. E., & Pilgrim, W.-C. (2002). Collective dynamics of liquid Hg investigated by inelastic X-ray scattering. *Journal of Non-Crystalline Solids*, *312–314*, 163–167. doi:10.1016/S0022-3093(02)01677-0

Inui, M., Hong, X., & Tamura, K. (2003). Local structure of expanded fluid mercury using synchrotron radiation: From liquid to dense vapor. *Physical Review B: Condensed Matter and Materials Physics*, *68*. 094108. doi:10.1103/PhysRevB.68.094108.

Inui, M., Ishikawa, D., Matsuda, K., Tamura, K., Tsutsui, S., & Baron, A. Q. R. (2005). Observation of fast sound in metal–nonmetal transition in liquid Hg. *Journal of Physics and Chemistry of Solids*, *66*(12), 2223–2229. doi:10.1016/j.jpcs.2005.09.021

Inui, M., Kajihara, Y., Ohara, K., Matsuda, K., Ishikawa, D., & Hosokawa, S. (2016). Critical Indices ν and γ Experimentally Obtained at the Liquid-Vapor Critical Point in Fluid Hg. *Journal of Physical Society of Japan*, 85, 035001. doi:10.7566/JPSJ.85.035001.

Inui, M., Matsuda, K., Ishikawa, D., Tamura, K. & Ohishi, Y. (2007). Medium-Range Fluctuations Accompanying the Metal-Nonmetal Transition in Expanded Fluid Hg. *Physical Review Letters*, *98*, 185504. doi:10.1103/PhysRevLett.98.185504.

Ishikawa, D., Inui, M., Matsuda, K., Tamura, K., Tsutsui, S., Baron, A. Q. R. (2004). Fast sound in expanded fluid Hg accompanying the metal-nonmetal transition. *Physical Review Letters*, *93*, 097801. doi:10.1103/PhysRevLett.93.097801.

Isshiki, M., Ohishi, Y., Goto, S., Takeshita, K., & Ishikawa, T. (2001). High-energy X-ray diffraction beamline: BL04B2 at SPring-8. *Nuclear Instruments & Methods in Physics Research. Section A, Accelerators, Spectrometers, Detectors and Associated Equipment*, *467–468*, 663–666. doi:10.1016/S0168-9002(01)00441-7

Itou, M., Hiraoka, N., Ohata, T., Mizumaki, M., Deb, A., Sakurai, Y., & Sakai, N. (2001). Present status of the Cauchois-type Compton scattering spectrometer at SPring-8. *Nuclear Instruments & Methods in Physics Research. Section A, Accelerators, Spectrometers, Detectors and Associated Equipment*, *467-468*, 1109–1112. doi:10.1016/S0168-9002(01)00592-7

Jüngst, S., Knuth, B., & Hensel, F. (1985). Observation of Singular Diameters in the Coexistence Curves of Metals. *Physical Review Letters*, *55*(20), 2160–2163. doi:10.1103/PhysRevLett.55.2160 PMID:10032064

Kimura, K., Matsuda, K., Hiraoka, N., Fukumaru, T., Kajihara, Y., Inui, M. & Yao, M. (2014). Inelastic x-ray scattering study of plasmon dispersions in solid and liquid Rb. *Physical Review B: Condensed Matter and Materials Physics*, *89*, 014206. doi:10.1103/PhysRevB.89.014206

Kimura, K., Matsuda, K. & Yao, M. (2013). Plasmon line width in liquid metals. *Journal of the Physical Society of Japan*, *82*, 115001. doi:10.7566/JPSJ.82.115001

Kittel, C. (1986). *Introduction to Solid State Physics* (6th ed.). New York: John Wiley and Sons.

Kohno, H., & Yao, M. (2002). Dynamic critical phenomena of mercury. *Journal of Physics Condensed Matter*, *14*(6), L171–L176. doi:10.1088/0953-8984/14/6/106

Kresse, G., & Hafner, J. (1997). Ab initio simulation of the metal/nonmetal transition in expanded fluid mercury. *Physical Review B: Condensed Matter and Materials Physics*, *55*(12), 7539–7548. doi:10.1103/PhysRevB.55.7539

Kukkonen, C. A., & Wilkins, J. W. (1979). Electron-electron scattering in simple metals. *Physical Review B: Condensed Matter and Materials Physics*, *19*(12), 6075–6093. doi:10.1103/PhysRevB.19.6075

Landau, L. D., & Zeldovich, J. (1943). On the relation between the liquid and the gaseous states of metals. *Acta Physico-Chimica URSS*, *18*, 194–196.

Maebashi, H., & Takada, Y. (2009). First-Principles Understanding of the Anomalous Structural Change in an Expanded Liquid Alkali Metal. *Journal of the Physical Society of Japan*, *78*, 053706. doi:10.1143/JPSJ.78.053706

Mahan, G. D. (2000). *Many-Particle Physics*. New York: Kluwer Academic/Plenum. doi:10.1007/978-1-4757-5714-9

Matsuda, K., Fukumaru, T., Kimura, K., Tamura, K., Katoh, M., Kajihara, Y., . . . Sakurai, Y. (2015). X-ray Compton scattering experiments for fluid alkali metals at high temperatures and pressures. AIP Conference Proceedings, 1673. doi:10.1063/1.4928258

Matsuda, K., Inui, M., & Tamura, K. (2006). Structural studies of expanded fluid rubidium up to the supercritical regions. *Science and Technology of Advanced Materials*, *7*(6), 483–489. doi:10.1016/j.stam.2006.07.003

Matsuda, K., Nagao, T., Kajihara, Y., Inui, M., Tamura, K., Nakamura, J., Kimura, K., Yao, M., Itou, M., Sakurai, Y. & Hiraoka, N. (2013). Electron momentum density in liquid silicon. *Physical Review B: Condensed Matter and Materials Physics*, 88, 115125. doi: 10.1103/PhysRevB.88.115125

Matsuda, K., Tamura, K. & Inui, M. (2007). Instability of the Electron Gas in an Expanding Metal. *Physical Review Letters*, *98*, 096401. doi:10.1103/PhysRevLett.98.096401

Matsuda, K., Tamura, K., Inui, M., Kajihara, Y., Nagao, T., Yao, M., & Sakurai, Y. et al. (2011). Synchrotron radiation studies of an expanded fluid alkali metal: From the structure towards the electronic state. *The European Physical Journal. Special Topics*, *196*(1), 95–107. doi:10.1140/epjst/e2011-01421-4

Matsuda, K., Tamura, K., Katoh, M., & Inui, M. (2004). Molybdenum cell for x-ray diffraction measurements of fluid alkali metals at high temperatures and high pressures. *The Review of Scientific Instruments*, *75*(3), 709–712. doi:10.1063/1.1646769

Mirzoev, A. A., Mirzoev, Jr., A. A., Sobolev, A. N., & Gelchinski, B. R. (2008). Analysis of the electronic structure of liquid rubidium by the methods of ab initio moleculardynamics, linear muffin-tin orbitals and recursion. *Journal of Physics: Condensed Matter*, *20*, 114104. doi:10.1088/0953-8984/20/11/114104

Mott, N. F. (1966). The electrical properties of liquid mercury. *Philosophical Magazine*, *13*(125), 989–1014. doi:10.1080/14786436608213149

Pattison, P., & Williams, B. (1976). Fermi surface parameters from the Fourier analysis of Compton profiles. *Solid State Communications*, *20*(6), 585–588. doi:10.1016/0038-1098(76)91066-8

Pfeifer, H. P., Freyland, W., & Hensel, F. (1979). Equation of state and transport data on expanded liquid rubidium up to 1700 °C and 400 bar. *Berichte der Bunsengesellschaft fur Physikalische Chemie*, *83*(3), 204–211. doi:10.1002/bbpc.19790830305

Pilgrim, W. C., Ross, M., Yang, L. H., & Hensel, F. (1997). Monatomic-molecular transition in expanded rubidium. *Physical Review Letters*, *78*(19), 3685–3688. doi:10.1103/PhysRevLett.78.3685

Redmer, R., & Warren, W. W. (1993). Magnetic-Susceptibility of Cs and Rb from the Vapor to the Liquid-Phase. *Physical Review B: Condensed Matter and Materials Physics*, *48*(20), 14892–14906. doi:10.1103/PhysRevB.48.14892 PMID:10008020

Reiter, G. F., Deb, A., Sakurai, Y., Itou, M., Krishnan, V.G., & Paddison, S. J. (2013). Anomalous Ground State of the Electrons in Nanoconfined Water. *Physical Review Letters*, *111*, 036803. doi: 10.1103/PhysRevLett.111.036803.

Sakai, N. (1987). Simulation of Compton Double Scatterings of Circularly Polarized γ-Rays by Magnetic Electrons. *Journal of the Physical Society of Japan*, *56*(7), 2477–2485. doi:10.1143/JPSJ.56.2477

Shimoji, M. (1977). *Liquid Metals An Introduction to the Physics and Chemistry of Metals in the Liquid State*. London: Academic Press.

Shülke, W. (2007). *Electron Dynamics by Inelastic X-Ray Scattering*. Oxford University Press.

Söderlind, P., & Ross, M. (2000). Rubidium at high pressure and temperature. *Journal of Physics Condensed Matter*, *12*(6), 921–931. doi:10.1088/0953-8984/12/6/314

Sturm, K., & Oliveira, L. E. (1981). Wave-vector-dependent plasmon linewidth in the alkali metals. *Physical Review B: Condensed Matter and Materials Physics*, *24*(6), 3054–3062. doi:10.1103/PhysRevB.24.3054

Tamura, K., & Hosokawa, S. (1991). X-ray diffraction measurements for expanded liquid mercury around the metal- nonmetal transition region. *Journal de Physique. IV*, *C5*, 39–44.

Tamura, K., & Hosokawa, S. (1998). Structural studies of expanded fluid mercury up to the liquid-vapor critical region. *Physical Review B: Condensed Matter and Materials Physics*, *58*(14), 9030–9038. doi:10.1103/PhysRevB.58.9030

Tamura, K., Inui, M., & Hosokawa, S. (1999). Energy-dispersive x-ray diffraction equipment for fluids at extreme conditions of high temperatures and high pressures. *The Review of Scientific Instruments*, *70*(1), 144–152. doi:10.1063/1.1149556

Urban, M., & Sadlej, A. J. (1995). Electronic structure and electric properties of the alkali metal dimers. *The Journal of Chemical Physics*, *103*(22), 9692–9704. doi:10.1063/1.469984

vom Felde, A., Sprösser-Prou, J., & Fink, J. (1989). Valence-electron excitations in the alkali metals. *Physical Review B: Condensed Matter and Materials Physics*, *40*(15), 10181–10193. doi:10.1103/PhysRevB.40.10181 PMID:9991563

Warren, W. W. Jr, Brennert, G. F., & El-Hanany, U. (1989). NMR investigation of the electronic structure of expanded liquid cesium. *Physical Review B: Condensed Matter and Materials Physics*, *39*(7), 4038–4050. doi:10.1103/PhysRevB.39.4038 PMID:9948738

Wigner, E. P. (1934). On the Interaction of Electrons in Metals. *Physical Review*, *46*(11), 1002–1011. doi:10.1103/PhysRev.46.1002

Wilson, A. H. (1931). The theory of electronic Semi-Conductors. *Proceedings of the Royal Society of London. Series A, Containing Papers of a Mathematical and Physical Character*, *133*(822), 458–491. doi:10.1098/rspa.1931.0162

Winter, R., Hensel, F., Bodensteiner, T., & Gläser, W. (1987). The static structure factor of Cesium over the whole liquid range up to the critical point. *Berichte der Bunsengesellschaft fur Physikalische Chemie*, *91*(12), 1327–1330. doi:10.1002/bbpc.19870911205

Ziman, J. M. (1961). A theory of the electrical properties of liquid metals. I: The monovalent metals. *Philosophical Magazine*, *6*(68), 1013–1034. doi:10.1080/14786436108243361

Ziman, J. M. (1979). *Principles of the Theory of Solids*. Cambridge, UK: Cambridge University Press.

KEY TERMS AND DEFINITIONS

Dynamic Structure Factor: The dynamic structure factor is a physical quantity that describes a time-dependent density-density correlation between particles in a material. This quantity that is proportional to the inelastic scattering cross section can be obtained by inelastic X-ray or Neutron scatterings.

Fluid Metals: Liquid or gaseous state (including the supercritical states above the critical temperature and pressure) of elemental metals or metallic alloys.

Inelastic X-Ray Scattering: A technique to investigate various excitations such as electronic charge and spin fluctuations, and atomic density fluctuations (phonons) in materials, by probing the energy and momentum variations between incident and scattered X-rays with an inelastic mode. In this chapter, 'inelastic X-ray scattering' usually means 'high-resolution inelastic X-ray scattering' to observe phonons in a material, while X-ray Compton scattering and X-ray Raman scattering are also inelastic X-ray scatterings.

Metal-Nonmetal Transition: A phenomenon that occurs owing to decreasing number of carriers (conduction electrons and holes), decreasing carrier mobility or group velocity in a material. In a fluid system, typical evidence of the transition is that high metallic electrical conductivity is reduced into a poor insulating one.

Small-Angle X-Ray Scattering: A technique for probing medium- and long-range structures of 10-1000 Å in length (lamellar structure, density fluctuations, etc.) in a material using elastic X-ray scattering. As the scattering angle much smaller than usual X-ray scattering is needed, a diffractometer specifically designed is utilized.

Static Structure Factor: The static structure factor $S(Q)$ is a physical quantity that depends on the average atomic configuration in a disordered material. It is defined as $S(Q)=I(Q)/Nf(Q)^2$, where Q, $I(Q)$, N, and $f(Q)$ are the modulus of a scattering wave vector, the intensity of scattered X-rays, the number of atoms and the atomic form factor, respectively. The pair distribution function is deduced by Fourier-transformation of $S(Q)$.

X-Ray Compton Scattering: A technique to observe the electron momentum density in a material, by probing the energy and momentum variations between incident and scattered X-rays with an inelastic mode.

X-Ray Diffraction: A technique for probing microscopic atomic arrangements (nearest neighbor distance, nearest neighbor coordination number, etc.) in a disordered material using elastic X-ray scattering. A diffraction pattern generated by scattered X-rays includes information on the atomic configuration in the material through interference effects even if the material has a periodic or non-periodic structure.

X-Ray Raman Scattering: A technique to investigate excitations by electronic charge and spin fluctuations in a material, by probing the energy and momentum variations between incident and scattered X-rays with an inelastic mode. The energy resolution for this purpose is not as high as the one for observing phonons with much smaller excitation energies.

Section 2

Supercritical Natural Circulation Flow and Its Applications

Chapters in this section are selected to discuss the convective behaviors and circulation flow heat transfer characteristics. Supercritical loop systems are very basic concept for various applications of effective energy conversion and thermal utilization processes. This section is then focused on the basic principles, mechanisms and applications in this aspect. The results and proposals in this section may contribute to related natural circulation system designs of supercritical fluids.

Chapter 5
Principles, Experiments, and Numerical Studies of Supercritical Fluid Natural Circulation System

Lin Chen
Tohoku University, Japan & Japan Society for the Promotion of Science (JSPS), Japan

ABSTRACT

Due to the unique thermal and transport properties, Supercritical natural circulation loop (NCL, or thermosyphon) has been proposed in many energy conversion systems. This chapter presents the principals of supercritical natural circulation loop system and its application challenges. A specially designed experimental prototype system is introduced and compared with numerical findings. The system is operated in wide range of high pressures in the critical region. It is found that in a supercritical NCL system, very high Reynolds number natural convection flow can be achieved only by simple heating and cooling. Thermal performance analysis and parameter effects are carried out along with the experimental development. The heat transfer dependency on operation and its mechanisms are also explained and summarized in this chapter. The comparison of experimental and numerical results contributes to better understanding of NCL stability phenomena and applications in energy systems.

BASIC DESIGN AND PRINCIPALS OF SUPERCRITICAL NATURAL CIRCULATION LOOPS (NCLS)

Thermosyphon (or called NCL: natural circulation loop) is one basic loop heat and mass transfer mechanism by fluid buoyancy forces under terrestrial conditions. Due to the absence of mechanical components, natural circulation provides a reliable way of heating/cooling and energy conversion from heat source to heat sink. Representative applications of NCL can be found in solar heating and cooling systems (Huang, 1980; Kalogirou, 2004; Yamaguchi et al., 2010), geothermal process (Kreitlow &

DOI: 10.4018/978-1-5225-2047-4.ch005

Reistad, 1978), nuclear power plant (Zvirin, 1981; Dimmick et al., 2007) and others. Zvirin (1981) and Grief (1998) have made general reviews on NCL systems.

However, the efficiency of NCL systems is still a problem. Water based NCL is now mostly utilized, for example in solar collectors, which has very low circulation flow rate and it takes a long time for heat accumulation (Kalogirou, 2004). Therefore, it is not preferable for high intensity and fast energy conversion processes. Also, for applications in higher temperature and pressure, possible two phase flow and system instabilities are found in NCL (Huang, 1980; Kalogirou, 2004; Yamaguchi et al., 2010). Due to the of density and loop friction variations when the thermosyphon loop heat input is varied, fluid 'density waves' or flow reversals are also found, both for single-phase and two-phase flows (Vijayan et al., 1995; Chatoorgoon, 2001; Kumar and Gopal, 2012). Since 1960s, Welander (1967) and others (Holman & Boggs, 1960) have begun to study the fluid dynamics and stability laws for such NCL systems. Later, a lot of experimental and analytical/numerical studies have been carried out to investigate the system behaviors (for normal fluids) under various input and geometric effects as briefly reviewed by Chen et al. (2010). For some studies, NCL power-flow rate curves are proposed to describe the stability threshold (Dimmick et al., 2002; Jain & Rizwan-uddin, 2008). However, it is later reported that system stability cannot be judged by single parameters due to the complex effects from specific system design and fluid properties (Zhang et al., 2010). Therefore, the NCL controlling parameter correlations are analytically developed by one-dimensional modeling of Vijayan et al. (2004). Also, stability maps have been proposed analytically by Cammarata et al. (2003) and others (Misale et al., 2007).

Although comprehensive studies were conducted by earlier researchers as discussed above, most of these studies were limited to using water as the working fluid operated under normal pressure conditions. Recently, there is an increasing interest in the study of near-critical/supercritical CO_2 based NCL (Yamaguchi et al., 2010; Dimmick et al., 2002; Chatoorgoon, 2001; Kumar & Gopal, 2012; Chen et al., 2010; Jain & Rizwan-uddin, 2008; Zhang et al., 2010). For high power systems, supercritical water (critical point: $Tc = 647.1$ K and $Pc = 22.06$ MPa) is proposed, and supercritical CO_2 (critical point: $Tc = 304.13$ K and $Pc = 7.378$ MPa) is also promising as it can operate under relative lower temperature and pressure compared with supercritical water. Indeed, supercritical CO_2 NCL systems have already been tested in solar heater (Yamaguchi et al., 2010) and proposed in new generation nuclear plant design and other energy conversion designs (Dimmick et al., 2012; Holman & Boggs, 1960; Chen et al., 2010; Jain & Rizwan-uddin, 2008; Zhang et al., 2010; Vijayan et al., 2004), where relative higher circulation and heat transfer efficiencies have been identified and optimized. Near the critical point, the fluid is characterized by the feature of large thermal and transport properties variations with even a very small change in temperature and/or pressure, as shown in Figure 1, Figure 2, Figure 3, and Figure 4. Near-critical fluid is very dense and highly expandable, which property can help produce effective natural convective flow in NCL with large density difference. Sometimes the critical diverges will also happen in the transition process in the near-critical region. The high specific heat and low viscosity of near-critical/supercritical CO_2 fluid may achieve high heat transfer performances when it is operated near the critical region. It has also been proved that compact and miniaturized system design can be achieved for supercritical fluids (Kumar & Gopal, 2009; Basu et al., 2008).

The basic system stability of supercritical CO_2 based NCL have been studied by several groups and the major concerns are system stability. For pressurized CO_2 based NCLs, steady-state analysis (Kumar & Gopal, 2009) and effect of system pressure on the steady state performance are numerically studied by Kumar and Gopal (2009). In the supercritical region, Chatoorgoon (2001) reported a non-linear numerical code to examine the supercritical flow stability in a single-channel natural-convection loop.

Figure 1. Variation of thermoproperties with temperature and pressure in the critical region: specific heat

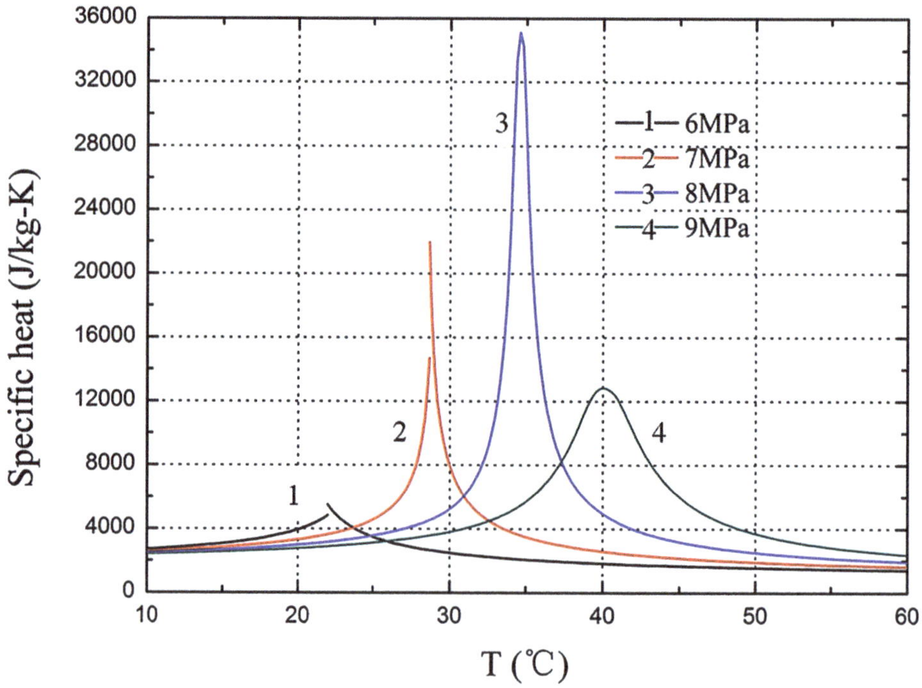

Figure 2. Density

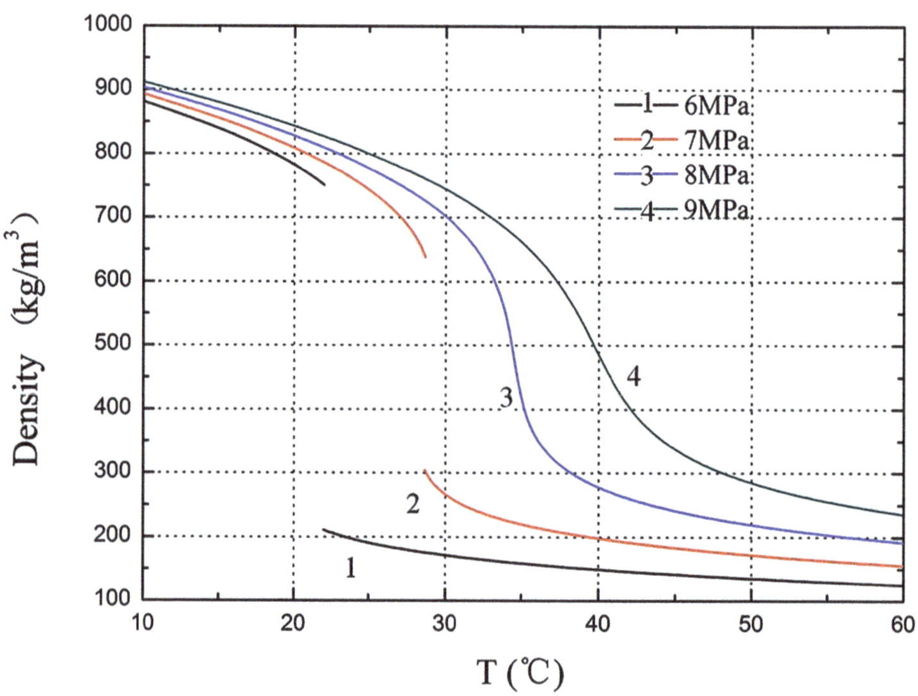

Figure 3. Thermal conductivity

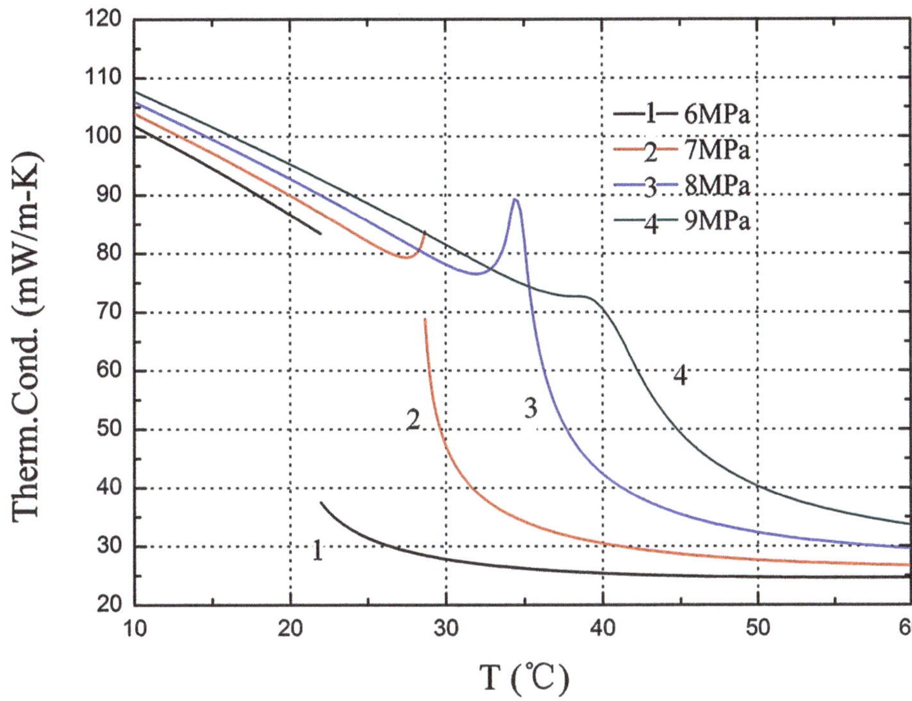

Figure 4. Viscosity

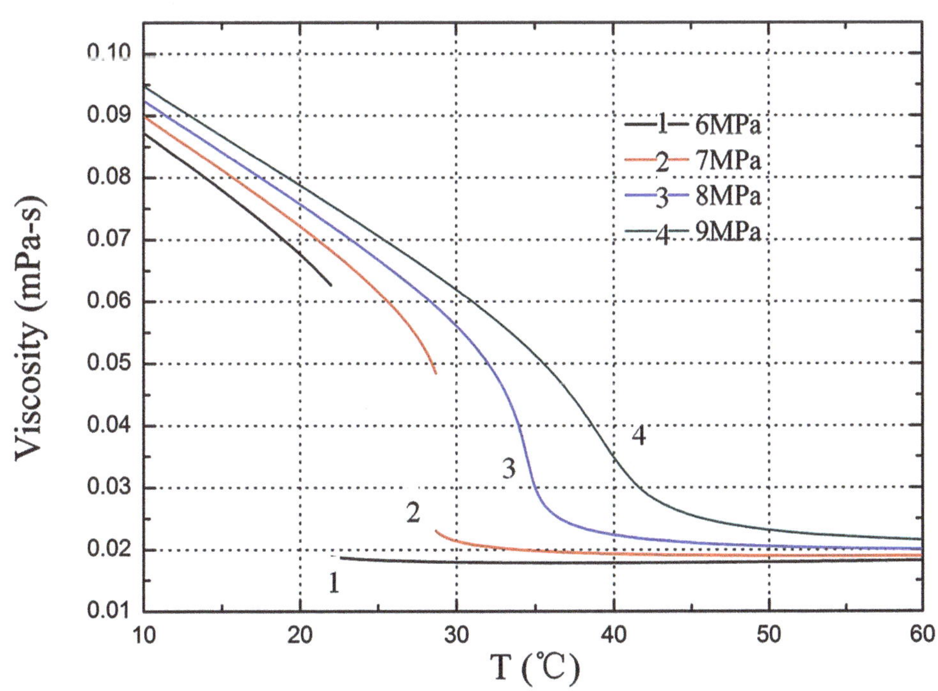

A theoretical stability criterion is also developed to verify the numerical prediction. Chatoorgoon et al. (2005) reported non-dimensional parameters governing flow stability boundary on supercritical CO_2. Chen and Zhang group (Chen et al., 2010; Zhang et al., 2010; Chen & Zhang, 2011; Chen et al., 2012) have developed two-dimensional numerical simulations of various effects on supercritical NCL stability and heat transfer. The above studies also briefly reviewed the recent developments in supercritical fluid based NCLs and the controlling parameters summarized as: heat input/output conditions, loop inclination, loop aspect ratio, loop diameter as well as gravity. In addition, the basic physics of supercritical NCL have been discussed by Swapnalee et al. (2012), Chatoorgoon group (Chatoorgoon, 2001; Chatoorgoon et al., 2005), Chen and Zhang group (Chen et al., 2010; Zhang et al., 2010; Chen & Zhang, 2011; Chen et al., 2012), Kumar and Gopal group (mainly sub-critical pressurized loop) (Kumar & Gopal, 2009; Yadav et al., 2012), and others. Ambrosini (2007) has reported the analogous between boiling and supercritical flows. Jain and Rizwan-uddin (2008) also numerically studied the supercritical flow instabilities and compared the results with two-phase flow instability. Yadav et al. (2012) numerically proposed the heat transfer correlation specifically for NCL heater and cooler, which is different for single horizontal heating or cooling conditions.

However, very few experimental investigations are found for supercritical CO_2 natural circulation systems. Shan et al. (2007) has reported one numerical and experimental study into supercritical natural convection in a reactor chamber and reported strong buoyancy effect inside; Yoshikawa et al. (2005) recently set up one (relative complex) supercritical CO_2 based natural circulation loop and reported the parameter fluctuations and flow rate curves; Jain 2005) has also compared numerical results with the specific supercritical water loop (at University of Wisconsin-Madson) and supercritical CO_2 loop (at Argonne National Laboratories). Kumar and Gopal (2012) made similar experimental model with that of Yoshikawa et al. (2005) (with curved pipe lines at heater and cooler), and the operation pressure ranges from 2.0 MPa to 5.2 MPa (in the sub-critical region). Also, Lomperski and Cho (2004) made experimental study on a rectangular supercritical CO_2 loop and no unstable flows were reported in the operation range considered. More recently in 2012, T'Joen and Rohde (2012) used Freon R23 (at 5.7 MPa) as a scaling fluid and studied the instability limit of NCL. Debrah et al. (2013) reported one supercritical water NCL loop (at 25 MPa) operated in very high heat flux region, where instabilities are identified. Indeed, very limited experimental data and operation ranges have been reported near-critical and supercritical NCLs and only two or three groups' data from 1960s can be referred for heat transfer correlation development (Swapnalee et al., 2012). Both the system stability and heat transfer have been strongly recommended for further experimental studies (Chen & Zhang, 2011; Chen et al., 2012; Swapnalee et al., 2012).

In recent years, the supercritical natural circulation system is still under intense discussions. For example, in continuation to previous developments, Sarkar and Basu (2017) have discussed the suitability of supercritical NCL operation designs and concluded the suitability by numerical methods. Numerical modeling is proved to be still a very critical method for the discussion of various kinds of NCLs and the unique behaviors of supercritical fluids (Archana et al., 2015). Experimental studies are still very difficult due to high pressure, while the information is critical for the verification and validation of numerical findings (Chen et al., 2013c, 2013f; Liu et al., 2016). Sakar et al. (2014) have made a recent review of the state-of-the-art on supercritical NCLs in nuclear application studies.

Motivated by the interest in thermal and hydraulic behaviors in supercritical thermosyphon, the present study is focused on the experimental verifications of basic stability and heat transfer behaviors. One rectangular supercritical mini-loop model is established and the effects of various input parameters as heat flux, initial charge amount, operation pressure and cooling temperature are discussed. It should also be noted that the stability behaviors of supercritical thermosyphon is strongly dependent on the heat transfer conditions. The coupled effects between density waves, flow fluctuations and heat transfer conditions play a key role in related systems (Jain & Rizwan-uddin, 2008; Chen & Zhang, 2011; Chen et al., 2012; Jain, 2005). The experimental results are also compared with previous studies, especially for stability condition and heat transfer correlations at the heater and cooler.

EXPERIMENTAL DISCUSSIONS

Experimental System Set-Ups

System Design and Realization

Basic experimental apparatus in the current study is designed as one rectangular circulation mini-loop with uniform inner diameter throughout. The established loop system is consisted of a CO_2 container, an electric heater (as heat source), a thermostats and counter-flow cooler (using water as secondary coolant), and a rectangular loop (including two adiabatic legs and two horizontal pipes, connected by four elbows), as shown in Figure 5 and Figure 6. The current system is one typical kind of simplified model for natural circulation systems (Dimmick et al., 2002; Zvirin, 1981; Grief, 1998; Vijayan et al., 1995;

Figure 5. NCL experimental system established in this study: schematic system design and measuring points

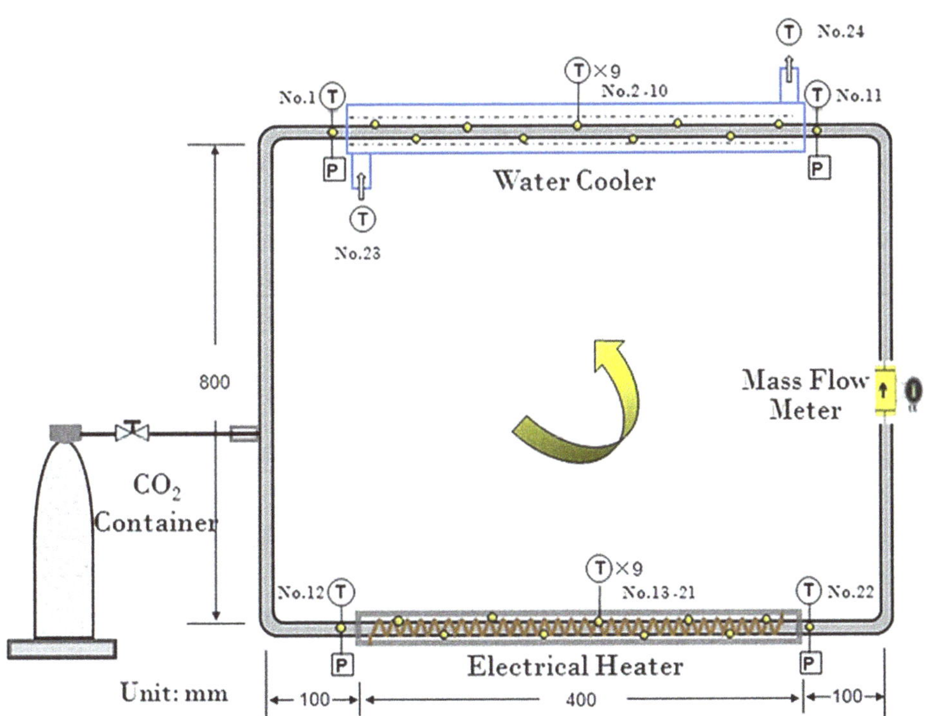

Figure 6. Experimental loop system established

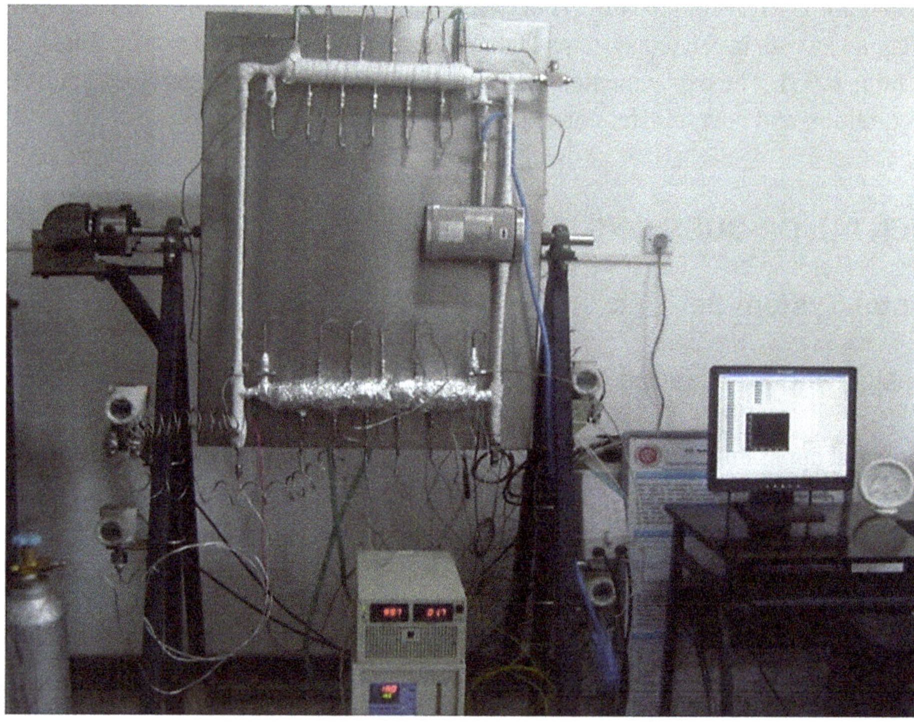

Chatoorgoon, 2001; Kumar and Gopal, 2012; Chen et al., 2010; Jain & Rizwan-uddin, 2008; Zhang et al., 2010; Vijayan et al., 2004; Cammarata et al., 2003; Misale et al., 2007; Kumar & Gopal, 2009a, 2009b; Chatoorgoon et al., 2005; Chen & Zhang, 2011; Chen et al., 2012).

The main loop, including the bends, is made of stainless steel (1Cr18Ni9Ti) with uniform inner diameter d = 8 mm throughout the circulation loop, the outer diameter D = 10 mm. The geometrical dimensions/parameters of loop, locations of thermocouples and pressure transducer and other parts can be seen in Figure 5 and Figure 6. The loop heater is placed at the mid-part of bottom horizontal pipe while the cooler is set at the mid-part of the upper one. For the heater part, the loop was inserted into a cooper cylinder with an outer diameter of 20 mm and length of 400 mm, as shown in Figure 5 and Figure 6. The copper cylinder was wrapped with electric resistance, with the resistance wires curved around the inner stainless pipe heating area. The small space between inner stainless pipe and outer heater was sealed with highly-thermo-conductive silica gel, which ensures good electric heating of the pipe.

The electric resistance was connected with DC current electrical source (HY-2500A-220) to heat the pipe, where the voltage (U) could be regulated from 0 to 220 V to control heat flux at the heater. The cooling loop is set at the upper horizontal loop section and it was established as a tube-in-tube counter-flow kind water cooler, the outer diameter of the cooler is 62.3 mm and the inner pipe can be cooled at constant temperature (T_c) ranged from 8.0 to 30.0 °C. Using LX-300 water thermostats and circulating system (with the water flow rate ranges from 0-15.0 L/min), the outlet and inlet cooling water temperature differences can be controlled below 0.1 °C. The whole experimental loop area is deeply insulated and the room temperature also controlled by an air-conditioner so as to minimize heat losses to the ambient (overall heat losses were estimated below 1.0%). The experimental loop with its current design can sustain a maximum pressure of 20.0 MPa and a maximum mass flow rate of 20.0 kg/h.

Measurement and Data Accumulation System

As shown in Figure 5 and Figure 6, some temperature and pressure measuring points are selected to record system parameters and one data collection system is also used in the current experimental loop.

1. T-type thermocouples are placed across the heater and cooler test the temperature behaviors across the heat exchange areas as shown in Figure 5. Four thermocouples (Noted as No. 1, 11, 12, 22) were placed in the middle of the pipe to measure the inlet ($T_{in.c}$) and outlet ($T_{out.c}$) temperatures of the cooling section and inlet ($T_{in.h}$) and outlet ($T_{out.h}$) temperatures of the heating section, respectively. Both of the cooling pipe and heating pipe were placed with nine uniformly spaced thermocouples (No. 2-10 and No. 13-21) to measure the wall outer face temperature of heat transfer pipes. Two of them (No. 23 and 24) were used to test the inlet and outlet temperature of the cooling water. All the thermocouples were calibrated in a constant temperature bath and the measurement error was estmated to be within ±0.2 °C.
2. Four pressure gauge transducer (Model HART 2600T, ABB) were symmetrically placed at the inlet and outlet of the cooling and heating pipe to measure the static operation pressure (p); The pressure measurement error is within ±0.075%.
3. One Coriolis-type mass flow meter (Model SITRANS FC 300, SIEMENS) was placed at the right vertical leg to measure the mass flow rate ($\dot{m}$), which is capable of recording 0-350 kg/h mass flow. The measurement error of mass flow rate is within ±0.1%.

Operation and Procedures

In addition, all the data were recorded by computer data collection system and Labview (Version 8.6) software was used for the in-time screening (the reading error for the data collection is within ±0.1%). A photograph of the experiment apparatus is shown in Figure 6. CO_2 fluid with a purity of 99.99% was fed to the test loop from the container. Due to the small size of the mini-loop, the charging quantity is relatively small and greatly dependent on the initial charge conditions. The real charging amount is decided by controlling the initial charging pressure (P_0) at fixed T_0 for the close loop.

Before operating of the system, the room temperature is first controlled by an air-conditioner to a certain value. Then all the testing apparatus and sensors should be checked and data acquisition system be started. All data are recorded once for each minute in the current experiment set-up. After that the NCL is injected and evacuated with pure CO_2 (99.99%) for several times and make sure a good sealing and pure fluid charge into the loop. High pressure charge can be very dangerous and special care should be paid to safety operation. After the charge and safety check, the cooling water thermostats and circulation system can be started; after steady water circulation is reached the DC heating system can be initiated. The heating flux, cooling flow temperature, charge amount can be adjusted for each runs to change the operation conditions. In the current discussion, each experiment runs were injected and started at identical initial temperature (T_0 = 25.0 °C). In the present study, the initial charging pressure ranges from 6.2 MPa to 10.0 MPa (with three typical conditions of P_0 = 6.5 MPa, 7.0 MPa, and 7.8 MPa mainly selected for discussion and comparison). The final steady operation pressures range from 7.77 MPa to 14.6 MPa, which are all above the critical point (i.e. in the supercritical region).

Data Procedures

As the focus of the current study is the basic flow stability and heat transfer in near-critical region, some parameters for analysis are firstly defined. However, less analysis for heat transfer behaviors are found for NCL systems, only recently Yadav et al. (2012) numerically proposed specific friction factors and heat transfer correlations for it. Previous analysis also used forced convection friction factor and heat transfer laws as it has been found that steady supercritical CO_2 loop can easily reach high Reynolds numbers flow (Yamaguchi et al., 2010; Chen & Zhang, 2011; Chen et al., 2012) and that should be very similar to forced convection in circular tubes. Accordingly, the current results will be compared with that of Yadav et al. (2012) and also other studies for circular tube supercritical heat transfer models (Liao & Zhao, 2002a, 2002b).

In the current design, the upper wall was cooled at constant temperature as has discussed formerly. And for the heater, uniform heat transfer to the pipe can also be achieved due to uniform heating of outer copper cylinder (with also enhanced the axial heat conduction along the pipe). Then the heating pipe was close to a constant temperature condition during the test. The wall temperatures along the cooling pipe and heating pipe for three typical test conditions are listed in Table 1 and Table 2, respectively. It can be seen that the maximum temperature difference along the wall is within 2.0 °C, which is very small compared with fluid temperature changes in the heater/cooler.

Based on the energy balance under a steady state, the heat transfer rate $\dot{Q}$, between the pipe wall and the CO_2 can be obtained from:

$$\dot{Q}_c = \dot{m}\left(h_{in,c} - h_{out,c}\right) \quad (1)$$

$$\dot{Q}_h = \dot{m}\left(h_{out,h} - h_{in,h}\right) \quad (2)$$

Table 1. Wall outer face temperature of the upper cooling pipe

No.	Experimental Conditions	Temperature Measured by Thermocouples No. 2-10 (°C)								
		2	3	4	5	6	7	8	9	10
1	P_0=6.5MPa,U=70V,T_c=15°C	20.2	19.9	20.3	20.1	20.1	20.3	19.8	20.2	20.1
2	P_0=7.0MPa,U=80V,T_c=21°C	22.1	22	22.4	22.3	22.2	22.5	21.9	22.7	22.6
3	P_0=7.8MPa,U=100V,T_c=18°C	19.9	19.9	20.2	20.1	20.1	20.5	19.7	20.4	20

Table 2. Wall outer face temperature of the lower heating pipe

No.	Experimental Conditions	Temperature Measured by Thermocouples No. 13-21 (°C)								
		13	14	15	16	17	18	19	20	21
1	P_0=6.5MPa,U=70V,T_c=15°C	47.0	47.7	46.3	46.1	46.1	47.1	45.9	46.4	47.2
2	P_0=7.0MPa,U=80V,T_c=21°C	57.8	58	57.3	57.6	57.6	59.8	57.3	58.1	58.4
3	P_0=7.8MPa,U=100V,T_c=18°C	74.9	74.8	74.3	74.3	74.8	76.6	76.1	76.5	76.8

where $\dot{m}$ is the mass flow rate, and h_{in} and h_{out} denote the enthalpy of CO_2 at the inlet and outlet temperature. The latter subscripts c and h represents to the upper cooling pipe and lower heating pipe, respectively. It could be found easily the symmetry in form for this couple of equation (Equation 1 and Equation 2). For the purpose of simplicity, the following equations would be written in a synthetically single form, rather than adding latter subscripts to differentiate the two pipes.

The bulk mean Reynolds number is defined as:

$$\mathrm{Re}_b = \frac{4\dot{m}}{\pi d \mu_b} \tag{3}$$

where μ_b is the viscosity of CO_2 at bulk mean temperature T_b. The bulk mean temperature is defined as:

$$T_b = \frac{T_{in} + T_{out}}{2} \tag{4}$$

Based on the measurements of the nine thermocouples under each horizontal heat transfer pipe, the wall outer face mean temperatures averaged over the entire length were obtained. A corresponding inner wall mean temperature of each horizontal heat transfer pipe T_w can then be calculated based on the wall thickness of the test pipe, the wall outer face mean temperature and the heat transfer power given by Equation 1 and Equation 2. Subsequently, the logarithmic mean temperature difference (LMTD) can be obtained for the two horizontal test pipe as:

$$LMTD = \left| \frac{(T_{out} - T_w) - (T_{in} - T_w)}{\ln\left(\frac{T_{out} - T_w}{T_{in} - T_w}\right)} \right| \tag{5}$$

The mean heat transfer coefficient is calculated by:

$$h = \frac{\dot{Q}}{A \times LMTD} \tag{6}$$

where A denotes the inner area of the two horizontal pipes.

The Nusselt number is defined as:

$$Nu_b = \frac{h_b d}{k_b} \tag{7}$$

where d is the pipe diameter, k_b is the thermal conductivity of CO_2 at bulk mean temperature T_b.

In addition, the system major uncertainties come from axial conduction of heater transfer area, room temperature control, mass flow rate measurement, temperature and pressure measurement and recording (as discussed in 2.1.1), heat transfer area measurement (estimated below 0.25%). Therefore, the estimated uncertainties for averaged Reynolds number and LMTD are ±0.3% and ±4.1%, respectively. And for heat transfer coefficient and averaged Nusselt number, the uncertainties are both around ±16.5% within ±3.0 °C from pseudo-critical point and around ±7.6% for most ranges out of that critical region.

Experimental Results on Stability Evolution

Basic experimental tests were carried out under the boundary and initial conditions as described in the previous section. For each experimental run, the system should be cooled to ambient temperature and carefully checked and injected as discussed in operation procedure part. In the current tests, system instabilities were identified for sub-critical flows due to two phase oscillations. For supercritical operation conditions, the flow is always stable under current heating and cooling settings. In this section, heat transfer is also focused for steady supercritical NCL flows. Detailed results and discussions can be found in the following parts.

System Stability Trends

As discussed, the initial charge pressure ranges from 6.2 MPa to 10.0 MPa in the current experiments. The flow conditions are found to be initial pressure dependent and from sub-critical to supercritical conditions, different stability situations are identified. Due to the large density differences in the near critical region (see Figure 1, Figure 2, Figure 3, and Figure 4), the charge pressure will greatly affect the charge amount and flow status during NCL operation. It is found that when the initial charge pressure is relatively lower from critical point, unstable flows are seen. As shown in Figure 7, mass flow and pressure fluctuations are found throughout the operation period with initial conditions of p_0 = 6.0 MPa, T_0 = 25.0 °C, and cooling water flow temperature T_c = 15.0 °C, heater power 182 W. Initial overshoots of flow rate are also reported for pressurized CO_2 NCL systems previously (Yoshikawa et al., 2005; Kumar and Gopal, 2008). Indeed, sub-critical with two-phase flow condition can be identified for the current unstable case as shown in Figure 7. Due to the large density differences of liquid and vapor in the loop, repetitive flow reversals are seen in Figure 7. However, with the increase of initial pressure, stability transition from oscillating flow to stable single-directional flow can be found. As shown in Figure 7 with initial pressure of 6.3 MPa, the flow changed from fluctuation mode to steady mode. Indeed, the transition point lies around the critical point of CO_2 fluid of 7.38 MPa as shown in Figure 8, which indicates that single-directional steady flow can be achieved when the operation pressure goes above the critical point.

For systems operating in the supercritical state, stable flows are seen with the current model conditions. The results of a representative experimental case (p_0 = 7.0 MPa, T_0 = 25.0 °C, U = 100V, T_c = 21.0 °C) is are plotted in Figure 9 and Figure 10. In Figure 9 and Figure 10, the inlet and outlet temperatures in the heater and cooler, and the average heat transfer wall temperatures are shown. It is seen in Figure 9 that the wall heating temperature gradually developed to a steady state at around 2500s, through which time the fluid changed from pressurized liquid state to supercritical state. No phase change happened during that time, so very small fluctuations are found and the temperature can go through gradual change, which is much different from the two-phase cases shown in Figure 7 and Figure 8. During the evolution, the temperature differences across the heater and cooler also gradual increased and reached final steady state

Figure 7. Evolution of NCL mass flow and pressure during sub-critical operation: initial p_0 = 6.0 MPa

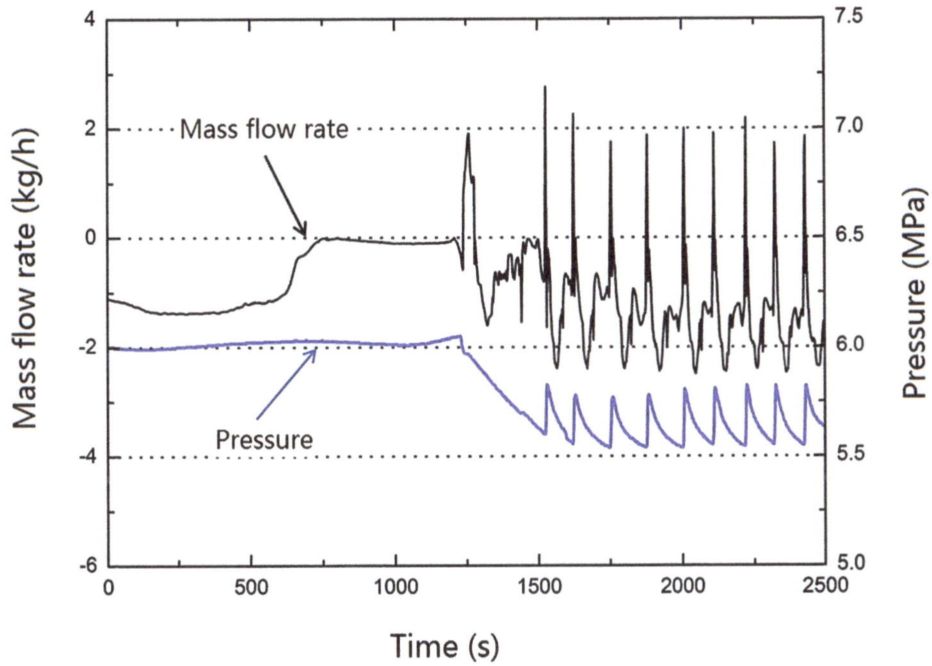

Figure 8. Initial p_0 = 6.3 MPa

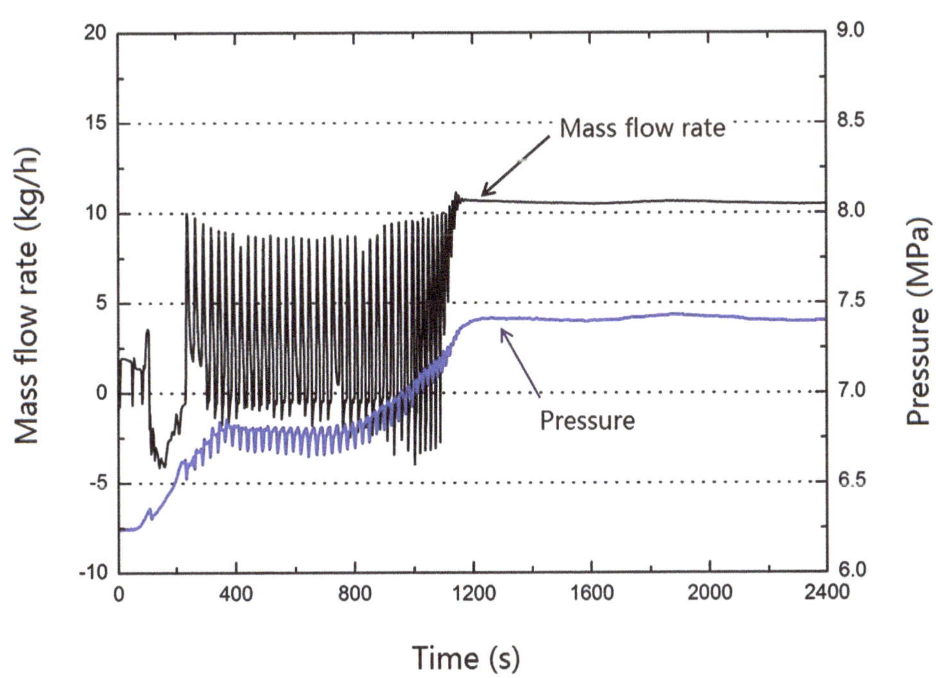

Figure 9. Evolution of fluid parameters during near-critical operation, initial p_0 = 7.0MPa: fluid and wall temperatures

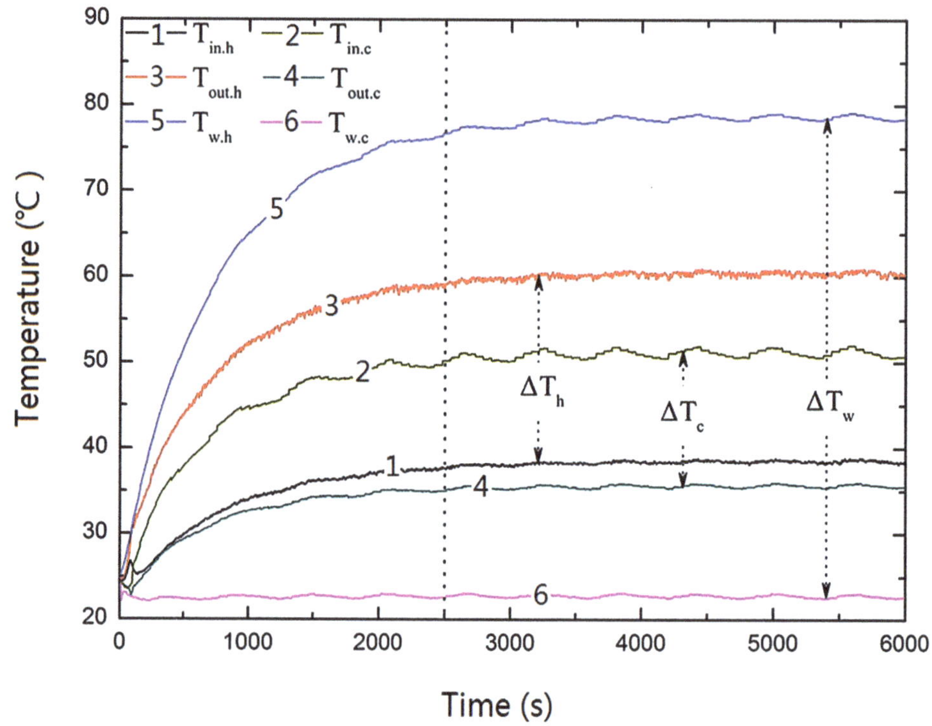

Figure 10. Mass flow rate and pressure

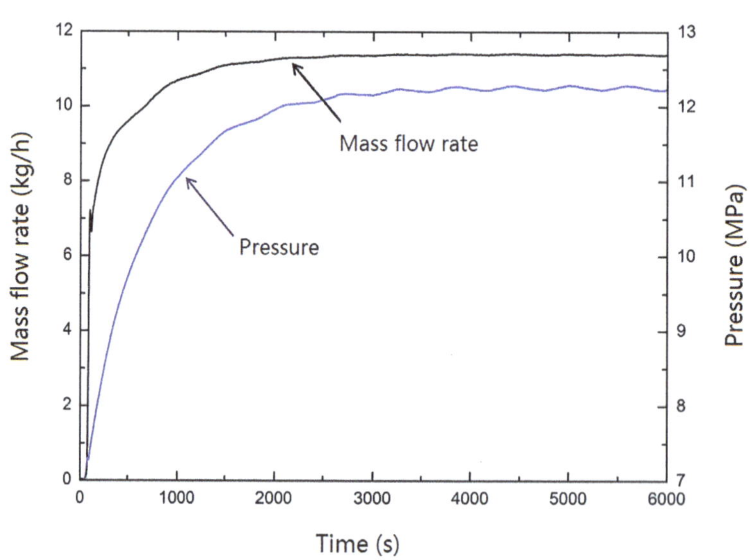

(which differences ΔT_h and ΔT_c are 22.0 °C and 15.0 °C, respectively). The above experimental results agreed with the numerical studies of Chen and Zhang (2011, 2014). Compared with related studies of supercritical Freon R23 (T'Joen & Rohde, 2012) and supercritical water (Debrah et al., 2013), where the system will become unstable for very high heat input (several kW), in the current operation range (50 W to 200 W) the system is always stable.

In addition, as shown in Figure 10, rapid initial mass flow starting up is seen for the current NCL, which is typical for supercritical fluid based NCL systems as also verified in numerical studies (Chen et al., 2010; Zhang et al., 2010; Chen & Zhang, 2011; Chen et al., 2012). Still, it should be noted that small periodic 'waves' are seen in the curves of Figure 9 and Figure 10 for steady operation region after 2500s, which is due to the sensitive properties near the supercritical states (as shown in Figure 1, Figure 2, Figure 3, and Figure 4). Such 'waves' are also found in supercritical NCL study of Yoshikawa et al. (2005). In the current study, system stability differences are very large for near-critical fluids. From 6.0 MPa to 7.0 MPa, as the fluid status will change from two phase state to pressurized liquid state and then to supercritical state. During those changes, the charge amount of 7.0 MPa case will be 3.5 time that of 6.0 MPa case, leading to much different behaviors for near-critical systems. In the experiments of Kumar and Gopal (2012) the operation pressure ranges from 2.0 MPa to 5.2 MPa, but the flow stability behaviors changes are relatively small. From the current results, special care should be paid to the charge and operation conditions for real system stability design.

The discussion of flow rate and power input level will also be useful for analysis of supercritical loops as discussed (Jain & Rizwan-uddin, 2008; Yoshikawa et al., 2005; Dimmick et al., 2002). In the current closed experimental system, the operation pressure will change with initial charge amount and power input as shown in Figure 7, Figure 8, Figure 9, and Figure 10. However, in the supercritical region operation in the current parameter range, the basic flow shows stable trend. Representative operation pressure and flow rate curve are plotted in Figure 11 (for initial 6.50 MPa pressure, cooling water temperature 17.2 °C and different power input levels). Also the pressure differences across the heater are also shown. It can be seen that mass flow rate will decrease with input power level (or steady operation pressure). The pressure difference increases with flow rate, which indicate higher losses for near-critical conditions.

Parameter Effects and Stability Map

In order to study the detailed stability parameter behaviors of stable cases (which is preferred for real system operation), three typical conditions of P_0 = 6.5 MPa, 7.0 MPa, and 7.8 MPa mainly selected for discussion and comparison in the following discussions. When initial experiment conditions are given, the final steady flow behaviors are determined by boundary heat transfer conditions, the fluid temperature ranges between the two wall temperatures (Chen et al., 2010; Chatoorgoon et al., 2005; Chen & Zhang, 2011; Chen et al., 2012; Swapnalee et al., 2012). The experimental system parameter contours (including operating pressure and Reynolds numbers) are shown in Figure 12, Figure 13, and Figure 14 for initial charging pressure of 6.5 MPa (which pressure indicates stable flow as discussed formerly). In Figure 12 the pressure contours are set under the two heat transfer wall temperatures at heater and cooler. It can be seen from Figure 12 that the pressure will increase with either of wall temperatures at heater or cooler and the pressure changes are relatively more sensitive to heater wall temperature $T_{w,h}$ than cooler wall temperature $T_{w,c}$ (see the gradient of pressure lines). The above effect may be due to the more sensitive changes of properties at higher temperature (at the cooler), as in the sub-critical cooler the percentage of properties changes can be much smaller for the good thermostatic cooler.

Figure 11. Operation pressure and flow rate evolution. Initial p_0 = 6.50 MPa, cooling water T = 17.2 °C and different power input levels.

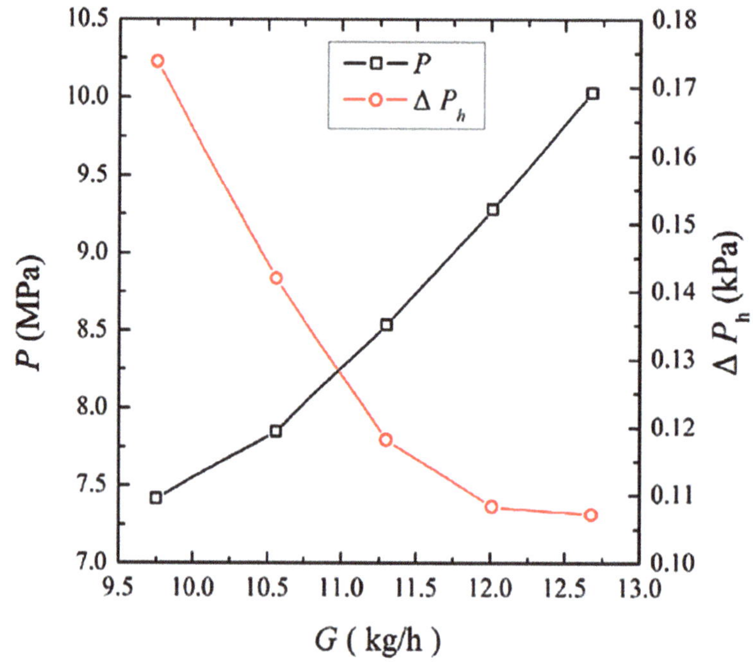

Figure 12. Operation pressure p contours vs lower wall cooling temperature $T_{w.c}$ and upper wall heating temperature $T_{w.h}$.

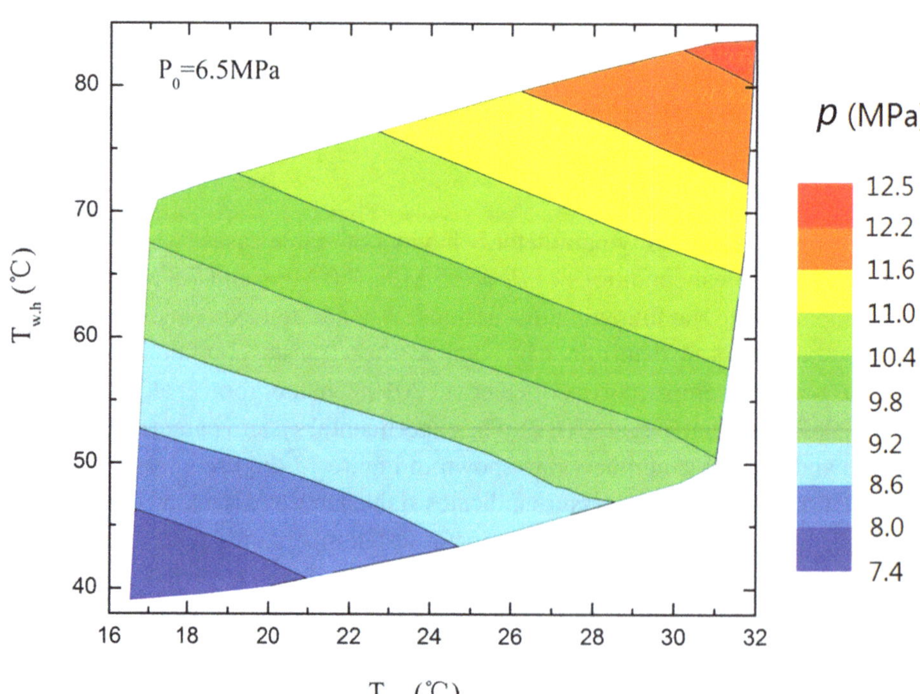

Figure 13. Bulk mean Reynolds number contours vs lower wall cooling temperature and upper wall heating temperature: Re at the heater

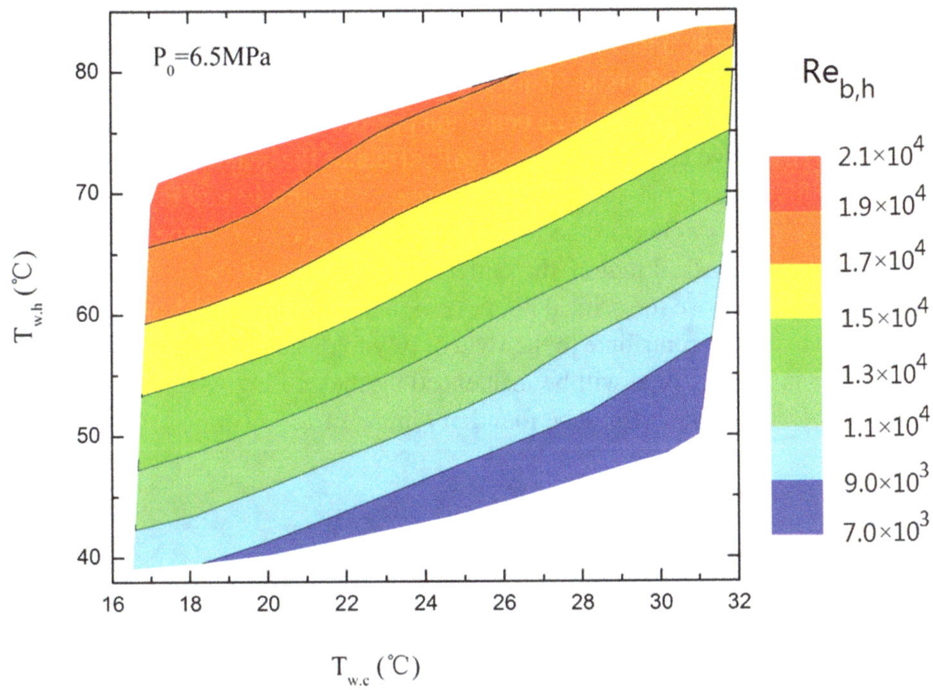

Figure 14. Re at the cooler

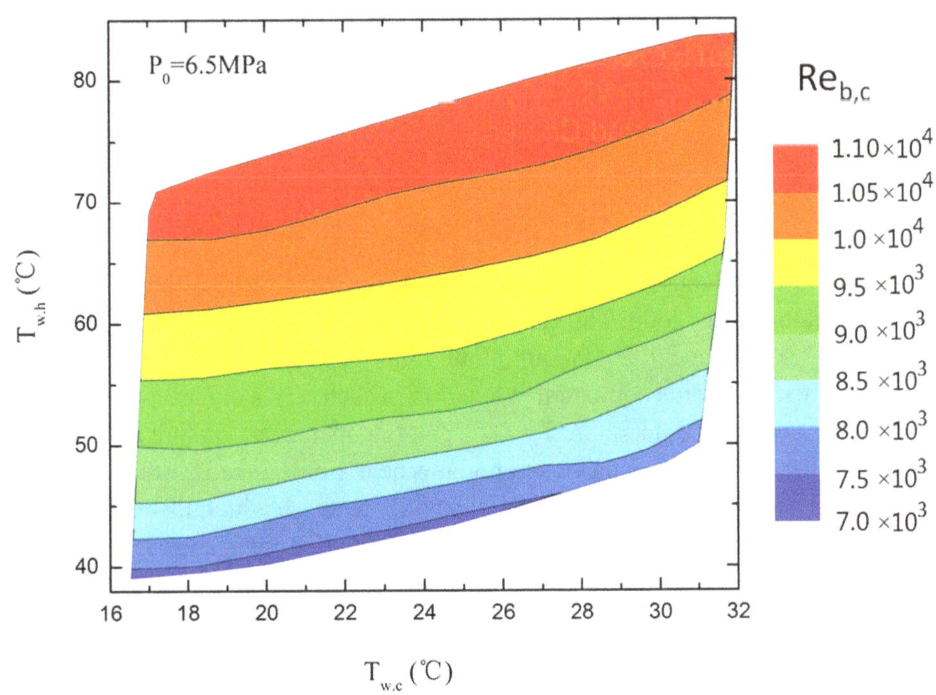

The evolutions of bulk mean Reynolds numbers defined at the heater and cooler are shown in Figure 13 and Figure 14, respectively. From Figure 13 and Figure 14, it is seen that the Reynolds numbers calculated across the heater is larger than ones for the cooler, due to the higher viscosity (see Figure 1, Figure 2, Figure 3, and Figure 4; under relatively lower temperature) found at the cooler. In Figure 13, it is seen that larger Reynolds numbers are found in the experiments with higher heating temperature $T_{w,h}$ and lower cooling temperature $T_{w,c}$, which condition can provide the larger driving buoyancy force. This difference between the two heat transfer pipes will influence the heat transfer behaviors (which will be discussed later in this chapter). Moreover, it can be seen in Figure 13 and Figure 14 that the heating wall temperature has a greater effect on the Reynolds number in cooling pipe (considering the curve gradient of Re). In the supercritical region, the outlet fluid viscosity of the heating pipe drops sharply as the temperature increase above the critical point. And then, the viscosity-dropped fluid circulates to the cooling section, which may contribute to the style of Reynolds number lines in Figure 13. Similarly, the Reynolds number in heating pipe will be influenced by the cooling wall temperature greater than the heating wall temperature, which could be found in Figure 14.

Experimental Results on Heat Transfer

Temperature Behaviors under Different Operation Pressures

Previous heat transfer analysis on NCL focused on the bulk mean temperature evolution of heater and cooler (Chen & Zhang, 2011; Chen et al., 2012). The experiment data of the two bulk temperatures in the heater and cooler under different initial pressures are plotted in Figure 15. In that figure, each symbol represents a steady experimental test. It is seen in Figure 15 that the variation of two temperatures behaves almost linearly for every run of P_0. With the increase of P_0, the difference between the two bulk temperatures increases, which condition has been found critical for the maintenance of stable flow circulation (Chen & Zhang, 2011; Chen et al., 2010).

Heat Transfer Rate for Heater and Cooler

Under the present experimental controlling condition, the flow circulation is determined by the heat input/remove of the two heat transfer pipes. Any change in the boundary controlling condition will affect the fluid temperature, and thus heat transfer of the two horizontal pipes is affected due to the flow circulation throughout the closed loop. The heat transfer coefficients for different initial charging pressures (p_0 = 6.5 MPa, 7.0 MPa and 7.8 MPa) are plotted in Figure 16, Figure 17, and Figure 18. Every two symbols together with the same comparative location in each series order represent one experiment case, characterized by two bulk temperature $T_{b.c}$, $T_{b.h}$, and the operation pressure p.

It can be seen that the heat transfer coefficient remains nearly unchanged for the cooler (due to stable temperature maintain) and for the heater the coefficient will decrease with bulk temperature (due to the decrease of thermal conductivity as shown in Figure 1, Figure 2, Figure 3, and Figure 4). The evolution trend in Figure 16, Figure 17, and Figure 18 agrees with the pressurized liquid CO_2 and supercritical cases of Yadav et al. (2012). It is reported by Liao and Zhao (2002a, 2002b) that the peak of heat transfer coefficient at each operation pressure locates where the T_b near the corresponding pseudocritical temperature T_{pc} under both heating and cooling condition. However, the condition for the current closed NCL shows stable behaviors at the cooler due to the nearly constant heat transfer condition. This indi-

Figure 15. Bulk mean temperature at the heater $T_{b,h}$ vs that at the cooler $T_{b,c}$ at different initial charging pressures.

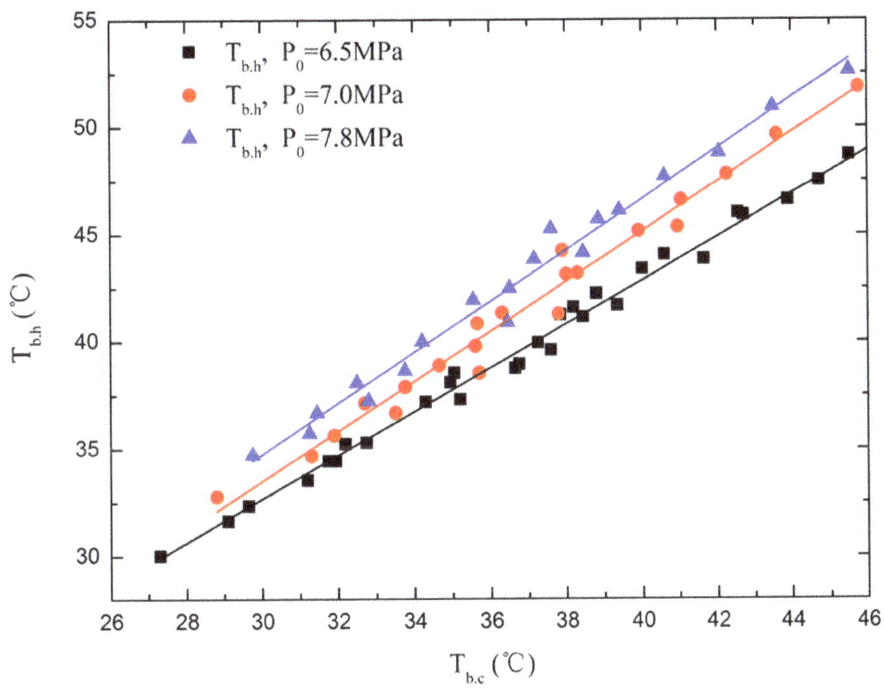

Figure 16. Variation of the mean heat transfer coefficient hh and hc with bulk mean temperature T_b for the heater and cooler: p_0 = 6.5 MPa

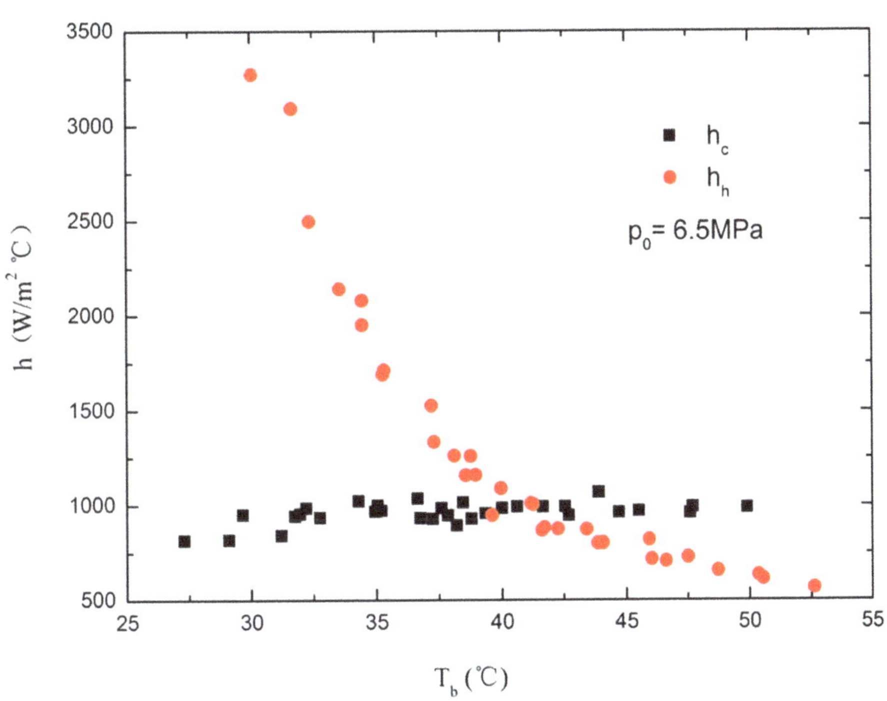

Figure 17. $p_0 = 7.0$ MPa

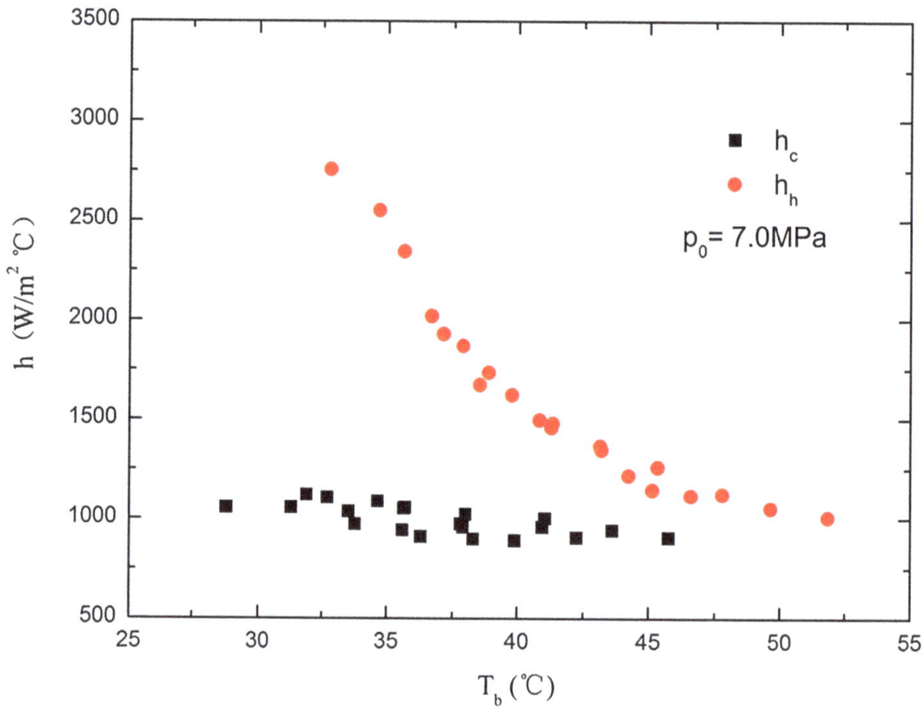

Figure 18. $p_0 = 7.8$ MPa

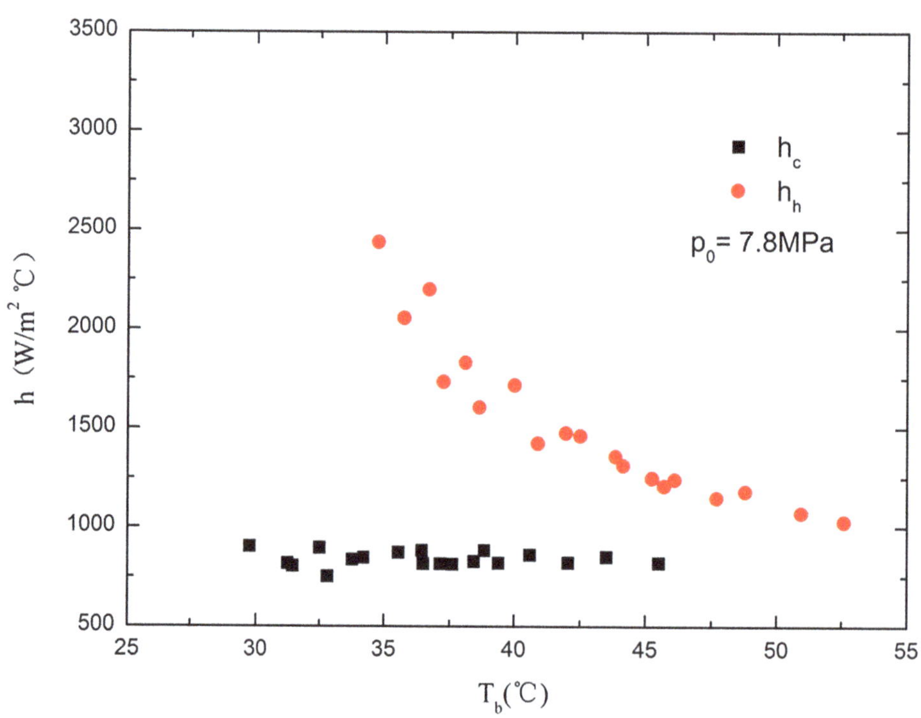

cates that the correlations of circular tubes may not be suitable for every case of NCL heater and cooler (Yadav et al., 2012). In addition, the coefficient at the heater will be relatively higher (and the curve steeper) for relatively lower initial pressure as shown in Figure 16, Figure 17, and Figure 18. Indeed, lower initial pressure will for the current case is nearer to the pseudo-critical point, where the highest peaks of specific heat and thermal conductivity are seen (as the trend can also be seen from Figure 1), therefore higher heat transfer coefficients (at the heater) are found in Figure 16. Also, pseudo-critical point will change with operation pressures, which may also affect the heat transfer behaviors as the peak properties will change accordingly (as shown in Figure 1, Figure 2, Figure 3, and Figure 4). Similar effects have also been reported in the summary of Yadav et al. (2012). Possible heat transfer deterioration phenomena can also be seen during this change. In particular, it is found in Figure 16 that h_h drops below h_c when the T_b exceed a steady state of $T_{b.c}$ = 37.8 °C, $T_{b.h}$ = 41.3 °C (corresponding p = 9.85 MPa), which indicates that deteriorated heat transfer happened. This information can be very important for trans-critical NCL system design and applications. Further studies are still needed to clarify the detailed trend of heat transfer evolutions.

Further Heat Transfer Analysis and Comparisons

The variations of Nusselt numbers for the heater at different p_0 are plotted in Figure 19. For the sake of eliminating the operation pressure effect, dimensionless temperature T_b/T_{pc} is adopted for performance comparisons. For all the experiment cases, the upper limit of dimensionless temperature T_b/T_{pc} is found to be below 1.0. Therefore, all cases were operated in the supercritical region with bulk temperatures below the pseudo-critical temperature at corresponding operation pressure. Near the pseudo critical point Cp and thermal conductivity both have high peaks (the pseudo-critical point will be dependent on the pressure as shown in Figure 1, Figure 2, Figure 3, and Figure 4). This is also one reason of always stable flow and different from the predictions constant pressure operation conditions in previous numerical simulations (Chen et al., 2010; Chen & Zhang, 2011). As discussed, due to the variations of the thermal physical properties, the Nu_b increases with T_b/T_{pc} as shown in see Figure 19. Indeed, this effect will be more evident at the heater when T_b is getting close to T_{pc} so the respective general slope of $Nu_{b.h}$ is found larger in Figure 19. The cooler Nusselt numbers only increase a little the increase of T_b/T_{pc}, due to the constant temperature conditions for each case and the values are around 110, 90 and 80 for p_0 = 6.5 MPa, 7.0 MPa and 7.8 MPa, respectively.

For different initial pressure conditions as shown in Figure 19, the general trend is similar to the evolution of heat transfer rate discussed in the former part. In previous numerical study of Chen and Zhang group (Chen & Zhang, 2011; Chen et al., 2012), Yadav et al. (2012) and others (Swapnalee et al., 2012; Vijayan et al., 2004), general trends are similar. But the correlations diverge for each specific condition studied, indicating the complexity of near-critical friction and heat transfer behaviors. Specifically, in the current study, strong pressure dependency is found in experiments. As shown in Figure 19, arger T_b/T_{pc} range is found for relative lower p_0 (with also relative larger Nu_b range). That can be ascribed to the smaller operation pressure when smaller charging mass is initially injected into the closed loop, and which brings about smaller corresponding T_{pc}. Also, slope increasing of the liner fit line of $Nu_{b.h}$ could be found as p_0 decrease, In addition, as the heat transfer enhancement effect near pseudo-critical temperature is relatively weak for cooler, so no obvious trend for the cooler are identified in the present study.

Figure 19. Variation of Nusselt numbers ($Nu_{b,h}$ and $Nu_{b,c}$) with bulk mean temperature at the heater and cooler

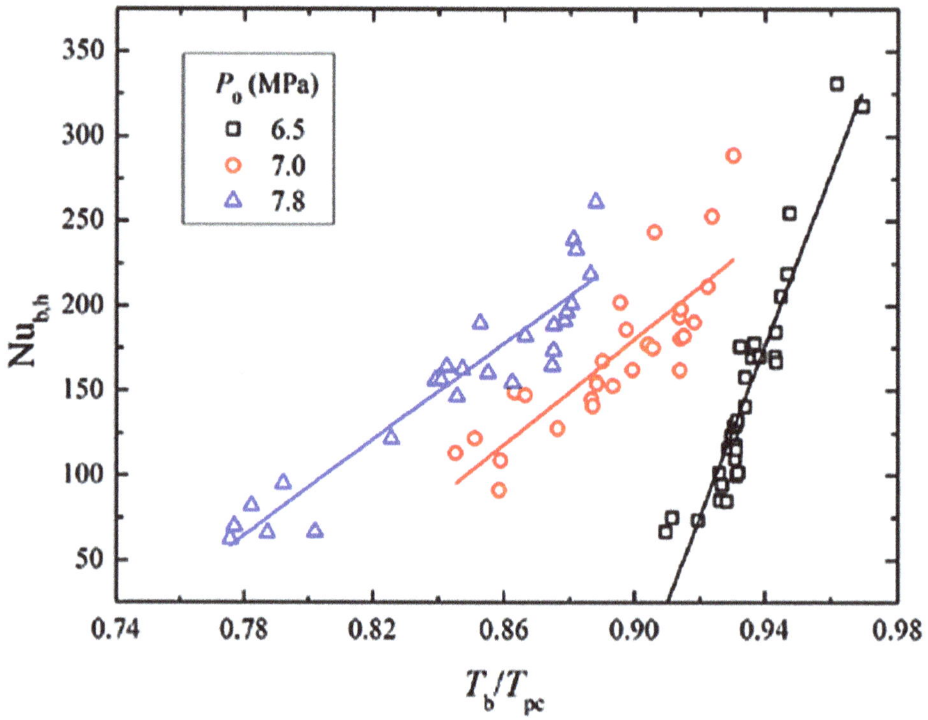

NUMERICAL DISCUSSIONS OF SUPERCRITICAL NCL

Numerical Model Formulations

Physical Model and Numerical Realization

The geometry of the thermosyphon loop is rectangular and vertically fixed with heat source in the lower side and heat sink in the upper side. Effect of pipe diameter on the system heat transfer behavior and flow are investigated in this study. As shown in Figure 20, the circulation loop is two-dimensional with diameter D. The geometry is $H\times L_0$ for vertical pipe and horizontal pipe. The middle parts of the upper and lower pipes (with length of L) are heating and cooling wall sections with prescribed temperature T_H and T_L respectively. Other pipes are adiabatic walls. The present model is also based on a solar collecting system using supercritical CO_2 as working fluid (Zhang & Yamaguchi, 2008) and has been tested and analyzed in previous studies (Chen et al., 2010; Zhang et al., 2010). Model pipe wall thickness is assumed zero. For convenience of analysis, cross section walls numbered from ① to ⑧ are set to monitor some important flow statistics as shown in Figure 20.

Figure 20. Schematic of the problem (thermosyphon) studied

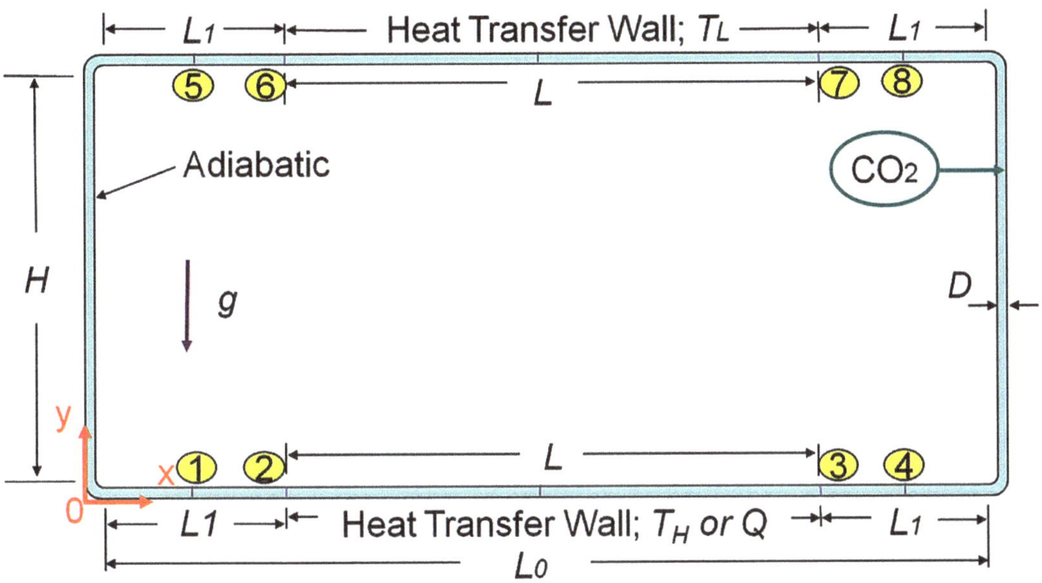

Controlling Equations and Methods

Due to the sensitive thermal physical properties of supercritical CO_2 fluid, fluid dynamics model with modified Navier-Stokes equations are solved together with continuity equation and system energy conservative equation on control volume. In the model Navier-Stokes equations are rewritten for in-compressible flow considering the relative small velocity (0~20m/s) in the loop and the large acoustic speed (10^2 m/s magnitude) of CO_2 fluid under current model conditions. Then the governing equations are written as following:

Continuity equation:

$$\frac{\partial \rho}{\partial t} + \nabla \cdot (\rho \vec{V}) = 0 \tag{8}$$

Navier-stokes equation for incompressible flow:

$$\frac{\partial \rho \vec{V}}{\partial t} + \nabla \cdot (\rho \vec{V}\vec{V}) = -\nabla P + \nabla \left[\mu (\nabla \cdot \vec{V} + \nabla \vec{V}^{\mathrm{T}}) \right] + \rho \vec{g} \tag{9}$$

Energy equation:

$$\frac{\partial \rho E}{\partial t} + \nabla \cdot (\rho \vec{V} E) = \nabla (\lambda \nabla T) + \beta T \frac{\partial P}{\partial t} + \Phi + \dot{Q}_W \tag{10}$$

where Φ represents the viscous dissipation function; and E is defined as:

$$E = \int_{T_0}^{T} C_p dT + \frac{V^2}{2} \tag{11}$$

T_0 is the reference value used in Equation 11.

As it is found in previous studies (Chen et al., 2010) the flow Reynolds number can be as high as 10^4, the flow can be turbulent. However, turbulent models for supercritical fluid are less developed/used in supercritical fluid flow studies and still under intense study (Yang et al., 2007). Therefore, in the present simulation a general Renormalization Group (RNG) $k-\varepsilon$ model is selected as the first step to introduce turbulent effect. The RNG $k-\varepsilon$ turbulent model is used together with an accurate description of fluid properties for dealing with near wall regions in a computational efficient means. This method has also been used in previous supercritical CO_2 turbulent flow studies, which shows much closer results to experimental data compared with other methods (Lisboa et al., 2010).

The fundamental equations for RNG $k-\varepsilon$ model are written below:

$$\rho \frac{Dk}{Dt} = \frac{\partial}{\partial x_i}\left(\alpha_k \mu_{eff} \frac{\partial k}{\partial x_i}\right) + G_k + G_b - \rho\varepsilon \tag{12}$$

$$\rho \frac{D\varepsilon}{Dt} = \frac{\partial}{\partial x_i}\left(\alpha_\varepsilon \mu_{eff} \frac{\partial \varepsilon}{\partial x_i}\right) + C_{1\varepsilon}\frac{\varepsilon}{k}\left(G_k + C_{3\varepsilon}G_b\right) - C_{2\varepsilon}\rho\frac{\varepsilon^2}{k} - R \tag{13}$$

where G_k represents the generation of turbulent kinetic energy due to the mean velocity gradients; G_b is the generation of turbulent kinetic energy due to buoyancy;

$G_b = -\vec{g}\frac{\mu_t}{\rho \Pr_t}\frac{\partial \rho}{\partial x_i}$ and $G_k = \mu_t S^2$ where S is modulus of the mean rate-of-strain tensor:

$S \equiv \sqrt{2S_{ij}S_{ji}}$ With the mean strain tensor given by:

$$S_{ij} = \frac{1}{2}\left(\frac{\partial u_i}{\partial x_i} + \frac{\partial u_j}{\partial x_i}\right)$$

and in the above equations:

$$R = \frac{C_\mu \rho \eta^3 \left(1 - \eta/\eta_0\right)}{1 + \beta\eta^3}\frac{\varepsilon^2}{k} \tag{14}$$

where $\eta \equiv Sk/\varepsilon$; $\eta_0 = 4.38$; $\beta = 0.012$; $C_\mu = 0.0845$;

Other parameters in the equations are as below:

$$\alpha_k = \alpha_\varepsilon = 1.393;\ \mu_t = \rho C_\mu \frac{k^2}{\varepsilon};\ C_{1\varepsilon} = 1.42;\ C_{2\varepsilon} = 1.68;\ C_{3\varepsilon} = \tanh\left|\frac{v}{u}\right|$$

where v is the where v is the component of the flow velocity parallel to the gravitational vector and u is the component of the flow velocity perpendicular to the gravitational vector.

For simulation of turbulent heat transfer, viscous heating and dissipation terms in Equation 10 are maintained except for the thermal conductivity is replaced by effective thermal conductivity $\lambda_{eff} = \alpha C_p \mu_{eff}$, where α can be calculated from RNG theory equations.

Methods and Simulation Details

In the present simulation, the NIST Standard Reference Database (REFPROP) Version 8.0 (2006) are referred to as primary source for thermal physical properties of CO_2 fluid at supercritical region. Here 9.0 MPa is set as the operation pressure and the thermal-physical properties of CO_2 are defined as functions of both temperature and pressure. For all the calculations in this study those properties are obtained from the NIST database and are arranged by piecewise-polynomial interpolation functions. The R-square of the fit for each physical property exceeds 0.995 and all the maximum relative differences between interpolated values and the values from NIST Database are small than 1%. This approach was also used in many related studies (Shan et al., 2007) with confident results.

For the detailed simulation procedure, an unsteady flow condition is applied. A structured non-uniform grid system is used to solve the discretized equations. As the large parameter gradients are possible near the pipe wall and the elbows, smaller and finer grids are set. The total number of mesh points is about 2.0×10^5 and in the testing this gridding system is able to provide independent behavior solutions and can provide information with sufficient accuracy. Finite-volume method is used to solve the governing equations with non-slip boundary condition. For numerical algorithm, the pressure-implicit with splitting of operators (PISO) algorithm is used to solve the coupling model between velocity and pressure as previous relevant studies (Chen et al., 2010; Chen et al., 2012). The momentum terms and energy terms in the governing equations are iterated with a secondary-order upwind scheme. The numerical model has been previously used and validated in a number of natural heat transfer studies (Zhang & Yamaguchi, 2008; Chen et al., 2010). Converge limitation is set at 10^{-3} for the residuals of velocity and continuity and for energy term the limitation is 10^{-6} for residuals of iteration.

Numerical Results and Discussion

Continued with our previous study (Chen et al., 2010; Zhang et al., 2010) and the established solar collector experimental model (Zhang & Yamaguchi, 2008), the geometric parameters in Figure 20 are set as: $L_0 = 5.0$ m, $L = 3.0$ m, $L_1 = 1.0$ m, $H = 2.0$ m; loop diameter D is set at 0.006 m and 0.015 m for comparison investigations; prescribed cooling walls temperature is maintained at $T_L = 298$ K while the heating walls (constant wall temperature for each case run) experience a wide temperature range: $T_H = 343$ K, 523 K, 823 K and 1023 K are simulated for each of the pipe diameter. Those heating wall temperatures then covers from normal water heater to solar collector, low grade heat re-collector, and even nuclear reactor, etc. Time promotion is started with zero velocity throughout the loop. For each run, initial temperature and pressure for CO_2 fluid are 298 K and 9.0 MPa.

Stability Trends and Transitions

Figure 7 and Figure 8 show the average flow velocity monitored on the plane in the middle of lower pipe. It can be seen that for 6 mm diameter group (see Figure 21) flow transition from periodic flow reversal to single direction flow happens when the heating wall temperature is 523 K. In fact, this phenomenon has been reported by Chen et al. (2010) in a former study. That is dominated by the second 'critical temperature' of supercritical CO_2 fluid, where the thermal physical properties begin linear change mode with temperature as shown in Figure 1, Figure 2, Figure 3, and Figure 4. The transition temperature can be identified by the thermosyphon mean fluid temperature (Chen et al., 2010; 2012). This effect is applied to mini loops. For larger loops the flow is always single direction regardless of heating wall temperature (see Figure 22). In Figure 22, 15 mm diameter cases are all stable flow after only several overshoots/fluctuations. It should be noted that in Figure 13, T_H = 823 K, case the negative velocity value is due to the random overshoot flow direction in the loop. For this study, positive velocity value shows the flow is clockwise direction and the verse.

From the comparison of Figure 21 and Figure 22 it can be seen that for larger loop (D = 15 mm) higher stable flow velocity can be achieved than small loop (D = 6 mm) group. Also the flow velocity will increase with the rise of heating wall temperature except for 343 K case. For 6 mm case, low viscosity leads to flow reversal and for 15 mm case that leads to high flow acceleration. Jain and Rizwan-uddi, (2008) also found very high velocity as 3 m/s to 4 m/s for supercritical CO_2 thermosyphon with pipe diameter of 7.484 cm and with this pipe diameter the flow is still one-directional flow. That's to say for large pipe supercritical flow seems to be more stable. Calculated Reynolds numbers are plotted in Figure 13. It is

Figure 21. Velocity monitored of cases for different heater temperatures: D = 6 mm group

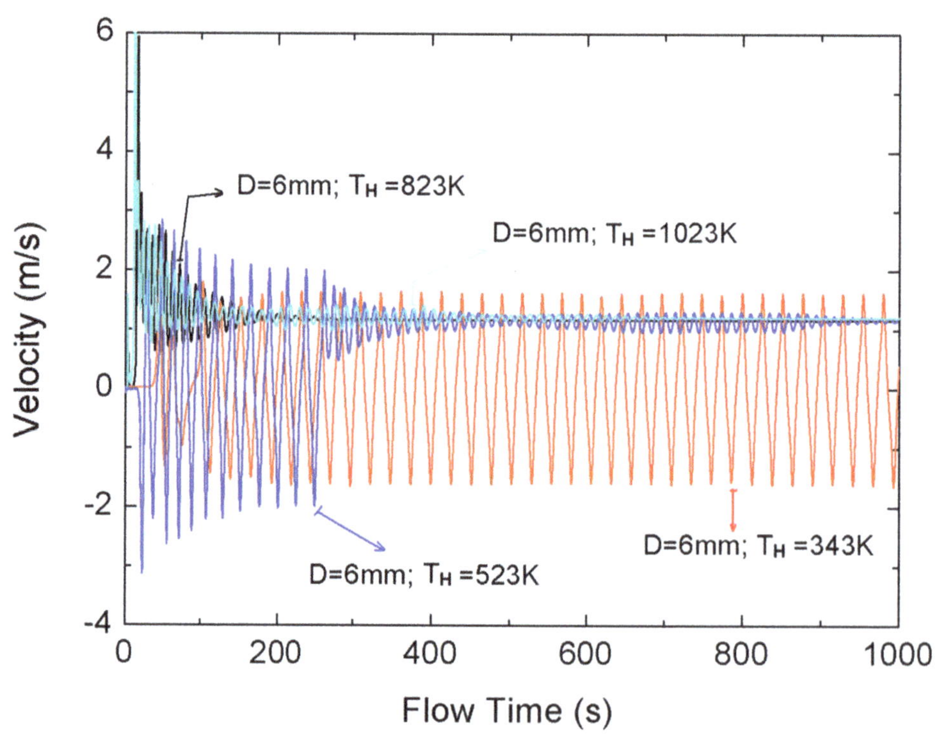

Figure 22. D = 15 mm group

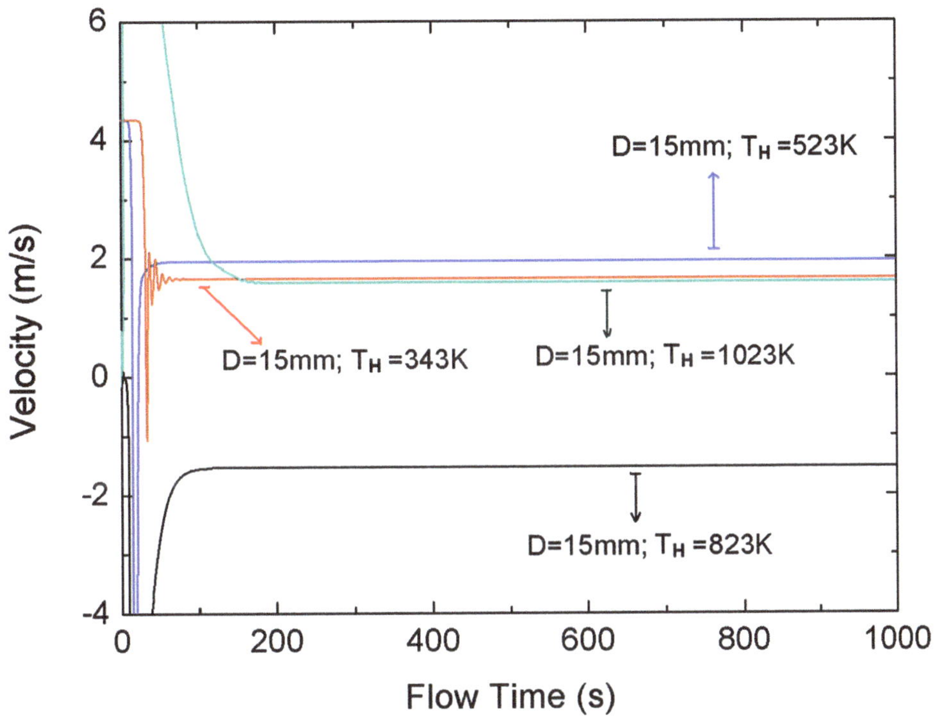

seen that Reynolds number will decrease with the increase of heating wall temperature and will increase with the increase of loop diameter. The great change of supercritical CO_2 Reynolds number shown in Figure 23 can be ascribed to the great change of fluid viscosity with temperature; also the increase of diameter. In addition, large Reynolds number around 10^4 to 10^5 is rarely seen in normal fluid like water and air flow driven by buoyancy. Also, the presented changes can partly be the results of the applied RNG turbulence model and the effect of transport parameters used in the model will clearly add to the approximation of the flow instability and heat transfer, the later one will be discussed in the next part.

Viewpoint of Heat Transportation at Cooler and Heater

The most interesting results in this simulation is the transition phenomena found for the effect of different heat source temperatures. For heating temperatures below 523K, the cases are all repetitive reversal flow. With the increase of heat source temperature, the reversal frequency also increase and velocity and temperature fluctuation magnitude enlarged. However, for the 523K heat source temperature case, the flow will experience a transition phase. For the beginning period the loop flow is repetitive reversal and the fluctuation magnitude gradually decrease, as shown in Figure 24 and Figure 25. Then at a critical condition the flow will change into single one directional flow. With higher heat source temperature the flow will be single one directional flow from the beginning.

The dependence of flow-pattern upon heating wall temperature could be ascribed to the temperature-sensitive thermal-physical properties of supercritical CO_2 fluid. Previous works on supercritical CO_2 flow mostly focused near its critical point around 304.13K. As discussed by Chen et al. (2010), due to the

Figure 23. Reynolds number of each case for different diameters

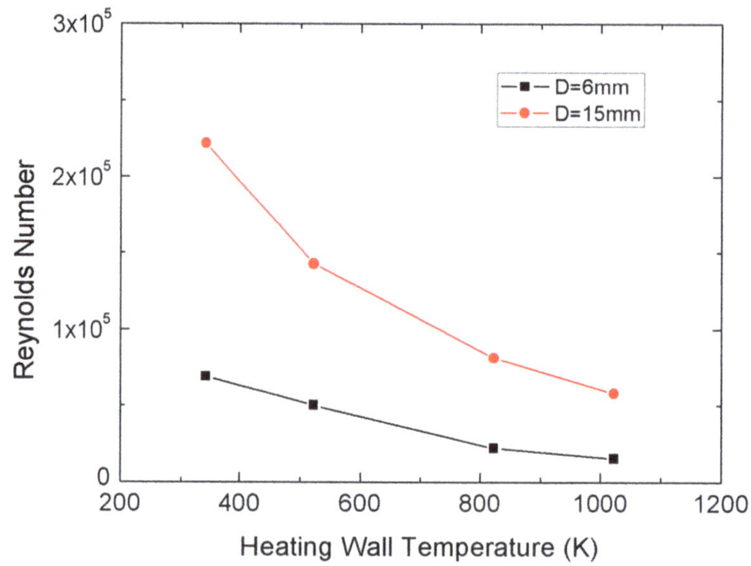

Figure 24. Variations of average flow velocity and temperature on cross section at 523K heat source temperatures; the flow transition is seen: velocity

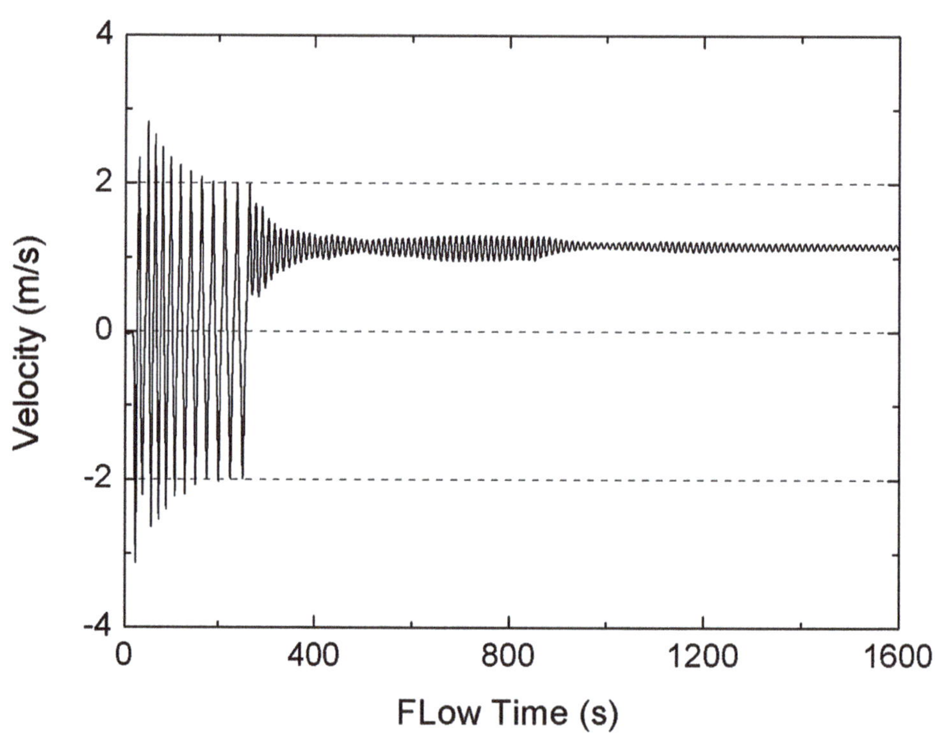

Figure 25. Temperature

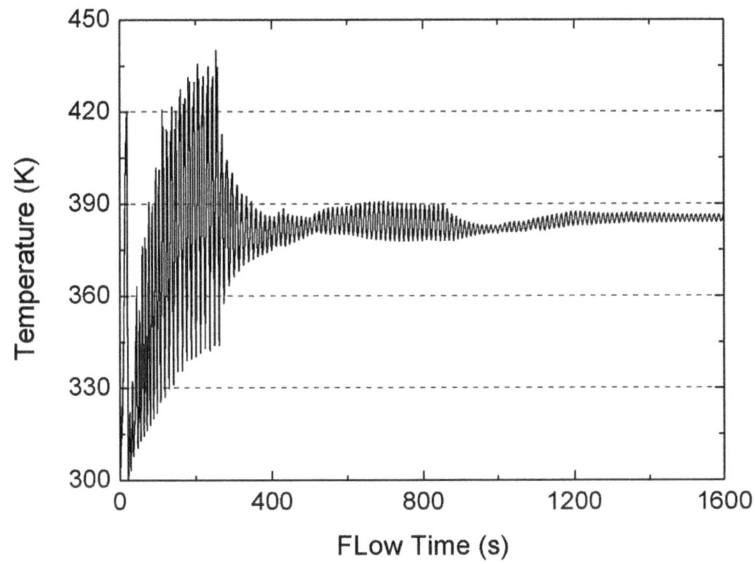

change of thermal conductivity, the fluid is not sufficiently cooled and 'density waves' are seen, which is also liable to flow reversals when this oscillation is accumulated with time. Then the fluid becomes both dense and with low-viscosity. Due to the strong buoyancy forces formed by large density changes in the rectangular loop, flow back happens. The basic mechanism is similar to that of two phase NCL by boiling systems (Linzer & Walter, 2003), but in supercritical fluid no phase change happens and heat transfer condition is more dominant (Chen et al., 2010).

In order to discuss the heating and cooling effect, the temperature difference across the heater and cooler are calculated and plotted in Figure 26 and Figure 27. The difference calculated for heater is the difference between average temperature on plane 2 and 3, for cooler is the difference between plane 6 and 7. For convenience the stable period from 700 s to 800 s are plotted. For 6 mm cases the temperature difference fluctuates for reversal flow case and for higher heating wall temperature cases the temperature difference becomes greater. For stable cases, both in Figure 26 and Figure 27, the T-difference of heater are always greater than that of cooler. The T-differences for 15 mm cases are always greater than those of 6 mm cases.

As the T-difference can represent the extent of heating or cooling efficiency, here it can be concluded that 15 mm loops perform better on stabilized heat transfer. Also, the curves in Figure 26 have some kind of periodic fluctuations with cases and no such curves are seen in Figure 27. The fluctuations are caused by the fluctuations of heat transfer performances at the cooler/heater, which will lead to unbalanced density wave with hot packet and cold packet. Similar phenomena are also found in the study of Vijayan et al. (1995) and Misale et al. (1999) and others. For more stable cases such fluctuations can be minimized. It can be concluded that both the heater and cooler of 15 mm cases reach a more stable condition. In another study of Jain and Rizwan-uddin (2008), supercritical CO_2 flow can still be stable when pipe diameter reaches 74 mm (much higher heat input is needed though).

From the present study, it is recommended that larger loop be used to achieve developed flow field and stable heat transfer condition. Such natural circulation system can guarantee the stable of system

Figure 26. Temperature differences across heater and cooler: D = 6 mm

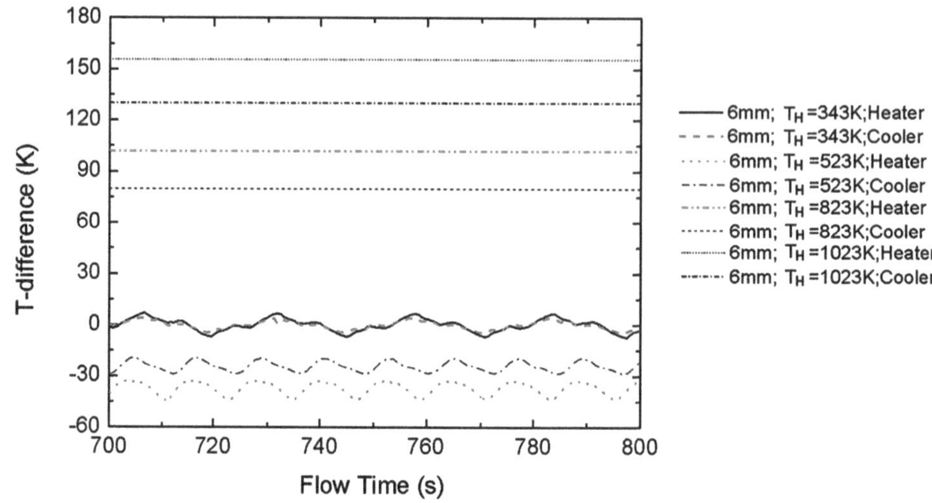

Figure 27. D = 15 mm

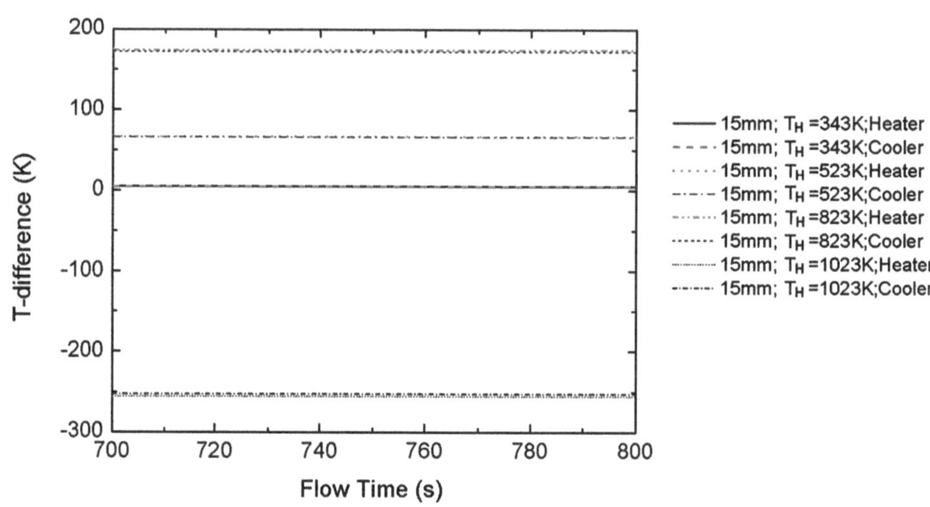

and low down the cost without pumps or valves used. Dostal et al. (2006) has explored supercritical water as working fluid in new generation reactors, and supercritical CO_2 can be one substitute choice as discussed. Other investigations have also made it especially useful for high temperature situations with more stable and self-induced single direction flow with high heat transportation efficiency (Jain and Rizwan-uddin, 2008) and control.

Comparison of Numerical Results and Experiments

For single-phase natural circulation Vijayan et al. (2004) have developed an analytical correlation between steady state Reynolds number and modified Grashof number in 1-Dimentional water cases. In the correlation geometric factors are also considered by introducing parameter N_G, which is the ratio of loop total length divided by loop diameter. The correlation can be expressed as following:

$$\mathrm{Re}_{ss} = C\left[\frac{\left(Gr_m\right)_{\Delta z}}{N_G}\right]^r \tag{15}$$

where Re_{ss} is steady state Reynolds number; (Gr_m) is modified Grashof number; geometric factor $N_G = L_f/D$; C and r are correlation parameters; in Vijayan's model C and r are predetermined by flow regimes of laminar, transition or turbulent flow. Recently in 2012, Swapnalee et al. (2012) developed the model for supercritical NCLs by dividing critical region according to the relationships between dimensionless density and enthalpy parameters, where C and r are experimentally identified for each parameter region.

In the present calculation, the supercritical Reynolds number is relatively higher than the predicted values from Vijayan's model developed for water (Vijayan, 2004) and also higher than the new correlation by Swapnalee et al. (2012). It is found the Vijayan model is silghtly lower that the present supercritical CO_2 based simulation, as shown in Figure 28 and Figure 29. In fact, for the supercritical fluid modeling in Vijayan's model, thermal and transport properties are generally fixed with little change to temperature and pressure, and leaving the changes of density and enthalpy for steady state condition. The plotted points in Figure 28 are only for steady state single direction flow cases. Though the general trends for both high Re group (Figure 28-near critical region according to Swapnalee and Vijiayan) and low Re group

Figure 28. Correlated plots for Reynolds number and modified Grosholf number of cases (comparison with Swapnalee and Vijayan's new correlation for supercritical fluids): high Re region

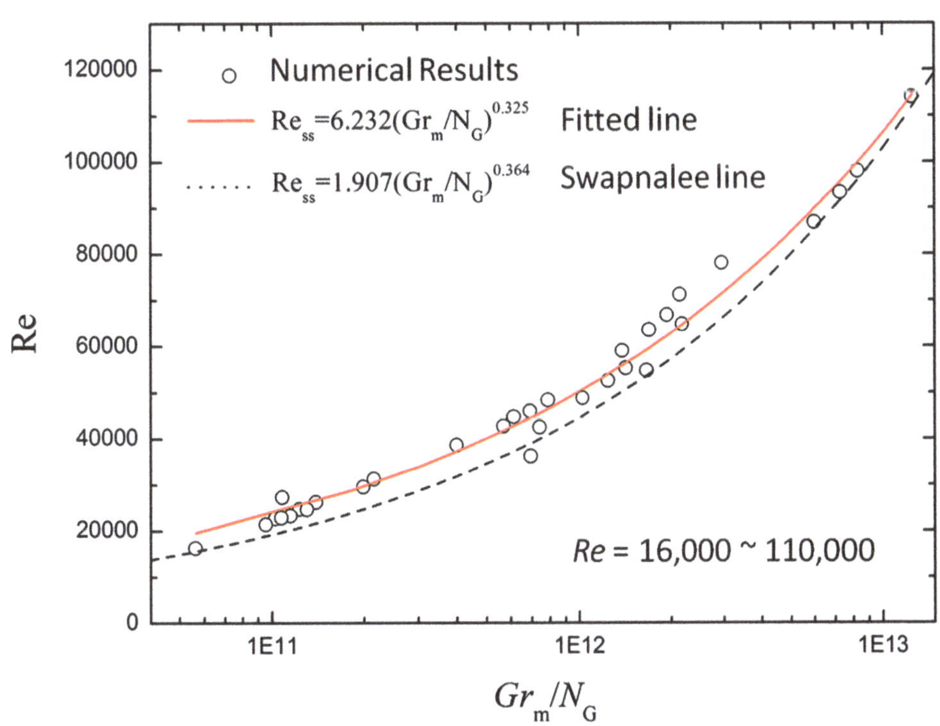

Figure 29. Low Re region

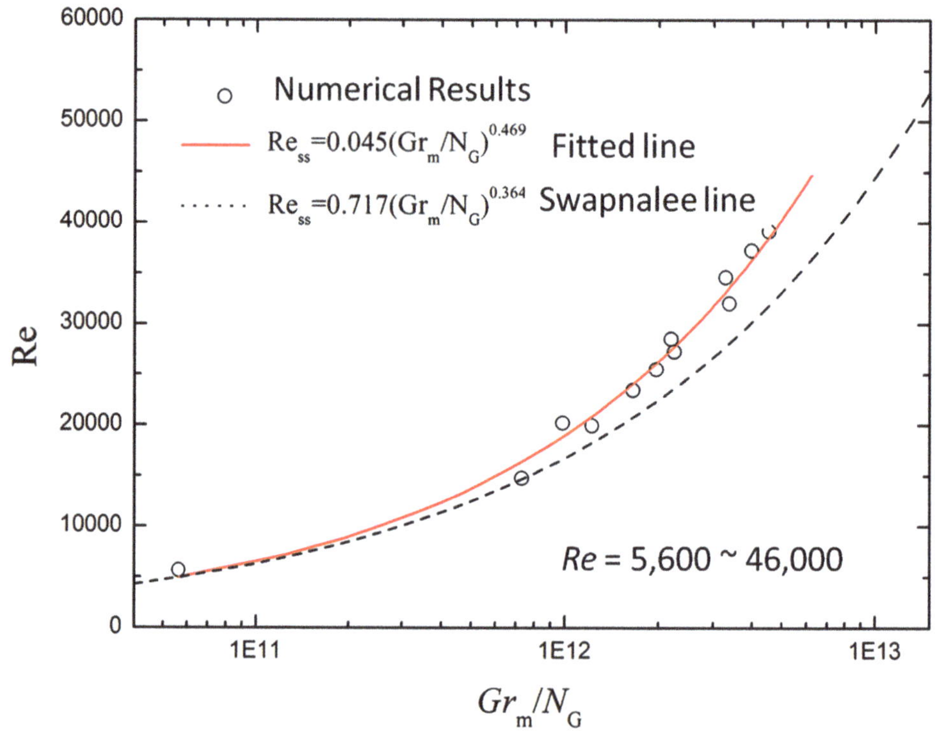

(Figure 29-relatively far from critical region according to Swapnalee and Vijiayan) are similar, relatively greater discrepancies are seen in the moderate Re transition region between Figure 28 and Figure 29. Also it should be noted that the current model is two dimensional (one dimentional for Swapnalee and Vijiayan's model) and the effect of loop diameter (as the flow field can be largely different (Chen et al., 2011). More work should be done to the further development the original correlation or new relations be set up in order to suit for non-Boussinesq feature of supercritical CO_2 in the future (as also discussed by Swapnalee et al. (2012). For further study, various data should be correlated as shown in this chapter thus to expand the feasibility of such correlation for the summary and predictions for mechanical designs.

Various Factors for the Flow and Heat Transfer Behaviors

Besides the above discussed factors of stability evolutions, there are several more important kinds that govern the basic trends of circulation stability. From a theoretical viewpoint, if the natural circulation system takes a symmetric geometric design, the stability solution may have three solutions: single direction flows (of clockwise and anti-clockwise) and the steady state solution. If a small perturbation occurs in the natural circulation system, the flow will soon tends to one-direction flows. However, if the fluid loop design has high legs, then it is possible for loop oscillations of flow and back-flows. That is a basic situation of flow instability. Another kind of flow instability is the flow rate fluctuations, which is due to the heating intensity and the cooling intensity: if the flow heat transfer of the two ends becomes unstable, the fluid will have high temperature end and low temperature end, which may induce density difference throughout the loop and have fluctuations in the flow rate. Further, the stability behavior of a natural circulation loop may be governed by several factors, which include the loop aspect ratio, the heating/cooling intensity, inclination angle of the loop, heater/cooler orientations, pipe diameter, and also the operation strategy of the loop system (Chatoorgoon, 2001; Cammarata, 2003; Chatoorgoon et al., 2005; Chen et al., 2010; Chen & Zhang, 2011; Sarkar et al., 2016). And if supercritical fluid is used for the circulation loop, as the fluid property is very sensitive to heat transfer, the stability conditions for the loop system will be much more complicated.

1. The heat transfer performances go up with that of flow rates. For fixed heat input, heat source location changes have no significant influence on Nu number. The effect of cooler heat transfer (and fluid paroperties) can be crucial for the heat transfer improvement and system stability behaviors (Chen and Zhang, 2011; Chen et al., 2013c).
2. The time to reach steady flow of cases will decrease with heat flux and the flow rates increases at the same time. The Re number ranges from 5×10^3 to 8.5×10^4, which indicates the effects of near critical low viscosity (Chen et al., 2010, 2012, 2013a).
3. About the heater/cooler orientation. Vertical heating cases are all stable in a wide range of heat flux conditions from 600 W/m^2 to 60000 W/m^2 due to the stable buoyancy force torques across the NCL; horizontal heating cases show unstable transitions and oscillations under natural circulation condition (Chen et al., 2014).
4. One-dimensional assumption and simplification of cross loop distribution of critical properties may affect the fitness of NCL flow and heat transfer correlations models. More comparison of numerical and experimental results is needed for future development.

Stability Nature Discussion: Channel Flows of Supercritical Fluids

Basic Model Set-Ups for Supercritical Channel Flows

This chapter is also focused on the trans-critical flow in mini-channels, in which the thermo-properties of CO_2 fluid vary dramatically. As discussed formerly, though there have been a lot of investigations of supercritical channel flows, the detailed information and mechanisms are still not clearly stated (Jackson, 2001; He et al., 2005; Jue et al., 2007; Chen et al., 2013a, 2013b, 2013c; Zhang et al., 2010; Chen et al., 2010; Zappoli, 2003; Carles, 2010). The model in the current study is the simple long, thin mini-channels, where the channel diameter, length, initial and boundary conditions can be changeable, as shown in Figure 30. Constant heat fluxes Q are applied from the boundary walls. By changing the inlet

Figure 30. The geometry of the channel flow studied in the present study

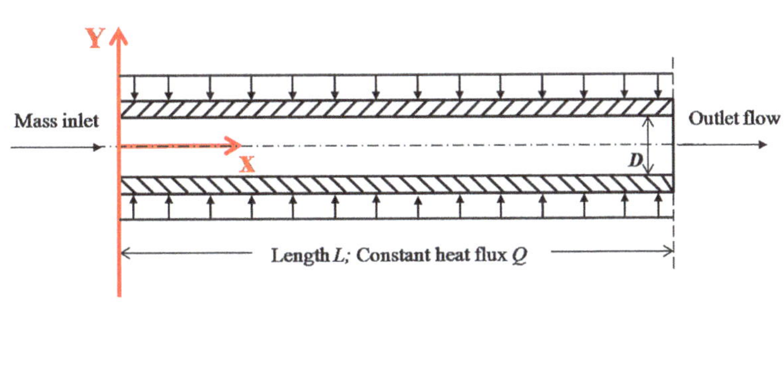

and boundary parameters, various conditions and model operations can be tested. The focus of this study is different from the previous studies, where the heat transfer parameters or correlations are summarized. However, this study would focus on the new phenomena inside detailed mini-channel trans-critical flows.

Indeed, trans-critical flow inside mini-channels, especially for the transition of pressurized fluid to near-critical state or supercritical state can be of critical importance for real application designs but are not well studied or explained previously (Carles, 2010; Chen 2013d, 2014; Pioro et al., 2004; Duffey & Pioro, 2005; Cheng et al., 2008; Adebivi & Hall, 1976; Zhang & Yamaguchi, 2008; Liao & zhao, 2002a, 2002b; Gupta et al., 2013; Zhao et al., 2014; Xi et al., 2014). For such transitions of state, the confinement of channel flow, the drastic changes of fluid properties, pressure and heat transfer coupling, turbulent effect and others all contribute to the complex basic problem, thus making it quite interesting problem to investigate. Also, there are a lot of related mini-sized applications and new energy systems intensely under development in recent years (Chen & Zhang, 2014; Zhang et al., 2014; Fernandez et al., 2010; Dostal et al., 2006; Zhang et al., 2006; Ochsner, 2008).

Stability Discussion and Multi-Peak Phenomena in Supercritical Heat Transfer

When $D = 0.4$ mm and $Re_{in} = 1000$, it is notable that the variation of flow characteristics resembles those of $D = 1.2$ mm, $Re_{in} = 5000$. Two peaks of HTC can also be found in Figure 31, which indicates that the formation of multiple HTC peaks may relate to the ratio of D to Re_{in}. Compared with the cases of $Re_{in} = 5000$, the HTCs in current condition are generally smaller because of lower velocity magnitudes. Peaks under 9.0 MPa are still lower than that under 7.4 MPa in current condition. Similar explanations also apply here, but with an additional condition that a smaller velocity magnitude has extended the flow time so that both extra peaks of heat transfer are less obvious here.

The multi-peak phenomena identified in the current study for relative high Reynolds number flows are less discussed in previous studies (Liao & Zhao, 2002a, 2002b). Indeed, the critical transitions under low Reynolds numbers of near-critical fluid flow in similar scale have been found recently (Chen et al., 2013a, 2013b, 2013e), where the unstable vortex flow and sensitive local temperature equilibrium phenomena are seen. For relative larger channels/loops, supercritical stability problem due to critical transitions are also found a big problem for circulation flow (Nikitin et al., 2006; Zhang et al., 2010; Chen et al., 2013c; Zappoli, 2003; Carles, 2010; Chen et al., 2013d, 2013f, 2013g, 2014; Kumar & Gopal, 2009; Debrah et al., 2013), where the major concerns are laid on the sensitive heating and cooling in the critical vicinity. Still, there are studies that reflected the stability transitions and 'density waves' of supercritical channel flows (Xi et al., 2014; Debrah et al., 2013; Chen et al., 2013a, 2013e, 2014; Chen & Zhang, 2014; Zhang et al., 2014), where the 'density wave' is nominated to be the main reason of instabilities. So here the multi-peaks phenomena indeed should be mentioned to have the same origin of physic background: the specific heat peak and density drops at the same time. Therefore, the sudden heat absorption and thermal expansion happens inside confined channels, which lead to the first peak; then the fast gas-like supercritical flow continues to expand and form turbulent heat transfer region just after a small interval. During such transition, deteriorated heat transfer is also possible, which may be very important for small scale heat transfer devices design, especially the local equilibrium process should be considered. In addition, there are still a lot of new problems in mini-channels with near-critical/supercritical fluids. Indeed, though nowadays more and more studies have explored into even much smaller scales, more studies should be conducted in this field to get detailed information.

Figure 31. The variation of flow characteristics to normalized position on the x axis when channel diameter D = 1.2 mm and Re_{in} = 5000. Wall heat flux Q is 50000 W/m². (a) The variation of local convective HTC (hx); (b) local Nusselt number Nux; (c) normalized bulk temperature T_b/T_{pc}.

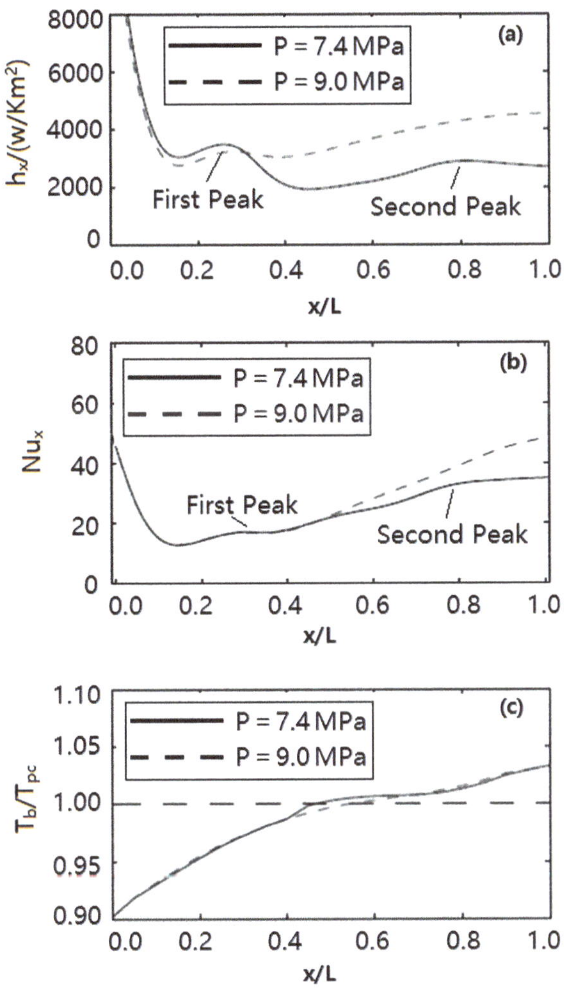

APPLICATIONS OF SUPERCRITICAL NCL

Solar Water Heaters

Based on the natural convection supercritical CO_2 solar water heater concept put forward in this study and previous designs (Yamaguchi et al., 2010), a new experimental prototype is established. As shown in Figure 32 and Figure 33, the new model has no pumping devices and is mainly consisted of two parts: the supercritical CO_2 solar collector circulation system and the water heating and control system. The supercritical CO_2 natural collector side has three parts: (1) evacuated tube solar collector panel; (2) supercritical CO_2 - water heat exchanger; (3) pressure control valve mounted between the heat exchanger outlet and evacuated tube collector inlet. In water side, inlet and outlet valves are mounted specifically and connected to the hot water tank.

Figure 32. Schematic diagram of the supercritical CO_2 natural convection based solar water heater: diagram of flow and system structure

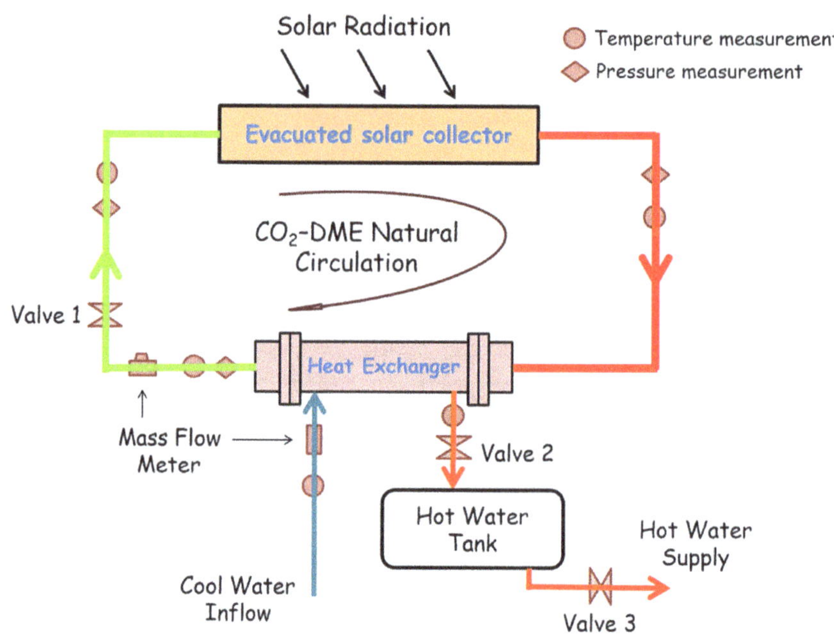

Figure 33. Side view of experimental prototype figure

In the tests, the solar collector water heater is tested with declination angle of 45° with respect to the horizontal level. During the operation, the system is firstly filled with nitrogen to drive out air in the circulation system. Then it is filled with supercritical CO_2 fluid and refilled for several times to make sure the collector is filled with pure CO_2 fluid as possible. After the injection, the water side flow valve is opened and solar collector is exposed to the sunlight. At the same time, the data collecting system begins.

Applications in energy conversion systems with supercritical natural circulation systems are proved possible. Due to the specific thermo-physical properties of supercritical CO_2 fluid, the system can be compactly made and achieve smooth convective flow with Reynolds number high than 6700. The CO_2 fluid temperature in the heat exchanger can be as high as 80 °C and a stable supply of hot water above 45°C is maintained. Basic explanations and seasonal performance discussions are made on this promising kind solar system.

Cycle Operation Pressure Reduction by CO_2-DME Mixture Working Fluids

Basics of CO_2-DME Mixture

One major concern of the current kind supercritical solar thermal conversion is the high operation pressure, which may bring about extra cost of manufacturing and maintenance. In the current experimental study, DME (dimethyl ether) is chosen to pre-mix with supercritical CO_2 with proper volume fraction, attempting to reduce/adjust the operation pressure and test the system performance of such kind of mixture working fluids. Indeed, DME is also one substitute working fluid of CFCs/HCFCs with its preferable compression/condensation properties. DME fluid is also of less corrosive and has good miscibility with other polar/non-polar fluid (like CO_2) but it is very flammable and explosive. In recent years, some groups have tried to mix DME with CO_2 as working fluid to reduce the high supercritical CO_2 operation pressure and in turn the CO_2 fluid help reduce the flammability of DME (Hasan et al., 2008). For example, Koyama et al. (2006) experimentally studied the behaviors of CO_2-DME (with 10% DME) mixture in heat pump systems, where high standard COP was maintained and considerably lower discharge pressure was achieved. Later, Onaka et al. (2007) reported that besides operation pressure reduction, CO_2-DME mixture based heat pump COP increases with DME mass fraction (in the range of 0% to 10% in their analysis). Indeed, the thermal-physical properties of refrigerant can be different when DME is introduced. The above studies referred to the increase of thermal conductivity and viscosity of CO_2-DME mixture as shown in Figure 34, Figure 35, Figure 36, and Figure 37 (data obtained from NIST reference package of fluid thermodynamic and transport properties REFPROP version 8.0; in Figure 34, Figure 35, Figure 36, and Figure 37 pure CO_2 and pure DME fluid properties are also plotted for comparison), which may play an important role in respective systems when compared with pure CO_2 or pure DME single fluid system.

The present study is one continuation of our previous experimental studies of trans-critical experimental study on CO_2 solar water heater (Zhang & Yamaguchi, 2008; Zhang et al., 2013) and preliminary natural experimental prototype solar water heater (Zhang & Yamaguchi, 2008; Yamaguchi et al., 2010; Zhang et al., 2013). In the present study DME is also proposed to be mixed into CO_2 fluid within the explosive

Figure 34. Variation of thermo-physical properties of the CO_2-DME mixture refrigerant with with different DME mass fractions at 8.0 MPa: density

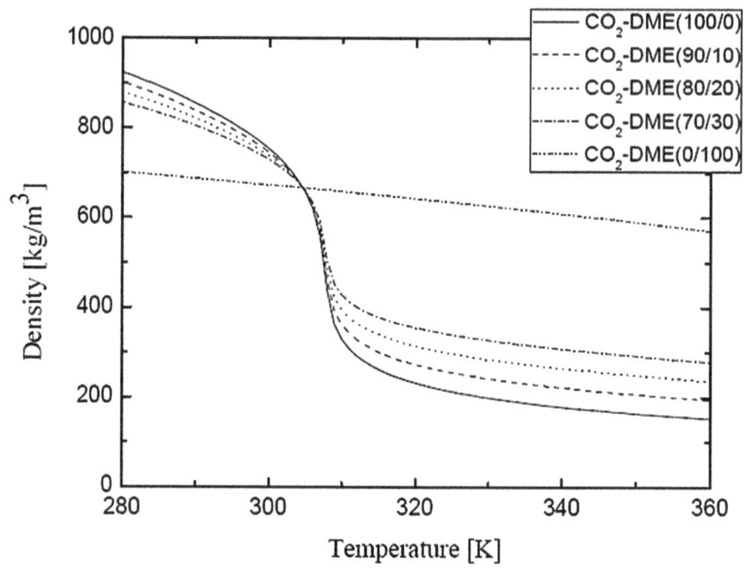

Figure 35. Specific heat

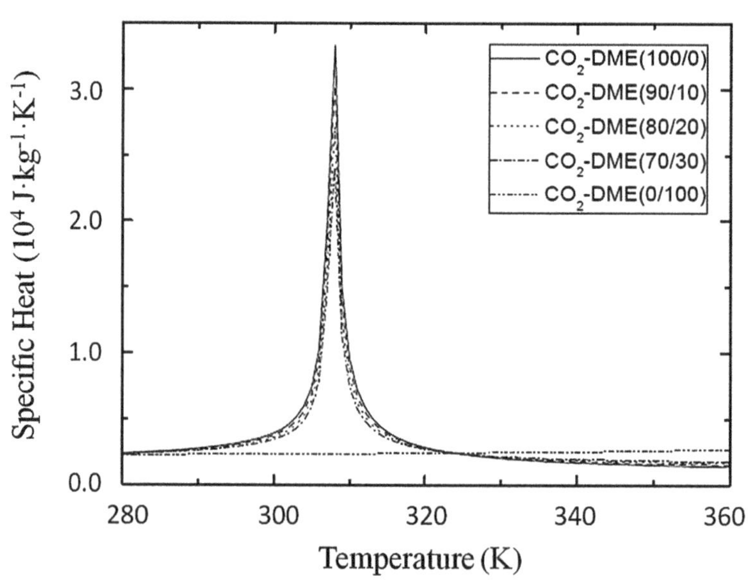

fraction limit thus to take advantages of both fluids. Respective experimental systems are designed to investigate the system behaviors when CO_2-DME mixture fluid is used in solar thermal conversion. The following parts will introduce the basic system design and system operation process; then the basic results on the pressure reduction effect of DME introduction; after that the system features/parameters are compared with previous pure CO_2 based tests and the seasonal efficiency, feasibility study as well as future development are also discussed into detail in this chapter.

Figure 36. Thermal conductivity

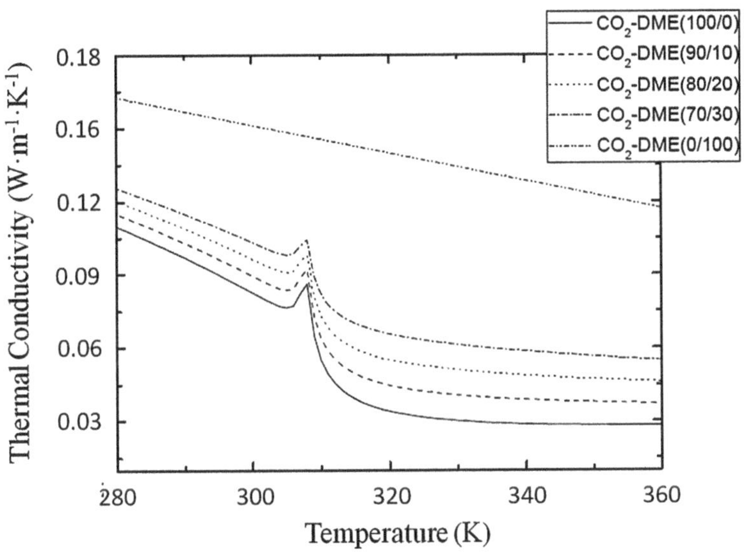

Figure 37. Viscosity

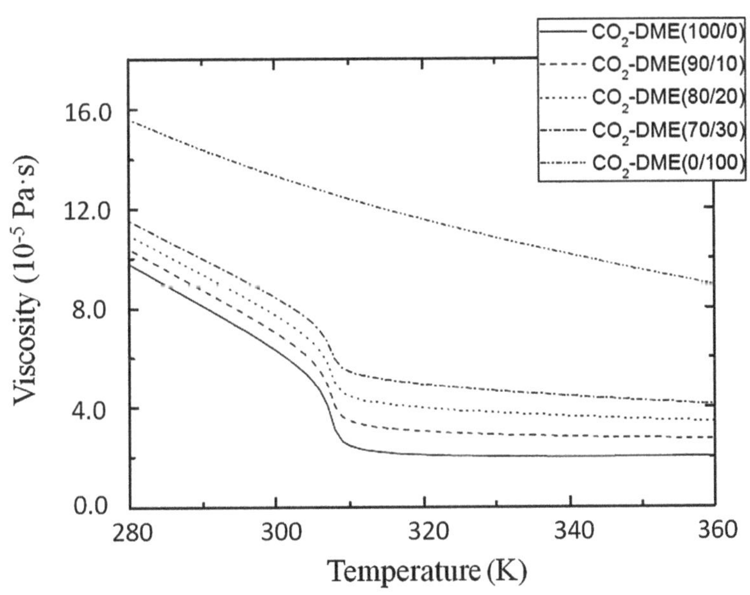

Pressure Effect of CO_2-DME Mixture Flow in the System

Based on the above discussion on pressure behaviors of supercritical CO_2-DME mixture, it is found that the pressure dependency can be very different from previous pure CO_2 system. It is known that the pressure changes to similar trend with solar radiation instead of monotonically increasing mode for pure CO_2 system (Zhang & Yamaguchi, 2008; Yamaguchi et al., 2010; Zhang et al., 2013). Due to the co-change mode and the pressure drop effect in the afternoon, the overall operation pressure is successfully reduced in CO_2-DME mixture based solar collector tests, assuring the mixture fluid as one possible choice in the future development.

The system pressures during summer season tests in 2010 are plotted against DME mass fraction in Figure 38. In Figure 38, the averaged operation pressure during the season tests are also added near the operation pressure points for each DME mass fraction. It can be seen from Figure 38 that the average operation pressure decreases with the increase of DME mass fraction: the averaged operation pressure for pure CO_2, 10% DME and 20% DME tests are 9.47 MPa, 8.95 MPa and 8.75 MPa, respectively. Therefore, the method of CO_2-DME mixture natural convection solar collector can be one choice to reduce the operation pressure of supercritical system, as also suggested in related studies (Hasan et al., 2008; Koyama et al., 2006; Onaka et al., 2007). However, DME mass fraction higher than 20% to 30% can be difficult for supercritical solar thermal system due to the explosive limit of DME fluid.

In addition, the introduction of DME into the supercritical CO_2 system also brings about new evolution trend of mass flow rate in the collector, which becomes more sensitive but still stable with variation of solar radiation due to the changes of thermal physical properties like thermal conductivity, viscosity, etc. Similar to previous studies, the difficulties of high precision parameter capturing and near-critical region analysis still need further developments (Zhang & Yamaguchi, 2008; Yamaguchi et al., 2010; Zhang et al., 2013; Chen & Zhang, 2014; Zhang et al., 2014). Also, though the pressure reductions have been achieved, the absolute operation pressure (higher than 8.0 MPa) is still relatively high. Future designs of similar systems should pay attention to those characteristics different from pure CO_2 based systems.

Collector Efficiency Variations with Different DME Mass Fraction

Besides the operation pressure reduction, system efficiency is also generally maintained at relatively high level when DME is introduced, as discussed in the former sections. It is essential that the current solar thermal conversion just by natural convective flow can have efficiencies generally above 40%, even 20% DME is mixed in the system. The collector efficiency variations with comprehensive parameter C for different DME mass fractions are plotted in Figure 39. It can be seen from Figure 39 that with the increase of DME mass fraction the distribution of collector efficiency is less expanded. Also for pure CO_2

Figure 38. Variations of the time-averaged pressures with the mass fraction of DME

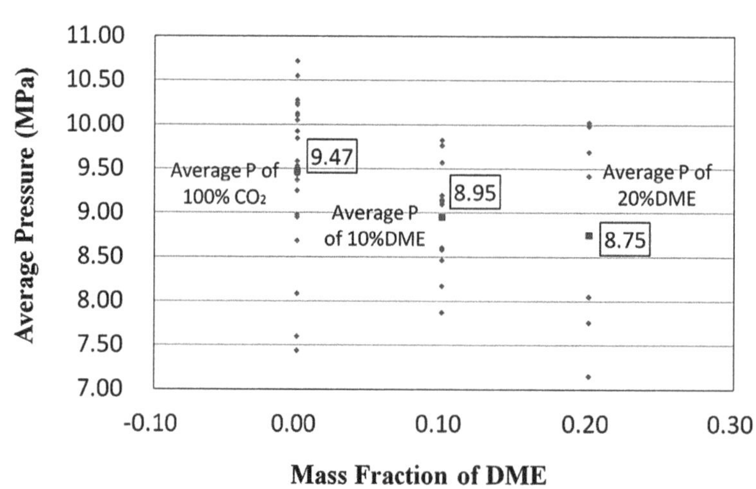

Figure 39. Variations of the measured collector efficiency with the comprehensive coefficient under three different mass fractions of DME, 0%, 10% and 20%. Every point shown in this figure represents a time-averaged value during one day.

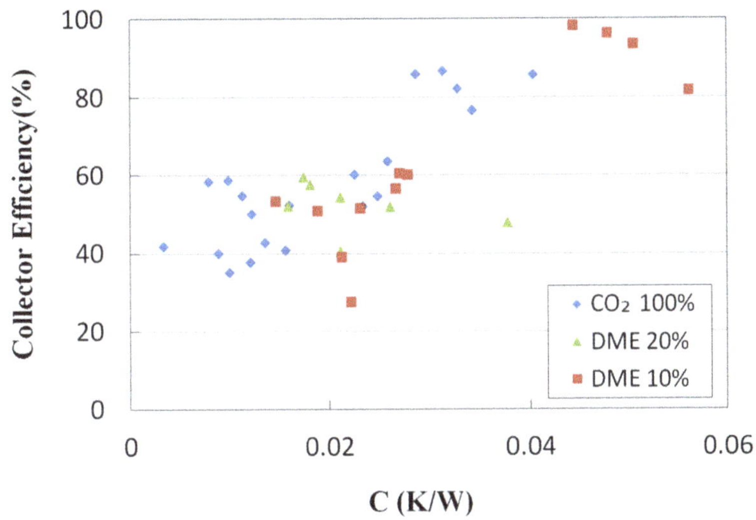

cases and 10% DME cases the efficiency generally increase with *C* parameter, while 20% DME cases do not show similar trend. It is also found in the previous discussion that though the pressure behavior is preferable and the temperature sensitivity and stability of flow are also good, higher DME percentage may make it difficult to maintain stable flow. Table 3 presents the seasonal tests of 2010 summer for average system efficiency of different DME mass fraction. From Table 3 the average efficiencies are around 50% and they seem less affected by the DME mass fraction, which indicates the effect of DME fraction on the current solar conversion tests is limited during the summer/autumn season.

Other Heating and Cooling Designs

Concept Design of Supercritical CO_2 Assisted Solar Li-Br Refrigeration

In the present study, a new solar cooling system is proposed using supercritical CO_2 at the collector side and by using double effect LiBr-H_2O absorption at the refrigeration side. By combining the advantages of both systems, experimental set ups of the supercritical fluid side are tested. For the LiBr-H_2O side, theoretical analysis and system performances are discussed. The system formation of the connected cycles and its basic features and feasibility are also included in the discussion. This study is one first trial on such combined cycles, so as to achieve a continuous effort of increasing collecting thermal temperature

Table 3. Collector efficiency under different DME fractions

Mass Fraction of DME	0	10%	20%
Collector Efficiency	51.7%	49.9%	50.8%

and enhancing solar cooling efficiency. It is hoped that this study can contribute to related application system designs and optimizations.

The schematic formulation of the proposed solar cooling system is shown in Figure 40. A closed supercritical carbon dioxide cycle (the solar heating) and a double effect lithium bromide-water cycle (the chiller) are thermally connected together. Based on the previous solar water heater design (Zhang & Yamaguchi, 2008; Chen & Zhang, 2014; Yamaguchi et al., 2010), the current developed solar collector side comprises an evacuated tubes array, a heat exchanger, heat recovery systems, feed pump and other auxiliary equipment. This solar heating system can heat supercritical carbon dioxide to high temperature (even higher than 120 °C), which enables it to power the chiller cycle effectively. In the supercritical collector side, the pressurized CO_2 fluid is first heated to high temperature inside the collector panel; then the fluid release heat to the generator of the refrigeration cycle side through one heat exchanger; after that the supercritical fluid temperature can be further decreased by additional heat recovery and utilization systems (for example, for water heating or drying); then the cooled fluid is pumped into the collector panel again to form continuous circulation flows.

The refrigeration cycle is introduced by a double effect series flow lithium bromide-water cycle, as shown in Figure 40. The right-hand side part of Figure 40 is the LiBr-water refrigeration cycle. The high-pressure generator (HPG) of the chiller cycle first obtains heat from the heat exchanger of the supercritical collector cycle; before that the low-pressure generator (LPG) is heated up by the water vapor generated in the high-pressure generator (with external energy supply still possible; from recovery systems). Since the low-pressure generator runs under relative lower temperature, it may also utilize the waste heat recovered from the supercritical solar collector cycle. And if the solar energy is not sufficient to power the cooling cycle, subsidiary energy sources (e.g. waste heat or geothermal energy) can also

Figure 40. Schematic diagram of the supercritical CO_2 based lithium bromide-water solar cooling system

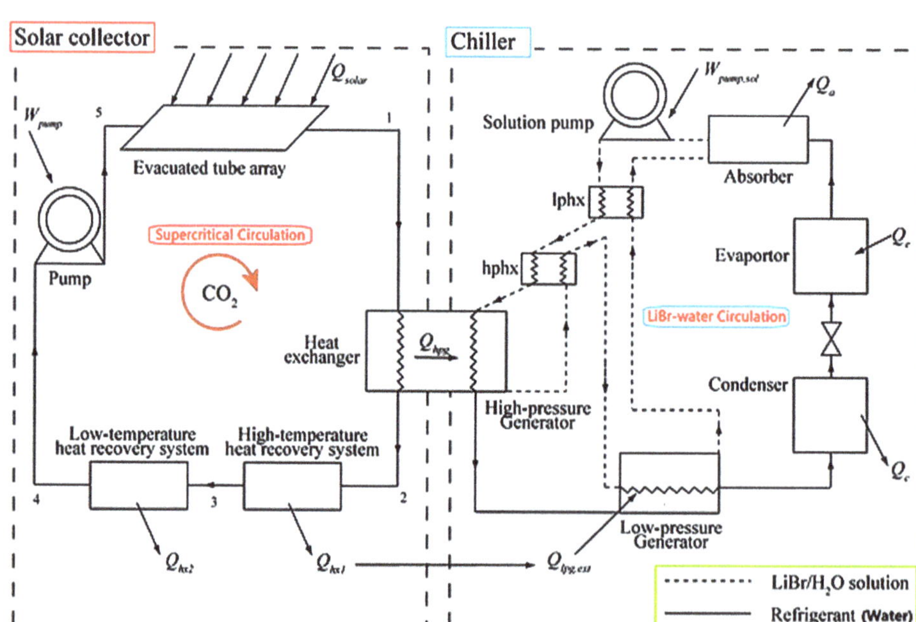

be used to power the LPG so as to maintain the circulation flows of the cycle. In Figure 40, the *lphx* stands for the low pressure heat exchanger while the *hphx* represents the high pressure heat exchanger.

The energy flow and fluid flow process are also marked in Figure 40. By continuous energy supply from the solar collector side, relatively stable LiBr-H_2O refrigeration capacity can be achieved. In the present study, the basic concept design and formation of the basic cycles are discussed into detail. For the solar collector part, one modified experimental prototype is established and tested under outdoor weather conditions. And for the LiBr-H_2O refrigeration cycle, theoretical analysis and model estimation are made by using the experimental data obtained in real conditions. This is the first trial of the designing with supercritical solar powered LiBr-H_2O refrigeration cycles.

Others Designs

Indeed, there are also many other kinds of energy systems and related applications by supercritical fluids. In nuclear engineering, the cooling system with a secondary refrigeration cycle can use the high temperature range supercritical fluid to achieve stable cooling. The basic system operation strategies and off-parameter design conditions, especially the sudden load changes or sudden increase/decrease of heating load, have been tested numerically in previous study of Chen et al. (2012). For high pressure turbines and energy production systems, supercritical fluids may also contribute as one possible choice in future. Recently, the supercritical fluid based power cycles (mainly by Rankine Cycles) have been intensely studies (Kim et al., 2016). Other related designs of coupled cooling/heating cycles have also been proposed in literature. Some of the related energy field applications will also be found in other chapters of this book. Though there is still a lot of work to do in this field, it is very promising for the development of supercritical systems.

FUTURE RESEARCH DIRECTIONS

The above results and discussions show the basic flow and heat transfer trend and applications of supercritical CO_2 thermosyphon, which is much different from traditional working fluid based systems. As trans-critical systems are often utilized due to the advantages of heat transfer, the current experimental results show that pressurized liquid or supercritical states should be preferable. The general heat transfer trends are similar to previous numerical predictions (Chen et al., 2010; Zhang et al., 2010; Chen et al., 2013d), but there are still a lot many new feature for future investigations, such as the near-critical region abnormalities, the enhancement and deteriorates, the scaling effect.

In the current experimental study, the dependence of heat transfer performances on pressure variations is verified. Also, the heating and cooling correlations can be dependent on system pressure evolutions during operation. Judging from the variations as discussed in the former part, correlations from NCL studies still need further justifications (Zhang et al., 2010; Chen et al., 2013d). How to correlate well the flow equations and the boundary heat transfer conditions is still a tough task for future studies.

For heat transfer applications, for example, in the solar collector or new generation nuclear cooling, the initial charge amount will not only affect the stability behavior, it will also affect the heat transfer behaviors as the pseudo-critical point will change with system pressure during operation. Possible heat transfer deteriorations are found in the current experiments, which may be critical for real applications when system pressure control is absent. It is hoped that future studies go deeper into critical natural convection systems. However, in the field of trans-critical/supercritical thermosyphon technologies, with its wide applications and complex behaviors, more in-depth studies and designs are still expected in this field.

CONCLUSION

This chapter presents systematic experimental and numerical investigations into the flow and heat transfer of CO_2 based natural circulation loop. Supercritical loop under natural circulation conditions are found to be one basic model for various energy system applications in industry. The application-related problems of supercritical fluid dynamic behaviors and stability evolutions are discussed into detail in this chapter. A high pressure supercritical natural circulation loop experimental apparatus is carefully established and operated first in this study. The system is operated from sub-critical region, near-critical region to supercritical region to test the stability and heat transfer characteristics.

The heat transfer temperature has proportional effects on the final operation pressure. The heating wall temperature has a greater effect on the Reynolds number in cooling pipe. The Nusselt numbers show increase with T_b/T_{pc} value and the overall heat transfer performance is found to be greatly dependent on the NCL operation pressure.In the numerical study, sophisticated modeling method is developed with detailed consideration of turbulence effect of supercritical fluids. Flow instabilities and transitions from unstable cases to single-directional stable flows are identified and compared between the experimental and numerical studies. The basic stability trend is found to be initial pressure dependent.

The variation of bulk temperatures behaves almost linearly for each run of initial charging pressure. With the increase of initial charging pressure, the difference between the heater and cooler temperatures will also increase, leading to stable flow circulation. The heat transfer coefficient of the cooling pipe changes a little with variation of bulk mean cooling temperature, and for the heater higher Nusselt number is found for conditions closest to the pseudo-critical condition. Main factors of supercritical natural circulation loops are discussed for the heating and cooling temperatures, inclination angle, aspect ratio, heating intensity, injection amount, pipe diameter, operation strategy, etc. Multi-peaks of heat transfer parameters are found in trans-critical cased, which governs the main heat transfer behaviours of the channel. The multi-peaks appear when the main stream temperature respectively reaches and leaves pseudo-critical vicinity.

Representative applications of supercritical NCL systems are introduced in this chapter. The system performance of several kinds of solar collector and generation discussion is based on theoretical analysis of the supercritical thermodynamic cycles as well as supercritical solar collector experimental data. Besides the well-known solar water systems by supercritical fluid, the introduction of DME successfully reduced the operation pressure and it is found that higher DME mass fraction leads to lower average operation pressure.

Future studies including experiments are still needed for the in-depth understanding of stability controlling factors and heat transfer behaviors of supercritical fluid based NCLs. Those studies will surely contribute to the future development of energy systems using supercritical fluids.

ACKNOWLEDGMENT

This chapter is a summary of the studies in natural circulation loops and the related application designs in recent years. The support of National Science Foundation of China and JSPS Overseas Researcher Grant (No. 16F16068) are gratefully acknowledged. The author is very thankful for the kind cooperation of Prof. X.R. Zhang and the assists in numerical/experimental analysis from Dr. Bili Deng, Mr. Bin Jiang, Ms. Jia Liu, Mr. Yimin Chen in Renewable Thermal Lab in College of Engineering, Peking University. Dr. Lin Chen is also very grateful for the discussion/suggestions Dr. Yuhui Cao in Department of Physics, University of Chinese Academic of Sciences, from Prof. Hiroshi Yamaguchi in Energy Conversion Research Center of Doshisha University (Japan) and Prof. Yuhiro Iwamoto in Nagoya Institute of Technology, and from Prof. Shigenao Maruyama, Prof. Atsuki Komiya and Prof. Junnosuke Okajima in Institute of Fluid Science, Tohoku University (Japan).

REFERENCES

Ambrosini, W. (2007). On the analogies in the dynamic behavior of heated channels with boiling and supercritical fluids. *Nuclear Engineering and Design*, *237*(11), 1164–1174. doi:10.1016/j.nucengdes.2007.01.006

Archana, V., Vaidya, A. M., & Vijayan, P. K. (2015). Numerical modeling of supercritical CO_2 natural circulation loop. *Nuclear Engineering and Design*, *293*, 330–345. doi:10.1016/j.nucengdes.2015.07.030

Basu, D. N., Bhattacharyys, S., & Das, P. K. (2008). Effect of geometric parameters on steady-state performance of single-phase NCL with heat loss to ambient. *International Journal of Thermal Sciences*, *47*(10), 1359–1373. doi:10.1016/j.ijthermalsci.2007.10.010

Cammarata, L., Fichera, A., & Pagano, A. (2003). Stability maps for rectangular circulation loops. *Applied Thermal Engineering*, *23*(8), 965–977. doi:10.1016/S1359-4311(03)00027-9

Carles, P. (2010). A brief review of the thermaophysical properties of supercritical fluids. *The Journal of Supercritical Fluids*, *53*(1-3), 2–11. doi:10.1016/j.supflu.2010.02.017

Chatoorgoon, V. (2001). Stability of supercritical fluid flow in a single-channel natural-convection loop. *International Journal of Heat and Mass Transfer*, *44*(10), 1963–1972. doi:10.1016/S0017-9310(00)00218-0

Chatoorgoon, V., Voodi, A., & Upadhye, P. (2005). The stability boundary for supercritical flow in natural-convection loops Part II: CO_2 and H_2. *Nuclear Engineering and Design*, *235*(24), 2581–2593. doi:10.1016/j.nucengdes.2005.06.004

Chen, L., Deng, B. L., Jiang, B., & Zhang, X. R. (2013d). Thermal and hydrodynamic characteristics of supercritical CO_2 natural circulation in closed loops. *Nuclear Engineering and Design*, *257*, 21–30. doi:10.1016/j.nucengdes.2013.01.016

Chen, L., Deng, B. L., & Zhang, X. R. (2013c). Experimental study of trans-critical and supercritical CO_2 natural circulation flow in a closed loop. *Applied Thermal Engineering*, *59*(1-2), 1–13. doi:10.1016/j.applthermaleng.2013.05.017

Chen, L., Deng, B. L., & Zhang, X. R. (2013f). Experimental investigation of CO_2 thermosyphon flow and heat transfer in the supercritical region. *International Journal of Heat and Mass Transfer, 64*, 202–211. doi:10.1016/j.ijheatmasstransfer.2013.03.077

Chen, L., & Zhang, X. R. (2011). Simulation of heat transfer and system behavior in a supercritical CO_2 based thermosyphon: Effect of pipe diameter. *Journal of Heat Transfer-Transactions of ASME, 133*(12), 2505–2513. doi:10.1115/1.4004434

Chen, L., & Zhang, X. R. (2014). Experimental analysis on a novel solar collector system achieved by supercritical CO_2 natural convection. *Energy Conversion and Management, 77*, 173–182. doi:10.1016/j.enconman.2013.08.059

Chen, L., Zhang, X. R., Cao, S., & Bai, H. (2012). Study of trans-critical CO_2 natural convection flow with unsteady heat input and its implications on system control. *International Journal of Heat and Mass Transfer, 55*(23-24), 7119–7132. doi:10.1016/j.ijheatmasstransfer.2012.07.027

Chen, L., Zhang, X. R., Deng, B. L., & Jiang, B. (2013g). Effects of inclination angle and operation parameters on supercritical CO_2 natural circulation loop. *Nuclear Engineering and Design, 265*, 895–908. doi:10.1016/j.nucengdes.2013.06.037

Chen, L., Zhang, X. R., & Jiang, B. (2014). Effects of heater orientations on the natural circulation and heat transfer in a supercritical CO2 rectangular loop. Journal of Heat Transfer-Transactions of ASME, 132.

Chen, L., Zhang, X. R., Okajima, J., & Maruyama, S. (2013a). Thermal relaxation and critical instability of near-critical fluid microchannel flow. *Physical Review E: Statistical, Nonlinear, and Soft Matter Physics, 87*(4), 043016. doi:10.1103/PhysRevE.87.043016 PMID:23679522

Chen, L., Zhang, X. R., Okajima, J., & Maruyama, S. (2013b). Abnormal microchannel convective fluid flow near the gas-liquid critical point. *Physica A, 398*, 10–24. doi:10.1016/j.physa.2013.11.002

Chen, L., Zhang, X. R., Okajima, J., & Maruyama, S. (2013e). Numerical simulation of near-critical fluid convective flow mixing in microchannels. *Chemical Engineering Science, 97*, 67–80. doi:10.1016/j.ces.2013.04.010

Chen, L., Zhang, X. R., Yamaguchi, H., & Liu, Z.-S. (2010). Effect of heat transfer on the instabilities and transitions of supercritical CO_2 flow in a natural circulation loop. *International Journal of Heat and Mass Transfer, 53*(19-20), 4101–4111. doi:10.1016/j.ijheatmasstransfer.2010.05.030

Cheng, L., Ribatski, G., & Thome, J. R. (2008). Analysis of supercritical CO_2 cooling in macro- and micro-channels. *International Journal of Refrigeration, 31*(8), 1301–1316. doi:10.1016/j.ijrefrig.2008.01.010

Debrah, S. K., Ambrosini, W., & Chen, Y. (2013). Discussion on the stability of natural circulation loops with supercritical pressure fluids. *Annals of Nuclear Energy, 54*, 47–57. doi:10.1016/j.anucene.2012.10.015

Dimmick, G. R., Chatoorgon, V., Khartabil, H. F., & Duffey, R. B. (2002). Natural-convection studies for advanced CANDU reactor concepts. *Nuclear Engineering and Design, 215*(1-2), 27–38. doi:10.1016/S0029-5493(02)00039-0

Dostal, V., Hejzlar, P., & Driscoll, M. J. (2006). The supercritical carbon dioxide power cycle: Comparison to other advanced power cycles. *Nuclear Technology, 154*, 283–301.

Duffey, R. B., & Pioro, I. L. (2005). Experimental heat transfer for supercritical carbon dioxide flowing inside channels (survey). *Nuclear Engineering and Design*, *235*(8), 913–924. doi:10.1016/j.nucengdes.2004.11.011

Fernandez, N., Hwang, Y., & Radermacher, R. (2010). Comparison of CO_2 heat pump water heater performance with baseline cycle and two high COP cycles. *International Journal of Refrigeration*, *33*(3), 635–644. doi:10.1016/j.ijrefrig.2009.12.008

Grief, R. (1998). Natural circulation loops. *Journal of Heat Transfer*, *110*(4b), 1243–1257. doi:10.1115/1.3250624

Gupta, S., Saltanov, E., Mokry, S. J., Pioro, I., Trevani, L., & McGillivray, D. (2013). Developing empirical heat-transfer correlations for supercritical CO_2 flowing in vertical bare tubes. *Nuclear Engineering and Design*, *251*, 116–131. doi:10.1016/j.nucengdes.2013.02.048

Hasan, M. M., Afroz, A. M., & Tsubaki, K. (2008). Heat Transfer Coefficients and Pressure Drops during In-tube Condensation of CO_2/DME Mixture Refrigerant. *International Journal of Refrigeration*, *29*(6), 76–80.

He, S., Jiang, P. X., Xu, Y. J., Shi, R. F., Kim, W. S., & Jackson, J. D. (2005). A computational study of convection heat transfer to CO_2 at supercritical pressures in a vertical mini tube. *International Journal of Thermal Sciences*, *44*(6), 521–530. doi:10.1016/j.ijthermalsci.2004.11.003

Holman, J. P., & Boggs, J. H. (1960). Heat transfer to Freon 12 near the critical state in a natural circulation loop. *Journal of Heat Transfer*, *82*(3), 221–226. doi:10.1115/1.3679913

Huang, B. J. (1980). Similarity theory of solar water heater with natural circulation. *Solar Energy*, *25*(2), 105–116. doi:10.1016/0038-092X(80)90465-X

Jackson, J. D. (2001). Some striking features of heat transfer with fluids at pressures and temperatures near the critical point. *Keynote Paper for International Conference on Energy Conversion and Application (ICECA)*, Wuhan, China.

Jain, P. K., & Rizwan-uddin, . (2008). Numerical analysis of supercritical flow instabilities in a natural circulation loop. *Nuclear Engineering and Design*, *238*(8), 1947–1957. doi:10.1016/j.nucengdes.2007.10.034

Jue, Y., Yoshiaki, O., Yuiki, I., Jie, L., & Jaewoon, Y. (2007). Numerical investigation of heat transfer in upward flow of supercritical water in circular channels and tight fuel rod bundles. *Nuclear Engineering and Design*, *237*(4), 420–430. doi:10.1016/j.nucengdes.2006.08.003

Kalogirou, S. A. (2004). Solar thermal collectors and applications. *Progress in Energy and Combustion Science*, *30*(3), 231–295. doi:10.1016/j.pecs.2004.02.001

Kim, M. S., Ahn, Y., Kim, B., & Lee, J. I. (2016). Study on the supercritical CO_2 power cycles for landfill gas firing gas turbine bottoming cycle. *Energy*, *111*, 893–909. doi:10.1016/j.energy.2016.06.014

Koyama, S., Takato, N., Kuwahara, K., Jin, D., Xue, J., & Miyara, A. (2006), Experimental study on the performance of a refrigerant mixture CO2/DME system. *Proceedings of JSRAE Annual Conference*, 133-136. (In Japanese)

Kreitlow, D. B., Reistad, G. M., Miles, C. R., & Culver, G. G. (1978). Thermosyphon models for d(ownhole heat exchanger application in shallow geothermal systems. *Journal of Heat Transfer*, *100*(4), 713–719. doi:10.1115/1.3450883

Kumar, K. K., & Gopal, M. R. (2009). Effect of system pressure on the steady state performance of a CO_2 based natural circulation loop. *Applied Thermal Engineering*, *29*(16), 3346–3352. doi:10.1016/j.applthermaleng.2009.05.010

Kumar, K. K., & Gopal, M. R. (2011). Experimental studies on CO_2 based single and two-phase natural circulation loops. *Applied Thermal Engineering*, *31*(16), 3437–3443. doi:10.1016/j.applthermaleng.2011.06.029

Liao, S. M., & Zhao, T. S. (2002a). Measurements of heat transfer coefficients from supercritical carbon dioxide flowing in horizontal mini/micro channels. *Journal of Heat Transfer-Transactions of ASME*, *124*(3), 413–420. doi:10.1115/1.1423906

Liao, S. M., & Zhao, T. S. (2002b). An experimental investigation of convection heat transfer to supercritical carbon dioxide in miniature tubes. *International Journal of Heat and Mass Transfer*, *45*(25), 5025–5034. doi:10.1016/S0017-9310(02)00206-5

Linzer, W., & Walter, H. (2003). Flow reversal in natural circulation systems. *Applied Thermal Engineering*, *23*(18), 2363–2372. doi:10.1016/S1359-4311(03)00214-X

Liu, G. X., Huang, Y. P., Wang, J. F., Lv, F., & Leung, L. K. H. (2016). Experiments on the basic behaviour of supercritical CO_2 natural circulation. *Nuclear Engineering and Design*, *300*, 376–383. doi:10.1016/j.nucengdes.2016.01.021

Lomperski, S., & Cho, D. (2004). Stability of a natural circulation loop with a fluid heated through the thermo dynamic pseudo critical point. *Proceedings of ICAPP'04*.

Misale, M., Frogheri, M., DAuria, F., Fontani, E., & Garcia, A. (1999). Analysis of single-phase natural circulation experiments by system codes. *International Journal of Thermal Sciences*, *38*(11), 977–983. doi:10.1016/S1290-0729(99)00106-4

Misale, M., Garibaldi, P., Passos, J. C., & Bitencourt, G. G. (2007). Experiments in a single-phase natural circulation mini-loop. *Experimental Thermal and Fluid Science*, *31*(8), 1111–1120. doi:10.1016/j.expthermflusci.2006.11.004

Nikitin, K., Kato, Y., & Ngo, L. (2006). Printed circuit heat exchanger thermal-hydraulic performance in supercritical CO_2 experimental loop. *International Journal of Refrigeration*, *29*(5), 807–814. doi:10.1016/j.ijrefrig.2005.11.005

NIST. (2006). *NIST Standard Reference Database-REFPROP*. Version 8.0. NIST.

Ochsner, K. (2008). Carbon dioxide heat pipe in conjunction with a ground source heat pump (GSHP). *Applied Thermal Engineering*, *28*(16), 2077–2082. doi:10.1016/j.applthermaleng.2008.04.023

Onaka, Y., Miyara, A., Tsubaki, K., & Koyama, S. (2007), Performance analysis on heat pump cycle of CO2/DME mixture refrigerant. *Proceedings of JSRAE Annual Conference*, 333-336. (in Japanese)

Pioro, I. L., Khartabil, H. F., & Duffey, R. B. (2004). Heat transfer to supercritical fluids flowing in channels - empirical correlations (survey). *Nuclear Engineering and Design*, *230*(1-3), 69–91. doi:10.1016/j.nucengdes.2003.10.010

Sarkar, M. K. S., & Basu, D. N. (2017). Numerical appraisal on the suitability of supercritical condition in natural circulation loop with isothermal boundary conditions. *International Journal of Thermal Sciences*, *111*, 30–40. doi:10.1016/j.ijthermalsci.2016.08.002

Sarkar, M. K. S., Tilak, A. K., & Basu, D. N. (2014). A state-of-the-art review of recent advances in supercritical natural circulation loops for nuclear applications. *Annals of Nuclear Energy*, *73*, 250–263. doi:10.1016/j.anucene.2014.06.035

Shan, X. Y., Schmidt, D. P., & Watkins, J. J. (2007). Study of natural convection in supercritical CO_2 cold wall reactors: Simulations and experiments. *The Journal of Supercritical Fluids*, *40*(1), 84–92. doi:10.1016/j.supflu.2006.03.023

Swapnalee, B. T., Vijayan, P. K., Sharma, M., & Pilkhwal, D. S. (2012). Steady state flow and static instability of supercritical natural circulation loops. *Nuclear Engineering and Design*, *245*, 99–112. doi:10.1016/j.nucengdes.2012.01.002

TJoen, C., & Rohde, M.TJoen. (2012). Experimental study of the coupled thermo-hydraulic-nertronic stability of a natural circulation HPLWR. *Nuclear Engineering and Design*, *242*, 221–232. doi:10.1016/j.nucengdes.2011.10.055

Vijayan, P. K., Austregesilo, H., & Teschendorff, V. (1995). Simulation of the unstable oscillatory behavior of single-phase natural circulation with repetitive flow reversals in a rectangular loop using the computer code ATHLET. *Nuclear Engineering and Design*, *155*(3), 623–641. doi:10.1016/0029-5493(94)00972-2

Vijayan, P. K., Bade, M. H., Saha, D., Sinha, R. K., & Venkat, R. V. (2004). A generalized flow correlation for single-phase natural circulation loops. *Proceedings of the 17th National and 6th ISHMT/ASME Heat and Mass Transfer Conference*.

Welander, P. (1967). On the oscillatory instability of a differentially heated fluid loop. *Journal of Fluid Mechanics*, *29*(01), 17–30. doi:10.1017/S0022112067000606

Xi, X., Xiao, Z., Yan, X., Xiong, T., & Huang, Y. (2014). Numerical simulation of the flow instability between two heated parallel channels with supercritical water. *Annals of Nuclear Energy*, *64*, 57–66. doi:10.1016/j.anucene.2013.09.017

Yadav, A. K., Gopal, M. R., & Bhattacharyya, S. (2012). CO_2 based natural circulation loops: New correlations for friction and heat transfer. *International Journal of Heat and Mass Transfer*, *55*(17-18), 4621–4630. doi:10.1016/j.ijheatmasstransfer.2012.04.019

Yamaguchi, H., Sawada, N., Suzuki, H., Ueda, H., & Zhang, X. R. (2010). Preliminary study on a solar water heater using supercritical carbon dioxide as working fluid. *Journal of Solar Energy Engineering-Transactions of ASME*, *132*(1), 101–106. doi:10.1115/1.4000350

Yang, J., Oka, Y., Ishiwatari, Y., Liu, J., & Yoo, J. (2007). Numerical investigation of heat transfer in upward flow of supercritical water in circular tubes and tight fuel rod bundles. *Nuclear Engineering and Design*, *237*(4), 420–430. doi:10.1016/j.nucengdes.2006.08.003

Yoshikawa, S., Smith, R. L. Jr, Inomata, H., Matsumura, Y., & Arai, K. (2005). Performance of a natural convection circulation system for supercritical fluids. *The Journal of Supercritical Fluids*, *36*(1), 70–80. doi:10.1016/j.supflu.2005.02.007

Zappoli, B. (2003). Near-critical fluid hydrodynamics. *Comptes Rendus. Mécanique*, *331*(10), 713–726. doi:10.1016/j.crme.2003.05.001

Zhang, X. R., Chen, L., & Yamaguchi, H. (2010). Natural convective flow and heat transfer of supercritical CO_2 in a rectangular circulation loop. *International Journal of Heat and Mass Transfer*, *53*(19-20), 4112–4122. doi:10.1016/j.ijheatmasstransfer.2010.05.031

Zhang, X. R., & Yamaguchi, H. (2008). An experimental study on evacuated tube solar collector using supercritical CO_2. *Applied Thermal Engineering*, *28*(10), 1225–1233. doi:10.1016/j.applthermaleng.2007.07.013

Zhang, X. R., Yamaguchi, H., Uneno, D., Fujima, K., Enomoto, M., & Sawada, N. (2006). Analysis of a novel solar energy powered Rankine cycle for combined power and heat generation using supercritical carbon dioxide. *Renewable Energy*, *31*(12), 1839–1854. doi:10.1016/j.renene.2005.09.024

Zhang, X. R., Zhang, Y. L., & Chen, L. (2014). Experimental study on solar thermal conversion based on supercritical natural convection. *Renewable Energy*, *62*, 610–618. doi:10.1016/j.renene.2013.08.025

Zhao, M., Gu, H. Y., & Cheng, X. (2014). Experimental study on heat transfer of supercritical water flowing downward in circular tubes. *Annals of Nuclear Energy*, *63*, 339–349. doi:10.1016/j.anucene.2013.07.003

Zvirin, Y. (1981). A review of natural circulation loops in pressurized water reactors and other systems. *Nuclear Engineering and Design*, *67*(2), 203–225. doi:10.1016/0029-5493(82)90142-X

KEY TERMS AND DEFINITIONS

Critical Diverges: The thermal and transport property varies and show great diverges process when a fluid status is near the critical region.

Heat Transfer Enhancement: The incensement of heat transport of a surface, body, apparatus. Usually it is a target to be achieved by improving the thermal and mechanical designs, or the choice of working fluids.

Natural Convection: The convection phenomena induced by buoyancy forces under gravity. Usually it indicates the ones different from forced convection under outer forces. In supercritical fluids, natural convection flow is strong as the fluid density changes are evident under temperature variations.

Stability Transition: The transitions among stable flow, quasi-stable flow, unstable flow, etc. It is critical for supercritical fluids for the definition of stability evolutions and changes.

Stability Map: The map that states the stability region, transition lines, and parameters for the system stability conditions.

System Design: The concept and construct of a targeted system, which may include the geometric design, fluid design, operation design, etc.

Solar Thermal: The utilization of solar energy by converting it into thermal kinds, which is usually achieved by a collector system.

Supercritical Fluid: The fluid state that above some critical temperature and pressure point, where the fluid will never be liquefied by pressurization process. The fluid status is special as the solubility, flow viscosity, thermal conductivity, compressibility will have large changes during the transitions.

Supercritical Nuclear Cooling: The application of supercritical fluids in nuclear cooling and heat transportation process, which is possible for the secondary loop circulation system. The concept of supercritical fluids based nuclear system is now one of the trends for future nuclear systems.

Supercritical Refrigeration: The concept of using supercritical fluids in refrigeration systems. As a working fluid, supercritical state connects high pressure region and the low pressure region (the gas phase) so as to achieve cooling process. Some designs also use the throttling process, or jet flow of sudden-expansion flow from supercritical status to achieve low temperature environment.

Supercritical Heat Exchanger: The heat exchanger specially designed for supercritical fluids, which may be relatively compact as the supercritical fluid heat transfer rate is relatively high and it is operated under high pressure.

Thermosyphon: Another name for natural circulation loop (NCL), which is more focused on the thermal transportation aspect of a loop system utilizing natural circulation working fluid as the basic transport mechanism.

APPENDIX: NOMENCLATURE

A: Area (m^2)
C_p: Specific heat at constant pressure ($J \cdot kg^{-1} \cdot K^{-1}$)
C_p: Specific heat at constant volume ($J \cdot kg^{-1} \cdot K^{-1}$)
C_μ, $C_{1\varepsilon}$,
$C_{2\varepsilon}$, $C_{3\varepsilon}$: Parameters in RNG model equations
d or D: Width/diameter (m, μm)
E: Energy ($J \cdot kg^{-1}$)
g: Gravitational acceleration ($m^2 \cdot s^{-1}$)
G_k: Turbulent kinetic energy due to velocity gradients
G_b: Turbulent kinetic energy due to buoyancy
Gr: Grashof number
Gr_m: Modified Grashof number
Gr_m^* : Modified Grashof number by pseudo-critical point values
h: Mean heat transfer coefficient, enthalpy ($W \cdot m^{-2} \cdot K^{-1}$, $J \cdot kg^{-1}$)
H: Length of vertical pipes
k: Thermal conductivity ($W \cdot m^{-1} \cdot K^{-1}$)
L: Length (m)
L_0: Total length of a horizontal pipe (m)
L_1: Adiabatic pipe length on horizontal pipe (m)
$LMTD$: Logarithmic mean temperature difference (K)
$\dot{m}$: Mass flow rate ($kg \cdot h^{-1}$)
N_G: Non-dimentional loop geometric parameter
Nu: Nusselt number
p: Pressure (Pa or MPa)
P: External surface force (Pa)
P: Pressure of the fluid (Pa or MPa)
Pr: Prandtl number
Pr_t: Turbulent Prandtl number
q: Heating power input (W)
$\dot{Q}$: Heat transfer rate (W)
R: Parameter in RNG model equations
T: Temperature, (°C or K)
Re: Reynolds number
t: Time (s)
u: Velocity ($m \cdot s^{-1}$)
U: Voltage (V)
$\overline{V}$: Dimensional velocity ($m \cdot s^{-1}$)
V: Non-dimensional velocity
x: x coordinate location
X: Dimensionless axial coordinate ($X=x/L_0$)
y: y coordinate location

Greek Symbols

ρ: Density (kg·m^{-3})
μ: Viscosity (kg·m^{-1}·s^{-1})
α: Thermal diffusivity; thermal conductivity parameter (m^2·s, -)
α_k, α_s: Parameters in RNG model
β: Thermal expansion coefficient; parameter in RNG model equations (K^{-1}, -)
ε: Turbulence dissipation rate (J·kg^{-1})
μ_t μ_{eff}: Viscosity parameter in RNG model
η: Parameter in RNG model
Φ: Dissipation function, $\left(\equiv\left(\bar{\tau}\cdot\nabla\right)\cdot\bar{V}\right)$
$\bar{\tau}$: Shear tensor, $\left(\equiv\begin{pmatrix}\tau_{xx} & \tau_{xy}\\ \tau_{yx} & \tau_{yy}\end{pmatrix}\right)$

Subscripts

0: Initial state
b: Bulk fluid
c: Upper cooling of the pipe
h: Upper heating of the pipe
in: Inlet
out: Outlet
pc: Pseudo-critical
r : Radial direction
ref: Reference value; bulk value
T: Total length
w: Wall
x: Specific axial location
c: Upper cooling of the pipe
h: Upper heating of the pipe
Δz: Vertical height

Chapter 6
Supercritical Natural Circulation Loop:
A Technology for Future Reactors

Dipankar Narayan Basu
Indian Institute of Technology Guwahati, India

Milan Krishna Singha Sarkar
Indian Institute of Technology Guwahati, India

ABSTRACT

Supercritical natural circulation loop is a compelling technology for cooling of modern nuclear reactors, which promises enhanced thermalhydraulic performance in a simple design. Being a new concept, related knowledge base is relatively thin and involves several conflicting theories and controversies. Present chapter summarizes the observation till date, starting from the very fundamentals. The phenomenon of natural circulation and suitability of supercritical medium as working fluid are discussed in details. Different methods of analyses, including analytical, simple 1-D numerical and multidimensional computational codes, as well as experimental, are elucidated. A comprehensive discussion is presented about the effect of various geometric and operating parameters on the system behavior, from both thermalhydraulic and stability point of view. Finally, a few recommendations are included about the operation of such loops and future direction of research.

INTRODUCTION

The term *circulation*, in engineering sense, refers to the transport of a fluid through a closed circuit. When such a transmission takes place at the absence of any designated prime mover and solely because of a favorable density gradient across the loop under the influence of a body force, it is referred to as *natural circulation*. Gebhart (1973) employed the term *internal natural convection* to describe *flows arising in a body of fluid contained in a cavity or completely bounded by surfaces*. The most distinctive characteristics of natural circulation systems are their high sensitivity to the operative conditions and predisposition towards instability, predominantly due to the resilient pairing between flow and tem-

DOI: 10.4018/978-1-5225-2047-4.ch006

perature fields. Considerably lower circulation rate, in contrast to the loops with prime movers, ensures identical orders of momentum and viscous dissipations, with negligible inertial effect, and pivotal role of the prevailing body force field, commonly gravity. Accordingly, the flow field appears as an implicit function of numerous constituents encompassing geometry, boundary conditions and location, and hence is implausible to envisage without a comprehensive exploration.

Natural circulation loops (NCLs), despite the mathematical intricacy, proposes a convenient route of energy and species transport. The density differential can be accomplished either by introducing a lighter phase into the primary fluid or by modulating fluid temperature through complementary energy interactions with the surrounding in different segments of the flow path. The later contrives a proficient option of energy transport from a high-temperature source to a low-temperature sink, without them in direct contact. Warmer fluid from the source can rise to the sink owing to buoyancy, to dispense the accrued energy there, and return as a cooler medium, prepared to accumulate energy from the source again. Therefore, it is obligatory to place the sink at a higher elevation than the source to establish the favorable buoyancy field, as is shown in Figure 1, and that generally remains the only constraint for an NCL configuration. Such simplicity in construction to suit any physical silhouette and enhanced reliability due to the absence of rotating machinery have stimulated innumerable engineering innovations, ranging from gigantic power cycles, nuclear plants and automobiles, through domestic refrigerators, chemical processes and solar heaters, to miniature chip cooling, with undisputed success.

It is improbable to converge on any initiation period for commercial utilization of NCLs as heat transport systems. One of the pioneering application can be identified in early-1950s for turbine rotor cooling. Several arrangements have historically been proposed and developed with varying nature of the working medium, shape and imposed body force. A complete list of classification is shown in Figure 2. While the other factors are typical to a situation, the selection of operating fluid is generally governed by

Figure 1. Schematic representation of a generalized natural circulation loop, with sink mounted at a higher elevation than the source

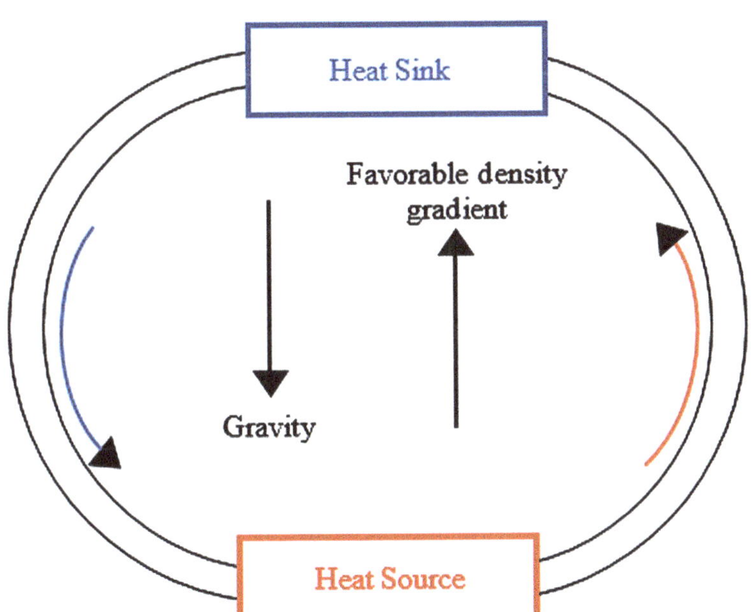

Figure 2. Classification of natural circulation loops based on various criteria

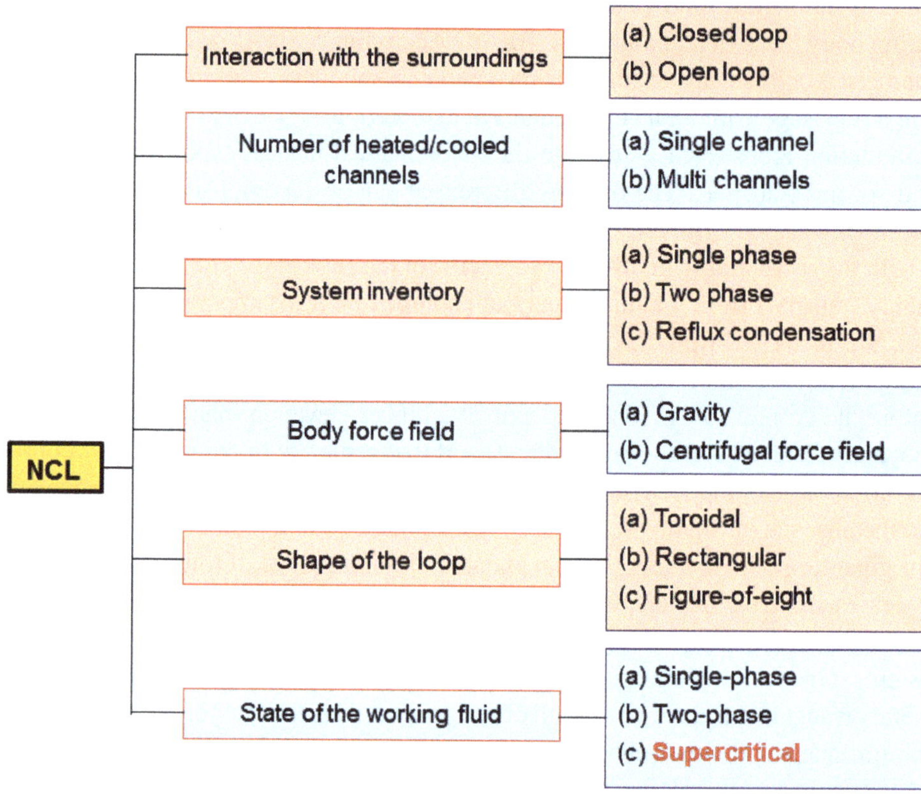

the operation convenience and range of parameters explored. Single-phase, mostly liquid-based, loops are favored in PWRs, solar heaters and electronic cooling applications; whereas two-phase systems, comprising distinct boiling and condensing sections, are prevalent in power cycles, refrigerators and heat pipes. Both are well-explored devices and under scrutiny since the very inception of the concept of NCL. The supercritical NCL, however, is a relatively new concept, with the pioneering research paper being publicized only in late 1990s. Significant count of theoretical and experimental studies have followed over next one and half-decades, only to propose several contradicting theories about the system performance, on both thermalhydraulic and stability aspects, leaving a reasonably thin knowledge base. Reported observations not being in consensus makes it very difficult for the beginners to grasp the initial concepts, which prepares the backdrop of the present chapter. Following segment will discuss the suitability of supercritical fluid as the working medium in NCL and also narrate the development. The thermalhydraulic behavior will be analyzed thereafter, followed by discussion on stability aspects, the objective being gradually introducing a reader with basic engineering knowledge into the topic of supercritical NCLs.

BACKGROUND

In order to exploit the enormous energy potential of nuclear fission, modern reactor designs are increasingly adopting higher temperature levels, to expand the thermal efficiency limit. While the single- and two-phase systems are constrained by the saturation temperature criterion, advances in material research allows sustained operation at high temperatures, approaching the borders of 800°C or even higher, and also at higher pressures. That encouraged the scientists to explore the supercritical state and led to the conceptual development of the supercritical water reactor (SCWR), which is an integral constituent of the generation-IV initiative of reactor design. While the US reference design of SCWR primarily adopts forced-flow channels, a few schemes have also recommended embracing NCL as the primary coolant circulation mode, such as the high-temperature CANDU reactor (Bushby et al., 2000) and B-500 SKDI design (Silin et al., 1993). Implementation of natural circulation for reactor core cooling during normal, as well as accidental, condition promises a significant gain in thermodynamic efficiency (~44%). The once-through design also eliminates bulky components like separating drum, dryer and recirculation channels, warranting a compact structure. The possibility of employing a variety of fluid to suit the requisite temperature range makes supercritical loop a lucrative technology.

Single-phase liquid is characterized by large thermal conductivity compared to vapor and quite regularized heat transfer behavior. A boiling loop, on the contrary, can exhibit substantial volumetric expansion, yielding considerably large flow rates. Supercritical fluid can integrate the advantages of both the mediums. While the change in density around the pseudocritical point is comparable to liquid-to-vapor phase change, Prandtl number exhibits a peak there, registering excellent heat transport characteristics as a consequence. Accordingly, the supercritical NCLs (SCNCLs) are characterized by large flow rates and significant heat transfer coefficient, and are anticipated to demonstrate traits of both single- and two-phase NCLs. Vijayan et al. (2010) reported the steady-state behavior of SCNCL to be similar to the two-phase loops, while its stability characteristics followed the single-phase version. However, the drastic variation in the thermophysical properties of any supercritical fluid is confined to a small region around the pseudocritical point, typically covering a temperature span of about 30°C, with fairly monotonous representation away from the pseudocritical temperature (T_{pc}). Consequently, the benefits of supercritical fluid in an NCL can be harnessed only by operating around that particular zone and hence the performance of SCNCL can deviate a lot depending on the regime of operation.

Following the conceptual development of CANDU-X reactor around 1998, which was proposed to operate with supercritical water at 25 MPa pressure level as coolant in both forced and natural circulation mode, Chatoorgoon (2001) was the first to propose an analytical model of SCNCL to envisage the stability and thermalhydraulic characteristics. While most of the conclusions from their simplified approach has been found to veer from real-world, their study served as an excellent launching pad for research on SCNCL. Several semi-analytical and computational studies followed, with limited number of experimental efforts as well. While the initial efforts concentrated mostly on the overall performance, gross discrepancy between the theoretical and experimental results were observed. That shifted the focus towards the local behavior of supercritical fluid in different segments of the loop, which has consequently been found to play a major role on the overall behavior. More on this issue will be elaborated in the following section.

ONE-DIMENSIONAL ANALYSES

Due to the absence of any exclusive driving mechanism, fluid motion in any NCL is instigated following the development of the density gradient, which is the consequence of energy diffusion from the interacting boundaries to the static bulk fluid. Zvirin (1985) hypothesized the motion in any NCL as a combination of four flow instabilities, namely,

- Development of local Rayleigh-Bénard cells beyond a critical Rayleigh number.
- Onset of global motion at higher powers owing to the rupture of such cells.
- Fluctuations in steady flow with growing amplitude of oscillations.
- Multiple steady-states leading to metastable instability.

While the operation of NCL is initiated by the favorable instability of the second kind, the operator always needs to be mindful about the appearance of the third and fourth varieties. Accordingly, the flow and heat transfer analyses of such loops are mandatorily accompanied by stability appraisal and SCNCLs are no exceptions. In fact, the rapid property variation of supercritical fluid over certain range of temperature has added a new dimension to the concerned stability behavior. Therefore, present section will discuss both the thermalhydraulic and stability aspects simultaneously.

Analytical Model

It is extremely difficult to simulate the supercritical heat transfer in a simple analytical framework, primarily because of the highly nonlinear relationship between enthalpy and temperature, and so other thermophysical properties, around the pseudocritical point. Chatoorgoon (2001) assumed point heat source and sink in a rectangular structure, in order to circumvent enthalpy distribution along heated and cooled walls. Noting that the traditional two-phase loops are unstable along the negative slope of the internal characteristics curve (pressure drop versus flow rate), he postulated the stability threshold to be the solution of

$$\frac{\partial G}{\partial Q} = 0 \tag{1}$$

where G refers to the steady-state mass flux and Q is the imposed heater power. Validity of this particular approximation cannot be substantiated and hence this criterion can solely be viewed as a first-hand estimate.

The one-dimensional momentum conservation equation can be expressed as,

$$\frac{\partial G}{\partial t} + \frac{\partial}{\partial s}\left(\frac{G^2}{\rho}\right) = -\frac{\partial p}{\partial s} - \rho g \cos\beta - \frac{f}{2d}\left(\frac{G^2}{\rho}\right) \tag{2}$$

Here the inclination β is zero for vertically upward motion and f denotes the effective friction factor, which is dependent on fluid viscosity. Being a temperature-dependent property, viscosity, and hence the friction factor, varies along the flow direction indicated by the coordinate s. Adopting the simplified configuration deliberated by Chatoorgoon (2001) with equal strength of the source and sink and assuming adiabatic boundaries elsewhere, fluid properties can be lumped separately over three segments, as is shown in Figure 3. Subsequently integrating the steady-state version of Equation 2 over the flow path and noting the pressure drop to be zero over the loop, a solution can be obtained as follows.

$$G^2 = \frac{2gdH\left(\rho_1 - \rho_2\right)}{\frac{f_1L_1 + f_3L_3}{\rho_1} + \frac{f_2L_2}{\rho_2}} \tag{3}$$

H being the height of the loop and the terms with subscripts 1, 2 and 3 refers to the parameters at the corresponding segments. Therefore, depending on the magnitude of the geometrical variables, flow rate can attain a peak, before declining with further increase in power. A typical profile is shown in Figure 4 following the configuration of Chatoorgoon (2001), which is consistent with the concomitant physics. Initial increase in power is expected to enhance the enthalpy difference across the source and cause a rise in driving buoyancy, thereby leading to the flow enhancement. The frictional resistance, however, also keep on increasing with the flow rate and becomes the dominating factor beyond a certain limit. That explains the reduction in flow at higher powers.

Applying the postulate of Equation 1 and approximating density as power-law function of enthalpy, the maximum mas flux (G_m), which is proposed to correspond to the stability threshold, can be correlated as,

Figure 3. Simplified geometry of open NCL with point heat source and sink
Considered for the analytical model following Chatoorgoon (2001).

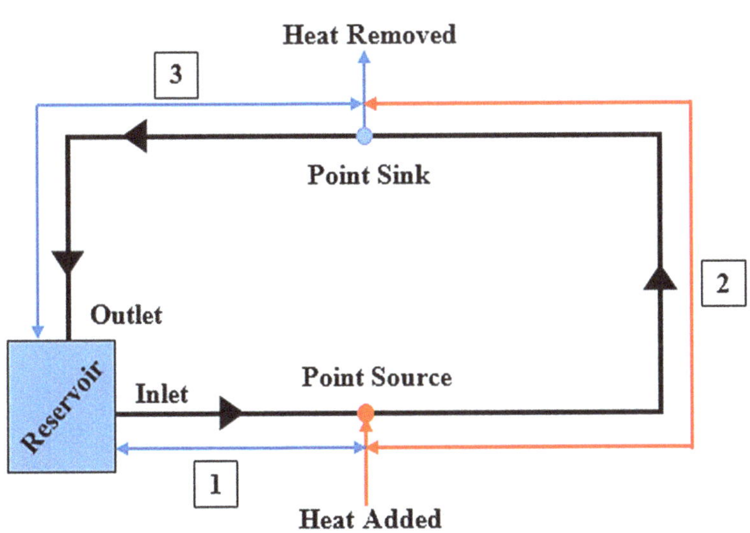

Figure 4. Typical mass flow rate variation with power for an SCNCL

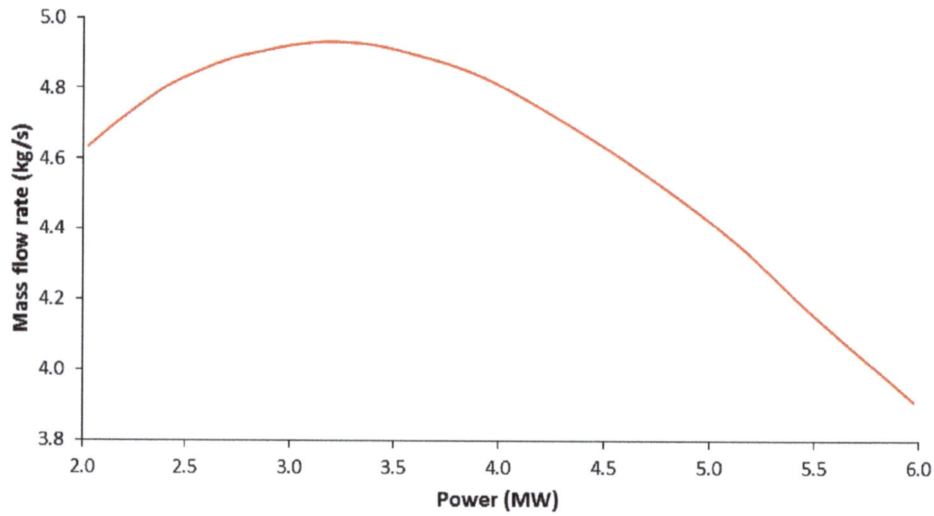

$$\left(\frac{G_m}{\rho_2}\right)^2 = \frac{2gdH}{f_2 L_2} = \xi^2 \tag{4}$$

Corresponding source power (Q_b), termed as the *bounding power*, can therefore be estimated by simple energy balance across the heater as,

$$Q_b = G_m A \Delta h_h \tag{5}$$

Here A is the cross-sectional area of the channel, considered to be uniform, and Δh_h is the enthalpy rise experienced by the fluid during its passage through the heater. At power values lower than Q_b, i.e., on the positive slope of the flow-power profile, any increase in flow at fixed power will lower the enthalpy at heater exit and hence cause a rise in density thereof. That would tend to suppress the imposed perturbation itself, reverting the system back to steady-state. At higher powers ($> Q_b$), however, such enhancement in the outlet density would amplify the disturbance, guiding towards instability. Fluid density in the hot arm (ρ_2) is inversely proportional to the corresponding enthalpy (h_2) and hence can be viewed to have similar relationship with source power. Therefore flow-ρ_2 characteristic can also be employed in lieu of the flow-power profile and any state point on the positive slope there can be designated as unstable.

For the sake of generalization, requisite parameters can be non-dimensionalized as,

$$G_m^* = \frac{G_m}{\xi \rho_1},$$
$$Q_b^* = \frac{Q_b}{A G_m h_1}, \tag{6}$$
$$R^* = \frac{\rho_2}{\rho_1}$$

Above definitions can be utilized to compare the performance amongst dissimilar configurations and also to generate generalized stability map, which can be defined as the locus of the stability threshold. A comparison of working fluids was attempted by Chatoorgoon et al. (2005) extending this framework to loops with distributed source and sink and the stability response of CO_2 was reported to be similar to water, which is an important observation. The critical properties for CO_2 (7.377 MPa and 304.1 K) are far easier to achieve in contrast to water (22.064 MPa and 647.1 K) and therefore CO_2 can be employed as a lab-scale model fluid to simulate water.

The analytical approach is quite simple and can provide a closed-form solution, but with several limitations. In an effort to simplify the mathematical model, a semi-lumped approach has been followed, thereby completely neglecting the fluid property variation within heater, even with the distributed sources. The postulate of Chatoorgoon presented in Equation 1 was also proposed more in hindsight and without any physical justification. Use of an equation-of-state (EOS) sans pressure-dependence is definitely not the correct choice for supercritical fluid, which exhibits significant compressibility, particularly for pressures just above the critical value. The analytical solution, therefore, better be viewed more as a starting estimate for more-involved techniques.

The configuration discussed so far is of an open loop with continuous inflow and outflow, allowing knowledge of inlet conditions. However, such a geometry is quite impracticable in reactor cooling and closed loops (Figure 5) are universally preferred, where the state of fluid at no location can be comprehended a priori, necessitating iterative solution and rendering the analytical approach invalid. Adoption of the computational techniques is more judicial here, which is discussed in the next section.

Finite Difference Computation: Steady-State Thermalhydraulics

It is a well-established fact that the frictional and heat transfer relations commonly employed for subcritical fluids are often not sufficient for supercritical ones. Heat transfer coefficient may suffer change of the order of two to three decimals with variation in the level of prevailing temperature, and therefore the choice of concerned correlation must depend on circumstances. Property relations are usually complicated functions of measurable quantities. For example, Span and Wagner (1996) defined Helmholtz energy of supercritical CO_2 as polynomial of density and temperature, and other properties as derivatives of the same. It is impossible to account for so many factors in the simplified structure of lumped-parameter approach. On the contrary, numerical procedures like finite difference method can conveniently incorporate such intricacies by dividing the entire computational domain into large number of segments and treating the condition in each independently.

The time-dependent equations for conservation of mass, momentum and energy for any fluid in 1-D can be summarized as,

Figure 5. Schematic view of a rectangular (a) open loop and (b) closed loop

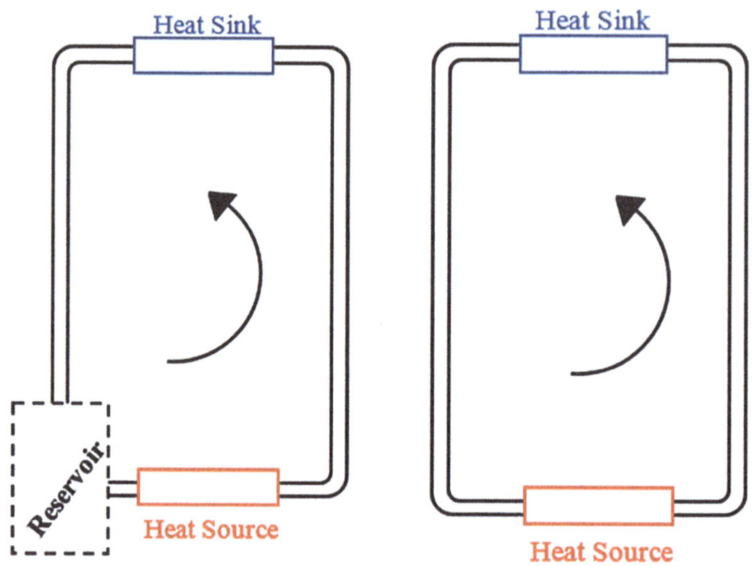

$$\frac{\partial \rho}{\partial t}+\frac{\partial G}{\partial s}=0 \tag{7a}$$

$$\frac{\partial G}{\partial t}+\frac{\partial}{\partial s}\left(\frac{G^2}{\rho}\right)=-\frac{\partial p}{\partial s}-\rho g\cos\beta-C_k\left(\frac{G^2}{\rho}\right) \tag{7b}$$

$$\frac{\partial}{\partial t}\left[\rho\left(h+\frac{u^2}{2}\right)\right]+\frac{\partial}{\partial s}\left[G\left(h+\frac{u^2}{2}\right)\right]-Gg\cos\beta=q''' \tag{7c}$$

Here the coefficient C_k characterizes frictional, as well as, the minor losses, such as at the bends of the loop. If the minor losses are neglected, Equation 7b reduces to Equation 2. The changes in kinetic and potential energies can be ignored in most of the loops, considering the large amount of heat exchange associated with reactor cooling, thereby simplifying Equation 7c as well. Contributions like fluid conduction and viscous dissipation are also omitted from Equation 7c. Proper EOS must also be considered, alongside the boundary conditions. For an open loop, both ends of the loop can be assumed to be connected to a large reservoir with specified pressure and temperature. That allows known inlet and outlet pressures and inlet temperature as boundary conditions, along with the source- and sink-side information. For a closed loop, however, state at any location is an unknown and can be appraised only through iterations.

In order to realize the steady-state solution, transient terms can be dropped from Equations 7a–c, leading to uniform mass flux throughout the loop, as $G_{ss}\neq G(s)$. Total loop needs to be divided into m-number of small cells, with node points located at the cell edges, as is shown in Figure 6, and the first and $(m+1)^{th}$ node coinciding to indicate any specific location, such as the inlet plane for an open loop. All thermophysical properties are assumed to be lumped at the nodes, facilitating property variation only between the nodes. Conservation equations are integrated between two neighboring nodes to convert the partial differential equations into algebraic ones. For example, discretized energy equation for the $(i+1)^{th}$ node, omitting kinetical and potential energy changes, can be presented as,

$$\left(h_{ss}\right)_{i+1} = \left(h_{ss}\right)_i + \frac{q_i''' + q_{i+1}'''}{2G_{ss}} \tag{8}$$

Similarly, the discretized momentum equation can be used to estimate local pressure $\left(p_{ss}\right)_{i+1}$ and the local density can subsequently be calculated through EOS, noting $\rho = \rho\left(p, h\right)$. The entire calculation, however, must follow an iterative approach, initiated by a guessed mass flux and knowledge of enthalpy/temperature at the first node to allow forward spatial marching. The later item is a known quantity for an open loop, whereas needs to be approximated in closed systems, only to be corrected by evaluating enthalpy profile at all nodes and confirming the continuity of temperature at the reference plane. Final constraint is provided by the zero-pressure-drop criterion over the loop, which provides the desired convergence of mass flux. A flow chart detailing the steady-state solution procedure for an open loop is presented in Figure 7.

Figure 6. Typical nodalization adopted for 1-D finite-difference simulation of SCNCL

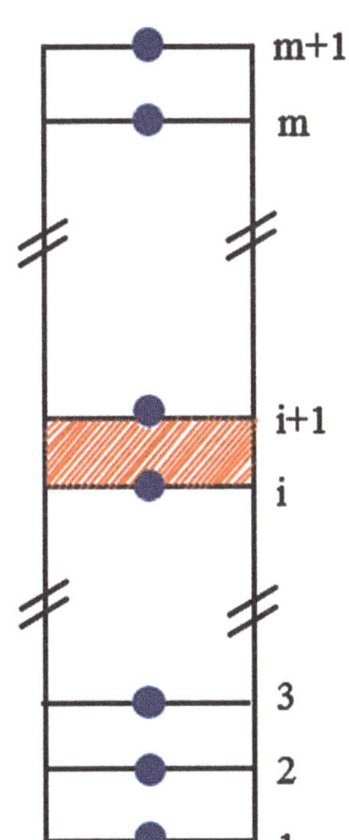

Figure 7. An example of the flow of numerical control during steady-state simulation of SCNCL following finite-difference method

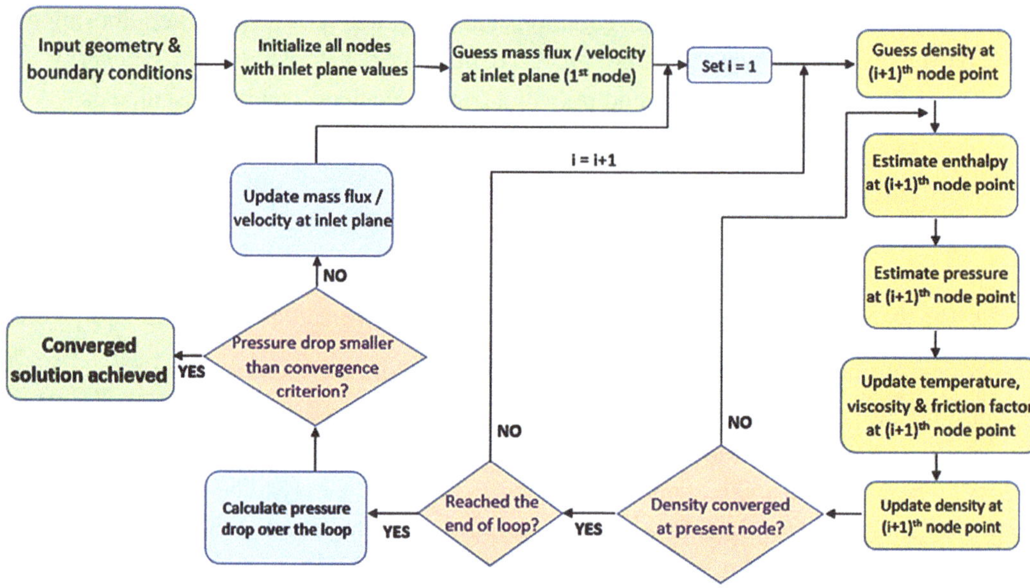

The precision of steady-state solution may depend on both the adopted discretization scheme and meshing technique. While the dependence on the former is generally weak, it is mandatory to ensure the grid-independent nature of the profiles by sequentially increasing the number of nodes in the domain and comparing the corresponding observations. Jain and Rizwan-udding (2008) found a grid spacing of 0.1 m to be sufficient for a CO_2-loop having 32 m of length at 8 MPa system pressure, as any further reduction in spacing was found to have insignificant impact. Due to the large-scale property variation, optimum number of nodes varies with the fluid and even with the operating condition for the same fluid. Therefore it is suggested to check for the effect of mesh refinement with any change in simulation parameters. The choice of EOS is also important and, as the properties at each node are handled individually in a finite-difference framework, most updated property relations can expediently be integrated into the numerical scheme. NIST database, incorporating several refrigerants and hydrocarbons, is a popular example. Use of exclusive relations, such as the one proposed by Span and Wagner (1996) for CO_2, is also not uncommon.

The nature of flow-power profile predicted through numerical simulation is, in general, quite similar to the one shown in Figure 4, albeit with possible quantitative deviation. Chatoorgoon (2001) reported a reduction in the bounding power corresponding to the peak flow rate with the analytical approach. In other words, the lumped-parameter approach produces a more conservative estimate. Vijayan et al. (2010) also reached a similar conclusion, noting that the reduction in flow rate corresponded to the pseudocritical region, which experiences a rapid deterioration in thermal expansion coefficient. A general correlation was proposed by Swapnalee et al. (2012) for estimating the steady-state flow rate by relating dimensionless density and dimensionless enthalpy in a sigmoidal curve form as,

$$\rho^* = A + \frac{B}{1 + exp\left(1.518h^* + 0.5689\right)} \tag{9}$$

$$\rho^* = \frac{\rho}{\rho_{pc}} \; h^* = \beta_{pc}\left(h - h_{pc}\right)$$

where, $A = 0.15704$ and $B = 2.4785$. Combining the cyclic integral of the momentum equation with energy balance and non-dimensionalizing the concerned form, their final relation was expressed as,

$$Re_{ss} = C\left[\frac{Gr_m}{N_G}\right]^r \tag{10}$$

Here N_G is a geometrical parameter (= L / d for an uniform-diameter loop) and experimentally-fitted values are used for the constants C and r.

Numerical Stability Appraisal

Towards numerical evaluation of stability characteristics, both the linear and non-linear methods are in use. While the former adopts the frequency-domain approach mostly to identify the stability threshold, the other one operates in the time-domain and is capable of assessing the nature of fluctuations beyond the stability margin. Linear approach is generally easier to use and computationally more economical, as it can configure the stability map within a shorter span of time. It, however, is not capable of differentiating between different forms of instabilities, which necessitates the nonlinear approach.

The linear approach requires the transient version of the conservation equations to be perturbed over the steady-state and linearized assuming the perturbation to be infinitesimally small. Eigenvalues are estimated from the resultant set of linear equations. Real parts of the eigenvalues reveal the stability behavior. If any of the eigenvalues possess a positive real part, the system is unstable, whereas a stable system will have only negative real parts. Purely imaginary eigenvalues characterize a neutrally stable loop. Common tools such as Nyquist plots and root-locus technique are frequently employed alongside the linear approach. SUCLIN developed by Sharma et al. (2010) is a good example of the linear stability code. They also elucidated the possible limitations of this approach, as the extent of unstable zone was under-predicted. It is not possible to impose perturbation on all fluid properties and also on constitutive relations employed for friction factor and heat transfer coefficient. SUCLIN also neglected the perturbation in specific volume due to pressure oscillations, in order to facilitate linearization. Still such a code can provide a prompt estimate about the approximate stability margin and relative influences of different loop variables, which can further be refined through the nonlinear method.

Nonlinear stability analysis concerns the transient solution of the complete set of relevant conservation equations, such as the one presented in Equations 7a-c, united with suitable EOS and correlations, over a desired period of time. It is commonly initiated with the knowledge of the steady-state solution. A small perturbation is imposed on any significant variable, such as the flow rate or velocity at a certain location, and the resultant temporal evaluation is monitored following a time-marching. The imposed disturbance affects the variables in the direction of motion with time and, owing to the loop geometry, eventually modifies itself, thereby dynamically altering the very reason of its appearance and activating a self-correcting mechanism. The disturbances are found to subside with time for a stable system, returning the loop to its initial operating point. An unstable loop is characterized by oscillation growth and possible reversal in the flow direction. Neutrally-stable behavior, epitomized through oscillations of constant amplitude beyond a period of initial development, corresponds to the stability threshold.

Time-domain approach is computer-intensive and more time consuming, as simulation over a significant period of time is essential before any conclusion can be drawn about the nature of system stability. So, it is quite common to proceed with frequency-domain codes for realizing the stability map, whereas the time-domain codes are reserved for more rigorous appraisal of the nature of oscillations at the unstable points. As both the methods are initiated from the same set of conservation equations, they are expected to yield similar qualitative predictions, but may have quantitative disparity, as was reported by Sharma et al. (2010). The nonlinear approach allows the incorporation of more practical factors, such as the advanced property relations, complicated constitutive relations and intricate system geometry, and hence are expected to lead towards more feasible perspective of the loop.

The accuracy of the nonlinear approach, however, is strongly reliant on the choice of time-step. As a time-marching procedure is followed, it is indispensable to have a converged flow and thermal field at every time step, employing the solution available for the previous step as a possible guess. A larger time-step may introduce numerical diffusion into the scheme, thereby artificially stabilizing the oscillations and portraying an erroneous stability response. Tilak and Basu (2015) presented a very good assessment on the impact of such incorrect choice for geometry similar to Chatoorgoon (2001). Their observations are presented in Figure 8. As is evident, for the same power input, 0.1 s time-step predicts highly stabilized behavior, whereas a choice of 0.0575 s leads to unstable fluctuations. A choice of even smaller time-step was found to produce larger oscillation growth, accordingly giving the same qualitative prediction, but at the expense of radical increase in the simulation time expended. Choice of time-step is also influenced by the selected grid spacing and the prevailing boundary conditions, as a larger flow rate may require closely-spaced nodes and smaller time-step. Jain and Rizwan-uddin (2008) suggested a Courant number criterion of close to unity, along with an absolute pressure convergence of 10^{-6} MPa, as a possible guideline. Their FIASCO code opposed the observations from the SPORTS code developed by Chatoorgoon (2005) about the positioning of the stability threshold. Both the analytical and numerical results of Chatoorgoon (2005) indicated the neutrally-stable state to converge with the maxima on flow-power profile with 95% accuracy, which is in contrary to the practice. Jain and Rizwan-uddin (2008) attributed such error to the dissipative and dispersive effect introduced by wrong selection of the time-step.

The role of adopted numerical scheme on the stability prediction has found significant interest for single-phase NCLs (Basu et al., 2014). Several schemes were proposed, mostly to reduce the diffusive impact, and use of low-diffusion schemes was suggested in general. Use of explicit upwind scheme with detailed spatial discretization or second-order technique was also recommended, besides the effort to minimize the truncation error. Unfortunately, such concentrated effort for SCNCL is lacking, as per the open literature. Adoption of implicit scheme, with control-volume-based discretization in space and forward-marching in time, is found in codes like SPORTS and FIASCO. Nonlinear code NOLSTA developed by Sharma et al. (2010) employed upwind discretization. Lack of reliable experimental data makes it difficult to comment on the success of any particular scheme, unless a comprehensive comparative analysis is accessible.

Figure 8. An example of the impact of time-step on the transient prediction of a rectangular SCNCL
Tilak & Basu, 2015.

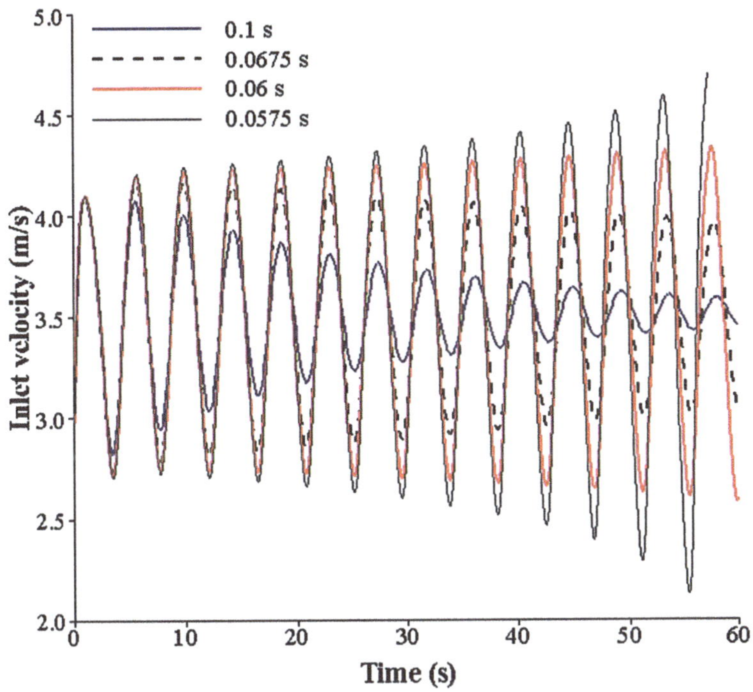

MULTIDIMENSIONAL ANALYSES

One-dimensional models can provide a decent insight into the functionalities of an SCNCL and flow features. However, the heat transfer characteristics can strongly be influenced by the local behavior. Only the fluid present within a thin layer in the vicinity of the heated wall can attain the wall temperature, whereas the bulk near centerline remains well below that. Considering the large density variation experienced by the supercritical fluid, near-wall fluid can have noticeably lower density than the bulk, generating strong local buoyancy and subsequently setting up an upward local motion. The reverse is true at the sink, with cooler and denser fluid near the wall and net downward movement within a cross-section. Sample temperature contours and velocity vectors at a typical heated channel in demonstrated in Figure 9. As a consequence, to such local phenomenon, fluid near the upper wall of the sink and lower wall of the source directly energize the heat and mass transfer into the main stream, leading to enhanced heat transfer at those locations. On the contrary, heat transfer is restricted at the lower wall of the cooler and upper wall of the heater. Zhang et al. (2010) presented the temporal variation in wall Nusselt number at four specific locations and the same is reproduced in Figure 10. Here wall 1 and 2 corresponds to the upper and lower wall of the cooler, whereas 3 and 4 refers to the upper and lower walls of the heater. It is clearly evident the heat transfer at wall 1 and 4 are consistently higher than others, substantiating the above discussion and hence highlighting the need of multidimensional analyses for SCNCL. While the one-dimensional assumption completely ignores such local variations, it yields more of an average prediction, as the property values at any cross-section can be viewed to be area-averaged. Therefore, when the energy conservation is satisfied, it is possible to have a reasonably accurate estimation of flow-power

Figure 9. Typical examples of the steady-state (a) temperature contours and (b) velocity vectors at the horizontal heated section of a rectangular CO2-based loop, showing the asymmetric nature

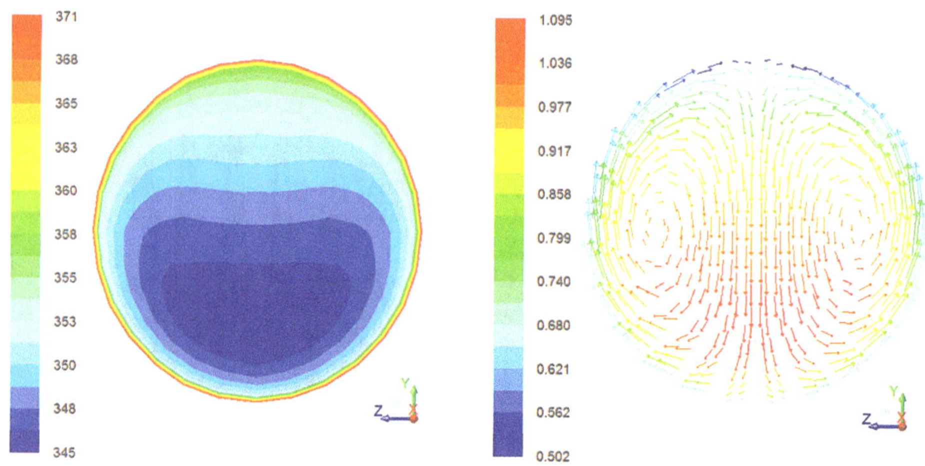

Figure 10. Temporal evolution of the average Nusselt number at four different location of a rectangular loop
Zhang et al., 2010.

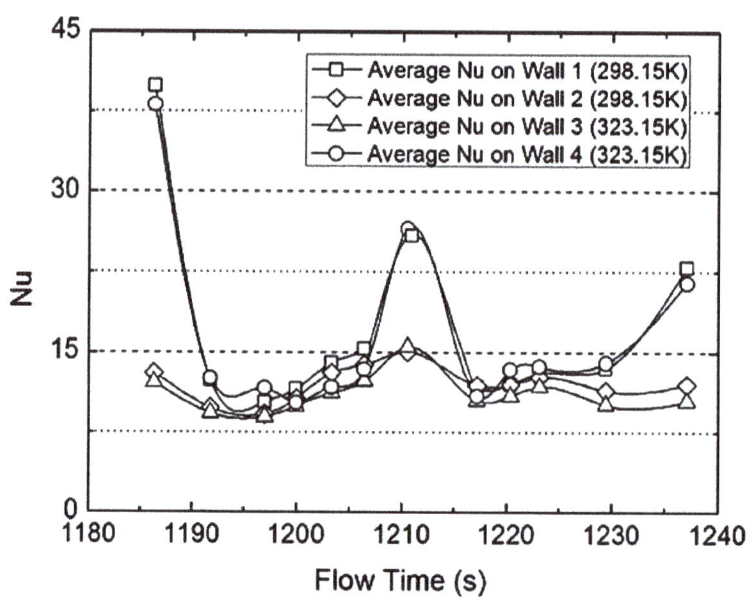

characteristic. However, heat transfer being a local phenomenon, depending precisely on the interaction between the wall and microlayer adjacent to it, coupled with the rapid property variation close to the pseudocritical point, multidimensional analysis is mandatory.

A 3-D analysis allows to use to complete form of conservation equations, thereby eliminating the need of constitutive relations and removing one possible source of inaccuracy from the numerical framework. The normal gradients of velocity and temperature at the walls can predict the shear stress and heat transfer coefficient respectively, which is not possible in 1-D models, where parameters are considered to vary only along the flow direction. Incompressible assumption is frequently imposed, considering the low velocity range in SCNCLs, compared to the acoustic speeds to common supercritical fluids. The set of governing equations are summarized below.

$$\frac{\partial \rho}{\partial t} + \nabla \cdot \left(\rho \vec{u}\right) = 0 \tag{11a}$$

$$\frac{\partial}{\partial t}\left(\rho \vec{u}\right) + \nabla \cdot \left(\rho \vec{u}\vec{u}\right) = -\nabla p + \nabla\left[\mu\left(\nabla \cdot \vec{u}\right)\right] + \rho \vec{g} \tag{11b}$$

$$\frac{\partial}{\partial t}\left(\rho E\right) + + \nabla \cdot \left(\rho \vec{u} E\right) = \nabla\left(\lambda \nabla T\right) + \beta T \frac{dp}{dt} + \Phi + q''' \tag{11c}$$

Here Φ designates viscous dissipation, referring to irreversible conversion of frictional work to thermal energy, and E is total energy given by,

$$E = \int_{T_0}^{T} C_p dT + \frac{u^2}{2} \tag{12}$$

T_0 is a suitably-chosen reference temperature, popular choice being sink or ambient temperature. Potential energy component is neglected in this definition.

Flow Reynolds number associated with SCNCLs can be of the order of 10^4, thereby necessitating the adoption of a suitable turbulence model. While the choice of turbulence model differs based on the nature of the problem for single-phase systems, hardly any attempt has been reported in literature on the impact of the same on SCNCLs. RNG $k - \varepsilon$ model is the preferred choice till date, which has been found to provide reasonable prediction. But a comparative assessment to judge the performance of other models is necessary and can be a topic for future research. Such a model must also be accompanied by accurate EOS. Sarkar and Basu (2015) integrated NIST reference database into their computational platform, allowing the estimation of all requisite properties as functions of local pressure and temperature at every iteration step, which definitely is the most comprehensive approach. To save some computational effort, certain groups have defined properties as piecewise-polynomial functions, extracted from the NIST database (Chen et al., 2012a). Such simplification was found to introduce less than 1% error in property prediction and hence can be adhered with. It may, of course, fail to give an exact representation of the local phenomena.

For the solution of such complicated system of equations, use of system codes like ANSYS-Fluent is almost universally-accepted. It is a finite volume based solver, which allows unstructured meshing over the domain under scrutiny. The perfection of the mesh system is important to have correct prediction. Accordingly, the mesh properties like orthogonal quality and skewness needs to be checked prior to simulation. It is pertinent to mention here that the axisymmetric assumption for the circular geometry is better being avoided, as the temperature and velocity profiles are not azimuthally symmetric (Figure 9).

Grid-independency check is an obligatory requirement, which can be ensured by progressively increasing the number of meshes and comparing the corresponding solutions, till the desired level of satisfaction. For a loop with circular cross-section, the same can be achieved by systematically adding either radial layers or azimuthal rays. A total number of 326836 nodes were employed by Sarkar and Basu (2015) to simulate a loop length of 7.4 m. Similarly, a loop of 5.4 m length was represented by 2×10^5 number of meshes by Chen et al. (2012a). Noting the massive mesh requirement, associated governing equations including turbulence and involved EOS, an idea about the computational resource requirement can be clearly comprehended. Such obligation can exponentially increase during a transient simulation and hence adaptive is preferred, with variable time-step. Zhang et al. (2010) reported a range of 10^{-3}–0.1 s, based on the convergence criteria.

EXPERIMENTAL OBSERVATIONS

Exhaustive experimentation is of paramount importance in any thermofluid application and even more so for intricate systems, such as SCNCLs. Experimental data, while revealing critical information on related thermalhydraulic phenomenon, can be employed either for retrofitting constants in semi-analytic relations, such as in Equation 10, or for judging the suitability of closure laws and turbulence models in computational simulations. The asymmetric nature of temperature and velocity distribution makes it even more significant to perform experiments on supercritical loops, to acquire thorough understanding of concomitant transport phenomena. Unfortunately, the volume of reliable experimental data on SCNCLs is quite limited, mainly owing to the concern of managing high pressure and temperature in a laboratory environment. In order to deal with supercritical water, system must maintain a pressure well-above 22.1 bar, while the pseudocritical temperature lies in the range of 650–670 K. It is an extremely demanding combination, without proper training and industrial framework. Therefore, it is a trend to employ fluids with more moderate critical point parameters at the laboratories. R134a, the refrigerant, has a reasonably low critical pressure (~4.06 MPa), whereas CO_2 is characterized by a very low critical temperature (~304.1 K). Critical pressure of CO_2 is about 7.4 MPa, which can comfortably be achieved as well. Accordingly, R134a and CO_2 have gained popularity as alternate supercritical fluids, despite water being the natural choice for reactor cooling.

Instrumentation can be another serious concern for SCNCLs. No intrusive measurement is allowed inside an NCL structure, to avoid any additional pressure losses and also to preserve the local buoyancy development, where the structure of wall plays an important role. It is quite common for any NCL to have the primary measurement restricted only to temperatures, with the flow rate being calculated from energy balance across the heater. That approach is reasonable for single-phase loops, but may introduce significant error in supercritical ones because of the large temperature variation in a single cross-section, the precision being dependent on both the employed instrument and the method of measurement. Consequently, conflicting experimental observations have been reported, with substantial disparity in conclusions.

An important work was reported by Sharma et al. (2010, 2012), based on the test facility developed at Bhabha Atomic Research Centre (BARC), India, with supercritical CO_2 as the working medium. They observed unstable oscillations with lower secondary-side flow rate and with heat inlet temperature close to the critical point. Swapnalee et al. (2012) also observed instability around the pseudocritical region, which was typified as density-wave or excursive type. Yu et al. (2011), however, postulated that Ledinegg instability cannot appear in SCNCLs, through experiments on a water-based loop. Operating within a pressure range of 24.2–25.2 MPa for water, Chen et al. (2012b) observed unstable fluctuations around the maxima of the flow-power curve, most of which was identified to be of dynamic-type. In their follow-up study (Chen et al., 2013a), however, possibility of static instability was hinted around the same location. Hence the contradiction is very much apparent, calling for more comprehensive experimentation with improved measurement and better appraisal of the concerned results.

PARAMETRIC EFFECTS

Considering the application of SCNCL in reactor cooling, it is desirable to have superior heat transfer characteristics over the entire spectrum of operating conditions. Heat transfer coefficient depends on the flow rate, as well as the thermophysical properties. Prandtl number of any supercritical fluid exhibits a peak at the pseudocritical point (Figure 11), while thermal conductivity suffers a decline from liquid-like to vapor-like level. Accordingly, the heat transfer behavior is strongly governed by the prevailing temperature range, with possibility of both enhancement and deterioration under certain circumstances. As the flow in SCNCL is determined by interplay between friction and buoyancy, and both these forces are influenced by several geometric and operating parameters, it is important to systematically analyze the role of each such variable. Owing to its inherent advantage of reasonably accurate prediction at moderate computational cost, 1-D finite difference method is the most preferred approach for such parametric appraisal, particularly for the evaluation of flow-power characteristics. The heat transfer behavior,

Figure 11. Variation of Prandtl number and volumetric expansion coefficient with temperature for CO2 at 8 MPa pressure

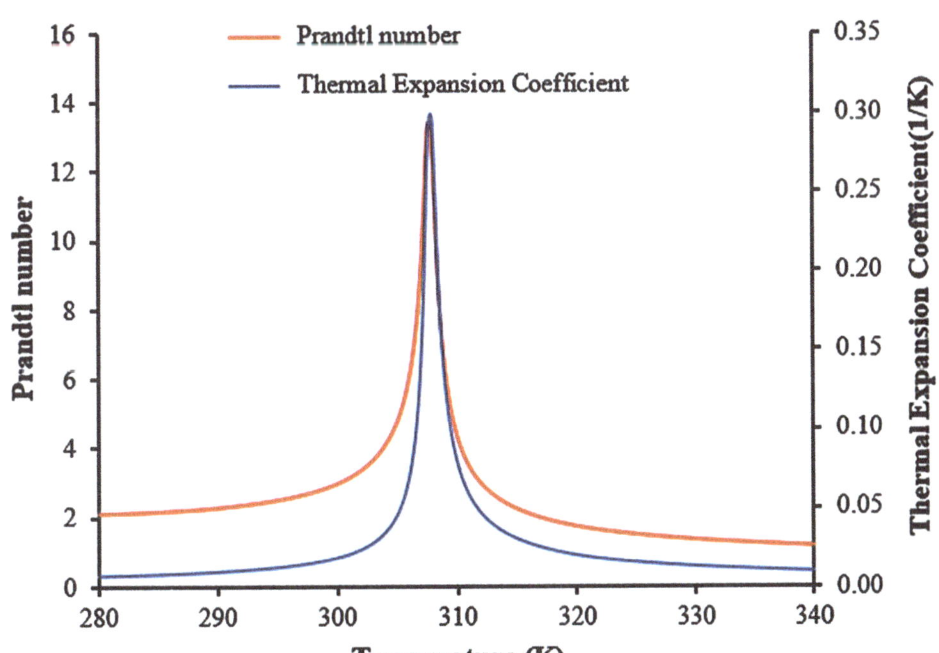

however, is strongly influenced by the multidimensional effects, as discussed before, and hence most of the concerned conclusions are drawn either through CFD simulations or experiments. The role of some of these factors is discussed below.

Temperature Difference Between Source and Sink

This is probably the most crucial factor, which strongly influences the steady-state thermalhydraulics. For single-phase NCLs, buoyancy shows proportionate increase with source-to-sink temperature differential, because of the enhanced density difference across the source. But the drastic property variation around the pseudocritical point makes the case of SCNCLs more intriguing. Because of the peak in volumetric expansion at T_{pc} (Figure 11), fluid experiences large buoyancy, if it is allowed to cross T_{pc} within the source.

Fluid viscosity is also quite low beyond the pseudocritical zone, which can contribute towards large flow rate and hence higher rate of heat transfer. If the sink temperature, however, is well below T_{pc}, viscous resistances in the cold arm will be reasonably high, deciding the flow rate accordingly. On the contrary, when the sink temperature itself is higher than T_{pc}, change in fluid density on heat addition will be significantly low, resulting in smaller flow rate, despite gas-like viscosity level and regardless of the source temperature. Similar effect can also be achieved by imposing source temperature lower than T_{pc}. Figure 12 illustrates the variation in mass flow with sink temperature for a constant source temperature of 371 K in a CO_2-based rectangular loop. As the lowest fluid temperature crosses T_{pc}, the decline in flow is very much apparent.

Therefore, even a large temperature differential does not warrant high flow rate, depending rather on the positioning of the pseudocritical temperature relative to the imposed values. T_{pc} itself is a function

Figure 12. Typical variation in mass flow rate with sink temperature for a CO2-based rectangular loop with constant source temperature of 371 K and 8 MPa pressure

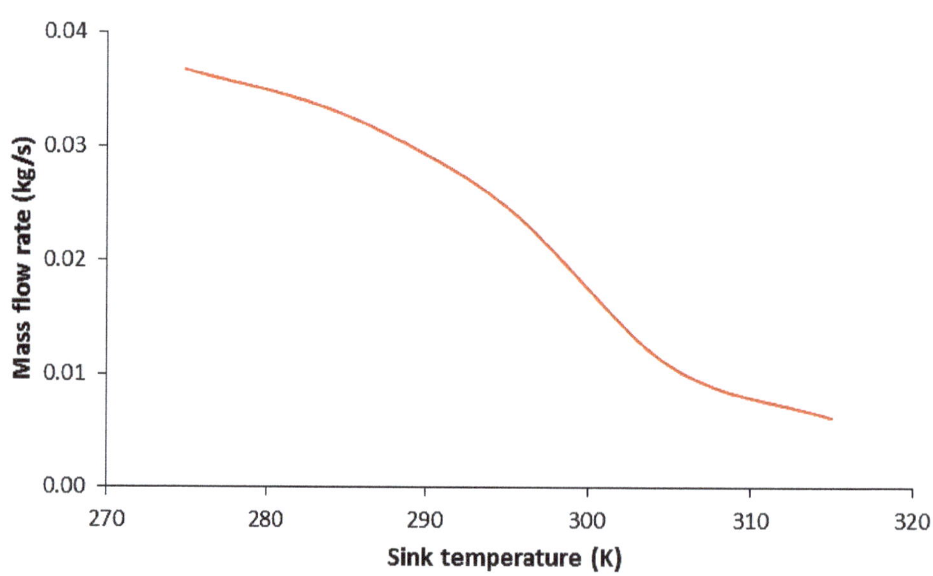

of the pressure, allowing operation with the enhanced flow rate till larger sink temperature at elevated pressure levels. Hence it can be concluded that an increase in system pressure or lowering in sink temperature is essential to facilitate large flow through an SCNCL, while the source temperature needs to be well-above the pseudocritical limit.

Heat transfer coefficient is expected to follow the mass flow profile, which implies a deterioration at higher sink temperatures. As reported by Yadav et al. (2012), cooler-side heat transfer coefficient increases with operating pressure and produces a peak with rise in sink temperature, which closely follows the appearance of the pseudocritical point. Their results are reproduced in Figure 13. It is very much evident that the order of heat transfer coefficient can be comparable to any boiling loop under certain situations. This observation is consistent with the experimental findings of Tokanai et al. (2010).

Input Heat Flux/Power

For a heat flux supported boundary condition, role of input power is quite similar to the source temperature. However, it is not possible to predict the largest fluid, and hence wall, temperature only from the value of power and without a proper analysis, unlike temperature-coupled loop, where the highest temperature involved is that of the source itself. Largest temperature in power-coupled loops is expected to appear at the heater exit and generally depends on the nature of fluid, system pressure and power level. For a specified sink, when the fluid is allowed to cross the pseudocritical limit inside the heater, large density differential is available between the vertical arms. It is quite likely to have fluid temperature well-above T_{pc} in hot leg and well-below that in the other, leading to large buoyancy and

Figure 13. Combined effect of sink temperature and system pressure on the heat transfer coefficient
Yadav et al., 2012.

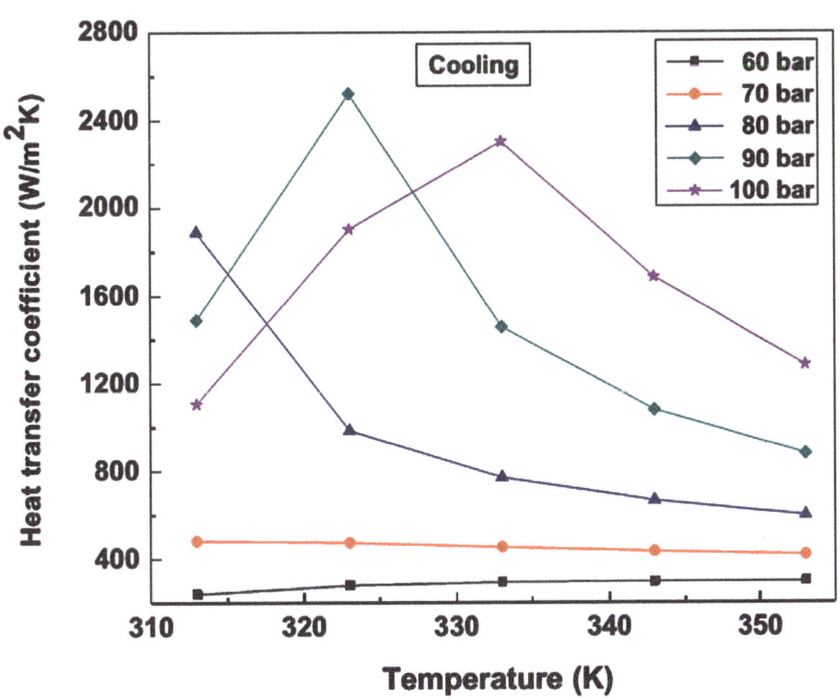

hence substantial flow rate. A large mass flow also ensures small temperature rise across the heater. But, at larger powers, lowest temperature may remain above T_{pc}, resulting in a gas-like representation of the fluid throughout the loop. Concerned loop behavior is similar to a single-phase NCL, with weak buoyancy field and small flow rate. This combination of high power and low flow leads to significant rise in fluid temperature at the heater exit. Heat transfer coefficient follows the profile of flow rate and, therefore, suffers a similar drop in its magnitude beyond a certain power range. Sarkar and Basu (2015) identified this phenomenon as a natural circulation version of *heat transfer deterioration (HTD).*

The variation in mass flow rate and heat transfer coefficient with power for a CO_2-based SCNCL is reproduced from their work in Figure 14. Such a drastic decline in heat transfer coefficient definitely suggests a failure in the thermal interaction between the solid wall and adjacent fluid layers, which is the indicative phenomenon of HTD for forced-flow channel. Of course, such deterioration here is instigated by the fall in mass flow rate and hence can also be viewed as a flow-induced HTD. Loops are found to be more vulnerable to such undesirable consequences with greater sink temperature and for pressure just-above the critical value. Appearance of HTD can be categorized as a limiting operating point for SCNCLs with heat flux boundary condition, as imposing power levels beyond this limit can raise the wall temperature by 300 K or higher, which can be detrimental from material point of view.

Figure 14. Variation in (a) mass flow rate and (b) heat transfer coefficient with power and sink temperature for a CO2-based rectangular loop with 8 MPa pressure, demonstrating the appearance of HTD

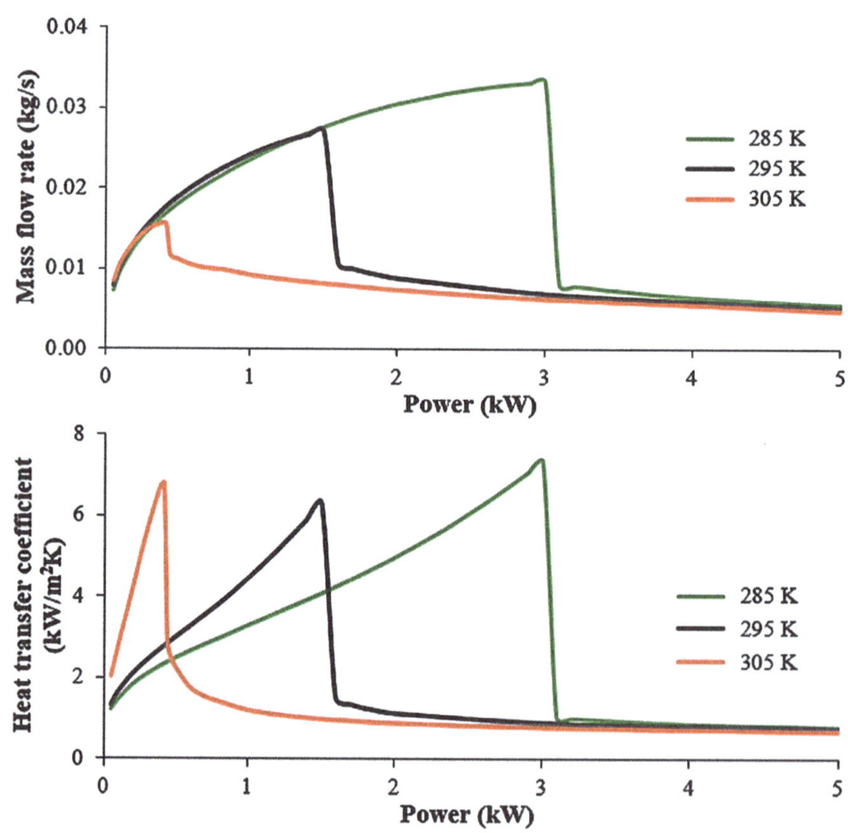

Orientation of Heater and Cooler

Buoyancy being the driving force in NCL, it can also be viewed to provide the destabilizing effect, with friction acting as the restraining one. Therefore, any factor which leads to an enhancement in buoyancy can also introduce larger instability into the system, and the relative positioning of the source and sink in the geometry can have serious impact on the same. For a specific combination of boundary conditions and a particular location of the sink, largest buoyancy can be achieved by allowing the fluid with the largest temperature the longest vertical rise. Therefore, horizontal orientation of the heater in the rectangular configuration is expected to be the most destabilizing one and the same has been observed through both numerical and experimental analyses. Numerical study of Chen et al. (2013b) considered four different heater positions, with the cooler mounted at the middle of the top horizontal arm of a rectangle. Both the vertical configurations produced stable results, with near-identical values of Nusselt number over the entire range of heat flux simulated. Between the two horizontal configurations, however, considerable deviation was observed at lower heat fluxes, with a monotonous phase-lag between Nusselt number values at higher heat fluxes. Unstable transitions and oscillations were also observed with horizontal heaters. Similar conclusions were inferred by Swapnalee et al. (2012) as well, as they identified the symmetric horizontal-heater-horizontal-cooler (HHHC) configuration to be the most unstable one with repeated flow reversals. It is not possible to predict the flow direction in a symmetric loop a priori, as flow is equally probable to be in either-way, with the eventual direction being reliant on the mode of flow initiation and minor irregularities, such as the wall roughness and nature of pipe bends. Therefore, such loops are more prone towards flow reversal and instability. On the contrary, asymmetric orientation ensures a pre-defined flow direction, resulting in more regularized stability behavior. It must be mentioned here that the loop orientation has been found to influence the flow direction more prominently, in contrast to the heat transfer and stability aspects. More concentrated studies are necessary to draw a better appraisal.

Tube Diameter

Heat transfer surface area is directly proportional to the tube diameter, whereas the frictional resistances exhibit the inverse relationship. Therefore large-diameter tube necessitates higher heat flux to maintain similar flow rate and temperature distribution. Substantial rise in Nusselt number was reported by Chen and Zhang (2011) with increase in diameter and at lower wall temperatures. Their observation, of course, was based upon only two different diameter values, which are quite close to each other, and hence better not to consider as a universal one. The qualitative effect of increase in diameter is to enhance both mass flow rate and heat transfer. Reduction in friction, with no significant change in buoyancy is likely to induce greater instability in the system. But such a conclusion awaits experimental validation.

The same is applicable for the length as well. While an increase in the horizontal dimension enhances the length of the total flow path and hence the friction, any rise in the vertical length reinforces both buoyancy and friction. Therefore, it is difficult to visualize the influence of loop dimensions on system performance without thorough experimental data.

Inclination Angle

Introducing inclination to the vertical arms directly modulate the effective gravitational force on them and hence reduce the developed buoyancy by a factor of $\cos\theta$, θ being the angle to vertical. Therefore, the inclined loop experiences lower mass flow rate in comparison to the vertical one, operating under the same set of parameters. Consequently, it shows an inferior heat transfer performance, characterized by reduction in Nusselt number with increased inclination. Generally, the effect of inclination is less pronounced at higher heat fluxes. At lower powers, however, rapid deterioration in heat transfer has been observed through experiments on inclined loops. The basic objective behind the inclination is to stabilize SCNCLs, due to the reduced-gravity effect and resultant lowering in buoyancy. However, no tangible experimental data is available till date to establish this particular concept.

SOLUTIONS AND RECOMMENDATIONS

Following the comprehensive discussion on thermalhydraulic and stability aspects of supercritical loops and methodologies for corresponding experimental and theoretical analyses, several recommendations can be made about the design and operation of SCNCLs. While the dearth of reliable experimental data is a concern and several guidelines emerging from theoretical studies lack experimental verification, quite a few conceptions are universally followed in modern research and application, and some of them are summarized here.

Existence of maxima in flow-power curve has been observed both numerically and experimentally, similar to two-phase NCLs. While it was postulated by early research to correspond to the stability threshold, that idea was discarded later. Instead the power corresponding to maxima can act as an important design guideline from thermalhydraulic point of view. For loops with heat flux controlled boundary, alike nuclear reactors, the maxima can also correspond to the peak in heat transfer coefficient and the initiation of HTD. Any increase in power can be accompanied by a drastic rise in temperature level and fall in flow rate. Therefore, it is recommended to restrict the operation till the HTD appearance, which can be delayed by lowering sink temperature and increasing system pressure. It is also suggested to control the input power so as to allow the fluid to cross the pseudocritical point within the heater, as that will ensure a large density differential across the source and hence a substantial flow rate, leading to improved heat transfer.

The stability behavior of the system can be controlled either by design-level modification in the geometry or by runtime modulation of the operating variables. Several suggestions can be proposed to expand the stable zone of operation such as,

- Decrease in diameter and increase in frictional length.
- Shorter source-to-sink vertical distance.
- Inclination of the adiabatic arms to vertical.
- Higher operating pressure and lower sink temperature.

Computational solution is the most preferred approach for both thermalhydraulic evaluation and stability appraisal. While the 1-D finite-difference models are easy to formulate and computationally less expensive, they mostly lead to approximate flow characteristics and are generally insufficient to infer on heat transport behavior. Therefore, the use of finite-volume-based multidimensional simulation is recommended. Considering the associated resource requirement, it is important to optimize any such model, both in terms of the mesh structure and employed numerical scheme. Choice of turbulence model is an important factor, which can seriously impact the prediction, particularly with higher heat flues. Therefore, it is suggested to compare the prediction with several combinations of meshes and turbulence models, before selecting one for further analyses. Use of system codes such as ANSYS-Fluent is probably the better option.

Property estimation plays an extremely crucial role in SCNCL calculations, as the entire thermalhydraulics is controlled by the sharp property variation around the pseudocritical zone. Hence incorporation of modern and highly-involved EOS is a must. Most of the initial studies employed simplified property relations in 1-D framework, commonly to produce the trends. However, it is necessary to introduce advanced EOS in 3-D analyses, as only that can mimic the local behavior and hence yield more realistic prediction.

FUTURE RESEARCH DIRECTIONS

From the preceding discussion, it is very much evident that, despite the enormous potential as a reactor cooling technology, the biggest concern with SCNCLs is the lack of experimentation. As has been outlined earlier, principal reasons behind the same are the difficulty in handling supercritical conditions and issues with measuring tools. With the advancement in material research, it is now possible to afford suitable materials, which can sustain operation over long-duration with high pressure and temperature, at reasonable cost. Therefore, systematic experimentation with well-instrumented test facility is the need of the hour. It is pleasing to see the increase in the number of experimental publications in the present decade. Large facilities are available now in research organizations, such as BARC, India and CIAE, China, along with some institutional laboratories. Importance of experiment will only grow, as newer application areas of SCNCLs will be explored, to expand its scope beyond nuclear fields.

The limited amount of experimental data available as of now focuses primarily on the thermalhydraulic characterization, frequently producing a correlation for flow rate or heat transfer coefficient as the end-product. On the contrary, appraisal of stability behavior is quite scarce and also not in consensus across different studies, as has been pointed out earlier. Therefore, future experimental research is expected to focus more on the stability characterization. It is very common to employ popular subcritical correlations for friction factor and heat transfer coefficient during SCNCL analysis, which may lead to wrong prediction and hence, when applied to practice, grave consequences. It is extremely important to develop correlations exclusively for supercritical NCLs to facilitate modeling in future. It is a proven fact for single-phase loops that the laminar-to-turbulent transition appears at a very low Reynolds number, necessitating the adoption of unique relations and modeling methodologies. The same also applies for SCNCLs, where the presence of strong local buoyancy effect makes the hydrodynamic behavior even more complicated. Therefore, the thermalhydraulic research on SCNCL must produce similar relations prior to industrial recognition.

While the 3-D modeling has gained popularity in last few years, it is mostly restricted to steady-state cases and hence to heat transfer predictions, predominantly owing to the substantial computational resource requirement. Smaller time-step is a necessity for precise stability evaluation, which can induce enormous load on a computational framework involving meshes of the order of 10^5. However, as has been discussed, simplified 1-D models, despite being very efficient with nonlinear transient calculations, ignores the local factors and can produce only an averaged prediction. With the escalating growth in computational capabilities, it is expected that bulk of the future numerical research will be targeted towards more-involved multidimensional stability appraisal. A suitable combination of advanced experimentation and exhaustive multidimensional simulation can help the knowledge base on SCNCL to mature and lead way towards widespread industrial application.

CONCLUSION

Supercritical NCL is offers very promising technology for cooling of nuclear reactors, as it facilitates fluid transport without using any prime movers and hence confiscate any possibility of drive-related accidents. Despite the potential, it is a relatively new technology, conceptual origin of which goes back only by one-and-half decade. Decent number of theoretical, as well as quite a few experimental, research studies has been reported on SCNCLs. While several crucial thermalhydraulic phenomenon have been acknowledged, and explained, there are significant controversies as well, on the nature of stability response in particular. The limited knowledge base makes it very difficult for the practicing engineers to design and fabricate one such system to suit a real-life application. This particular chapter has been initiated to fill such void and to provide a complete platform to the beginners for understanding the technology.

The chapter introduces the concept of NCL to start with and elucidates the advantages of supercritical medium as the working fluid. Different methods of analyses have been discussed in detail. 1-D finite-difference-based models are the most preferred approach for both steady-state characterization and transient evaluation. Thermalhydraulic and stability aspects are almost inseparable for any NCL. Attempt is made to discuss the modeling methodologies for both separately. Different aspects of multidimensional model, with the help of commercial codes, have also been presented. Some aspects of the experimental analyses have also been detailed. Dependence of thermalhydraulic and stability characteristics on several geometric and operating parameters of the loop has been elucidated, combining both the experimental and numerical observations. Finally, a few recommendations have been provided about SCNCL analyses and some topics as the subject of future research has been identified.

REFERENCES

Basu, D. N., Bhattacharyya, S., & Das, P. K. (2014). A review of modern advances in analyses and applications of single-phase natural circulation loop in nuclear thermal hydraulics. *Nuclear Engineering and Design*, *280*, 326–348. doi:10.1016/j.nucengdes.2014.09.011

Bushby, S. J., Dimmick, G. R., Duffey, R. B., Spinks, N. J., Burrill, K. A., & Chan, P. S. W. (2000). Conceptual designs for advanced, high-temperature CANDU reactors.*Proceedings of the SCR-2000*.

Chatoorgoon, V. (2001). Stability of supercritical fluid flow in a single-channel natural-convection loop. *International Journal of Heat and Mass Transfer*, *44*(10), 1963–1972. doi:10.1016/S0017-9310(00)00218-0

Chatoorgoon, V., Voodi, A., & Upadhye, P. (2005). The stability boundary for supercritical flow in natural-convection loops Part II: CO_2 and H_2. *Nuclear Engineering and Design*, *235*(24), 2581–2593. doi:10.1016/j.nucengdes.2005.06.004

Chen, L., Zhang, X., Cao, S., & Bai, H. (2012a). Study of trans-critical CO_2 natural convective flow with unsteady heat input and its implications on system control. *International Journal of Heat and Mass Transfer*, *55*(23-24), 7119–7132. doi:10.1016/j.ijheatmasstransfer.2012.07.027

Chen, L., Zhang, X., & Jiang, B. (2013b). Effects of heater orientations on the natural circulation and heat transfer in a supercritical CO_2 rectangular loop. *Journal of Heat Transfer*, *136*(5), 052501. doi:10.1115/1.4025543

Chen, L., & Zhang, X.-R. (2011). Simulation of heat transfer and system behavior in supercritical CO_2 based thermosyphon: Effect of pipe diameter. *Journal of Heat Transfer*, *133*(12), 2505–2513. doi:10.1115/1.4004434

Chen, Y., Zhao, M., Yang, C., Bi, K., & Du, K. (2012b). An experiment on flow and heat transfer characteristics in natural circulation of supercritical water.*Proceedings of 3rd China-Canada Joint Workshop on Supercritical-Water-Cooled Reactors (CCSC-2012).*

Chen, Y., Zhao, M., Yang, C., Bi, K., & Du, K. (2013a). Experiment of heat transfer of supercritical water in natural circulation with different diameters of heated tubes.*Proceedings of 6th International Symposium on Supercritical Water-cooled Reactors (ISSCWR6).*

Gebhart, B. (1973). Natural circulation flows and stability. In J. P. Hartnett & T. F. Irvine Jr., (Eds.), *Advances in heat transfer* (Vol. 9, pp. 273–348). New York: Academic Press.

Jain, P. K., & Rizwan-uddin, . (2008). Numerical analysis of supercritical flow instabilities in a natural circulation loop. *Nuclear Engineering and Design*, *238*(8), 1947–1957. doi:10.1016/j.nucengdes.2007.10.034

Sarkar, M. K. S., & Basu, D. N. (2015). Working regime identification for natural circulation loops by comparative thermalhydraulic analyses with three fluids under identical operating conditions. *Nuclear Engineering and Design*, *293*, 187–195. doi:10.1016/j.nucengdes.2015.07.046

Sharma, M., Pilkhwal, D. S., Vijayan, P. K., Saha, D., & Sinha, R. K. (2012). Steady state behavior of natural circulation loops operating with supercritical fluids for open and closed loop boundary conditions. *Heat Transfer Engineering*, *33*(9), 809–820. doi:10.1080/01457632.2012.646920

Sharma, M., Vijayan, P. K., Pilkhwal, D. S., Saha, D., & Sinha, R. K. (2010). Linear and nonlinear stability analysis of a supercritical natural circulation loop. *Journal of Engineering for Gas Turbines and Power*, *132*(10), 102904. doi:10.1115/1.4000342

Silin, V. A., Voznesensky, V. A., & Afrov, A. M. (1993). The light water integral reactor with natural circulation of the coolant at supercritical pressure B-500 SKDI. *Nuclear Engineering and Design*, *144*(2), 327–336. doi:10.1016/0029-5493(93)90148-3

Span, R., & Wagner, W. (1996). A new equation of state for carbon dioxide covering the fluid region from the triple-point temperature to 1100 K at pressures up to 800 MPa. *Journal of Physical and Chemical Reference Data*, *25*(6), 1509–1596. doi:10.1063/1.555991

Swapnalee, B. T., Vijayan, P. K., Sharma, M., & Pilkhwal, D. S. (2012). Steady state flow and static instability of supercritical natural circulation loops. *Nuclear Engineering and Design*, *245*, 99–112. doi:10.1016/j.nucengdes.2012.01.002

Tilak, A. K., & Basu, D. N. (2015). Computational investigation of the dynamic response of a supercritical natural circulation loop to aperiodic and periodic excitations. *Nuclear Engineering and Design*, *284*, 251–263. doi:10.1016/j.nucengdes.2014.12.028

Tokanai, H., Ohtomo, Y., Joriguchi, H., Harada, E., & Kuriyama, M. (2010). Heat transfer of supercritical CO_2 flow in natural convection circulation system. *Heat Transfer Engineering*, *31*(9), 750–756. doi:10.1080/01457630903500924

Vijayan, P. K., Sharma, M., Pilkhwal, D. S., Saha, D., & Sinha, R. K. (2010). A comparative study of single-phase, two-phase and supercritical natural circulation in a rectangular loop. *Journal of Engineering for Gas Turbines and Power*, *132*(10), 102913. doi:10.1115/1.4000866

Yadav, A. K., Ram Gopal, M., & Bhattacharyya, S. (2012). CFD analysis of a CO_2 based natural circulation loop with end heat exchangers. *Applied Thermal Engineering*, *36*, 288–295. doi:10.1016/j.applthermaleng.2011.10.031

Yu, J., Che, S., Li, R., & Qi, B. (2011). Analysis of Ledinegg flow instability in natural circulation at supercritical pressure. *Progress in Nuclear Energy*, *53*(6), 775–779. doi:10.1016/j.pnucene.2011.04.001

Zhang, X. R., Chen, L., & Yamaguchi, H. (2010). Natural convective flow and heat transfer of supercritical CO_2 in a rectangular circulation loop. *International Journal of Heat and Mass Transfer*, *53*(19-20), 4112–4122. doi:10.1016/j.ijheatmasstransfer.2010.05.031

Zvirin, Y. (1985). The instability associated with the onset of motion in a thermosyphon. *International Journal of Heat and Mass Transfer*, *28*(11), 2105–2111. doi:10.1016/0017-9310(85)90104-8

KEY TERMS AND DEFINITIONS

Equation-of-State: A functional relation to estimate a dependent thermophysical property as a function of two independent properties.

Finite-Difference Method: A numerical scheme of converting partial differential equations to equivalent algebraic equations.

Friction Factor: The ratio of shear stress to associated kinetic head.

Heat Transfer Coefficient: The ratio of heat transported in convective mode to the driving temperature difference.

Natural Circulation Loop: A closed circuit of fluid transport, where flow is realized at the absence of any driving mechanism.

Perturbation: An infinitesimally-small disturbance introduced on any flow parameter.

Pseudocritical Temperature: The temperature value corresponding to the maximum specific heat at a specific supercritical pressure.

Stability Analysis: A method of identifying the nature of fluid response against any disturbance in the flow system under steady-state.

Stability Threshold: An operating point appearing directly on the stability map. Neutrally-stable state.

Thermalhydraulic: Combined fluid flow and heat transfer characteristics.

Chapter 7
Application of Supercritical Carbon Dioxide for Solar Water Heater

Yuhiro Iwamoto
Nagoya Institute of Technology, Japan

Hiroshi Yamaguchi
Doshisha University, Japan

ABSTRACT

For supercritical CO_2, a small change in temperature or pressure can result in large change in density, especially in the state close to the critical point. The large change in density can easily induce the natural convective flow. In this chapter, a solar water heater using supercritical CO_2 which is originally designed and constructed will be introduced. The solar water heater is a closed loop system with main components of an evacuated solar collector and a heat exchanger. The working fluid of CO_2 is naturally driven by the large change in density with absorbing and transporting heat in the solar collector. And the heat energy (hot water) is produced by exchanging the transferred heat with water in the heat exchanger. This chapter will describe the typical system operation and performance at different season and climates.

INTRODUCTION

In recent years, the global warming, atmosphere pollution and energy depletion have become critical issues in the world. The main factor leading these issues is that our life strongly depends on the consumption of fossil fuels such as coal and crude oil. Larger amount of fossil fuels become consumed with rapid economic development and growth of population. In order to establish the sustainable energy society without the strong-dependency of fossil fuels, it is important to develop environmentally friendly renewable energy systems which exhaust no greenhouse effect gas. Solar energy, wind energy, biomass and so on have attracted much attention as renewable energy. Among them, the amount of solar energy is rather large comparing other renewable energies. The amount of solar energy striking the Earth in one hour

DOI: 10.4018/978-1-5225-2047-4.ch007

is 4.3×10^{20} J/h, which value exceeds all energy consumed on the planet in one year of 4.1×10^{20} J/year. Therefore, the solar energy is the one of promising environmentally friendly and sustainable energy, and a lot of research and development of the renewable energy system using solar energy have been promoted.

There are different methods for utilizing the solar resource such as solar electricity, solar thermal energy, and solar-derived fuel from biomass. Among these solar resources, solar thermal energy can be efficiently recovered into other energies (Zhang et al., 2005; Zhang et al., 2006; Zhang et al., 2007; Zhang et al., 2008). Solar thermal utilization system can be categorized into low-temperature solar thermal systems, which may not involve sunlight concentration, and high-temperature solar thermal systems, which generally require concentration of sunlight. Low-temperature solar thermal systems have the potential to supply a significant number of households and commercial buildings with heating, cooling, and refrigeration. Solar water heater is one of the most promising low-temperature solar thermal systems (Davidson et al., 2002; Chen et al., 2003; Greffet et al., 2002; Mancini et al., 2003; Kalogirou, 2003). During recent years, a lot of studies were carried out in the field of solar water heater (Budihardjeo & Morrison, 2009; Dubey & Tiwari, 2008; Anderson & Morrison, 2007; Esen & Esen, 2005; Morrison et al., 2005; Morrison et al, 2004; Dahl & Davidson, 1997; Moson & Davidson, 1995; Rosengarten et al.), which are mainly for the solar water heaters of water-in-glass evacuated tube type.

In addition, natural refrigerants have attracted much attentions as working fluid in energy conversion systems in a sense of preventing the global warming. Among the natural refrigerants, the interest in utilizing CO_2 has been considerably increased for the past decades. Although CO_2 is responsible for over 60% of the greenhouse effect, it has good aspects of being environmentally friendly refrigerant and being alternative working fluid in energy conversion systems. The fluoro carbons are widely used as working fluid in energy conversion systems such as air conditioner. However, the fluoro carbon has very high global warming point and is high responsible for the greenhouse effect. CO_2 has great potential as a refrigerant, because CO_2 has low toxicity, nonflammable and low global warming point which value is 1/1300 comparing a fluoro carbon of R134a. Additionally, in recent years, supercritical fluid have been welcomed in many advanced energy conversion systems. Supercritical fluid is a substance at a temperature and pressure above its thermodynamic critical point. It can diffuse through solids like gas and dissolve materials like liquid. The critical temperature of CO_2 is low of 31.1 °C. This low critical point enables the large variation of the thermophysical properties with small changes in temperature or pressure. Especially, the density dramatically decreases when the phase of CO_2 changes from the liquid state to the supercritical state. This large change in the density can easily induce natural convective flow due to the effect of the buoyancy. This natural convective flow can be applied to the solar water heater.

Recycling CO_2 is expected to contribute the reduction of the greenhouse effect, and CO_2 capture and storage technologies (Lorentzen, 1990; Lorentzen & Pettersen, 1992; Hashimoto & Saikawa, 2004; Kim et al., 2004; Neksa et al., 1998; Neksa, 2002; Stene, 2005) have been rapidly developed. Based on the above mentioned motivations, a new solar water heater using supercritical CO_2 has been proposed (Yamaguchi et al., 2010). The new solar water heater can produce thermal energy as hot water from solar thermal energy without any electric power consumption. The solar water heater is comprised of two loops: one is a supercritical CO_2 circulation loop and another is a water loop. The two loops are coupled with each other by a heat exchanger. CO_2 filled in the closed loop absorbs solar thermal energy in an evacuated solar collector and is driven by the buoyancy force generated by the phase change of CO_2 from the liquid state to the supercritical state. The CO_2 transfer heat to the heat exchanger and exchange the heat with water in the water loop. After passing through the heat exchanger, the supercritical CO_2 becomes the liquid state and come back to the evacuated solar collector. The water absorbs heat from the supercritical CO_2 in the heat exchanger and its temperature increases. Then the heated water flows into the hot water tank, which can be directly used. Because the working fluid of CO_2 is operated by the natural convective flow and transfer heat, the proposed solar water heater is novel energy conversion system using supercritical CO_2 without any electric power consumption to recovery solar thermal energy to thermal energy (hot water).

In this chapter, the newly designed and constructed solar water heater using supercritical CO_2 will be introduced, and the typical system operation and performance will be described.

SUPERCRITICAL CO_2

In recent years, natural refrigerants have attracted much attention as working fluid in energy conversion systems. Table 1 lists the properties of typical natural refrigerants and fluoro carbons, concerning the Ozone Depletion Potential (ODP), the Global Warming Potential (GWP) and the safeness evaluation carried by ASHRAE Standard Group. CO_2 is the most environmentally friendly refrigerant among the refrigerant and is ranked "A1" as safeness, which means that CO_2 has low toxicity and nonflammable. Especially, it should be noted that the GWP is significantly low comparing fluoro carbons, for instance, the value of GWP is 1/1300 comparing R134a which is commercially spread and used in energy conversion systems in the world. Figure 1 shows the Pressure-Enthalpy diagram of CO_2. The critical temperature and pressure are 31.1 °C and 7.37 MPa, respectively. CO_2 become the supercritical state at above these critical points. Interestingly, the critical temperature is low in a room temperature range. This feature enables the utilization of CO_2 as working fluid in energy conversion system with low-grade heat sources. Figure 2 shows the variation of the thermal physical properties of CO_2 depending on the temperature at the fixed pressure of 9.0 MPa. The density significantly decreases with the temperature rising, changing the phase from the liquid state to the supercritical state. This large variation of the density enables the induction of the strong natural convective flow due to the effect of the buoyancy force. Other thermal physical properties also dramatically vary close to the critical point. In this region, the Prandl number ($Pr = \mu C_p / \lambda$) become very high value. This implies that the heat transfer is enhanced.

Table 1. ODP, GWP and Safeness evaluated by ASHRAE Standard Group of major refrigerants

Refrigerant		Ozone Depletion Potential (ODP)	Global Warming Potential (GWP)	Safeness
Fluoro carbons	R22	0	1700	A1
	R134a	0	1300	A1
	R404A	0	3850	A1
	R407C	0	1370	A1
	R410A	0	1370	A1
	R507A	0	3900	A1
Natural refrigerant	NH_3	0	<1	B2
	C_3H_8	0	3	A3
	C_4H_{10}	0	3	A3
	CO_2	0	1	A1

Figure 1. Pressure and enthalpy diagram of CO_2

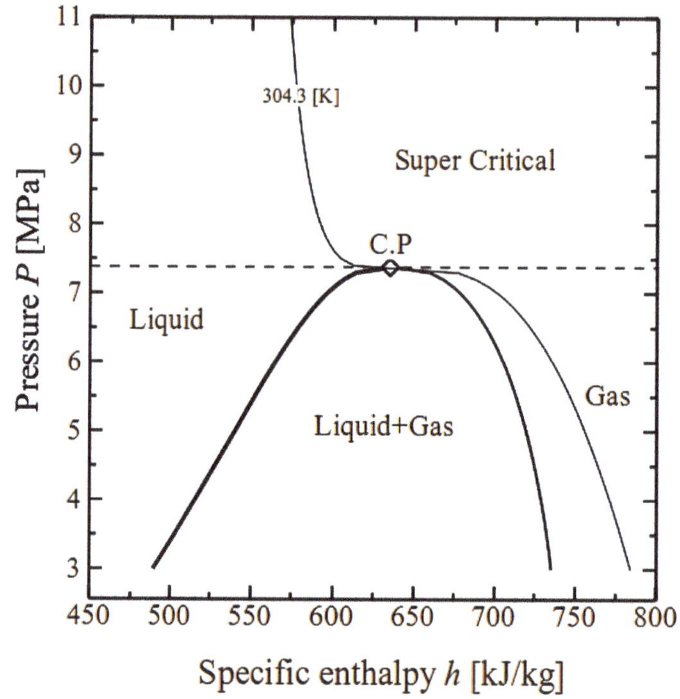

Figure 2. Thermal physical properties of CO_2 with varying the temperature at the pressure of 9 MPa

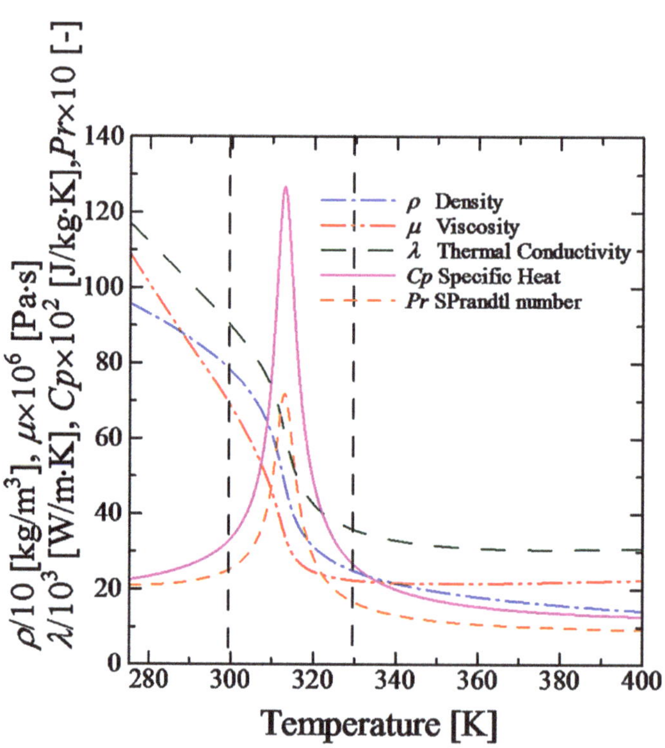

With considering the above mentioned features of CO_2, CO_2 is a good candidate as working fluid in energy conversion systems with low-grade heat source existing around our life. And one of the promising application of the supercritical CO_2 is a "Solar Water Heater".

EVACUATED SOLAR COLLECTOR

Evacuated solar collector is the heart of the solar water heater using supercritical CO_2. Its characteristics play an important role in the successful operation of the solar water heater. The evacuated solar collector used in this study is originally designed and constructed to be suitable with the unique supercritical CO_2 properties and the solar water heater. Figure 3 depicts the evacuated solar collector used. The collector has great properties of heat absorption and thermal insulation. The collector is mainly composed of a glass envelop, inner glass tube with selective coating, Aluminum fin and stainless U-tube. The annular space between the glass envelop and inner glass tube is arranged to be vacuum for keeping high thermal insulation. The inner glass tube is coated with a selective solar absorber coating. The coating has a nice optical property that absorbs a wide range of wave from visible to infrared ray radiation. The solar absorbance and the emissivity at the temperature of 373 K are 0.927 and 0.193, respectively. The stainless U-tube is inserted in the inner tube and Aluminum fin. The absorbed heat is conducted through the inner glass tube wall as well as the Aluminum fin, and it can be removed well by CO_2 running in the present study, in the stainless U-tube. The maximum operating pressure is as high as 12.5 MPa with considering the high critical pressure of CO_2. In the present experimental setup, the evacuated solar collector of 1.93 m2 (glass area) is used, and the other details of the utilized collector can be seen in refs. (Zang et al., 2005; Zang et al., 2007; Zang et al., 2008).

Figure 3. Schematic diagram of evacuated solar collector utilized in the present experimental setup system

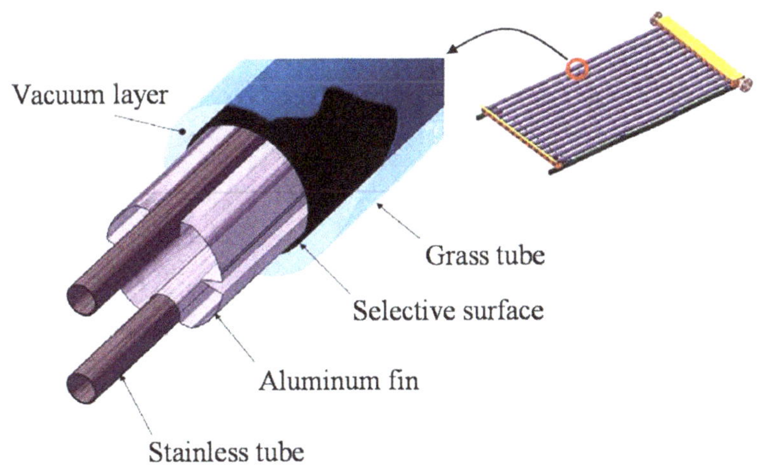

SOLAR WATER HEATER USING SUPERCRITICAL CO_2

As described in the section of supercritical CO_2, large variation of the density can induce the strong natural convective flow. And supercritical CO_2 has a great potential as a heat carrier, because the Prandl number is greatly enhanced closed to the critical point. Based on these unique characteristics, a solar water heater using supercritical CO_2 has been originally designed, constructed and tested in the present work. Figure 4 and Figure 5 show the picture of actual designed and constructed prototype experimental setup (Figure 4) and the schematic diagram presenting the details of the construction (Figure 5). The system is mainly composed of two loops: one is a supercritical CO_2 circulation loop and another is a water loop. The two loops are coupled with each other by a heat exchanger. The CO_2 closed loop is a simple configuration with an evacuated solar collector, a heat exchanger and a flow meter. Before the system operation, CO_2 exists in the liquid state in the system at approximately 6 MPa and 20 °C, (These values are dependent on season and weather). When the solar collector starts absorbing solar thermal energy, the temperature and the pressure of CO_2 increase and the fluid status will be in the supercritical region at above the critical point. The large difference of density between the liquid sate and the supercritical state induces the buoyancy force which becomes the driving force of CO_2, and then CO_2 naturally flows with absorbing solar thermal energy in the collector. The supercritical CO_2 transfers heat to the heat exchanger. The heat is exchanged with the water in the water loop. After passing through the exchanger, the supercritical CO_2 is cooled down and become the liquid state again. The liquefied CO_2 flows back the evacuated solar collector. In the water loop, cool water is fed into the heat exchanger and absorbs heat from the supercritical CO_2. Then the heated water flows into the hot water tank and is stored there.

For the aim of controlling the system operation, the valve 1 and 2 are embedded into the CO_2 loop and the water loop to adjust the mass flow rates of CO_2 and water. The cool water fed into the heat exchanger is directly supplied from a city water tank. The whole system is designed to be the maximum operating pressure as high as 10.0 MPa with considering the high critical pressure of CO_2. In the pres-

Figure 4. Picture of actual designed and constructed prototype experimental setup

Figure 5. Schematic diagram of the experimental system of the solar water heater using supercritical CO_2 as working fluid

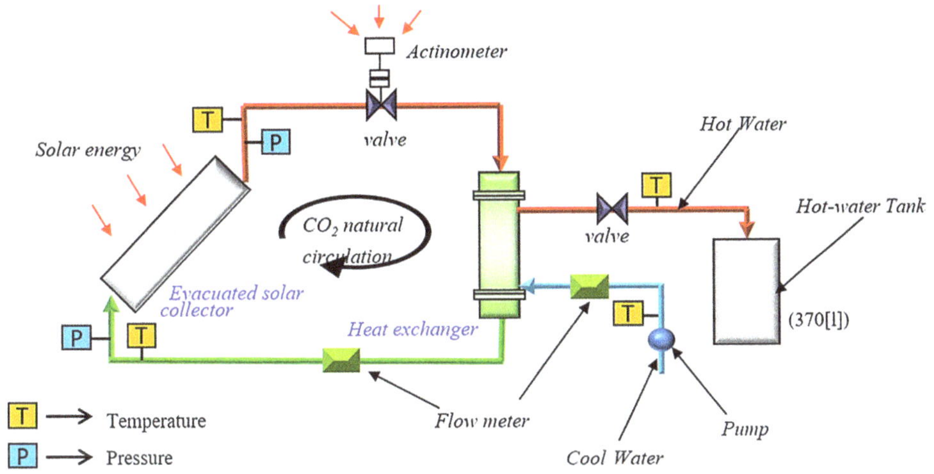

ent study, the solar collector is set declined with an angle of 45 degree with respect to the horizontal level. In order to evaluate the system performance, thermocouples, pressure gauges and flow meters are installed into the system (see Figure 5). The other details of the experimental setup can be seen in Ref. (Yamaguchi et al., 2010).

TYPICAL SYSTEM OPERATIN AND PERFORMANCE

The typical system operation and performance will be presented in this section. The experiment was carried out on typical sunny day of autumn, 3rd November, in Kyotanabe city, Kyoto, Japan. In the test day, the average and highest outside temperatures are 18.1 °C and 25.8 °C, respectively. The average solar radiation is 526 W. The surface of the solar collector is faced to the south direction. The experiment starts from 10 o'clock and the mass flow rate of the water is adjusted with the valve 2 to produce the hot water at the temperature of 50 °C. The experiment is finished when the 50 °C hot water cannot be available. CO_2 of 3.5 kg is filled in the system.

Figure 6 shows the variations of the solar radiation and the CO_2 mass flow rate with time elapsing. In the test day of 3rd, November, the solar radiation increases and decreases with the peak at noon. Before 13 o'clock, the weather is pretty good with clear blue sky. After 13 o'clock, the solar radiation begins fluctuating. This is because clouds shut out the sun light. It is observed that the CO_2 starts driving after approximately 30 minutes from when the experiment is started, namely solar thermal energy is supplied to the solar collector. There is a time lag of the solar radiation and the CO_2 mass flow rate increases. This is because, as mentioned above, the natural convective flow is induced by the effect of the buoyancy, which is generated when the phase of CO_2 changes from the liquid state to the supercritical state. Therefore, some time to supply enough heat energy to change the phase is necessary. Once the CO_2 starts circulating in the system, the mass flow rate is found to rapidly increase, and then become stable. Interestingly, the mass flow rate is stable even when the solar radiation fluctuates due to the effect of the clouds. This is due to the good characteristics of the evacuated solar collector and supercritical CO_2. The designed

Figure 6. Variations of the measured data with the test time, which include solar radiation and CO_2 flow rate

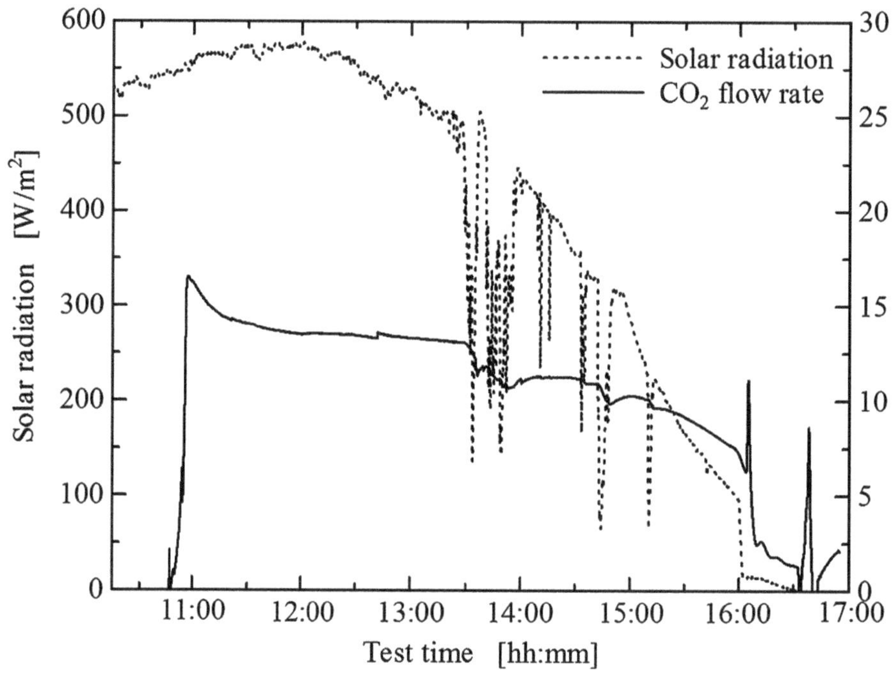

collector can absorb visible radiation as well as infrared ray radiation. Therefore, the collector can keep absorbing solar thermal energy even when the sun light is shut out by the clouds. Additionally, small supply of solar thermal energy can maintain the natural convective flow of CO_2 in the collector. For supercritical fluids, even very small change in temperature can result in large density variations when it is close to the critical point. In this experiment, the maximum Reynolds number of 6500 is observed, which is rather high value for natural convective solar water heater.

The variations of the temperatures and pressures at the inlet and outlet of the evacuated solar collector during the test time are plotted in Figure 7. And based on the measured temperatures and pressures, the pressure and enthalpy diagram is depicted in Figure 8. The value of enthalpy is calculated by using a program package for thermophysical properties of fluids-National Institute of Standards and Technology (NIST) reference fluid thermodynamic and transport properties (REFPROP 8.0). In Figure 8, the clockwise direction represents the time elapsing. Before the system operating, the temperatures and pressures both at the inlet and outlet are approximately 20 °C and 6 MPa. Based on these value, the state of CO_2 at the inlet and out let are gas-liquid two-phase. After the experiment starts, CO_2 begins absorbing solar thermal energy in the evacuated solar collector, and then the pressure in the whole system and the temperature at the outlet of the collector increase and become stable. The pressures are found to reaches at approximately 9 MPa. The temperature reaches the maximum value of 80 °C. These pressures and temperature are clearly over the critical point, and, as shown in Figure 8, the phases of CO_2 at the inlet and the outlet of the collector become the liquid state and the supercritical state, respectively. This result indicates that the solar water heater operates successfully in liquid-supercritical state system. After 4 o'clock, when the sun starts setting, the temperatures and pressures goes down with changing the phase into the initial state of gas-liquid two-phase.

Figure 7. Variations of the measured data with the test time, which include CO_2 temperature and CO_2 pressure

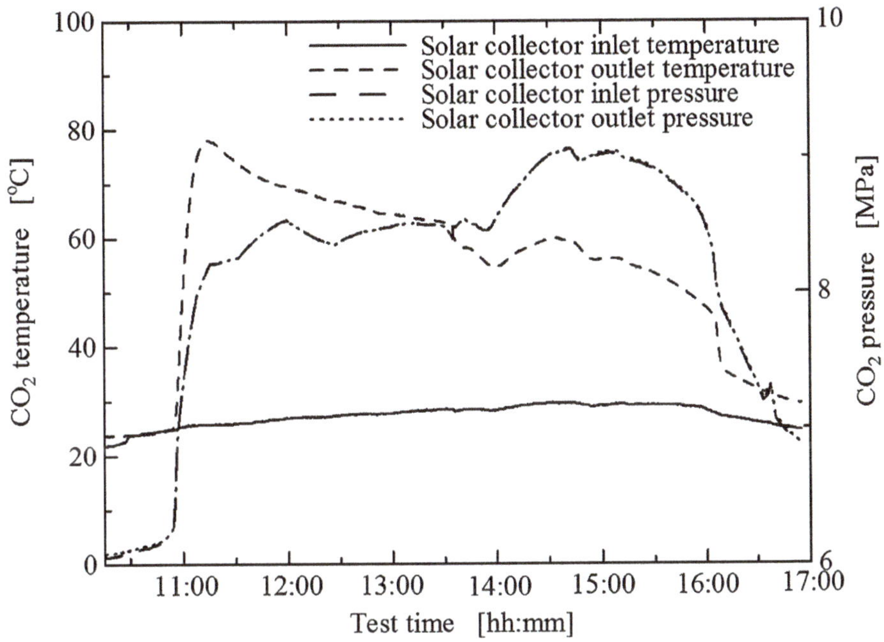

Figure 8. CO2 fluid states in the solar collector during the test period, presented in CO_2 p-h diagram. The green line represents the CO_2 fluid state at the collector inlet and the blue one shows the CO_2 fluid at the collector outlet, respectively.

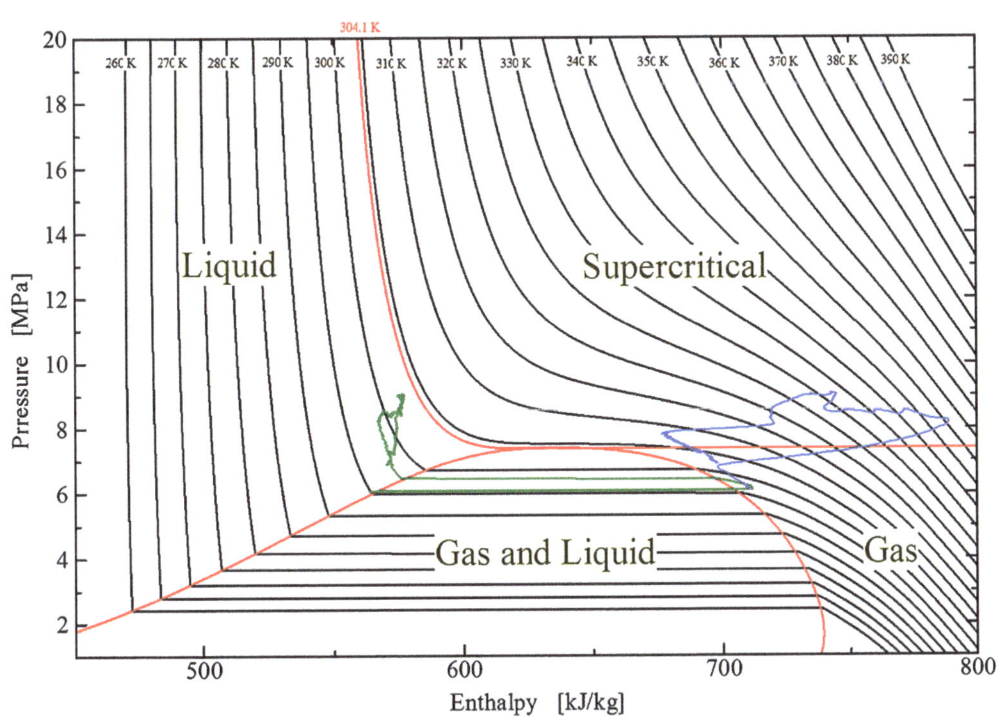

Figure 9 shows the variations of the temperatures of water in the water loop at the inlet and outlet of the heat exchanger with the test time. The variation of the mass flow rate of the water is also plotted in the figure. Here, it should be noted that the water is directly fed into the system from the city water tank. Therefore, the temperature depends on season and weather. Further, the mass flow rate of water is adjusted to make the temperature at the outlet of the heat exchanger be 50 °C. The cool water fed from the city water tank absorbs heat from supercritical CO_2 in the heat exchanger and the temperature of the water increase. The hot water with the temperature of 50 °C is found to be available stably during the system operation. After 15 o'clock, the temperature at the outlet slightly drops with the decrease of the solar radiation.

The most interesting aspect the readers have would be how much solar thermal energy can be collected in the evacuated solar collector and recovered into thermal energy as hot water. Figure 10 shows the variations of the collected heat quantity and the recovered heat quantity with the test time. These quantities are calculated based on the measured quantities of the temperatures and the pressures by the following equations:

$$Q_C = m_{CO_2}\left(h_{CO} - h_{CI}\right) \tag{1}$$

$$Q_H = m_W c_p \left(T_{WO} - T_{WI}\right) \tag{2}$$

Figure 9. Variations of the measured water flow rate and water temperature with the test time

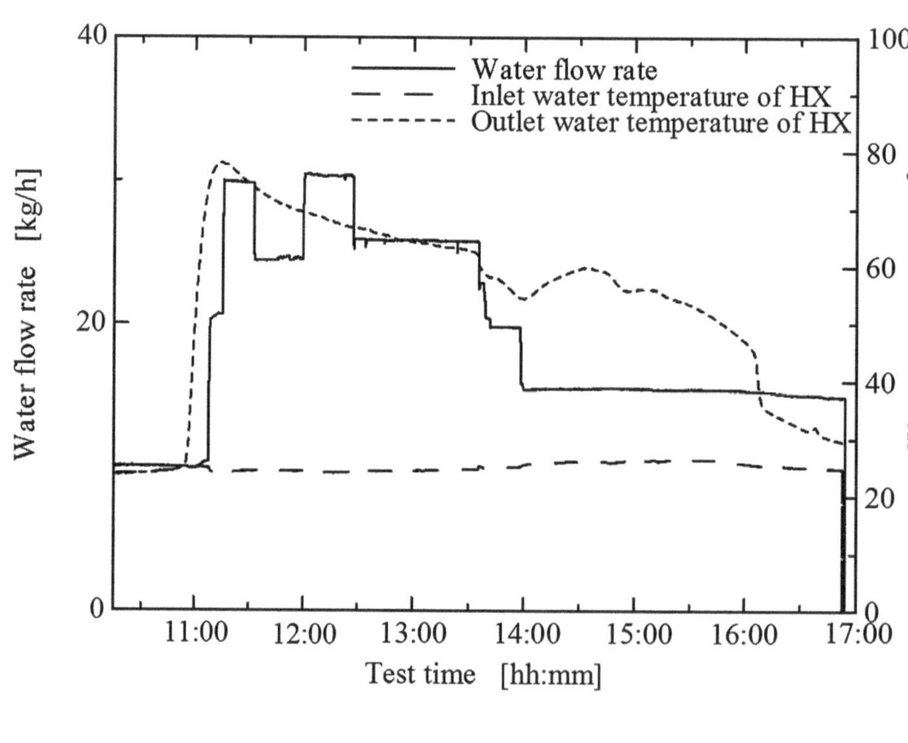

Figure 10. Variations of the performance parameters with the test time, which include collected heat quantity and recovered heat quantity.

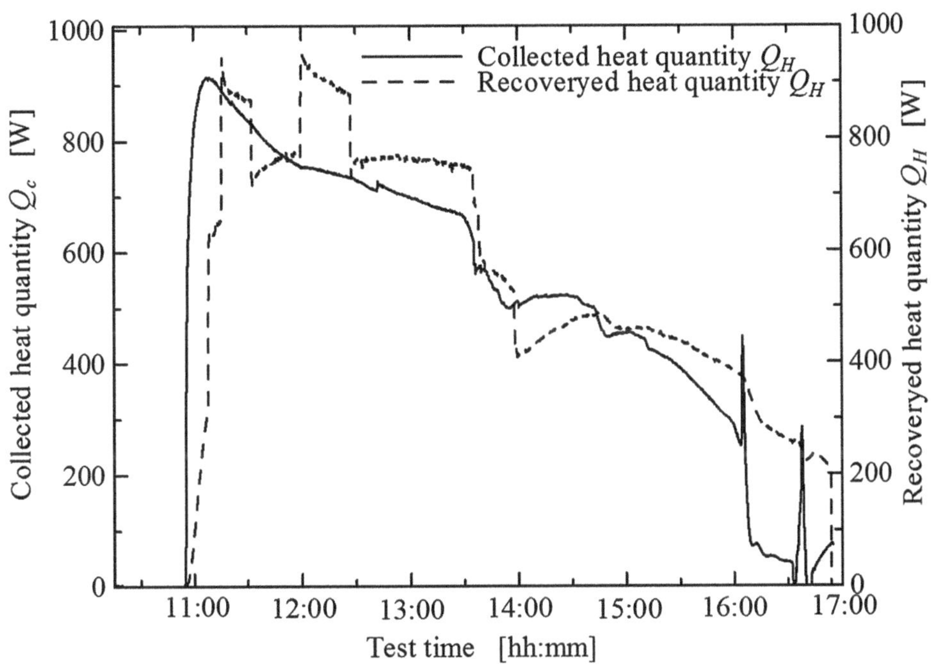

where, m_{CO2} and m_W are the mass flow rates of CO_2 and water, respectively, c_p the specific heat, h_{CO} and h_{CI} the enthalpy of CO_2 at outlet and inlet of the solar collector, respectively, and T_{WO} and T_{WI} the temperature of water at the outlet and inlet of the heat exchanger, respectively. As shown in Figure 10, as soon as the experiment starts, the collected and the recovered heat quantities are zero. This is because some time is necessary for CO_2 to absorb the solar thermal energy and to start running by the effect of the buoyancy force. After CO_2 is driven, both of the quantities rapidly increase. It is observed that there is a time lag between the startups, and sometimes, the recovered heat quantity becomes over the collected heat quantity. It may be physically explained as follows. Solar radiation heats the CO_2 fluid at one time. Because this heating process and corresponding heat transfer need to take some time period to be finished, the temperature change of the CO_2 fluid will occur later than that time of the solar radiation heating. After both of the quantities reach the maximum value, they gradually decrease with the decrease of the solar radiation. The heat recovery efficiency η_{REC} is determined based on the obtained heat quantities by the following equations:

$$\eta_{REC} = \frac{\int Q_H dt}{\int Q_C dt} \tag{3}$$

Based on the measured collected heat quantity and the recovered heat quantity during the test time, the heat recovery efficiency η_{REC} is 83.6%. In this experiment, the evacuated solar collector of 1.93 m^2

is used and the hot water at the temperature 50 °C of 104 l is available. These values are rather high comparing the conventional solar water heater, for instance the solar water heaters of water-in-glass evacuated tube type, which has generally the heat recovery efficiency of about 50%.

INFLUENCE OF CO_2 STATE ON SYSTEM PERFORMANCE

In the present solar water heater, the CO_2 is driven by the buoyancy force generated by the large variation of the density near the critical point. Therefore, the states of CO_2 at the inlet and the outlet of the evacuated solar collector strongly influence on the system performance. In order to investigate the effect, two experiments with different conditions: one is the liquid-supercritical state condition, in which the state of CO_2 at the inlet of the collect is liquid state and the outlet the supercritical state, and another is the supercritical-supercritical state condition, in which both of the states of CO_2 at the inlet and the outlet of the collector are the supercritical state, were conducted. Table 2 lists the experimental conditions with respected to the outside average and highest temperatures, the radiation time and the average solar radiation, carried out on 22nd, November and 14th, June. The condition of 22nd, November is a typical climate of winter, and one of 14th, June is summer in Japan.

Based on the measured temperatures and pressures, the enthalpies at the inlet and outlet of the evacuated solar collector are calculated and plotted in Figure 11. Figure 11 shows pressure-enthalpy diagram. The red line represents the data of the typical summer day of 14th, June, and the blue line represents the data of the typical winter day of 22nd, November. The left side and right edges of each solid-lines represent the pressure and enthalpy values at the inlet and outlet of the evacuated solar collector, respectively. As shown in Figure 11, the liquid-supercritical state system in the evacuated solar collector is formed in case of the typical winter day. On the other hand, in the case of typical summer day, the state of the CO_2 at the inlet of the collector is the supercritical state, and the supercritical-supercritical state system is found to be formed. This difference may be occurred by the effect of the amount of solar radiation and the temperature of feed-water in the heat exchanger. The amount of the solar radiation in summer is larger than that in winter. And the temperature of the feed-water in summer is higher than that in winter. In this experiment, the flow rate of the feed-water is adjusted to produce the hot water at temperature of 50 °C. Therefore, the CO_2 fed into the heat exchanger is not cooled well to change its phase to liquid state. The heat recovery efficiency in summer is found to be lower value of 49.8% comparing with one in winter of 85.8%. This large difference of the efficiency is clearly due to the effect of the state of CO_2 at the inlet of the collector. If the system operates in the liquid-supercritical state system, a large amount of the solar thermal energy can be absorbed in the collector due to the effect of the phase change of CO_2.

Table 2. Experimental conditions with respected to the outside average and highest temperatures, the radiation time and the average solar radiation, carried out on a typical winter day (22nd November) and a typical summer day (14th June)

Date	Outside Average Temperature [°C]	Outside Highest Temperature [°C]	Solar Radiation Time [h]	Average Solar Radiation [W]
22nd, November	7.7	14.3	9.0	418
14th, June	23.3	30.0	10.6	825

Figure 11. CO_2 fluid states in the solar collector, presented in CO_2 p-h diagram. The blue line represents the data on a typical winter day (22nd November) and the red line shows the data on a typical summer day (14th June)

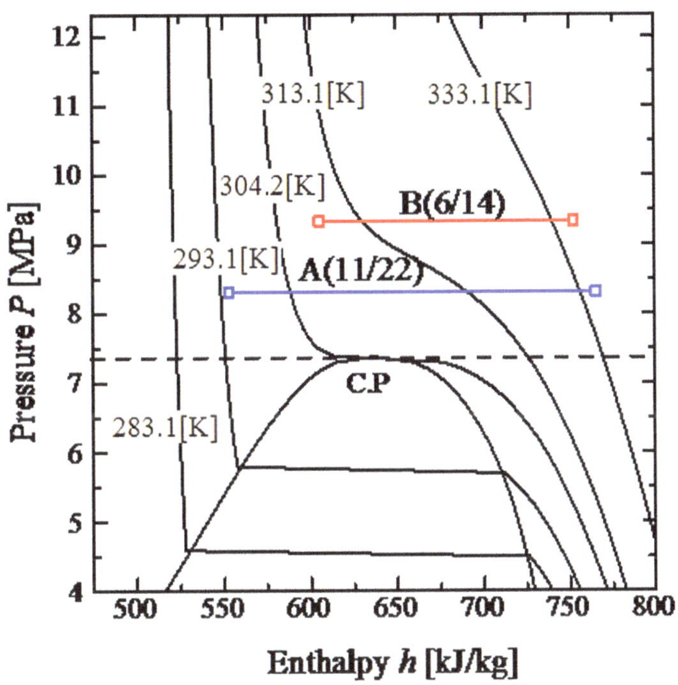

On the other hand, in the case of the supercritical-supercritical state system operation, no phase change deteriorates the heat transfer. However, the value of the heat recovery efficiency of 49.8% is still high value, which is nearly equal to the conventional solar water heater.

The system efficiency clearly depends on season. Figure 12 organizes the data of the heat recovery efficiency at various temperature condition of the feed-water. As shown in this figure, it is found that the heat recovery efficiency at the temperature bellow 25 °C is high value. This is because the liquid-supercritical state system is created in the evacuated solar collector, and CO_2 efficiently absorbs solar thermal energy with changing its phase. On the other hand, the efficiency is found to be lower value in the case of the temperature of the feed-water over 25 °C. This is due to the supercritical-supercritical state operation.

MAXIMUM TEMPERATURE OF HOT WATER

Other advantage of new solar water heater is that the system can produce higher temperature water compering the conventional one. It is well known that the maximum temperature of the hot water created in the conventional solar water heater is limited as approximately 40 °C in winter. The proposed system can produce rather higher temperature water even in winter. Because the critical temperature of CO_2 is low of 31.1 °C, the temperature keep increasing with absorbing the solar thermal energy after the phase changes into the supercritical state. And exchanging heat between the CO_2 and water in the heat exchanger can result in the generation of higher temperature water.

Figure 12. Heat recovery efficiency depending of feed-water temperature

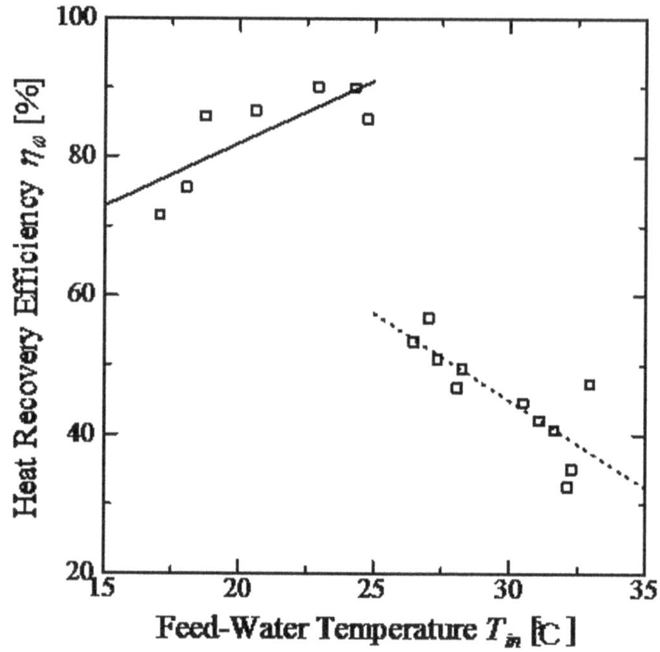

Figure 13, Figure 14, and Figure 15 show the temperature of the CO_2 at the inlet and outlet of the evacuated solar collector in different season of summer (Figure 13), autumn (Figure 14) and winter (Figure 15). The temperatures of the water at the inlet and outlet of the heat exchanger are also plotted in the figure. The outside average temperatures in summer, autumn and winter are 30.0, 20.5, 6.0 °C, respectively. In the present study, in order to investigate the maximum temperature of the hot water in the different season, the mass flow rate is set to be 10 kg/h. All of the season, the temperatures rapidly increase with the sun rinsing, and then reach the peak value. After that, the temperatures gradually drop with the decrease of the solar radiation. The maximum temperature of generated hot water at each seasons of summer, autumn and winter are observed as 78.0, 74.2 and 69.5 °C, respectively. All of these value are rather high comparing with one of the conventional solar water heater. Interestingly, the very high temperature of 69.5 °C is observed even in winter.

FUTURE RESEARCH DIRECTIONS

In this chapter, the typical operation and performance of new solar water heater using supercritical CO_2 were introduced. The main advantages of the new solar water heater are as follows:

- A small change in the temperature of CO_2 can result in a large variation of the density. And it induce strong natural convective flow in the system.
- A wide range of solar thermal energy from visible to infrared ray radiation can be absorbed by the originally designed evacuated solar collector.
- The above results imply that the system can operate not only in sunny day but also cloudy day

Figure 13. Variation of the measured CO_2 and water temperature with test time on a summer day (3rd, August)

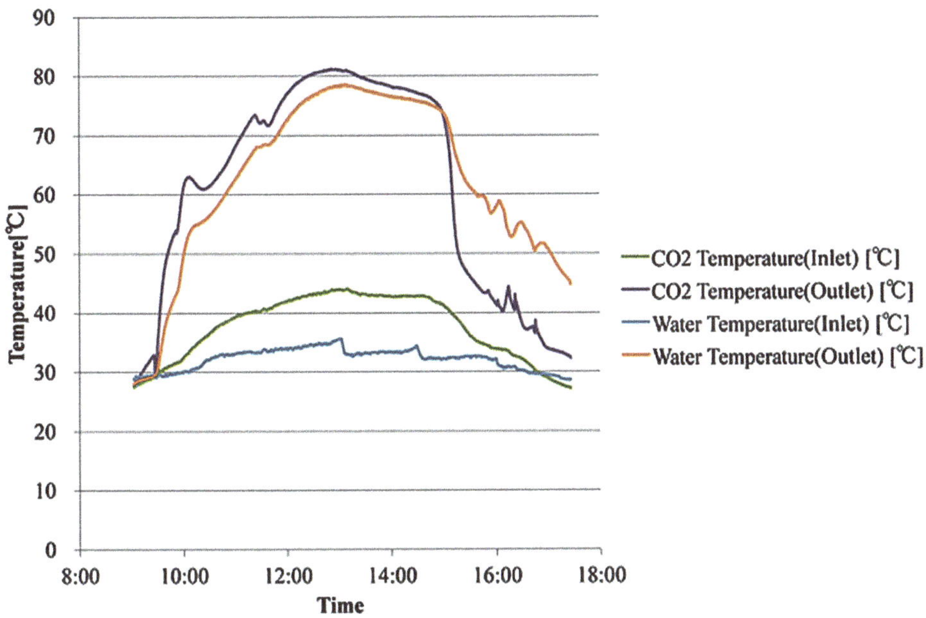

Figure 14. An autumn day (1st, October)

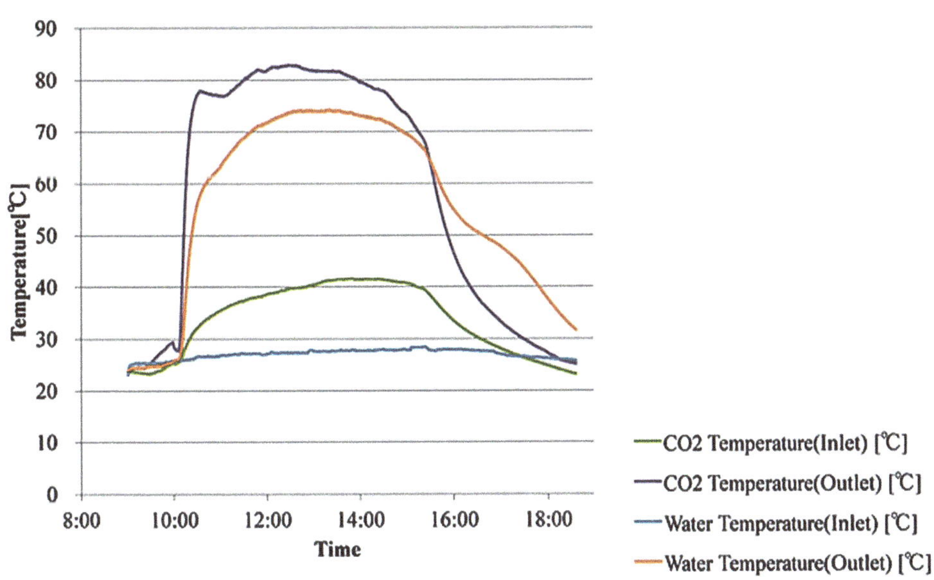

Figure 15. A winter day (4th, April)

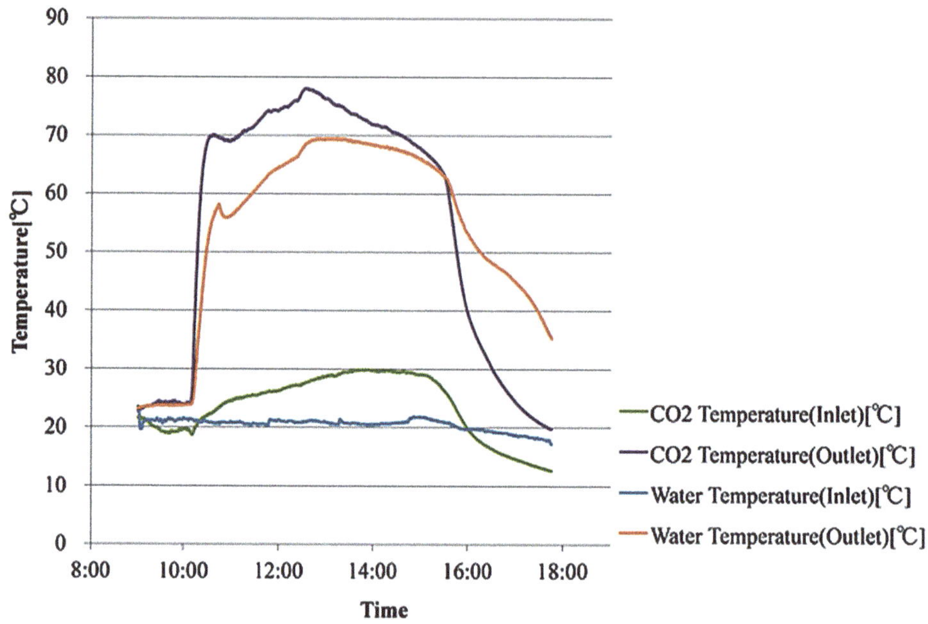

- The liquid-supercritical state condition created in the evacuated solar collector enhances the heat recovery efficiency. Even if the supercritical-supercritical state is formed in the collector, the efficiency is nearly equal to conventional one.
- The heat recovery efficiency is rather higher than that of conventional one.
- Rather high temperature water can be produced. Even in winter, hot water at temperature over 50 °C is stably available.

Although the proposed system has above mentioned great advantages and actually high performance data have been obtained, i.e. high Reynolds number of 6500, high temperature water of 78.0 °C and the high heat recovery efficiency of 85.8%, these values were obtained without any optimization of the system. The system operation and performance are influenced by many factors, i.e. season, weather, the solar radiation, the outside temperature, the mass of CO_2, the temperature of the feed-water, the mass flow rate of the feed-water, the inclination angle of the evacuated solar collector and so on. Figure 16 shows the organized data with respect to the average solar radiation, the outside temperature and the heat collection efficiency for two years. In the first year, the efficiency is found to be relatively low comparing the second year. In the first year, the experiments were conducted without any optimizing experimental parameter in order to investigate the basic operation and performance of the system. On the other hand, in the second year, it is found that the efficiency is greatly enhanced in some day of November comparing the same season of the first year. This is because the experimental parameters, i.e., the filled mass of CO_2 and the mass flow rate of the feed-water, are optimized for the season and weather. The best optimized condition and parameters have not been revealed. The experiment under many different conditions should be conducted, and the optimized condition and parameter should be found experimentally as well as theoretically.

Additionally, in recent years, due to the extensive attentions for renewable energy system, the roof then becomes a competitive field. The competitive device with the solar water heater is the solar photovoltaic. Interestingly, the proposed solar water heater can absorbs the solar thermal energy form visible to infrared ray radiation, which means that this system could be operated in a cloudy day or in facing the north direction. Further, because very strong natural convective flow is induced in the system, the evacuated solar collector can be placed on the wall with vertical setting. As the future work, the investigation of the system performance with varying the angle of the evacuated solar collector is very interested and this investigation will develop the new application field of the solar water heater.

CONCLUSION

For the supercritical CO_2, the small change in the temperature or the pressure can result in the large change in the density, which can easily induce the natural convective flow. CO_2 is responsible for the greenhouse effect, but is an environmentally friendly refrigerant which has a great potential as a heat transport carrier. A new solar water heater using supercritical CO_2 has been proposed. It is a kind of renewable energy system and can recycle CO_2, which means that it can contribute the reduction of the greenhouse gas. The system can recover solar thermal energy into heat energy (hot water). In this chapter, the originally designed and constructed solar water heater using supercritical CO_2 was introduced.

Figure 16. Data with respect to the average solar radiation, the outside temperature and heat collection efficiency at all days when the experiment carried out for two years

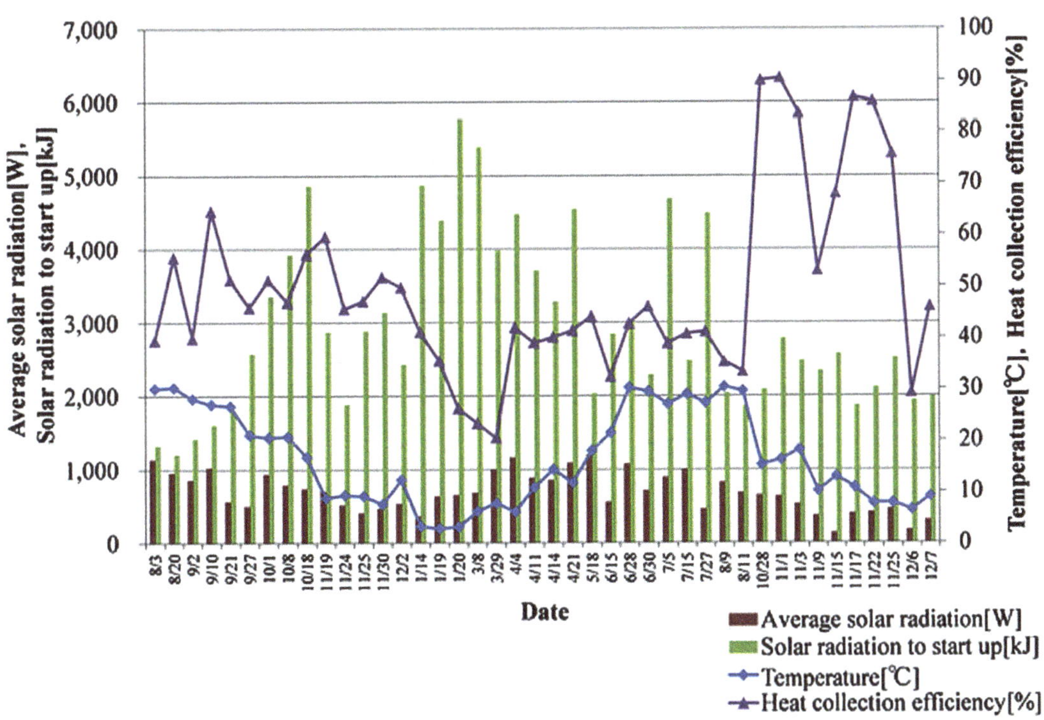

And the typical system operation and performance were described. The obtained results show that the proposed solar water heater can stably operate every seasons and weathers, because the evacuated solar collector can absorb the solar thermal energy in a wide range from visible to infrared ray radiation. The system performance is strong influenced by the seasons and weather, namely, the solar radiation and the outside temperature. The fluid state at the inlet of the evacuated solar collect is a key factor to determine the system efficiency. When the state is liquid, a large natural convective flow is induced in the evacuated solar collector, and the CO_2 can absorb the large amount of solar thermal energy with changing the state from liquid to supercritical. The maximum heat recovery efficiency of 85.8% is observed in autumn, which value is rather higher comparing the conventional solar water heater. Additionally, the system can supply high temperature hot water of 78.0°C.

REFERENCES

Anderson, T. N., & Morrison, G. L. (2007). Effect of Load Pattern on Solar-Boosted Heat Pump Water Heater Performance. *Solar Energy*, *81*(11), 1386–1395. doi:10.1016/j.solener.2007.01.018

Budihardjo, I., & Morrison, G. L. (2009). Performance of Water-in-Glass Evacuated Tube Solar Water Heaters. *Solar Energy*, *83*(1), 49–56. doi:10.1016/j.solener.2008.06.010

Chen, G., Dresselhaus, M. S., Fleurial, J.-P., & Caillat, T. (2003). Recent Developments in Thermoelectric Materials. *International Materials Reviews*, *48*(1), 45–66. doi:10.1179/095066003225010182

Dahl, S. D., & Davidson, J. H. (1997). Performance and Modeling of Thermosyphon Heat Exchangers for Solar Water Heaters, *ASME. Journal of Solar Energy Engineering*, *119*(3), 193–200. doi:10.1115/1.2888018

Davidson, J. H., Mantell, S. C., & Jorgensen, G. (2002). *Status of the Development of Polymeric Solar Water Heating Systems*. In D. Y. Goswami (Ed.), *Advances in Solar Energy* (Vol. 15, pp. 149–186). Boulder, CO: American Solar Energy Society.

Dubey, S., & Tiwari, G. N. (2008). Thermal Modeling of a Combined System of Photovoltaic Thermal PV/T Solar Water Heater. *Solar Energy*, *82*(7), 602–612. doi:10.1016/j.solener.2008.02.005

Esen, M., & Esen, H. (2005). Experimental Investigation of a Two-Phase Closed Thermosyphon Solar Water Heater. *Solar Energy*, *79*(5), 459–468. doi:10.1016/j.solener.2005.01.001

Greffet, J. J., Carminati, R., Joulain, K., Mulet, J. P., Mainguy, S., & Chen, Y. (2002). Coherent Emission of Light by Thermal Sources. *Nature*, *416*(6876), 61–64. doi:10.1038/416061a PMID:11882890

Hashimoto, K., & Saikawa, M. (2004). Trend Report of Development of Residential CO2 Heat Pump Hot Water Heater in Japan. *Proceedings for the JSME Ninth Power Energy Technical Symposium*, 425–430.

Kalogirou, S. A. (2004). Solar Thermal Collectors and Applications. *Progress in Energy and Combustion Science*, *30*(3), 231–295. doi:10.1016/j.pecs.2004.02.001

Kim, M. H., Pettersen, J., & Bullard, C. W. (2004). Fundamental Process and System Design Issues in CO_2 Vapor Compression Systems. *Progress in Energy and Combustion Science*, *30*(2), 119–174. doi:10.1016/j.pecs.2003.09.002

Lorentzen, G. (1990). *Trans-Critical Vapour Compression Cycle Device*. International Patent Publication, Patent No. WO 90/07683.

Lorentzen, G., & Pettersen, J. (1992). New Possibilities for Non-CFC Refrigeration. *Proceedings of the IIR International Symposium on Refrigeration, Energy and Environment*, 147–163.

Mancini, T., Heller, P., Butler, B., Osborn, B., Wolfgang, S., Vernon, G., & Moreno, J. et al. (2003). Dish-Stirling Systems: An Overview of Development and Status, *ASME. Journal of Solar Energy Engineering*, *125*(2), 135–151. doi:10.1115/1.1562634

Mason, A. A., & Davidson, J. H. (1995). Measured Performance and Modeling of an Evacuated-Tube, Integral-Collector-Storage Solar Water Heater, *ASME. Journal of Solar Energy Engineering*, *117*(3), 221–228. doi:10.1115/1.2847803

Morrison, G. L., Budihardjo, I., & Behnia, M. (2004). Water-in-Glass Evacuated Tube Solar Water Heaters. *Solar Energy*, *76*(1–3), 135–140. doi:10.1016/j.solener.2003.07.024

Morrison, G. L., Budihardjo, I., & Behnia, M. (2005). Measurement and Simulation of Flow Rate in a Water-in-Glass Evacuated Tube Solar Water Heater. *Solar Energy*, *78*(2), 257–267. doi:10.1016/j.solener.2004.09.005

Nekså, P. (2002). CO_2 Heat Pump Systems. *International Journal of Refrigeration*, *25*(4), 421–427. doi:10.1016/S0140-7007(01)00033-0

Nekså, P., Rekstad, H., Zakeri, G. R., & Schiefloe, P. A. (1998). CO_2-Heat Pump Water Heater: Characteristics, System Design and Experimental Results. *International Journal of Refrigeration*, *21*(3), 172–179. doi:10.1016/S0140-7007(98)00017-6

Rosengarten, G., Morrison, G., & Behnia, M. (1999). A Second Law Approach to Characterising Thermally Stratified Hot Water Storage With Application to Solar Water Heaters, *ASME. Journal of Solar Energy Engineering*, *121*(4), 194–200. doi:10.1115/1.2888166

Stene, J. (2005). Residential CO_2 Heat Pump System for Combined Space Heating and Hot Water Heating. *International Journal of Refrigeration*, *28*(8), 1259–1265. doi:10.1016/j.ijrefrig.2005.07.006

Yamaguchi, H., Sawada, N., Suzuki, H., Ueda, H., & Zhang, X. R. (2010). Preliminary Study on a Solar Water Heater Using Supercritical Carbon Dioxide as Working Fluid. *Journal of Solar Energy Engineering*, *132*(1), 011010. doi:10.1115/1.4000350

Zhang, X. R., & Yamaguchi, H. (2008). An Experimental Study on Evacuated Tube Solar Collector Using Supercritical CO_2. *Applied Thermal Engineering*, *28*(10), 1225–1233. doi:10.1016/j.applthermaleng.2007.07.013

Zhang, X. R., Yamaguchi, H., Fujima, K., Enomoto, M., & Sawada, N. (2005). A Feasibility Study of CO_2-Based Rankine Cycle Powered by Solar Energy, *JSME Int. J. Ser. B*, *48*, 540–547.

Zhang, X. R., Yamaguchi, H., & Uneno, D. (2007). Experimental Study on the Performance of Solar Rankine System Using Supercritical CO_2. *Renewable Energy*, *32*(15), 2617–2628. doi:10.1016/j.renene.2007.01.003

Zhang, X. R., Yamaguchi, H., Uneno, D., Fujima, K., Enomoto, M., & Sawada, N. (2006). Analysis of a Novel Solar Energy Powered Rankine Cycle for Combined Power and Heat Generation Using Supercritical Carbon Dioxide. *Renewable Energy*, *31*(12), 1839–1854. doi:10.1016/j.renene.2005.09.024

KEY TERMS AND DEFINITIONS

Collected Heat Quantity: Quantity with respect to heat absorbed in the evacuated solar collector.

Evacuated Solar Collector: The heart of the solar water heater using supercritical CO_2. The collector has great properties of heat absorption and thermal insulation.

Fluoro Carbon: A kind of refrigerant, which has great potential as heat transfer carrier and is commercially used as working fluid in energy conversion system. But it has high Global Warming Point.

Heat Recovery Efficiency: A non-dimensional parameter representing the system performance of the solar water heat. It is a ratio between the collected heat quantity and the recovered heat quantity.

Natural Refrigerant: A kind of refrigerant, which exists in nature and is expected as alternative refrigerant with fluoro carbon. Because it is environmentally friendly heat transfer carrier with low Global Warming Point.

Recovered Heat Quantity: Quantity with respect to heat supplied from the solar water heater.

Solar Water Heater: A kind of renewable energy system, which absorbs solar thermal energy and supply heat energy as hot water.

Chapter 8
Heat Transfer and Fluid Flow of Supercritical Fluids in Advanced Energy Systems

Hongzhi Li
Xi'an Thermal Power Research Institute, China

Yifan Zhang
Xi'an Thermal Power Research Institute, China

ABSTRACT

This chapter aims to clarify the supercritical fluids thermal hydraulics characteristics including heat transfer, pressure drops and flow instabilities for the purpose of accurate design and efficient safe operation of advanced energy systems. At first, the convection heat transfer models considering the effect of nonlinear properties and the effect of buoyancy and acceleration have been provided and discussed. Secondly, the hydraulic resistance models for supercritical fluids have been selected and suggested for different conditions. Thirdly, the published models for supercritical flow instabilities based on four different regional partitions are summarized and clarified. At last, two typical case studies have been provided to further intuitively elaborate the thermal hydraulics of supercritical fluids within the advanced energy systems.

INTRODUCTION

In the past decades, supercritical fluids, due to their highly efficient heat transfer effectiveness, have been worldwide extensively concerned to be used in many promising advanced energy conversion and power systems such as supercritical carbon dioxide power cycles configured to operate with a variety of heat sources, GEN IV supercritical water nuclear power technology and supercritical hydrocarbon fuels (e.g. kerosene and methane etc.) rocket and supersonic combustion scramjet engines. It is of great importance to figure out the supercritical fluids thermal hydraulics characteristics including heat transfer, pressure drops and flow instabilities for the purpose of accurate design and efficient safe operation of these advanced energy systems.

DOI: 10.4018/978-1-5225-2047-4.ch008

This chapter is started with an introduction on general background and characteristics of supercritical fluids. In section 1, the general backgrounds and associated thermal hydraulics key issues of several advanced energy systems using supercritical fluids as working fluid will be introduced. In addition, the nonlinear thermal physical properties such as specific heat, density, thermal conductivity and viscosity play an important role in determination of supercritical fluids heat transfer and fluid flow characteristics, thus the general thermal physical properties of supercritical fluids will be illustrated and analyzed.

The section 2 will focus on the heat transfer issues of supercritical fluids. The forced convection heat transfer models considering the effect of non-uniform and nonlinear properties in both spatial and temporal dimensions, the mixed convection heat transfer models considering the effect of buoyancy, the effect of acceleration on the heat transfer performance under high heat fluxes will be discussed.

The section 3 will focus on the hydraulic resistance issues of supercritical fluids. Accurate prediction of pressure loss must be well known for design purposes. Total pressure drop in supercritical systems consist of typically four components: frictional pressure loss, local flow obstructions, acceleration effects due to density changes over length, and gravitational pressure effects. For supercritical fluids, the frictional factor is significantly different from that of the conventional single phase fluids. In this section, several correlations will be selected and suggested for different conditions based on an extensive literature review of hydraulic resistances of supercritical fluids.

The section 4 will focus on the flow instabilities issues of supercritical fluids. Although there is no distinct liquid or gas phases exist, the physical properties of supercritical fluids also change sharply with temperature, in the neighboring region of the pseudo-critical temperature point. It means that the supercritical fluids with high density and low density may flow through the heated channel alternately. This phenomenon may finally lead to the flow instabilities of supercritical fluids, which is similar to the density wave oscillation of two-phase flows at subcritical pressures. Since the flow instabilities of supercritical fluids will lead to heat transfer deterioration and equipment damage, it has been the hot topic in the study of the safe use of supercritical fluids. In this section, the experimental and theoretical work carried out by various investigators over a period of several years will be summarized. Then, the published models for supercritical flow instabilities based on four different regional partitions are summarized and compared in this chapter, the differences and characteristics of these models will be clarified.

The section 5 will provide two case studies to further intuitively elaborate the thermal hydraulics of supercritical fluids within the advanced energy systems. The first one will focus on the heat transfer of supercritical CO_2 at prototypic printed circuit heat exchangers, which could be used for advanced supercritical CO_2 power cycles. The second one will discuss the stability boundaries obtained based on different partition methods and the advantages and disadvantages of each model are analyzed in detail, which will provide technical support for the design of efficient and safe ultra-supercritical steam boilers.

BACKGROUND

With the increasing concern about the fast rise in the energy needs and environmental issues such as air pollution, global warming and ozone depletion, we need to develop a highly efficient, relatively low-cost power conversion system with the minimal environmental impacts by the development of the new concept and advanced power cycles as well as by the refinement of the existing conventional power cycles. There are two main advantages to using a supercritical fluid: First, above the critical pressure no boiling will occur, eliminating the need for condensers, and situations where critical heat flux and burn out could occur. This makes for simplified plant design using fewer components, and therefore, reduced plant costs. Supercritical fluids make an excellent choice for an advanced power cycle. At high temperatures and low pressures, relative to the critical point, it behaves as a gas, while at low temperatures and high pressures it resembles a liquid. This results in low compressor work while maintaining gas like behavior in the rest of the cycle. However, due to the complexity in the property variations, there may be some drawbacks with supercritical fluids in terms of heat transfer, if buoyancy induced heat transfer deterioration occurs in the heat exchangers.

The performance of the various heat exchangers plays a significant role in the cycle efficiency. The heat transfer and fluid flow behavior of supercritical fluids is significantly different from that of a constant property fluid. The radical change in physical properties near the critical point may limit the heat exchange rate of the heat exchangers due to the presence of a pinch point inside the components. Therefore, it is of great interest and necessity to study such heat transfer and fluid flow regime and to evaluate the local heat transfer coefficient and flow regime throughout the components, which is of great importance in heat exchanger design and cycle optimization.

Comprehensive literature reviews of earlier studies on heat transfer and fluid flow characteristics of supercritical fluids can be found in Jackson (2013), Kurganov, Zeigarnik and Maslakova(2012) and Pioro and Duffey(2006). A large amount of previous works and more recent experimental and numerical studies concerning supercritical water, carbon dioxide and cryogenic methane have demonstrated that conventional forced convection heat transfer and frictional factor correlations fail to accurately predict the heat transfer and pressure loss of supercritical fluids with deviations larger than 60% from the experimental data, especially near the pseudo-critical point, indicating that due to rapid changes in thermal physical properties of supercritical fluids near the pseudo-critical point, the heat transfer and fluid flow regime at supercritical pressures could be significantly different from that of constant property fluid. This disparity necessitates an intensive study on developing better physically improved heat transfer and flow resistance models for supercritical fluids. This is also the major objective of this chapter.

GENERAL CHARACTERISTICS OF SUPERCRITICAL FLUIDS

A supercritical fluid can be practically defined as a fluid that is at a pressure above its critical pressure. Based upon the temperature of the fluid it may resemble a gas or a liquid, but always remains a single phase. The heat transfer and fluid flow regime at supercritical pressures could be significantly influenced by the thermal physical properties which present drastic variations within the vicinity of the pseudo-critical temperature. Take supercritical carbon dioxide as an example (Li, Zhang, Zhang, Yao, Kruizenga, Anderson, 2016). It can be seen from Figure 1 and Figure 2 that the specific heat exhibits a peak near the pseudo-critical point. The density as well as the thermal conductivity and the dynamic viscosity decrease dramatically within a very narrow temperature range, especially near the critical point. Besides, the thermal conductivity has a local peak near the critical point. However, the changes of the thermal physical properties become less pronounced with an increase in pressure, which would in turn bring differences in heat transfer features at different supercritical pressures.

Previous studies revealed that the effectiveness of heat transfer at supercritical pressures could be significantly influenced by the ratio of heat flux to mass flux. At high mass flux with relatively low heat flux the heat transfer coefficient could be enhanced near the pseudo-critical temperature due to the large value of specific heat capacity under such conditions. However, the enhancement of heat transfer

Figure 1. Thermal physical properties of supercritical CO_2: density and specific heat capacity

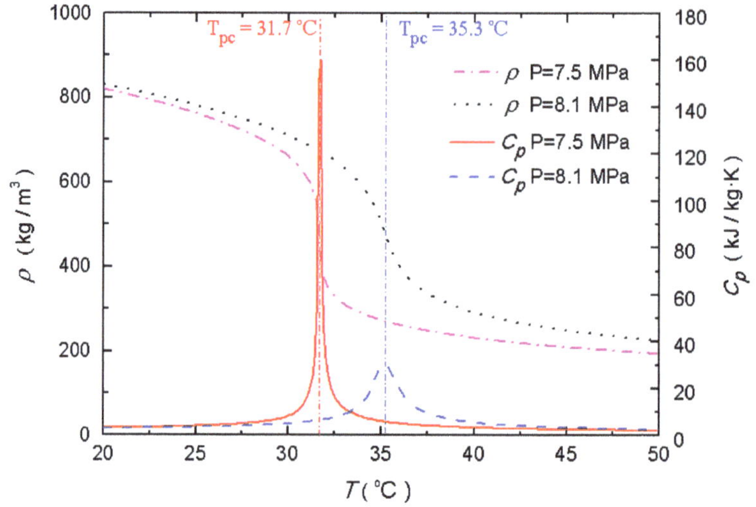

Figure 2. Thermal conductivity and dynamic viscosity

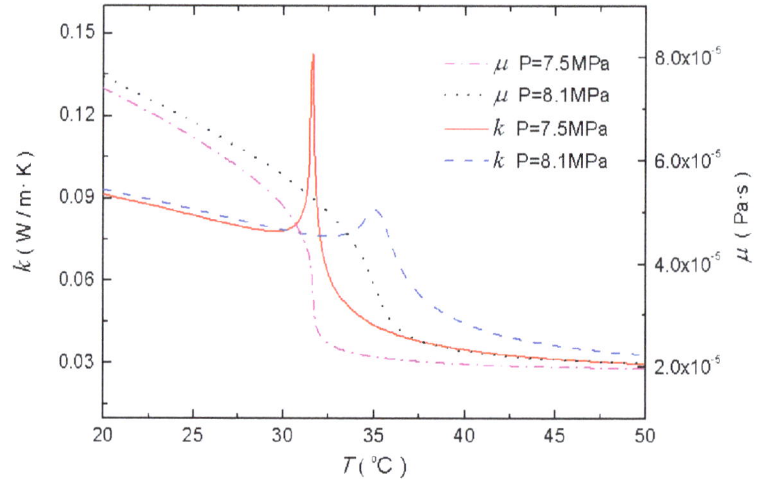

can be reduced with the increase of the heat flux as a result of strong non-uniform and non-linear of thermal physical and transport properties in the boundary layers. When the ratio of the heat flux to mass flux goes to a considerable high level, the heat transfer augmentation would be totally inhibited or even deteriorated due to turbulence production being attenuated and flow laminarization near the heated wall by virtue of buoyancy or flow acceleration. In addition, unlike the identical heat transfer behaviors between the heating and cooling modes for constant property fluid flow, in the pseudo-critical region certain apparent distinctions could be observed due to the reverse distributions of the thermo-physical properties in the boundary layer. This interesting and complicated fluid flow and heat transfer features will be present in the flowing sections.

HEAT TRANSFER

Forced Convection Heat Transfer Model for Fluids with Variable Properties

Satisfactory analytical methods have not yet been developed due to the difficulties in dealing with the non-uniform and non-linear properties variations, especially in the turbulent flows. Therefore, generalized empirical correlations based on experimental data are used for predictions of the heat transfer coefficient at supercritical pressures.

The classical Dittus-Boelter correlation, Equation 1, which has been used intensively as a basis for modified supercritical heat transfer correlations, was proposed to be used in a comparison with the experimental and computational data.

$$Nu_b = 0.023\,\mathrm{Re}_b^{0.8}\,\mathrm{Pr}_b^{n} \tag{1}$$

$n = 0.3$ for heating of fluid, and $n = 0.4$ for cooling of fluid.

Here, $Nu_b = \dfrac{hd}{k_b}$ is the Nusselt number, $h = \dfrac{T_w - T_b}{q_w}$ is the heat transfer coefficient, T_w and T_b are the wall and the bulk temperatures, q_w is the wall heat flux per unit area, d is the hydraulic diameter, and k_b is the heat conductivity at bulk temperature. $Re_b = \dfrac{Gd}{\mu_b}$ is the Reynolds number and $Pr_b = \dfrac{\mu_b C_{p,b}}{k_b}$ is the Prandtl number of the bulk fluid. μ_b is the viscosity at bulk temperature, G is the mass flux per unit area and $C_{p,b}$ is the heat capacity of the bulk fluid.

A majority of previous studies have demonstrated that conventional single phase heat transfer correlations are no longer valid for supercritical fluids since the effect of the strong non-uniformity of the properties on heat transfer effectiveness cannot be simply evaluated using bulk temperature. In this case, the temperature variation in the near-wall region is of importance since the properties are highly sensitive to temperature. In other words, small changes in temperature would result in large changes in properties. Thus, the best way to reflect the influence of the varying properties is to integrate the properties within the boundary layer. However, this integration method is difficult to implement in practical applications because the temperature profile in the boundary layer is not readily obtained from experiments. One practical approach is that the correlations can be evaluated at a simple linearly averaged temperature between wall and bulk, which is known as the film temperature. Another competing method is to employ suitable powers of wall-to-bulk property ratios to account for the effect of the variable properties. As a matter of fact, a variety of existing corrclations (Jackson, 2013; Liao & Zhao, 2002; IIuai, Koyama, & Zhao, 2005; Kruizenga, Li, & Anderson, 2012) for predicting heat transfer of supercritical fluids basically apply the above two methods.

Jackson's correlation (Jackson, 2013), Equation 2, has been extensively proved to work well for predicting supercritical fluid heat transfer experiments because it employs wall-to-bulk property ratios to take account of the variations of properties in the radial direction.

$$Nu_b = 0.0183\,\mathrm{Re}_b^{0.82}\,\mathrm{Pr}_b^{0.5}\left(\frac{\rho_w}{\rho_b}\right)^{0.3}\left(\frac{\bar{C}_p}{C_{p,b}}\right)^{n} \tag{2}$$

where the exponent n is (T in Kelvin)

$n = 0.4$ for $T_b < T_w < T_{pc}$ and for $1.2T_{pc} < T_b < T_w$;

$n = 0.4 + 0.2\left(\frac{T_w}{T_{pc}} - 1\right)$ for $T_b < T_{pc} < T_w$;

$n = 0.4 + 0.2\left(\frac{T_w}{T_{pc}} - 1\right)\left[1 - 5\left(\frac{T_b}{T_{pc}} - 1\right)\right]$ for $T_{pc} < T_b < 1.2T_{pc}$

In these equations, the index b refers to the bulk and w to the wall conditions, the index pc means 'pseudo-critical'.

In general, they perform better than conventional single phase correlations and could capture the qualitative trend of heat transfer coefficients, but quantitatively in the near pseudo-critical regions they show substantial differences from each other. This disparity necessitates a further work to develop a better physically improved correlation for supercritical fluids.

A PDF-Based Physically Improved Heat Transfer Model

In many engineering practices Reynolds-averaged rule can be applied for turbulent energy equation, where the instantaneous temperature T could be partitioned into a time-averaged value $\bar{T}$ and a fluctuation value $T^{'}$ as in Equation 3:

$$T = \bar{T} + T^{'} \tag{3}$$

Then, an essential question arises. Whether a property evaluated at mean temperature also equals the value calculated from averaging all the properties evaluated at every instantaneous temperature within the same interval? This is true for most of the engineering turbulent flow cases since the properties could be assumed constant or they are linearly dependent on the temperature. However, for turbulent flow at supercritical pressures, especially in the pseudo-critical region, this method yields considerable error as the properties are highly nonlinear dependent on temperature. For example, as shown in Figure 1, the specific heat capacity exhibits a maximum in the vicinity of pseudo-critical region. As Equation 4 shows, specific heat evaluated at mean temperature does not equal to the calculated from temporal averaging at every instantaneous temperature. Therefore, the work on development of an effective method to get physically reasonable time-averaged properties plays an essential role in accurately predicting the heat transfer behavior at the supercritical pressures.

$$C_p(\bar{T}) \neq \frac{1}{\Delta t}\int_t^{t+\Delta t} C_p(T)dt \tag{4}$$

The features of the turbulent flow suggest that the instantaneous variables are not deterministic but continuously random. This means the instantaneous temperature, on which the physical and transport properties depend, could not be readily obtained in a deterministic manner. In other words, it is impractical to measure every instantaneous temperature and properties. A possible statistical solution coming into the authors' minds is the Probability Density Function (PDF) for describing the distributions of the random instantaneous temperatures. In probability theory and statistics, the PDF of the continuous random variables describes the relative frequencies of the different values for that random variable. The normal distribution which is often used to describe, at least approximately, any variable that tends to cluster around the mean value, thus is very appropriate for describing the distributions of the instantaneous temperature away from their mean value.

The physically reasonable time-averaged properties were generated by employing the idea of the probability density function of the instantaneous temperature which is given by the Equation 5

$$\varphi(T,\bar{T}) = \frac{1}{\sqrt{2\pi\sigma^2}} e^{-\frac{(T-\bar{T})^2}{2\sigma^2}} \tag{5}$$

where the value of the probability density function $\varphi(T,\bar{T})$ is between 0 and 1, T is the instantaneous temperature, $\bar{T}$ is the time-averaged temperature and the parameter σ ,which is the variance describing how far values lie from the mean, could be regarded as a measure of intensity of turbulent temperature. As can be seen from Figure 3, which shows the PDF distribution with mean value $\bar{T} = 30$ and various σ, the graph of the probability density function is bell-shaped, with a peak at the mean. The underlying physical ground is that the probability of the instantaneous temperature presenting at the mean temperature achieves the highest and the probability occurring at other temperature would decrease with the increase of its distance away from the mean. Another interesting observation is that the shape of PDF would become lower and broader with the increase of the parameter σ. The intensity of the turbulent temperature depends on certain imposed conditions e.g. mass flux, heat flux, so that the parameter σ may need to be changed according to the different working conditions.

Since we did not measure the instantaneous temperature, the value of the intensity of turbulent temperature could not be theoretically derived or accurately measured by the experiments, so that the accurate value of the parameter σ could not be obtained. As a compromise, we assume, for a given position of the cross section, the time-averaged temperature $\bar{T}$ equals to the bulk temperature T_b, the local instantaneous temperatures fluctuated in the vicinity of the local bulk temperature but could not exceed the local wall temperature. In other words, the local instantaneous temperatures lie in the range from ($T_b - |T_w - T_b|$) to ($T_b + |T_w - T_b|$).

In statistics, the probability that a normal deviate lies in the range $\bar{T} - n\sigma$ and $\bar{T} + n\sigma$ is given by Equation 6. The distributions of PDF and tolerance intervals were illustrated in Figure 4, It can be seen that when n=3, about 99.7% of values drawn from a normal distribution are within three standard de-

Figure 3. Distributions of PDF with various σ and $\bar{T} = 30$ °C

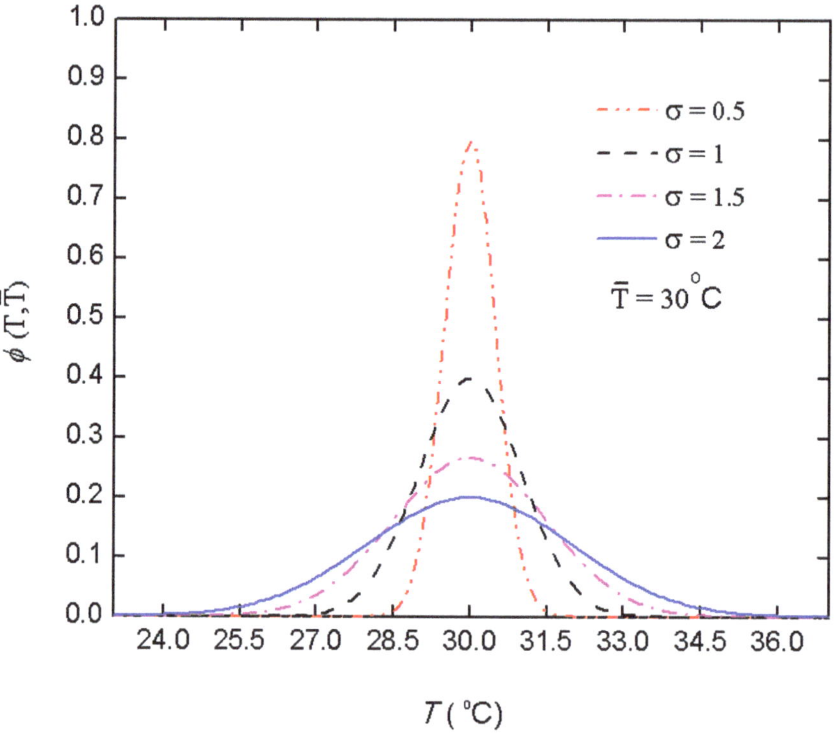

Figure 4. Distributions of PDF and tolerance intervals with $\sigma = 1$ and $\bar{T} = 30$ °C

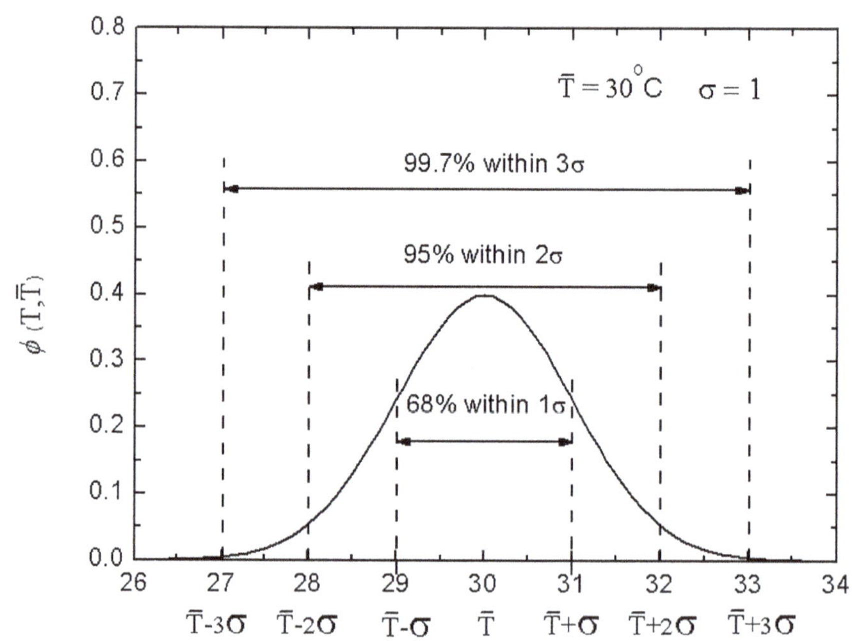

viations (3σ) away from the mean. This fact is known as the the "3σ rule" (Li, Zhang, Zhang, Yao, Kruizenga, Anderson, 2016).

$$F(\bar{T}+n\sigma)-F(\bar{T}+n\sigma)=erf\left(\frac{n}{\sqrt{2}}\right) \tag{6}$$

Here, F(x) is the cumulative distribution function and erf(x) is error function.

In the present study, based on the above assumptions and 3σ rule, we could get a relatively reasonable value of σ according to Equation 7

$$\sigma=\frac{\left|T_w-T_b\right|}{3} \tag{7}$$

Once the parameter σ is determined, the probability density function $\varphi(T,\bar{T})$, as a shaping function, could serve as a weighting factor to time-average the instantaneous specific heat capacity and get the PDF-based time-averaged bulk specific heat $C_{p,b,pdf}$ as described by Equation 8. This PDF-based time-averaged model could also be applied to all the other physical and transport properties such as density, dynamic viscosity and thermal conductivity.

$$C_{p,b,pdf}=\int_{-\infty}^{+\infty}C_p(T)\varphi(T,\bar{T})dT=\int_{T_b-3\sigma}^{T_b+3\sigma}C_p(T)\varphi(T,T_b)dT \tag{8}$$

Figure 5, Figure 6, Figure 7, and Figure 8 show PDF-based time-averaged properties of supercritical CO_2 with various σ at the pressure of 7.5MPa. The original properties, indicated by $\sigma=0$, are also shown as a reference. Compared to the rapid changes of original properties over a relatively narrow band of temperature, the PDF-based time-averaged properties exhibit a similar but milder trend in the vicinity of the pseudo-critical temperature (T_{pc}= 31.7°C at P=7.5MPa). Specifically, the dramatic maximum of specific heat as well as the local peak of thermal conductivity at the pseudo-critical temperature is significantly reduced with the increase of the parameter σ. From this point of view, the parameter σ could be interpreted as a "smoothing" factor. The variation behavior of PDF-based time-averaged properties becomes smoother as the "smoothing" factor being enhanced.

A semi-empirical physically improved new correlation evaluated by PDF-based time-averaged properties for forced convection heat transfer of CO_2 at supercritical pressures is obtained as Equation 9:

$$Nu_{b,pdf}=CRe_{b,pdf}^{m}Pr_{b,pdf}^{n}\left(\frac{\rho_w}{\rho_{b,pf}}\right)^p\left(\frac{\bar{C}_{p,pdf}}{C_{p,b,pdf}}\right)^q \tag{9}$$

Based on the authors' previous experimental and numerical data (Li, Zhang, Zhang, Yao, Kruizenga, Anderson, 2016), the values of the empirical parameters C, m, n, p, q were shown in Table 1 for heating and cooling mode supercritical CO_2 heat transfer performance, respectively.

Figure 5. PDF-based time-averaged thermal properties of CO_2 with various σ at 7.5MPa: specific heat capacity

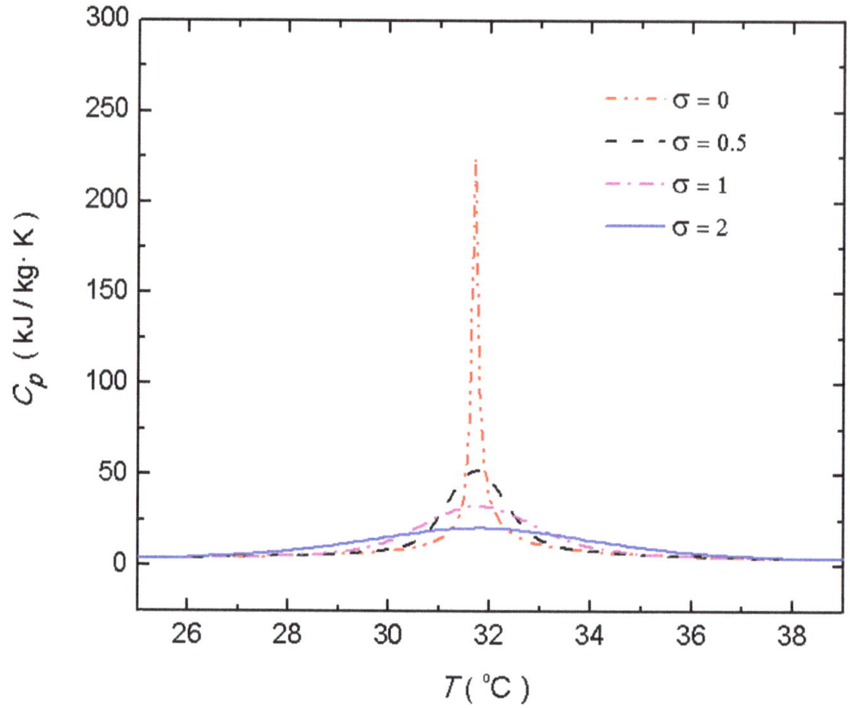

Figure 6. Density

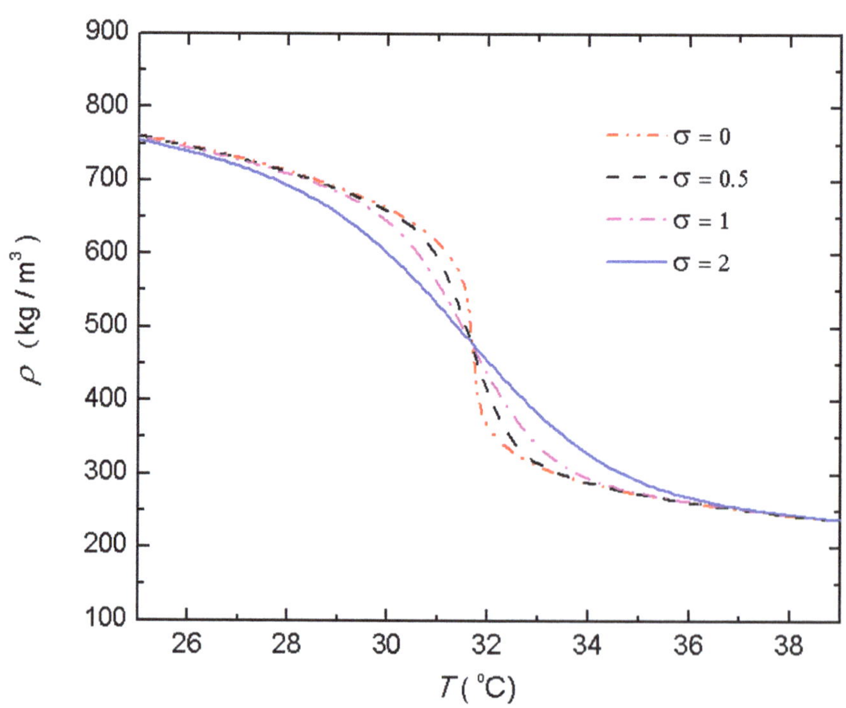

Figure 7. Thermal conductivity

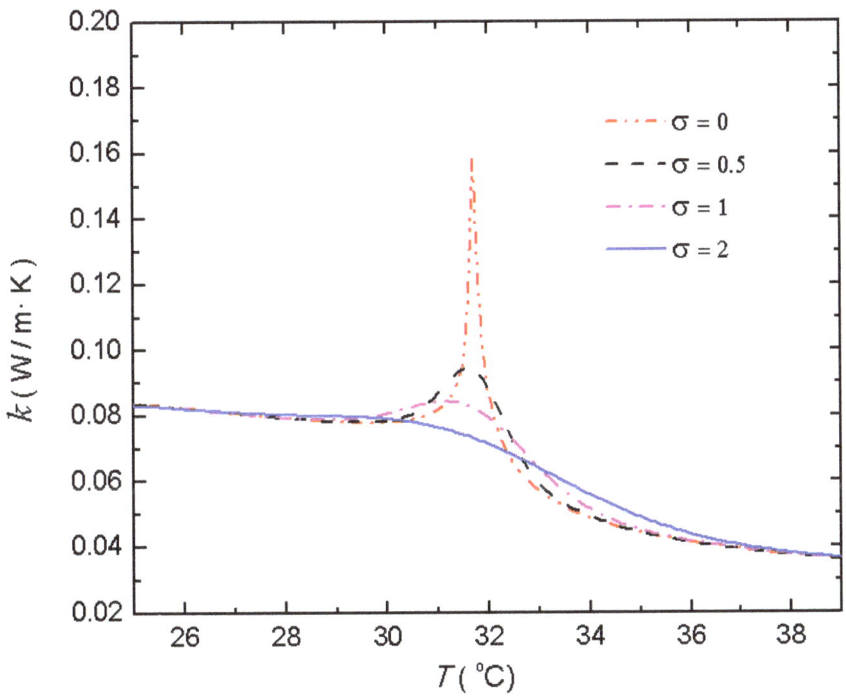

Figure 8. Dynamic viscosity

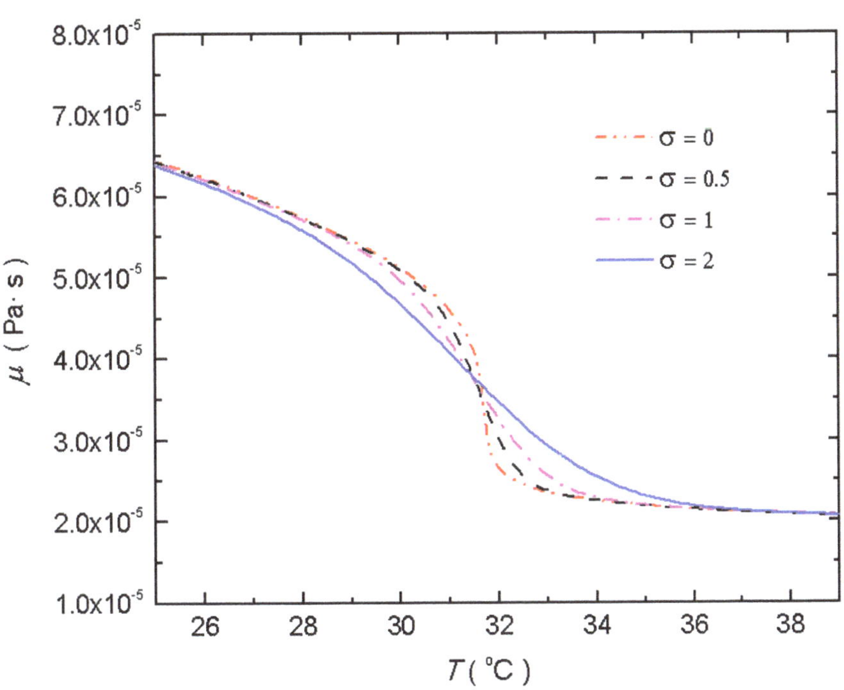

Table 1. The values of the parameters for the PDF-based supercritical CO_2 empirical correlation

	C	m	n	p	q
Heating	0.0409	0.681	0.650	0.223	0.397
Cooling	0.0781	0.662	0.455	0.392	0.334

To look into the new correlation, the single phase base ($CRe^{m}_{b,pdf}Pr^{n}_{b,pdf}$) equation ensures a higher accuracy for forced convection heat transfer, as at the region far from the pseudo-critical temperature, the property ratios multiplier approaches one and the PDF-based time-averaged properties will also have no effect. Thus, under this situation the new developed correlation would be turned back to the single phase correlation. But near the pseudo-critical region, the property ratio modification terms would contribute to account for the variation of property distribution in the cross section, especially in the boundary layer. This could be regarded as a modification in spatial dimensions and would lead to an increase in the accuracy of prediction in the region near the pseudo-critical region. In addition, PDF-based time-averaged properties aim to represent the effect of instantaneous temperature fluctuations and in turn the strongly nonlinear dependent property fluctuations. This could be regarded as a modification in temporal dimension and would contribute to collapsing the data to its average value.

This section provides a general method how to get a physically improved heat transfer correlation, which can also be used for developing other supercritical fluids PDF-based correlations.

Buoyancy Effect

There has been a large number of experimental and numerical studies performed investigating the effect of buoyant forces on the heat transfer. In the buoyancy-aided case such as upward flow with heating, starting at conditions where the buoyancy is negligible small indicates that heat transfer is fully turbulent, as the buoyant force increases there is a relatively sharp deterioration in heat transfer, which can cause Nusselt numbers to fall to nearly 50% of predicted forced turbulent equations. However, above a certain threshold of buoyancy influence, a recovery or even enhancement in the heat transfer effectiveness occurs, which indicates the buoyancy dominated the flow regime. In contrast, for the buoyancy opposed case such as downward flow with heating, a systematic improvement in heat transfer effectiveness is observed for conditions when buoyancy becomes significant.

This buoyancy effect has been studied by several investigators with numerical models for better understanding the physical behavior. Several authors have developed non-dimensional number groups, derived from wall shear stress modifications, to be used as correlating parameters for regions where buoyancy or acceleration are important.

Jackson (2013) derived the term Bo, Equation 10, to evaluate the buoyancy,

$$Bo = \frac{Gr_b}{\mathrm{Re}_b^{2.625}\,\mathrm{Pr}_b^{0.4}} \tag{10}$$

where Gr_b is the Grashof number defined by differences between wall and bulk density:

$$Gr_b = \frac{g\left(\rho_b - \rho_w\right)d^3}{\rho_b \upsilon_b^2} \tag{11}$$

Kim and Kim (2010) used many of the same assumptions as Jackson, but instead chose to include a Grashof number based on heat flux:

$$Bu = \frac{Gr_q}{\mathrm{Re}_b^{3.425}\,\mathrm{Pr}_b^{0.8}}\left(\frac{\mu_w}{\mu_b}\right)\left(\frac{\rho_b}{\rho_w}\right)^{0.5} \tag{12}$$

$$Gr_q = \frac{g\beta_b q_w^{"} d^4}{k_b \upsilon_b^2} \tag{13}$$

By comparing Jackson's and Kim's correlating parameters it is clear that they are very similar. This same procedure can be done for acceleration and will be presented in the next section.

Acceleration Effect

The mass flow rate should keep constant at all axial positions, thus due to heat being received by the fluid the bulk temperature rises, the bulk density falls and in turn the bulk velocity increases, i.e. the fluid accelerates. An additional pressure difference must be imposed to cause this acceleration and in the boundary layer region the gradient of shear stress across the flow has to undergo an adjustment to balance the excess pressure gradient. Hence, it follows that the shear stress falls more quickly with distance from the wall than it would in the absence of any flow acceleration. Consequently, its value in the laminar sub-layer region is lower than that for a steady fully developed flow and the production of turbulence is reduced and the effectiveness of heat transfer is impaired. The higher the heat flux the greater is the impairment of heat transfer.

Jackson (2013) proposed a non-dimensional number Ac as Equation 14 to reflect the effect of thermally-induced acceleration flow.

$$Ac_b = \frac{\beta_b q_w^{"} d}{k_b\,\mathrm{Re}_b^{1.63}\,\mathrm{Pr}_b} \tag{14}$$

This correlation takes the earlier Jackson correlation and then solves it iteratively with the acceleration parameter and variable property terms.

A similar term was derived by Kim and Kim (2010) and was of very similar form:

$$Ac = \frac{q_w^{"}\beta_b}{GC_{p,b}\,\mathrm{Re}_b^{0.625}}\left(\frac{\mu_w}{\mu_b}\right)\left(\frac{\rho_b}{\rho_w}\right)^{0.5} \tag{15}$$

It can be concluded that while good, necessary progress has been made in semi-empirical correlations, more work needs to be done to better refine and tune the correlations for practical use.

HYDRAULIC RESISTANCE

General Model for Total Pressure Drop

In general, the total pressure drop in supercritical systems consist of typically four components: frictional pressure loss, local flow obstructions, acceleration effects due to density changes over length, and gravitational pressure effects. These effects can be calculated according to the Equation 16:

$$\Delta P = \Delta P_f + \Delta P_a + \Delta P_l + \Delta P_g \tag{16}$$

where ΔP is the total pressure drop, ΔP_f is the pressure drop due to frictional resistance, which is defined as Equation 17

$$\Delta P_f = f \frac{L}{D} \frac{G^2}{2\rho} \tag{17}$$

where f is the friction factor, G is the mass flux, L is the length of test section, D is the hydraulic diameter, and ρ is the bulk fluid density.

ΔP_a is the pressure drop due to acceleration of flow defined as Equation 18

$$\Delta P_a = G^2 \left(\frac{1}{\rho_{out}} - \frac{1}{\rho_{in}} \right) \tag{18}$$

here the density is designated by the inlet and outlet conditions.

ΔP_l is the pressure drop due to local flow obstructions defined as Equation 19

$$\Delta P_l = \zeta \frac{G^2}{2\rho} \tag{19}$$

ζ is the friction factor due to local obstructions.

ΔP_g is the pressure drop due to gravity defined as Equation 20

$$\Delta P_g = \pm g \left(\frac{i_{out}\rho_{out} + i_{in}\rho_{in}}{i_{out} + i_{in}} \right) L \sin\theta \tag{20}$$

g is the acceleration due to gravity, i is the enthalpy (inlet or outlet), L is the test section length, and θ is the angle from horizontal. Depending on the experimental setup several of these terms may be neglected.

Correlations of Frictional Factor for Supercritical Fluids

The pressure loss due to friction is the most uncertain of the pressure loss terms, since an empirical correlation needs to be used to determine the total pressure loss.

Pioro and Duffey (2006) performed an extensive literature review study on hydraulic resistances of several supercritical fluids with various Reynolds numbers, flow orientations, and tube diameters.

The Blasius correlation, Equation 21, is assumed to predict fully developed turbulent flows in smooth pipes with Reynolds numbers less than 105. When compared to experimental data for supercritical water, CO_2, and R-22 it was found to predict most values within ±20%. However, this correlation does not take the surface roughness into consideration.

$$f_{Blasius} = \frac{0.316}{\mathrm{Re}^{0.25}} \tag{21}$$

The use of the explicit Colebrook and White friction factor has been more reliable correlation, as the surface roughness effects are included.

$$f_{CW} = \left(-2\log\left(\frac{\Delta / d}{3.7} + \frac{2.51}{\mathrm{Re}\sqrt{f_{CW}}}\right)\right)^{-2} \tag{22}$$

where Δ is the surface roughness. This correlation has been found to match closely with the well-known Moody friction factor. Unfortunately, in most of the literature the roughness was not reported, making it difficult to evaluate the role of surface roughness in frictional pressure losses for supercritical fluids.

FLOW INSTABILITY

General Considerations

The flow instabilities may cause oscillations of flowing and thermal parameters of the equipment and systems, such as the boiler water wall in thermal power plants, the steam generators in nuclear power plants and the heat exchangers in chemical engineering, etc. Undesirable oscillations may result in the boiling crisis, and even induce the thermal fatigue of the pipes and the disrepair of the equipment. Hence, the research on flow instabilities is significant and necessary.

It is well known that the two-phase flow instability is one of the most common flow instabilities, because of the phase change and physical properties of the subcritical fluid, and much effort had been made to study the two-phase flow instabilities (e.g., Brauner & Moalem, 1992; Fukuda & Kobori, 1979).

Considering there is no distinct liquid or gas phases exist, are there any flow instabilities of the supercritical fluids. In fact, the physical properties of supercritical fluids also change sharply with temperature, in the neighboring region of the pseudo-critical temperature point. It means that the supercritical fluids with high density and low density may flow through the heated channel alternately. This phenomenon may finally lead to the flow instabilities of supercritical fluids, which is similar to the density wave oscillation of two-phase flows at subcritical pressures. Since the flow instabilities of supercritical fluids will also lead to heat transfer deterioration and equipment damage, it has been the hot topic in the study of the safe use of supercritical fluids.

The study of flow instabilities at supercritical pressure is more difficult than that at subcritical pressure, because of the complexities of physical properties of supercritical fluids and the limitation of testing technology. However, with the development in science and technology, more and more researches have been made for the flow instabilities of supercritical fluids in the recent decades. The main research methods are still experimental study and theoretical study.

Theoretical study is an effective and convenient access to investigate the flow instabilities of supercritical fluids. Xiong, Yan, Xiao, Li, Huang and Yu (2013) set up an in-house code by using time domain method to study the flow instabilities of supercritical fluids in parallel channels. The results showed that the inlet temperature had a non-monotonic effect on the threshold power, and the threshold power was roughly proportional to the mass flow rate. Zuber (1966) established a two region model for studying DWO of SCW in a single tube by using the frequency domain method. The demarcation point between the two regions was the pseudo critical point. According to the model of Zuber, influences of pressure, mass flux and inlet subcooling could be obtained conveniently. Later, a three region model was presented by Zhao (2005), which contained heavy fluid region, heavy & light fluid mixture region and light fluid region. The three region model had a better accuracy than the two region model. Zhang, Li, Li, Wang, Zhang and Lei (2015) proposed a new model to study the density wave instabilities of supercritical water flowing in tubes, according to the frequency domain method. In this model, a new regional partition of supercritical water is proposed for better dealing with the great variations of physical properties of supercritical water in the neighborhood region of pseudo critical point, based on the analyses of the physical properties. Many other calculation studies are also been conducted by lots of scholars (e.g., Antoni & Dumaz, 2003; Gómez, Class, Lahey & Schulenberg, 2008; Su, Feng, Zhao, Tian, Su & Qiu, 2013; Yi, Koshizuka & Oka, 2004)

Experimental study is another effective access to investigate the flow instabilities. However, it is difficult and complex to study the density wave instability of water under supercritical pressure according to experiments, due to the difficult experimental conditions under supercritical pressures. Hence, experimental studies on the flow instabilities of supercritical fluids are very limited, and only few of relevant experiments can be found in literature. An experimental study on the flow instabilities of supercritical water had been performed by Xiong, Yan, Xiao, Li, Huang and Yu (2012) in two parallel channels with pressures from 23 to 25MPa. The length of the heated pipes was 3000 mm, and the inner diameter was 6mm, and the outer diameter equaled to 11 mm. Based on the experimental phenomena, the way of defining the onset of flow instability of supercritical water in parallel channels was proposed. Zhang, Wang, Gu, Qiao, Shi and Luo (2013) studied the flow instabilities of supercritical water in a square channel at near-critical and supercritical pressures. The square channel is composed of a 10 mm×2.5 mm circular pipe and a 15mm×15mm square pipe. Joen and Rohde (2011) studied the flow instabilities at supercritical pressure using Freon R23 on a scaled version test facility, and made the relevant stability maps.

Models of Flow Instabilities for Supercritical Fluids

Research methods and models of flow instabilities for subcritical fluids are relatively mature, such as Time domain method, Frequency domain method and Homogenous flow model, Drift-flux model, and the the two-phase flow instabilities can be predicted well by these models and methods. These models and methods can also be employed to study the supercritical flow instabilities, if the flow region in tube at the supercritical pressure could be divided into three parts just like that at the subcritical pressure.

Hence,many relevant efforts had been carried out in decades. Four different regional partitions of supercritical water (SCW) were proposed and published (e.g., Zuber,1966; Antoni & Dumaz, 2003; Zhao, 2005; Zhang, Li, Li, Wang, Zhang & Lei, 2015), respectively. These regional partitions are summarized and compared in this chapter.

In the study by Zuber (1966), a two-region model was made and employed for the stability analyses of SCW in tubes. Supercritical water in the whole flow region was considered as two parts, i.e., gas like region and liquid like region. The demarcation point between these two regions was the pseudo critical point (the point with the maximum specific heat C_p). The upstream was defined as the liquid like region, and the downstream was defined as the gas like region. Detailed partition method of SCW from Zuber can be shown as Figure 9.

In the study by Antoni and Dumaz (2003), a three-region model was made and employed for the stability analyses of SCW in tubes. Supercritical water in the whole flow region was considered as three parts. A specific region of SCW was defined as "two-phase mixture", and the demarcation points of three regions were the pseudo critical point and a point with "latent heat" of 400kJ/kg.

The detailed regional partition from Antoni and Dumaz is shown in Figure 9B_2 represents the pseudo critical point and $\boldsymbol{A}_2$ represents the point with "latent heat" of 400kJ/kg. The two points divide the whole flow region into three parts, i.e., heavy fluid region (H-F), heavy & light fluid mixture region (H-L-M) and light fluid region (L-F).

In the study by Zhao (2005), a three-region model was made and employed for the stability analyses of SCW in tubes. The whole flow region was also divided as H-F, H-L-M and L-F. The two demarcation points of the three-region model were named as ***A*** and ***B***, which represented Pseudo Saturated Water (PSW) and Pseudo Saturated Steam (PSS), respectively. The calculation equations of the temperatures of ***A*** and ***B*** were written as Equation 23 and Equation 24 respectively (Zhao, 2005).

$$T_A = 350\,°\mathrm{C} \tag{23}$$

$$T_B = n_2 + \left(\frac{P - n_3}{n_1}\right)^{0.5} \tag{24}$$

where,

n_1 = 0.10192970039326E-02,
n_2 = 0.57254459862746E+03,
n_3 = 0.13918839778870E+02, and
P = The system pressure.

The detailed regional partition of SCW from Zhao can be shown as Figure 9

The flow region of SCW is also divided into three region (H-F, H-L-M and L-F) (Zhang, Li, Li, Wang, Zhang & Lei, 2015), shown in Figure 9. The correlations of T_{PSW} and T_{PSS} are shown as Equation 25 and Equation 26.

$$T_{PSW} = m_{PSW} + m'_{PSW} \cdot \left(P \cdot 10^{-6}\right) + m''_{PSW} \cdot \left(P \cdot 10^{-6}\right)^2 \tag{25}$$

$$T_{PSS} = m_{PSS} + m'_{PSS} \cdot \left(P \cdot 10^{-6}\right) + m''_{PSS} \cdot \left(P \cdot 10^{-6}\right)^2 \tag{26}$$

where,

m_{PSW} = 309.50661
m'_{PSW} = 1.26935
m''_{PSW} = 0.01174
m_{PSS} = 279.44318
m'_{PSS} = 8.5502
m''_{PSS} = -0.07014

Other physical properties of SCW at pseudo critical water(PSW) and pseudo critical steam (PSS), such as h_{PSW} and h_{PSS}, can also be determined based on the value of temperature and pressure. Then, based on the energy conservation equations, the values of Z_{PSW} and Z_{PSS} can be calculated according to Equation 27 and Equation 28. It should be pointed out that the height of inlet is defined as 0 here.

$$Z_{PSW} = \frac{\rho_{in} u_{in} A' \left(h_{PSW} - h_{in}\right)}{q'S} \tag{27}$$

$$Z_{PSS} = \frac{\rho_{in} u_{in} A' \left(h_{PSS} - h_{in}\right)}{q'S} \tag{28}$$

where, Z_{PSW} is the height of PSW, Z_{PSS} is the height of PSS, ρ_{in} is the density at inlet, u_{in} is the inlet velocity, q' is the surface heat flux, S is the heated perimeter, A' is the flow area, h_{PSW} is the specific enthalpy at PSW and h_{PSS} is the specific enthalpy at PSS.

To show the detailed regional partition differences directly, the above mentioned regional partitions are all expressed in Figure 9. $A_1(B_1)$ is the demarcation line of the regional partition of SCW from Zuber, (1966), A_2,B_2 are the demarcation lines of the regional partition of SCW from Antoni and Dumaz (2003), A_3,B_3 are that from Zhao (2005), and A_4,B_4 are that from Zhang, Li, Li, Wang, Zhang and Lei (2015).

From Figure 9, the four regional partitions in different models express great differences of the ranges of the detailed regions, and this may lead to great different results. To further discuss the differences of

Figure 9. Comparison of different regional partitions of SCW in different models

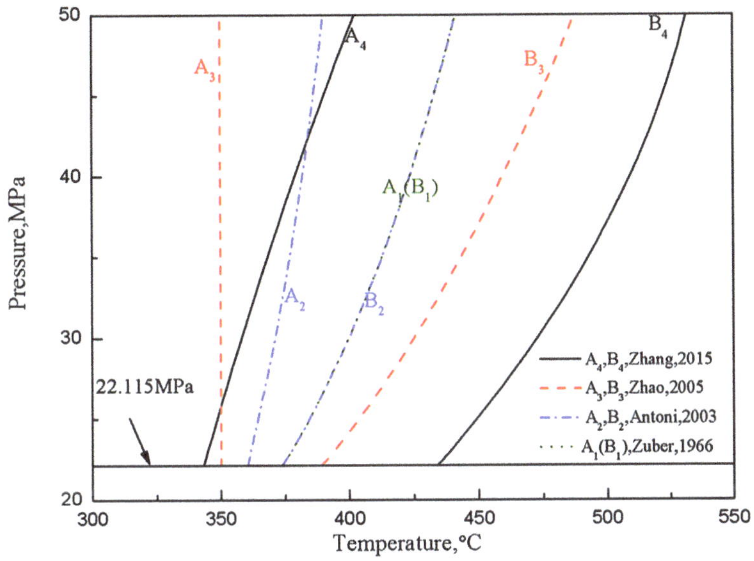

these models and illustrate the advantages of the present model, the regional partitions of SCW from different models are compared in the diagrams with physical properties variation tendencies, at different pressures, 23, 26, 32 and 40MPa, shown in Figure 10, Figure 11, Figure 12, and Figure 13.

According to the regional partition of SCW from Zuber shown in Figure 10, Figure 11, Figure 12, and Figure 13, the whole region of SCW is divided into two parts, and both H-F and L-F all contain somewhat parts where the coefficient of volume expansion of SCW changes sharply. Hence, considering the definitions of H-F and L-F, the great variations of the coefficient of volume expansion and density near the pseudo critical point are ignored, and the partition method from Zuber is not so accurate. This problem has be improved partly in the partition method from Antoni, shown in Figure 10, Figure 11, Figure 12, and Figure 13, a so called H-L-M is proposed specially to represent the region near the pseudo critical point. However, the so called H-L-M only contains part of the region near the pseudo critical point, and the coefficient of volume expansion of SCW also changes sharply in the low temperature part of L-F. In view of the disadvantages of the two previous partition methods, Zhao (2005) provides his partition method. As shown in Figure 10, Figure 11, Figure 12, and Figure 13, the partition method from Zhao is more suitable than that from Zuber and Antoni, because the major part of the region near the pseudo critical point is included in H-L-M.

However, analyzing the partition method from Zhao in detail, there is still two factors not so suitable. Firstly, T_{PSW} is defined as a constant, equaling to 350°C. According to Figure 10, Figure 11, Figure 12, and Figure 13, it is obvious that the region near the pseudo critical point at different pressures does not have the same starting point. It means that T_{PSW} may change with the pressure, and defining T_{PSW} as a constant is not so suitable. Secondly, the temperature of the right demarcation point between H-L-M and L-F is somewhat low, and this question is more obvious especially at high pressures, such as 32 and 40MPa shown in Figure 12 and Figure 13. Hence, the coefficient of volume expansion of SCW still changes greatly in the low temperature part of L-F, and that is not so suitable with the definition of L-F.

The two problems in the partition method of SCW from Zhao have been solved in the study of Zhang, Li, Li, Wang, Zhang. and Lei (2015), because the three regions are defined in a new way based on the detailed analyses of physical properties of SCW. It is easily observed from Figure 10, Figure 11, Figure 12, and Figure 13that H-L-M in the present model matches with the region near the pseudo critical point better than that in the previous models. This partition method may finally make the calculation model more accurate in the study on flow instabilities of SCW.

Figure 10. Comparison at 23MPa

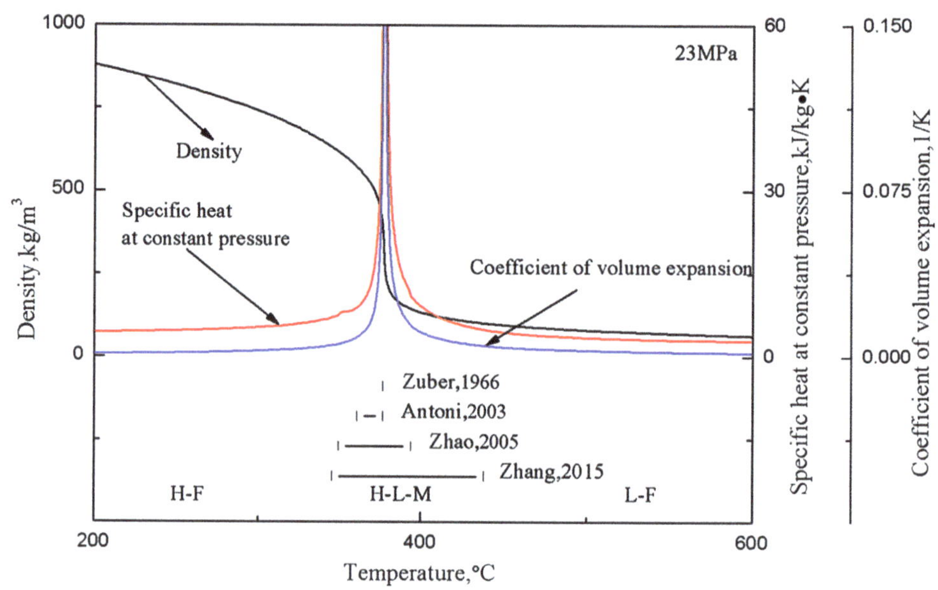

Figure 11. Comparison at 26MPa

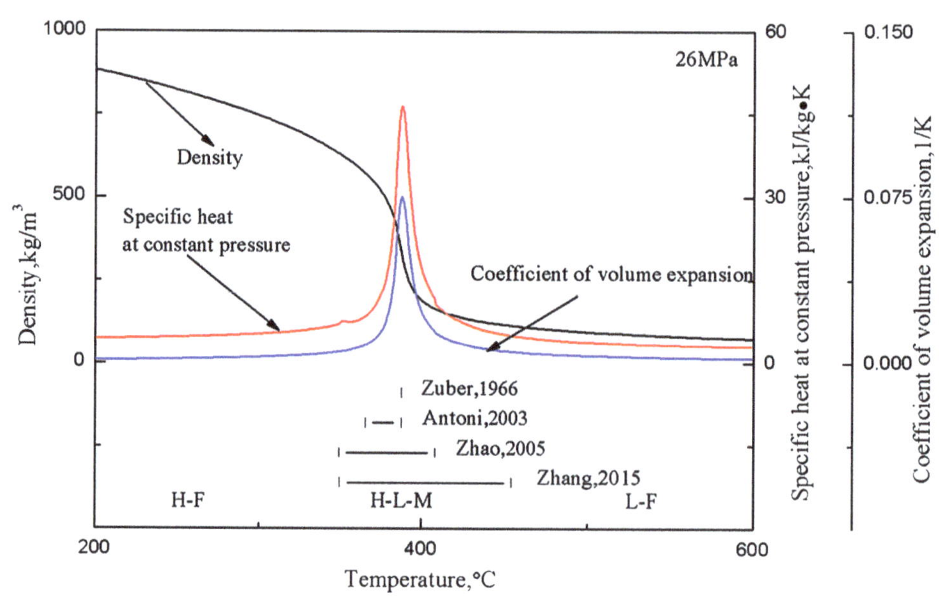

Figure 12. Comparison at 32MPa

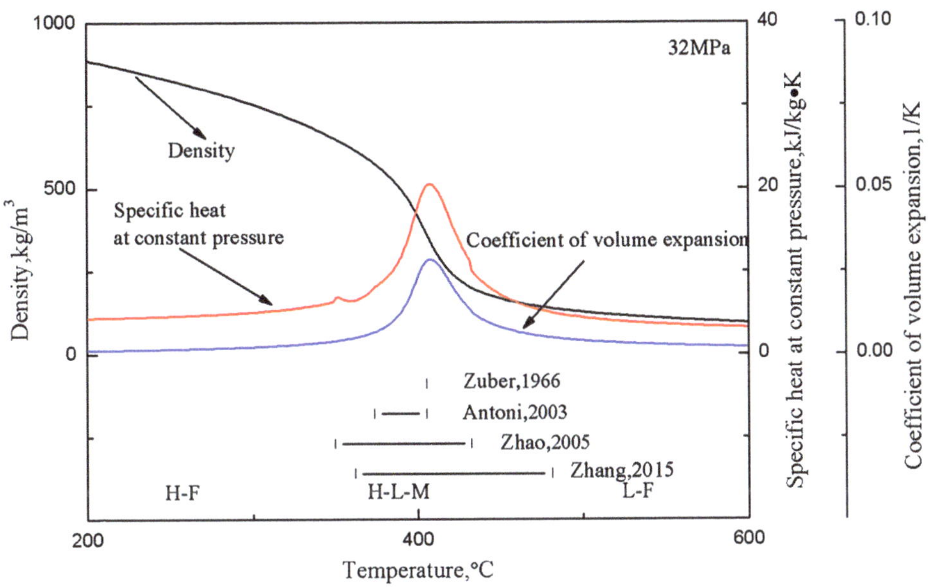

Figure 13. Comparison at 40MPa

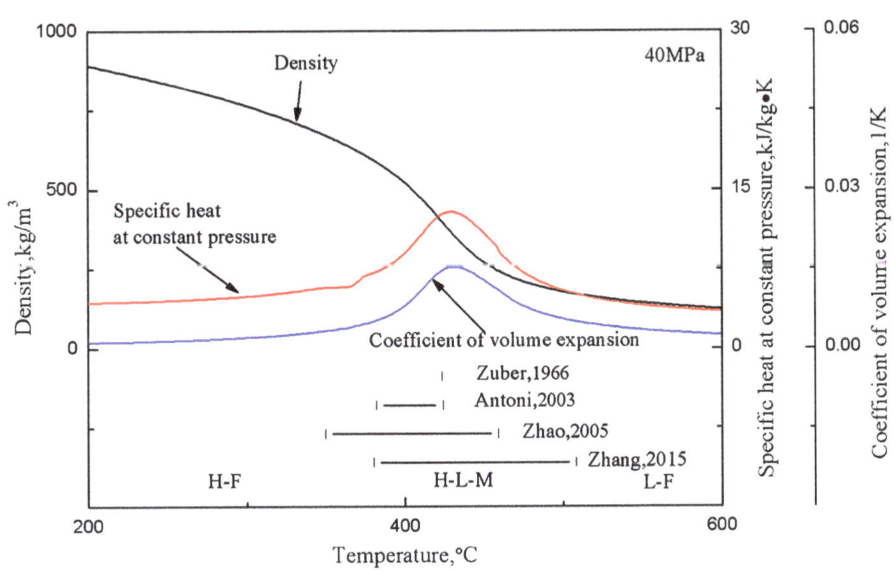

CASE STUDIES

Heat Transfer of Supercritical CO_2 at Prototypic Printed Circuit Heat Exchangers

In this section, the heat transfer performance of supercritical CO_2 at prototypic printed circuit heat exchangers is discussed for better understanding the heat transfer models in the section 2.

Figure 14 shows a schematic diagram of the experimental systems used for the investigation on the supercritical CO_2 heat transfer. The test facility and the detailed descriptions of the experimental apparatuses can be found in the authors' previous studies (Li, Kruizenga, & Anderson, 2011; Kruizenga, Li, & Anderson, 2012). A large amount of experiments were performed at steady-state conditions. Primary operational parameters included the inlet pressures of 7.5-8.5MPa, mass fluxes ranged from 160 kg/m^{2}s to 760 kg/m^{2}s, the average heat fluxes of 20kW/m^2 to 90kW/m^2 and inlet temperatures from 5°C to 95 °C with a bulk temperature rise or fall across the test section varying from less than 2°C to a maximum of 60°C.

The PCHE test section consists of several pieces: the heat exchanger plate, the mating plate, and the water blocks. The heat exchanger plate is designed to be interchangeable with other prototypic designs, with flow paths fabricated by chemically etching designs into type 316 stainless steel. The channel configuration used in the present study consists of nine semi-circular, parallel channels of 1.17mm hydraulic diameter. Thermocouples are implanted into the metal wall to measure the axial variation in wall temperature. Each thermocouple hole is 1mm in diameter and 33mm deep, spaced axially by 50mm. Attached to the top of the mating plate and bottom heat exchanger plate are ten individual aluminum water blocks used to provide heating or cooling, while also measuring the local heat removal by performing an energy balance on each block. The inlet and outlet temperatures, and the volumetric flow rate of water for each block are measured. The local heat flux and the heat transfer coefficient are

Figure 14. Schematic diagram of the experimental system

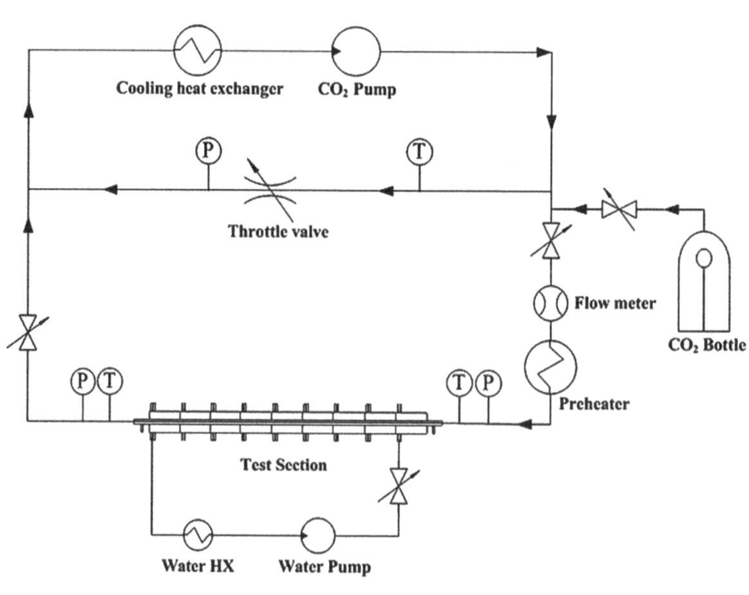

calculated as discussed in the authors' previous paper (Li, Kruizenga, & Anderson, 2011; Kruizenga, Li, & Anderson, 2012). The whole piping is thermally insulated to minimize the heat loss.

ANSYS FLUENT 13.0, the commercially available CFD software, was employed to model the experiments. In virtue of symmetry, the computational domain involves one of the nine semi-circular parallel channels inside the experiment test section. A piecewise linear interpolation method is used to best approximate the actual outer temperature boundary conditions in the simulation according to the experimentally measured wall temperature data. These temperature profiles are implemented by employing the User Defined Functions (UDF) in the ANSYS FLUENT. Thus, the wall temperature in contact with the fluid is calculated by the coupling of the convective and conductive heat transfer equations. The experimental wall temperature boundary conditions could be easily replaced with the constant heat flux boundary conditions for supplementary studies beyond experiments operational limits. The inlet and outlet boundary conditions are specified as mass flux inlet and pressure outlet, respectively, while non-wall faces are specified as symmetry boundaries.

A complete set of governing equations for mass, momentum, energy, and turbulence are discretized by the finite volume method. In the present study, the Shear Stress Transport (SST) k-ω turbulence model, which blends the robustness and independence of the k-ε model for the bulk flow and the accurate formulation of the Wilcox modified k-ω model in the near wall region. The QUICK scheme is used for approximating the convection terms in the governing equations for an improved accuracy. The coupling between the pressure and the velocities is implemented by the SIMPLEC algorithm. The temperature and pressure dependent properties of CO_2 are calculated by linking the NIST Standard Reference Database to the FLUENT solver. The convergence criterion for temperature is that the residual is less than 10^{-7} and for the other variables the residuals are less than 10^{-5}. The computational grids were generated using Gambit, which was packaged with the FLUENT software. The computational domain was divided into several small regions so that the hexahedral structured mesh could be generated in every small region. Mesh refinement in radial direction, near the wall, ensures property gradients are properly resolved. The whole computational domain was discretized with a total of $\sim 5\times10^5$ grid cells by non-uniform grids with the grids of the heat transfer regions being finer and those in the extension domains being coarser. To meet the requirements of the turbulence near-wall models, slight changes in mesh spacing are needed to ensure that the values of the non-dimensional distance (y+) at the wall-adjacent nodes are always less than 1. The specific numerical modeling methods and the data reduction to determine the local bulk temperature, the heat transfer coefficient and the local heat flux could be found in the authors' previous work (Kruizenga, Li, & Anderson, 2012).

The classical Dittus-Boelter (DB) correlation, Equation 1, was employed in a comparison with the current experimental and computational data. As can be seen from Figure 15 and Figure 16, The DB correlation performs a disappointed work on predicting the heat transfer behaviors in the pseudo-critical region either in the heating mode or in the cooling mode. To better understand this, the Nusselt numbers of various working conditions determined by the FLUENT simulations were normalized by the DB correlation and were plotted against to the normalized bulk temperature. It could be seen from Figure 15 and Figure 16, the normalized Nusselt number approximately equals one in the regions far from the pseudo-critical point, which indicates the DB correlation predicts fairly well outside the pseudo-critical region. However, within the pseudo-critical region, the DB correlation dramatically either overestimates the heat transfer coefficient in the heating mode as shown in Figure 15, or underestimates the heat transfer coefficient in the cooling mode as shown in Figure 16. This is mainly because the DB correlation has no properties ratio correction terms taking account of the properties distribution in the radial direction and consequently no ability to identify the differences between the heating and cooling modes heat transfer. Thus it exhibits almost identical heat transfer behaviors between the heating and cooling modes while, in fact, as discussed above the heat transfer effectiveness in the cooling mode was proved to be much better than that in the heating mode.

Jackson's Nusselt correlation, Equation 2, which employs wall-to-bulk properties ratios to take account of the variations of properties in the radial direction, was also be used for comparison with our experimental and computational data. As can be seen from Figure 15 and Figure 16, the Jackson correlation prediction does exhibit different behaviors between the heating and cooling mode heat transfer by taking advantage of its property ratio correction terms. However, the heat transfer coefficients near the pseudo-critical temperature were still overestimated by Jackson correlation, although this overestimation is less than that of DB correlation. As the instantaneous properties are highly nonlinear dependent on temperature, the properties evaluated at time-averaged temperature yield considerable error.

The PDF-based correlation, Equation 9, employs the property ratio modification terms to account for the variation of properties distributions in the cross section, especially in the turbulence boundary layer. This can be regarded as a modification in spatial dimension and contribute to an increase in the accuracy of prediction. More impressively, the new physically improved correlation employs the PDF-based properties modification to take into account the effect of instantaneous temperature fluctuations and the dependent strongly nonlinear property fluctuations. This could be regarded as a modification in temporal dimension and lead to collapsing the data to its average true value. As can be seen from Figure 15 and Figure 16, the newly developed correlation did quite a promising job that it reproduce the experimental and computational data at almost all of the working conditions in both heating and cooling modes.

Figure 15. Comparisons of the computational data with predictions of the DB, Jackson and PDF-based correlations versus T_b/T_{pc}: cooling

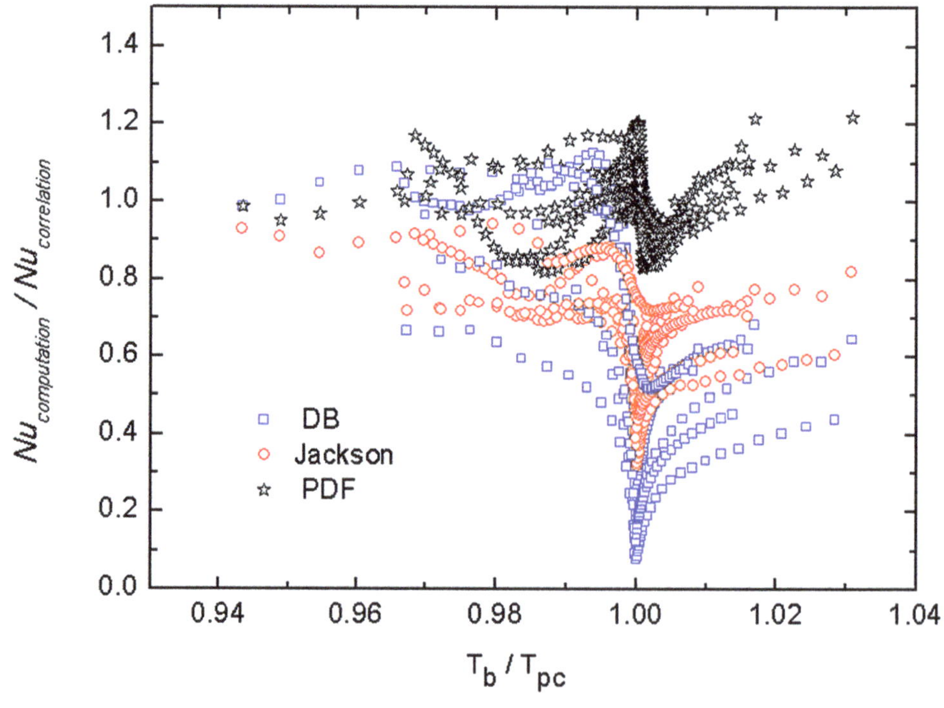

Figure 16. Heating

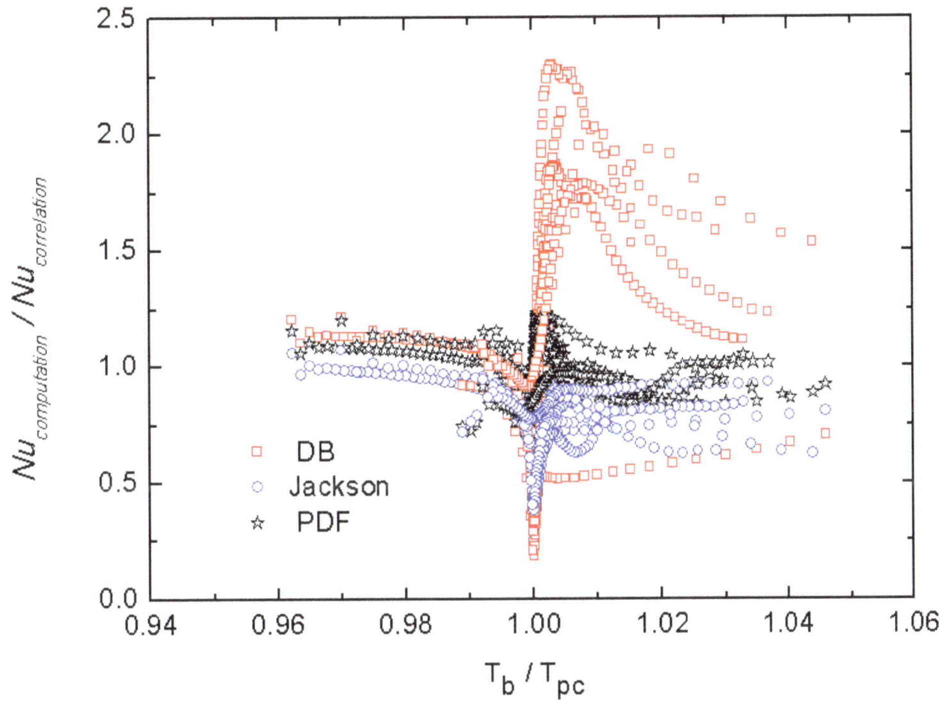

Density Wave Instabilities of Supercritical Water Flowing in Heated Tubes

In this case, density wave instabilities of supercritical water flowing in the vertically upward heated tubes are studied by frequency domain method.

According to the above mentioned partition methods, SCW flows in a tube can be simplified to a three region model. Figure 17 shows the schematic diagram of the three region model. The heating section is divided into three parts, which is named as H-F, H-L-M and L-F from inlet to outlet respectively. Z_{PSW} and Z_{PSS} are the interfaces which represent PSW and PSS respectively, and the values of Z_{PSW} and Z_{PSS} are considered to be influenced by the perturbations in the process of the dynamical calculation.

In order to analyze density wave instabilities of supercritical water in tubes, the following assumptions are proposed, (1) The perturbations of density of SCW in heavy fluid region are ignored; (2) SCW in light fluid region is assumed as ideal gas; (3) Differential pressure from inlet to outlet is constant; (4) Heat flux is uniform in the axial direction.

The details of the conservation equations are expressed as follows.

Equations for H-F

The density of SCW in H-F is assumed as constant, equaling to that at the PSW. The conservation equations can be described by Equations 29-31.

Figure 17. Schematic diagram of the present model

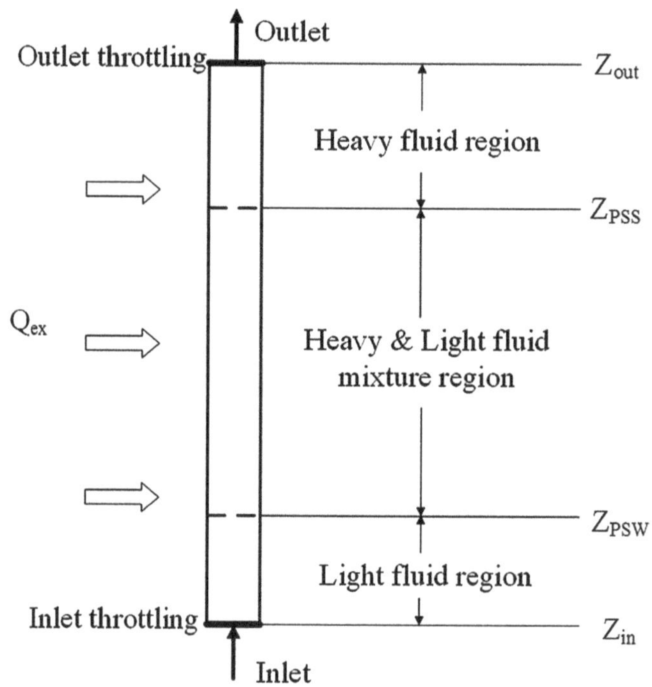

$$\frac{\partial G_{hr}}{\partial Z} = 0 \tag{29}$$

$$\rho_{PSW}\frac{\partial h_{hr}}{\partial t} + \frac{\partial G_{hr}\cdot h_{hr}}{\partial Z} = \frac{Q}{A'} \tag{30}$$

$$\frac{\partial G_{hr}}{\partial t} + \frac{1}{\rho_{PSW}}\frac{\partial G_{hr}^2}{\partial Z} = -\frac{\partial P_{hr}}{\partial Z} - \rho_{PSW}g - \frac{\lambda_{hr}}{2D}\frac{G_{hr}^2}{\rho_{PSW}} \tag{31}$$

where, G_{hr} is the mass flux of a particle in H-F, *Z* is the height, *t* is time, h_{hr} is the specific enthalpy of a particle in H-F, A' is the flow area, P_{hr} is the pressure of a particle in H-F, *g* is gravity acceleration, λ_{hr} is the coefficient of frictional resistance in H-F, *D* is the internal diameter of the tube and ρ_{PSW} is the density of SCW at PSW.

Equations for H-L-M

Equations 32-34 are the conservation equations for H-L-M.

$$\frac{\partial \rho_{mr}}{\partial t} + \frac{\partial G_{mr}}{\partial Z} = 0 \tag{32}$$

$$\frac{\partial \rho_{mr} h_{mr}}{\partial t} + \frac{\partial G_{mr} \cdot h_{mr}}{\partial Z} = \frac{Q}{A'} \tag{33}$$

$$\frac{\partial G_{mr}}{\partial t} + \frac{\partial G_{mr}^2 / \rho_{mr}}{\partial Z} = -\frac{\partial P_{mr}}{\partial Z} - \rho_{mr} g - \frac{\lambda_{mr}}{2D} \frac{G_{mr}^2}{\rho_{mr}} \tag{34}$$

where, ρ_{mr} is the density of a particle in H-L-M, G_{mr} is the mass flux of a particle in H-L-M, Z is the height, t is time, h_{mr} is the specific enthalpy of a particle in H-L-M, A' is the flow area, P_{mr} is the pressure of a particle in H-L-M, g is gravity acceleration, λ_{mr} is the coefficient of frictional resistance in H-L-M and D is the internal diameter of the tube.

In this region, the density, ρ_{mr}, employed in the process of the dynamical calculation is considered as a dynamic parameter, which is not only the function of height (**Z**) but also the function of time (***t***), and the value of the density is influenced by the small perturbations of other parameters.

Equations for L-F

Equations 35-37 are the conservation equations of an arbitrary particle in L-F.

$$\frac{\partial \rho_{lr}}{\partial t} + \frac{\partial G_{lr}}{\partial Z} = 0 \tag{35}$$

$$\frac{\partial \rho_{lr} h_{lr}}{\partial t} + \frac{\partial G_{lr} \cdot h_{lr}}{\partial Z} = \frac{Q}{A'} \tag{36}$$

$$\frac{\partial G_{lr}}{\partial t} + \frac{\partial G_{lr}^2 / \rho_{lr}}{\partial Z} = -\frac{\partial P_{lr}}{\partial Z} - \rho_{lr} g - \frac{\lambda_{lr}}{2D} \frac{G_{lr}^2}{\rho_{lr}} \tag{37}$$

where, ρ_{lr} is the density of a particle in L-F, G_{lr} is the mass flux of a particle in L-F, Z is the height, t is time, h_{lr} is the specific enthalpy of a particle in L-F, A' is the flow area, P_{lr} is the pressure of a particle in L-F, g is gravity acceleration, λ_{lr} is the coefficient of frictional resistance in L-F and D is the internal diameter of the tube.

In this region, SCW is assumed as idea gas, and the physical properties of SCW can be calculated based on the ideal gas state equation.

Equation Transformation and Feedback Relations

Based on the fundamental equations, the characteristic equation and open-loop transfer function can be obtained by frenquency domain method. The detailed derivation process is complex, and it is not shown in this book (Detailed derivation process can be found in the literature (Zhang, Li, Li, Lei, & Wang, 2014; Zhang, Li, Li, Wang, Zhang. & Lei, 2015).

Finally, the characteristic equation in the form of equation 38 and the open-loop transfer function in the form of Equation 39 can be got.

$$F(s) = 1 + G(s)H(s) = 0 \tag{38}$$

$$G(s)H(s) = \frac{1}{C_6' \cdot s} \cdot \frac{C_1' \cdot s^2 + C_2' \cdot s + C_3'}{C_3' \cdot s^2 + C_4' \cdot s + C_5'} \tag{39}$$

Based on Nyquist stability criterion, the necessary and sufficient conditions of the system's stability is:

$$N_d = P_0 \tag{40}$$

where, N_d is the number of anticlockwise encirclements of the point (-1, j_0) by Nyquist locus, P_0 is the total number of the poles (poles of the open-loop transfer function) in the right-half plane.

According to this necessary and sufficient conditions of the system's stability, the stability of density wave instabilities of supercritical water can be predict easily.

Stability Boundary Obtained Based on Different Partition Methods

The experiments used for verification and comparison in this book were conducted by Khabensky and Gerliga (1995). A long heated pipe (12.6m high, 0.01m internal diameter) was employed as the test section, and the pressure was at 23 and 23.5 MPa. Table 2 shows the detailed working conditions and results of stability boundaries.

The models based on above mentioned partition methods are used to calculate the stability boundaries of the cases in Table 2, and the results are compared with the experimental ones.

Figure 18, Figure 19, Figure 20, and Figure 21, respectively, show the comparisons of the calculated results and the published experimental results (Khabensky & Gerliga, 1995), and the boundary N_{TPC} is used to represent the stability situations. According to the definition of N_{TPC}", the relationship between N_{TPC} and q' is liner. Hence, in the dimensionless analyses, it is reasonable to employ the boundary N_{TPC} to represent the stability situations instead of the boundary heat flux, which is widely used to represent the demarcation point between the stable region and the unstable region (Zhang, Li, Li, Lei, & Wang, 2014). Hence, a working condition with the N_{TPC} lower than the boundary N_{TPC} is stable, and that with the N_{TPC} greater than the boundary N_{TPC} is unstable, and the working condition with the N_{TPC} equaling to the boundary N_{TPC} is the critical working condition (Zhang, Li, Li, Lei, & Wang, 2014).

Table 2. Cases for the verification of the present model

Case	Pressure (MPa)	Inlet Enthalpy (kJ/kg)	Inlet Flow Rate (kg/(m²s))	Experimental Boundary Heat Flux (kW/m²)	Experimental Boundary N_{TPC}
1	23	1200	180	81.3	2.937
2	23	1200	150	69.5	3.013
3	23	1590	300	116.0	2.515
4	23	840	135	69.5	3.348
5	23.5	1000	150	116.0	5.399
6	23.5	1000	100	58.0	4.049

Khabensky & Gerliga, 1995.

In Figure 18, Figure 19, Figure 20, and Figure 21, the x axis is the experimental N_{TPC}, and the y axis is the calculated N_{TPC}. The data points in Figures 18, Figure 19, Figure 20, and Figure 21, respectively, are the boundary N_{TPC} got from different models, based on the six cases shown in Table 2. From Figure 18, all the 6 calculation results are greater than the experiment ones, and only one of the 6 cases owns a calculation result with relative error less than 20%, according to dynamic calculation by using the model from Zuber (1966). The primary reason of such relative errors is that the regional partition of SCW are somewhat rough, and the great variations of physical properties of SCW in the neighborhood region of the pseudo critical point are neglected. According to Figure 19, the calculation results of the model from Antoni and Dumaz (2003) are a little more accurate than that of the model from Zuber (1966). It is proved that the three region model is more suitable than the two region one for the study on density wave instability of SCW in tubes, because a special region in the three model, H-L-M, is defined to deal with the great variations of physical properties of SCW in the near region of the pseudo critical point. However, from Figure 19, there are only two calculation results own the relative errors less than 20%, and the greatest relative error of the calculation result is even 60% approximately. Hence, the calculation results of the model from Antoni and Dumaz (2003) are also not so accurate. According to Figure 20, the relative errors of the calculation results express a better accuracy than the foregoing ones, and there are three cases own the calculation results with relative errors less than 20%, based on the dynamic calculation by using the model from Zhao (2005). Shown as Figure 21, the calculation results by the model from Zhang, Li, Li, Wang, Zhang. and Lei (2015) are more accurate than that from the previous models, and four cases own the calculation results with relative errors less than 20%, and all the relative errors of the calculation results from this model is not greater than 26%.

SOLUTIONS AND RECOMMENDATIONS

This chapter provides basic physical mechanisms and practical solutions for better understanding the supercritical fluids thermal hydraulics characteristics including heat transfer, pressure drops and flow instabilities for the purpose of accurate design and efficient safe operation of advanced energy systems.

Figure 18. Stability boundary obtained based on different partition methods
Model from Zuber (1966).

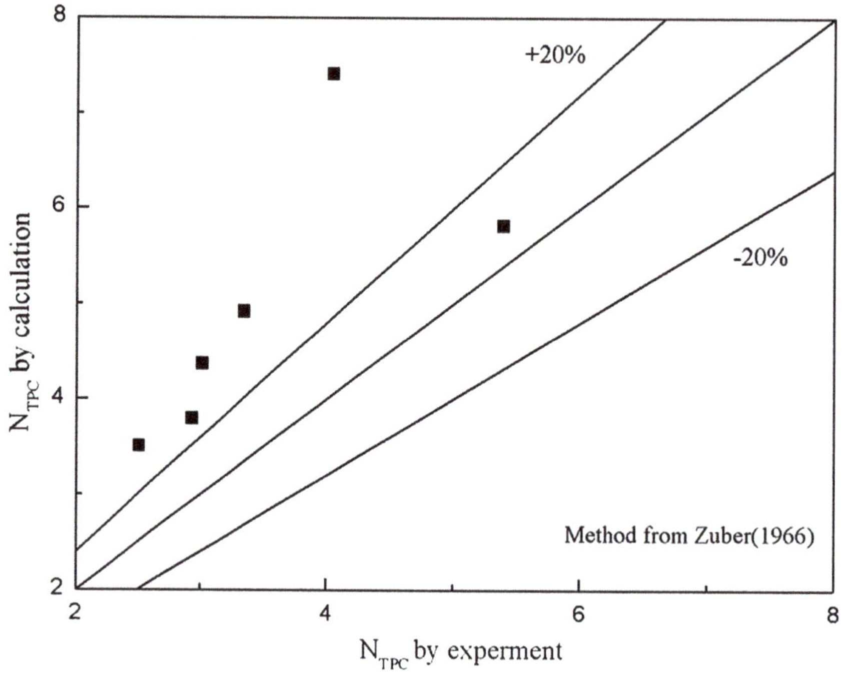

Figure 19. Model from Antoni & Dumaz (2003)

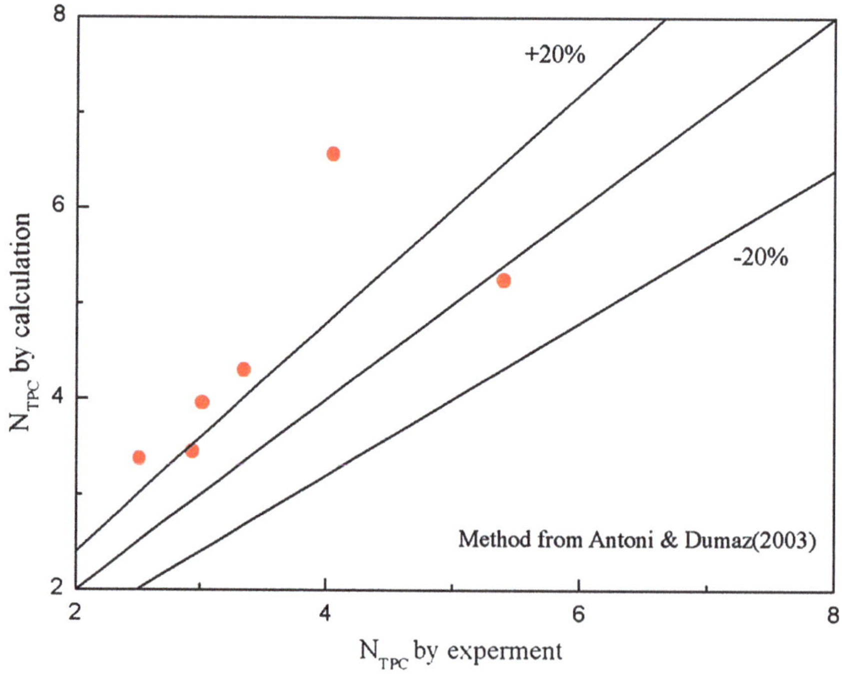

Figure 20. Model from Zhao (2005)

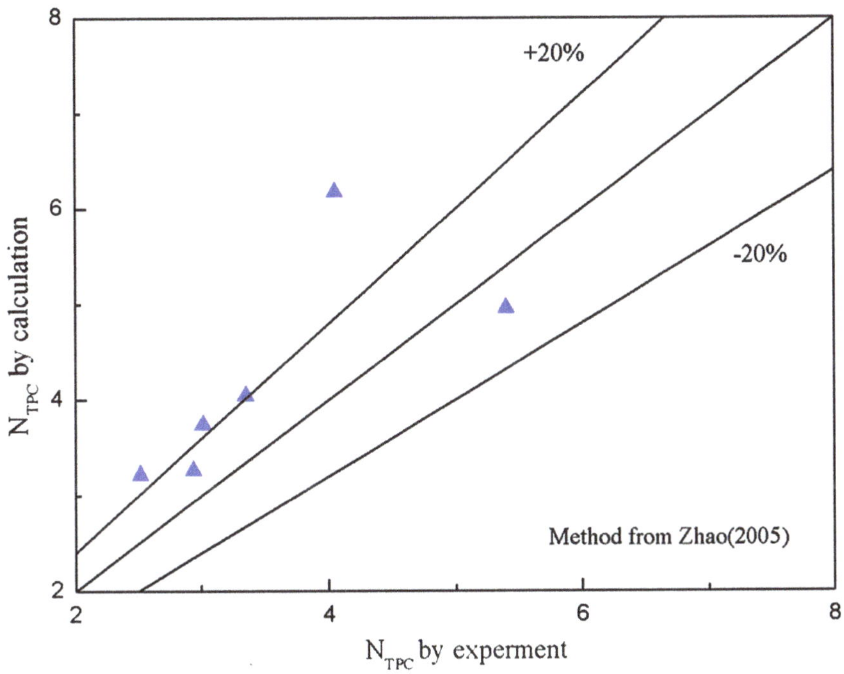

Figure 21. Model from Zhang (2015)

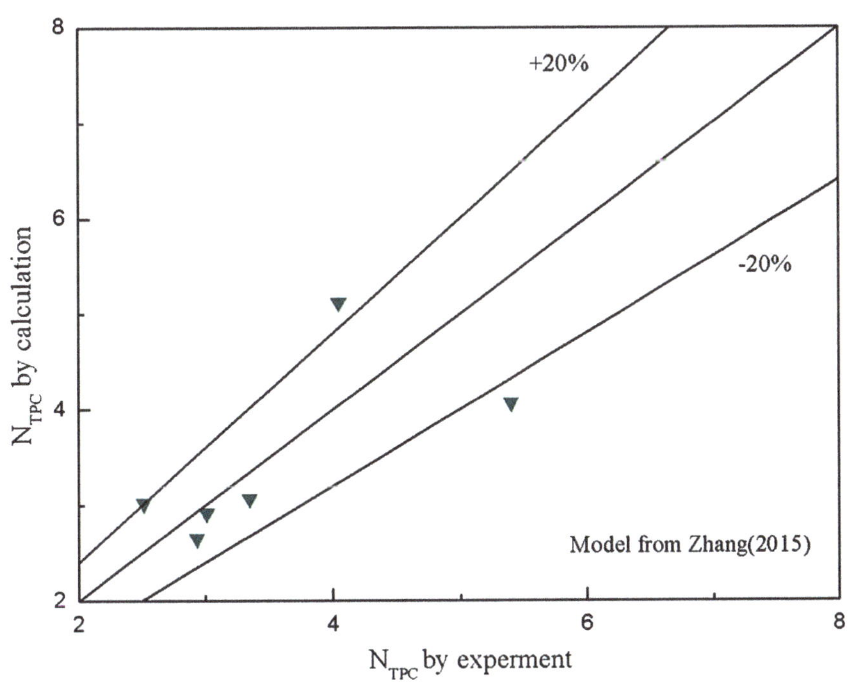

FUTURE RESEARCH DIRECTIONS

Further in-depth study on the consistency of the improved correlations with different supercritical fluids and the underlying physical mechanisms of the thermal hydraulic models will be imperative and underway.

CONCLUSION

The supercritical fluids thermal hydraulics characteristics including heat transfer, pressure drops and flow instabilities are different from those of the single phase fluids and very important for accurate design and efficient safe operation of advanced energy systems.

The newly developed heat transfer correlation employs PDF-based time-averaged properties accounting for the effect of instantaneous temperature fluctuations and the dependent strongly nonlinear property fluctuations in temporal dimension, in conjunction with the property ratio correction terms to account for the effect of variable properties in spatial dimensions, which could better predict the heat transfer performance of the supercritical fluid.

The use of the explicit Colebrook and White friction factor has been more reliable correlation, as the surface roughness effects are included. This correlation has been found to match closely with the well-known Moody friction factor.

According to the suitable partition method of supercritical water, the models for two-phase flow instabilities can be improved and employed in the study of the supercritical water instabilities. The four published partition method of supercritical water are compared with each other, and that form Zhang (2015) is recommended.

REFERENCES

Antoni, O., & Dumaz, P. (2003, January). Preliminary calculations of a supercritical light water reactor concept using the CATHARE code.*Proceedings of ICAPP*, 4-7.

Brauner, N., & Moalem, M. D. (1992). Stability analysis of stratfied liquid-liquid flow. *International Journal of Multiphase Flow*, *18*(1), 103–121. doi:10.1016/0301-9322(92)90009-6

Fukuda, K., & Kobori, T. (1979). Classification of two-phase flow instability by density wave oscillation model. *Journal of Nuclear Science and Technology*, *16*(2), 95–108. doi:10.1080/18811248.1979.9730878

Gómez, T. O., Class, A., Lahey, R. Jr, & Schulenberg, T. (2008). Stability analysis of a uniformly heated channel with supercritical water. *Nuclear Engineering and Design*, *238*(8), 1930–1939. doi:10.1016/j.nucengdes.2007.10.031

Gu, H., Li, H., Wang, H., & Luo, Y. (2013). Experimental investigation on convective heat transfer from a horizontal miniature tube to methane at supercritical pressures. *Applied Thermal Engineering*, *58*(1-2), 490–498. doi:10.1016/j.applthermaleng.2013.04.049

Huai, X. L., Koyama, S., & Zhao, T. S. (2005). An experimental study of flow and heat transfer of supercritical carbon dioxide in multi-port mini channels under cooling conditions. *Chemical Engineering Science*, *60*(12), 3337–3345. doi:10.1016/j.ces.2005.02.039

Jackson, J. D. (2013). Fluid flow and convective heat transfer to fluids at supercritical pressure. *Nuclear Engineering and Design*, *264*(11), 24–40. doi:10.1016/j.nucengdes.2012.09.040

Khabensky, V. B., & Gerliga, V. A. (1995). Hydrodynamic flow instabilities in power equipment components. CBS Publisher & Distributors.

Kim, D. E., & Kim, M. H. (2010). Experimental study of the effects of flow acceleration and buoyancy on heat transfer in a supercritical fluid flow in a circular tube. *Nuclear Engineering and Design*, *240*(10), 3336–3349. doi:10.1016/j.nucengdes.2010.07.002

Kruizenga, A., Li, H., & Anderson, M. (2012). Supercritical carbon dioxide heat transfer in horizontal semicircular channels. *Jouranl of Heat Transfer-Transactions of the ASME*, *134*(8), 081802-1–10.

Kurganov, V. A., Zeigarnik, Y. A., & Maslakova, I. V. (2012). Heat transfer and hydraulic resistance of supercritical-pressure coolants. Part I: Specifics of thermophysical properties of supercritical pressure fluids and turbulent heat transfer under heating conditions in round tubes (state of the art). *International Journal of Heat and Mass Transfer*, *55*(11-12), 3061–3075. doi:10.1016/j.ijheatmasstransfer.2012.01.031

Li, H., Kruizenga, A., Anderson, M., Corradini, M., Luo, Y., Wang, H., & Li, H. (2011). Development of a new forced convection heat transfer correlation for CO2 in both heating and cooling modes at supercritical pressures. *International Journal of Thermal Sciences*, *50*(12), 2430–2442. doi:10.1016/j.ijthermalsci.2011.07.004

Li, H., Wang, H., Luo, Y., Gu, H., Shi, X., Chen, T., & Zhu, Y. et al. (2009). Experimental investigation on heat transfer from a heated rod with a helically wrapped wire inside a square vertical channel to water at supercritical pressures. *Nuclear Engineering and Design*, *239*(10), 2004–2012. doi:10.1016/j.nucengdes.2009.05.011

Li, H., Zhang, Y., Zhang, L., Yao, M., Kruizenga, A., & Anderson, M. (2016). PDF-based modeling on the turbulent convection heat transfer of supercritical CO_2 in the printed circuit heat exchangers for supercritical CO_2 Brayton cycle. *International Journal of Heat and Mass Transfer*, *98*(7), 204–218. doi:10.1016/j.ijheatmasstransfer.2016.03.001

Liao, S. M., & Zhao, T. S. (2002). Measurements of heat transfer coefficients from supercritical carbon dioxide flowing in horizontal mini/micro channels. *Jouranl of Heat Transfer-Transactions of the ASME*, *124*(3), 413–420. doi:10.1115/1.1423906

Pioro, I. L., & Duffey, R. B. (2006). *Heat transfer and hydraulic resistance at supercritical pressures in power engineering applications*. New York, NY: ASME Press.

Pioro, I. L., Khartabil, H. F., & Duffey, R. B. (2004). Heat transfer to supercritical fluids flowing in channels - empirical correlations (survey). *Nuclear Engineering and Design*, *230*(1-3), 69–91. doi:10.1016/j.nucengdes.2003.10.010

Su, Y., Feng, J., Zhao, H., Tian, W., Su, G., & Qiu, S. (2013). Theoretical study on the flow instability of supercritical water in the parallel channels. *Progress in Nuclear Energy*, *68*(1), 169–176. doi:10.1016/j.pnucene.2013.06.005

T'Joen, C. T., & Rohde, M. (2011, August). Experimental study on the stability of a natural circulation driven super-critical water cooled reactor. *22nd International Symposium on Transport Phenomena, Delft*, The Netherlands.

Xiong, T., Yan, X., Huang, S., Yu, J., & Huang, Y. (2013). Modeling and analysis of supercritical flow instability in parallel channels. *International Journal of Heat and Mass Transfer*, *57*(2), 549–557. doi:10.1016/j.ijheatmasstransfer.2012.08.046

Xiong, T., Yan, X., Xiao, Z., Li, Y., Huang, Y., & Yu, J. (2012). Experimental study on flow instability in parallel channels with supercritical water. *Annals of Nuclear Energy*, *48*(2), 60–67. doi:10.1016/j.anucene.2012.05.018

Yi, T. T., Koshizuka, S., & Oka, Y. (2004). A linear stability analysis of supercritical water reactors,(I) thermal-hydraulic stability. *Journal of Nuclear Science and Technology*, *41*(12), 1166–1175. doi:10.1080/18811248.2004.9726345

Zhang, L., Wang, H., Gu, H., Qiao, S., Shi, X., & Luo, Y. (2012, October). Experimental investigations on flow instabilities in a forced circulation loop at near-critical and supercritical pressures.*AIP Conference Proceedings*, 1547, 312.

Zhang, Y., Li, H., Li, L., Lei, X., & Wang, T. (2014). Study on two-phase flow instabilities in internally-ribbed tubes by using frequency domain method. *Applied Thermal Engineering*, *65*(1-2), 1–13. doi:10.1016/j.applthermaleng.2013.12.069

Zhang, Y., Li, H., Li, L., Wang, T., Zhang, Q., & Lei, X. (2015). A new model for studying the density wave instabilities of supercritical water flows in tubes. *Applied Thermal Engineering*, *75*(C), 397–409. doi:10.1016/j.applthermaleng.2014.09.029

Zhao, J. (2005). *Stability analysis of supercritical water cooled reactors. Doctor of Philosophy in Nuclear Science and Engineer*. Massachusetts Institute of Technology.

Zuber, N. (1966, May). *An analysis of thermally induced flow oscillations in the near-critical and supercritical thermodynamic region*. Final Report Nas8-11422. NASA.

KEY TERMS AND DEFINITIONS

Boundary Heat Flux: The boundary of the unstable region. Fluid instabilities will occuer in the heating tubes with a higher heat flux than Boundary Heat Flux.

Probability Density Function: In probability theory, a probability density function (PDF), or density of a continuous random variable, is a function that describes the relative likelihood for this random variable to take on a given value.

Pseudo Critical Steam: In the three region model of supercritical water instabilities, Pseudo Critical Steam is a new definition, similar as the saturated steam at subcritical pressures. It is the upper limit of the heavy & light fluid mixture region.

Pseudo Critical Temperature: For supercritical fluids, the temperature, for a given pressure, at which the specific heat exhibits a maximum.

Pseudo Critical Water: In the three region model of supercritical water instabilities, Pseudo Critical Water is a new definition, similar as the saturated water at subcritical pressures. It is the lower limit of the heavy & light fluid mixture region.

Section 3
Supercritical Thermal Conversion of Different Scales

In this section, multi-scale analysis is discussed for supercritical and near-critical fluid systems: from simple channels to complex heat transfer system designs. The confinement and scaling effect of supercritical fluid heat transfer flow is focused in this section. The knowledge obtained by major researchers in this field will contribute to various applications of supercritical fluid based heating, cooling, microscale transportation, reaction, etc.

Chapter 9
Heat Transfer in Supercritical Fluids:
Going to Microscale Times and Sizes

Sergey B. Rutin
Institute of Thermal Physics, Ural Branch, Russian Academy of Sciences, Russia

Aleksandr D. Yampol'skiy
Institute of Thermal Physics, Ural Branch, Russian Academy of Sciences, Russia

Pavel V. Skripov
Institute of Thermal Physics, Ural Branch, Russian Academy of Sciences, Russia

ABSTRACT

Results of experimental study of non-stationary heat transfer in supercritical fluids, which were obtained using the method of controlled pulse heating of low-inertia wire probe, are discussed. The aim of this study was to clarify the peculiarities of heat conduction mode at significant heat loads. A threshold decrease in the "instant" heat transfer coefficient, the more pronounced the closer the pressure value to critical pressure, has been found, as well as the absence of impact of the isobaric heat capacity peak known from stationary measurements on the experimental results. These results give new insights into selection of the operating pressure of supercritical heat transfer agent. Small time and spatial scale in the experiments (units of millisecond and units of micrometer) in combination with high-power heat release (up to 20 MW/m^2) makes it possible to associate the results with the behavior of boundary layer region of heat transfer agent.

INTRODUCTION

The study of heat transfer in supercritical fluids (SCFs) has a long history (see, for example, Anisimov, 2011; Berche, Henkel, & Kenna 2009; Kurganov, Zeigarnik, & Maslakova, 2012; Pioro, & Duffey, 2007) and is closely associated with the task of increasing the efficiency of thermoengineering devices. Application of heat transfer agents at supercritical conditions can significantly improve the thermal efficiency

DOI: 10.4018/978-1-5225-2047-4.ch009

of thermoengineering units. However, insufficient understanding of the physical processes occurring in the flow presents an obstacle for the use of, for example, supercritical water as a heat transfer agent in the nuclear reactors (Kurganov et al., 2014; Pioro, & Duffey, 2007). The fundamental problems of heat and mass transfer in SCFs appear to be unresolved. It is curious that at the same time, supercritical fluids have long been successfully used as working fluids and hundreds of power plants in the world are working on supercritical water for decades. It is obvious that the requirements for manageability and predictability of the supercritical water behavior under conditions of extreme thermal loads are significantly higher in nuclear power industry than in the conventional thermal power industry. Therefore, the transfer of experience accumulated in the conventional thermal power industry to the nuclear power industry proved to be a non-trivial task.

In the opinion of the authors, of note are two important issues. First, there is no theoretical model that would be able to describe all heat transfer modes that were experimentally observed. Secondly, the majority of experimental studies of heat transfer in SCFs and measurements of the thermophysical properties of substances at supercritical conditions have been performed by stationary or quasi-stationary methods. Taking into account the extremely high convective instability of SCFs and gravitational sensitivity of parameters in the near-critical region, the known pattern of transfer phenomena in SCFs is not complete. In other words, there is a shortage of research methods for such complex objects as SCFs.

This motivated development of the specialized techniques for studying heat transfer, operating at small characteristic sizes and times. Moreover, these techniques allow one to set up experiments under conditions of high heat flux densities (Rutin & Skripov, 2012, 2013a, 2013b), as shown below. This approach made it possible to virtually avoid the influence of convection and gravity on the experimental results and obtain the data on conductive heat transfer mode in the course of high-power heat release, which are of interest for both fundamental and applied aspects. The most important results obtained at small characteristic sizes and times can be formulated as follows. First, the effect of threshold decrease in the heat transfer intensity was revealed in course of a fast transition between compressed liquid and supercritical fluid states along the isobar. The effect was more pronounced at pressures (p) close to the critical pressure (p_c). Second, for all investigated substances (Rutin & Skripov, 2013b, 2013c, 2014; Rutin, Volosnikov, & Skripov, 2015), the revealed effect was observed in the range of reduced pressures from 1 to $3p/p_c$ and completely disappeared in the vicinity of pressure $p/p_c = 3$. Third, it was found that the peaks of isobaric heat capacity and excess thermal conductivity, which are known from stationary measurements, do not affect the experimental results.

BACKGROUND

The topic of heat transfer phenomena in SCF is so extensive that its comprehensive coverage within a single chapter is not feasible. Different groups of researchers studied the supercritical heat transfer phenomena at different times have formed their own views of the problem. In this sense, our research team is not an exception. Therefore, the discussion presented in this section reflects the opinion formed in the authors' team about both retrospective and prospects of studies of heat transfer in SCFs. The authors are aware that almost all questions relating to SCFs are debatable; one just needs to go into the sufficient details. Take as example such a seemingly simple question: what is a supercritical fluid, and at what state parameters is it possible to consider it as a gas-like or a liquid-like? The views of various researchers on this issue are essentially different. Follow the practitioners of heat transfer, such as a point of view presented in the monograph by Pioro and Duffey (2007), the following phase diagram can be depicted, see Figure 1.

In this diagram, the region of existence of a supercritical fluid corresponds to the condition $T > T_c$ and $p > p_c$. Continuation of the saturation line designated as "pseudo-critical line" separates "liquid-like, high-density" and "gas-like, low-density" states of SCF. The pseudo-critical line shown in the diagram is a generator of the isobaric heat capacity maxima in the supercritical region of parameters. In fact, in this approach the pseudo-critical line appears as something commonly accepted and not requiring further explanation. The same line separates the liquid-like and gas-like states of SCF.

Actually, the question regarding the parameters at which SCF can be considered as a liquid- or gas-like, is not so simple and unambiguous. For example, in the work of Gorbaty and Bondarenko (2007) devoted to the structure of supercritical water, the authors emphasize that even under normal conditions there is no common point of view on water structure. Based on a large array of experimental data, they conclude that at a supercritical pressure, water is a micro-heterogeneous mixture of gas-like and liquid-like molecular configurations quickly transforming from one to another. Also, the attempts to construct different kinds of lines continuing the saturation line in the subcritical region were challenged (Gorbaty and Bondarenko, 2007).

There is another view regarding this issue (Brazhkin, Fomin, Lyapin, Ryzhov, & Trachenko, 2012; Brazhkin & Trachenko, 2012). Following these authors, it is possible to present the phase diagram shown in Figure 2. As seen in Figure 2, there is a large region of parameters corresponding to the condition $T > T_c$ and $p > p_c$, in which the substance is in a stable crystalline phase! It is pointed out that at supercritical parameters near the melting line, SCF should most closely match the liquid-like state. The main idea

Figure 1. Fragment of the phase diagram of water
Pioro & Duffey, 2007.

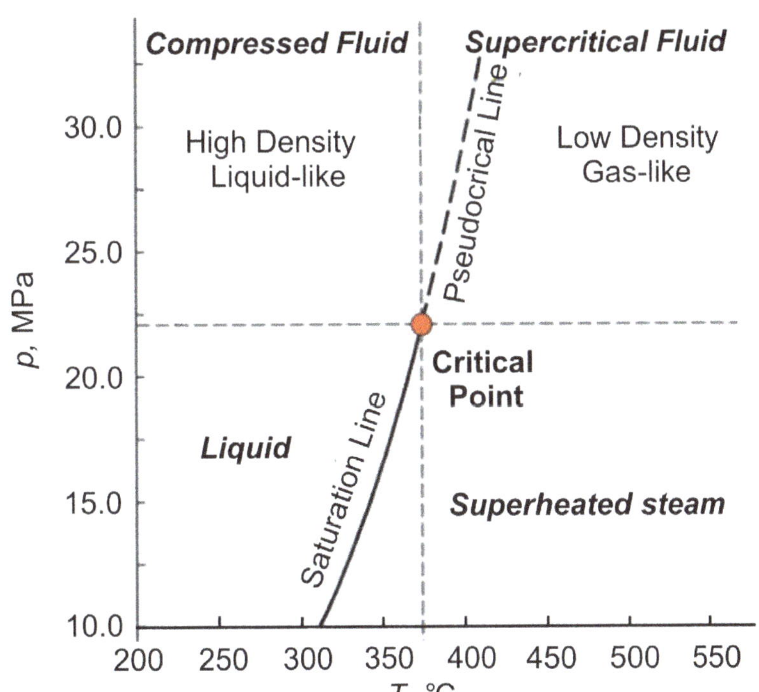

Figure 2. Phase diagram of a substance in a wide range of the pressure
Brazhkin et al., 2012.

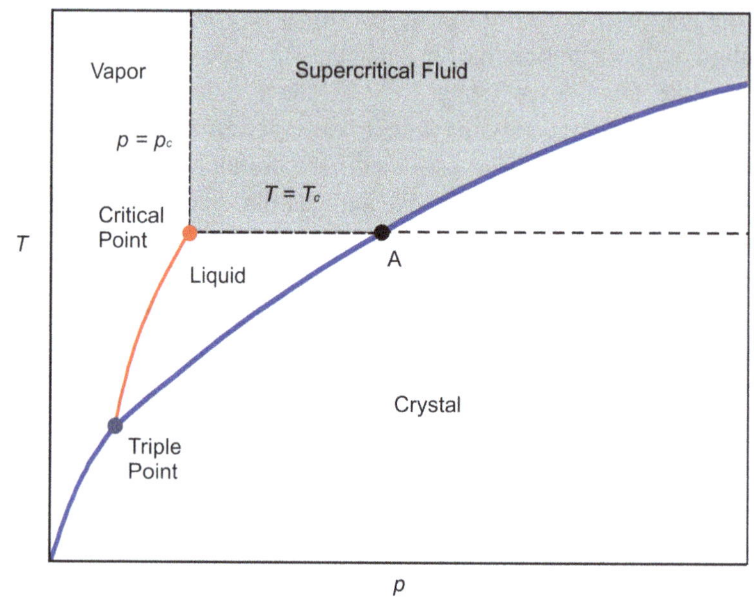

of these works (Brazhkin et al., 2012; Brazhkin & Trachenko, 2012) is to substantiate the existence of a transition zone between the conventionally liquid-like and gas-like states of SCF by the criterion of change in the diffusion mechanism from a vibrational-hopping, inherent to fluid, to a ballistic-collisional that is characteristic of the gas. Brazhkin et al. (2012) propose to call the transition zone separating the change in the diffusion mechanism the Frenkel line.

The issue of the physical nature of supercritical fluids also remains open. There is substantial number of experimental investigations which point to the "two-phase" or "two-component" nature of the substance in supercritical state. One of the "phases" has a higher density and can be considered a liquid-like. Another phase has a lower density and can be considered a gas-like. The known phenomenon of critical opalescence directly indicates the SCF structuring. Of course, the term "structure" is not fully applicable to fluids. It is appropriate to speak of a complex continuously varying system in which there is the near-ordering with its own characteristic times and a set of energetically favorable configurations.

In the authors' opinion, the most successful attempt of theoretical description of supercritical fluids was made by Chaikina (2011; 2012). In these studies, it has been shown that the large fluctuations of density that give rise to the fluctuations of collective velocity and temperature of the fluid correspond to the critical state; the energy stored in the fluctuations of collective velocity when approaching the critical state is almost comparable to the thermal energy of the system. In these works, the opinion regarding two competing basic ideas designed to describe the physical picture of the critical state by means of a drop gas or a fractal cluster has been also expressed. The choice was made in favor of the latter.

Let now discuss the supercritical heat transfer. Study of the supercritical heat transfer has more than fifty-year history (see, for example, Guildner, 1958; Schmidt, 1960; Sengers, 1962; Skripov & Potashev, 1962). Experimental works on this subject can be divided into the study of heat exchange at the free and the forced convection (Pioro & Duffey, 2007). The idea of applying supercritical fluids as heat transfer agents in power-generation industry predetermined a shift of interest to the experiments on the forced convective heat transfer. In this regard, there are relatively few works concerned with heat transfer in SCF at free convection. The reason is obvious: the low heat flux densities that are registered in experiments on the free convective heat transfer (from 1 to 10 kW/m^2) are too low for the power-generation applications. The main result of the experimental work on heat transfer at free and forced convection can be formulated as follows: the presence of the peak of heat transfer coefficient, which is the most pronounced near the critical point and decreases as the parameters are shifted deeper in the supercritical region, can be considered as an established fact. Notably, the line of heat transfer maximum coincides with the line of the isobaric heat capacity maximum. It is generally accepted that it is the heat capacity maximum that provides the maximum of the heat transfer coefficient. Other factors, such as the volume expansion coefficient, are also mentioned while explaining the heat transfer maximum in the near vicinity of critical point.

The behavior of the heat transfer coefficient maximum at forced convection has a number of important features. At the fixed values of heat flux density and mass flux rate, a decrease in the peak height and the shift of its position toward the region of higher temperatures are observed as the pressure increases. Such a behavior is identical to the behavior observed in experiments on free convective heat transfer. A completely different picture is observed when the pressure and mass flux are fixed in the experiments, and the heat flux density is a variable parameter. In this case, while increasing the heat flux density, the heat transfer peak decreases rapidly (see, for example, Wang, Bi, & Leung, 2016; Wang, H., Wang, W., Bi, & Wang, L. 2015; Mori, Kaida, Ohno, Yoshida, & Hamamoto, 2012) and disappears completely, just at the values of interest to the power-generation industry.

It should be noted that the majority of the works on forced convective heat transfer is performed in a relatively close vicinity of the critical point. The logic leading researchers to the vicinity of the critical point is quite understandable. Indeed, the presence of the heat transfer peak was detected exactly in this region. It is unclear, however, why no attention was paid to the rapid drop of the heat transfer peak with increasing the heat flux density, the most important parameter for power-generation industry. Despite the large number of works, the physical picture of the processes in a supercritical heat transfer agent flow remains unclear. This fact is a serious obstacle to the implementation of the idea of using supercritical water in nuclear power industry (Pioro & Duffey, 2007). The origin of some heat transfer modes found in experiments, in particular, modes with a deteriorated heat transfer (Kurganov et al., 2014) has remained not entirely clear. Objectively, there are no theories or even correlations that can predict the emergence of such regimes.

On the basis of the above discussion, our aims were defined as follows: (1) to elucidate the characteristic features of supercritical heat transfer, which accompanies the intense heat release in a compressed liquid, and (2) to provide the experimental justification for selecting the operating pressure of supercritical water for industrial heat transfer equipment, including mini-sized heat exchangers.

MAIN FOCUS OF THE CHAPTER

Experimental

A concept of the real-time control of heating of a thin wire probe underlies the techniques applied in our experiments. The idea turned out to be fruitful for carrying out comparative measurements under strictly specified heating conditions (Skripov, Smotritskiy, Starostin, & Shishkin, 2007). A platinum thermometry wire, 20 μm in diameter, was used as a probe. The characteristic thermal relaxation time of such a probe is on the order of 1 μs. Therefore, it can be assumed that the probe is virtually inertia-less at heating times t of the order of 10^{-3} s and one can carry out the heating modes that are confined in time and inaccessible to traditional methods. In addition, the probe surface area is relatively small at the working length from 1 to 2 cm. This makes it possible to obtain high values of heat flux density at a comparatively small electric power applied to the probe. The heat flux density in our experiments reached 4 MW/m^2 for organic liquids and 20 MW/m^2 for water.

The control procedure is based on the implementation of Proportional–Integral controller (shortly, *PI*-power controller) covered by the negative feedback, see Figure 3. The essence of a control algorithm is as follows. The power supplied to the probe $P_{out}(t)$ is proportional to the driving voltage $U_{in}(t)$:

$$P_{out}(t) = n \cdot U_{in}(t) \tag{1}$$

where n is a constant factor.

To simplify the interpretation of the results, the technique of constant power, $P_{out}(t) = \text{const}$, was chosen as the main heating mode in setting up the experiments. The technique of constant power assumes a predetermined repeatability of the value of a given power with high accuracy in a series of single pulses.

Figure 3. Schematic diagram of the experimental setup: 1 – Control pulse generator; 2 – PI-power controller; 3 – precise standard resistor; 4 – wire probe, connected with the standard resistor in series; 5 – analog-digital converter; 6 – computer

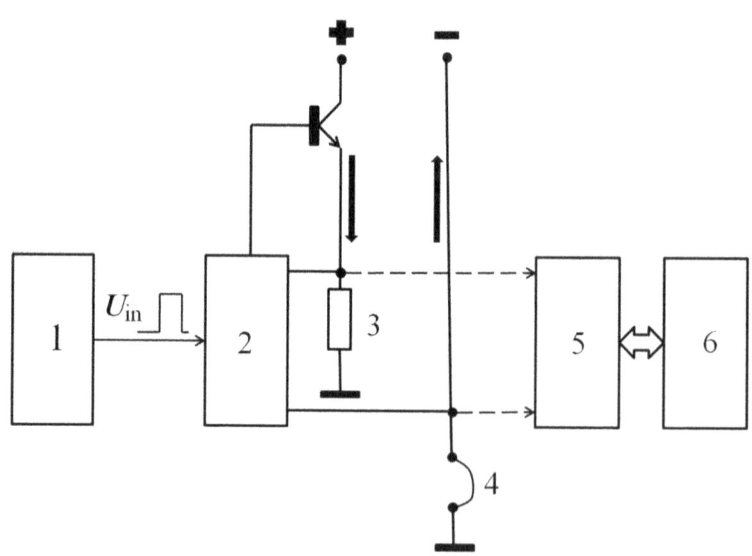

Changing any external parameters is performed during the gap between the pulses. In the case of SCFs, the varied parameter is the pressure. Why is the control of exactly power performed? Because the energy introduced in a probe–fluid system during the pulse is the same for any pulse of this series and can be attributed to the change in the external parameter change in the thermophysical properties and/or the presence (or absence) of phase and chemical transformations in the substance. This creates the basis for carrying out comparative experiments with respect to the chosen base value. Moreover, these relative measurements significantly decrease the errors associated with non-ideality of the probe (namely, edge effects) and non-zero heat capacity of the probe. In addition, in the mode of relative measurements, the probe surface area is excluded from consideration.

In the experiments, the evolution of the average probe temperature $T(t)$ at a given value of the heating power P is monitored. These data are sufficient to determine the heat flux density through the probe surface, which is an integral characteristic of the pulse experiment

$$q = (P - P_{Pt})/\pi \bullet d \bullet l \qquad (2)$$

and the heat transfer coefficient of the substance

$$k(t) = q(t)/\Delta T(t) \qquad (3)$$

at any time t. The reciprocal value

$$R_{\lambda}(t) = \Delta T(t)/q(t) \qquad (4)$$

is called the thermal resistance of a substance at a given heating mode (Rutin & Skripov, 2012). Here d and l are the probe diameter and its length, $\Delta T(t)$ is the temperature rise, $P_{Pt}(t)$ is the fraction of the applied power consumed for heating of the probe itself. In the chosen experimental conditions, P_{Pt} value did not exceed 10%. Ignoring the heat capacity of the probe, which is appropriate in the case under discussion, the task of finding the relative thermal resistance $R_{\lambda i}(t)/R_{\lambda 0}(t)$ is reduced to finding $\Delta T_i/\Delta T_0$ ratio for selected values of P and t. Here the indices "i" and "0" refer to a number of pressure values, one of which (indicated by subscript "0") is selected as the base value. The presented procedure has been described in more detail elsewhere (Rutin & Skripov, 2012, 2013a, 2013b). The substance under study is in an absolutely stable state before the start of the heating pulse. In the course of the experiment, cell with the fluid is placed in a thermostated pressure chamber at a temperature close to the ambient one. For clarity, the progress of experiment is shown schematically in Figure 4.

Because the heat transfer patterns observed in the experiment are similar for all fluids studied by the method of pulse heating, we restrict ourselves to the results of experiments on water. Note that water is the most common and accessible heat transfer agent. Therefore, the most detailed database of thermophysical properties is available for water. Figure 5 shows the experimental data for the evolution of the probe temperature and Figure 6 shows the calculation of the relative thermal resistance at different values of the reduced pressure (from 1.04 to 1.99). In the course of calculating the relative thermal resistance, the value of $p/p_c = 5$ was selected as the base pressure. Position of the base curve in the diagram $T(t; p/p_c = 5)$ was derived by extrapolating the experimental curves $T(t)$ shown in Figure 5.

Figure 4. Schematic representation of the experiment on a probe heating in the pressure–volume plane. Experiment starts at ambient temperature. Arrows show heating at three different values of the supercritical pressure.

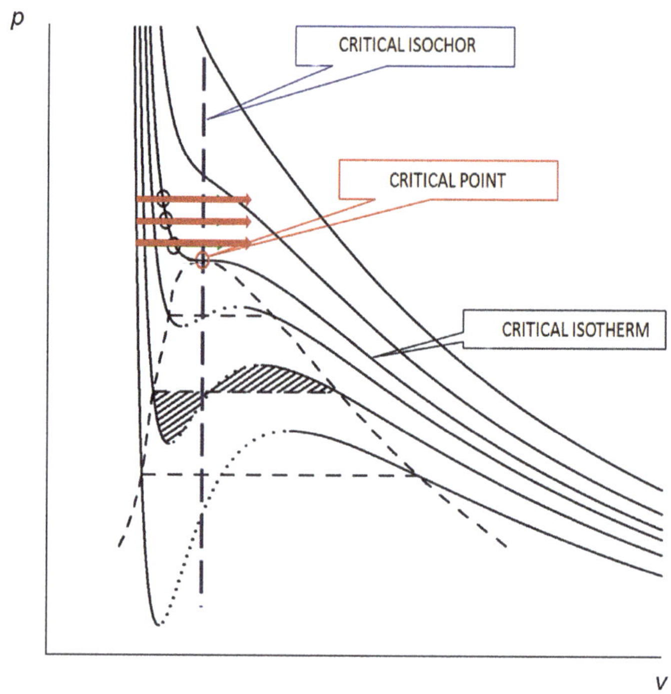

Here and below, the inserts show the actual pattern on a larger scale. Additional explanations are given in the text.

The process of extrapolation included the following steps. Firstly, the computer experiments on the probe heating (namely, $\Delta T(t)$) for $p/p_c = 1$, 2, and 5, respectively, were performed. The physical base for such experiments was the Fourier equation. The thermophysical properties of water that are required for such calculations were taken from the known data collection (Wagner, & Pruβb, 2002); the essential probe dimensions and heating power value were taken from the real experiment. Then a fitting function in the form of

$$y = a + b \cdot \sqrt{x + c} + d \cdot \sqrt[3]{x + e}\,,$$

which describes these curves quite well, was revealed. This function appeared to be suitable for both calculated and experimental curves. The coefficients *a*, *b*, *c*, *d*, *e* for three computer experiments ($p/p_c = 1$, 2, and 5) and two real experiments ($p/p_c = 1$ and 2) were found with the Least Squares method. Finally, the coefficients for function describing real experiment at $p/p_c = 5$ were derived from coefficients for $p/p_c = 2$ and ratio between coefficients for computer experiments at $p/p_c = 2$ and $p/p_c = 5$. The final formula for the base heating curve reconstruction ($p/p_c = 5$) on the $T(t)$ diagram has the following form:

$$y = 624.6 + 347.3 \cdot \sqrt{x + 0.01444} - 528.4 \cdot \sqrt[3]{x + 1.956} \qquad (5)$$

Figure 5. Characteristic heating curves at constant heating power mode in water along the isobar: probe temperature as a function of time

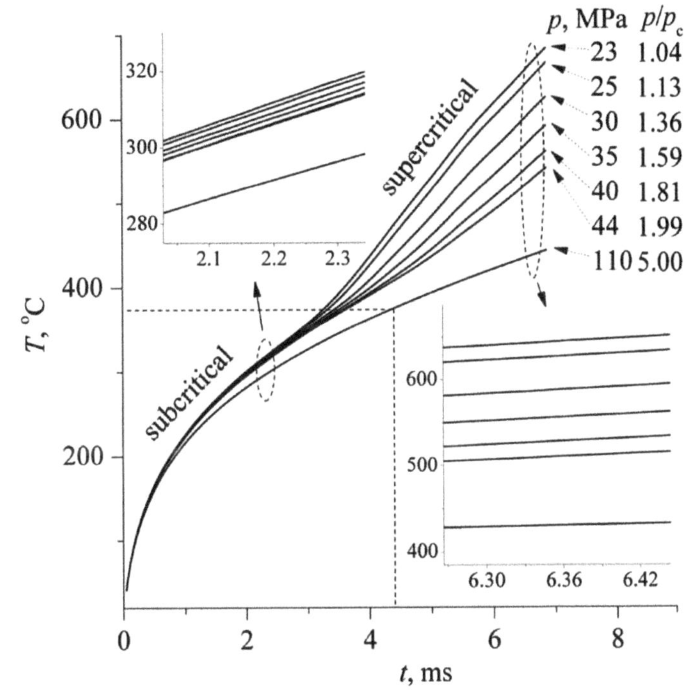

In the experiments, a sharp upward deviation of heating trajectories *T*(*t*) to the region of higher temperatures was found in the course of crossing the critical temperature vicinity. This result clearly indicates the threshold decrease in the heat transfer intensity in the course of isobaric approach to the region of supercritical temperatures. Indeed, the higher is the heating curve for a given power of heat release, the higher the corresponding value of the thermal resistance. This conclusion is illustrated by Figure 6.

Results and Discussion

In the experiments, the effect of isobaric heat capacity peak, which is known from stationary measurements, was not found. In this regard, the computer experiment on the probe heating in water was performed. The parameters of the probe and heating mode were taken from the real experiment. The properties of water, which are necessary for the temperature field calculation, were taken from the same data collection as discussed above. Time-dependent heat transfer equations were solved with the finite volume method. The calculation procedure for evolution of the probe temperature taking into account the thermophysical properties of the substance and the probe material has been reported in detail by Rutin, Smotritskiy, Starostin, Okulovsky, and Skripov (2013). The results of comparing experiment and calculation for the pressure of 23 MPa are shown in Figure 7. Similar results in more detail are discussed elsewhere (Rutin & Skripov, 2014).

A zone of local growth of heat transfer intensity is clearly seen on the calculated heating trajectory, see Figure 7. For clarity, this zone unambiguously associated with the isobaric heat capacity peak is marked by the dashed circle. No peculiarities associated with the said peak are observed in the experimental

Figure 6. Characteristic heating curves at constant heating power mode in water along the isobar: thermal resistance at different pressures ($R_{\lambda i}$) reduced to thermal resistance at the highest pressure ($R_{\lambda 0}$; $p/p_c = 5.0$) as a function of time

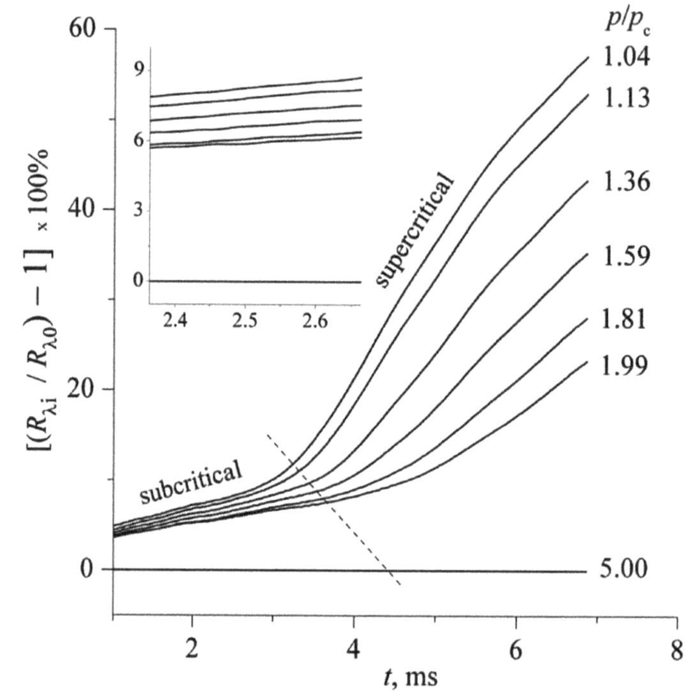

heating trajectory. An important point is that the characteristic time needed to cross the near-critical region is on the order of 10^{-4} s for the selected power value, while the heated layer providing response signal has a thickness of the order of 10^{-7} m, see Figure 8.

Because no influence of the isobaric heat capacity peak on the results of pulse experiments carried out under the constant power mode was detected, the following question arises. This result can be attributed to the fact that a key region of parameters generating the isobaric heat capacity peak is crossed too quickly. Indeed, the influence of the characteristic time of the experiment on the measured value of thermophysical property is discussed with respect to the measurements of heat capacity (see, for example, Anisimov, 1991). This issue can be resolved by increasing the characteristic measurement time. However, such an increase has a physically understandable limitation from above associated with the convection onset. In this regard, it is important to determine the characteristic time for the convection development in pulsed experiments.

To address these issues, the technique of temperature plateau has been applied. The main feature of this technique is that it allows stopping the process of the probe temperature rise for a specified period of time upon reaching the selected value. Entering the region of supercritical parameters is carried out in the same way as in the constant power mode (see Figure 4). The difference is that in the technique of temperature plateau, the probe is rapidly heated (during the order of 10 μs) to a given temperature value $T = T_{pl}$. This value is then maintained with high accuracy throughout the pulse. The length of the second (actual measurement) part of the pulse varied from a few to hundreds of milliseconds. A recorded value in the experiments with temperature plateau is the power needed to maintain a constant difference in

Figure 7. Characteristic heating curves at constant heating power mode in water along the isobar of 23 MPa: real experiment and computer simulation

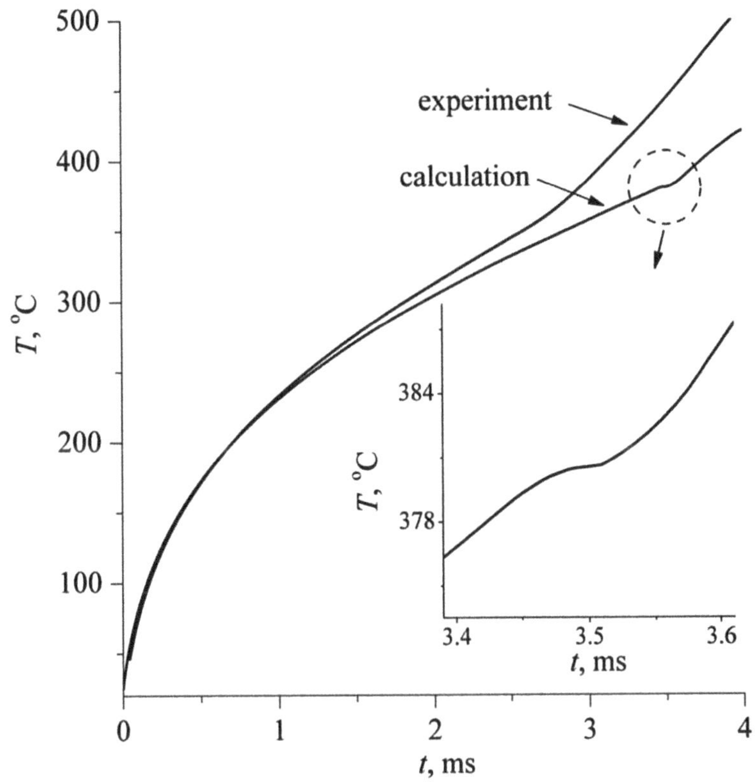

temperature of the probe and that of the substance volume. Pressure is a parameter in the experiment. The procedure for calculating the relative thermal resistance is the same as described for constant power technique. The procedure of detection of convection onset signs has been described elsewhere (Rutin, Volosnikov, & Skripov, 2015).

A comparative experiment involving both techniques was performed with isopropanol, see Figure 9. Within the second technique, a significant increase in the characteristic heating time was achieved (at the highest temperature of 350°C, the starting time of convection onset was about 30 ms). The result of comparison was surprising. Indeed, in the constant power method, the time of passage of the temperature region corresponding to the peak of isobaric heat capacity is appeared to be of the order of 300 μs, see Figure 8. In its turn, the second method made it possible to increase the characteristic heating time to dozens of milliseconds with a conventionally zero heating rate in the time interval corresponding to the temperature plateau. Nevertheless, the fact of threshold decrease in the heat transfer intensity when transferring the substance from the initially stable state to the supercritical one along the isobar was derived as well, see right part of Figure 9.

Indicated numbers correspond to the p_i/p_c values as follows—1: 6.12, 2: 5.71, 3: 5.31, 4: 4.90, 5: 4.49, 6: 4.08, 7: 3.67, 8: 3.27, 9: 2.86, 10: 2.45, 11: 2.04, 12: 1.63, 13: 1.22, 14: 1.02 (left)and 1: 6.0, 2: 5.10, 3: 4.08, 4: 3.06, 5: 2.04, 6: 1.63, 7: 1.55, 8: 1.47, 9: 1.39, 10: 1.31, 11: 1.22, 12: 1.14, 13: 1.06, 14: 1.02 (right).

Figure 8. Pulse heating of water along the isobar of 23 MPa: calculation of the temperature field in a direction normal to the probe surface for the temperature values of (T_c - 10 K, curve 1) and (T_c + 10 K, curve 2) at the constant heating power mode. The probe diameter is 20.8 μm.

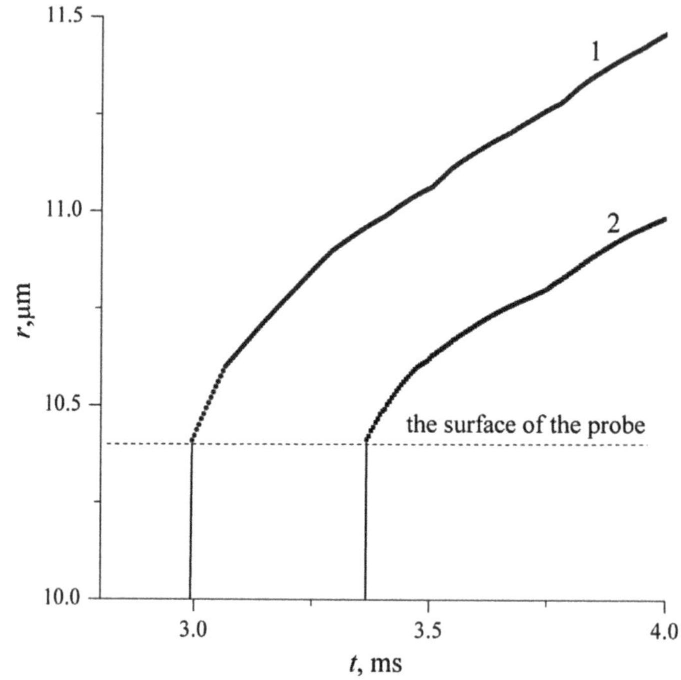

Based on the analysis of Figure 9, one can draw a conclusion about quantitative agreement of the effects of decreasing the heat transfer intensity in the critical temperature vicinity for both approaches discussed. It is evident that this phenomenon manifests itself in a wide range of variation of the characteristic times (with respect to the methods of high-power pulse heating of the probe in the substance). According to Rutin et al. (2015), it is inherent in the conduction mechanism of heat transfer in supercritical fluids in general.

SOLUTIONS AND RECOMMENDATIONS

Data for supercritical heat transfer obtained in our experiments made it possible formulation of a "road map" of the study. One of the next steps discussed below is aimed at implementation of the idea of the fourth generation nuclear reactors (Generation IV, sixth concept, as cited by Pioro & Duffey, 2007). Cooling of such reactor core is supposed to be carried out by supercritical water. As noted in Background, the logic of operating pressure selection was insufficiently substantiated for Generation IV. At close examination of this issue, the following contradiction arises. On one hand, there is a natural desire to make use of the known peak of the heat transfer coefficient. This leads to the selection of operating pressure in the vicinity of the critical pressure value. Conventional choice is 25 MPa (see Pioro & Duffey, 2007) or ~ $1.14p/p_c$. At the same time, the fact of collapse of the heat transfer peak with increasing the heat flux density toward the values of practical interest for power-generation is ignored. On the other

Figure 9. Pulse heating of isopropanol under the constant power mode (left) and temperature plateau mode (right): the change in thermal resistance at different pressures ($R_{\lambda i}$) reduced to thermal resistance at the highest pressure ($R_{\lambda 0}$; p/p_c = 6.0) vs. heating time (left) and vs. temperature T_{pl} (right)

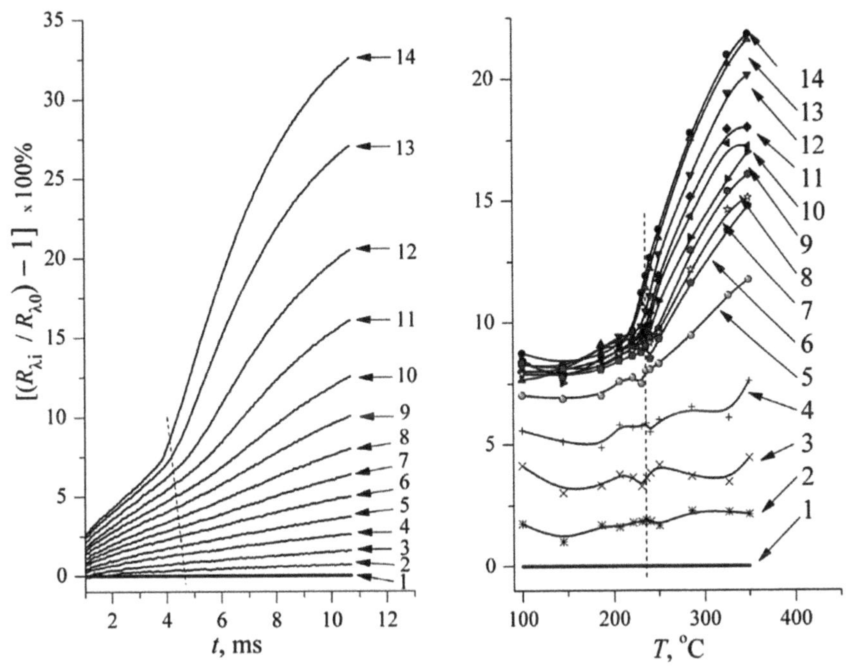

hand, selection of the temperature at an outlet of device indicates an intention to obtain the high thermal efficiency. However, the selected temperature proves to be considerably higher than the temperature region in which the heat transfer peak manifests itself. Also of note is that in the near-critical region of parameters, there is the drastic and non-monotonic change in thermophysical properties (TPPs) of a substance. This makes it almost impossible to perform engineering modeling of heat exchangers with supercritical heat transfer agent. Furthermore, the emergence of deteriorated heat transfer modes, flow oscillations, and some other phenomena detected in the near-supercritical region are directly related to a confined (with respect to temperature) change in TPPs in this area.

We anticipate an objection that choosing the higher values of the operating pressure will lead to a rise in price of Generation IV concepts. The answer to it is a well-known tendency to the increased operating parameters of supercritical water in the conventional thermal power industry (Wichtmann A., Deckers M., & Ulm, 2005). At present, the thermal power plant blocks with parameters called "ultra-supercritical" have been developed and successfully maintained. The mastered operating parameters are approaching 40 MPa (~1.82 p/p_c) and 700°C (~1.5 T/T_c). The process is accompanied by the development of appropriate structural steels capable of operating at these parameters for a long time. The authors believe that this trend is absolutely correct. The only thing missing is the target benchmark for selecting the operating pressure.

In this connection, let return to the heat transfer patterns obtained for a constant power mode in our experiments. By an example for isopropanol, let consider a change in the intensity of heat transfer from the probe to the environment in the subcritical range of parameters. Figure 10 shows the trajectories of the probe heating $T(t)$ at different subcritical pressures.

The inset of this graph clearly shows that the heating trajectories are slightly, but systematically bent downward with increasing the reduced pressure. This result indicates a slight decrease in the thermal resistance caused by small changes in TPPs as pressure (in subcritical region) changes. Sharp bends of heating trajectories are caused by the process of spontaneous boiling-up, the discussion of which is beyond the scope of this chapter. Note only that spontaneous boiling-up under conditions of maintaining a constant heating power leads to a sharp decrease in the trajectory of probe heating due to energy pumping out that is associated with the heat of the liquid–vapor phase transition. The line of spontaneous boiling-up temperatures with respect to pressure, by definition, is terminated at the critical point.

Let us consider the trajectory of probe heating in isopropanol in a wide range of supercritical pressures. Figure 11 shows heating trajectories and their derivatives with respect to time in the entire region of the probe temperature changes. There is a threshold decrease in the intensity of heat transfer in the near-supercritical region of pressure values, which almost completely degenerates when the pressure is increased up to $3p/p_c$.

Similarly, Figure 12 represents the same data at $T < T_c$, i.e., until crossing the critical temperature in the course of heating process. From Figure 12, one can see that the heating process does not contain any features until the moment of crossing the T_c value, and the heating trajectory gradually decreases in the course of pressure increase. The derivatives with respect to heating trajectories also do not show any abnormal behavior. If one returns to Figure 10, it becomes clear that in the region of subcritical temperatures the heat transfer process is identical for both subcritical and supercritical pressures (with

Figure 10. Characteristic heating curves at constant heating power mode in isopropanol. The reduced pressure ($p/p_c < 1.0$) serves as a parameter. Additional explanations are given in the text.

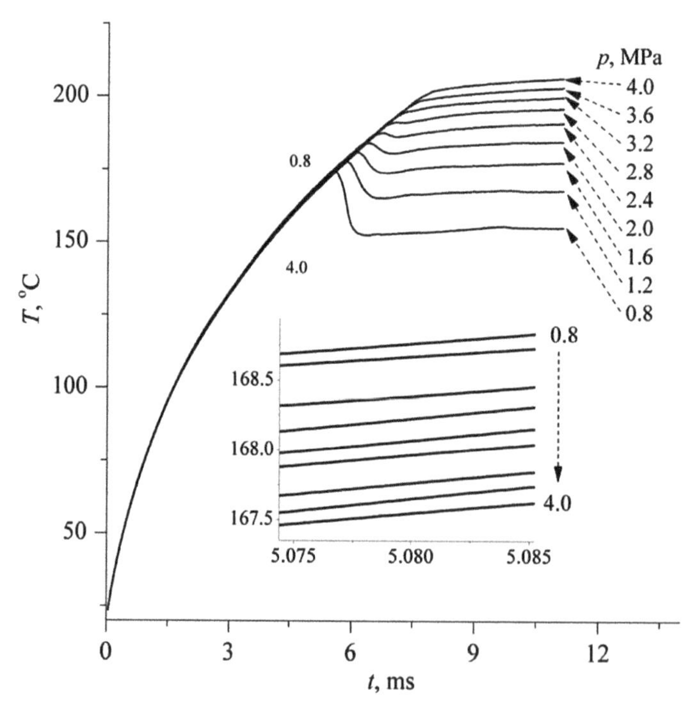

Figure 11. Characteristic heating curves at constant heating power mode in isopropanol. The reduced pressure ($p/p_c > 1.0$) serves as a parameter. Here and below, its values are the same as in Figure 9.

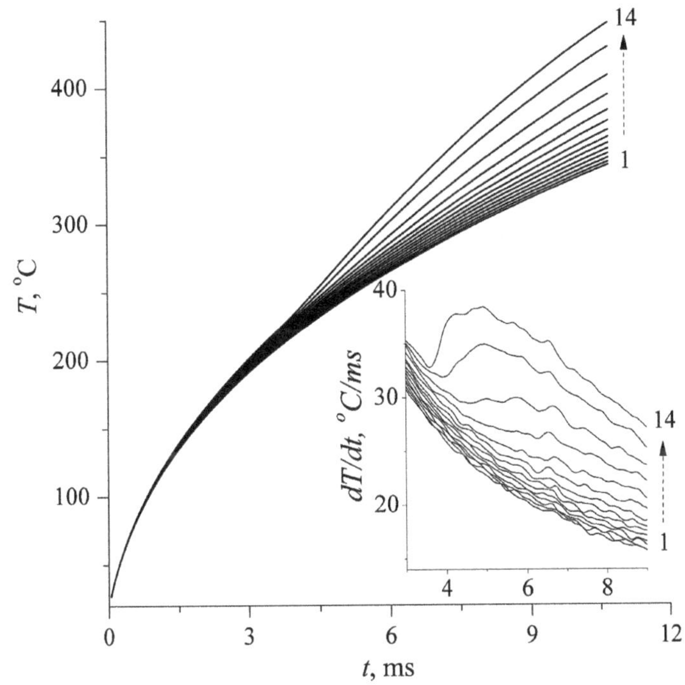

Figure 12. Characteristic heating curves at constant heating power mode in isopropanol at subcritical ($T < T_c$) region. The reduced pressure ($p/p_c > 1.0$) serves as a parameter

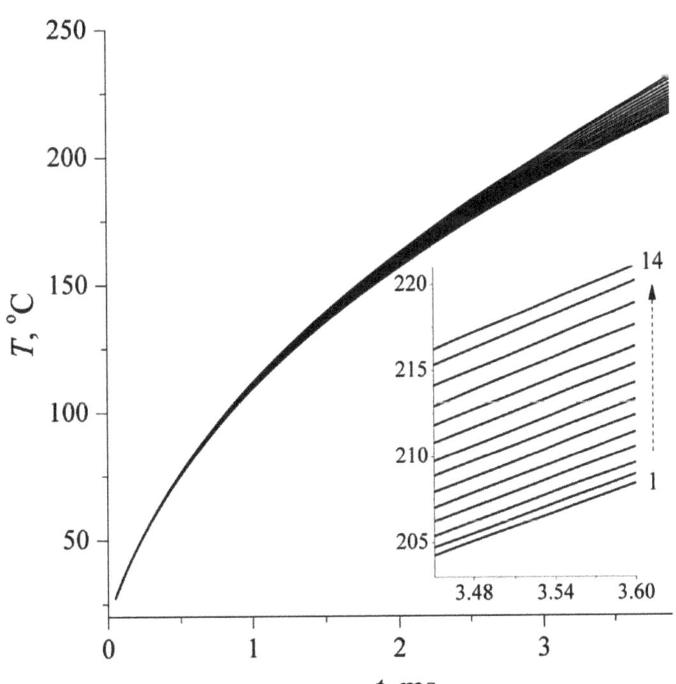

the exception of the process of spontaneous boiling-up at $p < p_c$). The heat transfer patterns observed in the experiment are characterized by a slight decrease in heating trajectories with increasing pressure. This can be explained by low compressibility of the fluid and consequently a small change in its TPPs as the pressure changes.

The most interesting thing is that such a behavior was also found directly in the supercritical region of parameters. Figure 13 shows the course of the heating trajectories according to data shown in Figure 11, but at the values of reduced pressure over $3p/p_c$. It is easy to notice that the behavior of data corresponding to sufficiently deep entering into the region of supercritical pressure, as shown in Figure 13 (see also the curves 1-8 in the left part of Figure 9), does not differ from the behavior of the data presented in the entire range of pressures at temperatures below the critical value (Figure 12) and at subcritical pressures (Figure 10).

Such an obvious similarity of heating trajectories gives justification to propose the target benchmarks for selecting a value of operating pressure in industrial heat transfer equipment, including Generation IV concepts, in the range of 2.5÷3.0 p/p_c. Let emphasize that the above data were obtained under conditions of high heat loads (up to 20 MW/m^2 for water) and in a scale of small characteristic sizes and times. In the actual heat exchange devices, the limiting heat transfer factor is the boundary layer; more precisely, its viscous sublayer. Viscous sublayer remains substantially immobile and provides transfer of thermal energy from the walls to the heat transfer agent. Let assign the results obtained in experiments to the boundary layer region, which is quite justified, taking into account the spatial scale of our experiments. It can be stated that the thermal resistance of supercritical fluid in the near-supercritical region exceeds the thermal resistance at $3p/p_c$ by dozen's percent, see Figure 9.

Figure 13. Pulse heating of isopropanol at a deep penetration into supercritical (p/p_c >3) region: probe temperature as a function of heating time

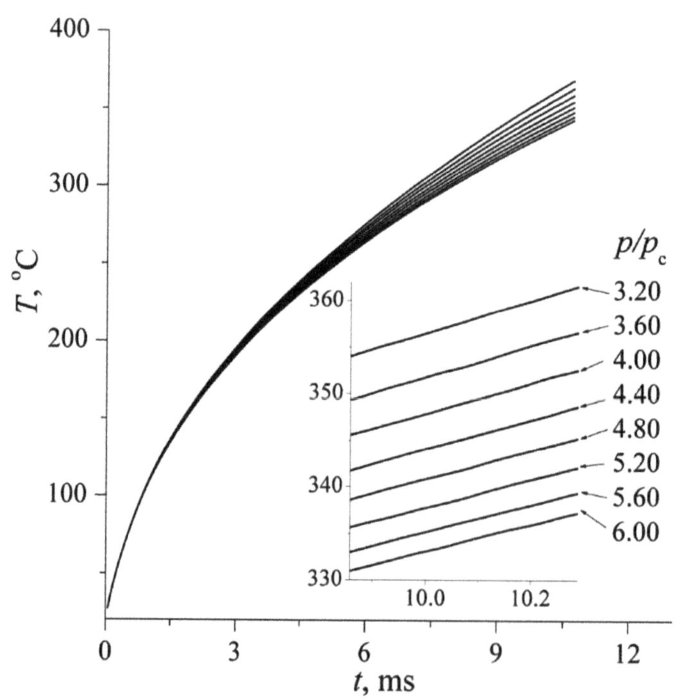

Additional arguments in favor of the choice of higher pressure values (as compared to those built-in the Generation IV concepts) can be obtained using the known data on water TPPs. Figure 14 shows graphs of the thermal effusivity of water for some values of supercritical pressure. A glance at Figure 14 is enough to understand that at temperatures of interest to the thermal power industry, namely, 600÷700°C, the thermal effusivity values at the chosen pressure values are significantly different.

Realizing that the distance from a thin wire probe to the industrial heat transfer equipment is large enough, the authors propose to return to the lab experiments on the forced convective heat transfer under conditions of sufficiently high heat loads. Statement of this problem is formulated in the most concise form by Rutin and Skripov (2016). The aims of this study are to compare the limiting heat flux densities for water, in which the heat transfer process remains stable, for the pressure of 25, 44, and 66 MPa; to find out the presence (or absence) of deteriorated heat transfer modes and the conditions for the emergence of such modes. It is also important to check the possibility of modeling the heat transfer intensity for the pressures of 2÷3 p/p_c in a way similar to calculations for incompressible fluid. This move would enable to simplify the engineering calculation of heat exchange equipment in the future. The authors believe that the probability of obtaining a positive result in experiments on the forced convective heat transfer is sufficiently high.

Figure 14. Thermal effusivity $b = (\lambda \rho c_p)^{0.5}$ of water at different pressures (b_i, 44 MPa, curve 1; 66 MPa, curve 2) reduced to thermal effusivity of water at the lowest pressure (b_0, 25 MPa) as a function of temperature.

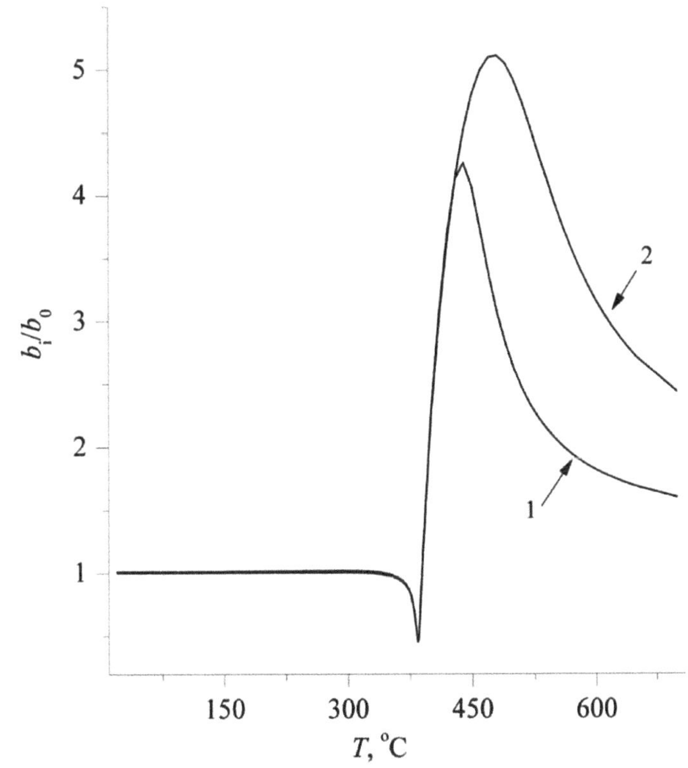

FUTURE RESEARCH DIRECTIONS

In order to completely solve the applied problems of heat transfer in supercritical water it is necessary to perform several stages of experimental work. As was shown above, the heat transfer patterns in the short-term wire probe experiments are almost identical for different fluids. Nevertheless, it is necessary to perform a systematic study of the heat transfer specifically in water. Water is the most demanded and the most common heat transfer agent. Moreover, the most complete data on TPPs in a wide range of parameters, including the supercritical state parameters are available specifically for water. All these data were obtained under stationary conditions. Nevertheless, their availability would allow a quantitative assessment, which is not always possible for other fluids. Experiments will be performed in both the constant power mode and the mode of temperature plateau in the range of reduced pressures from 1 to $3p/p_c$.

The data obtained on intensity of heat transfer, including data on the characteristic times of convection onset, will make it possible to justify the benchmark for selecting the working pressure of supercritical water for industrial heat transfer equipment. The authors have already suggested the research teams involved in the experimental study of supercritical heat transfer under conditions of forced convection, to perform comparative experiments in water in a wide range of supercritical pressures, up to the value of $3p_c$ (Rutin, & Skripov, 2016). An analysis of the available scientific publications on this topic leads to the conclusion that they are focused primarily on the description of supercritical heat transfer in the near-supercritical region of parameters. In this regard, the abovementioned proposal is still valid, and the authors are ready to cooperate in solving this problem.

CONCLUSION

Application of short-term techniques for studying the supercritical heat transfer allowed obtaining new interesting results that are able to significantly influence the choice of the direction of future works on this subject. The idea of using controlled pulse heating of a wire probe proved to be very fruitful, and the results presented in this work are far from exhausting the possibilities of this approach. Short time of thermal relaxation of the probe made it possible to implement heating regimes that are not available in the traditional hot-wire methods (Assael, Antoniadis, & Wakeham, 2010), and to identify the patterns of heat transfer at small characteristic times and sizes, and, importantly, at the high heat flux densities. Within this approach, the impact of the two main factors (namely, convection and gravity) complicating the study of heat conduction in SCFs is proven to be almost eliminated. As a result, the effect of threshold decrease in the heat transfer intensity was revealed in course of a fast transition from compressed liquid to supercritical fluid state along the isobar. The final solution of the problem of coexistence of this effect with the peaks of thermophysical properties in the near-critical region, which are known from stationary measurements, has not been found so far.

Data obtained in the pulse experiments gave grounds for the assumption that, under conditions of high heat flux densities, thermal resistance of the boundary layer of supercritical fluid is a limiting factor for heat transfer. The stronger the impact of this factor is expressed, the closer the pressure value to critical pressure. In this context, the expansion of the region of variation of supercritical pressure values in the experiments towards their increase appears to be timely. The authors believe that the results of such study will provide the answer to the question of expediency of transition to higher working pressure values.

ACKNOWLEDGMENT

The study was partially supported by the Russian Foundation for Basic Research, project no. 16-08-00381-a and by the Presidium of the UB RAS Complex Program (project no. 15-20-2-18). Also, technical support from R.G. Ismagilov, G.L. Levit, and D.V. Volosnikov is very much appreciated.

REFERENCES

Anisimov, M. A. (1991). *Critical Phenomena in Liquids and Liquid Crystals*. Philadelphia, PA: Gordon & Breach Science Publishers.

Anisimov, M. A. (2011). Letter to the editor: Fifty years of breakthrough discoveries in fluid criticality. International Journal of Thermophysics, 32, 2001–2009.

Assael, M. J., Antoniadis, K. D., & Wakeham, W. A. (2010). Historical evolution of the transient hot-wire technique. *International Journal of Thermophysics*, *31*(6), 1051–1072. doi:10.1007/s10765-010-0814-9

Berche, B., Henkel, M., & Kenna, R. (2009). Critical phenomena: 150 years since Cagniard de la Tour. Journal of Physical Studies, 13(3).

Brazhkin, V. V., Fomin, Yu. D., Lyapin, A. G., Ryzhov, V. N., & Trachenko, K. (2012). Two liquid states of matter: A dynamic line on a phase diagram. *Physical Review E: Statistical, Nonlinear, and Soft Matter Physics*, *85*(3), 031203. doi:10.1103/PhysRevE.85.031203 PMID:22587085

Brazhkin, V. V., & Trachenko, K. (2012). What separates a liquid from a gas. *Physics Today*, *65*(11), 68–69. doi:10.1063/PT.3.1796

Chaikina, Yu. A. (2011). Distinctive features of random local motions in critical and supercritical fluids. *Russian Journal of Physical Chemistry B*, *5*(7), 1116–1129. doi:10.1134/S1990793111070128

Chaikina, Yu. A. (2012). Development of finite fluctuations in local density, velocity, and temperature for real fluids. *Russian Journal of Physical Chemistry B*, *6*(7), 834–843. doi:10.1134/S1990793112070020

Gorbaty, Yu. E., & Bondarenko, G. V. (2007). Water in the supercritical state. *SCF-TP Online, 2*(2) 5–19. Retrieved from http://scf-tp.ru/articles/2007_02/index.html

Guildner, L. A. (1958). The thermal conductivity of carbon dioxide in the region of the critical point. *Proceedings of the National Academy of Sciences of the United States of America*, *44*(11), 1149–1153. doi:10.1073/pnas.44.11.1149 PMID:16590326

Kurganov, V. A., Zeigarnik, Yu. A., & Maslakova, I. V. (2012). Heat transfer and hydraulic resistance of supercritical-pressure coolants. Part I: Specifics of thermophysical properties of supercritical pressure fluids and turbulent heat transfer under heating conditions in round tubes (state of the art). *International Journal of Heat and Mass Transfer*, *55*(11-12), 3061–3075. doi:10.1016/j.ijheatmasstransfer.2012.01.031

Kurganov, V. A., Zeigarnik, Yu. A., & Maslakova, I. V. (2014). Heat transfer and hydraulic resistance of supercritical pressure coolants. Part IV: Problems of generalized heat transfer description, methods of predicting deteriorated heat transfer; empirical correlations; deteriorated heat transfer enhancement; dissolved gas effects. *International Journal of Heat and Mass Transfer*, *77*, 1197–1212. doi:10.1016/j.ijheatmasstransfer.2014.06.014

Mori, H., Kaida, T., Ohno, M., Yoshida, S., & Hamamoto, Y. (2012). Heat transfer to a supercritical pressure fluid flowing in sub-bundle channels. *Journal of Nuclear Science and Technology Journal of Nuclear Science and Technology*, *49*(4), 373–383. doi:10.1080/00223131.2012.669240

Pioro, I. L., & Duffey, R. B. (2007). *Heat transfer and hydraulic resistance at supercritical pressures in power engineering applications*. New York, NY: ASME Press. doi:10.1115/1.802523

Rutin, S. B., & Skripov, P. V. (2012). Apparatus for studying heat transfer in nanofluids under high-power heating. *Journal of Engineering Thermophysics*, *21*(2), 144–153. doi:10.1134/S1810232812020063

Rutin, S. B., & Skripov, P. V. (2013a). Investigation of not fully stable fluids by the method of controlled pulse heating. 1. Experimental approach. *Thermochimica Acta*, *562*, 70–74. doi:10.1016/j.tca.2013.03.030

Rutin, S. B., & Skripov, P. V. (2013b). Heat transfer under supercritical parameters of pulse-heated liquid. *Russian Journal of Physical Chemistry B*, *7*(8), 943–949. doi:10.1134/S1990793113080101

Rutin, S. B., & Skripov, P. V. (2013c). Heat transfer in supercritical fluids under pulse heating regime. *International Journal of Heat and Mass Transfer*, *57*(1), 126–130. doi:10.1016/j.ijheatmasstransfer.2012.10.027

Rutin, S. B., & Skripov, P. V. (2014). Heat transfer under high-power heating of liquids. 2. Transition from compressed to supercritical water. *International Journal of Heat and Mass Transfer*, *79*, 526–531. doi:10.1016/j.ijheatmasstransfer.2014.08.043

Rutin, S. B., & Skripov, P. V. (2016). Controlled high-power heat release as a tool to selecting working pressure for supercritical water. *Journal of Engineering Thermophysics*, *25*(2), 166–173. doi:10.1134/S1810232816020028

Rutin, S. B., Smotritskiy, A. A., Starostin, A. A., Okulovsky, Yu. S., & Skripov, P. V. (2013). Heat transfer under high-power heating of liquids. 1. Experiment and inverse algorithm. *International Journal of Heat and Mass Transfer*, *62*, 135–141. doi:10.1016/j.ijheatmasstransfer.2013.02.060

Rutin, S. B., Volosnikov, D. V., & Skripov, P. V. (2015). Heat transfer under high-power heating of liquids. 3. Threshold decrease of heat conduction in supercritical region. *International Journal of Heat and Mass Transfer*, *91*, 1–6. doi:10.1016/j.ijheatmasstransfer.2015.07.001

Schmidt, E. (1960). Warmetransport durch naturliche konvektion in stiffen bei kritischem zustand. *International Journal of Heat and Mass Transfer*, *1*(1), 92–101. doi:10.1016/0017-9310(60)90012-0

Sengers, J. V. (1962). *Thermal conductivity measurements at elevated gas densities including the critical region* (Unpublished thesis). Universiteit van Amsterdam, Amsterdam, The Netherlands.

Skripov, P. V., Smotritskiy, A. A., Starostin, A. A., & Shishkin, A. V. (2007). A method of controlled pulse heating: Applications. *Journal of Engineering Thermophysics*, *16*(3), 155–163. doi:10.1134/S1810232807030071

Skripov, V. P., & Potashev, P. P. (1962). Heat exchange with carbon dioxide along the supercritical isobars under free convection. [in Russian]. *Journal of Engineering Physics*, *5*(2), 30–34.

Wagner, W., & Prußb, A. (2002). The IAPWS formulation 1995 for the thermodynamic properties of ordinary water substance for general and scientific use. *Journal of Physical and Chemical Reference Data*, *31*(2), 387–535. doi:10.1063/1.1461829

Wang, H., Bi, Q., & Leung, L. K. H. (2016). Heat Transfer from a 2 × 2 Wire-Wrapped Rod Bundle to Supercritical Pressure Water. *International Journal of Heat and Mass Transfer*, *97*, 486–501. doi:10.1016/j.ijheatmasstransfer.2016.02.036

Wang, H., Wang, W., Bi, Q., & Wang, L. (2015). Experimental study of heat transfer and flow resistance of supercritical pressure water in a SCWR sub-channel. *The Journal of Supercritical Fluids*, *100*, 15–25. doi:10.1016/j.supflu.2015.02.011

Wichtmann, A., Deckers, M., & Ulm, W. (2005). *Ultra-supercritical steam turbine turbosets – best efficiency solution for conventional steam power plants*. Paper presented at International Conference on Electrical Engineering, Kunming, China.

KEY TERMS AND DEFINITIONS

Compressed Liquid: A liquid under supercritical pressure and subcritical temperature.

Critical Point: Critical point is the end point of liquid-vapor coexistence line. The properties of liquid and vapor become identical in the critical point.

Near-Critical Region: The region where the properties of a substance manifest characteristic anomalies.

Saturation Line: A line of equilibrium coexistence of liquid and vapor phases. Triple point and critical point serve as a start and end points, respectively.

Supercritical Fluid: A substance under supercritical pressure and supercritical temperature.

Thermal Effusivity: The essential thermophysical parameter with respect to heat transfer under non-stationary regime.

Ultra-Supercritical Parameters: Values of pressure and temperature exceeding 1.5 times their critical values.

Chapter 10
Study on Cooling Heat Transfer of Supercritical Carbon Dioxide Applied to Transcritical Carbon Dioxide Heat Pump

Chaobin Dang
University of Tokyo, Japan

Eiji Hihara
University of Tokyo, Japan

ABSTRACT

Understanding the heat transfer characteristics of supercritical fluids is of fundamental importance in many industrial processes such as transcritical heat pump system, supercritical water-cooled reactor, supercritical separation and supercritical extraction processes. This paper addresses recent experimental, theoretical and numerical studies on cooling heat transfer of supercritical CO_2. A systematic study on heat transfer coefficient and pressure drop of supercritical CO_2 was carried out at wide ranges of tube diameter, mass flux, heat flux, temperature and pressure. Based on the understanding of temperature and velocity distributions at cross-sectional direction provided by the numerical simulation, a new prediction model was proposed, which agreed well with the experimental results. In addition, the effect of lubricating oil was also discussed with the focus on the change in flow pattern and heat transfer performance of oil and supercritical CO_2.

INTRODUCTION

Re-Discovering of Carbon Dioxide Refrigerant

Carbon dioxide, hereafter CO_2, was initially applied as a refrigerant in late 1800s because it was proven safe and highly efficient. It was commonly used on board ship till it was substituted by CFCs (chlorofluorocarbons) in 1930s because the latter was operated at lower system pressures. Things have changed

DOI: 10.4018/978-1-5225-2047-4.ch010

since 1990s. Due to two treatments on environmental protection, i.e. the Montreal Protocol and Kyoto Protocol, carbon dioxide has been re-discovered and is expected to be a promising refrigerant in applications such as hot water heater, automotive A/C, and low temperature refrigerators.

Lorentzen (1995) compared several common alternative refrigerants. In contrast with CFCs and HCFCs, the alternatives, such as ammonia, hydrocarbons, and CO_2, have an ODP (Ozone Depletion Potential) of zero and a negligible GWP (Global Warming Potential). As for HFC134a, although its ODP is zero, it has a GWP as high as 1200 (100 year) or 3100 (20 year). With respect to the safety of "old" refrigerants, only CO_2 can compete with the non-flammable HFCs. Although CO_2 is also listed as a greenhouse gas, it is because of the large amounts emitted from many industrial applications. In comparison with HFCs, the GWP of CO_2 is negligible when CO_2 is used as a refrigerant. Therefore, the use of CO_2 as a refrigerant has major benefits of being environmentally benign and safety.

When CO_2 is applied to conventional vapor compression cycle, it shows significantly different characteristics comparing with other refrigerants. Due to its low critical temperature ($Tc_p = 31.1$°C, $Pc_p = 7.38$ MPa), the CO_2 heat pump cycle has to be operated trans-critically when the ambient temperature is near or higher than the critical temperature. In this case, the heat absorbing process takes place at subcritical pressure whereas the heat rejection takes place at supercritical state. This kind of trans-critical cycle is initially proposed by Lorentzen and his coworkers for automotive air conditioning and hot water heat pump systems (Figure 1).

Due to the temperature glide in the gas cooler at supercritical pressure, the temperature profiles of the CO_2 and the secondary fluid can be advantageously adapted in order to minimize heat transfer loss and hence improve energy efficiency. In addition, high discharge temperature of coolant about 90°C-100°C can be obtained without increasing the refrigerant side pressure and temperature so much. Thus an obvious preferable application of CO_2 heat pump could be hot air or water production.

Although the high vapor pressure requires the redesign for more durable compressor and other parts, the high pressure and low viscosity of CO_2 may lead to a small pressure loss. Furthermore, the pressure ratio of CO_2 heat pump is quite low, only 2-3, compared to the value of 4-5 in conventional vapor compression cycle. Higher compressor efficiency can be achieved for the CO_2 trans-critical cycle.

In Japan, since the commercialization of the "Eco Cute" in 2001, the production of CO_2 heat pump water heaters has increased steadily. According to JRAIA (The Japan Refrigeration and Air Condition-

Figure 1. Temperature-entropy chart for a trans-critical process of hot water heat pump
Lorentzen, 1995.

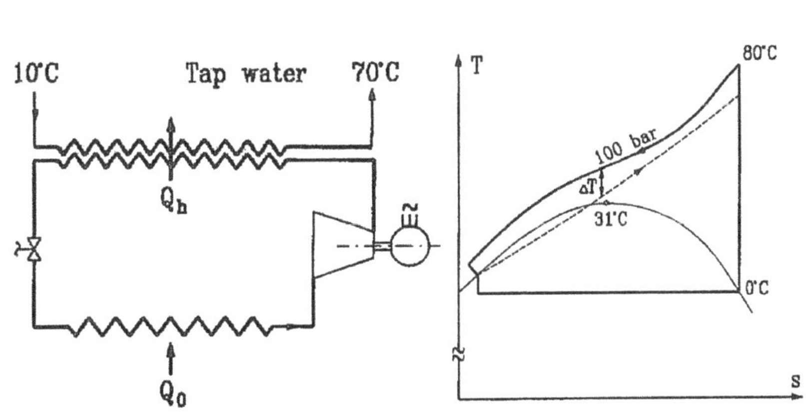

ing Industry Association), the total installations in Japan achieved 5 million units by March 2016. The coefficient of performance (COP) of the heat pump unit increased from 3.46 in its first generation units to 4.90 in those produced toward the end of 2007. See Figure 2.

In conclusion, the CO_2 technology is very attractive as it is environmentally benign and safe. Properly designed systems for such applications as hot water supplier and automotive A/C have been proved competitive to the former CFC-based and ozone depleting system. Main difficulty of design and optimization of CO_2 heat pump is related to the lack of information about the flow and heat transfer at supercritical pressure in cooling process, this paper addresses recent experimental, theoretical and numerical studies on cooling heat transfer of the supercritical CO_2, including the effect of lubricating oil with the focus on the change in flow pattern and heat transfer performance of oil and supercritical CO_2 at different operation conditions.

Figure 2. Shipment of Eco cute in Japan and the transition of COP: (a) shipment in Japan; (b) COP variations

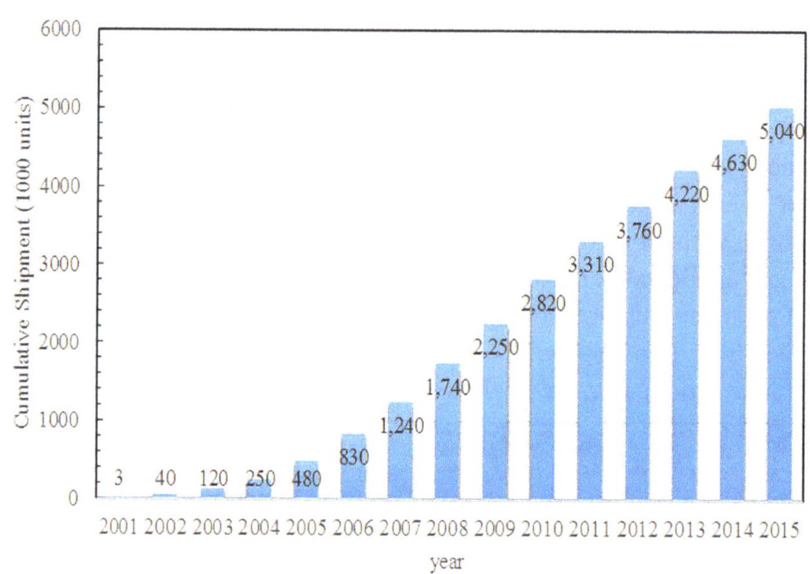

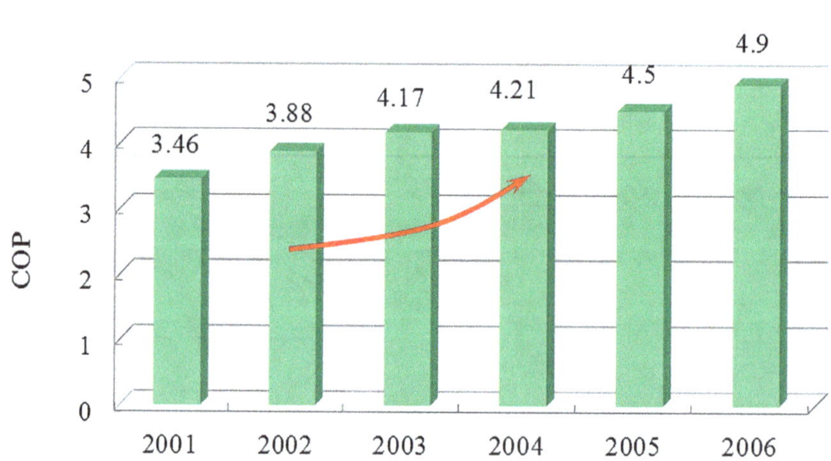

Thermophysical Properties

Thermophysical properties affecting flow and heat transfer of supercritical fluid significantly include specific heat (c_p), density (ρ), thermal conductivity (k) and dynamic viscosity (μ). Figure 3 shows an example of the variations of thermophysical properties with temperature at pressure of 8 MPa. Thethermophysical properties are calculated according to property software Refprop 6.01 of NIST. The most significant feature of the temperature dependence of thermophysical property is that c_p takes a maximum value at the pseudocritical temperature. At pressure of 8 MPa, c_p at the pseudocritical temperature (34.52℃) is 29.6 kJ/kgK, which is more than 10 times larger than that at 20℃. In addition, c_p at the pseudocritical temperature of 9 MPa decreases about 12.0 kJ/kgK, about half of that at 8 MPa, while at pressure of 10 MPa, a further smaller value of c_p at pseudocritical temperature can be seen.

Near the pseudocritical temperature, ρ and μ decrease sharply with the increase of temperature. Although k also decreases with temperature, it shows a local maximum near the pseudocritical temperature. The existence of the thermal conductivity spike was long questioned, because eliminating all fluid movement in conductivity measurement under supercritical pressure is difficult. However, lots of detailed experiments of the thermal conductivity had proved the existence of this anomalous characteristic of the thermal conductivity at pseudocritical temperature (Sengers, 1965).

Heat Transfer Characteristics at Supercritical Pressure

For constant property fluid flow, the Gnielinski's equation (1976) is well known in the prediction of heat transfer coefficient,

$$Nu = \frac{\left(f/8\right)\left(\mathrm{Re}-1000\right)\mathrm{Pr}}{1.07+12.7\sqrt{f/8}\left(\mathrm{Pr}^{2/3}-1\right)} \tag{1}$$

Figure 3. Thermophysical properties of supercritical carbon dioxide near the pseudocritical point at pressure of 8 MPa

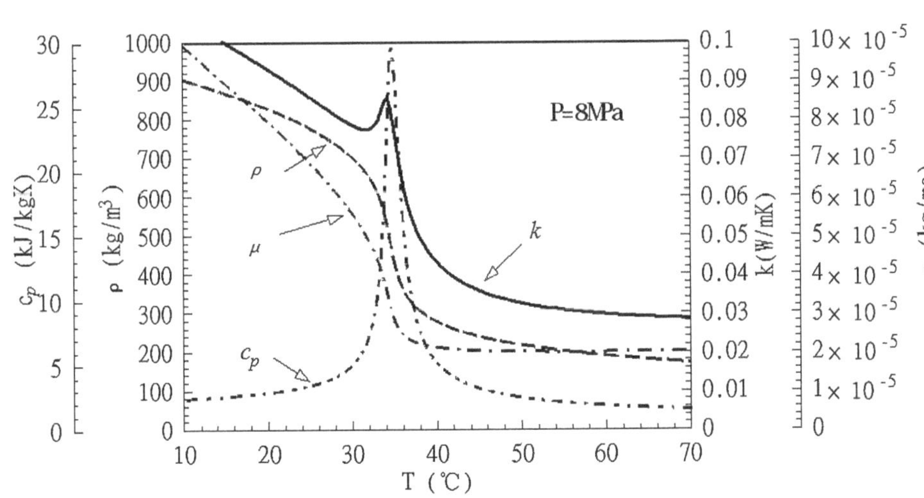

where f is friction factor and is calculated from Filonenko's equation,

$$f = \left[1.82\log_{10}(\mathrm{Re}) - 1.64\right]^{-2} \tag{2}$$

This model, together with many other similar models, is proposed on the assumption that the fluid properties remain constant throughout the flow field. In order to apply these models to heat transfer problems with temperature-dependent properties variations, usually they are revised in a property ratio method to take into consideration of the properties distribution in cross section.

Heat transfer of temperature-dependent-property problem has been well investigated (Kays and Crawford, 1980). Usually it involves a different form for liquids and for gases. For liquids, c_p and k are relatively independent of temperature, while μ decreases greatly with temperature, thus the viscosity ratio is commonly selected as a correction factor to take into account the effect of variation of properties with temperature on the heat transfer performance.

$$Nu = Nu_m\left(\frac{\mu_w}{\mu_m}\right)^a,\ f = f_m\left(\frac{\mu_w}{\mu_m}\right)^b \tag{3}$$

where the subscript "m" refers to an appropriate mean temperature, and "w" refers that the properties being calculated at wall temperature.

For gases, c_p varies only slightly with temperature, but μ and k increase to about the 0.8 power of the absolute temperature. Furthermore, ρ varies inversely with the first power of the absolute temperature, thus the temperature ratio can be conveniently used in the correlation,

$$Nu = Nu_m\left(\frac{T_w}{T_m}\right)^a,\ f = f_m\left(\frac{T_w}{T_m}\right)^b \tag{4}$$

By means of this simple scheme, the investigation of temperature-dependent-property fluid flow is to find out a proper expression of mean temperature and proper values of exponents "a" and "b".

At supercritical pressure, since all the thermophysical properties show drastic variations with both the temperature and pressure, this simple correlation scheme is no longer valid. To take into consideration of properties variations, all the properties have to be properly included in the correlation. Furthermore, a universal mean temperature or reference temperature is difficult to be defined.

COOLING HEAT TRANSFER CHARACTERISTICS OF SUPERCRITICAL CARBON DIOXIDE

Experimental Apparatus

To investigate the characteristics of flow and heat transfer of supercritical CO_2 under cooling condition, a series of measurements were carried out by using both a pump cycle to investigate the heat transfer characteristics of pure carbon dioxide, and a heat pump cycle in order to study the effect of lubricating oil on the heat transfer characteristics. Figure 4 shows the schematic drawing of the pump cycle and heat pump cycle.

The test section was a tube in tube counter flow heat exchanger with CO_2 flowed inside the inner tube and coolant (water) flowed along the outside annular passage. Figure 5 shows the schematic diagram of the test section, together with a photograph of the test section with insulation. The inner tube was a horizontally located smooth copper tube and the outer tube was made of acrylic resin. Four different inner tubes were prepared for the experiments, with inner diameters of 1, 2, 4 and 6 mm. The length of the test sections was 0.5 m, and the thickness of the tubes was 1 mm. The distances of annular passages

Figure 4. Schematic diagram of the experimental cycle: (a) experimental cycle for pure carbon dioxide; (b) experimental cycle for carbon dioxide and oil mixture

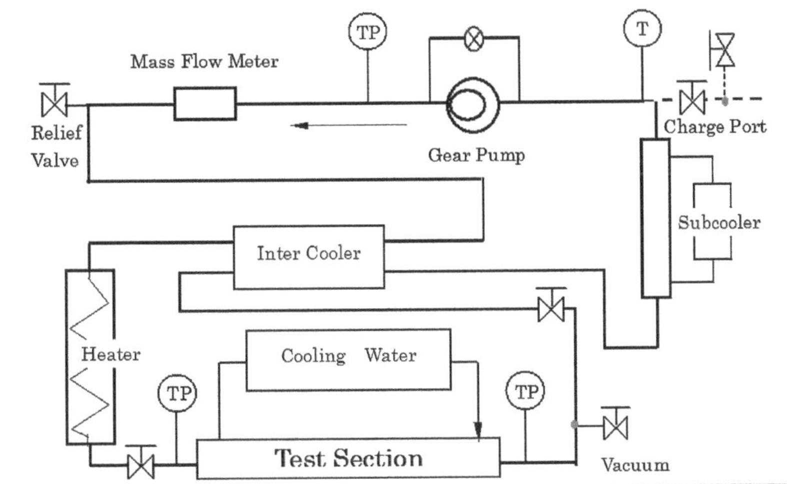

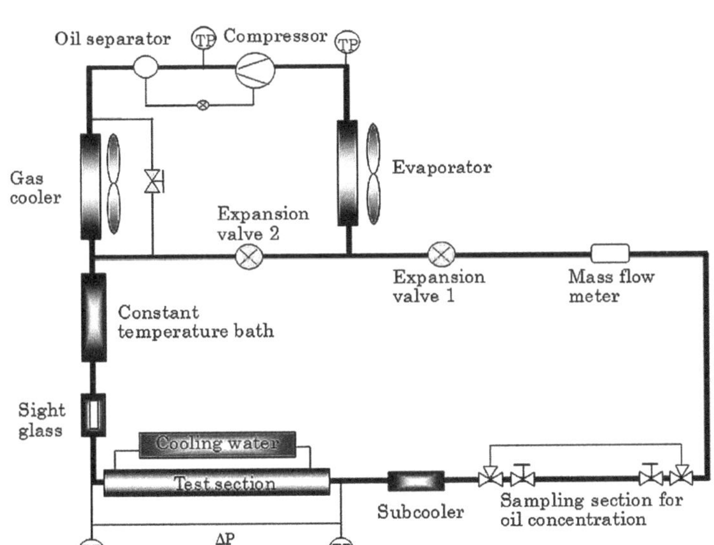

between outer and inner tubes were set to be 1.5, 2, 2 and 3 mm, respectively. The test sections were covered with 25 mm thick polystyrene insulator.

The experiments were conducted at different CO_2 mass flux, heat flux conditions. At each test condition, inlet temperature of the test section was changed from 20ºC to 70ºC to obtain a full understanding over the pseudocritical area.

Inner diameter of the test tubes ranged between 1 and 6 mm. The upper limit of tube diameter was selected to accord with practical application since currently most of heat exchangers are composed with tubes of 4 to 6 mm in diameter. While the lower limit of tube diameter was selected to accord with the tendency of applying small size tube in CO_2 gas cooler. The tube, smaller than 1 mm ID, was not included in this measurement because of the difficulty in measurement.

Upper limit of mass flux was selected according to the ability of the circulation pump. Furthermore, for different tube diameter conditions, the mass flux was selected to keep the product $G \times d$ the range of 1.2 ~2.4 kg/ms. The influence of heat flux on heat transfer coefficient has been proved significant for a middle size tube, but negligible for a small size tube. Four different heat fluxes from 6 to 33 kW/m^2 were scheduled for 6 mm tube in order to investigate the effect of heat flux. While for other three diameter tubes, measurements were conducted at one heat flux Q_w=12 kW/m^2. Three different pressure conditions 8, 9 and 10 MPa were tested to investigate the effect of pressure on heat transfer coefficient and pressure drop.

The heat transfer coefficient was calculated from heat flux and temperature difference, and the pressure drop was calculated from difference of the measured pressures at inlet and outlet of the test section:

$$\alpha = \frac{Q_w}{\Delta T} \quad (5)$$

where the heat flux was calculated from heat exchange rate of waterside.

Both the heat exchange rates calculated from waterside and from CO_2 side were compared, and the deviations were found smaller than 10% under most experimental conditions. Only when the average temperature of the test section is close to the pseudocritical point, a relatively large deviation of about 15% was obtained.

LMTD (Logarithmic Mean Temperature Difference) between the wall and CO_2 was used to calculate the average temperature difference. Detailed discuss of the validity of LMTD method can be found on dang and Hihara (2004a).

Experimental Results and Discussion

Some examples of experimental results are compared to represent the effect of mass flux, pressure, heat flux and tube diameter on heat transfer coefficient and pressure drop. To make the comparison clear, the experimental results are represented together with results of numerical calculation. The numerical calculations are conducted by solving incompressible Navier-Stokes equations. A turbulence model proposed by Jones-Launder (1972) is applied, with turbulent Prandtl number set to be 0.9.

Figure 5. Schematic diagram and photograph of the test section: (a) test section design; (b) picture of the test section

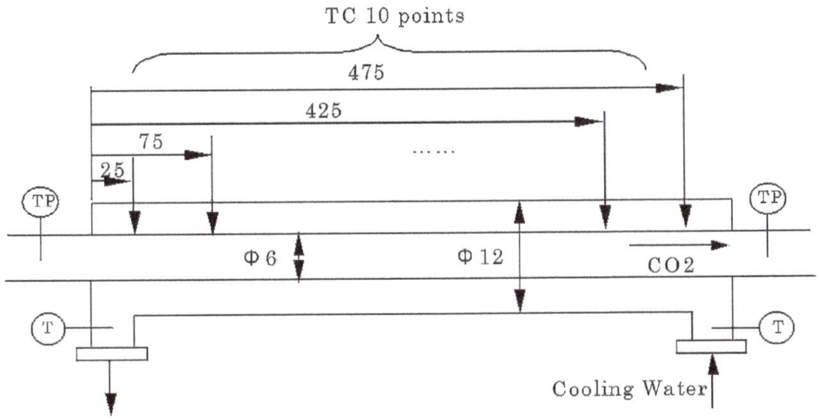

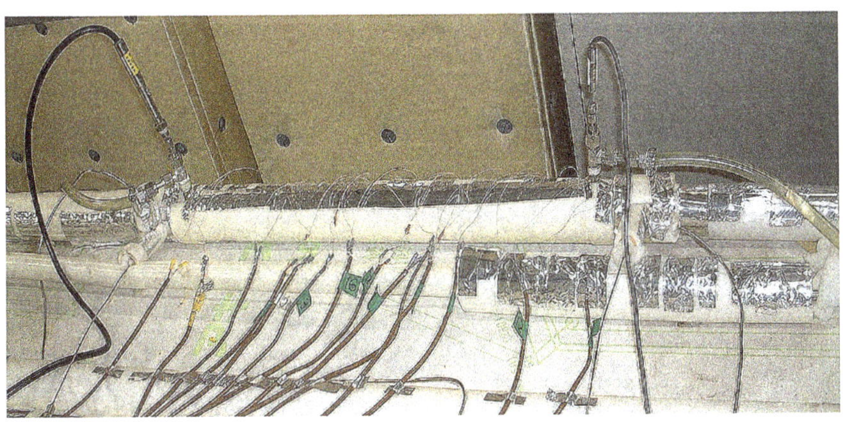

Heat Transfer Coefficient

1. Effect of Mass Flux

In Figure 6, the effect of mass flux on heat transfer coefficient is illustrated. The comparisons are carried out for two mass fluxes (G = 200 kg/m^{2}s and 400 kg/m^{2}s) in a 6 mm ID tube with T_{in} = 75°C, P_{in} = 8 MPa and Q_w = 12 kW/m^2. In Figure 6(a), the comparison of heat transfer coefficient is conducted against the bulk temperature, while in Figure 6(b), it is compared against film temperature. Here the film temperature is defined as an average of the bulk temperature and wall temperature.

Measured data are plotted in Figure 6 with a horizontal error bar representing the temperature change from the inlet to the outlet of the test section, and with a vertical error bar representing the uncertainty of measurement. Results of numerical calculation are illustrated in solid lines. A dashed line with T_m denotes the location of pseudocritical temperature. At pressure of 8 MPa, it is 34.52°C. On the whole, the experimental data agrees well with the numerical calculation results.

Similar to the effect of mass flux on heat transfer coefficient of constant property fluid flow, with increase of mass flux, the heat transfer coefficient increases due to the intensification of the turbulent diffusion. When the mass flux increases twice, the shape of heat transfer coefficient remains unchanged while the heat transfer coefficient value increases about 1.85 times.

For a fixed mass flux, the heat transfer coefficient increases during cooling until a maximum is reached at bulk temperature a little higher than the pseudocritical temperature. After that, the heat transfer coefficient changes to a descent with further decrease of temperature, and approaches to a constant value at the temperature far lower than the pseudocritical temperature. Temperature change from the inlet to the outlet of test section decreases with the temperature approaching the pseudocritical temperature, which is due to the increase in specific heat, since the specific heat reaches a maximum at the pseudocritical temperature.

In Figure 6(b), the heat transfer coefficients are illustrated against film temperature. Locations of the peak for both experimental and numerical calculation results agree well with the location of pseudocritical temperature, which implies that the film temperature can be used as reference temperature conveniently in developing heat transfer model.

Figure 6. Effect of mass flux on the heat transfer coefficient for d=6mm, P=8MPa, Q_w=12kW/m², G=200,400kg/m²s: (a) comparison with bulk temperature; (b) comparison with film temperature

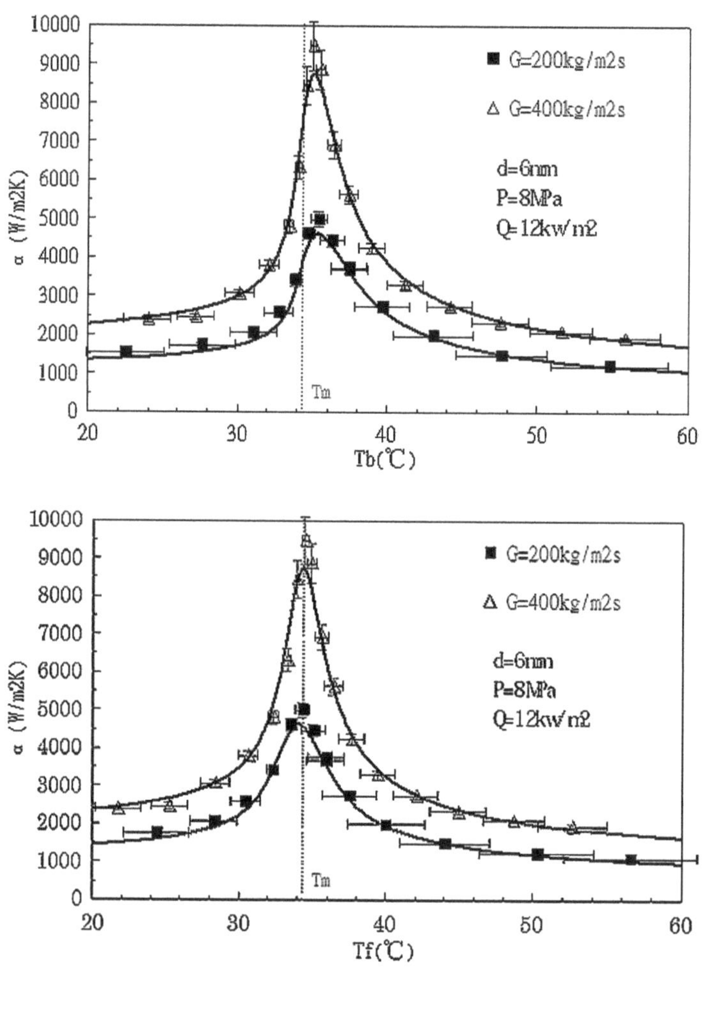

2. Effect of Pressure

Figure 7 shows measured heat transfer coefficients for $d = 4$ mm, $T_{in} = 75°C$, $G = 800$ kg/m²s and $Q_w = 12$ kW/m² at different inlet pressure of $P = 8$, 9 and 10 MPa. With film temperature used as reference temperature, the location of the peak of heat transfer coefficient is coincident well with the position of pseudocritical temperature. In the temperature area far lower or higher than the pseudocritical temperature, the heat transfer coefficients at different pressures take similar values. But the maximum heat transfer coefficient varies with pressure significantly. The maximum value decreases from 17 to 10 kW/m²K when pressure increases from 8 to 9 MPa, while at pressure of 10 MPa, the maximum value of heat transfer coefficient decreases to 7 kW/m²K.

3. Effect of Heat Flux

In Figure 8, the heat transfer coefficients at four different heat fluxes of 6, 12, 24 and 33 kW/m² are compared for $d = 6$ mm, $T_{in} = 75°C$, $G = 200$ kg/m²s and $P_{in} = 8$ MPa. Unlike the heat transfer process of constant property fluid, where heat flux has no or only a weak influence on the heat transfer coefficient, at supercritical pressure, heat flux has large influence on the distribution of properties in the radial direction, thus affects the flow and heat transfer process significantly.

Firstly, the peak values of heat transfer coefficients at different heat fluxes are compared. As describe above, a large heat transfer coefficient is mainly caused by the large specific heat at pseudocritical temperature. With an increase of heat flux, the difference between wall temperature and bulk temperature increases, due to the decrease in average specific heat. Thus, the heat flux seems to have a negative effect on the peak value of the heat transfer coefficient. At the same time, the decrease of wall temperature corresponds to an increase of thermal conductivity and an increase of dynamic viscosity at near wall

Figure 7. Effect of pressure on the heat transfer coefficient, d = 4 mm, G = 800 kg/m²s, Q_w = 12 kW/m², P = 8, 9, 10 MPa

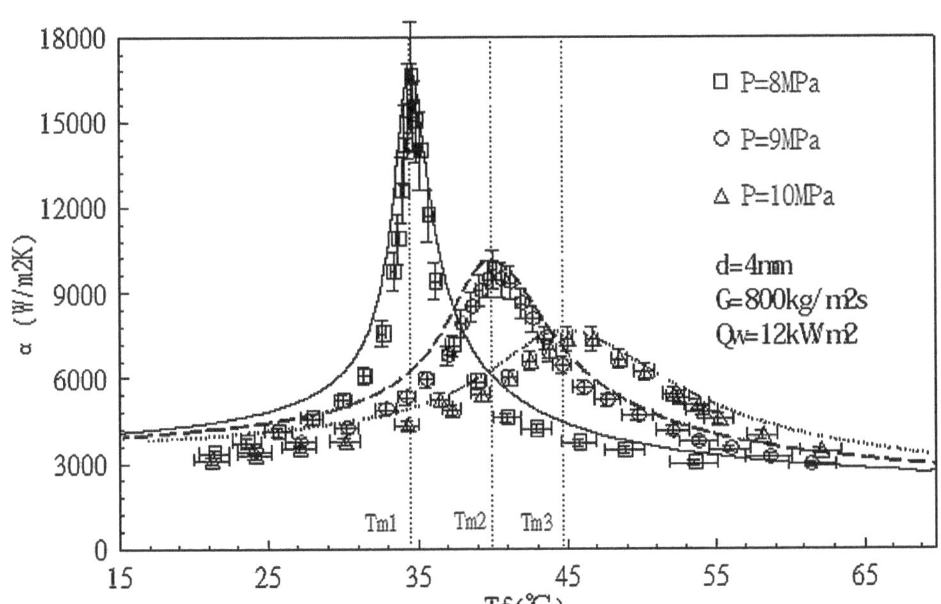

Figure 8. Comparison of heat transfer coefficient at different heat fluxes: d = 6 mm, T_{in} = 75°C, G = 200 kg/m²s, P = 8 MPa, Q_w = 6-33 kW/m²

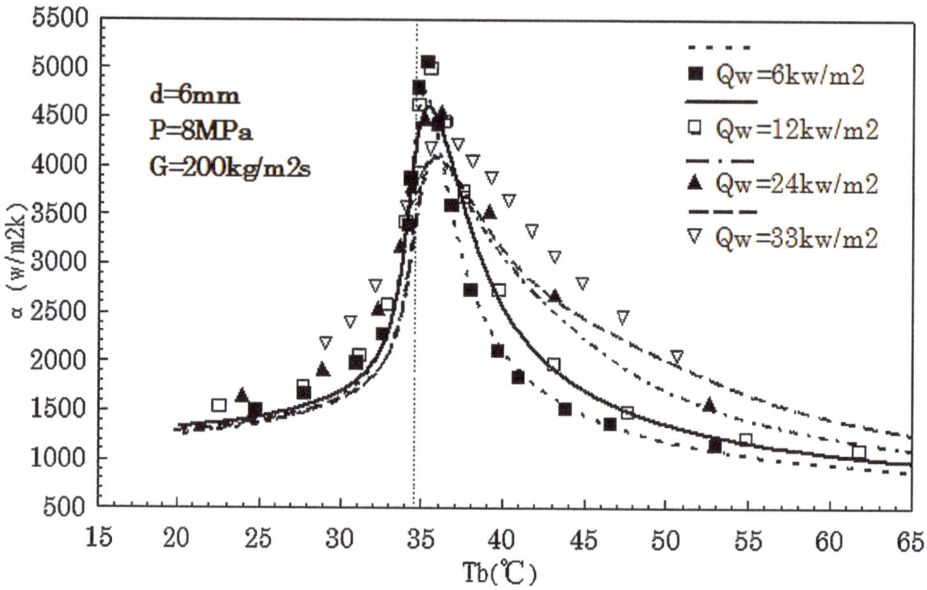

region. The former is a positive factor for the heat transfer coefficient while the latter is a negative factor. From the comparisons in Figure 8, it seems that the effect of specific heat is dominant in pseudocritical area, the maximum value of heat transfer coefficient decreases slightly with the increase of heat flux.

For the temperature region apart from pseudocritical temperature, the effect of heat flux on the heat transfer coefficient is rather complex. As shown in Figure 8, at bulk temperature lower than the pseudocritical temperature, the heat transfer coefficient seems independent of the heat flux, whereas at high temperature side, with the increase of heat flux, the heat transfer coefficient increases significantly.

A simple explanation can be made by applying a two-layer model to investigate the effect of property distribution on heat transfer coefficient. The boundary layer is divided into a viscous sublayer and a turbulent core region. Qualitatively, heat conduction is assumed dominant inside the viscous sublayer, where thermal conductivity and specific heat are two important properties to determine the magnitude of heat conduction; in the turbulent core region, turbulent convection is assumed dominant, and viscosity and specific heat are important properties. Since the heat transfer coefficients at different heat fluxes are compared at the same bulk temperature, attributions of the turbulent convection in turbulent core area to the heat transfer coefficient can be assumed having similar magnitudes, thus the difference is mainly determined by the difference of heat conduction process in the near wall region.

When the heat transfer coefficient is compared at bulk temperature lower than the pseudocritical temperature, a higher heat flux corresponds to a lower wall temperature, thus results in an increase of thermal conductivity and a decrease of specific heat at the near wall region. Since effects of the two properties on heat transfer coefficient are in reverse, they may offset each other and result in a weak dependence on heat transfer coefficient. On the contrary, at bulk temperature higher than the pseudocritical temperature, with the decrease of wall temperature, both the thermal conductivity and the specific heat increase, therefore the heat transfer coefficient is enhanced.

4. Effect of Tube Diameter

In order to investigate the effect of tube diameter on heat transfer, usually the Nusselt number is compared at the same Reynolds numbers. For supercritical fluid flow, as described above, using different reference temperature to calculate the Nusselt number from the measured heat transfer coefficient gives different result. To avoid the complexity of defining reference temperature, an effective heat transfer coefficient, which is defined as multiplying heat transfer coefficient by tube diameter, is used for the comparison at different tube diameters.

Figure 9 shows the effect of tube diameter on effective heat transfer coefficient. Comparison is made for four different diameters of 1 to 6 mm and $G \times d = 2.4\mathrm{kg/ms}$, with bulk temperature as reference temperature. From the results of numerical calculation, at temperature lower than the pseudocritical temperature, calculated effective heat transfer coefficient seems independent of tube diameter, while at temperature higher than the pseudocritical temperature, with increase of tube diameter, the effective heat transfer coefficient increases slightly. But due to the uncertainty of measurement, this tendency cannot be seen clearly from measured data. The distribution of thermodynamic properties in the radial direction again is the key factor to understand the variations of effective heat transfer coefficient with tube diameter. Since the comparison is made at the same bulk temperatures, a larger tube diameter corresponds to a smaller heat transfer coefficient and a relatively lower wall temperature. At temperature lower than the pseudocritical temperature, with the increase of tube diameter, thermal conductivity at near wall region increases, at the same time specific heat at near wall region decreases. The two reverse tendencies may compensate each other. As a result, the effective heat transfer coefficient seems independent of the tube diameter. On the contrary, at temperature higher than the pseudocritical temperature, with increase of tube diameter, both the thermal conductivity and the specific heat at the near wall region increase, and thus those correspond to a higher effective heat transfer coefficient.

Figure 9. Effect of tube diameter on heat transfer coefficient: $G\times d = 2.4$ kg/ms, $P = 9$ MPa, $Q_w = 12$ kW/m^2

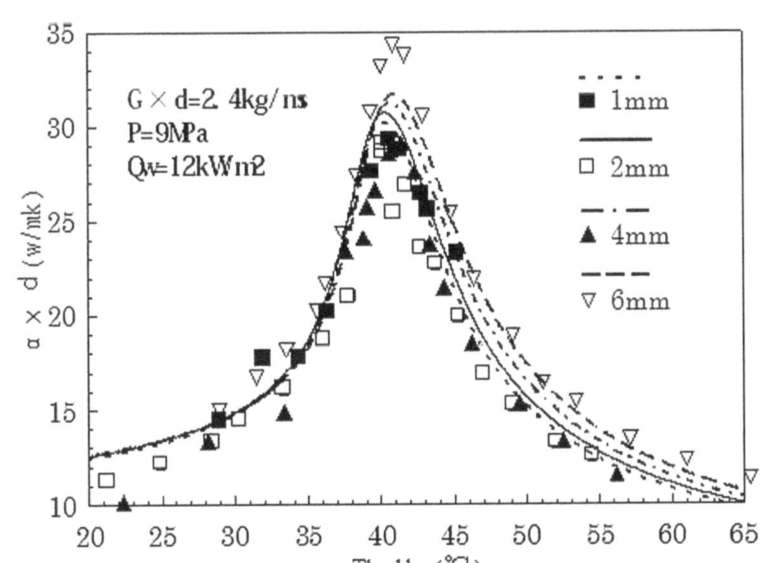

Pressure Drop

Due to the small dynamic viscosity of supercritical CO_2, in present experiment, only the pressure drops in 1 and 2mm tubes were measured, where the pressure drop means frictional pressure drop.

In Figure 10, the measured pressure drops are compared with the results of numerical calculation at different mass fluxes. The horizontal error bar represents the temperature change from the inlet to the outlet of the test section, and the vertical error bar represents the uncertainty of measurement. Although the uncertainty of measurement is a little large compared with the measured pressure drop, with the assistance of numerical calculation, the effect of mass flux and pressure on pressure drop can be seen clearly.

As shown in Figure 10, a larger mass flux corresponds to a larger pressure drop, which is due to the increase of Reynolds number. At a given pressure and mass flux, the pressure drop increases with the increase of CO_2 bulk temperature due to the increase of average velocity and decrease of density. When the bulk temperature is lower than the pseudocritical temperature, only a slight increase of pressure drop is observed with increase of temperature, whereas a sharp increase can be seen when the temperature approaches the pseudocritical temperature. After that, the pressure drop increases continuously with a lower gradient than that at pseudocritical temperature, but larger than that at lower temperature.

Prediction of Heat Transfer Coefficient and Pressure Drop

Main difficulty of predicting heat transfer coefficient of supercritical fluid flow seems to be concentrated on how to properly estimate the effect of thermophysical properties distributions, since the effects of pressure, heat flux, and tube diameter on heat transfer coefficient are all related to the distribution of properties in radial direction. Obviously, it is difficult to include all these factors in a simple correlation,

Figure 10. Effect of pressure on pressure drop: d = 1 mm, L = 0.5 m, G = 1200 kg/m²s, Q_w = 12 kW/m²

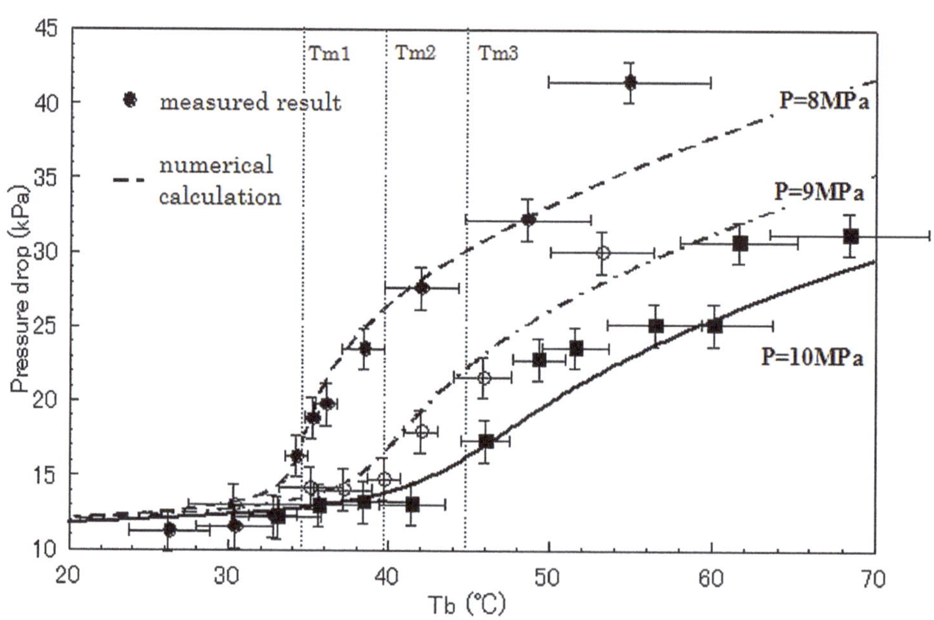

therefore some simplifications based on the understanding of properties distributions in radial direction and their influences on heat transfer coefficient are necessary.

In order to apply the correlation into the variable property fluid flow, usually two methods are employed, i.e. the property ratio method and the reference temperature method. Although the property ratio method is effective in single-phase flow with a slight property variation with temperature, it generally fails to give reasonable prediction for supercritical fluid due to the drastic variation of properties with temperature and pressure. From the comparison of several existing correlations in heat transfer prediction, the only one model that gives a relatively good prediction (Petrov and Popov 1988) is proposed by means of reference temperature method. Here, a new correlation is proposed to use the reference temperature method through the following procedure:

1. Due to the fact that the heat transfer resistance is concentrated mainly on viscous sublayer near the wall, the properties inside viscous sublayer are important in the prediction of the heat conduction process and near wall flow, the thermal conductivity and friction factor in correlation are calculated from the properties at film temperature (T_f).
2. At a temperature lower than the pseudocritical temperature, when the bulk temperature is used as reference temperature, the heat transfer coefficient can be seen independent of the heat flux and tube diameter (Figure 8 and Figure 9). This leads to a simplification that the same correlation form as constant property fluid flow can be employed.
3. At a temperature higher than the pseudocritical temperature, the heat transfer coefficient varies with heat flux and tube diameter, which requires the introduction of average property in the radial direction. Here, an integrated specific heat is introduced.
4. In pseudocritical region, the film temperature was found adequate in the calculation of Prandtl number.

Following the above procedure, the proposed correlation is composed as

$$\alpha = Nu\frac{k_f}{d} \tag{6}$$

$$Nu = \frac{(f_f/8)(\mathrm{Re}_b - 1000)\,\mathrm{Pr}}{1.07 + 12.7\sqrt{f_f/8}(\mathrm{Pr}^{2/3} - 1)} \tag{7}$$

$$\bar{c}_p = \frac{h_b - h_w}{T_b - T_w} \tag{8}$$

$$\mathrm{Pr} = \begin{cases} \mathrm{Pr}_1 = \bar{c}_p \cdot \mu_b / k_b, \text{ for } \mathrm{Pr}_1 \geq \mathrm{Pr}_2 \\ \mathrm{Pr}_2 = \bar{c}_p \cdot \mu_f / k_f, \text{ for } \mathrm{Pr}_2 \geq \mathrm{Pr}_3 \\ \mathrm{Pr}_3 = c_{pb} \cdot \mu_b / k_b, \text{ for } \mathrm{Pr}_3 > \mathrm{Pr}_2 \end{cases} \tag{9}$$

$$\mathrm{Re}_b = Gd / \mu_b \tag{10}$$

$$\mathrm{Re}_f = Gd / \mu_f \tag{11}$$

$$f_f = \left[1.82\log_{10}(\mathrm{Re}_f) - 1.64\right]^{-2} \tag{12}$$

where subscript "*b*" represents bulk temperature, and "*f*" represents film temperature.

Accuracy of the proposed correlation is examined from the comparisons with experimental results, about 93% of 458 valid measurement results are found within the ±20% limits. In Figure 11, detailed comparisons of the proposed correlation with experimental results are represented. The correlation describes the effect of mass flux and effect of pressure successfully.

In this experiment, only the pressure drop in 1 and 2 mm tubes were measured. Under these conditions, the Filonenko's equation was found giving the similar prediction as Petrov and Popov's model. In case of a middle size tube is concerned, due to the fact that the actual pressure drop value is almost negligible, the small error of about 10% to 20% of Filonenko's equation is considered acceptable.

Peak Value of Heat Transfer Coefficient at Pseudocritical Temperature

It has been demonstrated from both the experimental results and numerical simulation results that for the cooling heat transfer of turbulent supercritical fluid flow, the heat transfer coefficient takes maximum value when the film temperature equals to the pseudocritical temperature. In Figure 12, the heat transfer coefficient is illustrated against bulk temperature, together with the radial distance of pseudocritical point from the wall. For each bulk temperature, the dimensionless distance from the wall y^+ where the temperature takes pseudocritical temperature is shown.

If the three layers model is applied, from the wall, the boundary layer can be divided into viscous sublayer, buffer layer and turbulent core region. Because there is no information about the boundaries of the three layers for a variable property fluid, the same boundaries as that for constant property fluid are applied, i.e. the boundary between viscous sublayer and buffer layer is assumed that $y^+=5$, and the boundary between buffer layer and turbulent core region is $y^+=30$.

In cooling process, the pseudocritical point firstly appears at the wall. With the further decrease of bulk temperature, the pseudocritical point moves from viscous sublayer toward buffer layer. The heat transfer coefficient takes a maximum when the pseudocritical point locates at boundary of the viscous sublayer and the buffer layer. After the buffer layer, the pseudocritical point moves towards the turbulent core region and the heat transfer coefficient decreases rapidly.

From the basic equation of turbulent heat transfer:

$$q = -\rho c_p(\alpha + \alpha_t)\frac{dT}{dy} \tag{13}$$

The temperature difference ΔT between wall and centerline can be calculated as

Figure 11. Comparison of proposed correlation with experimental data: (a) for 4mm tube; (b) for 2mm tube

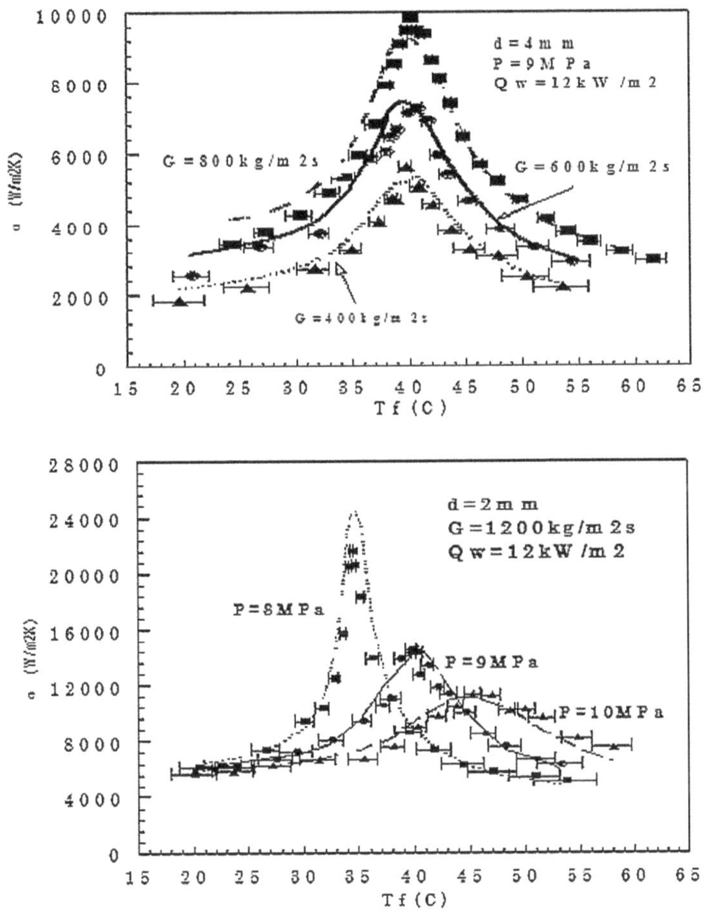

$$\Delta T = (T_c - T_w) = \int_0^R \frac{-q}{k + \rho C_p a_t} dy \tag{14}$$

where α_t is the eddy diffusivity for heat.

The temperature difference between the wall and centerline can be calculated from the sum of the temperature changes in the three layers.

$$\begin{aligned} \Delta T &= \Delta T_v + \Delta T_b + \Delta T_t \\ &= \int_0^5 \frac{-q}{k + \rho c_p \alpha_t} dy^+ + \int_5^{30} \frac{-q}{k + \rho c_p \alpha_t} dy^+ + \int_{30}^{r_w^+} \frac{-q}{k + \rho c_p \alpha_t} dy^+ \\ &\approx \int_0^5 \frac{-q}{k} dy^+ + \int_5^{30} \frac{-q}{k + \rho c_p \alpha_t} dy^+ + \int_{30}^{r_w^+} \frac{-q}{\rho c_p \alpha_t} dy^+ \end{aligned} \tag{15}$$

where ΔT_v, ΔT_b and ΔT_t represent the temperature change in viscous sublayer, buffer layer and turbulent core region, respectively. In viscous sublayer, the term of turbulent eddy diffusion is treated negligible compared to molecular diffusion. In the turbulent core region, the molecular diffusion is neglected because of a high eddy diffusion effect. But in buffer layer, both the turbulent diffusion and the molecular diffusion must be considered.

At the constant heat flux condition, a small ΔT corresponds to a large heat transfer coefficient. Influence of the location of pseudocritical point in the radial direction can described as follows:

1. **Pseudocritical Temperature Locates Inside the Viscous Sublayer:** The molecular diffusion is dominant in this area; thus, the large specific heat does not decrease the temperature change greatly. Consequently, the appearance of pseudocritical point in viscous sublayer does not affect the heat transfer coefficient significantly.
2. **Pseudocritical Temperature Locates Inside the Buffer Layer:** If the specific heat takes large value inside the buffer layer, the temperature difference of buffer layer decreases greatly. That is to say, the appearance of pseudocritical point in buffer layer will correspond to a high heat transfer coefficient.
3. **Pseudocritical Temperature Locates Inside the Turbulent Core Region:** The increase of specific heat will lead to a decrease of temperature difference in turbulent core region, thus corresponds to a high value of heat transfer coefficient. But due to the fact that the temperature difference at the turbulent core region is small compared to other two layers, the contribution of the decreased ΔT_t to the increase of heat transfer coefficient is limited.

On the whole, it seems that the heat transfer coefficient is apt to take a maximum value when the pseudocritical temperature point is located in the buffer layer. As shown in Figure 12, at the cross section of highest heat transfer coefficient, the location of pseudocritical temperature point is at y^+=5.5.

Figure 12. Relationship of heat transfer coefficient with location of pseudocritical point in radial direction; d = 6 mm, P = 8 MPa, G = 200 kg/m²s, Q_w = 12 kW/m²

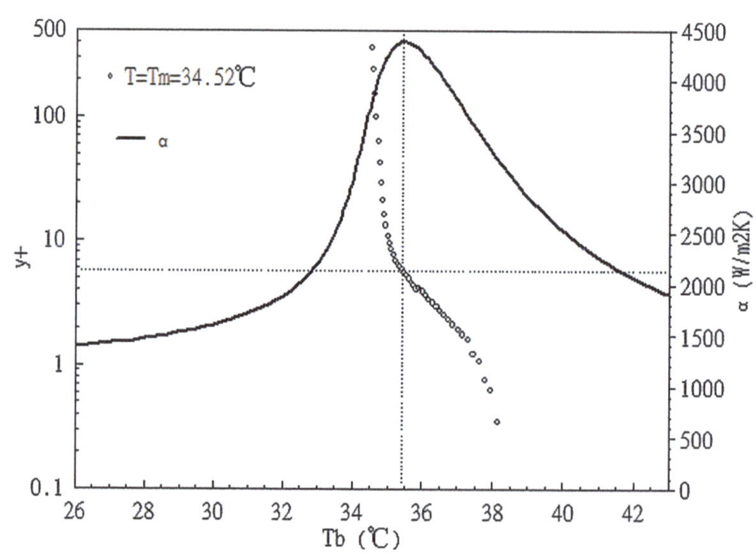

For the purpose of gas cooler application where the heat flux is not high, approximately the temperature at the buffer layer can be represented by film temperature. Thus, it is possible to predict if the heat transfer coefficient will reach a maximum approximately by using the film temperature as reference temperature.

Numerical Simulation of the Cooling Heat Transfer of Supercritical Fluids

Experimental measurements of in-tube cooling heat transfer of supercritical carbon dioxide show that the radial distribution of temperature and the corresponding thermophysical properties in the transverse cross-section contribute significantly to the variation in heat transfer coefficient. Due to a high operating pressure for supercritical fluids and a relatively small tube-diameter used in practical applications, direct measurements of velocity and temperature distributions are difficult. Numerical calculation seems to be the only practical way to provide such information. If the supercritical fluid flow is assumed in steady state and in local thermodynamic equilibrium, flow and heat transfer can be simulated by solving governing equations numerically with local properties (i.e., specific heat, density, thermal conductivity and viscosity) being considered. Major difficulty of the numerical calculation is proper selection of the turbulence model. Many numerical calculations for supercritical water or carbon dioxide have been reported adopting mixing length models or two-equation models. However, no direct comparison of different turbulence models being applied to supercritical fluids has been reported. Due to the drastic variation in thermophysical properties in both flow direction and radial direction, the turbulence structure might change greatly. Such a comparison is therefore necessary to obtain better knowledge of heat transfer characteristics of supercritical fluids.

In this study, four turbulence models were applied to the numerical calculation of heat transfer of supercritical fluid, and simulation results of heat transfer coefficient were compared with experimental results. The turbulence models being compared were a mixing length model by Bellmore and Reid (1983), called the BR model, and three low Reynolds number k-ε models (JL model by Jones and Launder (1972), LS model by Launder and Sharma (1973, and MK model by Myong and Kasagi (1990). Furthermore, the effect of turbulent Prandtl number Pr_t on heat transfer simulation was investigated by setting Pr_t to be constant and by calculating Pr_t using the model proposed by Myong et al. (1989).

Calculation Models

Governing Equations and Turbulence Models

Governing equations used to describe the forced convection heat transfer in a horizontal tube include time-averaged continuity, momentum and energy equations. The flow is assumed to be axisymmetric and steady state without considering the effect of buoyant force, which is not significant under present experimental conditions.

Continuity Conservation:

$$\frac{\partial}{\partial z}(\rho u)+\frac{1}{r}\frac{\partial}{\partial r}(r\rho v)=0 \tag{16}$$

Momentum Conservation:

$$\frac{\partial}{\partial z}(\rho uu)+\frac{1}{r}\frac{\partial}{\partial r}(r\rho vu)=-\frac{\partial P}{\partial z}+\frac{\partial}{\partial z}\left[(\mu+\mu_t)\frac{\partial u}{\partial z}\right]+\frac{1}{r}\frac{\partial}{\partial r}\left[(\mu+\mu_t)r\frac{\partial u}{\partial r}\right] \tag{17}$$

$$\frac{\partial}{\partial z}(\rho uv)+\frac{1}{r}\frac{\partial}{\partial r}(r\rho vv)=-\frac{\partial P}{\partial r}+\frac{\partial}{\partial z}\left[(\mu+\mu_t)\frac{\partial v}{\partial z}\right]+\frac{1}{r}\frac{\partial}{\partial r}\left[(\mu+\mu_t)r\frac{\partial v}{\partial r}\right]-\frac{2(\mu+\mu_t)v}{r^2} \tag{18}$$

Energy Conservation:

$$\frac{\partial}{\partial z}(\rho uh)+\frac{1}{r}\frac{\partial}{\partial r}(r\rho vh)=\frac{\partial}{\partial z}\left[c_p(\frac{\mu}{\Pr}+\frac{\mu_t}{\Pr_t})\frac{\partial T}{\partial z}\right]+\frac{1}{r}\frac{\partial}{\partial r}\left[c_p(\frac{\mu}{\Pr}+\frac{\mu_t}{\Pr_t})r\frac{\partial T}{\partial r}\right] \tag{19}$$

where turbulent viscosity μ_t is determined according to the turbulence model.

In this research, four turbulence models, namely, the BR model, JL model, LS model and MK model, were applied to supercritical fluid and the results of heat transfer coefficient were compared with experimental results of this research and those by Tanaka et al. (1967).

The BR model is a modified mixing length model proposed by Bellmore and Reid and applied by Lee and Howell to their simulation of supercritical water under heating conditions. In this model, density fluctuation with variation of enthalpy is considered.

$$\mu_t=\rho l_m^2\left|\frac{\partial u}{\partial y}\right|\left[1-\left(\frac{\beta}{c_p}\frac{l_m}{\Pr_t}\frac{\partial h}{\partial y}\right)-\left(\frac{\beta}{c_p}\frac{l_m}{\Pr_t}\frac{\partial h}{\partial y}\right)^2\right] \tag{20}$$

To apply this model in the near-wall region, damping function of Van Driest is applied.

$$l_m=l'\left[1-\exp\left(-y^+/26\right)\right] \tag{21}$$

where l' is the Prandtl mixing length, defined as:

$$l'=\begin{cases}0.44(R-r) & for\ 0.44(R-r)<0.09R\\ 0.09R & for\ 0.44(R-r)>0.09R\end{cases} \tag{22}$$

For the three two-equation low Reynolds number *k*-*ε* models, the turbulent viscosity is calculated from the turbulent energy *k* and the turbulent energy dissipation rate *ε*,

$$\mu_t=\rho c_\mu f_\mu k^2/\varepsilon \tag{23}$$

Transport equations for k and ε are:

$$\frac{\partial}{\partial z}(\rho u k)+\frac{1}{r}\frac{\partial}{\partial r}(r\rho v k)=\frac{\partial}{\partial z}\left[\left(\mu+\frac{\mu_t}{\sigma_k}\right)\frac{\partial k}{\partial z}\right]+\frac{1}{r}\frac{\partial}{\partial r}\left[\left(\mu+\frac{\mu_t}{\sigma_k}\right)r\frac{\partial k}{\partial r}\right]+\rho P_k-\rho\varepsilon-D \tag{24}$$

$$\frac{\partial}{\partial z}(\rho u \varepsilon)+\frac{1}{r}\frac{\partial}{\partial r}(r\rho v \varepsilon)=\frac{\partial}{\partial z}\left[\left(\mu+\frac{\mu_t}{\sigma_\varepsilon}\right)\frac{\partial \varepsilon}{\partial z}\right]+\frac{1}{r}\frac{\partial}{\partial r}\left[\left(\mu+\frac{\mu_t}{\sigma_\varepsilon}\right)r\frac{\partial \varepsilon}{\partial r}\right]+\frac{\rho\varepsilon}{k}\left(c_1 f_1 P_k-c_2 f_2\varepsilon\right)+E \tag{25}$$

The JL model is the original and simplest one among the low Reynolds number k-ε models. It was first applied to the prediction of wall boundary-layer flows in which streamwise accelerations were so severe that the boundary layer reverted partially towards laminar. Koshizuka et al. (1995) applied this model to the simulation of supercritical water under heating condition. The JL model reportedly describes well the behavior of supercritical water under both low and high heat flux conditions.

The LS model was derived from the JL model with a shift in the coefficients to be consistent with the optimum forms for free shear flows, and a change in damping function. This model has been widely used for both constant-property fluid and temperature-dependent property fluid, and found to predict well the heat transfer coefficient when the Prandtl number (Pr) is near unity (Patel et al. (1984), Mikielewicz et al. (2002)). In addition, both the JL model and the LS model have the principal advantage that the damping function is determined only by turbulent Reynolds number (Re_t), and thus only local values of properties are used.

The MK model was developed by Myong and Kasagi based on two characteristics length scales for the dissipation rate. As a result, this model correctly predicts the major turbulence quantities at the near-wall region and in the region far from the wall. Accompanied by a turbulent Prandtl number (Pr_t) model proposed by the same research group, this model is effective in a wide range of Pr and Re, $10^{-2} < Pr < 5 \times 10^4$ and $10^4 < Re < 5 \times 10^5$. The MK Pr_t model is:

$$Pr_t = 0.75+\frac{1.63}{\ln(1+Pr/\,0.0015)} \tag{26}$$

Table 1 lists the damping function f_μ and function f_2 of the three low Reynolds number k-ε models, and Table 2 lists the model constants and extra terms.

For the BR model, JL model, and LS model, Pr_t was set to be constant at 0.9, whereas for the MK model, the MK Pr_t model was used.

Initial and Boundary Conditions

In the comparison here, initial conditions were obtained from a preliminary calculation of fully developed flow of constant-property fluid, with thermophysical properties calculated at inlet temperature and pressure.

Boundary conditions were given as:

Table 1. Damping functions

Model	f_μ	f_2
JL	$\exp\left[-2.5/\left(1+R_t/50\right)\right]$	$1-0.3\exp\left(-R_t^2\right)$
LS	$\exp\left[-3.4/\left(1+R_t/50\right)^2\right]$	$1-0.3\exp\left(-R_t^2\right)$
MK	$\left[1-\exp\left(-y^+/70\right)\right]\left(1+3.45/R_t^{0.5}\right)$	$\left\{1-(2/9)\exp\left[-\left(R_t/6\right)^2\right]\right\}\left[1-\exp\left(-y^+/5\right)\right]^2$

Table 2. Model constants and extra terms

Model	c_1	c_2	σ_k	σ_ε	D	E
JL	1.55	2.0	1.0	1.3	$2\mu\left(\partial\sqrt{k}/\partial y\right)^2$	$\frac{2\mu\mu_t}{\rho}\left(\partial^2 u/\partial y^2\right)^2$
LS	1.44	1.92	1.0	1.3	$2\mu\left(\partial\sqrt{k}/\partial y\right)^2$	$\frac{2\mu\mu_t}{\rho}\left(\partial^2 u/\partial y^2\right)^2$
MK	1.4	1.8	1.4	1.3	0	0

$$u = v = k = \varepsilon = 0;\ T_w = T_1 + q_w y_1/k;\ \partial P/\partial r = 0 \text{ at } z > 0 \text{ and } r = R \quad (27)$$

$$v = 0;\ \partial\varphi/\partial r = 0,\ \varphi = u, T, h, P, k, \varepsilon \text{ at } z > 0 \text{ and } r = 0 \quad (28)$$

$$\partial\varphi/\partial z = 0,\ \varphi = u, v, T, h, P, k, \varepsilon \text{ at } z = L \quad (29)$$

Solution Procedure

Thermodynamic properties were calculated using the program of Refprop 7.0 (1997). Because the data exchange with this property program is very slow, first the properties were tabulated at the inlet pressure at temperature intervals of 0.0005°C. Then, during the iteration process, the fluid properties were updated by interpolating from the tables with temperature. Because a short tube (0.5 m in length) was used in the experiment, generally the pressure drop in the test section was negligible. During the numerical calculation, the properties were updated only as functions of local temperature from the table.

The governing general finite difference equations for continuity, momentum, energy and turbulence variables were solved iteratively with the updating of fluid properties and temperature. The iteration procedure was conducted using the SIMPLE algorithm (Patankar, 1980) until convergence was achieved. The convergence criteria were given for velocities and enthalpy as:

$$\left|\frac{\varphi^{i+1}-\varphi^{i}}{\varphi^{i}}\right|<10^{-3},\ \varphi=u,v,h \tag{30}$$

For stability of convergence, relaxation factors from 0.5 ~ 0.7 were introduced for velocities u, v, pressure P and turbulence variables k, ε. For enthalpy h, a large relaxation factor of 0.99 was found adequate to balance the stability and convergence speed. thermophysical properties were calculated directly from the updated temperature, without relaxation.

A staggered grid scheme was used in the computation to avoid producing a checkerboard pressure field. The grid number in the radial and axial directions was set to be 150 × 360, with the grid compressed near the wall to ensure that the dimensionless distance y^+ of the first node from the wall at the inlet was smaller than 0.1. In the axial direction, a uniform grid was adopted. Grid-independence of the calculation was examined by increasing grid number in both the radial and axial directions.

Comparison of Turbulence Models for Supercritical Fluid Flow

The difference in heat transfer characteristics between the cooling conditions and heating conditions is due to the reverse distribution of thermophysical properties in the radial direction. For example, when α at different heat fluxes is compared, α under heating conditions is generally much lower than that under cooling conditions. In addition, α decreases with increasing heat flux under heating conditions, whereas under cooling conditions, generally, α increases with increasing heat flux except at pseudocritical temperature T_m, where a slight decrease in heat transfer occurs. These results can be explained by considering the distribution of thermophysical properties in the radial direction, especially the effect of specific heat and thermal conductivity. Although the purpose of our study here was to evaluate turbulence models applied to the cooling process of supercritical carbon dioxide, it is preferable that the same turbulence model can be applied to both conditions. The comparison between the results of numerical calculation and experimental data under both heating and cooling conditions were therefore considered here.

Under Heating Conditions

The four turbulence models were applied to supercritical fluid flow, and were then compared with measured data for supercritical carbon dioxide under heating conditions by Tanaka et al. (1967). Figure 13 shows the comparison results for α at two different heat fluxes, q_w = 33.0 and 51.7 kW/m², for a heating tube with inner diameter d = 10 mm, pressure P = 7.848 MPa and mass flux G = 990 kg/m²s.

The α was calculated from q_w and temperature difference between wall temperature T_w and bulk temperature T_b as

$$\alpha = q_w/\left(T_b - T_w\right) \tag{31}$$

The T_b was calculated from the averaged enthalpy (h_b) of transverse cross-section, which was defined as

$$h_b = \frac{\int_0^R \rho u h \pi r dr}{\int_0^R \rho u \pi r dr} \tag{32}$$

Figure 13 shows that qualitatively all four turbulence models yielded similar tendencies in α with respect to T_b. At $T_b < T_m$, the BR model and JL model simulated well the measured α, whereas the MK model and LS model slightly underestimated α. At $T_b > T_m$, the BR model slightly overestimated the measured α, whereas the three other models simulated well the measured α. Near T_m, no clear trend was evident in the difference between the calculated α from the BR, JL, and MK models and the measured α, whereas the calculated α by the LS model was slightly lower, about 20%, than the measured α at q_w = 33 kW/m^2, increasing to 33% lower at a higher q_w of 51.7 kW/m^2.

Under Cooling Conditions

Figure 14 shows a comparison of calculated α with the measured α under cooling conditions for a middle-sized tube of d = 6 mm, P = 8 MPa, G = 200 kg/m^{2}s, and q_w = 6 kW/m^2. The BR model, JL model, and MK model simulated well the measured α, whereas the LS model underestimated the measured α. Furthermore, the LS model yielded α values lower than all values calculated using the three other models; for example, the maximum α predicted by the LS model was about 40% lower than that calculated using the JL model.

To investigate how the distribution of physical properties affects the simulation results for α by different turbulence models, f_μ and effective viscosity μ_{eff} in the near-wall region were calculated and then compared. Figure 15(a) shows that the asymptotic feature was $f_\mu = 0(y^{-1})$ for the MK model and $f_\mu = 0(y^0)$ for the JL and LS models. Figure 15(a) also shows that f_μ for the LS model was much smaller than that for JL model, which leading to a smaller μ_{eff} in the near-wall region (Figure 15b). When a fluid is cooled at supercritical pressure, the slow generation of turbulent energy due to the small μ_{eff} near the wall corresponds to a small α, resulting in a low temperature at the wall. The low temperature will increase the dynamic viscosity near the wall, thus further slowing the generation of turbulent energy. Due to such a positive feedback, the LS model significantly underestimates α of supercritical carbon dioxide under cooling conditions.

Figure 16 shows the comparison of measure α with those calculated using the JL model, MK model, and BR model at a relatively high q_w of 33 kW/m^2. The LS model is not included in the comparison because a divergence in the temperature fluctuation occurred under such a large q_w condition. The figure shows that none of the three turbulence models simulated well the measured α over the entire T_b range. The BR model and MK model, however, simulated well the measured α over a wide T_b range except at T_m. The JL model simulated well the maximum measured α, but slightly underestimated the measured α at $T_b > T_m$.

Figure 13. Comparison of predicted heat transfer coefficient α predicted by different turbulence models and measured α for supercritical carbon dioxide under heating conditions. Dashed line denotes the pseudocritical temperature $T_{m.}$ (a) q_w=33.0kW/m²; (b) q_w=51.7kW/m².

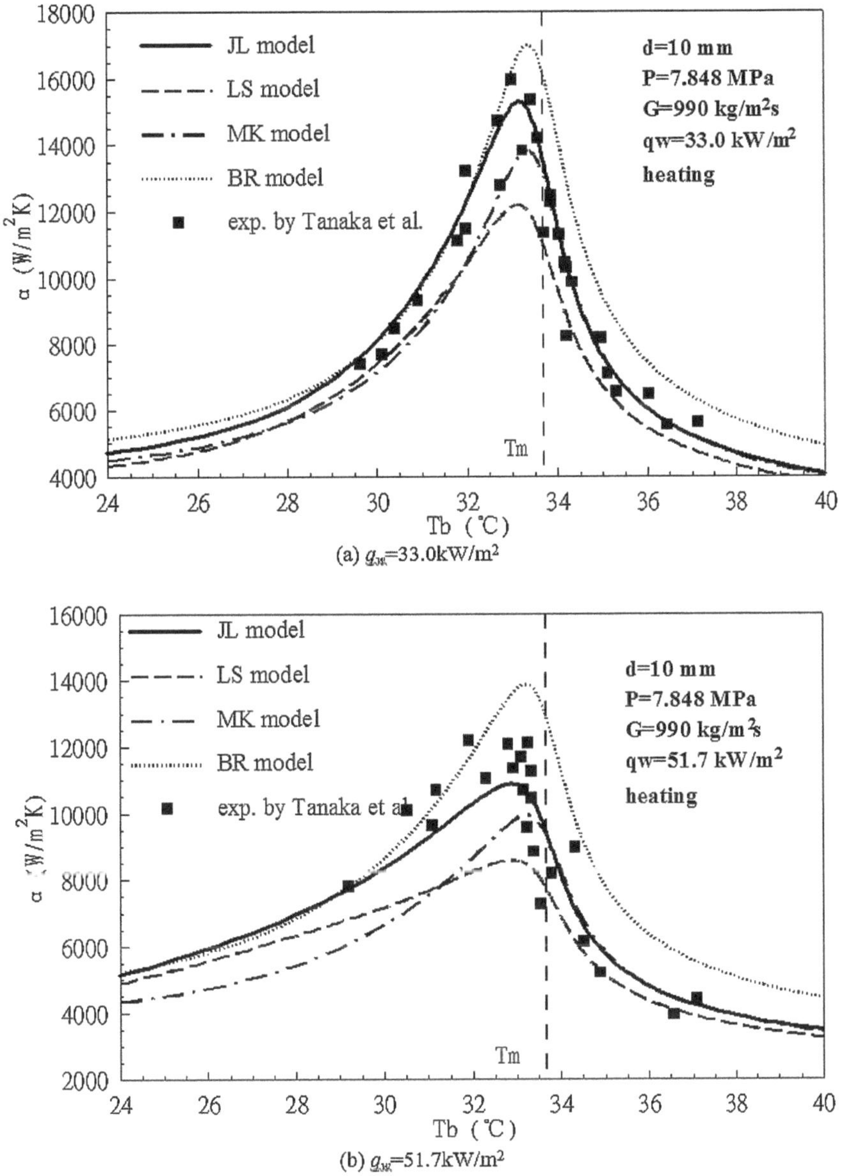

The effect of q_w on α was previously reported. Those experimental results show a decrease in maximum α with increasing q_w. Figure 17 shows the α calculated using the JL model and MK model at different q_w. Both turbulence models simulated well the measured α at $T_b > T_m$ and $T_b < T_m$. However, the JL model also simulated the tendency previously reported that the maximum α decreases with increasing q_w, whereas the MK model simulated the reverse tendency of this effect of q_w on α at T_b near T_m. The maximum α calculated using the BR model was apparently independent of q_w. These results clearly indicate that the JL model gives better estimation of α compared with the two other models.

Figure 14. Comparison of heat transfer coefficient α predicted by different turbulence models and measured α for supercritical carbon dioxide under cooling conditions at a relatively low heat flux q_w. Horizontal error bars represent temperature change from the inlet to the outlet of the test section, and vertical error bars represents the uncertainty in the measurement.

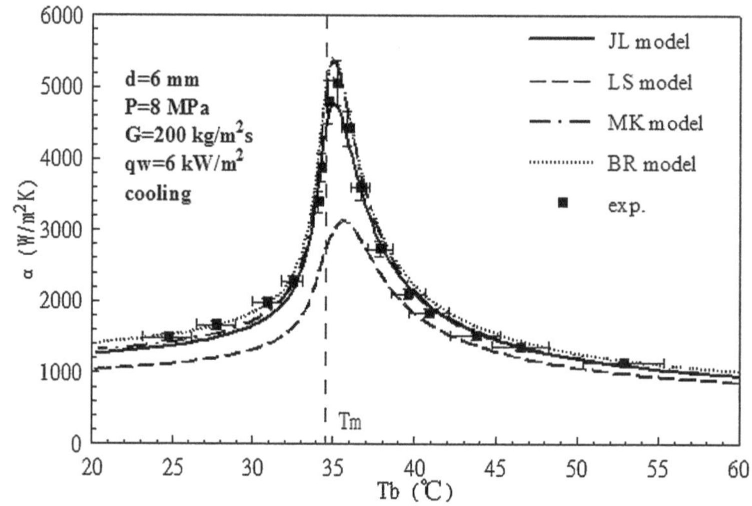

Non-Dimensional Distance y⁺ and Turbulent Prandtl Number Pr_t

The JL model and LS model have the principal advantage that f_μ is determined only by Re_t, no non-dimensional distance y^+ is used in the two models. For the BR model and MK model, y^+ is introduced in f_μ. In the above calculations, local properties were used to define y^+, as used by Satry and Schnurr (1975) in their simulation of supercritical water under heating condition:

$$y_1^+ = \frac{y\sqrt{\tau_w/\rho}}{\upsilon} \tag{33}$$

where y is distance for the wall, τ_w is wall shear stress, ρ is density and υ *is* kinematic viscosity.

Goldmann (1954) suggested that the following definition should be applied to consider the effect of variation in properties in radial direction:

$$y_2^+ = \int_0^y \frac{\sqrt{\tau_w/\rho}}{\upsilon} dy \tag{34}$$

This definition was used by Tanaka (1967) in the prediction of heat transfer of supercritical carbon dioxide and by Yoshida (2000) of supercritical water under heating conditions.

Equation 33 and 34, together with another y^+ model as follows,

Figure 15. Distribution of (a) damping functions f_μ and (b) effective viscosities μ_{eff} near the wall calculated by different turbulence models.

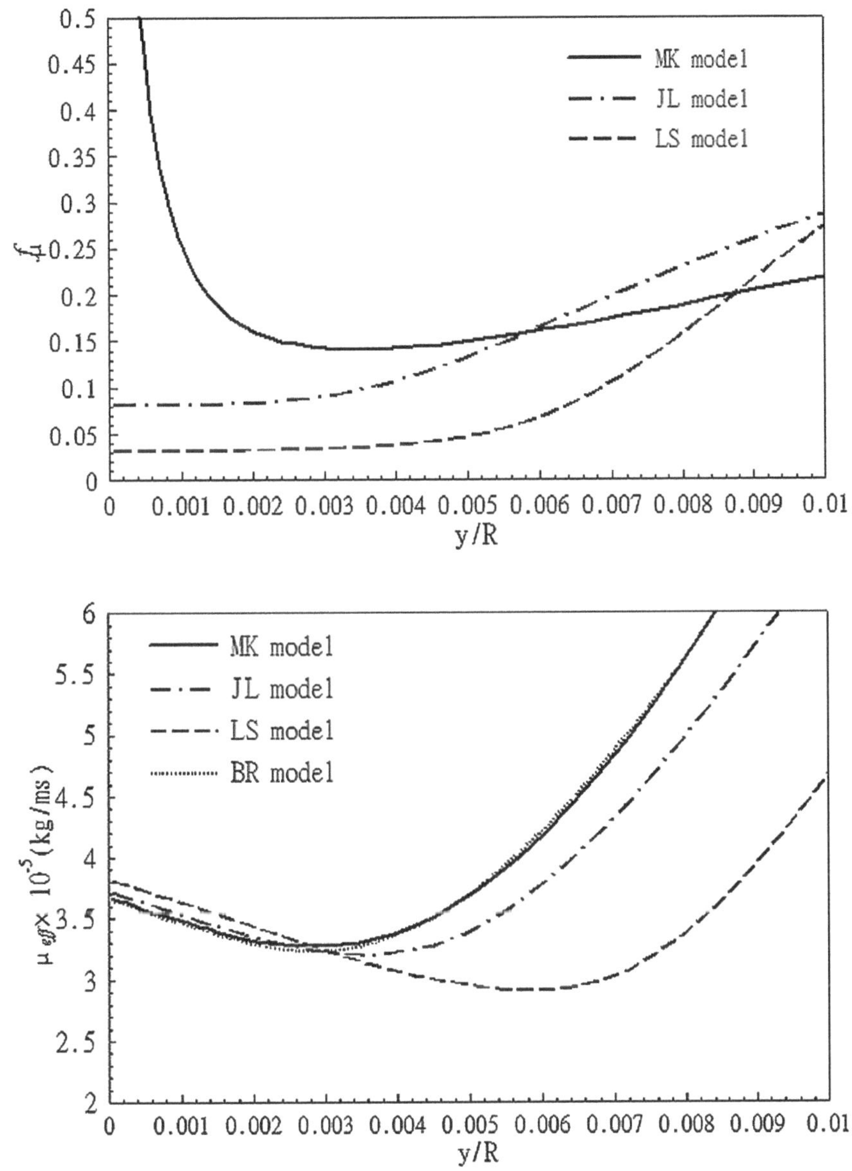

$$y_3^+ = \frac{y\sqrt{\tau_w/\rho_w}}{\upsilon_w} \tag{35}$$

were compared by Lee and Howell (1998) using the BR model for supercritical water under heating conditions. Differences in the predicted α by these three definitions of y^+ were found to be less than 3%.

Under cooling conditions, however, a different definition of y^+ will yield significantly different results. Figure 18 shows a comparison of α calculated by the MK model using the three definitions of y^+ for d

Figure 16. Comparison of predicted heat transfer coefficient α predicted by different turbulence models and measured α for supercritical carbon dioxide under cooling conditions at a relatively high heat flux q_w.

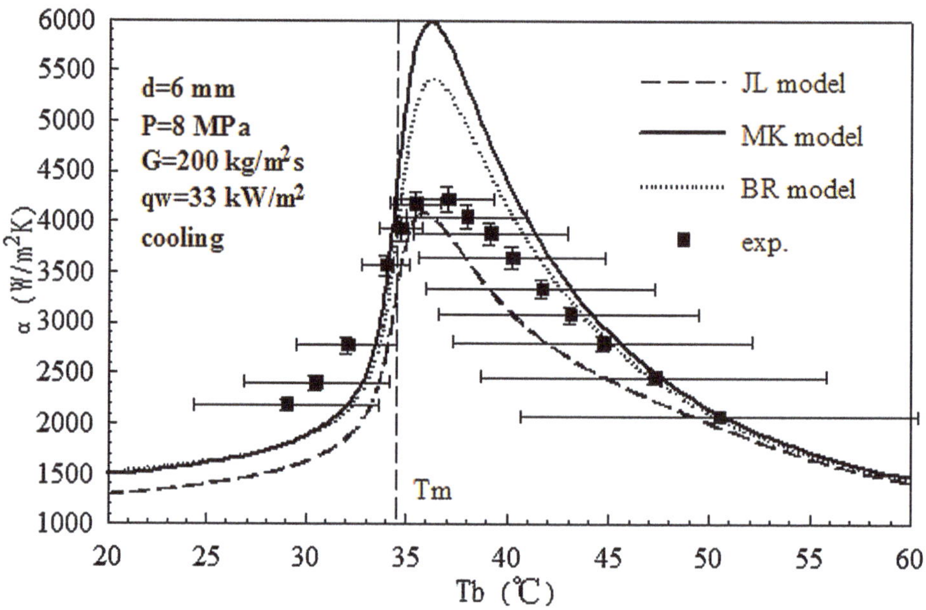

$= 6$ mm, $P = 8$ MPa, $G = 200$ kg/m²s and $q_w = 33$ kW/m². Calculated maximum α corresponding to y_2^+ was about 10% lower than that corresponding to y_1^+, but about 33% higher than that corresponding to y_3^+. This comparison again shows the effect of property distribution in near wall region.

Figure 19 shows the distributions of ρ, υ and their product $1/\sqrt{\rho\upsilon}$ in the near-wall region. Under cooling conditions, with decreasing y towards the wall, both ρ and υ increase. Form the definitions of y^+ it is clear that $y_1^+ > y_2^+ > y_3^+$ at the same y. This explains the comparison results shown in Figure 18 that $\alpha(y_1^+) > \alpha(y_2^+) > \alpha(y_3^+)$. Under heating conditions, however, ρ decreases, while υ increases with decreasing y. The product $1/\sqrt{\rho\upsilon}$ did not significantly change with increasing y. As a result, the calculated α under heating conditions was less sensitive to the definition of y^+ than that under cooling conditions.

Figure 20 shows the calculated α by the MK model and BR model using Equation 35 at different q_w. Compared with Figure 17(b) where the prediction result of the MK model shows a reverse effect of q_w on α, by considering the integrated effect of properties, both the MK model and the BR model show approximately the same tendencies with experimental results. However, the two models still slightly overestimated the maximum α.

To investigate the effect of turbulent Prandtl number Pr_t on α, we compared three Pr_t models, where Pr_t was set at a constant 0.9 and 1.0, and at Pr_t calculated by the MK Pr_t model. Figure 21 shows the α calculated using the three Pr_t models with the JL model. The range for Pr considered here was 1.5 to 12, and thus the Pr_t by the MK model ranged from 0.93 to 0.99. The figure shows that no significant difference in α calculated by the different Pr_t models. The maximum difference was only about 3% between α calculated using the MK Pr_t model and that using the constant-Pr_t model where $Pr_t = 0.9$, and that between the two constant Pr_t models was less than 4%. Compared to the experimental uncertainty of about 10%, the difference caused by the Pr_t model is negligible under the experimental conditions considered here.

Figure 17. Comparison of predicted heat transfer coefficient α at different heat flux q_w and measured α for supercritical carbon dioxide under cooling conditions. (a) JL model; (b) MK model.

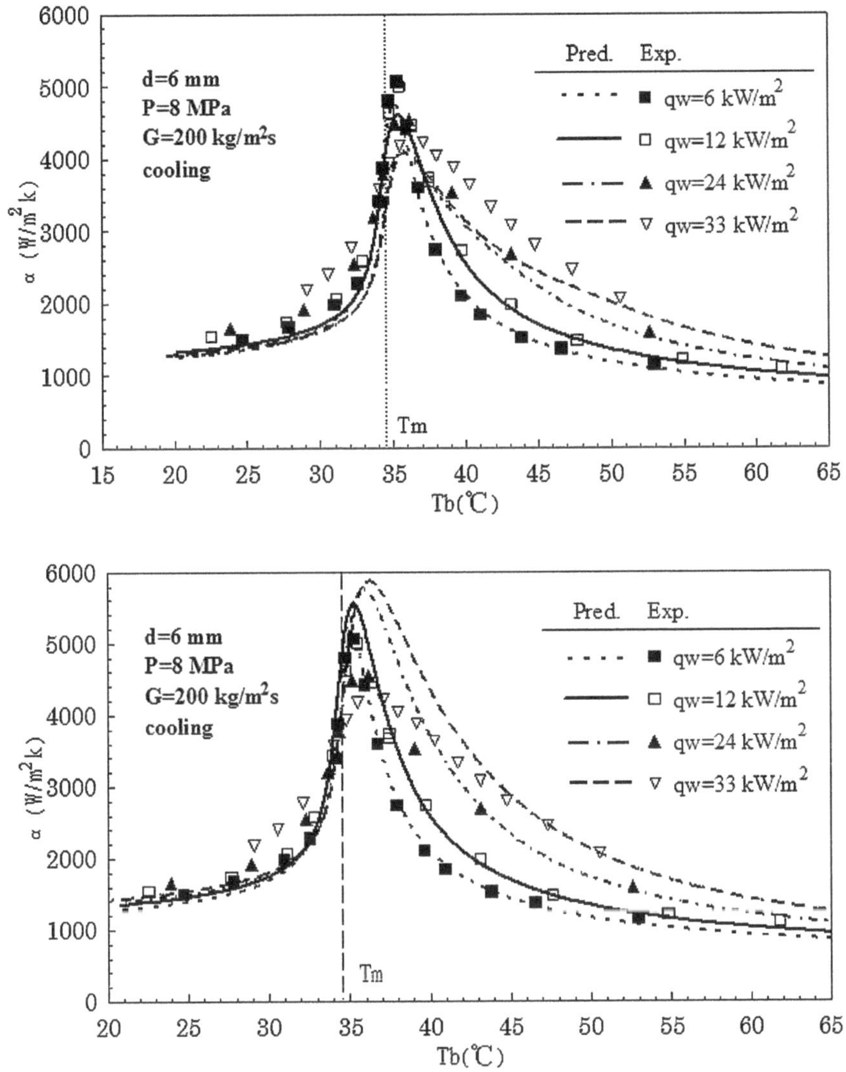

The above discussion focused on the heat transfer characteristics of supercritical CO_2 flowing in smooth tubes under fully developed turbulent condition. Further discussion of supercritical CO_2 under laminar condition or flowing in helically coiled tube can be found in Dang and Hihara (2010), Xu, et al (2016) and Wang, et al (2015).

EFFECT OF LUBRICATING OIL ON THE FLOW AND HEAT TRANSFER OF SUPERCRITICAL CARBON DIOXIDE

In an actual heat pump cycle, lubricating oil is commonly used in the compressor for lubrication, cooling, and sealing purposes. However, small amounts of lubricating oil may be discharged by the compressor

Figure 18. Comparison of heat transfer coefficient α calculated using different definitions of non-dimensional distance from the wall y^+

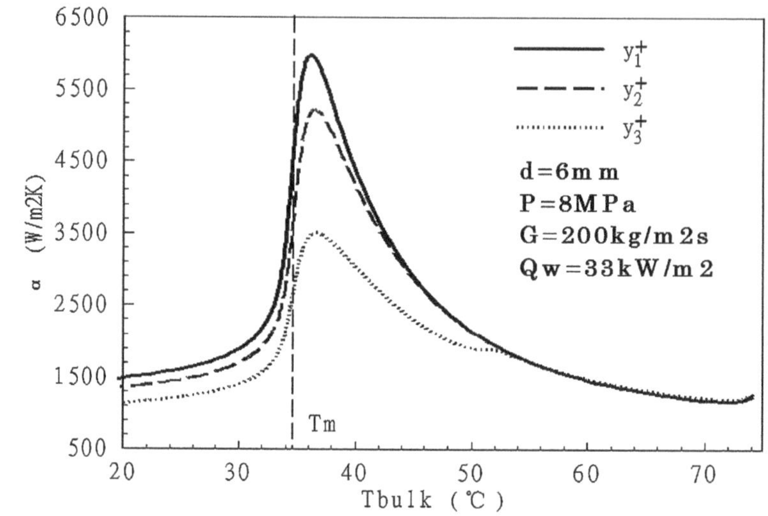

Figure 19. Distributions of density ρ, kinematic viscosity υ and $1/\sqrt{\rho\upsilon}$ at the near wall region. d = 6 mm, P = 8 MPa, G = 200 kg/m²s, q_w = 33 kW/m².

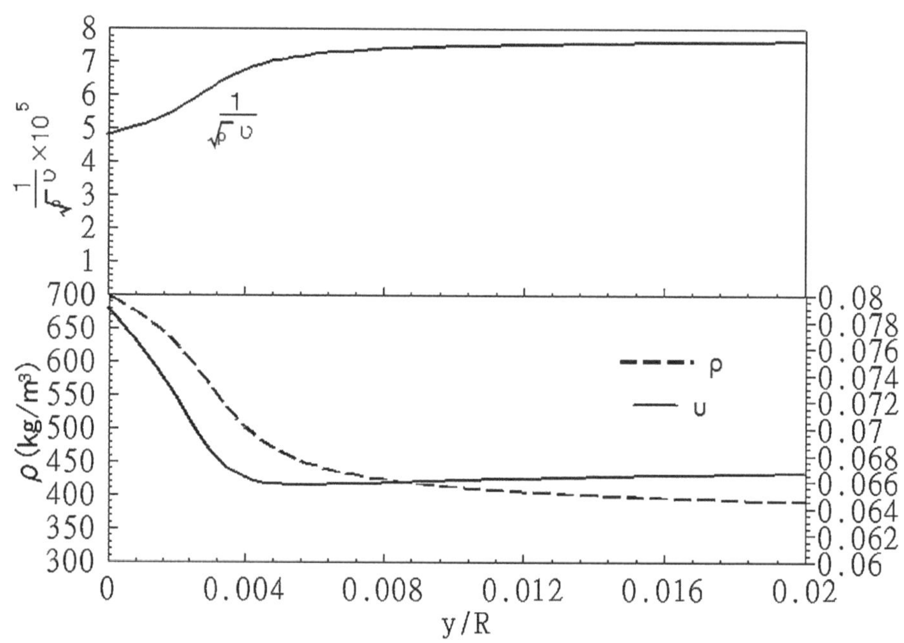

along with the refrigerant; this oil would flow through the heat exchangers and expansion devices and finally return to the compressor. Therefore, a mixture of the refrigerant and a small amount of oil, instead of pure refrigerant, flows inside the heat exchanger and exchanges heat with the environment. Due to the excellent lubricity and stability in the trans-critical CO_2 environment, PAG (polyalkylene glycol) type

Figure 20. Comparison of heat transfer coefficient α predicted at different heat flux q_w and measured α for supercritical carbon dioxide under cooling conditions. Non-dimensional distance y^+ calculated using Equation 35. (a) MK model; (b) BR model.

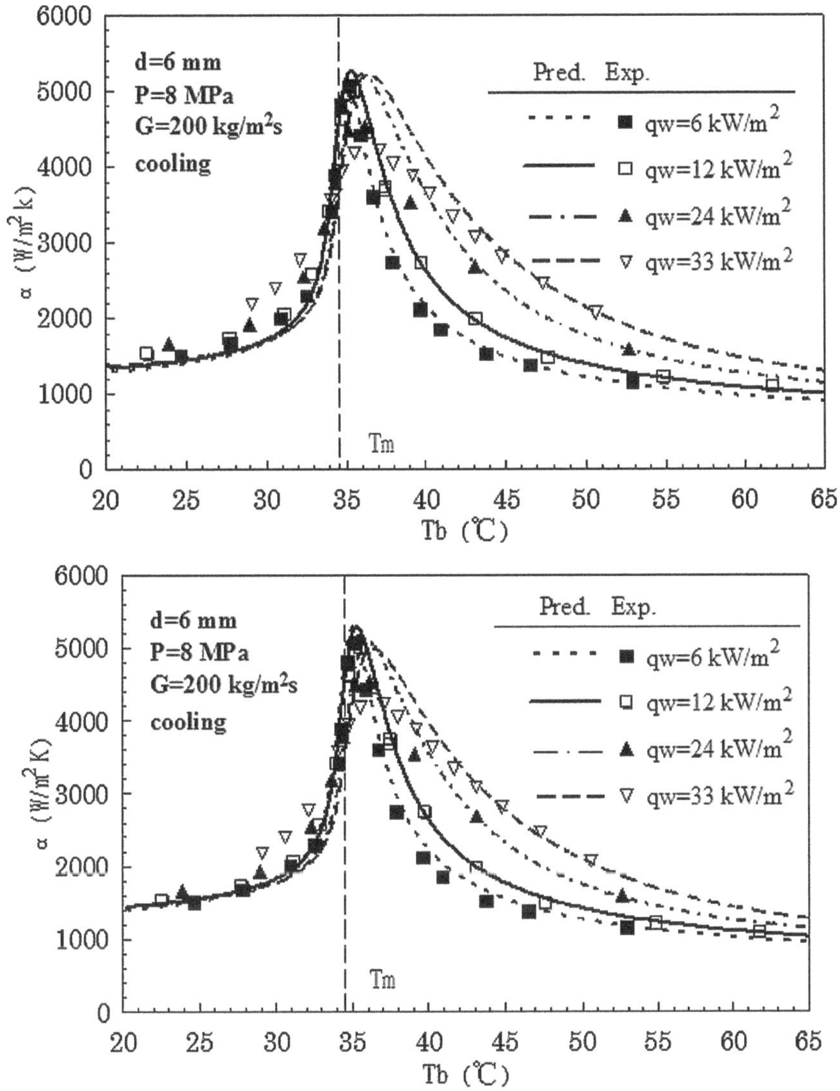

oils are widely applied in the CO_2 heat pump cycle (Kaneko et. al 2006). It is, therefore, important to understand the heat transfer performance of supercritical CO_2 with entrained PAG-type lubricating oils.

Two-Phase Flow Pattern of Supercritical Carbon Dioxide with Entrained PAG-Type Lubricating Oil in a Gas Cooler

Dang et al. (2007) have investigated the heat transfer characteristics of supercritical CO_2 with small amounts of entrained PAG oil. The experimental results showed a decrease in the heat transfer coefficient and an increase in the pressure drop with the entrainment of lubricating oil. The maximum reduction

Figure 21. Comparison of heat transfer coefficient α predicted using different turbulent Prandtl number (Pr_t) models for supercritical carbon dioxide under cooling conditions.

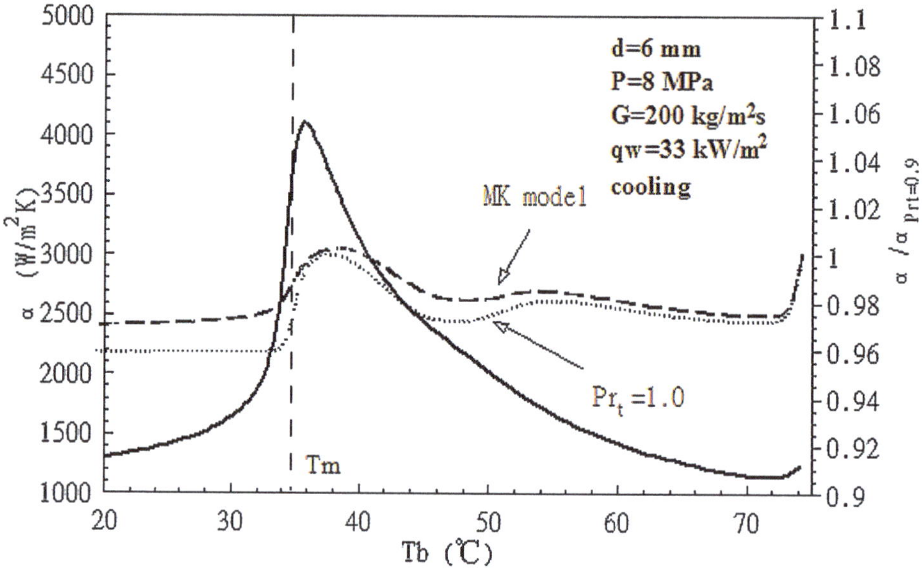

in the heat transfer coefficient was about 75%, which occurred in the vicinity of the pseudocritical temperature. In addition, the transition of the flow pattern was found to drastically affect the heat transfer performance. The changes in the thermophysical properties of supercritical CO_2 with the temperature and pressure are the main parameters used to predict the flow pattern and heat transfer performance of supercritical CO_2 with lubricating oil; moreover, a change in the solubility of CO_2 into PAG oil would only make the issue more challenging. To understand the experimental results of the heat transfer performance with oil entrainment, information concerning the flow pattern inside the gas cooler is of significant importance. However, mainly due to the difficulty in dealing with the high operation pressures, a systematic investigation of the flow pattern under wide experimental conditions is absent in the literature. The current available experimental results from open literature concerning the flow and heat transfer of supercritical CO_2 contaminated with PAG-type oil include the following: Yun et al (2006) measured the heat transfer of supercritical CO_2 with PAG type oil in microchannel with a diameter of 1.0 mm. The mass flux was varied from 200 to 400 $kgm^{-2}s^{-1}$ at heat flux of 20 kWm^{-2}, and the test section inlet pressures of CO_2 were changed from 8.4 to 10.4 MPa with inlet temperatures controlled from 40 to 80°C. By comparing the measured value at different oil concentration, they concluded that the average heat transfer coefficient decreased 9% for 2 wt%, and 20.4% for 4 wt%. Gao and Honda (2002) investigated the heat transfer characteristics of supercritical CO_2 with an oil concentration of 1% inside a 4.5 mm I.D. tube at a pressure of 7.7 MPa, and reported a maximum decrease of 40% in its heat transfer performance at pseudocritical temperature when compared to pure CO_2. Mori et al. (2002) observed the flow pattern of a CO_2–oil mixture inside a gas cooler with a 6 mm I.D. tube at a pressure of 9.5 MPa. Although the oil concentration was unknown, it was found that the oil separated from CO_2 and formed an oil layer near the inner wall. Subsequent measurements of the heat transfer coefficient revealed a decrease in heat transfer, which was explained by the heat resistance of the oil layer.

In order to clarify the heat transfer mechanism of supercritical CO_2 flowing with a small amount of entrained lubricating oil, visualization experiments have been conducted using two sapphire sight-glass tubes with I.D. of 2 and 6 mm. In this report, the flow pattern inside the gas coolers under different pressure, mass flux, and temperature conditions are analyzed, and the relationship between the flow pattern and the heat transfer performance is discussed.

Experimental Setup

A schematic drawing of the test loop has been shown in Figure 4(b), which consists of three main components—a main loop, test section, and oil sampling section. The main loop was a conventional CO_2 heat pump cycle for controlling the temperature and pressure at the inlet of the test section; an oil separator was installed at the exit of the compressor to control the oil concentration circulated in the loop. The mass flow rate of refrigerant was controlled by adjusting the rotation speed of the compressor and the opening of the expansion valves. The test section included a sapphire sight glass tube for visual observation of the refrigerant flow pattern and a tube-in-tube heat exchanger for the measurement of the heat transfer coefficients under various experimental conditions. The oil concentration was measured at the sampling section using the shut-down method. The test sections are those used for heat transfer coefficient measurement of pure carbon dioxide. The oil concentration was measured behind the heat transfer test section using a 4.35 mm I.D. tube with a length of 1.5 m.

The sight glass was made of sapphire in order to endure the high operation pressure of supercritical CO_2. The outside diameter of the two sight glasses were both 30 mm with inner diameters of 6 and 2 mm, respectively, therefore the wall thickness was 12 and 14 mm, respectively. The sight glasses were carefully assembled, and test results demonstrated that they could withstand high pressures of up to 20 MPa at an environmental temperature of 60°C. A sight window was assembled to enable the observation of the flow pattern inside the sapphire glass. The dimensions of the sight window were 10 × 58 mm for the 6 mm I.D. sight glass and 10 × 20 mm for the 2 mm I.D. sight glass. Figure 22 shows the details of the manufactured sight glasses and an enlarged view of the 2 mm I.D. sight glass.

The flow pattern inside the sight glass was recorded using a high-speed CCD camera (MEMRECAM fx-6000, NAC Corp.) with a maximum frame rate of 70000 fps (Frame per second). Because the CO_2–oil droplet entrained flow was much faster than the oil film, the record was repeated twice for the same experimental conditions using different frame rate. High-speed recording, 10000 or 6000 fps, was applied for the observation of the flow pattern over the bulk regions for the 2 mm and 6 mm tubes, respectively, and low-speed recording (500 fps) was adopted for the observation of the flow pattern in near-wall areas. The recorded images were viewed on a monitor and transferred to a computer simultaneously. From recorded continuous images, the velocity of oil droplets and oil film can be calculated therefore the slip ratio of the oil droplets to bulk CO_2 flow, as well as the slip ratio of oil film to bulk CO_2 flow, can be determined.

Experimental Conditions

Since PAG oil is partially miscible with CO_2 at the supercritical pressure, the flow pattern of the CO_2–oil mixture may vary with the oil concentration, mass flux, temperature, and pressure conditions. Table 3 denotes the experimental conditions. Flow visualizations were conducted using 2 and 6 mm I.D. tubes

Figure 22. Visual observation instruments

Sight glasses (6 mm I.D. and 2 mm I.D.)

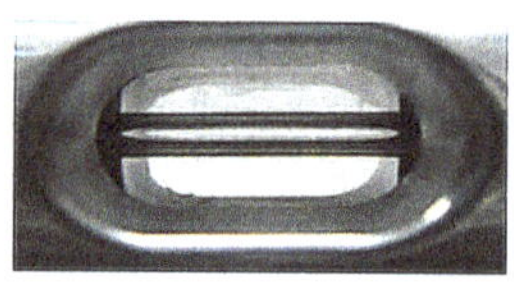

Enlarged view of sight glass (2 mm I.D.)

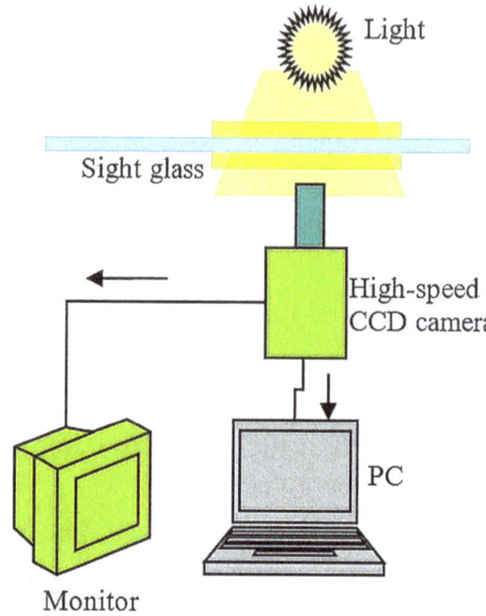

Experimental setup for visual observation

at pressures of 8 and 10 MPa, mass fluxes from 200 to 1200 kg/m^{2}s, and temperatures from 20 to 60°C. Observations of the change in the flow pattern with oil concentrations of 1 and 5 wt% were carried out under the above-mentioned experimental conditions.

Visual Observation Results

Classification of Flow Pattern

Figure 23 illustrates the observed flow pattern of supercritical CO_2 flowing with PAG oil under the above-mentioned experimental conditions. Here, "V" denotes supercritical CO_2, "D" denotes oil droplet, and "F" denotes oil film. The observed flow pattern includes the following: (a) mist flow (M), where a small amount of oil droplets flow with CO_2 and no oil film is observed; (b) annular-dispersed flow (AD), where both oil droplets and an oil film are observed; (c) annular flow (A), where no or few oil droplets are observed; (d) wavy flow (W), where the oil film only exists at the bottom of the cross-section; (e)

Table 3. Experimental conditions

Tube material	Copper
Tube I.D.	2 mm, 6 mm
Oil type	PAG100
Oil concentration	1 wt%, 5 wt%
Mass flux [kg/m^{2}s]	200~1200
Pressure [MPa]	8, 10

Figure 23. Classification of flow pattern: M: mist flow; AD: annular-dispersed flow; A: annular flow; W: wavy flow; WD: wavy-dispersed flow

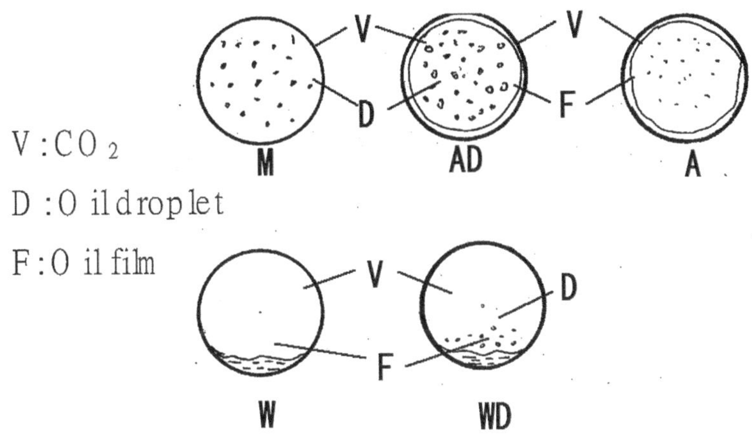

wavy-dispersed flow (WD), where oil droplets are observed flowing along with an oil film at the bottom of the cross-section.

Some examples of the high-speed camera images are shown in Figure 24, Figure 25, and Figure 26. The white stripe at the centerline is due to the refraction of incoming light. Since supercritical CO_2 is transparent, the white particles moving along with CO_2 are oil droplets. The flow pattern was found changing with the temperature, pressure, and oil concentration, as well as with the tube diameter.

Flow Pattern for 2 mm I.D. Tube

Figure 24 shows a comparison of the flow pattern for the 2 mm I.D. tube under different temperature, oil concentration, and pressure conditions. At a low temperature of 25°C and an oil concentration of 1 wt%, it is clear that the flow pattern is mist flow, with oil droplets flowing along with the bulk CO_2 at a slip ratio of about 0.7 in the flow direction; simultaneously, the droplets move up and down along the radial direction due to the disturbing velocity of the bulk CO_2 flow. The average diameter of the oil droplets ranges from 50 to 100 μm. Under this condition, no distinct oil film flowing along the inner wall is observed.

Since the solubility of CO_2 in the oil decreases with an increase in temperature, the viscosity and surface tension of the oil droplets increase with the temperature. As a result, the separated oil appears to adhere to the inner wall and forms an oil-rich layer at a high temperature, with 20 to 60 wt% CO_2 dissolved inside it; this layer is visible as stripes at a temperature of 30°C, as shown in Figure 5(a). With a further increase in temperature, the oil-rich layer becomes much thicker, and the possible movement of the oil droplets in the bulk region cannot be observed. The layer moves at a very low speed, which may lead to a decrease in the heat transfer coefficient due to its heat resistance.

When comparing the flows at oil concentrations of 1 and 5 wt% at the same temperature, the oil-rich layer can be seen clearly even at a low temperature of 25°C under the high-oil-concentration condition and, at the same time, the number of oil droplets flowing along with CO_2 in the bulk region also increases proportionally.

Comparing the flows under pressures of 8 and 10 MPa, no distinct difference is observed in the flow pattern under low- and high-temperature conditions. However, in the near-pseudocritical-temperature region, a sharp contrast in the flow patterns at different pressures is observed. When comparing the flow patterns at pressures of 8 and 10 MPa at 40°C and an oil concentration of 5%, it is seen that the oil-rich layer at 8 MPa is much thicker than that at 10 MPa and the movement of oil droplets is less obvious. The difference observed is due to the change in the pseudocritical temperature with pressure, which corresponds to the changes in the thermophysical properties of CO_2 and solubility of CO_2 into PAG oil; the pseudocritical temperature is 34.5 and 44.8°C at 8 and 10 MPa, respectively. At 40°C, the CO_2 flow is on the high-temperature side of the pseudocritical temperature when the pressure is 8 MPa, the density difference between CO_2 and the oil is large, and the solubility of CO_2 into the oil is low; consequently, the oil tends to separate from the bulk area and adhere to the inner wall. On the contrary, at pressure of 10 MPa, the CO_2 flow is on the low-temperature side of the pseudocritical temperature, the densities of CO_2 and the oil are close to each other, and the oil layer is easily transformed into oil droplets.

Flow Pattern for 6 mm I.D. Tube

Figure 25 shows the transition of the flow pattern with temperature for the 6 mm I.D tube at a low mass flux of 200 kg/m^{2}s. The flow pattern forms a wavy-dispersed or wavy

In Figure 26, the flow pattern for the 6 mm I.D. tube at a high mass flux of 800 kg/m^{2}s is shown. Unlike the flow pattern at a low mass flux of 200 kg/m^{2}s, the oil film can be seen formed at both the upper and bottom inner walls. This annular flow pattern is confirmed at different temperatures of 30 and 40°C. Obviously, the wavy flow is unstable at high mass flux due to the increased turbulent disturbance.

Figure 24. Flow visualization of CO_2 with entrained PAG oil; d = 2 mm, G = 800 kg/m^{2}s: (a) P = 8 MPa, x = 1%; (b) P = 8 MPa, x = 5%; (c) P = 10 MPa, x = 5%

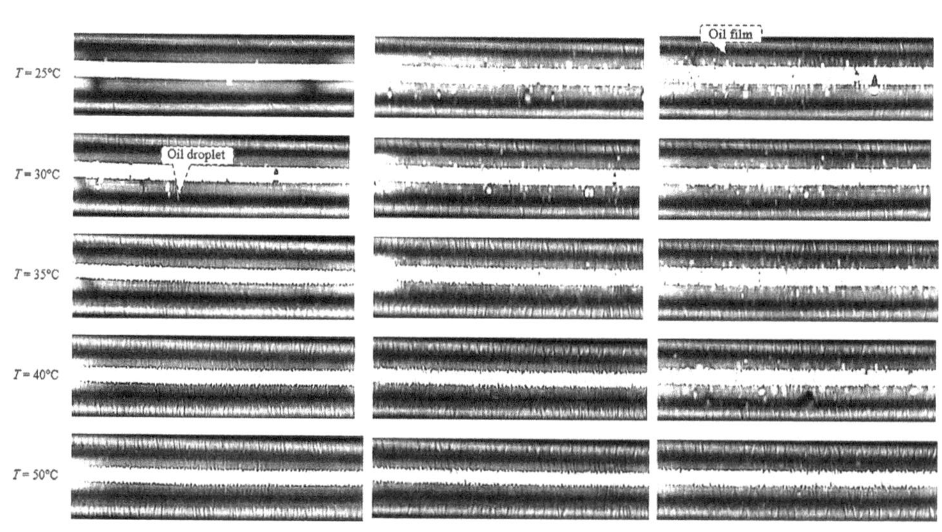

Figure 25. Comparison of flow pattern at different temperature for 6 mm I.D. tube; d = 6 mm, P = 8 MPa, G = 200 kg/m²s, x = 1%: (a) T = 25°C; (b) T = 30°C; (c) T = 35°C; (d) T = 40°C; (e) T = 45°C; (f) T = 50°C

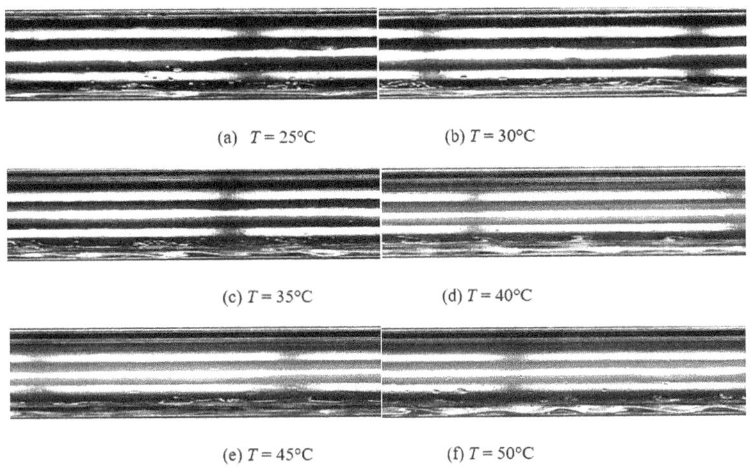

Figure 26. Comparison of flow pattern at different temperature for 6 mm I.D. tube; d = 6 mm, P = 8 MPa, G = 800 kg/m²s, x = 1%: (a) T = 30°C; (b) T = 40°C

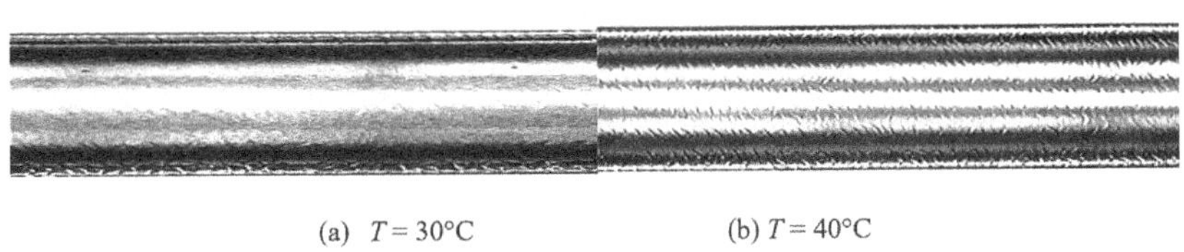

Comparison of Flow Pattern at Different Tube Diameters

Previous researches on flow pattern change are focus on tube diameter larger than 20 mm, the flow pattern at small sized tube less than 2 mm is not well investigated, and no much knowledge available for a two-phase flow with a large viscosity and surface tension like lubricating oil. The flow pattern map proposed for water-air or refrigerant were tested and the results showed that the previous flow map could not explain the flow pattern changed demonstrated in this research. Therefore, an attempt for the flow map prediction has been scheduled. However, the results are unavailable at present.

A comparison of the flow pattern at tube diameters of 2 and 6 mm tubes are presented in Figure 27. The visual observations were conducted at pressure of 8 MPa, mass flux of 800 kg/m²s, the temperature and oil concentration is 30°C and 1 wt%, respectively. It is obvious from comparison that the oil film for 2 mm tube is thicker than that of 6 mm tube, and the oil droplets is much larger with smaller number density. This result is due to the decreased turbulent disturbance with the decrease in tube diameter when compared under the same mass flux, temperature and pressure condition. The thicker oil film for small sized tube corresponds to a much more significant effect of heat resistance. Due to the small viscosity of CO_2 and high operating pressure, a small size tube is usually proposed to be used for CO_2 heat pump.

Figure 27. comparison of flow pattern at different tube diameters; P = 8 MPa, G = 800 kg/m²s, T = 30°C, x = 1%: (a) d = 2 mm; (b) d = 6mm

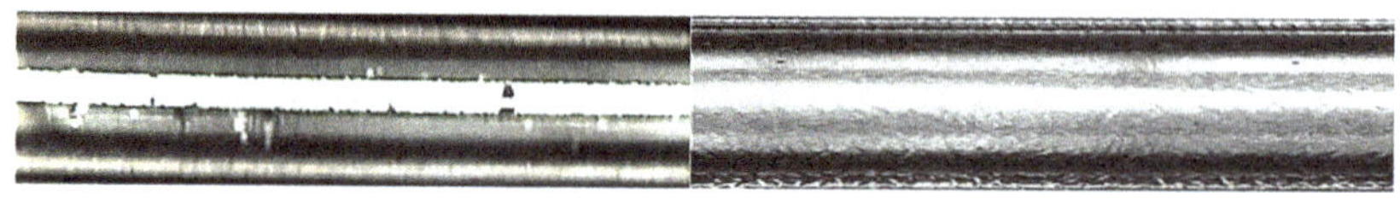

(a) d= 2 mm (b) d = 6 mm

However, the visual observation results imply that the formation of oil film and the corresponding heat transfer deterioration due to the introduction of lubricating oil should be taken in consideration to decide the optimum tube dimension.

DISCUSSION

The relationship between the flow pattern of CO_2–PAG oil and the heat transfer performance (Dang, et al. 2007) is shown in Figure 28 and Figure 29 for the 2 and 6 mm I.D. tubes, respectively.

In Figure 28, a comparison of the heat transfer coefficients and the flow pattern for the 2 mm I.D. tube at different temperatures are shown. The solid line denotes the heat transfer coefficient of pure CO_2. It is seen that the heat transfer coefficient decreases due to the introduction of PAG oil, and this decrease is most significant at the pseudocritical temperature. Comparing the curves at 30°C and 50°C, the drop in the heat transfer coefficient at 50°C is much larger than that at 30°C for an oil concentration of 5%. From the visual observation, it is found that the flow pattern at low temperature is mist flow with a considerable amount of oil flowing over the bulk area and a thin oil film. At high temperature, a thick oil film is observed, which corresponds to a sharp decrease in the heat transfer coefficient. This visual observation implies that the heat resistance due to the formation of the oil film is the main reason for the

Figure 28. Flow pattern of CO_2–oil and heat transfer coefficient for 2 mm I.D. tube

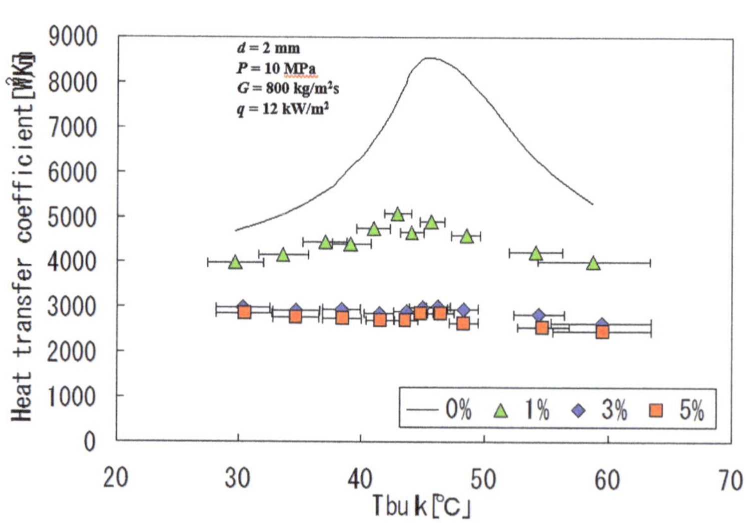

heat transfer deterioration. And the oil droplets flowing in the bulk region do not contribute significantly to the heat transfer deterioration.

Figure 29 shows a comparison of the heat transfer coefficient and flow pattern for the 6 mm I.D. tube at two mass fluxes, 200 and 800 kg/m^{2}s. The heat transfer coefficient decreases significantly at 800 kg/m^{2}s due to the heat transfer resistance of the oil film flowing along inner wall. However, the heat

Figure 29. Flow pattern of CO_2–oil (40°C) and heat transfer coefficient for 6 mm I.D. tube: (a) G = 200 kg/m^{2}s; (b) G = 800 kg/m^{2}s

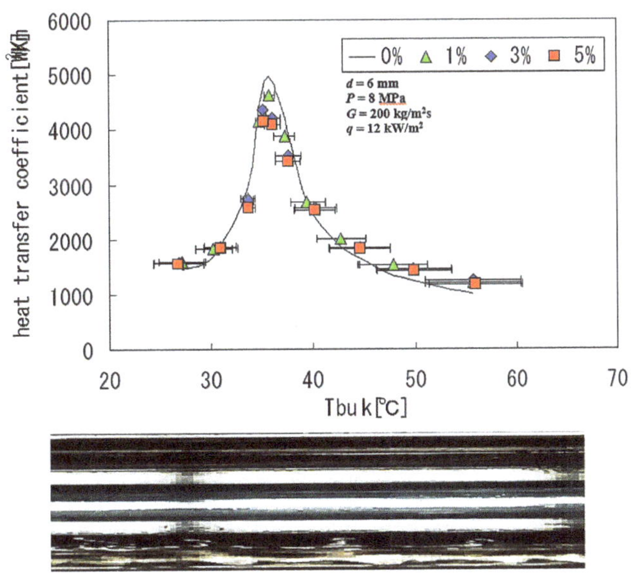

(a) G = 200 kg/m^{2}s

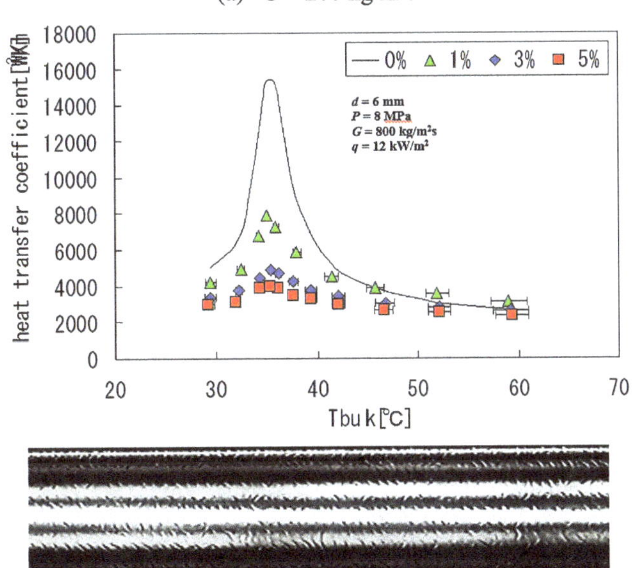

(b) G = 800 kg/m^{2}s

transfer coefficient does not drop significantly at a mass flux of 200 kg/m^{2}s, even if the oil concentration is increased to 5%. From visual observation, it is found that the flow at the low mass flux is wavy flow, and the oil film is only observed at the bottom of the tube. However, at the high mass flux, the oil film is formed over the entire inner wall, which leads to a significant drop in heat transfer performance.

It is evident again by comparing the heat transfer coefficient shown in Figure 28 and Figure 29 that although the heat transfer coefficient for a small sized tube is much higher than that of larger sized tube for pure CO_2, by introducing the partial miscible lubricating oil, much more significant deterioration of the heat transfer coefficient was observed for smaller sized tube than that of large sized tube, which is the results of the change in flow pattern since the oil film for small sized tube may much thicker than large sized tube. It is therefore revealed that the effect of lubricating oil should be considered in detail when the tube diameter is determined.

A prediction model based on neural network method was proposed to predict the heat transfer coefficient of supercritical CO_2 containing a small amount of PAG type oil (Dang & Hihara, 2012, Dai et al, 2015a). The heat transfer characteristics of CO_2 with other type of lubricating oil, PVE and ECP, was also investigated to reveal the effect of solubility of oil into supercritical CO_2 on the heat transfer performance (Dang, et al, 2012). Mixture of CO_2 with low global warming potential working fluids for heat pump water heater was also discuss to reduce the operating pressure and improve system performance (Dai, et al, 2015b).

CONCLUSION

In this chapter, a series of experimental and analytical studies on supercritical CO_2 fluid flow and heat transfer characteristics are conducted and explained.

1. The cooling heat transfer characteristics of supercritical CO_2 differ from constant property fluid due to the distribution of thermal properties along both the radial and longitudinal directions.
2. The heat transfer coefficient takes maximum value when the film temperature equals to the pseudocritical temperature.
3. A new heat transfer coefficient was proposed, which fit the experimental and numerical results well.
4. Presence of Lubricating oil may decrease the heat transfer performance sharply; however, the flow pattern is determined by many factors, including the tube diameter, oil concentration, temperature, pressure, and mass flux.
5. From the comparison of the flow pattern observations and the heat transfer measurements, it seems that the mechanism behind the heat transfer performance drop is mainly due to the formation of an oil film along the inner wall, and the oil droplets have no significant contribution to the heat transfer deterioration.

REFERENCES

Bellmore, C. P., & Reid, R. L. (1983). Numerical Prediction of wall temperature for near critical para-hydrogen in turbulent upflow inside vertical tubes. *Journal of Heat Transfer*, *105*(3), 22–29. doi:10.1115/1.3245618

Dai, B., Li, M., Dang, C., & Ma, Y. (2015a). Effects of lubricating oil on thermal performance of water-cooled carbon dioxide gas cooler. *Applied Thermal Engineering*, *80*, 288–300. doi:10.1016/j.applthermaleng.2015.01.071

Dai, B., Li, M., Dang, C., Tian, H., & Ma, Y. (2015b). Thermodynamic performance assessment of carbon dioxide blends with low GWP working fluids used for heat pump water heater. *International Journal of Refrigeration*, *56*, 1–14. doi:10.1016/j.ijrefrig.2014.11.009

Dang, C., & Hihara, E. (2004 a). In-tube cooling heat transfer of supercritical carbon dioxide, part 1: Experimental measurement. *International Journal of Refrigeration*, *27*(7), 736–747. doi:10.1016/j.ijrefrig.2004.04.018

Dang, C., & Hihara, E. (2004 b). In-tube cooling heat transfer of supercritical carbon dioxide, part 2: Comparison of turbulence models in numerical calculation. *International Journal of Refrigeration*, *27*, 748–760. doi:10.1016/j.ijrefrig.2004.04.017

Dang, C., & Hihara, E. (2010). Numerical study on in-tube laminar heat transfer of supercritical fluids. *Applied Thermal Engineering*, *13*(13), 1567–1573. doi:10.1016/j.applthermaleng.2010.03.010

Dang, C., & Hihara, E. (2012). Predicting the cooling heat transfer coefficient of supercritical CO_2 with a small amount of entrained lubricating oil by using the neural network method. *International Journal of Refrigeration*, *35*(4), 1130–1138. doi:10.1016/j.ijrefrig.2012.01.022

Dang, C., Hoshika, K., & Hihara, E. (2012). Effect of lubricating oil on the flow and heat transfer characteristics of supercritical carbon dioxide. *International Journal of Refrigeration*, *35*(5), 1410–1417. doi:10.1016/j.ijrefrig.2012.03.015

Dang, C., Iino, K., Fukuoka, K., & Hihara, E. (2007). Effect of lubricating oil on cooling heat transfer of supercritical carbon dioxide. *International Journal of Refrigeration*, *30*(4), 724–731. doi:10.1016/j.ijrefrig.2006.09.006

Dang, C., Iino, K., & Hihara, E. (2008). Study on two-phase flow pattern of supercritical carbon dioxide with entrained PAG-type lubricating oil in a gas cooler. *International Journal of Refrigeration*, *31*(7), 1265–1272. doi:10.1016/j.ijrefrig.2008.01.014

Dang, C., Iino, K., & Hihara, E. (2010). Effect of PAG-type lubricating oil on heat transfer characteristics of supercritical carbon dioxide cooled inside a small internally grooved tube. *International Journal of Refrigeration*, *33*(3), 558–565. doi:10.1016/j.ijrefrig.2009.12.010

Gao, L., & Honda, T. (2002). Experiments on heat transfer characteristics of heat exchanger for CO2 heat pump system. In *Proc. of the Asian Conference on Refrigeration and Air Conditioning*. Kobe, Japan: Japan Society of Refrigeration and Air Conditioning Engineers.

Gnielinski, V. (1976). New equations for heat and mass transfer in turbulent pipe and channel flow. *Chemical Engineering Journal.*, *16*(2), 359–368.

Goldmann, K. (1954). Heat transfer to supercritical water and other fluids with temperature-dependent properties. *Chemical Engineering Progress Symposium Series, 50*(11), 105-110.

Hashimoto, K., Tanahashi, M., & Tomita, M. (2002). Direct numerical simulation on the near wall heat transfer of carbon dioxide. *Proceeding of 5th IIR-Gustav Lorentzen Conference on Natural Working Fluids.* Guangzhou, Guangzhou, China: International Institute of Refrigeration.

Howell, J. R., & Lee, S. H. (1999). Convective heat transfer in the entrance region of a vertical tube for water near the thermodynamic critical point. *International Journal of Heat and Mass Transfer*, *42*(7), 1177–1187. doi:10.1016/S0017-9310(98)00263-4

Jackson, J. D., & Hall, W. B. (1979). Forced convection heat transfer to fluids at supercritical pressure. In S. Kakac & D. B. Spalding (Eds.), *Turbulent Forced Convection in Channels and Bundles* (pp. 563–611). New York: McGraw-Hill.

Jones, W. P., & Launder, B. E. (1972). The prediction of laminarization with a two-equation model of turbulence. *International Journal of Heat and Mass Transfer*, *15*(2), 301–314. doi:10.1016/0017-9310(72)90076-2

JRAIA. (2016). *About 'Eco Cute' Exceeding 5 million Units.* Retrieved from https://www.jraia.or.jp/product/heatpump/i_broke.html

Kaneko, M., Ikeda, H., Tokiai, T., Nagao, S., Suto, H., & Tamano, M. (2008). The development of PAG-PVE copolymer (ECP). *Proc. Int. Symp. on New Refrigerants and Environmental Technology*, 376-388.

Kaneko, M., Ikeda, H., Tokiai, T., Yoshii, A., & Suto, H. (2006). The development of PAG refrigeration lubricants for automotive A/C with CO2. JSAE Automotive Air-Conditioning Conference, 142-159.

Kays, W. M., & Crawford, M. E. (1980). *Convective heat and mass transfer*. New York: McGraw-Hill.

Koshizuka, S., Takano, N., & Oka, Y. (1995). Numerical analysis of deterioration phenomena in heat transfer to supercritical water. *International Journal of Heat and Mass Transfer*, *38*(16), 3077–3084. doi:10.1016/0017-9310(95)00008-W

Launder, B. E., & Sharma, B. I. (1974). Application of the energy dissipation model of turbulence to calculation of flow near a spinning disc. *Letters of Heat Mass Transfer.*, *1*(2), 131–138. doi:10.1016/0094-4548(74)90150-7

Lee, S. H., & Howell, J. R. (1998). Turbulent developing convective heat transfer in a tube for fluids near the critical point. *International Journal of Heat and Mass Transfer*, *41*(10), 1205–1218. doi:10.1016/S0017-9310(97)00217-2

Lorentzen, G. (1995). The use of natural refrigerants: A complete solution to the CFC/HCFC predicament. *International Journal of Refrigeration*, *18*(3), 190–197. doi:10.1016/0140-7007(94)00001-E

Mikielewicz, D. P., Shehata, A. M., Jackson, J. D., & McEligot, D. M. (2002). Temperature, velocity and mean turbulence structure in strongly heated internal gas flows - Comparison of numerical predictions with data. *International Journal of Heat and Mass Transfer*, *45*(21), 4333–4352. doi:10.1016/S0017-9310(02)00119-9

Mori, K., Onishi, J., Shimaoka, H., Nakanishi, S., & Kimoto, H. (2002). Cooling heat transfer characteristics of CO2 and CO2-oil mixture at supercritical pressure conditions. In *Proc. of the Asian Conference on Refrigeration and Air Conditioning*. Kobe, Japan: Japan Society of Refrigeration and Air Conditioning Engineers.

Myong, H. K., & Kasagi, N. (1990). A new approach to the improvement of k-ε turbulence model for wall-bounded shear flows. *JSME International Journal, Series II.*, *33*(1), 63–72.

Myong, H. K., Kasagi, N., & Hirata, M. (1989). Numerical prediction of turbulent pipe flow heat transfer for various Prandtl number fluids with the improved k-ε turbulence model. *JSME International Journal, Series II.*, *32*(4), 613–622.

Patankar, S. V. (1980). *Numerical heat transfer and fluid flow*. New York: McGraw Hill.

Patel, V. C., Rodi, W., & Shceuerer, G. (1984). Turbulence models for near-wall and low Reynolds number flows. A review. *AIAA Journal*, *23*(9), 1308–1319. doi:10.2514/3.9086

Petrov, N. E., & Popov, V. N. (1988). Heat transfer and hydraulic resistance with turbulent flow in a tube of water under supercritical parameters of state. *Thermal Engineering*, *35*(6), 577–580.

Pitla, S. S., Ramadhyani, S., & Groll, E. A. (2001). Convective heat transfer from in-tube flow of turbulent supercritical carbon dioxide: Part 1-numerical analysis. *International Journal HVAC&R Research.*, *7*(4), 345–366. doi:10.1080/10789669.2001.10391280

REFPROP. (n.d.). *NIST Standard Reference Database 23, Version 7.0*. NIST.

Sengers, J. V. (1965). Thermal conductivity and viscosity of simple fluids. *International Journal of Heat and Mass Transfer*, *8*(8), 1103–1116. doi:10.1016/0017-9310(65)90138-9

Tanaka, H., Nishiwaki, N., & Hirata, M. (1967). Turbulent heat transfer in vertical tubes at supercritical pressures. *Proc. J.S.M.E. Semi-International Symposium*, 127-134.

Van Driest, E. R. (1956). On turbulent flow near a wall. *Journal of the Aeronautical Sciences*, *23*(11), 1007–1011. doi:10.2514/8.3713

Wang, K., Xu, X., Wu, Y., Liu, C., & Dang, C. (2015). Numerical investigation on heat transfer of supercritical CO_2 in heated helically coiled tubes. *The Journal of Supercritical Fluids*, *99*, 112–120. doi:10.1016/j.supflu.2015.02.001

Xu, X., Wu, Y., Liu, C., Wang, K., Dang, C., & Bai, W. (2016). Experimental investigation on heat transfer characteristics of supercritical CO_2 cooled in horizontal helically coiled tube. *International Journal of Refrigeration*, *67*, 190–201. doi:10.1016/j.ijrefrig.2016.03.010

Yoshida, S., & Mori, H. (2000). Heat transfer to supercritical pressure fluids flowing in tubes. *Proceedings of SCR-2000*. Tokyo: The University of Tokyo.

Yun, R., Hwang, Y., & Radermacher, R. (2006). Gas cooler heat transfer and pressure drop characteristics of CO2/oil mixture in a microchannel. *Proc. of the 7th IIR-Gustav Lorentzen Conference on Natural Working Fluids*, 503-505.

KEY TERMS AND DEFINITIONS

Compressed Fluid: A fluid at a pressure above the critical pressure, but at a temperature below the critical temperature.

Heat Pump: A device that takes heat from one source and moves it to another location through electric or mechanical means. Heat pumps may be used either to heat or cool.

Heat Transfer Coefficient: The proportionality constant between the heat flux and the thermodynamic driving force for the flow of heat.

Pseudocritical Point: A point at a pressure above the critical pressure and at a temperature corresponding to the maximum value of the specific heat at this particular pressure.

Refrigerant: A substance or mixture, usually a fluid, used in a heat pump and refrigeration cycle. In most cycles it undergoes phase transitions from a liquid to a gas and back again.

Supercritical Fluid: A fluid at a pressure higher than the critical pressure and at a temperature higher than the critical temperature. However, in the present chapter, a term "supercritical fluid" includes both "supercritical fluid" and "compressed fluid".

Transcritical Cycle: A thermodynamic cycle where the working fluid goes through both subcritical and supercritical states.

Chapter 11
Supercritical Carbon Dioxide and Its Application to Rankine Cycle

Hiroshi Yamaguchi
Doshisha University, Japan

ABSTRACT

Supercritical CO_2 has been given much attention to be a working fluid in a power cycle due to its unique properties. The supercritical CO_2 solar Rankine cycle system was designed and developed by using the benefit of supercritical state of CO_2 to generate electric power and supply heat energy in environmentally friendly manner. The development of main components in the system are introduced and discussed particularly by focusing on the properties of CO_2 for obtaining higher performance. The properties of CO_2 in near critical region are also discussed in this chapter. Operating the power cycle in the supercritical region of CO_2 enhances the heat transfer in energy exchanging process and improves the cycle efficiency.

INTRODUCTION

The Global warming and climate change crisis have been a great concern in our human history. The main cause comes from the greenhouse gases (GHGs) emission produced by fossil fuel-powered generation facilities in such as the combustion process of a coal, oil or gas (Jacobson, 2009). The other significant cause also comes from burning of fossil fuels for driving vehicle, inefficient energy usage in homes and businesses operation. To deal with the problems, the prime importance is the greenhouse gas reduction with our highest efforts in science and technology.

1997 Kyoto Protocol, the world's first climate change treaty, is a binding agreement for reducing greenhouse gas emissions with 192 parties' agreement. The second commitment period (2013-2020) had been announced in 2012, and proposing the targeting cut of 20% GHG emission levels of 1990 by 2020. The Hydrofluorocarbons (HFCs) had been recommended and widely used in previous years as a working fluid in the industries, instead of other chemical working fluid, because HFC has no effect to the Ozone Layer Depletion (zero ODP). However, HFCs has very high Impact on Global Warming

DOI: 10.4018/978-1-5225-2047-4.ch011

Potential (high GWP), which directly causes global warming crisis. On the other hand, HFCs is predicted approximately 2 – 4% increase of the overall climate forcing impact by 2050 (Marley, 2001). Due to the fact mentioned above, in 2013, the international agreement of industrial countries to phase down HFC use under the Montreal Protocol has started, targeting 80% reduction of HFC by 2030 (Su, Fang, Li, Wu, Zhang, Xu, & Hu, 2015).

In the 21st session of the Conference of the Parties (COP), which was held in 2015, to the United Nation Framework Convention on Climate Change (UNFCCC), so called "Paris Agreement", also suggests to reduce the carbon emission at the earliest and handle the increasing of global average temperature to well below 2 °C (Victor and Kennel, 2014). To achieve the objective, the natural working fluid such as carbon dioxide (R-744 or CO_2), which is also listed in the required collective control in the Kyoto Protocol has recommended to utilize and use instead of HFCs. Taking in account of these reasons CO_2 itself is environmentally friendly, which is by definition 0 and 1 of ODP and GWP, respectively.

In the meaning of recycling and utilizing of GHGs, the interest to use CO_2 as a natural working fluid in the thermodynamic cycle has been greatly increased due to its high vapor pressure and large volumetric refrigeration capacity when compared to other working fluids. Moreover, the critical pressure and critical temperature of CO_2 are 7.38 MPa and 31.1 °C, respectively, which are very much lower than other working fluids as well. Thus, CO_2 can become as supercritical state easily even when it is operated at comparatively low temperature. The characteristics and properties of various working fluids are compared in Table 1 (Kim, Pettersen and Bullard, 2004). The other point of interest in CO_2 properties is that classified as non-flammable, non-toxic working fluid and also chemical inactive, which are safe to human health and the environment. From these reasons, many attempts have been made to replace ordinary working fluids to CO_2 in power cycle, in which the Rankine cycle system is also one of the efforts.

In the aim to achieve the targeted level of the global warming treaty and the world's sustainable development, a renewable energy system for electric power generation and thermal energy supply is essential. The new combined power/heat thermo-fluid cycle, so called supercritical CO_2 solar Rankine

Table 1. Characteristic of some representative working fluids

Properties	R-12	R-22	R-134a	R-407C	R-410A	R-717	R-209	R-744
ODP/GWP	1/8500	0.05/1700	0/1300	0/1600	0/1900	0/0	0/6	0/1
Flammability/toxicity	N/N	N/N	N/N	N/N	N/N	Y/Y	Y/N	N/N
Molecular mass (kg/kmol)	120.9	86.5	102	86.2	72.6	17	44.1	44
Critical pressure (MPa)	4.11	4.97	4.07	4.64	4.79	11.42	4.25	7.38
Critical temperature (°C)	11.2	96	101.1	86.1	70.2	133	96.7	31.1
Reduced pressure[a]	0.07	0.1	0.07	0.11	0.16	0.04	0.11	0.47
Reduced temperature[b]	0.71	0.74	0.73	0.76	0.79	0.67	0.74	0.9
Refrigerant capacity[c] (kJ/m^3)	2734	4356	2868	4029	6763	4382	3907	22,454

Kim et al., 2004.

R-12: dichlorodifluoromethane; R-22: chlorodifluoromethane; R-134a: tetrafluoroethane; R-407C: ternary mixture of difluoromethane/pentafluoroethane/tetrafluoroethane (23/25/52, %); R-410A: binary mixture of difluoromethane/pentafluoroethane (50/50, %); R-717: ammonia; R-290: propane; R-744: carbon dioxide.

[a]Ratio of saturation pressure at 0 °C to critical pressure.

[b]Ratio of 273.15 K (0 °C) to critical temperature in Kelvin.

[c]Volumetric refrigeration capacity at 0 °C.

cycle system (SRCS), which uses CO_2 as a working fluid with solar energy as the energy source, has been originally designed, constructed and tested by Zhang et al. since 2004 in Kyoto, Japan. The early stage of development of SRCS with the good acceptance will be described in this chapter. The unique properties of CO_2 give the high performance and effectiveness to the system that can avoid the limitation appeared in the Rankine cycle, such as using water as a working fluid. The original SRCS was tested in various weather conditions in Kyoto area, where the system works well in the transcritical cycle in test hours. Preliminary data for the electric and heat generation output were also gained with acceptable thermodynamic efficiency. Along with the introduction of SRCS, thermo-physical details of CO_2 and its supercritical state will be provided in this chapter. Moreover, the basic principle of supercritical CO_2 Rankine cycle is described and chief elements of the cycle are introduced and discussed in view of the future energy creation technique in SRCS.

POWER CYCLE SYSTEM

Thermodynamic cycle is composited with various thermodynamic processes to transfer heat and work in and outside of the cycle. Eventually, by giving heat input (as thermal energy) thermo-physical properties of working fluid in the cycle such as pressure, temperature and other variables are returned to the initial state after extracting work output. The concept of thermodynamic cycle has been accorded to the law of thermodynamic, which makes use of Rankine cycle for converting heat (as thermal energy) to the useful work as an ideal reference cycle, such as the steam power plant (Winterbone, 2004).

The thermodynamic cycle can be classified as two primary classes, which are power cycle or heat pump cycle. Normally, the power cycle is used to convert thermal energy (heat) input into mechanical work output, while the heat pump cycle uses mechanical work input to transfer low-thermal energy to high-thermal (temperature) energy.

The thermodynamic power cycles are widely used as a primary operation in heat engine, which propels the world nowadays, mostly in (electric) power generation system. In this chapter, the power generation cycles are only discussed and compared, namely the purposed systems are Rankine cycle and Supercritical Rankine cycle.

Rankine Cycle

Rankine cycle is one of the closed-loop power generation cycle, which is generally used in a thermal power plant (Wallace & Linning, 1970). The ideal Rankine cycle is a practical model of energy conversion from heat to mechanical work, which consists of four components and four processes, i. heating at constant pressure state (in boiler), ii. isentropic expansion (in turbine), iii. heat rejecting in constant pressure state (in condenser) and iv. isentropic compression (in feed pump), as shown in Figure 1, where Figure 1(a) is a schematic of the Rankine cycle with the four components with four representative cycle points of ①, ②, ③ and ④. Figure 1(b) is the cycle diagram plotted in T (temperature) and s (entropy); T-s diagram, and Figure 1(c) is the cycle diagram plotted in P (pressure) and h (enthalpy); P-h diagram. In Figure 1(b) and Figure 1(c), the four processes are indicated with the four representative points.

Figure 1. (a) Schematic diagram of Rankine cycle with four components; (b) T-s diagram; and (c) P-h diagram of Rankine cycle

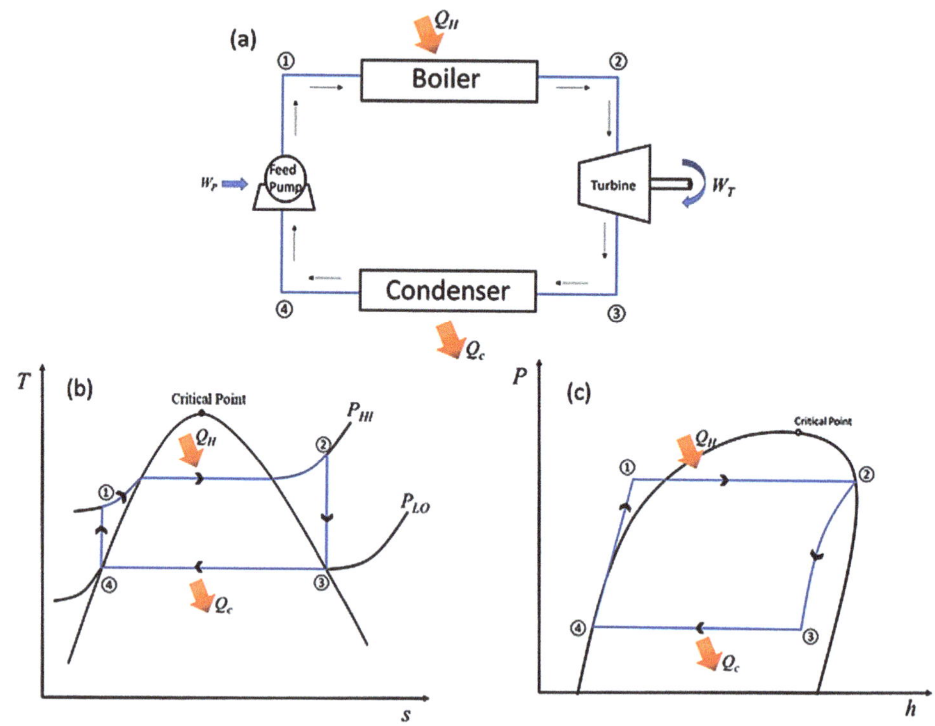

1. Constant pressure heat absorption (Q_H), (① - ②).
2. Power output (W_T) by isentropic expansion, (② - ③).
3. Constant pressure heat rejection (Q_C), (③ - ④).
4. Isentropic compression (Reversible adiabatic), (④ - ①).

The efficiency of Rankine cycle can be written as:

$$\eta_{\text{Rankine}} = \frac{W_T - W_P}{Q_H - Q_C} \quad (1)$$

The Rankine cycle gains greater thermal efficiency, in which the cycle provides two phase changes in the cycle process. In practical process of Rankine cycle, however, external power input W_P in #4; ④ - ① process is required to achieve the closed thermodynamic cycle as formulated in Equation 1.

SUPERCRITICAL RANKINE CYCLE

In selection of appropriate practical working substance, the working fluid, which has relatively low critical temperature and pressure, is to be used in the Rankine cycle when operating in supercritical region

Figure 2. Supercritical Rankine cycle: (a) T-s diagram; and (b) P-h diagram

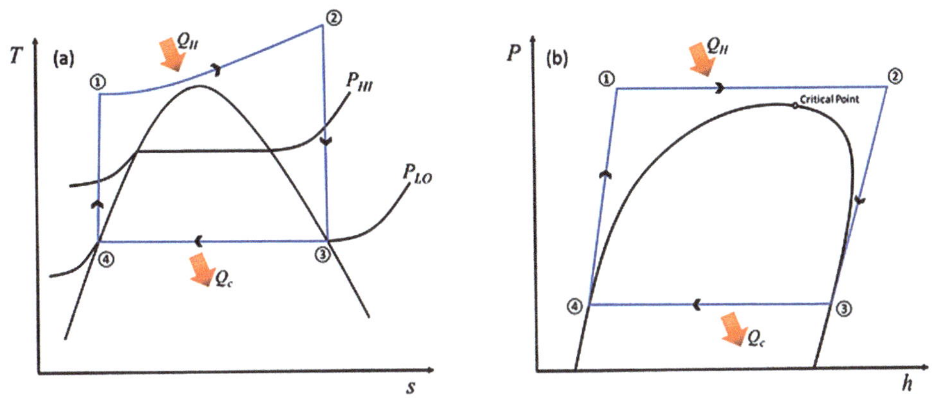

(Hough, 2009). The working fluid is compressed to achieve its supercritical state over critical point to obtain higher efficiency (with higher heat transfer mode in the heating process ① - ②) in supercritical Rankine cycle as shown in Figure 2(c) and Figure 2(b).

The diagram of supercritical Rankine cycle of *T-s* and *P-h* diagram are drawn in Figure 2(a) and Figure 2(b) respectively. Different from the ideal Rankine cycle as shown in Figure 1(b) and Figure 1(c), working fluid is heated and compressed to high pressure and temperature to reach the supercritical state, where high-mode phase change occurs in the process of heating ① - ②. Then, supercritical working fluid at ② passes through the turbine in adiabatic expansion process (② - ③) for power generation (W_T). As shown in Figure 2(a) and Figure 2(b), the supercritical Rankine cycle is the trans-critical cycle, where the critical point is included inside of the closed loop (cycle).

In consideration of realizing the supercritical Rankine cycle, which should be the most effective with efficient power generation cycle as described above, CO_2 would be the choice for the purpose in the present study.

GAS BEHAVIOR

Thus far, in an ordinary power cycle the ideal gas is usually agreed to explain the change of thermodynamic state in a power cycle. However, at the state of high temperature and high pressure, the behavior of real gas deviates significantly from the ideal gas (Wakefield, 1996).

For one mole of real gas;

$$\frac{PV}{RT} \rightarrow 1_{\text{lim it} P \rightarrow 0} \tag{2}$$

where P is pressure, V is volume, R is gas constant and T is temperature.

It can be seen that when the pressure of gas is approached to zero, the behavior of real gas will be similar to the ideal gas, with which the ideal gas law can be defined as

$$PV = RT \tag{3}$$

In extension of Equation 2, the deviation of real gas from the ideal gas can be considered by the compressibility factor (*Z*) by comparing the volume of 1 mole of real gas with the molar volume of an ideal gas at the same temperature and pressure condition. The compressibility factor can be defined as

$$Z = \frac{\text{molar volume of real gas at same } T \text{ and } P}{\text{molar volume of ideal gas at same } T \text{ and } P} = \frac{PV}{RT} \tag{4}$$

Figure 3 shows plots of the compressibility factor of various working fluid with higher pressure range. It can be obviously seen that the ideal gas law is not suitable to describe the behavior of real gas as the pressure being higher range. Hydrogen (H) and Helium (He), *Z* are greater than one for all pressure state, while other working fluids (Nitrogen (N_2) and Oxygen (O_2)), *Z* are lower than one at relatively lower pressure and greater than one at high pressure state, which means these working fluids are more compressible and less compressible at lower and higher pressure states, respectively. In case of CO_2, on the other hand, the pressure increases largely as *Z* decreases to minimum, the value of which is approximately 0.27 at critical temperature and pressure point (Ahmed, 2006). At higher pressure, *Z* is kept at above one, where CO_2 is no longer compressible. At this condition, specific volume becomes smaller, while the density becomes greater compared with *Z* being lower than one.

Figure 3. The compressibility factor (Z) of various working fluids with pressure

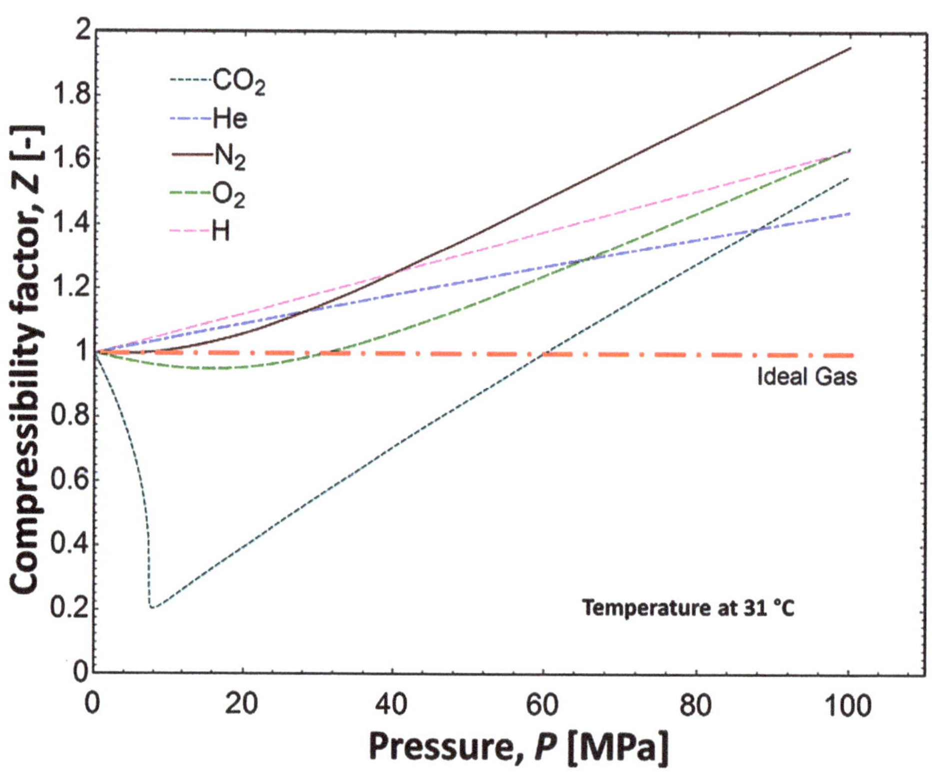

Figure 4. Enthalpy for temperature variation (a) 6 MPa and (b) 9 MPa

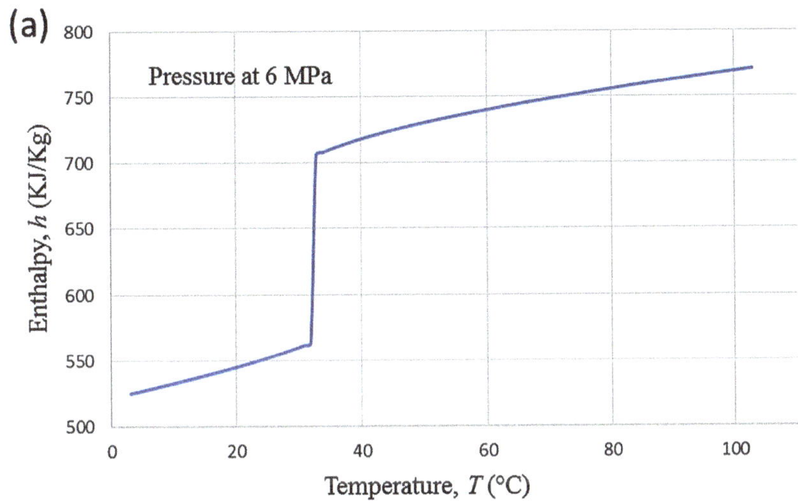

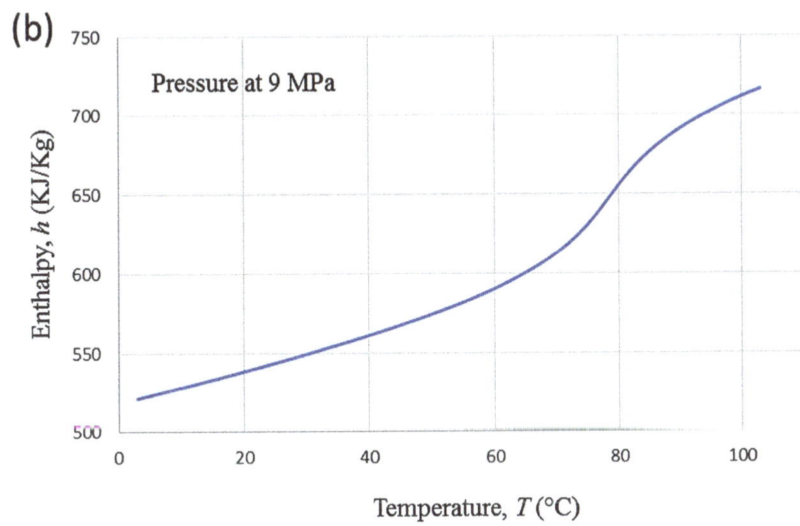

In a real gas, heat absorbing and emitting occurs when it undergoes two separate phase changes in an energy conversion process such as in an ordinary Rankine cycle. When the phase change process occurs in real gas, the internal energy and kinetic energy of the real gas are changed (Brown, 1980). The phase change of real gas refers to the molecular bond energy, while the temperature of the real gas is kept at constant. Figure 4(a) is a plot of the enthalpy with the temperature variation (constant pressure at 6 MPa), it is seen during the phase change, heat is added to the working fluid without increase in temperature as it generally observed in an ordinary Rankine cycle before superheating. However, in the supercritical Rankine cycle, the heat absorbing with enthalpy change (for temperature variation with constant pressure at 9 MPa) occurs as indicated in Figure 4(b), where CO_2 undergoes liquid to supercritical states (as this process is explained in details in later section), while heat emitting process follows the same as an ordinary phase change with constant temperature process.

As explained earlier, CO_2 used as a working fluid is operated at supercritical state in the Rankin cycle, in which properties are much deferent from ideal gas state (ideal gas as a working fluid). In the following section, the properties of CO_2 are discussed and clarified in details, in which the author focuses on the unique properties of CO_2 in its supercritical state.

CARBON DIOXIDE (CO_2) IN SUPERCRITICAL STATE

CO_2 has been regarded as a promising next-generation working fluid in various purposes due to its ecological and environmentally safely. Besides of high pressure and temperature supercritical state, which is used in the power generation cycle of Rankine cycle (in this study), it is mentioned here that a dry ice solid-gas state of CO_2 is also interesting phase for use in a heat pump cycle as a cooling/refrigeration mechanism, such as being investigated in CO_2 cascade refrigeration cycle system by Yamaguchi et al. (2008). The challenging heat pump system can achieve the temperature below -56.6 °C. The Moiler diagram (*P-h* diagram) of CO_2 is shown in Figure 5 (Yamaguchi et al. 2008), in which the critical point of CO_2 is marked (at temperature of 31.1 °C and pressure of 7.38 MPa), while the triple point is also marked at temperature of -56.6 °C and pressure of 0.518 MPa.

Supercritical fluid has much different properties compared to regular fluid in general, where the state of fluid is neither liquid state nor gas state (nor the mixture of the both), but it is a substance in a unique state, which exists above the critical point as showing in Figure 5. As also displayed in Figure 5, the pseudocritical line is the one that separates between liquid state and supercritical state. If a process at a constant pressure (such as described at $P = 9.0$ MPa) undertakes from (a) to (b), say the line (a) – (b), a crossing point at which the line (a) – (b) crosses the pseudocritical line, is so called the pseudocritical point.

Figure 5. Mollier (P-h) diagram of CO_2
Yamaguchi et al. 2008.

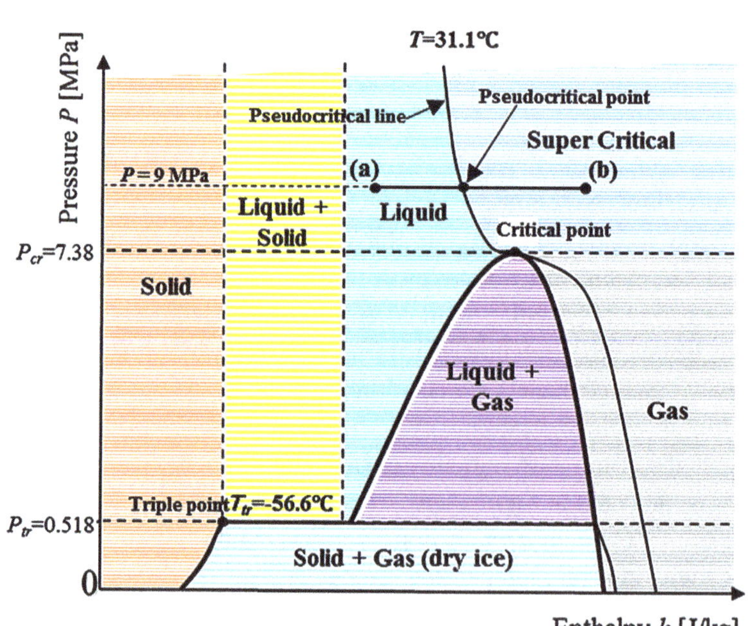

The supercritical state CO_2 at 9 MPa, which is a representative operation pressure in the present power cycle, system (SRCS) (Zhang, Chen, & Yamaguchi, 2010), has the remarkable energy transfer (heat transfer) characteristic in which its thermophysical properties exhibit rapid variation with a change in temperature, especially near the pseudocritical point (around 312 K) as indicated in Figure 6. It has the transitions in its properties, such as flow viscosity, thermal conductivity and density, as temperature is increased across the pseudocritical point at the given pressure, all the properties (except for c_p) decrease greatly. On the other hand, the specific heat (c_p) reaches the peak at pseudocritical point. This characteristic brings many different feature of heat transfer from the constant properties of ordinary fluids. It should be noted that, when temperature increases higher than the pseudocritical the region, CO_2 has slowly change in its properties, i.e. specific heat and density become smaller, while viscosity and thermal conductivity increase along the temperature variation due to the reason that supercritical CO_2 has highly compressible. The charge of representative properties of supercritical CO_2 at 9 MPa is shown in Figure 6, PROPATH (1997).

The forced convection heat transfer is a high mode of transferring the thermal energy of working fluid in the power cycle under consideration (the supercritical Rankine cycle), since the heat transfer efficiency in the forced convection heat transfer can be largely increased by operating the working fluid at a supercritical region. In the supercritical region, working fluid behaves much different, as seen in Figure 6, from an ordinary power cycle operation condition. The large change in the fluid properties

Figure 6. Various of thermophysical properties with temperature at 9MPa, density $\rho \times 10^{-2}(kg/m^3)$, specific heat $c_p \times 10^{-3}(J/Kg \cdot K)$, thermal conductivity $k \times 10^2(W/m \cdot K)$ and dynamic viscosity $\mu \times 10^5(Pa \cdot s)$ Zhang et al., 2010.

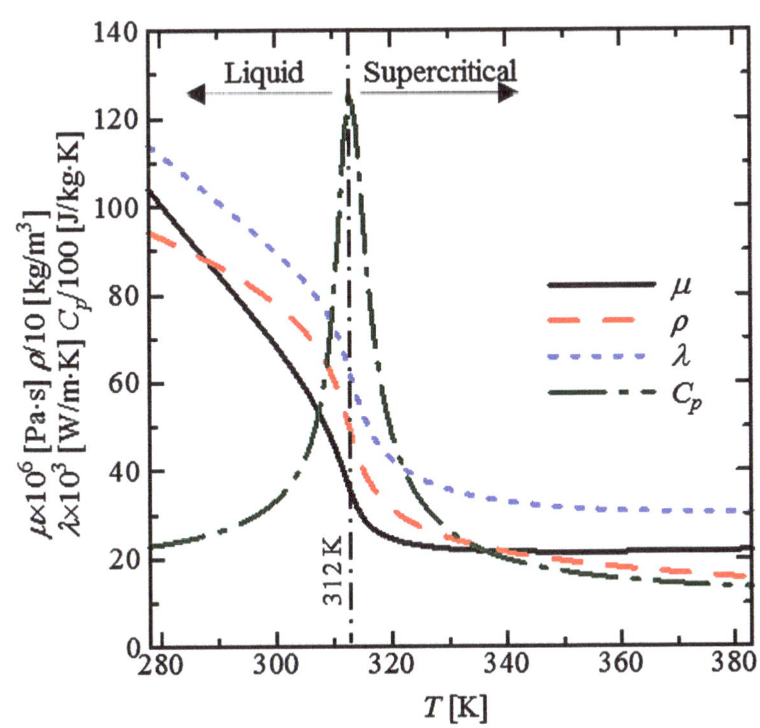

enhances the convection heat transfer coefficient of working fluid in the supercritical Rankine cycle, particularly the change in specific heat (c_p) as described below (Yunus & Afshin, 2011).

Figure 7 shows the temperature dependence of specific heat (c_p) of CO_2 at various pressures. It can be obviously seen that the specific heat is greatly increased at critical region (7 – 8 MPa) and at the critical pressure point (7.38 MPa), the specific heat goes much larger beyond the other values. By using CO_2 as a working fluid in the power cycle, the operation temperature and pressure is very important factors to achieve the highest output energy value as the specific heat is one of the big factors in heat transfer process. The Prandlt number (Pr) of CO_2, which is used to measure the heat transfer effectiveness in terms of fluid properties of the thermal conductivity and specific heat, is used to examine the convective heat transfer coefficient in the supercritical region. Similarly to Figure 7, in Figure 8, the temperature variation of Prandlt number of CO_2 at various pressures are plotted at supercritical region, the Prandlt number goes to high peak, in which it can be confirmed that the higher convective heat transfer coefficient of CO_2 at the pseudocritical point can be feasibly acclaimed.

SUPERCRITICAL CO_2 SOLAR RANKINE CYCLE SYSTEM (SRCS)

Similar to the supercritical Rankine cycle as described in Figure 2, the present SRCS is composed of four main components, namely evacuated solar collector, turbine, heat exchanger units and mechanical feed pump (or thermally driven pump) as displayed in Figure 9 (Zhang, Yamaguchi, Uneno, Fujima, Enomoto, & Sawada, 2006), where (a) is the schematic and (b) is the representative *P-h* diagram (cycle diagram). CO_2 contained in evacuated solar collector is heated by solar energy and turned into the super-

Figure 7. Specific heat (c_p) of CO_2 with various pressure

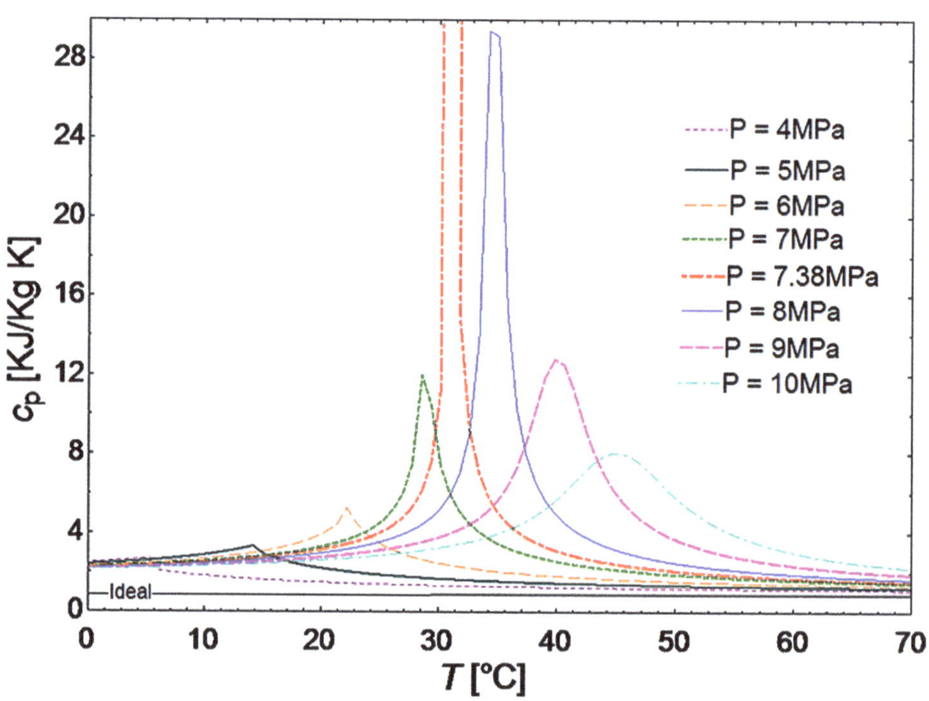

Figure 8. Prandtl number of CO_2 with various pressure

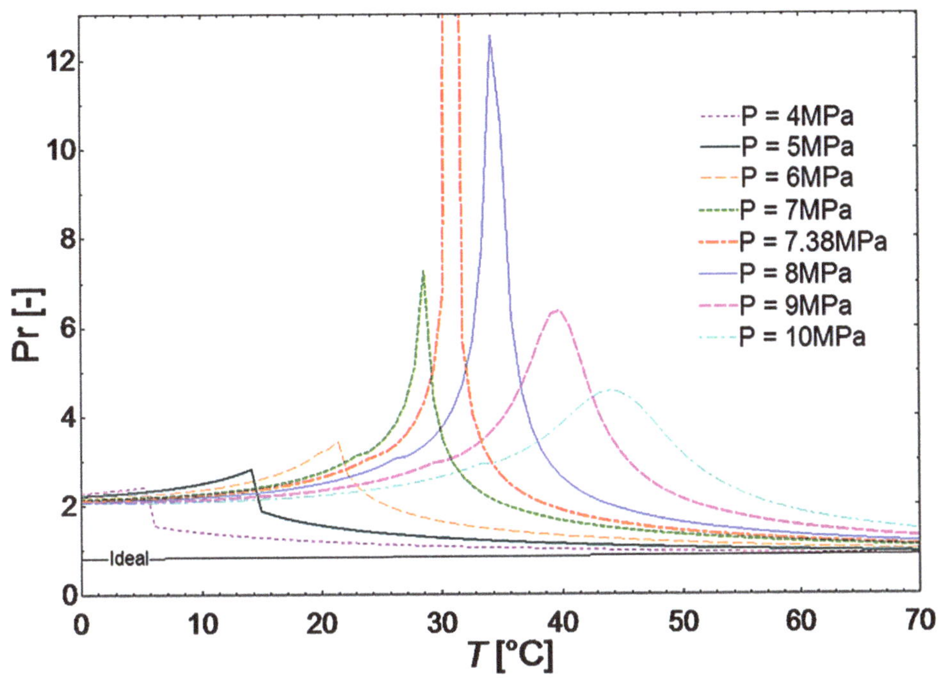

Figure 9. Supercritical CO_2 Rankine cycle system
Zhang et al., 2006.

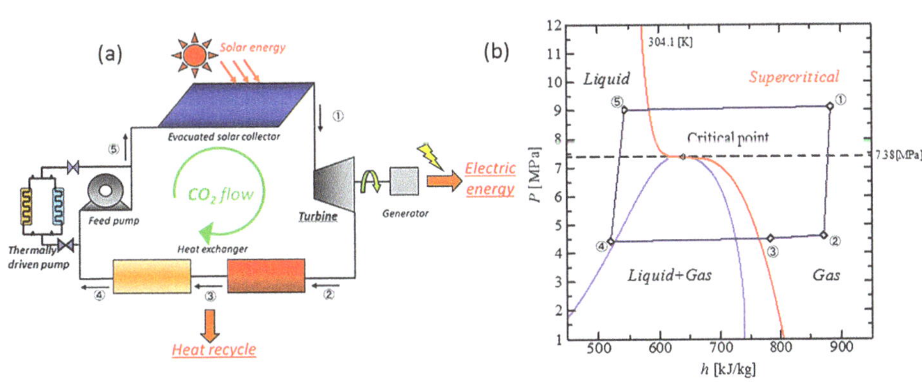

critical state. Owing to the expansion of supercritical CO_2 through a turbine, electric power is generated. After leaving from the turbine, waste heated in liquid CO_2 is absorbed by heat exchanger units and heat energy is supplied to appliances. Mechanical feed pump feeds the liquid CO_2 back to evacuated solar collector with high pressure state, and the cycle recommences. It can be seen in the cycle process of SRCS that both electric and thermal (heat) energy are generated at the same time from the system with environmentally friendly manner, namely from the solar energy.

The original SRCS was tested in the various weather conditions in Kyoto area installed at a roof top as seen in Figure 10, the system works well with the transcritical cycle in the test hours (Zhang et

Figure 10. Outlook of Supercritical CO_2 solar Rankine cycle system
Zhang et al., 2005.

al. 2006). The electric and heat generation output were recorded with an acceptable thermodynamic efficiency. Some details of the system components are provided in the following sections. Particular attention is given in the following section for the evacuated solar collector, which is the key component of the system. The loss due to the energy consumption and mechanical loss in the mechanical feed pump are discussed. As a solution for the problem of the mechanical feed pump, a new invention in pumping part, so called the thermally driven pump, is proposed and designed for the loss prevention. A designing view of supercritical CO_2 turbine of generating electric energy is also separately introduced in the following sections (Dostal, Driscoll, Hejzlar, & Todreas, 2002).

It is noted here that all thermo-physical properties of CO_2 used in this study is obtained from database PROPATH (1997) based on the actual experimental data of CO_2.

EVACUATED SOLAR COLLECTOR (NIU ET AL. 2011 & 2013)

The evacuated solar collector which was used as a heat absorption equipment in the SRCS, which consists of 15 units with total solar collective effective area of 1.43 m^2 per unit (Niu, Yamaguchi, Iwamoto, & Zhang, 2013). In order to obtain the highest performance of the SRCS, the optimal arrangement in 15 units of evacuated solar collector was studied as schematically displayed in Figure 11 (Niu et al. 2013), in which CO_2 mass flow rate at inlet of evacuated solar collector (⑤ in Figure 9(a) and Figure 9(b)), heat collection efficiency and temperature at outlet of evacuated solar collector were examined in the study. The evacuated solar collector was arranged in three arrangements, as shown in Figure 11, where (a), (b) and (c) are respectively; (a) parallel with each 5 units, (b) cascade with each 5 units in series and (c) 5 units only in series. The performance of SRCS with three independent arrangements of evacuated solar collector was evaluated by considering the heat collection amount (Q_c), CO_2 mass flow rate ($\dot{m}$) at

Figure 11. Three arrangements of evacuated solar collector
Niu et al., 2013.

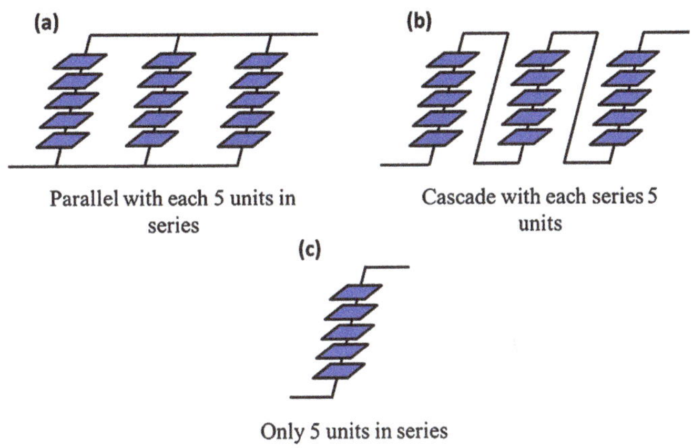

collector's inlet and temperature (T_I) at the outlet of evacuated solar collector. Also, the solar radiation intensity (I) is measured for the reference. Q_c is calculated with the enthalpy (h) of CO_2 between inlet and outlet of evacuated solar collector (as positioned in Figure 9(a) and Figure 9(b)) obtained from measured temperature and pressure by using PROPATH database (1997), the formulas used for the evaluation is as follows (Niu et al. 2013)

$$Q_c = \int \dot{m}(h_1 - h_4)dt \tag{5}$$

The thermal energy (heat) gain from solar radiation (Q_{SR}) was evaluated based on the total solar collective effective area and solar radiation intensity, which can be defined as

$$Q_{SR} = \int AIdt \tag{6}$$

For calculating the heat collection efficiency η_c can be found by

$$\eta_c = \frac{Q_c}{Q_{SR}} \tag{7}$$

The results of heat collection efficiency, solar radiation, CO_2 temperature at the outlet of evacuated solar collector and CO_2 mass flow rate at inlet of evacuated solar collector with optimum time for the SRCS with three arrangements of evacuated solar collector units were obtained and the performance was evaluated in a typical sunny day in winter season of Kyoto, Japan. Figure 12(a) and Figure 12(b) show the results obtained from evacuated solar collector with the arrangement of parallel with each 5 units in series. In Figure 12(a), the amount of solar radiation and heat collection efficiency were plotted for the operation time. In the morning period, the amount of solar radiation increases from 0.09 kW/m^2 at the beginning to a high value of 0.5 kW/m^2, and vice versa at noon till the zero at sunset. The heat

Figure 12. (a) Heat collection efficiency and solar radiation, and (b) CO_2 temperature at the outlet of collectors and CO_2 mass flow rate at the outlet of the feed pump against the operation time from 7:00 to 17:00 measured in the SRCS with the solar collector units having parallel arrangement with each 5 units in series
Niu et al., 2013.

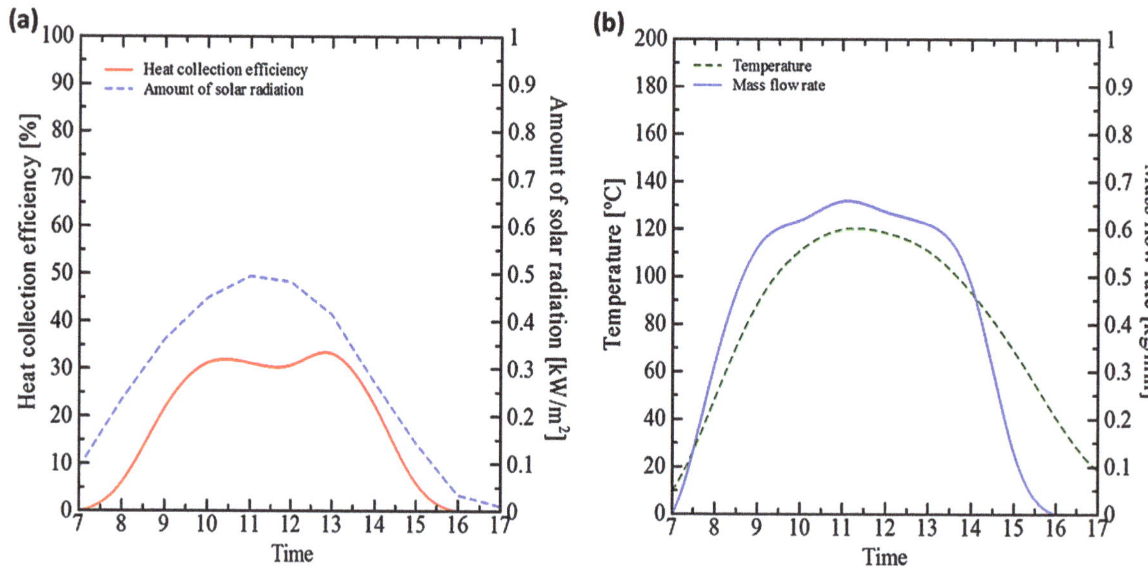

collection efficiency also increases from the beginning of operation and stays fairly stable at time 9:00 to 14:00, which is not so depended with the solar radiation at that time, and then decreases to zero at time around 16:00. Corresponding with Figure 12(a), Figure 12(b) shows CO_2 mass flow rate at inlet of evacuated solar collector achieved the highest value of approximately 0.6 kg/min during time 9:00 to 14:00. The CO_2 temperature at the outlet of evacuated solar collector is raised from the beginning from 10 °C to be the highest temperature of 110 °C at 11:30 and decreases to 20 °C at last measuring here.

Similar to Figure 12, the results of the SRCS with the cascade arrangement of evacuated solar collector with each series 5 units are shown in Figure 13. It can be found that the results in the cascade arrangement have similar behavior with the parallel arrangement (Figure 12). However, the cascade arrangement can be achieved higher value of heat collection efficiency (40%), CO_2 temperature at outlet of evacuated solar collector (154.5 °C) and CO_2 mass flow rate at inlet of evacuated solar collector (0.75 kg/min) compared with the parallel arrangement.

To investigate the effect of heat collection area, Figure 14(a) and Figure 14(b) show the results of SRCS with the arrangement of evacuated solar collector in only 5 units in series. The heat collection efficiency and CO_2 mass flow rate at outlet of evacuated solar collector are found much higher than in parallel arrangement (Figure 12) and cascade arrangement (Figure 13), in which the highest values are obtained as 90% and 0.85 kg/min, respectively. The reasons may be thought from the fact that the flow pathway of CO_2 in 5 series is shorter than other, thus the CO_2 mass flow rate is larger compared with other two arrangements with same amount of CO_2 in the system. Therefore, since the heat collection area is smaller than others, the fact leads the CO_2 temperature at the outlet of evacuated solar collector become lower than others with the highest value of 90 °C. From the results of three arrangements for the

Figure 13. (a) Heat collection efficiency and solar radiation, and (b) CO_2 temperature at the outlet of collectors and CO_2 mass flow rate at the outlet of the feed pump against the operation time from 7:00 to 17:00 measured in the SRCS with the solar collector units having cascade arrangement with each 5 units in series
Niu et al., 2013.

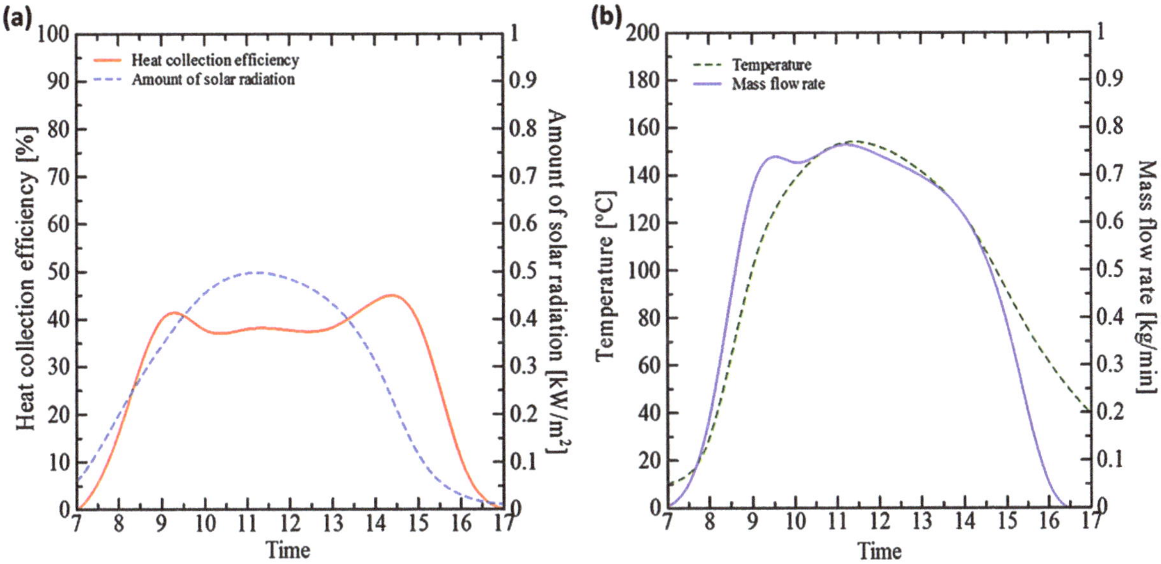

Figure 14. (a) Heat collection efficiency and solar radiation, and (b) CO_2 temperature at the outlet of collectors and CO_2 mass flow rate at the outlet of the feed pump against the operation time from 7:00 to 17:00 measured in the SRCS with the solar collector units having only 5 in series
Niu et al., 2013.

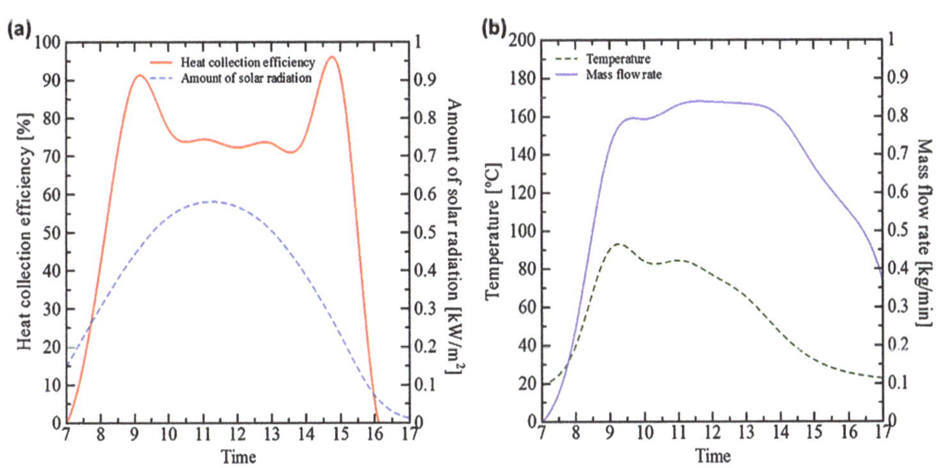

aim of electric power generation and heat utilization, it can be suggested that the cascade arrangement in series would give high performance in the CO_2 temperature and the CO_2 mass flow rate at outlet and inlet of evacuated solar collector, respectively.

The performance of evacuated solar collector is not only depended on the construction or collectors' arrangements itself, but also on the flow behavior of working fluid (CO_2) inside the evacuated solar collector, in which active forced convection heat transfer is taking place (as mentioned earlier). From the *P-h* (Moiler) diagram of SRCS (in Figure. 9(b)), it can be seen that the working fluid, CO_2, undergoes the state change from liquid to supercritical state. As mentioned earlier, CO_2 particularly near to the critical point, the supercritical state properties enhance strong convective flows, which improve heat transfer characteristic largely. In order to optimize operation condition of the supercritical CO_2 in the evacuated solar collector, the knowledge of local heat transfer characteristic of supercritical CO_2 inside the solar collector is necessary to verify in details (Niu, Yamaguchi, Zhang, Iwamoto, & Hashitani, 2011)

However, the difficulty arises to the direct measurement of the evacuated solar collector, due to the complex channel geometry of the collector and system path, where CO_2 is enclosed in the pressurized state.

In order to obtain fundamental data, the test section of a simple geometry is used instead of evacuated solar collector to simulate the supercritical CO_2 flow experimentally, the schematic diagram of the experimental facility is shown in Figure 15. The detailed structure of the test section ① is shown in Figure 16, in which a stainless tube heated by silicon rubber heater to simulate the solar radiation is provided in the basic Rankine cycle. The heat transfer coefficient in the heating test section at different operation condition of the system was investigated in the study. The local heat transfer coefficient of supercritical CO_2 (h_x) in heating tube (test section) can be evaluated by measuring temperature at the inner wall of tube surface (T_{IW}), temperature at inlet of heating tube (T_{IN}) and input heat flux of heating tube (Q_r). h_x is thus expressed by

Figure 15. Schematic diagram of the experimental facility
Niu et al., 2011.

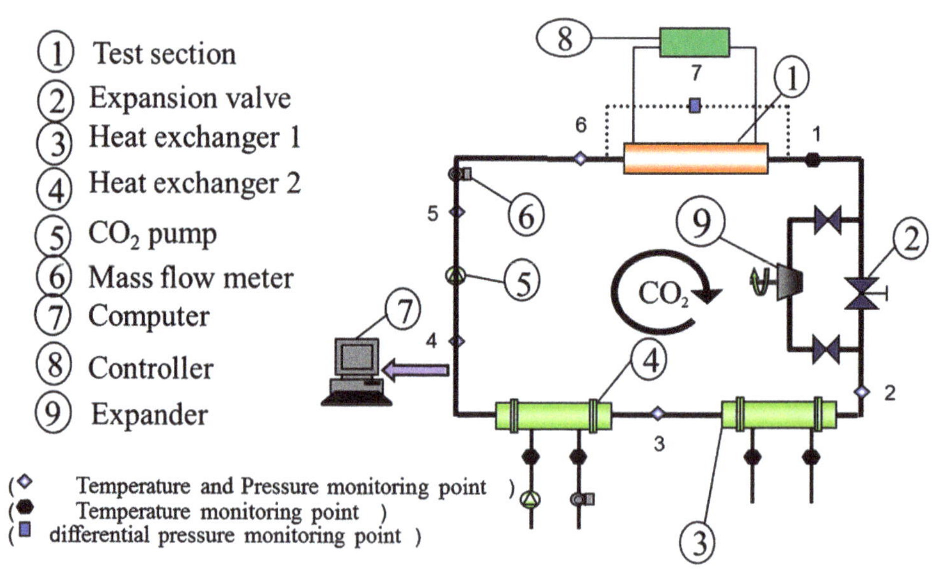

Figure 16. Test section
Niu et al., 2011.

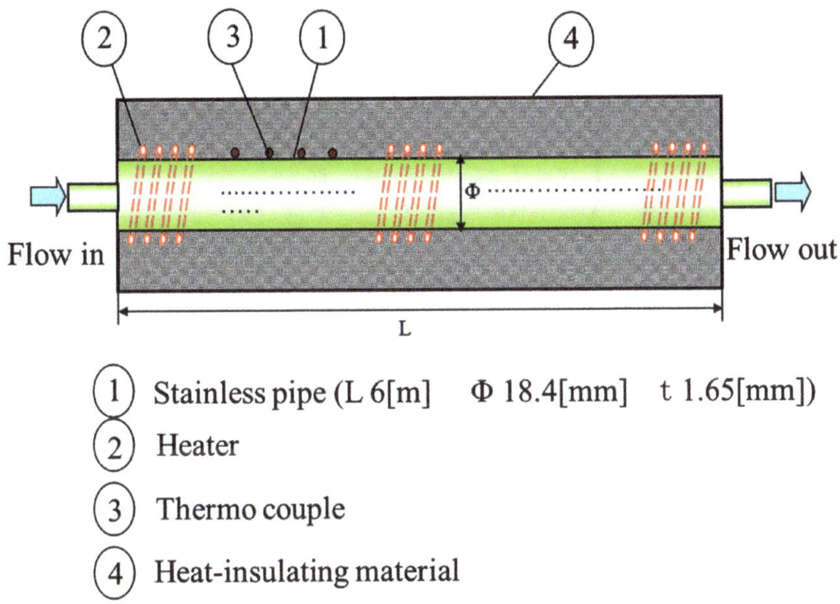

$$h_x = \frac{Q_r}{T_{IW} - T_{IN}} \tag{8}$$

where T_{IW} can be defined as

$$T_{IW} = T_{OW} - \frac{Q_r d_1}{2\lambda_{SUS}} \ln\left(\frac{d_2}{d_1}\right) \tag{9}$$

and where T_{OW} is outer wall temperature of the test section (positioned ③ in Figure 16), λ_{SUS} is the thermal conductivity of the stainless heating tube, and d_1 and d_2 are inner and outer diameter of heating tube, respectively.

In Figure 17, representative variation of local heat transfer coefficient of the liquid to the supercritical state in the tube is plotted. It is seen that the bulk temperature (T_b) is increased along the tube, and reaches the critical temperature through the entrance region, in which it keeps stable in the middle of tube. Then T_b increases above the critical temperature (T_{pc}) at the outlet region. Heat transfer coefficients of CO_2 (h_x) quickly decreases at the inlet region along the heating tube, but keeps at high value due to higher Prandtl number (Pr) (high heat transfer mode) and reaches the first peak at the peak point of Prandtl number, $x/L = 0.3$. The second peak is found at $x/L = 0.7$, and it might be caused by the heat transfer with hydrodynamic interaction phenomena (Shiralkar & Griffith, 1969)

To confirm the effect of pressure and mass flow rate to the heat transfer coefficient, meticulous tests for CO_2 mass flow rate ($\dot{m}$) of 19 kg/m^{2}s and 38 kg/m^{2}s with three diffidence inlet pressure (7.8, 8.6 and 10.4 MPa) were conducted and results are presented in Figure 18(a) and Figure18(b), respectively.

Figure 17. Variation of local heat transfer coefficient h_x, of the CO_2 along the heated tube at $Q_r = 3.6 \times 10^3$ W/m² and $\dot{m} = 25$ kg/m²s
Niu et al. 2011.

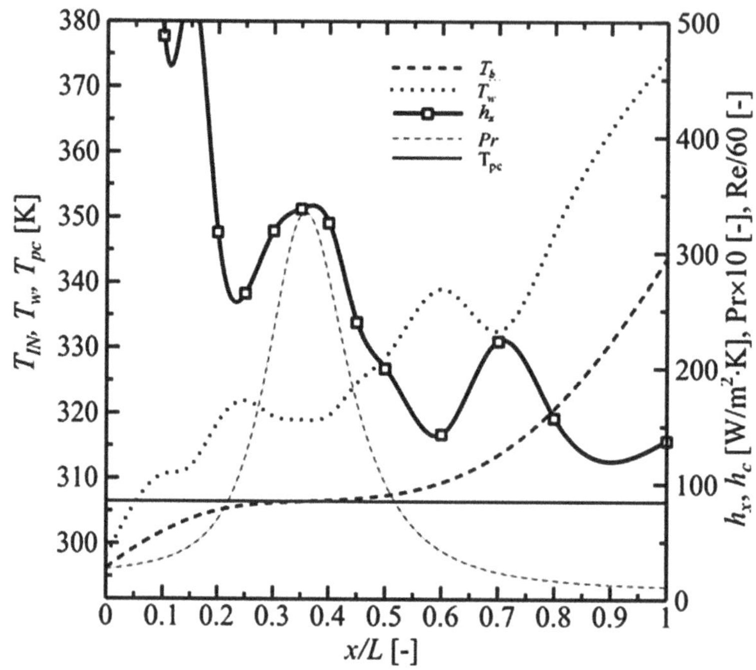

From both results in Figure 18(a) and (b), the trend of heat transfer coefficient is found to be the same as Figure 17. The heat transfer coefficient decreases when pressure increase, the highest values were found at pressure of 7.6 MPa. From the results of the fundamental tests (Figure 17 and Figure 18), it was found the general trend that the increasing of mass flow rate improves the heat transfer coefficient of supercritical CO_2 in the heating tube. It is further suggested that the higher mass flow rate and the inlet pressure near the critical pressure give the higher value of heat transfer coefficient with twin peaks (the first exists one at pseudocritical point and the second one exists at down steam close to the exist) in the actual evacuated solar collector (test section). The appearance of the second peak seems the general trend and would be caused by a hydrodynamic interaction of same degree. No detailed explanation of the cause has been clarified yet.

SUPERCRITICAL CO_2 TURBINE (CHEN ET AL. 2015 & PUMANERATKUL ET AL. 2015)

In the SRCS, turbine is one of the key components in terms of electric power generation. However, at present, there is no specific compact turbine for (supercritical CO_2) available in commercial market yet. The supercritical CO_2, which is used as a working fluid in SRCS, is highly compressible with large density and viscosity variation. The specific design of the turbine for supercritical CO_2 is necessarily required for obtaining high performance electric power generation from SRCS. The centrifugal turbine, which is intended to be used in the present SRCS, was investigated experimentally and analytically for

Figure 18. Heat transfer coefficients at different pressures at constant K and (a) $\dot{m}$ = 19 kg/m²s and (b) $\dot{m}$ = 38 kg/m²s
Niu et al., 2011.

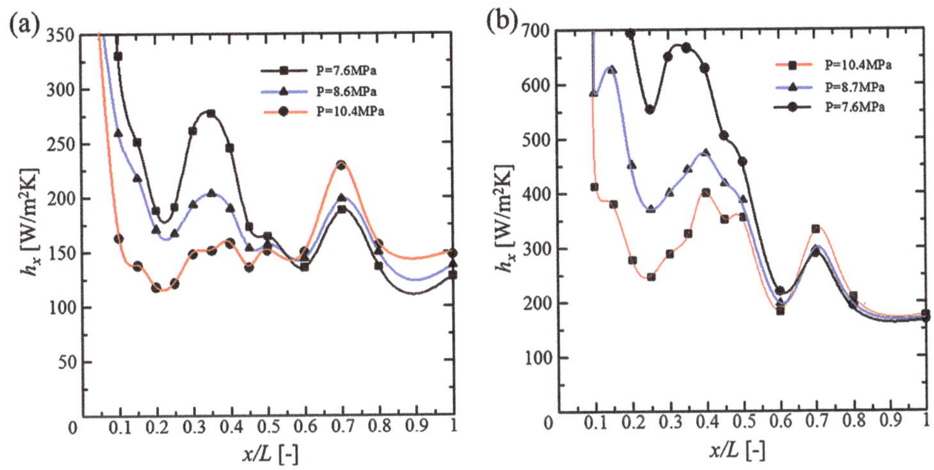

development of a specific functional design purpose (Chen, Yamaguchi, Zhang, & Niu, 2015). The detailed schematic of the prototype turbine and nozzle geometry are shown in Figure 19(a) and Figure 19(b), respectively. As shown in Figure 19(a), the shell of prototype turbine ④ is made of stainless steel SUS304 for enduring the high pressure in the operation, and which is also designed to avoid leakage of high pressure supercritical CO_2. The magnet coupling ③ consists of four magnets, which are made of neodymium with two on each side. The dimension and magnetic flux density of each magnet are ø10×20 mm and 560 mT, respectively. The detail geometry of the test nozzle is shown in Figure 19(b), the length of connecting part is 0.25 inch and the test nozzle is connected with the flow cycle path from the heating tube (test section) or SRCS evacuated solar collector. The guide vane, which is normally used in ordinary turbine for adjust the flow angle of fluid is not necessary in this first step of turbine design, the nozzle is used instead. The various diameter of nozzle from 0.5 mm to 1.0 mm were used with the length up to 40 mm (due to the limit of turbine design).

In order to give the efficientness and good performance of the turbine as shown in Figure 19(a), the nozzle design, to which detailed knowledge of flow phenomena is necessary and essential, is the prime importance.

A numerical analysis was conducted herewith experiment for the nozzle design, and those are introduced in this section. The governed equations of compressible (one dimensional) flow are solved numerically (by the finite difference method), where Fanno flow is assumed to examine the flow in a microtube (nozzle) as follows (Yamaguchi, 2008)

$$\frac{d\rho}{\rho} + \frac{du}{u} = 0 \tag{10}$$

$$\frac{du}{u} + \frac{1}{kM^2}\frac{dp}{p} + \frac{c_f}{2}\left(\frac{4}{d_n}\right)dx = 0 \tag{11}$$

Figure 19. Detailed schematic of: (a) the prototype turbine; and (b) test nozzle
Chen et al., 2015.

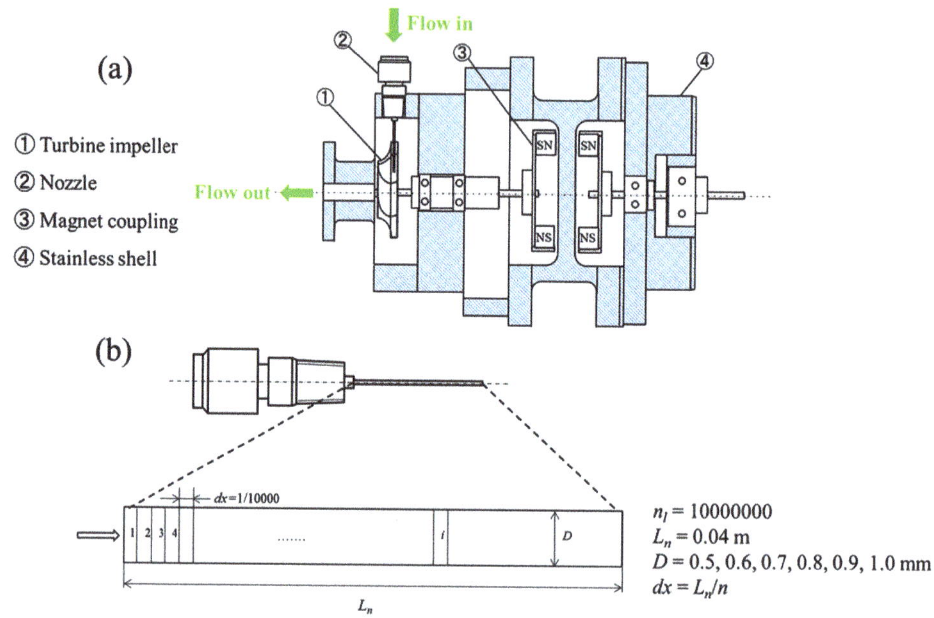

$$\frac{dT}{T} + \left(k-1\right) M^2 \frac{du}{u} = 0 \tag{12}$$

where u, P, ρ and T are velocity, pressure, density and temperature of CO_2 flow, respectively. M and γ is Mach number and specific heat ratio, respectively. It is noted that all CO_2 properties at given states are derived from the database PROPATH (1997).

The relationship between Fanning friction coefficient and Darcy friction coefficient (λ) can be described as (assumed CO_2 as gas state; there was no theory or experimental correlation is available at the present time)

$$\lambda = 4f \tag{13}$$

where $\lambda = 0.3164\times Re^{-0.25}$ at the Reynolds number $3.0\times10^3<Re<10^5$, and $\lambda = 0.0032+0.221\times Re^{0.237}$ at the Reynolds number $10^5<Re<10^6$ (Yamaguchi, 2008).

The pressure (P) can be obtained from the equation state as

$$P = \rho RTZ \tag{14}$$

with Compressibility factor for CO_2 (Z) is based on Angus EOS (Angus et al. 1976) as

$$Z = 1 + \frac{\rho}{\rho_c}\sum_{i=1}^{6}\sum_{j=1}^{9} a_{ij}\left(\frac{\rho}{\rho_c} - 1\right)^i\left(\frac{T_c}{T} - 1\right)^j \tag{15}$$

where $\rho_c = 468$ kg/m^3 and $T_c = 304.2$ K, which are CO_2 critical density and critical pressure, respectively, the coefficients a_{ij} of Angus EOS for CO_2 is tabulated in Table 2.

To calculate Mach number in the nozzle, the speed of sound (w_s) can be derived from

$$w_s^2 = \left(\frac{c_p}{c_v}\right)\left(\frac{\partial P}{\partial \rho}\right)_T \tag{16}$$

and

$$\left(\frac{\partial P}{\partial \rho}\right)_T = RT \times \left[1 + 2\omega\sum_{i=0}^{6}\sum_{j=0}^{9} a_{ij}\left(\tau - 1\right)^j\left(\omega - 1\right)^{i-1}\right] \tag{17}$$

where the parameters are defined as, $\omega = \rho/\rho_c$ and $\tau = T_c/T$.

The following set of equations (Yamaguchi, 2008) are resultantly derived to solve with the finite difference technique, defining Mach number as $M = u/w_s$

$$\frac{du}{u} = \frac{kM^2\psi}{1 - M^2}\frac{4c_f}{d_n}dx, \psi = 1 + \frac{k-1}{2}M^2 \tag{18}$$

Table 2. Coefficients for a_{ij}

i	j: 0	1	2	3	4	5	6
0	-0.725854437	-1.68332974	0.259587221	0.376945574	-0.67075537	-0.871456126	-0.149156928
1	0.447869183	1.26050691	5.96957049	15.4645885	19.4449475	8.64880497	0
2	-0.172011999	-1.83458178	-4.61487677	-3.81121926	3.6171349	4.92265552	0
3	0.004463049	-1.76300541	-11.1436705	-27.8215446	-27.168572	-6.42177872	0
4	0.255491571	2.37414246	7.50925141	6.61133318	-2.4266321	-2.57944032	0
5	0.05946673	1.16974683	7.4370641	15.0646731	9.57496845	0	0
6	-0.14796001	-1.69233071	-4.68219937	-3.13517448	0	0	0
7	0.013671044	-0.10049233	-1.63653806	-1.87082988	0	0	0
8	0.039228458	0.441503812	0.88674197	0	0	0	0
9	-0.01198721	-0.084605195	0.046456437	0	0	0	0

Thermodynamic Tables Project, 1976.

$$\frac{d\rho}{\rho} = -\frac{kM^2\psi}{1-M^2}\frac{4c_f}{d_n}dx \tag{19}$$

$$\frac{dp}{p} = -\frac{kM^2\left[1+\left(k-1\right)M^2\right]}{2\left(1-M^2\right)}\frac{4c_f}{d_n}dx \tag{20}$$

$$\frac{dM}{M} = \frac{kM^2\left\{\left(k-1\right)M^2+2\right\}}{1-M^2}\frac{c_f}{d_n}dx \tag{21}$$

The choked flow in the nozzle of the turbine was found to occur as Mach number equals to one along the nozzle length, and in effect it limits the CO_2 flow, and resultantly the efficiency of turbine (Miller, 1983) The turbine nozzle with various diameter were examined for designing purpose in the study. The results are shown in Figure 20. It is found that the choked flow occurred at the diameter of 0.5 mm to 0.6 mm in this particular nozzle design. By 0.5 mm, total choked flow occurs at the beginning of the entrance region around $L = 0.015$ m, while for 0.6 mm, the choked flow can be found occurring around 0.016 m of the nozzle length. It is suggested that the nozzle diameter bigger than 0.6 mm is appropri-

Figure 20. Variation of Mach number along the turbine nozzle
Chen et al. 2015.

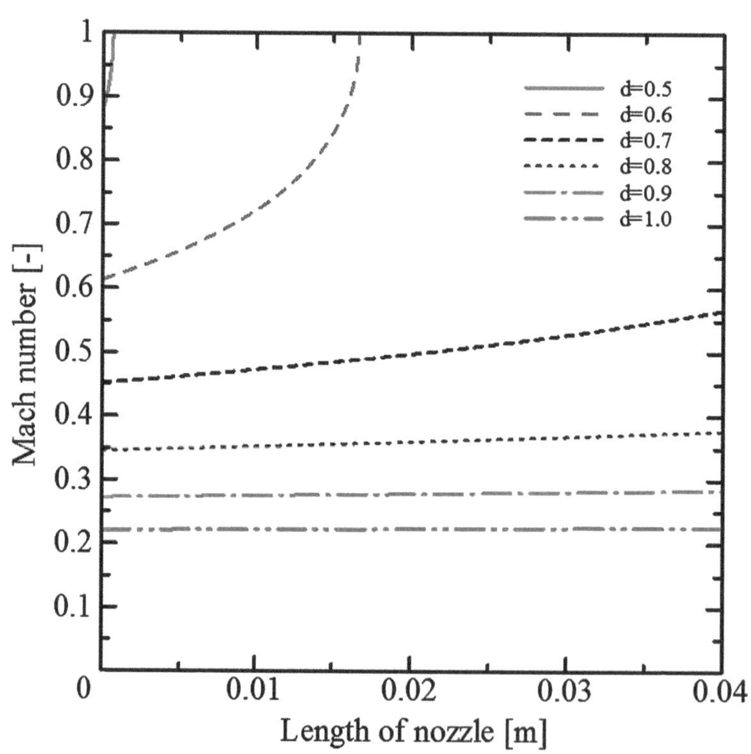

ate for designing this specific supercritical turbine with the specific rotor mode as adapted in the study (Chen et al. 2015) It is mentioned here that the approach adopted in the study is valid for any specific design of turbine in this type in general.

It is further examined from the analysis of the specific supercritical turbine, where the density, pressure and velocity of CO_2 in nozzle were calculated for a sake of examining performance characteristic, with the representative nozzle diameter of 1.0 mm with the nozzle length up to 40 mm, as shown in Figure 21. It is found that the velocity increases almost linearly along the nozzle length as the density and pressure decreasing, indicating that CO_2 flow properties in the nozzle are almost linear and Mach number of CO_2 flow shows slightly increasing along the nozzle (as verified in Figure 20).

Based on the numerical study, which was applied to the actual test machine (turbine), Figure 22 shows experimental results of actual turbine rotational speed with designed three nozzle length (25, 29 and 33 mm). As seen in Figure 22, the rotational speed is substantially increased by increasing nozzle length as well as the pressure. The rotational speed of turbine is directly concerned with electric power generation from the SRCS and in the study more than 3000 rpm (with free of load) of the test turbine was achieved for the nozzle of d_n = 1.0 mm with the length of maximum of 33 mm. From the prime study, a fundamental design point of the specific supercritical CO_2 turbine was gained for a compact centrifugal rotor (Pumaneratkul, Iwamoto, & Yamaguchi, 2015).

Figure 21. Numerical result of CO_2 thermal physical property in nozzle (d_n = 1.0 mm)
Pumaneratkul et al. 2015.

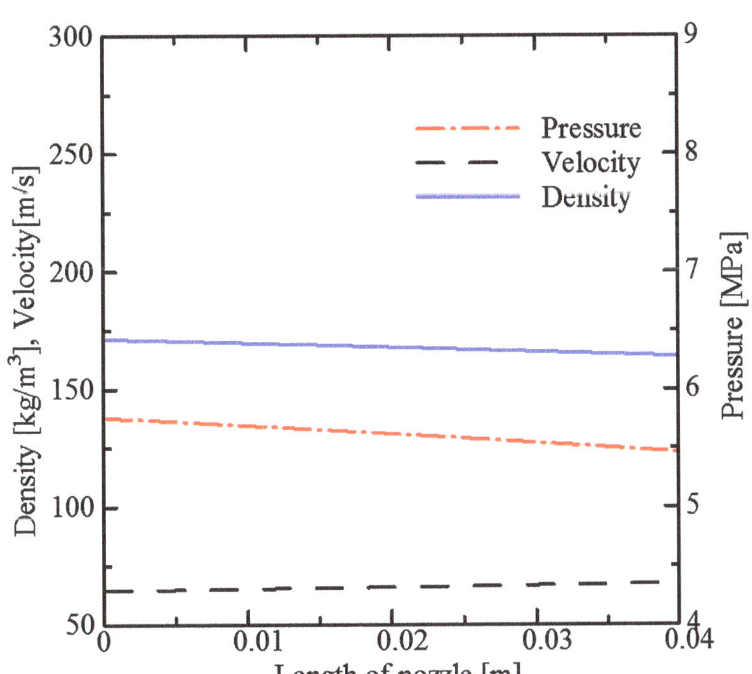

Figure 22. Experimental results of rotational speed of turbine for nozzles of 25, 29 and 33 mm (d_n = 1.0 mm)
Pumaneratkul et al. 2015.

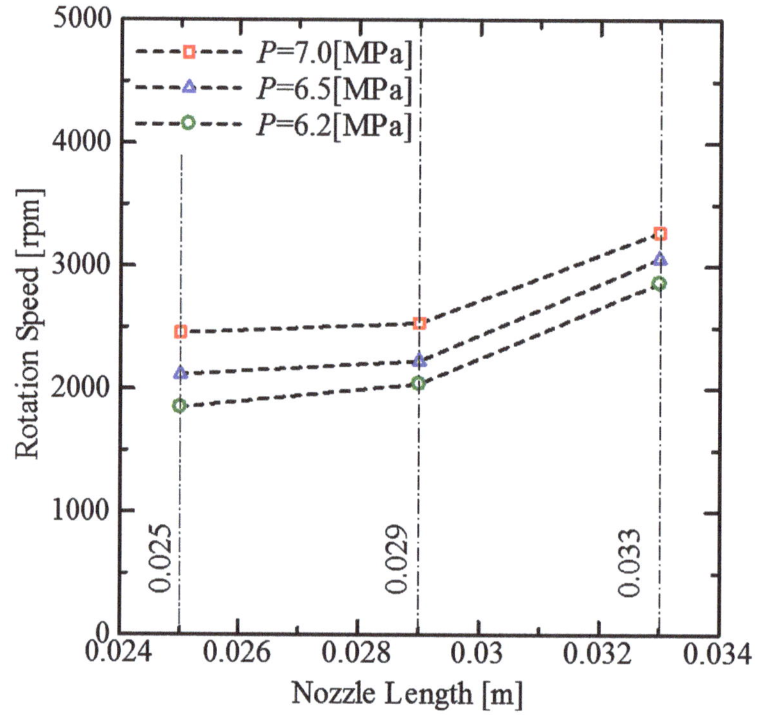

THERMALLY DRIVEN PUMP (YAMAGUCHI ET AL. 2013, PUMANERATKUL ET AL. 2015 & KUWAHARA ET AL. 2011)

In the typical SRCS, mechanical feed pump is used to feed liquid CO_2 back to evacuated solar collector at high pressure state. However, the mechanical feed pump consumed very high rate of energy and also has large mechanical loss, which decreases the total efficiency of the system (Yamaguchi, Zhang, Niu, & Hashitani, 2013). The thermally driven pump was newly introduced to be used for the replacement of the mechanical feed pump under developing stage of SRCS. The detailed of thermally driven pump is shown in Figure 23(a).

As shown in Figure 23(a), the thermally driven pump is a refrigerant (liquid CO_2)-circulation pump, which is composed of two expansion tanks. The substantial pressure difference between thermally driven pump and outside of the pump (evacuated solar collector and heat exchanger units) are generated by using hot and cool water (secondary circulation fluid) for heating and cooling CO_2 inside the tank. The pressure difference drives CO_2 without any mechanical movement and electric power required. Because of the superiority, a prototype thermally driven pump was newly installed and tested to be utilized in the actual Rankine cycle instead of the mechanical feed pump. The working principle of thermally driven pump are described below (referring 1 - 4 operation process as displayed in Figure 23(b)):

Figure 23. Schematic of (a) novel-concept thermally driven pump in test set-up and (b) its working principle
Yamaguchi et al. 2013.

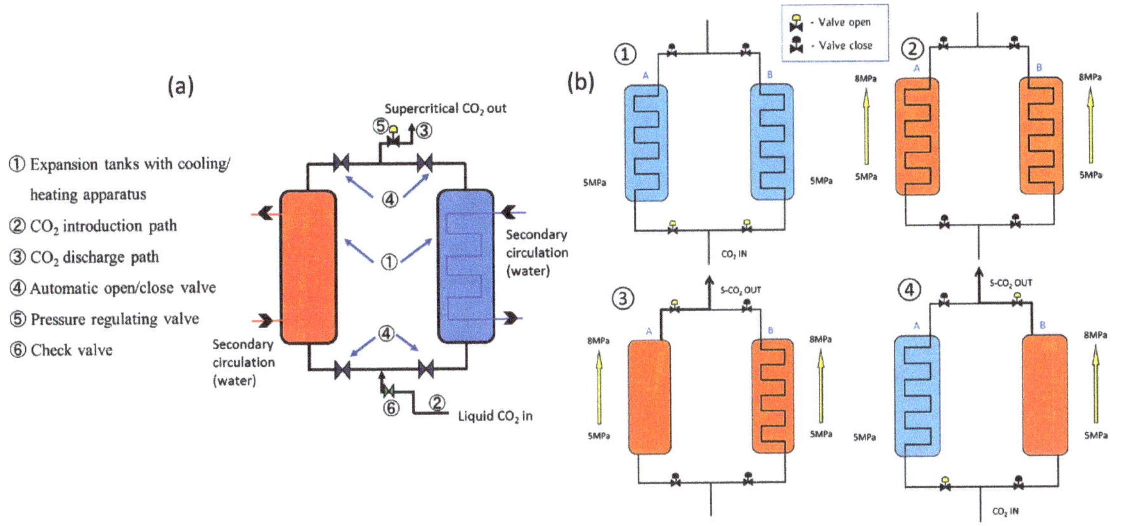

1. When lower valves are opened, low pressure CO_2 (5 MPa) flows from heat exchanger by the pressure difference.
2. When lower valves are closed, hot water is heated CO_2 to high pressure (5 MPa to 8 MPa), CO_2 turns into supercritical state.
3. When upper valve of tank A is opened, supercritical CO_2 (8 MPa) flows to evacuated solar collector by the pressure difference. Then cold water is used to cool down tank A (where pressure is reduced to lower than 5 MPa).
4. When upper valve of tank B is opened to let supercritical CO_2 flow, lower valve of tank A is opened to let low pressure CO_2 (5 MPa) flow in the tank. And cycle recommences.

The experimental results were obtained and representative data are displayed for the outlet flow from the thermally driven pump as it was installed in the CO_2 supercritical Rankine cycle for a trail test. The resultant pressure and temperature at the exit point of thermally driven pump are shown in Figure 24 and 25 (Pumaneratkul, Iwamoto, Niu, Kuwahara, & Yamaguchi, 2015), respectively. In Figure 24, the outlet flow (mass flow rate) of CO_2 can be found as high as 60 kg/h at the peak point. However, as observed the peak point is occurred in only very short period of time, due to the nature of the function of the thermally driven pump (batch processing) and the interval time between each peak is found to be quite long due to the nature of the operation. Because of these reasons, the CO_2 outlet flow from thermally driven pump is not continuous. The pressure and temperature of CO_2 at the exit point of thermally driven pump from the same experimental are also plotted in Figure 25. It is found that, at the peak points, both temperature and pressure are achieved to be in the supercritical region. These feeding mechanism (as shown in Figure 24 and Figure 25) improves the performance of SRCS, because CO_2 turns to be supercritical state before entering the evacuated solar collector. This phenomenon can be confirmed by the *P-h* (Mollier) diagram of SRCS with thermally driven pump as indicated in Figure 26 (Yamaguchi et al. 2013), the large enthalpy (Δh) occurs between state ③ to ④. In which CO_2 is directly turned into

Figure 24. CO_2 mass flow rate in the thermally driven pump
Pumaneratkul et al. 2015.

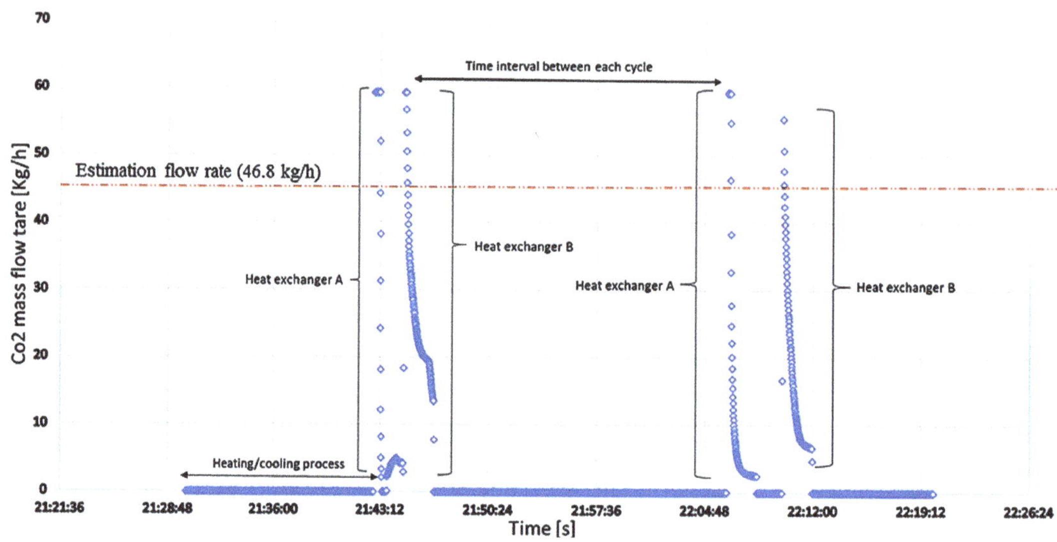

Figure 25. Pressure and temperature of CO_2 in thermally driven pump
Pumaneratkul et al. 2015.

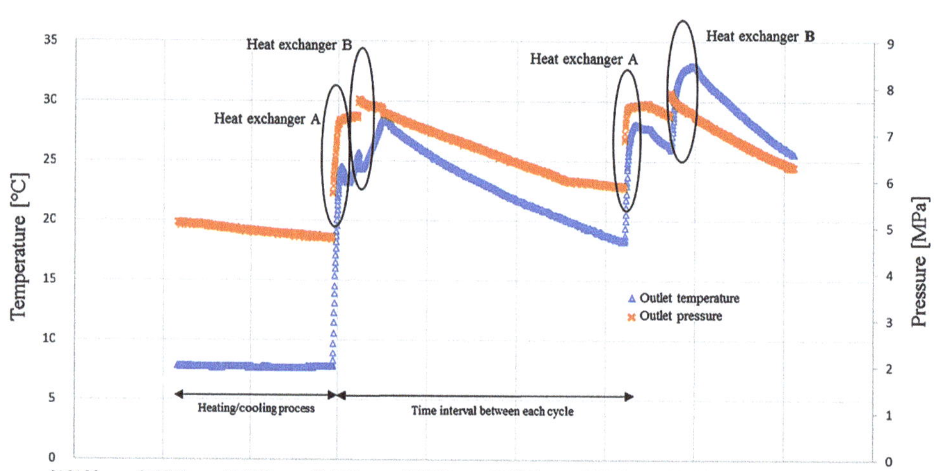

the supercritical state from low pressure state without passing to the high pressure state, which does not occur in ordinary SRCS operation process (Figure 9(b)).

To deal with the problem in non-continuous flow of CO_2 with thermally driven pump, which is directly related to the performance of the system itself, the simple one-dimension quasi flow model is considered to investigate the flow of the heat exchanging process in the thermally driven pump. The basic design of thermally driven pump dependent on designing a spiral flow pathway, where unique properties of supercritical CO_2 flow passes. By the working principle, the elevated enthalpy Δh with associated mass flow rate enhances the performance of pumping process of the pump. The flow of CO_2 is

Figure 26. P-h diagram of supercritical CO_2 Rankine cycle system with thermally driven pump
Yamaguchi et al. 2013.

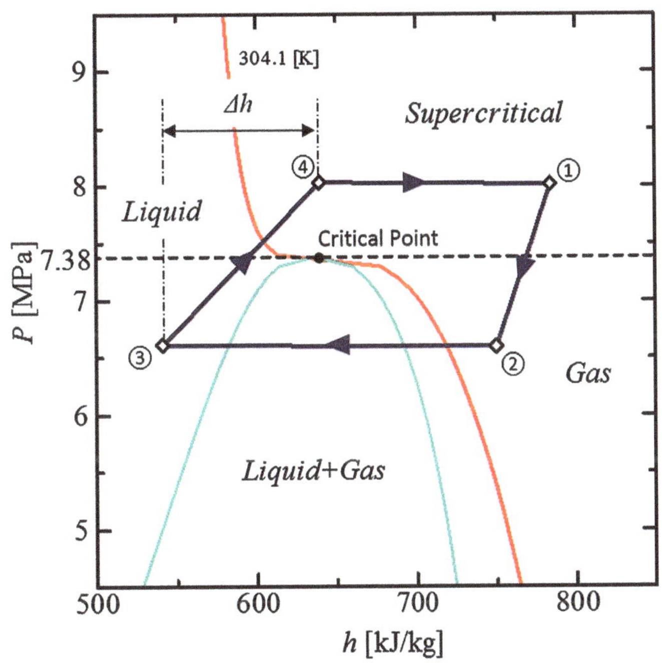

assumed to be laminar flow and the mass flow rate (by the thermally driven pump) can be approximated by the following manner.

The energy conservation equation, which relates moving fluid's pressure (*P*), density (ρ), velocity (v), height (y), and also enthalpy (h) and friction loss (f) in the fluid flow in the tube, can be written by the one-dimensional energy equation from inside it the thermally driven pump (Point 1 in Figure 27) to outlet of the thermally driven pump (Point 2 in Figure 27) as

$$P_1 + \frac{1}{2}\rho_1 v_1^2 + \rho_1 g y_1 + h_1 = P_2 + \frac{1}{2}\rho_2 v_2^2 + \rho_2 g y_2 + h_2 + f \quad (22)$$

From the process of thermally driven pump, stagnant flow (at Point 1) having zero velocity exists ($v_1 = 0$) at to the vale opening, while fluid passing outlet point of pipe (at Point 2) has velocity as v_2. The Equation 22 becomes as

$$\frac{1}{2}\rho_2 v_2^2 = \Delta P + \Delta h - \rho_2 g y_2 - f \quad (23)$$

where the friction loss (f) can be combined with the friction loss in the swirling flow (White, 1932) and the loss due to change in mass of two phase flow regime (Kakak & Ishil, 1983).

The result obtained from the analytical estimation (by Equation 23) gives CO_2 mass flow rate as approximately 30% (at maximum flow rate) lower from the experimental result, i.e. 46.8 kg/h and 60

Figure 27. Quasi-static flow model of thermally driven pump

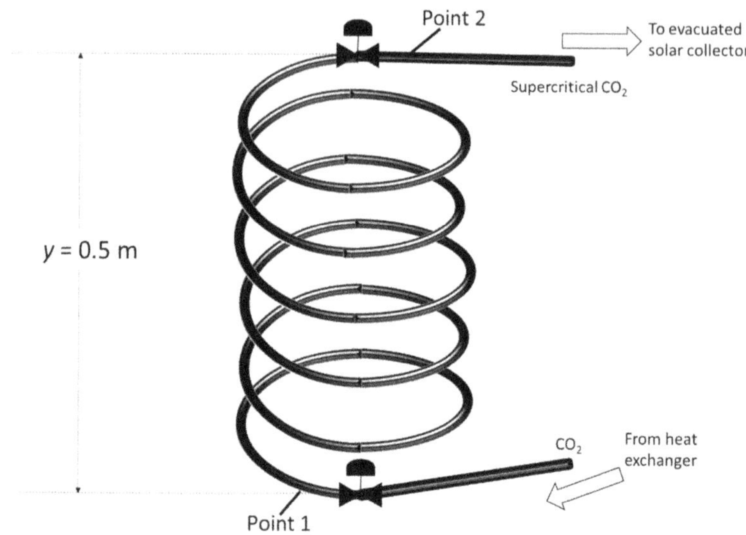

kg/h (shown in Figure 24), respectively. This is due to the fact that the energy conservation equation is still not sufficient (only first order energy balance equation) to estimate more accurate supercritical CO_2 mass flow rate in thermally driven pump. The further flow model would be proposed in future study. It must be further commented from the primary analysis, that the key parameters of CO_2 outlet flow from thermally driven pump are the pressure difference ΔP and the enthalpy difference (the elevated enthalpy Δh) between inside the pump (Point 1) and outlet of the pump (Point 2). ΔP and Δh have big influence on the CO_2 mass flow rate of the thermally driven pump, as approximately 90% in total. On the other hand, the parameters such as height and friction loss have very small influence to the CO_2 mass flow rate. In future analysis, the parameters for accurate estimation of pressure P and enthalpy h have to be considered as the first priority to development further advanced thermally driven pump.

To achieve the aim of replacing mechanical feed pump with thermally driven pump, it is to be expected that the efficiency of the cycle using thermally driven pump would be higher than using mechanical feed pump, because there should not be any mechanical loss occurred in the pumping process. To confirm, the objective system efficiency of the total Rankine system, by either using mechanical feed pump and using thermally driven pump, were compared based on two conditions; the ideal condition, where the efficiency of feed pump (η_P) was set at 100% and practical condition, where efficiency of feed pump (η_P) was set at 50% (Kuwahara, Niu, Yamaguchi, Iwamoto, & Zhang, 2011). In the initial state of development, power for pumping and heating water in thermally driven pump are ignored. The efficiencies are calculated as

$$\eta_{thf} = \frac{\left(h_1 - h_2\right) - \left(h_5 - h_4\right) / \eta_P}{h_1 - h_4} \tag{24}$$

$$\eta_{tht} = \frac{\left(h_1 - h_2\right)}{h_1 - h_4} \tag{25}$$

where h is specific enthalpy on each position of process in the system (in Figure 9), h_1 - h_2, h_4 - h_3, h_1 - h_3 and h_1 - h_4 are the energy (enthalpy) drop through the turbine, and net energy (enthalpy) gain by the system with the thermally driven pump and mechanical feed pump, respectively. As displayed in Figure 28(a) and Figure 28(b) for the pump efficiency in ideal condition and practical condition, respectively, it is clearly seen that the system efficiency of cycle using thermally driven pump is higher than that using mechanical feed pump in both ideal and practical conditions. The system efficiency when using thermally driven pump at P_1 = 11 MPa is about 3.5% higher compared with that using mechanical feed pump (Kuwahara et al. 2011). Moreover, it can be mentioned that the system thermal efficiency of both thermally driven pump and mechanical feed pump are increased with the $P_{1.}$ In future development of the thermally driven pump, a transient model (with the multi-dimensional dynamic equations) are required for the energy transfer and exchanging process of CO_2 flow.

CONCLUSION

The supercritical CO_2 solar Rankine cycle system is developed to generate electric and heat energy with environmentally friendly manner by combining the solar energy and natural working fluid CO_2. The system can be regarded as one of the solution for tackling Global Warming and climate change issue.

In the typical working condition of CO_2, the energy conversion performance is too low to be used as a working fluid in power cycle compared with other working fluid. However, where CO_2 is operated in supercritical region, the rate of heat transfer is found much higher and resultantly the system would be able to achieve high performance characteristics. This is the main motivation for developing the supercritical CO_2 solar Rankine cycle system which can achieve the excellent system performance, adapting unique properties of supercritical CO_2.

In the development of components in SRCS to date, it is summarized as

Figure 28. Comparisons of the system efficiencies of the loop system with thermally driven and feed pumps in (a) ideal and (b) practical conditions, respectively
Kuwahara et al. 2011.

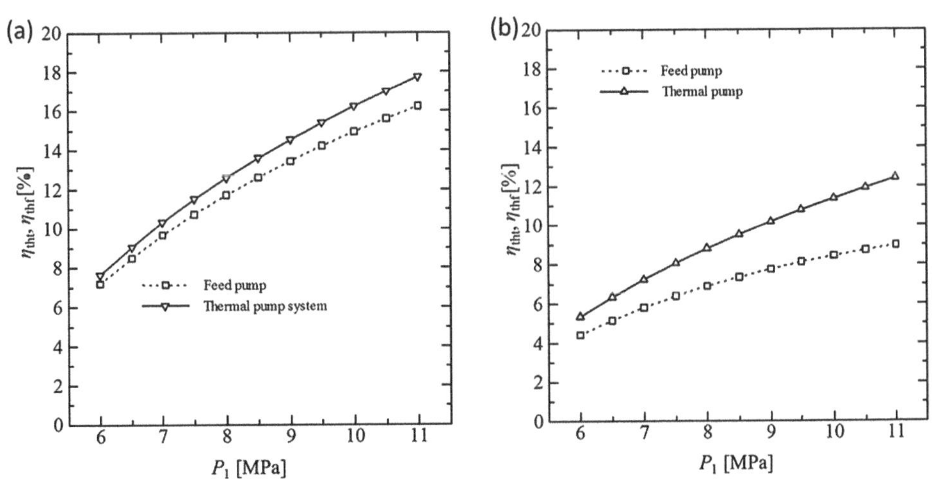

- **Evacuated Solar Collector:** The performance can be improved by the flow passage arrangement and flow condition of supercritical CO_2. The heat transfer is enhanced with the unique properties of supercritical CO_2.
- **Turbine:** Newly designed specific turbine for supercritical state CO_2 was introduced in the developing system of SRCS for gaining good performance in electric power generation, avoiding the choked flow in the test nozzle to prevent the decrease in efficiency of turbine.
- **Thermally Driven Pump:** Mechanical feed pump is replaced by thermally driven pump for avoiding the mechanical loss and electric power requirement in the feeding operation. SRCS with thermally driven pump gives higher thermal efficiency than ordinary SRCS with mechanical feed pump. However, the outlet flow of thermally driven pump is not continuous. For further improvement of the device the key operating parameter are the pressure difference and the enthalpy difference to shorter the discontinuous time interval.

REFERENCES

Ahmed, T. (2006). *Reservoir engineering handbook*. Gulf Professional Publishing.

Brown, A. (1980). Engineering thermodynamics work and heat transfer. *Rogers, GFC and Mayhew, YR, Longman, 1980*, 667.

Chen, M. F., Yamaguchi, H., Zhang, X. W., & Niu, X. D. (2015). Performance analyses of a particularly designed turbine for a supercritical CO2-based solar Rankine cycle system. *International Journal of Energy Research*, *39*(13), 1819–1827. doi:10.1002/er.3299

Dostal, V., Driscoll, M. J., Hejzlar, P., & Todreas, N. E. (2002, January). A supercritical CO2 gas turbine power cycle for next-generation nuclear reactors. In *10th International Conference on Nuclear Engineering* (pp. 567-574). American Society of Mechanical Engineers. doi:10.1115/ICONE10-22192

Hough, S. (2009). *Supercritical Rankine Cycle*. University of Idaho. Technical Paper. ME-517. Retrieved from http://www. if. uidaho. edu/~ gunner/ME443-543/HW/rankine.pdf

Jacobson, M. Z. (2009). Review of solutions to global warming, air pollution, and energy security. *Energy & Environmental Science*, *2*(2), 148–173. doi:10.1039/B809990C

Kakac, S., & Ishil, M. (1983). *Advances in two-phase flow and heat transfer fundamentals and applications volumes I and II*. Academic Press.

Kim, M. H., Pettersen, J., & Bullard, C. W. (2004). Fundamental process and system design issues in CO_2 vapor compression systems. *Progress in Energy and Combustion Science*, *30*(2), 119–174. doi:10.1016/j.pecs.2003.09.002

Kuwahara, T., Niu, X. D., Yamaguchi, H., Iwamoto, Y., & Zhang, X. R. (2011). Performance study of supercritical CO_2-based solar Rankine cycle system with a novel-concept thermally driven pump. *Proceedings of the Institution of Mechanical Engineers. Part A, Journal of Power and Energy*, *225*(4), 413–419. doi:10.1177/0957650910396420

Miller, R. W. (1983). *Flow measurement engineering handbook*. Academic Press.

Morley, J. (2001). *The Future of HFCs as Refrigeration Working fluids*. Hemel Hempstead, UK: Dupont Fluoroproducts.

Niu, X. D., Yamaguchi, H., Iwamoto, Y., & Zhang, X. R. (2013). Optimal arrangement of the solar collectors of a supercritical CO_2-based solar Rankine cycle system. *Applied Thermal Engineering*, *50*(1), 505–510. doi:10.1016/j.applthermaleng.2012.08.004

Niu, X. D., Yamaguchi, H., Zhang, X. R., Iwamoto, Y., & Hashitani, N. (2011). Experimental study of heat transfer characteristics of supercritical CO_2 fluid in collectors of solar Rankine cycle system. *Applied Thermal Engineering*, *31*(6), 1279–1285. doi:10.1016/j.applthermaleng.2010.12.034

PROPATH Group. (1997). *PROPATH: a program package for thermophysical properties*. Author.

Pumaneratkul, C., Iwamoto, Y., Niu, X. D., Kuwahara, T., & Yamaguchi, H. (2015). Characteristics of thermally driven pump in supercritical CO_2 solar Rankine cycle system. *International Journal of Mechanical and Production Engineering*, *3*(4), 1–6.

Pumaneratkul, C., Iwamoto, Y, Yamaguchi, H. (2015). Nozzle test to Supercritical CO_2 turbine in Rankine cycle system. *International Journal of Advances in Science and Technology*, 167-174.

Shiralkar, B. S., & Griffith, P. (1969). Deterioration in heat transfer to fluids at supercritical pressure and high heat fluxes. *Journal of Heat Transfer*, *91*(1), 27–36. doi:10.1115/1.3580115

Su, S., Fang, X., Li, L., Wu, J., Zhang, J., Xu, W., & Hu, J. (2015). HFC-134a emissions from mobile air conditioning in China from 1995 to 2030. *Atmospheric Environment*, *102*, 122–129. doi:10.1016/j.atmosenv.2014.11.057

Thermodynamic Tables Project. (1976). International Thermodynamic Tables of the Fluid State: Carbon Dioxide. Pergamon Press.

Victor, D. G., & Kennel, C. F. (2014). Climate policy: Ditch the 2 C warming goal. *Nature*, *514*(7520), 30–31. doi:10.1038/514030a PMID:25279903

Wakefield, C. B. (1996). *Non ideal gas behavior (No. CONF-960220--)*. Edmond, OK: University of Central Oklahoma.

Wallace, J. F., & Linning, A. W. (1970). *Basic engineering thermodynamics*. Pitman Publishing Ltd.

White, C. M. (1932). Friction Factor and Its Relation to Heat Transfer. *Transactions of the Institution of Chemical Engineers*, *18*, 66–86.

Winterbone, D. E. (1997). *Advanced thermodynamics for engineers*. London: Arnold.

Yamaguchi, H. (2008). *Engineering fluid mechanics* (Vol. 85). Springer Science & Business Media.

Yamaguchi, H., Zhang, X. R., & Fujima, K. (2008). Basic study on new cryogenic refrigeration using CO 2 solid–gas two phase flow. *International Journal of Refrigeration, 31*(3), 404-410.

Yamaguchi, H., Zhang, X. R., Niu, X. D., & Hashitani, N. (2013). A novel thermally driven pump and its test in a supercritical CO_2 loop system. *International Journal of Energy Research*, *37*(11), 1331–1338. doi:10.1002/er.2934

Yunus, C. A., & Afshin, J. G. (2011). *Heat and Mass Transfer: Fundamentals and Applications*. Academic Press.

Zhang, X. R., Chen, L., & Yamaguchi, H. (2010). Natural convective flow and heat transfer of supercritical CO_2 in a rectangular circulation loop. *International Journal of Heat and Mass Transfer, 53*(19), 4112–4122. doi:10.1016/j.ijheatmasstransfer.2010.05.031

Zhang, X. R., Yamaguchi, H., Fujima, K., Enomoto, M., & Sawada, N. (2005). A feasibility study of CO_2-based Rankine cycle powered by solar energy. *JSME International Journal. Series B, Fluids and Thermal Engineering, 48*(3), 540–547. doi:10.1299/jsmeb.48.540

Zhang, X. R., Yamaguchi, H., Uneno, D., Fujima, K., Enomoto, M., & Sawada, N. (2006). Analysis of a novel solar energy-powered Rankine cycle for combined power and heat generation using supercritical carbon dioxide. *Renewable Energy, 31*(12), 1839–1854. doi:10.1016/j.renene.2005.09.024

KEY TERMS AND DEFINITIONS

Compressibility Factor: A ratio between the specific volume of real gas with the specific volume of ideal gas at the same temperature and pressure condition.

Environment Friendly: A process or product that is not harmful to the environment.

Equation of State (EOS): An equation that expresses the interconnection between the pressure, temperature and volume of a substance in thermodynamic system.

Numerical Analysis: A study for approximation solution by using mathematical method, which are considered to be used with competitional method.

Reynolds Number: The dimensionless quantity that shows the proportion of inertial force and viscous force, which indicates the type of fluid flow.

Supercritical Fluid: A state that temperature and pressure of substance above its critical point, where the distinct liquid and gas phase do not contain.

Thermodynamic Cycle: The process or arrangement, in which the working fluid goes through the cyclic process that involves transfer of heat and work in/out of the system.

APPENDIX: NOMENCLATURE

A: Area (m^2)
C_f: Fanning fiction coefficient (-)
c_p: Constant pressure specific heat (kJ/Kg·K)
c_v: Constant volume specific heat (kJ/Kg·K)
d: Diameter (m)
f: Fiction factor (-)
g: Gravity force (m^2/s)
h: Enthalpy (kJ/Kg):
h_x: Local heat transfer coefficient (W/m^2 K):
I: Solar radiation intensity (kW/m^2)
k: Specific heat ratio (-)
L: Tube length (m)
$\dot{m}$: Mass flow rate (kg/s)
M: Mach number (-)
P: Pressure (MPa)
P_{HI}: High pressure state (MPa)
P_{LO}: Low pressure state (MPa)
Pr: Prandtl number (-)
Q_C: Heat rejection (kW)
Q_H: Heat absorption (kW)
Q_c: Heat collection of evacuated solar collector (kW)
Q_r: Input heat flux per unit area (W/m^2)
Q_{SR}: Heat produced by solar radiation (kW)
R: Gas constant (-)
Re: Reynolds number (-)
s: Entropy (kJ/kg K)
T: Temperature (°C)
T_b: Bulk temperature (°C)
T_c: Critical temperature (°C)
T_w: Tube wall temperature (°C)
t: Time (sec.)
u: Speed (m/s)
V: Volume (m^3)
W_p: Pump work (kW)
W_T: Turbine work (kW)
x: Position on the tube (m)
y: Height (m)
Z: Compressibility factor (-)

Greek Letters

η: Efficiency (%)
$\eta_{Rankine}$: Rankine cycle efficiency (%)
η_{P}: Pump efficiency (%)
η_{tht}: System thermal efficiency with thermally driven pump (%)
η_{thf}: System *t*hermal efficiency with mechanical feed pump (%)
ρ: Density (kg/m^3)
ρ_c: Critical density (kg/m^3)
λ: Thermal conductivity (W/m K)
γ: Specific heat ratio (-)

Subscripts

1-5: Positions ① – ⑤ shown in Figure 9

Acronyms

CFC: Chlorofluorocarbon
GWP: Global warming potential
OPD: Ozone depletion potential
GHG: Greenhouse gas

Chemical Compounds

CO_2: Carbon dioxide
He: Helium
N_2: Nitrogen
O_2: Oxygen
H: Hydrogen

Chapter 12
Supercritical Fluids and Their Applications in Power Generation

Huijuan Chen
GE Global Research Center, USA

Ricardo Vasquez Padilla
Southern Cross University, Australia

Saeb Besarati
SolarReserve, USA

ABSTRACT

Supercritical fluids have been studied and used as the working fluids in power generation system for both high- and low-grade heat conversions. Low-grade heat sources, typically defined as below 300 °C, are abundantly available as industrial waste heat, solar thermal, and geothermal, to name a few. However, they are under-exploited for power conversion because of the low conversion efficiency. Technologies that allow the efficient conversion of low-grade heat into mechanical or electrical power are very important to develop. First part of this chapter investigates the potential of supercritical Rankine cycles in the conversion of low-grade heat to power, while the second part discusses supercritical fluids used in higher grade heat conversion system. The selection of supercritical working fluids for a supercritical Rankine cycle is of key importance. This chapter discusses supercritical fluids fundamentals, selection of supercritical working fluids for different heat sources, and the current research, development, and commercial status of supercritical power generation systems.

INTRODUCTION

Currently, two-thirds of the world's electricity demand is met by non-renewable fossil fuels which has led to serious environmental problems and a widespread energy crisis. In trying to limit the emissions from the electricity generating sector, new energy resources as well as radically new technologies should

DOI: 10.4018/978-1-5225-2047-4.ch012

be developed and/or current technologies be improved so that the power output per unit of pollution is reduced.

Renewable energy sources, such as solar thermal and geothermal, and vast amounts of industrial waste heat are potentially promising energy sources capable, in part, to meet the world electricity demand. However, the above mentioned energy sources are available largely at moderate temperatures. The conventional steam Rankine cycle works only efficiently at above 300 °C, and the conversion efficiency becomes uneconomically low for the low-grade heat sources conversion. In this context, developing other technologies that allow the efficient conversion of low-grade heat into mechanical or electrical power is of great significance.

The objective of this chapter is to discuss the potential of using supercritical fluids to convert low and higher grade thermal energy, for which, discussion of the desired properties of the working fluid candidates, performance of the supercritical cycles, and supercritical cycle optimization were carried out. Other objectives of this work include screening the working fluids for supercritical Rankine cycles, comparing the performance of supercritical Rankine cycles with other thermodynamic cycles, and suggesting some feasible applications of the supercritical Rankine cycle.

BACKGROUND

In a supercritical thermodynamic power cycle, the working fluid is compressed to its supercritical pressure and heated to supercritical state directly, bypassing the two-phase vaporization that a conventional Rankine cycle would have. Professor Martyn Poliakoff from University of Nottingham demonstrated the instant transformation of CO_2 from liquid to its supercritical state very clearly in a video ("Supercritical fluids - YouTube," n.d.).

For low-grade heat conversion, various thermodynamic cycles such as the organic Rankine cycle, supercritical Rankine cycle, Kalina cycle, Goswami cycle, and trilateral flash cycle have been proposed and studied. Although there are broad claims of 15–50% more power output for the same heat input for Kalina cycles relative to organic Rankine cycles, data from actual cycles in operation and simulations under identical conditions of ambient temperature and cooling systems showed that the difference in performance is only 3% in favor of Kalina cycle(DiPippo, 2004). However, the organic Rankine cycle is much less complex and need less maintenance. The focus of this chapter regarding low-grade heat conversion is a derivative of organic Rankine cycle, which is supercritical Rankine cycle. Supercritical Rankine cycle has been reported to generally have higher efficiency than an organic Rankine cycle, especially when the heat source is sensible heat (Chen et al., 2011).

A conceptual configuration and a P–h diagram of a supercritical Rankine cycle are shown in Figure 1(a) and Figure 1(b). In this conceptual configuration, the working fluid is pumped above its critical pressure (a->b), and then heated isobarically from liquid directly to supercritical vapor (b->c); the supercritical vapor is expanded in the turbine to extract mechanical work (c->d); after expansion, the fluid is condensed in the condenser by dissipating heat to a heat sink (d->a); the condensed liquid is then pumped to the high pressure again, which completes the cycle. The major difference between a subcritical and a supercritical Rankine cycle lies in the heating process of the working fluid as seen in Figure 2. In a supercritical Rankine cycle, the working fluid is heated directly from the liquid state into the supercritical state, bypassing the two-phase region (b->c in Figure 2). By bypassing the isothermal boiling process, the supercritical Rankine cycle allows the working fluid to have a better thermal match

Figure 1. A typical supercritical Rankine cycle
Chen et al., 2011.

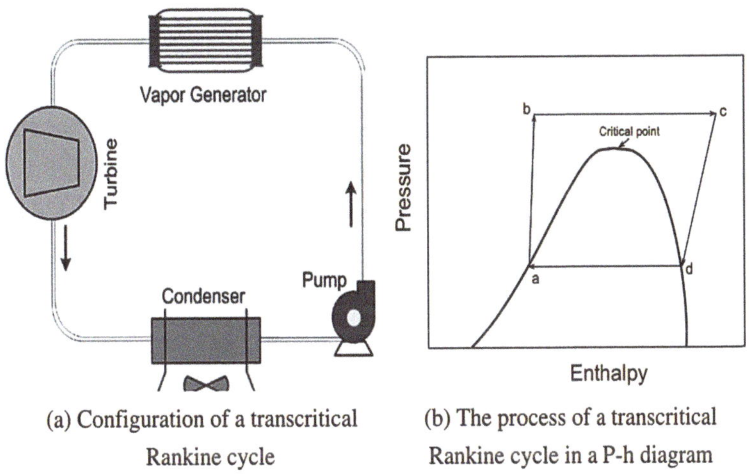

Figure 2. The isobaric heating process of a fluid under supercritical and subcritical pressures
Chen et al., 2011.

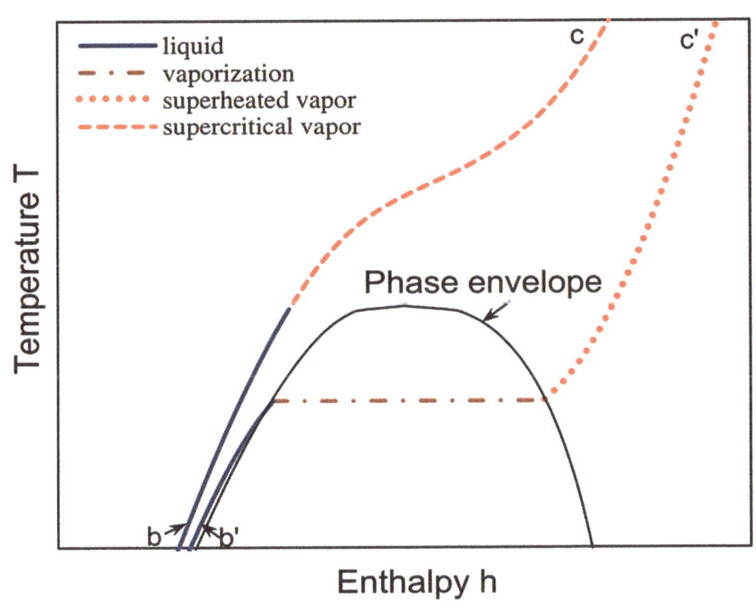

with the heat source, resulting in less exergy loss. Furthermore, by avoiding the boiling process, the configuration of the heating system is potentially simplified. Figure 1 shows the configuration and process of a CO_2 supercritical Rankine cycle in a T–s diagram.

The choice of a working fluid is of key importance for a supercritical Rankine cycle. Carbon dioxide (CO_2), being abundant, non-flammable, non-toxic and inexpensive, has been extensively studied as a supercritical working fluid by a number of researchers. Zhang et al. (X. R. Zhang, Yamaguchi, Uneno, et al., 2006; X. R. Zhang, Yamaguchi, Fujima, Enomoto, & Sawada, 2007, 2006; X.-R. Zhang, Yamaguchi, & Uneno, 2007) indicates the thermal efficiency of a CO_2-based supercritical Rankine cycle to be 8.0–11.4% depending on the working conditions. Chen et al. (Yang Chen, 2006; Y. Chen, Lundqvist, & Platell, 2005) found that under the same thermodynamic mean heat rejection temperature, a CO_2-based supercritical power cycle gives a slightly higher power output than a R123-based organic Rankine cycle. Wang et al. (Wang, Sun, Dai, & Ma, 2010) and Cayer et al. (Cayer, Galanis, & Nesreddine, 2010) did parametric studies on the thermodynamic performance and exergy destruction in a CO_2-based supercritical Rankine cycle to find that parameters such as turbine inlet pressure and temperature have significant effects on the exergy efficiency of the supercritical CO_2 power cycle. Beside CO_2, hydrocarbons (Ochs & O'Connor, 2006) and refrigerants (Chacartegui, Sánchez, Muñoz, & Sánchez, 2009; Etemoglu, 2008; Karellas & Schuster, 2008; Larjola, 1995; McMahan, Klein, & Reindl, 2007) have also been studied as working fluids in supercritical Rankine cycles, and the results showed that the thermal efficiency could be improved by 10-20%, compared to the same working fluids used in the conventional organic Rankine cycle.

For high-grade heat conversion, the major thermodynamic cycles using supercritical fluids are supercritical CO_2 cycles and supercritical steam cycles. For the case of Concentrated Solar Power (CSP) systems, Turchi et al. (2010) carried out a comprehensive analysis on predicting the future price of molten salt solar power tower in the U.S market. The main factors that can alleviate the cost were summarized as: Reducing the{Citation} cost of the heliostats, increasing the operating temperature and employing more efficient power cycles, and including thermal storage with the systems to increase the capacity factor. By taking all these factors into consideration, the Levelized Cost of Electricity (LCOE) of this technology is expected to drop to 10-13 cents USD/kWh over the next ten years, and it has the potential to become competitive with natural gas combined cycle systems.

In an attempt to make CSP price competitive with conventional power, the U.S. Department of Energy (DOE) announced an initiative that aimed at reducing the total costs and reaching the target cost of 6 cents/kWh by 2020. Besides revolutionary improvements in numerous aspects of the solar power

Figure 3. Thermal efficiency of different power cycles vs. heat source temperature
Wright et al., 2011.

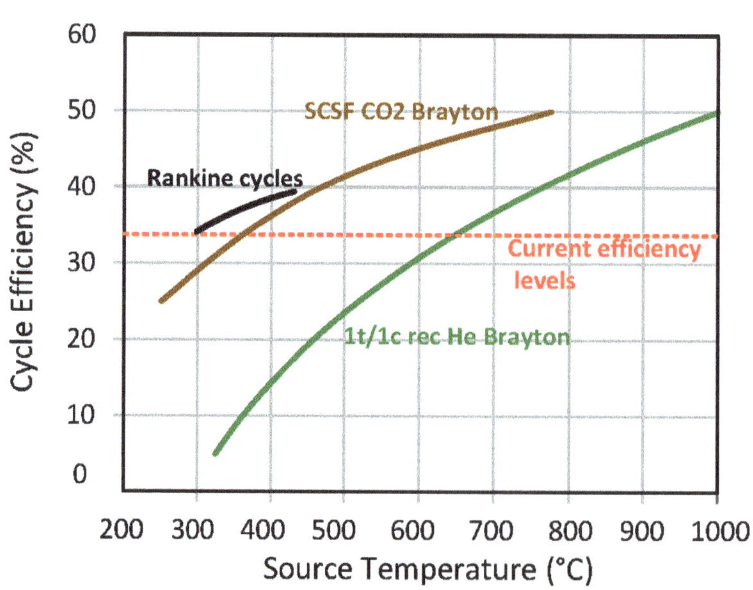

Figure 4. Relative size of the components in S-CO_2, helium and steam power cycles
Wright et al., 2011; Musgrove et al., 2014.

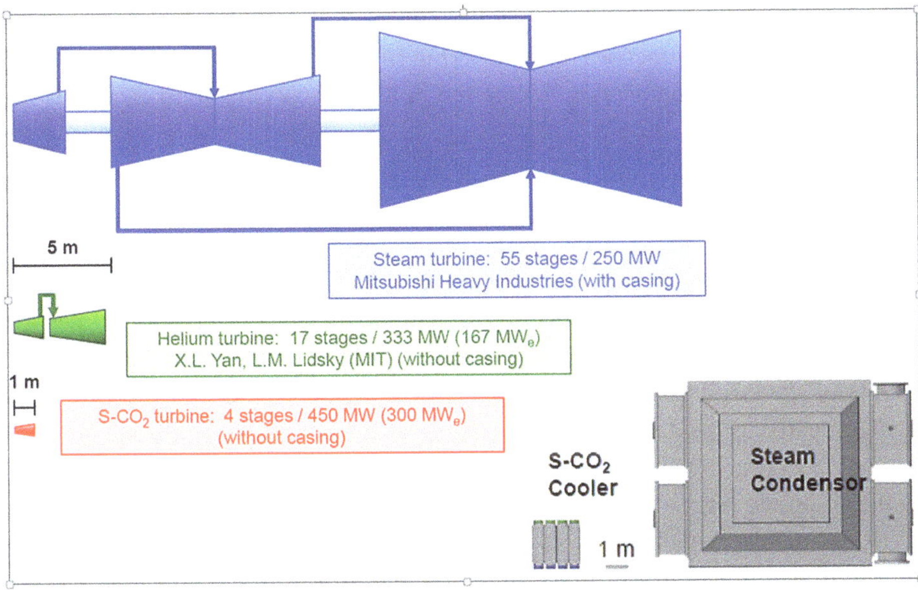

tower plants, new generations of power cycles are required to generate power at higher efficiencies, and lower costs compared with conventional steam cycles in order to meet this target. Supercritical carbon dioxide (S-CO_2) power cycle has emerged as one of the alternatives that have been proposed for the future CSP plants. Figure 3 compares the performance of the S-CO_2 cycle with the steam Rankine and helium Brayton cycles (Wright et al., 2011). As can be seen, S-CO_2 cycle has efficiency benefit over steam when the heat source temperature is greater than 420°C. In addition, the S-CO_2 cycle can reach 50% efficiency at a much lower heat source temperature compared to helium. The high efficiency of the cycle is due to low compression work, as the density of CO_2 increases substantially around the critical point. The density of CO_2 around the critical point is comparable with the liquids, which requires low pumping power. On the other hand, higher density of the working fluid means smaller power conversion components. Figure 4 compares the size of a S-CO_2 turbine with a helium and a steam turbine.

The size of a S-CO_2 cooler and a steam condenser are also compared in this Figure. As can be clearly seen, the sizes of the components of the S-CO_2 cycle are much smaller than the other cycles.

WORKING FLUID PROPERTIES AND SELECTION CRITERIA

For supercritical Rankine cycles, the working fluid affects system efficiency, operating conditions, environmental impact and economic viability. The literature shows extensive analyses and comparisons among different working fluids. However, most of the comparisons were conducted under certain predefined temperature conditions and used only a few working fluids (Chacartegui et al., 2009; Chen et al., 2006; Dai, & Gao, 2009; Delgado-Torres & García-Rodríguez, 2007; Hung et al., 1997; Hung, 2001; Madhawa et al., 2007; Mago et al. 2008; Nowak et al., 2008; Quoilin, 2007; Tchanche et al., 2009; Yamamoto et al. 2001). The claims for best working fluids and the cycle with highest efficiencies may not hold true under other operating conditions and among other working fluids. In reality, a working fluid must not only have the necessary thermo-physical properties that match the application but also possess adequate chemical stability in the desired temperature range. Instead of suggesting a best working fluid, the thermos-physical properties of the working fluids and selection criteria are discussed. The aim is to provide guidance for the users to locate the potential working fluid candidates for different cycles at various conditions.

Types of Working Fluids

As described before, the working fluid in a supercrical Rankine cycle would expand at supercritical state to drive a turbine and generate power. In order to differentiate the expansion process, the working fluids are described are "dry", "Isentropic" or "wet" fluid based on the slope of the saturation vapor curve. The slope (dT/ds) of the saturation vapor curve of a working fluid in a T–s diagram can be positive (e.g. isopentane), negative (e.g. R22) or near infinity (e.g. R11), and the fluids are accordingly called "wet", "dry" and "isentropic" fluids. Since the value of dT/ds leads to infinity for isentropic fluids, the inverse of the slope, (i.e. ds/dT,) is used to express how "dry" or "wet" a fluid is. The inverse of the slop ds/dT is defined as ξ in (Chen et al., 2010), and then if $\xi > 0$: a dry fluid (e.g. pentane), $\xi \approx 0$: an isentropic fluid (e.g. R11), and $\xi < 0$: a wet fluid (e.g. water). Figure 5 shows the three types of fluids in a T–s diagram.

Typical turbine does not allow large percentage of liquid accruing during the expansion in order not to danger it. Therefore, wet fluids like water and carbon dioxide usually need to be superheated in order to avoid large percentage of liquid occurring during the expansion, while many organic fluids, which are dry or isentropic, do not need superheating because the expansion exit ends up in the superheat vapor region. That is a major characteristic of dry or isentropic working fluids. However, if the fluid is "too

Figure 5. Three types of working fluids: dry, isentropic, and wet
Chen et al., 2010.

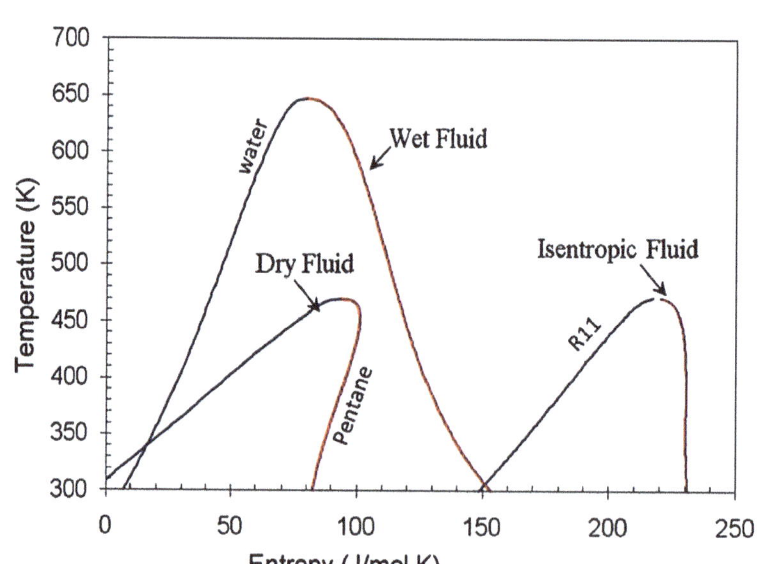

Figure 6. T–s diagram shows a dry fluid and a wet fluid used in supercritical cycles. (a) Pentane as the working fluid. (b) Propyne as the working fluid
Chen et al., 2010.

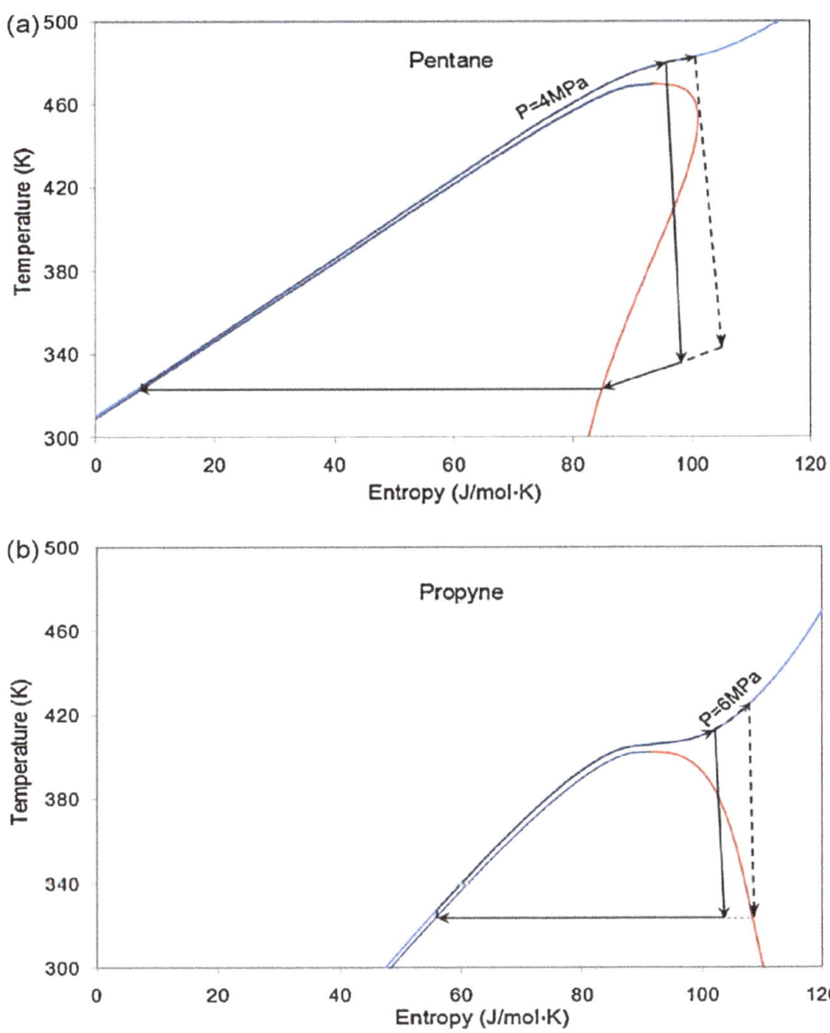

dry,'' the expanded vapor will leave the turbine with substantial ''superheat'', which is a waste and adds to the cooling load in the condenser. In this case, the cycle efficiency can be increased using this superheat to preheat the liquid after it leaves the feed pump and before it enters the boiler.

Figure 6 shows a dry fluid, propyne, and a wet fluid pentane used in supercritical Rankine cycles. If the expansion is carried out such that it does not go into the two-phase region (the dashed lines in Figure 6(a) and Figure 6(b), dry fluids may leave the turbine with substantial amount of superheat, which adds to the burden for the condensation process or a recovery system is needed. Wet fluids, on the other hand, will need higher turbine inlet temperature to avoid two-phase region but there is less concern about desuperheating after the expansion. If the process is allowed to pass through the two-phase region (the solid lines in Figure 6), the dry fluid can still leave the turbine at superheated state, while the wet fluid stays in the two-phase region at the turbine exit. Bakhtar et al.(Bakhtar et al., 2006; Bakhtar et al., 1999;

Bakhtar & Mahpeykar, 1997; Bakhtar et al., 1997; Bakhtar et al., 1997) found that for a wet fluid, such as, water, the fluid first subcools and then nucleates to become a two-phase mixture. The formation and behavior of the liquid in the turbine create problems that would lower the performance of the turbine. For dry fluids, Goswami et al. (1991) and Demuth(1981, 1983) found that only extremely fine droplets (fog) were formed in the two-phase region and no liquid was actually formed to damage the turbine before it started drying during the expansion. Demuth also found that the turbine performance should not degrade significantly as a result of the turbine expansion process passing through and leaving the moisture region if no condensation occurs. Meanwhile, potential gains in the net fluid effectiveness on the order of 8% can be achieved resulting from the selection of the turbine expansion process whose equilibrium states pass through the two-phase region. To this end, dry fluids may serve better than wet fluids in supercritical states if the turbine expansion involves two-phase region.

Influence of Latent Heat, Density and Specific Heat

Maizza and Maizza (1996) suggested that high latent heat, high density and low liquid specific heat are preferable for organic or supercritical organic Rankine cycle, as a fluid with a high latent heat and density absorbs more energy from the source in the evaporator and thus reduces the required flow rate, the size of the facility, and the pump consumption. However, Yamamoto et al. (Yamamoto et al., 2001) suggested that low latent heat is better because the saturated vapor at the turbine inlet would provide the best operating condition. In order to obtain a comprehensive conclusion, Chen and her colleagues conducted a theoretical analysis by deriving the expression of the enthalpy change through the turbine (Chen et al., 2010). They found that fluids with higher latent heat give higher unit work output when the temperatures and other parameters are defined. Fluids with higher density need smaller equipment setup for same power production. In brief, working fluids with high density, low liquid specific heat and high latent heat are expected to give high turbine work output.

Critical Points of the Working Fluids

Condensation is a necessary process in supercritical Rankine cycle. The design condensation temperature is normally above 300 K in order to reject heat to the ambient; therefore, fluids like methane with critical temperatures far below 300 K are out of consideration because of the difficulty in condensing. On the other hand, the critical point of a fluid considered as the working fluid of a supercritical Rankine cycle should not be too high to overpass with the available heat source. The critical point of a working fluid, being the peak point of the fluid saturation line in a T–s diagram, suggests the proper operating temperature range for the working fluid of liquid and vapor forms, and the critical temperature is an important data for fluid selection. Another important thermodynamic property is the freezing point of the fluid, which must be below the lowest operating temperature in the cycle. The fluid must also work in an acceptable pressure range. Very high pressure or high vacuum has a tendency to impact the reliability of the cycle or increase the cost.

Stability of the Fluids and Compatibility with Materials in Contact

Unlike water or carbon dioxide, organic fluids usually suffer chemical deterioration and decomposition at high temperatures(Zhang et al., 2006). The maximum operating temperature is thus limited by the chemical stability of the working fluid. Additionally, the working fluid should be noncorrosive and compatible with engine materials and lubricating oil. Calderazzi and Paliano(1997) studied the thermal stability of R-134a, R-141b, R-13I1, R-7146 and R-125 associated with stainless steel as the container material. Andersen and Bruno (Andersen & Bruno, 2005) presented a method to assess the chemical stability of potential working fluids by ampule testing techniques. The method allows the determination of the decomposition reaction rate constant of simple fluids at the temperatures and pressures of interest. Researchers at General Electric conducted experimental studies of cyclopentane and cyclopentene(Ginosar et al., 2011). For Cyclepentane, they found that it would likely to have acceptable stability under the envisioned conditions. Same conclusion was obtained by researchers from Idaho National Laboratories (Calderazzi & di Paliano, 1997). For cyclopentene, although theoretically it can achieve higher efficiency under the same condition, obvious degradation occurs at 300°C and therefore it is not suggested to be used in cycles that operate at around or above 300°C. Details of the tests can be found from the report (Zia et al., 2013).

Environmental Aspects and Safety

As we learn more about their environmental impacts, some organic working fluids have been phased out, or being phased out due to their ozone depletion potential (ODP), global warming potential (GWP), or atmospheric lifetime (ALT). The ODP and GWP represent substance's potential to contribute to ozone degradation and globe warming. Over the years, working fluids such as R-11, R-12, R-113, R-114, and R-115 have been phased out and fluids such as R-21, R-22, R-123, R-124, R-141b and R-142b are being phased out in 2020 or 2030. Main while, alternative fluids are being found and applied. The alternatives are expected to retain the attractive properties and avoid their adverse environmental impact. The most promising candidates are still found among fluids containing fluorine and carbon atoms. The inclusion of one or more hydrogen atoms in the molecule, results in it being largely destroyed in the lower atmosphere by naturally occurring hydroxyl radical, ensuring that little of the fluid survives to enter the stratosphere (Crook & Engineers, 1994).

One can use ASHRAE refrigerant safety classification as a good indicator of the fluid's level of danger. Generally, characteristics like noncorrosive, non-flammable, and non-toxic are expected. But depending on the operating condition, not all of them are always practically satisfying or critically necessary. Many fluids, like R-601, are considered flammable but this is not a problem if there is no ignition source around. However, auto ignition is a problem, in particular for longer alkanes at temperatures above 200°C. The maximum allowable concentration and the explosion limit should also be considered during fluid selection.

In addition, the availability and cost of the working fluids are also among the considerations when selecting working fluids in production. Traditional refrigerants used in organic Rankine cycles might be expensive. This cost could be reduced by a more massive production of those refrigerants, or by the use of low cost hydrocarbons.

To sum up, the selection of working fluids is a complicated process. A fluid will be evaluated from its thermodynamic and physical properties, stability and compatibility, environmental impacts, safety, availability and cost upon its being considered as the working fluid of a supercritical Rankine cycle. The next section is a discussion of the fluids discovered and proposed for supercritical Rankine cycle.

FLUID CANDIDATES AND THEIR PROPERTIES

Fluid Candidate Screening

More than 50 working fluids have been suggested in the literature, among which some have been phased out due to the environmental concerns, and some are not practical for application, such as methane due to its properties.

Chen et al. (2010) screened and summarized 35 potential working fluid candidates for supercritical Rankine cycles. Although aspects other than the thermodynamic properties were pointed out in the last section, they are better discussed in specific engineering designs. For the screened fluids, important thermodynamic properties such as molecular weight, critical temperature, critical pressure, critical density, vapor specific heat capacity, latent heat, and *ds/dT* were calculated and listed for reference (see Table 1). The fluids' vapor specific heat capacity C_p, the latent heat L, and the ratio *ds/dT* are functions of temperature, and the values given in the table are based on 320K (46.8°C) for those whose critical temperatures are above 320K (46.8°C), because 320K (46.8°C) is the approximate design temperature for condensation. For those fluids whose critical temperatures are below 320 K (46.8°C), it is assumed that the condensation is designed to be at 290 K (16.8°C), and the calculation is based on that.

Table 1 can suggest how well the substance serves as a working fluid in an organic Rankine cycle or a supercritical Rankine cycle. The molecular weight suggests the density of the fluid; the critical point suggests the possible operating temperature and pressure range; the turbine work output can be estimated from Equation 6 provided in (Chen et al., 2010)with the value of vapor specific heat capacity (C_p) and the latent heat (L); inverse of the saturated vapor slope parameter ξ describes the fluid type and suggests the effectiveness of superheating.

There is no best fluid that meets all the criteria discussed in previous sections for heat sources with different temperatures. Compromise must be made when selecting the fluids. Among all the criteria and concerns, the authors find that the critical temperature and the ξ value are important parameters that suggest which type of cycle a fluid may serve and the applicable operating temperature of the fluid. Thereby, T– ξ charts are introduced in paper (Chen et al., 2010). The screened 35 fluids were distributed in the charts with their critical temperature and ξ value in Figure 7, from which the critical temperature and the type of each working fluid can be easily found out. Water, toluene, benzene and ammonia are labeled in Figure 7; the remaining fluids are shown in a close-up look in Figure 8. The fluids are divided into 5 groups based on their locations in the T– ξ chart and discussed in the following.

Fluid Discussion

Fluids Water, Ammonia, Benzene and Toluene

It can be observed from Figure 7 that water is located in the upper left corner of the chart, which indicates it is the wettest fluid and has the highest critical temperature among all the fluids plotted. These characteristics make it unsuitable for low temperature heat conversion. However, in high temperature heat conversion applications, such as coal and nuclear power plants, supercritical Rankine cycle using water as the working fluid is typically seen. Ammonia as a deep wet fluid with ξ value of -10.48 J/kg•K^2, needs superheating when used in a Rankine cycle. Ammonia is not recommended in supercritical Rankine cycles, since the critical pressure (11.33MPa) is relatively high. Meanwhile, ammonia is

Table 1. Properties of the screened working fluids

ASHRAE Number	Name	Molecular Weight	*Tc* (K)	*Pc*(MPa)	Vapor *Cp* (J/kg K)	Latent Heat *L* (kJ/kg)	ξ (J/kgK2)
R-21	Dichlorofluoromethane	102.92	451.48	5.18	339.85	216.17	0.78
R-22	Chlorodifluoromethane	86.47	369.30	4.99	1069.13	158.46	1.33
R-23*	Trifluoromethane	70.01	299.29	4.83	3884.02	89.69	6.49
R-32	Difluoromethane	52.02	351.26	5.78	2301.61	218.59	4.33
R-41*	Fluoromethane	34.03	317.28	5.90	3384.66	270.04	7.20
R-116*	Hexafluoroethane	138.01	293.03	3.05	4877.91	30.69	5.54
R-123	2,2-Dichloro-1,1,1-trifluoroethane	152.93	456.83	3.66	738.51	161.82	0.26
R-124	2-Chloro-1,1,1,2-tetrafluoroethane	136.48	395.43	3.62	908.70	132.97	0.26
R-125	Pentafluoroethane	120.02	339.17	3.62	1643.89	81.49	1.08
R-134a	1,1,1,2-Tetrafluoroethane	102.03	374.21	4.06	1211.51	155.42	0.39
R-141b	1,1-Dichloro-1-fluoroethane	116.95	477.50	4.21	848.37	215.13	0.00
R-142b	1-Chloro-1,1-difluoroethane	100.50	410.26	4.06	1036.52	185.69	0.00
R-143a	1,1,1-Trifluoroethane	84.04	345.86	3.76	1913.97	124.81	1.49
R-152a	1,1-Difluoroethane	66.05	386.41	4.52	1456.02	249.67	1.14
R-170*	Ethane	30.07	305.33	4.87	5264.72	223.43	8.28
R-218	Octafluoropropane	188.02	345.02	2.64	1244.87	58.29	0.45
R-227ea	1,1,1,2,3,3,3-Heptafluoropropane	170.03	375.95	3.00	1013.00	97.14	0.76
R-236ea	1,1,1,2,3,3-Hexafluoropropane	152.04	412.44	3.50	973.69	142.98	0.76
R-245ca	1,1,2,2,3-Pentafluoropropane	134.05	447.57	3.93	1011.26	188.64	0.60
R-245fa	1,1,1,3,3-Pentafluoropropane	134.05	427.20	3.64	980.90	177.08	0.19
HC-270	Cyclopropane	42.08	398.30	5.58	1911.81	366.18	1.54
R-290	Propane	44.10	369.83	4.25	2395.46	292.13	0.79
R-C318	Octafluorocyclobutane	200.03	388.38	2.78	896.82	93.95	1.05
R-3-1-10	Decafluorobutane	238.03	386.33	2.32	928.83	77.95	1.32
FC-4-1-12	Dodecafluoropentane	288.03	420.56	2.05	884.25	86.11	1.56
R-600	Butane	58.12	425.13	3.80	1965.59	336.82	1.03
R-600a	Isobutane	58.12	407.81	3.63	1981.42	303.44	1.03
R-601	Pentane	72.15	469.70	3.37	1824.12	349.00	1.51
R-717	Ammonia	17.03	405.40	11.33	3730.71	1064.38	10.48
R-718	Water	18.00	647.10	22.06	1943.17	2391.79	17.78
R-744*	Carbon dioxide	44.01	304.13	7.38	3643.72	167.53	8.27
R-1270	Propene	42.08	365.57	4.66	2387.36	284.34	1.77
	Propyne	40.06	402.38	5.63	2100.54	431.61	1.87
	Benzene	78.11	562.05	4.89	1146.72	418.22	0.70
	Toluene	92.14	591.75	4.13	1223.90	399.52	0.21

Chen et al., 2010.

Note: *The critical temperature of the fluid is below 320 K, and the data is given based on 290 K.

Figure 7. Distribution of the screened 35 working fluids in T– ξ chart
Chen et al., 2010.

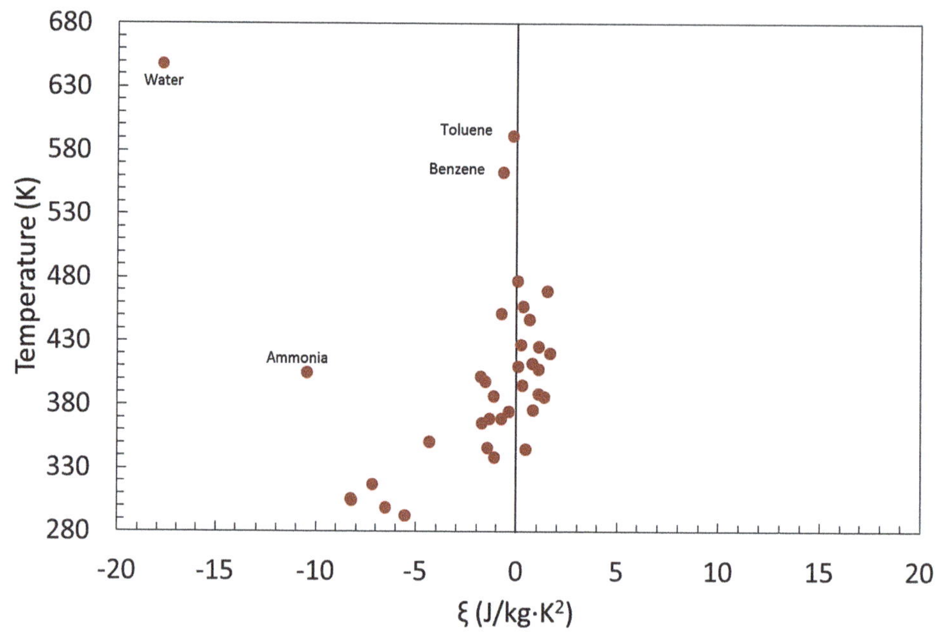

Figure 8. Close-up look of the distribution of the remaining 31 working fluids in T– ξ chart
Chen et al., 2010.

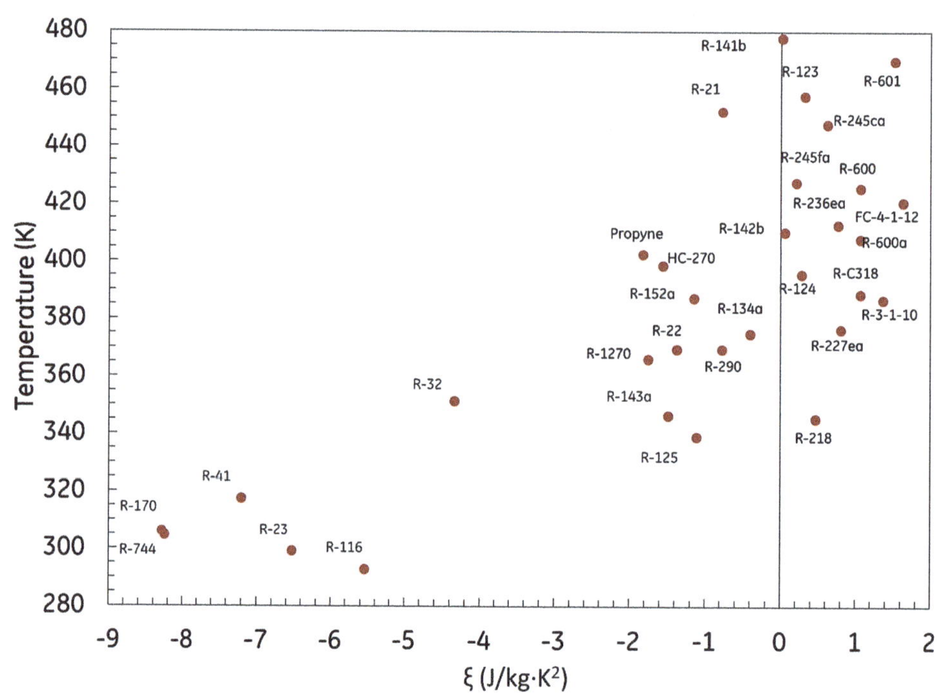

highly hydrophilic, and the ammonia–water solution is corrosive, limiting the materials that may be used. Benzene and toluene are considered as isentropic fluids with relatively high critical temperatures, which are desirable characteristics for supercritical Rankine cycles. At the same time, benzene and toluene are chemically stable in these potential operating conditions(Andersen & Bruno, 2005).

Fluids R-170, R-744, R-41, R-23, R-116, R-32, R-125 and R-143a

From Figure 8 it can be observed that fluids R-170, R-744, R-41, R-23, R-116, R-32, R-125 and R-143a are wet fluids with low critical temperatures and reasonable critical pressures (from Table 1), which are desirable characteristics for supercritical Rankine cycles. Carbon dioxide (R-744) and R134a have been studied in supercritical Rankine cycles in the literature. Among these fluids, R-170, R-744, R-41, R-23 and R-116 have critical temperatures below 320 K, which require low condensing temperatures, not achievable under many circumstances. The critical temperatures of R-32, R-125 and R-143a are above 320 K, so the design of condensers for these fluids is not a big concern. Provided other aspects are satisfied, R-32, R-125 and R-143a could be promising working fluids for supercritical Rankine cycle.

Fluids Prozpyne, HC-270, R-152a, R-22 and R-1270

Propyne, HC-270, R-152a, R-22 and R-1270 are wet fluids with relatively high critical temperatures (Figure 8). They can be applied in supercritical Rankine cycles if the temperature profile of the heat source meets the requirements. However, propyne, HC-270 (cyclopropane) and R-1270 (propene) are not normally used in their supercritical state due to the stability concerns. Propyne, HC-270 and R-1270 have relatively low molecular weight (Table 1). Applying these fluids implies a larger system size compared to those fluids with higher molecular weight.

Fluids R-21, R-142b, R-134a, R-290, R-141b, R-123, R-245ca, R-245fa, R-236ea, R-124, R-227ea, R-218

This group of fluids can be considered isentropic fluids (Figure 8). They can be applied in supercritical Rankine cycle when the temperature of the heat source meets their temperature requirements. Since the isentropic expansion would not cause wet fluid problems, substantial heat beyond critical temperature is typically not necessary with these fluids. Among these fluids, R-141b, R-123, R-21, R-245ca, R-245fa, R-236ea and R-142b have critical temperature above 400 K, making them more likely to be used for heat source with relatively higher temperature. The remaining fluids in this group can be used for low-temperature heat sources.

Fluids R-601, R-600, R-600a, FC-4-1-12, R-C318, R-3-1-10

Fluids R-601, R-600, R-600a, FC-4-1-12, R-C318, R-3-1-10 are considered dry fluids as it can be seen from Figure 8. Based on the analysis in (Chen et al., 2010), further heating beyond critical temperature has a negative effect on the cycle efficiency with dry fluids, substantial heating beyond critical temperature is not recommended from the efficiency stand point of view. The decision on which fluids could be used may be based on how the operating temperature is tailored to cope with the heat source temperature profile.

Zeotropic Mixture of Fluids

Besides the pure fluids discussed above, there are researches of using zeotropic mixture as the working fluids of supercritical Rankine cycle. A zeotropic mixture is a mixture of fluids that do not have the same vapor phase and liquid phase composition at any point of the vapor–liquid equilibrium state. The major difference between a pure fluid and a zeotropic mixture fluid in a supercritical Rankine cycle is that during the condensation process, a zeotropic mixture creates a thermal glide that allows it to have better thermal match with its cooling fluid. This thermal glide can potentially improve the condensation effectiveness and improve system efficiency. Chen. et al. (2011) studied the performance of R134a/R32 zeotropic mixture based supercritical Rankine cycle and compared it with pure R134a based organic Rankine cycle to find that 10-30% efficiency improvement can be achieved by using zeotropic mixture in supercritical Rankine cycle. Heberle et al. (2012) studied the zeotropic mixture fluids of isobutane/isopentane and R227ea/R245fa used in subcritical Rankine cycles and found that due to a glade match of temperature profiles in the condenser and evaporator, the system efficiency can increase up to 15%. Liu. et al. (2014) presented a method to determine the optimal ORC condensation pressure when using a zeotropic mixture. Various theoretical researches have shown the benefit of using zeotropic mixture in organic Rankine cycle or supercritical Rankine cycle and more experimental studies need to be conducted to verify it in reality.

SUPERCRITICAL POWER CYCLES FOR CONCENTRATED SOLAR THERMAL SYSTEMS

In order to be competitive with conventional power plants, Concentrated Solar Power (CSP) Systems need further development on different aspects, such as advanced the use of more efficient power cycles and solar components, increasing deployment, reducing manufacturing cost and integration of thermal energy storage to improve dispatchable power on demand (Ma & Turchi, 2011). Central solar receiver systems have received great attention in CSP applications, since it is possible to achieve high temperatures (Tian & Zhao, 2013), which leads to higher thermal cycle efficiencies. Recently, the Supercritical carbon dioxide (S-CO_2) Brayton power cycle has been proposed as a power cycle for a central solar receiver system due to the following advantages (Cassetti & Colombo, 2013; Ma & Turchi, 2011; Persichilli et al., 2012; C. Turchi et al., 2009):

- Non-toxic and non-flammable (Group A1)(ASHRAE, 2013).
- Lower capital cost compared to other thermodynamic cycles.
- High efficiencies at high temperatures.
- CO_2 can be used as direct heat transfer fluid (HTF) in the solar receiver.
- Compactness and less complexity than Rankine cycles (Turchi et al., 2009).
- Scale advantage for modular S-CO_2 tower receiver/generator in terms of maintenance and cost.
- Compatibility with thermal energy storage.
- Better temperature glide, since CO_2 is a single phase during heating and cooling process.

In the 1960s, Feher(1968) proposed a supercritical thermodynamic power cycle by using CO_2 as a working fluid. The Feher's cycle achieved a thermal efficiency of ~55% under ideal conditions and due to its thermodynamic performance it was proposed as a compact and portable electric power generator. In 2004, Dostál (2004) proposed that the S-CO_2 Brayton cycle was suitable for nuclear applications.

Dostál et al. (2006) and Hejzlar et al. (Hejzlar et al., 2006) showed that the S-CO_2 recompression Brayton cycle is more efficient than helium Brayton cycles operating at the same turbine inlet temperature and has smaller thermal and turbomachinery components. Kulhánek and Dostál (2011) performed thermodynamic simulations for different S-CO_2 Brayton cycle configurations; they found that partial cooling and recompression S-CO_2 Brayton cycle have the highest thermal efficiency under ideal operating conditions. Bae et al. (2015) came to the same conclusion, except for those cases when turbomachinery with high efficiency is used. (Moisseytsev et al., 2008)

Moisseytsev et al. (2008a) studied different S-CO_2 Brayton configurations: reheating, double recompression cycle and intercooling. They concluded that the new configurations did not show a significant improvement with respect to the recompression cycle. Garg et al. (2013) compared the thermal performance of S-CO_2 Brayton cycle with transcritical Brayton cycle in the gas region; they found that S-CO_2 Brayton cycle has higher efficiency than the transcritical configuration at the same turbine inlet temperature and the specific work output of S-CO_2 cycle is 5 times higher than the same cycle operating below critical conditions.

As mentioned above, the S-CO_2 Brayton cycle was initially proposed for nuclear applications, however due to its high operating temperature it was later proposed as an alternative power cycle for CSP systems. According to Table 1, carbon dioxide has moderate critical properties which make it adaptable to most source and sink temperatures. Therefore, the S-CO_2 power cycle can be utilized in fossil fuel, CSP, geothermal, and nuclear power plants. Figure 9 shows the ranges of operating temperature for each of these power plants and the efficiencies that can be obtained using S-CO_2 cycle. According to the figure, in CSP plants efficiencies in the range of 43% to 54% can be obtainable under wet cooling conditions. However, CSP plants are usually located in the areas where water resources are limited; therefore, dry cooling may be preferred over wet cooling. Under dry cooling conditions, close to 50% efficiency is still achievable, which is consistent with the framework of the DOE SunShot program.

Figure 9. Operating temperature range of different power plants and the efficiencies that can be obtained using S-CO_2 cycle
Sargent and Lundy LLC, 2003.

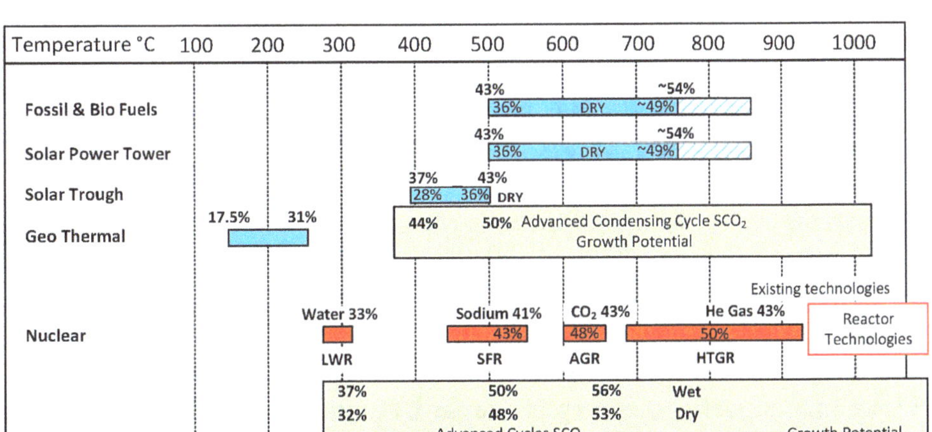

Turchi et al. (2012a), Turchi et al. (2013a) carried out thermodynamic simulations of S-CO_2 Brayton cycles by including reheating and dry cooled condensers for different configurations: Simple, Recompression, Partial Cooling and Recompression with Main-Compression Intercooling. They found that S-CO_2 Brayton cycles achieve higher thermal efficiency than conventional steam Rankine cycles operating at the same turbine inlet temperature and at temperatures above 650 °C all configurations, except the simple configuration, achieved approximately the same thermal efficiency. Turchi et al. (2013a) also showed that adding reheat increases the thermal efficiency of S-CO_2 Brayton cycles when no pressure drop is assumed. Dunham and Iverson (2014) reviewed different power cycles for CSP systems applications and studied the overall performance of the power cycles integrated with a solar receiver; they found that the recompression configuration can achieve higher efficiencies than the other configurations at 30 MPa and turbine inlet temperatures above 600 °C. Neises and Turchi (2014) studied the thermal performance of simple, recompression and partial cooling S-CO_2 Brayton cycles for CSP applications with different high pressures and recuperator conductances. Their results showed that partial cooling achieved higher efficiencies compared to recompression configuration, however for high heat exchanger conductance (>10 MW/K), recompression cycle achieved similar thermal efficiencies. Besarati and Goswami (2014) proposed an improvement of the S-CO_2 Brayton cycle by adding a bottoming Organic Rankine Cycle (ORC). Their results showed an increase of 3-7% points over the thermal cycle efficiency and the best thermal performance was obtained in the recompression cycle. Ho et al. (2014) studied the integration of S-CO_2 Brayton cycle with solar central receivers by using different solar receiver type and cycle layouts (direct and indirect configuration). They concluded that the tubular receiver design is suitable for both cycle layouts. Iverson et al. (2013) presented a S-CO_2 Brayton cycle facility at Sandia National Laboratory. The thermodynamic cycle produces around 250 kWe and it is able to operate under transient conditions similar to those expected in a CSP plant.

Dynamic simulations of S-CO_2 Brayton cycles (Singh, Kearney, & Manzie, 2013; Singh, Miller, Rowlands, & Jacobs, 2013) have been also performed to study the stability of the cycle and operating conditions near to its critical point and evaluate control strategies (Moisseytsev & Sienicki, 2008a, 2008b; Carstens et al., 2008; Moisseytsev & Sienicki, 2011; Moisseytsev & Sienicki, 2012; Conboy et al., 2013). Part-load performance and off-design analysis for simple and recompression S-CO_2 Brayton cycles have also been studied (Dyreby et al., 2013). The results have shown that high efficiencies can be achieved by S-CO_2 Brayton cycles even operating under part-load and with high ambient temperatures.

Some experimental facilities of small scale (Conboy et al., 2012; Wright et al., 2008; Wright et al., 2009; Wright & Pickard, 2009; Garg et al., 2014) have been built to evaluate the thermal performance of simple and recompression S-CO_2 Brayton cycles and develop turbomachinery (Kim et al., 2014; Lee et al., 2012; Pecnik et al., 2012) suitable for the operating conditions proposed. In addition, the experimental work has been useful in investigating problems during startup and shutdown of S-CO_2 Brayton cycles (Clementoni et al., 2014; Kimball & Clementoni, 2012) and validating thermodynamic models (Pasch, et al., 2014).

Figure 10 shows the current status of different power cycles that can be used in the CSP plants. Despite lower operating temperatures compared with others, steam Rankine cycle is the only power cycle that is used in all the commercial power plants. S-CO_2 cycle is still under development for pilot studies. S-CO_2 combined cycle (CC) is another alternative that can be considered.

A CSP plant using S-CO_2 as the HTF and the working fluid with a single-tank thermal storage system is shown in Figure 11. Considering the compact size of s-CO_2 turbomachinery, modular power generation in the receiver without energy storage has recently been proposed (Turchi et al., 2012).

Figure 10. Current status of different power cycles to be used in CSP plants
Musgrove et al., 2014.

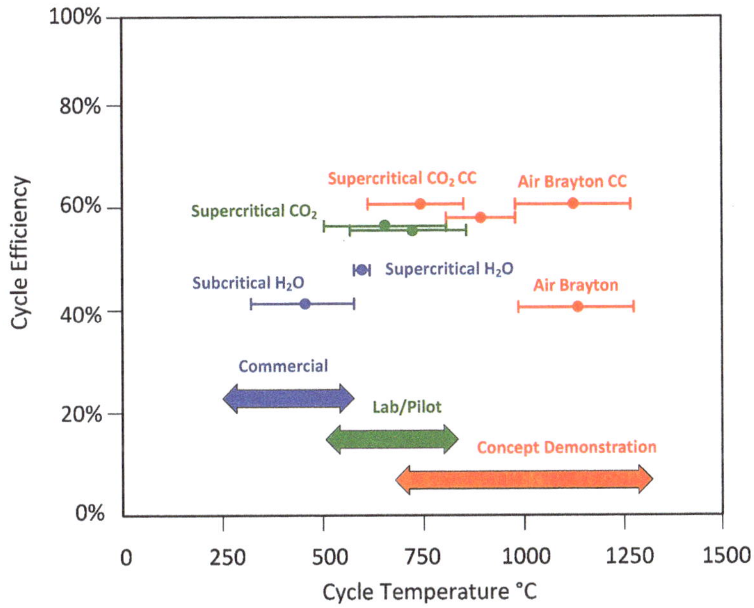

Figure 11. Direct s-CO_2 concept in CSP plant with thermal storage
Turchi et al., 2012.

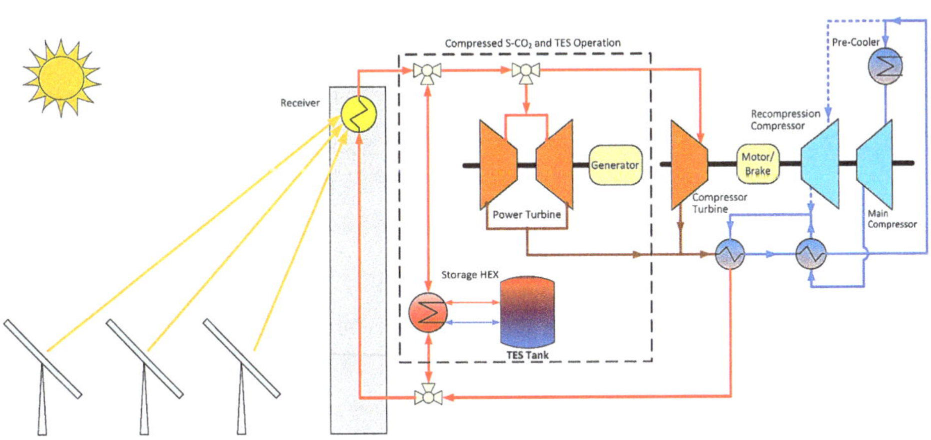

Table 2 shows the turbine size, shaft speed, and CO_2 mass flow rate for power ratings of 0.3, 3 and 300 MW. As can be seen, at the 3MW power rate, the turbine wheel diameter is only 15 cm with a speed of 50,000 rpm. Therefore, it is possible to place the power cycle inside the tower (Figure 12).

In other words, using this technology the power can be generated inside the tower right after the receiver and the fluid does not need to flow through long pipes. Hence, the system is more compact, and the pressure drop and heat loss are less, which consequently leads to higher efficiency and lower cost.

Despite all the positive aspects mentioned in this introduction, there are some uncertainties about utilization of s-CO_2 in the CSP plants. The main concerns are the high pressure of the fluid and lack of experience in operating closed loop Brayton cycles.

Table 2. CO_2 turbine size at different power rates

Power Rate (MW)	Turbine Wheel Diameter (m)	Desired Shaft Speed (RPM)	CO_2 Flow (kg/sec)
0.3	0.04	125,000	3.5
3	0.15	50,000	35
300	1.5	3,600	3500

C. S. Turchi, Ma, & Dyreby, 2012b.

Figure 12. Schematic of modular tower concept using s-CO_2
Turchi et al., 2012b.

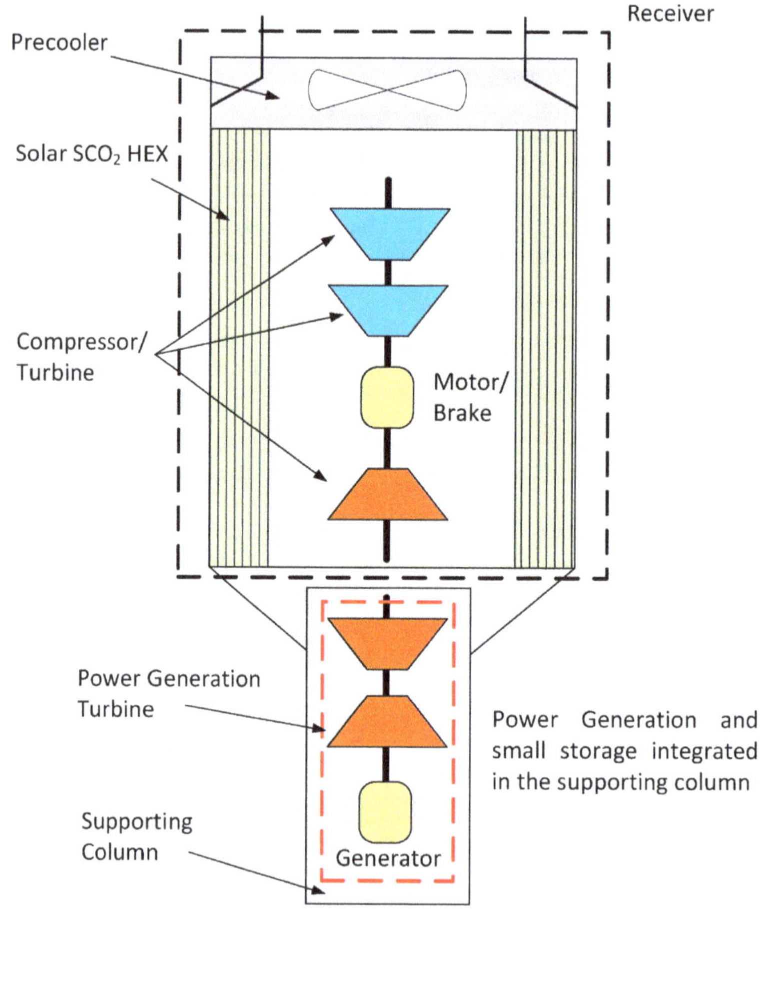

Characteristics of S-CO_2 Around the Critical Point

The critical point of CO_2 is 7.38 MPa and 30.98°C (314.13 K). Around the critical point, CO_2 is not an ideal gas, and its behavior is very sensitive to the pressure and temperature. In other words, the fluid properties vary significantly around the critical point.

Figure 13 shows the density variations of CO_2 at different operating conditions. As can be seen, the density is very high close to the critical point and comparable to liquids. Therefore, the compression work is considerably reduced if carbon dioxide enters the compressor close to the critical condition, which is the main advantage of S-CO_2 over the air Brayton cycle.

Wright et al. (2011) compared the density of S-CO_2 in a closed loop Brayton cycle with that of water. At the specified condition, the density of the CO_2 at the inlet of the compressor is 60% of the density of water, which results in low compression power requirement.

In addition to the density, other properties of CO_2 also change drastically around the critical point. As can be seen from Figure 14, the thermal conductivity of CO2 maximizes close to the critical point reaching 148.95 (mW)/(m-K) at 305 K. According to REFPROP (Lemmon et al., 2013), the thermal conductivity of water at 305K is 618.41 (m W)/(m-K) . At the atmospheric pressure and the same temperature, the thermal conductivity of air is given as 26.355 (mW)/(m-K).

Figure 15 shows how the specific heat of CO_2 changes close to the critical point. The large variations in specific heat affect the recuperator design in the power cycle. It is known that for a certain operating condition, a pinch-point exists in the recuperator. The pinch-point is the location where the temperature difference between the hot and cold streams is the lowest. As the specific heat varies radically with the changes in the pressure and the temperature, the temperature difference between the fluids varies widely within the recuperator. Consequently, the pinch-point location may be found somewhere along the recuperator, not at the inlet or the outlet. Therefore, more detailed analysis of the temperature profiles is necessary to evaluate the performance of the recuperator. Moreover, the recuperator size and efficiency are directly affected by the operating pressure. Therefore, unlike many other recuperators working with ideal gases such as helium where the temperature difference is almost constant and only dependent on

Figure 13. Variations of CO_2 density at different temperatures and pressures
Taken from (Mostaghim, 2014).

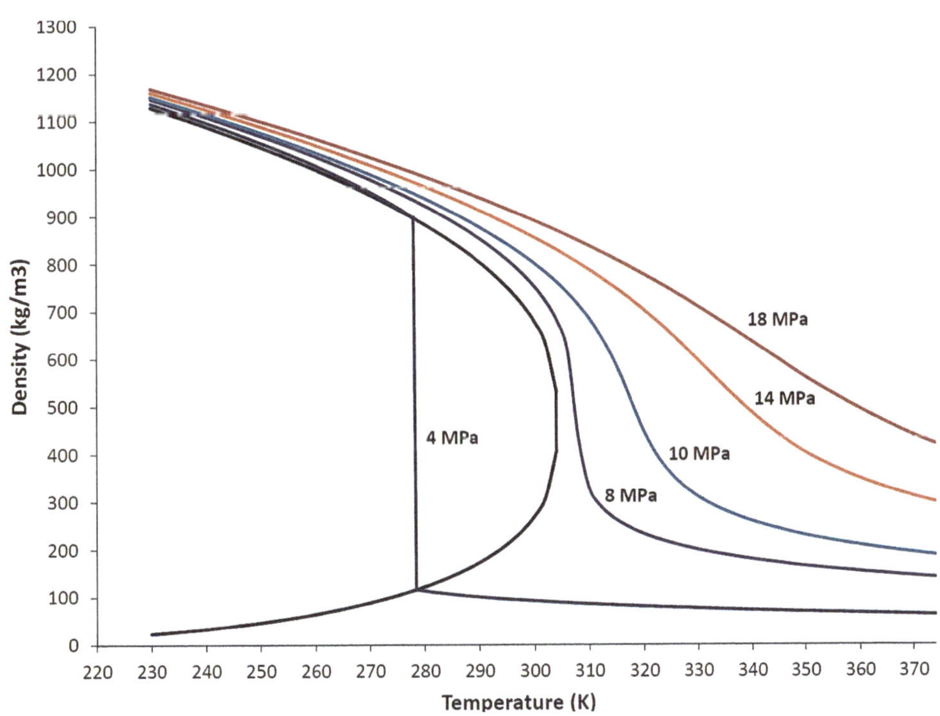

Figure 14. Variations of CO_2 thermal conductivity at different operating conditions
Taken from (Mostaghim, 2014).

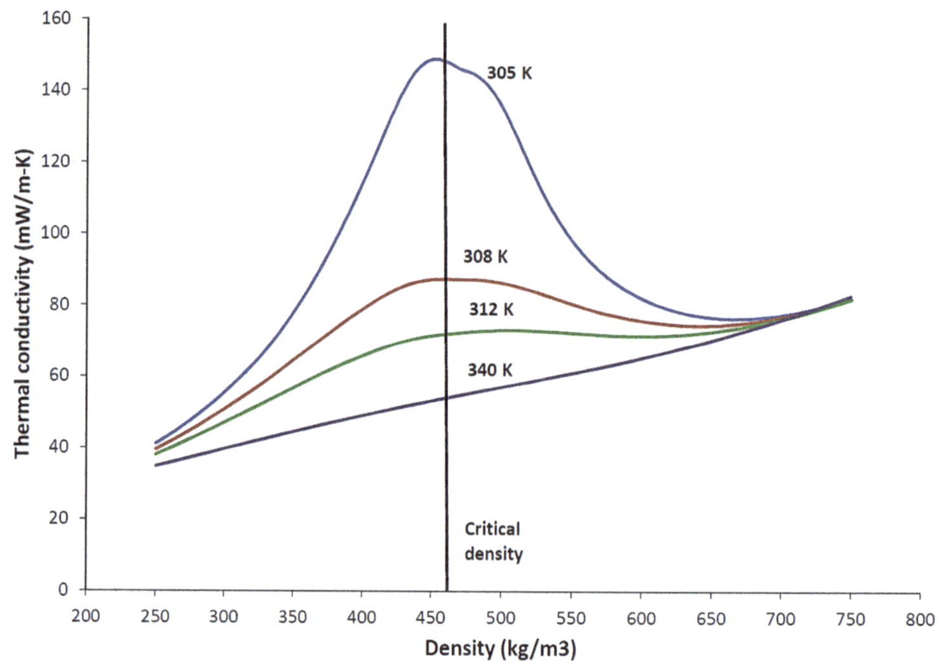

Figure 15. Variations of the specific heat close to the critical temperature at different pressures
Taken from (Mostaghim, 2014).

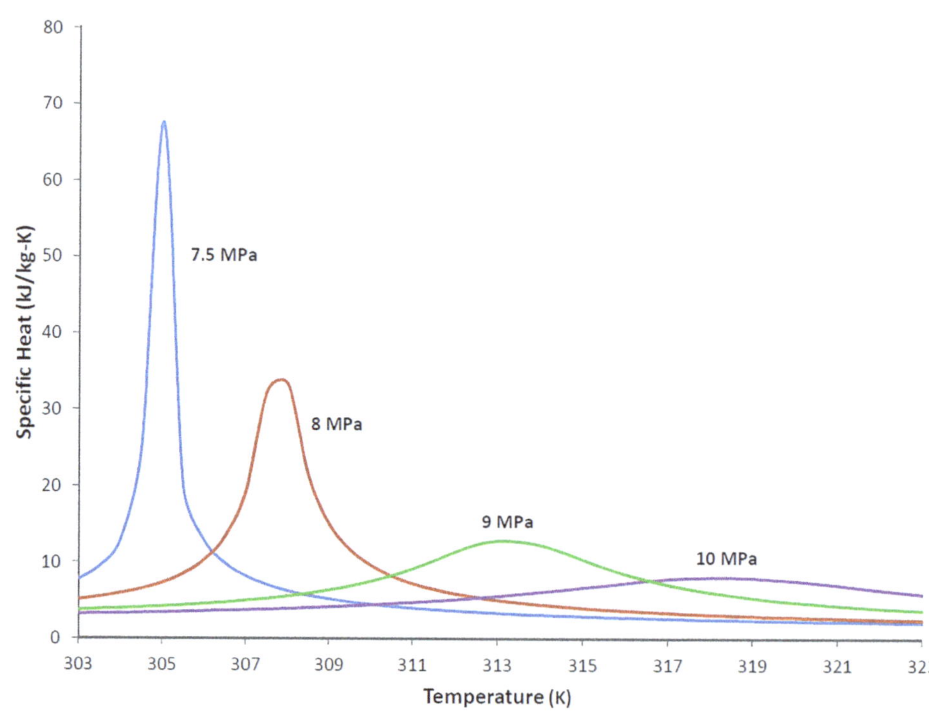

the pressure ratio and the temperatures, the operating pressure is also important and has to be optimally determined. In addition, the high specific heat of the CO_2 close to the critical point requires high mass flow rate cooling water in the precooler which increases the parasitic losses (Dostál, 2004).

S-CO_2 Cycle Configurations

There are usually four closed loop configurations of S-CO_2 Brayton cycles (Padilla et al., 2015). These configurations are based on modifications proposed to improve the thermal efficiency of S-CO_2 Brayton cycle (Moisseytsev et al., 2008b): multiple recompression, intercooling between compressor and reheating. The description of each configuration is presented below. Figure 16 shows the typical temperature-entropy diagrams for the four S-CO_2 Brayton cycle layouts with reheating at a turbine inlet temperature of 700 °C.

Simple Cycle (Simple)

The simple configuration is based on a simple Brayton cycle incorporating a recuperator (Figure 17). In this cycle, CO_2 at high temperature and high pressure (state 1) is expanded at intermediate pressure through a primary turbine (first stage, T1). The exit stream (state 2) is reheated at high temperature (state 3) and subsequently expanded at lower pressure through a secondary turbine (T2).

Figure 16. Temperature entropy diagram for different S-CO_2 Brayton cycle configurations. Operating conditions: Turbine inlet temperature of 700 °C, cycles high pressure 25 MPa

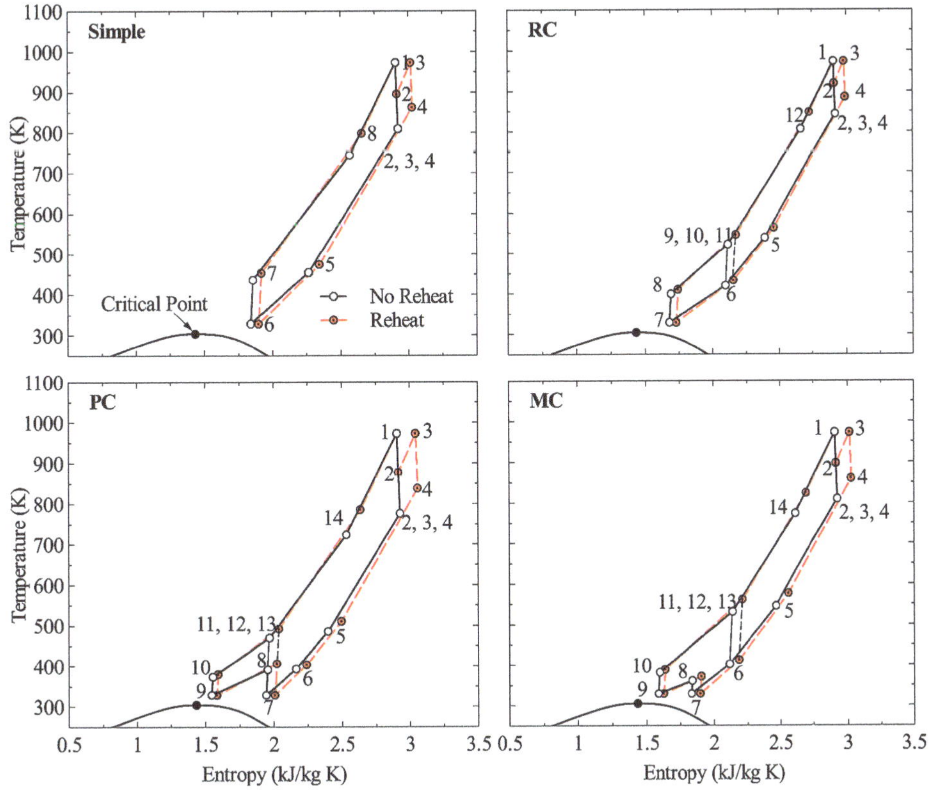

A recuperator (HTR, High thermal regenerator) is used to recover energy from the exit stream (state 4) leaving the secondary turbine and preheat the stream that is going to the solar receiver (state 8). In order to increase the pressure (state 6), the stream leaving the regenerator (state 5) is cooled (state 6) prior to the compressor (C1). Part of the work produced by the turbines is used to drive the compressor (C1).

Recompression Cycle (RC)

Due to the rapid variation of thermophysical properties of CO_2 near critical conditions (Dostál, 2004) the Simple cycle configuration is limited by temperature pinch point problems in the HTR due to the difference in the heat capacity rate between the hot and cold side (C. S. Turchi, Ma, Neises, & Wagner, 2013b). Therefore, a S-CO_2 recompression Brayton cycle configuration (Figure 18) is proposed. In this configuration, two recuperators are used (LTR, Low thermal recuperator and HTR, high thermal recuperator). The stream leaving the LTR (state 6) is split into two streams; the first stream enters the main compressor (C1) and the second stream is diverted to the recompressor, which is a secondary compressor operating at the exit pressure and temperature of the low thermal recuperator (Padilla et al., 2015). This split flow reduces cold fluid capacitance and therefore avoids pinch point problems. The stream leaving the main compressor (state 8) is preheated in the LTR and then mixed with the stream (state 10) leaving from the recompressor (C2). The mixed stream (state 11) recovers energy though the HTR prior to the solar receiver. In this configuration, part of the work produced by the turbines is also used to drive both compressors (C1 and C2).

Figure 17. S-CO_2 Simple Brayton cycle configurations with reheat
Adapted from (Padilla et al., 2015).

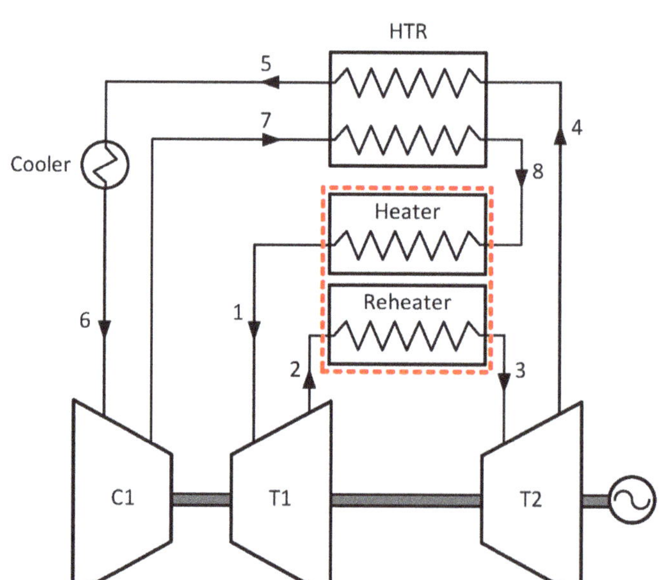

Figure 18. S-CO_2 Recompression Brayton cycle configurations with reheat
Adapted from (Padilla et al., 2015).

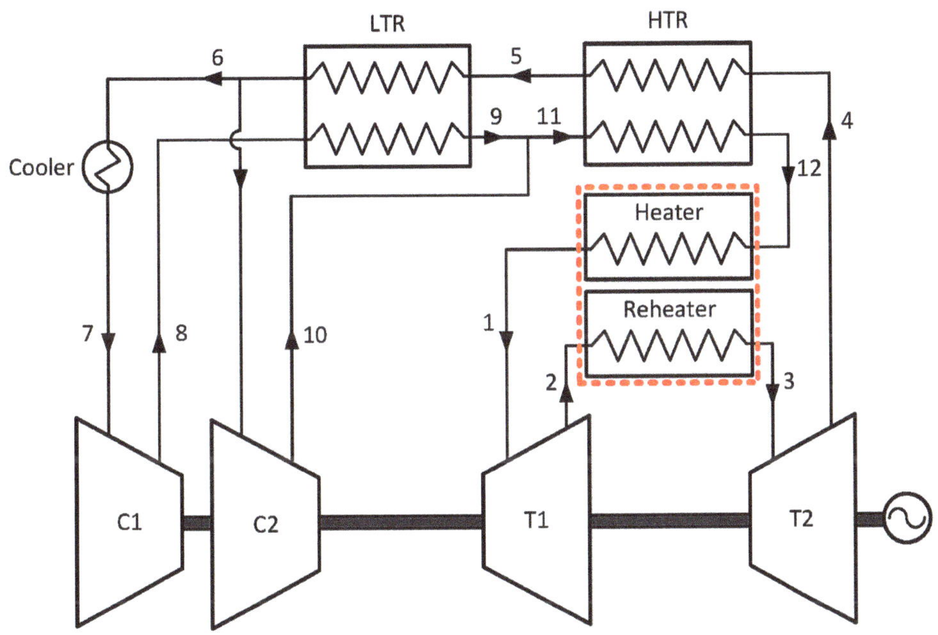

Partial Cooling Cycle (PC)

In order to reduce the work input during the compression process, intercooling between different compression stages can be incorporated. In the partial cooling configuration (Figure 19) the stream exiting from the low thermal recuperator (LTR) is cooled (state 6) and compressed (state 8) in the first stage of the main compressor (C1).

Then, the stream (state 8) is split into two; one stream is cooled (state 9) and recompressed to the high pressure of the cycle (state 10) before is preheated (state 11) in the low thermal recuperator (LTR), whereas the other stream is directly compressed at the high pressure of the cycle (state 12) and mixed with the stream exiting the LTR (stream 11). The mixed stream (state 13) recovers heat in the high thermal recuperator (HTR) prior to entering the heater, with the remaining flow path similar to the recompression cycle described previously.

Recompression with Main Compression Intercooling (MC)

This configuration (Figure 20) is similar to the recompression cycle, except that an intermediate intercooling stage is introduced in the main compression process (C1 and C2).

Other alternative configurations have also proposed (Clifford K. Ho, 2015). In addition, Padilla et al. (2016) presented an alternative configuration of a S-CO_2 Brayton cycle configurations without reheat integrated with a ejector, which reduces the pressure at the solar receiver without having a significant impact on the thermal efficiency as compared to the convectional supercritical CO_2 Brayton cycles (Figure 21).

Figure 19. S-CO_2 Partial Cooling Brayton cycle configurations with reheat
Adapted from (Padilla et al., 2015).

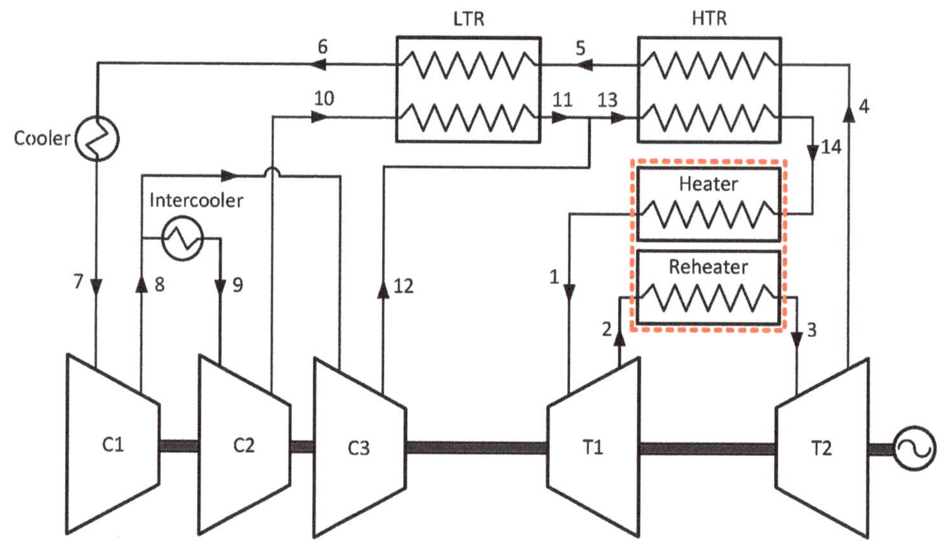

Figure 20. S-CO_2 Recompression with Main Compression Intercooling Brayton cycle configurations with reheat
Adapted from (Padilla et al., 2015).

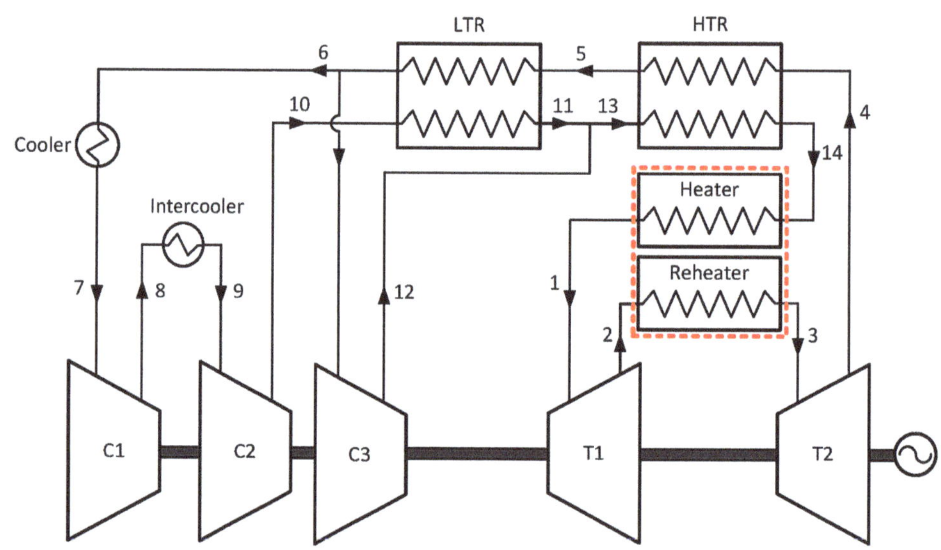

FUTURE RESEARCH DIRECTIONS

There has been extensive research published on supercritical fluids used in thermodynamic cycles for power conversion. On the low-grade heat conversion side, working fluids selection and system optimization are continuously seen in recent publications. However, in order for the supercritical cycles to be used for real power generation systems, more headwear design and experimental work need to be done.

Figure 21. S-CO_2 Simple Brayton cycle configurations assisted by ejector

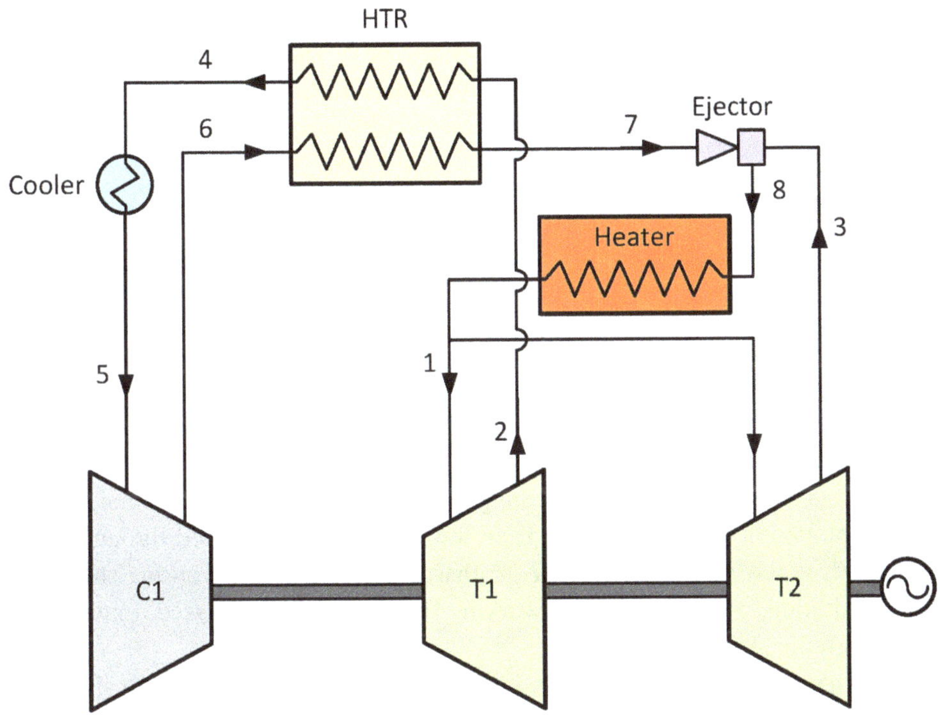

Specifically, questions such as how the transition from liquid to supercritical fluid impacts the design of the heat exchanger, what are the major design specifications required for a turbine to expand a supercritical fluid, and how to design the condensation process to make the best use of its property if the working fluid is a zeotropic fluid need to be answered and validated through experiments. Zeotropic mixture as a working fluid for supercritical Rankine cycle have great potential as it can potentially improve the exergy efficiency of the heating and cooling processes of the cycle. Potential studies on zeotropic fluids for supercritical Rankine cycle may include making the proper zeotropic mixture for specific applications, such as the thermal glide created by the zeotropic mixture fluid during the condensation process would allow good thermal matching with its cooling agent. Some organic fluids have stability and compatibility issues at relatively high temperate, so validating the fluids being used in the cycle can sustain the operating condition with experimental data is very important. Eventually, the power cycle needs to make economic sense, therefore, technical-economic analysis of the supercritical Rankine cycles would have values on the practicality side.

On the high-grade heat conversion side, S-CO_2 Brayton cycle has the potential to be the next power cycle for nuclear and solar concentrated systems. Extensive research has been published for S-CO_2 Brayton cycle in the area of thermo-economic optimization, heat exchangers, turbines and transient analysis and off-design operating conditions. One of the key issues with this power cycle is the design of an appropriate turbine for the high temperature and pressure conditions; therefore more research needs to be carried out in this area to find a suitable turbine for this cycle. In addition, due to the uncertainties associated with the small scale of the proposed experimental setups, it is necessary to build, in a short term, a S-CO_2 Brayton cycle in a commercial scale to show the advantage of this power cycle and find the right balance between the investment and operational costs and its thermal efficiency.

For the case of concentrated solar thermal systems, there are many challenges associated with the integration of the S-CO_2 Brayton cycles with the solar system, thermal storage and alternative heat transfer fluids. Direct and indirect configurations have been proposed based on the operating conditions of CO_2 at the high pressure side. In addition, development of experimental setups to integrate S-CO_2 Brayton cycles with solar towers is necessary to study the transient effect associated with the intermittence of solar radiation.

CONCLUSION

This chapter discussed supercritical working fluids and their applications in power cycles for both high- and low- grade heat conversion. The properties of the working fluids play vital role in the cycle performance, therefore, impact of the working fluid candidates' thermos-physical properties such as types of working fluids, influences of latent heat, density and specific heat, and the effectiveness of superheating were discussed in detail. Among all the fluids suggested for low-grade heat conversion, 35 fluids were screened out, and plotted in the $T-\xi$ charts. Then the fluids were discussed through grouping based on their distributions in the $T-\xi$ charts. The $T-\xi$ charts can be used for working fluid selection for a specific cycle. In addition, zeotropic mixture working fluid is briefly discussed. Extended information can be found through the references provided.

For high-grade heat conversion, the theoretical and experimental results have showed that S-CO_2 Brayton cycle is a promising option for central solar receivers systems due to high thermal efficiencies compared to conventional steam Rankine cycles. Although there are many challenges, researchers have focused on finding a balance between the investment and operational costs of S-CO_2 Brayton cycle and the effect of the heat source (solar, nuclear, etc.) on its optimal operating conditions and thermal efficiency.

REFERENCES

Andersen, W. C., & Bruno, T. J. (2005). Rapid Screening of Fluids for Chemical Stability in Organic Rankine Cycle Applications. *Industrial & Engineering Chemistry Research, 44*(15), 5560–5566. https://doi.org/10.1021/ie050351s

Bae, S. J., Lee, J., Ahn, Y., & Lee, J. I. (2015). Preliminary studies of compact Brayton cycle performance for Small Modular High Temperature Gas-cooled Reactor system. *Annals of Nuclear Energy, 75*, 11–19. doi:10.1016/j.anucene.2014.07.041

Bakhtar, F., Henson, R. J. K., & Mashmoushy, H. (2006). On the Performance of a Cascade of Turbine Rotor Tip Section Blading in Wet Steam. Part 5: Theoretical Treatment. *Proceedings of the Institution of Mechanical Engineers. Part C, Journal of Mechanical Engineering Science, 220*(4), 457–472. doi:10.1243/09544062JMES151

Bakhtar, F., & Mahpeykar, M. R. (1997). On the performance of a cascade of turbine rotor tip section blading in nucleating steam Part 3: Theoretical treatment. *Proceedings of the Institution of Mechanical Engineers. Part C, Journal of Mechanical Engineering Science, 211*(3), 195–210. doi:10.1243/0954406971521773

Bakhtar, F., Mashmoushy, H., & Buckley, J. R. (1997). On the performance of a cascade of turbine rotor tip section blading in wet steam Part 1: Generation of wet steam of prescribed droplet sizes. *Proceedings of the Institution of Mechanical Engineers. Part C, Journal of Mechanical Engineering Science, 211*(7), 519–529. doi:10.1243/0954406971521908

Bakhtar, F., Mashmoushy, H., & Jadayel, O. C. (1997). On the performance of a cascade of turbine rotor tip section blading in wet steam Part 2: Surface pressure distributions. *Proceedings of the Institution of Mechanical Engineers. Part C, Journal of Mechanical Engineering Science, 211*(7), 531–540. doi:10.1243/0954406971521917

Bakhtar, F., Rassam, S. Y., & Zhang, G. (1999). On the performance of a cascade of turbin rotor tip section blading in wet steam Part 4: Droplet measurements. *Proceedings of the Institution of Mechanical Engineers. Part C, Journal of Mechanical Engineering Science, 213*(4), 343–353. doi:10.1243/0954406991522301

Besarati, S. M., & Goswami, D. Y. (2014). Analysis of Advanced Supercritical Carbon Dioxide Power Cycles With a Bottoming Cycle for Concentrating Solar Power Applications. *Journal of Solar Energy Engineering, 136*(1), 010904–010904.

Calderazzi, L., & di Paliano, P. C. (1997). Thermal stability of R-134a, R-141b, R-13I1, R-7146, R-125 associated with stainless steel as a containing material. *International Journal of Refrigeration, 20*(6), 381–389. doi:10.1016/S0140-7007(97)00043-1

Carstens, N. A., Vilim, R. B., Hejzlar, P., & Driscoll, M. J. (2008). Dynamic modeling of the S-CO2 recompression cycle. In *International Conference on Advances in Nuclear Power Plants, ICAPP 2008* (Vol. 2, pp. 784–798).

Cassetti, G., & Colombo, E. (2013). Minimization of local impact of energy systems through exergy analysis. *Energy Conversion and Management, 76*, 874–882. doi:10.1016/j.enconman.2013.08.044

Cayer, E., Galanis, N., & Nesreddine, H. (2010). Parametric study and optimization of a transcritical power cycle using a low temperature source. *Applied Energy, 87*(4), 1349–1357. doi:10.1016/j.apenergy.2009.08.031

Chacartegui, R., Sánchez, D., Muñoz, J. M., & Sánchez, T. (2009). Alternative ORC bottoming cycles FOR combined cycle power plants. *Applied Energy, 86*(10), 2162–2170. doi:10.1016/j.apenergy.2009.02.016

Chen, H., Goswami, D. Y., Rahman, M. M., & Stefanakos, E. K. (2011). A supercritical Rankine cycle using zeotropic mixture working fluids for the conversion of low-grade heat into power. *Energy, 36*(1), 549–555. doi:10.1016/j.energy.2010.10.006

Chen, H., Goswami, D. Y., & Stefanakos, E. K. (2010). A review of thermodynamic cycles and working fluids for the conversion of low-grade heat. *Renewable & Sustainable Energy Reviews, 14*(9), 3059–3067. doi:10.1016/j.rser.2010.07.006

Chen, H., Yogi Goswami, D., Rahman, M. M., & Stefanakos, E. K. (2011). Energetic and exergetic analysis of CO_2- and R32-based transcritical Rankine cycles for low-grade heat conversion. *Applied Energy, 88*(8), 2802–2808. doi:10.1016/j.apenergy.2011.01.029

Chen, Y. (2006). *Novel cycles using carbon dioxide as working fluid: new ways to utilize energy from low-grade heat sources*. KTH, Stockholm, Sweden. Retrieved from https://www.diva-portal.org/smash/get/diva2:10578/FULLTEXT01.pdf

Chen, Y., Lundqvist, P., Johansson, A., & Platell, P. (2006). A comparative study of the carbon dioxide transcritical power cycle compared with an organic rankine cycle with R123 as working fluid in waste heat recovery. *Applied Thermal Engineering*, *26*(17–18), 2142–2147. doi:10.1016/j.applthermaleng.2006.04.009

Chen, Y., Lundqvist, P., & Platell, P. (2005). Theoretical research of carbon dioxide power cycle application in automobile industry to reduce vehicles fuel consumption. *Applied Thermal Engineering*, *25*(14–15), 2041–2053. doi:10.1016/j.applthermaleng.2005.02.001

Clementoni, E. M., Cox, T. L., & Sprague, C. P. (2014). Startup and Operation of a Supercritical Carbon Dioxide Brayton Cycle. *Journal of Engineering for Gas Turbines and Power*, *136*(7), 71701. doi:10.1115/1.4026539

Conboy, T., Pasch, J., & Fleming, D. (2013). Control of a Supercritical CO 2 Recompression Brayton Cycle Demonstration Loop. *Journal of Engineering for Gas Turbines and Power*, *135*(11), 111701. doi:10.1115/1.4025127

Conboy, T., Wright, S., Pasch, J., Fleming, D., Rochau, G., & Fuller, R. (2012). Performance characteristics of an operating supercritical CO_2 Brayton cycle. *Journal of Engineering for Gas Turbines and Power*, *134*(11), 111703. doi:10.1115/1.4007199

Crook, A. W., & Engineers, I. of E. (1994). Profiting from Low-grade Heat: Thermodynamic Cycles for Low-temperature Heat Sources. IET.

Dai, Y., Wang, J., & Gao, L. (2009). Parametric optimization and comparative study of Organic Rankine Cycle (ORC) for low grade waste heat recovery. *ResearchGate*, *50*(3), 576–582. https://doi.org/10.1016/j.enconman.2008.10.018

Delgado-Torres, A. M., & García-Rodríguez, L. (2007). Preliminary assessment of solar organic Rankine cycles for driving a desalination system. *Desalination*, *216*(1), 252–275. doi:10.1016/j.desal.2006.12.011

Demuth, O. J. (1981). Analyses of Mixed Hydrocarbons Binary Thermodynamic Cycles for Moderate Temperature Geothermal Resources. *ResearchGate, 2*. Retrieved from https://www.researchgate.net/publication/253167221_Analyses_of_Mixed_Hydrocarbons_Binary_Thermodynamic_Cycles_for_Moderate_Temperature_Geothermal_Resources

Demuth, O. J. (1983). *Preliminary assessment of condensation behavior for hydrocarbon-vapor expansions which cross the saturation line near the critical point* (No. EGG-M-00283; CONF-830812-54). EG and G Idaho, Inc. Retrieved from http://www.osti.gov/geothermal/biblio/5673705-preliminary-assessment-condensation-behavior-hydrocarbon-vapor-expansions-which-cross-saturation-line-near-critical-point

DiPippo, R. (2004). Second Law assessment of binary plants generating power from low-temperature geothermal fluids. *Geothermics*, *33*(5), 565–586. doi:10.1016/j.geothermics.2003.10.003

Dostál, V. (2004). *A supercritical carbon dioxide cycle for next generation nuclear reactors*. Massachusetts Institute of Technology.

Dostál, V., Hejzlar, P., & Driscoll, M. J. (2006). The supercritical carbon dioxide power cycle: Comparison to other advanced power cycles. *Nuclear Technology*, *154*(3), 283–301.

Dunham, M. T., & Iverson, B. D. (2014). High-efficiency thermodynamic power cycles for concentrated solar power systems. *Renewable & Sustainable Energy Reviews*, *30*(0), 758–770. doi:10.1016/j.rser.2013.11.010

Dyreby, J. J., Klein, S. A., Nellis, G. F., & Reindl, D. T. (2013). Modeling Off-Design and Part-Load Performance of Supercritical Carbon Dioxide Power Cycles. In *ASME Turbo Expo 2013: Turbine Technical Conference and Exposition. Volume 8: Supercritical CO_2 Power Cycles; Wind Energy; Honors and Awards* (Vol. 8, p. V008T34A014). ASME. https://doi.org/10.1115/GT2013-95824

Etemoglu, A. B. (2008). Thermodynamic Evaluation of Geothermal Power Generation Systems in Turkey. *Energy Sources, Part A: Recovery, Utilization, and Environmental Effects*, *30*(10), 905–916. https://doi.org/10.1080/15567030601082589

Feher, E. G. (1968). The supercritical thermodynamic power cycle. *Energy Conversion*, *8*(2), 85–90. doi:10.1016/0013-7480(68)90105-8

Garg, P., Kumar, P., Dutta, P., Conboy, T., & Ho, C. (2014). Design of an Experimental Test Facility for Supercritical CO2 Brayton Cycle. In *ASME 2014 8th International Conference on Energy Sustainability collocated with the ASME 2014 12th International Conference on Fuel Cell Science, Engineering and Technology* (p. V001T05A005). ASME.

Garg, P., Kumar, P., & Srinivasan, K. (2013). Supercritical carbon dioxide Brayton cycle for concentrated solar power. *The Journal of Supercritical Fluids*, *76*, 54–60. doi:10.1016/j.supflu.2013.01.010

Ginosar, D. M., Petkovic, L. M., & Guillen, D. P. (2011). Thermal Stability of Cyclopentane as an Organic Rankine Cycle Working Fluid. *Energy & Fuels*, *25*(9), 4138–4144. doi:10.1021/ef200639r

Goswami, D. Y., Hingorani, S., & Mines, G. (1991). A Laser-Based Technique for Particle Sizing to Study Two-Phase Expansion in Turbines. *Journal of Solar Energy Engineering*, *113*(3), 211–218. doi:10.1115/1.2930495

Heberle, F., Preißinger, M., & Brüggemann, D. (2012). Zeotropic mixtures as working fluids in Organic Rankine Cycles for low-enthalpy geothermal resources. *Renewable Energy*, *37*(1), 364–370. doi:10.1016/j.renene.2011.06.044

Hejzlar, P., Dostal, V., & Driscoll, M. J. (2006). A supercritical CO2 cycle- a promising power conversion system for Generation IV reactors. In *Proceedings of the 2006 International Congress on Advances in Nuclear Power Plants, ICAPP'06* (Vol. 2006, pp. 722–731).

Ho, C. K. (2015). *Cost and Performance Tradeoffs of Alternative Solar-Driven Supercritical CO_2 Power Cycles (presentation). (No. SAND2015-4055PE). Sandia National Laboratories*. Albuquerque, NM: SNL-NM.

Ho, C. K., Conboy, T., Ortega, J., Afrin, S., Gray, A., Christian, J. M., ... Wani, P. (2014). High-Temperature Receiver Designs for Supercritical CO_2 Closed-Loop Brayton Cycles. In *ASME 2014 12th International Conference on Fuel Cell Science, Engineering and Technology* (Vol. 1, p. V001T02A003). ASME. https://doi.org/10.1115/ES2014-6328

Hung, T.-C. (2001). Waste heat recovery of organic Rankine cycle using dry fluids. *ResearchGate, 42*(5), 539–553. https://doi.org/10.1016/S0196-8904(00)00081-9

Hung, T. C., Shai, T. Y., & Wang, S. K. (1997). A review of organic rankine cycles (ORCs) for the recovery of low-grade waste heat. *Energy*, *22*(7), 661–667. doi:10.1016/S0360-5442(96)00165-X

Iverson, B. D., Conboy, T. M., Pasch, J. J., & Kruizenga, A. M. (2013). Supercritical CO_2 Brayton cycles for solar-thermal energy. *Applied Energy*, *111*(0), 957–970. doi:10.1016/j.apenergy.2013.06.020

Karellas, S., & Schuster, A. (2008). Supercritical Fluid Parameters in Organic Rankine Cycle Applications. *International Journal of Thermodynamics*, *11*(3), 101–108. https://doi.org/10.5541/ijot.1034000217

Kim, S. G., Lee, J., Ahn, Y., Lee, J. I., Addad, Y., & Ko, B. (2014). CFD investigation of a centrifugal compressor derived from pump technology for supercritical carbon dioxide as a working fluid. *The Journal of Supercritical Fluids*, *86*, 160–171. doi:10.1016/j.supflu.2013.12.017

Kimball, K. J., & Clementoni, E. M. (2012). Supercritical Carbon Dioxide Brayton Power Cycle Development Overview. In *ASME Turbo Expo 2012: Turbine Technical Conference and Exposition. Volume 5: Manufacturing Materials and Metallurgy; Marine; Microturbines and Small Turbomachinery; Supercritical CO_2 Power Cycles* (Vol. 5, p. 931). ASME. https://doi.org/10.1115/GT2012-68204

Kulhánek, M., & Dostál, V. (2011). Thermodynamic Analysis and Comparison of Supercritical Carbon Dioxide Cycles. *Proceedings of Supercritical CO_2 Power Cycle Symposium.*

Larjola, J. (1995). Electricity from industrial waste heat using high-speed organic Rankine cycle (ORC). *International Journal of Production Economics*, *41*(1), 227–235. doi:10.1016/0925-5273(94)00098-0

Lee, J., Lee, J. I., Ahn, Y., & Yoon, H. (2012). Design Methodology of Supercritical CO_2 Brayton Cycle Turbomachineries. In *ASME Turbo Expo 2012: Turbine Technical Conference and Exposition. Volume 5: Manufacturing Materials and Metallurgy; Marine; Microturbines and Small Turbomachinery; Supercritical CO_2 Power Cycles* (Vol. 5, p. 975). ASME. https://doi.org/10.1115/GT2012-68933

Lemmon, E. W., Huber, M. L., & McLinden, M. O. (2013). *NIST Standard Reference Database 23: Reference Fluid Thermodynamic and Transport Properties-REFPROP, Version 9.1.* NIST.

Liu, Q., Duan, Y., & Yang, Z. (2014). Effect of condensation temperature glide on the performance of organic Rankine cycles with zeotropic mixture working fluids. *Applied Energy*, *115*, 394–404. doi:10.1016/j.apenergy.2013.11.036

Ma, Z., & Turchi, C. (2011). Advanced supercritical carbon dioxide power cycle configurations for use in concentrating solar power systems. *Proceedings of Supercritical CO_2 Power Cycle Symposium.*

Madhawa Hettiarachchi, H. D., Golubovic, M., Worek, W. M., & Ikegami, Y. (2007). Optimum design criteria for an Organic Rankine cycle using low-temperature geothermal heat sources. *Energy*, *32*(9), 1698–1706. doi:10.1016/j.energy.2007.01.005

Mago, P. J., Chamra, L. M., Srinivasan, K., & Somayaji, C. (2008). An examination of regenerative organic Rankine cycles using dry fluids. *Applied Thermal Engineering*, *28*(8–9), 998–1007. doi:10.1016/j.applthermaleng.2007.06.025

Maizza, V., & Maizza, A. (1996). Working fluids in non-steady flows for waste energy recovery systems. *Applied Thermal Engineering*, *16*(7), 579–590. doi:10.1016/1359-4311(95)00044-5

McMahan, A., Klein, S. A., & Reindl, D. T. (2007). A Finite-Time Thermodynamic Framework for Optimizing Solar-Thermal Power Plants. *Journal of Solar Energy Engineering*, *129*(4), 355–362. doi:10.1115/1.2769689

Moisseytsev, A., & Sienicki, J. et al. (2008a). *Performance improvement options for the supercritical carbon dioxide Brayton cycle. Argonne National Laboratory (ANL). Funding organisation.* USDOE Office of Science. doi:10.2172/935094

Moisseytsev, A., Sienicki, J., & others. (2008b). *Performance improvement options for the supercritical carbon dioxide Brayton cycle*. Academic Press.

Moisseytsev, A., & Sienicki, J. (2011). *Investigation of plant control strategies for the supercritical CO_2 Brayton cycle for a sodium-cooled fast reactor using the plant dynamics code. Argonne National Laboratory (United States). Funding organisation.* USDOE Office of Nuclear Energy, Science and Technology.

Moisseytsev, A., & Sienicki, J. J. (2008a). Controllability of the supercritical carbon dioxide brayton cycle near the critical point. In *International Conference on Advances in Nuclear Power Plants, ICAPP 2008* (Vol. 2, pp. 799–809).

Moisseytsev, A., & Sienicki, J. J. (2008b). Transient accident analysis of a supercritical carbon dioxide Brayton cycle energy converter coupled to an autonomous lead-cooled fast reactor. *Nuclear Engineering and Design*, *238*(8), 2094–2105. doi:10.1016/j.nucengdes.2007.11.012

Moisseytsev, A., & Sienicki, J. J. (2012). Dynamic simulation and control of the S-CO_2 cycle: From full power to decay heat removal. In *Advances in Thermal Hydraulics 2012* (pp. 52–60). ATH.

Mostaghim Besarati, S. (2014). *Analysis of Advanced Supercritical Carbon Dioxide Power Cycles for Concentrated Solar Power Applications*. Graduate Theses and Dissertations.

Musgrove, G., Conboy, T. M., & Wilkes, J. (2014). *Fundamentals of Supercritical CO_2*. Presented at the ASME Turbo Expo 2014, Düsseldorf, Germany.

Nowak, W., Borsukiewicz-Gozdur, A., & Stachel, A. A. (2008). Using the low-temperature Clausius–Rankine cycle to cool technical equipment. *Applied Energy*, *85*(7), 582–588. doi:10.1016/j.apenergy.2007.09.001

Ochs, T. L., & O'Connor, W. K. (2006, March 7). *Energy recovery during expansion of compressed gas using power plant low-quality heat sources*. Retrieved from http://www.google.com/patents/US7007474

Padilla, R. V., Soo Too, Y. C., Benito, R., & Stein, W. (2015). Exergetic analysis of supercritical CO_2 Brayton cycles integrated with solar central receivers. *Applied Energy*, *148*, 348–365. doi:10.1016/j.apenergy.2015.03.090

Padilla, R. V., Too, Y. C. S., Benito, R., McNaughton, R., & Stein, W. (2016). Thermodynamic feasibility of alternative supercritical CO_2 Brayton cycles integrated with an ejector. *Applied Energy*, *169*, 49–62. doi:10.1016/j.apenergy.2016.02.029

Pasch, J., Conboy, T., Fleming, D., Carlson, M., & Rochau, G. (2014). Steady State Supercritical Carbon Dioxide Recompression Closed Brayton Cycle Operating Point Comparison With Predictions. In *ASME Turbo Expo 2014: Turbine Technical Conference and Exposition* (p. V03BT36A008). ASME. doi:10.1115/GT2014-25777

Pecnik, R., Rinaldi, E., & Colonna, P. (2012). Computational Fluid Dynamics of a Radial Compressor Operating With Supercritical CO_2. In *ASME Turbo Expo 2012: Turbine Technical Conference and Exposition. Volume 5: Manufacturing Materials and Metallurgy; Marine; Microturbines and Small Turbomachinery; Supercritical CO_2 Power Cycles* (Vol. 5, p. 985). ASME. https://doi.org/10.1115/GT2012-69640

Persichilli, M., Kacludis, A., Zdankiewicz, E., & Held, T. (2012). *Supercritical CO_2 Power Cycle Developments and Commercialization: Why SCO_2 can Displace Steam*. Presented at the Power-Gen India & Central Asia.

Quoilin, S. (2007, June). *Experimental study and modeling of a low temperature Rankine Cycle for small scale cogeneration*. Université de Liège. Retrieved from http://orbi.ulg.ac.be/handle/2268/1273

Sargent and Lundy LLC. (2003). *Assessment of Parabolic Trough and Power Tower Solar Technology Cost and Performance Forecasts*. DIANE Publishing.

Singh, R., Kearney, M. P., & Manzie, C. (2013). Extremum-seeking control of a supercritical carbon-dioxide closed Brayton cycle in a direct-heated solar thermal power plant. *Energy*, *60*, 380–387. doi:10.1016/j.energy.2013.08.001

Singh, R., Miller, S. A., Rowlands, A. S., & Jacobs, P. A. (2013). Dynamic characteristics of a direct-heated supercritical carbon-dioxide Brayton cycle in a solar thermal power plant. *Energy*, *50*, 194–204. doi:10.1016/j.energy.2012.11.029

Supercritical fluids - YouTube. (n.d.). Retrieved October 27, 2016, from https://www.youtube.com/watch?v=yBRdBrnIlTQ

Tchanche, B. F., Papadakis, G., Lambrinos, G., & Frangoudakis, A. (2009). Fluid selection for a low-temperature solar organic Rankine cycle. *Applied Thermal Engineering*, *29*(11–12), 2468–2476. doi:10.1016/j.applthermaleng.2008.12.025

Tian, Y., & Zhao, C.-Y. (2013). A review of solar collectors and thermal energy storage in solar thermal applications. *Applied Energy*, *104*, 538–553. doi:10.1016/j.apenergy.2012.11.051

Turchi, C., Ma, Z., & Dyreby, J. (2009). Supercritical CO2 for application in concentrating solar power systems. *Proceedings of Supercritical CO_2 Power Cycle Symposium*.

Turchi, C., Mehos, M., Ho, C. K., & Kolb, G. J. (2010). *Current and Future Costs for Parabolic Trough and Power Tower Systems in the Us Market: Preprint (No. NREL/CP-5500-49303). National Renewable Energy Laboratory*. Golden, CO: NREL.

Turchi, C. S., Ma, Z., & Dyreby, J. (2012a). Supercritical Carbon Dioxide Power Cycle Configurations for Use in Concentrating Solar Power Systems. In *ASME Turbo Expo 2012: Turbine Technical Conference and Exposition. Volume 5: Manufacturing Materials and Metallurgy; Marine; Microturbines and Small Turbomachinery; Supercritical CO_2 Power Cycles* (Vol. 5, pp. 967–973). ASME. https://doi.org/10.1115/GT2012-68932

Turchi, C. S., Ma, Z., & Dyreby, J. (2012b). Supercritical carbon dioxide power cycle configurations for use in concentrating solar power systems. In *ASME Turbo Expo 2012: Turbine Technical Conference and Exposition* (pp. 967–973). San Diego, CA: American Society of Mechanical Engineers. doi:10.1115/GT2012-68932

Turchi, C. S., Ma, Z., Neises, T., & Wagner, M. (2012). Thermodynamic Study of Advanced Supercritical Carbon Dioxide Power Cycles for High Performance Concentrating Solar Power Systems. In *ASME Turbo Expo 2012: Turbine Technical Conference and Exposition* (pp. 375–383). San Diego, CA: American Society of Mechanical Engineers. doi:10.1115/ES2012-91179

Turchi, C. S., Ma, Z., Neises, T. W., & Wagner, M. J. (2013a). Thermodynamic Study of Advanced Supercritical Carbon Dioxide Power Cycles for Concentrating Solar Power Systems. *Journal of Solar Energy Engineering*, *135*(4), 41007. doi:10.1115/1.4024030

Turchi, C. S., Ma, Z., Neises, T. W., & Wagner, M. J. (2013b). Thermodynamic Study of Advanced Supercritical Carbon Dioxide Power Cycles for Concentrating Solar Power Systems. *Journal of Solar Energy Engineering*, *135*(4), 041007–041007. doi:10.1115/1.4024030

Wang, J., Sun, Z., Dai, Y., & Ma, S. (2010). Parametric optimization design for supercritical CO_2 power cycle using genetic algorithm and artificial neural network. *Applied Energy*, *87*(4), 1317–1324. doi:10.1016/j.apenergy.2009.07.017

Wright, S. A., Conboy, T. M., & Rochau, G. E. (2011). Supercritical CO2 Power Cycle Development Summary at Sandia National Laboratories. In *International Conference on ORC Power Systems Delft*.

Wright, S. A., Fuller, R., Noall, J., Radel, R., Vernon, M. E., & Pickard, P. S. (2008). Supercritical CO2 brayton cycle compression and control near the critical point. In *International Conference on Advances in Nuclear Power Plants, ICAPP 2008* (*Vol. 2*, pp. 810–819).

Wright, S. A., & Pickard, P. S. (2009). Supercritical CO2 test loop operation and first test results. In *International Congress on Advances in Nuclear Power Plants 2009, ICAPP 2009* (Vol. 1, pp. 351–360). Atomic Energy Society of Japan.

Wright, S. A., Pickard, P. S., Vernon, M. E., Radel, R. F., & Fuller, R. (2009). Description and test results from a supercritical CO2 Brayton cycle development program. In *7th International Energy Conversion Engineering Conference*. doi:10.2514/6.2009-4607

Yamamoto, T., Furuhata, T., Arai, N., & Mori, K. (2001). Design and testing of the Organic Rankine Cycle. *Energy*, *26*(3), 239–251. doi:10.1016/S0360-5442(00)00063-3

Zhang, X. R., Yamaguchi, H., Fujima, K., Enomoto, M., & Sawada, N. (2006). AIP Conference Proceedings: Vol. 832. *Experimental Performance Analysis of Supercritical CO2 Thermodynamic Cycle Powered by Solar Energy* (pp. 419–424). AIP Publishing. doi:10.1063/1.2204534

Zhang, X. R., Yamaguchi, H., Fujima, K., Enomoto, M., & Sawada, N. (2007). Theoretical analysis of a thermodynamic cycle for power and heat production using supercritical carbon dioxide. *Energy*, *32*(4), 591–599. doi:10.1016/j.energy.2006.07.016

Zhang, X.-R., Yamaguchi, H., & Uneno, D. (2007). Experimental study on the performance of solar Rankine system using supercritical CO_2. *Renewable Energy*, *32*(15), 2617–2628. doi:10.1016/j.renene.2007.01.003

Zhang, X. R., Yamaguchi, H., Uneno, D., Fujima, K., Enomoto, M., & Sawada, N. (2006). Analysis of a novel solar energy-powered Rankine cycle for combined power and heat generation using supercritical carbon dioxide. *Renewable Energy*, *31*(12), 1839–1854. doi:10.1016/j.renene.2005.09.024

Zia, J., Sevincer, E., Chen, H., Hardy, A., Wickersham, P., Kalra, C., … Vandeputte, T. (2013). *High-potential Working Fluids for Next Generation Binary Cycle Geothermal Power Plants* (Report). Retrieved from http://digital.library.unt.edu/ark:/67531/metadc846464/m1/14/

KEY TERMS AND DEFINITIONS

Brayton Cycle: A thermodynamic cycle that describes the workings of a constant pressure heat engine. Modern gas turbine engines and airbreathing jet engines follow the Brayton cycle to generate electricity.

Concentrated Solar Power: A system that generates power by using mirrors or lenses to concentrate a large area of sunlight, or solar thermal energy, onto a small area. Electricity is generated when the concentrated light is converted into heat to drive a power cycle.

Exergy: In thermodynamics, the exergy of a system is the maximum useful work possible during a process that brings the system into equilibrium with a heat reservoir. In contrast to energy, exergy accounts for the irreversibility of a process due to increase in entropy.

Recuperator: A special purpose counter-flow energy recovery heat exchanger.

Solar Receiver: A solar receiver collects heat by absorbing sunlight. This device is used in capturing solar radiation.

Supercritical Fluid: A supercritical fluid is a fluid at a temperature and pressure above its critical point, where distinct liquid and gas phases do not exist.

Supercritical Rankine cycle: A supercritical Rankine cycle is a thermodynamic process where heat is transferred to a fluid at a pressure above its critical point and the fluid is transformed into supercritical phase before being expanded in a turbine that generates power. The spent vapor is condensed to liquid and recycled back through the cycle.

Zeotropic Mixture: A zeotropic mixture is a mixture of fluids that do not have the same vapor phase and liquid phase composition at any point of the vapor–liquid equilibrium state and because of that, a zeotropic mixture does not have a critical point that a pure fluid has, instead, it has a "critical line".

Section 4

Supercritical Power Generation Systems and Green Energy Harvesting

Chapters in this section are mainly focused on power generation systems assisted by supercritical fluids and supercritical cycles. Based on the current understanding of supercritical transport mechanism (heat, mass, chemical reaction, etc.), system scale analysis is focused for the discussion of power efficiency and generation system designs. The compact and high efficiency in power engineering, chemical engineering, thermal engineering as well as nuclear engineering are found attractive for scientists and engineers in this field. Still, the contribution of green energy harvesting and other related supercritical energy systems are also included.

Chapter 13
Application of Supercritical Pressures in Power Engineering

Igor Pioro
University of Ontario Institute of Technology, Canada

Mohammed Mahdi
University of Ontario Institute of Technology, Canada

Roman Popov
University of Ontario Institute of Technology, Canada

ABSTRACT

SuperCritical Fluids (SCFs) have unique thermophyscial properties and heat-transfer characteristics, which make them very attractive for use in power industry. In this chapter, specifics of thermophysical properties and heat transfer of SCFs such as water, carbon dioxide and helium are considered and discussed. Also, particularities of heat transfer at SuperCritical Pressures (SCPs) are presented, and the most accurate heat-transfer correlations are listed. SuperCritical Water (SCW) is widely used as the working fluid in the SCP Rankine "steam"-turbine cycle in fossil-fuel thermal power plants. This increase in thermal efficiency is possible by application of high-temperature reactors and power cycles. Currently, six concepts of Generation-IV reactors are being developed, with coolant outlet temperatures of 500°C~1000°C. SCFs will be used as coolants (helium in GFRs and VHTRs; and SCW in SCWRs) and/or working fluids in power cycles (helium; mixture of nitrogen (80%) and helium [20%]; nitrogen, and carbon dioxide in Brayton gas-turbine cycles; and SCW "steam" in Rankine cycle).

INTRODUCTION

SuperCritical Fluids (SCFs) are not a human invention. Actually, compressed fluid (i.e., fluid at pressures above the critical one, but at temperatures below the critical one) - water exists in oceans below the depth of ~2.2 km, which corresponds to the critical pressure of water - 22.064 MPa. If there is an underwater volcano with molten lava at this depth then water can reach temperatures above the critical one (373.95°C) and becomes SCF.

DOI: 10.4018/978-1-5225-2047-4.ch013

Also, nature has been processing minerals in aqueous solutions at near or above the critical point of water for billions of years (Levelt & Sengers, 2000). In the late 1800s, scientists started to use this natural process in their labs for creating various crystals. During the last 50 - 60 years, this process, called hydrothermal processing (operating parameters: water pressure from 20 to 200 MPa and temperatures from 300 to 500ºC), has been widely used in the industrial production of high-quality single crystals (mainly gem stones) such as sapphire, tourmaline, zircon and others.

First works dedicated to specifics of thermophysical properties of fluids and heat transfer at Super-Critical Pressures (SCPs) started as early as the 1930s. E. Schmidt and his associates (Schmidt, 1960; Schmidt et al., 1946) investigated free-convection heat transfer of fluids at the near-critical point with the application to a new effective cooling system for turbine blades in jet engines. They found that free-convection Heat Transfer Coefficient (HTC) at the near-critical state was quite high and decided to use this advantage in single-phase thermosyphons with an intermediate working fluid at the near-critical point.

In the 1950s, the idea of using SuperCritical Water (SCW) appeared to be rather attractive for SC "steam" generators. At SCPs, there is no liquid-vapour-phase transition; due to that there is no such phenomenon as critical heat flux or dryout. Only within a certain range of parameters a deterioration of heat transfer may occur. The main objective of operating "steam" generators at SCPs was to increase the total thermal efficiency of a power plant. Work in this area was mainly done in the USA, former USSR, Germany and some other countries in the 1950s - 1980s. Currently, use of SCW in coal-fired power plants is the largest application of fluid at SCPs in industry.

At the end of the 1950s and the beginning of the 1960s, some studies were conducted to investigate a possibility of using SCFs in nuclear reactors (Pioro & Duffey, 2007) with the same objective as in the thermal-power engineering, i.e., to increase thermal efficiency of water-cooled-reactors Nuclear Power Plants (NPPs). Several concepts of nuclear reactors using SCW as a coolant were developed in the USA, former USSR and some other countries. However, this idea was abandoned for almost 30 years and regained momentum in the 1990s.

At the end of 20th century, leading nuclear-power countries started to look into a possibility to develop next generation nuclear-power reactors. And in 2001, under the leadership of the U.S. Department of Energy's (DOE) Office of Nuclear Energy, Science and Technology a framework for international co-operation in research and development for the next generation of nuclear-energy systems has been launched as the Generation-IV International Forum (GIF) Initiative, which was originally signed by Argentina, Brazil, Canada, France, Japan, Republic of Korea, South Africa, the United Kingdom (UK) and the United States (US). Later on, Switzerland (2002), Euratom (2003), the People's Republic of China (2006), the Russian Federation (2006), and Australia (2016) have joined the GIF.

GIF experts have evaluated about 130 reactor concepts and selected six reactor technologies for further research and development. These technologies include (listed in the alphabetical order): 1) Gas-cooled Fast Reactor (GFR), 2) Lead-cooled Fast Reactor (LFR), 3) Molten Salt Reactor (MSR), 4) Supercritical Water-cooled Reactor (SCWR), 5) Sodium-cooled Fast Reactor (SFR), and 6) Very High Temperature Reactor (VHTR).

All Generation-IV nuclear-reactor concepts are high-temperature reactors with reactor coolant outlet temperatures ranging from 500°C and up to 1000°C. These reactors will use SCFs as reactor coolants (helium in GFRs and VHTRs; SCW in SCWRs) and / or working fluids in power cycles (helium; mixture of nitrogen (80%) and helium (20%); nitrogen; and carbon dioxide in Brayton gas-turbine cycles; and SCW in Rankine "steam"-turbine cycle). Also, SCP carbon dioxide and refrigerants can be used in power cycles of geothermal plants.

SPECIFICS OF THERMOPHYSICAL PROPERTIES AND HEAT TRANSFER AT SCPS

The most widely used SCFs in power engineering are water and after that carbon dioxide, helium and refrigerants (Pioro (Ed.), 2016; Pioro & Duffey, 2007). Quite often, carbon dioxide and refrigerants were considered as modeling fluids and have been used instead of water due to significantly lower critical pressures and temperatures (see Table 1), which decreases the complexity and costs of thermalhydraulic experiments. Therefore, knowledge of thermophysical-properties specifics at critical and supercritical pressures is very important for the safe and efficient use of SCFs in power industry.

Definitions of Terms and Expressions Related to Critical and Supercritical Regions

Prior to a general discussion on specifics of thermophysical properties at critical and supercritical pressures, it is important to define special terms and expressions used at these conditions. For a better understanding of these terms and expressions their definitions (based on the following sources: Pioro (Ed.) (2016); Cengel & Boles (2015), Pioro & Duffey (2007)) are listed below together with complementary Table 1 and Figure 1 through Figure 13.

- **Compressed Fluid:** The fluid at a pressure above the critical pressure, but at a temperature below the critical temperature (see Figures 1 - 3).
- **Critical Point (Also Called a Critical State):** The point in which the distinction between the liquid and gas (or vapour) phases disappears, i.e., both phases have the same temperature, pressure and specific volume or density. The *critical point* is characterized with the phase-state parameters: T_{cr}, P_{cr} and v_{cr} (or ρ_{cr}), which have unique values for each pure substance (see Table 1 and Figures 1 - 4).
- **Fluid:** A substance that continually deforms (flows) under an applied shear stress. Fluids are a subset of the phases of matter and include liquids, gases, plasmas, and, to some extent, plastic solids. Fluids are substances that have zero shear modulus or, in simpler terms, a fluid is a sub-

Table 1. Basic properties of supercritical fluids for power engineering

Fluid	Chemical Formula	Molar Mass	Temperatures				Critical Point			
			Triple Point		Normal Boiling		T_{cr}		P_{cr}	ρ_{cr}
		kg/kmol	K	°C	K	°C	K	°C	MPa	kg/m³
Carbon Dioxide	CO_2	44.010	216.59	-56.56	194.69	-78.46	304.13	30.98	7.377	467.60
Helium	He	4.003	2.18	-270.97	4.22	-268.93	5.20	-267.95	0.228	72.57
Nitrogen	N_2	28.013	63.15	-210.00	77.36	-195.79	126.19	-146.96	3.396	313.30
R-134a	CH_2FCF_3	102.03	169.85	-103.30	247.08	-26.07	374.21	101.06	4.059	511.90
Water	H_2O	18.015	273.16	0.01	373.12	99.97	647.10	373.95	22.064	322.00

stance, which cannot resist any shear force applied to it. Although the term "fluid" includes both the liquid and gas phases, in common usage, "fluid" is often used as a synonym for "liquid", with no implication that gas could also be present.

- **Near-Critical Point:** Actually a narrow region around the critical point, where all thermophysical properties of a pure fluid exhibit rapid variations (see Figure 5b).
- **Overheated Vapor:** The vapor at pressures below the critical pressure, and at temperatures above the saturation temperature, but below the critical temperature (see Figures 1 - 3).
- **Pseudocritical Line:** The line, which consists of pseudocritical points (see Figures 1 - 3).
- **Pseudocritical Point (Characterized with P and T_{pc}):** The point at a pressure above the critical pressure and at a temperature ($T_{pc} > T_{cr}$) corresponding to the maximum value of specific heat at this particular pressure (see Figures 6 and 7).
- **Saturated Liquid:** The liquid that is about to vaporize (see Figures 1 - 3).
- **Saturated-Liquid Line:** The saturated-liquid states connected by the line that meets the saturated-vapor line in the critical point, forming a dome (see Figures 1 - 3).
- **Saturated Liquid-Vapor Mixture (Wet Region):** The mixture of the liquid and vapor phases that coexist in equilibrium (see Figures 1 - 3).
- **Saturated Liquid-Vapor Mixture Region:** All the states that involve both the liquid and vapor phases in equilibrium and are located under the dome (see Figures 1 - 3).
- **Saturated Vapor:** The vapor that is about to condense (see Figures 1 - 3).
- **Saturated-Vapor Line:** The saturated-vapor states connected by the line that meets the saturated-liquid line at the critical point, forming a dome (see Figures 1 - 3).
- **Saturation Pressure (P_{sat}):** The pressure at which a pure substance changes phase at a given temperature (see Figures 1 - 3).
- **Saturation Temperature (T_{sat}):** The temperature at which a pure substance changes phase at a given pressure (see Figures 1 - 3).
- **Subcooled Liquid:** Has a temperature less than the saturation temperature corresponding to the pressure (see Figures 1 - 3).
- **Supercritical Fluid:** The fluid at pressures and temperatures that are higher than the critical pressure and critical temperature (see Figures 1 - 3). However, in the present Chapter, the term *supercritical fluid* usually includes both terms - a *supercritical fluid* and *compressed fluid.*
- **Supercritical "Steam":** Actually supercritical water, because at supercritical pressures fluid is considered as a single-phase substance (see Figure 1). However, this term is widely (and incorrectly) used in the literature in relation to supercritical-"steam" generators and turbines.
- **Supercritical "Vapor":** Actually, supercritical fluid, because at supercritical pressures fluid is considered as a single-phase substance (see Figures 2 and 3).
- **Superheated Steam:** The steam at pressures below the critical pressure, but at temperatures above the critical temperature (see Figure 1).
- **Superheated Vapor:** The vapor at pressures below the critical pressure, but at temperatures above the critical temperature (see Figures 2 and 3).
- **Triple Point:** The state at which all three phases coexist in equilibrium (see Table 1).

Figure 1. Thermodynamics diagrams for water: (a) pressure - temperature and (b) temperature - specific entropy. For better understanding what means "High Density (liquid like)" and "Low Density (gas like)", see Figures 5b and 6

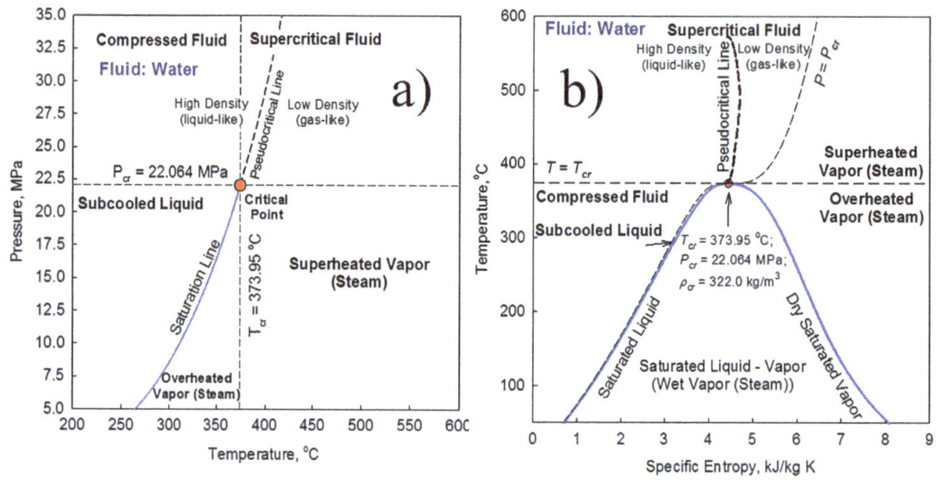

Figure 2. Thermodynamics diagrams for carbon dioxide: (a) pressure - temperature and (b) temperature - specific entropy. For better understanding what means "High Density (liquid like)" and "Low Density (gas like)", see Figure 7

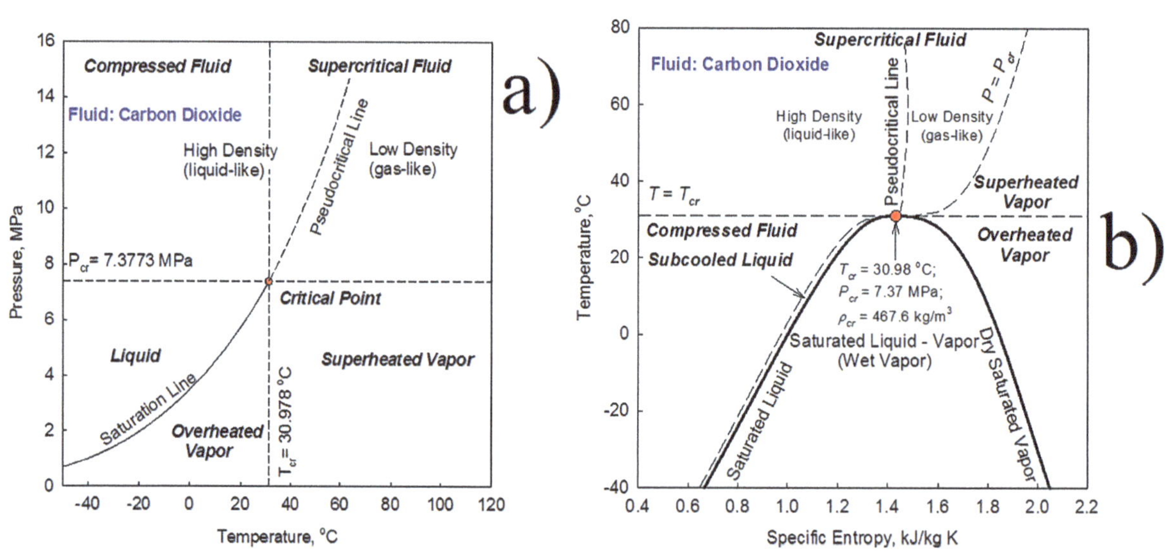

All thermophysical properties of fluids presented in this Chapter were calculated through the NIST REFPROP software (2013). NIST REFPROP software Version 9.1 includes 121 pure fluids, 5 pseudo-pure fluids (such as air), and mixtures with up to 20 components and calculates the following properties within a wide range of temperatures and pressures including subcritical, critical and supercritical ones: Temperature, pressure, density, energy, enthalpy, entropy, specific heats at constant pressure and volume, sound speed, compressibility factor, Joule-Thomson coefficient, quality, 2nd and 3rd virial coefficients,

Figure 3. Thermodynamics diagrams for helium: (a) pressure - temperature and (b) temperature - specific entropy

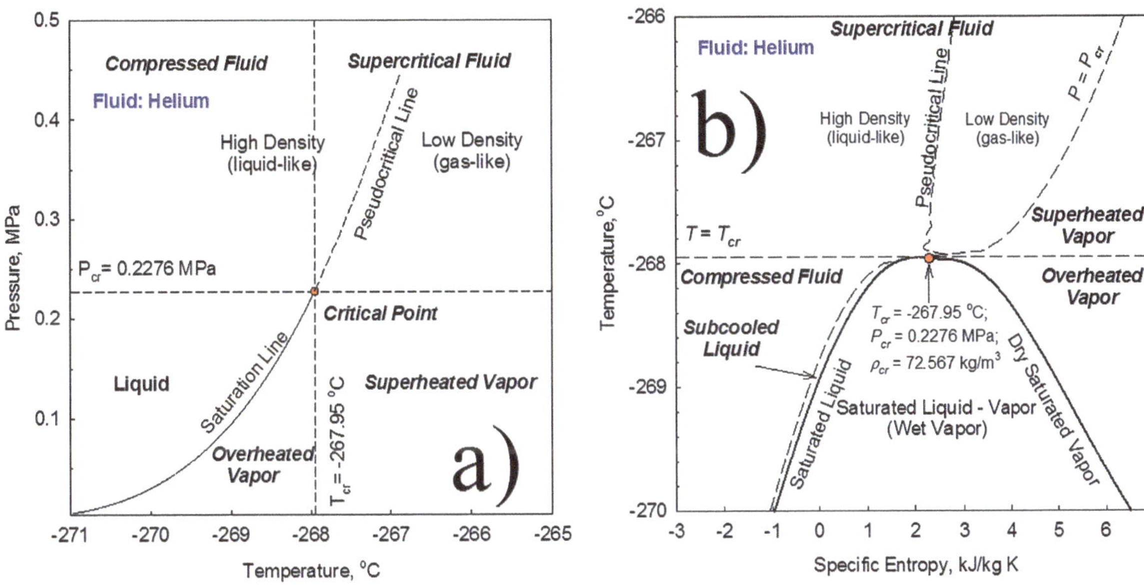

Figure 4. Transition of carbon dioxide (CO_2) through various phases: (a) - (d) Two-phase conditions (moving along saturation line towards critical point; (e) Transition through critical point; and (f) Supercritical CO_2 (single phase)
Courtesy of L. Trevani and D. McGillvray, UOIT.

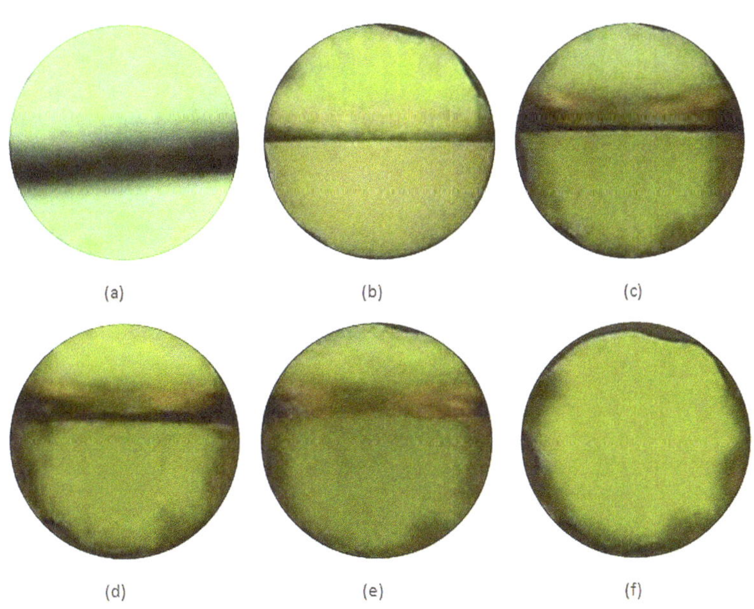

2nd and 3rd acoustic virial coefficients, Helmholtz energy, Gibbs energy, heat of vaporization, fugacity, fugacity coefficient, chemical potential, *K* value, molar mass, B12, thermal conductivity, viscosity, kinematic viscosity, thermal diffusivity, Prandtl number, surface tension, dielectric constant, gross and net heating values, isothermal compressibility, volume expansivity, isentropic coefficient, adiabatic compressibility, specific heat input, exergy, Gruneisen, critical-flow factor, excess values, dp/dr, d^2p/dr^2, dp/dT, dr/dT, dp/dr, d^2p/dr^2. Also, some uncertainties are provided for various properties. Based on the abovementioned, NIST REFPROP (2013) software is the most comprehensive source for thermophysical, thermodynamic and other properties of fluids (liquids, vapors, and gases).

In addition, the NIST has the mini-REFPROP program, which is a sample version of the full REFPROP program and is meant for use as a teaching tool in the introduction of thermodynamics to students. It contains a limited number of pure fluids (water, CO_2, R-134a, nitrogen, methane, propane, hydrogen, and dodecane) and also, allows mixture calculations of nitrogen with methane for teaching Vapor-Liquid Equilibrium (VLE). Therefore, this version is for free Internet download.

$$\left(\frac{P}{P_{cr}}\right)_{water} = \left(\frac{P}{P_{cr}}\right)_{CO_2} .$$

Figures 8 - 13 show graphs with properties' profiles (density, thermal conductivity, specific heat, specific enthalpy, dynamic viscosity, and Prandtl number) vs. temperature for water at 15 MPa (Pressurized Water Reactor (PWR) conditions, shown here just for reference purposes) and selected SCFs (carbon dioxide, helium, nitrogen, nitrogen (80%) - helium (20%) mixture, and water) at pressures corresponding to those of outlet of reactors or inlet to power cycles.

Additional graphs with properties' profiles vs. temperature for the following supercritical fluids:

Figure 5. Profiles of Density vs. Temperature at various pressures for water. (a) subcritical pressures; (b) critical and supercritical pressures

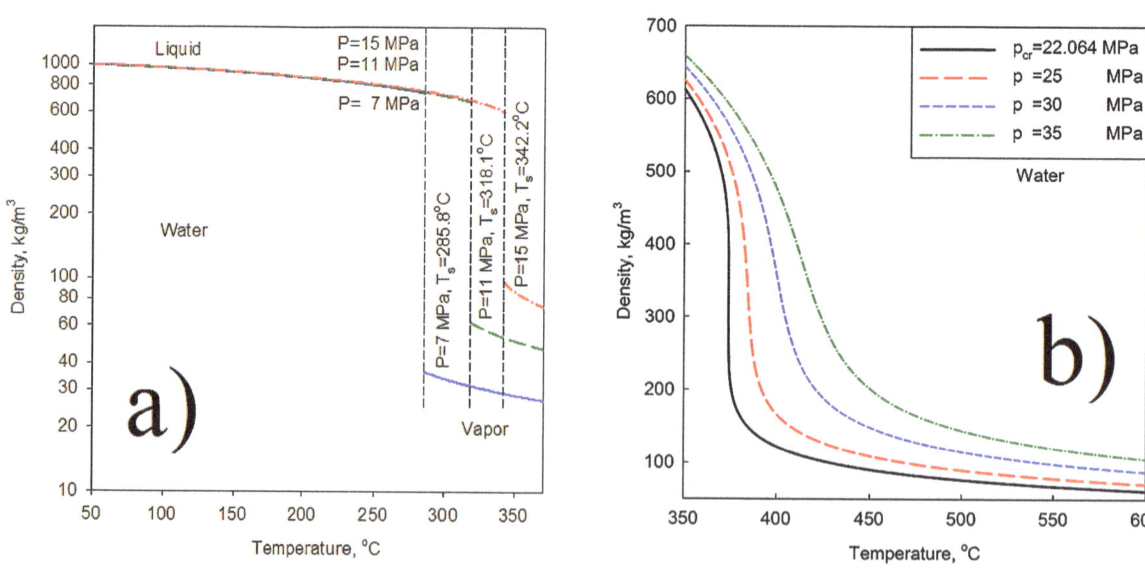

Figure 6. Profiles of Thermal Conductivity, Specific Heat, Density, and Dynamic Viscosity vs. Temperature: Supercritical water at pressure of 25 MPa. Pseudocritical region is about ±25°C around pseudocritical point.
Based on NIST (2013); prepared by Eu. Saltanov, UOIT.

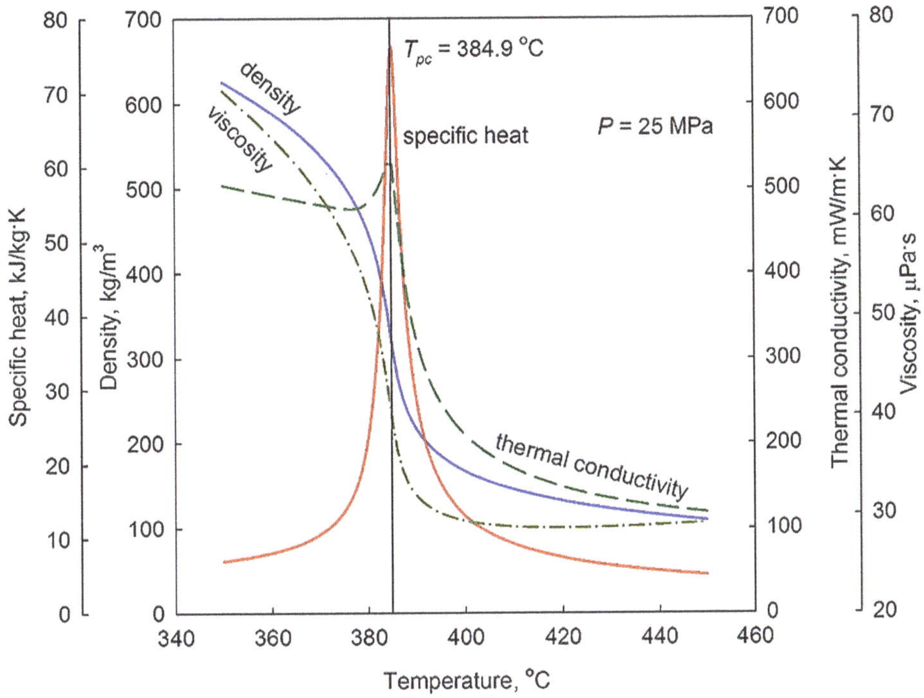

- Water, carbon dioxide, and helium can be found in Handbook of Generation IV Nuclear Reactors (Pioro (Ed.), 2016).
- Water, carbon dioxide, and R-134a and their comparison can be found in Gupta et al. (2013a).
- Water and R-12 can be found in Pioro and Mokry (2011a).
- Water, carbon dioxide, helium and R-134a can be found in Pioro and Duffey (2007).

Also, more detailed information on nuclear-reactor coolants, their properties, applications and literature sources on this topic and additional information on application of SCPs in thermal- and nuclear-power engineering can be found in Handbook of Generation IV Nuclear Reactors (Pioro (Ed.), 2016); IAEA-TECDOC-1746 (2014); Schulenberg & Starflinger (2012); Pioro (2011); Oka et al. (2010); and Pioro & Duffey (2007).

In general, crossing the pseudocritical line from left to right is quite similar to crossing the saturation line from liquid into vapor (see Figures 5b - 7). The major difference in crossing these two lines is that all changes (even drastic variations) in thermophysical properties at critical and supercritical pressures are continuous and gradual, and take place within a certain temperature range. On the contrary, at subcritical pressures there are properties discontinuity on the saturation line: one value for liquid and another for vapor. Therefore, fluids at critical and SC pressures behave as single-phase substances. However, it should be pointed out that below critical / pseudocritical temperatures (see Figures 1 - 3) SCF behave

Figure 7. Profiles of Thermal Conductivity, Specific Heat, Density, and Dynamic Viscosity vs. Temperature: Supercritical carbon dioxide at pressure of 8.4 MPa. Pseudocritical region is about ±25°C around pseudocritical point. Carbon-dioxide pressure of 8.4 MPa corresponds to water pressure of 25 MPa
Based on scaling law (Pioro & Duffey, 2007); Based on NIST (2013); prepared by Eu. Saltanov, UOIT.

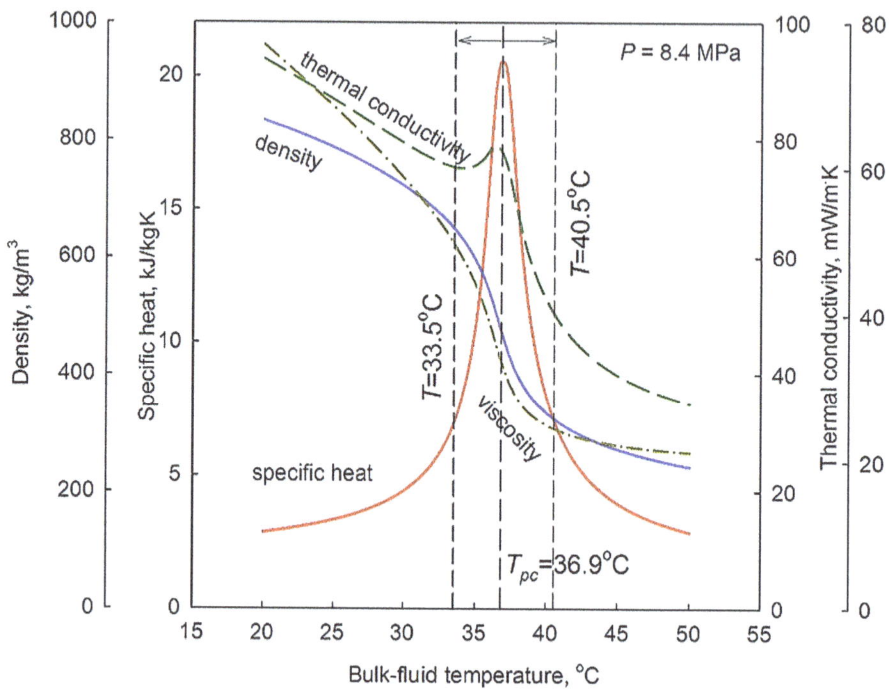

as "liquid-like" substances and above critical / pseudocritical temperatures - as "gas-" or "vapor-like" substances (see Figures 5b - 7).

The general trends of various properties near the critical and pseudocritical points can be illustrated on a basis of those of water (see Figures 5b - 13). Figure 5 shows variations in density of water at three subcritical pressures (Figure 5a): 1) 7 MPa, which usually corresponds to pressure inside modern Boiling Water Reactors (BWRs), Advanced BWRs (ABWRs), and in Rankine steam cycle of majority light-water-cooled NPPs; 2) 11 MPa, which usually corresponds to pressure inside modern Pressurized Heavy Water Reactors (PHWRs); and 3) 15 MPa, which usually corresponds to pressure inside modern PWRs; and critical (P_{cr} = 22.064 MPa) and three SCPs (P = 25.0, 30.0, and 35.0 MPa) (Figure 5b).

At critical and SC pressures a fluid is considered as a single-phase substance in spite of the fact that all thermophysical properties undergo significant changes within the critical (see Figure 5b) and pseudocritical regions (see Figures 5b - 7). Near the critical point, these changes are dramatic (see Figure 5b). In the vicinity of pseudocritical points, with an increase in pressure, these changes become less pronounced (Figure 5b).

Also, it can be seen that properties such as density (Figures 5b - 8) and dynamic viscosity (Figures 6, 7, and 12) undergo a significant drop (near the critical point this drop is almost vertical) within a very narrow temperature range, while specific enthalpy undergo a sharp increase (Figure 11). Volume expansivity, specific heat (Figures 6, 7, and 10), thermal conductivity (Figures 6, 7, and 9) and Prandtl number (Figure 13) have peaks near the critical and pseudocritical points. The magnitude of these peaks decreases very quickly with an increase in pressure (Pioro (Ed.), 2016; Pioro & Mokry, 2011a; Pioro & Duffey, 2007). Also, "peaks" transform into "humps" profiles at pressures beyond the critical pressure. It should be noted that the dynamic viscosity (Figures 6, 7, and 12), kinematic viscosity and thermal conductivity (Figures 6, 7, and 9) undergo through their minimum right after critical and pseudocritical points.

Specific heat of water (as well as of other fluids) has a maximum value in the critical point (Pioro (Ed.), 2016; Pioro & Mokry, 2011a; Pioro & Duffey, 2007). The exact temperature that corresponds to the specific-heat peak above the critical pressure is known as the pseudocritical temperature (see Figures 1 - 3, 6, 7, and 10). For water at pressures approximately above 300 MPa and for carbon dioxide at pressures above 30 MPa a peak (here it is better to say a "hump") in specific heat almost disappears, therefore, term such as a *pseudocritical point* no longer exists (Pioro (Ed.), 2016; Pioro & Mokry, 2011a; Pioro & Duffey, 2007). The same applies to the *pseudocritical line*. It should be noted that peaks in the thermal conductivity and volume expansivity may not correspond to the pseudocritical temperature.

SPECIFICS OF HEAT TRANSFER AT SUPERCRITICAL PRESSURES

Definitions of Terms and Expressions Related to Critical and Supercritical Regions

Prior to a general discussion on specifics of heat transfer at critical and supercritical pressures, it is important to define special terms and expressions used at these conditions (Pioro, 2016; Pioro & Duf-

Figure 8. Profiles of Density vs. Temperature for water at 15 MPa (PWR conditions) and selected SCFs at pressures corresponding to those of outlet of reactors or inlet to power cycles

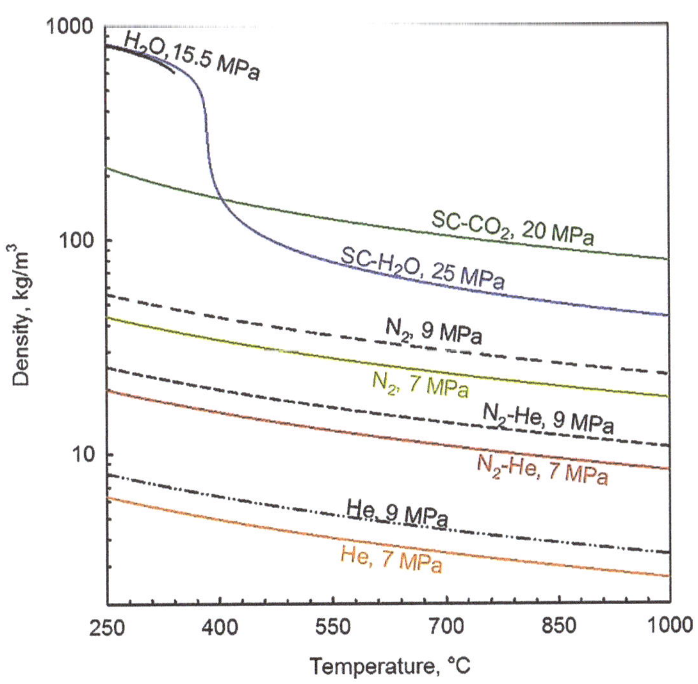

Figure 9. Profiles of Thermal Conductivity vs. Temperature for water at 15 MPa (PWR conditions) and selected SCFs at pressures corresponding to those of outlet of reactors or inlet to power cycles

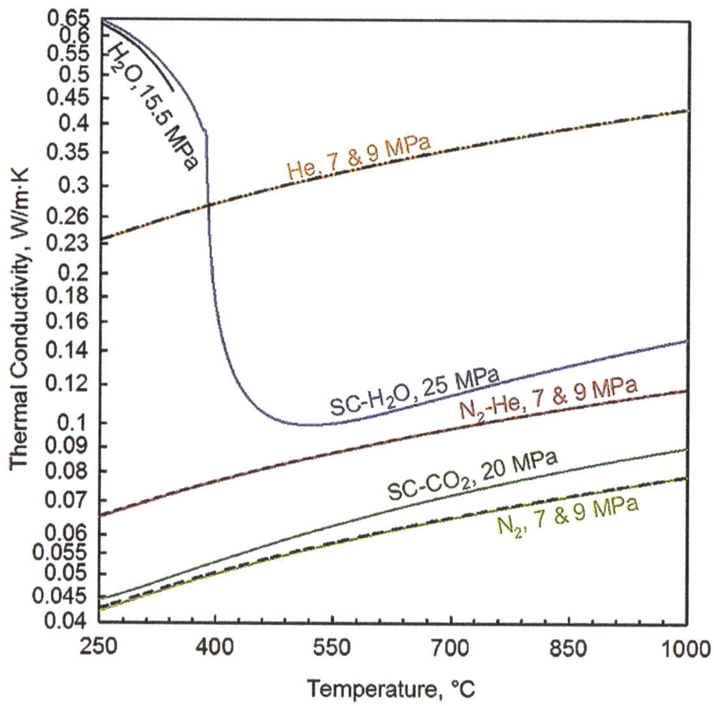

Figure 10. Profiles of Specific Heat vs. Temperature for water at 15 MPa (PWR conditions) and selected SCFs at pressures corresponding to those of outlet of reactors or inlet to power cycles

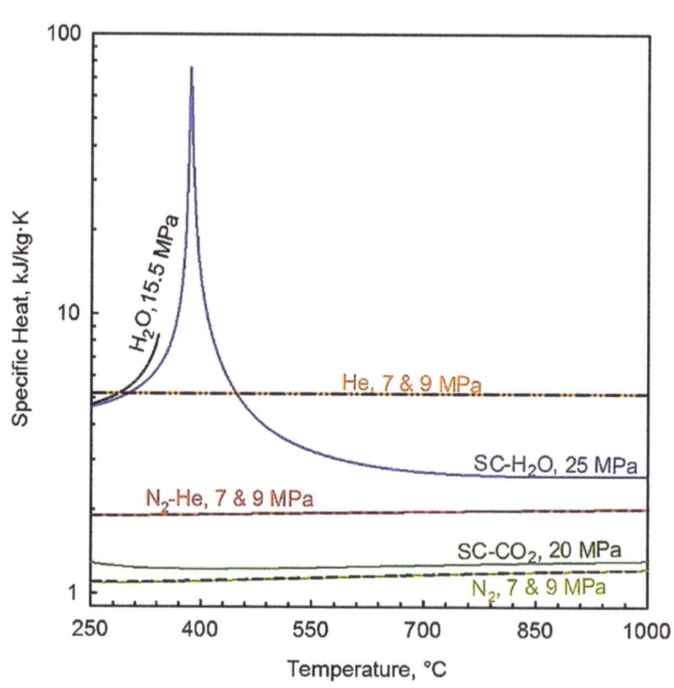

Figure 11. Profiles of Specific Enthalpy vs. Temperature for water at 15 MPa (PWR conditions) and selected SCFs at pressures corresponding to those of outlet of reactors or inlet to power cycles

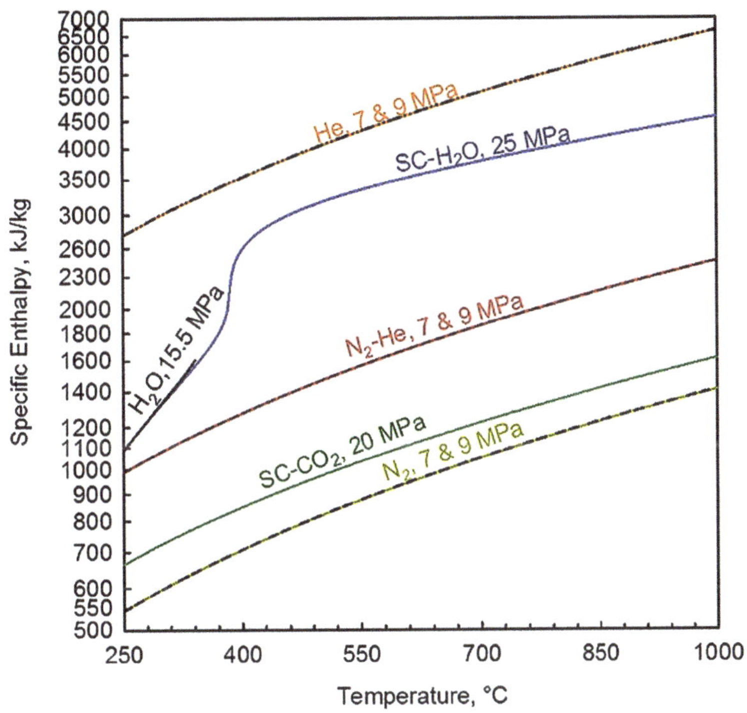

fey, 2007). For a better understanding of these terms and expressions their definitions are listed below together with complementary Figure 14.

- **Deteriorated Heat Transfer (DHT):** Characterized with lower values of the HTC compared to those at normal heat transfer; and hence, has higher values of wall temperature within some part of a test section or within the entire test section.
- **Improved Heat Transfer (IHT):** Characterized with higher values of the HTC compared to those at normal heat transfer; and hence, lower values of wall temperature within some part of a test section or within the entire test section. In our opinion, the improved heat-transfer regime or mode includes peaks or "humps" in the HTC near the critical or pseudocritical points.
- **Normal Heat Transfer (NHT):** Can be characterized in general with HTCs similar to those of subcritical convective heat transfer far from the critical or pseudocritical regions, when they are calculated according to the conventional single-phase Dittus-Boelter-type correlations: $\mathbf{Nu} = 0.023\ \mathbf{Re}^{0.8}\mathbf{Pr}^{0.4}$.
- **Pseudo-Boiling:** A physical phenomenon similar to subcritical-pressure nucleate boiling, which may appear at SCPs. Due to heating of an SCF with a bulk-fluid temperature below the pseudocritical temperature (high-density fluid, i.e., "liquid-like"), some layers near the heating surface may attain temperatures above the pseudocritical temperature (low-density fluid, i.e., "gas-like"). This low-density "gas-like" fluid leaves the heating surface in a form of variable density (bubble) volumes. During the pseudo-boiling, the HTC usually increases (IHT regime).

Figure 12. Profiles of Dynamic Viscosity vs. Temperature for water at 15 MPa (PWR conditions) and selected SCFs at pressures corresponding to those of outlet of reactors or inlet to power cycles

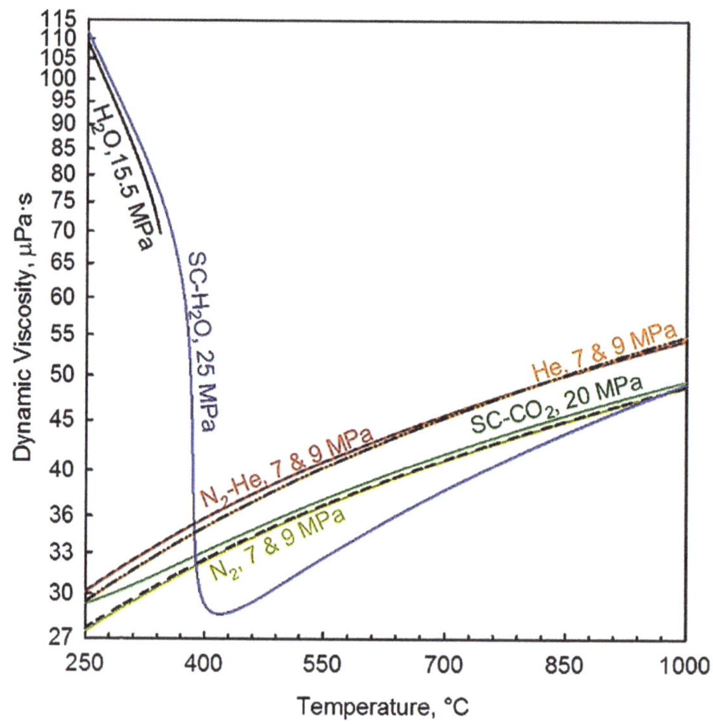

Figure 13. Profiles of Prandtl Number vs. Temperature for water at 15 MPa (PWR conditions) and selected SCFs at pressures corresponding to those of outlet of reactors or inlet to power cycles

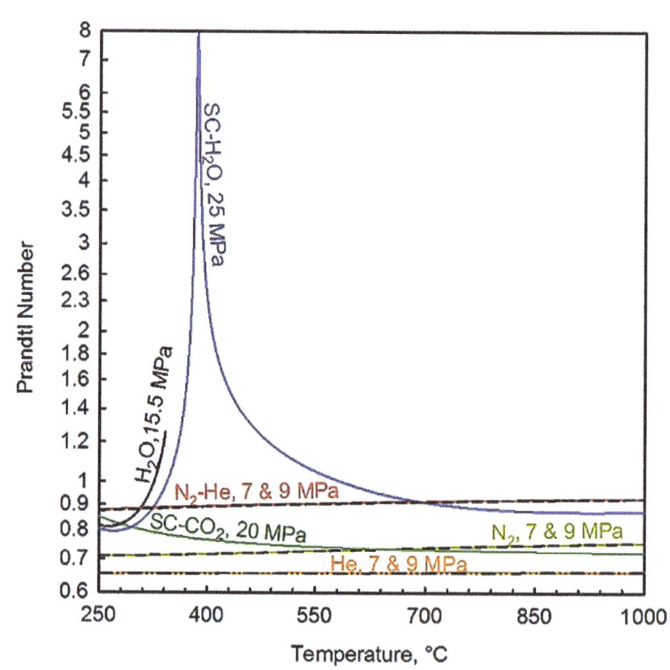

Figure 14. Temperature and Heat-Transfer-Coefficient (HTC) profiles along heated length of channels cooled with SCW: (a) vertical circular tube (data by Kirillov et al. (2003)): D = 10 mm and L_h = 4 m and (b) vertical 3-rod bundle with 4 helical ribs (Razumovskiy et al., 2016): D_{hy} = 2.4 mm; L_h = 0.485 m; heated elements OD = 5.2 mm; ribs height 0.6 mm and width 1 mm

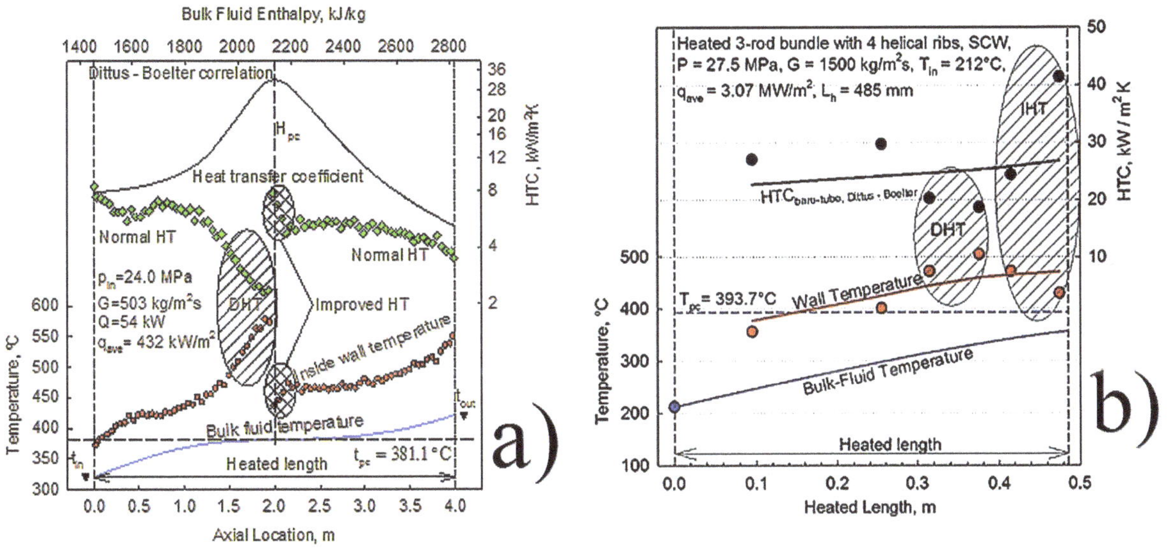

- **Pseudo-Film Boiling:** A physical phenomenon similar to subcritical-pressure film boiling, which may appear at SCPs. At pseudo-film boiling, a low-density fluid (a fluid at temperatures above the pseudocritical temperature, i.e., "gas-like") prevents a high-density fluid (a fluid at temperatures below the pseudocritical temperature, i.e., "liquidlike") from contacting ("rewetting") a heated surface. Pseudo-film boiling leads to the DHT regime.

Also, more detailed information on supercritical heat transfer can be found in Handbook of Generation IV Nuclear Reactors (Pioro (Ed.), 2016); IAEA-TECDOC-1746 (2014); Gupta et al. (2013a); Schulenberg & Starflinger (2012); Pioro (2011); Pioro & Mokry (2011b); Mokry et al. (2011); Oka et al. (2010); and Pioro & Duffey (2007).

The most common case in forced convection without phase change is turbulent flow of liquid or gas over / along a heat-transfer surface or inside a flow geometry. Usually, such type of heat transfer is based on the following form of a correlation: $\mathbf{Nu} = f(\mathbf{Re}^m; \mathbf{Pr}^n)$. The most widely used correlation for turbulent flow of liquid or gas at subcritical pressures is the Dittus-Boelter correlation (1985), as introduced by McAdams (1942) (for details, see Winterton (1998)):

$$\mathbf{Nu} = 0.023\mathbf{Re}^{0.8}\,\mathbf{Pr}^{0.4} \text{ (for heating)} \tag{1}$$

This correlation is valid within the following range: $\mathbf{Re} \geq 10,000$; $0.6 \leq \mathrm{Pr} \leq 160$; and $\left(L / D\right) \geq 10$. Reynolds number is $\mathbf{Re} = \frac{uD_{hy}}{v} = \frac{\rho u D_{hy}}{\mu} = \frac{GD_{hy}}{\mu}$ and all thermophysical properties are based on a bulk-fluid temperature. However, the conventional Dittus-Boelter correlation (Equation 1) does not include the effect of heat flux. Later, two non-dimensional ratios ($(\mu_b / \mu_w)^{0.14}$ or $(\mathbf{Pr}_b / \mathbf{Pr}_w)^{0.25}$) were proposed to take into account the effect of heat flux or effect of variations in thermophysical properties due to non-uniform temperature distribution in the cross section of a flow channel. However, in spite of these additional terms the conventional Dittus-Boelter correlation (Equation 1) is not applicable for SCFs within improved and deteriorated heat-transfer regimes (see Figure 14a and Figure 14b). Nevertheless, it shows reasonable HTC prediction outside the critical and pseudocritical regions. Therefore, the Dittus-Boelter correlation might be used for SC helium, nitrogen and their mixtures at temperatures way above critical or pseudocritical ones, when these fluids behave as compressed gases.

The majority of studies deal with heat transfer and hydraulic resistance of working fluids, mainly water, carbon dioxide and helium, in circular bare tubes (Pioro & Duffey, 2007). A limited number of studies were dedicated to heat transfer and pressure drop in annuli, rectangular-shaped channels and bundles.

In general, three major heat-transfer regimes (for their definitions, see the beginning of this Section) can be noticed at critical and SC pressures (for details, see Figure 14a and Figure14b) (Pioro & Duffey, 2007):

1. Normal Heat Transfer (NHT),
2. Improved Heat Transfer (IHT),
3. Deteriorated Heat Transfer (DHT).

Also, two special phenomena (for their definitions, see the beginning of this Section) may appear along a heated surface: 1) pseudo-boiling; and 2) pseudo-film boiling.

These heat-transfer regimes and special phenomena appear to be due to significant variations of thermophysical properties near the critical and pseudocritical points (see Figures 6, 7 and 14) and due to operating conditions.

Some researchers have suggested that variations in thermophysical properties near critical and pseudocritical points result in the maximum value of HTC. A magnitude of the peak in HTC decreases with increasing heat flux and pressure.

The pseudo-boiling phenomenon is thought to be due to large differences in fluid density below the pseudocritical point (high-density fluid, i.e., "liquid-like") and beyond (low-density fluid, i.e., "gas-like").

The process of pseudo-film boiling (i.e., low-density fluid prevents high-density fluid from "rewetting" a heated surface) is similar to film boiling, which occurs at subcritical pressures. Pseudo-film boiling leads to the DHT regime. However, the pseudo-film-boiling phenomenon may not be the only reason for the DHT.

Unfortunately, satisfactory analytical methods for practical prediction of forced-convection heat transfer at SCPs have not yet been developed due to the difficulty in dealing with steep property variations, especially, in turbulent flows and at high heat fluxes. Therefore, generalized correlations based on experimental data are used for HTC calculations at SCPs.

There are numerous correlations for convective heat transfer in circular tubes at SCPs (for details, see Zahlan et al. (2011; 2010) and Pioro and Duffey (2007)). However, an analysis of these correlations has shown that they are more or less accurate only within the particular dataset, which was used to derive the correlation, but show a significant deviation in predicting other experimental data.

In general, many of these correlations are based on the conventional Dittus-Boelter-type correlation (see Equation 1) in which the regular specific heat in Prandtl number is replaced with the cross-sectional averaged specific heat ($c_{p\ ave}$) or ($\bar{c}_p$) within the range of (T_w - T_b); $\left(\frac{H_w - H_b}{T_w - T_b}\right)$, J/kg K. In this case, the cross-sectional averaged Prandtl number ($\mathbf{Pr}_{ave}$) or ($\overline{\mathbf{Pr}}$) is $\left(\frac{\mu \bar{c}_p}{\rho}\right)$. Profiles of regular and cross-sectional-averaged specific heats and corresponding to that **Pr** and $\overline{\mathbf{Pr}}$ numbers along the heated channel of a flow channel are shown in Figure 15. Also, additional terms, such as: $\left(\frac{k_b}{k_w}\right)^k$; $\left(\frac{\mu_b}{\mu_w}\right)^m$; $\left(\frac{\rho_b}{\rho_w}\right)^n$; etc., can be added into correlations to account for significant variations in thermophysical properties within a cross section due to a non-uniform temperature profile, i.e., due to heat flux.

It should be noted that usually generalized correlations, which contain fluid properties at bulk and wall temperatures, require iterations to be solved, because there are two unknowns: 1) HTC and 2) wall temperature. Therefore, the initial wall-temperature value at which fluid properties will be estimated should be "guessed" to start iterations.

In general, experimental HTC values show just a moderate increase within the pseudocritical region. This increase depends on flow conditions and heat flux: higher heat flux - less increase.

Figure 15. Regular and averaged Prandtl number and specific-heat profiles for water along heated length of SCWR fuel channel: AHFP - Axial Heat Flux Profile. With increasing heat flux peaks in averaged specific heat and Prandtl number decrease and move upstream from pseudocritical point defined by bulk-fluid temperature
Mokry et al., 2008.

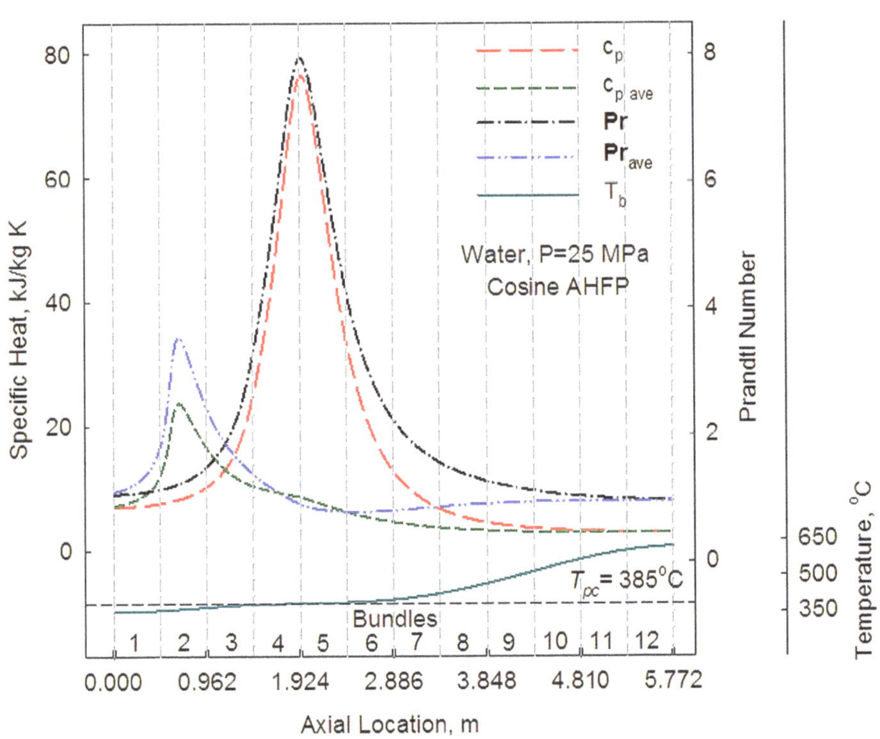

The majority of empirical correlations were proposed in the 1960s - 1970s (Pioro & Duffey, 2007), when experimental techniques were not at the same level (i.e., advanced level) as of today. Also, thermophysical properties of water have been updated since that time (for example, a peak in thermal conductivity in critical and pseudocritical points within a range of pressures from 22.1 to 25 MPa for water was not officially recognized until the 1990s).

Therefore, recently new correlations, based on a new set of heat-transfer data and the latest thermophysical properties of SCW (Mokry et al., 2011; Gupta et al. 2013a) and SC carbon dioxide (Gupta et al. 2013b; Saltanov et al., 2015) were developed and evaluated.

The following correlations (Equations 2 - 6) have been developed by Dr. I. Pioro with his students.

1. Correlation for SCW, which is based on bulk-fluid-temperature approach and cross-section-averaged specific heat (Mokry et al., 2011):

$$\mathrm{Nu_b} = 0.0061\ \mathrm{Re_b^{0.904}}\ \overline{\mathrm{Pr}}_\mathrm{b}^{-0.684}\left(\frac{\rho_w}{\rho_b}\right)^{0.564}. \tag{2}$$

Correlation (Equation 2) is the most accurate heat-transfer correlation for SCW forced convection compared to other 14 heat-transfer correlations: Uncertainty ±25% for HTC values and ±15% for wall temperatures (this statement is based on IAEA TECDOC (2014) and papers by Zahlan et al. (2011; 2010)). This correlation was verified within the following operating conditions: Water, upward flow, vertical bare tubes with inside diameter 3 - 38 mm, pressure 22.8 - 29.4 MPa, mass flux 200 - 3000 kg/m^2s, and heat flux 70 - 1250 kW/m^2. Also, Equation 2 showed very good predictions for subcritical liquid (second by an RMS error after the Gnielinski (1976) correlation) and the best for superheated steam compared to other four correlations (based on Zahlan et al. (2010; 2011)).

2. Correlation for SCW, which is based on wall-temperature approach and cross-section-averaged specific heat (Gupta et al., 2013a):

$$\boldsymbol{Nu}_w = 0.0033\mathbf{Re}_w^{0.941}\ \overline{\mathbf{Pr}}_w^{-0.764}\left(\frac{\mu_w}{\mu_b}\right)^{0.398}\left(\frac{\rho_w}{\rho_b}\right)^{0.156} \tag{3}$$

Correlation (Equation 3) is also the most accurate heat-transfer correlation for SCW forced convection compared to other 14 heat-transfer correlations (the uncertainty is the same as that for Equation 2) (this statement is based on paper by Zahlan et al. (2011)). And this correlation was verified within the same operating conditions as those for Equation 2.

However, it should be noted that both correlations: Equation 2 and Equation 3; as well as many other SCFs heat-transfer correlations, can predict only HTC values at NHT and IHT regimes. Unfortunately, there are no reliable correlations to predict HTCs at the DHT regime.

3. Nevertheless, the following empirical correlation for SCW was proposed for calculating the minimum heat flux at which the DHT regime appears in forced convection in bare vertical tubes with upward flow (Mokry et al., 2011):

$$q_{dht} = -58.97 + 0.745 \cdot G \text{, kW/m}^2. \tag{4}$$

Correlation (Equation 4) is valid within the following range of experimental parameters: Water, upward flow, vertical bare tube with inside diameter 10 mm, pressure 24 MPa, mass flux 200 – 1500 kg/m²s, and bulk-fluid inlet temperature 320 – 350°C: Uncertainty ±15% for the DHT heat flux.

4. Correlation for SC carbon dioxide, which is based on wall-temperature approach and cross-section-averaged specific heat (Gupta et al., 2013b):

$$\boldsymbol{Nu}_w = 0.0038\mathbf{Re}_w^{0.957}\,\overline{\mathbf{Pr}}_w^{-0.14}\left(\frac{\rho_w}{\rho_b}\right)^{0.84}\left(\frac{k_w}{k_b}\right)^{-0.75}\left(\frac{\mu_w}{\mu_b}\right)^{-0.22} \tag{5}$$

Correlation (Equation 5) is the heat-transfer correlation for SC carbon-dioxide forced convection: Uncertainty ±30% for HTC values and ±20% for wall temperatures. This correlation was verified within the following operating conditions: Carbon dioxide, upward flow, vertical bare tube with inside diameter 8 mm; pressure 7.6 - 8.8 MPa, mass flux 700 - 3170 kg/m²s, heat flux 10 - 615 kW/m², inlet temperature 20 - 40°C, and outlet temperature 30 - 225°C.

5. The following empirical correlation for SC carbon dioxide was proposed for calculating the minimum heat flux at which the DHT regime appears in forced convection in bare vertical tubes with upward flow (Saltanov et al., 2015):

$$q_{dht} = 66.81 + 0.18 \cdot G \text{, kW/m}^2. \tag{6}$$

Correlation (Equation 6) is valid within the following range of experimental parameters: Carbon dioxide, upward flow, vertical bare tube with inside diameter 8 mm, pressure 7.6 – 8.8 MPa, mass flux 890 – 2990 kg/m²s, and bulk-fluid inlet temperature 20 – 35°C. The RMS error is about 9%.

Correlations (Equation 2, Equation 3, and Equation 5) can be applied not only for circular tubes, but for any other flow geometries through using the hydraulic-equivalent diameter $\left(D_{hy} = \frac{4 \cdot A_{fl}}{p_{wet}}\right)$ instead of an inside-tube diameter in Nusselt and Reynolds numbers. Moreover, these three correlations can be used for bundle flow geometries as a conservative approach, because in bundles HTCs are usually higher than those in bare tubes due to various appendages, grids, endplates, spacers, etc., which turbulize flow and enhance heat transfer.

Information on the total pressure drop in forced convection can be found in Pioro (Ed.) (2016); Pioro and Duffey (2007); and Pioro et al. (2004).

SUPERCRITICAL-PRESSURE FOSSIL-FUEL THERMAL POWER PLANTS

It is well-known from thermodynamics (Cengel & Boles, 2015) that thermal efficiency of power plants depends primarily on temperature and, also, on pressure of the working fluid at the turbine inlet. At the end of 1950s, subcritical pressure fossil-fuel thermal power plants have reached gross[1] thermal efficiencies of about 36 - 42% at operating temperatures of 530 - 560°C and pressures of 15 - 17 MPa. However, further increase in thermal efficiencies through rising temperatures and corresponding to that pressures were limited by fast decreasing critical heat flux at boiling closer to the critical pressure of water (P_{cr} = 22.064 MPa) (Dragunov et al., 2015).

Therefore, it was decided to move operating conditions of thermal power plants (mainly coal-fired) to SCPs. Figure 16 shows a typical scheme of such plant.

In the 1950s, in the former USSR the Podol'sk Machine-Building plant manufactured several small experimental SC "steam" generators for research institutions such as: *(i)* "ВТИ" ("VTI") - All-Union Heat Engineering Institute (Moscow) with "steam" parameters of 29.4 MPa and 600°C, *(ii)* "ЦКТИ" ("TsKTI") - Central Boiler-Turbine Institute by Polzunov (St.-Petersburg) and *(iii)* Kiev Polytechnic Institute with "steam" parameters of 39 MPa and 700°C.

However, the implementation of SC power-plant "steam" generators in the former USSR started with units of 300-MW_{el} capacities (Ornatskiy et al. 1980). Two leading Russian manufacturing plants: "ТКЗ" ("TKZ") - Taganrog Power-Plant Steam Generator's Manufacturing plant (Taganrog, Ukraine) and Machine-Building plant (Podol'sk, Russia) developed and manufactured first units with the assistance of research institutes such as "ЦКТИ" ("TsKTI") and "ВТИ" ("VTI").

Power-plant "steam" generator "ТПП-110" ("TPP-110") manufactured by the "ТКЗ" ("TKZ") in 1961 was the first industrial unit operating at SC conditions in the former USSR, and was used at a coal-fired power plant. A total of six units were put into operation.

The 300-MW_{el} power-"steam"-generating units have the following characteristics:

- **"Steam" Capacity, t/h:** 950 – 1000.
- **Pressure (Primary "Steam"), MPa:** 25.
- **Temperature (Primary "Steam"), °C:** 545 – 585.
- **Pressure (Secondary Steam), MPa:** 3.5 - 3.9.
- **Feed-Water Temperature, °C:** 260 – 265.

The next stage in further development of SC "steam" generators in the former USSR involved an increase in their capacity to 500 MW_{el} and 800 MW_{el}. In 1966, the first 1000-MW_{el} Ultra-SuperCritical (USC) plant started its operation in Kashira with the primary "steam" pressure of 30.6 MPa, and primary and reheat temperatures of 650 and 565°C, respectively. In modern designs of SC units, the capacity was upgraded to 1200 MW_{el} (Ornatskiy et al. 1980). This, one of the largest in the world, SC power-generating unit operates with a single-shaft turbine at the Kostroma district power plant and is a gas-oil-fired "steam" generator.

Parameters of largest former USSR supercritical-pressure turbines are listed in Table 2. Typical 600-MW_{el} SC coal-fired power plant detailed schematic is shown in Figure 17, and simplified *T-s* diagrams of SC and subcritical-pressure power plants - in Figure 18 and Figure 19, respectively, for comparison purposes. Data in Table 3 (see Row No. 1) show that internal needs of a plant is about 8.2% from the

Figure 16. Typical scheme of supercritical-pressure coal-fired thermal power plant: 1) Cooling tower; 2) Cooling-water pump; 3) Transmission line (3-phase); 4) Step-up transformer (3-phase); 5) Electrical generator (3-phase); 6) Low-pressure steam turbine; 7) Condensate pump; 8) Surface condenser; 9) Intermediate-pressure steam turbine; 10) Steam control valve; 11) High-pressure steam turbine; 12) Deaerator; 13) Feedwater heater; 14) Coal conveyor; 15) Coal hopper; 16) Coal pulverizer; 17) Boiler steam drum; 18) Bottom-ash hopper; 19) Superheater; 20) Forced-draught (draft) fan; 21) Reheater; 22) Combustion-air intake; 23) Economiser; 24) Air preheater; 25) Precipitator; 26) Induced-draught fan; and 27) Flue-gas stack

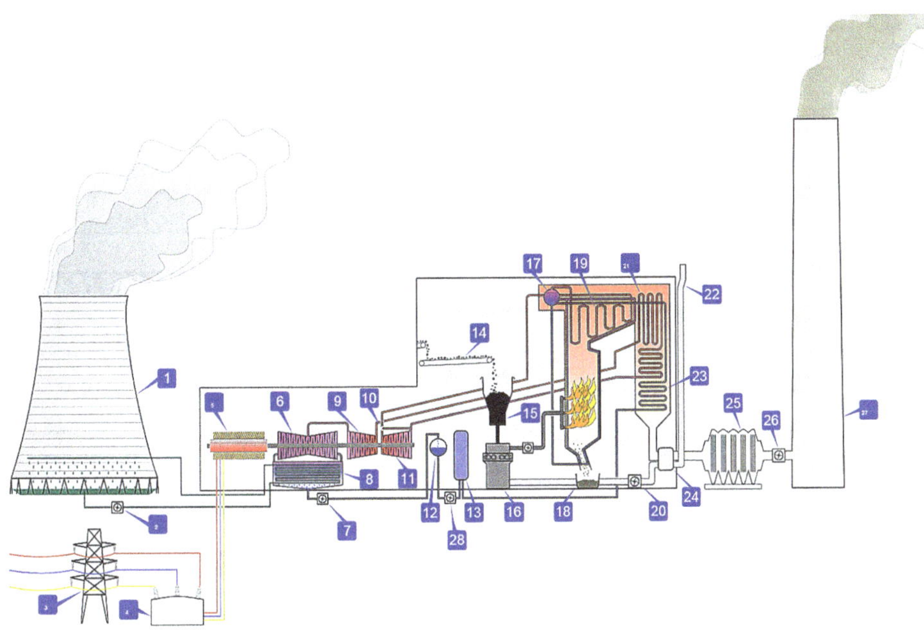

gross power. Based on that values of gross thermal efficiencies in rows Nos. 2 and 3 have been estimated (shown in Italic style).

In the early 1950s, the development work on SC "steam" generators started in the USA. The first SC "steam" generator was put into operation at the Philo Plant of American Electric Power in 1957. Major USA manufacturers of power-plant "steam" generators such as Babcock & Wilcox, Combustion Engineering, Inc., Foster Wheeler, and others were involved in the development and manufacturing of SC units. The supercritical units found their application at capacities from 400 and to 1380 MW_{el}.

US power "steam"-generating units have the following averaged characteristics (Ornatskiy et al. 1980):

- **"Steam" Capacity, t/h:** 1110 - 4440
- **Pressure (Primary "Steam"), MPa:** 23 - 26
- **Temperature (Primary "Steam"), °C:** 538 - 543
- **Temperature (Secondary Steam), °C:** 537 - 551

The characteristics of two SC units built by "Babcock & Wilcox" are listed below (Ornatskiy et al. 1980).

Table 2. Parameters of largest former USSR supercritical-pressure turbines

Parameters	K-1200-240	K-800-240	K-800-240*
Power, MW_{el} (max power)	1200 (1380)	800 (850)	800 (835)
Main Steam			
Pressure, MPa	23.5	23.5	23.5
Temperature, °C	540	540	560
Max Flow Rate through High Pressure (HP) Turbine, t/h	3950	2650	2500
Reheat Steam			
Pressure, MPa	3.5	3.2	3.4
Temperature, °C	540	540	565
No. of Steam Extractions	9	8	8
Outlet Pressure, kPa	3.6	3.4	2.9
Cooling Water			
Temperature, °C	12	12	12
Flow Rate, m^3/h	108,000	73,000	85,000
Feedwater Temperature, °C	274	274	270
Turbine Layout			
No. of Cylinders	5	5	6
No. of HP Cylinders	1	1	-
No. of Intermediate Pressure (IP) Cylinders	2	2	-
No. of Low Pressure (LP) Cylinders	2	2	-
Turbine Mass and Dimensions			
Total Mass, t	1900	1300	1600
Total Length, m	48	40	40
Total Length with Electrical Generator, m	72	60	46
Average Diameter of HP Turbine, m	3.0	2.5	2.5

Grigor'ev and Zorin, 1982.
* Double-shaft turbine

Power plant "steam" generator put into operation at the "Paradise" power plant (USA) in 1969 (for 1130 MW_{el} unit):

- **"Steam" Capacity, t/h:** 3630
- **Pressure (Primary "Steam"), MPa:** 24.2
- **Temperature (Primary "Steam"), °C:** 537
- **Steam Capacity (Secondary Steam), t/h:** 2430
- **Pressure (Secondary Steam), MPa:** 3.65
- **Temperature (Secondary Steam), °C:** 537
- **Feed-Water Temperature, °C:** 288

Figure 17. Tom'-Usinsk 600-MW$_{el}$ supercritical-pressure coal-fired power plant detailed schematic: Single-reheat-regenerative cycle
Kruglikov et al., 2009.
Cond P - Condensate Pump; CP - Circulation Pump; Cyl - Cylinder; GCHP - Gas Cooler of High Pressure; GCLP - Gas Cooler of Low Pressure; H - Heat exchanger (feedwater heater); and TDr - Turbine Drive

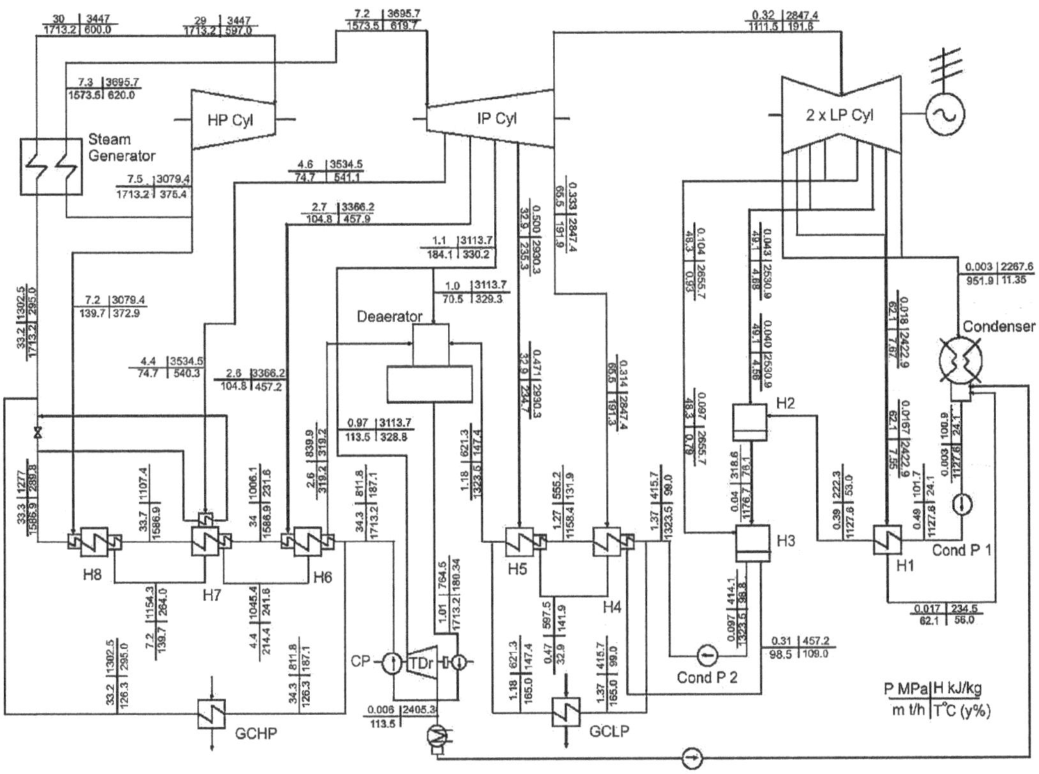

Power-plant "steam" generators put into operation at the "Emos" (1973) and "Gevin" (1974 - 1975) power plants (USA) (for 1130 MW$_{el}$ units):

- **"Steam" Capacity, t/h:** 4438
- **Pressure (Primary "Steam"), MPa:** 27.3
- **Temperature (Primary "Steam"), °C:** 543
- **Steam Capacity (Secondary Steam), t/h:** 3612
- **Pressure (Secondary Steam), MPa:** 4.7
- **Temperature (Secondary Steam), °C:** 538
- **Feed-Water Temperature, °C:** 291

In Japan, the first SC "steam" generator (600 MW$_{el}$) was commissioned in 1967 at the Anegasaki plant. Nowadays, many thermal power plants are equipped with SC "steam" generators and turbines. Major parameters of selected Hitachi, Ltd. SC power plants (turbines) are listed in Table 4, and photos of a SC turbine - in Figure 20.

Figure 18. Simplified T-s diagram of the Tom'-Usinsk 600-MW_{el} supercritical-pressure Rankine steam cycle with single reheat: Gross thermal efficiency ~48.6% (for details, see Table 3)

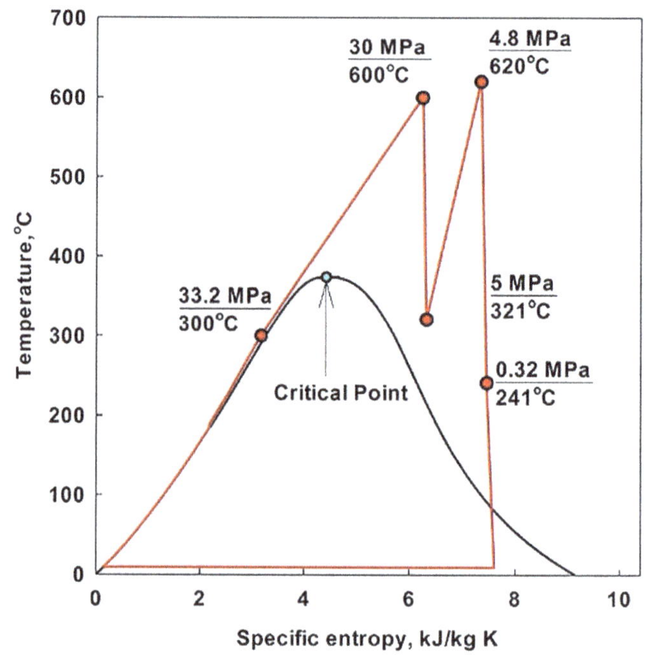

Figure 19. Simplified T-s diagram of typical subcritical-pressure Rankine steam cycle with single reheat: Gross thermal efficiency ~42%
Based on Dragunov et al., 2015.

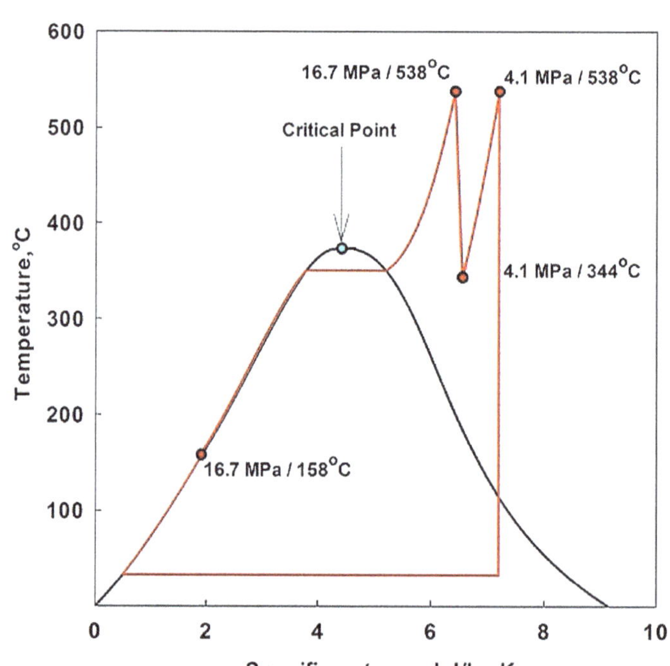

Table 3. Selected parameters of SC coal-fired power plants

No.	Power		"Steam" Inlet Parameters			Feed Water	Steam in Condenser	Thermal Efficiency, %	
	MW_{el}		$P_{primary}$	T, °C		T, °C	P, kPa	Net	Gross
	Net	Gross	MPa	Primary	Reheat				
1	450	490	24.1	566	566	-	5	-	-
2	660	-	28	600	600	290-295	5	43.5	*47.5*
3	660	-	30	600	620	300	3	44.6	*48.6*

Based on EPCOR (2007) - row no. 1 and Kruglikov et al. (2009) - rows nos. 2 and 3.

In China, the first 600-MW_{el} SC power plant was put into operation in 1992: Operating pressure - 24.2 MPa and temperatures - 538°C/566°C (main/reheat). In the first phase (more than 10 years), SC power plants were mainly imported from abroad and have three various levels of power: 350, 600 and 900 MW_{el}. In 2000, the Chinese government decided to develop the Chinese SC technology based on a 2 × 600 MW_{el} SC power-plant project. The first domestically designed SC power plant started its commercial operation in 2004 and has a pressure of 24.2 MPa and temperatures of 566°C/566°C (main/reheat). In November of 2006, a new level of SC parameters was achieved with starting of operation of 1000-MW_{el} thermal power plant with a pressure of 26.2 MPa and temperatures of 600°C/600°C (main/reheat).

More details on history of SC power plants can be found in Pioro and Duffey (2007) and (Ornatskiy et al. 1980).

Currently, so-called, "Ultra-SuperCritical (USC) boilers / turbines" are now being researched and deployed worldwide. Using single / double-steam reheat and advanced high-temperature blade materials, the turbine inlet temperature is being extended to 625°C at pressures of up to 38 MPa with overall efficiencies then approaching 51 - 55%. Figure 21 shows the 906-MW_{el} unit of Boxberg lignite-fired power plant in Germany equipped with USC parameters steam turbine.

In general, the gross thermal efficiencies of modern thermal power plants with subcritical-parameters steam generators are about 36 - 42%, but reaches 45 - 50% or even higher with SC parameters (i.e., with a "steam" pressure of 23.5 - 26 MPa and inlet turbine temperatures within the range of 535 - 585°C, the thermal efficiency is about 45 - 50% or even higher at ultra-supercritical parameters (25 - 38 MPa and 600 - 625°C). From the thermodynamics point of view it is clear that higher temperatures and eventually pressures - higher thermal efficiency. However, current advances in power plants are limited by materials either by their mechanical strength at high temperatures or by their cost.

The major driving force for all advances in thermal power plants is directed towards increasing thermal efficiency in order to reduce operating fuel costs and minimize specific emissions. Typical ranges of thermal efficiencies of modern thermal and nuclear power plants[2] are listed in Table 2 for reference purposes and can reach up to 62% in the combined-cycle mode (Pioro (Ed.), 2016).

Today, the major sources for electricity generation in the world (see Figure 22a) are: 1) Coal-fired power plants (39.9%); 2) Natural-gas-fired power plants (22.6%); 3) "Large" hydro-power plants (17.2%); and 3) Nuclear power from various reactor designs (11.2%). The remaining 9.2% of the electrical energy is generated using oil (4.2%) and the rest (5%) - with biomass, wind, geothermal, and solar energy in selected countries. In addition, energy sources, such as wind and solar and some others, like wave-power, are intermittent from depending on Mother Nature. Therefore, coal-fired power plants (many of which are SCP plants) are the major source of electricity generation in the world! Moreover, the most populated

Table 4. Major parameters of selected Hitachi, Ltd. supercritical-pressure turbines

First Year of Operation	Power Rating MW_{el}	Pressure MPa(g)	T_{main}/T_{reheat} °C
2011	495	24.1	566/566
2010	809	25.4	579/579
	790	26.8	600/600
2009	1000	25.0	600/620
	1000	25.5	566/566
	600	24.1	600/620
2008	1000	24.9	600/600
	887	24.1	566/593
	677	25.5	566/566
2007	1000	24.9	600/600
	870	25.3	566/593
2006	600	24.1	566/566
2005	495	24.1	566/566
2004	700	24.1	538/566
2003	1000	24.5	600/600
2002	700	25.0	600/600
1998	1000	24.5	600/600
1994	1000	24.1	538/566
1992	700	24.1	538/566
1991	600	24.1	538/566
1989	1000	24.1	538/566
	700	24.1	538/566
1984 and 1985	600	24.1	538/538
1983	700	24.1	538/538
	600	24.1	538/566
	350	24.1	538/566
1981	500	24.1	538/538
1979	600	24.1	538/566
1977	1000	24.1	538/566
	600	24.1	538/552/566*
1975	450	24.1	538/566
1974	500	24.1	538/566
1973	600	24.1	538/552/566*
	450	24.1	538/566
1971 and 1972	600	24.1	538/566

Mokry et al., 2008.

* Double-steam-reheat-cycle turbines. Analysis of the data in Table 4 shows that usually SC turbines have a single reheat.

Figure 20. Steam turbine (1 HP, 1 IP and 2 double-flow LP cylinders, tandem arrangement) of SCP coal-fired power plant (Japan): 700 MW_{el}, TC4F-43, 3000 rpm, "steam" parameters - 25.0 MPa, and 600°C (primary) / 600°C (reheat)
Courtesy and copyright by Hitachi, Ltd.

a)

b)

Figure 21. 906-MW_{el} unit of Boxberg lignite-fired power plant in Germany equipped with SSP5-6000 steam turbine and cooled through cooling tower (Siemens press photo; copyright Siemens AG, Munich/ Berlin, Germany). Ultra-SuperCritical (USC) parameters steam turbine - inlet primary / reheat steam pressures - temperatures: 28 MPa - 600°C / 6 MPa - 610°C; 1 HP, 1 IP and 2 double-flow LP cylinders, 4 casings, and tandem arrangement. Feed Water Heaters (FWHs) - 9 stages; 3 HP Preheaters (FWPHs) with 1 de-superheater; 5 plate-type LP FWHs, and 2 × 50% electrical motor-driven FW pumps. Cooling tower - natural-draft wet-cooling tower. Electrical generator - 2-pole 3000 rpm / 50 Hz; 99% efficiency; water / hydrogen cooled; hydrogen at 0.6 MPa (improved cooling compared to that with atmospheric air by a factor of 14). Net-plant thermal efficiency (LHV-basis) is about 46% (gross thermal efficiency (estimated value) ~50%). CO_2 emissions is well-below 800 g/kWh.

Table 5. Typical gross thermal efficiencies of various modern thermal and nuclear power plants (power plants are listed from high thermal efficiencies to lower ones)

No.	Power Plant	Gross Efficiency %
1	Combined-cycle power plant (combination of Brayton gas-turbine cycle (fuel - natural or Liquefied Natural Gas (LNG); combustion-products parameters at the gas-turbine inlet: $T_{in} \approx 1650°C$) and Rankine steam-turbine cycle (steam parameters at the turbine inlet: $T_{in} \approx 620°C$ ($T_{cr} = 374°C$)).	Up to 62
2	SCP coal-fired thermal power plant (new plants) (Rankine-cycle "steam" inlet turbine parameters: $P_{in} \approx 25$ - 38 MPa ($P_{cr} = 22.064$ MPa), $T_{in} \approx 540$ - 625°C ($T_{cr} = 374°C$) and $T_{reheat} \approx 540$ - 625°C).	Up to 55
3	Subcritical-pressure coal-fired thermal power plant (older plants) (Rankine-cycle steam: $P_{in} \approx 17$ MPa, $T_{in} \approx 540°C$ ($T_{cr} = 374°C$) and $T_{reheat} \approx 540°C$).	Up to 40-42
4	Carbon-dioxide-cooled reactor (Advanced Gas-cooled Reactor (AGR) NPP (Generation III, current fleet) (reactor coolant - carbon dioxide: $P \approx 4$ MPa and $T_{in} / T_{out} \approx 290 / 650°C$; secondary Rankine-cycle steam: $P_{in} \approx 17$ MPa ($T_{sat} \approx 352°C$) and $T_{in} \approx 560°C$ ($T_{cr} = 374°C$))	Up to 42
5	Liquid-Metal Fast-Breeder Reactor (LMFBR) or Sodium-cooled Fast Reactor (SFR) NPP (Generation III, IV, currently just two reactors – BN[3]-600 and BN-800 operate in Russia) (reactor coolant - liquid sodium: $P \approx 0.1$ MPa and $T_{max} \approx 500$ - 550°C; secondary Rankine-cycle steam: $P_{in} \approx 14$ MPa ($T_{sat} \approx 337°C$) and $T_{in} \approx 505°C$ ($T_{cr} = 374°C$)).	Up to 40
6	Pressurized Water Reactor (PWR) NPP (Generation III+, to be implemented within next 1 - 10 years) (reactor coolant - light water: $P \approx 16$ MPa ($T_{sat} = 347°C$) and $T_{out} \approx 327°C$; secondary Rankine-cycle steam: $P_{in} \approx 7.8$ MPa and $T_{in} = T_{sat} \approx 293°C$).	Up to 36-38
7	PWR NPP (Generation III, current fleet) (reactor coolant - light water: $P \approx 16$ MPa ($T_{sat} = 347°C$) and $T_{in} / T_{out} \approx 290 / 325°C$; secondary Rankine-cycle steam: $P_{in} \approx 7.2$ MPa and $T_{in} = T_{sat} \approx 288°C$).	32-36
8	Boiling Water Reactor (BWR) NPP (Generation III, current fleet) (reactor coolant - light water; direct cycle; steam parameters at the turbine inlet: $P_{in} \approx 7.2$ MPa and $T_{in} = T_{sat} \approx 288°C$). ABWR NPP (Generation III+) has approximately the same thermal efficiency.	~34
9	Light-water Graphite-moderated Reactor (LGR) NPP (boiling reactor, pressure-channel design) NPP (Generation II and III, current fleet) (reactor coolant - light water; direct cycle; steam parameters at the turbine inlet: $P_{in} \approx 6.6$ MPa and $T_{in} = T_{sat} \approx 282°C$).	~32
10	Pressurized Heavy Water Reactor (PHWR) NPP (Generation III, current fleet) (reactor coolant - heavy water: $P_{in} \approx 11$ MPa; $P_{out} \approx 10$ MPa ($T_{sat} = 311°C$) and $T_{in} / T_{out} \approx 265 / 310°C$; secondary Rankine-cycle steam (light water): $P_{in} \approx 4.6$ MPa and $T_{in} = T_{sat} \approx 259°C$).	~32

Based on data from Pioro (Ed.) (2016).

countries such as China, India, and USA rely heavily on coal-generated electricity (see Figure 22b, c, and d): 1) China (80%); 2) India (72%); and 3) USA (38.7%) (for additional information on this topic, see Pioro (Ed.) (2016) and Pioro and Duffey (2015)).

One of the important features of coal-fired power plants including SCP plants is their ability to follow up with fast changing demand for electricity during a day (see Figure 23) (Pioro (Ed.) (2016); Pioro & Duffey (2015)).

Despite the advances in SCP thermal power-plants design and operation worldwide, they are still considered as not of minimum environmental impact due to significant carbon-dioxide emissions[4] and air pollution as a result of the combustion process (Pioro (Ed.), 2016). In addition, coal-fired power-plants produce also virtual mountains of slag and ash, and other gas emissions may contribute to acid rains. Therefore, more "environmentally friendly", but still large-capacity and high-reliability sources for electricity generation are required. One of such sources is nuclear power.

Figure 22. Electricity generation by source in the world and selected countries: (a) World: Population 7,470 millions; Electrical Energy Consumption (EEC) 19,710 TW h/year or 310 W/Capita; Human Development Index (HDI) 0.711 or HDI Rank 103; (b) China: Population 1,385 millions; EEC 5,523 TW h/year or 461 W/Capita; HDI 0.727 or HDI Rank 90; (c) India: Population 1,334 millions; EEC 865 TW h/year or 79 W/Capita; HDI 0.609 or HDI Rank 130; (d) USA: Population 325 millions; EEC 3,832 TW h/year or 1,360 W/Capita; HDI 0.915 or HDI Rank 8; Renewables (6.9%): Wind (4.4%); Biomass (1.7%); Geothermal (0.4%); and Solar (0.4%)

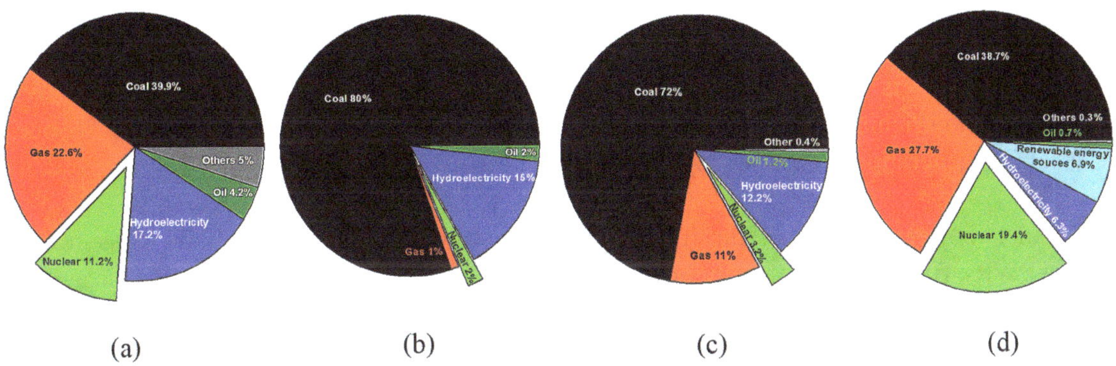

Figure 23. Power generated (a) and capacity factors (b) of various energy sources in Ontario (Canada) on June 19, 2012

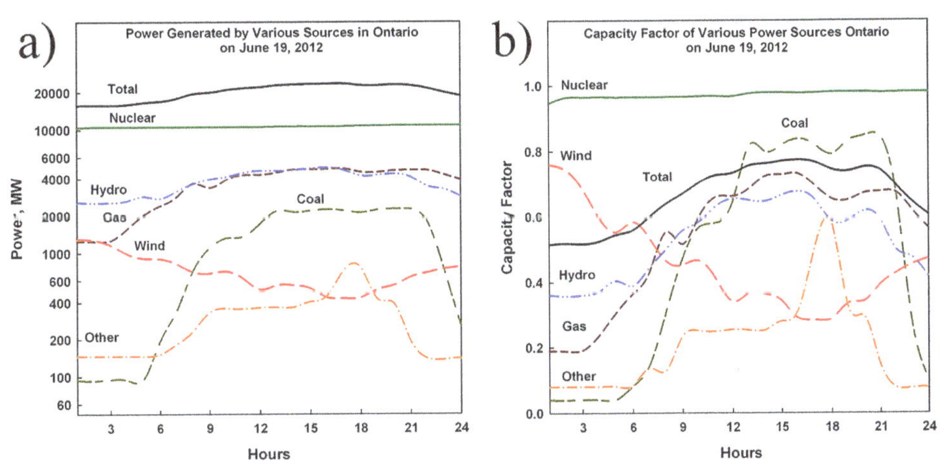

APPLICATION OF SUPERCRITAL FLUIDS AND PRESSURES IN GENERATION-IV NUCLEAR-POWER PLANTS

Current Nuclear-Power Reactors and NPPs

Current fleet of nuclear-power reactors consist of 436 reactors of Generation-II and III (Nuclear News, 2016; Pioro (Ed.), 2016): 1) 280 PWRs; 2) 78 BWRs; 3) 48 PHWRs; 4) 14 AGRs; 5) 15 LGRs; and 6) 1 LMFBR or SFR (also, for details on operating conditions of these reactors, see Table 5).

Major advantages of nuclear power are:

- High capacity factors are achievable, often in excess of 90% with long operating cycles, making the units suitable for semi-continuous base-load operation, alongside intermittent windmills backed by gas peaking plants;
- Essentially negligible operating emissions of carbon dioxide into atmosphere compared to alternate thermal power plants; And
- Relatively small amount of fuel required (for example, a 500-MW_{el} coal-fired SCP power plant requires 1.8 million ton of coal annually, but a fuel load into a 1300-MW_{el} Pressurized Water Reactor (PWR) is 115 t (3.2% enrichment[5]) or into a 1330-MW_{el} Boiling Water Reactor (BWR) - 170 t (1.9% enrichment)). Therefore, this source of energy is considered as the most viable one for electrical generation for the next 50 - 100 years.

All nuclear-power reactors operating in the world as per March of 2016 use subcritical-pressure reactor coolants: PWRs, BWRs, and LGRs - light water; PHWRs - heavy water (moderator is also heavy water at atmospheric pressure); AGRs - carbon dioxide; and SFR - liquid or molten sodium. Power cycle for all these reactors is subcritical-pressure Rankine steam-turbine cycle (for details on operating conditions, see Table 5). Therefore, thermal efficiencies of modern NPPs range from 30 - 32 to 42% (Table 5). Moreover, all AGR NPPs (the most efficient NPPs as of today with the thermal efficiency up to 42%) will be shut down within next 10 - 15 years, and there are no plans to build new ones.

Analysis of data in Table 5 shows that current NPPs with water-cooled reactors (~97% of all nuclear-power reactors) have significantly lower thermal efficiencies (30 - 36%) compared to those achieved in thermal industry (combined-cycle power plants up 62% gross thermal efficiency and SCP fossil-fuel plants - up to 55%).

In spite of all current advances in nuclear power, NPPs have the following deficiencies: 1) Have relatively low thermal efficiencies, especially, water-cooled-reactors NPPs (up to 1.6 times lower than that for modern advanced thermal power plants (see Table 5)); 2) Generate radioactive wastes; 3) Risk of radiation release during severe accidents; and 4) Production of nuclear fuel is not an environment-friendly process. Therefore, all these deficiencies should be addressed in next generation nuclear reactors and NPPs.

Generation-IV Nuclear-Reactor Concepts and NPPs

Therefore, in 2001, a group of countries, including Argentina, Brazil, Canada, France, Japan, Republic of Korea, South Africa, the United Kingdom (UK) and the United States (US), has initiated an international collaboration (Generation-IV International Forum (GIF): to develop the next generation nuclear reactors (Generation-IV reactors). Later on, Switzerland (2002), Euratom (2003), the People's Republic of China (2006), the Russian Federation (2006), and Australia (2016) have joined this Initiative. The ultimate goal of developing such reactors is an increase in thermal efficiencies of NPPs from 30 - 36% to 45 - 50% and even higher. This increase in thermal efficiency would result in a higher generation of electricity compared to current Light-Water-Reactor (LWR) technologies per 1 kg of uranium.

The GIF Program has narrowed design options of nuclear reactors to six concepts (see Tables 6 and 7, and Figures 24 - 37). These concepts are (listed in the following order - helium-cooled reactors, molten metals-and molten salt-cooled reactors, and SCW-cooled reactor): 1) Very High Temperature

Reactor (VHTR) (Figure 24); 2) Gas-cooled Fast Reactor (GFR) (Figure 28); 3) Sodium-cooled Fast Reactor (SFR) (Figure 32); 4) Lead-cooled Fast Reactor (LFR) (Figure 35); 5) Molten Salt Reactor (MSR) (Figure 36); and 6) Supercritical Water-cooled Reactor (SCWR) (Figure 37). Table 6 lists basic reference parameters of Generation-IV nuclear reactors/concepts including their possible power cycles.

All Generation-IV nuclear-reactor concepts are high-temperature reactors with reactor-coolant outlet temperatures ranging from 500°C and up to 1000°C. These reactors will use SCFs as reactor coolants (helium in GFRs and VHTRs; SCW in SCWRs) and / or working fluids in power cycles (helium, mixture of nitrogen (80%) and helium (20%), nitrogen, and carbon dioxide in Brayton gas-turbine cycles. These nuclear-reactor concepts differ one from each other in terms of their design, neutron spectrum, coolant, moderator, operating temperatures and pressures. A brief description of each Generation-IV nuclear-reactor concept and possible power cycles has been provided below.

In general, all reactors have certain limits, which must be accounted for in their designs, two of which are: 1) maximum sheath temperature and 2) maximum fuel centerline temperature. Maximum sheath or cladding temperature depends on materials used, diameter and wall thickness of a sheath, and limits

Table 6. Basic reference parameters of Generation-IV nuclear reactors/concepts (for details see Figures 24 - 37)

Reactor	Neutron Spectrum	Reactor Coolant	Moderator	Operating Range		Reactor Cycle	Power Cycle	P_{in}	T_{in}
				P	T				
				MPa	°C			MPa	°C
VHTR	Thermal	Helium	Graphite	7	490-1000	Direct	Brayton He	7	1000
						Indirect	Brayton $N_{2(80\%)}$-$He_{(20\%)}$		
							Combined		
GFR	Fast	Helium	-	9	490-850	Direct	Brayton He	9	850
						Indirect	Brayton $N_{2(80\%)}$-$He_{(20\%)}$		
							Combined		
SFR	Fast	Sodium	-	~0.1	370-550	Indirect	Rankine (Brayton SC CO_2)*	~14	~505
LFR	Fast	Lead or LBE	-	~0.1	550-800 (420-540)**	Indirect	Rankine (Rankine SCP)*	~17 (24.5)	~505 (520)
MSR	Epi-thermal	Sodium fluoride with dissolved uranium	Graphite	~0.1	T_{out}=700-800	Indirect	Brayton or combined	-	-
MSFR	Fast		-	~0.1	T_{out}=700-800	Indirect			
SCWR	Thermal	Water		~25	300-625	Direct	Rankine SCP	~25	~625
		Water	Heavy water			Indirect			
	Fast	Water	-		300-625	Direct			
						Indirect			

Based on Handbook of Generation IV Nuclear Reactors (Pioro (Ed.), 2016).

Explanations to Table 6: *(…) - possible back-up design; **(…) - current possible design; LBE - Lead-Bismuth Eutectic; P_{in} and T_{in} - inlet parameters to steam or gas turbine; T_{out} - maximum outlet-reactor temperatures.

Table 7. Estimated ranges of thermal efficiencies (gross) of Generation-IV NPP concepts (shown here just for reference purposes; NPPs are listed from high thermal efficiencies to lower ones)

No.	Nuclear Power Plant	Thermal Eff., %
1	Very High Temperature Reactor (VHTR) NPP (reactor coolant – helium: P=7 MPa and T_{in}/T_{out}=640/1000°C; primary power cycle – direct Brayton gas-turbine cycle; possible back-up – indirect Brayton cycle(s), combined cycle or Rankine steam cycle) (for details, see Table 8 and Figures 24 - 27).	≥55
2	Gas-cooled Fast Reactor (GFR) NPP (reactor coolant – helium: P=9 MPa and T_{in}/T_{out}=490/850°C; primary power cycle – direct Brayton gas-turbine cycle; possible back-up – indirect Brayton cycle(s), combined cycle or Rankine steam cycle) (for details, see Table 9 and Figures 28 - 31).	≥50
3	SuperCritical Water-cooled Reactor (SCWR) NPP (one of Canadian concepts; reactor coolant – light water: P=25 MPa and T_{in}/T_{out}=350/625°C (T_{cr}=374°C); direct cycle; high-temperature steam superheat: T_{out}=625°C; possible back-up - indirect supercritical-pressure Rankine "steam" cycle with high-temperature steam superheat) (for details, see Table 14 and Figure 37).	45-50
4	Molten Salt Reactor (MSR) NPP (reactor coolant – sodium-fluoride salt with dissolved uranium fuel: T_{out}=700-800°C; primary power cycle – indirect supercritical-pressure carbon-dioxide Brayton gas-turbine cycle; possible back-up – supercritical-pressure Brayton gas-turbine cycle with other working fluids or combined cycle) (for details, see Table 13 and Figure 36).	~50%
5	Lead-cooled Fast Reactor (LFR) NPP (Russian design Brest-300: reactor coolant – liquid lead: P≈0.1 MPa and T_{in}/T_{out}=420/540°C; primary power cycle – indirect subcritical- (~18 MPa) or supercritical-pressure Rankine "steam" cycle: P_{in}≈24.5 MPa (P_{cr}=22.064 MPa) and T_{in}/T_{out}=340/520°C (T_{cr}=374°C); high-temperature steam superheat; possible back-up in some other countries – indirect supercritical-pressure carbon-dioxide Brayton gas-turbine cycle) (for details, see Table 12 and Figure 35).	~43
6	Sodium-cooled Fast Reactor (SFR) NPP (Russian design BN-600: reactor coolant – liquid sodium (primary circuit): P≈0.1 MPa and T_{in}/T_{out}=380/550°C; liquid sodium (secondary circuit): T_{in}/T_{out}=320/520°C; primary power cycle – indirect Rankine steam cycle: P_{in}≈14.2 MPa (T_{sat}≈337°C) and $T_{in\ max}$=505°C (T_{cr}=374°C); steam superheat: P≈2.45 MPa and T_{in}/T_{out}=246/505°C; possible back-up in some other countries - indirect supercritical-pressure carbon-dioxide Brayton gas-turbine cycle) (for details, see Tables 10 and 11, and Figures 32 - 34).	~42-43.6

Based on Handbook of Generation IV Nuclear Reactors (Pioro (Ed.), 2016).

maximum temperatures and pressures inside a reactor. Maximum fuel centerline temperature should be well below a melting temperature of fuel, for example, for uranium dioxide (UO_2), as the most common current nuclear fuel, the melting temperature is about 2850°C, but the industry accepted limit for safe and reliable operation of a reactor is 1850°C for the fuel centerline temperature, i.e., 1000°C below the melting point. Also, in the same way as for modern thermal power plants, materials (their mechanical and other properties, and their cost) used in NPP power cycles limit current level of maximum thermal efficiencies.

Very High Temperature Reactor (VHTR)

Very High Temperature Reactor (VHTR) (see Figure 24 and Table 8) is a thermal-neutron-spectrum reactor (Pioro & Kirillov, 2013). In the VHTR, graphite and helium have been chosen as the moderator and coolant. The inlet and outlet temperatures of the coolant are 640 and 1000°C, respectively, at a pressure of 7 MPa (US DOE, 2002). Due to such high outlet temperatures, the thermal efficiency of VHTR can be about 50% or even higher. At the beginning, the ultimate purpose of this nuclear-reactor concept was the co-generation of hydrogen through high-temperature electrolysis. However, recently, the VHTR is considered as a multipurpose nuclear-power reactor, which can generate electricity, co-generate hydrogen through high-temperature electrolysis or thermo-chemical cycles, supply high-temperature process heat, support desalination and district heating (Pioro (Ed.), 2016).

In general, the US DOE supports research on several Generation-IV reactor concepts including High Temperature Gas-cooled Reactor (HTGR) for the Next Generation Nuclear Plant (NGNP).

Originally, for a VHTR and GFR the direct Brayton helium-turbine cycle was considered (see Figure 28). However, due to various technical problems indirect cycles are currently under consideration including combined cycles. Figures 25 - 27 show such power cycles for the VHTR concept. Figure 25 shows the indirect primary Brayton gas-turbine cycle with SCP nitrogen-helium mixture as the working fluid and indirect secondary subcritical-pressure Rankine steam-turbine cycle. Figure 26 shows the indirect primary Brayton gas-turbine cycle with SCP carbon dioxide as the working fluid and indirect secondary SCP Rankine turbine cycle also with carbon dioxide as the working fluid. Figure 27 shows the *T-s* diagram of this cycle.

Gas-Cooled Fast Reactor (GFR)

Gas-cooled Fast Reactor (GFR) (see Table 9 and Figure 28) is a fast-neutron-spectrum reactor, which can be used for the production of electricity and co-generation of hydrogen through thermochemical cycles or high-temperature electrolysis (Pioro & Kirillov, 2013). The coolant is helium with inlet and outlet temperatures of 490 and 850°C, respectively. The net-plant efficiency is about 48% with the

Figure 24. Very High Temperature Reactor (VHTR): SCP helium-cooled, graphite-moderated, thermal-neutron-spectrum reactor with core outlet temperature 900-1000°C (shown only with hydrogen co-generation)
Courtesy of Generation IV International Forum.

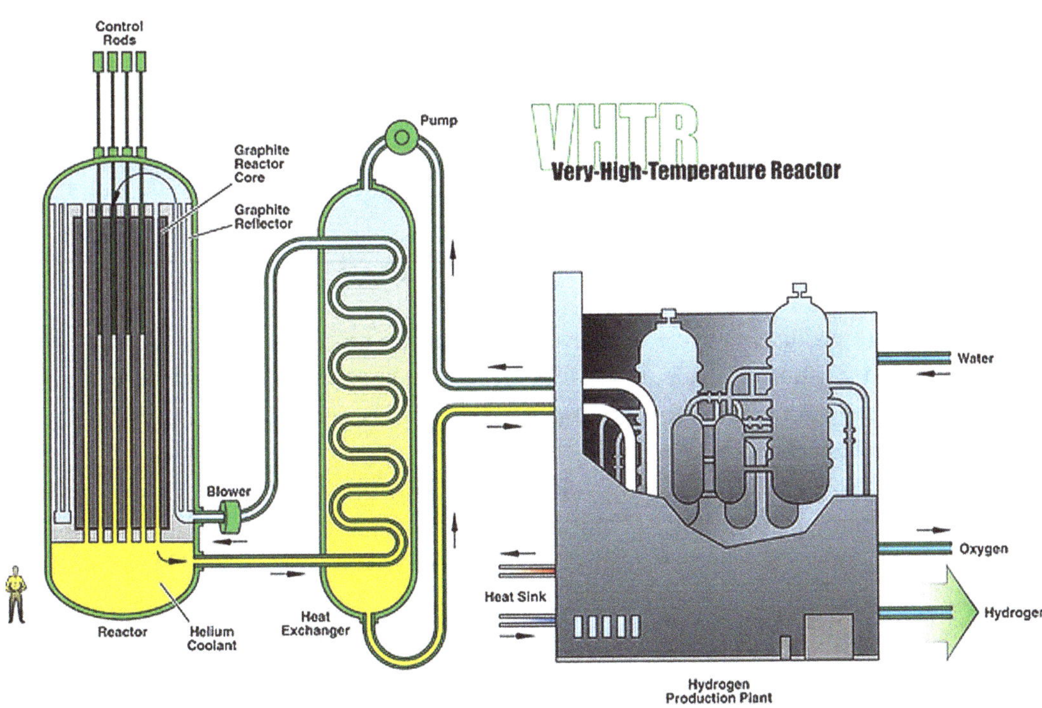

Table 8. Key-design parameters of Very High Temperature Reactor (VHTR) concept

Reactor Parameter	Unit	Reference Value
Reactor power	MW_{th}	600
Average power density	MW_{th}/m^3	6-10
Coolant inlet/outlet temperatures	°C	640/1000
Coolant/Massflow rate	kg/s	Helium/320
Reference fuel compound	-	ZrC-coated particles in pins or pebbles
Net-plant efficiency	%	>50

Pioro & Kirillov, 2013.

Figure 25. Simplified schematic of VHTR NPP (reactor coolant - helium at 5 MPa) with indirect combined cycle (Primary - Brayton gas-turbine cycle (working fluid - mixture of nitrogen and helium at 5 MPa) and Secondary - Rankine steam-turbine cycle) and hydrogen cogeneration
Based on schematic from Gauthier et al. (2004).

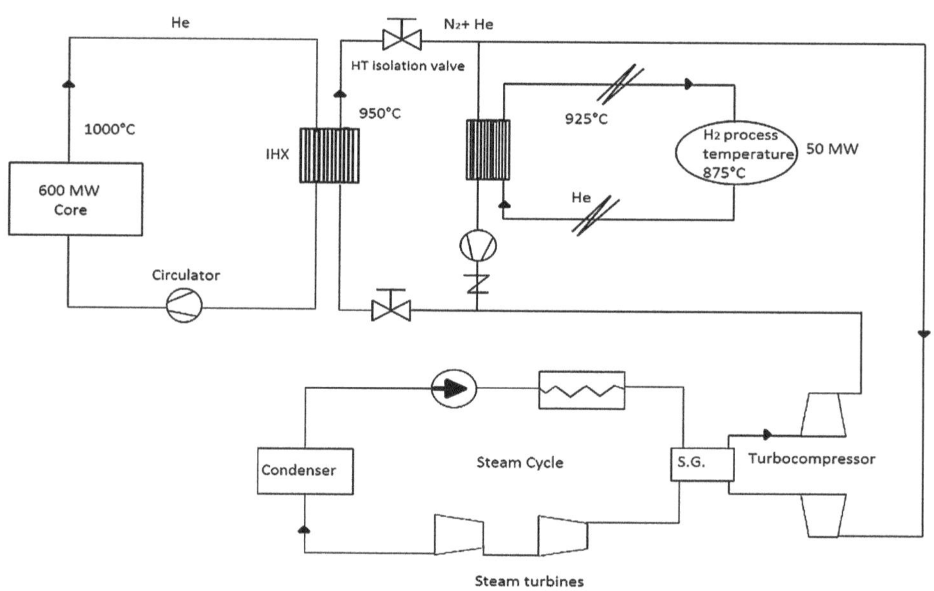

direct Brayton helium-gas-turbine cycle. However, due to some problems with implementation of the direct Brayton helium-gas-turbine cycle, the indirect Rankine steam cycle or even indirect supercritical carbon-dioxide Brayton gas-turbine cycle are also considered. The indirect cycles will be linked to the GFR through heat exchangers.

Figures 29 and 31 show possible power cycles for the GFR concept. Figure 29 shows the indirect primary Brayton gas-turbine cycle with SCP nitrogen-helium mixture as the working fluid and indirect secondary subcritical-pressure Rankine steam-turbine cycle. Figure 30 shows the *T-s* diagram of this cycle. Figure 31 shows the indirect primary Brayton gas-turbine cycle with SCP nitrogen as the working fluid and indirect secondary SCP Rankine turbine cycle with carbon dioxide as the working fluid.

Figure 26. Simplified schematic of VHTR NPP (reactor coolant - helium at 7 MPa) with indirect combined cycle (Primary - SCP Brayton gas-turbine cycle (working fluid - carbon dioxide at ~20 MPa) and Secondary - SCP Rankine cycle (working fluid - carbon dioxide at ~21 MPa)). T-s diagram of this cycle is shown in Figure 27
Based on schematic from Bae et al. (2014).

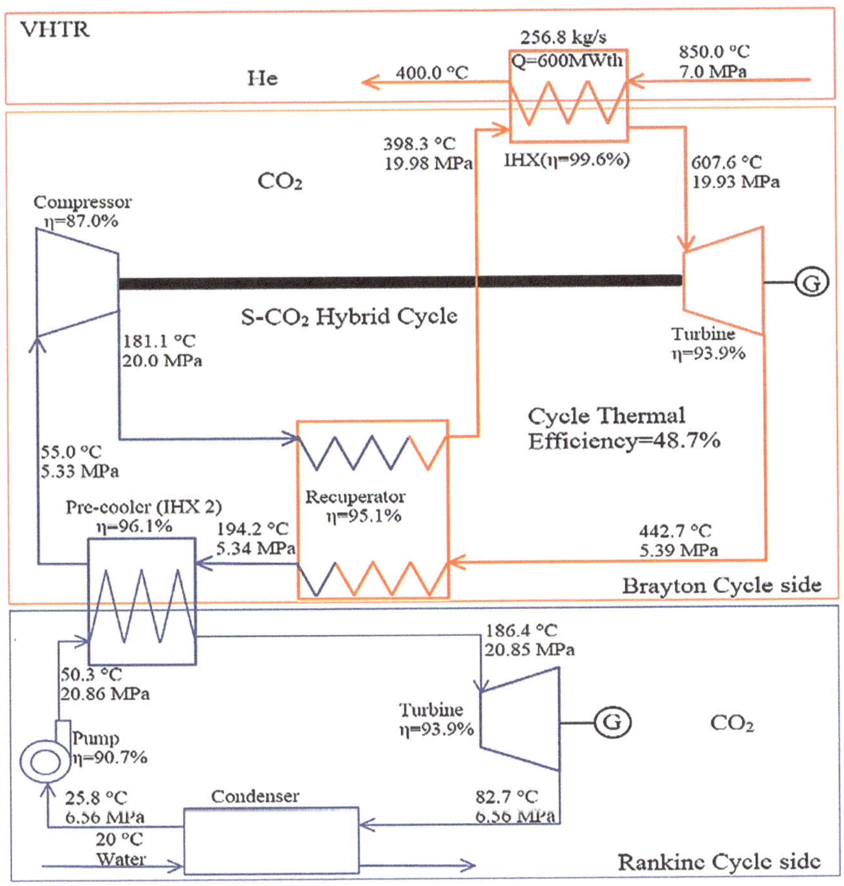

Sodium-Cooled Fast Reactor (SFR)

Similar to GFR, SFR (see Figure 32 and Table 10) is a fast-neutron-spectrum reactor (Pioro & Kirillov, 2013). The main objectives of SFR are the management of high-level radioactive wastes and generation of electricity. SFR uses liquid sodium as a reactor coolant with an outlet temperature between 530 and 550°C at the atmospheric pressure. The primary choices of fuel for SFR are oxide and metallic fuels. The SFR concept is also on the priority list of the US DOE.

Currently, SFR is the only one Generation-IV reactor concept implemented in the power industry. Russia and Japan are leaders within this area. In particularly, Russia operates two SFRs (BN-600 and BN-800) at the Beloyarsk NPP (Table 11). Japan has operated SFR at the Monju NPP some time ago. In Russia and Japan the SFRs are connected to the subcritical-pressure Rankine steam-turbine cycle through heat exchangers or steam generators (see Figure 32).

Figure 27. T-s diagram of VHTR NPP indirect combined power cycle with SCP carbon dioxide in Brayton and Rankine cycles. For details, see schematic in Figure 26.
Based on diagram from Bae et al. (2014).

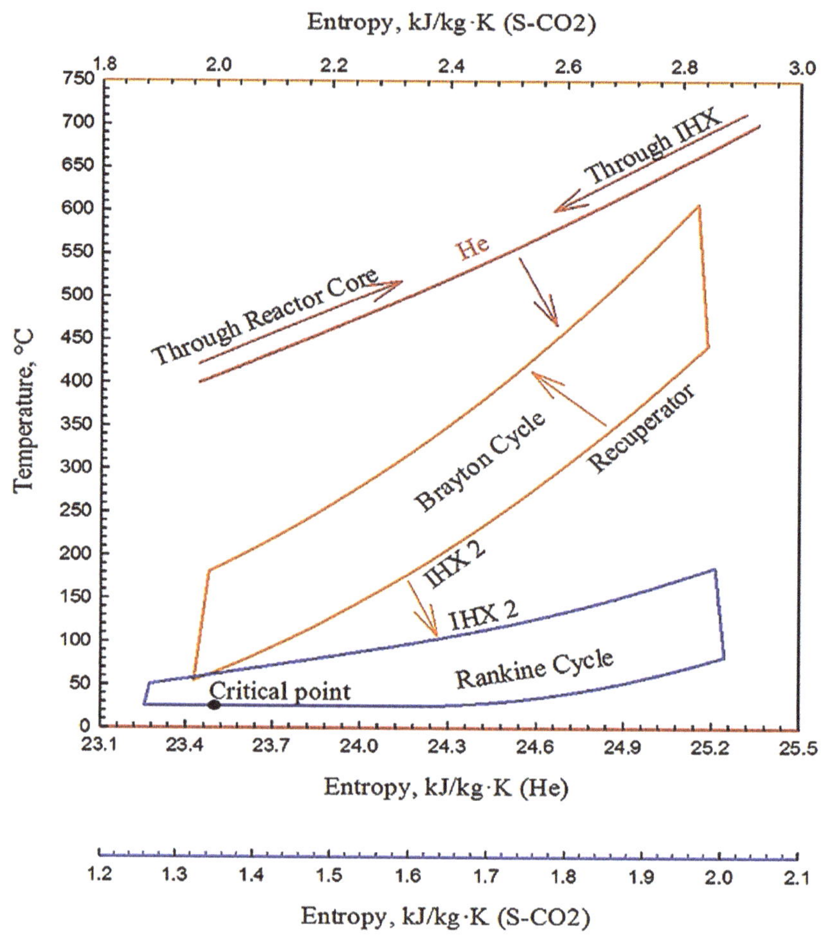

However, in the US and some other countries a supercritical carbon-dioxide Brayton gas-turbine cycle (see Figures 33 and 34) is considered as the power cycle for future SFRs, because carbon dioxide and sodium are considered to be more compatible than water and sodium. In general, sodium as the alkali metal is a highly-reactive substance. It may ignite spontaneously in air and reacts violently with water creating sodium hydroxide and hydrogen, which ignites spontaneously. The ignition temperature of sodium in air depends on the area of surface exposed. Sodium vapor ignites at room temperature; droplets - at about 121.1°C (250°F); an agitated pool - at 204.4°C (400°F). Therefore, special precautions should be taken for safe operation of this type reactor. One of them is the triple-flow circuit with the intermediate sodium loop between the reactor coolant (primary sodium) and water as the working fluid in the power cycle.

Figure 28. Gas-cooled Fast Reactor (GFR): SCP helium-cooled, fast-neutron-spectrum reactor with closed fuel cycle and outlet temperature about 850°C (shown with direct SCP Brayton helium-turbine power cycle)
Courtesy of Generation IV International Forum.

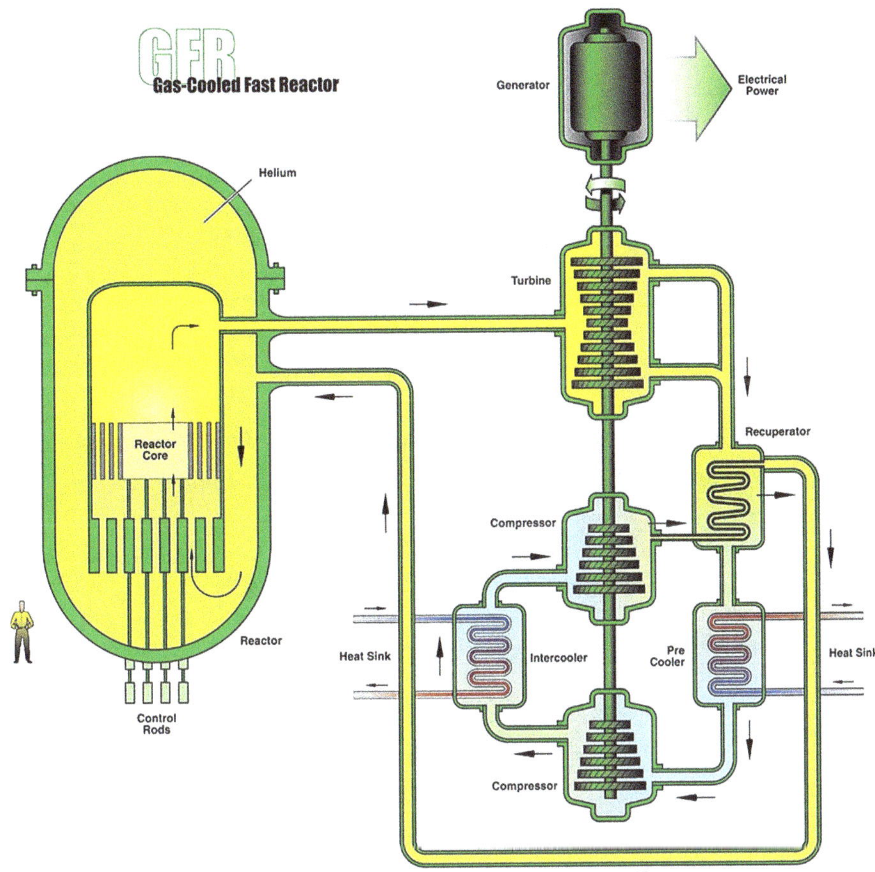

Table 9. Key-design parameters of Gas-cooled Fast Reactor (GFR) concept

Reactor Parameters	Unit	Reference Value
Reactor power	MW_{th}	600
Coolant inlet/outlet temperatures	°C	490/850
Pressure	MPa	9
Coolant massflow rate	kg/s	320
Average power density	MW_{th}/m^3	100
Reference fuel compound	-	UPuC/SiC (70/30%) with about 20% Pu
Net-plant efficiency	%	48

Pioro & Kirillov, 2013.

Figure 29. Simplified schematic of GFR (reactor coolant - helium at 7 MPa) with indirect combined cycle (Primary - SCP Brayton gas-turbine cycle (working fluid - mixture of nitrogen and helium at 6.5 MPa) and Secondary - Rankine steam-turbine cycle (at 15 MPa)). T-s diagram of this cycle is shown in Figure 30.
Based on schematic from Poette et al. (2013).

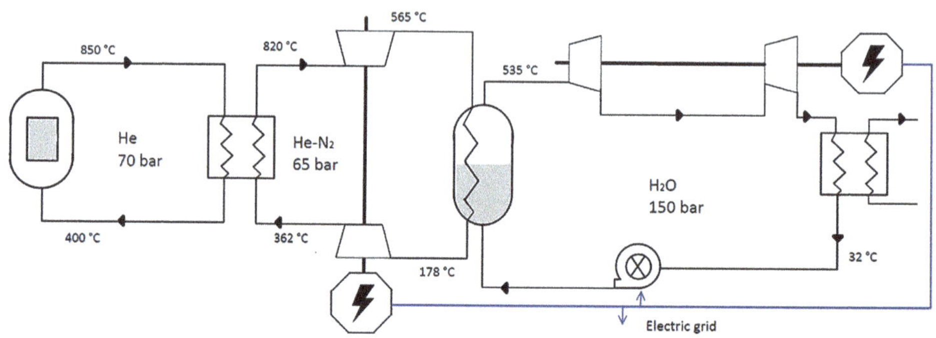

Figure 30. T-s diagram of GFR indirect combined power cycle with SCP nitrogen-helium mixture in Brayton cycle and Rankine steam cycle. For details, see schematic in Figure 29.
Based on data from Poette et al. (2013).

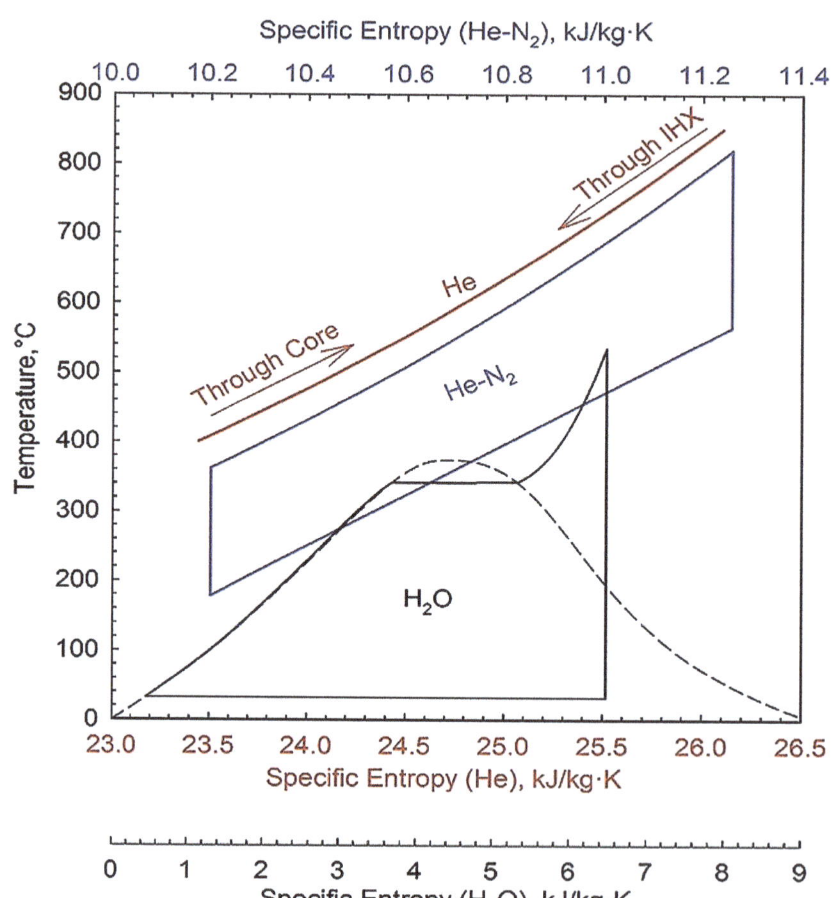

Figure 31. Simplified schematic of GFR (reactor coolant helium at 7 MPa) with indirect Brayton cycles (Primary - SCP Brayton gas-turbine cycle (working fluid - nitrogen at 6.5 MPa) and Secondary - SCP Brayton cycle (working fluid - carbon dioxide at 15 MPa)) and hydrogen co-generation
Based on schematic from Bae et al. (2014).

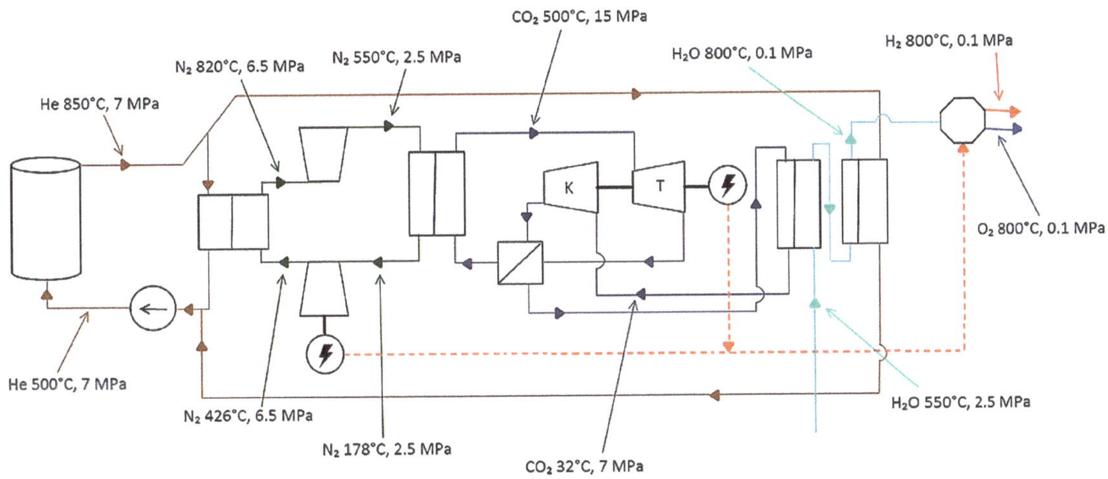

Figure 32. Sodium-cooled Fast Reactor (SFR): Molten-sodium-cooled, fast-neutron-spectrum reactor with closed fuel cycle and outlet temperatures within 500 - 550°C (shown pool-type reactor with indirect subcritical-pressure Rankine steam-turbine power cycle)
Courtesy of Generation IV International Forum.

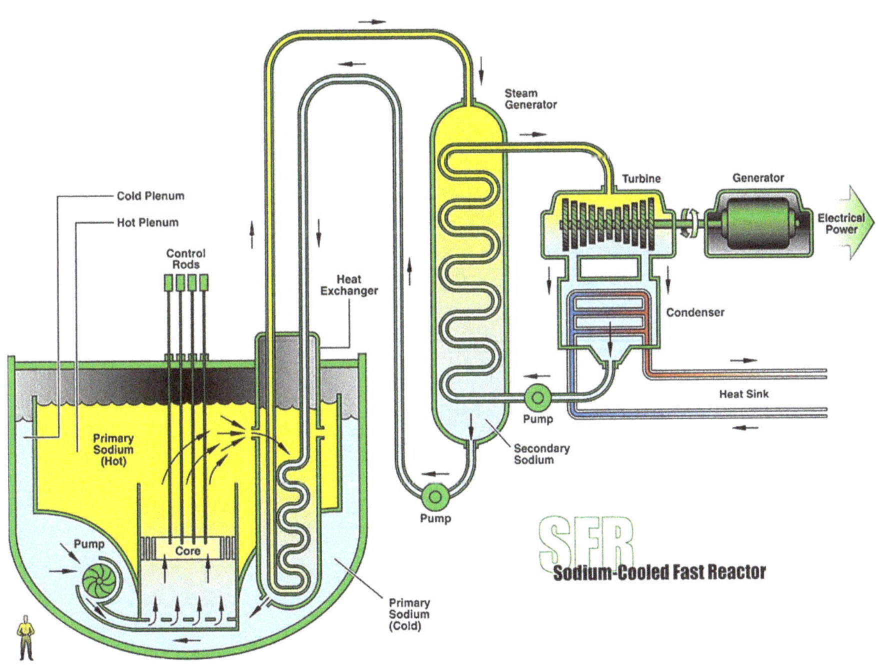

Table 10. Key-design parameters of generic SFR concept

Reactor Parameter	Unit	Reference Value
Reactor power	MW_{th}	1000-5000
Thermal efficiency	%	40–42%
Coolant	-	Sodium
Coolant melting/boiling temperatures	°C	98/883
Coolant density at 450°C	kg/m^3	844
Pressure inside reactor	MPa	~0.1
Coolant maximum outlet temperature	°C	530-550
Average power density	MW_{th}/m^3	350
Reference fuel compound	-	Oxide or metal alloy
Cladding	-	Ferritic or ODS* ferritic
Average burnup	GWD**/MTHM***	~150-200

Pioro & Kirillov, 2013.
* ODS - Oxide Dispersion Strengthened
** GWD is GigaWatt-Days (1 GWd = 8.64 x 1013 J)
*** MTHM - Metric Tonne of Heavy Metal

Table 11. Key-design parameters of Russian SFRs

No.	Parameter	BN-600*	BN-800**	BN-1200***
1	Thermal power, MW_{th}	1470	2100	2800
2	Electrical power, MW_{el}	600	880	1220
3	Basic components: No of turbines × type No of generators × type	3 × K-200-130 3 × ТГВ-200-М	1 × K-800-130 1 × ТЗВ-800-2	1 × K-1200-160 1 × ТЗВ-1200-2
4	Pressure vessel Diameter, m Height, m	12.86 12.60	12.96 14.82	16.9 20.72
5	No of heat-transfer loops	3	3	4
6	T of reactor coolant: sodium, primary loop - T_{in}/T_{out}, °C	377/550	354/547	410/550
7	T of intermediate coolant: sodium, secondary loop - T_{in}/T_{out}, °C	328/518	309/505	355/527
8	T of power-cycle working fluid: water/steam - T_{in}/T_{out}, °C	240/505	210/490	275/510
9	P at steam-generator outlet, MPa	13.7	14.0	17.0
10	Scheme of steam reheat with	Sodium	Steam	Steam
11	Basic unchangeable components service term, years	30	40	60
12	NPP thermal efficiency (gross), %	42.5	41.9	43.6
13	NPP thermal efficiency (net), %	40.0	38.8	40.5

Pioro & Kirillov, 2013.
* BN-600 – the only one long-term operating SFR in the world (Beloyarsk NPP); commercial start – 1981.
** BN-800 – the only second operating SFR in the world (Beloyarsk NPP), commercial start - 2016.
*** BN-1200 – concept of future SFR.

Figure 33. Simplified schematic of SFR (reactor coolant - sodium at ~0.1 MPa) with intermediate loop (heat-transfer fluid - sodium at ~0.1 MPa) and indirect cycle (SCP Brayton gas-turbine cycle (working fluid - carbon dioxide at 20 MPa): Comp. - Compressor; HTR - High Temperature Recuperator; IHX - Intermediate Heat eXchanger; LTR - Low Temperature Recuperator. T-s diagram of this cycle is shown in Figure 34.
Based on schematic from Moisseytsev and Sienicki (2009).

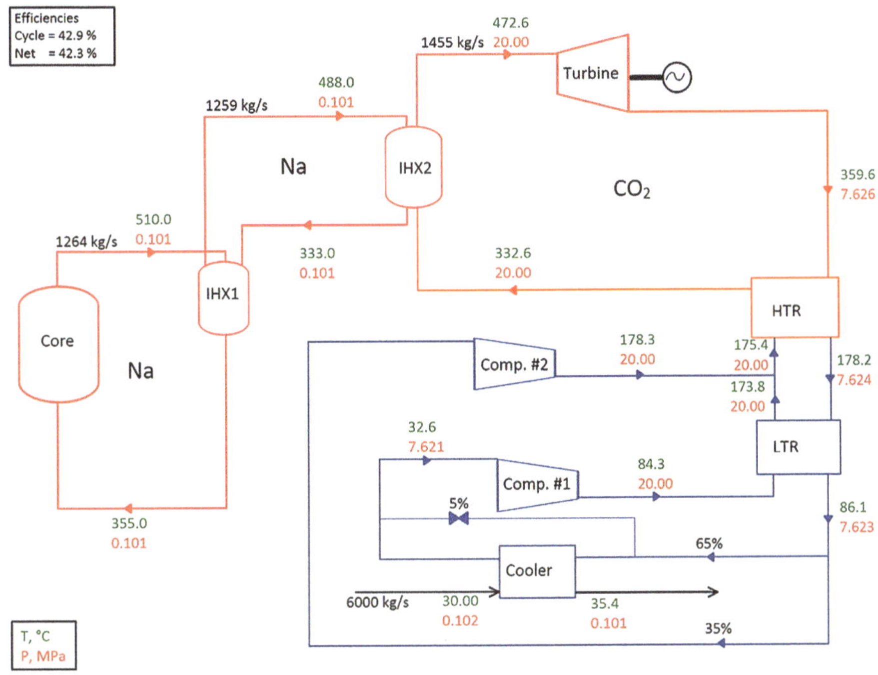

Lead-Cooled Fast Reactor (LFR)

LFR (see Figure 35 and Table 12) is also a fast-neutron-spectrum reactor, which uses lead or lead-bismuth as the reactor coolant (Pioro & Kirillov, 2013). The outlet temperature of the coolant is about 550°C (but can be as high as 800°C) at the atmospheric pressure. The primary choice of fuel is a nitride fuel. The SCP carbon-dioxide Brayton gas-turbine cycle has been chosen as a primary choice for the power cycle in US and some other countries, while subcritical (superheated steam pressure ~18 MPa) or SCP (SC "steam" pressure 24.5 MPa) Rankine steam-turbine cycles are considered as primary choices in Russia (see Table 12).

Molten Salt Reactor (MSR)

MSR (see Figure 36 and Table 13) is a thermal-neutron-spectrum reactor, which uses a molten fluoride salt with dissolved uranium while the moderator is made of graphite (Pioro & Kirillov, 2013). The inlet temperature of the coolant (e.g., fuel-salt mixture) is 565°C while the outlet temperature reaches 700°C. However, the outlet temperature of the fuel-salt mixture can be even increased to 850°C, when the co-

Figure 34. T-s diagram of SFR indirect power cycle with SCP carbon-dioxide Brayton cycle. For details, see schematic in Figure 33.
Based on data from Moisseytsev and Sienicki (2009).

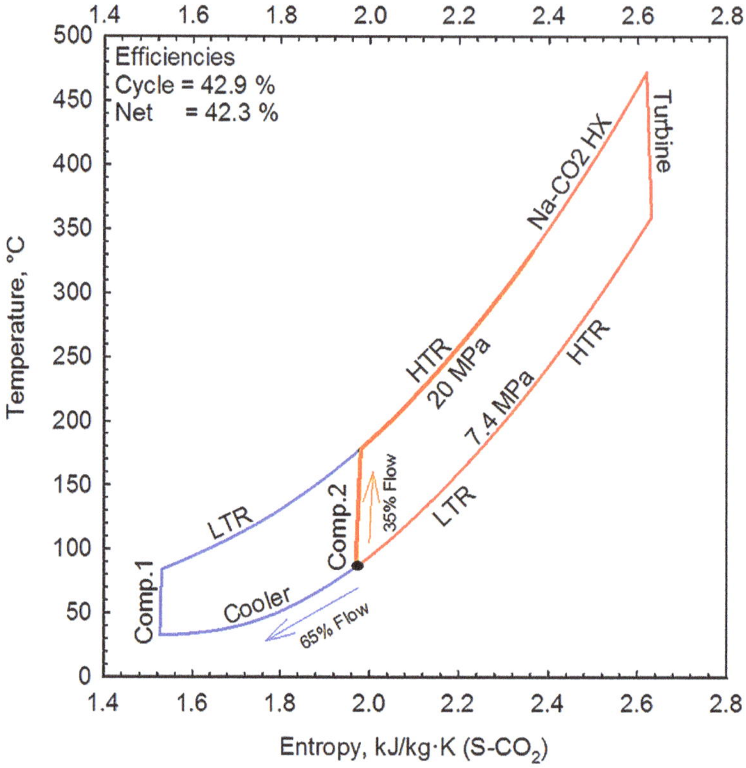

generation of hydrogen is considered as an option. The thermal efficiency of the plant is between 45 and 50%. The MSR concept is also on the priority list of the US DOE.

MSR NPP will have possibly indirect SCP Brayton gas-turbine cycle(s) as it is shown in Figure 36.

SuperCritical Water-Cooled Reactor (SCWR)

The design of SCWRs is seen as the natural and ultimate evolution of today's conventional water-cooled nuclear reactors (Pioro and Kirillov, 2013):

- Modern PWRs operate at pressures of 15 - 16 MPa.
- BWRs are the once-through or direct-cycle design, i.e., steam from a nuclear reactor is forwarded directly into a turbine.
- Some experimental reactors used nuclear steam reheat with outlet steam temperatures well beyond the critical temperature, but at pressures below the critical pressure (Saltanov and Pioro, 2011). And
- Modern supercritical-pressure turbines, at pressures of about 25 MPa and inlet temperatures of about 600°C, operate successfully at coal-fired thermal power plants for more than 50 years (for details, see Section in this Chapter: SCP Fossil-Fuel Thermal Power Plants).

Figure 35. Lead-cooled Fast Reactor (LFR): Molten-lead-cooled, fast-neutron-spectrum reactor with closed fuel cycle and outlet temperatures within 500 - 550°C (shown with indirect SCP Brayton gas-turbine power cycle)
Courtesy of Generation IV International Forum.

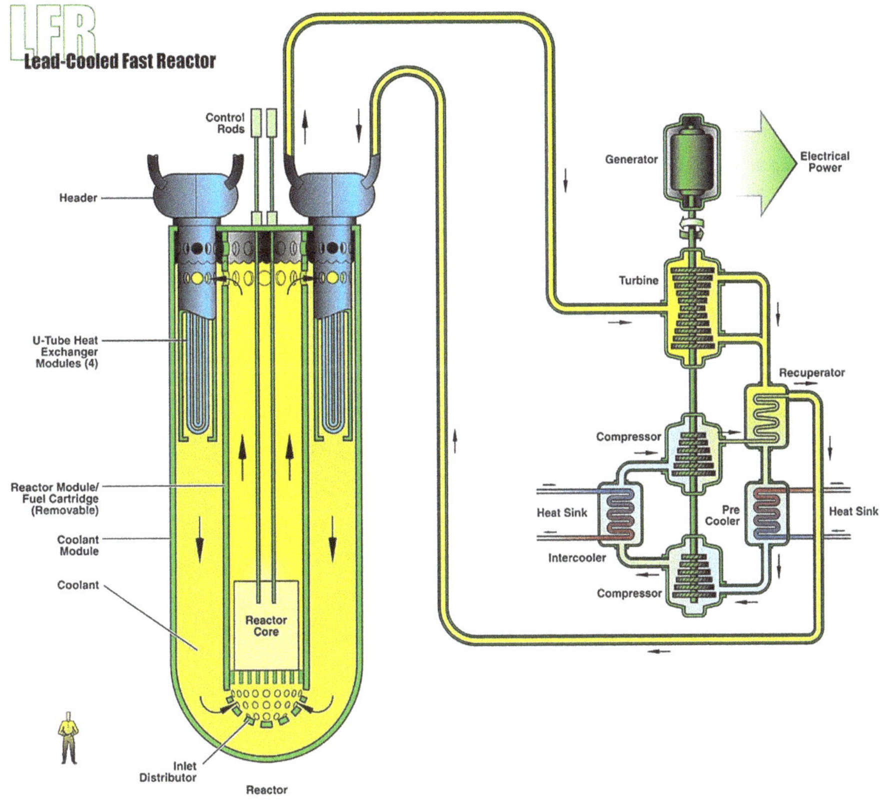

In general, SCWRs can be classified based on a pressure boundary, neutron spectrum and/or moderator (Pioro (Ed.), 2016; IAEA-TECDOC-1746, 2014; Schulenberg & Starflinger, 2012; Pioro, 2011; Oka et al., 2010; and Pioro & Duffey, 2007). In terms of the pressure boundary, SCWRs are classified into two categories, a) Pressure Vessel (PV) SCWRs (see Figure 37), and b) Pressure Tube (PT) or Pressure Channel (PCh) SCWRs. The PV SCWR requires a pressure vessel with a wall thickness of about 50 cm in order to withstand supercritical pressures. Table 14 lists general operating parameters of modern PV-SCWR concepts. On the other hand, the core of a PT SCWR consists of distributed pressure channels, with the wall thickness of about 10 mm, which might be oriented vertically or horizontally, analogous to CANDU and RBMK reactors, respectively.

The main objectives of using SCW in nuclear reactors are: 1) Increasing the efficiency of modern nuclear-power plants, which is currently 30 - 36%, to approximately 45 - 50%, 2) decreasing operational and capital costs by eliminating steam generators, steam separators, steam dryers, etc.; and 3) possible co-generation of hydrogen through thermo-chemical cycles.

In general, mainly thermal-spectrum SCWRs are currently under development worldwide. However, several concepts of fast-spectrum SCWRs are also considered (Oka et al., 2010).

Table 12. Key-design parameters of LFRs planned to be built in Russia

Reactor Parameter	Unit	Brest-300	Brest-1200
Reactor power (thermal/electrical)	MW	700/300	2800/1200
Thermal efficiency	%	43	
Primary coolant	-	Lead	
Coolant melting/boiling temperatures	°C	328/1743	
Coolant density at 450°C	kg/m^3	10,520	
Pressure inside reactor	MPa	~0.1	
Coolant inlet/outlet temperatures	°C	420/540	
Coolant massflow rate	t/s	40	158
Maximum coolant velocity	m/s	1.8	1.7
Fuel	-	UN+PuN	
Fuel loading	t	16	64
Term of fuel inside reactor	years	5	5–6
Fuel reloading per year	-	1	
Core diameter/height	m / m	2.3/1.1	4.8/1.1
Number of fuel bundles	-	185	332
Fuel-rod diameter	mm	9.1; 9.6; 10.4	
Fuel-rod pitch	mm	13.6	
Maximum cladding temperature	°C	650	
Steam-generator pressure	MPa	24.5	
Steam-generator inlet/outlet temperatures	°C	340/520	
Steam-generator capacity	t/s	0.43	1.72
Term of reactor	years	30	60

Based on NIKIET data.

SCWR is the only one Generation-IV concept, which never has been tested as an experimental, research or transportation reactor. However, subcritical- and supercritical-pressure Rankine steam-turbine cycles are the only cycles to be used in Generation-IV NPPs, which were proven for many years in thermal power industry to be effective and reliable. In general, the major current problem for advances in the SCWR development is reliability of materials to be used inside reactor. Compatibility of SCW with materials has been successfully resolved in SCP fossil-fuel thermal power plants long time ago, but there are no data on materials, which can withstand highly aggressive medium such as SCW, high temperatures and pressures, and high neutron flux. Therefore, more research is needed, especially, within this area.

CONCLUSION

- Electricity generation in the world is mainly based on coal-fired power plants (~40% of the total electricity generated) many of which are supercritical-pressure plants. Therefore, use of supercritical water in Rankine power cycle is the largest current application of supercritical fluids industry.

Figure 36. Molten Salt Reactor (MSR): Molten-salt-cooled reactor with outlet temperatures within 700 - 800°C (shown with indirect SCP Brayton gas-turbine power cycle)
Courtesy of Generation IV International Forum.

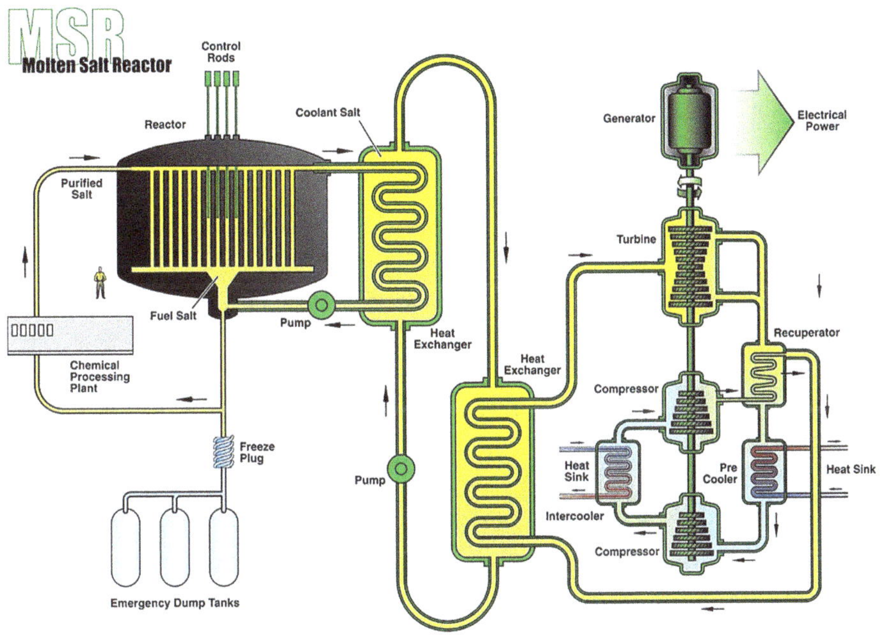

Table 13. Key-design parameters of MSR concept

Reactor Parameters	Unit	Reference Value
Reactor power	MW_{el}	1000
Net thermal efficiency	%	44-50
Average power density	MW_{th}/m^3	22
Fuel-salt inlet/outlet temperatures	°C	565/700 (800)
Moderator	-	Graphite
Neutron-spectrum burner	-	Thermal-Actinide

Pioro & Kirillov, 2013.

Modern supercritical-pressure coal-fired power plants have reached gross thermal efficiency up to 55% compared to older subcritical-pressure coal-fired power plants (42 - 43%). Therefore, it is clear that moving from subcritical pressures to supercritical ones can significantly increase thermal efficiency of the Rankine steam cycle.

- Current level of operating pressures and maximum temperatures in supercritical-pressure fossil-fuel power plants is: P = 23.5 - 38 MPa and T = 535 - 625°C, respectively. The vast majority of supercritical-pressure turbines are single-reheat ones with secondary subcritical-pressure, but superheated steam parameters: P = 3 - 8 MPa and T = 535 - 625°C. However, further increase in

Figure 37. SuperCritical Water-cooled Reactor (SCWR): Supercritical water-cooled, thermal-neutron-spectrum reactor with outlet temperatures within 510 - 625°C (shown with direct SCP Rankine "steam"-turbine power cycle)
Courtesy of Generation IV International Forum.

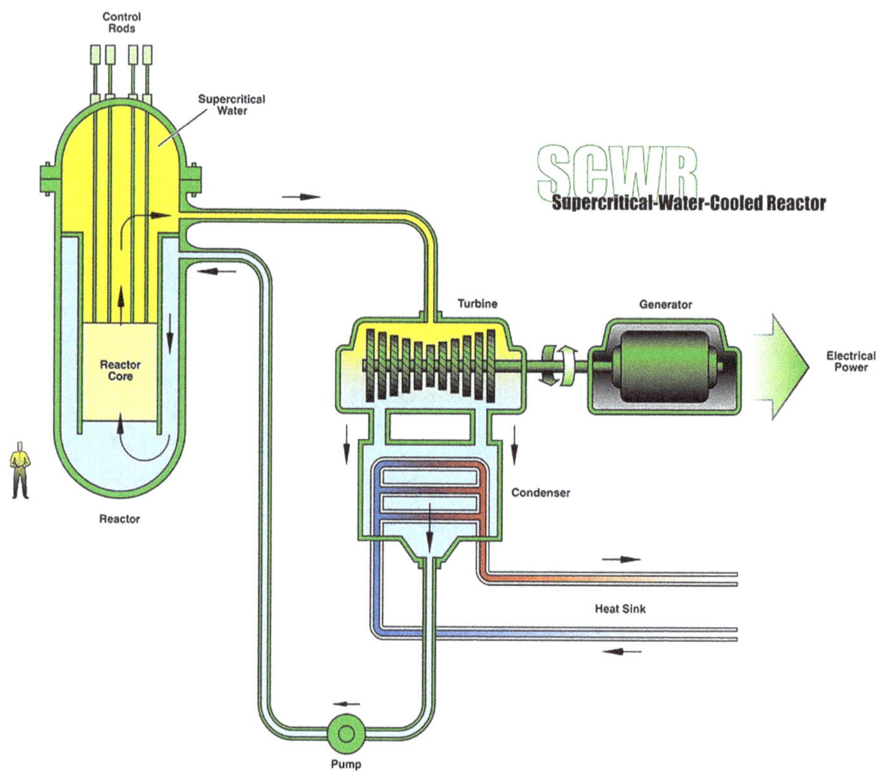

Table 14. Selected concepts of Pressure-Vessel SuperCritical Water-cooled Reactors (PV SCWRs)

Parameters	Unit	Pressure-Vessel SCWR Concepts		
Country	–	**Russia**		**USA**
Spectrum	–	**Thermal**	**Fast**	**Thermal**
Power electrical	MW	1500	1700	1600
Thermal efficiency	%	34	44	45
Pressure	MPa	25	25	25
Coolant inlet/outlet temperatures	°C	280/550	280/530	280/500
Massflow rate	kg/s	1600	1860	1840
Core height/diameter	m/m	3.5/2.9	4.1/3.4	4.9/3.9
Fuel	–	UO_2	MOX	UO_2
Enrichment	$\%_{wt}$	–	–	5
Maximum cladding temperature	°C	630	630	–
Moderator	–	H_2O	–	H_2O

Pioro & Duffey, 2007.

operating temperatures and pressures and corresponding to that in thermal efficiencies of supercritical-pressure fossil-fuel power plants are limited by strength of materials and their cost.

- Despite the advances in supercritical-pressure thermal power-plants design and operation worldwide, they are still considered as not of minimum environmental impact due to significant carbon-dioxide emissions and air pollution as a result of the combustion process. In addition, coal-fired power-plants produce also virtual mountains of slag and ash, and other gas emissions may contribute to acid rains. Therefore, more "environment friendly", but still large-capacity and high-reliability sources for electricity generation is required. One of such sources is nuclear power. As of today, this source of energy is considered as the most viable one for base-load electrical generation for the next 50-100 years.
- Currently, 436 nuclear-power reactors generate about 11% of the total electricity in the world. All current Nuclear Power Plants (NPPs) use indirect or direct subcritical-pressure Rankine steam-turbine power cycles. Therefore, their thermal efficiencies are within 30 - 36% for water-cooled nuclear-reactors NPPs (97% of all reactors) and 40-42% for sodium-cooled and carbon-dioxide-cooled reactors NPPs.
- Therefore, new generation or Generation-IV nuclear-power reactors and NPPs are required. These NPPs must have thermal efficiencies close to those of modern thermal power plants, i.e., within the range of at least 40 – 50%, and improved safety measures and designs in order to be built in the nearest future. Six concepts are considered currently as future Generation-IV nuclear-power reactors. These concepts are: 1) Very High Temperature Reactor (VHTR); 2) Gas-cooled Fast Reactor (GFR); 3) Sodium-cooled Fast Reactor (SFR); 4) Lead-cooled Fast Reactor (LFR); 5) Molten Salt Reactor (MSR); and 6) Supercritical Water-cooled Reactor (SCWR).
- All Generation-IV nuclear-reactor concepts are high-temperature reactors with reactor-coolant outlet temperatures ranging from 500°C and up to 1000°C. These reactors will use supercritical fluids as reactor coolants (helium in GFRs and VHTRs; SCW in SCWRs) and / or working fluids in power cycles (helium, mixture of nitrogen (80%) and helium (20%), nitrogen, and carbon dioxide in Brayton gas turbine cycle; and supercritical water/"steam" in Rankine cycle). Due to that thermal efficiencies of Generation-IV NPPs are expected to be within the range of 40 - 50% and even high higher for helium cooled reactors NPPs.

Also, supercritical-pressure carbon dioxide and refrigerants can be used in power cycles of geothermal plants.

REFERENCES

Bae, S. J., Ahn, Y., & Lee, J. I. (2014). Preliminary Study of the Supercritical CO2 Hybrid Cycle for the HTGR Application. *Transactions of the Korean Nuclear Society Spring Meeting*.

Cengel, Y. A., & Boles, M. A. (Eds.). (2015). *Thermodynamics - An Engineering Approach* (8th ed.). New York, NY: McGraw-Hill Companies, Inc.

Dittus, F. W., & Boelter, L. M. K. (1985). Heat Transfer in Automobile Radiators of the Tubular Type. *International Communications in Heat and Mass Transfer*, *12*(1), 3–22. doi:10.1016/0735-1933(85)90003-X

Dragunov, A. Saltanov, E., Pioro, I., Kirillov, P., & Duffey, R. (2015). Power Cycles of Generation III and III+ Nuclear Power Plants. ASME Journal of Nuclear Engineering and Radiation Science, 1(2).

EPCOR Genesee Generating Station Phase 3. Project Background. (2007). *CIV E 240. Technical Communications*. W. Kindzierski, Dept. Civil and Environmental Engineering.

Gauthier, J., Brinkmann, G., Copsey, B., & Lecomte, M. (2004). ANTARES: The HTR/VHTR Project at Framatome ANP. *Proceedings of the 2nd International Topical Meeting on High Temperature Reactor Technology (HTR 2004)*.

Gnielinski, V. (1976). New equation for heat and mass transfer in turbulent pipe and channel flow. *International Chemical Engineering*, *16*, 359–366.

Grigor'ev, V. A., & Zorin, V. M. (Eds.). (1982). *Thermal and Nuclear Power Plants*. Moscow, Russia: Energoatomizdat Publishing House. (In Russian)

Gupta, S., Saltanov, E., Mokry, S. J., Pioro, I., Trevani, L., & McGillivray, D. (2013a). Developing Empirical Heat-Transfer Correlations for Supercritical CO_2 Flowing in Vertical Bare Tubes. *Nuclear Engineering and Design*, *261*, 116–131. doi:10.1016/j.nucengdes.2013.02.048

Gupta, S., Saltanov, E., & Pioro, I. (2013b). Heat-Transfer Correlation for Supercritical Carbon Dioxide Flowing in Vertical Bare Tubes. *Proceedings of the 21st International Conference on Nuclear Engineering (ICONE-21)*. doi:10.1115/ICONE21-16453

IAEA TECDOC Series. (2014). IAEA-TECDOC-1746. Author.

Kirillov, P. L., Lozhkin, V. V., & Smirnov, A. M. (2003). Investigation of borders of deteriorated regimes of a channel at supercritical pressures (In Russian). State Scientific Center of Russian Federation Institute of Physics and Power Engineering by the name of A.I. Leypunskiy.

Kruglikov, P. A., Smolkin, Y. V., & Sokolov, K. V. (2009). Development of engineering solutions for thermal scheme of power unit of thermal power plant with supercritical parameters of steam. *Proceedings of the International Workshop "Supercritical Water and Steam in Nuclear Power Engineering: Problems and Solutions"*. (in Russian)

Levelt Sengers, J. M. H. L. (2000). Supercritical fluids: Their properties and applications. In Supercritical Fluids. Kluwer Academic Publishers.

McAdams, W. H. (1942). *Heat Transmission* (2nd ed.). New York, NY: McGraw-Hill.

Moisseytsev, A., & Sienicki, J. (2009). Investigation of Alternative Layouts for the Supercritical Carbon Dioxide Brayton Cycle for a Sodium-Cooled Fast Reactor. *Proceedings of the Supercritical CO_2 Power Cycle Symposium*. doi:10.1016/j.nucengdes.2009.03.017

Mokry, S., Naidin, M., & Baig, F. (2008). Conceptual thermal-design options for pressure-tube SCWRs with thermochemical co-generation of hydrogen. *Proceedings of the 16th International Conference on Nuclear Engineering (ICONE-16)*. doi:10.1115/ICONE16-48313

Mokry, S., Pioro, I. L., Farah, A., King, K., Gupta, S., Peiman, W., & Kirillov, P. (2011). Development of Supercritical Water Heat-Transfer Correlation for Vertical Bare Tubes. *Nuclear Engineering and Design*, *241*(4), 1126–1136. doi:10.1016/j.nucengdes.2010.06.012

NIST (National Institute of Standards and Technology). (2013). *NIST Standard Reference Database 23*. NIST Reference Fluid Thermodynamic and Transport Properties - REFPROP. Ver. 9.1. Technology Administration, U.S. Department of Commerce: http://www.nist.gov/srd/nist23.cfm

Nuclear News. (2016). Publication of the American Nuclear Society (ANS).

Oka, Y., Koshizuka, S., Ishiwatari, Y., & Yamaji, A. (Eds.). (2010). *Super Light Water Reactors and Super Fast Reactors*. New York, NY: Springer. doi:10.1007/978-1-4419-6035-1

Ornatskiy, A. P., Dashkiev, Yu. G., & Perkov, V. G. (Eds.). (1980). *Supercritical Steam Generators*. Kiev, Ukraine: Vyshcha Shkola Publishing House. (In Russian)

Pioro, I. (2011). The Potential Use of Supercritical Water-Cooling in Nuclear Reactors. Chapter. In S. B. Krivit, J. H. Lehr, & Th. B. J. Kingery (Eds.), *Nuclear Energy Encyclopedia: Science, Technology, and Applications*. Hoboken, NJ: Wiley & Sons. doi:10.1002/9781118043493.ch28

Pioro, I. (Ed.). (2016). Handbook of Generation IV Nuclear Reactors. Duxford, UK: Elsevier - Woodhead Publishing (WP).

Pioro, I., & Duffey, R. (2015). Nuclear Power as a Basis for Future Electricity Generation. ASME Journal of Nuclear Engineering and Radiation Science, 1(1).

Pioro, I., Duffey, R., & Dumouchel, T. (2004). Hydraulic Resistance of Fluids Flowing in Channels at Supercritical Pressures (Survey). *Nuclear Engineering and Design*, *231*(2), 187–197. doi:10.1016/j.nucengdes.2004.03.001

Pioro, I., & Kirillov, P. (2013). *Generation IV Nuclear Reactors as a Basis for Future Electricity Production in the World. In Materials and Processes for Energy: Communicating Current Research and Technological Developments* (pp. 818–830). Formatex Research Center.

Pioro, I., & Mokry, S. (2011a). Thermophysical Properties at Critical and Supercritical Conditions, Chapter in book. In A. Belmiloudi (Ed.), *Heat Transfer. Theoretical Analysis, Experimental Investigations and Industrial Systems* (pp. 573–592). Rijeka, Croatia: INTECH. doi:10.5772/13790

Pioro, I., & Mokry, S. (2011b). Heat Transfer to Fluids at Supercritical Pressures, Chapter in book. In A. Belmiloudi (Ed.), *Heat Transfer. Theoretical Analysis, Experimental Investigations and Industrial Systems* (pp. 481–504). Rijeka, Croatia: INTECH. doi:10.5772/13792

Pioro, I. L., & Duffey, R. B. (2007). *Heat Transfer and Hydraulic Resistance at Supercritical Pressures in Power Engineering Applications*. New York, NY: ASME Press. doi:10.1115/1.802523

Poette, C., Guedeney, P., & Stainsby, R. (2013). *Gas Cooled Fast Reactors: Recent Advances and Prospects, French Alternative Energies and Atomic Energy Commission (CEA)*. FR13 Conference, Paris, France.

Razumovskiy, V. G. Pis'mennyi, E. N., Sidawi, K., Pioro, I. L., & Koloskov, A. E. (2016). Experimental Heat Transfer in an Annular Channel and 3-Rod Bundle Cooled with Upward Flow of Supercritical Water. ASME Journal of Nuclear Engineering and Radiation Science, 2(1).

Saltanov, E., Pioro, I., Mann, D., Gupta, S., Mokry, S., & Harvel, G. (2015). Study on Specifics of Forced-Convective Heat Transfer in Supercritical Carbon Dioxide. ASME Journal of Nuclear Engineering and Radiation Science, 1(1).

Saltanov, E., & Pioro, I. (2011). World Experience in Nuclear Steam Reheat. In *Nuclear Power: Operation, Safety and Environment.* INTECH. Retrieved from http://www.intechopen.com/books/nuclear-power-operation-safety-and-environment/world-experience-in-nuclear-steam-reheat

Schmidt, E. (1960). Wärmetransport durch natürliche konvection in Stoffen bei Kritischen zustand. *International Journal of Heat and Mass Transfer, 1*(1), 92–101. doi:10.1016/0017-9310(60)90012-0

Schmidt, E., Eckert, E., & Grigull, V. (1946). *Heat transfer by liquids near the critical state.* AFF Translation, No. 527. Air Materials Command.

Schulenberg, Th., & Starflinger, J. (2012). *High Performance Light Water Reactor. Design and Analyses.* KIT.

Winterton, R. H. S. (1998). Where did the Dittus and Boelter equation come from? *International Journal of Heat and Mass Transfer, 41*(4-5), 809–810. doi:10.1016/S0017-9310(97)00177-4

Zahlan, H., Groeneveld, D., & Tavoularis, S. (2010). Look-Up Table for Trans-Critical Heat Transfer. *Proceedings of the 2nd Canada-China Joint Workshop on Supercritical Water-Cooled Reactors (CCSC-2010).*

Zahlan, H., Groeneveld, D. C., Tavoularis, S., Mokry, S., & Pioro, I. (2011). Assessment of supercritical heat transfer prediction methods. *Proceedings of the 5th International Symposium on SCWR (ISSCWR-5).*

KEY TERMS AND DEFINITIONS

Brayton Cycle: Thermodynamic power cycle based on gas turbine. Widely used in gas-fired thermal power plants (consists of compressor, combustion chamber (source of heat) and gas turbine connected to electrical generator; open cycle; working fluid: combustion products). Might be used in Generation-IV (next generation) nuclear power plants (consists of reactor (source of heat), (heat exchanger for indirect cycle), gas turbine connected to electrical generator, recuperator, heat exchanger (cooler) to reject heat, compressors, and intercooler; closed cycle; working fluids: supercritical-pressure helium (direct and indirect cycles), nitrogen (indirect cycle), nitrogen-helium mixture (indirect cycle), and carbon dioxide (indirect cycle)).

Critical Point (Also Called a Critical State): The point in which the distinction between the liquid and gas (or vapour) phases disappears, i.e., both phases have the same temperature, pressure and specific volume or density. The critical point is characterized with the phase-state parameters: T_{cr}, P_{cr} and v_{cr} (or ρ_{cr}), which have unique values for each pure substance.

Deteriorated Heat Transfer (DHT): Characterized with lower values of the heat transfer coefficient compared to those for normal heat transfer; and hence, has higher values of wall temperature within some part of a test section or within the entire test section.

Improved Heat Transfer (IHT): Characterized with higher values of the heat transfer coefficient compared to those for normal heat transfer; and hence, lower values of wall temperature within some part of a test section or within the entire test section. In our opinion, the improved heat-transfer regime or mode includes peaks or "humps" in the heat transfer coefficient near the critical or pseudocritical points.

Normal Heat Transfer (NHT): Characterized in general with heat transfer coefficients similar to those of subcritical convective heat transfer far from the critical or pseudocritical regions, when they are calculated according to the conventional single-phase Dittus-Boelter-type correlations: $Nu = 0.023\ Re^{0.8}Pr^{0.4}$.

Pseudocritical Point (Characterized with P and T_{pc}): The point at a pressure above the critical pressure and at a temperature ($T_{pc} > T_{cr}$) corresponding to the maximum value of specific heat at this particular pressure.

Rankine Cycle: thermodynamic power cycle based on steam turbine. Widely used in fossil-fuel thermal power Plants (consists of steam generator installed inside combustion chamber or flue-gases duct (source of heat), steam turbine connected to electrical generator, condenser, circulation pump, feedwater heaters; working fluid: subcritical- or supercritical-pressure water/steam). Widely used in all current nuclear power plants (subcritical-pressure cycle). Might be used in Generation-IV nuclear power plants (consists of reactor (source of heat), (heat exchanger for indirect cycle), steam turbine connected to electrical generator, condenser, circulation pump, feedwater heaters; working fluid: subcritical- or supercritical-pressure water/steam (direct and indirect cycles) or supercritical-pressure carbon dioxide (indirect cycle)).

Supercritical "Steam": Actually supercritical water, because at supercritical pressures fluid is considered as a single-phase substance. However, this term is widely (and incorrectly) used in the literature in relation to supercritical-"steam" generators and turbines.

Supercritical Fluid: The fluid at pressures and temperatures that are higher than the critical pressure and critical temperature. However, in the present Chapter, the term supercritical fluid usually includes both terms - a supercritical fluid and compressed fluid.

Superheated Steam: Steam at pressures below the critical pressure, but at temperatures above the critical temperature.

ENDNOTES

1 Gross efficiency of a unit during a given period of time is the ratio of the gross electrical energy generated by the unit to the energy consumed during the same period by the same unit. The difference between gross and net efficiencies is internal needs for electrical energy of a power plant, which might be not so small (5% or even more).

2 Currently, all nuclear-power reactors (as per March of 2016, there are in total 436 reactors in the world) are connected to direct (ABWRs, BWRs, and Light-water Graphite-moderated Reactors (LGRs)) or indirect (all other types of reactors) subcritical-pressure Rankine steam cycle.

3 BN - Fast Sodium reactor (in Russian abbreviations).

4 For example, the largest in the world 5,780-MW_{el} Taichung coal-fired power plant (Taiwan) is the world's largest emitter of carbon dioxide with over 40 million tons per year.

5 Natural uranium consists of ~99.3% U^{238}, 0.7% U^{235}, and < 0.01% U^{234}. 3.2% enrichment means 3.2% of U^{235} in the uranium fuel.

APPENDIX: NOMENCLATURE

A: Area, m^2

c_p: Specific heat at constant pressure, J/kg K

$\bar{c}_p$: Averaged specific heat within the range of (T_w - T_b); $\left(\frac{H_w - H_b}{T_w - T_b}\right)$, J/kg K

D: Inside diameter, m

D_{hy}: Hydraulic-equivalent diameter, m

G: Mass flux, kg/m^{2}s; $\left(\frac{m}{A_{fl}}\right) = \rho\, u$

H: Specific enthalpy, J/kg

h: Heat transfer coefficient, W/m^{2}K

k: Thermal conductivity, W/m K

L: Heated length, m

m: Mass-flow rate, kg/s; $\left(\rho \cdot V\right)$

P, p: Pressure, Pa

p_{wet}: Wetted perimeter, m

Q: Heat-transfer rate, W

q: heat flux, W/m^2; $\left(\frac{Q}{A_h}\right)$

s: Specific entropy, J/kg K

T, t: Temperature, ºC

u: Axial velocity, m/s

V: Volume-flow rate, m^3/kg or volume, m^3

v: Specific volume, m^3/kg; $\left(\frac{1}{\rho}\right)$

Greek Letters

α: Thermal diffusivity, m^2/s; $\left(\frac{k}{c_p \cdot \rho}\right)$

η: thermal efficiency

μ: Dynamic viscosity, Pa·s

ρ: Density, kg/m^3

υ: Kinematic viscosity, m^2/s; $\left(\frac{\mu}{\rho}\right)$

Non-Dimensional Numbers

Nu: Nusselt number; $\left(\frac{h \cdot D_{hy}}{k}\right)$

Pr: Prandtl number; $\left(\frac{\mu \cdot c_p}{k}\right) = \left(\frac{\upsilon}{\alpha}\right)$

$\overline{\Pr}$: Cross-sectional average Prandtl number within the range of (T_w - T_b); $\left(\frac{\mu \cdot \overline{c}_p}{k}\right)$

Re: Reynolds number; $\left(\frac{G \cdot D_{hy}}{\mu}\right)$

Subscripts or Superscripts

ave: Average
b: Bulk-fluid
cr: Critical
dht: Deteriorated heat transfer
el: Electrical
fl: Flow
h: Heated
hy: Hydraulic equivalent
in: Inlet
out: Outlet or outside
pc: Pseudocritical
sat: Saturation
th: Thermal
w: Wall
wt: Weight

Abbreviations and Acronyms

ABWR: Advanced Boiling Water Reactor
AGR: Advanced Gas-cooled Reactor
AHFP: Axial Heat Flux Profile
BN: Fast Sodium (reactor) (in Russian abbreviations)
BREST: Fast REactor with Lead Coolant (in Russian abbreviations)
BWR: Boiling Water Reactor
CANDU: CANada Deuterium Uranium (reactor)
DHT: Deteriorated Heat Transfer
DOE: Department Of Energy (USA)
Ed.: Editor

EEC: Electrical Energy Consumption
EU: European Union
G: Generator
(g): gauge (pressure)
GFR: Gas Fast Reactor
GIF: Generation-IV International Forum
HDI: Human Development Index
HP: High Pressure
HT: Heat Transfer
HTC: Heat Transfer Coefficient
HTGR: High Temperature Gas-cooled Reactor
IHT: Improved Heat Transfer
IHX: Intermediate Heat eXchanger
IP: Intermediate Pressure
IPPE: Institute for Physics and Power Engineering
LBE: Lead-Bismuth Eutectic
LFR: Lead-cooled Fast Reactor
LGR: Light-water Graphite-moderated Reactor
LMFBR: Liquid-Metal Fast-Breeder Reactor
LP: Low Pressure
Ltd: Limited
LWR: Light-Water Reactor
MOX: Mixed OXide (fuel)
MSR: Molten Salt Reactor
NGNP: Next Generation Nuclear Plant
NHT: Normal Heat Transfer
NIST: National Institute of Standards and Technology (USA)
NPP: Nuclear Power Plant
OD: Outside Diameter
PHWR: Pressurized Heavy Water Reactor
PT: Pressure Tube
PV: Pressure Vessel
PWR: Pressurized Water Reactor
REFPROP: REFerence PROPerties
RBMK: Reactor of Large Capacity Channel type (in Russian abbreviations)
RMS: Root-Mean Square (error)
S: Supercritical
SC: SuperCritical
SCF: SuperCritical Fluid
SCP: SuperCritical Pressure
SCW: SuperCritical Water
SCWR: SuperCritical Water-cooled Reactor
SFR: Sodium-cooled Fast Reactor

S.G.: Steam Generator
"ТКЗ" ("TKZ"): Taganrog Power-Plant Steam Generator's Manufacturing plant (Taganrog, Ukraine)
"ЦКТИ" ("TsKTI"): Central Boiler-Turbine Institute by Polzunov (St.-Petersburg, Russia)
USC: Ultra-SuperCritical
UOIT: University of Ontario Institute of Technology
US: United States (of America)
USA: United States of America
USSR: Union of Soviet Socialist Republics
VHTR: Very High Temperature Reactor
"ВТИ" ("VTI"): All-Union Heat Engineering Institute (Moscow, Russia)

Chapter 14
Energy Conversion Using the Supercritical Steam Cycle

Thomas Schulenberg
Karlsruhe Institute of Technology, Germany

ABSTRACT

A supercritical steam (or Rankine) cycle is used today for more most of the new coal-fired power plants. More recently, it has been proposed as well for future water-cooled nuclear reactors to enhance their efficiency and to reduce their costs. This chapter provides the technical background explaining this technology. Some criteria for boiler design and operation, like drum or once-through boiler design, fixed or sliding pressure operation and coolant mixing, are discussed in general to explain the particular challenges of supercritical steam cycles. Examples of technical solutions are given for two large-scale applications: a coal-fired power plant and a supercritical water-cooled reactor, both producing around 1000 MW electric power.

INTRODUCTION

Coal fired power plants are among the most prominent examples of energy conversion using a supercritical steam cycle. They were the result of a continuous development process, starting from sub-critical Rankine cycles in the early 20th century, which were fired with moving grates, up to supercritical steam cycles in the 1990s, which were fired with pulverized coal. Termuehlen and Emsperger (2003), who took part in this development as senior engineering experts at SIEMENS, report about the history of these coal fired power plant and summarize the status reached by the end of the century.

In the 1990s, SIEMENS took significant effort to increase power and efficiency and, simultaneously, to decrease the specific plant erections costs per kW installed power, as reported by Kefer and Seiter (1996). The supercritical steam cycle became the standard for all new built power plants and the specific costs of new power plants built around 2010 were finally down at around 1100 €/kW, including flue gas cleaning systems for dust, NOx and SO_2.

DOI: 10.4018/978-1-5225-2047-4.ch014

Figure 1. Power and efficiency of coal fired power plants in Germany

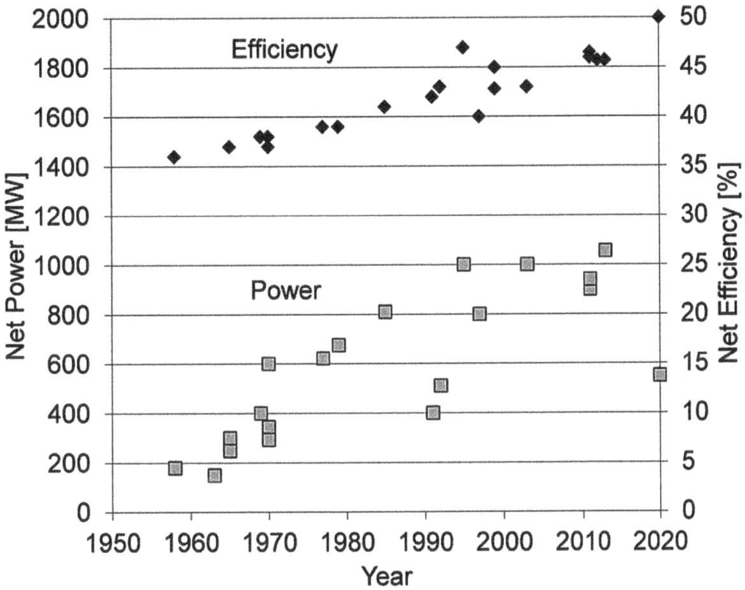

The history of power and efficiency of power plants built in Germany since the 1950s is shown exemplarily in Figure 1. Within 50 years, the power was increased by a factor of 5, and the net efficiency by more than 10% points. The most important driver for power and efficiency improvements was the live steam temperature, i.e. the steam temperature at the inlet of the high pressure turbine, as shown in Figure 2. It is exceeding now 600°C with a potential to reach 700°C in near future, enabling 50% net efficiency.

Figure 2. Net efficiency of coal fired power plants, shown in Figure 1, increasing with live steam temperature

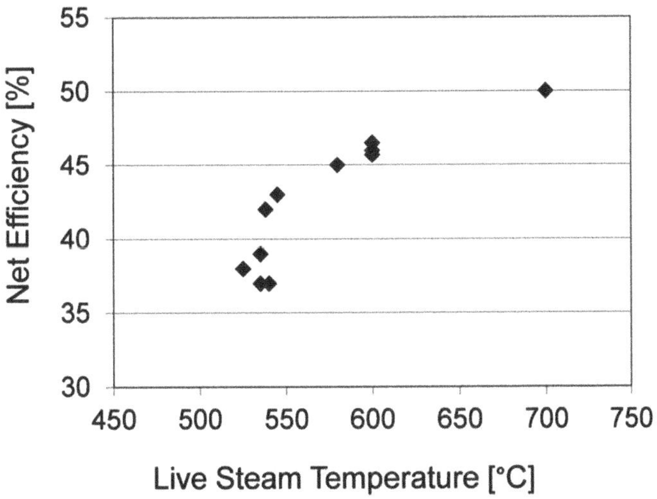

Among the latest power plants in Germany, using such technologies are:

- Unit 8 of Rheinhafendampfkraftwerk (RDK) Karlsruhe, with 912 MW electric power, firing hard coal with more the 46% efficiency, using live steam of 600°C at 275 bar and reheat steam of 620°C
- Unit 9 of GKM Mannheim, with 911 MW electric power, firing hard coal with 46.4% efficiency, using live steam of 600°C at 285 bar and reheat steam of 610°C
- Datteln Unit 4, with 1055 MW electric power, firing hard coal with more the 45% efficiency, using live steam of 600°C at 275 bar and reheat steam of 620°C
- Hamburg Moorburg, a twin unit with 1654 MW, firing hard coal with more the 46% efficiency, using live steam of 600°C at 276 bar and reheat steam of 620°C
- Niederaussem Unit K, with 1000 MW electric power, firing brown coal with more than 43% efficiency, using live steam of 580°C at 290 bar and reheat steam of 600°C
- Neurath Units F and G, two units with 1100 MW electric power each, firing brown coal with more than 43% efficiency, using live steam of 600°C at 275 bar and reheat steam of 620°C
- Boxberg Unit R with 675 MW electric power each, firing brown coal with 43.7% efficiency, using live steam of 600°C at 285 bar and reheat steam of 620°C

A coal fired power with 700°C live steam temperature, reaching 50% net efficiency, was planned to be built in Wilhelmshaven by 2020, but plans were postponed recently because coal is still too cheap to justify the increased costs of nickel-base alloys needed for this future technology.

These supercritical coal fired power plants will be used in this chapter as examples explaining the technology of supercritical steam cycles in general. Of course, a similar steam cycle can also be used to convert any other kind of thermal power with high temperature. A particular challenge is the application to nuclear power plants, which will be used to discuss general problems that might occur when using supercritical steam cycles for other applications.

THE RANKINE CYCLE

Since more than 100 years, the thermal power from coal combustion has been converted to electric power using the Rankine steam cycle. The working medium has always been water as liquid or steam. The physical properties of water and steam can be taken from steam tables, which are nowadays available in electronic form. The steam cycle is usually plotted in the temperature-entropy (T-s) diagram to illustrate the thermodynamic process. The example shown in Figure 3 shows lines of constant pressure p with solid lines, as well as the lines of saturated liquid and saturated steam with dashed lines.

Around 1900, a Rankine cycle required just a feedwater pump, an economizer to preheat the feedwater to the boiling temperature, an evaporator, a superheater and a condenser, as sketched in Figure 4. Plotted into the T-s-diagram, Figure 5, we find the inlet and outlet of the feedwater pump close together, and the endpoints of other components as vertices of the steam cycle. Ideally, the turbine appears as a vertical line in this diagram, meaning that a steam expansion without flow losses would be isentropic, but a real steam turbine at that time had just 60% efficiency, which produced a lot of entropy instead, so that a real turbine expansion would appear as a skewed line.

The net power of the steam cycle, which is the turbine power minus the pump power, can rather be read from the enthalpy-entropy (h-s) diagram, Figure 6. For each state, the enthalpy can be determined

Figure 3. Temperature-entropy (T-s) diagram of water

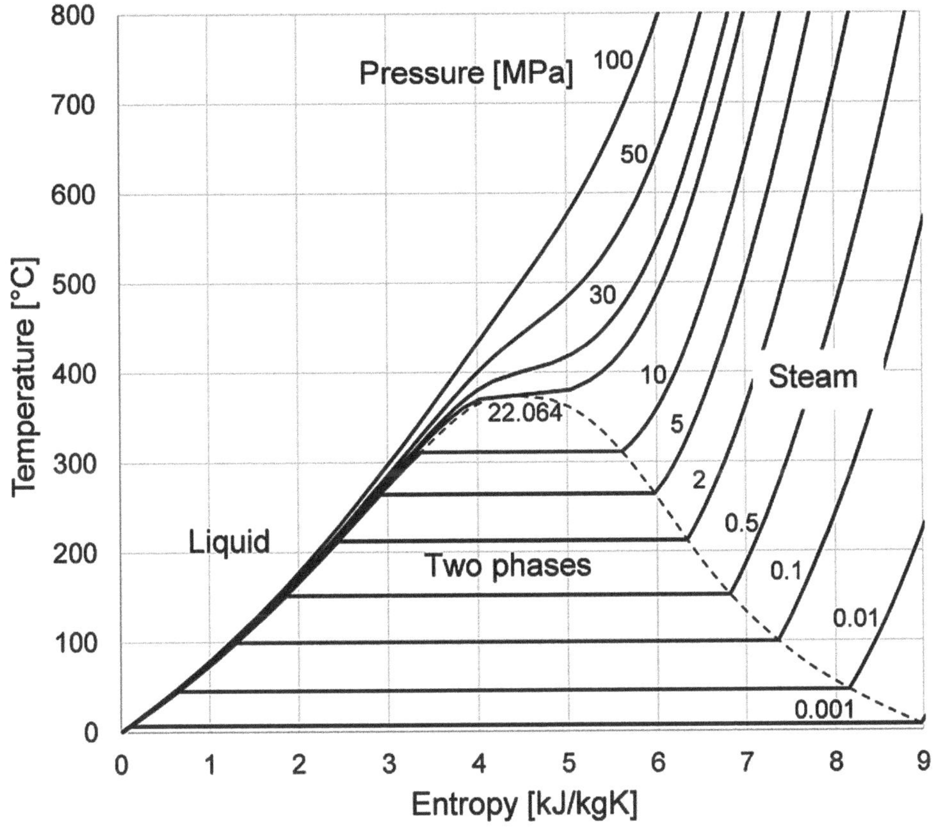

Figure 4. Major components of a simple Rankine cycle

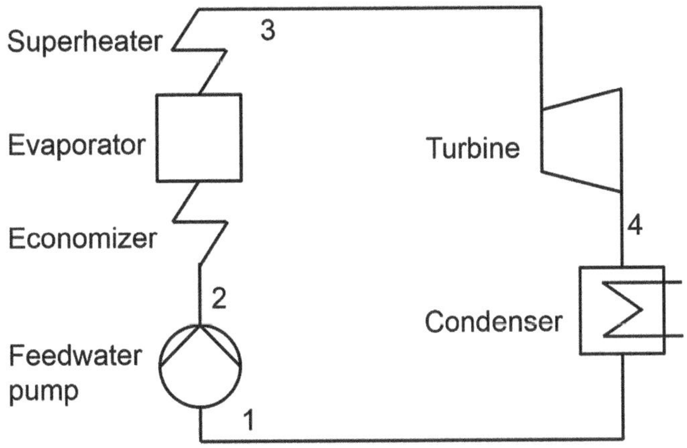

from data of the *T-s*-diagram using the steam table. The turbine power equals the enthalpy difference between turbine inlet and outlet, h_3-h_4, multiplied by the steam mass flow, and the pump power equals the enthalpy difference between the pump outlet and inlet, h_2-h_1, multiplied with the liquid mass flow. The heat, which is added to the feedwater in the economizer, in the evaporator and in the superheater

Figure 5. Ideal, simple cycle Rankine cycle in the T-s-diagram

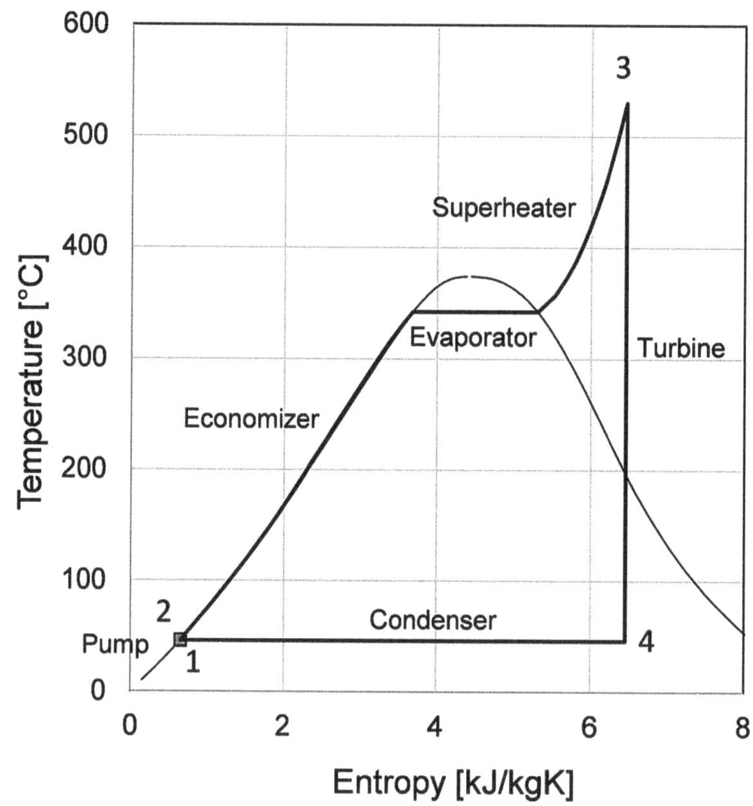

Figure 6. Ideal, simple Rankine cycle in the h-s-diagram

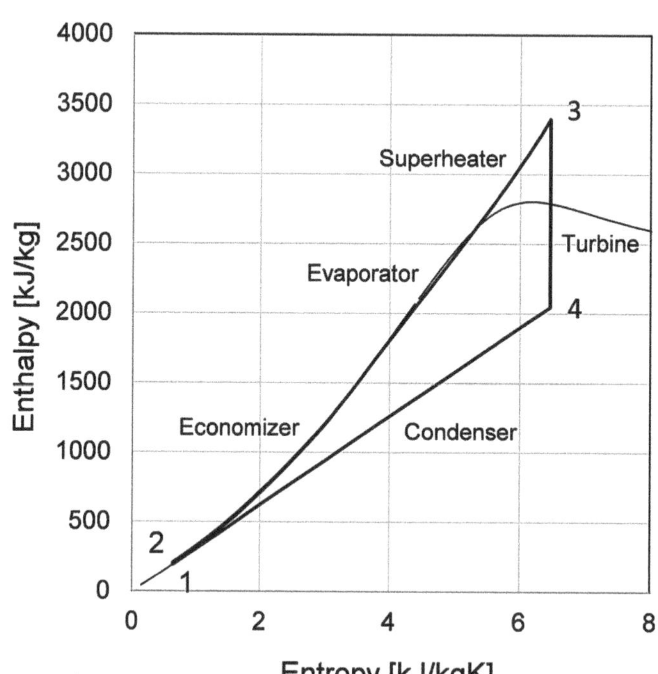

equals the enthalpy difference between points 3 and 2, multiplied with the mass flow of water through these components. Assuming the same mass flow through all components of the steam cycle, the thermal efficiency η can be determined as

$$\eta = \frac{h_3 - h_4 - \left(h_2 - h_1\right)}{h_3 - h_2} \quad (1)$$

In 1990, a simple Rankine cycle could reach ideally a thermal efficiency of around 34%, but the real net efficiency of the power plant was just 15% due to poor component efficiencies.

Since then, the net efficiency has significantly been improved to more than 46% by the following measures, as reported by Bald and Heusinger (1996):

- The feedwater is mostly preheated by steam, which is extracted from the steam turbines, and only the final heat-up step to saturation (boiling) conditions is provided by the flue gas in the economizer.
- The single steam cycle is replaced by a high pressure (HP), intermediate pressure (IP) and low pressure (LP) turbine, with a reheat of steam between the HP turbine outlet and the IP turbine inlet.
- The steam temperature at the inlet of the HP turbine (live steam) and at the inlet of the IP turbine has been increased to 600°C or more, once boiler steels with 9 to 12% Cr became available for pipes and headers, having enough creep strength and corrosion resistance at this temperature. Austenitic stainless steels are taken today for the hottest superheater are reheater tubes, with a temperature limit of 650°C. Ni-base alloys for even higher temperatures are already available today with a penalty, however, in their significant higher costs and their lower fatigue strength, such that a live steam temperature of 700°C or more is not yet considered to be economic.
- Turbine efficiencies have been improved e.g. by an advanced turbine airfoil design to more than 90%.

The simplified flow scheme of a typical advanced coal fired power plant in shown in Figure 7. Feedwater lines and steam lines are shown in black, steam extractions for feedwater preheating in gray. The following data refer to a power plant with a net power of 1000 MW. Starting from the condenser outlet, the feedwater pressure is first increased by the condensate pumps to ~3 MPa to allow heat-up without the risk of boiling. The first feedwater preheater, taking low pressure steam from an extraction point of the LP turbine just upstream the last turbine stage, is usually inserted into the neck of the condenser, thus minimizing the length of the large steam supply lines to it. Up to 4 further regenerative preheaters, taking steam which is extracted from the LP turbines at further upstream positions, are heating up the feedwater to 180°C. Two of these LP preheaters are shown in Figure 7 as examples. The extracted steam mass flow needed for each preheat stage can be determined from an energy balance of the condensing steam giving its heat to the feedwater mass flow.

The deaerator at 2.5MPa condenses superheated steam from the IP turbine outlet in a counter-current flow with sub-cooled feedwater such that non-condensable gases remain, which are released from there to the atmosphere. The deaerator is serving simultaneously as another pre-heat stage. The large feedwater tank of the deaerator is needed as an expansion vessel, allowing the feedwater pump to work

Figure 7. Advanced coal fired power plant with regenerative feedwater preheaters and reheater

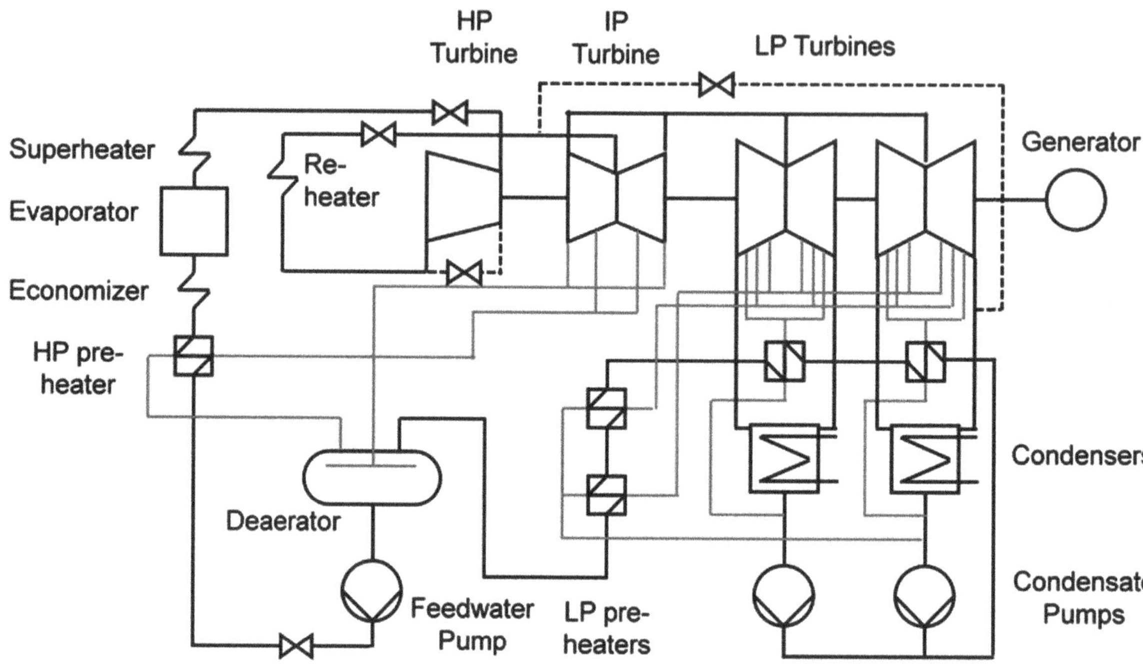

independently from the condensate pumps. The feedwater pumps on the lowest floor of the power plant are increasing the pressure to around 30 MPa. Up to three further HP preheaters, of which just one is shown here, are increasing the feedwater temperature to 250°C or more, taking steam which is extracted from the IP turbine.

Economizer, evaporator and superheater are heat exchangers of the steam generator, using radiative and convective heat of the flue gas. After expansion in the HP turbine, the steam is forwarded again to the steam generator to be reheated. The reheated steam is supplied to a dual flood IP turbine and from there, without further reheat, directly to the LP turbines. The number of LP turbines needed for the power plant is not only determined by the steam mass flow, but even more by the condenser pressure, which is in turn a function of the temperature of the cooling water. All turbines are coupled to drive a joint, water-cooled generator.

This advanced Rankine cycle is shown in the *T-s*-diagram in Figure 8. The condensate leaving the condenser at point 1 is increased in pressure by the condensate pumps, increasing the temperature by less than 1°C. Three preheat steps at low pressure, which are shown in this example, are first de-superheating the extracted steam and are condensing it then on the outside of U-tubes, warming up the feedwater inside. At point 2, a deaerator with a feedwater tank is using steam, which is extracted from the IP turbine, to separate non-condensable gases from the steam and to preheat the feedwater simultaneously as a preheat stage 4. It is followed by a feedwater pump, rising the pressure to 30 MPa, which causes a little step of 6°C in the *T-s*-diagram. A high pressure preheater using steam with even higher pressure of the IP turbine is finally increasing the feedwater temperature to more than 250°C. Note that the feedwater cannot be preheated by low temperature flue gas in a coal fired power plant, as the flue gas has to preheat the combustion air instead.

Figure 8. Advanced, supercritical Rankine cycle in the T-s-diagram

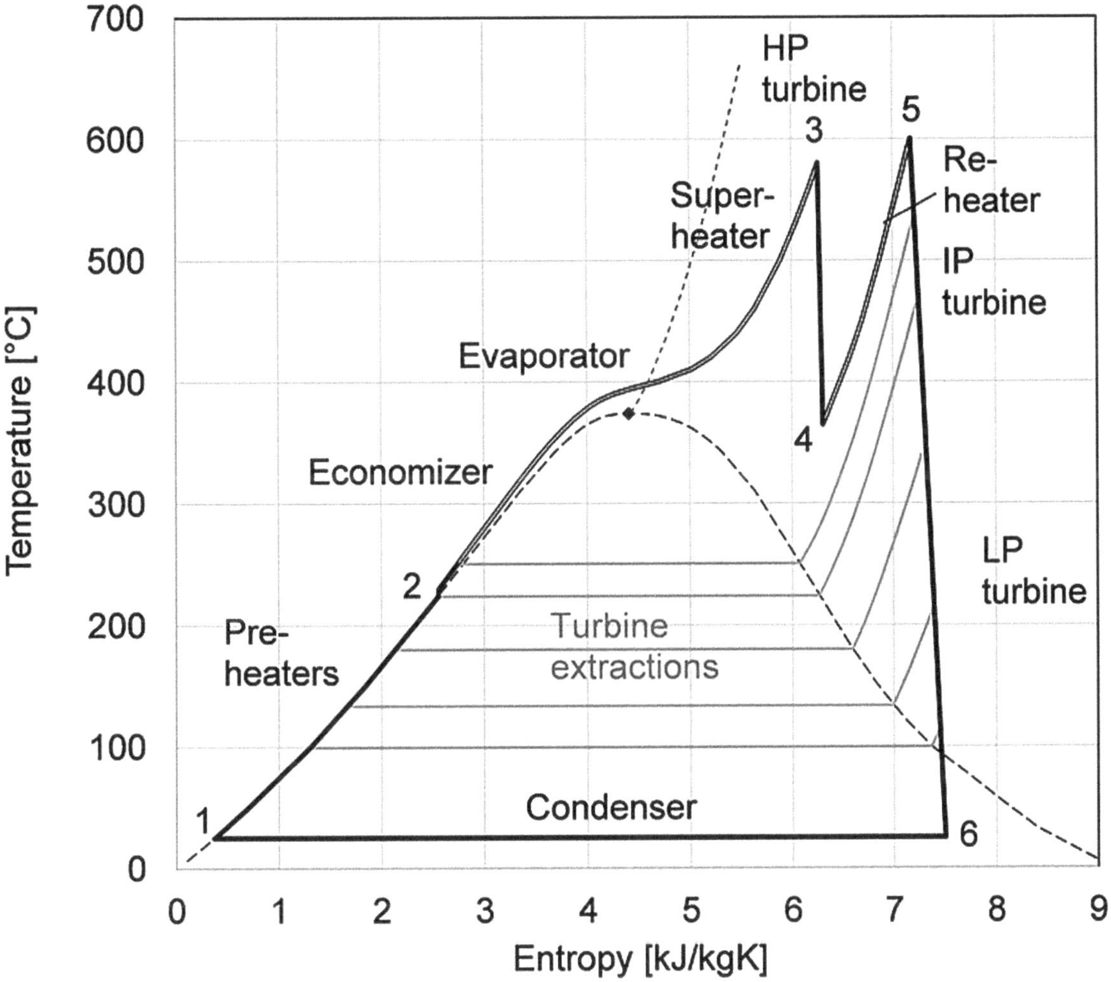

Such high pressure is already supercritical, meaning that the pressure has exceeded the critical pressure of water at 22.064 MPa. Evaporation at supercritical pressure is a continuous process with single-phase heat transfer, and we cannot differ exactly anymore between economizer, evaporator and superheater. These terms are used, however, similarly here. The evaporator is meant as the heat exchanger crossing the pseudo-critical line with a continuous transition from liquid like to steam like properties, indicated with a dashed line in Figure 8.

The optimum live steam pressure, and thus the reason for the transition to supercritical pressure, can be understood by following the steam cycle backwards through the turbines. The steam quality (steam mass fraction) at the LP turbine outlet is ideally chosen at around 87%. A higher steam quality would result in a higher enthalpy at the LP turbine outlet, and thus in less turbine power at a given inlet enthalpy. A lower steam quality would cause unacceptable droplet erosions in the LP steam turbine. The LP turbine outlet condition, point 6 in Figure 8, can thus be considered as a fixed point in the *T*-*s*-diagram at a given condenser pressure.

Following the cycle backwards now along the IP and LP turbine expansion line, we reach a temperature of 600°C at point 5, which is considered in this example as the creep limit of the reheater tube material, at a pressure of 6 MPa. The flowpath backwards through the reheater, taking a pressure drop of 0.1 MPa into account and keeping enough margin from the saturation line, brings us to the HP turbine outlet at 6.1 MPa and 365°C, point 4. From there, backwards along the HP turbine expansion line up to 580°C, brings us to the HP turbine inlet at 26.5 MPa, point 3. Note that the HP turbine inlet temperature is usually 20°C colder than the IP turbine inlet temperature at the same creep limit of superheater tube material, as the thicker HP tube walls allow only 20°C colder inside wall temperatures at same average wall temperature.

The high pressure at the HP turbine inlet is thus an inevitable consequence of the HP turbine inlet temperature, and the transition from subcritical to supercritical pressure became unavoidable at a live steam temperature beyond 560°C. The high turbine inlet temperature, in turn, is one of the major reasons for the efficiency improvements of advanced coal fired power plants, as shown in Figure 2.

STEAM GENERATOR DESIGN

The steam generator, often called "boiler" for short, is a key component of the supercritical steam cycle. It comprises the economizer, the evaporator, the superheaters and the reheaters. In a coal fired power plant with pulverized coal combustion, the evaporator and the first superheater form a tube wall, housing the combustion zone in the lower half and the 2nd and higher superheater as well as the reheater as horizontal tube bundles in the upper half. The economizer is a horizontal tube bundle heat exchanger on top of the superheaters, where the flue gas has already been cooled down below ~500°C.

Evaporator Design and Operation

As long as the steam pressure was subcritical, the evaporator has been designed as a drum boiler, sketched in Figure 9. Only around 20% of the feedwater was evaporated, avoiding dryout and film boiling, and cyclone separators inside the drum were separating the two-phase flow out of the evaporator into saturated steam, supplied to the superheater, and saturated liquid, which was returned to the evaporator inlet through the downcomer. In case of low heat flux to the evaporator tubes, the recirculating flow could be driven by natural convection. In others cases, a recirculation pump was installed at the lowest point of the downcomer, controlling the steam quality at the evaporator outlet. A valve in the feedwater supply line was controlling the liquid level inside the drum.

It is obvious that such a drum boiler cannot work at supercritical pressure, as liquid and steam are a single phase which cannot be separated. Instead, water must be fully evaporated and superheated without any flow recirculation, as sketched in Figure 10. Such a once-through boiler, often denoted as Benson boiler honoring its inventor, can also be operated at subcritical pressure, but a dryout of the liquid film at the evaporator tube walls becomes unavoidable then. The heat flux must be kept sufficiently low and the mass flux inside the tubes sufficiently high to stay within the material creep limits under post-dryout conditions. Therefore, the evaporator tubes are usually inclined as a spiral tubing instead of vertical, reaching a higher mass flux compared with vertical tubes at same pitch.

Figure 9. Natural convection drum boiler design for subcritical pressure

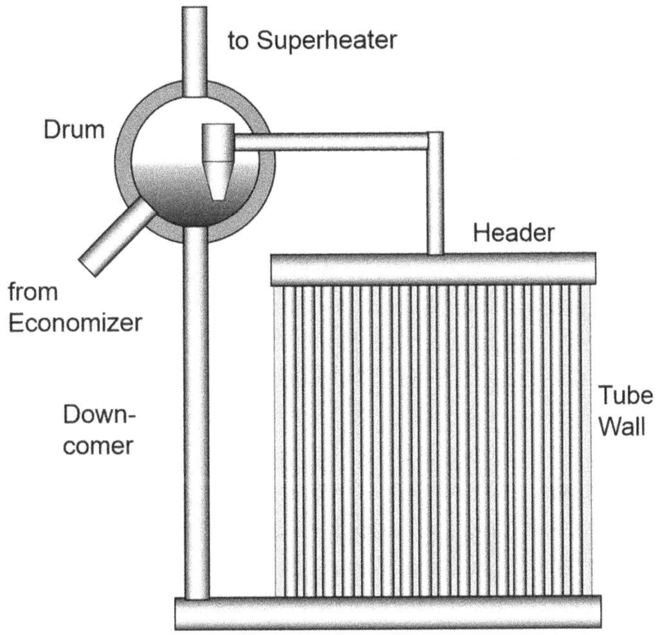

Figure 10. Once-through boiler design with spiral tubing for sub- and supercritical pressure

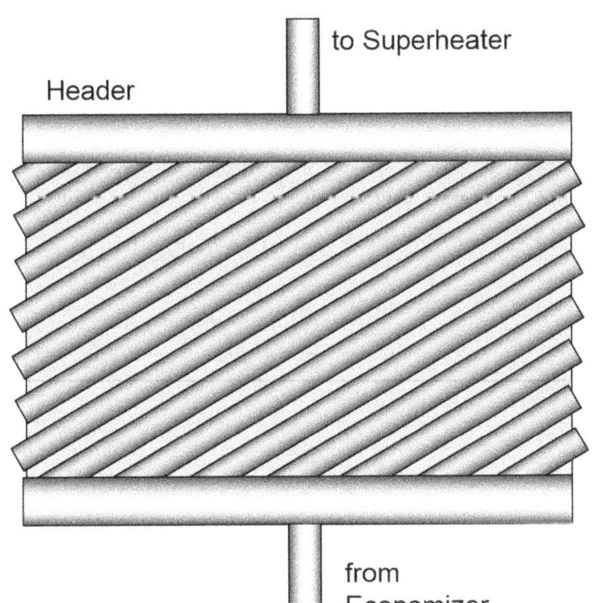

Though being a material challenge, the once-through boiler has an economic advantage: the large and thick-walled drum was expensive and needed to be warmed-up slowly to minimize thermal fatigue, whereas the once-through boiler can be heated-up faster and it is equipped only with a small recirculation system with separators and recirculation pump for start-up.

Today, all supercritical coal fired power plants are designed with once-through boilers. The tube wall of the evaporator is designed as a spiral tubing with an inclination of 60° to the vertical, providing a mass flux of around 2500 kg/m^{2}s at full power. The heat flux is kept below 400 kW/m^2 at full power to keep the material temperature below ~550°C under post-dryout conditions. The evaporator can thus be operated at supercritical and at subcritical pressure. Like with drum boilers, however, film boiling is always avoided; it would distroy the tube wall. A valve in the feedwater supply line is controlling the steam quality at the evaporator outlet.

With drum boilers as well as with once-through boilers, the pressure at evaporator outlet is controlled by the turbine governor valve at the turbine inlet. A sliding pressure operation, as indicated in Figure 11, is the preferred operation mode of the once-through boiler in the upper power range. Trying to keep the evaporator outlet enthalpy constant, the mass flow through the evaporator tubes must change almost proportionally with power. According to Stodola's ellipse law (Stodola, 1927), the pressure drop of the steam turbines varies almost proportional with mass flow. Thus, if the turbine governor valve is kept wide open, the evaporator outlet pressure is increasing proportional with power. Such sliding pressure operation is minimizing the power consumption of the feedwater pump and is thus maximizing efficiency at lower power. Consequently, the evaporator is operated at supercritical pressure only above 80% power, and at subcritical pressure in the remaining power range.

The sliding pressure, however, cannot be extended to zero power. Instead, a minimum mass flow and a minimum pressure must be provided to avoid flow instabilities and overheating at low power. Therefore, below ~40% power as well as during start-up from cold conditions, the water leaving the evaporator outlet is not fully evaporated anymore, and steam and liquid must be separated to avoid flow instabilities in the superheater. For this purpose, a few cyclone separators are foreseen at the evaporator outlet. Liquid

Figure 11. Sliding pressure operation of a once-through boiler

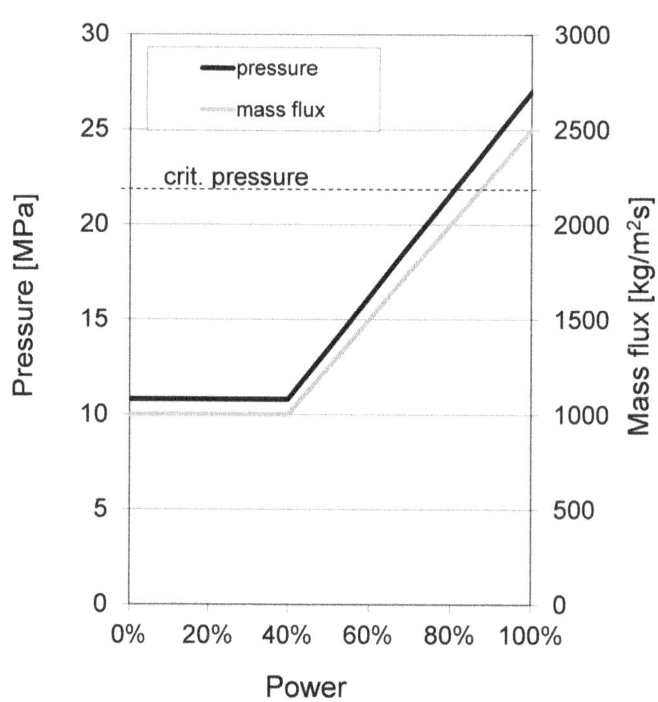

water separated there is collected and resupplied to the evaporator inlet header via a small recirculation pump. The total mass of this start-up system, however, is small compared with a high pressure drum, such that the fast cycling capabilities and the economic advantages are still kept.

Heat Transfer Challenges

While supercritical water is an excellent coolant, the change of heat transfer during sliding pressure operation is a challenge for the evaporator design. The problem shall illustrated by a transient heat transfer test of Köhler and Hein (1986), who measured the wall temperatures of an electrically heated evaporator tube at supercritical and at sub-critical pressure. The tube was made of stainless steel 1.4981 with a length of 6.1 m, an inner diameter of 14 mm and a wall thickness of 2 mm. A uniform heat flux of 619 kW/ m^2 and a mass flux of 2000 kg/m^2s were kept constant during the test, as well as the inlet temperature of 347°C. Starting from a supercritical pressure of 24.7 MPa, the pressure was reduced with ~30 kPa/s, passing the critical pressure at time 73 s. Once a pressure of 19.1 MPa was reached, the pressure was kept constant. The pressure history of this experiment is plotted in Figure 12. The measured wall temperatures are plotted in Figure 13 versus the coolant enthalpy, which increased linearly with the tube length.

As long as the pressure was supercritical, the tube was well cooled, reaching a maximum wall temperature of around 400°C. As soon as the pressure became subcritical, those parts of the tube which had been colder than the critical temperature (374°C) were wetted, but the temperature of other parts increased rapidly to almost 550°C due to film boiling. Staying at 19.1 MPa for around 10 min, the hot zones were cooling down by axial conduction through the tube wall and they were wetted again as soon as the wall temperature became colder than the minimum film boiling temperature. This phenomenon, called conduction controlled rewetting during transient depressurization, has been described in detail by Schulenberg and Raque (2014). The test demonstrates that transient heat transfer at near-critical pressure, and not the supercritical heat transfer at full power, is limiting the allowable heat flux to 400 kW/m^2 at maximum.

Figure 12. Pressure history of the boiler tube experiment of Köhler and Hein (1986)

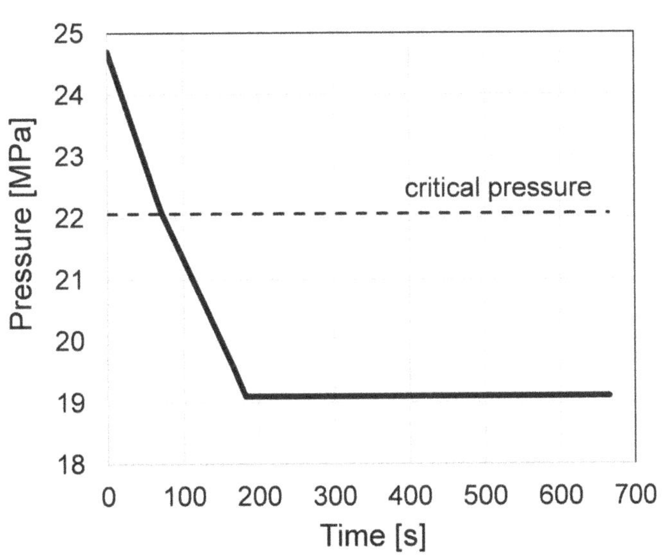

Figure 13. Transient tube temperature during depressurization of an evaporator tube from supercritical pressure
Köhler and Hein, 1986.

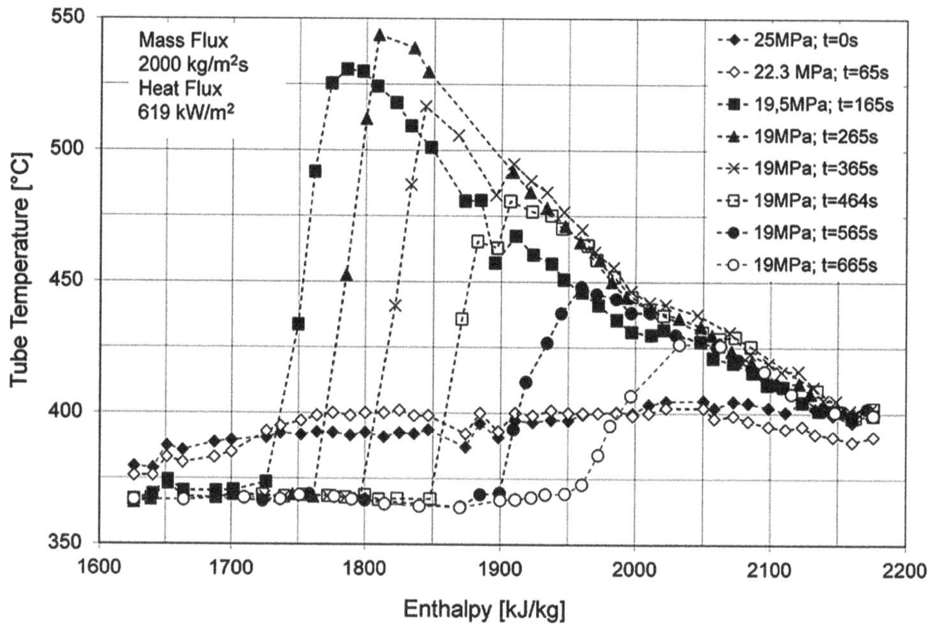

Superheater and Reheater Design

Superheater and reheater tube bundles are designed as horizontal U-tube bundles arranged in the upper boiler house, except the first superheater, which is designed as the vertical tube wall of the upper boiler house. A typical arrangement of a pulverized coal fired power plant is shown in Figure 14. Superheaters are shown in dark gray, reheaters in light gray, and economizers in black. Heat from the flue gas is transferred to the tube bundles by radiation and convection. The tube wall is hanging from a steel frame, which is erected first to pull up all components. The tube wall of the first superheater has to carry the weight of the evaporator tube wall underneath. The ash hopper at the bottom of the tube wall is thus expanding downwards by almost 1 m when the boiler is warming up. Such a tower design became the most economical solution for pulverized coal combustion. It can reach a total height of more than 120 m for a power plant with net electric power of 1000 MW.

Superheaters and reheaters must be designed such that the hottest tube is still colder than the creep and corrosion limit of the tube material. The steam temperature variation, however, can be significant, due to different radiation exposure to the flame, flue gas flow non-uniformities, larger or smaller ash layers, or corrosion effects. Unfortunately, those tubes in a parallel tube bundle, which are receiving the highest heat input, cause the largest pressure drop of the steam flow inside, which decreases the mass flow and increases the local heat-up of steam even more.

The challenge of superheater design is thus to reach a maximum average steam outlet temperature at a given peak steam temperature, in order to maximize efficiency while minimizing material costs. As a non-uniform heat-up cannot be avoided in general, intensive steam mixing is needed inside the headers, which are foreseen between each heat-up step. Moreover, feedwater can be injected into the connecting

Figure 14. Typical arrangement of superheaters and reheaters in the boiler house of a coal fired power plant

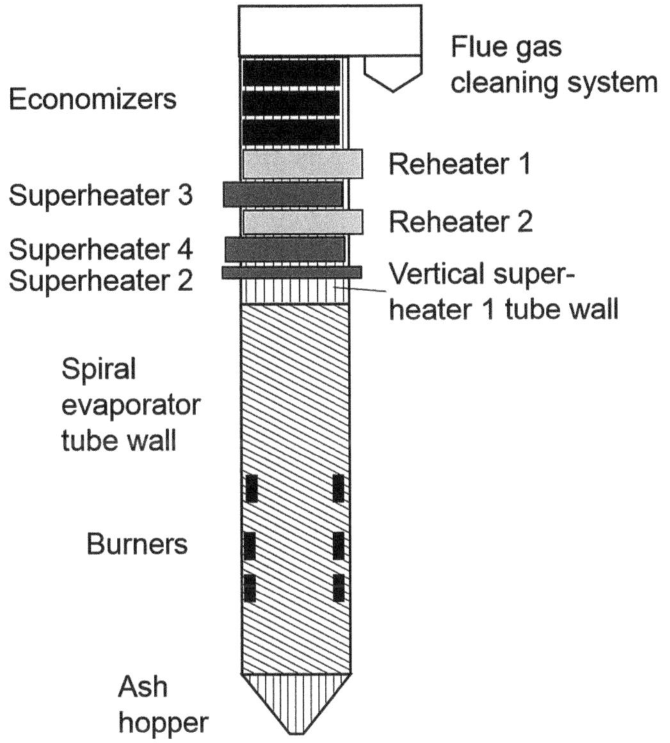

pipes between headers to reduce peak temperatures. A typical arrangement of mixing headers and injectors of a superheater is sketched in Figure 15. Steam from different side walls of the first superheater is mixed in joint headers and different mixing combinations are used in the following stage. This mixing scheme is repeated also in stage 3 and 4 of the superheater as well as in the two reheat stages. As a result, peak and average temperature differ finally by less than 10°C, enabling best use of the tube material.

Figure 15. Mixing in headers (gray) of a superheater and feedwater injection for peak temperature control

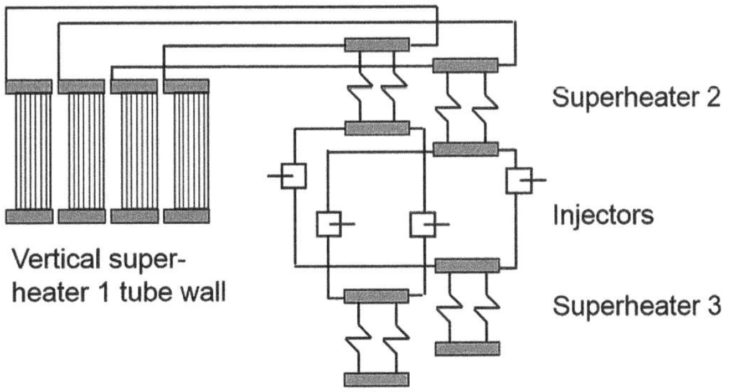

SUPERCRITICAL STEAM TURBINE DESIGN

The steam turbine train consists of a high pressure (HP) turbine, an intermediate pressure (IP) turbine and one or more low pressure (LP) turbines, which are all aligned on the same shaft and drive a common generator. They are rotating with synchronous speed of 50 Hz or 60 Hz, depending on the grid frequency. Plane radial (or “journal”) bearings are needed between each of these turbines, and only a single axial (or “thrust”) bearing is allowed, usually placed between the HP and the IP turbine to minimize clearance of these turbines. A horizontal split of the IP and LP turbines allows to remove each of these turbine rotors once the upper half of the turbine casing has been opened. Steam supply lines are installed preferably in the lower half, if possible, to avoid that steam flanges need to be disconnected when the turbine casing is opened. The horizontal split, however, causes local bending stresses so that the casing needs twice the wall thickness. Therefore, the HP turbine casing, which has the strongest wall thickness anyway, is designed without this horizontal split and the rotor can only be pulled out axially, after the entire HP turbine has been taken out and turned vertically.

HP steam is supplied through two thick-walled steam lines symmetrically from both sides to the HP turbine. A shut-down valve and a turbine governor valve (control valve) are foreseen on each steam line. They are operated with a hydraulic piston which opens the valve against a spring. Steam from the HP turbine outlet returns back to the reheater of the boiler, and the reheated steam is supplied to the IP turbine from top and bottom through two IP steam lines. Both IP steam lines need a shut-down valve and a control valve each like the HP steam lines. Bypass lines between the HP lines and the IP lines, as well as bypass lines between the IP lines and the condenser, blow down the steam into the condenser in case of a turbine trip or during the start-up process, when the boiler and the turbines are warmed up. They are indicated with dashed lines in Figure 7.

The condensers are placed underneath the LP turbines, and the entire LP turbine casing serves as the steam line from the LP turbine outlet to the condenser.

The IP and LP turbines are designed symmetrically to balance the axial thrust, which minimizes the size of the axial bearing, with the inlet from the center and the outlets to both sides. The reheated IP steam is supplied through diagonal vanes to each side. The vanes hold a thermal shield around the rotor to avoid that hot steam impinges on the rotor. The shield avoids thus local thermal stresses and thermal fatigue with change of steam temperature. The steam extractions on both sides are at different positions between the vanes, providing steam for different preheaters. Each rotor is usually designed as a monolithic steel block with grooves for the rotor blades. Vanes are installed in a separate vane carrier as an inner casing, and not in the outer casing, to ease radial and axial alignment with the rotor.

Labyrinth seals on both ends of the rotor shaft minimize steam leakage. Gland steam needs to be supplied to the inner glands, to avoid air ingress when the turbine is cold. The outer gland is kept at a pressure below ambient pressure to suck in some ambient air rather than releasing steam to the outside. Air and steam taken from there are separated in a gland steam condenser.

DYNAMICS OF A SUPERCRITICAL STEAM CYCLE

The major control loops of a once-through steam cycle in a coal fired power plant are:

- The thermal power and thus also the generator power are controlled by the coal feeder. The air ratio in the furnace is controlled by air flaps.
- In a once-through boiler, the mass flow of steam is equal to the mass flow of feedwater, which is controlled by the valve or speed of the feedwater pumps. As power, live steam enthalpy and mass flow are correlated, the mass flow of the feedwater pump is also responsible for the live steam temperature at a given thermal power. It can further be corrected by feedwater injection into the superheater, indicated in Figure 15. The live steam pressure is controlled by the HP turbine control valve or by the HP turbine bypass valve.
- The reheat temperature is controlled by feedwater injection into the reheater. The reheat pressure is controlled by the IP turbine control valve or by the IP turbine bypass valve.
- The condensate pumps are controlling the liquid level in the hotwell of the condensers. The condensate pressure is the saturation pressure, given by the temperature of the cooling water, as long as non-condensable gases have been evacuated from the condenser.

A closer look at the start-up procedure of a coal fired power plant shall illustrate the plant dynamics. Figure 16 sketches mass flows, pressure, temperatures, turbine speed and generator power during a cold start. Prior to the start-up phase indicated there, the cooling systems had to be started, the lube oil systems for bearings had to be started, fans and pumps were started, an auxiliary boiler had to produce gland steam, and the furnace had to be purged by combustion air for 12 min. As long as the boiler is cold, the air preheater in the flue gas duct cannot warm-up the combustion air to dry the coal. Instead, the boiler needs to be started with an oil flame as the coal flame cannot be ignited yet. This oil flame, however, is already evaporating some water of the steam cycle and the steam pressure rises to around 40% within 1 hour. The first steam, which is produced, is needed to warm up the steam lines and the steam valves. After around 1 hour, the turbines can be accelerated to full speed by slightly opening the steam turbine valves and the turbines continue to warm up for one more hour. Most of the steam, however, goes still through the turbine bypass lines, which are controlling the pressure. The steam turbine valves are still controlling the turbine speed. While warming up the turbines, the first coal is injected into the furnace, but the oil flame is still needed to keep the coal burning.

Around 2 hours after first ignition, the turbine generator is synchronized and the turbine control valves are further opened. The generator is producing now about 10% power and the speed control of the turbines is no more required. Around 2.5 hours after first ignition, the turbine bypass lines are closed and the coal mass flow is increased gradually, still keeping a small oil flame for flame stabilization. Sliding pressure operation starts once 40% power has been reached. The live steam temperature is still keeping some margin from the design temperature, even when full power is reached, to avoid thermal fatigue of steam turbines. It takes more than 5 hours after first ignition before the power plant can run with full efficiency at design conditions.

If the power plant is shut down for a day or just for some hours, it can start again much faster. Again, the boiler must be started with an oil flame, but coal can already be added after around 15 min. Warm-up of the turbines is not required, as they are thermally insulated and kept warm by auxiliary steam. As they do not cool down significantly during 8 hours, the generator can by synchronized right after the turbine has been accelerated to full speed, less than 30 min after first ignition. The power can increase much faster and full power design conditions are reached already 70 to 80 min after first ignition.

The complex dynamic interaction between fuel systems, combustion and flue gas systems with the steam cycle can be modelled with APROS, a dynamic simulation software developed by VTT and Fortum,

Figure 16. Start-up of a cold supercritical coal fired power plant

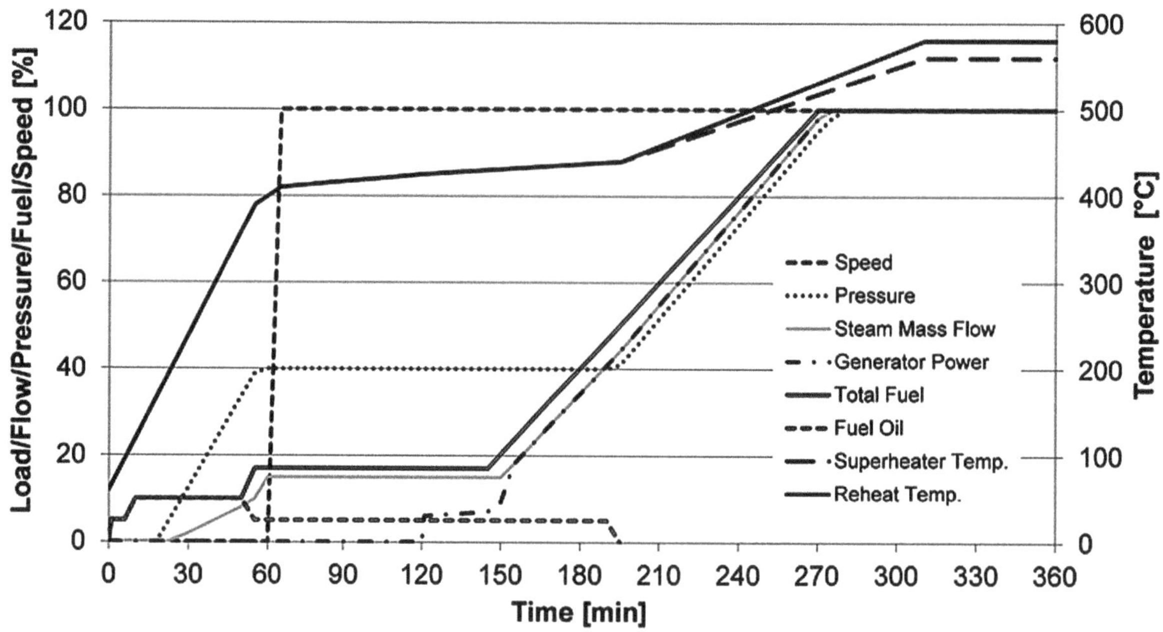

Finland. Starkloff et al. (2015) describe the application of this code to simulate the response of a coal fired power plant with a once-through boiler to step changes in the load setpoint. Details of component models in APROS for this application are further outlined by Kuronen et al. (2016). Results confirm that the once-through boiler can react faster to load changes than the former drum boilers, which had to limit pressure gradients of the live steam to avoid excessive thermal stresses in the drum.

SUPERCRITICAL WATER COOLED REACTORS

The most common water cooled nuclear reactor type is the Pressurized Water Reactor (PWR), in which liquid water is heated up in a closed primary system at ~15 MPa from ~280 to ~320 °C. Within this primary system, a steam generator transfers heat to a secondary system, in which water at 7 to 8 MPa is evaporated and supplied as saturated steam to a steam turbine. The steam generator is a natural convection drum boiler without superheater in the terminology of this chapter.

More economical is the Boiling Water Reactor (BWR), which generates steam for the HP turbine directly inside the reactor core. The reactor pressure vessels comprises thus a forced convection drum boiler with economizer, evaporator, separators and steam driers, and recirculation pumps. Again, a superheater is not included in this steam cycle. Like with the PWR, the steam is dried and reheated after leaving the HP turbine, using steam which has been extracted from the main steam line upstream of the HP turbine inlet. Such reheat is needed to meet the steam quality target of ~87% at the outlet of the LP turbine. A BWR with superheater (called HDR) had been built in Germany in the 1960s, but the plans were given up already after the first commissioning tests. Schulenberg and Starflinger (2012) report about more details.

The net efficiency of nuclear power plants with a saturated steam cycle is around 34% to 36% only, depending on the temperature of the cooling water. The advantages of supercritical coal fired power plants had motivated therefore the Generation IV International Forum to look for similar options for future light water reactors with superheated steam at supercritical pressure, called Supercritical Water Cooled Reactors (SCWR). The higher steam enthalpy could enable a direct, once-through steam cycle such that neither steam generators nor steam separators and dryers would be required, and even primary coolant pumps could be omitted. Moreover, steam turbines and reheaters could be significantly smaller than today. The power plant, therefore, should have less plant erection costs, while the thermal efficiency would even be higher – a clear economic advantage. As supercritical water does not boil with a phase change, a boiling crisis would physically be excluded, and superheated steam could be produced without excessive hot spots in the core. Moreover, as coal fired power plants with supercritical steam conditions have been operated since more than 20 years now, this nuclear plant concept could benefit from proven design of turbines, feedwater pumps and most other components of the steam cycle, except the steam generator.

The steam cycle of such SCWR is shown in Figure 17. Concepts described by Oka et al. (2010) and by Schulenberg and Starflinger (2012) were considering a live steam temperature of ~500 °C, leaving more margin for hot spots than in a coal fired power plant. Like with the BWR, the HP steam is supplied directly to the HP turbine. It is not reheated inside the steam generator after leaving the HP turbine, but in a heat exchanger with HP steam, which is extracted from the main steam line.

Parameters of this steam cycle are plotted into the *T-s*-diagram in Figure 18. With a live steam temperature of 500°C only, a supercritical pressure would not have been required yet from steam cycle analyses, and a subcritical steam cycle would have given a similar net efficiency. The reason for the supercritical pressure is rather due to the excellent heat transfer, allowing a high heat flux and thus a high power density in the reactor core.

Reactor Core Design Concepts

Schulenberg and Starflinger (2012) could show that all steam cycle components of a 1000 MW power plant, except the steam generator and the reheater, were quite similar indeed to those of coal fired power plants. The reactor core as the steam generator of this system, however, differs significantly from a steam generator of a coal fired power plant because of the following, generic issues:

Figure 17. Supercritical steam cycle of a Supercritical Water Cooled Reactor (SCWR)

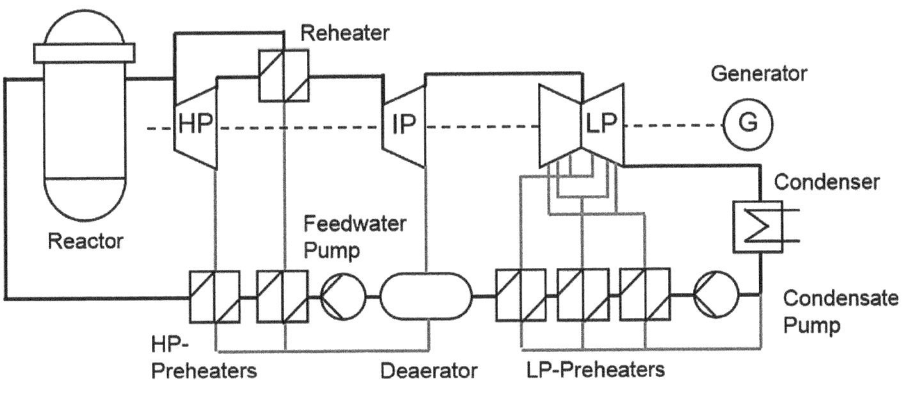

Figure 18. Supercritical steam cycle of an SCWR in the T-s-diagram

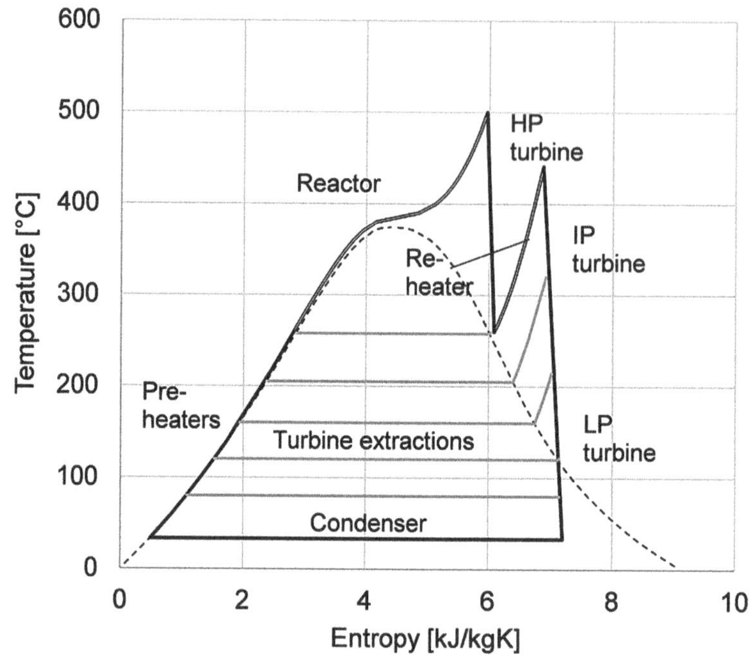

- The heat flux of fuel rods in a PWR or BWR is > 1500 kW/m^2, and a similar heat flux is planned for the SCWR. Such heat flux, however, exceeds the heat flux limit of 400 kW/m^2 of the steam generator in a coal fired power plant by far. Therefore, the reactor can only be operated at supercritical pressure, except at very low power, e.g. for residual heat removal. A sliding pressure operation cannot be allowed.
- The reactor core of an SCWR does not allow any coolant mixing between fuel assemblies. Any water cooled reactor with thermal neutron spectrum, like the PWR, BWR or SCWR, requires water not only as coolant, but also as moderator. Steam or superheated supercritical water, however, does not provide much moderation due to its low density, such that the BWR or SCWR fuel assemblies must be housed in assembly boxes, leaving enough liquid water in the gaps between them for moderation. The assembly boxes, however, prevent any coolant mixing between assemblies, and severe hot spots must be expected in the superheaters of an SCWR.

The second issue can be solved, in principle, by coolant mixing outside the reactor core and by multiple heat-up steps through the core. The reactor concepts, which have been studied so far, can thus be classified into the following generic concepts, which are sketched in Figure 19:

1. A single-pass design is used for all BWRs. The coolant, as well as moderator water inside the gaps between fuel assemblies, enters the reactor core at the bottom and leaves it on top. No mixing between fuel assemblies is foreseen.
2. A two-pass design has been chosen by Oka et al. (2010) for their Super Light Water Reactor concept. Coolant of the economizer at the core periphery, shaded in dark gray, as well moderator water for

Figure 19. General core design concepts for SCWRs

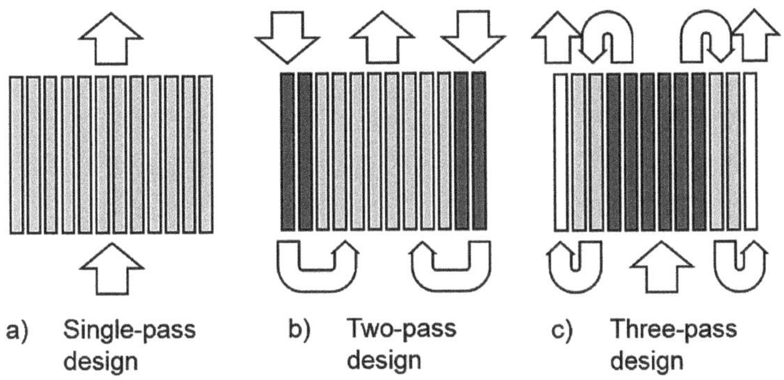

all fuel assemblies, is running downwards. The coolant is mixed then underneath the reactor core and is superheated afterwards in the central fuel assemblies, shaded in light gray.

3. A better interstage mixing is possible with a three-pass core design, which has been chosen by Schulenberg and Starflinger (2012) for the High Performance Light Water Reactor (HPLWR). Coolant is used first as moderator water through assembly gaps, through additional water rods inside assemblies and in the reflector surrounding the core like in an economizer. It is running upwards then through the evaporator assemblies in the central part of the core, shaded in dark gray, is mixed above the reactor core, is superheated with a downward flow in assemblies around the evaporator, shaded in light gray, mixed again underneath the core, and is finally heated up to 500°C in average in the second superheater at the core periphery, shaded in white.

The three-pass core design of the HPLWR, designed for a net electric power of 1000 MW and a net efficiency of 43.5%, could reach a difference between peak coolant temperature and average coolant temperature of 100 °C at the core outlet, which is better than the coolant temperature distribution of the two-pass core design, but still much worse than the temperature distribution of a steam generator of a coal fire power plant. The peak steam temperature of the HPLWR, with its average live steam temperature of 500°C, is comparable with the one of a coal fired power plant with 600°C live steam temperature.

The complexity of the reactor core, however, is increasing with each heat-up step. Mixing plena must be foreseen above and underneath the reactor core, and the core design optimization process, with its interaction of neutron physics, fuel burn-up and thermal-hydraulics, is getting more constraints. Nevertheless, the reactor design concept of the HPLWR, shown in Figure 20, does not differ too much from a conventional PWR. Three assemblies of the reactor core are shown exemplarily, characterizing the evaporator assemblies with upward flow, the first superheater assemblies with downward flow, and the second superheater assemblies with upward flow again. Mixing plena are foreseen above the core and underneath the core support plate, consisting each of a separate inner and outer mixing plenum. The upper mixing plenum can be removed after extraction of the steam outlet pipes for replacement of fuel assemblies. The feedwater inlet has been equipped for this concept with a passive backflow limiter to minimize coolant losses in case of a break of the feedwater line. Control rods inserted from top and the control rod guide assembly are similar to those of PWRs. Schulenberg and Starflinger (2012) provide more details of this design concept.

Figure 20. Concept of the HPLWR
Schulenberg & Starflinger, 2012.

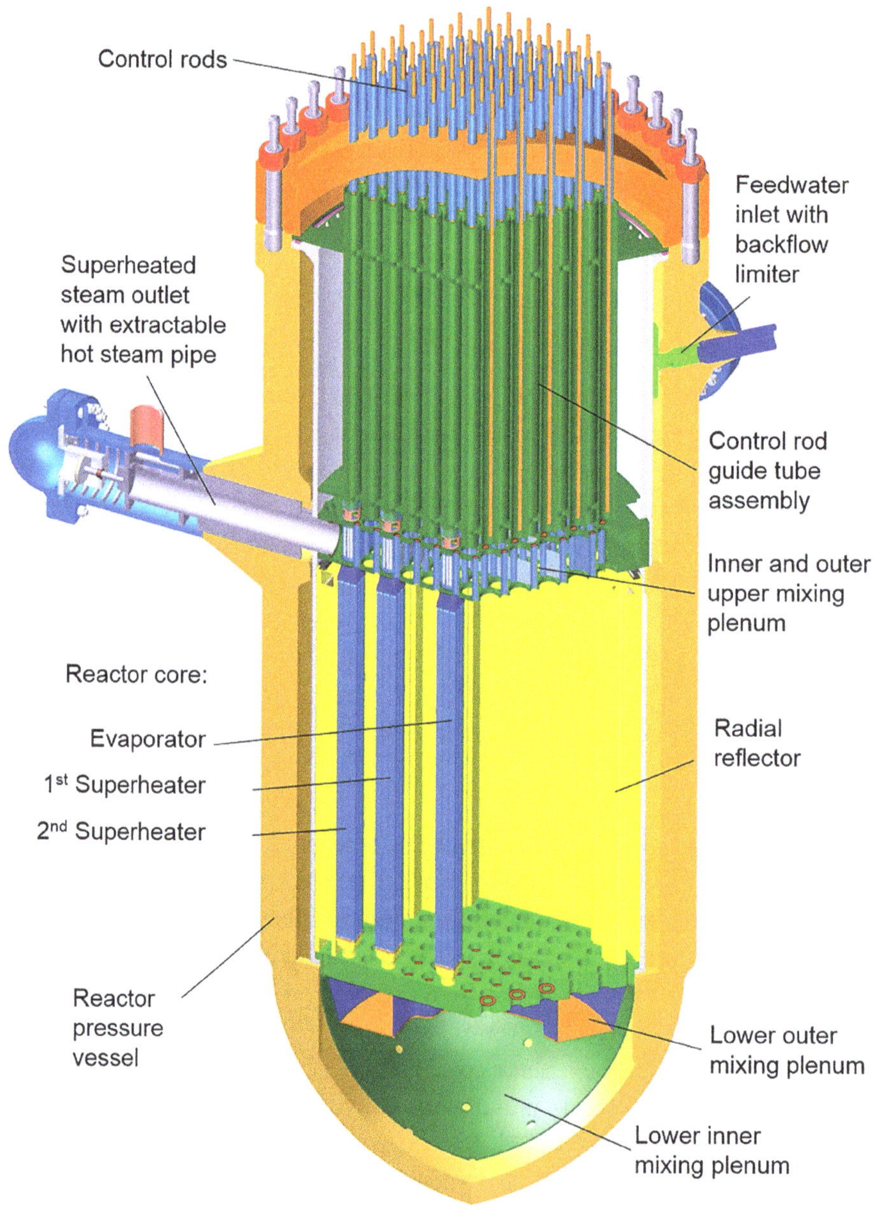

Numerical Simulation of the Supercritical Steam Cycle

The steam cycle dynamics of the HPLWR have been simulated numerically by Schlagenhaufer et al. (2009). Using the system code APROS again, they modelled the once-through steam cycle and the safety systems of this power plant to predict the system performance during start-up and shut-down, as well as the response of the safety systems for several accident scenarios. The steam cycle control was assuming:

- Power control by positioning the control rods of the reactor;
- Live steam temperature at the reactor outlet by the speed of the feedwater pumps;
- Live steam pressure control by the turbine governor valve;
- Reheat temperature control by a valve controlling the mass flow of extracted steam to the reheater.

Different from coal fired power plants, the live steam pressure and temperature were kept constant between 50% and 100% of full power to maximize heat transfer in the reactor core and to optimize the thermal efficiency of the steam cycle. A start-up system with a battery of cyclone separators was required, however, during start-up to warm-up the feedwater, to bypass the cold turbine and to increase pressure. The start-up system could also be used to minimize thermal stresses of reactor components during shut-down.

CONCLUSION

The supercritical steam cycle is an improved Rankine cycle at supercritical pressure, aiming at higher steam cycle efficiency and lower specific costs. It had been developed continuously from conventional, subcritical steam cycles. Coal fired power plants are using such supercritical steam cycle already since the 1990s. Transition to supercritical pressure became necessary with increasing live steam temperature to meet the target steam quality at the outlet of the low pressure turbine. The proven drum boiler design needed to be changed to a once-through boiler, as water and steam cannot be separated anymore at supercritical pressure. The most common operation mode is the sliding pressure operation, where the system pressure varies proportionally with power, except in the low power range. The most challenging heat transfer problem occurs then at the transition from supercritical to subcritical pressure, such that the evaporator heat flux must be limited to less than 400 kW/m^2.

As a different application of the supercritical steam cycle, a conceptional design study of a supercritical water cooled reactor is shown in this chapter. It is comparable with the supercritical Rankine cycle of a coal fired power plant except the steam generator, which is replaced by a nuclear reactor core. Reheat of steam cannot be provided inside the reactor core, but a small mass flow of high pressure steam is extracted instead to superheat the intermediate pressure steam. Recent design studies of a supercritical steam cycle for nuclear power plants are assuming a fixed, supercritical pressure, avoiding overheating at subcritical pressure, which allows a significantly higher heat flux. Such reactors have never been built up to now and more development effort and tests would be required before such reactors can be licensed, in particular for the nuclear reactor core. A particular design challenge will be coolant mixing inside the superheater, for which different concepts exist.

REFERENCES

Bald, A., & Heusinger, K. (1996, April). Energieerzeugung mit fortschrittlichen Dampfkraftwerken entlastet die Umwelt. *SIEMENS Power Journal*, 5-11.

Kefer, V., & Seiter, C. (1996, April). Ein 2x660-MW-Kohlekraftwerk mit niedrigen Life-Cycle-Kosten. *SIEMENS Power Journal*, 12-15.

Köhler, W., & Hein, D. (1986). *Influence of the wetting state of a heated surface on heat transfer and pressure loss in an evaporator tube, NUREG/IA-0003*. US-NRC.

Kuronen, J., Hotti, M., & Tuuri, S. (2016), Modelling and dynamic simulation of cyclically operated pulverized coal-fired power plant. *9th EUROSIM Congress on Modelling and Simulation.*

Oka, Y., Koshizuka, S., Ishiwatari, Y., & Yamaji, A. (2010). *Super Light Water Reactors and Super Fast Reactors*. Springer. doi:10.1007/978-1-4419-6035-1

Schlagenhaufer, M., Starflinger, J., & Schulenberg, T. (2009). Plant control of the High Performance Light Water Reactor. *Proc. Global 2009.*

Schulenberg, T., & Starflinger, J. (2012). *High performance light water reactor, design and analyses.* KIT Scientific Publishing.

Starkloff, R., Alobaid, F., Karner, K., Epple, B., Schmitz, M., & Boehm, F. (2015). Development and validation of a dynamic simulation model for a large coal-fired power plant. *Applied Thermal Engineering, 91*, 496–506. doi:10.1016/j.applthermaleng.2015.08.015

Stodola, A. (1927). *Steam and Gas Turbines*. New York: McGraw-Hill.

Termuehlen, H., & Emsperger, W. (2003). *Clean and efficient coal-fired power plants*. New York: ASME Press. doi:10.1115/1.801942

KEY TERMS AND DEFINITIONS

Dryer: An assembly of parallel plates, separating small droplets from a steam in a zig-zag-flow.
Economizer: Heat exchanger heating up feedwater to boiling temperature.
Feedwater: Water supplied to the steam generator.
Fuel Assembly: An assembly of fuel rods containing UO_2 fuel, to be installed inside the reactor core.
Heat Flux: Ratio of heat provided to a heat exchanger to heat transfer surface area.
Live Steam: High pressure steam to be provided to the high-pressure turbine.
Mass Flux: Ratio of mass flow through a heat exchanger tube to tube cross section.
Mass Flux: Ratio of mass flow through a heat exchanger tube to tube cross section.
Reactor Core: Part of the nuclear reactor which is producing heat by nuclear fission.
Saturated Steam: Steam at boiling temperature; not superheated.
Separator: Vertical tubes with a swirl flow, separating liquid water from steam by centrifugal forces.

Chapter 15
The Basic Thermal Hydraulic Issues of Applying Supercritical Fluid to Nuclear Reactors

Xiao Yan
Nuclear Power Institute of China, China

Xi Sui
Nuclear Power Institute of China, China

Jinguang Zang
Nuclear Power Institute of China, China

Yanping Huang
Nuclear Power Institute of China, China

Ting Xiong
China Nuclear Energy Industry Corp., China

Zejun Xiao
Nuclear Power Institute of China, China

ABSTRACT

This chapter is mainly focused on illustrating some introductory progress on thermal hydraulic issues of supercritical water, including heat transfer characteristics, pressure loss characteristics, flow stability issues and numerical method. These works are mainly performed in Nuclear Power Institute of China (NPIC) these years, to give a basic idea of elementary but important topics in this area. An analytical method was proposed up to predict the heat transfer coefficient and friction coefficient based on the two-layer wall function. Flow instability experiments have been carried out in a two parallel channel system with supercritical water, aiming to provide an up-to-date knowledge of supercritical flow instability phenomena and initial validation data for numerical analysis. An in-house code has been developed in NPIC in order to better utilize and further expand the experimental results on supercritical flow instability. At last, some future research directions are suggested for reference.

INTRODUCTION

The Supercritical Water-Cooled Reactor (SCWR) is a high temperature, high pressure water cooled reactor that operates above the thermodynamic critical point (374°C, 22.1 MPa) of water. The reactor outlet temperature can be as high as 500°C which can increase the thermal efficiency up to 40% or even more. Supercritical water is a single phase fluid without two-phase interfaces, leading to simplifica-

DOI: 10.4018/978-1-5225-2047-4.ch015

tion of the primary system, no need for steam generator, pressurizer and main circulation pump of the PWRs, and no need for the inner circulation pumps, steam separator and dryer of BWRs. As for the safety advantages, due to the characteristics of the supercritical water, no phase change would occur in the SCWR core under nominal conditions, therefore there are no DNB risks like in the PWRs (See 2015 GIF Annual Report).

Gaps exist in applying supercritical fluid into nuclear reactors. Among them, the heat transfer, pressure loss and flow stability are some typical and key thermal hydraulic issues to be solved. The significant changes in the thermal physical properties may have big influence on the flow resistance characteristics. When the wall temperature is greater than the pseudocritical point and bulk fluid temperature is below, large variation of fluid property will happen across the boundary layer. In such situation, the wall heat transfer coefficient will be dependant on the heat flux. As the increase of heat flux, the wall heat transfer coefficient is decreased. The friction coefficient will not depend solely on the bulk Reynolds number, but also bonds tightly with the physical property of the boundary layer. Experiments of supercritical water needed to be performed and data needed to be collected and compiled to develop the suitable empirical or semi-empirical correlations. Meanwhile, the theoretical analysis is relatively few, attributed to the fact of the difficulty in dealing with the complex variation of the physical properties.

The large density difference between the core inlet and outlet raises doubts about the possible occurrences of flow instabilities similar to those observed in BWR. As the licensing of SCWR will probably require, at a minimum, demonstration of the ability to predict the onset of instabilities, it is necessary to understand the instability phenomena in SCWR, to identify the important variables affecting these phenomena, and ultimately to generate the maps identifying the stable operating conditions of different SCWR designs.

Numerical methods are now playing an important part in investigating the flow and heat transfer behaviors of SCW under different flowing conditions. Experimental data obtained in the corresponding tests could be used to verify the validity of different numerical tools, including the system analysis code and CFD code. These methods could be used to investigate the flow instability characteristics under different working conditions. Detailed distributions of both the flow parameters and the fluid properties in regions near the heated wall can be obtained on the basis of numerical results to help understand the complexity and distinct features of the flow and heat transfer of the SCW, especially in the case with deteriorated heat transfer.

In this chapter, four sections would be organized to illustrate some introductory progress on thermal hydraulic issues of supercritical water. These sections are:

Section I: The heat transfer characteristics of supercritical water.
Section II: The pressure loss characteristics of supercritical water.
Section III: The flow stability issues of supercritical water in parallel channels.
Section IV: Modeling and analysis of supercritical flow instability in parallel channels.

For the first and second sections, an analytical method was proposed up to predict the heat transfer coefficient and friction coefficient based on the two layer wall function. Such kind method was verified with experimental data obtained in NPIC. It could provide a perspective to explain the unusual flow and heat transfer behavior related to supercritical water. For the third section, some research would be described on flow instability in two parallel channels with supercritical water. Parametric studies show that the flow becomes more stable with increasing pressure or decreasing inlet temperature in the range of present experiments, and the mechanisms have been discussed compared to that for two-phase flow. The stability boundaries are illustrated in a two-dimensional plane using two dimensionless parameters proposed for supercritical flow.

For the fourth section, an in-house code has been developed applying time-domain approach to better utilizing and further expanding the experimental results. The numerical code is described in detail and validated by the experimental results. Based on the validated code, the geometrical simplification is further discussed and effects of system parameters on the flow instability are numerically analyzed.

The thermal hydraulics of SCWR is a broad area covering every aspect or process of transferring energy from the core to the electricity generated equipment. Some monographs have already given a detailed description of this area, such as Oka et al. (2010), Pioro and Duffey (2007), IAEA (2014) etc.

This chapter is mainly focused on relative research work performed in NPIC these years, aimed to give a basic idea of elementary but important topics in this area.

BACKGROUND

The Supercritical Water-Cooled Reactor (SCWR) nuclear energy system is a combination of the advanced nuclear reactor technology and the updated supercritical fossil boiler technology. Compared with the current water-cooled reactor NPPs whose average plant efficiency is only around 30% due to use of obsolete subcritical turbine-generator system, the SCWR has significant economic and technical advantages. The Generation IV International Forum selected SCWR as one of the six most promising Generation IV nuclear energy systems in 2002 after technical investigation and comparisons among more than 100 initiative reactor types for several years.

The Supercritical Water-Cooled Reactor (SCWR) uses the supercritical water as the coolant. Supercritical water is mainly characterized by its special thermal physical properties which vary strongly, especially near the pseudo critical line. Figure 1 shows the thermal physical properties of water at 24MPa. The specific heat capacity (Cp) of water presents a peak within a very narrow range in the vicinity of the pseudo-critical temperature. Density (ρ), thermal conductivity (λ) and dynamic viscosity (μ) decrease sharply near the pseudo-critical point. Such dramatic variations of thermal physical properties have a significant influence on heat transfer characteristics of supercritical water.

In general terms, the conceptual designs of SCWRs can be grouped into two main categories: pressure vessel concepts proposed first by Japan and then by a Euratom partnership and by China more recently, and a pressure tube concept proposed by Canada, generically called the Canadian-SCWR. Table 1 summarizes the primary design parameters for each concept. A number of these parameters are still evolving with better understanding through the design process. Other than the specifics of the core design, these concepts have many similar features (e.g. outlet pressure and temperatures, thermal neutron spectra, steam cycle options, materials, etc.). Therefore, the R&D needs for each reactor type are common; this enables collaborative research to be pursued.

Take the Chinese SCWR conceptual design CSR1000 developed by Nuclear Power Institute of China (NPIC) for example (NPIC, 2016). The primary circuit of CSR1000 is a direct-cycle system consisting of a two-pass, thermal neutron reactor cooled and moderated by supercritical light water, two

Figure 1. The thermal physical properties of water at 24MPa

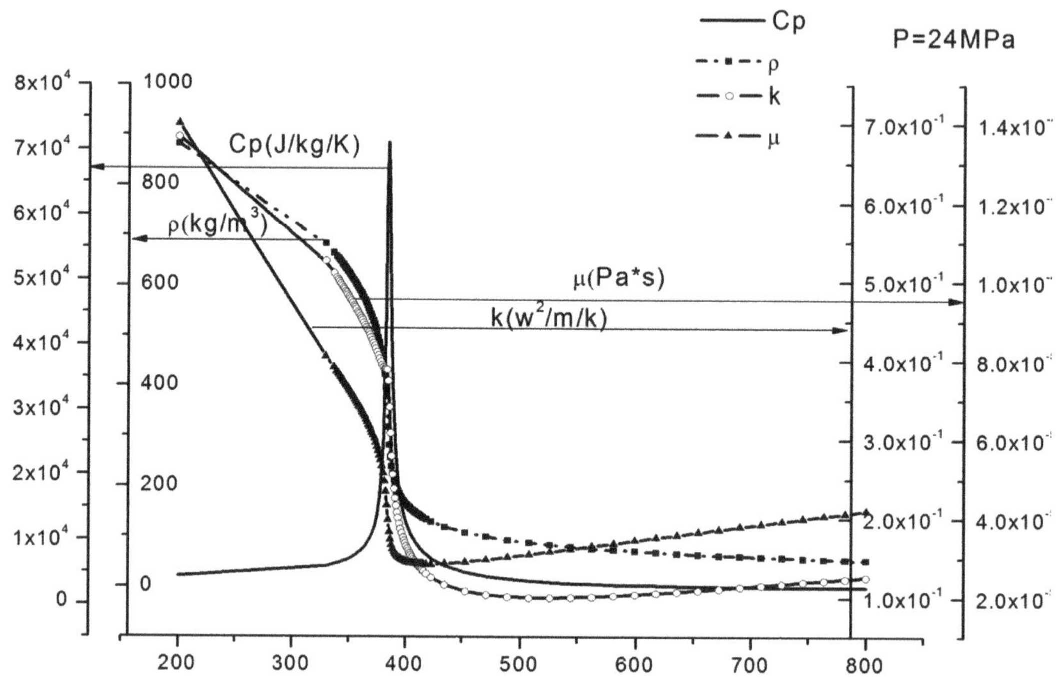

Table 1. Summary of design parameters for some SCWR concepts

Parameters	Unit	Canadian SCWR	HPLWR	JSCWR	Super Fast Reactor	CSR1000
Country	-	**Canada**	**EU**	**Japan**		**China**
Reactor Type	-	**PT**	**RPV**	**RPV**	**RPV**	**RPV**
Spectrum	-	**Thermal**	**Thermal**	**Thermal**	**Fast**	**Thermal**
Thermal Power	MW_{th}	2540	2300	4039	1602	2300
Thermal efficiency	%	48	43.5	42.7	~44	43.5
Pressure	MPa	25	25	25	25	25
Tin coolant	°C	350	280	290	280	280
Tout coolant	°C	625	500	510	508	500
Flow rate	kg/s	1512	1179	2105	820	1190
Active core height	M	5.0	4.2	4.2	2	4.2
Equiv. core diameter	m	~5.5	3.8	3.34	1.86	3.4
Fuel	-	Pu-Th	UO_2	UO_2	MOX	UO_2
Cladding material	-	SS	316SS	310SS	SS	310S
Drod	mm	7/12.4	8	7	5.5	9.5
Pitch	mm	vary	9.44	-	6.55	10.5
Tmax cladding	°C	850	620	700	643	650
Moderator	-	D_2O	H_2O	H_2O	-/ZrH	H_2O

primary loops connected with supercritical turbine and feedwater pumps etc. The primary circuit is also interfaced with passive safety features. The core thermal power is 2300MWt with a calculated system thermal efficiency 43.5%, leading to system output electrical power around 1000MWe. The primary circuit operates at 25.0 MPa. The feedwater temperature is 280°C, and the average core outlet coolant temperature is about 500°C.

The CSR1000 reactor core consists of 157 fuel assemblies, 156 FAs of 1st, 2nd, 3rd cycle fuels and one 4th cycle FA. The core coolant flow rate around 1190 kg/s, is significantly lower than those of current LWRs since the enthalpy rise in the core is much higher.

The two-pass core flow pattern is arranged (Figure 2). Fifty-seven FAs are located in the center, signed as "I" style fuel assembly; the rest lie in periphery of the core, marked as "II" style fuel assembly. In order to meet refueling requirement, the structure of "I" style fuel assemblies must be the same as that of "II" style fuel assemblies. Considering two-pass core arrangement, moderator water passage and coolant passage must be separated from each other. In addition, moderator water and coolant should have enough flow area, and there is no interference with structure among the different components of fuel assembly.

Figure 2. Two-pass core design

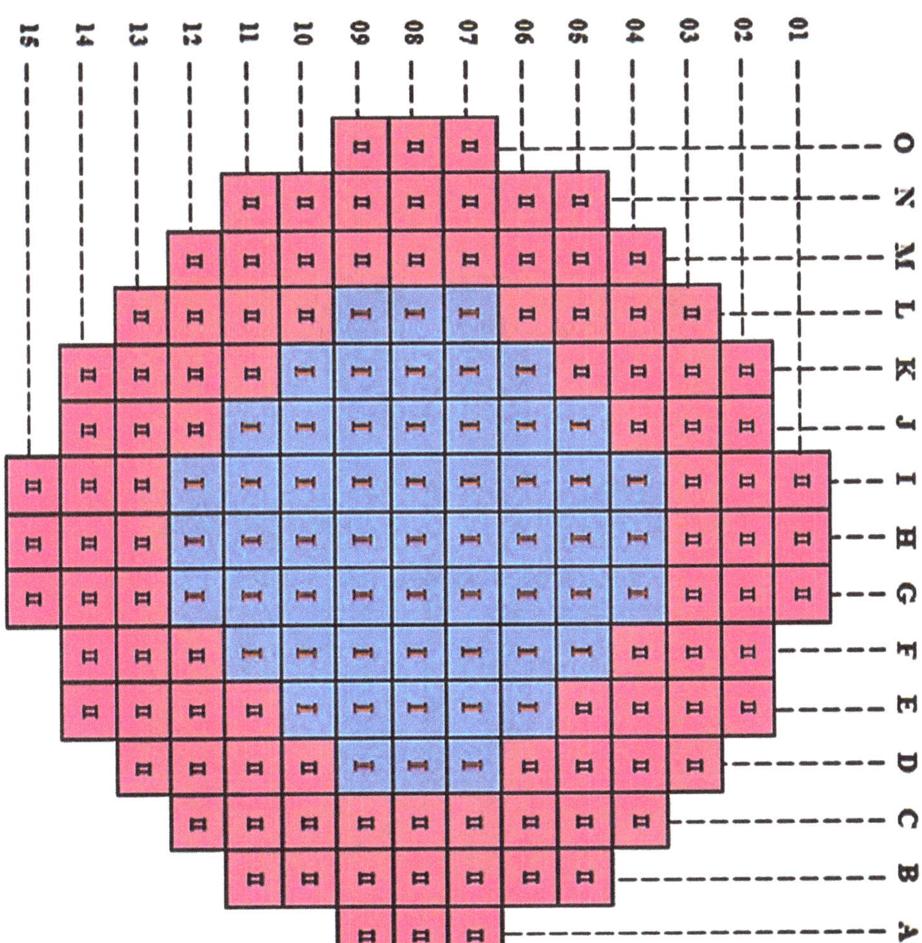

In order to simplify structural design and obtain more uniform moderation, the standard fuel assembly cluster is composed of 4 square sub-assemblies as shown in Figure 3, and each of them consists of 56 fuel rods and a square water (moderator) rod in the center surrounded by a square channel box, and the cross-plate control rods similar to that of BWRs are used. The fuel rods contain UO2 pellets like PWR fuels in the modified stainless-steel cladding. In the heat-insulated water rod, lower temperature water flows downward to keep enough moderation in the core. A 9×9 square arrangement for fuel rods in each subassembly is adopted, while central moderator box takes up 5×5 fuel rod cells. A dual wire is wrapped around each fuel rod, leaving a clearance of 0.1 mm between the wire and the fuel rods in order to mainly space the fuel rods and to enhance the heat transfer among the fuel rods. The outer length of inner moderator box is 52.5 mm, with a wall thickness of 0.8 mm; while the assembly box has an outer size of 98.5 mm and a wall thickness of 2.0 mm with ZrO2 filling whose thickness is 1.0mm. The cluster side length is 218 mm, close to a PWR size, which is convenient for refueling process. The boxes and cladding materials are 310S stainless steel, which shows good performance under supercritical water condition in the tests performed by NPIC and other researchers. The cross control rod with a width of 8mm is located at the center of 4 moderator box, which is inserted into the cross channel from the top of assembly.

The fuel assembly cluster has a total length of 5.9 m, which is composed by fuel rods, top/bottom nozzle, adaptor plates, assembly boxes, nozzle connectors, moderator boxes, moderator tubes, and etc(Figure 4). In order to strengthen the laterally support and connect subassemblies into a cluster, 4 grids with a height

Figure 3. 2×2 assembly cluster of CSR1000

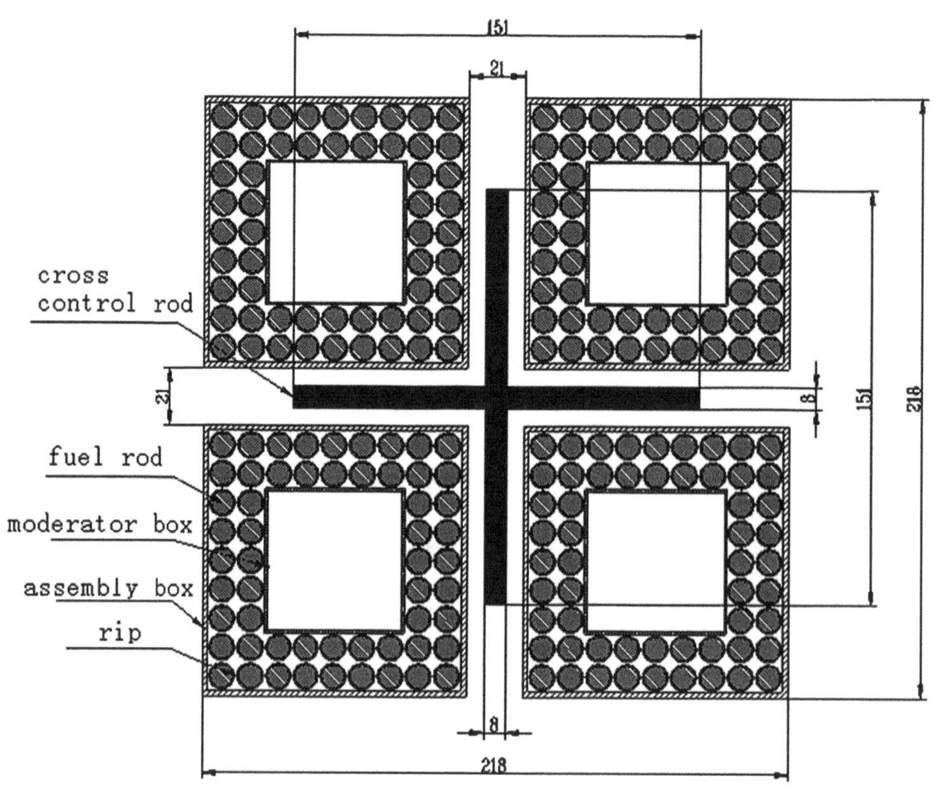

of 30 mm are designed. Combining four sub-assemblies to form an assembly cluster allows reducing the number of individual control rod drives to similar numbers as in a PWR. A preliminary choice of mature fuel design on PWRs is adopted, with a diameter of 9.5 mm and a cladding wall thickness of 0.57 mm. In order to accommodate more fission gas and reduce the central temperature of fuel pellets, the annular fuel pellet are used, with an outer diameter of 8.192 mm and a central hole of 2.0 mm.

Besides, the CSR1000 main technical parameters, core design, key structure design and systems design, etc. were carefully studied along with the material R&D and thermal hydraulic experimental R&D. The design and tests were closely integrated and the internal results feedback optimized the design.

Figure 4. Illustration of fuel assembly cluster

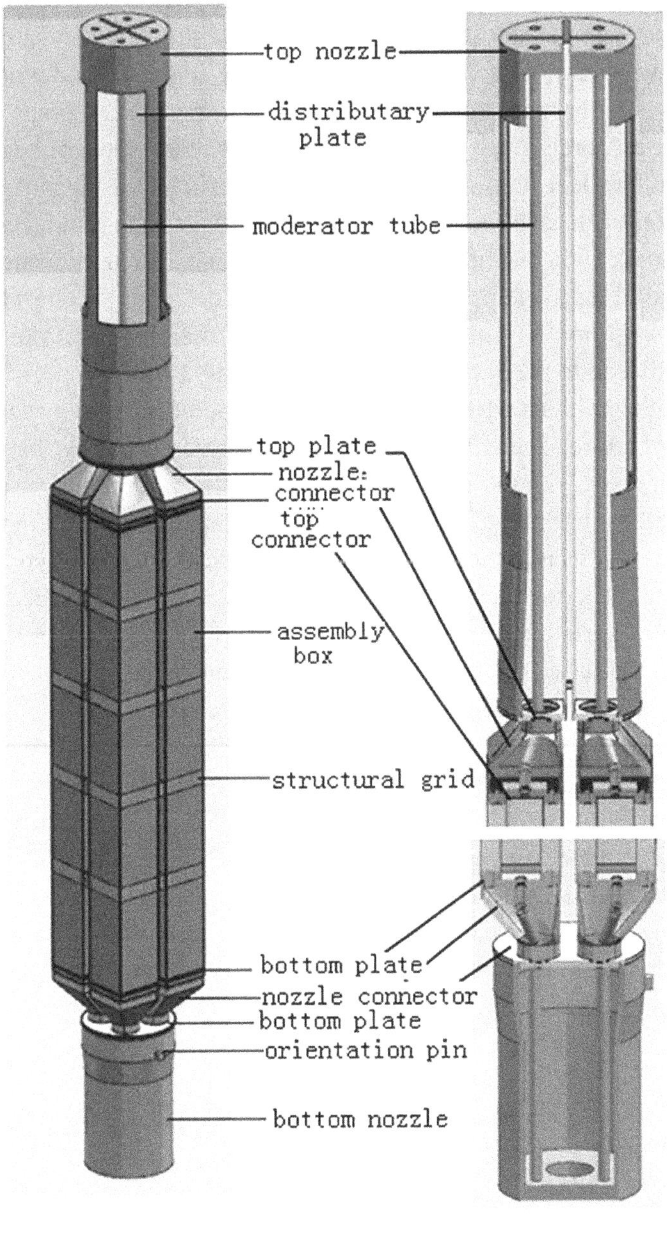

THE HEAT TRANSFER CHARACTERISTICS OF SUPERCRITICAL WATER

Heat Transfer Introduction

The development of supercritical water reactors requires more accurate predictions of water and steam heat transfer coefficients over a wide range of operating conditions. Many experiments have been performed to investigate the forced convective heat transfer phenomena, together with the numerical simulations and theoretical analysis. Yamagata et al. (1972), Swenson et al.(1965), Bishop et al.(1964), Shitsman (1963) etc. carried out experiments in the earlier years in last 60's. In recent years, more and more activities are being devoted to the R&D development of SCWR, such as Licht et al. (2008), Mokry et al. (2010), Misawa et al. (2009) and Zahlan et al. (2011). Pioro and Duffey (2007), Cheng and Schulenberg (2001), OKa et al. (2010) gave a review of ongoing research about these activities. Pioro et al. (2011) assessed the work related with the heat transfer and pressure drop characteristics at supercritical pressures based on the experiments of water and carbon dioxide. Most of the studies have found that the heat transfer near the pseudocritical point has unusual behaviors where the specific heat capacity arrives at its maximum value. In some cases, at relatively high mass fluxes, an enhancement of heat transfer occurs in the pseudocritical region at low heat flux whose heat transfer coefficient is bigger than the value predicted by subcritical correlations. But as the heat flux increases, its magnitude is reduced. In other cases, when the mass flow rate is small, while the heat flux is relatively large, heat transfer deterioration may occur, in companion with local wall peak temperature. This unusual behavior is believed to be related with the sharp variation of the fluid property in the neighborhood of pseudocritical point where small change in fluid temperature will lead to big variation in fluid properties. This is especially possible in a heated tube under supercritical conditions when the pseudocritical point is just between the wall temperature and the bulk fluid temperature. The crossing of the pseudocritical line may have big influence on the velocity and temperature profile in the wall boundary layer. At least, the physical property couldn't assume to be constant, and the traditional law of the wall may be dependent on the physical properties.

Although many correlations have been proposed up to predict such unusual heat transfer behaviors as mentioned above, the theoretical analysis is relatively few, attributed to the fact of the difficulty in dealing with the complex variation of the physical properties. In this chapter, a method of predicting the heat transfer coefficient based on the new thermal function of the wall was proposed up and could provide some explanations to the strange heat transfer characteristics.

The Thermal Law of the Wall for Varied Property Fluid

Generally, two kinds of mechanisms dominate the heat transfer behaviors in the boundary layer: the molecule diffusion and the turbulent fluctuation. And according to each of their contributions, two zones are distinguished: the molecule heat conduction layer and the turbulent heat diffusion layer. Based on the boundary layer assumption, Boussinesq hypothesis and the Reynolds analogy rule, a new thermal function of the wall could be derived in account of the fluid property variation across the boundary layer.

The general energy conservation equation:

$$\frac{\partial \overline{\rho h}}{\partial t} + \frac{\partial \overline{\rho u_i h}}{\partial x_i} = -\frac{\partial}{\partial x_i}\left(-k\frac{\partial t}{\partial x_i} + \overline{\rho u_i' h'}\right) + \frac{\partial \bar{\rho}}{\partial t} + u_i\frac{\partial \bar{\rho}}{\partial x_i} + \left(\tau_{ij} - \overline{\rho u_i' u_j'}\right)\frac{\partial \overline{u_i}}{\partial x_i} \tag{1}$$

In steady equilibrium two dimensional heated boundary flows, the transient term, kinetic energy dissipation term and the pressure drop term are neglected. According to the boundary layer assumption, the flow direction scale was far larger than the wall normal direction scale. So the axial gradient of some quantities would be ignored in contrast with the normal direction gradients. The thermal boundary equation was obtained as below:

$$\rho u \frac{\partial h}{\partial x} + \rho v \frac{\partial h}{\partial y} = \frac{\partial}{\partial y}\left(k \frac{\partial t}{\partial y} - \rho \overline{v'h'}\right) \tag{2}$$

The Reynolds heat flux term $\overline{v'h'}$ could be modeled directly with the Boussinesq eddy viscous hypothesis:

$$\rho \overline{v'h'} = \rho c_p \overline{v't'} = -\rho c_p \varepsilon_H \frac{\partial t}{\partial y} \tag{3}$$

Near the boundary layer, the convection term was small according to the Cuette flow assumption.

$$0 = \frac{\partial}{\partial y}\left[\rho c_p \left(\alpha + \varepsilon_H\right) \frac{\partial t}{\partial y}\right] \tag{4}$$

Integrate the above expression from 0 at the wall to distance y away from the wall.

$$q_0 = \left[-\rho c_p \left(\alpha + \varepsilon_H\right) \frac{\partial t}{\partial y}\right] \tag{5}$$

where q_0 stands for the wall heat flux, t is temperature, and ε_H is the eddy diffusivity of heat transfer, α is the molecule diffusivity.

In the molecule dominated zone, the heat flux transported by eddy fluctuation is negligible, so Equation 5 could be simplified as such:

$$q = -\rho c_p \alpha \frac{\partial t}{\partial y} \tag{6}$$

Through some mathematical operations and integration from the wall to the distance y from the wall, the final thermal law of the wall in laminar conduction region is written below.

$$t_1^+ = \int_0^{y^+} \mathrm{Pr}\, dy^+ \tag{7}$$

The non-dimensional temperature was defined as below, in addition with the wall friction temperature $T_{\tau l}$ and average value of ρc_p in laminar viscous region.

$$t_1^+ = -\int_{t_w}^{t} \frac{u_\tau \rho c_p dt}{q} = -\int_{t_w}^{t} \frac{dt}{T_\tau} = \frac{t_w - t}{T_{\tau 1}} \tag{8}$$

$$T_{\tau 1} = \frac{q\left(t_w - t\right)}{-\int_{t_w}^{t} u_\tau \rho c_p dt} = \frac{q}{u_\tau \left\langle \rho c_p \right\rangle_1} \tag{9}$$

$$\left\langle \rho c_p \right\rangle_1 = \frac{\int_{t_w}^{t} u_\tau \rho c_p dt}{t_w - t} \tag{10}$$

In turbulent dominated zone, the laminar conduction can be neglected in comparison with the turbulent transport effects. Equation 5 can be simplified as such:

$$q_0 = -\rho c_p \varepsilon_H \frac{\partial t}{\partial y} \tag{11}$$

According to the Reynolds analogy assumption, there exists a simple similarity between the momentum exchange and the energy exchange:

$$\varepsilon_H = \frac{\varepsilon_M}{\mathrm{Pr}_t} \tag{12}$$

Pr_t is the turbulent Prandtl number, and is in the range of 0.7 and 0.9 based on the measurements. Here Pr_t is assumed to be 0.85.

ε_M is the eddy viscosity which is a kind of a measure of the space size of the turbulent eddies. According to the research in next section, an expression considering the water property variation along the wall normal direction was proposed.

$$\varepsilon_M = \kappa y^+ \left(\frac{\rho_w}{\rho}\right)^{1/2} \tag{13}$$

Assume that the division line between the laminar conduction layer and the turbulent transported layer is at the place of y^+ equal to y_l^+. Combine all the assumptions above, and integrate the equation from the division line, the final expression in turbulent heat transfer region is written below:

$$t^{+} = \frac{\mathrm{Pr}_t}{\kappa}\ln\frac{y^{+}}{y_l^{+}} + t_l^{+} = \frac{\mathrm{Pr}_t}{\kappa}\ln\frac{y^{+}}{y_l^{+}} + \mathrm{Pr}\, y_l^{+} \tag{14}$$

The non-dimensional temperature was defined below with the definition of wall friction temperature:

$$\tau_2^{+} = \int_{y_l^{+}}^{y^{+}} \mathrm{Pr}_t \left(\frac{\rho}{\rho_w}\right)^{1/2} \frac{1}{\kappa y^{+}} dy^{+} \tag{15}$$

$$t_2^{+} = \int_{t_1}^{t} \frac{-\rho c_p u_\tau}{q} dt = -\int_{t_1}^{t} \frac{dt}{T_\tau} = \frac{t_1 - t}{T_{\tau 2}} \tag{16}$$

$$T_{\tau 2} = \frac{q}{u_\tau \left\langle \rho c_p \right\rangle_2} \tag{17}$$

The new thermal function of the wall has generally the same form as those under subcritical conditions, such as Kays et al. (2005), White (1974). However, it is more complicated due to the fact that big variations of fluid property of supercritical water could largely change the actual heat transfer behavior near the wall. It is hard to get a direct explicit algebraic form as the traditional one. The fluid property will govern the heat transfer capability and the temperature profile will react in return.

In subcritical conditions, the fluid property variation is expected to be small, the two wall friction temperature $T_{\tau 1}$ and $T_{\tau 2}$ will have generally the same value. An explicit form of the thermal function of the wall could be obtained:

$$t^{+} = \mathrm{Pr}\, y^{+} \text{ when } y^{+} < y_l^{+} \tag{18}$$

$$t^{+} = \frac{\mathrm{Pr}_t}{\kappa}\ln\frac{y^{+}}{y_l^{+}} + t_l^{+} = \frac{\mathrm{Pr}_t}{\kappa}\ln\frac{y^{+}}{y_l^{+}} + \mathrm{Pr}\, y_l^{+} \text{ when } y^{+} \geq y_l^{+} \tag{19}$$

The Heat Transfer Correlation Derivation

The thermal law of the wall provides an expression of describing the temperature profile through the boundary layer and this will lead to a development of heat transfer correlation after integrated from wall to the centerline of fluid channel, roughly paralleling the friction correlation derivation. Due to the complexity of the thermal law of the wall, some assumptions were made in order to get an algebraic equation.

Assume that the fluid temperature in center line of the pipe satisfies with the turbulent thermal law of the wall and is equal to t_c:

$$t_c = t_1 - T_{\tau 2} \int_{y_1^+}^{r_0^+} \mathrm{Pr}_t \left(\frac{\rho}{\rho_w} \right)^{1/2} \frac{1}{\kappa y^+} dy^+ \tag{20}$$

t_1 is the temperature on the division line between the laminar conduction layer and the turbulent transport layer. It satisfies the thermal law of the wall in laminar conduction layer:

$$t_1 = t_w - T_{\tau 1} \int_0^{y_1^+} \mathrm{Pr}\, dy^+ \tag{21}$$

Combine Equation 20 and Equation 21, then:

$$\frac{t_w - t_c}{T_{\tau 2}} = \frac{T_{\tau 1}}{T_{\tau 2}} \int_0^{y_1^+} \mathrm{Pr}\, dy^+ + \int_{y_1^+}^{r_0^+} \mathrm{Pr}_t \left(\frac{\rho}{\rho_w} \right)^{1/2} \frac{1}{\kappa y^+} dy^+ \tag{22}$$

In circular heated flow, big temperature gradient exists in the wall boundary region but the boundary layer is very thin in comparison with the whole pipe diameter so it contributes little portion to the average fluid temperature. Assume that the centerline temperature is the same as the average flow temperature.

$$\frac{t_w - t_c}{T_{\tau 2}} \approx \frac{t_w - t_m}{T_{\tau 2}} = \frac{t_w - t_m}{\dfrac{q}{\left\langle \rho c_p \right\rangle u_\tau}} = \frac{t_w - t_m}{\dfrac{q}{\left\langle \rho c_p \right\rangle_2 u_b \sqrt{\dfrac{c_f}{2}} \sqrt{\dfrac{\rho_b}{\rho_w}}}} = \frac{\mathrm{Re} \left\langle \mathrm{Pr} \right\rangle_2 \sqrt{\dfrac{c_f}{2}} \left(\dfrac{\rho_b}{\rho_w} \right)^{1/2}}{Nu} \tag{23}$$

$$Nu = \frac{\mathrm{Re} \left\langle \mathrm{Pr} \right\rangle_2 \sqrt{\dfrac{c_f}{2}} \left(\dfrac{\rho_b}{\rho_w} \right)^{1/2}}{\dfrac{T_{\tau 1}}{T_{\tau 2}} \displaystyle\int_0^{y_1^+} \mathrm{Pr}\, dy^+ + \int_{y_1^+}^{r_0^+} \mathrm{Pr}_t \left(\frac{\rho}{\rho_w} \right)^{1/2} \frac{1}{\kappa y^+} dy^+} = \frac{\mathrm{Re} \left\langle \mathrm{Pr} \right\rangle_2 \sqrt{\dfrac{c_f}{2}} \left(\dfrac{\rho_b}{\rho_w} \right)^{1/2}}{\dfrac{T_{\tau 1}}{T_{\tau 2}} \left\langle \mathrm{Pr} \right\rangle_1 y_1^+ + \dfrac{\left\langle \mathrm{Pr}_t \rho^{1/2} \right\rangle_2}{\rho_w^{1/2}} \dfrac{1}{\kappa} \ln \dfrac{r_0^+}{y_1^+}} \tag{24}$$

$\left\langle \cdots \right\rangle_1$ stands for the average operator in the laminar conduction region, and $\left\langle \cdots \right\rangle_2$ stands for the average operator in the turbulent transport region. There are also many unknown parameters in the aforementioned expressions. So some assumptions had to be made in order to get an explicit form of heat transfer correlation. Assume that the wall temperature is the characteristic temperature for laminar conduction layer and the turbulent transport region will cover the range of wall temperature to the average bulk fluid temperature.

$$T_{\tau 1} = \frac{q}{u_\tau \left\langle \rho c_p \right\rangle_1} \approx \frac{q}{u_\tau \rho_w c_{pw}} \tag{25}$$

$$T_{\tau 2} = \frac{q}{u_\tau \left\langle \rho c_p \right\rangle_2} \approx \frac{q}{u_\tau \rho_b c_p} \tag{26}$$

$$\left\langle \mathrm{Pr} \right\rangle_1 \approx \mathrm{Pr}_w = \frac{\rho_w c_{pw} v_w}{k_w} \tag{27}$$

$$\left\langle \mathrm{Pr} \right\rangle_2 \approx \overline{\mathrm{Pr}} = \frac{\rho_b \overline{c_p} v_b}{k_b} \tag{28}$$

$$\left\langle \rho^{1/2} \right\rangle_2 = \rho_b^{1/2} \tag{29}$$

Substitute the expressions above into Equation 24:

$$Nu - \frac{\mathrm{Re}\,\overline{\mathrm{Pr}}\sqrt{\dfrac{c_f}{2}}}{\overline{\mathrm{Pr}}\dfrac{\mu_w}{\mu_b}\dfrac{k_b}{k_w}\left(\dfrac{\rho_b}{\rho_w}\right)^{1/2} y_1^+ + \mathrm{Pr}_t \dfrac{1}{\kappa} \ln \dfrac{r_0^+}{y_1^+}} \tag{30}$$

$$r_0^+ = \frac{u_\tau r_0}{v} = \sqrt{\frac{c_f}{2}}\left(\frac{\rho_b}{\rho_w}\right)^{1/2} \frac{\mathrm{Re}}{2} \tag{31}$$

According to literature Zang et al. (2012), such relationship between the centerline velocity and the distance away from the wall were obtained:

$$u_{c,t}^+ = y_0^+ = \frac{1}{\kappa}\int_{y_0^+}^{y^+} \frac{dy^+}{y^+} = \frac{1}{\kappa}\ln\left(\frac{y^+}{y_0^+}\right) \tag{32}$$

$$u_\tau = \sqrt{\frac{c_f}{2}}\sqrt{\frac{\rho_b}{\rho_w}}u_m \tag{33}$$

$$u_{c,t}^+ = \int\left(\frac{\rho}{\rho_w}\right)^{1/2}\frac{du}{u_\tau} \sim \left\langle\left(\frac{\rho}{\rho_w}\right)^{1/2}\right\rangle_2 \frac{u_c}{u_\tau} \sim \left(\frac{\rho_b}{\rho_w}\right)^{1/2}\frac{u_m}{u_\tau} = \sqrt{\frac{2}{c_f}} \tag{34}$$

So, Equation 30 can be further converted into such form:

$$Nu = \frac{\mathrm{Re}\,\overline{\mathrm{Pr}}\frac{c_f}{2}}{\left(\overline{\mathrm{Pr}}\frac{\mu_w}{\mu_b}\frac{k_b}{k_w}\left(\frac{\rho_b}{\rho_w}\right)^{1/2} y_1^+ - \mathrm{Pr}_t y_0^+ + \frac{\mathrm{Pr}_t}{\kappa}\ln\frac{y_0^+}{y_1^+}\right)\sqrt{\frac{c_f}{2}} + \mathrm{Pr}_t} \tag{35}$$

The velocity in the centerline of the flow channel is assumed to have little difference with the average velocity.

$\overline{\mathrm{Pr}}$ is the average Prandtl number and is defined as below:

$$\overline{Pr} = \frac{\mu_b \overline{c}_p}{k_b} = \frac{\mu_b}{k_b}\frac{\left(h_w - h_b\right)}{\left(T_w - T_b\right)} \tag{36}$$

κ is Karman constant, y_0^+ stands for the laminar momentum layer thickness, and y_l^+ stands for the laminar thermal layer thickness.

In isothermal conditions, the fluid property variation along the wall normal direction doesn't exist. So the heat transfer correlation can be further simplified into such form:

$$Nu = \frac{\mathrm{Re}\,\mathrm{Pr}\frac{c_f}{2}}{\left(\mathrm{Pr}\, y_1^+ - \mathrm{Pr}_t y_0^+ + \frac{\mathrm{Pr}_t}{\kappa}\ln\frac{y_0^+}{y_1^+}\right)\sqrt{\frac{c_f}{2}} + \mathrm{Pr}_t} \tag{37}$$

The Gnielinski correlation which has similar theory background has been verified by many experiments. Its form is like this:

$$Nu = \frac{\left(\mathrm{Re} - 1000\right)\mathrm{Pr}\frac{c_f}{2}}{1 + 12.7\left(\mathrm{Pr}^{2/3} - 1\right)\sqrt{c_f / 2}} \tag{38}$$

with its suitable range: $2300 \le \mathrm{Re} \le 5 \times 10^6$, $0.5 \le \mathrm{Pr} \le 2000$. The friction coefficient is calculated with McAdams equation:

$$f = 0.184\,\mathrm{Re}^{-0.2} \tag{39}$$

The comparison between Equation 37 and Equation 38 could see that two equations have very similar forms and even the same correlation coefficients when $y_0^+ = y_1^+ = 12.7$, $\mathrm{Pr}_t = 1.0$. The main difference is that the Gnielinski correlation has adjusted the Re term to (Re-1000) term, and adds the Pr term with 2/3 exponent.

The supercritical water experiments suggest that the heat transfer coefficient could be represented by the subcritical correlation in the region far away from the pseudocritical point where the fluid property variation is not that strong. In order to extend the subcritical equation to be applied into supercritical conditions and keep the consistent form in theory, the supercritical equation was first decided in such form.

$$Nu = \frac{\mathrm{Re}\,\overline{\mathrm{Pr}}\frac{c_f}{2}}{12.7\left(\frac{\mu_w}{\mu_b}\frac{k_b}{k_w}\left(\frac{\rho_b}{\rho_w}\right)^{1/2}\overline{\mathrm{Pr}}^{2/3} - 1\right)\sqrt{\frac{c_f}{2}} + 1} \tag{40}$$

The main advantage of such correlation is that when the wall fluid property and the bulk fluid property don't have much difference, the equation above could be converted into Gnielinski form. So the correlations in two different regions could be united together.

Compared with Heat Transfer Experimental Data

The heat transfer correlation of supercritical water has two distinct differences in comparison with that of subcritical water. First, subcritical correlation always imports the fluid Prandtl number as the characteristic quantity, while the average Prandtl number was frequently applied in supercritical water. Many experiments have proved this conclusion that the introduction of the average Prandtl number would largely improve the data fitting accuracy. The derivation process also supports this conclusion. It is the average Prandtl number across the boundary layer that dominates the heat transfer ability in supercritical water. Second, the supercritical heat transfer correlation has to take account of the fluid property variation factor besides the traditional subcritical form. These fluid property factors include the specified heat transfer capacity, dynamic viscosity, density, thermal conductivity, et al. The new heat transfer correlation reflects the impact of those property factors.

The new heat transfer correlation was compared with NPIC (Nuclear Power Institute of China) supercritical water experiments with 3-meter headed length, 6mm diameter. The operating matrix is below: mass flux 602~1231 kg/m²/s, heat flux 355~882kW/m², operating pressure 23MPa~25MPa. Through the comparison with experiments, it is found that the thermal conductivity, dynamic viscosity and density ratio in the first term do not take any parts in the correlation if a proper frictional coefficient is selected. So the final supercritical form is this:

$$Nu = \frac{\mathrm{Re}\,\overline{\mathrm{Pr}}\,\frac{c_f}{2}}{12.7\left(\overline{\mathrm{Pr}}^{2/3} - 1\right)\sqrt{\frac{c_f}{2}} + 1} \tag{41}$$

The Popov (1978) correlation is adopted to calculate the frictional coefficient.

$$c_f = \frac{1}{4}\frac{1}{\left(1.82\log_{10}\mathrm{Re}_b - 1.64\right)^2}\left(\frac{\rho_w}{\rho_b}\right)^{0.4}\left(\frac{\mu_w}{\mu_b}\right)^{0.2} \tag{42}$$

It has to pointed that Petukhov and Kirillov (1958) also proposed a correlation with the same as Equation 41, but using the frictional correlation without material factor correction:

$$c_f = \frac{1}{4}\frac{1}{\left(1.82\log_{10}\mathrm{Re}_b - 1.64\right)^2} \tag{43}$$

Figure 5 shows the different performances of the two correlations the original Gninelinski correlation (Equation 38) and the modified Gninelinski correlation (Equation 41), through comparing them with the experimental data. The software REFPROP developed by NIST (2007) was used to calculated the thermophysical properties of water. For the original Gninelinski correlation, the prediction accuracy is poor. Most of the predicted values stay beyond the 25% percent range of the experimental data. However, for the modified correlation, 92.5% data points lie within ±25% of the experimental data and 96.3% data points lie within ±30% of the experimental data. The new heat transfer correlation was not adjusted with experimental data but shows satisfactory results which proves the physical validity of this method.

According to the derivation process above, the Dittus-Boelter equation could also be extend to supercritical conditions. First, turn the bulk fluid Prandtl number into average Prandtl number; second, add the fluid property variation factor to account for the differences between the wall temperature and bulk fluid temperature.

The normal Dittus-Bolter equation:

$$Nu = 0.023\,\mathrm{Re}^{0.8}\,\mathrm{Pr}^{0.4} \tag{44}$$

The modified form of Dittus-Bolter equation:

Figure 5. Comparisons of new heat transfer correlation with experimental data

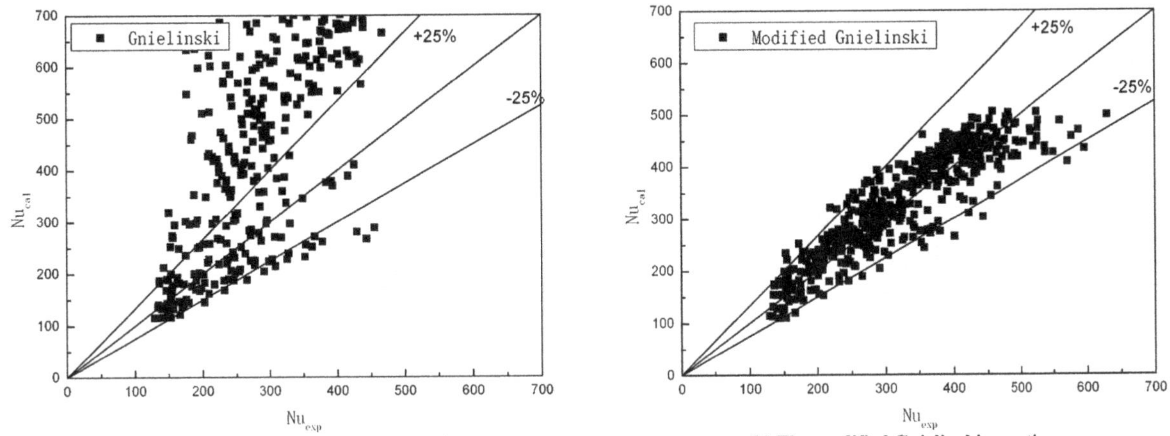

(a) The normal Gnielinski equation **(b) The modified Gnielinski equation**

$$Nu = 0.023\,\mathrm{Re}^{0.8}\,\overline{\mathrm{Pr}}^{0.4}\left(\frac{\rho_w}{\rho_b}\right)^{n} \quad (45)$$

After fitted with the experimental data, the exponent n is equal to 0.35. So the modified Dittus-Bolter equation is this:

$$Nu = 0.023\,\mathrm{Re}^{0.8}\,\overline{\mathrm{Pr}}^{0.4}\left(\frac{\rho_w}{\rho_b}\right)^{0.35} \quad (46)$$

Figure 6 gives the comparison of experimental data with those calculated by the normal Dittus-Bolter equation and modified Dittus-Bolter equation. The normal Dittus-Bolter equation over predicts the experimental data seriously especially when Nusselt number gets increased. Only 20.5% of predicted values lie within 25% of experimental data, 24.4% of predicted values lie within 30% of experimental data. After modified, its prediction accuracy is largely improved. About 90.7% of predicted values lie within 25% of experimental data, 96.1% of predicted values lie within 30% of experimental data. This method provides some guidelines to develop new heat transfer correlation under supercritical conditions and could build a bridge between subcritical region and supercritical region.

Many researchers have also proposed their heat transfer correlations based on their experimental data.

Bishop et al.(1964) equation:

$$Nu_x = 0.0069\,\mathrm{Re}_x^{0.9}\,\overline{\mathrm{Pr}}_x^{0.66}\left(\frac{\rho_w}{\rho_b}\right)^{0.43}\left(1 + 2.4\frac{D}{x}\right) \quad (47)$$

The application range is listed below. Pressure:22.8-27.6 MPa, Bulk fluid temperature: 282-527°C, mass flow rate 651-3662kg/m²/s, the heat flux: 0.31-3.46 MW/m².

Figure 6. Comparisons of new heat transfer correlation with experimental data

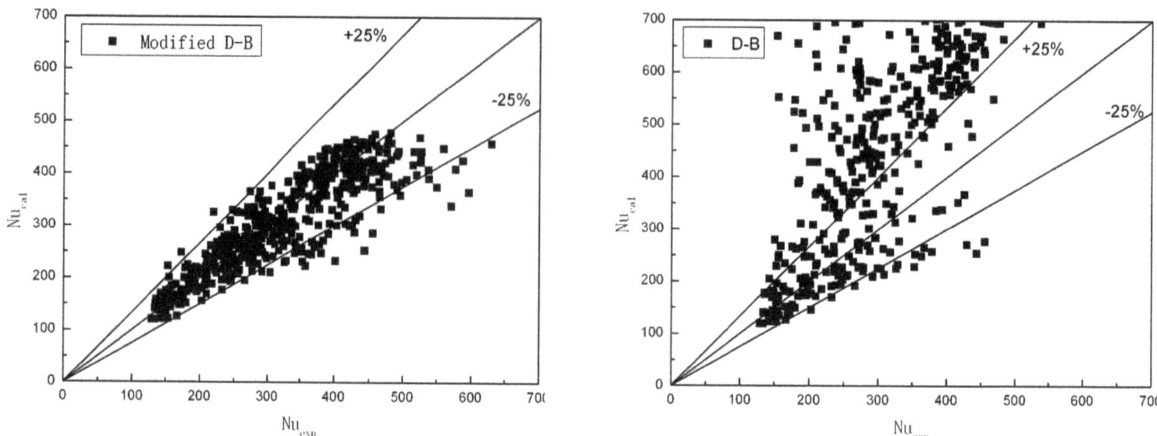

(a) The normal Dittus-Bolter equation **(b) The modified Dittus-Bolter equation**

Jackson (2002) equation:

$$Nu = 0.0183\,\mathrm{Re}_b^{0.82}\,\mathrm{Pr}_b^{0.5}\left(\frac{\rho_w}{\rho_v}\right)^{0.3}\left(\frac{\overline{c}_p}{c_{pb}}\right)^{n} \tag{48}$$

The exponent n is a piecewise function in different temperature range:

$$n = 0.4 \qquad T_b < T_w < T_{pc} \; and \; 1.2T_{pc} < T_b < T_w$$

$$n = 0.4 + 0.2\left(\frac{T_w}{T_{pc}} - 1\right) \qquad T_b < T_{pc} < T_w$$

$$n = 0.4 + 0.2\left(\frac{T_w}{T_{pc}} - 1\right)\left[1 - 5\left(\frac{T_b}{T_{pc}} - 1\right)\right] \qquad T_{pc} < T_b < 1.2T_{pc} \; and \; T_b < T_w$$

Swenson equation (see Swenson et al.,1965):

$$\frac{hD}{\kappa_w} = 0.00459\left(\frac{DG}{\mu_w}\right)^{0.923}\left(\frac{H_w - H_b}{T_w - T_b}\frac{\mu_w}{\kappa_w}\right)^{0.613}\left(\frac{\rho_w}{\rho_b}\right)^{0.23} \tag{49}$$

With the following range of flow and operating parameters:P=22.8-41.4MPa, G=542- 2150kg/m^2/s, Tb=75-576°C, Tw=93-649°C, D=9.42mm.

Mokry (see Mokry et al., 2011; Mokry et al., 2010) prosposed such equation based on the experimental dataset obtained in supercritical water flowing upward in a 4-m long vertical bare tube with 10-mm inner diameter.

$$Nu_b = 0.0061\,\mathrm{Re}_b^{0.904}\,\overline{\mathrm{Pr}}_b^{-0.684}\left(\frac{\rho_w}{\rho_b}\right)^{0.564} \tag{50}$$

With the following range of operating parameters: P≈24MPa, Tin=320-350°C, Tout=380-406°C, G=200-1500kg/m²/s, q=70-1250kW/m², D=10mm.

Figure 7 shows the comparisons of experimental Nusselt numbers with kinds of correlations. The Bishop Equation lies closely with the experimental data except that some data points are a bit larger; The Jackson equation over predicts the Nusselt numbers while the Swenson and Mokry equation under predicts the Nusselt numbers. However, the Mokry equation shows compact distribution of the predicted values.

Figure 7. Comparisons of different heat transfer correlations with experimental data

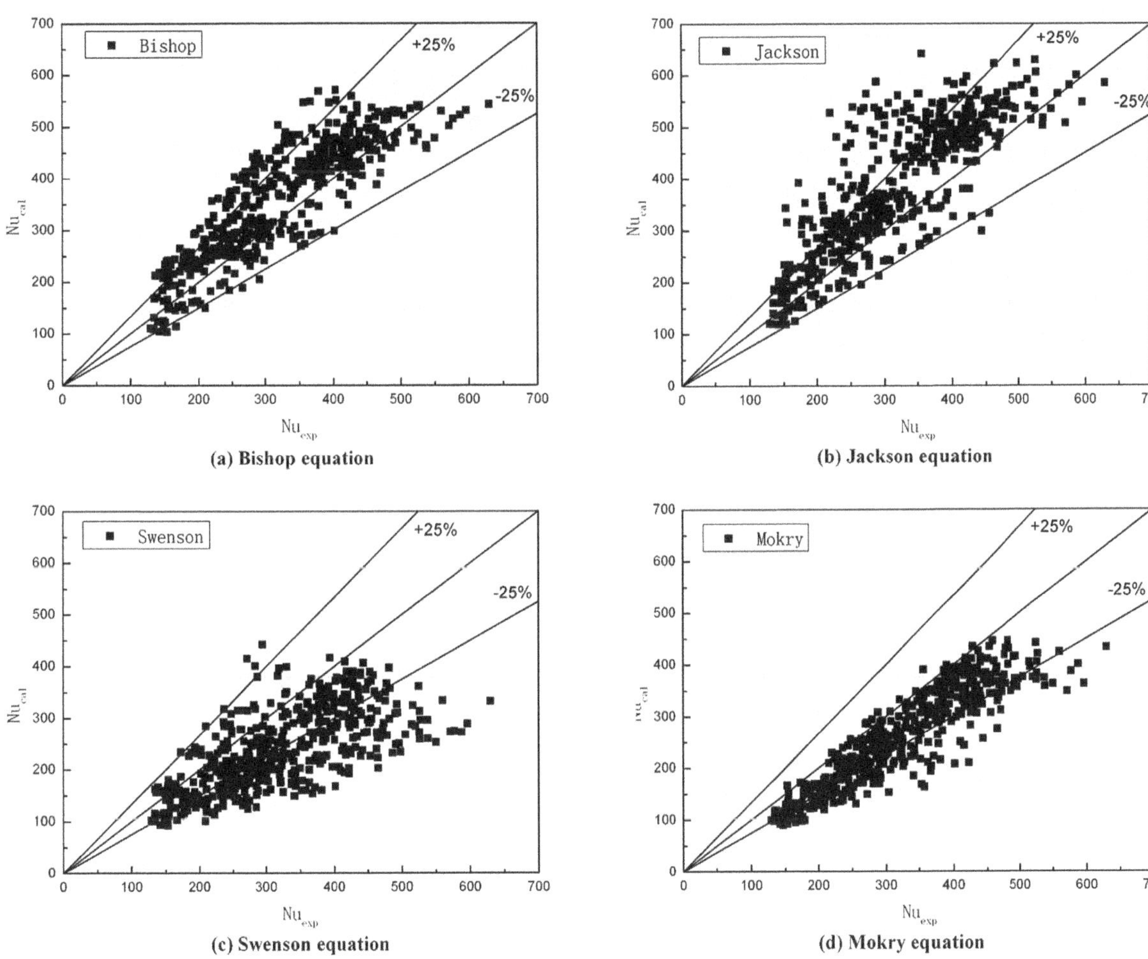

THE PRESSURE LOSS CHARACTERISTICS OF SUPERCRITICAL WATER

Pressure Loss Introduction

Many studies have been carried out in the past on turbulent flow in different geometry shapes under subcritical conditions. Hartnett (1962) developed a kind of analytical prediction of friction factor for non-circular geometry by integration the law of the wall over the cross section. Rehme (1973) proposed a simple method for the prediction of friction factors in non-circular channels if only the geometry factor of the pressure drop relationship for laminar flow is known. The proposed method was tested with experimental results with respect to non-circular channels such as triangular shaped ducts, eccentric annuli, rod bundles and provided a good description of the experimental data. Lee (1995) applied the similar methods to an infinite rod array with a low pitch to diameter by taking account of the variation of local wall shear stress. Su (2002) followed the same way, but presented the friction factor equation in a different form which is similar to that in a circular pipe. They further extended the methods to rough surface and prediction of Nusselt number in turbulent convection flow. In Su (2002)'s work, the one layer wall function theory was adopted. It shows good agreement with Cheng & Todreas method in high Reynolds number. But as Reynolds is getting small, the discrepancy is increased. The reason is believed to be the lack of considering viscous effect in the viscous layer. Bae (2011) et al derived the turbulent geometry parameters for a rod bundle in consideration with the influences of the channel wall and the local shear stress distribution. For a typical geometry of a rod bundle, each type of subchannel could be further subdivided into basic elements by the zero stress line. So the turbulent geometry parameters and the friction factors for a rod bundle could be obtained by combining the geometry parameters of these subchannels.

Compared with the majority of studies under subcritical conditions, relatively few studies are devoted to applying wall function method to investigating hydraulic resistance of supercritical water. Most studies have been performed by means of experiments in circular tube (Pioro (2004)) and obtained experimental correlations by data fitting. The experimental results show that the friction factors of supercritical water were largely dependent on the thermal physical properties.

In this chapter, a method based on two-layer wall function theory instead of single layer was given. Moreover, this new method could be applied to supercritical water and had good prediction accuracy with the experiments. The basic idea is to adopt the two-layer wall function law with Van Direst transformation that could account for the fluid property variation along the wall normal direction. After integrated through the cross section, the average fluid velocity is obtained and also the friction coefficient.

The Velocity Law of the Wall for Varied Property Fluid

In order to extend the new frictional correlation to supercritical water, a modified wall function based on Van Direst transformation was adopted. It has the similar form as the traditional wall function, but with different definitions of dimensionless velocity. The expression is below:

$$\begin{cases} u_1^+ = y^+, & y^+ \le y_l^+ \\ u_2^+ = \dfrac{1}{\kappa}\ln y^+ + b, & y^+ > y_l^+ \end{cases} \tag{51}$$

Here κ is Karman constant and generally to be 0.41. b is constant and always to be 5.0. The subscript number 1 stands for the viscous region, while subscript number 2 stands for the turbulent core region. y_l^+ is the division line between the laminar viscous layer and the turbulent core region and according to W.M.Kays (2005), the value of 10.8 is selected. If the reader is interested in the detailed process of this new wall function, please refer to Fu (2010) or Zang (2012). The two dimensionless velocities are defined below.

$$u_1^+ = \int_0^u \frac{\mu}{\mu_w}\frac{du}{u_\tau} = \frac{u}{u_{\tau 1}} = \frac{u}{\phi_l u_\tau}, u_2^+ = \int_{u_0}^u \left(\frac{\rho}{\rho_w}\right)^{1/2} \frac{du}{u_\tau} = \frac{u}{u_{\tau 2}} = \frac{u}{\phi_t u_\tau} \tag{52}$$

$$u_{\tau 1} = \phi_l u_\tau, u_{\tau 2} = \phi_t u_\tau \tag{53}$$

$$\phi_l = \frac{u}{\int_0^u \frac{\mu}{\mu_w} du} = \frac{\mu_w}{\mu_{bl}}, \phi_t = \frac{u}{\int_{u_0}^u \left(\frac{\rho}{\rho_w}\right)^{1/2} du} = \left(\frac{\rho_w}{\rho_{bt}}\right)^{1/2} \tag{54}$$

The new dimensionless velocity is more complicated to account for the fluid property variation through the boundary layer. In order to be consistent with the traditional wall function form, the fluid property factor is attributed to the frictional velocity with ϕ_l and ϕ_t. These two factors stand for the average viscous and density value in laminar and turbulent region, respectively. The μ_{bl} is the average viscous value in the boundary laminar region and ρ_{bt} is the average density value in the boundary turbulent region. They are both calculated with the velocity as the weight factor. Based on the wall function, the velocity profile could be summarized below:

$$u = \begin{cases} u_{\tau 1} y^+, & y^+ \le y_l^+ \\ u_{\tau 2}\left(\frac{1}{\kappa}\ln y^+ + b\right), & y^+ > y_l^+ \end{cases} \tag{55}$$

As for the special characteristics of supercritical water, two material factors were proposed up to account for the variation of fluid property across the laminar viscous layer and turbulent layer.

$$\phi_l = \frac{u}{\int_0^u \frac{\mu}{\mu_w} du} = \frac{\mu_w}{\mu_{bl}}, \phi_t = \frac{u}{\int_{u_0}^u \left(\frac{\rho}{\rho_w}\right)^{1/2} du} = \left(\frac{\rho_w}{\rho_{bt}}\right)^{1/2} \tag{56}$$

Here μ_{bl} and ρ_{bt} stands for the average dynamic viscosity in the laminar viscous layer and average density in the turbulent layer, respectively. It is assumed that the characteristic temperature in the laminar viscous layer is the wall temperature and the turbulent region spans from the wall temperature at the upper limit of viscous layer to the bulk fluid temperature. So the average density in this region should be integrated from the wall temperature to the bulk fluid temperature. The three-order Simpson algorithm is applied. So:

$$\mu_{bl} = \mu_w \tag{57}$$

$$\rho_{bt} = \rho\left(T_w, T_b\right) = \frac{1}{6}\left(\rho\left(T_b\right) + \rho\left(T_w\right) + 4 \times \rho\left(T_{mid}\right)\right) \tag{58}$$

Here T_{mid} is the temperature at the average enthalpy of bulk fluid temperature and wall temperature. Based on such substitution, it is found that the material factor in the laminar region is equal to one. This is out of general expectation. It is tended to believe that the friction coefficient is related to the viscosity. Since the temperature would have impact on the viscosity, the variation of viscosity would naturally influence the friction coefficient. However, such effect is cancelled by this assumption. The key fluid property factor for the friction coefficient is the fluid density.

$$\phi_l = \frac{\mu_w}{\mu_{bl}} = 1 \tag{59}$$

$$\phi_t = \left(\frac{\rho_w}{\rho_{bt}}\right)^{1/2} \tag{60}$$

We have to mention that some assumptions were made to achieve this conclusion. It is not always the same case to eliminate the impact of viscosity. For example, if we assume that the laminar viscous region spans from the wall temperature to the bulk fluid temperature and the turbulent region is characterized only by bulk fluid temperature, both the viscosity and density would take effects in such case. This is also our initial opinion at early stage (Zang (2012)). In such case, it predicts smaller friction coefficient than the experimental data. So we have to adjust our assumptions to fit the data.

The Frictional Correlation Derivation

The regular flow channel could be divided into a combination of the elementary units according to its symmetric line. The elementary unit can be further decomposed of simple analogous-annular channel at the zero shear stress line. For example, the cir-annular can be decomposed of two parts: the inner cir-annular and outer cir-annular. The square-annular have eighth symmetry and its elementary unit can also be decomposed of inner and outer parts. For the typical geometry of an infinite rod bundle of square array, the basic subchannel is eighth symmetric. There is no need of further decomposition for this geometry. The three typical geometries are shown in Figure 8.

Figure 8. Schematic diagram of a typical channel and the corresponding subchannels: (a) circle annular; (b) square annular; (c) rod bundle of square array

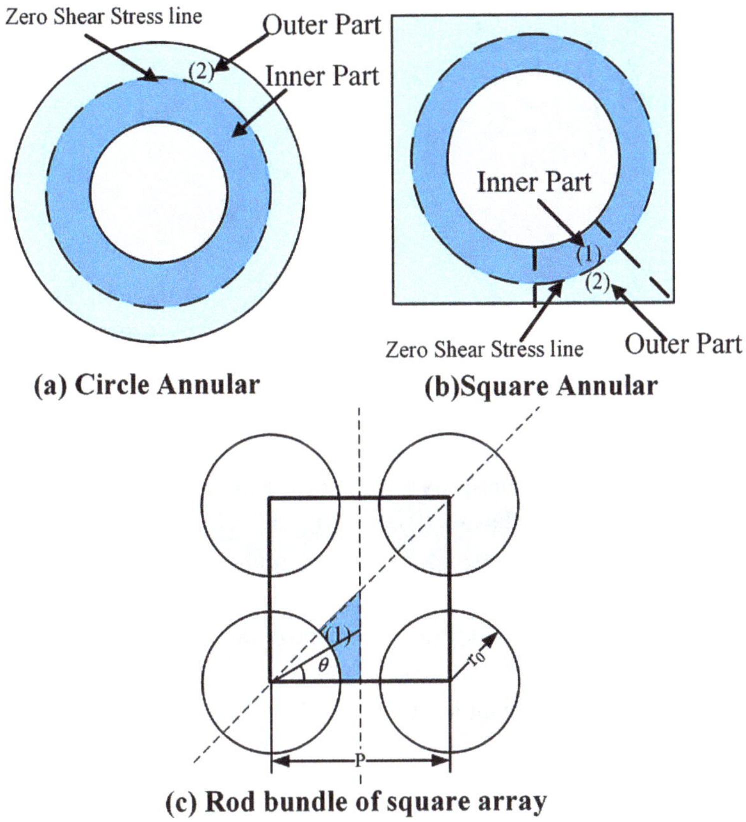

Each of these elementary shapes can be considered as analogous-annular with inner radius $r_1(\theta)$ and outer radius $r_2(\theta)$. The integration of physical function *f* can be directly written in a general form:

$$\int_0^{\theta_{\max}} \int_{r_1}^{r_2} f(r) \times r dr d\theta = \int_0^{\theta_{\max}} \int_{y_1}^{y_2} f(y) \times r(y) \frac{dr}{dy} dy d\theta \tag{61}$$

Here *y* is the distance to the nearest wall taken from the periphery of the rod outward in radial direction; *r* is the distance to the circle center. As the wall function is always described by an expression of the wall distance, we make a substitution of r with r(y) and dr with dr/dy*dy.

For the inner part (Figure 9a), the inner radius is $r_0(\Theta)$, the relationship between r and y is:

$$r(y) = r_0(\theta) + y \tag{62}$$

For the outer part (Figure 9b), the outer radius is $r_0(\Theta)$, the relationship between r and y is:

$$r(y) = r_0(\theta) - y \tag{63}$$

Figure 9. Schematic diagram of inward integration and outward integration: (a) inward; (b) outward

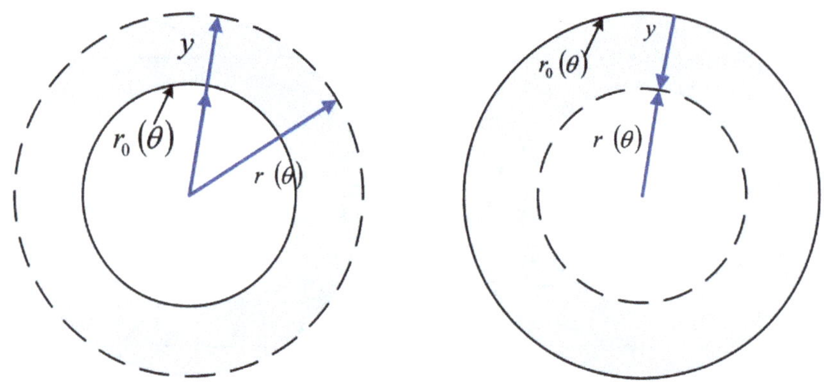

The sign of plus or minus in the expressions stands for the inner part or the outer part. For example, for the circular tube with radius r_0, we have $r_1(\theta)=0$, $r_2(\theta)=r_0$, $r(y)=r_0-y$;For the inner part of circular-annulus with inner radius r_0, we have $r_1(\theta)=r_0$, $r(y)=r_0+y$., and $r_2(\theta)$ is the zero shear stress line at approximately middle position; For a rod bundle in a square array with pitch *P* and rod radius r_0, we have $r_1(\theta)=r_0$,$r(y)=r_0+y$,$r_2(\theta)=P/2/\cos(\theta)$.θmax defines the shape of the rod bundle. It is equal to π/4 for square array and π/6 for hexagonal array.

For non-circumferential symmetry channel, the wall shear stress varies along the wall periphery at different angles. It is possible to write the frictional velocity in such form:

$$u_\tau = \overline{u}_\tau F\left(\theta\right) \tag{64}$$

Here the function F(θ) describes the variation profile of wall shear stress. Several functions have been proposed up by some researchers. This function depends on the geometry shape and detailed geometry parameters. In this chapter, the F(θ) function refers to Bae and Park (2011)'s work.

So for arbitrary analogous annular channel, we obtain the mean flow velocity by integration the two-layer wall function over the channel area A.

$$\begin{aligned} u_m &= \frac{1}{A}\int_0^{\theta_{\max}}\int_0^{p_0} u\left(r_0 \pm y\right)dyd\theta = \frac{1}{A}\frac{\nu^2}{\overline{u}_\tau^2}\int_0^{\theta_{\max}}\int_0^{p_0^+} u\left(r_0^+ \pm y^+\right)dy^+d\theta \\ &= u_{\tau 1}\frac{1}{A}\frac{\nu^2}{\overline{u}_\tau^2}\int_0^{\theta_{\max}}\int_0^{y_l^+} y^+F^2\left(\theta\right)\left(r_0^+ \pm y^+\right)dy^+d\theta \\ &+ u_{\tau 2}\frac{1}{A}\frac{\nu^2}{\overline{u}_\tau^2}\int_0^{\theta_{\max}}\int_{y_l^+}^{p_0^+}\left[\frac{1}{\kappa}\ln\left[y^+F\left(\theta\right)\right]+b\right]\left(r_0^+ \pm y^+\right)F\left(\theta\right)dy^+d\theta \\ &= u_{\tau 1}\alpha_1 + u_{\tau 2}\alpha_2 \end{aligned} \tag{65}$$

Here α_1 and α_2 are dimensionless velocity profile parameters.

$$A = \frac{\nu^2}{\overline{u}_\tau^2} \int_0^{\theta_{\max}} \int_0^{p_0^+} \left(r_0^+ \pm y^+\right) dy^+ d\theta \tag{66}$$

$$\begin{aligned} \alpha_1 &= \frac{1}{A} \frac{\nu^2}{\overline{u}_\tau^2} \int_0^{\theta_{\max}} \int_0^{y_l^+} y^+ F^2\left(\theta\right)\left(r_0^+ \pm y^+\right) dy^+ d\theta \\ \alpha_2 &= \frac{1}{A} \frac{\nu^2}{\overline{u}_\tau^2} \int_0^{\theta_{\max}} \int_{y_l^+}^{p_0^+} \left[\frac{1}{\kappa} \ln\left[y^+ F\left(\theta\right)\right] + b\right]\left(r_0^+ \pm y^+\right) F\left(\theta\right) dy^+ d\theta \end{aligned} \tag{67}$$

According to the definition of local friction coefficient, the average frictional velocity and average flow velocity can be related:

$$u_\tau = \sqrt{\frac{c_f}{2}} \sqrt{\frac{\rho_b}{\rho_w}} u_m = \sqrt{\frac{c_f}{2}} \sqrt{\frac{\rho_b}{\rho_w}} \left(u_{\tau 1}\alpha_1 + u_{\tau 2}\alpha_2\right) \tag{68}$$

So the final calculation expression for the local friction coefficient is:

$$\sqrt{\frac{2}{c_f}} = \sqrt{\frac{\rho_b}{\rho_w}} \left(\phi_l \alpha_1 + \phi_t \alpha_2\right), c_f = \frac{2}{\frac{\rho_b}{\rho_w}\left(\phi_l \alpha_1 + \phi_t \alpha_2\right)^2} \tag{69}$$

For complex geometry shape and velocity profile, it is not easy to get an explicit algebraic form, so we just keep the original expression without further substitution. The numerical method is used for practical applications.

The expression above actually constitutes a general method of obtaining friction coefficient based on the wall function. If a more general wall function is given that could account for other factors, it is also feasible to derive a more general frictional correlation.

Comparison with Pressure Loss Experimental Data

The experimental data was obtained in circular pipe at supercritical conditions in Nuclear Power Institute of China (NPIC). Figure 10 shows one of the experimental data to explain the special characteristics of friction coefficient under supercritical conditions. From the Figureure, it is found that the friction coefficient c_f behaves like 'V' shape along the increase of the fluid enthalpy. c_f falls down to its minimum value at the position with fluid enthalpy approximately equal to 2000kJ/kg. The 'V' shape profile is attributed to the variation of fluid property across the boundary layer to the bulk fluid region. Figure 11 presents the variation curve of fluid temperature, wall temperature and their responding densities along the flow direction, in correspondence with Figure 10. The fluid temperature increases slowly and has a nearly flat region near the pseudocritical point. Accordingly, the fluid density drops down gradually. In contrast, the wall temperature rises quickly and crosses the pseudocritical point near the inlet. So the

Figure 10. Friction coefficient profile along the pipe

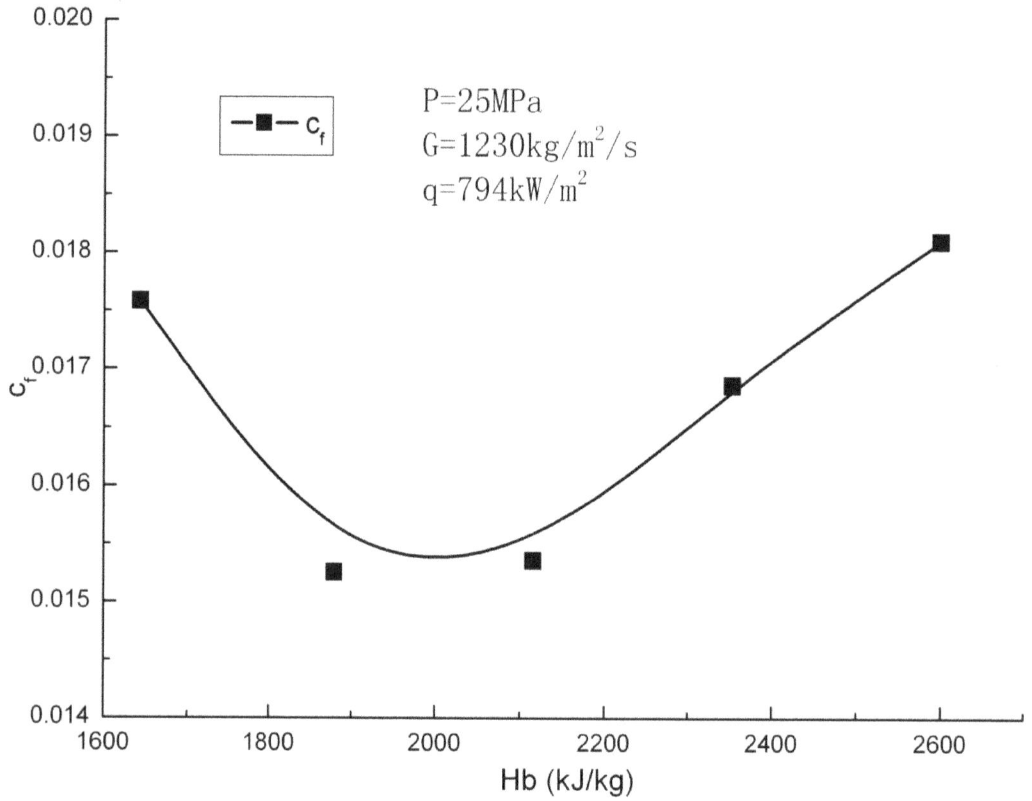

density near the wall falls down to its valley and remains at an even stage. The boundary layer takes main part in determining the momentum transfer. The variation of fluid property in this region would have direct impact on the friction coefficient.

Figure 12 presents the variations of the material factors: density ratio and viscosity ratio along the flow direction, also with the friction coefficient. From analysis above, the key fluid property is the density and viscosity seems to have little effect. But for comparison, we presented them both. From

Figure 12, it is found that the density ratio and viscosity ratio resemble the similar 'V' shape as c_f curve. The density ratio profile is closer to that of c_f curve which could explain one aspect of the reason that density ratio is important and cf resembles 'V' shape.

A more comprehensive comparison of Equation 69 with experimental data was made based on the experiments performed in NPIC. The operation parameters of the experimental data are listed here: mass flux 602~1231 kg/m²/s, heat flux 355~882kW/m², operating pressure 23MPa~25MPa. The comparison result is shown in Figure 13. About 82.7% of predicted values lie within the error of ±20%. About 92.6% of predicted values lie within ±25% and 97.5% within ±30%.

The comparisons of experimental data with other correlations (Pioro, 2007) were also performed:

Mikheev (1956):

Figure 11. The profile of wall temperature, fluid temperature and their corresponding densities

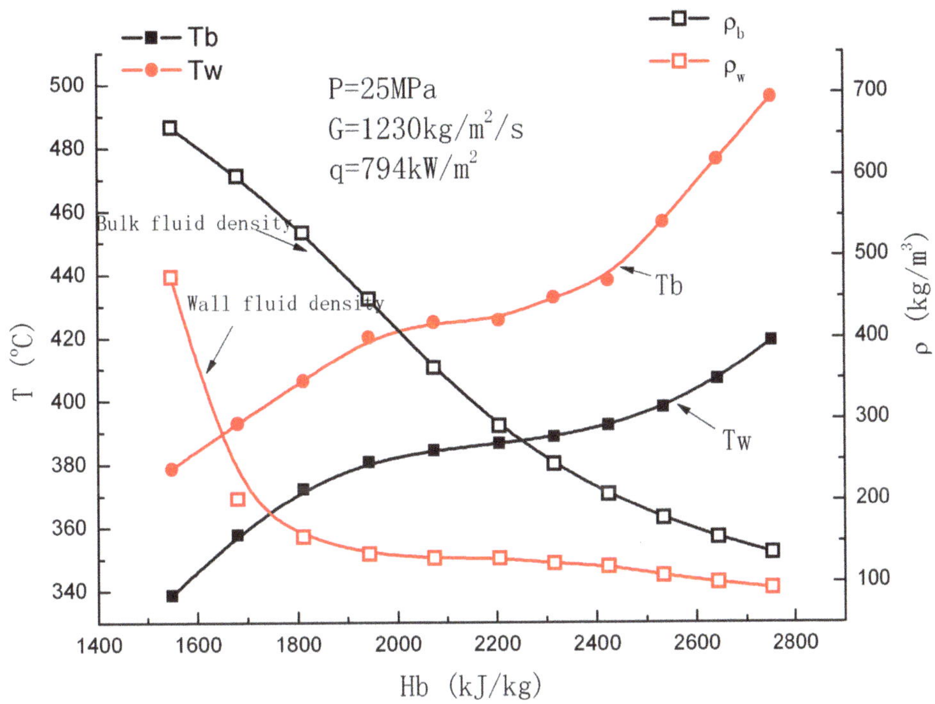

Figure 12. The Friction coefficients and the material factors

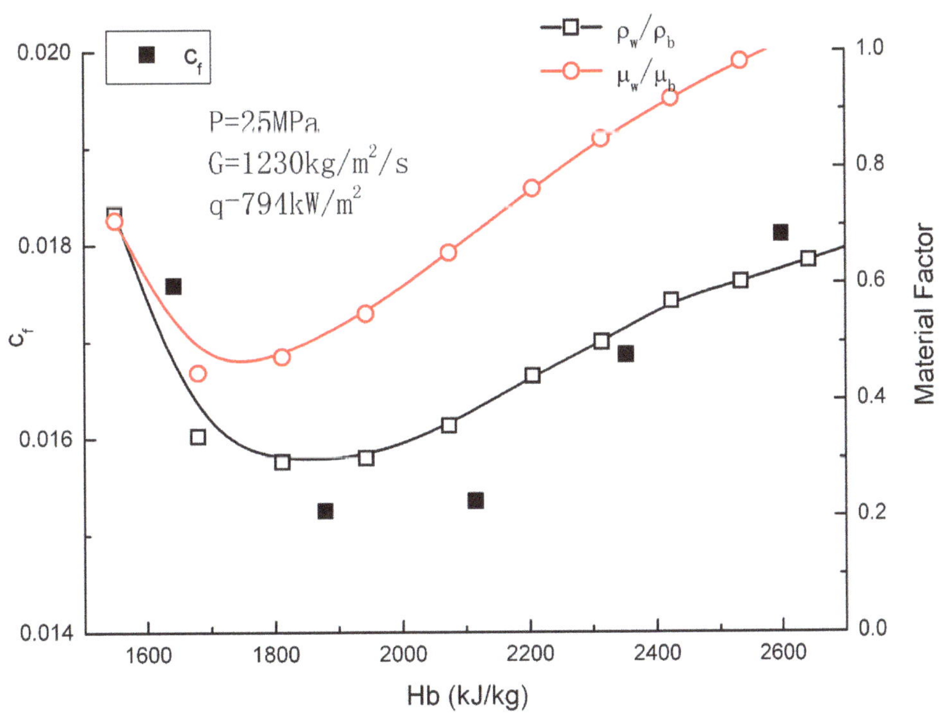

Figure 13. The comparison with experimental data of NPIC

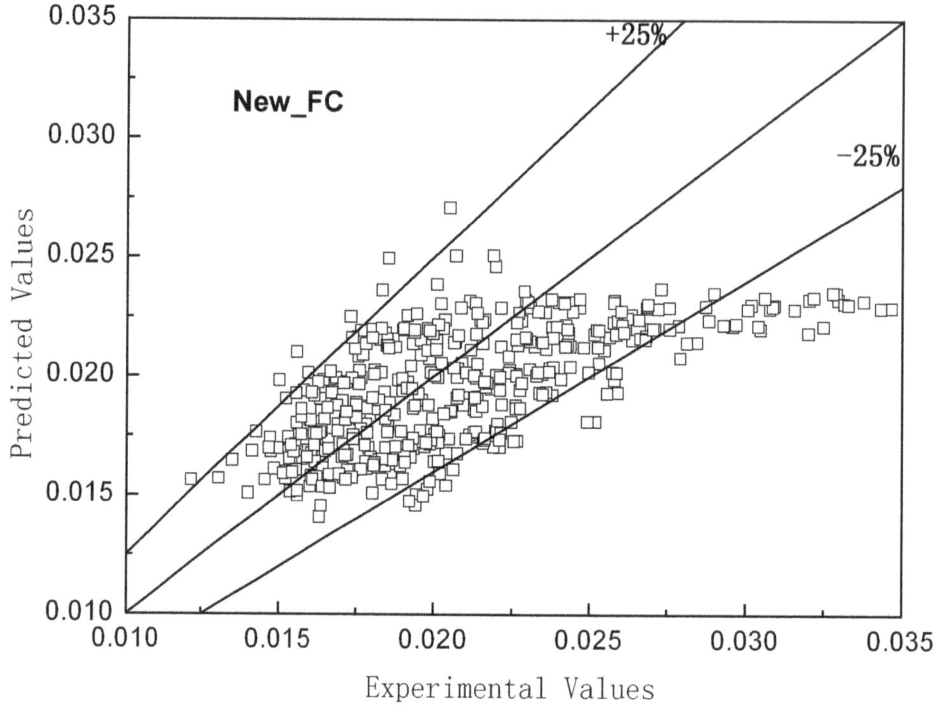

$$f = f_{iso}\left(\frac{\mathrm{Pr}_w}{\mathrm{Pr}_b}\right)^{\frac{1}{3}} \tag{70}$$

Popov (1978):

$$f = f_{iso}\left(\frac{\rho_w}{\rho_b}\right)^{0.4}\left(\frac{\mu_w}{\mu_b}\right)^{0.2} \tag{71}$$

Tarasova and Leont'ev:

$$f = f_{iso}\left(\frac{\mu_w}{\mu_b}\right)^{0.22} \tag{72}$$

Kirillov (1990):

$$f = f_{iso}\left(\frac{\rho_w}{\rho_b}\right)^{0.4} \tag{73}$$

The f_{iso} is the isothermal friction coefficient calculated by the Colebrook correlation.

Figure 14 presents the comparison of various correlations with experimental data. The Colebrook correlation is developed for isothermal flow. It doesn't account for the damping effect of material factor, so it predicts larger value than the experiments. Tarasova et al adopted the dynamic viscosity as the correction factor and the improvement is very limited. Mikheev used the Prandtl number to correct the

Figure 14. The comparison of different correlations with experimental data: (a) Colebrook correlation; (b) Tarasova correlation; (c) Mikheev Correlation; (d) Kirillov correlation; (e) Popov correlation

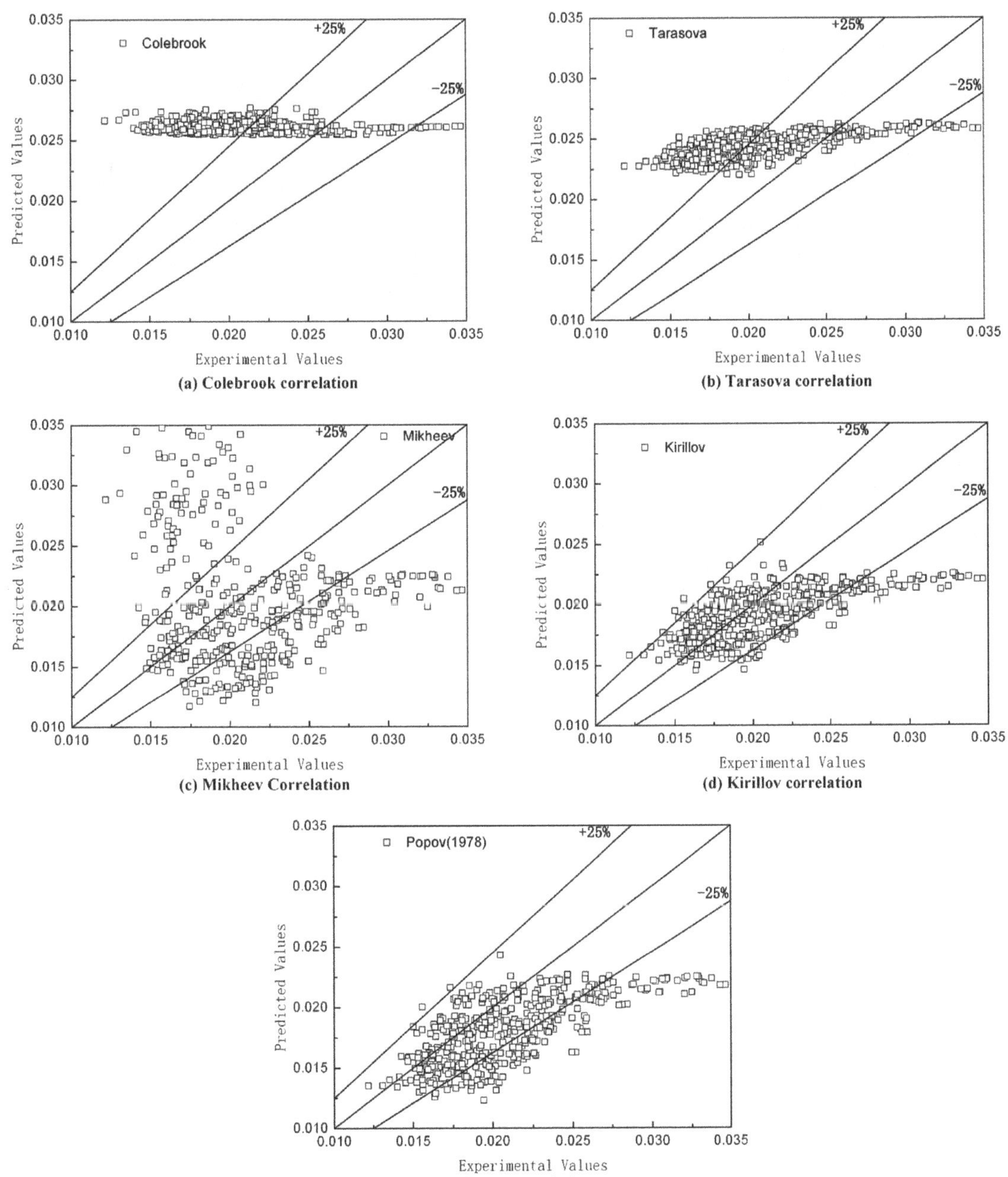

friction coefficient. From the analysis above, the Prandtl number is not the key parameter, so the prediction scatters a lot. Both Kirillov and Popov correlations adopted the density ratio as the correction factor and they showed good agreement with experimental data.

THE FLOW STABILITY ISSUES OF SUPERCRITICAL WATER IN PARALLEL CHANNELS

Instability Introduction

The reference SCWR design is a direct cycle system operating at 25.0 MPa with core inlet and outlet temperature of 280 and 500 °C. The coolant density hence decreases from about 780 kg/m^3 at the core inlet to about 90 kg/m^3 at the core outlet, which raises doubts about the possible occurrences of flow instabilities similar to those observed in BWR, such as Ledinegg instability, density wave oscillations and coupled neutronic thermal-hydraulic instabilities, as discussed in the Generation IV roadmap. The instabilities are highly undesirable in nuclear reactor systems, as they may result in failure of control mechanism, lead to fatigue damage of reactor components and affect the heat transfer characteristics. As the licensing of SCWR will probably require, at a minimum, demonstration of the ability to predict the onset of instabilities, it is necessary to understand the instability phenomena in SCWR, to identify the important variables affecting these phenomena, and ultimately to generate the maps identifying the stable operating conditions of different SCWR designs (Ji et al., 2001).

Studies on flow instabilities of boiling channels have been carried out for decades aimed at the investigation of basic phenomena and modeling techniques (Ishii and Zuber, 1970; Boure et al., 1973; Kakac and Bon, 2008), making the phenomena of two-phase flow instabilities rather clearly understood matters. Despite the fact that coolant in SCWR core will experience remarkable density changes similar to that in BWR, essential differences exist between these two systems. In the former case the rather sudden change in fluid density occurs in a single phase fluid, while in the latter interfaces are formed, with the appearance of an additional phase owing to boiling. As there is still no affirmative understanding of the phase change effect on the dynamic behaviour, it is reasonable to doubt the occurrence of dynamic instabilities as well as the mechanism driving the possible dynamic behaviour in the supercritical system. Moreover, hydraulic resistance of supercritical fluids is believed to be highly influenced by the significant thermophysical properties variations and high heat flux (Pioro et al., 2004), thus making investigations on the static or dynamic instabilities more essential.

Theoretical investigations of flow stability of supercritical fluids go back to the work of Zuber (1966). Straight pipe flow in a once through system was considered and conclusions were drawn that supercritical flow instability would be similar to two-phase flow instability, possessing both the excursive and oscillatory nature. Various types of geometrical configurations with supercritical fluids have been studied in recent years.

Chatoorgoon (2001) investigated the stability of supercritical water flow in a single channel under natural circulation conditions using a non-linear numerical code, and later examined the derived non-dimensional parameters governing supercritical flow instability boundary through a numerical experiment comprising 94 simulated cases (Chatoorgoon et al., 2005). Yu et al. (2011) studied the Ledinegg flow instability in natural circulation at supercritical pressure using a numerical analysis code verified by experimental results, and reached the conclusion that there would be no Ledinegg instabilities occurring at supercritical pressure in the loop.

Linear stability analysis in the frequency domain was carried out by Yi et al. (2004) aimed at the system stability of SCWR. One dimensional conservation equations were numerically discretized and linearised, and thermal-hydraulic stability was analysed using decay ratio. Parametric studies showed that the thermal-hydraulic stability may be improved by increasing orifice pressure drop coefficient, by decreasing the power to flow rate ratio, or by decreasing the inlet temperature. Cheng and Yang (2008) introduced a point-hydraulic model to establish a criterion for predicting the onset of a self-sustaining flow oscillation in a closed cooling system. Four dimensionless numbers were clarified and stability maps were derived.

Based on the classical phase change and sub-cooling numbers adopted in the case of boiling channels, Ambrosini and Sharabi (2006) proposed two dimensionless parameters for the analysis of stability in heated channels with supercritical fluids. The adopted parameters were proved to result in dimensionless density independent on dimensionless density, thus giving generality to the predictions of unstable conditions obtained as a function of dimensionless parameters. Also the dynamic behaviour of heated channels with boiling and supercritical fluids were numerically studied and compared (Ambrosini, 2007), showing a very similar qualitative and quantitative behaviour between the two systems. Numerical results showed that density wave oscillations and Ledinegg instabilities may occur in supercritical fluid systems, but the stability thresholds were far from presently proposed supercritical reactor core operating conditions.

Parallel channels with supercritical fluid were studied applying in-house numerical code (Chatoorgoon et al., 2007; Chatoorgoon, 2008; Ortega Gómez et al., 2007) and system code such as Relap5 (Ambrosini, 2009). Both horizontal and vertical channels were analyzed, as well as symmetric and asymmetric channels. Different dimensionless numbers were used to obtain the stability boundary maps.

As can be seen, though stability analysis of supercritical fluid has been performed widely, few experimental results are available to verify the knowledge of flow instability phenomena at supercritical pressure. Researches have been carried out applying different geometrical models and numerical techniques or other theoretical approaches; nevertheless validation of the conclusions still plays a key role in supercritical flow instability investigation. In the present chapter, experiments were carried out in order to offer an up-to-date knowledge of flow instability phenomena in two parallel channels with supercritical water. Parametric effects on flow instability are discussed and stability boundaries are presented.

Experiments

Experimental Facility and Measurements

A supercritical water loop has been constructed in Nuclear Power Institute of China (NPIC) to investigate various thermal hydraulic issues of supercritical water. The schematic diagram of the experimental loop is shown in Figure 15. Distilled and de-ionized water from the water tank is driven through a filter by a plunger pump, of which an accumulator is placed at the outlet to eliminate the mass flow rate fluctuations. Before flowing into the test section the working fluid is pre-heated in a pre-heater. Therefore the inlet temperature of the test section can be adjusted to desired values. The pre-heater is heated electrically by an AC power supply with maximum heating capability of 200 kW, while the test section is heated by a DC power supply with maximum heating capability of 600 kW. The water from the test section mixes with that from the bypass line in a mixing plenum. After being cooled by a heat-exchanger, the mixed

Figure 15. Schematic diagram of the experimental loop

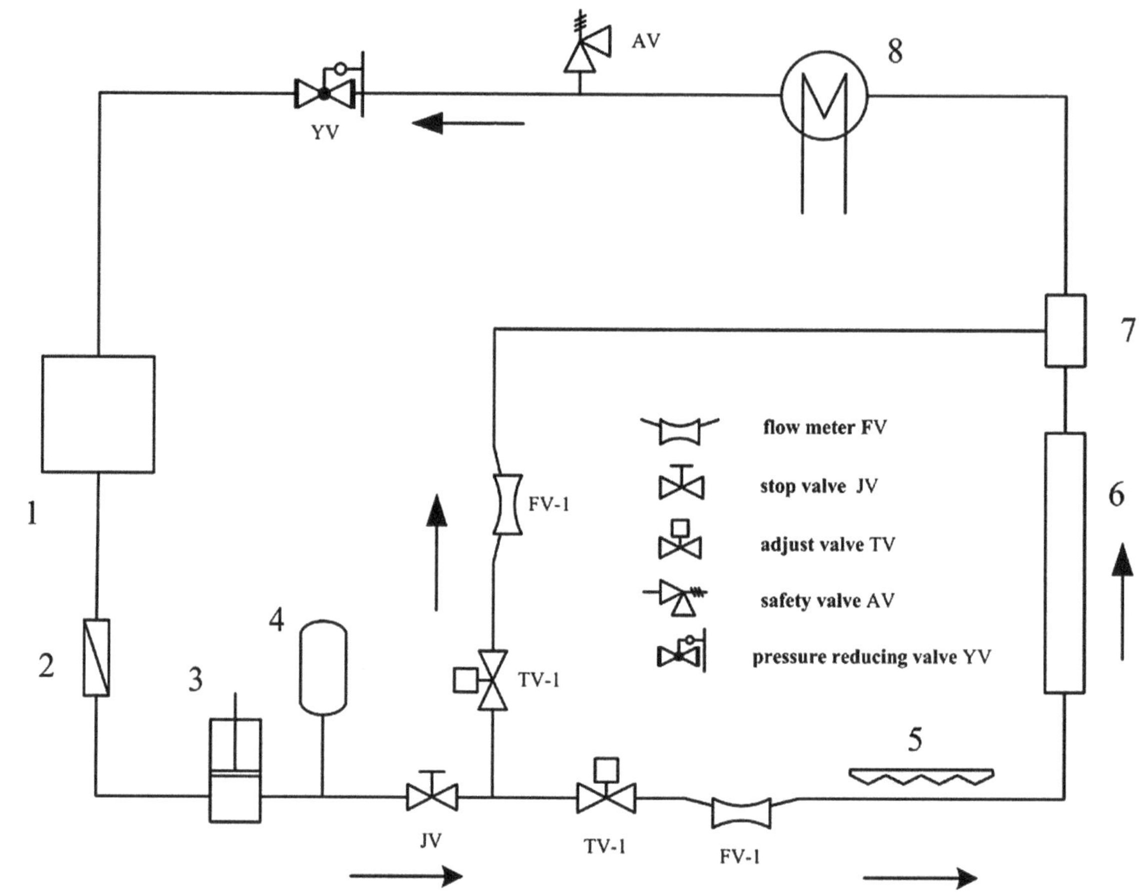

1. water tank 2. filter 3. plunger pump 4. accumulator
5. pre-heater 6. test section 7. mixing plenum 8. heat-exchanger

water passes through a pressure reducing valve and flows back into the water tank. The mass flux through the test section is controlled by adjusting the valves either in the test section line or in the bypass line. The operating limit of the system is 30 MPa.

The test section is a two-parallel-channel system consisting of a lower plenum, flow meters, entrance sections, heated sections, riser sections and an upper plenum. The two heated sections are INCONEL Alloy pipes with inner and outer diameters of 6 and 11 mm, while the heated length is 3000 mm. The application of INCONEL Alloy pipes ensures that the heat flux along the axis is uniform, since the resistance changes little with the temperature. Orifices are placed at the outlet of the heated sections. The whole piping is thermally insulated to minimize the heat loss. Figure 16 presents the schematic diagram of the test section.

Figure 16. Schematic diagram of the test section

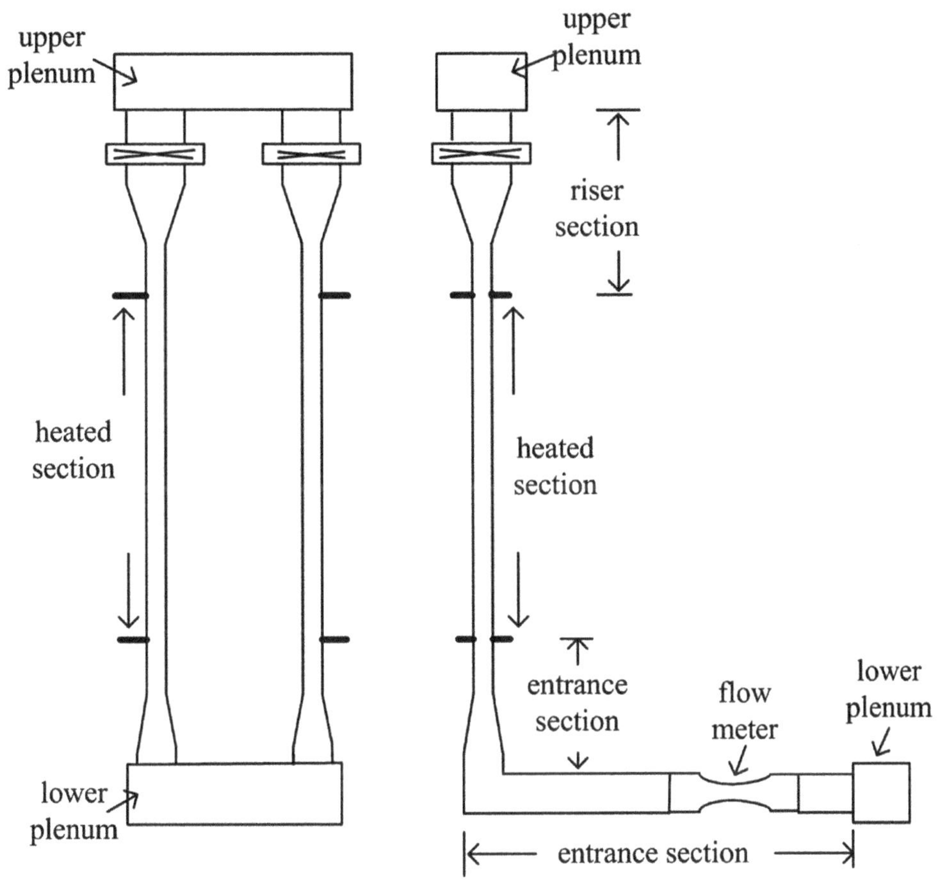

Fluid temperature at the inlet and the outlet of the test section were measured by sheathed N-type thermocouples. 14 sheathed N-type thermocouples have been spaced along each heated pipe at 7 different axial locations to measure outlet wall temperature, identify heat transfer deterioration and measure the wall temperature near the deterioration segment. Pressure measurements were made at the lower plenum, inlets of the heated sections, outlets of the heated sections and upper plenum. Heat input applied to the system was measured by means of current and voltage readings.

High frequency measurement was required for mass flow rate in the experiments, since the inlet flow oscillation was used as a reference to indentify the onset of parallel instability. A prior verification test was conducted to ensure the high dynamic response of flow meters. The pressure of the test was 25 MPa. The mass flow rate was manually adjusted to oscillate between two values with sufficiently high frequency. Figure 17 presents the result of the verification test. As can be seen, the maximum and minimum steady-state values are approximately 230kg/h and 125kg/h, respectively. And the flow meter could capture the maximum and minimum values during flow oscillation without amplitude losses, which indicates the dynamic response of the flow meter is acceptable.

All the signals of measured parameters were collected and recorded by a high-speed data acquisition system.

Figure 17. Result of dynamic response test

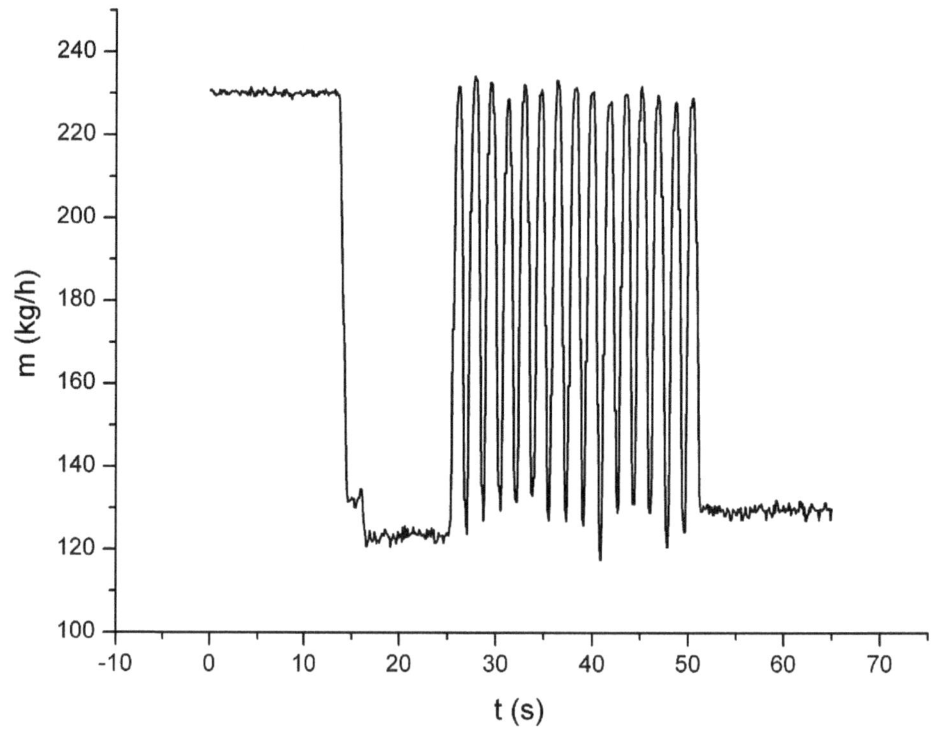

Experimental Procedure and Conditions

Before starting the main experiments, steady-state tests were conducted to check a proper working of the instrumentation and to evaluate the heat loss from the test section. Experimental values of pressure drop along the channel for isothermal flow were compared to empirical correlations, aiming to ensure the conformity and accuracy of the measured data for pressure drop and mass flow rate. Subsequently heat balance tests were carried out. Fluid enthalpy rises through both of the two-parallel channels were calculated separately and compared with each other to insure symmetrical heat supply for the parallel system. The total fluid enthalpy rise was compared with the total DC heating power. The heat loss was 5% or so, which was taken into consideration while determining the stability boundaries in this chapter.

Flow instability experiments have been performed for different thermal hydraulic conditions. The system pressure of the test section was varied between 23 MPa and 26 MPa. The mass flux of each heated section was in the range between 600 kg/m^{2}s and 800kg/m^{2}s. The inlet temperature range was between 180 ^{o}C and 260 ^{o}C, which has been predicted and modified according to real-time experimental results. During the experiments the system pressure, mass flux and inlet fluid temperature of the test section were firstly adjusted to the desired values, and heat flux was increased to the initial value forecasted by numerical methods. Then the heat flux was increased gradually to observe whether flow instabilities occur or not. The heat flux has been increased at a rate slow enough to make the process quasi steady state. Meanwhile careful attention was paid to temperature variations. If the maximum wall temperature or fluid temperature reached the limit, which is determined by the strength of the pipes, without occurrence of flow instability, the working condition was changed for another circumstance.

Physical Approximation

As well known, the ratio of inlet and outlet pressure drops to distributed friction is relevant in determining unstable conditions. Therefore the physical model should be illustrated specifically as long as the experimental data are used for validation. In present experimental facility, the entrance section (that is, the section from the lower plenum to the entrance of the heated section) and the riser section (that is, the section from the exit of the heated section to the upper plenum) are comprised of several throttles, bend tubes and pipes in various diameters, thus making the geometrical configuration complicated for numerical analysis. Nevertheless, in order to make the experimental data available for validating the boundaries obtained by general numerical studies, the geometry is simplified on the basis of the calculated pressure drops as well as the dimensions of the test section.

A proposed geometrical model is illustrated in Figure 18. The simplification is made by changing the entrance and the riser sections. That is to say, pipes in various inner diameters are substituted with one pipe, which has the same inner diameter with the heated section, and local pressure drops are assumed to be concentrated at special orifices. Key parameters in the geometrical model include the inlet (outlet) orifice pressure drop coefficients, k_{in} (k_{out}), and the length of the entrance (riser) sections.

Figure 18. A proposed geometrical model

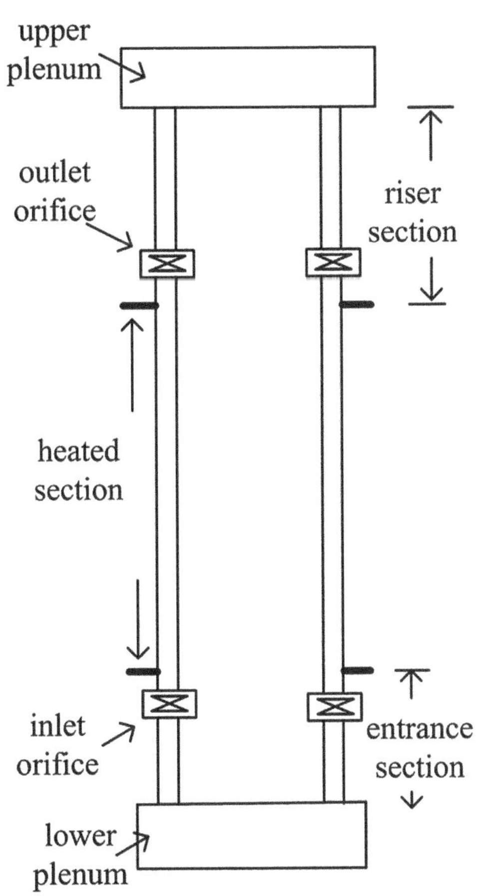

During the experiments, two pressure drops were measured for each channel of the two-parallel system, i.e. pressure drop from the lower plenum to the entrance of the heated section, ΔP_{in}, and that from the exit of the heated section to the upper plenum, ΔP_{out}. ΔP_{in} or ΔP_{out} is composed of friction pressure drop, local pressure drop and gravitational pressure drop:

$$\Delta P = \Delta P_f + \Delta P_l + \Delta P_g \tag{74}$$

where

$$\Delta P_f = \sum f_r \frac{l}{D} \frac{G^2}{2\rho} \tag{75}$$

$$\Delta P_l = \sum k \frac{G^2}{2\rho} \tag{76}$$

$$\Delta P_g = \rho g H \tag{77}$$

The gravitational pressure drop is determined by the height of the entrance or riser section and not affected by various kinds of throttles or bend tubes. Therefore, to simulate the gravitational pressure drop, the lengths of entrance (riser) sections in the simplified geometry can be set exactly the same with the height of entrance (riser) sections in the experiments, and that is probably the best choice.

Subsequently, the friction pressure drop in the simplified geometry can be calculated by applying appropriate friction factor. Here the following correlation (Haaland, 1983) has been used:

$$f_r = \left[-1.8 \log \left(\left(\frac{\varepsilon}{3.7D} \right)^{1.11} + \frac{6.9}{\mathrm{Re}} \right) \right]^{-2} \tag{78}$$

where $\varepsilon = 1.5 \times 10^{-6} m$ (for hard-drawn pipes).

Then the inlet (outlet) orifice pressure drop coefficients, k_{in} (k_{out}), can be calculated by deducting the gravitational pressure drop and frictional pressure drop from ΔP_{in} (ΔP_{out}).

In the experiments, the heights of the entrance and riser sections were 200 and 350 mm, respectively. And the calculated values of inlet (outlet) orifice pressure drop coefficients exhibit certain differences between the two channels, that is:

$$k_{in,1} = 5.40;\ k_{in,2} = 5.50;\ k_{out,1} = 4.93;\ k_{out,2} = 6.46 \tag{79}$$

Thus, the parameters in the simplified geometry are all quantified.

Results and Discussion

Evolution of Flow Rates

Figure 19 illustrates the evolution of mass flow rates with increasing heat flux in the parallel-channel system. The pressure was kept constant at supercritical pressure, and the heat flux was increased gradually according to the experimental procedure. The evolution can be divided into four stages as follows:

1. The flow rates of the two channels kept broadly symmetrical when heat flux and fluid temperature were relatively low. The outlet fluid temperatures of the two channels were quasi-symmetric, nevertheless one was indeed higher.
2. The flow rates were redistributed between the two channels, leading to be more asymmetrical. The redistribution occurred when the two outlet fluid temperatures were in the vicinity of pseudo-critical point and the higher one started rising beyond the point.
3. The flow rates of both channels started to fluctuate irregularly, which was probably induced by the channel with lower flow rate and higher outlet fluid temperature. The time-average flow rate of each channel remained constant.
4. The flow rates of the two channels started oscillating out of phase as the heat flux reached the stability boundary.

Figure 19. Evolution of mass flow rates

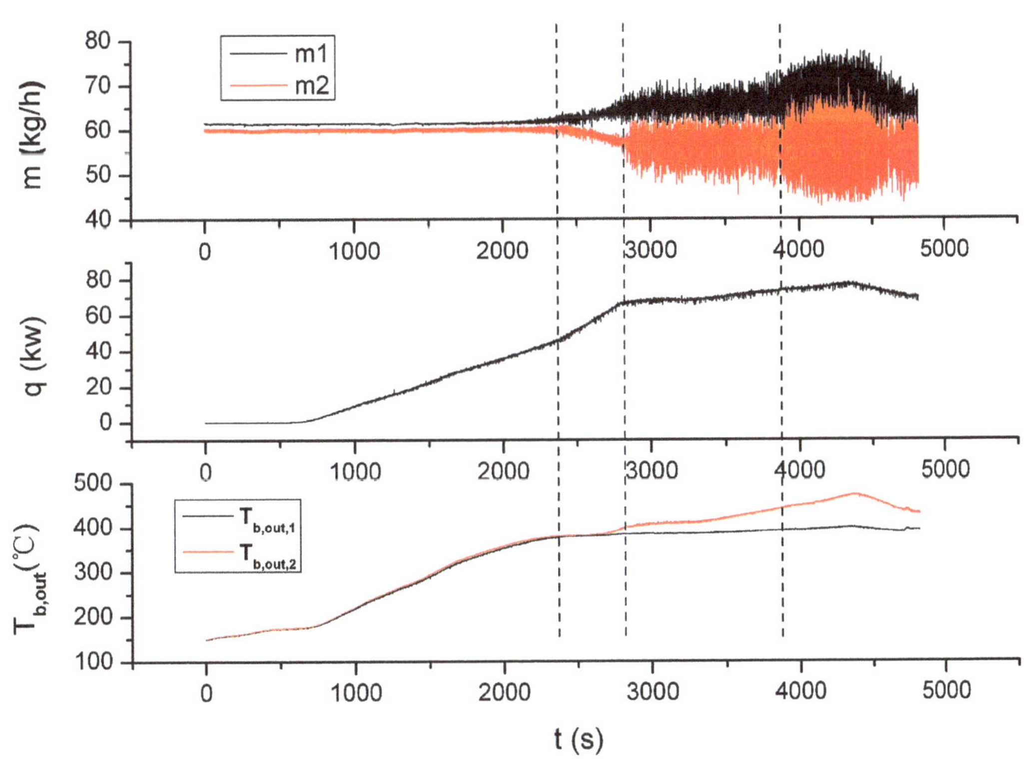

The redistribution of mass flow rates in stage 2 can be interpreted by the supercritical water property variation and different resistance characteristics between the two channels. As indicated in Equation 79, the outlet orifice pressure drop coefficients exhibited a slight difference, namely $K_{out,1}<K_{out,2}$. That caused slight differences in flow rates ($G_1>G_2$) as well as outlet fluid temperatures ($T_{b,out,1}<T_{b,out,2}$) before stage 2. Accordingly, $T_{b,out,2}$ was always larger than $T_{b,out,1}$ during the temperature rise. When $T_{b,out,1}$ reached the pseudo-critical point, $T_{b,out,2}$ had already got beyond the point. Owing to the high specific heat in the pseudo-critical region (Figure 20), $T_{b,out,2}$ increased faster than $T_{b,out,1}$ considering the symmetrical heat supply. Hence the temperature difference widened. Meanwhile, as a result of sharp density change near the pseudo-critical temperature, outlet fluid densities differed obviously between the two channels ($\rho_1>\rho_2$). Considering the outlet local pressure drop shown in Equation 76, G_2 would decrease to a new value, while G_1 increased in order to satisfy the boundary condition, i.e. equal pressure drop and constant total mass flow rate. The flow rate redistribution ended when the outlet fluid temperatures both got across the pseudo-critical point and density variations with temperature became relatively gentle.

As a result of the redistribution, the little asymmetry caused by different resistance characteristics was enlarged. The asymmetry plays a significant role in determining the feasibility of flow instability experiments and appears to be affected mainly by two factors, i.e. the fluid temperature and total mass flow rate, based on the experience of the experiments. The effect of fluid temperature could be interpreted by Equation 76 and Equation 79. In the experiments the main difference between the two channels lay in the outlet orifice pressure drop coefficients, which directly influenced the outlet local pressure drop. When the temperature was low, fluid with high density at the outlet led to a small ratio of outlet local pressure drop to overall pressure drop along each channel. Thus the difference of outlet local pressure drops between the two channels exhibited a negligible effect on flow rate distribution. On the contrary, when the temperature got relatively higher, the outlet local pressure drop increased as a fraction of overall pressure drop. In that way, the asymmetry of outlet local pressure drop led to asymmetry of flow rates between the channels.

The same explanation can be applied to the effect of total mass flow rate on the asymmetry. The ratio of outlet local pressure drop to overall pressure drop increased with increasing flow rate. Hence the asymmetry became larger at a higher total mass flow rate. As a matter of fact, in the experiments when the total mass flow rate was 120 kg/h, there was an individual flow rate difference of 65 kg/h versus 55 kg/h after the flow rate redistribution. And the relative difference became larger with a total mass flow rate of 160 kg/h, which was 90 kg/h versus 70 kg/h.

Therefore, considering a higher total mass flow rate, the outlet temperatures of the two channels would differ more widely if the same ratio of heat flux to total mass flow rate is considered. This means the maximum temperature would reach higher. With the limit of the same critical wall temperature, the ratio of heat flux to total mass flow rate must be reduced in the experiments. That would make the occurrence of parallel instability more difficult, according to related theoretical research or two-phase flow instability experiences. Unstable circumstances presented in this chapter are all at an approximate total mass flow rate of 120 kg/h.

Onset of Oscillations

As the preliminary experimental investigation on supercritical flow instability, it was sometimes difficult to discriminate whether an oscillation is "unstable" or "stable". Inspired by the earlier work done on boiling systems, the onset of supercritical flow oscillations in parallel channels is defined (shown

Figure 20. Specific heat of water

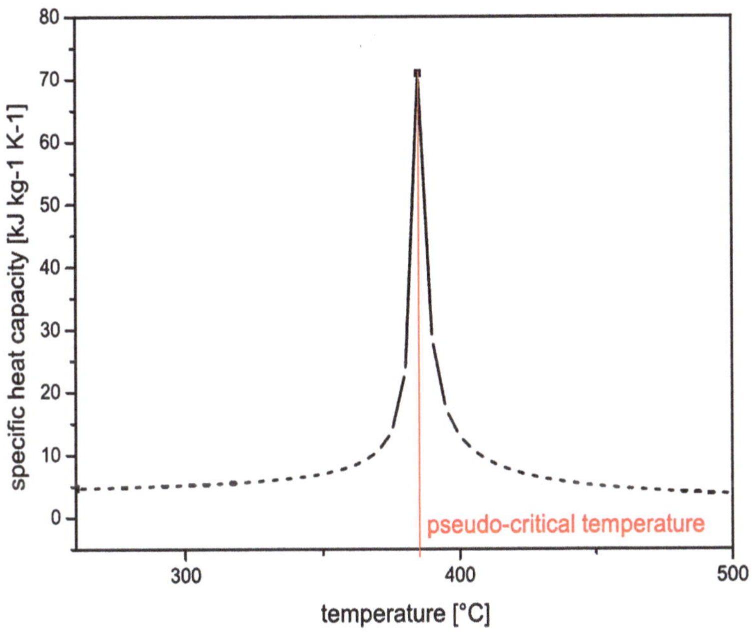

in Figure 21). Short-life transients were disregarded and only the sustained out-of-phase oscillations accompanied by evident amplitude enlargement were considered. System parameters at the onset were obtained and processed to define the stability boundary.

Parametric Effects on Instability

Different sets of experiments were conducted in order to get a primary understanding of the parametric effects on supercritical flow instability. Owing to the practical constraints, which basically determined the feasibility of acquiring desired instability phenomena for supercritical water, limited instability boundary

Figure 21. The way of defining the onset of flow oscillations in parallel channels (a) and (b)

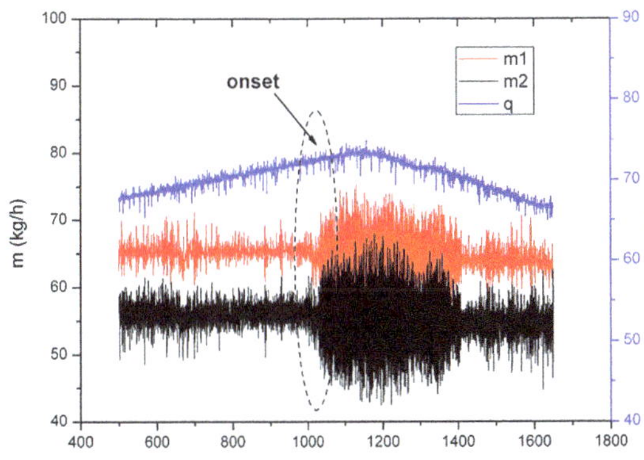

(a)

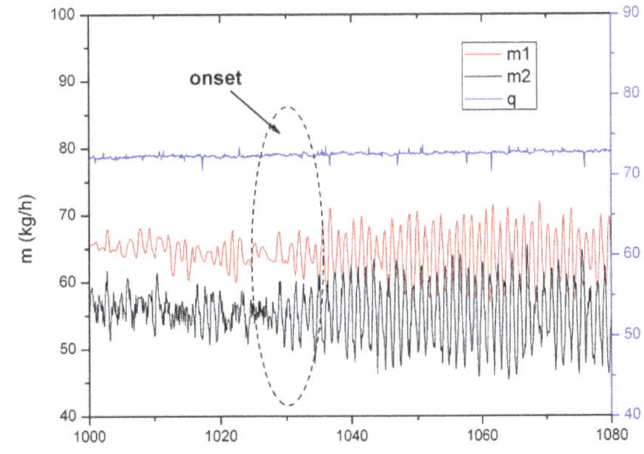

(b)

points were obtained. Nevertheless, boundaries corresponding to various pressures and inlet temperatures were somehow adequate to primarily illustrate the parametric effects. The effects are presented below.

Effect of Pressure

For the parallel channels with supercritical water, an increase in system pressure was found to have a stabilizing effect while other system parameters were kept constant. Figure 22 shows the influence of pressure on threshold heat flux for three inlet temperatures. The threshold heat flux is the individual heat flux per channel corresponding to the stability boundary. The results show that the threshold heat flux increases while the pressure is increased from 23 MPa to 25 MPa, and no instability phenomena were observed when the pressure was 26 MPa, limited by the same critical wall temperature in the experiments. Such stabilizing effect of increasing pressure on flow instability is well-known for two-phase flow. In two-phase flow, increasing system pressure reduces the void fraction and decreases the two-phase flow friction and momentum pressure drops, and thus stabilizes the system (Boure et al., 1973).

Similar mechanism of such pressure effect can be interpreted for supercritical water. In SCWR, the main concern about the possible occurrence of flow instabilities is caused by the sudden change of water density near the pseudo-critical point, which results in large density variation across the core. As shown in Figure 23, the density variation in the vicinity of pseudo-critical point becomes steeper at lower system pressure, and the change across the whole temperature range is relatively larger. This characteristic somehow explains the stabilizing effect of increasing pressure on supercritical flow, though the mechanism of supercritical flow instability needs to be researched further.

Figure 22. Effect of system pressure on threshold power

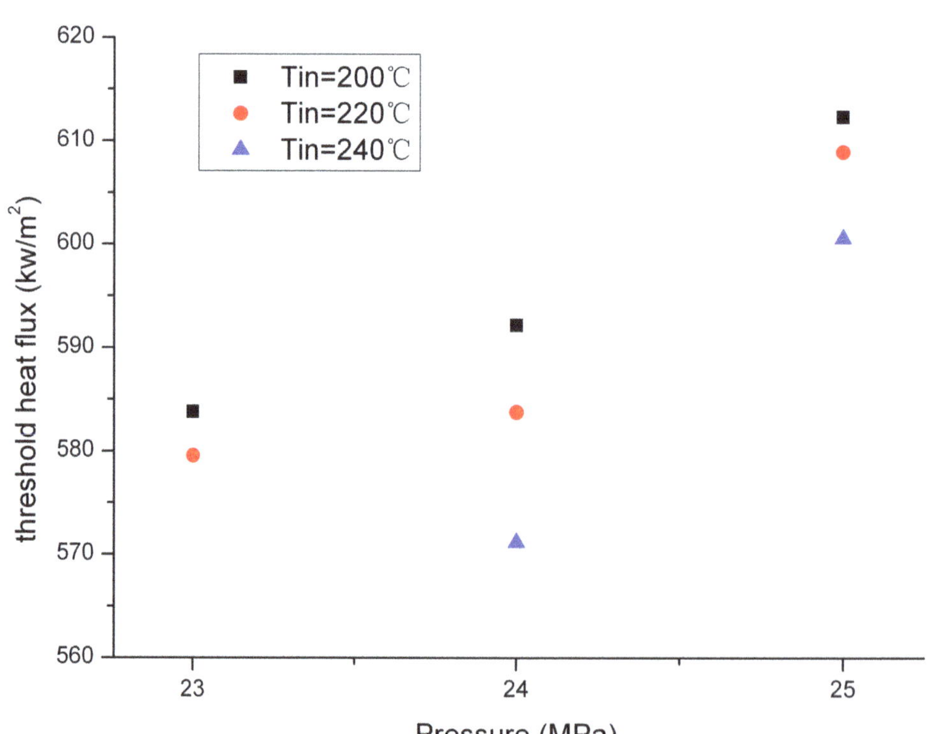

Figure 23. Density variations near the pseudo-critical points

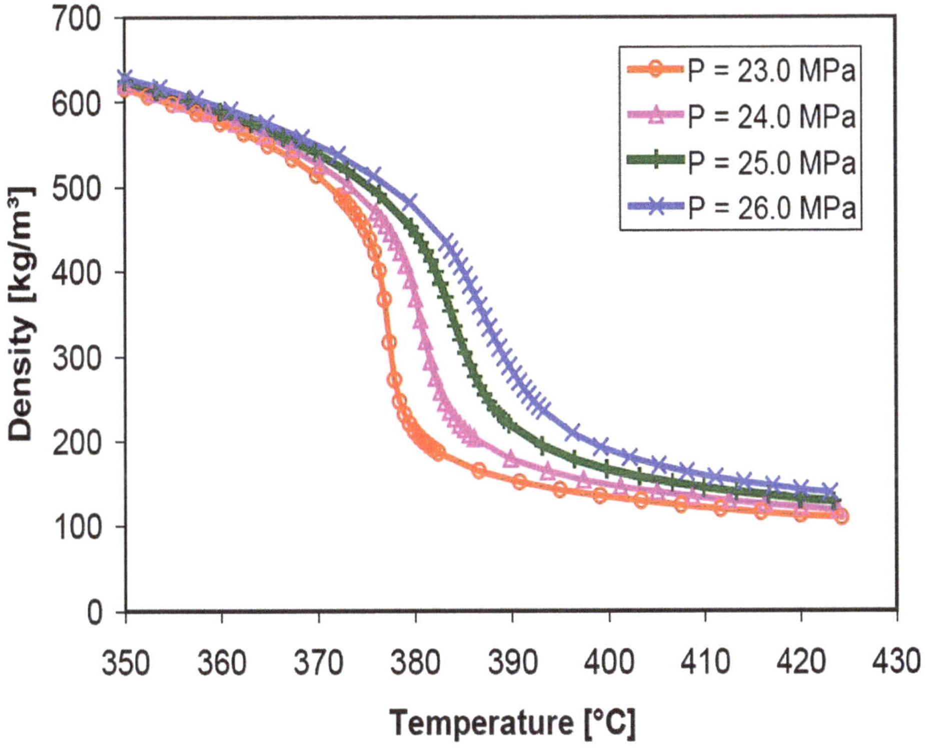

Effect of Inlet Temperature

Figure 24 presents the effect of inlet fluid temperature on the threshold heat flux. The threshold heat flux decreases while the inlet temperature increases from 180 °C to 240 °C, which means the system is more unstable. In spite of the decreasing heat flux, outlet fluid temperature as well as the maximum wall temperature corresponding to the stability boundary increases, thus making the instability phenomena more difficult to be observed in the experiments at an inlet temperature of 260 °C.

Ambrosini (2007) obtained a stability map for the heated channel with supercritical fluid by a linearised simplified program and RELAP5. The results indicate that for a particular heat flux and flow rate there is a definite range of inlet temperature within which the flow would oscillate and above or below this it is stable. Similar finding has been presented for two-phase flow instability. That is, for small degrees of subcooling, increasing inlet subcooling destabilizes the flow due to significant incremental change of transit time, whereas for high degrees of subcooling the effect is stabilizing due to enlargement of non-boiling length and reduction of void fraction (Boure et al., 1973). Though the geometry used by Ambrosini (2007) is different from the parallel channels presented here, the non-monotonic effect of inlet temperature similar to two-phase flow reveals that there is a possible chance of encountering an inlet temperature, at which the threshold heat flux exhibits a minimum, in parallel systems with supercritical water. As indicated by Figure 24, the temperature of that minimum point is probably above 240 °C. Considering the higher temperature at the outlet for both the fluid and the wall, confirming the existence of that non-monotonic effect needs further improvement on experimental techniques.

Figure 24. Effect of inlet temperature on threshold power

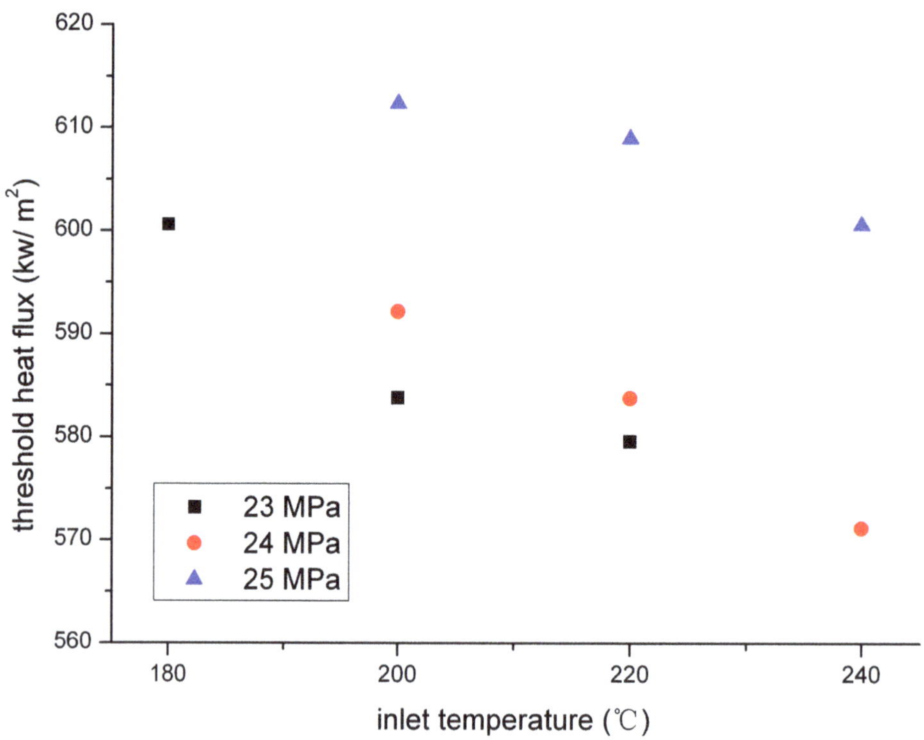

Stability Boundaries

To reduce the number of independent parameters and better define the stability boundary, dimensionless groups have been introduced for two-phase flow instability analysis. Two dimensionless parameters, namely subcooling number, N_{sub}, and phase change number, N_{pch}, are used to illustrate the stability boundary in a two-dimensional plane:

$$N_{sub} = \frac{h_f - h_{in}}{h_{fg}} \frac{\nu_{fg}}{\nu_f} \tag{80}$$

$$N_{pch} = \frac{q_0'' \Pi_h L}{\rho_f w_{in} h_{fg} A} \frac{\nu_{fg}}{\nu_f} \tag{81}$$

As for supercritical water, several studies have been carried out on the derivation of new dimensionless parameters in recent years. In this preliminary experimental investigation, dimensionless parameters introduced in Ambrosini and Sharabi (2006) are used. By replacing the boiling point adopted in two-phase flows with the pseudo-critical point to define the reference status in supercritical fluids, following substitution is performed:

$$\frac{\nu_{fg}}{h_{fg}}\frac{1}{\nu_f} \rightarrow \left(\frac{\partial \nu}{\partial h}\right)_{p,PC}\frac{1}{\nu_{pc}} = \frac{\beta_{pc}}{C_{P,pc}} \tag{82}$$

Hence, two dimensionless numbers are obtained:

$$N_{SPC} = \frac{\beta_{pc}}{C_{p,Pc}}(h_{pc} - h_{in}) \tag{83}$$

$$N_{TPC} = \frac{\beta_{pc}}{C_{p,Pc}}(h_{exit} - h_{in}) \tag{84}$$

where the sub-pseudo-critical number, N_{SPC}, and true trans-pseudo-critical number, N_{TPC}, are introduced to replace the subcooling number and phase change number, respectively.

All unstable points were collected and the pressures, inlet temperatures, mass flow rates and heat fluxes corresponding to the stability boundaries were translated into N_{TPC} and N_{SPC}, as shown in Figure 25. As limited boundary points are available, it is still not completely clear if the two dimensionless numbers are really successful to collapse the different pressure data in a similar trend. Nevertheless, the results could serve as a basis for further investigation.

Figure 25. Stability boundaries

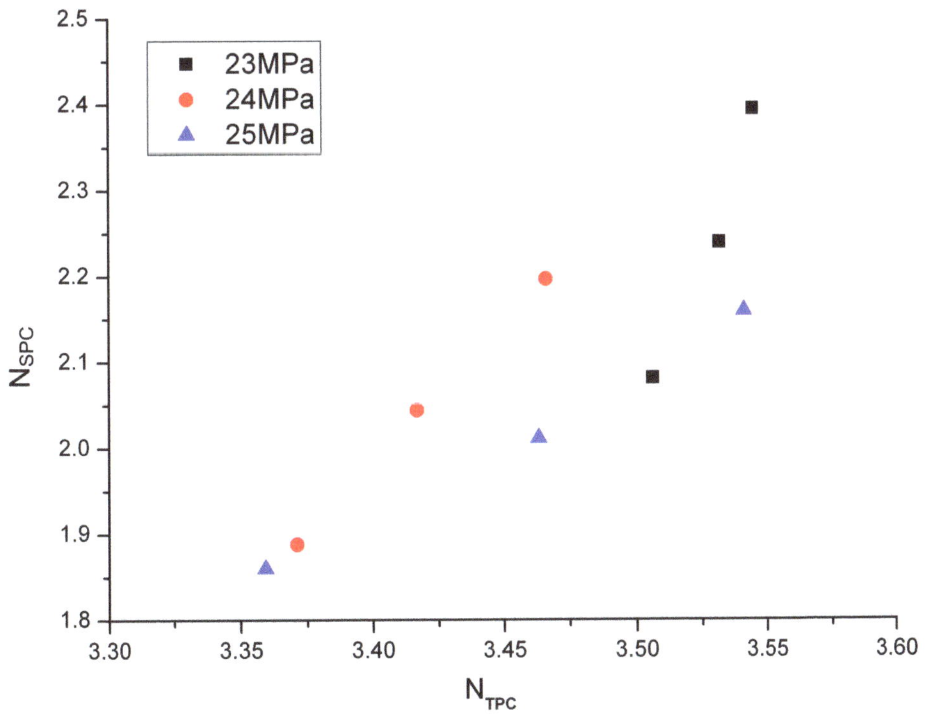

MODELING AND ANALYSIS OF SUPERCRITICAL FLOW INSTABILITY IN PARALLEL CHANNELS

Introduction

Solution methods for stability analysis of supercritical flow are mainly developed from those for two-phase flow and can be divided into two kinds, including frequency-domain method (Yi et al. 2004, Tin et al. 2004, Zhao et al. 2005) and time-domain method (Chatoorgoon 2001, 2008; Podowski 2010). The former one is based on linearized equations those are converted to transfer functions by Laplace transform, while the latter one is based on some kind of numerical discretization and integration in time (Ambrosini 2008). Furthermore, commercial software (such as Relap5 and Fluent) has been used to predict the dynamic behavior in the heated channel with supercritical fluid (Ambrosini, 2007, 2009; Sharabi, 2008a, 2008b).

Despite the fact that several numerical methods have been applied in supercritical flow stability analysis, there is a complete lack of experimental validation due to the difficulties related to the very high pressure and temperature for supercritical water. Scaling procedures have been proposed (Marcel 2009; Rohde 2011) to downscale the supercritical water loop for experimental studies by using fluid-to-fluid modeling, so that a proper fluid (e.g. Freon R-23) could be used in the test facility to simulate the SCWR with considerably reduced pressure, temperature and power.

In order to better utilizing and further expanding the experimental results, an in-house code have been developed applying time-domain approach. In this chapter the geometrical modeling of the test section in the experiments is firstly proposed. Then the numerical code is described in detail and validated by the experimental results. Based on the validated code, the geometrical simplification is further discussed. Finally, effects of system parameters on the flow instability are numerically analyzed.

System and Physical Model

Physical Approximation

According to experience of two-phase flow instability and relevant theoretical research on supercritical flow instability, the ratio of inlet and outlet pressure drops to distributed friction probably plays an important role in determining stability boundaries. Thus the geometrical conFigureuration must be modeled accurately when the experimental data are used to verify the numerical predictions.

As shown in Figure 15 the test section is comprised of several throttles, bend tubes and non-uniform cross-section-area pipes, somehow complicated for numerical modeling. A simplification of the physical model is illustrated in Figure 26, namely Model A in this chapter. The main difference between Model A (see Figure 26) and the test section (see Figure 15) is the local pressure drop distribution along the entrance and riser sections. For the test section the local pressure drops are caused by the flow meters, orifices and non-uniform cross-section-area pipes, while for Model A the local pressure drops are assumed to be concentrated at the special orifices and those at other places are artificially neglected. This simplification is reasonable considering the fact that no heat power is applied to the entrance and riser sections.

Figure 26. Schematic diagram of Model A

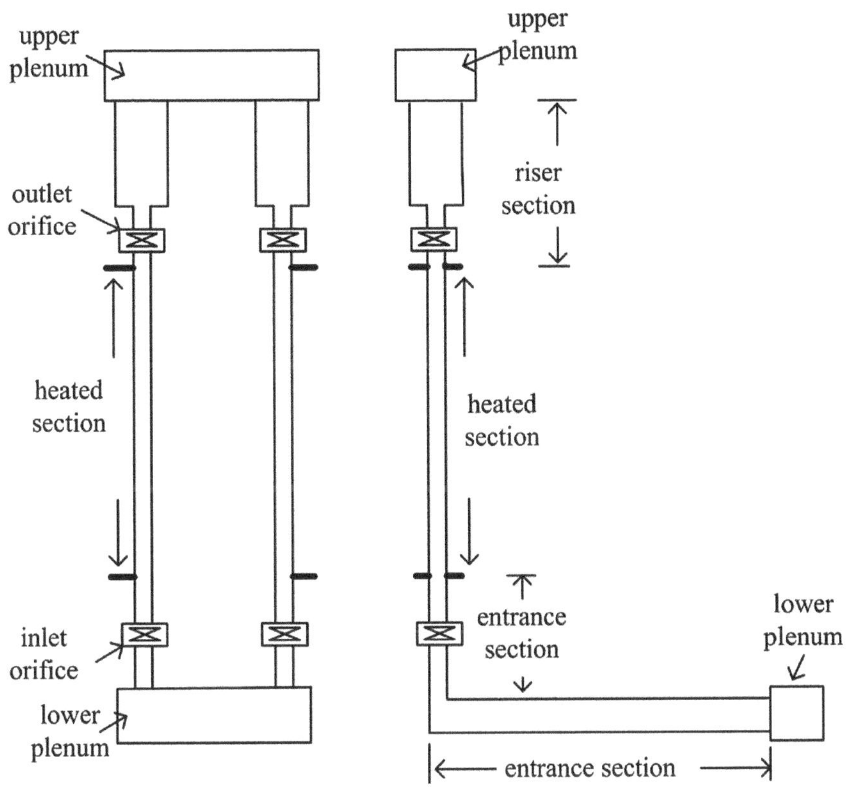

Though available for the numerical analysis, Model A might be still complicated. Generally speaking, two other geometrical models have been considered in the flow stability analysis, which are also adopted and discussed in this chapter.

1. **Model B Illustrated in Figure 27:** This model leaves out the horizontal pipes and pipes in various diameters. As shown in Figure 27, the two-parallel system consists of entrance sections, heated sections and riser sections in the same flow area. The length of the entrance sections for Model B is the same with the height of the vertical entrance sections in Figure 15. The length of the riser sections remains the same. Local pressure drops are assumed to be concentrated at special orifices.
2. **Model C Illustrated in Figure 28:** This model shows another typical two-parallel-channel system considered in flow instability analysis. The entrance and riser sections are eliminated, while local pressure drops are assumed to be concentrated at special orifices.

Parameters Quantification

For each of the above geometrical models, the inlet and outlet orifice pressure drop coefficients are unknown but significant, which essentially determine the inlet and outlet pressure drops. To quantify the pressure drop coefficients, special measurements have been performed in the experiments for the inlet pressure drops ΔP_{in} (pressure drop from lower plenum to the entrance of heated section) and outlet

Figure 27. Schematic diagram of Model B

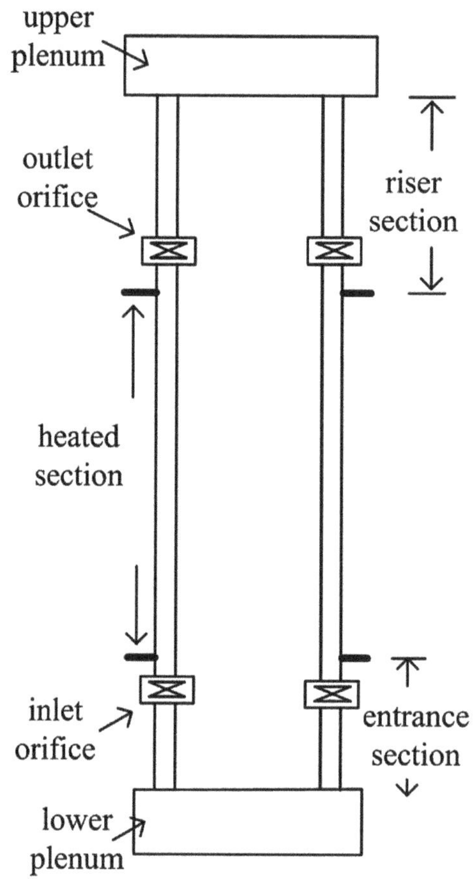

pressure drops ΔP_{out} (pressure drop from the exit of heated section to upper plenum) at steady states. ΔP_{in} or ΔP_{out} is composed of friction pressure drop, local pressure drop and gravitational pressure drop:

Finally, the inlet (outlet) orifice pressure drop coefficients k_{in} (k_{out}), can be calculated by deducting the gravitational pressure drop and frictional pressure drop from ΔP_{in} (ΔP_{out})

The way of determining and quantifying the key parameters seems inappropriate for Model C, as the entrance and riser sections are eliminated and accordingly the gravitational pressure drops have been "translated" into local pressure drops. In section 4, the numerical results for different geometrical models would be compared and discussed. Owing to the minimum as well as reasonable approximations, Model A could serve as a benchmark case to assess the applicability and reliability of numerical code comparing with experimental data. And the comparison with other two geometrical models aims at the study of geometrical simplification in flow instability analysis.

Numerical Approach

A numerical code has been developed in NPIC in order to satisfy the need for stability analysis of supercritical flow in various geometrical models (Figure 26, Figure 27, and Figure 28). The code solves the mass, momentum and energy equations by applying time-domain approach. Steady-state equations are

Figure 28. Schematic diagram of Model C

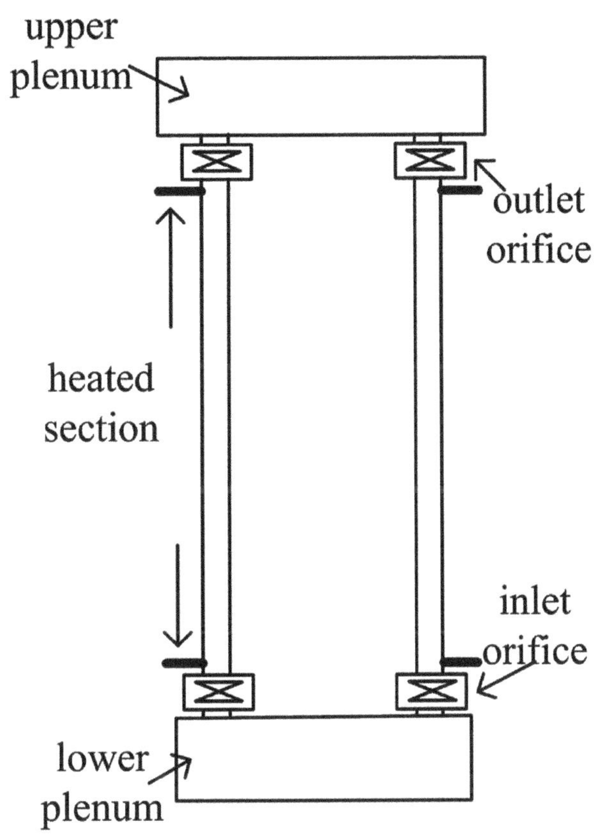

firstly solved. Then a small perturbation is imposed on the steady results to get the dynamic response of the system. If the disturbance grows in time and yields diverging flow oscillations, the steady-state system is considered to be unstable. On the other hand, if the disturbance leads to decaying oscillations to original steady state, then the system is considered to be stable. The stability boundary refers to the working condition when the disturbance yields sustained flow oscillations (neither diverge nor converge). In the present chapter, power level at the stability boundary is called the threshold power. That is to say, the system is unstable if the heat power is higher than the threshold power, while the system is stable if the heat power is lower.

Equation System

Despite the fact that the fluid density changes remarkably near the pseudo-critical region, there is no phase change for fluid at supercritical pressure. Therefore the basic mass, momentum and energy equations for one-dimensional channel flow are:

$$\frac{\partial \rho}{\partial t} + \frac{\partial (\rho u)}{\partial z} = 0 \tag{85}$$

$$\frac{\partial\left(\rho u\right)}{\partial t}+\frac{\partial\left(\rho u^{2}\right)}{\partial z}+\frac{\partial P}{\partial z}+C\rho u^{2}+\rho g\sin\theta=0 \tag{86}$$

$$\frac{\partial\left(\rho h\right)}{\partial t}+\frac{\partial\left(\rho uh\right)}{\partial z}=\frac{q_{l}}{A} \tag{87}$$

where C in the momentum equation is given by:

$$C=\frac{f_{r}}{2D}+\frac{K_{l}}{2}\delta_{d}(z-z_{k}) \tag{88}$$

and the friction factor is given by Equation 78.

Since the test section is not of uniform cross-sectional area, integration is carried out over the cross-sectional area perpendicular to the flow direction. Subsequently the following equations are obtained.

Mass:

$$A\frac{\partial\rho}{\partial t}+\frac{\partial M}{\partial z}=0 \tag{89}$$

Momentum:

$$\frac{\partial M}{\partial t}+\frac{\partial}{\partial z}\left(\frac{M^{2}}{\rho A}\right)+A\frac{\partial P}{\partial z}+C\frac{M^{2}}{\rho A}+\rho gA\sin\theta=0 \tag{90}$$

Energy:

$$A\frac{\partial\left(\rho h\right)}{\partial t}+\frac{\partial\left(Mh\right)}{\partial z}=q_{l} \tag{91}$$

The set of mass, momentum and energy equations is closed by the equation of state for the supercritical water:

$$\rho=f(P,h) \tag{92}$$

Discretization

Discretization of the governing equations employs control-volume-based discretization in space and backward-difference scheme in time. Integration of the conservation equations from spatial grid point (i) to (i+1) and forward-difference approximation for the time derivative yield the following equations:

Mass:

$$A\left(\frac{\rho_{i+1}^{j+1}-\rho_{i+1}^{j}+\rho_{i}^{j+1}-\rho_{i}^{j}}{2\Delta t}\right)+\frac{M_{i+1}^{j+1}-M_{i}^{j+1}}{\Delta z}=0 \tag{93}$$

Momentum:

$$\frac{M_{i+1}^{j+1}-M_{i+1}^{j}+M_{i}^{j+1}-M_{i}^{j}}{2\Delta t}+\frac{1}{A\Delta z}\left[\left(\frac{M^2}{\rho}\right)_{i+1}^{j+1}-\left(\frac{M^2}{\rho}\right)_{i}^{j+1}\right]+A\frac{\left(P_{i+1}^{j+1}-P_{i}^{j+1}\right)}{\Delta z}$$
$$+\frac{C_{i}^{j+1}}{2A}\left[\left(\frac{M^2}{\rho}\right)_{i}^{j+1}+\left(\frac{M^2}{\rho}\right)_{i+1}^{j+1}\right]+\frac{\rho_{i+1}^{j+1}+\rho_{i}^{j+1}}{2}gA\sin\theta=0 \tag{94}$$

Energy:

$$A\frac{\left(\rho h\right)_{i+1}^{j+1}-\left(\rho h\right)_{i+1}^{j}+\left(\rho h\right)_{i}^{j+1}-\left(\rho h\right)_{i}^{j}}{2\Delta t}+\frac{\left(Mh\right)_{i+1}^{j+1}-\left(Mh\right)_{i}^{j+1}}{\Delta z}=q_l \tag{95}$$

State:

$$\rho_{i+1}^{j+1}=f\left(P_{i+1}^{j+1},h_{i+1}^{j+1}\right) \tag{96}$$

where C_i^{j+1} is given by

$$C_i^{j+1}=\frac{f_{ri}^{j+1}}{2D}+\frac{K_l}{2}\delta_d(z-z_k) \tag{97}$$

and f_{ri}^{j+1} is given by

$$f_{ri}^{j+1}=\left[-1.8\log\left(\left(\frac{\varepsilon}{3.7D_H}\right)^{1.11}+\frac{6.9}{\mathrm{Re}_i^{j+1}}\right)\right]^{-2} \tag{98}$$

$$\mathrm{Re}_i^{j+1}=\frac{M_i^{j+1}D}{A\mu_i^{j+1}} \tag{99}$$

Steady-State Equations

By letting Δt be infinity in the above transient equations, the steady-state equations are produced:

$$M_{i+1} = M_i \tag{100}$$

$$\frac{1}{A\Delta z}\left[\left(\frac{M^2}{\rho}\right)_{i+1} - \left(\frac{M^2}{\rho}\right)_i\right] + A\frac{(P_{i+1} - P_i)}{\Delta z} + \frac{C_i}{2A}\left[\left(\frac{M^2}{\rho}\right)_i + \left(\frac{M^2}{\rho}\right)_{i+1}\right] + \frac{\rho_{i+1} + \rho_i}{2} gA\sin\theta = 0 \tag{101}$$

$$\frac{\left(Mh\right)_{i+1} - \left(Mh\right)_i}{\Delta z} = q_l \tag{102}$$

The steady-state equations are solved to produce initial conditions for the transient analysis. In addition, static pressure drop versus flow characteristics can be obtained for particular power level and inlet temperature. Therefore the Ledinegg instability can be studied, which however is not included in this chapter.

Boundary Conditions

For the physical models studied in this chapter, boundary conditions are given below:

1. Total mass flow rate of the two channels is given, i.e. $M_{total} = constant$.
2. Inlet temperature is given, i.e. $T_{in} = constant$.
3. Inlet pressure is given, i.e. $P_{in} = constant$.
4. Outlet pressures of the two channels are equal, i.e. $P_{out,1} = P_{out,2}$.

Solution Algorithm

The algorithm employed to solve the governing equations is based on the solution method developed by Chatoorgoon (1986), but changes specifically for two-channel system of non-uniform cross-section area. An overview of the solution algorithm follows below.

1. Steady-state equations are firstly solved to get initial values of ρ, M, P, h for both channels; subsequently a small perturbation is imposed on the system parameters and steps (2) ~ (4) are repeated for transient solution at the next time step level, i.e. j=0, 1, 2, 3 …
2. Inlet mass flow rate of one channel, M_1^{j+1}, is guessed; Accordingly, the inlet mass flow rate of the other channel is obtained by deducting M_1^{j+1} from M_{total}; Step (3) are performed for each channel individually
3. Steps (a) ~ (d) are repeated from inlet to outlet, i.e. i=1 to (n-1).
   a. The value of ρ_{i+1}^{j+1} is guessed and the mass Equation 93is solved for M_{i+1}^{j+1}

b. The momentum Equation 94 is solved for P_{i+1}^{j+1}

c. The energy Equation 95 is solved for h_{i+1}^{j+1}

d. Using P_{i+1}^{j+1} and h_{i+1}^{j+1}, an improved value of density, $\rho_{i+1}^{\prime j+1}$, is obtained by the state Equation 96. Steps (b) ~ (d) are repeated until a desired convergence of density is reached. The improved value $\rho_{i+1}^{\prime j+1}$ can be served as an updated guess for ρ_{i+1}^{j+1}.

4. After Step (3), the outlet pressures of both channels are obtained. The boundary condition $P_{out,1}$ = $P_{out,2}$ is checked. If the relative difference $\frac{\left|P_{out,1} - P_{out,2}\right|}{P_{out,1}}$ is within the specified tolerance limit, then the guess in Step (2) is correct and the solution moves to next time step level. Otherwise a "Regula-Falsi method" is employed to iterate on M_1^{j+1} until the boundary condition is satisfied.

In Step (1), the solution algorithm for the steady-state equations is similar to that listed above, thus not mentioned specifically here.

Results and Discussion

Validation

For verification of numerical predictions, flow stability boundaries calculated by the in-house code are compared with the experimental results. As indicated in section 2.2, Model A is employed due to the minimum as well as reasonable approximations.

To ensure independence of numerical results from temporal and spatial grid, a grid refinement study should be performed beforehand.

Temporal Grid Refinement

Effect of different time steps on flow stability boundaries is shown in Figure 29. For this temporal grid refinement study, the spatial grid size is set to be 0.05m. The system pressure is chosen to be 25 MPa and the inlet temperature is 220 °C. Threshold power is obtained at a total flow rate of 120 kg/h.

As shown in Figure 29, the time step does in fact influence the stability boundary. Reducing the time step size from 0.3 s to 0.02 s leads to a decrease of threshold power from 71.8 kW to 61.2 kW, indicating the system becomes more unstable. This is probably due to the dissipative and dispersive effects induced by the large time steps. Further reduction of the time step size to 0.005 s has no effect on the threshold power, suggesting that a time step size of 0.02 s is adequate for the analysis.

Moreover, to ensure independence of the temporal grid refinement from spatial grid size, a further study is performed when spatial grid size is set to be 0.025m. The results agree with that shown in Figure 29.

Spatial Grid Refinement

Similar to the temporal grid refinement study, the effect of spatial grid size on stability boundaries is analyzed. The results show that the threshold power remains the same when the spatial grid size is reduced from 0.05 m to 0.0125 m, revealing the spatial grid size does not have a significant effect on the stability boundary. Thus, a spatial grid size of 0.05 m is chosen.

Figure 29. Effect of time step on the threshold power

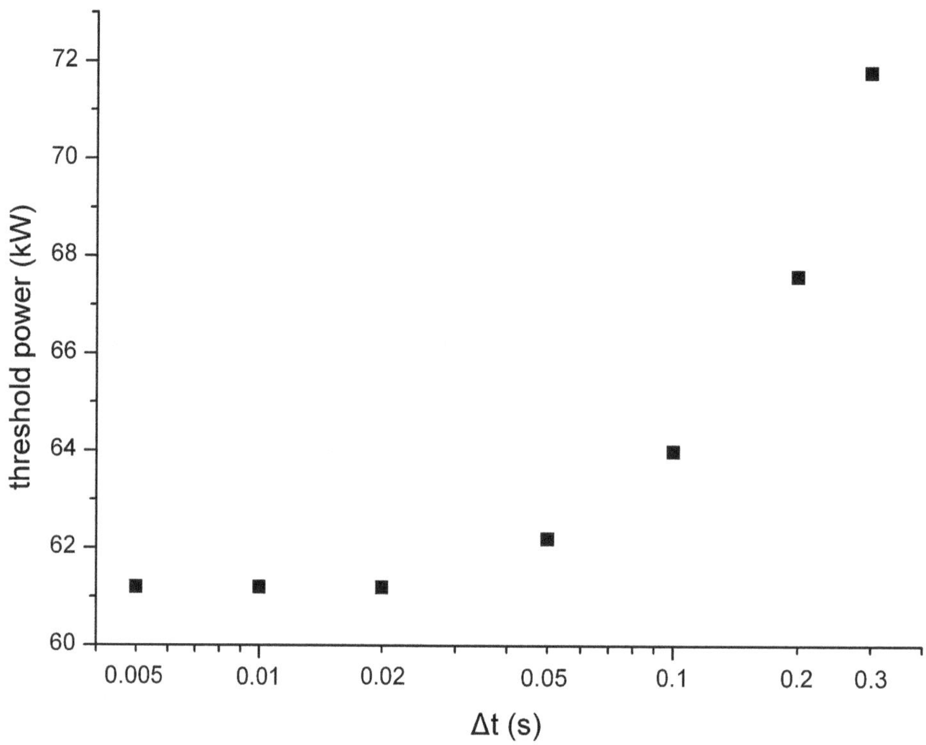

Comparison with Experiments

As proved by the grid refinement study, a time step size of 0.02 s and a spatial grid size of 0.05 m are chosen in the numerical analysis, which is sufficiently small to eliminate the possible numerical diffusion artificially induced by inappropriate grid sizes. System parameters including the system pressure, inlet temperature and total flow rate are set to be the same with the unstable points obtained in the experiments. Then the heat flux corresponding to the stability boundary is calculated. To illustrate the comparison in one two-dimensional plane, the system parameters are translated into dimensionless numbers N_{SPC} and N_{TPC}.

Figure 30 shows the comparison between the calculated and experimental N_{TPC}. The relative errors from -3% to -12%, though all the points show a conservative trend, indicate that the numerical code is capable to predict well the stability boundaries. The threshold powers in experiments are all higher than those in calculations. That might be induced by the heat capacity in the INCONEL Alloy pipes, which inhibits the flow oscillation.

Comparison Between Geometrical Models

In this section the geometrical simplification is discussed, for the purpose of making the experimental results universally and readily available for numerical analysis of supercritical stability. Also, due to the fact that experimental facility is generally complicated for numerical modeling, this study could

Figure 30. Comparison between the calculated and experimental results

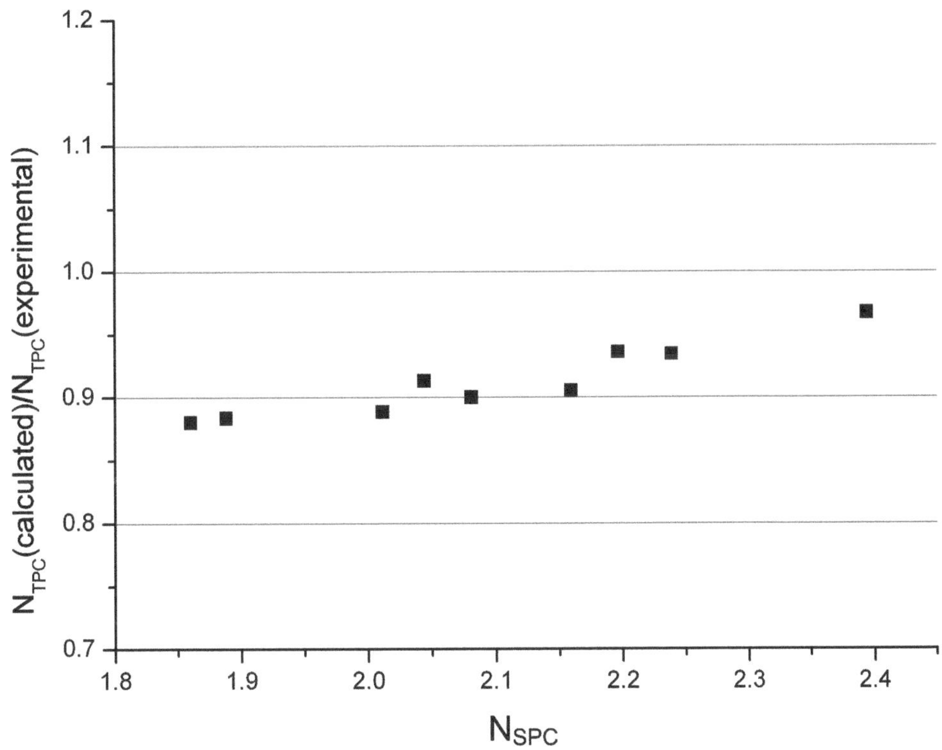

provide experience for the subsequent investigation when dealing with combination and comparison of experimental and numerical results.

Since being validated by the experimental results, the numerical code is applied to analyze the flow stability in each geometrical model proposed in section 2.1 (see Figure 26, Figure 27 and Figure 28). Similar to Model A discussed above, grid refinement studies are performed for Model B and Model C. Then the stability boundaries can be obtained and compared. For this analysis the total mass flow rate is set to be 120 kg/h and the system pressure is 25 MPa. The threshold heat fluxes are calculated for various inlet temperatures, as presented in Figure 31. It is observed that the boundaries for Model B agree well with that for Model A, while Model C deviates substantially.

The deviation of Model C confirms the speculation that the entrance and riser sections cannot be eliminated with respect to numerical modeling of flow stability. Indeed, if the gravitational pressure drop is "translated" into the local pressure drop, as a consequence of applying Model C, the inlet or outlet overall pressure drop in the calculation (ΔP_c) cannot be equal to that in the experiments (ΔP_e) for all thermal-hydraulic conditions. The difference between ΔP_c and ΔP_e is given by

$$\Delta P_c - \Delta P_e = \Delta P_{l,c} - \left(\Delta P_{fr,e} + \Delta P_{l,e} + \Delta P_{g,e}\right) = \left[k_{l,c} - (k_{fr,e} + k_{l,e} + k_{g,e})\right]\frac{G^2}{2\rho} \tag{103}$$

where both of $k_{l,c}$ and $k_{l,e}$ are constant. $k_{fr,e}$ represents the effect of friction pressure drop and changes slightly with the Reynolds number. $k_{g,e}$ represents the effect of gravitational pressure drop and can be given by

Figure 31. Comparison of boundaries between the geometrical models

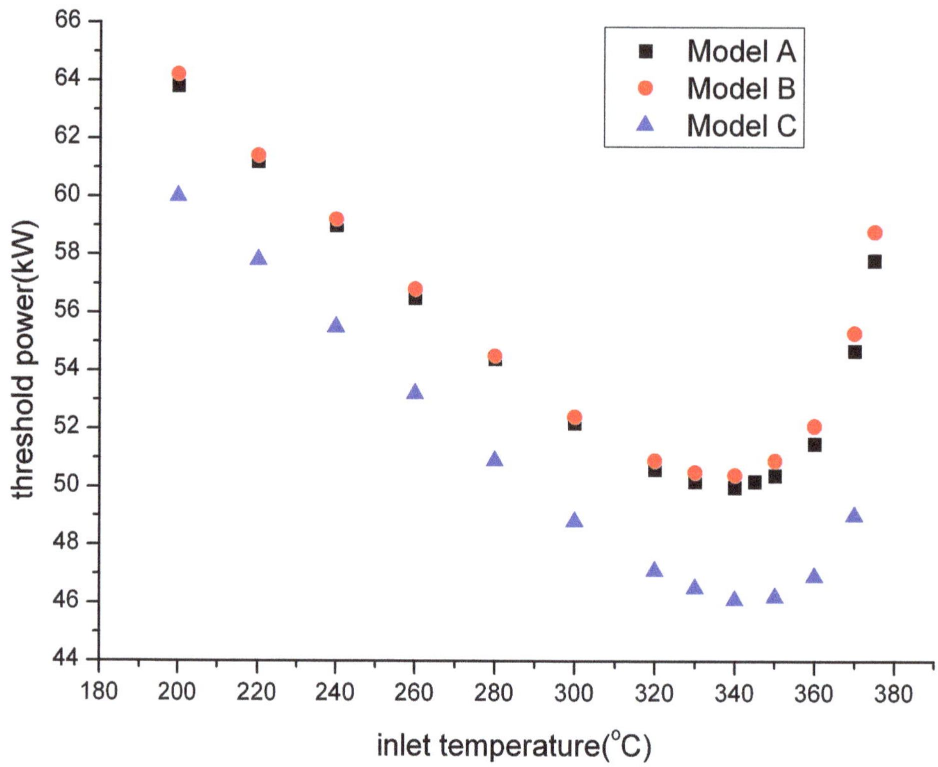

$$k_{g,e} = \frac{2\rho^2 gH}{G^2} \tag{104}$$

We can see that $k_{g,e}$ is considerably affected by the thermal-hydraulic conditions (G or ρ). If ΔP_c equals to ΔP_e at specific steady states, which is the basic rule when quantifying the pressure drop coefficients, ΔP_c differs remarkably from ΔP_e at some other thermal-hydraulic conditions during transient flow. In that way, Model C is incapable of simulating the test section.

Conversely, we can draw the conclusion once again that Model A is exactly suitable for the simulation, with respect to the effect of inlet or outlet pressure drop. The friction pressure drop distribution and gravitational pressure drop in Model A approximately equal those in the test section, thus

$$\Delta P_c - \Delta P_e = \Delta P_{l,c} - \Delta P_{l,e} = (k_{l,c} - k_{l,e})\frac{G^2}{2\rho} \tag{105}$$

where both of $k_{l,c}$ and $k_{l,e}$ are constant. ΔP_c and ΔP_e change simultaneously with the thermal-hydraulic conditions, and they can be kept equal all the time.

As for Model B, the gravitational pressure drop is equivalent to that in the test section due to the same height of entrance or riser section. Thus we have

$$\Delta P_c - \Delta P_e = (\Delta P_{fr,c} + \Delta P_{l,c}) - (\Delta P_{fr,e} + \Delta P_{l,e}) = [(k_{fr,c} - k_{fr,e}) + (k_{l,c} - k_{l,e})]\frac{G^2}{2\rho} \tag{106}$$

where both of $k_{l,c}$ and $k_{l,e}$ are constant. $k_{fr,c}$ and $k_{fr,e}$ represent the effect of friction pressure drop, and can be given by

$$k_{fr} = \sum f_r \frac{l}{D} \tag{107}$$

$k_{fr,c}$ and $k_{fr,e}$ change simultaneously with the Reynolds number, and the change is rather slight compared with $k_{l,e}$ for this experiment. Thus the change in ($k_{fr,c}$ -$k_{fr,e}$) is negligible. That is to say, ΔP_c and ΔP_e can be kept approximately equal all the time. Hence Model B is suitable for simulating the test section.

In the subsequent analysis, Model B is applied instead of Model A due to the relatively simple and common geometrical configuration. The entrance and riser sections were 200 and 350 mm, respectively. And the inlet (outlet) orifice pressure drop coefficients exhibit certain differences between the two channels, shown in Equation 79.

Parametric Effects on Stability

System parameters such as pressure, inlet temperature and mass flow rate may have a significant impact on flow instability, as indicated by correlative studies for two-phase flow. Considering the problems encountered in the experiments, the effects of inlet temperature and total mass flow rate are numerically analyzed. For the results presented below, Model B is applied as the geometrical model.

Effects of Inlet Temperature

The effects of inlet temperature on the threshold heat flux can be seen in Figure 31. For a relatively small inlet temperature, the threshold power decreases with increasing inlet temperature, indicating the system is more unstable, whereas for a relatively large inlet temperature the effect is stabilizing. Such non-monotonic effect of inlet temperature on flow stability is well-known for two-phase flow, which shows there is a definite range of inlet temperature within which the flow would oscillate and above or below this it is stable. The similar effect of inlet temperature reveals that the mechanisms of flow instability are probably alike between the supercritical flow and two-phase flow.

Moreover, to provide advice and experience for the feasibility and safety of experiments, the maximum bulk temperatures (i.e. outlet bulk temperatures) at stability boundary points are obtained. The working condition is consistent with that in Figure 31 (the total mass flow rate is set to be 120 kg/h and the system pressure is 25 MPa) .Despite the fact that the threshold power exhibits a decreasing trend at small inlet temperature (see Figure 31), Figure 32 shows that $T_{b,max}$ increases with increasing inlet temperature at a wide range. Here $T_{b,max}$ corresponds to the outlet bulk temperature at steady state when threshold power is applied to the system. The result indicates that the increase of inlet temperature would make the flow instability experiments more difficult.

Figure 32. Effect of inlet temperature on the outlet temperature

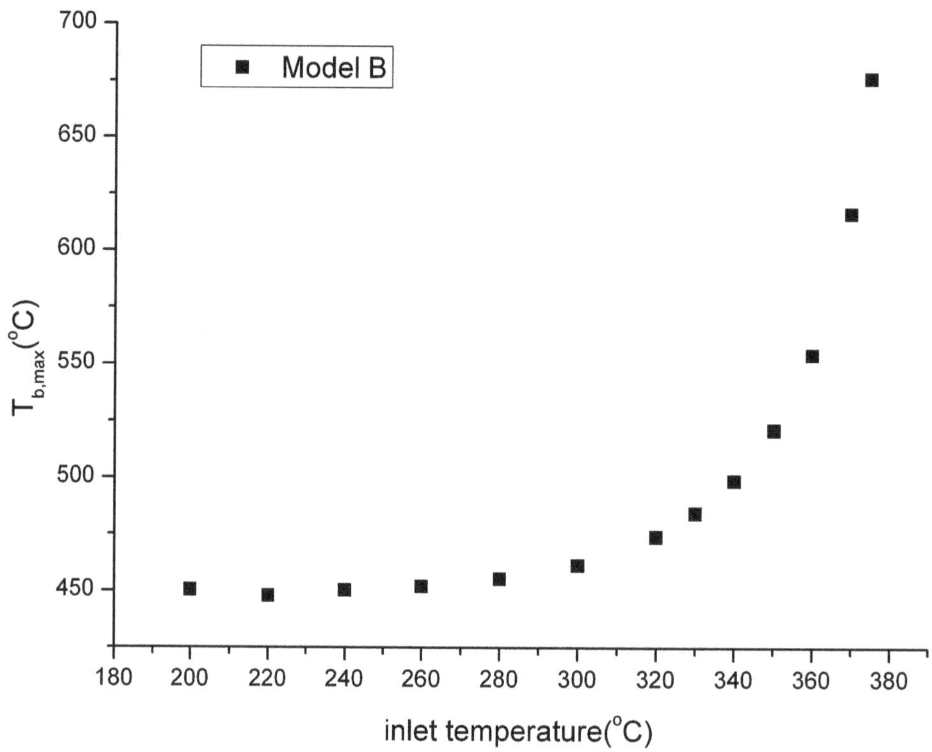

Effects of Total Flow Rate

The effect of mass flow rate on the threshold power has been indicated, in fact, by the dimensionless parameters proposed by Ambrosini and Sharabi (2006), where the form of Q/M represents the proportional relationship between the heat power and mass flow rate.

Different from the single channel applied by Ambrosini and Sharabi (2006), parallel-channel system is considered in this chapter. For the results presented below, the system pressure is 25 MPa and the inlet temperature is 220 °C. The effects of total flow rate are shown in Figure 33. For comparison, a symmetrical geometrical model is also applied and the results are presented in Figure 34. As for the symmetrical model, the inlet and outlet orifice pressure drop coefficients of both channels are set to be 5.5. The results show that the threshold power is roughly proportional to the total mass flow rate, no matter the system is symmetrical or not.

Another finding from the comparison is that the maximum bulk temperature has a different trend with increasing total flow rate. Due to the proportional relationship between the threshold power and mass flow rate, $T_{b,max}$ (corresponding to the outlet bulk temperature at steady state when threshold power is applied) remains almost the same for the symmetrical model (Figure 34). Nevertheless, for the asymmetrical model $T_{b,max}$ increases simultaneously with the total mass flow rate (Figure 33). This is caused by the asymmetrical distribution of flow rate between the two channels. As can be seen in Figure 35, the variance of individual mass flow rate widens with increasing total flow rate, from a relative difference

Figure 33. Effect of total flow rate on the threshold power and outlet temperature (asymmetrical)

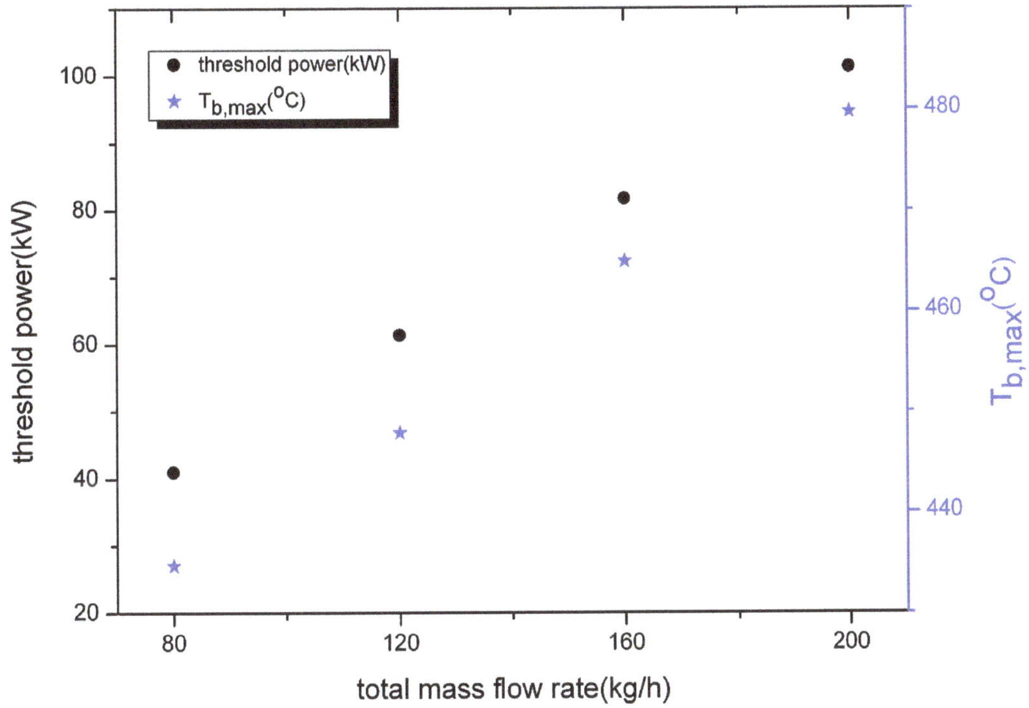

Figure 34. Effect of total flow rate on the threshold power and outlet temperature (symmetrical)

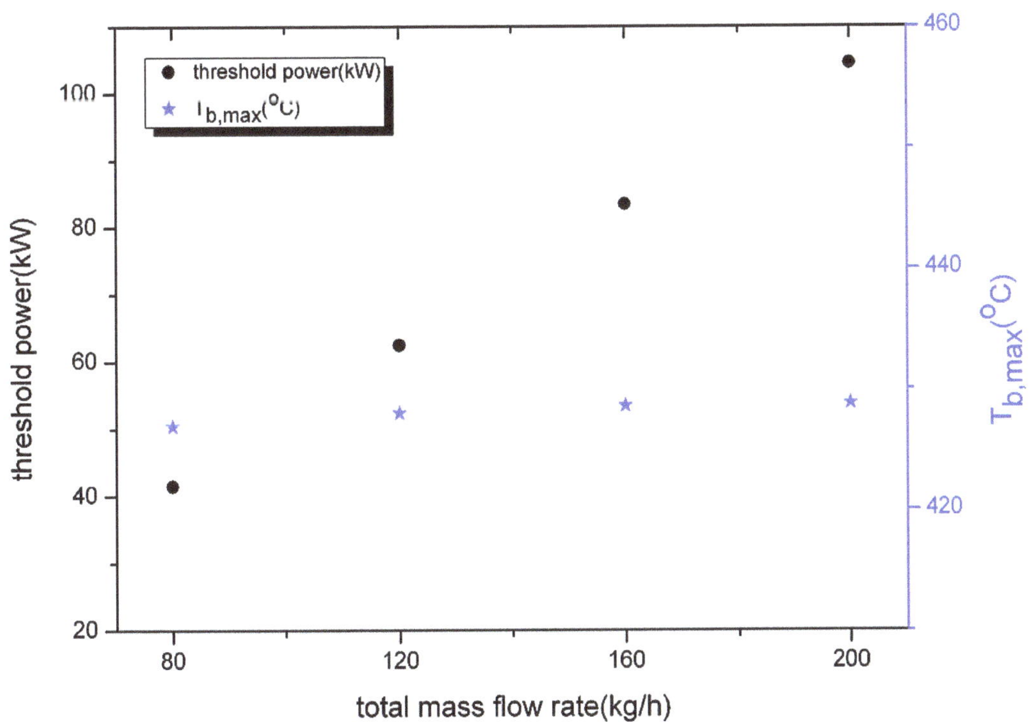

of 7.8% to 34.2%. Thus the outlet bulk temperature in the channel with lower flow rate increases. This finding indicates that it would be more difficult to perform an experiment at higher total mass flow rate, since the two-parallel-channel system is practically asymmetrical.

CFD Simulation Method

Besides the in-house code analysis above, the CFX code is used to analyze flow instability characteristics in two heated channels with supercritical water. Numerical models which could affect the onset of flow instability are analyzed systematically and proper numerical models are selected. From the sensitivity analysis, it is found that numerical models in the CFX code have a great influence on the onset of flow instability in two heated parallel channels with supercritical water. But up to now there are not enough unsteady experimental data for model validation, so it is difficult to make a decision that which numerical model is the best one. From the experience obtained by CFD engineers, it shows that higher order of the difference scheme and shorter time step will contribute to a more accurate result. Moreover PISO method is appropriate for the unsteady calculation and SST model is suitable for the heat transfer simulation under supercritical condition (Liu et al., 2013). But from a time saving perspective, the standard $k-\varepsilon$ model and 0.1s time step are still adopted incorporating with PISO method and second order up wind scheme in the following numerical simulations.

The numerical result is compared with experiment data, which shows a good agreement on the threshold power. The geometrical models in the experiment and numerical simulation are totally the same. But it should be noticed that due to manufacture and installation error of test section in the experiment, the

Figure 35. Flow rate distributions for different total flow rates (asymmetrical)

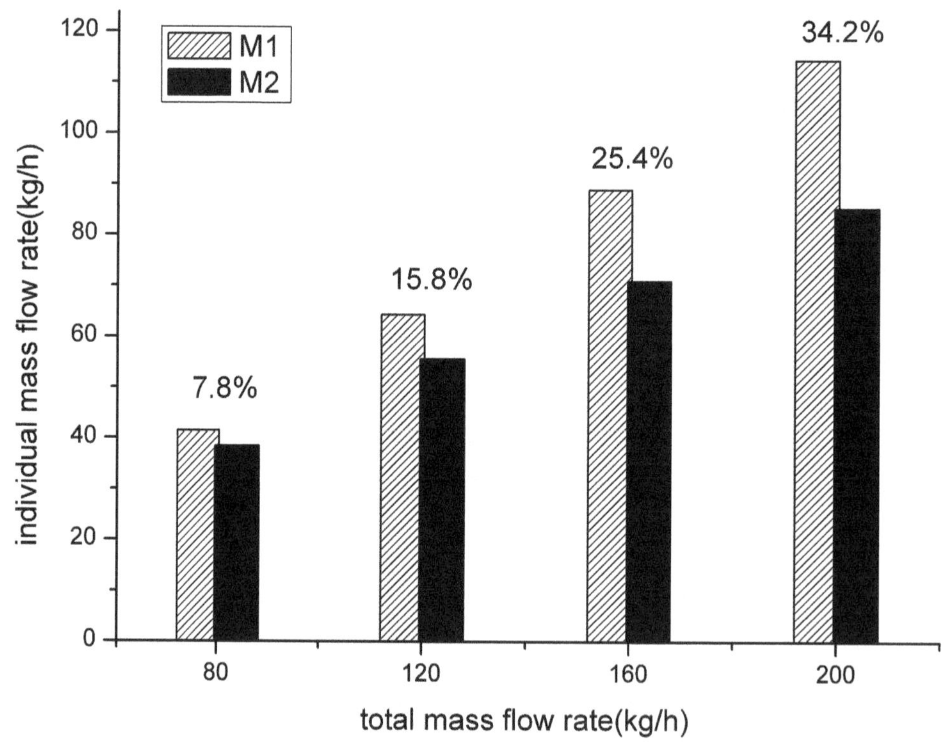

inlet and outlet pressure loss coefficients are different in each channel. So the pressure jump model is used in the numerical simulation to adjust the inlet and outlet pressure loss coefficients, which is shown in the following equation:

$$K = C \cdot n \tag{108}$$

Finally, the inlet and outlet pressure loss coefficients are the same in the experiment and numerical simulation

Table 2 shows the comparison among 1D numerical results obtained in last section, 3D numerical results and experiment data. Figure 36 shows the relative error of threshold power. It should be mentioned here that the 1D code is based on 1D conservation equations of mass, momentum and energy. Due to the simplification of physical models, the empirical correlations must be adopted to calculate heat transfer and frictional resistance, which is totally different from 3D code.

From Table 2 it can be found that the influence of inlet temperature on the instability boundary in experiment and numerical simulations are totally the same under different system pressure. With the increase of inlet temperature, the threshold power will decrease consequently and the decrease of threshold power in two kinds of numerical simulations is larger than the experiment. However, the influence of system pressure is complicated which is shown in Figure 37. With the increase of system pressure, the threshold power of experiment and 1D simulation increases monotonically, on the contrary the 3D result shows it will decrease first and then increase slightly.

Moreover it can be found that 1D result is absolutely more unstable especially when the inlet temperature and system pressure are high, the max relative between 1D result and experiment data error is about 12%. While 3D result shows a better agreement with experiment data, the max relative error is about 9%. Considering the two assumptions made above, the error between experiment and 3D calculation is acceptable.

Although there is a good agreement between experiment and 3D calculation on the threshold power, it is found the characteristics of oscillation are different, especially the period of oscillation. Figure 38 shows the oscillation in case3. It is interesting to find out that in the experiment there is a sustained out of phase oscillation with fixed amplitude when flow instability occurs. But in 3D numerical simulation

Table 2. Comparison between experiment and numerical simulations

Case	System Pressure (MPa)	Inlet Temperature (K)	Inlet Mass Flow Rate (kg/s)	Threshold Power (kw) Experiment	Threshold Power (kw) 1D Code	Threshold Power (kw) CFX Code
1	23	453	0.0342	67.9	65.7	69.8
2	23	473	0.0333	66.0	61.7	67.9
3	23	493	0.0333	65.6	59.0	65.0
4	24	473	0.0333	67.0	62.7	65.0
5	24	493	0.0333	66.0	60.3	62.8
6	24	513	0.0331	64.6	57.1	59.9
7	25	473	0.0328	69.3	62.7	65.6
8	25	493	0.0333	68.9	61.2	63.6
9	25	513	0.0339	67.9	59.8	61.6

Figure 36. Comparison among 1D result, 3D result and experiment data

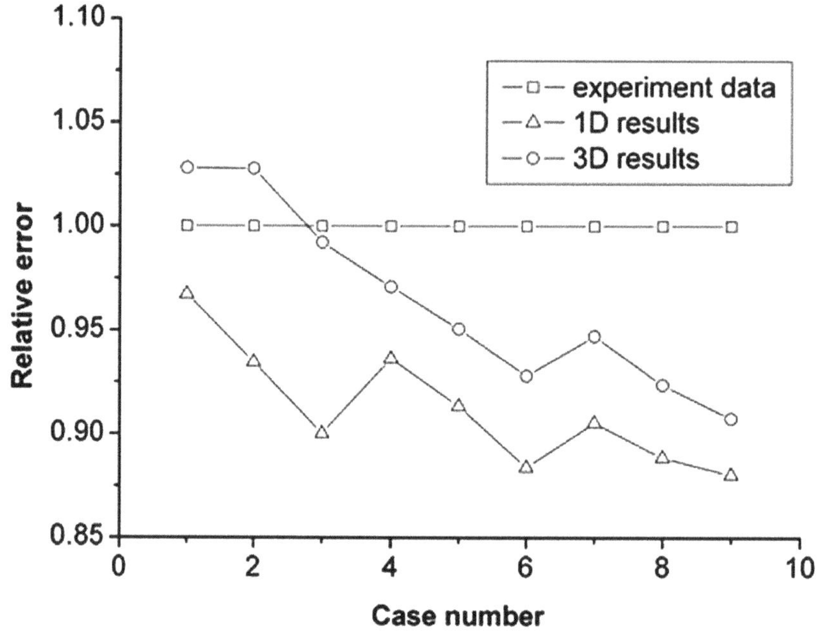

there is no sustained oscillation with fixed amplitude. If the heating power is larger than or equal to threshold power, the oscillation will diverge. On the contrary, if the heating power is less than threshold power, the system is absolutely stable.

Moreover it is found the period of oscillation in numerical simulation is much longer than experiment. Although the oscillation is diverged in numerical simulation, the period of oscillation is nearly unchanged which is equal to 4.5s. But in the experiment it is much shorter which is equal to 1s. This could be caused by the boundary conditions in 3D simulation where the random oscillation of total inlet mass flow rate is ignored. It is thought that the random oscillation of total inlet mass flow rate could bring a perturbation to the parallel channels and make the characteristics of oscillation different from numerical simulation. To make a better comparison between experiment and numerical simulation, two ways are feasible in the future. The first one is taking the random oscillation of inlet mass flow rate into consideration in numerical simulation. Another one is improving the experimental technique and making the total mass flow rate to be constant.

FUTURE RESEARCH DIRECTIONS

The design criteria for SCWR are based on the cladding temperature limit for normal operation and trip analyses. Experimental data on heat transfer and pressure drop are crucial in establishing this limit accurately. The SCWR may be susceptible to dynamic instability due to the sharp variation in fluid properties (such as density) at the vicinity of the critical point. This instability may lead to high cladding temperature in the fuel prematurely impacting on the operating and safety margins. In support of the design and operation of the reactor safety (or relief) valve and the automatic depressurization system,

Figure 37. Influence of system pressure (M_{int} =0.0333kg/s, T_{int} =220°C)

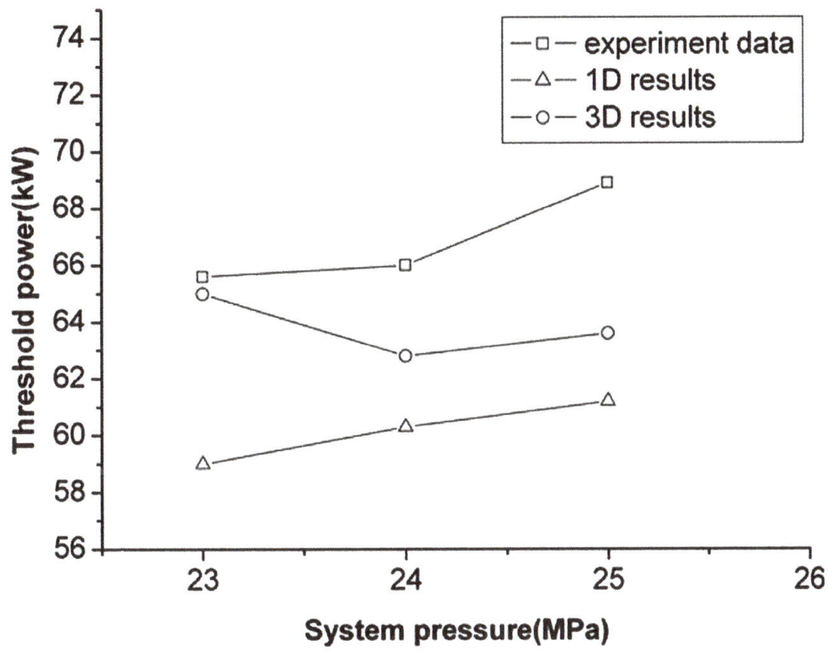

Figure 38. Oscillation of outlet mass flow rate M_{in1} in experiment and 3D simulation (case3, Q_e =65.6kw, Q_n =65kw)

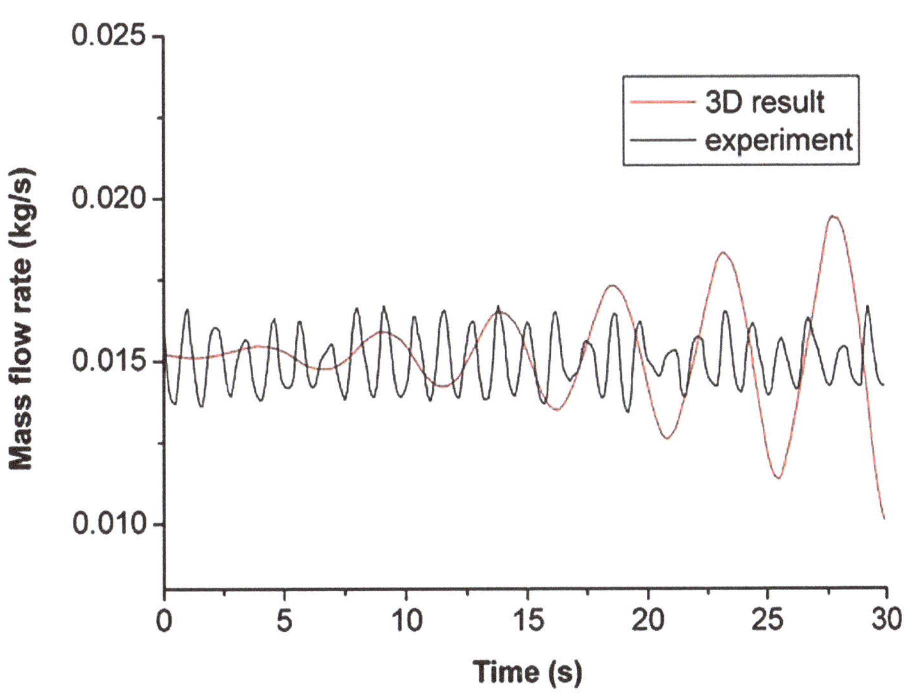

the critical (or choked) flow characteristic must be established at supercritical conditions since current information has been obtained at subcritical conditions. This established characteristic is also required in the analysis of a postulated large-break loss-of-coolant accident event.

According to the 2016-2020 Updated Project Plan for Thermal-Hydraulics and Safety Area drafted by SCWR PROJECT MANAGEMENT BOARD, the main technological task in the next few years covers the following aspects:

Experimental Heat transfer and pressure drop characteristics, including various geometry shapes, such as tubes, annuli, bundle; various coolant types, from water to Freon, to CO_2, and scaling method between different fluids; various heat transfer enhance structures, such as spacer or wire wraps. It has to point that, the more precise and refined temperature measurement technique is in an urgent need to get a full resolution of the temperature profile for the rod bundle.

- **Critical Flow:** There is a lack of data for critical (or choked) flow at supercritical conditions since LWRs (for which most critical flow work has been performed in the past) operate at subcritical pressures. Critical (or choked) flow phenomena are of great importance in designing/operating the reactor safety/relief valves and the automatic depressurization system, as well as in the analysis of LOCA events. Blowdown experiments are needed in a dedicated small-scale apparatus consisting of a pressure tank with discharge nozzles of different shape, size and length. Direct experimental measurements of the temperature and pressure along the discharge nozzles, and of the void fraction and flow rate at the nozzle outlet should be obtained. These data will enable accurate benchmark of existing critical-flow models and, if needed, development of new ones.
- **Instability:** Large variation of coolant density in the axial direction and strong coupling of the neutronic and thermal-hydraulic behavior makes the system theoretically susceptible to dynamic oscillations. Susceptibility to dynamic instability has been evaluated only for 1D single-channel steady-state operating conditions. Parallel-channel (out-of-phase) instabilities as well as instabilities that might arise during start-up or power transients have not been investigated.
- **System Code:** Existing water-reactor thermal design and safety analysis codes (including CFD codes) require modification, verification, and validation for applications to the SCWR. Hydraulics parameters and heat-transfer prediction methods derived from various databases must be incorporated into these codes. Routine for evaluating fluid properties must be updated to cover SC conditions. Validation of these codes against transient data obtained at the integral facility shall be performed once these data become available. A pre-test simulation may also be of interest to support the test loop design and establish test conditions. The extent of the international collaboration may be limited for this task because code selection for these analyses is different in each organization. Code-to-code benchmarking perhaps is of the most interest.
- **CFD Modeling:** The CFD has been playing an important role in analyzing local flow and thermal behavior of fluid in complicated geometries as well as conventional geometries and its usage in nuclear industry is continuously expanding. A numerical analysis of a flow and thermal field in subchannels and an estimation of heat transfer coefficient is required to support experiments as well as to provide accurate correlations as an input to the design and analysis codes. To prevent the fuel damage boundary of heat transfer deterioration, a mild version of boiling crisis, should be predicted as accurately as possible. The direct simulation by CFD code of the normal heat transfer as well as the deteriorated heat transfer is expected to provide valuable information in this regard.

- **Sub-Channel Analysis:** Sub-channel analysis is a key determinant on fuel operating temperatures. Existing sub-channel analysis codes are still to be improved in regards to the incorporation of physical properties of water at supercritical pressure and their robustness in various conditions. The results of calculation will determine whether turbulence promoting, or sub-channel coolant mixing features are required with supercritical water coolant.

CONCLUSION

This chapter is mainly focused on illustrating some introductory progress on thermal hydraulic issues of supercritical water, including heat transfer characteristics, pressure loss characteristics, flow stability issues and numerical method. These research work were performed in NPIC these years, aimed to give a basic idea of elementary but important topics in this area.

An analytical method was proposed to predict the heat transfer coefficient based on the two layer thermal wall function. The thermal wall function was first derived according to the boundary layer theory and some other assumptions. From the derivation process of the new heat transfer correlation, it could be seen that it is better to use the average Prandtl number as the characteristic quantity to describe the integration effect of the fluid property variation. This is in accordance with the experimental conclusions. Based on this method, the Gnielinski correlation and the Dittus-Bolter correlation were extended to supercritical water. The comparisons with the supercritical heat transfer experiments of NPIC shows good agreement of this method.

A general method based on two-layer wall function theory was set up which could be applied into different flow geometry, into supercritical water conditions. The two-layer wall function consist both the laminar viscous layer and the turbulent layer in the boundary layer. According to the relationship between the average velocity and friction velocity, the friction correlation could be derived. Based on the new wall function modified by Van Direst transformation, the new frictional correlation could be extended to supercritical water. The friction coefficient under supercritical conditions is different from that of subcritical water. One of the peculiarities is the 'V' shape profile of friction curve. It is contributed by the variation of fluid property across the boundary layer. It is found that the density ratio is the key parameter in influencing the friction coefficient. The prediction of the new correlation is in good agreement with the experimental data which verifies the physical correct of this method.

Flow instability experiments have been carried out in a two-parallel-channel system with supercritical water, aiming to provide an up-to-date knowledge of supercritical flow instability phenomena and initial validation data for numerical analysis. With the increase of the heat flux, the evolution of flow rates could be divided into four stages, quasi-symmetrical flow, asymmetrical flow caused by redistribution, irregular fluctuation, and parallel flow instability. The asymmetry of parallel channels is believed to be of great significance considering the feasibility of performing flow instability experiments with supercritical water, which provides certain advice and experience for further improvement of experimental apparatus and techniques. System parameters corresponding to the stability boundaries were recorded and translated into dimensionless parameters. Parametric studies show that the flow becomes more stable with increasing pressure or decreasing inlet temperature in the range of present experiments. It is suggested that experimental techniques be improved based on the experience provided in this chapter; thus more validation data could be applied to flow instability analysis of supercritical water.

An in-house code has been developed in order to better utilize and further expand the experimental results on supercritical flow instability. The numerical code is a one-dimensional code that can be applied to various geometrical configurations. The results show that the entrance and riser sections cannot be eliminated with respect to numerical modeling of flow stability, and a relative simple as well as common geometrical model is proposed for the purpose of making the experimental results universally and readily available for numerical analysis of supercritical stability. Also, this study is helpful and interesting from the point of view of combination and comparison of experimental and numerical results. Parametric effects on supercritical flow instability are numerically analyzed.

According to the update progress of thermal hydraulics these years, the main technological task in the next few years were suggested, including experimental heat transfer and pressure drop with more precise and refined measurement techniques, critical flow experiments and prediction methods, instability experiments and methods, the evaluation and improvement of system code and CFD code applied into supercritical water.

REFERENCES

Ambrosini, W. (2007). On the analogies in the dynamic behaviour of heated channels with boiling and supercritical fluids. *Nuclear Engineering and Design*, *237*(11), 1164–1174. doi:10.1016/j.nucengdes.2007.01.006

Ambrosini, W. (2008). Thermal-hydraulic stability limits. *International Students Workshop on High Performance Light Water Reactors*, Karlsruhe, Germany.

Ambrosini, W. (2009). *Continuing assessment of system and CFD codes for heat transfer and stability in supercritical fluids.4th international symposium on Supercritical Water-Cooled Reactors*, Heidelberg, Germany.

Ambrosini, W., & Sharabi, M. (2006). Dimensionless parameters in stability analysis of heated channels with fluids at supercritical pressures.*Proceedings of 14th International Conference on Nuclear Engineering (ICONE 14)*. doi:10.1115/ICONE14-89862

Bae, J.H., & Park, J. H. (2011). Analytical prediction of turbulent friction factor for a rod bundle. *Annals of Nuclear Energy, 38*, 348-357.

Bishop, A.A., Sandberg, R.O., & Tong, L.S. (1964). *Forced Convection heat transfer to water at near-critical temperature and supercritical pressures.* Westinghouse Electric Corporation WCAP-2056.

Boure, J. A., Bergles, A. E., & Tong, L. S. (1973). Review of two-phase flow instability. *Journal of Nuclear Engineering and Design*, *25*(2), 165–192. doi:10.1016/0029-5493(73)90043-5

Chatoorgoon, V. (1986). A simple thermalhydraulic stability code. *Nuclear Engineering and Design*, *93*, 51–67. doi:10.1016/0029-5493(86)90194-9

Chatoorgoon, V. (2001). Stability of supercritical fluid flow in a single-channel natural-convection loop. *International Journal of Heat and Mass Transfer*, *44*(10), 1963–1972. doi:10.1016/S0017-9310(00)00218-0

Chatoorgoon, V. (2008). Static instability in supercritical parallel-channel systems.*Proceedings of the 16th International Conference on Nuclear Engineering (ICONE 16).*

Chatoorgoon, V. (2008). Supercritical flow stability in horizontal channels. *Nuclear Engineering and Design*, *238*(8), 1940–1946. doi:10.1016/j.nucengdes.2007.10.033

Chatoorgoon, V., Shah, M., & Duffey, R. (2007). *Linear predictions of supercritical flow instability in parallel channels.* 3rd Int. Symposium on SCWR – Design and Technology, Shanghai, China.

Chatoorgoon, V., Voodi, A., & Fraser, D. (2005). The stability boundary for supercritical flow in natural convection loops part I: H_2O studies. *Nuclear Engineering and Design*, *235*(24), 2570–2580. doi:10.1016/j.nucengdes.2005.06.003

Cheng, X., & Schulenberg, T. (2001). *Heat Transfer at Supercritical Pressures-Literature Review and Application to an HPLWR.* Institut fur Kern-und Energietechnik.

Cheng, X., & Yang, Y. H. (2008). A point-hydraulics model for flow stability analysis. *Nuclear Engineering and Design*, *238*(1), 188–199. doi:10.1016/j.nucengdes.2007.06.012

Clauser, F. H. (1956). The turbulent buoundary layer. *Advances in Applied Mechanics*, *4*, 1–51. doi:10.1016/S0065-2156(08)70370-3

Fu, D. X., Ma, Y. W., & Li, X. L. (2010). *Direct numerical simulation to compressible turbulence (Chinese).* Beijing: Science Publisher.

Generation IV International Forum. (2015). *2015 GIF Annual Report.* Printed by the OECD Nuclear Energy Agency for the Generation IV International Forum.

Generation IV International Forum. (2016). *Updated Project Plan (2016-2020) On Thermal hydraulics And Safety For The International Research And Development of The Supercritical Water-Cooled Reactor Nuclear Energy System. Version 01.* Author.

Haaland, S. E. (1983). Simple and explicit formulas for the friction factor in turbulent flow. *Journal of Fluids Engineering*, *103*(1), 89–90. doi:10.1115/1.3240948

Hartnett, J. P., Koh, J. C. Y., & McComas, S. T. (1962). A comparison of predicted and measured friction factors for turbulent flow through rectangular ducts. *Trans.ASME J. Heat Transfer*, *84*(1), 82–88. doi:10.1115/1.3684299

IAEA Nuclear power technology development section. (2014). *Heat Transfer Behaviour and Thermohydraulics Code Testing for Supercritical Water Cooled Reactors (SCWRs).* Printed by the IAEA in Austria. IAEA-TECDOC-1746.

Ishii, M., & Zuber, N. (1970). Thermally induced flow instabilities in two-phase mixtures.*Proceedings of the Fourth International Heat Transfer Conference, 5.*

Jackson, J. D. (2002). *Consideration of the heat transfer properties of supercritical pressure water in connectin with the cooling of advanced nuclear reactors.* Paper presented at the 13th Pacific Basin Nuclear Conference, Shenzhen City, China.

Jackson, J.D., & Hall W.B. (1979a). *Forced convection heat transfer to fluids at supercritical pressure. Turbulent forced convection in channels and bundles*. New York: Hemisphere Publishing Corporation.

Jackson, J. D., & Hall, W. B. (1979b). Influences of buoyancy on heat transfer to fluids flowing in vertical tubes under turbulent conditions.Turbulent forced convection in channels and bundles, (pp. 613-640). Hemisphere Publishing Corporation.

Ji, S., Shirahama, H., Koshizuka, S., & Oka, Y. (2001). Stability analysis of supercritical pressure light water-cooled reactor in constant pressure operation. *Proc. of the Ninth International Conference on Nuclear Engineering (ICONE9).*

Kakac, S., & Bon, B. (2008). A review of two-phase flow dynamic instabilities in tube boiling systems. *International Journal of Heat and Mass Transfer*, *51*(3-4), 399–433. doi:10.1016/j.ijheatmasstransfer.2007.09.026

Kays, W. M., Crawford, M. E., & Weigand, B. (2005). *Convective heat and mass transfer New York*. McGraw-Hill.

Lee, K. B. (1995). Analytical prediction of subchannel friction factor for infinite bare rod square and triangular arrays of low pitch to diameter ratio in turbulent flow. *Nuclear Engineering and Design*, *157*(1-2), 197–203. doi:10.1016/0029-5493(95)00996-P

Licht, J., Anderson, M., & Corradini, M. (2008). Heat transfer to water at supercritical pressures in a circular and square annular flow geometry. *International Journal of Heat and Fluid Flow*, *29*(1), 156–166. doi:10.1016/j.ijheatfluidflow.2007.09.007

Marcel, C. P., Rohde, M., Masson, V. P., & Van der Hagen, T. H. J. J. (2009). Fluid-to-fluid modeling of supercritical water loops for stability analysis. *International Journal of Heat and Mass Transfer*, *52*(21-22), 5046–5054. doi:10.1016/j.ijheatmasstransfer.2009.03.022

Mentes, A., Kakac, S., Veziroglu, T.N. et al. (1989). Effect of inlet subcooling on two-phase flow oscillations in a vertical boiling channel. *Waerme- und Stoffuebertragung, 24*, 25-36.

Misawa, T., Nakatsuka, T., & Yoshida, H. (2009). *Heat Transfer Experiments and Numerical Analysis of Supercritical Pressure Water in Seven-rod Test Bundle*. Paper presented at the 13th International Topical Meeting on Nuclear Reactor Thermal Hydraulics (NURETH-13) Kanazawa City, Ishikawa Prefecture, Japan.

Mokry, S., Pioro, I. L., Kirillov, P., & Gospodinov, Y. (2010). Supercritical-water heat transfer in a vertical bare tube. *Nuclear Engineering and Design*, *240*(3), 568–576. doi:10.1016/j.nucengdes.2009.09.003

Nuclear Power Institute of China. (2016). *Status report – Chinese Supercritical Water-Cooled Reactor (CSR1000)*. Retrieved from https://aris.iaea.org

Oka, Y., Koshizuka, S., Ishiwatari, Y., & Yamaji, A. (2010). Super Light Water Reactors and Super Fast Reactors. New York: Springer.

Oka, Y., & Koshizuka, S. (2000). Design concept of once-through cycle supercritical-pressure light water cooled reactors.*Proceedings of SCR-2000*. The University of Tokyo.

Ortega Gómez, T., Class, A., Lahey, R. T., Jr., & Schulenberg, T. (2007). *Stability analysis of heated flow channels with supercritical water.* 3rd Int. Symposium on SCWR-Design and Technology, Shanghai, China.

Pioro, I. L., & Duffey, B. (2007). *Heat transfer and hydraulic resistance at supercritical pressures in power-engineering applications*. ASME Press. doi:10.1115/1.802523

Pioro, I. L., Duffey, R. B., & Dumouchel, T. J. (2004). Hydraulic resistance of fluids flowing in channels at supercritical pressures (survey). *Nuclear Engineering and Design*, *231*(2), 187–197. doi:10.1016/j.nucengdes.2004.03.001

Podowski, M. Z., & Gallaway, T. (2010). Stability analysis of Generation IV supercritical water reactors. In *Technical Meeting on Heat Transfer*. Pisa, Italy: Thermal-Hydraulics and System Design for Supercritical Water.

NIST Reference Fluid Thermodynamic and Transport Properties–REFPROP. (2007). *Ver. 8.0 NIST Standard Reference Database 23*. U.S. Dept. of Commerce.

Rehme, K. (1973). Simple method of predicting friction factors of turbulent flow in non-circular channels. *International Journal of Heat and Mass Transfer*, *16*(5), 933–950. doi:10.1016/0017-9310(73)90033-1

Rohde, M., Marcel, C. P., TJoen, C., Class, A. G., & van der Hagen, T. H. J. J. (2011). Downscaling a supercritical water loop for experimental studies on system stability. *International Journal of Heat and Mass Transfer*, *54*(1-3), 65–74. doi:10.1016/j.ijheatmasstransfer.2010.09.063

Sharabi, M. B., Ambrosini, W., & He, S. (2008). Prediction of unstable behaviour in a heated channel with water at supercritical pressure by CFD models. *Annals of Nuclear Energy*, *35*(5), 767–782. doi:10.1016/j.anucene.2007.09.019

Sharabi, M. B., Ambrosini, W., & He, S. (2008). Transient 3D stability analysis of SCWR rod bundle subchannels by a CFD code. *Proceedings of the 16th International Conference on Nuclear Engineering (ICONE16)*. doi:10.1115/ICONE16-48701

Shitsman, M. E. (1963). Impairment of the heat transmission at supercritical pressures. *High Temperature*, *1*, 237–244.

Su, K., & Freire, A. P. S. (2002). Analytical prediction of friction factors and Nusselt numbers of turbulent forced convection in rod bundles with smooth and rough surfaces. *Nuclear Engineering and Design*, *217*(1-2), 111–127. doi:10.1016/S0029-5493(02)00155-3

Swenson, H. S., Carver, J. R., & Kakarala, C. R. (1965). Heat transfer to supercritical water in smooth-bore tubes. *Journal of heat transfer, Transactions of the ASME, Series C*, *87*, 1289-1292. doi:10.1115/1.3689139

White, F. M. (1974). *Viscous Fluid Flow New York*. McGraw-Hill.

Xi, X., Xiao, Z. J., Yan, X., Li, Y., & Huang, Y. (2013). The axial power distribution validation of the SCWR fuel assembly with coupled neutronics-thermal hydraulics method. *Nuclear Engineering and Design*, *258*, 157–163. doi:10.1016/j.nucengdes.2013.01.031

Xi, X., Xiao, Z. J., Yan, X., Li, Y., & Huang, Y. (2014). An experimental investigation of flow instability between two heated parallel channels with supercritical water. *Nuclear Engineering and Design*, *278*, 171–181. doi:10.1016/j.nucengdes.2014.06.034

Xi, X., Xiao, Z. J., Yan, X., Xiong, T., & Huang, Y. (2014). Numerical simulation of the flow instability between two heated parallel channels with supercritical water. *Annals of Nuclear Energy*, *64*, 57–66. doi:10.1016/j.anucene.2013.09.017

Xiong, T., Yan, X., Huang, S. F., Yu, J., & Huang, Y. (2013). Modeling and analysis of supercritical flow instability in parallel channels. *International Journal of Heat and Mass Transfer*, *57*(2), 549–557. doi:10.1016/j.ijheatmasstransfer.2012.08.046

Xiong, T., Yan, X., Xiao, Z. J., Li, Y., Huang, Y., & Yu, J. (2012). Experimental study on flow instability in parallel channels with supercritical water. *Annals of Nuclear Energy*, *48*, 60–67. doi:10.1016/j.anucene.2012.05.018

Yamagata, K., Nishikawa, K., Hasegawa, S., Fujii, T., & Yoshida, S. (1972). Forced convective heat transfer to supercritical water flowing in tubes. *International Journal of Heat and Mass Transfer*, *15*(12), 2575–2593. doi:10.1016/0017-9310(72)90148-2

Yi, T. T., Koshizuka, S., & Oka, Y. (2004). A linear stability analysis of supercritical water reactors, (I) thermal–hydraulic stability. *Journal of Nuclear Science and Technology*, *41*(12), 1166–1175. doi:10.1080/18811248.2004.9726345

Yi, T. T., Koshizuka, S., & Oka, Y. (2004). A linear stability analysis of supercritical water reactors, (II) thermal–hydraulic stability. *Journal of Nuclear Science and Technology*, *41*(12), 1176–1186. doi: 10.1080/18811248.2004.9726346

Yu, J., Che, S., Li, R., & Qi, B. (2010). *Instability analysis of natural circulation at supercritical pressure.* The 8th International Topical Meeting on Nuclear Thermal-Hydraulics, Operation and Safety (NUTHO-8), Shanghai, China.

Zang, J. G., Yan, X., & Huang, S. F. (2012). *Analytical Prediction of Turbulent Friction Factor In Circular Pipe Under Supercritical Conditions.* Paper presented at the ICONE20POWER2012, Anaheim, CA. doi:10.1115/ICONE20-POWER2012-55159

Zang, J. G., Yan, X., Huang, S. F., Zeng, X., Li, Y., Huang, Y., & Yu, J. (2014). A general method of developing frictional formulas under supercritical conditions and in different geometries. *Annals of Nuclear Energy*, *65*, 262–271. doi:10.1016/j.anucene.2013.10.026

Zang, J. G., Yan, X., Huang, S. F., Zeng, X., Li, Y., Huang, Y., & Yu, J. (2014). A method of extending subcritical heat transfer formulas to supercritical conditions. *Nuclear Engineering and Design*, *266*, 186–193. doi:10.1016/j.nucengdes.2013.11.043

Zhao, J. Y. (2005). *Stability analysis of supercritical water cooled reactors* (PhD thesis). Massachusetts Institute of Technology.

Zuber, N., (1966). *An analysis of thermally induced flow oscillations in the near-critical and supercritical thermodynamic region*. Final Report Nas8-11422. NASA.

KEY TERMS AND DEFINITIONS

Critical Point: The point where the distinction between the liquid and gas phases disappears, i.e., both phases has the same temperature, pressure and volume. The critical point could also be called as critical state.

Deteriorated Heat Transfer: The heat transfer performance characterized with lower values of the wall heat transfer coefficient compared to those at the normal heat transfer.

Flow Instability: A kind of dynamic phenomena with its flow fluctuates regularly or irregularly, leading to other parameters fluctuate accordingly and could bring about damage to the system or equipment.

Near-Critical Point: The narrow region around the pseudocritical point where all the thermo physical properties of a pure fluid exhibit rapid variations.

Pseudocritical Point: The point at a pressure above the critical pressure and at a temperature where the specific heat capacity reaches its maximum value at this particular pressure.

Supercritical Fluid: Normally, the fluid with its pressures and temperatures higher than the critical pressure and critical temperature. However, in this chapter, the supercritical fluid includes both the temperature below and above the critical temperature as long as the pressure is above the critical pressure.

Wall Function: A kind of theory based on the wall boundary method to describe the near wall velocity and temperature profile.

APPENDIX 1: THERMAL HYDRAULIC TEST FACILITIES

Two test facilities were designed and constructed in NPIC for SCW T/H experiments; one is a small SCW mechanism loop (SSWT) (Figure 39) and the other is a large scale SCW T/H loop (LSWT) (Figure 40). With the two facilities, the flow and heat transfer experiments have been completed in various flow geometries, circular pipe, circle-annuli, square-annuli and small 2×2 rod bundles. A SCW test database has been compiled covering large range of operating conditions, which serves to develop experimental correlations and assess computer codes for the SCWR design.

Table 3. Design parameters of the test facilities

	SSWT	LSWT
Design Pressure, MPa	30.0	30.0
Design Temperature, °C	550	550
Max flow rate, m^3/h	5.0	30.0
Max Power, MW	1.2	5.0

Figure 39. Picture of the small scale SCW test facility (SSWT)

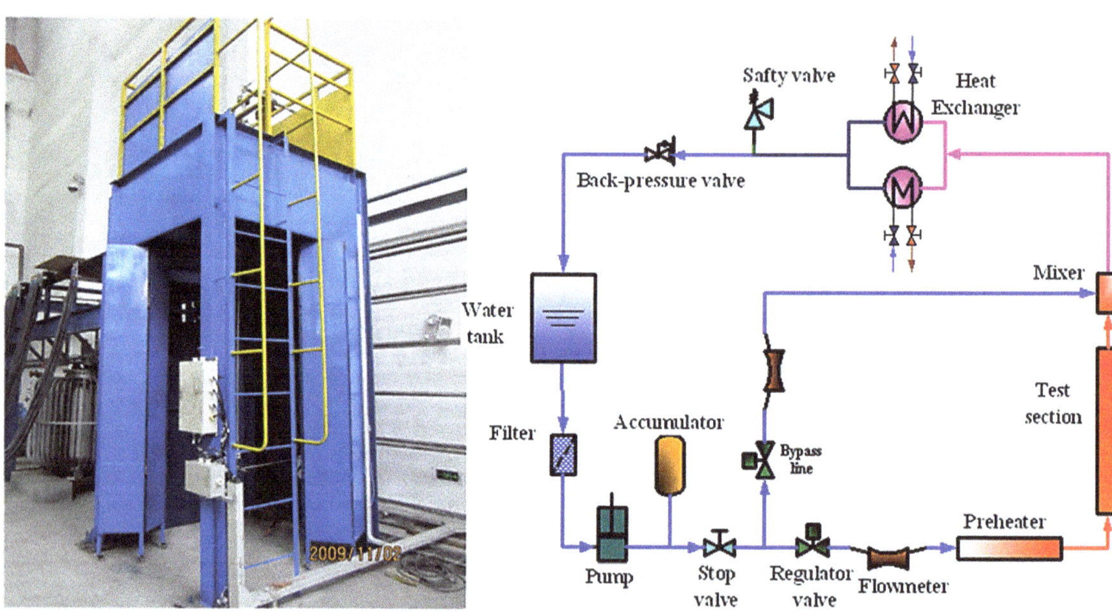

Figure 40. Picture of the large scale SCW test facility (LSWT)

APPENDIX 2: NOMENCLATURE

A: Flow area (m^2)
A,G:* Geometry parameters
B: Constant in wall function
$B_t(\mathrm{Pr})$: The function of Prandtl number
c_f: Fanning frictional coefficient
D: Tube diameter (m)
d_h: Hydraulic equivalent diameters (m)
f: Darcy-Weisbach frictional coefficient
$f(y)$: Function of y
$F(\theta)$: Function of θ
g: Gravity acceleration (m s^{-2})
G: Mass flux (kg m^{-2} s)
H_b: Bulk fluid Enthalpy (kJ kg^{-1})
k: Roughness height (m)
Nu: The Nusselt number
P: Pitch of rod (m) / Pressure (MPa)
Pr: The Prandtl number
$\overline{\mathrm{Pr}}$: The Average Prandtl number
$\langle \mathrm{Pr} \rangle_1$, $\langle \mathrm{Pr} \rangle_2$: The Average Prandtl number in laminar conduction and turbulent layer
q: Heat flux (kW m^{-2})

r : Radius (m)
r_0^+ : Dimensionless distance of r_0
Re : The Reynolds number
r_0 : Pipe radius
r_0^+ : Non-dimensional pipe radius
u_τ : Friction velocity
u, v : Velocity component in flow direction and normal direction
u', v' : Fluctuation velocity component in flow direction and wall normal direction
$u_{c,l}^+$: Non-dimensional velocity in viscous sublayer in compressed flow
$u_{c,t}^+$: Non-dimensional velocity in turbulent sublayer in compressed flow
u_m : Average velocity
y : Distance from the wall
y^+ : Non-dimensional distance from the wall
y_l^+ or y_t^+ : Non-dimensional distance of the upper bound of the laminar conduction layer
y_0^+ : Non-dimensional distance of the upper bound of the viscous layer
t : Temperature
t_w : Wall temperature
t_1 : Temperature at the upper bound of the laminar conduction layer
t_c : Temperature at the centerline of the pipe
Pr_t : Turbulent Prandtl number
c_p : Specific heat capacity
t_1^+, t_2^+ : Dimensionless temperature in laminar and turbulent region
$T_{\tau 1}, T_{\tau 2}$: Friction temperature in laminar and turbulent region
k : Thermal conductivity
h : Fluid enthalpy
$F(\theta)$: Function of θ

Greek Symbols

α : Molecular thermal diffusivity
α_1, α_2 : Velocity profile parameters
β : Pressure gradient parameter
β_1, β_2 : The geometry parameters
δ_1 : The displacement thickness of momentum boundary layer (m)
ε : Roughness height
ε_M : Turbulent viscosity
ε_H : Turbulent viscosity of heat transfer
φ_l : The material factor in the laminar layer

φ_t : The material factor in the turbulent layer
κ : Von Karman constant (=0.41)
θ : The angle position (rad)
τ : Shear stress (N m^{-2})
τ_w : Wall shear stress (N m^{-2})
μ : Dynamic viscosity (kg m^{-1} s^{-1})
μ_t : Turbulent viscosity (kg m^{-1} s^{-1})
μ_w : Dynamic viscosity evaluated at the wall temperature (kg m^{-1} s^{-1})
μ_{bl} : Dynamic viscosity in the viscous sublayer (kg m^{-1} s^{-1})
μ_b : Bulk average molecular viscosity (kg m^{-1} s^{-1})
ρ : Fluid Density (kg m^{-3})
ρ_{bt} : Fluid density in the turbulent sublayer (kg m^{-3})
ρ_{aver} or $\overline{\rho}$: The bulk average density (kg m^{-3})
ρ_w : Fluid density evaluated at the wall temperature (kg m^{-3})
τ_w : Wall shear stress (N m^{-2})
ν : Momentum viscosity (m^2 s)

Subscript

aver or *m*: Bulk average
b: Bulk
bl or 1: Viscous sublayer
bt or 2: Turbulent sublayer
I: In
iso : Isothermal
l: Laminar
o: Out
pc: Pseudocritical
t : Turbulent
w : Wall

Chapter 16
Supercritical Fluids as a Tool for Green Energy and Chemicals

Maša Knez Hrnčič
University of Maribor, Slovenia

Darija Cör
University of Maribor, Slovenia

Željko Knez
University of Maribor, Slovenia

ABSTRACT

Hydrothermal conversion of biomass is a promising technology for the conversion of biomass into biofuels and biobased chemicals. This chapter is focused on the waste biomass conversion for production of biofuels and chemicals by applying sub- and supercritical fluids. One of the biggest disadvantages in biomass conversion by SCF is the extremely high energy requirement for heating the media above the water critical point (374 °C, 221 bar). The idea behind the recent research is to reduce the operating temperature and energy requirements by processing biomass with water at much higher pressures. The importance of knowledge on behavior of multicomponent systems at elevated pressures and temperatures is underlined. Methods, developed by the authors of this chapter for determination of thermodynamic and transport properties for multicomponent systems of different solid compounds and supercritical fluid under extreme conditions are described. Future perspective of hydrothermal technology as a tool to obtain advanced materials and the possible scope for future research is also discussed.

INTRODUCTION

Large amounts of waste biomass produced offer a great opportunity to convert this volume of plant material and animal waste to a huge amount of energy. The conversion of biomass to biofuels and biobased chemicals has attracted a lot of attention recently, largely due to the environmental and socio-economic problems associated with the use of fossil fuels. In recent time, many novel technologies have been introduced for conversion of biomass to energy and chemicals. Biomass can be converted to energy

DOI: 10.4018/978-1-5225-2047-4.ch016

by biochemical and thermochemical processes. The two main thermochemical conversion processes to produce biofuel from biomass are the pyrolysis process and the hydrothermal process as promising technologies for the conversion of biomass into biofuels and biobased chemicals. In recent years, hydrothermal energy was accepted as a possible energy resource with a high potential already in the beginning of the previous century. Hydrothermal conversion of biomass has consequently attracted considerable attention. This chapter will therefore focus on the waste biomass conversion for production of biofuels (liquid or gaseous) and biochemicals (furfurals, organic acids, aldehydes, ketones) by applying sub- and supercritical fluids at pressures up to 1000 bar. Operating at high pressures may result in processes that require much lower amounts of energy and maybe also in new processes which are capable of performing new types of reactions that produce completely new bio-based products.

The production of value-added products from waste biomass, such as waste from agriculture, food or forestry industry, will be mostly considered. A topic of the intense research is to reduce the operating temperature and energy requirements by processing biomass with water at much higher pressures or adding supercritical CO_2 to the reaction mixture. Nevertheless, biomass conversion processes at high-temperature and pressures exceeding 300 bar are, for now, a completely new aspect of high-pressure technology for biomass conversion and literature regarding the reactions taking place in such systems practically does not exist. There are also several advantages of supercritical fluid processing under "real" high pressure like new product formulations and economic production of high value products due to the fact that the solubility of several dense gases in low soluble substances increases with increasing pressure and temperature. Most of the commercial supercritical water processes today are limited to pressures, where the complete solubility of water and CO_2 is not achievable. By elevating pressure, the total miscibility of water and CO_2 is already achieved at considerably lower temperatures. Another possibility is lowering the energy barrier by the addition of catalysts. Such treatment of biomass would be a completely new process. However, the data on the mechanisms and kinetics of reactions are scarce. Additionally, the total yields of the final products are practically unknown for these unconventionally high-pressures and temperatures for the multi-component systems.

Considering the above mentioned facts, it is highly important to investigate how of water and CO_2 in such a system would affect the biomass conversion reactions and what products would form, how different ratios and conditions affect the conversion and if it is possible to perform the reactions that normally occur above critical point at milder conditions. This knowledge will open completely new possibilities in biomass conversion, which could also lead to new types of reactions and products. Fundamental data obtained from such research could therefore revolutionize the concept of biomass conversion facilities or could open routes to completely new types of process intensification. Limitations of processes operating at the mentioned high temperatures and unconventionally high pressures are their investment costs that are significantly higher compared to ambient pressure technologies. Nevertheless, many recent studies have suggested that classical large-scale reactors needed for biomass conversion are maybe not necessary for processes involving supercritical water as reaction medium. It has been proven, that reaction kinetics of near-critical water and supercritical water with organic compounds are extremely fast and that long contact times should therefore be avoided.

BACKGROUND

The ability to produce solvent-free products and handle high viscous materials makes the SCFs technologies unique. Such novel technologies are emerging, for instance for particle design, sterilization, and separation of enantiomers (Brunner, 2005). Particulate products with the desired characteristics can be obtained (Kiran, 1994). The development of a process technology comprises the application of fundamentals to processes and has been a topic of an intense research in the past decades (Brennecke & Eckert, 1989; de Swaan Arons & Diepen, 1963).

Phase equilibria and reaction kinetics, the verification of process steps, as for example in specific reactors, and the design of process sequences to produce a product (energy) from raw materials are the main hindrances concerning the SCF applications. Some important bioconversions have decreased interest today, but other conversions gain interest or will gain interest, so that the fundamentals will somehow have to find its way. Detailed investigations on the basic thermodynamical and transport data like phase equilibria, density, viscosity, dielectric constant and diffusion coefficient have to be carried out to obtain the data fundamental for the design of a process in order to fulfil consumer and economic requirements (Brunner, 2015). Supercritical fluids have been widely applied in several processes developed to the commercial scale; – from pharmacy, food sciences to the textile industry, but however, pilot plant and industrial scale applications in the field of energy are still in its infancy. Solubility of the substance in SC or SubC fluid is the main parameter controlling the feasibility of the process. Determination of solubility of several compounds in SCFs is already reported in scientific literature. However, contributions reporting phase equilibria of systems with CO_2 are offered in particular (Fonseca, Dohrn, & Peper, 2011). Knowledge about phase equilibria and thermodynamic data is essential to design a process with a well-considered union between economic feasibility and safety.

A wide literature review showed that different authors suggest various, mainly well-established methods, to investigate thermodynamic data of binary systems of substances in presence of SCF (Aionicesei, Škerget, & Knez, 2008; Corma, García, & Leyva, 2005; Dindar & Kiran, 2002; Gourgouillon, Avelino, Fareleira, & Da Ponte, 1998; Gutiérrez, Rodríguez, Gracia, De Lucas, & García, 2013; Markočič, Škerget, & Knez, 2011; Mukherjee, Bhattacharyya, & Bagchi, 2002; Pantoula & Panayiotou, 2006). Those methods are either time consuming or require expensive, advanced equipment. Additionally, most of the methods may only be used for pure compounds, and may not be adopted to the diversity of the systems occurring in industry that include materials of a complex composition like biomass, bio oils, etc. There are only few articles dealing with the development of new methods (Calvignac, Rodier, Letourneau, Almeida dos Santos, & Fages, 2010; Calvignac, et al., 2010; Elvassore, Vezzù, & Bertucco, 2005; Knez Hrnčič, Markočič, Trupej, Škerget, & Knez, 2014). Solubility data, density, viscosity, dielectric constant, diffusion coefficient and interfacial tension certainly play a key role in a design of an efficient fractionation process of components of pyrolysis oil that are a product of fast pyrolysis or liquefaction of biomass and are dark brown organic liquids, which are chemically a complex mixture and/or emulsion of water and degradation products of lignin. and also open the innovative routes to completely new types of process intensification.

MAIN FOCUS OF THE CHAPTER

Issues, Controversies, Problems: Environmental Issues Relating Energy Consumption

The importance of the problem regrading environmental issues reflects also in the field of science popularization. Several conferences are organized worldwide. United Nations Conference on Climate Change was organized in France in the beginning of December 2015. Energy demand is still increasing and the use of fossil fuels still remains the principle energy source which may lead to uncontrolled net global warming and to an increased number of extreme climate events (Paris COP21, 2015).

In the past SCFs were used mainly as solvents in extraction processes such as decaffeination of coffee and tea, spices extraction, extraction of antioxidants from (Knez, Škerget, & Knez Hrňič, 2010), oil from seeds etc..., however in recent time technologies using SCFs as a processing media received new challenges. SCFs present a promising new alternative as a media for biomass conversion to green chemicals and energy. Currently over 80% of energy demand is still met by fossil fuels (oil, gas and coal) but these resources are finite. Beside the impact on the environment, the price of crude oil is fluctuating. However, fossil fuels are still the main source of energy and chemicals. Research on alternative sources of chemicals and fuel and production technologies has therefore attracted a great deal of attention (Barnes, 2015). Several investigations are recently performed in the field of supercritical fluids as a tool for energy production (Brunner, 2015). Biomass conversion processes applying sub- critical water and supercritical water are promising basis for processing of various intermediates to be used for energy and chemical purposes. Water as a reaction media opens new possibilities compared to "dry processes", allowing lower reaction temperatures.

From Sustainable Energy Resources to Energy and Fuel

Nowadays, there is a growing interest in the development of alternative technological processes with minimized environmental impact, such as reduced energy consumption, less toxic residues, better use of by-products and also better quality and safety of the final products (Knez et al., 2014). Currently, chemicals and energy are mainly derived from fossil fuels, what is challenged by the rapidly declined reserves, fluctuating price of crude oil and impact on the environment. Research on alternative sources of chemicals and fuel and production technologies has therefore attracted a great deal of attention Feng, Cheng, Yuan, Leitch, & Xu, 2013). Biomass energy is derived from distinct organic matter that can re-grow over a relatively short period of time compared with the hundreds of millions of years that it took for fossil fuels to form (Kullaiah Byrappa & Masahiro Yoshimura, 2013). Much effort is therefore oriented to the utilization of waste biomass for production of energy and chemicals. Figure 1 presents hierarchy of waste management as advised by the European Union (Wasteshipment, 2016). Residues produced by the agricultural and food industries contain valuable functionalized molecules such as flavonoids, waxes, biopolymers or fatty acids. Food supply chain waste has a high potential to be utilized as a feedstock for electricity generation (i.e. through anaerobic digestion), fuel generation (i.e. conversion of cellulosic biomass to bioethanol) or the production of animal feed (Pfaltzgraff, Cooper, Budarin, & Clark, 2013). Waste biomass and biomass of nature and agricultural cultivation are valuable organic reservoirs of raw materials and must be used in accordance with their organic composition (Kamm, Gruber, & Kamm, 2007).

Figure 1. Hierarchy of waste management as advised by the European Union
Wasteshipment, 2016.

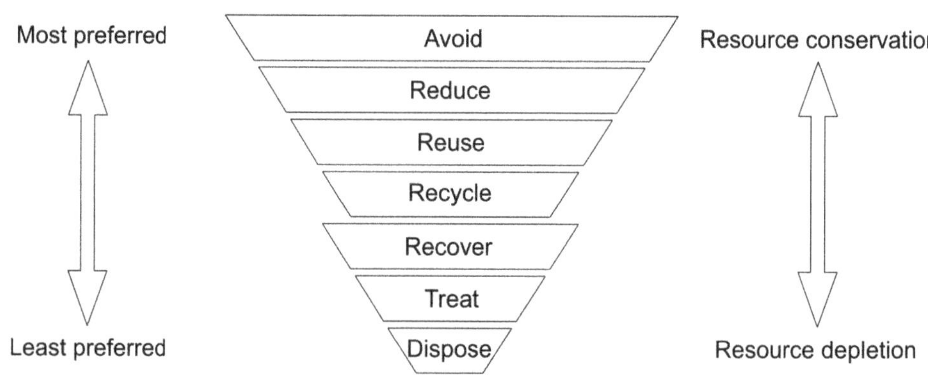

Anyhow, harvesting biomass to produce energy may result in reduced soil productivity by depletion of carbon and nutrients and may therefore not be sustainable (Woolf, Amonette, Street-Perrott, Lehmann, & Joseph, 2010). By biomass pyrolysis waste products may be utilized (Woolf et al., 2010).

The vital difference between biomass and fossil fuels as energy sources is the time scale. If it is managed on a sustainable basis, biomass is harvested as part of constantly replenished plants or crops in which the new growth in the re-plantation takes up CO_2 from the atmosphere at the same time as it is released by an energy conversion process (e.g. combustion) of the previous harvest. This maintains a closed carbon cycle with no net increase in atmospheric CO_2 levels, in contrast to that of fossil fuels, and therefore biomass is considered as a renewable energy resource. The large amounts of waste biomass, which is a renewable resource, cause serious problems when not being processed, which is a severe problem facing especially the countries of the third world that do not have strong regulatory instruments to control the environmental challenges.

Biofuels are liquid or gaseous fuels that are predominantly produced from biomass for transport sector applications. As biofuels are renewable, sustainable, carbon neutral and environmentally benign, they have been proposed as promising alternative fuels for gasoline and diesel engines (Hoogwijk et al., 2003). For different biomass types as a renewable resource there are a number of conversion technologies available (Cartmell et al., 2006). Primarily, the options are divided into two groups, namely thermal and chemical conversion. In general, the conversion technologies may release the energy directly, in the form of heat or electricity, or may be converted to another form, such as liquid biofuel (bioethanol, biodiesel), combustible biogas (H_2, CH_4) or charcoal-like solids (biochar). While for some classes of biomass resources there may be a number of usage options, for others there may only be one appropriate technology (Cabeza et al., 2015). Optimization of biomass supply chains could help in commercializing this sustainable energy source by reducing costs (Shabani, Akhtari, & Sowlati, 2013).

Conversion of Waste Biomass to Energy

Nevertheless, biomass derived energents have several pros and cons. Besides, several obstacles that have to be overcome to convert waste biomass to biomass fuels. Finding new methods to convert biomass energy resources to usable energy resources is inevitable (Mazaheri, Lee, Bhatia, & Mohamed, 2010; Van Bennekom et al., 2009) since there are several issues, such as some technical challenges, facing biodiesel production via transesterification, which include long residence times, high operating costs and energy consumption and on the other hand low production efficiency. Problems facing biodiesel production have to be solved in order to make the process more efficient and economical. Many efforts are proposed to convert biomass to bio fuel (Kruse, 2011). Green revolution focuses on the development of biodiesel production based on the use of non-catalytic esterification under high pressure and temperature, or with supercritical methanol (Glisic & Skala, 2009; Kesić et al., 2012; Qiu, Liu, & Riffat, 2011). General approach is to process the biomass in an aqueous phase. The specific implementation of this approach of interest is supercritical water gasification (SCWG), which involves the conversion of organic compounds to gaseous products (H_2, CO, CO_2, and CH_4) via reactions in and with water at a temperature and pressure exceeding the thermodynamic critical point (T_c = 374 oC and P_c = 221 bar) and therefore knowledge about the behavior of these gases with liquid products, obtained by fast pyrolysis of biomass under different conditions of temperature and pressure is essential in designing the separation processes following the conversion reactions. Figure 2 presents phase diagram of water with area of application for hydrothermal treatment. HT pretreatment could not only improve the mechanical dewatering of parent materials, but also the fuel quality, such as the carbon content and energy density (Zhao, Shen, Ge, Chen, & Yoshikawa, 2014). Anyhow, heating value of biomass depends on its composition. For instance, dry woody biomass consists of cellulose, hemicelluloses, lignin and ash. Its heating value can therefore be estimated from the heating value and weight fraction of each constituent.

Overview of High Pressure Processes for Biomass Treatment

Several applications have been proposed in the past years using hot compressed water as a reaction media, e.g. destruction of hazardous materials, synthesis of inorganic and organic materials, recovery of

Figure 2. Phase diagram of water with area of application for hydrothermal treatment
T_c - critical temerature, P_c - critical pressure.

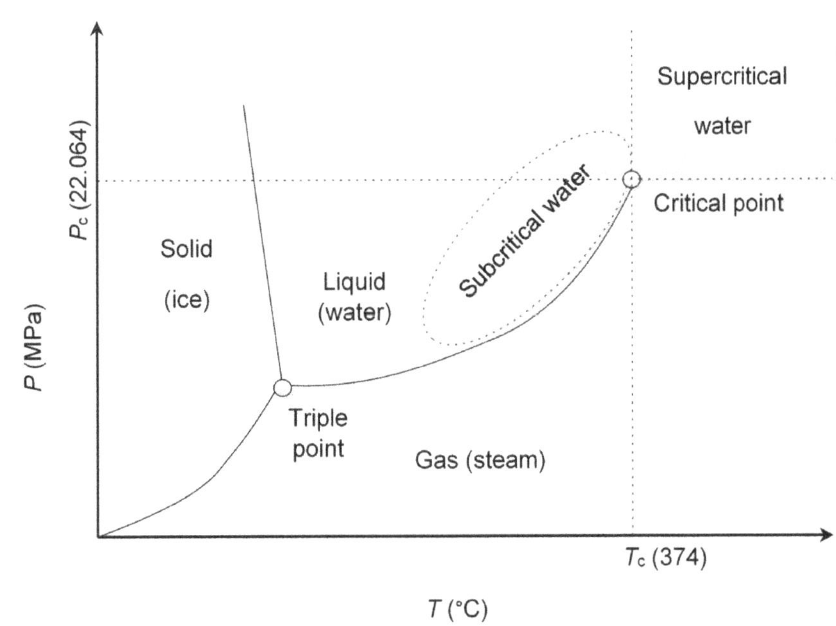

valuables from wastes recycling of polymers and energy recovery of biomass (Cabeza et al., 2015; Jain, Balasubramanian, & Srinivasan, 2016). There are certain drawbacks regarding utilization of waste biomass materials and forestry residues as sources of energy due to their uneven and troublesome properties and characteristics. Several beneficiation technologies which basically improve the physical and thermal characteristics of the original biomass materials by changing shape, size, size, distribution, density, water content, etc. Amongst these processes, the most important are: sizing and shaping, separation, drying, dewatering, densification (Baling, Pelletization and Briquetting) and torrefaction (Sugathapala, 2013). In many situations, more than one of the above types is required.

Generally, HT processes could be divided into four main processes, as demonstrated in Figure 3: carbonization (HTC), aqueous phase reforming (APR), liquefaction (HTL), and gasification (HTG).

An ancillary advantage of the uncongenially high processing conditions is that product streams are completely sterilized with respect to any possible pathogens including biotoxins, bacteria or viruses (Peterson et al., 2008). Hydrothermal carbonization is normally carried out at moderate conditions (250°C and 20 bar) "Hydro char" as the main product can be generated at even higher temperatures (250-800°C) based on wet pyrolysis. Reaction was tested with several types of biomass (Kruse, Funke, & Titirici, 2013; Liu, Quek, Kent Hoekman, & Balasubramanian, 2013; Titirici, Thomas, & Antonietti, 2007). When performing the process under moderate temperature, reaction rate is lower, normally it takes some hours. However, when applying high temperature, the reactive gasses are produced parallel with other pyrolysis products (bio-oils) resulting in interesting carbon structures like: nanotubes, films, microspheres, advance carbon materials.

The most widely investigated hydrothermal reaction over the past decade has been hydrothermal liquefaction involving conversion of waste biomass into chemicals and biofuels under medium temperature and high pressure (280-370°C, 100-250 bar). Water should be kept in a liquid state. The main product is water-insoluble bio-oil (called also bio-crude) with small quantities of water soluble products like

Figure 3. Applications of biomass thermochemical conversion processes
Kamm et al., 2007.

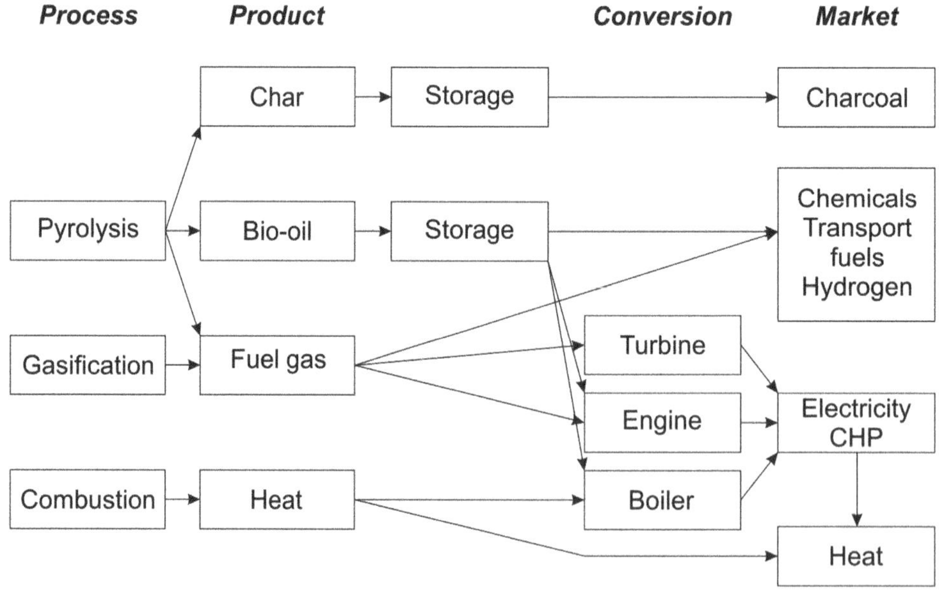

char (solid residue) and light gases (e.g. H_2, CO, CO_2, CH_4). Both, liquefaction and gasification can be minimized at a comparably low temperatures ranging from 160 to 250 °C in order to increase the yield of solid biofuel (Huber, Iborra, & Corma, 2006). Produced fuel has a high C/H ratio beside a higher stability and energy content. Various types of biomass have been subjected to the conversion concept, e.g., agricultural residues (Toor, Rosendahl, & Rudolf, 2011), forest residues as well as sewage sludge (Akizuki, Fujii, Hayashi, & Oshima, 2014). Three main types of supercritical water gasification are reported, depending on the temperature range: catalyzed near-critical gasification to methane and aqueous phase reforming (Pavlovič, Knez, & Škerget, 2013).Catalyzed near-critical gasification is conducted under lower near critical temperature range at 300-500°C with addition of heterogeneous catalysts, which is increase selectivity to CH_4-rich gas product and carbon dioxide (Devi, Ptasinski, & Janssen, 2003; Yoshida, Oshima, & Matsumura, 2004). The reaction mechanisms are intensively investigated for different catalysts. Pt-based catalysts exhibit high H_2 selectivity but low conversion of the alcohols. Ni-based catalysts exhibit high activity and conversion but low H_2 selectivity and yield (Chen, Li, Chen & Yan, 2015). The reaction takes place at relatively mild reaction conditions within temperature 220-250°C and pressures typical of 15-50 bar. One of the most important factors of an economic analysis is the expected cost of the end product or products and the total energy efficiency of waste biomass conversion process with SCFs.

SOLUTIONS AND RECOMMENDATIONS

Techniques, applying supercritical fluids as a processing media certainly have a high potentiality for future large-scale biofuel production, especially for biodiesel production from waste oil and waste fat due to a number of advantages that include fast kinetics, high fuel production rate, ease of continuous operation and in some cases even elimination of the necessity of catalysts. The sub-chapter presents high pressure processes for treatment of biomass as a perspective concept for production of energy and chemicals as well as emphasizes the importance of fundamental studies of phase equilibria of substances in the presence of supercritical fluids. Main product of hydrothermal liquefaction, which is one of the most investigated process for converting waste biomass into chemicals and biofuels, is water-insoluble bio-oil (called also bio-crude) with small quantities of water soluble products like char (solid residue) and light gases (e.g H_2, CO, CO_2, CH_4). Bio oil, commonly termed as pyrolysis oil, chemically consist of a complex mixture and/or emulsion of water and degradation products of lignin (e.g. guaiacols, catechols, syringols. vanillins), cellulose (such as levoglucosan, dehydrated sugars, di-sugars, furancarboxaldehydes), and hemicellulose (such as acetic acid, formic acid). Their composition varies depending on the type of biomass feedstock and conditions of pyrolysis process. As these free-flowing organic liquids are composed of numerous highly oxygenated components, which are reactive, a suitable way to stabilize the oil (and allowing production of hydrocarbons) should be considered. A possible solution may be in hydrotreating (Venderbosch, Ardiyanti, Wildschut, Oasmaa, & Heeres, 2010). Hydrogenation of bio oil yield products less prone to produce carbon during oxidation process. Water content of the bio-oil may differ considerably. Organic acids form esters with olefins and esters and water with alcohols, aldehydes form oligomers and resins. With water, aldehydes form hydrates. Combination of aldehydes and alcohols is suitable to form hemiacetals, or acetals and water. Products in thermodynamic equilibrium with the reactants are formed, which means that a change in temperature or relative amounts of water and other reactive compounds will upset the equilibrium and initiate compositional changes. Some reactions may form resins or polyolefins and may be irreversible under likely bio-oil storage conditions. Another important issue is air oxidation that tends to form more acids and reactive peroxides that catalyse the polymerization of unsaturated compounds. In a complex, real bio-oil, the number of possible chemical reactions is very high, especially considering the large number of complex oligomers. The reactions between the oligomers are responsible for nearly all the viscosity increases observed during aging of bio-oil (Diebold, 2000).

Several research groups performed experiments and obtained mixtures of gaseous and liquid products. Phase equilibrium data for ternary system bio oil/ diesel/CO_2 and bio oil/tail water/ CO_2 was determined in the frame of our previous research (Knez Hrnčič, Venderbosch, Škerget, & Knez, 2013a). Data for the system H_2/pyrolysis oil and H_2/hydrogenated pyrolysis oil was reported by Knez Hrnčič and co-authors (Knez Hrnčič, Venderbosch, Škerget, Ilić, & Knez, 2013b).

Hydrogen is expected to become an important fuel in the long-term since in combination with fuel cells it offers the opportunity of an intrinsically clean energy supply. Compared with the conventional methods, SCWG process is characterized by its high reaction efficiency and H_2 selectivity. Water plays a role of reaction medium. A high moisture content material can be applied (Wang, Wang, Zhao, Guo, & Guo, 2016). Compared with conventional biomass gasification technologies, supercritical water gasification possesses a number of advantages which are derived from the unique thermodynamic and fluid dynamic properties of water in supercritical state. Dielectric constant of water is besides being a function of temperature, a strong function of pressure. At normal temperature and pressure, dielectric constant of water is relatively large, approximately 80. The main reason the strong effect of hydrogen bond in water. With increasing temperature and pressure, dielectric constant of water decreases and is around 5 at critical point. Several benefits compared to conventional gasification procedures include high thermochemical conversion rate, high pressure products that are easy for future transportation and usage, opportunities for carbon capture, sequestration and storage, as well as to further pure hydrogen production via a further steam-methane reforming process.

During the past two decades, over 100 journal papers have been published regarding different aspects of heterogeneous catalysis in the SCWG process. These studies include both commercial as well as catalysts tailored for gasification in SCW (Wang et al., 2016). A variety of real biomasses including lignocellulosic biomass from different sources, sewage sludge, chicken manure, food wastes, algae, and fermentation residue have been successfully gasified in SCW (Azadi & Farnood, 2011; Matsumura et al., 2005). Performance of a gasification process is generally influenced by operating pressure and temperature, different catalysts and feedstock type and concentration, catalyst and loading, and the contact time between the catalyst and the feed.

Several obstacles are present on the way of scaling-up of a biomass gasification process. High energy requirements are present already in the first step of the process. Water must be heated from ambient to supercritical conditions and above. Another drawback is the highly aggressive nature of water, which is even more evident in combination with an oxidant, leading to corrosion related issues. In addition to the above mentioned drawbacks, salt deposition on the reactor walls as a result of very low solubility of inorganic compounds at supercritical conditions should be considered (Vadillo, Sánchez-Oneto, Portela, & Martínez de la Ossa, 2013). These are the main hindrances that should be removed to enable a wider exploitation of biomass conversion processes applying water in supercritical state on industrial (Kamm et al., 2007). In continuous processes, pumpability of the biomass slurry is a problem. As the dry matter content of the biomass slurry exceeds the pumpability limits, several clogging problems may appear. On the other hand, a high dry matter content contributes to a higher gas production which increases the energy efficiency and the profitability of the plant (Yakaboylu, Harinck, Smit, & de Jong, 2015). A detailed investigation of phase equilibrium for the systems gas/water and gas mixtures/water was carried out to provide the fundamental data for the design and operation of the pilot plant and dissemination of the concept to the industrial scale (Knez Hrnčič, Markočič, Kramberger, & Knez, 2011; Knez Hrnčič et al., 2013b; Kramberger, Markočič, & Knez, 2013; Markočič, Kramberger, van Bennekom, Heeres, Vos, & Knez, 2013). In the frame of the past research on reforming in supercritical water, a unit for the conversion of organic compounds to gaseous products was constructed (Figure 4). The main goal of the research was to increase the content of H_2 and CO and to reduce the hydrocarbon concentration, application of catalysts was proposed. Experimental conditions presented here covered a wide range of reaction temperature and types of catalysts. Reactions were performed at high pressure in the range of temperatures between 100°C and 600°C. Results were interpreted on a common basis in terms of carbon conversion and hydrogen and methane yields. No significant change in the composition of the substrate over the entire range of the investigated conditions has been observed. The possible explanation may be that all the catalyst reacted already with the surface of the reactor even before the reaction took place. There were only small changes in the concentration of the gases, which depended on the working pressure, during the experiments performed at a constant temperature of 750°C.

In Table 1, some applications of gasification as pilot scale and semi-industrial scale processes are summarized. Comparison between hydrothermal liquefaction (HTL), hydrothermal carbonization (HTC) and hydrothermal gasification (HTG) processes in last 5 years is presented in the Table 2.

Reaction mechanisms and kinetics of the process are another issues that have to be addressed. Due to the sufficient heat energy all of the biomass feed, including char, is normally converted to gasification products in a single pass through a gasifier system (Sugathapala, 2013). The overall combustion reaction is taking place in the combustion zone. The air introduced contains, besides oxygen, water vapor and inert gases such as nitrogen and argon. The inert gases are normally assumed to be non-reactive with the

Figure 4. High pressure - high temperature reactor for catalytic reforming of methane

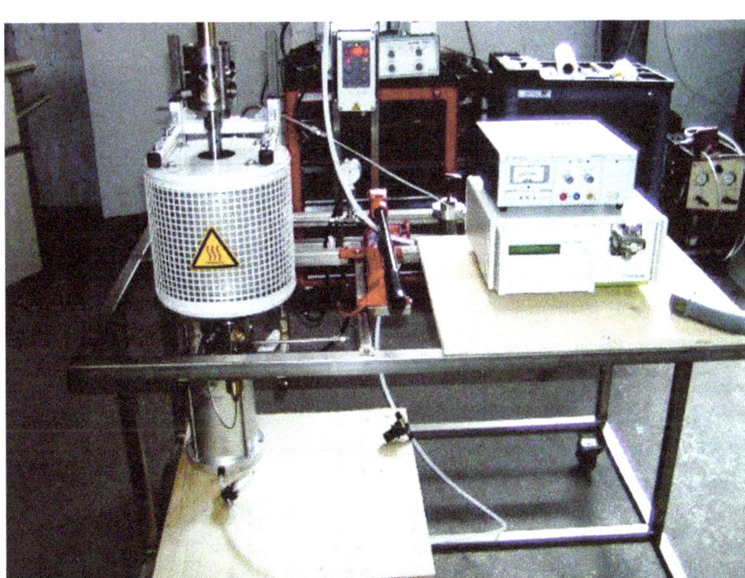

Table 1. Applications of the concept as pilot scale and semi-industrial scale processes in the last years

Plant	Brief Description	Reaction Conditions	Presence of Catalyst	Process Type	
woody biomass gasification	A high temperature air-blown gasification scale-up model for woody biomass gasification. The down-flow gasifier was used. In large-scale units, the cold gas has ability to achieve 80% efficiency in the air-blown gasification. The cold gas efficiency was over 80%, and the heating value of the produced gas was 6.6 MJ/ m_N^3 .	Feeding rate 1000 kg/h, input air temperature 427 °C. (Maximum achievable temperature is over 1100 °C the O/C is from 1.26 to 1.84).	-	Continuous	(Kobayashi et al., 2009)
CHP Agnion Biomasse Heizkraftwerk Pfaffenhofen	TRL 4-5 - Laboratory Component Testing - Component Prototype Development Pilot CHP - Combinate heat and power technology The Heatpipe-Reformer developed by agnion is an innovative plant for production of renewable energy. Production of high-quality syngas.	Input: Waste Wood (80,000.000 t/y) Output: syngas (32.500 MWth) power (electricity) (6.100 MWel) at temperatures around 800°C – to form a gas mixture made up of hydrogen and carbon monoxide	-	Continuous	(Bioenergy, 2015)
Skive CHP plant	TRL 9 - Full scale commercial deployment A bubbling fluidized bed (BFB) gasifier is used to produce gas from wood-based biomass. This gas is then used in of reciprocating engines in a combined heat and power (CHP) application.	wood pellets (100.000 t/d) power (electricity) (6.000 MWel) heat (20.000 MWth) pressurized, bubbling fluidized bed, bed material dolomite, operated 850°C	monolith catalysts	Continuous fully automated	(Bioenergy, 2015)
Urbas Calvello	A combustible gas, wood gas, is drawn from wood through a thermochemical processes. The raw gas is then separated of dust and tars through a filtering system. Such cleaned gas is afterward used to produce combined heat and power through a gas engine and generator.	Input: wood output: chips power (electricity) (0.199 MWel)	-	-	(Bioenergy, 2015)
Vaskiluodon Voima Biomass Gasification Plant	The biomass feedstock is dried in a belt dryer and gasified in a large CFB-gasifier. Produces biogas from wood, comprising mainly of forest residue, to generate electricity and provide district heating to the local community.	Input: biomass substituting coal Output: power (electricity) (140.000 MW)	-	Continuous Fully automated	(Bioenergy, 2015)
Synthesis Tembec Chemical Quebec	TRL 6-7 Component Prototype Demonstration-Pilot Plant	Input: spent sulphite liquor feedstock Output: ethanol (13,000.000 t/y)	-	Continuous	(Bioenergy, 2015)
FICFB Oberwart	TRL-9 Full-Scale Commercial Deployment High electrical efficiency (almost 30% with combined gas engine and ORC process) No problematic solid residues (only ash, carbon content) Length of District Heating System ca. 5200 m Temperature District Heating System VL 95°C / RL 65°C	Input: wood chips (8.700 MW) Output: - power (electricity) (2.800 MWel) - heat (4.100 MWth)	-	Continuous	(Bioenergy, 2015)

continued on following page

Table 1. Continued

Plant	Brief Description	Reaction Conditions	Presence of Catalyst	Process Type	
Synthesis Enerkem Sherbrooke	Pilot plant TRL 4-5 Gasification based process technology that transforms sorted municipal solid waste (MSW) and residues from the forest and agricultural industries into transportation fuels, high value chemicals and electricity. Steps: 1.feedstock pre-treatmant, 2. Gasification, 3. Synthetic gas conditioning, 4. Conversion into liquid fuel	Input: municipal solid waste, wood chips, treated wood, sludge, petroleum coke, spent plastics and wheat straw Output: ethanol (375.000 t/y), methanol (475.000 m^3/y), syngas	-	Continuous	(Bioenergy, 2015)
Synthesis Demonstration Plant, Westbury	TRL 6-7 Demonstration Pilot Plant	Input: waste wood, clean wood, saw mill residues (12,000.000 t/y) Output: FT liquids (0.568 m^3/h)	-	Continuous	(Bioenergy, 2015)
CHP Urbas Eberndorf	TRL 9 Full Scale Commercial Power / CHP technology A combustible gas, wood gas, is drawn from wood through thermochemical processes. The raw gas is then separated of dust and tars through a filtering system. Advantage: wood gas cogeneration requires no intermediate medium thus resulting in a higher electrical efficiency throughout the entire system.	Input: waste wood Output: power (electricity) (0.300 MWel)	-	Continuous	(Bioenergy, 2015)
GoBiGas	TRL 6-7 Demonstration Fuel synthesis The gasification technology is based on indirect gasification, which is supllemented by gas upgrading and SNG synthesis.	Input: Wood pellets (6.500 t/h), Forest residues (6.500 t/h) Output: SNG (20.000 MW), heat (5.000 MWth)	-	Continuous	(Bioenergy, 2015)
Holzgasanlage 2 Kaeser Gasel	TRL 9 Full-scale plant commercial Power / CHP technology; CHP gasifier unit with connection to heating device and power.	Input: wood chips (133.000 kg/h) Output: power (electricity) (0.140 MWel), heat (0.240 MWth)	-	Continuous	(Bioenergy, 2015)
BioSynergi CHP demonstration plant	TRL 6-7 Demonstration Pilot Plant Staged Open Core gasifier fuelled with forest wood chips (45-50% moisture on wet basis), as a part of a Internal combustion engine based CHP plant. The annual generation at the plant is expected to be 2,000 MWh electricity and 5,000 MWh heat.	Input: Wood chips Output: power (electricity) (0.300 MWel), heat (0.750 MWth)	-	Continuous	(Bioenergy, 2015)
Holzverstromungsanlage Bucher Escholzmatt	Full-scale Plant includes: feedstock and handling devices with heat and power gasifier unit, connection to district heating and power.	Input: wood chips (4,700.000 m3/y) Output: power (electricity) (0.130 MWel), heat (0.260 MWth)	-	Continuous	(Bioenergy, 2015)
Gasification plant fuelled with olive industry wastes	The energy recovery from olive industry wastes. The pilot plant contains a downdraft gasifier, gas cooling-cleaning stage and spark ignition engine with an adjusted carburettor. Cold gas efficiencies in the range of 70.7–75.5% were reached. The calorific value of the producer gas was 4.8 and 5.4 MJ kg^{-1}. The electric and CHP efficiency: 15% and almost 50%. Phases and main chemical reactions in a downdraft gasifier: Olive oil industry wastes, drying, pyrolysis, combustion, reduction, producer gas.	Operating temperature: 1050–1100 °C Outlet temperature: 400–500 °C Gas caloric value: > 4.5 MJ Nm^{-3}	-	Continuous	(Vera, Jurado, Margaritis, & Grammelis, 2014)

continued on following page

Table 1. Continued

Plant	Brief Description	Reaction Conditions	Presence of Catalyst	Process Type	
A two stage fluid bed plasma gasification plant	RDF-refused derived fuel from urban solid waste can be converted to clean syngas using conventional Fiber Bragg Grating system (FBG) tied with a plasma converter. Simultaneously converting ash-type components in the RDF into a stable vitrified product. The results show that the slag from the plasma converter unit can be assumed as fully recyclable. 90% of ash-type constituents in RDF are recuperated as a stable vitrified product.	Feed rate of the solids at rates of up to 50–60 kg/h plasma converter: 950–1150 °C FBG: 700–800 °C	-	Continuous	(Materazzi, Lettieri, Mazzei, Taylor, & Chapman, 2015)
Gasification of dried sewage sludge	Special rotary dryer produce particles, which differ significantly in size, geometry, porosity and weight. Three lines were planned to process the entire mass of 5.4 t/a ds6 with a water content of around 3–8%. Gas is blended with biogas from sludge fermentation and utilized in a gas engine or combustion chamber to produce heat.	Gasification temperature: 850–900 °C Installed power: 2.2 MW_{th} Type of dryer: Rotary dryer Electrical consumption: 75 kW Cold gas efficiency:70%	-	Continuous	(Judex, Gaiffi, & Burgbacher, 2012)
Japanese cedar wood (up-draft gasifier)	Gasification of cedar wood in a bench-scale externally heated updraft gasifier. Around 30-50 vol.% of H_2 can be produced from cedar wood gasification.	Feedstock: particle size of 1–2 mm Operating temperature: 650 °C – 950 °C Lower heating value of gas: 1–33.2 $mJNm^{-3}$	-	-	(Aljbour & Kawamoto, 2013)
Rice straw gasification (up-draft gasifier)	Rice straw gasification in an atmospheric fluidized-bed gasifier. A high quality syngas (10% H_2, 18% CO and 4% CH_4) with low tar content was produced.	Operating temperature: 700 °C – 850 °C higher heating value (HHV) of the produced gas: 5.1 $MJ(Nm)^{-3}$ cold gas efficiency: 52%	-	-	(Calvo, Gil, Otero, Morán, & García, 2012)
Willow and other agriculture residue-dry grains (up-draft gasifier)	The gasification has been performed using circulating fluidized bed (CFB) gasifier. The averaged concentration of H_2 obtained from willow and dry grains was around 28 vol. % and 20 vol. % respectively.	Feed rate: max. 20 kg/h Operating temperature: 800 °C–820 °C cold gas efficiency: 50 -70% power range: 60 kW	-	-	(Meng, de Jong, Fu, & Verkooijen, 2011)
gasification of pine woodchips (up-draft gasifier)	Gasification in bench-scale circulating fluidized-bed (CFB) gasifier. Steam was used as a fluidized and gasifying agent. Gas composition CO (27–40 vol.%), H_2 (22–27 vol.%), CH_4 (7–9 vol.%), CO_2 (39–42 vol.%)	Feed rate: 5 kg/h Operating temperature: 700-900 °C Power range: 15 kW higher heating value (HHV) of the produced gas: 17 $MJ(Nm)^{-3}$	-	-	(Ngo et al., 2011)
Bagasse gasification	Gasification was carried out in an intensified auto-thermal air-blown downdraft gasifier and at atmospheric pressure. The system consists of a gasifier, an air blower, a gas clean up system with two cyclones, a water scrubber and an ash collector. Very low tar content between 0.376–0.40 g Nm^{-3} in the gas.	Operating temperature: 1040 °C Power range: 50 kW	-	-	(Jordan & Akay, 2012)
Hazelnut shells gasification	Cracked hazelnut shells size between 5 and 10 mm. The gasifier consists of an ignition unit, a cyclone, a gas cleaning system, a measurement/ control system and data recording units. Gas which is produced by the gasifier enters the cyclone and then it is divided into two streams. Gas composition: H_2(13 vol.%), CO (23 vol.%), CO_2 (11 vol.%), CH_4 (4 vol.%) with low tar content.	10 kg feeding capacity Operating temperature: 1000 °C Lower heating value of gas (LHV): 1 MJ/Nm^3 - 5 MJ/ Nm^3 Power range: 45 kW	-	batch	(Olgun, Ozdogan, & Yinesor, 2011)

continued on following page

Table 1. Continued

Plant	Brief Description	Reaction Conditions	Presence of Catalyst	Process Type	
Biomass gasification	Pine wood blocks gasification unit consists of The gasification system consists of air intake equipment, a downdraft gasifier, ash chamber, gas cleaning system, roots blower, flow meters and off-gas burner. First the catalyst is uniformly distributed on the grate. Then the biomass is placed on charcoal. A specified (Lv et al., 2007) quantity of hot water is added into steam generating room. Afterwards feedstock in the gasifier is ignited. The gas contain around 70 vol.% of H_2 and CO. Biomass oxygen/steam gasification has its advantage for improving the hydrogen yield in less energy consumption.	Lower heating value (LHV) of fuel gas: 11.11 MJ/Nm^3 Hydrogen yield: 45.16 g H_2/kg	charcoal	batch	(Kirkels & Verbong, 2011)
Biomass gasification technology using integrated catalytic hot gas cleaning	A lab-scale pilot gasification facility of advanced gasification technology for production of high quality gas using innovative hot gas cleaning catalyst. In this process the tar content in the product gas could be reduced and the large amount of the aromatic ring systems in tar have been reformed and/or cracked. The increase contents of H_2 and CO and decrease of CH_4 and CO_2 was noticed. Gas composition: H_2 (54.2.%), CO (20 vol.%), CO_2 (6.7 vol.%), CH_4 (18.8 vol.%) with low tar content.	Feed rate: 4.4 kg/h gasifier reactor temperasture: 880 °C catalytic reactor temperature: 800°C tar content: below 83 mgNm^{-3}	Char-supported catalysts (woody chips soaked in 0.2 M $FeCl_3$ and dried)	batch	(Dong et al., 2013)
Plant of coal and supercritical water gasification system driven by concentrated solar energy	Gasification plant using concentrated solar energy system consists of: nitrogen bottle, feedstock tank, feeder, solar receiver/reactor, heat exchanger, cooler, filter; regulator, separator, pump, tanks, toroidal surface heliostat with the two axis spinning-elevation sun tracking, concentrator. Biomass glucose, corn meal and wheat stalk were gasified in the system for the production of hydrogen-rich gas.	Feedstock:1.03 t/h Maximal temperature and pressure:800 °C and 400 bar, respectively. Maximal concentrated solar power:163 kW. The maximal hydrogen production molar fraction: 55%	-	Continuous	(L. Guo, Jin, & Lu, 2015)
Solar receiver/reactor for hydrogen production	Hydrogen production with biomass gasification in supercritical water by using original solar hydrogen reactor. Gasification of glucose and corncob mixed with carboxymethyl cellulose mixture (CMC) was carried out to generate hydrogen-rich fuel gas. System consists of: feedstocks tank, pre-heater, high-pressure feeder, SCW reactor, heat exchanger, cooler, regulators, separators, pumps, valves, tanks, flow meter, multi-dishes concentrator.	The maximal throughput capacity was 16 kg/h. Feedstock temperature: 200–250 °C The maximal hydrogen production volume fraction: 55% Pressure: 240 bar Temperature: 543°C-676°C Feedstock: - Glucose concentration: 0.1 M, 0.2 M, 0.3 M, 0.4 M 2 wt% CMC + 1.5 wt% corncob, 2 wt% CMC + 3 wt% corncob, 2 wt% CMC + 3 wt% corncob + 0.5 wt% KOH, 2 wt% CMC + 5 wt% corncob	without or KOH	-	(Liao, Guo, Lu, & Zhang, 2013)
A transpiring-wall supercritical water oxidation reactor	SCWO pilot plant with a transpiring wall reactor including five streams (feed, oxygen, air, transpiring water and cooling water) are introduced into the reactor. Reactor is used to treat organic and recover energy. A total organic carbon (TOC) removal above 99% is reached even at a feed temperature of 369 °C.	Feed rate: 8–17 kg/h Reaction temperature: 476-725 °C	-	Continuous	(Zhang et al., 2012)

continued on following page

Table 1. Continued

Plant	Brief Description	Reaction Conditions	Presence of Catalyst	Process Type	
Hydrogen production from glycerol	Glycerol was used for hydrogen production using continuous flow tubular reactor which consists of: reactor, water tanks, pumps, cooler, preheater, mixer, reactor, regulators measuring equipment. The feedstock (glycerol solution) was first mixed with preheated water for avoiding side reactions. Hydrogen yield in relation to catalysts was in the following order: NaOH > Na_2CO_3>KOH > K_2CO_3 Hydrogen-rich gas products - hydrogen molar fraction was 52.0–68.9%.	Reaction temperature: 445-600 °C Pressure: 250 bar residence time: 3.9-9.0 s.	alkali catalysts	Continuous	(Guo et al., 2012)

Table 2. Comparison between hydrothermal liquefaction (HTL), hydrothermal carbonization (HTC) and hydrothermal gasification (HTG) processes in last 5 years

Type of Biomass Conversion	Material	Temperature (°C)	Yield (%)	HHV - High Heating Value (MJ/kg)	Energy Recovery (%)	Ref.
HTL	microalgae	350	38–64[a]	25–30	60-78	(Biller & Ross, 2011)
HTL	microalgae	300-350	27-47[a]	32-37.5	-	(Biller et al., 2012)
HTL	microalgae	300	66[a]	33-37	-	(Faeth, Valdez, & Savage, 2013)
HTL	sewage sludge	300	-	14	-	(Fonts, Gea, Azuara, Ábrego, & Arauzo, 2012)
HTL	manures	300	-	10–20	-	(Derek R Vardon et al., 2011)
HTL	lignocellulosics	300	35[a]	12-20	64	(Umeki, Yamamoto, Namioka, & Yoshikawa, 2010; L. Wang, Shahbazi, & Hanna, 2011; Zeton et al., 2014)
HTL	barley straw	280-400	35[a]	17	55	(Zhu, Toor, Rosendahl, & Chen, 2014)
HTL	cellulose	275-320	5.8 – 34[a]	-	-	(Yin & Tan, 2012)
HTL	algae	350	38-63.6[a]	-	-	(Elliott et al., 2013)
HTL	algae	300	24–45[a]	35–37	-	(Derek R. Vardon, Sharma, Blazina, Rajagopalan, & Strathmann, 2012)
HTL	brown macro-algae	350	9.8-17.8[a]	32-34	-	(Anastasakis & Ross, 2015)
HTL	cornelian cherry stones	250-300	28[a]	28.35	-	(Akalın, Tekin, & Karagöz, 2012)
HTL	*Litsea cubeba* seed	250–350	56.9[a]	40.8	-	(Wang et al., 2013)
HTL	microbial yeast biomass	300	56.4[a]	37	83.05	(Jena et al., 2015)

continued on following page

Table 2. Continued

Type of Biomass Conversion	Material	Temperature (°C)	Yield (%)	HHV - High Heating Value (MJ/kg)	Energy Recovery (%)	Ref.
HTL	microalgae	200-300	50-70[a]	36.9	101.8	(Zhang, Chen, Zhang, Luo, & Zhang, 2013)
HTL	wastewater culture algae	300	49[a]	33.3	-	(Chen et al., 2014)
HTC	lignocellulosic biomass	250	35-38[a]	28.4-29.2	-	(Xiao, Shi, Xu, & Sun, 2012)
HTC	woody and herbaceous biomass	175-295	10-12[a]	28–30	60-70	(Hoekman, Broch, Robbins, Zielinska, & Felix, 2013)
HTC	corn Stover	200-260	-	15.6-24.1	-	(Reza, Lynam, Uddin, & Coronella, 2013)
	miscanthus			20.6-22		
	switch grass			19-21		
	rice hull			20.3-24.6		
HTC	coconut fiber	150-375	35[b]	-	-	(Liu & Balasubramanian, 2012)
HTC	eucalyptus leaves		28[b]			
HTC	coconut fibers	250	45.3[b]	28.4	67.0	(Liu & Balasubramanian, 2014)
HTC	eucalyptus leaves		46.4[b]	25.8	63.3	
HTC	slash pine	235-275	53.6-69.7[c]	22.8-28.56	-	(Hoekman et al., 2014)
HTC	air dried sludge	200	-	14.37-15.09	-	(He, Giannis, & Wang, 2013)
HTC	palm oil empty fruit bunch	150-350	49-76[b]	>27	68-78	(Parshetti, Kent Hoekman, & Balasubramanian, 2013)
HTC	sugar	180-250 180-250	47.5-55.7[b]	25.4-25.5	-	(Fiori, Bassoa, Castelloa, & Baratierib, 2014)
	grape seeds		62.3-80.3[b]	26.4-29.8		
HTC	walnut shell	190–230	36.3-48.6[b]	23.2-29.3	50.2-57.5	(Román, Nabais, Laginhas, Ledesma, & González, 2012)
	sunflower stem		29.2-39.7[b]	24.5-28.9	50.4-58.9	
HTC	lignocellulosic biomass	190-260	47-8-83,5[c]	20.2-25.9	-	(Kambo & Dutta, 2014)
HTC	agricultural residues	180-220	34.1-81.7[b]	21.5-28.3	36.6-102.7	(Oliveira, Blöhse, & Ramke, 2013)
HTC	lignocellulosic biomass	180–260	37–95[b]	20–31	54–98	(Mäkelä, Benavente, & Fullana, 2015)
HTC	different biomasses	200-280	16.5-35.7[c]	21.11-26.15	24.7-58.23	(Gao et al., 2016)

continued on following page

Table 2. Continued

Type of Biomass Conversion	Material	Temperature (°C)	Yield (%)	HHV - High Heating Value (MJ/kg)	Energy Recovery (%)	Ref.
HTC	lignocellulosic biomass	200-275	42-86[c]	-	61.9-95.5	(Lynam, Reza, Yan, Vásquez, & Coronella, 2015)
HTC	pulp and paper waste	180–260	-	-	69-95	(Mäkelä, Benavente, & Fullana, 2016)
HTC	coniferous biomass	180-250	50-72[c]	21.95-28.36	72.5-79	(Sermyagina, Saari, Kaikko, & Vakkilainen, 2015)
HTC	olive mill waste	200-250	-	26.4-32.5	-	(Benavente, Calabuig, & Fullana, 2015)
	canned artichoke waste	200-250		16.8-30.26		
	orange juice waste	200-250		18.2-30.3		
HTC	miscanthus biomass	190-260	42.8-83.5	26–30	66–74	(Kambo & Dutta, 2015)
HTG	lignin alkali	300–600	12.0-54.9[d] 18.0-61.4[e]	-	-	(Güngören Madenoğlu, Sağlam, Yüksel, & Ballice, 2016)
	cellulose		21.1-66.2[d] 29.9-81.2[e]			
HTG	marine biomass	300–600	21-80[d] 62.51[f]	-	-	(Deniz et al., 2015)
HTG	xylose	400-600	24.4-65.9[d] 30-80.9[e]	-	-	(Gökkaya, Saglam, Yuksel, & Ballice, 2016)
HTG	wood residue	500-600	45-60[e] 20-26[f]	-	-	(Üremek Cengiz, Eren, Sağlam, Yüksel, & Ballice, 2016)
HTG	phenol	400-600	62-75[d] 76-80[e]	-	-	(Selvi Gökkaya, Saglam, Yüksel, & Ballice, 2015)
HTG	biodisel	300-600	8-60[f]	-	-	(Cengiz et al., 2015)
HTG	microalgae	600-650	57-98[d]	-	-	(Patzelt et al., 2015)
HTG	biomass	400-650	10-70[f]			(Castello & Fiori, 2015)
HTG	microalgae	650	-	32.4-33.5	-	(Möbius, Boukis, Hindersin, & Kerner, 2015)

[a]– oil yield dry basis
[b]– hydrochar yield ($kg_{hydrochar}/kg_{dry\ feed}$)
[c]– mass yield
[d]– gas yield without catalyst
[e]– gas yield with catalyst
[f]– hydrogen yield
HTL – hydrothermal liquefaction
HTC – hydrothermal carbonization
HTG – hydrothermal gasification

fuel constituents. However, the water vapor reacts with the hot carbon according to the heterogeneous water gas reaction ($CO + H_2O \Leftrightarrow CO_2 + H_2$). In the subsequent reaction, CO_2 is reduced to CO according to the Boudouard reaction. Salts, usually in the biomass, act as catalyst/promoter for the water-gas shift reaction, producing up to half of the hydrogen (Kruse, Bernolle, Dahmen, Dinjus, & Maniam, 2010). Another important reaction taking place in the reduction zone is the water shift reaction (Sugathapala, 2013). Instead of a deeper understanding of the reaction paths occurring during gasification, thermodynamic calculations were employed (Schmieder et al., 2000). Thermodynamic equilibrium calculations indicate that methane is preferably formed at lower temperatures, whereas at higher temperatures, hydrogen is the main product, besides CO_2 (Waldner & Vogel, 2005). Lower heating rates usually promote the formation of refractory compounds, resulting in more methane (rather than hydrogen) production. A possible solution to lower the high energy demand characteristic for the process could be the application of pressures exceeding conventional ones, up to 1000 bar. After exceeding the critical temperature (374 °C) at a pressure of 250 bar C_p of water decreases with further increase of temperature. The subsequent conclusion is that heating water at higher pressures to near critical conditions would result in a less energy intensive process.

Numerous non- conventional working fluids have been suggested for application in energy production processes due to their unique thermal and transport properties. The thermo-physical property change of the fluids is significant, in the high temperature and /or high pressure region (Yu, Chen, Zhao, Li, & Zhang, 2015). There are many advantages when using CO_2 supercritical cycles instead of classical systems: the compression ratio is low, so compressor efficiency is relatively high; volumetric refrigeration capacity is high, which helps reducing the size of the systems. Namely, in the proximity of critical point, the CO_2 fluid is very much expandable/compressible, the density and thermal conductivity change to be near one order of magnitude lower as temperature approaches the critical point. At the same time the Prandtl number and specific heat also form high peaks (Chen, Zhang, Okajima, Komiya, & Maruyama, 2016). When in a certain reaction system SCFs are applied as a reaction media, the particular thermodynamic and transport data have to be considered. Investigations of basic thermodynamic and transport data like phase equilibria, density, viscosity, dielectric constant and diffusion coefficients are of fundamental importance to overcome many uncertainties associated with the parameters of the extreme supercritical systems, the methods for incorporating risk and uncertainty into process design and optimization.

In the frame of our past research, several gravimetric methods have been developed or modified to carry out the measurements at elevated pressures and temperatures:

- **Determination of Density of the Binary Systems:** Density of some gases increases significantly at certain conditions and thus the buoyancy effect when applying the magnetic suspension balance method can be well expected. Consequently, the sample is lifted and the measurement cannot be performed. For these systems, the external balance method presents an alternative to the Magnetic Suspension Balance (Figure 5).
- **Determination of the Viscosity of Binary Systems:** Experimental viscosity data at high pressure and high temperature was measured for multi-component mixtures. The measurements were conducted in a HPHT reactor (Figure 6).
- **Investigation of Interfacial Tension of Binary Systems:** A novel technique for predicting the interfacial tension based on the capillary rise phenomenon for the gas saturated compounds has been proposed for performing interfacial tension tests at high pressure and high temperature conditions (Figure 7).

Figure 5. Magnetic suspension balance

Figure 6. a) NWA view cell with variable volume b) modification of system allowing viscosity measurements c) a sample of Polyethylene glycol (PEG) 1500 after addition of CO_2

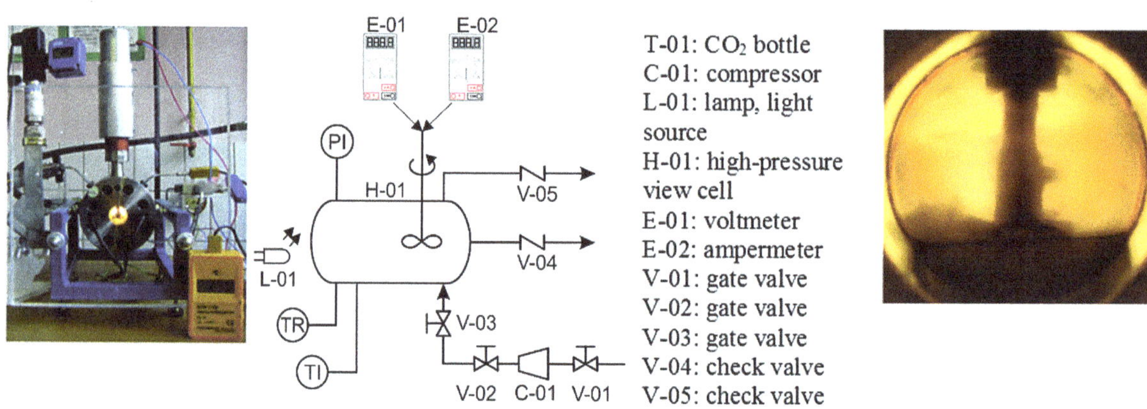

a) b) c)

To reduce hydrocarbon concentration and to increase the content of H2 and CO in the product, catalysts are introduced to the reaction mixture. If the aim of the reaction was the efficient gasification of biomass, alkali and metal catalysts were used. The alkali catalysts (NaOH, KOH, Na_2CO_3, $KHCO_3$, K2CO_3) (Cantero, Álvarez, Bermejo, & Cocero, 2015) significantly increased the H_2 and CO_2 yields and decreases CO yield by promoting the water-gas shift reaction (Guo et al., 2010). As heterogeneous catalysts, transition metals (Ru, Pd, Ni, Cu, Co) on different supports (TiO_2, Al_2O_3, ZrO_2, $CeZrO_2$, SiO_2) were used to increase biomass conversion and the yield of gas formation (Guan et al., 2014; Mehrani, Barati, Tavasoli, & Karimi, 2015). For the SCW reforming of biomass metals are much more attractive than alkalis. Their catalytic activity and stability of is influenced by the type of the support. Promoters such as Na, K, Mg or Cs may be also added to further improve catalyst activity and stability, and to favor the water-gas shift reaction. Activated carbon was also proposed as heterogeneous catalyst for enhancing biomass gasification in SCF (Ding, Azargohar, Dalai, & Kozinski, 2014). It was found that carbon gasification, the water-gas shift reaction and methanation reaction are favored by the presence

Figure 7. a) A coordinate measuring Machine Zeiss UMC 850 for determination of the radius of capillaries b) a scheme of the measuring system for determination of the interfacial tension

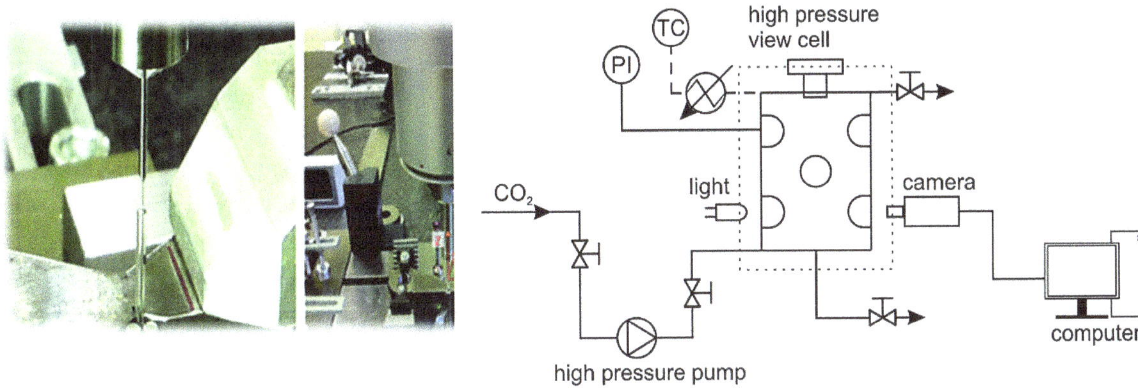

a) b)

of activated carbons. The main drawback is that activated carbon is prone to chemisorption of the intermediates which decreases the surface area, thus affecting its activity.

Corrosion is another issue that should be solved before considering construction materials. Unfortunately, it is almost impossible to prevent corrosion, however it is possible to control it. selection and the amount of corrosion inhibitor used depend on the acid type and its strength, the steel type, the desired protection time, and the expected temperature (Brondel et al., 1994). The main drawback considering corrosion inhibitors is that they are effective only for a particular metallic material in a certain environment (Finšgar & Merl, 2014). Already a very small change in the composition of a solution or alloy can significantly change the inhibition efficiency. Because the mechanism of how corrosion inhibitors work is usually not known and in spite of some proposed models for forecasting corrosion inhibition effectiveness, empirical experimental testing is still unavoidable (Brondel et al., 1994; Finšgar & Jackson, 2014). Inhibitors must be chosen after taking into account the nature and combinations of metals present, the nature of the corrosive environment, and the operating conditions: flow, temperature, and heat transfer. Recent studies have attempt to find a green (environmentally acceptable) corrosion inhibitor/inhibitor formulations such as plant extracts, essential oils and purified compounds (Znini et al., 2012).

FUTURE PERSPECTIVES

Applications of supercritical fluids certainly have a bright future. Particularly for the products for use in medicine, tissue engineering and nutrition from the point of the high value added products and treatment of excessive wastes generated by a wide range of processes, such as domestic and industrial wastewater treatment, wastes from oil and biodiesel industry, wastes from food, biomass processing and chemical industry. Combination of hydrolytic and bio-catalytic processes has proved to be a suitable method to convert very wet biomass and liquid streams to gases or other bio-based chemicals.

Integration of advanced processing concepts enable further application of residues from different industries with the principle goal to obtain value added products, avoiding the problems related to waste disposal. Limited supply of petroleum resources together with the ever-increasing demand for petroleum-based products has fueled the emergence of the bio economy. Global trends highlight the growth in fuel, energy, materials, chemicals and other products based on renewable feedstocks as economically viable alternatives to petroleum-based products. Therefore, there is no uncertainty that research needs and challenges that support the biorefinery concept, which is analogous to petroleum refinery, where instead of nonrenewable fossil fuels; renewable feedstocks are converted into chemicals, fuel, materials and other products. Thus, a biorefinery would integrate a variety of separation and conversion processes to produce multiple product streams from renewable feedstocks with little or no waste. Such developments in biorefinery approaches force us to consider integrated approaches to utilization of supercritical technology in biorefining where new approaches, such as energy recovery have key relevance. Namely, high temperatures involved in the process and the high energetic consumptions of these processes have been studied many times without an appropriate heat integration scheme. In design, knowledge of the reactions kinetics at different pressures and temperatures is of utmost importance and represents the fundamentals in the design of such process. The choice of biomass applied will dictate the operational conditions aiming to achieve a rapid conversion of polysaccharides with high saccharide production and low degradation products. The composition of biomass available varies considerably. Some natural materials contain relatively high amounts of complex substances that increase the resistance of the material to hydrolysis processes and require that conditions that are more aggressive. Lignocellulosic materials such as corn stover and sugar cane bagasse are some of them. Lignin is present in a complex structure with cellulose and hemicellulose that require high pressures and temperatures to carry out the complete hydrolysis. Thus the composition of the raw material is of fundamental importance in SubWH and SupWH processes, since it will determine (Brunner, 2015).

To overcome the obstacles between the research and application of the sustainable processing concepts in practice, more research is needed in order to understand the reaction mechanisms. Beside biomass treatment, hydrothermal technology has a great perspective for industrial scale production of materials, as they are part of the green chemistry. Particularly supercritical CO_2 and H_2O will have an important role in production of nano-particles and of nano-material owing to the multi-faceted advantages. The obtained products may be used as insulating materials or as catalysts. In such a process, initial material is dissolved in supercritical water in the first step. Then it is mixed with cold solutions to reduce the solvent power of the supercritical water and to achieve precipitation of the nano-materials. These are swept out with another flow of cold water and then separated from the water. Operation cost of such processes are relatively low due to the lower energy consumption and above all, the approach is less time consuming than the traditional processing methods (Brunner, 2015). Besides, the properties of the obtained products such as structure of electronic energy levels and transitions, electron affinity, conductivity, phase transition temperature, magnetic properties, melting points, etc. may be "tailored" since they depend on the nanoparticle size and shape, which is relatively easily controlled by varying the process conditions (Byrappa, 1991).

CONCLUSION

The chapter discussed hydrothermal processing of biomass to obtain bio-oil, bio-gases and bio-carbon with a highlight on current research trends that tend towards growth and implementation of new technologies which enable transition to "green" chemistry without further use of environmentally and health hazardous organic solvents as a basic step towards sustainable development. Such processing concepts comprise reuse of residues according to a basic concept of biorefineries. After extraction of source material, further conversion of residues (biomass) to a wide range of useful products, including second-generation biofuels, base chemicals, biospecial chemicals and bioefficient materials, is suggested. The main impetus for this conversion is on one way driven by a concern to the environment to reduce the usage of conventional solvents and energy. By the assessment of sustainable biomass potential and the evaluation of benefits of different bioenergy options, favorable conditions to set up a second-generation biofuel industry will be established. Several benefits are derived from the economic point of view; countries that are currently not able to produce second-generation biofuels could profit from new technology and market opportunities once the technology is commercially available.

"Process intensification" as a part of necessary sustainable development and "green" revolution uses also high pressure as a processing tool which enables production of goods with completely new specifications. Since higher pressures allow economically feasible processing, the future research will most probably lead towards investigation of the fundamental thermodynamic data pertaining to the conditions such processes.

Hydrothermal technique offers a great perspective of the future as an excellent possibility for processing of various product options from biomass degradation in hot compressed water. Hydrothermal processes are also a promising basis for the production of a variety of different intermediates to be used for energy and chemical purposes. As a reaction media water opens new perspectives compared to "dry processes", needing lower reaction temperatures. The polar character of water and the high reactivity of biomass with its many polar bonds in water offers an opportunity of processing at lower temperatures. Selectivity of water can be easily fine-tuned by slight changes in operating conditions – therefore different products with special properties can be obtained and by variation of pressure and temperature, production yields may be increased considerably. The advantages of hydrothermal technology are several and the waste biomass conversion for production of energy carriers are performed by applying supercritical water and supercritical water by the introduction of supercritical CO_2 to the binary systems supercritical water/biomass or supercritical water/biomass. The benefit of such processing reflects in lower energy consumption and production of specific products of a benefit. Current trends in hydrothermal materials processing are presented in relation to the concept of soft solution processing and above all, low energy consuming fabrication technique. Dissemination of the new approaches to the industrial scale processes, with the emphasis on waste biomass conversion for production of biofuels and bio-chemicals by applying supercritical fluids to improve the economic, environmental and energy efficiencies will play a significant role in the upstream and downstream processing in the oilfield industry. The challenge is not in the conversion of the feed to the product, but to achieve the product efficiently by understanding process backgrounds in terms of mass transfer, thermodynamic system characteristics, process optimization and scale up, which is still the main obstacle for a wider application of these technologies.

It is thus about the development of advanced technologies involving sub- and supercritical fluids which can provide efficient products, have low or even zero impact to the environment and do not require much energy.

Undoubtedly, much research is still needed and a close collaboration between scientists, industry and governmental agencies is required to make the large-scale application of such advances technologies for production of energy possible.

REFERENCES

Aionicesei, E., Škerget, M., & Knez, Ž. (2008). Measurement of CO_2 solubility and diffusivity in poly (l-lactide) and poly (d, l-lactide-co-glycolide) by magnetic suspension balance. *The Journal of Supercritical Fluids*, *47*(2), 296–301. doi:10.1016/j.supflu.2008.07.011

Akalın, M. K., Tekin, K., & Karagöz, S. (2012). Hydrothermal liquefaction of cornelian cherry stones for bio-oil production. *Bioresource Technology*, *110*, 682–687. doi:10.1016/j.biortech.2012.01.136 PMID:22342032

Akizuki, M., Fujii, T., Hayashi, R., & Oshima, Y. (2014). Effects of water on reactions for waste treatment, organic synthesis, and bio-refinery in sub- and supercritical water. *Journal of Bioscience and Bioengineering*, *117*(1), 10–18. doi:10.1016/j.jbiosc.2013.06.011 PMID:23867097

Aljbour, S. H., & Kawamoto, K. (2013). Bench-scale gasification of cedar wood – Part I: Effect of operational conditions on product gas characteristics. *Chemosphere*, *90*(4), 1495–1500. doi:10.1016/j.chemosphere.2012.08.029 PMID:22980961

Anastasakis, K., & Ross, A. B. (2015). Hydrothermal liquefaction of four brown macro-algae commonly found on the UK coasts: An energetic analysis of the process and comparison with bio-chemical conversion methods. *Fuel*, *139*, 546–553. doi:10.1016/j.fuel.2014.09.006

Azadi, P., & Farnood, R. (2011). Review of heterogeneous catalysts for sub-and supercritical water gasification of biomass and wastes. *International Journal of Hydrogen Energy*, *36*(16), 9529–9954. doi:10.1016/j.ijhydene.2011.05.081

Barnes, D. I. (2015). Understanding pulverised coal, biomass and waste combustion–a brief overview. *Applied Thermal Engineering*, *74*, 89–95. doi:10.1016/j.applthermaleng.2014.01.057

Benavente, V., Calabuig, E., & Fullana, A. (2015). Upgrading of moist agro-industrial wastes by hydrothermal carbonization. *Journal of Analytical and Applied Pyrolysis*, *113*, 89–98. doi:10.1016/j.jaap.2014.11.004

Biller, P., & Ross, A. B. (2011). Potential yields and properties of oil from the hydrothermal liquefaction of microalgae with different biochemical content. *Bioresource Technology*, *102*(1), 215–225. doi:10.1016/j.biortech.2010.06.028 PMID:20599375

Biller, P., Ross, A. B., Skill, S. C., Lea-Langton, A., Balasundaram, B., Hall, C., & Llewellyn, C. A. et al. (2012). Nutrient recycling of aqueous phase for microalgae cultivation from the hydrothermal liquefaction process. *Algal Research*, *1*(1), 70–76. doi:10.1016/j.algal.2012.02.002

Bioenergy, I. (2015). *Thermal Gasification of Biomass*. Task 33 Database. Retrieved May 05, 2016 from http://www.ieabioenergy.com/task/thermal-gasification-of-biomass/

Brennecke, J. F., & Eckert, C. A. (1989). Phase equilibria for supercritical fluid process design. *AIChE Journal. American Institute of Chemical Engineers*, *35*(9), 1409–1427. doi:10.1002/aic.690350902

Brondel, D., Edwards, R., Hayman, A., Hill, D., Mehta, S., & Semerad, T. (1994). Corrosion in the oil industry. *Oilfield Review, 6*(2), 4-18.

Brunner, G. (2005). Supercritical fluids: Technology and application to food processing. *Journal of Food Engineering*, *67*(1), 21–33. doi:10.1016/j.jfoodeng.2004.05.060

Brunner, G. (2015). Supercritical process technology related to energy and future directions–An introduction. *The Journal of Supercritical Fluids*, *96*, 11–20. doi:10.1016/j.supflu.2014.09.008

Byrappa, K. (1991). Hydrothermal growth of crystals. *Progress in Crystal Growth and Characterization of Materials*, *21*, R7–R10.

Byrappa, K., & Yoshimura, M. (2013). *Handbook of hydrothermal technology* (2nd ed.). Oxford, UK: William Andrew.

Cabeza, P., Queiroz, J. P. S., Criado, M., Jiménez, C., Bermejo, M. D., Mato, F., & Cocero, M. J. (2015). Supercritical water oxidation for energy production by hydrothermal flame as internal heat source. Experimental results and energetic study. *Energy*, *90*, 1584–1594. doi:10.1016/j.energy.2015.06.118

Calvignac, B., Rodier, E., Letourneau, J.-J., Almeida dos Santos, P. M., & Fages, J. (2010). Cocoa butter saturated with supercritical carbon dioxide: Measurements and modelling of solubility, volumetric expansion, density and viscosity. *International Journal of Chemical Reactor Engineering*, *8*(1). doi:10.2202/1542-6580.2191

Calvignac, B., Rodier, E., Letourneau, J.-J., Vitoux, P., Aymonier, C., & Fages, J. (2010). Development of an improved falling ball viscometer for high-pressure measurements with supercritical CO_2. *The Journal of Supercritical Fluids*, *55*(1), 96–106. doi:10.1016/j.supflu.2010.07.012

Calvo, L., Gil, M., Otero, M., Morán, A., & García, A. (2012). Gasification of rice straw in a fluidized-bed gasifier for syngas application in close-coupled boiler-gasifier systems. *Bioresource Technology*, *109*, 206–214. doi:10.1016/j.biortech.2012.01.027 PMID:22297044

Cantero, D. A., Álvarez, A., Bermejo, M. D., & Cocero, M. J. (2015). Transformation of glucose into added value compounds in a hydrothermal reaction media. *The Journal of Supercritical Fluids*, *98*, 204–210. doi:10.1016/j.supflu.2014.12.015

Cardenas-Toro, F. P., Forster-Carneiro, T., Rostagno, M. A., Petenate, A. J., Maugeri Filho, F., & Meireles, M. A. A. (2014). Integrated supercritical fluid extraction and subcritical water hydrolysis for the recovery of bioactive compounds from pressed palm fiber. *The Journal of Supercritical Fluids*, *93*, 42–48. doi:10.1016/j.supflu.2014.02.009

Cartmell, E., Gostelow, P., Riddell-Black, D., Simms, N., Oakey, J., Morris, J., & Pollard, S. J. et al. (2006). Biosolids a fuel or a waste? An integrated appraisal of five co-combustion scenarios with policy analysis. *Environmental Science & Technology*, *40*(3), 649–658. doi:10.1021/es052181g PMID:16509299

Castello, D., & Fiori, L. (2015). Supercritical water gasification of biomass: A stoichiometric thermodynamic model. *International Journal of Hydrogen Energy*, *40*(21), 6771–6781. doi:10.1016/j.ijhydene.2015.03.120

Cengiz, N. Ü., Yıldız, G., Sert, M., Selvi Gökkaya, D., Sağlam, M., Yüksel, M., & Ballice, L. (2015). Hydrothermal gasification of a biodiesel by-product crude glycerol in the presence of phosphate based catalysts. *International Journal of Hydrogen Energy*, *40*(43), 14806–14815. doi:10.1016/j.ijhydene.2015.08.097

Chen, G.-Y., Li, W.-Q., Chen, H., & Yan, B.-B. (2015). Progress in the aqueous-phase reforming of different biomass-derived alcohols for hydrogen production. *Journal of Zhejiang University*, *16*(6), 491–506. doi:10.1631/jzus.A1500023 PMID:26055913

Chen, L., Zhang, X.-R., Okajima, J., Komiya, A., & Maruyama, S. (2016). Numerical simulation of stability behaviors and heat transfer characteristics for near-critical fluid microchannel flows. *Energy Conversion and Management*, *110*, 407–418. doi:10.1016/j.enconman.2015.12.031

Chen, W.-T., Zhang, Y., Zhang, J., Yu, G., Schideman, L. C., Zhang, P., & Minarick, M. (2014). Hydrothermal liquefaction of mixed-culture algal biomass from wastewater treatment system into bio-crude oil. *Bioresource Technology*, *152*, 130–139. doi:10.1016/j.biortech.2013.10.111 PMID:24287452

Corma, A., García, H., & Leyva, A. (2005). Comparison between polyethylenglycol and imidazolium ionic liquids as solvents for developing a homogeneous and reusable palladium catalytic system for the Suzuki and Sonogashira coupling. *Tetrahedron*, *61*(41), 9848–9854. doi:10.1016/j.tet.2005.06.119

De Swaan Arons, J., & Diepen, G. (1963). Thermodynamic study of melting equilibria under pressure of a supercritical gas. *Recueil des Travaux Chimiques des Pays-Bas*, *82*(3), 249–256. doi:10.1002/recl.19630820308

Deniz, I., Vardar-Sukan, F., Yüksel, M., Saglam, M., Ballice, L., & Yesil-Celiktas, O. (2015). Hydrogen production from marine biomass by hydrothermal gasification. *Energy Conversion and Management*, *96*, 124–130. doi:10.1016/j.enconman.2015.02.048

Devi, L., Ptasinski, K. J., & Janssen, F. J. J. G. (2003). A review of the primary measures for tar elimination in biomass gasification processes. *Biomass and Bioenergy*, *24*(2), 125–140. doi:10.1016/S0961-9534(02)00102-2

Diebold, J. P. (2000). *A review of the chemical and physical mechanisms of the storage stability of fast pyrolysis bio-oils*. National Renewable Energy Laboratory.

Dindar, C., & Kiran, E. (2002). High-pressure viscosity and density of polymer solutions at the critical polymer concentration in near-critical and supercritical fluids. *Industrial & Engineering Chemistry Research*, *41*(25), 6354–6362. doi:10.1021/ie0108999

Ding, N., Azargohar, R., Dalai, A. K., & Kozinski, J. A. (2014). Catalytic gasification of glucose to H_2 in supercritical water. *Fuel Processing Technology*, *127*, 33–40. doi:10.1016/j.fuproc.2014.05.014

Dong, L., Asadullah, M., Zhang, S., Wang, X.-S., Wu, H., & Li, C.-Z. (2013). An advanced biomass gasification technology with integrated catalytic hot gas cleaning: Part I. Technology and initial experimental results in a lab-scale facility. *Fuel*, *108*, 409–416. doi:10.1016/j.fuel.2012.11.043

Elliott, D. C., Hart, T. R., Schmidt, A. J., Neuenschwander, G. G., Rotness, L. J., Olarte, M. V., & Holladay, J. E. et al. (2013). Process development for hydrothermal liquefaction of algae feedstocks in a continuous-flow reactor. *Algal Research*, *2*(4), 445–454. doi:10.1016/j.algal.2013.08.005

Elvassore, N., Vezzù, K., & Bertucco, A. (2005). Measurement and modeling of CO_2 absorption in poly (lactic-co-glycolic acid). *The Journal of Supercritical Fluids*, *33*(1), 1–5. doi:10.1016/j.supflu.2004.04.002

Faeth, J. L., Valdez, P. J., & Savage, P. E. (2013). Fast hydrothermal liquefaction of Nannochloropsis sp. to produce biocrude. *Energy & Fuels*, *27*(3), 1391–1398. doi:10.1021/ef301925d

Feng, S., Cheng, S., Yuan, Z., Leitch, M., & Xu, C. C. (2013). Valorization of bark for chemicals and materials: A review. *Renewable & Sustainable Energy Reviews*, *26*, 560–578. doi:10.1016/j.rser.2013.06.024

Finšgar, M., & Jackson, J. (2014). Application of corrosion inhibitors for steels in acidic media for the oil and gas industry: A review. *Corrosion Science*, *86*, 17–41. doi:10.1016/j.corsci.2014.04.044

Finšgar, M., & Merl, D. K. (2014). 2-Mercaptobenzoxazole as a copper corrosion inhibitor in chloride solution: Electrochemistry, 3D-profilometry, and XPS surface analysis. *Corrosion Science*, *80*, 82–95. doi:10.1016/j.corsci.2013.11.022

Fiori, L., Bassoa, D., Castelloa, D., & Baratierib, M. (2014). Hydrothermal carbonization of biomass: Design of a batch reactor and preliminary experimental results. *Chemical Engineering (Albany, N.Y.)*, 37.

Fonts, I., Gea, G., Azuara, M., Ábrego, J., & Arauzo, J. (2012). Sewage sludge pyrolysis for liquid production: A review. *Renewable & Sustainable Energy Reviews*, *16*(5), 2781–2805. doi:10.1016/j.rser.2012.02.070

Gao, Y., Yu, B., Wu, K., Yuan, Q., Wang, X., & Chen, H. (2016). Physicochemical, Pyrolytic, and Combustion Characteristics of Hydrochar Obtained by Hydrothermal Carbonization of Biomass. *BioResources*, *11*(2), 4113–4133. doi:10.15376/biores.11.2.4113-4133

Glisic, S., & Skala, D. (2009). The problems in design and detailed analyses of energy consumption for biodiesel synthesis at supercritical conditions. *The Journal of Supercritical Fluids*, *49*(2), 293–301. doi:10.1016/j.supflu.2008.12.011

Gökkaya, D. S., Saglam, M., Yuksel, M., & Ballice, L. (2016). Hydrothermal gasification of xylose: Effects of reaction temperature, pressure, and K_2CO_3 as a catalyst on product distribution. *Biomass and Bioenergy*, *91*, 26–36. doi:10.1016/j.biombioe.2016.04.013

Gourgouillon, D., Avelino, H., Fareleira, J., & Da Ponte, M. N. (1998). Simultaneous viscosity and density measurement of supercritical CO 2-saturated PEG 400. *The Journal of Supercritical Fluids*, *13*(1), 177–185. doi:10.1016/S0896-8446(98)00050-3

Guan, Q., Mao, T., Zhang, Q., Miao, R., Ning, P., Gu, J., & Chai, X.-S. et al. (2014). Catalytic gasification of lignin with Ni/Al_2 O_3–SiO_2 in sub/supercritical water. *The Journal of Supercritical Fluids*, *95*, 413–421. doi:10.1016/j.supflu.2014.10.015

Güngören Madenoğlu, T., Sağlam, M., Yüksel, M., & Ballice, L. (2016). Hydrothermal gasification of biomass model compounds (cellulose and lignin alkali) and model mixtures. *The Journal of Supercritical Fluids*, *115*, 79–85. doi:10.1016/j.supflu.2016.04.017

Guo, L., Jin, H., & Lu, Y. (2015). Supercritical water gasification research and development in China. *The Journal of Supercritical Fluids*, *96*, 144–150. doi:10.1016/j.supflu.2014.09.023

Guo, S., Guo, L., Cao, C., Yin, J., Lu, Y., & Zhang, X. (2012). Hydrogen production from glycerol by supercritical water gasification in a continuous flow tubular reactor. *International Journal of Hydrogen Energy*, *37*(7), 5559–5568. doi:10.1016/j.ijhydene.2011.12.135

Guo, Y., Wang, S., Xu, D., Gong, Y., Ma, H., & Tang, X. (2010). Review of catalytic supercritical water gasification for hydrogen production from biomass. *Renewable & Sustainable Energy Reviews*, *14*(1), 334–343. doi:10.1016/j.rser.2009.08.012

Gutiérrez, C., Rodríguez, J., Gracia, I., De Lucas, A., & García, M. (2013). Development of a strategy for the foaming of polystyrene dissolutions in $scCO_2$. *The Journal of Supercritical Fluids*, *76*, 126–134. doi:10.1016/j.supflu.2013.01.020

He, C., Giannis, A., & Wang, J.-Y. (2013). Conversion of sewage sludge to clean solid fuel using hydrothermal carbonization: Hydrochar fuel characteristics and combustion behavior. *Applied Energy*, *111*, 257–266. doi:10.1016/j.apenergy.2013.04.084

Hoekman, S. K., Broch, A., Robbins, C., Purcell, R., Zielinska, B., Felix, L., & Irvin, J. (2014). Process development unit (PDU) for hydrothermal carbonization (HTC) of lignocellulosic biomass. *Waste and Biomass Valorization*, *5*(4), 669–678. doi:10.1007/s12649-013-9277-0

Hoekman, S. K., Broch, A., Robbins, C., Zielinska, B., & Felix, L. (2013). Hydrothermal carbonization (HTC) of selected woody and herbaceous biomass feedstocks. *Biomass Conversion and Biorefinery*, *3*(2), 113–126. doi:10.1007/s13399-012-0066-y

Hoogwijk, M., Faaij, A., van den Broek, R., Berndes, G., Gielen, D., & Turkenburg, W. (2003). Exploration of the ranges of the global potential of biomass for energy. *Biomass and Bioenergy*, *25*(2), 119–133. doi:10.1016/S0961-9534(02)00191-5

Huber, G. W., Iborra, S., & Corma, A. (2006). Synthesis of Transportation Fuels from Biomass: Chemistry, Catalysts, and Engineering. *Chemical Reviews*, *106*(9), 4044–4098. doi:10.1021/cr068360d PMID:16967928

Jain, A., Balasubramanian, R., & Srinivasan, M. P. (2016). Hydrothermal conversion of biomass waste to activated carbon with high porosity: A review. *Chemical Engineering Journal*, *283*, 789–805. doi:10.1016/j.cej.2015.08.014

Jena, U., McCurdy, A. T., Warren, A., Summers, H., Ledbetter, R. N., Hoekman, S. K., & Quinn, J. C. et al. (2015). Oleaginous yeast platform for producing biofuels via co-solvent hydrothermal liquefaction. *Biotechnology for Biofuels*, *8*(1), 1. doi:10.1186/s13068-015-0345-5 PMID:26468320

Jordan, C. A., & Akay, G. (2012). Occurrence, composition and dew point of tars produced during gasification of fuel cane bagasse in a downdraft gasifier. *Biomass and Bioenergy*, *42*, 51–58. doi:10.1016/j.biombioe.2012.03.014

Judex, J. W., Gaiffi, M., & Burgbacher, H. C. (2012). Gasification of dried sewage sludge: Status of the demonstration and the pilot plant. *Waste Management (New York, N.Y.)*, *32*(4), 719–723. doi:10.1016/j.wasman.2011.12.023 PMID:22284442

Kambo, H. S., & Dutta, A. (2014). Strength, storage, and combustion characteristics of densified lignocellulosic biomass produced via torrefaction and hydrothermal carbonization. *Applied Energy*, *135*, 182–191. doi:10.1016/j.apenergy.2014.08.094

Kambo, H. S., & Dutta, A. (2015). Comparative evaluation of torrefaction and hydrothermal carbonization of lignocellulosic biomass for the production of solid biofuel. *Energy Conversion and Management, 105*, 746–755. doi:10.1016/j.enconman.2015.08.031

Kamm, B., Gruber, P. R., & Kamm, M. (2007). Biorefineries–Industrial Processes and Products Ullmann's Encyclopedia of Industrial Chemistry. Wiley-VCH Verlag GmbH & Co. KGaA.

Kesić, Ž., Lukić, I., Brkić, D., Rogan, J., Zdujić, M., Liu, H., & Skala, D. (2012). Mechanochemical preparation and characterization of CaO· ZnO used as catalyst for biodiesel synthesis. *Applied Catalysis A, General, 427*, 58–65. doi:10.1016/j.apcata.2012.03.032

Kiran, E. (1994). Polymer formation, modifications and processing in or with supercritical fluids. In Supercritical Fluids (pp. 541-588). Orono, ME: Springer Netherlands. doi:10.1007/978-94-015-8295-7_22

Kirkels, A. F., & Verbong, G. P. J. (2011). Biomass gasification: Still promising? A 30-year global overview. *Renewable & Sustainable Energy Reviews, 15*(1), 471–481. doi:10.1016/j.rser.2010.09.046

Knez, Ž., Markočič, E., Leitgeb, M., Primožič, M., Knez Hrnčič, M., & Škerget, M. (2014). Industrial applications of supercritical fluids: A review. *Energy, 77*, 235–243. doi:10.1016/j.energy.2014.07.044

Knez, Ž., Škerget, M., & Knez Hrňič, M. (2010). Principles of supercritical fluid extraction and applications in the food, beverage and nutraceutical industries. In S. Rizvi (Ed.), Separation, extraction and concentration processes in the food, beverage and nutraceutical industries (pp. 3-38). Elsevier

Knez Hrnčič, M., Markočič, E., Kramberger, B., & Knez, Ž. (2011). *Supercritical fluids as product and process solvents in emerging applications*. Paper presented at 13th European Meeting on Supercritical Fluids, Hague, The Netherlands.

Knez Hrnčič, M., Markočič, E., Trupej, N., Škerget, M., & Knez, Ž. (2014). Investigation of thermodynamic properties of the binary system polyethylene glycol/CO_2 using new methods. *The Journal of Supercritical Fluids, 87*, 50–58. doi:10.1016/j.supflu.2013.12.021

Knez Hrnčič, M., Venderbosch, R. H., Škerget, M., Ilić, L., & Knez, Ž. (2013b). Phase equilibrium data of hydrogen in pyrolysis oil and hydrogenated pyrolysis oil at elevated pressures. *The Journal of Supercritical Fluids, 80*, 86–89. doi:10.1016/j.supflu.2013.04.015

Knez Hrnčič, M., Venderbosch, R. H., Škerget, M., & Knez, Ž. (2013a). Observation of phase behavior for bio-oil+ diesel+ carbon dioxide and bio-oil+ tail water+ carbon dioxide system. *Journal of Chemical & Engineering Data, 58*(3), 648–652. doi:10.1021/je301146g

Kobayashi, N., Tanaka, M., Piao, G., Kobayashi, J., Hatano, S., Itaya, Y., & Mori, S. (2009). High temperature air-blown woody biomass gasification model for the estimation of an entrained down-flow gasifier. *Waste Management (New York, N.Y.), 29*(1), 245–251. doi:10.1016/j.wasman.2008.04.014 PMID:18653324

Kramberger, B., Markočič, E., & Knez, Ž. (2013). Phase equilibria of binary mixture of carbon monoxide and water at elevated temperatures and pressures. *Chemical Engineering Science, 99*, 77–80. doi:10.1016/j.ces.2013.05.048

Kruse, A. (2008). Supercritical water gasification. *Biofuels. Bioproducts and Biorefining*, *2*(5), 415–437. doi:10.1002/bbb.93

Kruse, A. (2011). Behandlung von Biomasse mit überkritischem Wasser. *Chemie Ingenieur Technik*, *83*(9), 1381–1389. doi:10.1002/cite.201100040

Kruse, A., Bernolle, P., Dahmen, N., Dinjus, E., & Maniam, P. (2010). Hydrothermal gasification of biomass: Consecutive reactions to long-living intermediates. *Energy & Environmental Science*, *3*(1), 136–143. doi:10.1039/B915034J

Kruse, A., Funke, A., & Titirici, M.-M. (2013). Hydrothermal conversion of biomass to fuels and energetic materials. *Current Opinion in Chemical Biology*, *17*(3), 515–521. doi:10.1016/j.cbpa.2013.05.004 PMID:23707262

Liao, B., Guo, L., Lu, Y., & Zhang, X. (2013). Solar receiver/reactor for hydrogen production with biomass gasification in supercritical water. *International Journal of Hydrogen Energy*, *38*(29), 13038–13044. doi:10.1016/j.ijhydene.2013.03.113

Liu, Z., & Balasubramanian, R. (2012). Hydrothermal Carbonization of Waste Biomass for Energy Generation. *Procedia Environmental Sciences*, *16*, 159–166. doi:10.1016/j.proenv.2012.10.022

Liu, Z., & Balasubramanian, R. (2014). Upgrading of waste biomass by hydrothermal carbonization (HTC) and low temperature pyrolysis (LTP): A comparative evaluation. *Applied Energy*, *114*, 857–864. doi:10.1016/j.apenergy.2013.06.027

Liu, Z., Quek, A., Kent Hoekman, S., & Balasubramanian, R. (2013). Production of solid biochar fuel from waste biomass by hydrothermal carbonization. *Fuel*, *103*, 943–949. doi:10.1016/j.fuel.2012.07.069

Lv, P., Yuan, Z., Ma, L., Wu, C., Chen, Y., & Zhu, J. (2007). Hydrogen-rich gas production from biomass air and oxygen/steam gasification in a downdraft gasifier. *Renewable Energy*, *32*(13), 2173–2185. doi:10.1016/j.renene.2006.11.010

Lynam, J. G., Reza, M. T., Yan, W., Vásquez, V. R., & Coronella, C. J. (2015). Hydrothermal carbonization of various lignocellulosic biomass. *Biomass Conversion and Biorefinery*, *5*(2), 173–181. doi:10.1007/s13399-014-0137-3

Mäkelä, M., Benavente, V., & Fullana, A. (2015). Hydrothermal carbonization of lignocellulosic biomass: Effect of process conditions on hydrochar properties. *Applied Energy*, *155*, 576–584. doi:10.1016/j.apenergy.2015.06.022

Mäkelä, M., Benavente, V., & Fullana, A. (2016). Hydrothermal carbonization of industrial mixed sludge from a pulp and paper mill. *Bioresource Technology*, *200*, 444–450. doi:10.1016/j.biortech.2015.10.062 PMID:26519695

Markočič, E., Kramberger, B., van Bennekom, J. G., Heeres, H. J., Vos, J., & Knez, Ž. (2013). Glycerol reforming in supercritical water; a short review. *Renewable & Sustainable Energy Reviews*, *23*, 40–48. doi:10.1016/j.rser.2013.02.046

Markočič, E., Škerget, M., & Knez, Ž. (2011). Solubility and diffusivity of CO_2 in poly (l-lactide)–hydroxyapatite and poly (d, l-lactide-co-glycolide)–hydroxyapatite composite biomaterials. *The Journal of Supercritical Fluids*, *55*(3), 1046–1051. doi:10.1016/j.supflu.2010.10.001

Materazzi, M., Lettieri, P., Mazzei, L., Taylor, R., & Chapman, C. (2015). Fate and behavior of inorganic constituents of RDF in a two stage fluid bed-plasma gasification plant. *Fuel*, *150*, 473–485. doi:10.1016/j.fuel.2015.02.059

Mazaheri, H., Lee, K. T., Bhatia, S., & Mohamed, A. R. (2010). Sub/supercritical liquefaction of oil palm fruit press fiber for the production of bio-oil: Effect of solvents. *Bioresource Technology*, *101*(19), 7641–7647. doi:10.1016/j.biortech.2010.04.072 PMID:20510608

Mehrani, R., Barati, M., Tavasoli, A., & Karimi, A. (2015). Hydrogen production via supercritical water gasification of bagasse using Ni–Cu/γ-Al2O3 nano-catalysts. *Environmental Technology*, *36*(10), 1265–1272. doi:10.1080/09593330.2014.984771 PMID:25387488

Meng, X., de Jong, W., Fu, N., & Verkooijen, A. H. M. (2011). Biomass gasification in a 100 kWth steam-oxygen blown circulating fluidized bed gasifier: Effects of operational conditions on product gas distribution and tar formation. *Biomass and Bioenergy*, *35*(7), 2910–2924. doi:10.1016/j.biombioe.2011.03.028

Möbius, A., Boukis, N., Hindersin, S., & Kerner, M. (2015). Cultivation and Hydrothermal Gasification of Microalgae Scenedesmus Obliquus–First Experimental Results. *Chemie Ingenieur Technik*, *87*(12), 1720–1725. doi:10.1002/cite.201500002

Mukherjee, A., Bhattacharyya, S., & Bagchi, B. (2002). Pressure and temperature dependence of viscosity and diffusion coefficients of a glassy binary mixture. *The Journal of Chemical Physics*, *116*(11), 4577–4586. doi:10.1063/1.1445747

Ngo, S. I., Nguyen, T. D. B., Lim, Y.-I., Song, B.-H., Lee, U.-D., Choi, Y.-T., & Song, J.-H. (2011). Performance evaluation for dual circulating fluidized-bed steam gasifier of biomass using quasi-equilibrium three-stage gasification model. *Applied Energy*, *88*(12), 5208–5220. doi:10.1016/j.apenergy.2011.07.046

Olgun, H., Ozdogan, S., & Yinesor, G. (2011). Results with a bench scale downdraft biomass gasifier for agricultural and forestry residues. *Biomass and Bioenergy*, *35*(1), 572–580. doi:10.1016/j.biombioe.2010.10.028

Oliveira, I., Blöhse, D., & Ramke, H.-G. (2013). Hydrothermal carbonization of agricultural residues. *Bioresource Technology*, *142*, 138–146. doi:10.1016/j.biortech.2013.04.125 PMID:23735795

Pantoula, M., & Panayiotou, C. (2006). Sorption and swelling in glassy polymer/carbon dioxide systems: Part I. Sorption. *The Journal of Supercritical Fluids*, *37*(2), 254–262. doi:10.1016/j.supflu.2005.11.001

Paris COP21. (2015). *Climate Change Conference*. Retrieved October 14, 2016 from http://www.cop21.gouv.fr/en/

Parshetti, G. K., Kent Hoekman, S., & Balasubramanian, R. (2013). Chemical, structural and combustion characteristics of carbonaceous products obtained by hydrothermal carbonization of palm empty fruit bunches. *Bioresource Technology*, *135*, 683–689. doi:10.1016/j.biortech.2012.09.042 PMID:23127830

Patzelt, D. J., Hindersin, S., Elsayed, S., Boukis, N., Kerner, M., & Hanelt, D. (2015). Hydrothermal gasification of Acutodesmus obliquus for renewable energy production and nutrient recycling of microalgal mass cultures. *Journal of Applied Phycology*, *27*(6), 2239–2250. doi:10.1007/s10811-014-0496-y

Pavlovič, I., Knez, Ž., & Škerget, M. (2013). Hydrothermal Reactions of Agricultural and Food Processing Wastes in Sub- and Supercritical Water: A Review of Fundamentals, Mechanisms, and State of Research. *Journal of Agricultural and Food Chemistry*, *61*(34), 8003–8025. doi:10.1021/jf401008a PMID:23848589

Peterson, A. A., Vogel, F., Lachance, R. P., Fröling, M., Antal, M. J. Jr, & Tester, J. W. (2008). Thermochemical biofuel production in hydrothermal media: A review of sub-and supercritical water technologies. *Energy & Environmental Science*, *1*(1), 32–65. doi:10.1039/b810100k

Pfaltzgraff, L. A., Cooper, E. C., Budarin, V., & Clark, J. H. (2013). Food waste biomass: A resource for high-value chemicals. *Green Chemistry*, *15*(2), 307–314. doi:10.1039/c2gc36978h

Qiu, G., Liu, H., & Riffat, S. (2011). Expanders for micro-CHP systems with organic Rankine cycle. *Applied Thermal Engineering*, *31*(16), 3301–3307. doi:10.1016/j.applthermaleng.2011.06.008

Reza, M. T., Lynam, J. G., Uddin, M. H., & Coronella, C. J. (2013). Hydrothermal carbonization: Fate of inorganics. *Biomass and Bioenergy*, *49*, 86–94. doi:10.1016/j.biombioe.2012.12.004

Román, S., Nabais, J. M. V., Laginhas, C., Ledesma, B., & González, J. F. (2012). Hydrothermal carbonization as an effective way of densifying the energy content of biomass. *Fuel Processing Technology*, *103*, 78–83. doi:10.1016/j.fuproc.2011.11.009

Schmieder, H., Abeln, J., Boukis, N., Dinjus, E., Kruse, A., Kluth, M., & Schacht, M. et al. (2000). Hydrothermal gasification of biomass and organic wastes. *The Journal of Supercritical Fluids*, *17*(2), 145–153. doi:10.1016/S0896-8446(99)00051-0

Selvi Gökkaya, D., Saglam, M., Yüksel, M., & Ballice, L. (2015). Supercritical water gasification of phenol as a model for plant biomass. *International Journal of Hydrogen Energy*, *40*(34), 11133–11139. doi:10.1016/j.ijhydene.2015.04.078

Sermyagina, E., Saari, J., Kaikko, J., & Vakkilainen, E. (2015). Hydrothermal carbonization of coniferous biomass: Effect of process parameters on mass and energy yields. *Journal of Analytical and Applied Pyrolysis*, *113*, 551–556. doi:10.1016/j.jaap.2015.03.012

Shabani, N., Akhtari, S., & Sowlati, T. (2013). Value chain optimization of forest biomass for bioenergy production: A review. *Renewable & Sustainable Energy Reviews*, *23*, 299–311. doi:10.1016/j.rser.2013.03.005

Sugathapala, A. G. T. (2013). *Technologies for Converting Waste Agricultural Biomass to Energy*. Osaka: Academic Press.

Titirici, M.-M., Thomas, A., & Antonietti, M. (2007). Back in the black: Hydrothermal carbonization of plant material as an efficient chemical process to treat the CO_2 problem? *New Journal of Chemistry*, *31*(6), 787–789. doi:10.1039/b616045j

Toor, S. S., Rosendahl, L., & Rudolf, A. (2011). Hydrothermal liquefaction of biomass: A review of subcritical water technologies. *Energy*, *36*(5), 2328–2342. doi:10.1016/j.energy.2011.03.013

Umeki, K., Yamamoto, K., Namioka, T., & Yoshikawa, K. (2010). High temperature steam-only gasification of woody biomass. *Applied Energy*, *87*(3), 791–798. doi:10.1016/j.apenergy.2009.09.035

Üremek Cengiz, N., Eren, S., Sağlam, M., Yüksel, M., & Ballice, L. (2016). Influence of temperature and pressure on hydrogen and methane production in the hydrothermal gasification of wood residues. *The Journal of Supercritical Fluids*, *107*, 243–249. doi:10.1016/j.supflu.2015.09.018

Vadillo, V., Sánchez-Oneto, J., Portela, J. R., & Martínez de la Ossa, E. J. (2013). Problems in supercritical water oxidation process and proposed solutions. *Industrial & Engineering Chemistry Research*, *52*(23), 7617–7629. doi:10.1021/ie400156c

Van Bennekom, J., Vos, J., Venderbosch, R., Paris Torres, M., Kirilov, V., Herres, H.J., … Penninger, J. (2009). *Supermethanol: reforming of crude glycerine in supercritical water to produce methanol for re-use in biodiesel plants*. Paper presented at 17th European Biomass Conference and Exhibition, Hamburg, Germany.

van de Beld, B., Elliott, D. C., Neuenschwander, G. G., & Kruse, A. (2005). Biomass gasification in near- and super-critical water: Status and prospects. *Biomass and Bioenergy*, *29*(4), 269–292. doi:10.1016/j.biombioe.2005.04.006

Vardon, D. R., Sharma, B., Scott, J., Yu, G., Wang, Z., Schideman, L., & Strathmann, T. J. et al. (2011). Chemical properties of biocrude oil from the hydrothermal liquefaction of Spirulina algae, swine manure, and digested anaerobic sludge. *Bioresource Technology*, *102*(17), 8295–8303. doi:10.1016/j.biortech.2011.06.041 PMID:21741234

Vardon, D. R., Sharma, B. K., Blazina, G. V., Rajagopalan, K., & Strathmann, T. J. (2012). Thermochemical conversion of raw and defatted algal biomass via hydrothermal liquefaction and slow pyrolysis. *Bioresource Technology*, *109*, 178–187. doi:10.1016/j.biortech.2012.01.008 PMID:22285293

Venderbosch, R. H., Ardiyanti, A. R., Wildschut, J., Oasmaa, A., & Heeres, H. J. (2010). Stabilization of biomass-derived pyrolysis oils. *Journal of Chemical Technology and Biotechnology (Oxford, Oxfordshire)*, *85*(5), 674–686. doi:10.1002/jctb.2354

Vera, D., Jurado, F., Margaritis, N. K., & Grammelis, P. (2014). Experimental and economic study of a gasification plant fuelled with olive industry wastes. *Energy for Sustainable Development*, *23*, 247–257. doi:10.1016/j.esd.2014.09.011

Waldner, M. H., & Vogel, F. (2005). Renewable production of methane from woody biomass by catalytic hydrothermal gasification. *Industrial & Engineering Chemistry Research*, *44*(13), 4543–4551. doi:10.1021/ie050161h

Wang, F., Chang, Z., Duan, P., Yan, W., Xu, Y., Zhang, L., & Fan, Y. et al. (2013). Hydrothermal liquefaction of Litsea cubeba seed to produce bio-oils. *Bioresource Technology*, *149*, 509–515. doi:10.1016/j.biortech.2013.09.108 PMID:24140857

Wang, L., Shahbazi, A., & Hanna, M. A. (2011). Characterization of corn stover, distiller grains and cattle manure for thermochemical conversion. *Biomass and Bioenergy*, *35*(1), 171–178. doi:10.1016/j.biombioe.2010.08.018

Wang, Y., Wang, S., Zhao, G., Guo, Y., & Guo, Y. (2016). Hydrogen production by partial oxidation gasification of a phenol, naphthalene, and acetic acid mixture in supercritical water. *International Journal of Hydrogen Energy*, *41*(4), 2238–2246. doi:10.1016/j.ijhydene.2015.12.115

Wasteshipment. (2016). Retrieved May, 10, 2016, from wasteshipment.com/waste-management

Woolf, D., Amonette, J. E., Street-Perrott, F. A., Lehmann, J., & Joseph, S. (2010). Sustainable biochar to mitigate global climate change. *Nature Communications*, *1*(5), 56. doi:10.1038/ncomms1053 PMID:20975722

Xiao, L.-P., Shi, Z.-J., Xu, F., & Sun, R.-C. (2012). Hydrothermal carbonization of lignocellulosic biomass. *Bioresource Technology*, *118*, 619–623. doi:10.1016/j.biortech.2012.05.060 PMID:22698445

Yakaboylu, O., Harinck, J., Smit, K., & de Jong, W. (2015). Supercritical water gasification of biomass: A literature and technology overview. *Energies*, *8*(2), 859–894. doi:10.3390/en8020859

Yin, S., & Tan, Z. (2012). Hydrothermal liquefaction of cellulose to bio-oil under acidic, neutral and alkaline conditions. *Applied Energy*, *92*, 234–239. doi:10.1016/j.apenergy.2011.10.041

Yoshida, T., Oshima, Y., & Matsumura, Y. (2004). Gasification of biomass model compounds and real biomass in supercritical water. *Biomass and Bioenergy*, *26*(1), 71–78. doi:10.1016/S0961-9534(03)00063-1

Yu, S.-C., Chen, L., Zhao, Y., Li, H.-X., & Zhang, X.-R. (2015). A brief review study of various thermodynamic cycles for high temperature power generation systems. *Energy Conversion and Management*, *94*, 68–83. doi:10.1016/j.enconman.2015.01.034

Zeton, P. V., Tews, I., Onarheim, K., Beckman, D., Zhu, Y., Solantausta, Y., & Snowden-Swan, L. et al. (2014). *Biomass Direct Liquefaction Options: TechnoEconomic and Life Cycle Assessment*. Richland: Pacific Northwest National Laboratory.

Zhang, F., Chen, S., Xu, C., Chen, G., Zhang, J., & Ma, C. (2012). Experimental study on the effects of operating parameters on the performance of a transpiring-wall supercritical water oxidation reactor. *Desalination*, *294*, 60–66. doi:10.1016/j.desal.2012.03.013

Zhang, J., Chen, W.-T., Zhang, P., Luo, Z., & Zhang, Y. (2013). Hydrothermal liquefaction of Chlorella pyrenoidosa in sub- and supercritical ethanol with heterogeneous catalysts. *Bioresource Technology*, *133*, 389–397. doi:10.1016/j.biortech.2013.01.076 PMID:23454385

Zhao, P., Shen, Y., Ge, S., & Yoshikawa, K. (2014). Energy recycling from sewage sludge by producing solid biofuel with hydrothermal carbonization. *Energy Conversion and Management*, *78*, 815–821. doi:10.1016/j.enconman.2013.11.026

Zhu, Z., Toor, S. S., Rosendahl, L., & Chen, G. (2014). Analysis of product distribution and characteristics in hydrothermal liquefaction of barley straw in subcritical and supercritical water. *Environmental Progress & Sustainable Energy*, *33*(3), 737–743. doi:10.1002/ep.11977

Znini, M., Majidi, L., Bouyanzer, A., Paolini, J., Desjobert, J.-M., Costa, J., & Hammouti, B. (2012). Essential oil of Salvia aucheri mesatlantica as a green inhibitor for the corrosion of steel in 0.5 MH_2SO_4. *Arabian Journal of Chemistry*, *5*(4), 467–474. doi:10.1016/j.arabjc.2010.09.017

KEY TERMS AND DEFINITIONS

Bio Based Chemicals: Sustainable, cost-advantaged, drop-in replacements and substitutes for petroleum-based chemicals that are used in variety of industrial applications.

Bio Oils: Dark brown organic liquids, which are chemically complex mixtures and/or emulsions of water and degradation products of lignin.

Biofuel: A fuel, such as biogas or biodiesel that is produced from renewable resources, especially plant biomass, vegetable oils, or treated municipal and industrial wastes.

High Pressure Phase Equilibrium: Equilibrium relations between phases at high pressure.

High Pressure View Cell: A reactor that allows both, observation through the windows and quick and precise thermodynamic measurements during experiments, carried out at elevated pressures by a precise volume control with measurement of the piston position in the cell. Pressure and temperature are maintained constant during entire measurement.

Hydrothermal Carbonization: Thermochemical conversion process, which usually takes place at moderate conditions (523 K and 2 MPa) where carbonaceous bio solid product named "hydro char" is obtained.

Hydrothermal Gasification: A process in which biomass waste without pre-drying reacts with water that acts not only as a solvent, but also as a reactant under subcritical and supercritical conditions to produce gaseous products, mainly CH_4, H_2, CO_2 and C_1-C_4 carbon gases.

Hydrothermal Liquefaction: A process for converting waste biomass into chemicals and biofuels, where the reaction is performed under medium temperature and high pressure (553-643 K, 10-25 MPa), sufficient to keep the water in a liquid state.

Chapter 17
Application of Supercritical Technologies in Clean Energy Production: A Review

Vladimir Anikeev
Institute of Catalysis, Siberian Branch, Russian Academy of Sciences, Russia

Khaled A. M. Gasem
University of Wyoming – Laramie, USA

Maohong Fan
University of Wyoming – Laramie, USA

ABSTRACT

In this chapter the authors seek to acquaint the reader with unusual properties of supercritical fluids, and how these properties are used for various applications in the synthesis of biofuels and the intensification of energy processes and thermal cycles. The review includes the studies of leading scientists from around the world related to effective and environmentally friendly methods and technologies of energy acquisition and conversion involving the synthesis of motor fuels, materials and chemicals, solar energy conversion and thermal cycle efficiencies using supercritical fluids. A special place in the chapter is occupied by studies of the transformation of vegetable oils, algae and wood in supercritical alcohols, esters and water to produce biodiesel fuels and hydrogen.

INTRODUCTION

The intensification of energy processes while preserving the quality of the environment is an integral component of global economic and social development. Energy and the technology of synthesizing motor fuels, hydrogen and chemicals occupy a special place among these processes. While there are many approaches and methods for achieving the above objectives, this review presents the results of research, practical applications and chemical processes carried out in supercritical fluids.

DOI: 10.4018/978-1-5225-2047-4.ch017

In this review, the authors seek to acquaint the reader with unusual properties of supercritical fluids, and how these properties are used for various applications in the synthesis of biofuels and the intensification of energy processes. The authors do not discuss in detail herein the structure of supercritical fluid nor its thermodynamic and thermal properties near the critical point, as these topics have been addressed in previously published papers (Schmidt et al., 1946; Eckert et al., 1996; Kiran, 2000; Sengers & Levelt, 2000).

Due to their unique properties, supercritical fluids (SCFs) offer an attractive environment for various types of chemical reactions and physical processes involving transformations in the underlying energy cycles. Many physical and chemical properties of SCFs lie between those of liquids and gases. SCFs have also unique solubility. While many chemical components are not soluble in fluid under normal conditions, these become almost completely soluble under supercritical conditions. Conversely, some substances (generally salts) that are readily soluble in liquid under normal conditions, they are not very soluble in this medium under supercritical conditions. Unlimited solubility effectively allows for a reduction in heat and mass limitations and increases the mass transfer rate, e.g., it significantly increases the rate of chemical reactions in catalyst grains.

The practical application of SCF technologies includes waste recycling and decomposition, biomass and coal conversion processes, synthesis of new materials, and synthesis of bio-fuels and hydrogen, among many others.

Among the chemical reactions performed in SCFs, the oxidation reactions carried out in supercritical water are of the greatest practical interest today. These form the basis of almost all known processes and technologies for processing different type of wastes, including the destruction of dissolved dangerous and harmful organic substances, biological wastes and military production (Harradine et al., 1993; Savage et al., 1995; Pisharody et al., 1996; Anikeet et al., 2006; Kritzer & Dinjus, 2001; Vadillo et al., 2014).

Technologies using supercritical conditions allow the conversion of wastes at a high rate and significantly lower temperature; therefore, non- NO_x, and SO_2, metal oxides require little or no energy supply. Organic waste components and oxygen dissolve almost completely in the supercritical solvent near its critical point, forming a single phase environment conducive to rapid oxidation of organics into CO_2, H_2O and N_2. In SCF in the presence of oxygen, practically all hydrocarbons are oxidized to CO_2 and water, and compounds containing nitrogen are oxidized to form N_2 or N_2O. If air is used as the oxidant, nitrogen under these conditions (temperature for supercritical water reaction requires slightly more than 380°C) is not oxidized, nor does NO_x form.

Because of their high diffusion coefficient and solubility of the reactants, supercritical solvents are attractive for heterogeneous catalytic reactions. They prevent coke formation and increase the rate of penetration of reagents into the pores of the catalyst pellet, thereby increasing its efficiency factor (Savage et al., 1995; Savage, 1999; Helwani et al., 2009; Semwal et al., 2011). The total rate of gas-liquid heterogeneous catalytic reactions, such as liquid phase oxidation reactions of organic and inorganic compounds, is strongly limited by the extremely low solubility of oxygen in the substrate within the diffusion inhibition, both in the gas and the liquid reagent. Considering the high solubility of gases in SC solvents, it can be assumed that the supercritical solvent in the presence of heterogeneous catalysts will not have any diffusion limitations. Furthermore, due to the high solubility of heavy hydrocarbons and coke, the supercritical reaction medium has a favorable effect on catalyst activity.

Studies of transformations (gasification) of biomass in supercritical fluids and solvents-including supercritical water to produce biofuel, hydrogen, low-molecular organic liquids, and polyhydric alcohols-have been intensively developed in the world in recent years (Mckendry, 2002; Calzavara et al., 2005;

Matsumura et al., 2005, 2006; Ni & Dyc, 2006; Peterson et al., 2008a; Kruse, 2009; Tan & Lee, 2011) . Thus, the synthesis of biodiesel from vegetable oils and fats by using lower alcohols as SCFs offers a promising substitute for conventional hydrocarbon fuels for diesel engines (Gerpen, 2005). The advantages of this method lie in the simplicity of the reactions in highly ecological processes that result in higher yields. Cellulose, lignin and triglycerides are the most common types of biomass suitable for such large-scale processing. A schematic overview of typical biomolecules for each of them is shown in Figure 1.

The need to conserve energy and preserve the environment requires the development of renewable sources to replace fossil fuels, including biofuels and other alternative sources of energy such as wind and solar. Such development will necessarily involve the use of new, highly efficient methods and principles.

The purpose of this review is to advance the union of modern research on energy and the environment with the use of the supercritical state of a solvent or working fluid (body). The review includes the studies of leading scientists from around the world focused on effective and environmentally friendly methods and technologies of energy acquisition and conversion involving the synthesis of fuels, materials and chemicals using supercritical fluids. Specifically, we will address the following topics:

- Synthesis of biofuels in supercritical fluids,
- Fischer-Tropsch synthesis in supercritical fluid reaction media,
- The use of supercritical fluids in the conversion and storage of thermal energy,
- Accumulation of concentrated solar energy,
- The use of supercritical fluids in nuclear power.

The authors do not attempt to include the greatest number of references on this subject, but instead limit their inquiry to the most critical studies.

Figure 1. Biomass feedstocks, including cellulose, triglycerides, and lignin
Peterson et al., 2008b.

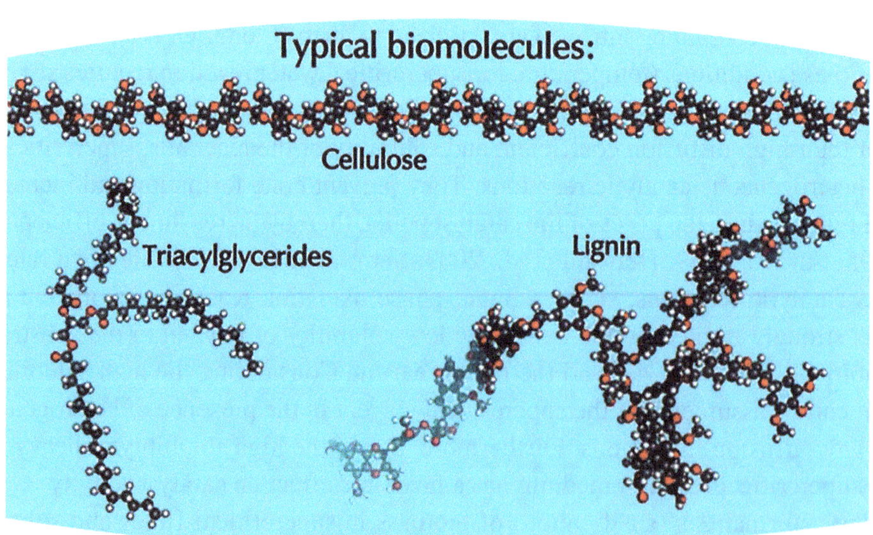

BIOFUELS: AN ENVIRONMENTALLY FRIENDLY ENERGY SOURCE

The constant increase in energy use in transportation and industry has led to increased pressure on natural resources. In addition, increased awareness of environmental issues such as global warming due to increased greenhouse gas emissions and the depletion of fossil fuels has led to the exploration of alternative energy sources and carriers. Biofuels derived from renewable resources of vegetable origin are among those considered seriously today.

Produced from renewable vegetable sources, biofuels are generally considered as either a replacement for or supplement to fossil fuels (Lovett et al., 2014). There are several types of biofuels, including bio-diesel, bio-ethanol, hydrogen, methane, methanol and others, particularly a number of monohydric alcohols. The raw material for the production of biofuels can be found in almost all crops, algae, and wild plants, including the waste products from processing these.

Among the above biofuels, the synthesis of biodiesel (methyl esters of fatty acids) receives the most attention today (Gerpen, 2005; Hoang et al., 2013). Biodiesel is today the most attractive fuel because it lacks sulfur, it has a high flash point, low content of aromatic compounds (American Society for Testing and Materials, 2002).

Many useful reviews published today devoted to analyzing the synthesis of biofuels in SCFs include investigations of bioethanol, biobutanol, biodiesel, biomethanol, pyrolysis oils, biogas, and hydrogen (Arai et al., 2009; Demirbas, 2009; Wen et al., 2009; Naik et al., 2010; Vyas et al., 2010; Deshpande et al., 2011; de Boer & Babri, 2011). Taken together, they offer a complete picture of the possibilities and advantages in the application of SCFs to the transformation of various types of biomass into biofuels generally, including hydrogen and biodiesel.

TECHNOLOGIES FOR BIOFUELS AND HYDROGEN PRODUCTION USING SUPERCRITICAL WATER BIOMASS CONVERSION

Biomass has been for a long time a major source of energy, and today still accounts for an estimated 10-14% of world energy consumption (Mckendry, 2002). Since biomass contains up to 95% water, biomass must be subjected to preliminary drying in conventional processes of gasification or liquefaction. To avoid this energy-consuming process, the biomass can be converted without prior drying in "hot" pressurized or supercritical water (Kruse et al., 2003; Toor et al., 2011).

Some of the earliest studies (Modell and Amin, 1978; Modell, 1982) show, first, that glucose and maple sawdust may be gasified in the water near its critical state in the presence of catalysts, promoting hydrogenation, reforming and catalytic cracking. Further, scientists from many countries have undertaken research aimed at gasification of biomass sub - or supercritical water (Elliott et al., 1994; Kruse et al., 2003; Williams and Onwudidi, 2005; Penninger et al., 2007; Resende et al., 2007; Resende et al., 2008; Toor et al., 2011).

Creating a universal method of converting biomass to the desired product faces the challenge of a large variety of biomaterials of various compositions and differing moisture content even within the same species. However, the use of sub- or supercritical water in biomass processing technology enables universal and cost-effective processes. Indeed, it is not necessary to dry the wet biomass. Moreover, water acts both as a reactant and the reaction medium in hydrolysis, dehydration, depolymerization and other reactions. Still, the assumption remains that the water in the liquid phase between the fibers of the biomass will, at the transition to the supercritical state, destroy the structure by increasing the material surface.

In recent years, the emphasis in the selection of the biomass type for processing into biofuel has shifted towards lignocellulosic biomass (non-food resources) and its primary processing products (lignin, hemicellulose). Lignocellulosic biomass, which is not a part of the starch-vegetable fiber material, is a renewable resource and can be considered as a potential raw material for the production of biofuels. The composition of lignocellulosic material varies depending on the plant species, growing conditions of the plants and the ages of their parts (Falkehag, 1975; Meister, 1996; Olsson et al., 2005)..

This section provides an overview of research on hydrothermal methods (into sub - and supercritical water) for treating biomass to produce liquid and gaseous fuels (methane, hydrogen, synthesis gas).

Biomass Gasification

Biomass is the fourth-largest energy source after coal, oil and natural gas. It is an attractive raw material for fuel and biomaterials for three main reasons. First, it is sourced from so-called carbon-neutral resources. Second, its conversion into other products does not increase greenhouse gases, while conventional incineration leads to emissions of some aromatic hydrocarbons, dioxins, furans, and heavy metals (Pasticakova, 2004). Third, for countries that do not possess significant natural mineral resources, it can be a major source of fuel and biomaterials.

As noted above, the high humidity of biomass requires its pre-drying for traditional processes of gasification or liquefaction. To avoid this, the biomass can be converted without prior drying in a "hot" pressurized or supercritical water (SCW) (Kruse et al., 2003; Matsumura & Minowa, 2004). In the absence of an oxidant, the biomass is rendered with a high degree of conversion in SCW into fuel gases (Antal, 1978; Sato et al., 2003). It is shown as well that at temperatures above 873K the gasification of biomass in SCW can produce H_2, CH_4, CO, CO_2 and small amounts of resins (Manarungson et al., 1993; Xu et al., 1996). The formation of solid carbon-containing products was not observed.

A different picture of biomass gasification takes place in subcritical water. Thus, at a pressure close to atmospheric pressure and temperatures higher than 463K, the biomass is reacted with steam to generate a large amount of tar and char; the gas phase contains higher hydrocarbons with small amounts of hydrogen and other light gases (Antal, 1982, 1983; Jakab et al., 1997).

Low-Temperature Gasification in Sub- and SCW

Typically, low-temperature gasification is carried out at temperatures in the range of 623-873K, and in some cases in the presence of heterogeneous catalysts (Matsumura et al., 2005). It should be noted that the use of heterogeneous catalysts in the gasification of solid biomass is almost unknown. However, it is well recognized that the deposited metal catalysts lose their activity during gasification of biomass near supercritical conditions as a result of oxidation of the active sites (Elliott et al., 1993). Silicon, and aluminum oxides, commonly used as carriers of active metal, also changes their properties (Shin et al., 2001). In this connection, new catalysts have to be created which include a combination of stable metals such as bimetallic ruthenium or nickel, and stable carriers such as zirconia or titania oxides.

Indeed, the TiO_2 carrier with deposited nanocrystal particles of ruthenium shows stable performance over two hours at a reaction temperature range of 523-773K and a pressure of 40 bar. Sodium carbonate was also used as an alkali catalyst in order to increase the rate of cellulose gasification (Minowa et al., 1998a, 1998b). Cellulose is converted in its presence at a considerably lower temperature (453K), showing that the alkaline catalyst lowers the temperature of the onset of degradation of the cellulose. This catalyst promotes the degradation of sugars, with a consequent increase in oil yield and gas.

Experiments using cellulose and glucose as a raw material led to the conclusion that the first reaction step in the conversion of biomass in the "hot" water is its hydrolysis (Kabyemela et al., 1999; Kruse et al., 2000), which is confirmed by the absence of the generation of gas, tar or charcoal in the reaction products. With an increase in temperature to 523K, cellulose and glucose decompose to produce gases, oil, coal, and water-soluble products, sugars in particular. Further increases in temperature lead to the complete degradation of cellulose; sugars and oils decompose to produce gases such as CO_2, H_2, CH_4, and simultaneously increase the yield of coal.

The High-Temperature Gasification of Biomass in SCW

The high-temperature gasification of biomass is generally carried out at temperatures above 780K. Due to the high reactivity of the biomass at these temperatures, a high gasification efficiency was reached. Products of primary biomass conversion, such as glycerol and glucose, may be gasified in the absence of catalysts. Conversion of glycerol in SCW at temperatures below 873K proceeds slowly, but with increasing temperature up to ~ 973K can achieve full conversion at glycerol concentrations below about 3 wt.%. Under these reaction conditions, the performance on H_2 and CO_2 increases sharply, whereas the concentration of CO is reduced. These results indicate a great contribution to a water vapor shift reaction at temperatures above 873K.

Since the reaction temperature is significantly above critical levels, pressure of the process does not significantly affect the composition of the biomass gasification products or their performance. However, the initial concentration of raw materials has a significant impact on performance.

The use of catalysts significantly accelerates the gasification of biomass. Thus, gasification of glucose and polyphenolic compounds is accelerated in the presence of KOH, $KHCO_3$ and Na_2CO_3 (Kruse et al., 2000).. Also, the content of H_2 increases in the reaction products and CO is reduced due to the water vapor (gas) shift reaction (Elliott et al., 1983). Gasification of biomass in the presence of $KHCO_3$ in SCW increases the concentration of water-soluble products, produces less coke, decreases the concentration of furfural, and increases the amount of phenol.

Taking into account the shortcomings of water-soluble salts as catalysts, heterogeneous catalysts such as dioxide zirconia ZrO_2 have been studied. The ability of ZrO_2 to accelerate the gasification of glucose and cellulose in SCW has been demonstrated (Watanabe et al., 2002). Similar results were obtained for the partial oxidative gasification of lignin (Watanabe et al., 2003).

The Selective Production of Hydrogen and Synthesis Gas

Hydrogen and synthesis gas (a mixture of hydrogen and carbon monoxide, and often some carbon dioxide) are among the most popular compounds for the energy and chemical industries, and are almost completely obtained from fossil resources, primarily natural gas. At the same time, their sourcing from renewable alternative resources is not discounted (Balat & Kirtay, 2010).

Typically, hydrogen and synthesis gas are produced by the gasification of biomass in sub- and supercritical water (Yu et al., 1993; Antal et al., 2000; Schmieder et al., 2000; Kruse et al., 2003; Lu et al., 2006; Ni et al., 2006; Resende et al., 2007, 2008; Susanti et al., 2014). Traditionally, these transformations are conducted in autoclave reactors or flow tubular reactors. In the first version, gasification processes are inefficient, requiring more time to restart the reactor of solid biomass. In carrying out the gasification of solid biomass in a tubular flow reactor in continuous regime, there is a serious problem of its sealing. Moreover, reactor and lines hammered solid (coal ash) and heavy tars during gasification. This leads to the need to stop the process in order to remove solids, which ultimately reduces the practicality, desirability and energy efficiency of the process (Matsumura & Minowa, 2004).

To solve these problems, methods for periodically removing solid deposits have been proposed (Antal et al., 2000), including optimization of process parameters (Lu et al., 2006) and performing the gasification of biomass in SCW in a fluidized bed reactor (Li et al., 2004; Matsumura & Minowa, 2004).However, despite the attractiveness of the fluidized bed gasification of solid biomass in SCW, the tough conditions (high pressure and temperature) make it difficult to conduct research or develop practical applications.

In their review, Guo et al. (2010), demonstrated the role of supercritical water and catalysts in hydrogen-producing reactions. It has been shown that reactions in supercritical water have many advantages over conventional gaseous and liquid reaction methods. Special attention has been paid to the use of homogenous and heterogeneous catalysts in the production of hydrogen. On the basis of published data, it was shown that alkali metal catalysts (Na_2CO_3, $KHCO_3$, K_2CO_3, NaOH, etc.) are effective in reactions of biomass gasification in SCW. Compared with homogeneous catalysts, heterogeneous catalysts have the advantages of high selectivity, recyclability, and are environmentally friendly, etc.

Lee and coworkers (2002) have shown that the water-gas shift reaction plays an important role in H_2 production. Specifically, in studying glucose gasification in SCW, they observed (see Figure 2) that when the temperature is over 650°C, CO concentration decreases and H_2 production is enhanced due to the

Figure 2. Variation of product distribution with reactor temperature in supercritical water gasification (glucose 0.6 M pressure 28 MPa and reactor residence time 30 s)

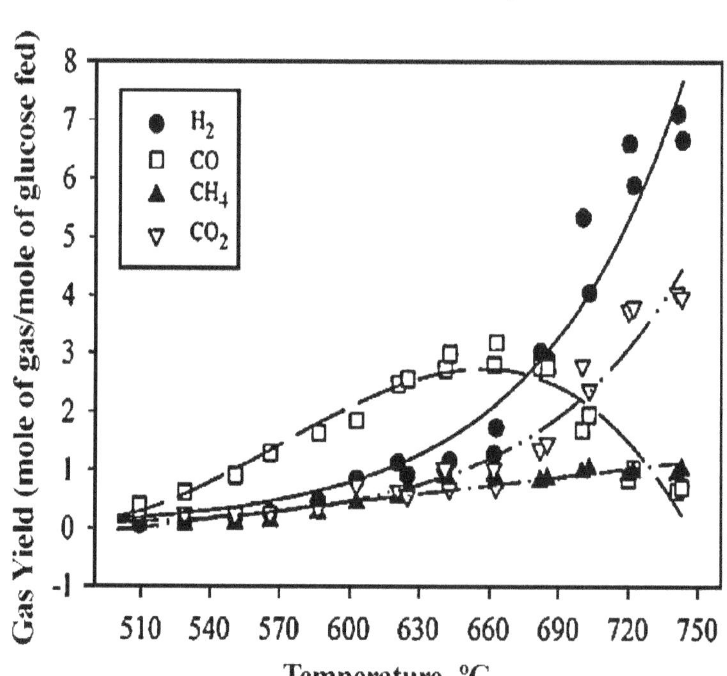

beginning of the water-gas shift reaction. Examining glycerol gasification in SCW with Ru/Al_2O_3 catalyst, Byrd and Gupt (2008), argue that CO yield is lower and H_2 yield higher than their equilibrium values.

Finally, experimental studies of hydrogen production by biomass gasification in SCW using concentrated solar energy were conducted in (Chen et al., 2010). The experimental results confirmed the technical feasibility of the gasification of biomass in SCW with concentrated solar energy, and validated some advantages of the thermo-chemical process. However, the main differences between the concentrated solar energy and conventional heating methods have not been determined.

Biomass Liquefaction Using Supercritical Fluids Towards Bio-Fuel

Hydrothermal treatment of biomass, in particular wood, is used for its high-temperature gasification (HTG) in order to generate the fuel gas, liquefying wood for chemical and hydrothermal carbonization (HC) (Minami & Saka, 2003). The special properties of supercritical water allow it to be used in these processes as the reaction medium, solvent, reagent and catalyst at the same time. Liquefaction of the feedstock during heating and receiving light organic compounds is the purpose of the first three processes. These compounds can be further converted into gaseous products, liquid fuel or other chemicals. The latter process can be used to produce charcoal.

Under ideal conditions of the HTG and using liquefied feed, a solid raw material is completely diluted in one step. However, under real conditions, the transformation may still lead to the formation of undesired products such as heavy tars and ash.

The liquefaction of wood was typically studied experimentally in the reactors - autoclaves under a hydrothermal gasification process (Vogel et al., 2007). The results of these experiments were used to further optimize the process through simulations (Mosteiro-Romero et al., 2014).

Woody biomass is the raw material with the most highly individual properties, depending on the type, place of growth, weather conditions and other factors. Due to such a variety of wood properties, averages of these properties are often used in the studies. As a result, it is assumed that the wood consists of about 40 50% cellulose, 20 30% hemicellulose, 20-28% of lignin, and smaller amounts of other compounds (Yu et al., 2008).

The liquefaction of woody biomass is traditionally achieved as a result of its pyrolysis and hydrolysis. Hydrolysis of biomass, typically in the presence of homogeneous catalysts, requires water splitting to form two or more substances. Wood hydrolysis, which normally processes its individual components in the presence of catalysts, is well known and does not require discussion (Toor et al., 2011).

Degradation of organic material occurs in the pyrolysis of wood in the absence of oxidizers, resulting in the formation of volatile compounds and charcoal. Pyrolysis of wood is used for the conversion of biomass to fuels, but its effectiveness is highly dependent on the moisture content of the feedstock. The transformation of wood into fuel and charcoal by pyrolysis is also a well-studied process (Di Blasi, 2008).

BIODIESEL AND BIOFUELS SYSTHESIS FROM VEGETABLE OILS IN SUPERCRITICAL LOWER ALCOHOLS AND ETHERS

The liquefaction of wood biomass in supercritical alcohols is a highly promising process for the production of liquid fuels and useful chemicals. The resulting liquid products can be directly used as liquid fuels, as alcohol itself is a good fuel.

Biodiesel Synthesis From Vegetable Oils in Supercritical Lower Alcohols

Methanol, ethanol and 2-propanol can be used as lower alcohols. It is expected that at temperatures over 350°C, all of the lignocellulosic components of biomass (cellulose, hemicellulose and lignin) will be converted to about 90-95% liquid compounds. It is important to examine the effectiveness of each kind of alcohol on the composition and rate of biomass liquefaction, as well as to study the conversion of various types of biomass. According to data in the literature (Ishikawa & Saka, 2001; Minami & Saka, 2003; Poudel & Oh, 2012), it is possible to conduct the liquefaction of more than 95 wt.% of wood biomass in supercritical alcohols in approximately 30 minutes. Nevertheless, the products with high molecular weight (~ 10 000 - 30 000) can be formed in the liquid phase at such a residence time.

Biodiesel is an alternative fuel for various types of power plants and is synthesized in chemical reactions such as transesterification of vegetable oils with lower alcohols. The products of the chemical reaction are fatty acid methyl esters (biodiesel) that represent an alternative non-toxic fuel for various kinds of power systems (Srivastava & Prasad, 2000; Knothe, 2001; Demirbas, 2003).

Emissions of carbon dioxide, hydrocarbons, carcinogens and soot have been reduced substantially due to the absence in biodiesel of aromatic hydrocarbons and sulfur compounds, as well as its low oxygen content (~ 11 wt. %) compared to diesel fuel derived from crude oil (Peterson & Hustrulid, 1998; Sheehan et al., 1998; US Environmental Protection Agency, 2002).

Traditionally, chemical transesterification or the esterification reaction of vegetable oils with lower alcohols is carried out in the presence of homogeneous catalysts such as alkali and acid. Feedstock and alcohol should not contain water for reactions catalyzed by alkali, as the water alters the reaction in the direction of saponification. The resulting soap reduces the yield of fatty acid methyl esters (FAME) and leads to problems of FAME and glycerol separation from the soapy water. Moreover, the alkali-catalyzed feedstock must have a low content of free fatty acids (FFA) (Ma & Hanna, 1999; Fukuda et al., 2001), which react with alkali catalysts to produce soap, which in turn suppresses the formation of FAME (Canakci & Gerpen, 2001). Unfortunately, many vegetable oils contain high amounts of free fatty acids. Provided its additional input to the reactor in order to compensate for its consumption in the reaction of saponification, the transesterification reaction can be carried out in the presence of an alkaline catalyst only when the content of free fatty acids in a vegetable oil is not more than 5%.

To meet synthesized biodiesel basic standards such as ASTM D6751 (American Society for Testing and Materials, 2002; En, 2003) requires, first, the separation of glycerol and ether, followed by washing and drying the ester. These processes are required for the neutralization of any catalyst residues and removal of formed soap.

Acid-Catalyzed Transesterification of Triglycerides

Low rate of reaction and necessity of acid separation from the reaction products is one disadvantage of processes for catalyzed transesterification oil using acids. Despite the fact that processes for the transesterification of vegetable raw catalyzed homogeneous acid catalysts are less sensitive to the presence of free fatty acids, their content in the raw material should not exceed about 1%.

One of the newer and most promising areas for effective and relatively cheap methods of synthesizing biofuels and chemicals from plant material as a waste byproduct of the food processing industry is to carry out reactions in supercritical solvents (Saka & Kusdiana, 2001; Kusdiana & Saka, 2001, 2004). First, conducting transesterification reactions of vegetable oils and fats using supercritical lower alcohols does not require the use of homogeneous catalysts, thus reducing demands on the quality of raw materials. Second, such a process can significantly reduce the residence time of the reaction mixture required to achieve 90-98% conversion of the feedstock to several tens of minutes, thereby enabling the use of a flow-type reactor that can dramatically increase the productivity of the process. This in turn allows discharge of the waste water, obtained in consequence of washing the reaction products to remove homogeneous alkali or acid catalysts. Finally, all of the above advantages in combination allow for significant reduction in the cost of biodiesel.

Application of the heterogeneous catalyst and flow system allows not only an increase in the transformation rate at a lower temperature (90-180 °C), but also reduces the alcohol/oil ratio on inlet to the reactor. Reduction of the reaction temperature makes it possible to further significantly reduce the yield of by-products, greatly improving the quality of the biodiesel. There are some studies in the literature aimed at carrying out the transesterification reaction of vegetable oil in a flow reactor in the presence of a heterogeneous catalyst in supercritical lower alcohols (Demirbas, 2007; Wang & Yang, 2007; Shama et al., 2011).

Biodiesel Synthesis From Vegetable Oils in Supercritical Ethers

The use of methyl acetate instead of methanol for supercritical synthesis of glycerol-free biodiesel from vegetable oils is a relatively new process, and its study is very limited in the literature (Saka & Isayama, 2009; Saka et al., 2010; Tan et al., 2010). In these works, methyl acetate has been tested on oils with different fatty acid compositions. The process was also applied to waste oil with a higher free fatty acid (FFA) content. The results demonstrate that oil composition does not significantly influence the biodiesel yield.

The simple one-step transesterification reaction of a triglyceride with methyl acetate is shown in the scheme in Figure 3. In fact, the transesterification of fatty acid triglycerides with methyl acetate usually proceeds in three steps via sequential substitution of one or two alkyl groups of fatty acid by hydroxyl groups to form di and monoglycerides, which can be present in the reaction products together with FFA, until three fatty acid methyl esters (FAME) (Anikeev et al., 2012) and one molecule of triacetin are formed. The production of triacetin is advantageous not only because it avoids the production of glycerol, but also because it has been demonstrated that mixtures of FAME and triacetin obtained by this process are suitable as biodiesel fuel (Saka & Isayama, 2009). Triacetin is itself a valuable compound that can be used in the cosmetic and food industry or as an additive to petroleum-derived fuels (Melero et al., 2007; Galan et al., 2009).

Figure 3. Reaction stoichiometry for transesterification reaction between trygliceride and methyl acetate to produce fatty acid methyl esters (FAME) and triacetin

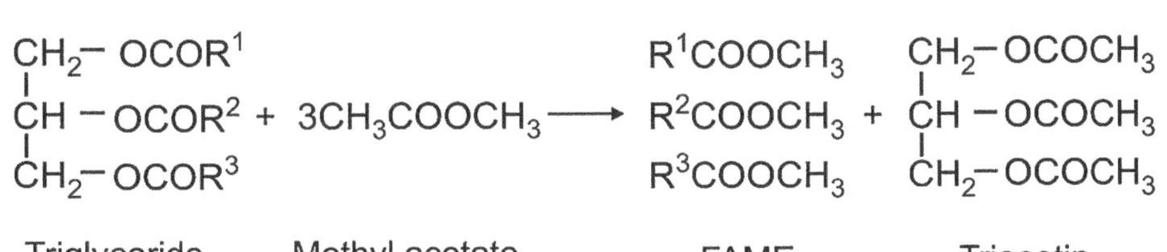

The esterification reaction of free fatty acids and methyl acetate, as shown in the scheme in Figure 4, generates one molecule each of FAME and acetic acid.

The influence of temperature, pressure and the molar ratio of reactants was studied in (Saka et al., 2010) by the transesterification of oils with supercritical methyl acetate. It has been shown that all oils achieved complete conversion after 50 min at 345^{0}C and 20 MPa with the methyl acetate:oil molar ratio equal to 42:1.

Kinetic of transesterification reactions of four oils with supercritical methyl acetate was studied by Saka et al., (2010) in a mixed batch reactor. Pseudo-first order equations used for modeling and the apparent rate constants and activation energies for all of the investigated vegetable oils are reported in Table 1.

The conversion of rapeseed oil and oleic acid with supercritical methyl acetate to FAME and triacetin (TA) in a flow tubular reactor was investigated by Tan et al., (2010) without catalyst. The results of these studies are shown in Figure 5 and Figure 6. It is shown that the transesterification reaction of triglycerides with methyl acetate can proceed under supercritical conditions, generating FAME and triacetin, Figure 5. The FAME yield after supercritical processing of oleic acid with methyl acetate is shown also in Figure 6.

Biodiesel and Fuel Synthesis From Algae

Algae are the fastest growing plant material and can be an important source of biomass. They can be produced on an industrial scale in artificial ponds, by photo-bioreactors, or extracted directly from the sea. According to various estimates, about 50% of their weight is comprised of oils. Therefore, along with vegetable oils from crops, algae can be used for the synthesis of biofuels, including biodiesel.

It is known that terrestrial plants are not effective in absorbing solar energy. Even fast-growing crops can convert solar energy into biomass at an average annual rate of only ~ 1 W / m^2 (Li et al., 2008). At the same time, the productivity of microalgae and photosynthetic microorganisms can be 50 times greater

Figure 4. Esterification reaction between FFA and methyl acetate to produce FAME and acetic acid

$$\underset{\text{Fatty acid}}{RCOOH} + \underset{\text{Methyl acetate}}{CH_3COOCH_3} \longrightarrow \underset{\text{FAME}}{RCOOCH_3} + \underset{\text{Acetic acid}}{CH_3COOH}$$

Table 1. The rate coefficients and activation energies of transesterification reaction of vegetable oils and supercritical methyl acetate at 20 MPa (with methyl acetate: oil molar ratio = 42:1)

Temperature (^{0}C)	Soybean Oil, k (1/s)	Sunflower Oil, k (1/s)	Jatropha Curcas Oil, k (1/s))	Waste Soybean Oil, k (1/s)
300	1.62 x 10^{-6}	2.43 x 10^{-6}	1.77 x 10^{-6}	1.79 x 10^{-6}
315	1.87 x 10^{-5}	2.04 x 10^{-5}	1.85 x 10^{-5}	1.95 x 10^{-5}
323	4.68 x 10^{-5}	4.93 x 10^{-5}	4.91 x 10^{-5}	4.98 x 10^{-5}
330	1.03 x 10^{-4}	1.08 x 10^{-4}	1.07 x 10^{-4}	1.07 x 10^{-4}
338	2.61 x 10^{-4}	2.63 x 10^{-4}	2.67 x 10^{-4}	2.69 x 10^{-4}
345	5.11 x 10^{-4}	5.08 x 10^{-4}	4.54 x 10^{-4}	5.28 x 10^{-4}
Activation energy (kJ/mol)	373	349	364	369

Figure 5. Yields of FAME and BDF (FAME and triacetin) from rapeseed oil as treated in supercritical methyl acetate at various temperatures in 20 MPa (methyl acetate/oil = 42/1 (mol/mol))
Saka 7 Isayama, 2009.

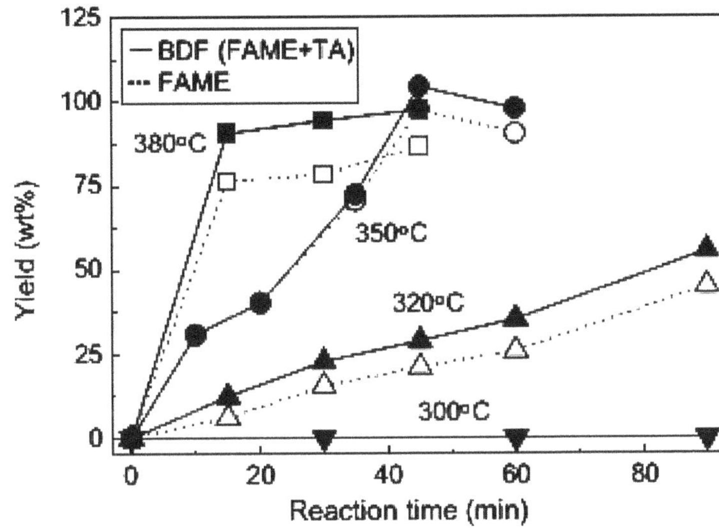

Figure 6. Yield of FAME from oleic acid as treated in supercritical methyl acetate at various temperatures in 20 MPa (methyl acetate/oleic acid = 14/1 (mol/mol))
Saka & Isayama, 2009.

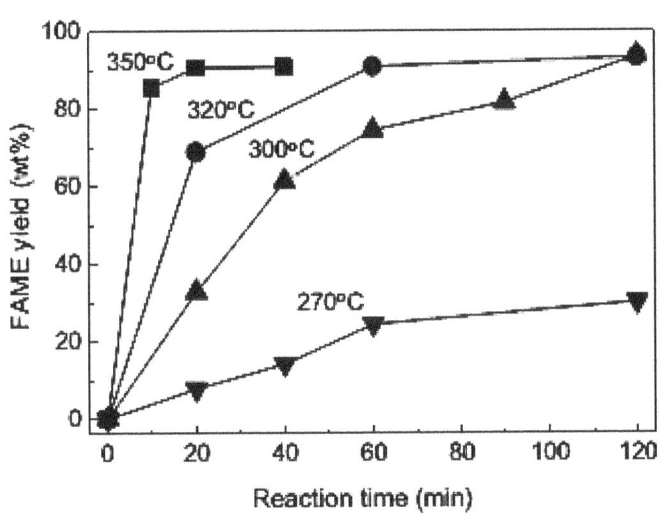

than leaves or grass (Demirbas, 2006). Microalgae grow in both marine and freshwater environments and, due to their simple cell structure and being in an aqueous medium in the environment of nutrients, are usually more efficient in converting solar energy into biomass.

In principle, the water resources of the Earth are not restricted to the sustainable cultivation of microalgae, which can be used to produce a wide range of advanced biofuels and biologically active compounds (Schenk et al., 2008; Huesemann & Benemann, 2009). Microalgae have several advantages (Li et al.,

2008; Schenk et al., 2008; Chen et al., 2009) in comparison with conventional sources of biomass: (1) culturing algae does not compete with agricultural land; (2) the algae can gather in batches throughout the year, providing their continuous supply; (3) per unit of area and per unit of fuel, less water is used than for the growth of other biomass; (4) algae can be cultivated in sewage or brackish water, thus significantly reducing the use of fresh water; (5) algae can be used for the disposal of industrial flue gases containing CO_2; and (6) growing algae facilitates the removal of nitrogen and phosphorus from the water and the absorption of CO_2 from the atmosphere.

Using conventional methods, biofuels were obtained from algae using costly and multi-step methods (Chisti, 2007; Singh and Olsen, 2011; Lam and Lee, 2012; Meleterova et al., 2015; Montingelli et al., 2015) comprised of drying and grinding the algae, extracting their oils with chemical solvents, and then converting this oil into biodiesel using homogeneous catalysts. A state-of-the-art review of the thermochemical conversion technologies of microalgae biomass into fuels is presented in (Chen et al., 2015).

The use of supercritical solvents allows for the transformation of algae to, for example, biodiesel in one step without the use of homogeneous catalysts (Patil et al., 2011). At the same time, the application of heterogeneous catalysts (Pd/C, Pt/C, Ru/C, Ni/SiO_2-Al_2O_3, CoMo/γ-Al_2O_3 and zeolites) (Duan & Savage, 2011) for converting Nannochloropsis sp. algae through liquefaction in subcritical water allowed investigators to significantly accelerate the process of obtaining crude bio-oil. These studies are among the first to use heterogeneous catalysts for processing algae in subcritical water.

Research studies on using supercritical CO_2 to extract microalgae lipids for biodiesel production has been explored recently (Halim et al., 2011). In this study, lipids from wet-paste Chlorococcum sp. biomass were extracted using supercritical CO_2 with a lipid yield of 7.1% at a critical temperature of 60°C, critical pressure of 30 MPa, and extraction time of 80 min. Since supercritical CO_2 is a nonpolar solvent, the presence of water in the wet biomass acts as a natural *polar* co-solvent, thus facilitating the extraction of polar lipids and improving total lipid yield.

As noted above, in addition to water and CO_2, lower alcohols (e.g., ethanol (Reddy et al., 2014) or methanol (Patil et al., 2011; Cao et al., 2013)) are used as chemical supercritical solvents for biodiesel from algae. In the case of ethanol, the authors of investigated the effects of the algae/ethanol ratio (varied from 1:6 to 1:15 wt./vol.) on biodiesel (fatty acid ethyl esters - FAEE) yield at the constant reaction temperature 260^{0}C and reaction time of 20 min (Reddy et al., 2014). As shown in Figure 7, the fatty acid ethyl esters yield increased with an algae/ ethanol ratio up to 1:9, and then decreased with an increasing algae/ethanol ratio. The yield of FAEE increases steadily with increasing temperature and reaction time. To conclude this section, we should pay attention to the study of the economic aspects of biofuel from algae in supercritical fluids undertaken by Brownbridge and coworkers (Brownbridge et al., 2014).

FISCHER-TROPSCH SYNTHESIS IN SUPERCRITICAL FLUID REACTION MEDIUM

The Fischer-Tropsch (FT) synthesis, directed mainly to the production of C_4-C_{15} olefins, is of considerable interest for the production of heavy hydrocarbons, including analogues of gasoline (Dry et al., 1981; Bukur et al., 2997). FT reactions are carried out in the gas or liquid phase. FT synthesis is carried out in the gas phase in reactors with a fixed or moving bed catalyst at temperatures of 525-575K, pressure of

Figure 7. FAEE yield against (a) algae to ethanol (wt/vol.) ratio, (b) reaction time and (c) reaction temperature
Reddy et al., 2014.

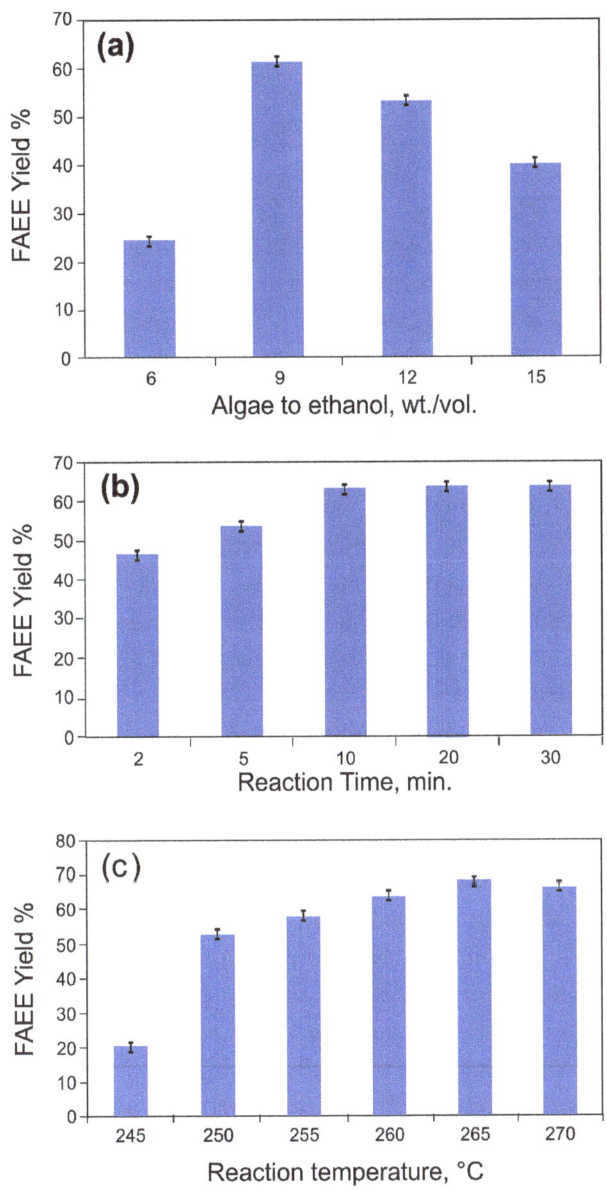

14-20 atm, and the initial ratio of H_2 / CO of 0.67 to 6.0. Liquid-phase FT synthesis is generally carried out in slurry reactors, which are multi-phase systems of a solid catalyst, liquid product and gas phase reactants. Increasing selectivity to olefin yield and the elimination of its inherent shortcomings (Dry et al., 1981; Yokota and Fujimoto, 1991, 1993; Fan et al., 1992; Bukur et al., 1997) is an urgent task for the FT process nowadays.

Review of the achievements in the research and implementation of the FT synthesis since 1950 is presented in detail by Saxena (Saxena, 1995). Particular attention here is paid to the problems of hydrodynamics, heat and mass transfer, and the mathematical modeling of FT synthesis in slurry reactors.

Along with the improvement of existing technologies and catalysts for FT synthesis, in recent years there has been increased interest in the study of FT synthesis under supercritical conditions Abbaslou et al., 2009), i.e., when the reaction is carried out in a solvent which, at the temperature and pressure of the reaction, is close to its critical state.

Propane or n-hexane is most often used as such a solvent in the FT synthesis process (Fan et al., 1992; Yokota and Fujimoto, 1993; Lang et al., 1995; Bukur et al., 1997). Due to the high solubility of the heavy hydrocarbons in the supercritical fluid, the diffusion processes are accelerated significantly in the catalyst grains, thereby suppressing secondary hydrogenation processes, isomerization and further condensation of products. As a result, increases in the reaction rate, the selectivity of the process towards a-olefins, and the duration of the catalyst life take place.

A detailed review of recent studies on the use of supercritical solvents in FT synthesis is presented by Abbaslou and coworkers (Abbaslou et al., 2009). This study compares differences in the activity of the catalysts, reaction selectivity to the yield of CH_4 and CO_2, the distribution of olefins and hydrocarbons, stability of catalysts and heat exchange for FT synthesis in the supercritical solvent, and FT synthesis in the gaseous phase. The effects of temperature, pressure, type of solvent, and the supercritical solvent ratio/synthesis gas to yield the desired products are also discussed. From an analysis of the literature data (Abbaslou et al., 2009), it was oncluded that the unique properties of supercritical fluid may improve the activity and selectivity of the catalyst, due to the higher rate of heat and mass transfer compared to FT synthesis in the gaseous phase.

Taking into account that the research of FT synthesis in supercritical fluid has been conducted only recently, there are many problems remaining to be solved. Among these are a reasonable choice of solvent, providing the optimal combination of supercritical conditions of the reaction medium to the operating conditions of the process, the study of the thermodynamics and kinetics of the reaction, and the implementation of optimization schemes of the FT, among others. Calculations and mathematical modeling-in particular, calculations of thermodynamics-are important in addressing some of these issues.

In most cases, the non-ideal properties of the reaction medium in the supercritical state appears near its critical point, which in turn affects the rate of chemical reactions and equilibrium constants of reversible reactions. The critical state of the mixture is generally regarded as the highest degree of non-ideality (Krichevskii, 1952; Savage et al., 1995; Sandler, 1999). Thus, the influence of non-ideality of the reaction mixture on the reaction kinetics of FT synthesis was first studied by Yermakova et al. (2000). Supercritical propane was chosen as the solvent. They demonstrated the possibility of a substantial increase in the rate of reaction under supercritical conditions, as well as determining the total yield of C_2 + olefins, including the target C_5-C_{11} fraction, as compared to the base case, which was carried out without the solvent.

Using this process as an example, Yermakova and coworkers (Yemakova et al., 2000; Yemakova & Anikeev, 2000) have discussed some of the common problems encountered in simulation reactors and process schemes for FT synthesis.

APPLICATION OF SUPERCRITICAL FLUIDS IN THE CONVERSION AND STORAGE OF THERMAL ENERGY

Heat in conventional power plants and cycles is transferred from the source to the working medium through the heat exchange surface. Either liquids, gases, metals or molten salts may be used as the working medium in a cycle.

In systems with an accumulation of thermal energy, the thermal energy is stored in the form of the internal energy of the coolant. As a rule, the heat accumulating - emitting loops and devices must meet the following requirements:

- High specific energy storage density,
- A good heat exchange between the coolant and the heat source,
- Mechanical and chemical stability of the material of the coolant,
- Chemical compatibility of coolant and materials of heat exchangers,
- The complete reversibility of the absorption/heat generation cycles,
- Long-time operation,
- Low cost of the device.

A traditional heat storage cycle consists of a heat receiver, two containers (1 and 2) housing the coolant, heat exchangers and a heat accumulator. The cold coolant from container 1 flows into the thermal energy receiver, where it is heated (vaporizes, melts or undergoes chemical conversion) by increasing its internal energy. The coolant then flows into container 2 or the heat accumulator. The coolant transfers the heat into the heat storage material in the accumulator and then enters container 1. In the absence of a separate heat accumulator in the cycle, the heated heat-transfer fluid in reservoir 2 can be stored for a while. When the accumulated heat energy is needed by the consumer, the hot heat-transfer fluid is passed through a heat exchanger where it is cooled, giving heat to the consumer, and it goes back into container 1.

In reality, the cycles for conversion and energy storage appear much more difficult; however, this does not matter for explaining the advantages of using supercritical fluid as a coolant. The density of thermal energy storage into the internal energy of the heat carrier determines its thermal properties. In a number of studies, highly conductive materials such as graphite (Bauer et al., 2006; Sari & Karaipekli, 2009; Cheng et al., 2010) and nanoscale additives (Khodadadi & Hosseinzadeh, 2007; Hosseinzadeh et al., 2012) were added to the heat carrier to affect desirable changes (mainly increases in thermal conductivity and specific heat).

The use of supercritical fluids for the conversion and storage of thermal energy can significantly increase the rate and density of stored energy. As mentioned earlier, this is due to a sharp increase in the specific heat of the fluid, heat- and mass transfer coefficients, and compressibility of the fluid in the vicinity of the critical parameters.

The advantage of a cycle with a supercritical heat carrier is demonstrated as energy accumulation step, and on step its release. The carrier (usually an organic liquid) is heated by an external heat source, it changes phase and goes into the supercritical region. Here, because the supply of heat occurs at a practically constant pressure (due to the high compressibility of the fluid) and a constant temperature (due to the increased heat capacity of the fluid), there is additional energy storage. In other words, the supercritical coolant allows the same amount of energy to accumulate at greater density than with other mediums. All of the above benefits are realized by the allocation of heat from the coolant to the consumer.

The second important point in determining the efficiency of thermal storage systems with supercritical coolant is the amount of heat transfer from the wall of the receiver to the heat carrier, and vice versa. Supercritical fluids, with high values of specific heat and thermal conductivity, provide significantly higher values of heat transfer coefficients compared to those of other carriers.

Supercritical Rankine Cycle

An Organic Rankine Cycle (ORC) uses organic working fluid instead of the water-steam of a Clausius Rankine Cycle (CRC). The main advantages of an ORC compared with a CRC are the possibility of the choice of working fluid providing maximum thermal efficiency of the cycle. A detailed review of an ORC's thermal cycle efficiencies for various working fluids for sub- and supercritical parameters is given in several papers (Saleh et al., 2007; Chen et al., 2010).

The advantage of Supercritical Rankine Cycle (SRC) compared with CRC and ORC lies in the absence of phase transformations during the heating process and related restrictions. The critical point of organic fluids is reached at lower pressures and temperatures compared with water. Therefore, supercritical fluid parameters are easily realized.

Several authors (Duan et al., 2000; Chen et al., 2010) analyzed about 35 working fluids for the ORC and SRC, and investigated the influence of fluid thermodynamic and physical properties on cycle performance. Both cycles were reviewed for low-grade heat conversion into power. It is shown that the fluid properties play an important role in cycle effectiveness. Working fluids with high density and high latent heat provide high unit turbine work output.

Chen et al. (2011) proposed a Supercritical Rankine Cycle (SRC) using zeotropic mixture working fluids for the conversion of low-grade heat into power and analyzed its efficiency. A zeotropic mixture of R134a and R32 (0.7/0.3, mass fraction) was used as the working fluid of the SRC, while the pure R134a is considered as the working fluid of the ORC. The thermal and energy efficiencies of both cycles were analyzed using computational methods.

The thermal efficiencies and network outputs of both cycles were studied for the turbine inlet temperature of 393 - 473 K and average ambient (condensing) temperature of 309.5K. The calculated thermal efficiencies of the ORC and the SRC are shown in Figure 8. It was found that thermal efficiency of the ORC using pure R134a was 9.70 -10.13%, while that of the SRC using the zeotropic mixture was a significantly higher - 10.77-13.35%.

Accumulation of Concentrated Solar Energy

Despite a slight decline in prices for natural mineral resources, interest in the use of alternative and renewable energy sources remains attractive, with energy from solar radiation leading the way. However, the effective integration of solar systems into the existing energy grid requires the use of energy storage equipment. This is because, despite the frequency of solar radiation during the day, peak solar activity does not coincide with peak energy consumption; consequently, energy storage must be used during peak solar activity, enabling the transfer of accumulated energy in the network at the peak of consumption (Tamme et al., 1991). This requirement for energy storage does not apply only to solar energy, as there is a need to smooth peak loads through the cycles of accumulation and release of energy in all energy systems.

Figure 8. Thermal efficiencies of the R134a-based organic Rankine cycle (ORC) and the zeotropic mixture-based Supercritical Rankine Cycle (SRC)

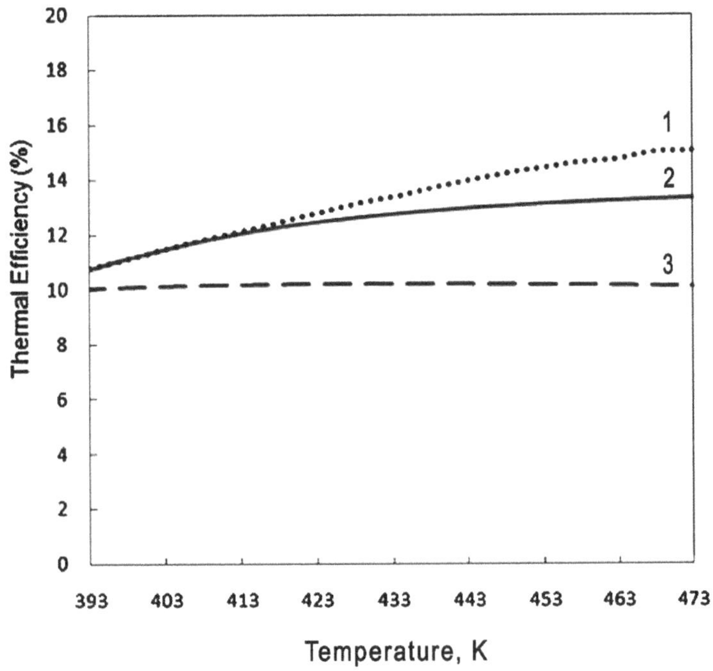

The past 80-90 years constitute the peak of interest in the research and application of energy from concentrated solar radiation. During this period, various methods and storage devices has been proposed, including devices based on thermo-chemical cycles (Anikeev, 1987; Anikeev et al., 1987, 1990). However, methods for the conversion and storage of solar energy using supercritical fluids have most actively been studied only recently (Ganapathi & Wirz, 2012; Lakeh et al., 2013).

Ganapathi and Wirz (2012), studying the perspective of using cumulative thermal cycles for the conversion of concentrated solar energy, have analyzed the possibility of using several dozen organic liquids as supercritical carriers. Among these, glycerol and naphthalene satisfy the required properties. The authors compared the specific energy storing properties for supercritical glycerin (720 kJ/kg), naphthalene (541 kJ/kg) and molten salts ($NaNO_3$, KNO_3) (141 kJ/kg). (Note that the specific energy of storing "hot" water is 418 kJ/kg.) As such, the cost of supercritical-fluid technology may be significantly lower compared to commonly used technology based on molten salts. Preliminary studies (Ganapathi & Wirz, 2012) have shown that the specific cost of the cycle using supercritical naphthalene as a coolant is $ 3/kWh compared with $ 25- 50/kWh using molten salts.

The Use of Supercritical Fluids in Nuclear Power

First appearing in the mid-1960s, the idea of using supercritical water was successfully implemented with the creation of steam generators as turbines for thermal power stations, thus significantly increasing their thermal efficiency. Besides increasing the efficiency of the cycle, the supercritical conditions of the working fluid result in the absence of phase transitions and, consequently, the absence of so-called critical heat flux (Pioro & Pioro, 1997; Pioro, 2014).

Water-cooled reactors with a thermal efficiency of about 30-35%, carbon dioxide gas-cooled reactors with efficiency up to 42%, and liquid sodium-cooled reactors with efficiency up to 40% are used in the operation of nuclear power plants. Due to a number of advantages, the use of supercritical working fluid in the power cycles of nuclear power plants is under active consideration.

The use of cycles with a supercritical working medium will allow the next generation of reactors to increase their efficiency up to ~ 50%. Supercritical steam Rankine cycles (Pioro & Duffey, 2007) and gas turbine Brayton cycles with supercritical carbon dioxide may be considered as such cycles.

CONCLUSION

The benefits of SCF in their application in various fields of energy, including the production of biofuels and thermal energy storage, has been amply demonstrated based on the results of studies in recent decades. The review discusses the transformation of vegetable oils, algae and wood in supercritical alcohols, esters and water to produce biodiesel fuels and hydrogen. Current issues of the solar energy conversion and storage with the use of the working fluid in a supercritical state, as well as the heat cycles in transforming nuclear energy considered also. The main problems, which it is advisable to solve to SCF technology played an important role in creating a sustainable development of the energy sector, are identified in the review. Recommendations that can improve the efficiency of energetic processes and cycles with the SCFs participation in terms of energy consumption, materials and equipment costs were made.

REFERENCES

Abbaslou, R. M. M., Mohammadzadeh, J. S., & Dalai, A. K. (2009). Review on Fischer-Tropsch synthesis in supercritical media. *Fuel Processing Technology*, *90*(7-8), 849–856. doi:10.1016/j.fuproc.2009.03.018

American Society for Testing and Materials. (2002). *Standard Specification for Biodiesel Fuel (b100) Blend Stock for Distillate Fuels, Designation D6751-02*. West Conshohocken, PA: ASTM International.

Anikeev, V. I. (1987). Experimental study of solar energy conversion by means of a thermochemical cycle. *Solar technology (Geliotechnika)*, *2*, 3–8.

Anikeev, V. I., Belobrov, N. S., & Piterkin, R. N. (2006). Results of testing the plant for supercritical water oxidation of nitroglycerin and diethylene glycol dinitrate. *I&ECR*, *45*, 7977–7981.

Anikeev, V. I., Kuzmin, V. A., & Bobrova, I. I. (1987). Experimental study of the solar energy catalytic device based on a closed thermochemical cycle. *Reports of the USSR Academy of Sciences*, *293*, 1427–1432.

Anikeev, V. I., Parmon, V. N., Kirillov, V. A., & Zamaraev, K. I. (1990). Theoretical and fundamental studies of solar catalytic power plants based on reversible reactions with participation of methane and syn gas. *International Journal of Hydrogen Energy*, *15*(4), 275–285. doi:10.1016/0360-3199(90)90048-4

Anikeev, V. I., Stepanov, D. A., & Yermakova, A. (2012). Thermodynamics of phase and chemical equilibrium in the processes of biodiesel fuel synthesis in sub- and supercritical methanol. *I&ECR*, *51*, 4783–4796.

Antal, J. M. Jr, Allen, S. G., Schulman, D., Xu, X., & Divilio, R. J. (2000). Biomass gasification in supercritical water. *Industrial & Engineering Chemistry Research*, *39*(11), 4044–4053. doi:10.1021/ie0003436

Antal, M. J. (1978). *Energy from Biomass and Wastes* (D. L. Klass, Ed.). Chicago, IL: IGT.

Antal, M. J. (1982). *Solar Energy* (K. W. Boer & J. A. Duffie, Eds.). New York: Plenum Press.

Antal, M. J. (1983). Effect of reactor severity on the gas-phase pyrolysis of celulose - and kraft lignin-derived volatile matter. *Industrial & Engineering Chemistry Product Research and Development*, *22*(2), 366–375. doi:10.1021/i300010a039

Arai, K., Smith, R. L. Jr, & Aida, T. M. (2009). Decentralized chemical processes with supercritical fluid technology for sustainable society. *The Journal of Supercritical Fluids*, *47*(3), 628–636. doi:10.1016/j.supflu.2008.11.008

Balat, H., & Kirtay, E. (2010). Review hydrogen from biomass present scenario and future prospects. *International Journal of Hydrogen Energy*, *35*(14), 7416–7426. doi:10.1016/j.ijhydene.2010.04.137

Bauer, T., Tamme, R., Christ, M., & Ottinger, O. (2006). PCM-graphite composites for high temperature thermal energy storage. *Proceedings of ECOSTOCK the 10th International Conference on Thermal Energy Storage*.

Brownbridge, G., Azadi, P., Smallbone, A., Bhave, A., Taylor, B., & Kraft, M. (2014). The future viability of algae-derived biodiesel under economic and technical uncertainties. *Bioresource Technology*, *151*, 166–173. doi:10.1016/j.biortech.2013.10.062 PMID:24220544

Bukur, D. B., Lang, X., Akgerman, A., & Feng, Z. (1997). Effect of Process Conditions on the Olefin Selectivity during Conventional and Supercritical Fischer-Tropsch Synthesis. *Industrial & Engineering Chemistry Research*, *36*(7), 2580–2587. doi:10.1021/ie960507b

Byrd, A. J., Pant, K. K., & Gupt, R. B. (2008). Hydrogen production from glycerol by reforming in supercritical water over Ru/Al_2O_3 catalyst. *Fuel*, *87*(13-14), 2956–2965. doi:10.1016/j.fuel.2008.04.024

Calzavara, Y., Joussot-Dubicn, C., Boissonnct, G., & Sarradc, S. (2005). Evaluation of biomass gasification in supercritical water process for hydrogen production. *Energy Conversion and Management*, *46*(4), 615–631. doi:10.1016/j.enconman.2004.04.003

Canakci, M., & Gerpen, J. V. (2001). Biodiesel production from oils and fats with high free fatty acids. *Transactions of the ASAE. American Society of Agricultural Engineers*, *44*(6), 1429–1436. doi:10.13031/2013.7010

Cao, H., Zhang, Z., Wu, X., & Miao, X. (2013). Direct biodiesel production from wet microalgae biomass of chlorella pyrenoidosa through in situ transesterification. BioMed Research International.

Chen, C. Y., Zou, J. H., Zhang, S. Y., Zaitlin, D., & Zhu, L. H. (2009). Strigolactones are a new-defined class of plant hormones which inhibit shoot branching and mediate the interaction of plant-AM fungi and plant-parasitic weeds. *Science in China. Series C, Life Sciences*, *52*(8), 693–700. doi:10.1007/s11427-009-0104-6 PMID:19727586

Chen, H., Goswami, D. Y., Rahman, M. M., & Stefanakos, E. K. (2011). A supercritical Rankine cycle using zeotropic mixture working fluids for the conversion of low-grade heat into power. *Energy*, *36*(1), 549–555. doi:10.1016/j.energy.2010.10.006

Chen, H., Goswami, D. Y., & Stefanakos, E. K. (2010). A review of thermodynamic cycles and working fluids for the conversion of low-grade heat. *Renewable & Sustainable Energy Reviews*, *14*(9), 3059–3067. doi:10.1016/j.rser.2010.07.006

Chen, J., Lu, Y., Guo, L., Zhang, X., & Xiao, P. (2010). Hydrogen production by biomass gasification in supercritical water using concentrated solar energy: System development and proof of concept. *International Journal of Hydrogen Energy*, *35*(13), 7134–7141. doi:10.1016/j.ijhydene.2010.02.023

Chen, W. H., Lin, B. J., Huang, M. Y., & Chang, J. S. (2015). Thermochemical conversion of microalgal biomass into biofuels: A review. *Bioresource Technology*, *184*, 314–327. doi:10.1016/j.biortech.2014.11.050 PMID:25479688

Cheng, W. I., Zhang, R., Xie, K., & Wang, J. (2010). Heat conduction enhanced shape-stabilized paraffin/HDPE composite PCMs by graphite addition: Preparation and thermal properties. *Solar Energy Materials and Solar Cells*, *94*(10), 1636–1642. doi:10.1016/j.solmat.2010.05.020

Chisti, Y. (2007). Biodiesel from microalgae. *Biotechnology Advances*, *25*(3), 294–306. doi:10.1016/j.biotechadv.2007.02.001 PMID:17350212

de Boer, K., & Bahri, P. A. (2011). Supercritical methanol for fatty acid methyl ester production: A review. *Biomass and Bioenergy*, *35*(3), 983–991. doi:10.1016/j.biombioe.2010.11.037

Demirbas, A. (2003). Biodiesel fuels from vegetable oils via catalytic and noncatalytic supercritical alcohol transesterifications and other methods: A survey. *Energy Conversion and Management*, *44*(13), 2093–2109. doi:10.1016/S0196-8904(02)00234-0

Demirbas, A. (2006). Biodiesel production via SCF method and biodiesel fuel characteristics. *Energy Conversion and Management*, *47*(15-16), 2271–2282. doi:10.1016/j.enconman.2005.11.019

Demirbas, A. (2007). Biodiesel from sunflower oil in superctitical methanol with calcium oxide. *Energy Conversion and Management*, *48*(3), 937–941. doi:10.1016/j.enconman.2006.08.004

Demirbas, A. (2009). Biofuels securing the planets future energy needs. *Energy Conversion and Management*, *50*(9), 2239–2249. doi:10.1016/j.enconman.2009.05.010

Deshpande, P., Kulkarni, K., & Kulkarni, A. D. (2011). Supercritical fluid technology in biodiesel production: A review. *Chemistry and Materials Research*, *1*, 27–32.

Di Blasi, C. (2008). Modeling chemical and physical processes of wood and biomass pyrolysis. *Progress in Energy and Combustion Science*, *34*(1), 47–90. doi:10.1016/j.pecs.2006.12.001

Dry, M. E. (1981). The Fischer-Tropsch synthesis. In Catalysis-Science and Technology. Springer-Verlag.

Duan, P., & Savage, P. (2011). Hydrothermal Liquefaction of a Microalga with Heterogeneous Catalysts. *Industrial & Engineering Chemistry Research*, *50*(1), 52–61. doi:10.1021/ie100758s

Duan, Y. Y., Shi, L., Sun, Q., Zhu, M. S., & Han, L. Z. (2000). Hermodynamic roperties of trifluoroiodomethane (CF_3I). *International Journal of Thermophysics*, *21*(2), 393–404. doi:10.1023/A:1006683529436

EN14214 European Committee for Standardization (CEN), EN 14214. (2003). *Automotive fuels - fatty acid methyl esters (FAME) for diesel engines - requirements and test methods*. Author.

Eckert, C. A., Knutson, B. L., & Debenedetti, P. G. (1996). Supercritical fluids as solvents for chemical and materials processing. *Nature*, *383*(6598), 313–318. doi:10.1038/383313a0

Elliott, D. C., & Sealock, L. J. (1983). Aqueous catalyst systems for the water-gas shift reaction. 1. Comparative catalyst studies. *Industrial & Engineering Chemistry Product Research and Development*, *22*(3), 426–431. doi:10.1021/i300011a008

Elliott, D. C., Sealock, L. J., & Bake, E. G. (1994). Chemical processing in high pressure aqueous environments. 3. Batch reactor process development experiments for organics destruction. *Industrial & Engineering Chemistry Research*, *33*(3), 558–565. doi:10.1021/ie00027a012

Elliott, D. C., Sealock, L. J., & Baker, E. G. (1993). Chemical processing in high-pressure aqueous environments. 2. Development of catalysts for gasification. *Industrial & Engineering Chemistry Research*, *32*(8), 1542–1548. doi:10.1021/ie00020a002

Falkehag, S. I. (1975). Synthesis of phenolic polymer. *Applied Polymer Symposia*, *28*, 247–257.

Fan, L., Yokota, K., Hanakata, Y., & Fujimoto, K. (1992). Supercritical phase Fischer-Tropsch synthesis: Catalyst pore-size effect. *AIChE Journal. American Institute of Chemical Engineers*, *38*(10), 1639–1648. doi:10.1002/aic.690381014

Fukuda, H., Kondo, A., & Noda, H. (2001). Biodiesel fuel production by transesterification of oils. *Journal of Bioscience and Bioengineering*, *92*(5), 405–416. doi:10.1016/S1389-1723(01)80288-7 PMID:16233120

Galan, M. I., Bonet, J., Sire, R., Reneaume, J. M., & Plesu, A. E. (2009). From residual to useful oil:revalorization of glycerine from the biodiesel synthesis. *Bioresource Technology*, *101*(15), 3775–3778. doi:10.1016/j.biortech.2009.01.066 PMID:19269813

Ganapathi, G. B., & Wirz, R. (2012, July). High density thermal energy storage with supercritical fluids. *Proceedings of the ASME 6th International Conference on Energy Sustainability ES2012*, 1-9. doi:10.1115/ES2012-91008

Gerpen, J. V. (2005). Biodiesel processing and production. *Fuel Processing Technology*, *86*(10), 1097–1107. doi:10.1016/j.fuproc.2004.11.005

Guo, Y., Wang, S. Z., Xu, D. H., Gong, Y. M., Ma, H. H., & Tang, X. Y. (2010). Review of catalytic supercritical water gasification for hydrogen production from biomass. *Renewable & Sustainable Energy Reviews*, *14*(1), 334–343. doi:10.1016/j.rser.2009.08.012

Halim, R., Gladman, B., Danquah, M. K., & Webley, P. A. (2011). Oil extraction from microalgae for biodieselproduction. *Bioresource Technology*, *102*(1), 178–185. doi:10.1016/j.biortech.2010.06.136 PMID:20655746

Harradine, et al.. (1993). Oxidation chemistry of energetic material in SCW. *Haz. Waste Mat.*, *10*, 223.

Helwani, Z., Othman, M. R., Aziz, N., Kim, J., & Fernando, W. J. N. (2009). Fernando,W.J.N. Solid heterogeneous catalysts for transesterification of triglycerides with methanol: A review. *Applied Catalysis A, General*, *363*(1-2), 1–10. doi:10.1016/j.apcata.2009.05.021

Hoang, D., Bensaid, S., & Saracco, G. (2013). Supercritical fluid technology in biodiesel production. *Green Process Synth.*, *2*, 407–425.

Hosseinizadeh, S. F., Darzi, A. A., & Tan, F. L. (2012). Numerical investigations of unconstrained melting of nano-enhanced phase change material (NEPCM) inside a spherical container. *International Journal of Thermal Sciences*, *51*, 77–83. doi:10.1016/j.ijthermalsci.2011.08.006

Huesemann, M. H., & Benemann, J. R. (2009). Biofuels from microalgae: Review of products, processes, and potential, with special focus on Dunaliella sp. In A. Ben-Amotz, J. E. W. Polle, & D. V. Subba Rao (Eds.), *The Alga Dunaliella: Biodiversity, Physiology, Genomics, and Biotechnology*. Science Publishers. doi:10.1201/b10300-20

Ishikawa, Y., & Saka, S. (2001). Chemical conversion of cellulose as treated in supercritical methanol. *Cellulose (London, England)*, *8*(3), 189–195. doi:10.1023/A:1013170020469

Jakab, E., Liu, K., & Meuzelaar, H. L. C. (1997). Thermal decomposition of wood and cellulose in the presence of solvent vapors. *Industrial & Engineering Chemistry Research*, *36*(6), 2087–2095. doi:10.1021/ie960335i

Kabyemela, B. M., Adschiri, T., Malaluan, R. M., & Arai, K. (1999). Glucose and Fructose Decomposition in Subcritical and Supercritical Water: Detailed Reaction Pathway, Mechanisms, and Kinetics. *Industrial & Engineering Chemistry Research*, *38*(8), 2888–2895. doi:10.1021/ie9806390

Khodadadi, J. M., & Hosseinzadeh, S. F. (2007). Nanoparticle-enhanced phase change materials (NEPCM) with great potential for improved thermal energy storage. *International Communications in Heat and Mass Transfer*, *34*(5), 534–543. doi:10.1016/j.icheatmasstransfer.2007.02.005

Kiran, E. (2000). Supercritical Fluids. Kluwer Academic Publishers. doi:10.1007/978-94-011-3929-8

Knothe, G. (2001). Historical perspectives on vegetable oil-based diesel fuels. *Inform (Silver Spring, Md.)*, *12*, 1103–1107.

Krichevskii, I. R. (1952). *Phase equilibria in solutions at high pressures*. Moscow: Goskhimizdat.

Kritzer, P., & Dinjus, E. (2001). An assessment of supercritical water oxidation (SCWO): Existing problems, possible solutions and new reactor concepts. *Chemical Engineering Journal*, *83*(3), 207–214. doi:10.1016/S1385-8947(00)00255-2

Kruse, A. (2009). Hydrothermal biomass gasification. *The Journal of Supercritical Fluids*, *47*(3), 391–399. doi:10.1016/j.supflu.2008.10.009

Kruse, A., Henningsen, T., Sinag, A., & Pfeiffer, J. (2003). Biomass gasification in supercritical water: Influence of the dry matter content and the formation of phenols. *Industrial & Engineering Chemistry Research*, *42*(16), 3711–3717. doi:10.1021/ie0209430

Kruse, A., Meier, D., Rimbrecht, P., & Schacht, M. (2000). Gasification of pyrocatechol in supercritical water in the presence of potassium hydroxide. *Industrial & Engineering Chemistry Research*, *39*(12), 4842–4848. doi:10.1021/ie0001570

Kusdiana, D., & Saka, S. (2001). Kinetics of transesterification in rapeseed oil to biodiesel fuels as treated in supercritical methanol. *Fuel*, *80*(5), 693–698. doi:10.1016/S0016-2361(00)00140-X

Kusdiana, D., & Saka, S. (2004). Effects of water on biodiesel fuel production by supercritical methanol treatment. *Bioresource Technology*, *91*(3), 289–295. doi:10.1016/S0960-8524(03)00201-3 PMID:14607489

Lakeh, R. B., Lavine, A. S., Kavehpour, Ganapathi, G.B., & Wirz, R. (2013, July). Effect of natural conversion on thermal energy storage in supercritical fluids. *Proceedings of the ASME 2013 7th International Conference on Energy Sustainability ES2013*. doi:10.1115/ES2013-18079

Lam, M. K., & Lee, K. T. (2012). Microalgae biofuels: A critical review of issues, problems and the way forward. *Biotechnology Advances*, *30*(3), 673–690. doi:10.1016/j.biotechadv.2011.11.008 PMID:22166620

Lang, X., Akgerman, A., & Bukur, D. B. (1995). Steady state Fischer-Tropsch synthesis in supercritical propane. *Industrial & Engineering Chemistry Research*, *34*(1), 72–77. doi:10.1021/ie00040a004

Lee, I.-G., Kim, M.-S., & Ihm, S.-K. (2002). Gasification of glucose in supercritical water. *Industrial & Engineering Chemistry Research*, *41*(5), 1182–1188. doi:10.1021/ie010066i

Li, J., Zhao, J. L., Xu, L. J., Zhou, L. G., Li, X. L., & Wang, J. G. (2008). Endophytic fungi from rhizomes of Paris polyphylla var. yunnanensis. *World Journal of Microbiology & Biotechnology*, *24*(5), 733–737. doi:10.1007/s11274-007-9531-3

Li, X. T., Grace, J. R., Lim, C. J., Watkinson, A. P., Chen, H. P., & Kim, J. R. (2004). Biomass gasification in a circulating fluidized bed. *Biomass and Bioenergy*, *26*(2), 171–193. doi:10.1016/S0961-9534(03)00084-9

Lovett, A., Sunnenberg, G., & Dockerty, T. (2014). The availability of land for perennial energy crops in Great Britain. *Global Change Biology Bioenergy*, *6*(2), 99–107. doi:10.1111/gcbb.12147

Lu, Y. J., Guo, L. J., Ji, C. M., Zhang, X. M., Hao, X. H., & Yan, Q. H. (2006). Hydrogen production by biomass gasification in supercritical water: A parametric study. *International Journal of Hydrogen Energy*, *31*(7), 822–831. doi:10.1016/j.ijhydene.2005.08.011

Ma, F., & Hanna, M. A. (1999). Biodiesel production: A review. *Bioresource Technology*, *70*(1), 1–15. doi:10.1016/S0960-8524(99)00025-5

Manarungson, S., Mok, W. S. L., & Antal, M. J. (1993). *Advances in Thermochemical Biomass Conversion* (A. V. Bridgwater, Ed.). London: Blackie Academic and Professional.

Matsumura, Y., & Minowa, T. (2004). Fundamental design of a continuous biomass gasification process using a supercritical water fluidized bed. *International Journal of Hydrogen Energy*, *29*(7), 701–707. doi:10.1016/j.ijhydene.2003.09.005

Matsumura, Y., Minowa, T., Potic, B., Kersten, S., Prins, W., Vanswaaij, W., & Kruse, A. et al. (2005). Biomass gasification in near- and super-critical water: Status and prospects. *Biomass and Bioenergy*, *29*(4), 269–292. doi:10.1016/j.biombioe.2005.04.006

Matsumura, Y., Sasaki, M., Okuda, K., Takami, S., Ohara, S., Umetsu, M., & Adschiri, T. (2006). Supercritical water treatment of biomass for energy and material recovery. *Combustion Science and Technology*, *178*(1-3), 509–536. doi:10.1080/00102200500290815

McKendry, P. (2002). Energy production from biomass (Part 1): Overview of biomass. *Bioresource Technology*, *83*(1), 37–46. doi:10.1016/S0960-8524(01)00118-3 PMID:12058829

Meister, J. J. (1996). Chemical modification of lignin. In Chemical modification of lignocellulosic materials. Marcel Dekker Inc.

Melero, J. A., Van Grieken, R., Morales, G., & Paniagua, M. (2007). Acid mesoporous silica for the acetylation of glycerol: Synthesis of bioadditives to petrol fuel. *Energy & Fuels*, *21*(3), 1782–1791. doi:10.1021/ef060647q

Meleterova, Y., Kostanek, F., Rouskova, M., Matejkova, M., Kostanek, P., & Solcova, O. (2015). Microalgae for bioenergy: Key technology nodes. *TheScientificWorldJournal*, *2015*, 1–6. doi:10.1155/2015/597618

Minami, E., & Saka, S. (2003). Comparison of the decomposition behaviors of hardwood and softwood in supercritical methanol. *Journal of Wood Science*, *49*(1), 73–78. doi:10.1007/s100860300012

Minowa, T., Fang, Z., Ogi, T., & Varhegyi, G. (1998b). Decomposition of cellulose and glucose in hot-compressed water under catalyst-free conditions. *Journal of Chemical Engineering of Japan*, *31*(1), 131–134. doi:10.1252/jcej.31.131

Minowa, T., Zhen, F., & Ogi, T. (1998a). Cellulose decomposition in hot-compressed water with alkali or nickel catalyst. *The Journal of Supercritical Fluids*, *13*(1-3), 253–259. doi:10.1016/S0896-8446(98)00059-X

Modell, M. (1982). *Processing methods for the oxidation of organics in supercritical water*. US Patent 4,338,199.

Modell, M., & Amin, S. I. (1978). *Gasification process*. US Patent 4113446.

Montingelli, M. E., Tedesco, S., & Olabi, A. G. (2015). Biogas production from algal biomass: A review. *Renewable & Sustainable Energy Reviews*, *43*, 961–972. doi:10.1016/j.rser.2014.11.052

Mosteiro-Romero, M., Vogel, F., & Wokaun, A. (2014). Liquefaction of wood in hot compressed water. Part1 - Experimental results. *ChemicalEngineering Science*, *109*, 111–122. doi:10.1016/j.ces.2013.12.038

Naik, S. N., Goud, V. V., Rout, P. K., & Dalai, A. K. (2010). Production of first and second generation biofuels: A comprehensive review. *Renewable & Sustainable Energy Reviews*, *14*(2), 578–597. doi:10.1016/j.rser.2009.10.003

Ni, M., & Dyc, L. et al.. (2006). An overview of hydrogen production from biomass. *Fuel Processing Technology*, *87*(5), 461–472. doi:10.1016/j.fuproc.2005.11.003

Ni, M., Leung, D. Y. C., Leung, M. K. H., & Sumathy, K. (2006). An overview of hydrogen production from biomass. *Fuel Processing Technology*, *87*(5), 461–472. doi:10.1016/j.fuproc.2005.11.003

Olsson, L., & Jorgensen, H. et al.. (2005). Bioethanol production from lignocellulosic material. In S. Dumitriu (Ed.), *Polysachharides: structural diversity and functional versatility* (pp. 957–993). New York: Marcel Dekker.

Pastircakova, K. (2004). Determination of trace metal concentrations in ashes from various biomass materials. *Energy Educ. Sci. Technol.*, *13*, 97–104.

Patil, P. D., Gude, V. G., Mannarswamy, A., Deng, S., Cooke, P., Munson-McGee, S., & Nirmalakhandan, N. et al. (2011). Optimization of direct conversion of wet algae to biodiesel under supercritical methanol conditions. *Bioresource Technology*, *102*(1), 118–122. doi:10.1016/j.biortech.2010.06.031 PMID:20591655

Penninger, J. M. L., Maass, G. J. J., & Rep, M. (2007). Compressed hydrogen-rich fuel gas (CHFG) from wet biomass by reforming in supercritical water. *International Journal of Hydrogen Energy*, *32*(10-11), 1472–1476. doi:10.1016/j.ijhydene.2006.10.026

Peterson, A. A., Vogel, F., Lachance, P. R., Fröling, M., Antal, M. J. Jr, & Tester, J. W. (2008b). Thermochemical biofuel production in hydrothermal media: A review of sub- and supercritical water technologies. *Energy Environ. Sci.*, *1*(1), 32–65. doi:10.1039/b810100k

Peterson, A. A., Vogel, F., Lachance, R. P., Fröling, M., Antal, M. J. Jr, & Tester, J. W. (2008a). Thermochemical biofuel production in hydrothermal media: A review of sub- and supercritical water technologies. *Energy Environ Sci.*, *1*(1), 32–65. doi:10.1039/b810100k

Peterson, C., & Hustrulid, T. (1998). Carbon cycle for rapeseed oil biodiesel fuels. *Biomass and Bioenergy*, *14*(2), 91–101. doi:10.1016/S0961-9534(97)10028-9

Pioro, I. (2014). Application of supercritical pressure in power engineering. Specifics of thermophysical properties and forced - convective heat transfer. In V. Anikeev & M. Fan (Eds.), *Supercritical Fluid Technology for Energy and Environmental Applications* (pp. 201–234). Elsevier B.V. doi:10.1016/B978-0-444-62696-7.00011-3

Pioro, I., & Duffey, R. B. (2007). *Heat Transfer and Hydraulic Resistance at Supercritical Pressures in Power Engineering Applications*. New York: ASME Press. doi:10.1115/1.802523

Pioro, L. S., & Pioro, I. (1997). *Industrial Two-Phase Thermosyphons*. New York: Begell House.

Pisharody, S. A., Fisher, J. W., & Abraham, M. A. (1996). Supercritical water oxidation of solid particulates. *Industrial & Engineering Chemistry Research*, *35*(12), 4471–4478. doi:10.1021/ie960269k

Poudel, J., & Oh, S. C. (2012). A kinetic analysis of wood degradation in supercritical alcohols. *Industrial & Engineering Chemistry Research*, *51*(12), 4509–4514. doi:10.1021/ie200496b

Reddy, H. K., Muppaneni, T., Patil, P. D., Ponnusamy, S., Cooke, P., Schaub, T., & Deng, S. (2014). Direct conversion of wet algae to crude biodiesel under supercritical ethanol conditions. *Fuel*, *115*, 720–726. doi:10.1016/j.fuel.2013.07.090

Resende, F. L. P., Fraley, S. A., Berger, M. J., & Savage, P. E. (2008). Non catalytic Gasification of Lignin in Supercritical Water. *Energy & Fuels*, *22*(2), 1328–1334. doi:10.1021/ef700574k

Resende, F. L. P., Neff, M. E., & Savage, P. E. (2007). Noncatalytic gasification of cellulose in supercritical water. *Energy & Fuels*, *21*(6), 3637–3643. doi:10.1021/ef7002206

Saka, S., Campanelli, P., Banchero, M., & Manna, L. (2010). Synthesis of biodiesel from edible, non-edible and waste cooking oils via supercritical methyl acetate transesterification. *Fuel*, *89*(12), 3675–3682. doi:10.1016/j.fuel.2010.07.033

Saka, S., & Isayama, Y. (2009). A new process for catalyst-free production of biodiesel using supercritical methyl acetate. *Fuel*, *88*(7), 1307–1313. doi:10.1016/j.fuel.2008.12.028

Saka, S., & Kusdiana, D. (2001). Biodiesel fuel from rapeseed prepared in supercritical methanol. *Fuel*, *80*(2), 225–231. doi:10.1016/S0016-2361(00)00083-1

Saleh, B., Koglbauer, G., Wendland, M., & Fischer, J. (2007). Working fluids for low-temperature organic Rankine cycles. *Energy*, *32*(7), 1210–1221. doi:10.1016/j.energy.2006.07.001

Sandler, S. I. (1999). *Chemical and Engineering Thermodynamics*. John Wiley & Sons, Inc.

Sari, A., & Karaipekli, A. (2009). Preparation, thermal properties and thermal reliability of palmitic acid/expanded graphite composite as form-stable PCM for thermal energy storage. *Solar Energy Materials and Solar Cells*, *93*(5), 571–576. doi:10.1016/j.solmat.2008.11.057

Sato, S., Lin, S. Y., Suzuki, Y., & Hatano, H. (2003). Hydrogen production from heavy oil in the presence of calcium hydroxide. *Fuel*, *82*(5), 561–567. doi:10.1016/S0016-2361(02)00328-9

Savage, P. E. (1999). Organic chemical reactions in supercritical water. *Chemical Reviews*, *99*(2), 603–621. doi:10.1021/cr9700989 PMID:11848994

Savage, P. E., Gopalan, S., Mizan, T. I., Martino, C. J., & Brock, E. E. (1995). Reactions at supercritical conditions: Application and Fundamentals. *AIChE Journal. American Institute of Chemical Engineers*, *41*(7), 1723–1778. doi:10.1002/aic.690410712

Saxena, S. C. (1995). Bubble column reactors and Fischer-Tropsch synthesis. *Catalysis Reviews. Science and Engineering*, *37*(2), 227–309. doi:10.1080/01614949508007096

Schenk, P. M., Thomas-Hall, S. R., Stephens, E., Marx, U., Mussgnug, J. H., Posten, C., & Hankamer, B. et al. (2008). Second generation biofuels: High-efficiency microalgae for biodiesel production. *Bioenergy Res.*, *1*(1), 20–43. doi:10.1007/s12155-008-9008-8

Schmidt, E., Eckert, E., & Grigull, V. (1946). *Heat transfer by liquids near the critical state, AFF Translation, No. 527*. Dayton, OH: Air Materials Command, Wright Field.

Schmieder, H., Abeln, J., Boukis, N., Dinjus, E., Kruse, A., Kluth, M., & Schacht, M. et al. (2000). Hydrothermal gasification of biomass and organic wastes. *The Journal of Supercritical Fluids*, *17*(2), 145–153. doi:10.1016/S0896-8446(99)00051-0

Semwal, S., Arora, A. K., Badoni, R. P., & Tuli, D. K. (2011). Biodiesel production using heterogeneous catalysts. *Bioresource Technology*, *102*(3), 2151–2161. doi:10.1016/j.biortech.2010.10.080 PMID:21106371

Sengers, J. M. H., & Levelt, L. (2000). Supercritical fluids: Their properties and applications. Supercritical Fluids. Springer.

Sharma, Y., Singh, B., & Korstad, J. (2011). Latest developments on application of heterogenous basic catalysts for an efficient and eco friendly synthesis of biodiesel: A review. *Fuel*, *90*(4), 1309–1324. doi:10.1016/j.fuel.2010.10.015

Sheehan, J., Camobreco, V., Duffield, J., Graboski, M., & Shapouri, H. (1998). *Life cycle inventory of biodiesel and petroleum diesel for use in an urban bus, NREL/SR-580-24089*. National Renewable Energy Laboratory.

Shin, H. Y., Matsumoto, K., Higashi, H., Iwai, Y., & Arai, Y. (2001). Development of a solution model to correlate solubilities of inorganic compounds in water vapor under high temperatures and pressures. *The Journal of Supercritical Fluids*, *21*(2), 105–110. doi:10.1016/S0896-8446(01)00089-4

Singh, A., & Olsen, S. I. (2011). A critical review of biochemical conversion, sustainability and life cycle assessment of algal biofuels. *Applied Energy*, *88*(10), 3548–3555. doi:10.1016/j.apenergy.2010.12.012

Srivastava, A., & Prasad, R. (2000). Triglycerides-based diesel fuels. *Renewable & Sustainable Energy Reviews*, *4*(2), 111–133. doi:10.1016/S1364-0321(99)00013-1

Susanti, R. F., Kim, J., & Yoo, K. (2014). *Supercritical water gasification for hydrogen production: current status and prospective of high temperature operation.*In V. Anikeev & M. Fan (Eds.), *Supercritical Fluid Technology for Energy and Environmental Applications* (pp. 111–138). Elsevier B.V.

Tamme, R., Taut, U., Streuber, C., & Kalfa, H. (1991). Energy storage development for solar thermal processes. *Solar Energy Materials*, *24*(1-4), 386–396. doi:10.1016/0165-1633(91)90077-X

Tan, K. T., & Lee, K. T. (2011). A review on supercritical fluids (SCF) technology in sustainable biodiesel production: Potential and challenges. *Renewable & Sustainable Energy Reviews*, *15*(5), 2452–2456. doi:10.1016/j.rser.2011.02.012

Tan, K. T., Lee, K. T., & Mohamed, A. R. (2010). A glycerol-free process to produce biodiesel by supercritical methyl acetate technology: An optimization study via response surface methodology. *Bioresource Technology*, *101*(3), 965–969. doi:10.1016/j.biortech.2009.09.004 PMID:19773156

Toor, S., Rosendahl, L., & Rudolf, A. (2011). Hydrothermal liquefaction of biomass: A review of subcritical water technologies. *Energy*, *36*(5), 2328–2342. doi:10.1016/j.energy.2011.03.013

US Environmental Protection Agency. (2002). *A comprehensive analysis of biodiesel impacts on exhaust emissions*. Draft Technical Report, EPA420- P-02-001.

Vadillo, V., Sánchez-Oneto, Portela, J.R., & Martínez de la Ossa, E.J. (2014). Supercritical water oxidation for wasterwater destruction with energy recovery. In Supercritical Fluid Technology for Energy and Environmental Applications. Elsevier B.V.

Vogel, F., Waldner, M., Rouff, A., & Rabe, S. (2007). Synthetic natural gas from biomass by catalytic conversion in supercritical water. *Green Chemistry*, *9*(6), 616–619. doi:10.1039/b614601e

Vyas, A. P., Verma, J. L., & Subrahmanyam, N. (2010). A review on FAME production processes. *Fuel*, *89*(1), 1–9. doi:10.1016/j.fuel.2009.08.014

Wang, L., & Yang, J. (2007). Transesterification of soybean oil with nano-MgO or not in supercrtical and subcritical methanol. *Fuel*, *86*(3), 328–333. doi:10.1016/j.fuel.2006.07.022

Watanabe, M., Inomata, H., & Arai, K. (2002). Catalytic hydrogen generation from biomass (glucose and cellulose) with ZrO_2 in supercritical water. *Biomass and Bioenergy*, *22*(5), 405–410. doi:10.1016/S0961-9534(02)00017-X

Watanabe, M., Inomata, H., Osada, M., Sato, T., Adschiri, T., & Arai, K. (2003). Catalytic effects of NaOH and ZrO_2 for partial oxidative gasification of n-hexadecane and lignin in supercritical water. *Fuel*, *82*(5), 545–552. doi:10.1016/S0016-2361(02)00320-4

Wen, D., Jiang, H., & Zhang, K. (2009). Supercritical fluids technology for clean biofuel production. *Progress in Natural Science*, *19*(3), 273–284. doi:10.1016/j.pnsc.2008.09.001

Williams, P. T., & Onwudili, J. (2005). Composition of products from the supercritical water gasification of glucose: A model biomass compound. *Industrial & Engineering Chemistry Research*, *44*(23), 8739–8749. doi:10.1021/ie050733y

Xu, X., Matsumura, Y., Stenberg, J., & Antal, M. J. (1996). Carbon-catalyzed gasification of organic feedstocks in supercritical water. *Industrial & Engineering Chemistry Research*, *35*(8), 2522–2530. doi:10.1021/ie950672b

Yermakova, A., & Anikeev, V. (2000). Thermodynamic calculations in modeling of multiphase processes and reactors. *Industrial & Engineering Chemistry Research*, *39*(5), 1453–1472. doi:10.1021/ie9905761

Yermakova, A., Anikeev, V., & Froment, G. (2000). Influence of non-ideality of reaction mixture on the Fischer-Tropsch reaction rate in supercritical solvent.[in Russian]. *Theoretical Foundation of Chemical Technology*, *34*(2), 203–211.

Yokota, K., & Fujimoto, K. (1991). Supercritical-Phase Fischer-Tropsch Synthesis Reaction. 2. The Effective Diffusion of Reactant and Products in the Supercritical-Phase. *Industrial & Engineering Chemistry Research*, *30*(1), 95–100. doi:10.1021/ie00049a014

Yokota, K., & Fujimoto, K. (1993). Supercritical-phase Fischer-Tropsch synthesis reaction. 3. Extraction capability of supercritical fluids. *Fuel*, *70*(8), 989–994. doi:10.1016/0016-2361(91)90056-G

Yu, D., Aihara, M., & Antal, J. M. (1993). Hydrogen production by steam reforming glucose in supercritical water. *Energy & Fuels*, *7*(5), 574–577. doi:10.1021/ef00041a002

Yu, Y., Lou, X., & Wu, H. (2008). Some recent advances in hydrolysis of biomass in hot- compressed water and its comparisons with other hydrolysis methods. *Energy & Fuels*, *22*(1), 46–60. doi:10.1021/ef700292p

Compilation of References

Accary, G., Raspo, I., Bontoux, P., & Zappoli, B. (2005). Rayleith-Benard and Schwarzschild instability in a supercritical fluid. *Advances in Space Research*, *36*(1), 11–16. doi:10.1016/j.asr.2005.02.082

Ambrosini, W., & Sharabi, M. (2006). Dimensionless parameters in stability analysis of heated channels with fluids at supercritical pressures.*Proceedings of 14th International Conference on Nuclear Engineering (ICONE 14)*. doi:10.1115/ICONE14-89862

Ambrosini, W. (2008). Thermal-hydraulic stability limits. *International Students Workshop on High Performance Light Water Reactors*, Karlsruhe, Germany.

Ambrosini, W. (2009). *Continuing assessment of system and CFD codes for heat transfer and stability in supercritical fluids.4th international symposium on Supercritical Water-Cooled Reactors*, Heidelberg, Germany.

Ameel, T. A., Warrington, R. O., Wegeng, R. S., & Drost, M. K. (1997). Miniaturization technologies applied to energy systems. *Energy Conversion and Management*, *38*(10-13), 969–982. doi:10.1016/S0196-8904(96)00127-6

Amiroudine, S., & Zappoli, B. (2003). Piston Effect induced thermal oscillations at the Rayleigh-Benard threshold in supercritical ^{3}He. *Physical Review Letters*, *90*(10), 105303. doi:10.1103/PhysRevLett.90.105303 PMID:12689007

Antal, M. J. (1983). Effect of reactor severity on the gas-phase pyrolysis of celulose - and kraft lignin-derived volatile matter. *Industrial & Engineering Chemistry Product Research and Development*, *22*(2), 366–375. doi:10.1021/i300010a039

Antoni, O., & Dumaz, P. (2003, January). Preliminary calculations of a supercritical light water reactor concept using the CATHARE code.*Proceedings of ICAPP*, 4-7.

Aoki, N., Umei, R., Yoshida, A., & Mae, K. (2011). Design method for micromixers considering influence of channel confluence and bend on diffusion length. *Chemical Engineering Journal*, *167*(2-3), 643–650. doi:10.1016/j.cej.2010.08.084

Asinari, P. (2005). Numerical prediction of turbulent convective heat transfer in mini/micro channels for carbon dioxide at supercritical pressure. *International Journal of Heat and Mass Transfer*, *48*(18), 3864–3879. doi:10.1016/j.ijheatmasstransfer.2005.03.028

Bae, S. J., Ahn, Y., & Lee, J. I. (2014). Preliminary Study of the Supercritical CO2 Hybrid Cycle for the HTGR Application. *Transactions of the Korean Nuclear Society Spring Meeting*.

Basu, D. N., Bhattacharyya, S., & Das, P. K. (2014). A review of modern advances in analyses and applications of single-phase natural circulation loop in nuclear thermal hydraulics. *Nuclear Engineering and Design*, *280*, 326–348. doi:10.1016/j.nucengdes.2014.09.011

Beckman, E. J. (2004). Supercritical and near-critical CO_2 in green chemical synthesis and processing. *The Journal of Supercritical Fluids*, *28*(2-3), 121–191. doi:10.1016/S0896-8446(03)00029-9

Brownbridge, G., Azadi, P., Smallbone, A., Bhave, A., Taylor, B., & Kraft, M. (2014). The future viability of algae-derived biodiesel under economic and technical uncertainties. *Bioresource Technology*, *151*, 166–173. doi:10.1016/j.biortech.2013.10.062 PMID:24220544

Bushby, S. J., Dimmick, G. R., Duffey, R. B., Spinks, N. J., Burrill, K. A., & Chan, P. S. W. (2000). Conceptual designs for advanced, high-temperature CANDU reactors.*Proceedings of the SCR-2000.*

Cengel, Y. A., & Boles, M. A. (Eds.). (2015). *Thermodynamics - An Engineering Approach* (8th ed.). New York, NY: McGraw-Hill Companies, Inc.

Chatoorgoon, V. (2008). Supercritical flow stability in horizontal channels. *Nuclear Engineering and Design*, *238*(8), 1940–1946. doi:10.1016/j.nucengdes.2007.10.033

Chen, J., Lu, Y., Guo, L., Zhang, X., & Xiao, P. (2010). Hydrogen production by biomass gasification in supercritical water using concentrated solar energy: System development and proof of concept. *International Journal of Hydrogen Energy*, *35*(13), 7134–7141. doi:10.1016/j.ijhydene.2010.02.023

Chen, L., Deng, B. L., & Zhang, X. R. (2013). Experimental Investigation of CO_2 thermosyphon Flow and Heat Transfer in the Supercritical Region. *International Journal of Heat and Mass Transfer*, *64*, 202–211. doi:10.1016/j.ijheatmasstransfer.2013.03.077

Chen, L., Zhang, X. R., Cao, S. M., & Bai, H. (2012). Study of trans-critical CO_2 natural convection flow with unsteady heat input and its implications on system control. *International Journal of Heat and Mass Transfer*, *55*(23-24), 7119–7132. doi:10.1016/j.ijheatmasstransfer.2012.07.027

Chen, L., Zhang, X. R., Okajima, J., Komiya, A., & Maruyama, S. (2016). Numerical simulation of stability behaviors and heat transfer characteristics for near-critical fluid microchannel flows. *Energy Conversion and Management*, *110*, 407–418. doi:10.1016/j.enconman.2015.12.031

Chen, L., Zhang, X. R., Yamaguchi, H., & Liu, Z. S. (2010). Effect of heat transfer on the instabilities and transitions of supercritical CO_2 flow in a natural circulation loop. *International Journal of Heat and Mass Transfer*, *53*(19-20), 4101–4111. doi:10.1016/j.ijheatmasstransfer.2010.05.030

Chen, L., & Zhang, X. R. (2014). Experimental Analysis on a Novel Solar Collector System achieved by Supercritical CO_2 Natural Convection. *Energy Conversion and Management*, *77*, 173–182. doi:10.1016/j.enconman.2013.08.059

Chen, W. H., Lin, B. J., Huang, M. Y., & Chang, J. S. (2015). Thermochemical conversion of microalgal biomass into biofuels: A review. *Bioresource Technology*, *184*, 314–327. doi:10.1016/j.biortech.2014.11.050 PMID:25479688

Chen, Y., Zhao, M., Yang, C., Bi, K., & Du, K. (2012b). An experiment on flow and heat transfer characteristics in natural circulation of supercritical water.*Proceedings of 3rd China-Canada Joint Workshop on Supercritical-Water-Cooled Reactors (CCSC-2012).*

Chen, Y., Zhao, M., Yang, C., Bi, K., & Du, K. (2013a). Experiment of heat transfer of supercritical water in natural circulation with different diameters of heated tubes.*Proceedings of 6th International Symposium on Supercritical Water-cooled Reactors (ISSCWR6).*

Cheng, L., & Thome, J. R. (2009). Cooling of microprocessors using flow boiling CO_2 in a micro-evaporator: Preliminary analysis and performance comparison. *Applied Thermal Engineering*, *29*(11-12), 2426–2432. doi:10.1016/j.applthermaleng.2008.12.019

Cheng, X., & Yang, Y. H. (2008). A point-hydraulics model for flow stability analysis. *Nuclear Engineering and Design*, *238*(1), 188–199. doi:10.1016/j.nucengdes.2007.06.012

Chisti, Y. (2007). Biodiesel from microalgae. *Biotechnology Advances*, *25*(3), 294–306. doi:10.1016/j.biotechadv.2007.02.001 PMID:17350212

Demirbas, A. (2007). Biodiesel from sunflower oil in superctitical methanol with calcium oxide. *Energy Conversion and Management*, *48*(3), 937–941. doi:10.1016/j.enconman.2006.08.004

Demirbas, A. (2009). Biofuels securing the planets future energy needs. *Energy Conversion and Management*, *50*(9), 2239–2249. doi:10.1016/j.enconman.2009.05.010

Duan, P., & Savage, P. (2011). Hydrothermal Liquefaction of a Microalga with Heterogeneous Catalysts. *Industrial & Engineering Chemistry Research*, *50*(1), 52–61. doi:10.1021/ie100758s

Duan, Y. Y., Shi, L., Sun, Q., Zhu, M. S., & Han, L. Z. (2000). Hermodynamic roperties of trifluoroiodomethane (CF_3I). *International Journal of Thermophysics*, *21*(2), 393–404. doi:10.1023/A:1006683529436

Ducoulombier, M., Colasson, S., Haberschill, P., & Tingaud, F. (2011). Charge reduction experimental investigation of CO_2 single-phase flow in a horizontal micro-channel with constant heat flux conditions. *International Journal of Refrigeration*, *34*(4), 827–833. doi:10.1016/j.ijrefrig.2011.01.013

Eckert, C. A., Knutson, B. L., & Debenedetti, P. G. (1996). Supercritical fluids as solvents for chemical and materials processing. *Nature*, *383*(6598), 313–318. doi:10.1038/383313a0

Elliott, D. C., Sealock, L. J., & Baker, E. G. (1993). Chemical processing in high-pressure aqueous environments. 2. Development of catalysts for gasification. *Industrial & Engineering Chemistry Research*, *32*(8), 1542–1548. doi:10.1021/ie00020a002

Falk, F. L., & Commenge, J. M. (2010). Performance comparison of micromixers. *Chemical Engineering Science*, *65*(1), 405–411. doi:10.1016/j.ces.2009.05.045

Falkehag, S. I. (1975). Synthesis of phenolic polymer. *Applied Polymer Symposia*, *28*, 247–257.

Fukuda, H., Kondo, A., & Noda, H. (2001). Biodiesel fuel production by transesterification of oils. *Journal of Bioscience and Bioengineering*, *92*(5), 405–416. doi:10.1016/S1389-1723(01)80288-7 PMID:16233120

Fukuda, K., & Kobori, T. (1979). Classification of two-phase flow instability by density wave oscillation model. *Journal of Nuclear Science and Technology*, *16*(2), 95–108. doi:10.1080/18811248.1979.9730878

Gebhart, B. (1973). Natural circulation flows and stability. In J. P. Hartnett & T. F. Irvine Jr., (Eds.), *Advances in heat transfer* (Vol. 9, pp. 273–348). New York: Academic Press.

Gerpen, J. V. (2005). Biodiesel processing and production. *Fuel Processing Technology*, *86*(10), 1097–1107. doi:10.1016/j.fuproc.2004.11.005

Gómez, T. O., Class, A., Lahey, R. Jr, & Schulenberg, T. (2008). Stability analysis of a uniformly heated channel with supercritical water. *Nuclear Engineering and Design*, *238*(8), 1930–1939. doi:10.1016/j.nucengdes.2007.10.031

Halim, R., Gladman, B., Danquah, M. K., & Webley, P. A. (2011). Oil extraction from microalgae for biodieselproduction. *Bioresource Technology*, *102*(1), 178–185. doi:10.1016/j.biortech.2010.06.136 PMID:20655746

Hirschfelder, J. O., Curtiss, C. F., & Bird, R. B. (1954). *Molecular Theory of Gases and Liquids*. New York: Wiley & Sons Inc.

Huai, X. L., Koyama, S., & Zhao, T. S. (2005). An experimental study of flow and heat transfer of supercritical carbon dioxide in multi-port mini channels under cooling conditions. *Chemical Engineering Science*, *60*(12), 3337–3345. doi:10.1016/j.ces.2005.02.039

Ishii, M., & Zuber, N. (1970). Thermally induced flow instabilities in two-phase mixtures.*Proceedings of the Fourth International Heat Transfer Conference, 5.*

Ishikawa, Y., & Saka, S. (2001). Chemical conversion of cellulose as treated in supercritical methanol. *Cellulose (London, England)*, *8*(3), 189–195. doi:10.1023/A:1013170020469

Jackson, J. D. (2013). Fluid flow and convective heat transfer to fluids at supercritical pressure. *Nuclear Engineering and Design*, *264*(11), 24–40. doi:10.1016/j.nucengdes.2012.09.040

Jakab, E., Liu, K., & Meuzelaar, H. L. C. (1997). Thermal decomposition of wood and cellulose in the presence of solvent vapors. *Industrial & Engineering Chemistry Research*, *36*(6), 2087–2095. doi:10.1021/ie960335i

Johnson, B. K., & Prudhomme, P. K. (2003). Chemical processing and micromixing in confined impinging jets. *AIChE Journal. American Institute of Chemical Engineers*, *49*(9), 2264–2282. doi:10.1002/aic.690490905

Kabyemela, B. M., Adschiri, T., Malaluan, R. M., & Arai, K. (1999). Glucose and Fructose Decomposition in Subcritical and Supercritical Water: Detailed Reaction Pathway, Mechanisms, and Kinetics. *Industrial & Engineering Chemistry Research*, *38*(8), 2888–2895. doi:10.1021/ie9806390

Kakac, S., & Bon, B. (2008). A review of two-phase flow dynamic instabilities in tube boiling systems. *International Journal of Heat and Mass Transfer*, *51*(3-4), 399–433. doi:10.1016/j.ijheatmasstransfer.2007.09.026

Kim, D. E., & Kim, M. H. (2010). Experimental study of the effects of flow acceleration and buoyancy on heat transfer in a supercritical fluid flow in a circular tube. *Nuclear Engineering and Design*, *240*(10), 3336–3349. doi:10.1016/j.nucengdes.2010.07.002

Klein, H., & Feuerbacher, B. (1987). Gravity influence on thermal relaxation near the critical point. *Physics Letters. [Part A]*, *123*(4), 183–187. doi:10.1016/0375-9601(87)90701-8

Knothe, G. (2001). Historical perspectives on vegetable oil-based diesel fuels. *Inform (Silver Spring, Md.)*, *12*, 1103–1107.

Koo, J., & Kleinstreuer, C. (2005). Analysis of surface roughness effects on heat transfer in micro-conduits. *International Journal of Heat and Mass Transfer*, *48*(13), 2625–2634. doi:10.1016/j.ijheatmasstransfer.2005.01.024

Krichevskii, I. R. (1952). *Phase equilibria in solutions at high pressures*. Moscow: Goskhimizdat.

Kruse, A., Henningsen, T., Sinag, A., & Pfeiffer, J. (2003). Biomass gasification in supercritical water: Influence of the dry matter content and the formation of phenols. *Industrial & Engineering Chemistry Research*, *42*(16), 3711–3717. doi:10.1021/ie0209430

Kruse, A., Meier, D., Rimbrecht, P., & Schacht, M. (2000). Gasification of pyrocatechol in supercritical water in the presence of potassium hydroxide. *Industrial & Engineering Chemistry Research*, *39*(12), 4842–4848. doi:10.1021/ie0001570

Kruse, A. (2009). Hydrothermal biomass gasification. *The Journal of Supercritical Fluids*, *47*(3), 391–399. doi:10.1016/j.supflu.2008.10.009

Kumar, V., Paraschivoiu, M., & Nigam, K. D. P. (2011). Single-phase fluid flow and mixing in microchannels. *Chemical Engineering Science*, *66*(7), 1329–1373. doi:10.1016/j.ces.2010.08.016

Kurganov, V. A., Zeigarnik, Y. A., & Maslakova, I. V. (2012). Heat transfer and hydraulic resistance of supercritical-pressure coolants. Part I: Specifics of thermophysical properties of supercritical pressure fluids and turbulent heat transfer under heating conditions in round tubes (state of the art). *International Journal of Heat and Mass Transfer*, *55*(11-12), 3061–3075. doi:10.1016/j.ijheatmasstransfer.2012.01.031

Kuronen, J., Hotti, M., & Tuuri, S. (2016), Modelling and dynamic simulation of cyclically operated pulverized coal-fired power plant. *9th EUROSIM Congress on Modelling and Simulation.*

Kusdiana, D., & Saka, S. (2001). Kinetics of transesterification in rapeseed oil to biodiesel fuels as treated in supercritical methanol. *Fuel, 80*(5), 693–698. doi:10.1016/S0016-2361(00)00140-X

Kusdiana, D., & Saka, S. (2004). Effects of water on biodiesel fuel production by supercritical methanol treatment. *Bioresource Technology, 91*(3), 289–295. doi:10.1016/S0960-8524(03)00201-3 PMID:14607489

Lam, M. K., & Lee, K. T. (2012). Microalgae biofuels: A critical review of issues, problems and the way forward. *Biotechnology Advances, 30*(3), 673–690. doi:10.1016/j.biotechadv.2011.11.008 PMID:22166620

Lang, X., Akgerman, A., & Bukur, D. B. (1995). Steady state Fischer-Tropsch synthesis in supercritical propane. *Industrial & Engineering Chemistry Research, 34*(1), 72–77. doi:10.1021/ie00040a004

Lee, I.-G., Kim, M.-S., & Ihm, S.-K. (2002). Gasification of glucose in supercritical water. *Industrial & Engineering Chemistry Research, 41*(5), 1182–1188. doi:10.1021/ie010066i

Lee, K. B. (1995). Analytical prediction of subchannel friction factor for infinite bare rod square and triangular arrays of low pitch to diameter ratio in turbulent flow. *Nuclear Engineering and Design, 157*(1-2), 197–203. doi:10.1016/0029-5493(95)00996-P

Li, H., Zhang, Y., Zhang, L., Yao, M., Kruizenga, A., & Anderson, M. (2016). PDF-based modeling on the turbulent convection heat transfer of supercritical CO_2 in the printed circuit heat exchangers for supercritical CO_2 Brayton cycle. *International Journal of Heat and Mass Transfer, 98*(7), 204–218. doi:10.1016/j.ijheatmasstransfer.2016.03.001

Li, J., Zhao, J. L., Xu, L. J., Zhou, L. G., Li, X. L., & Wang, J. G. (2008). Endophytic fungi from rhizomes of Paris polyphylla var. yunnanensis. *World Journal of Microbiology & Biotechnology, 24*(5), 733–737. doi:10.1007/s11274-007-9531-3

Li, X. T., Grace, J. R., Lim, C. J., Watkinson, A. P., Chen, H. P., & Kim, J. R. (2004). Biomass gasification in a circulating fluidized bed. *Biomass and Bioenergy, 26*(2), 171–193. doi:10.1016/S0961-9534(03)00084-9

Licht, J., Anderson, M., & Corradini, M. (2008). Heat transfer to water at supercritical pressures in a circular and square annular flow geometry. *International Journal of Heat and Fluid Flow, 29*(1), 156–166. doi:10.1016/j.ijheatfluidflow.2007.09.007

Lu, Y. J., Guo, L. J., Ji, C. M., Zhang, X. M., Hao, X. H., & Yan, Q. H. (2006). Hydrogen production by biomass gasification in supercritical water: A parametric study. *International Journal of Hydrogen Energy, 31*(7), 822–831. doi:10.1016/j.ijhydene.2005.08.011

Ma, F., & Hanna, M. A. (1999). Biodiesel production: A review. *Bioresource Technology, 70*(1), 1–15. doi:10.1016/S0960-8524(99)00025-5

Matsumura, Y., & Minowa, T. (2004). Fundamental design of a continuous biomass gasification process using a supercritical water fluidized bed. *International Journal of Hydrogen Energy, 29*(7), 701–707. doi:10.1016/j.ijhydene.2003.09.005

Matsumura, Y., Sasaki, M., Okuda, K., Takami, S., Ohara, S., Umetsu, M., & Adschiri, T. (2006). Supercritical water treatment of biomass for energy and material recovery. *Combustion Science and Technology, 178*(1-3), 509–536. doi:10.1080/00102200500290815

McAdams, W. H. (1942). *Heat Transmission* (2nd ed.). New York, NY: McGraw-Hill.

McKendry, P. (2002). Energy production from biomass (Part 1): Overview of biomass. *Bioresource Technology, 83*(1), 37–46. doi:10.1016/S0960-8524(01)00118-3 PMID:12058829

Melero, J. A., Van Grieken, R., Morales, G., & Paniagua, M. (2007). Acid mesoporous silica for the acetylation of glycerol: Synthesis of bioadditives to petrol fuel. *Energy & Fuels*, *21*(3), 1782–1791. doi:10.1021/ef060647q

Minami, E., & Saka, S. (2003). Comparison of the decomposition behaviors of hardwood and softwood in supercritical methanol. *Journal of Wood Science*, *49*(1), 73–78. doi:10.1007/s100860300012

Minowa, T., Fang, Z., Ogi, T., & Varhegyi, G. (1998b). Decomposition of cellulose and glucose in hot-compressed water under catalyst-free conditions. *Journal of Chemical Engineering of Japan*, *31*(1), 131–134. doi:10.1252/jcej.31.131

Minowa, T., Zhen, F., & Ogi, T. (1998a). Cellulose decomposition in hot-compressed water with alkali or nickel catalyst. *The Journal of Supercritical Fluids*, *13*(1-3), 253–259. doi:10.1016/S0896-8446(98)00059-X

Mokry, S., Pioro, I. L., Kirillov, P., & Gospodinov, Y. (2010). Supercritical-water heat transfer in a vertical bare tube. *Nuclear Engineering and Design*, *240*(3), 568–576. doi:10.1016/j.nucengdes.2009.09.003

Montingelli, M. E., Tedesco, S., & Olabi, A. G. (2015). Biogas production from algal biomass: A review. *Renewable & Sustainable Energy Reviews*, *43*, 961–972. doi:10.1016/j.rser.2014.11.052

Morini, G. L. (2004). Single-phase convective heat transfer in microchannels: A review of experimental results. *International Journal of Thermal Sciences*, *43*(7), 631–651. doi:10.1016/j.ijthermalsci.2004.01.003

Naik, S. N., Goud, V. V., Rout, P. K., & Dalai, A. K. (2010). Production of first and second generation biofuels: A comprehensive review. *Renewable & Sustainable Energy Reviews*, *14*(2), 578–597. doi:10.1016/j.rser.2009.10.003

Olsson, L., & Jorgensen, H. et al.. (2005). Bioethanol production from lignocellulosic material. In S. Dumitriu (Ed.), *Polysachharides: structural diversity and functional versatility* (pp. 957–993). New York: Marcel Dekker.

Onuki, A. (2004). *Phase Transition Dynamics*. Cambridge University Press.

Ornatskiy, A. P., Dashkiev, Yu. G., & Perkov, V. G. (Eds.). (1980). *Supercritical Steam Generators*. Kiev, Ukraine: Vyshcha Shkola Publishing House. (In Russian)

Patil, P. D., Gude, V. G., Mannarswamy, A., Deng, S., Cooke, P., Munson-McGee, S., & Nirmalakhandan, N. et al. (2011). Optimization of direct conversion of wet algae to biodiesel under supercritical methanol conditions. *Bioresource Technology*, *102*(1), 118–122. doi:10.1016/j.biortech.2010.06.031 PMID:20591655

Penninger, J. M. L., Maass, G. J. J., & Rep, M. (2007). Compressed hydrogen-rich fuel gas (CHFG) from wet biomass by reforming in supercritical water. *International Journal of Hydrogen Energy*, *32*(10-11), 1472–1476. doi:10.1016/j.ijhydene.2006.10.026

Peterson, C., & Hustrulid, T. (1998). Carbon cycle for rapeseed oil biodiesel fuels. *Biomass and Bioenergy*, *14*(2), 91–101. doi:10.1016/S0961-9534(97)10028-9

Pioro, I. L., & Duffey, R. B. (2006). *Heat transfer and hydraulic resistance at supercritical pressures in power engineering applications*. New York, NY: ASME Press.

Pioro, I. L., & Duffey, R. B. (2007). *Heat Transfer and Hydraulic Resistance at Supercritical Pressures in Power Engineering Applications*. New York, NY: ASME Press. doi:10.1115/1.802523

Pioro, I., Duffey, R., & Dumouchel, T. (2004). Hydraulic Resistance of Fluids Flowing in Channels at Supercritical Pressures (Survey). *Nuclear Engineering and Design*, *231*(2), 187–197. doi:10.1016/j.nucengdes.2004.03.001

Pioro, I., & Mokry, S. (2011a). Thermophysical Properties at Critical and Supercritical Conditions, Chapter in book. In A. Belmiloudi (Ed.), *Heat Transfer. Theoretical Analysis, Experimental Investigations and Industrial Systems* (pp. 573–592). Rijeka, Croatia: INTECH. doi:10.5772/13790

Pioro, I., & Mokry, S. (2011b). Heat Transfer to Fluids at Supercritical Pressures, Chapter in book. In A. Belmiloudi (Ed.), *Heat Transfer. Theoretical Analysis, Experimental Investigations and Industrial Systems* (pp. 481–504). Rijeka, Croatia: INTECH. doi:10.5772/13792

Pioro, L. S., & Pioro, I. (1997). *Industrial Two-Phase Thermosyphons*. New York: Begell House.

Pisharody, S. A., Fisher, J. W., & Abraham, M. A. (1996). Supercritical water oxidation of solid particulates. *Industrial & Engineering Chemistry Research*, *35*(12), 4471–4478. doi:10.1021/ie960269k

Poudel, J., & Oh, S. C. (2012). A kinetic analysis of wood degradation in supercritical alcohols. *Industrial & Engineering Chemistry Research*, *51*(12), 4509–4514. doi:10.1021/ie200496b

Reddy, H. K., Muppaneni, T., Patil, P. D., Ponnusamy, S., Cooke, P., Schaub, T., & Deng, S. (2014). Direct conversion of wet algae to crude biodiesel under supercritical ethanol conditions. *Fuel*, *115*, 720–726. doi:10.1016/j.fuel.2013.07.090

Rehme, K. (1973). Simple method of predicting friction factors of turbulent flow in non-circular channels. *International Journal of Heat and Mass Transfer*, *16*(5), 933–950. doi:10.1016/0017-9310(73)90033-1

Resende, F. L. P., Fraley, S. A., Berger, M. J., & Savage, P. E. (2008). Non catalytic Gasification of Lignin in Supercritical Water. *Energy & Fuels*, *22*(2), 1328–1334. doi:10.1021/ef700574k

Resende, F. L. P., Neff, M. E., & Savage, P. E. (2007). Noncatalytic gasification of cellulose in supercritical water. *Energy & Fuels*, *21*(6), 3637–3643. doi:10.1021/ef7002206

Rosa, P., Karayiannis, T. G., & Collins, M. W. (2009). Single-phase heat transfer in microchannels: The importance of scaling effects. *Applied Thermal Engineering*, *29*(17-18), 3447–3468. doi:10.1016/j.applthermaleng.2009.05.015

Rostami, A. A., Mujumdar, A. S., & Saniei, N. (2002). Flow and heat transfer for gas flowing in microchannels: A review. *Heat and Mass Transfer*, *38*(4-5), 359–367. doi:10.1007/s002310100247

Saka, S., & Isayama, Y. (2009). A new process for catalyst-free production of biodiesel using supercritical methyl acetate. *Fuel*, *88*(7), 1307–1313. doi:10.1016/j.fuel.2008.12.028

Saka, S., & Kusdiana, D. (2001). Biodiesel fuel from rapeseed prepared in supercritical methanol. *Fuel*, *80*(2), 225–231. doi:10.1016/S0016-2361(00)00083-1

Saleh, B., Koglbauer, G., Wendland, M., & Fischer, J. (2007). Working fluids for low-temperature organic Rankine cycles. *Energy*, *32*(7), 1210–1221. doi:10.1016/j.energy.2006.07.001

Sarkar, M. K. S., & Basu, D. N. (2015). Working regime identification for natural circulation loops by comparative thermalhydraulic analyses with three fluids under identical operating conditions. *Nuclear Engineering and Design*, *293*, 187–195. doi:10.1016/j.nucengdes.2015.07.046

Sato, S., Lin, S. Y., Suzuki, Y., & Hatano, H. (2003). Hydrogen production from heavy oil in the presence of calcium hydroxide. *Fuel*, *82*(5), 561–567. doi:10.1016/S0016-2361(02)00328-9

Savage, P. E., Gopalan, S., Mizan, T. I., Martino, C. J., & Brock, E. E. (1995). Reactions at supercritical conditions: Application and Fundamentals. *AIChE Journal. American Institute of Chemical Engineers*, *41*(7), 1723–1778. doi:10.1002/aic.690410712

Schmidt, E., Eckert, E., & Grigull, V. (1946). *Heat transfer by liquids near the critical state, AFF Translation, No. 527*. Dayton, OH: Air Materials Command, Wright Field.

Schulenberg, Th., & Starflinger, J. (2012). *High Performance Light Water Reactor. Design and Analyses*. KIT.

Semwal, S., Arora, A. K., Badoni, R. P., & Tuli, D. K. (2011). Biodiesel production using heterogeneous catalysts. *Bioresource Technology*, *102*(3), 2151–2161. doi:10.1016/j.biortech.2010.10.080 PMID:21106371

Sharabi, M. B., Ambrosini, W., & He, S. (2008). Transient 3D stability analysis of SCWR rod bundle subchannels by a CFD code. *Proceedings of the 16th International Conference on Nuclear Engineering (ICONE16)*. doi:10.1115/ICONE16-48701

Sharma, M., Pilkhwal, D. S., Vijayan, P. K., Saha, D., & Sinha, R. K. (2012). Steady-state behavior of natural circulation loops operating with supercritical fluids for open and closed loop boundary conditions. *Heat Transfer Engineering*, *33*(9), 809–820. doi:10.1080/01457632.2012.646920

Sharma, M., Vijayan, P. K., Pilkhwal, D. S., Saha, D., & Sinha, R. K. (2010). Linear and nonlinear stability analysis of a supercritical natural circulation loop. *Journal of Engineering for Gas Turbines and Power*, *132*(10), 102904. doi:10.1115/1.4000342

Shin, H. Y., Matsumoto, K., Higashi, H., Iwai, Y., & Arai, Y. (2001). Development of a solution model to correlate solubilities of inorganic compounds in water vapor under high temperatures and pressures. *The Journal of Supercritical Fluids*, *21*(2), 105–110. doi:10.1016/S0896-8446(01)00089-4

Shitsman, M. E. (1963). Impairment of the heat transmission at supercritical pressures. *High Temperature*, *1*, 237–244.

Silin, V. A., Voznesensky, V. A., & Afrov, A. M. (1993). The light water integral reactor with natural circulation of the coolant at supercritical pressure B-500 SKDI. *Nuclear Engineering and Design*, *144*(2), 327–336. doi:10.1016/0029-5493(93)90148-3

Singh, A., & Olsen, S. I. (2011). A critical review of biochemical conversion, sustainability and life cycle assessment of algal biofuels. *Applied Energy*, *88*(10), 3548–3555. doi:10.1016/j.apenergy.2010.12.012

Span, R., & Wagner, W. (1996). A new equation of state for carbon dioxide covering the fluid region from the triple-point temperature to 1100 K at pressures up to 800 MPa. *Journal of Physical and Chemical Reference Data*, *25*(6), 1509–1596. doi:10.1063/1.555991

Srivastava, A., & Prasad, R. (2000). Triglycerides-based diesel fuels. *Renewable & Sustainable Energy Reviews*, *4*(2), 111–133. doi:10.1016/S1364-0321(99)00013-1

Starkloff, R., Alobaid, F., Karner, K., Epple, B., Schmitz, M., & Boehm, F. (2015). Development and validation of a dynamic simulation model for a large coal-fired power plant. *Applied Thermal Engineering*, *91*, 496–506. doi:10.1016/j.applthermaleng.2015.08.015

Steinke, M. E., & Kandlikar, S. G. (2006). Single-phase liquid friction factors in microchannels. *International Journal of Thermal Sciences*, *45*(11), 1073–1083. doi:10.1016/j.ijthermalsci.2006.01.016

Stodola, A. (1927). *Steam and Gas Turbines*. New York: McGraw-Hill.

Su, Y., Feng, J., Zhao, H., Tian, W., Su, G., & Qiu, S. (2013). Theoretical study on the flow instability of supercritical water in the parallel channels. *Progress in Nuclear Energy*, *68*(1), 169–176. doi:10.1016/j.pnucene.2013.06.005

Tamme, R., Taut, U., Streuber, C., & Kalfa, H. (1991). Energy storage development for solar thermal processes. *Solar Energy Materials*, *24*(1-4), 386–396. doi:10.1016/0165-1633(91)90077-X

Tan, K. T., Lee, K. T., & Mohamed, A. R. (2010). A glycerol-free process to produce biodiesel by supercritical methyl acetate technology: An optimization study via response surface methodology. *Bioresource Technology*, *101*(3), 965–969. doi:10.1016/j.biortech.2009.09.004 PMID:19773156

Tan, K. T., & Lee, K. T. (2011). A review on supercritical fluids (SCF) technology in sustainable biodiesel production: Potential and challenges. *Renewable & Sustainable Energy Reviews*, *15*(5), 2452–2456. doi:10.1016/j.rser.2011.02.012

Termuehlen, H., & Emsperger, W. (2003). *Clean and efficient coal-fired power plants*. New York: ASME Press. doi:10.1115/1.801942

Tilak, A. K., & Basu, D. N. (2015). Computational investigation of the dynamic response of a supercritical natural circulation loop to aperiodic and periodic excitations. *Nuclear Engineering and Design*, *284*, 251–263. doi:10.1016/j.nucengdes.2014.12.028

Vijayan, P. K., Sharma, M., Pilkhwal, D. S., Saha, D., & Sinha, R. K. (2010). A comparative study of single-phase, two-phase and supercritical natural circulation in a rectangular loop. *Journal of Engineering for Gas Turbines and Power*, *132*(10), 102913. doi:10.1115/1.4000866

Vogel, F., Waldner, M., Rouff, A., & Rabe, S. (2007). Synthetic natural gas from biomass by catalytic conversion in supercritical water. *Green Chemistry*, *9*(6), 616–619. doi:10.1039/b614601e

Vyas, A. P., Verma, J. L., & Subrahmanyam, N. (2010). A review on FAME production processes. *Fuel*, *89*(1), 1–9. doi:10.1016/j.fuel.2009.08.014

Wang, L., & Yang, J. (2007). Transesterification of soybean oil with nano-MgO or not in supercrtical and subcritical methanol. *Fuel*, *86*(3), 328–333. doi:10.1016/j.fuel.2006.07.022

Wang, Q., Guan, Y. X., Yao, S. J., & Zhu, Z. Q. (2011). Controllable preparation and formation mechanism of BSA microparticles using supercritical assisted atomization with an enhanced mixer. *The Journal of Supercritical Fluids*, *56*(1), 97–104. doi:10.1016/j.supflu.2010.12.002

Watanabe, M., Inomata, H., & Arai, K. (2002). Catalytic hydrogen generation from biomass (glucose and cellulose) with ZrO_2 in supercritical water. *Biomass and Bioenergy*, *22*(5), 405–410. doi:10.1016/S0961-9534(02)00017-X

Watanabe, M., Inomata, H., Osada, M., Sato, T., Adschiri, T., & Arai, K. (2003). Catalytic effects of NaOH and ZrO_2 for partial oxidative gasification of n-hexadecane and lignin in supercritical water. *Fuel*, *82*(5), 545–552. doi:10.1016/S0016-2361(02)00320-4

Wen, D., Jiang, H., & Zhang, K. (2009). Supercritical fluids technology for clean biofuel production. *Progress in Natural Science*, *19*(3), 273–284. doi:10.1016/j.pnsc.2008.09.001

Winterton, R. H. S. (1998). Where did the Dittus and Boelter equation come from? *International Journal of Heat and Mass Transfer*, *41*(4-5), 809–810. doi:10.1016/S0017-9310(97)00177-4

Xiong, T., Yan, X., Xiao, Z., Li, Y., Huang, Y., & Yu, J. (2012). Experimental study on flow instability in parallel channels with supercritical water. *Annals of Nuclear Energy*, *48*(2), 60–67. doi:10.1016/j.anucene.2012.05.018

Xu, X., Matsumura, Y., Stenberg, J., & Antal, M. J. (1996). Carbon-catalyzed gasification of organic feedstocks in supercritical water. *Industrial & Engineering Chemistry Research*, *35*(8), 2522–2530. doi:10.1021/ie950672b

Yadav, A. K., Ram Gopal, M., & Bhattacharyya, S. (2012). CFD analysis of a CO_2 based natural circulation loop with end heat exchangers. *Applied Thermal Engineering*, *36*, 288–295. doi:10.1016/j.applthermaleng.2011.10.031

Yamagata, K., Nishikawa, K., Hasegawa, S., Fujii, T., & Yoshida, S. (1972). Forced convective heat transfer to supercritical water flowing in tubes. *International Journal of Heat and Mass Transfer*, *15*(12), 2575–2593. doi:10.1016/0017-9310(72)90148-2

Yi, T. T., Koshizuka, S., & Oka, Y. (2004). A linear stability analysis of supercritical water reactors, (II) thermal–hydraulic stability. *Journal of Nuclear Science and Technology*, *41*(12), 1176–1186. doi:10.1080/18811248.2004.9726346

Yi, T. T., Koshizuka, S., & Oka, Y. (2004). A linear stability analysis of supercritical water reactors,(I) thermal-hydraulic stability. *Journal of Nuclear Science and Technology*, *41*(12), 1166–1175. doi:10.1080/18811248.2004.9726345

Yokota, K., & Fujimoto, K. (1991). Supercritical-Phase Fischer-Tropsch Synthesis Reaction. 2. The Effective Diffusion of Reactant and Products in the Supercritical-Phase. *Industrial & Engineering Chemistry Research*, *30*(1), 95–100. doi:10.1021/ie00049a014

Yu, D., Aihara, M., & Antal, J. M. (1993). Hydrogen production by steam reforming glucose in supercritical water. *Energy & Fuels*, *7*(5), 574–577. doi:10.1021/ef00041a002

Yu, J., Che, S., Li, R., & Qi, B. (2011). Analysis of Ledinegg flow instability in natural circulation at supercritical pressure. *Progress in Nuclear Energy*, *53*(6), 775–779. doi:10.1016/j.pnucene.2011.04.001

Yu, Y., Lou, X., & Wu, H. (2008). Some recent advances in hydrolysis of biomass in hot- compressed water and its comparisons with other hydrolysis methods. *Energy & Fuels*, *22*(1), 46–60. doi:10.1021/ef700292p

Zappoli, B., & Carles, P. (1995). Thermoacoustic nature of the critical speeding-up. *European Journal of Mechanics - B/ Fluids*, *14*, 41–65.

Zappoli, B. (2003). Near-critical fluid hydrodynamics. *Comptes Rendus. Mécanique*, *331*(10), 713–726. doi:10.1016/j.crme.2003.05.001

Zhang, X. R., Chen, L., & Yamaguchi, H. (2010). Natural convective flow and heat transfer of supercritical CO_2 in a rectangular circulation loop. *International Journal of Heat and Mass Transfer*, *53*(19-20), 4112–4122. doi:10.1016/j.ijheatmasstransfer.2010.05.031

Zhang, X. R., Yamaguchi, H., & Uneno, D. (2007). Experimental study on the performance of solar Rankine system using supercritical CO_2. *Renewable Energy*, *32*(15), 2617–2628. doi:10.1016/j.renene.2007.01.003

Zhang, X. R., & Yamaguchi, H. (2008). An experimental study on evacuated tube solar collector using supercritical CO_2. *Applied Thermal Engineering*, *28*(10), 1225–1233. doi:10.1016/j.applthermaleng.2007.07.013

Zhang, Y., Li, H., Li, L., Lei, X., & Wang, T. (2014). Study on two-phase flow instabilities in internally-ribbed tubes by using frequency domain method. *Applied Thermal Engineering*, *65*(1-2), 1–13. doi:10.1016/j.applthermaleng.2013.12.069

Zhang, Y., Li, H., Li, L., Wang, T., Zhang, Q., & Lei, X. (2015). A new model for studying the density wave instabilities of supercritical water flows in tubes. *Applied Thermal Engineering*, *75*(C), 397–409. doi:10.1016/j.applthermaleng.2014.09.029

Zhao, C. X., He, L. Z., Qiao, S. Z., & Middelberg, A. P. J. (2011). Nanoparticle synthesis in microreactors. *Chemical Engineering Science*, *66*(7), 1463–1479. doi:10.1016/j.ces.2010.08.039

Zvirin, Y. (1985). The instability associated with the onset of motion in a thermosyphon. *International Journal of Heat and Mass Transfer*, *28*(11), 2105–2111. doi:10.1016/0017-9310(85)90104-8

American Society for Testing and Materials. (2002). *Standard Specification for Biodiesel Fuel (b100) Blend Stock for Distillate Fuels, Designation D6751-02*. West Conshohocken, PA: ASTM International.

van der Waals. (1873). *Over de Continuiteit van den Gas- en Voleistoftoestand*. Leiden, The Netherlands: Ed. Sijthoff.

Abbaslou, R. M. M., Mohammadzadeh, J. S., & Dalai, A. K. (2009). Review on Fischer-Tropsch synthesis in supercritical media. *Fuel Processing Technology*, *90*(7-8), 849–856. doi:10.1016/j.fuproc.2009.03.018

Anikeev, V. I., Kuzmin, V. A., & Bobrova, I. I. (1987). Experimental study of the solar energy catalytic device based on a closed thermochemical cycle. *Reports of the USSR Academy of Sciences*, *293*, 1427–1432.

Anikeev, V. I., Stepanov, D. A., & Yermakova, A. (2012). Thermodynamics of phase and chemical equilibrium in the processes of biodiesel fuel synthesis in sub- and supercritical methanol. *I&ECR*, *51*, 4783–4796.

Anikeev, V. I. (1987). Experimental study of solar energy conversion by means of a thermochemical cycle. *Solar technology (Geliotechnika)*, *2*, 3–8.

Antal, J. M. Jr, Allen, S. G., Schulman, D., Xu, X., & Divilio, R. J. (2000). Biomass gasification in supercritical water. *Industrial & Engineering Chemistry Research*, *39*(11), 4044–4053. doi:10.1021/ie0003436

Balat, H., & Kirtay, E. (2010). Review hydrogen from biomass present scenario and future prospects. *International Journal of Hydrogen Energy*, *35*(14), 7416–7426. doi:10.1016/j.ijhydene.2010.04.137

Beysens, D., Chatain, D., Nikolayev, V. S., Ouazzani, J., & Garrabos, Y. (2010). Possibility of long-distance heat transport in weightlessness using supercritical fluids. *Physical Review E: Statistical, Nonlinear, and Soft Matter Physics*, *82*(6), 061126. doi:10.1103/PhysRevE.82.061126 PMID:21230663

Blood, P. J., Denyer, J. P., Azzopardi, B. J., Poliakoff, M., & Lester, E. (2004). A versatile flow visualization technique for quantifying mixing in a binary system: Application to continuous supercritical water hydrothermal synthesis (SWHS). *Chemical Engineering Science*, *59*(14), 2853–2861. doi:10.1016/j.ces.2004.04.021

Boure, J. A., Bergles, A. E., & Tong, L. S. (1973). Review of two-phase flow instability. *Journal of Nuclear Engineering and Design*, *25*(2), 165–192. doi:10.1016/0029-5493(73)90043-5

Brauner, N., & Moalem, M. D. (1992). Stability analysis of stratfied liquid-liquid flow. *International Journal of Multiphase Flow*, *18*(1), 103–121. doi:10.1016/0301-9322(92)90009-6

Bukur, D. B., Lang, X., Akgerman, A., & Feng, Z. (1997). Effect of Process Conditions on the Olefin Selectivity during Conventional and Supercritical Fischer-Tropsch Synthesis. *Industrial & Engineering Chemistry Research*, *36*(7), 2580–2587. doi:10.1021/ie960507b

Calzavara, Y., Joussot-Dubien, C., Boissonnet, G., & Sarrade, S. (2005). Evaluation of biomass gasification in supercritical water process for hydrogen production. *Energy Conversion and Management*, *46*(4), 615–631. doi:10.1016/j.enconman.2004.04.003

Canakci, M., & Gerpen, J. V. (2001). Biodiesel production from oils and fats with high free fatty acids. *Transactions of the ASAE. American Society of Agricultural Engineers*, *44*(6), 1429–1436. doi:10.13031/2013.7010

Carles, P. (2010). A brief review of the thermophysical properties of supercritical fluids. *The Journal of Supercritical Fluids*, *53*(1-3), 2–11. doi:10.1016/j.supflu.2010.02.017

Chatoorgoon, V., Voodi, A., & Fraser, D. (2005). The stability boundary for supercritical flow in natural convection loops part I: H_2O studies. *Nuclear Engineering and Design*, *235*(24), 2570–2580. doi:10.1016/j.nucengdes.2005.06.003

Chatoorgoon, V. (1986). A simple thermalhydraulic stability code. *Nuclear Engineering and Design*, *93*, 51–67. doi:10.1016/0029-5493(86)90194-9

Chen, C. Y., Zou, J. H., Zhang, S. Y., Zaitlin, D., & Zhu, L. H. (2009). Strigolactones are a new-defined class of plant hormones which inhibit shoot branching and mediate the interaction of plant-AM fungi and plant-parasitic weeds. *Science in China. Series C, Life Sciences*, *52*(8), 693–700. doi:10.1007/s11427-009-0104-6 PMID:19727586

Chen, L., Zhang, X. R., Okajima, J., & Maruyama, S. (2013). Thermal relaxation and critical instability of near-critical fluid microchannel flow. *Physical Review E: Statistical, Nonlinear, and Soft Matter Physics*, *87*(4), 043016. doi:10.1103/PhysRevE.87.043016 PMID:23679522

Cheng, W. I., Zhang, R., Xie, K., & Wang, J. (2010). Heat conduction enhanced shape-stabilized paraffin/HDPE composite PCMs by graphite addition: Preparation and thermal properties. *Solar Energy Materials and Solar Cells*, *94*(10), 1636–1642. doi:10.1016/j.solmat.2010.05.020

Cheng, X., & Schulenberg, T. (2001). *Heat Transfer at Supercritical Pressures-Literature Review and Application to an HPLWR*. Institut fur Kern-und Energietechnik.

Clauser, F. H. (1956). The turbulent buoundary layer. *Advances in Applied Mechanics*, *4*, 1–51. doi:10.1016/S0065-2156(08)70370-3

Cui, H. H., & Li, Z. H. (2004). Flow characteristics of liquids in micro-tubes driven by a high pressure. *Physics of Fluids*, *16*(5), 1803–1810. doi:10.1063/1.1691457

de Boer, K., & Bahri, P. A. (2011). Supercritical methanol for fatty acid methyl ester production: A review. *Biomass and Bioenergy*, *35*(3), 983–991. doi:10.1016/j.biombioe.2010.11.037

Demirbas, A. (2003). Biodiesel fuels from vegetable oils via catalytic and noncatalytic supercritical alcohol transesterifications and other methods: A survey. *Energy Conversion and Management*, *44*(13), 2093–2109. doi:10.1016/S0196-8904(02)00234-0

Demirbas, A. (2006). Biodiesel production via SCF method and biodiesel fuel characteristics. *Energy Conversion and Management*, *47*(15-16), 2271–2282. doi:10.1016/j.enconman.2005.11.019

Deshpande, P., Kulkarni, K., & Kulkarni, A. D. (2011). Supercritical fluid technology in biodiesel production: A review. *Chemistry and Materials Research*, *1*, 27–32.

Di Blasi, C. (2008). Modeling chemical and physical processes of wood and biomass pyrolysis. *Progress in Energy and Combustion Science*, *34*(1), 47–90. doi:10.1016/j.pecs.2006.12.001

Dimmic, G. R., Chatoorgoon, V. V., Khartabil, H. F., & Duffey, R. B. (2002). Natural-convection studies for advanced CANDU reactor concepts. *Nuclear Engineering and Design*, *215*(1-2), 27–38. doi:10.1016/S0029-5493(02)00039-0

Elliott, D. C., Sealock, L. J., & Bake, E. G. (1994). Chemical processing in high pressure aqueous environments. 3. Batch reactor process development experiments for organics destruction. *Industrial & Engineering Chemistry Research*, *33*(3), 558–565. doi:10.1021/ie00027a012

Elliott, D. C., & Sealock, L. J. (1983). Aqueous catalyst systems for the water-gas shift reaction. 1. Comparative catalyst studies. *Industrial & Engineering Chemistry Product Research and Development*, *22*(3), 426–431. doi:10.1021/i300011a008

Fan, L., Yokota, K., Hanakata, Y., & Fujimoto, K. (1992). Supercritical phase Fischer-Tropsch synthesis: Catalyst pore-size effect. *AIChE Journal. American Institute of Chemical Engineers*, *38*(10), 1639–1648. doi:10.1002/aic.690381014

Flockhart, S. M., & Dhariwal, R. S. (1998). Experimental and numerical investigation into the flow characteristics of channels etched in <100> silicon. *ASME Journal of Fluid Engineering*, *120*(2), 291–295. doi:10.1115/1.2820647

Gnielinski, V. (1976). New equation for heat and mass transfer in turbulent pipe and channel flow. *International Chemical Engineering*, *16*, 359–366.

Grigor'ev, V. A., & Zorin, V. M. (Eds.). (1982). *Thermal and Nuclear Power Plants*. Moscow, Russia: Energoatomizdat Publishing House. (In Russian)

Gu, H., Li, H., Wang, H., & Luo, Y. (2013). Experimental investigation on convective heat transfer from a horizontal miniature tube to methane at supercritical pressures. *Applied Thermal Engineering*, *58*(1-2), 490–498. doi:10.1016/j.applthermaleng.2013.04.049

Haaland, S. E. (1983). Simple and explicit formulas for the friction factor in turbulent flow. *Journal of Fluids Engineering*, *103*(1), 89–90. doi:10.1115/1.3240948

Helmholtz, H. (1868). Über discontinuierliche Flüssigkeits-Bewegungen[On the discontinuous movements of fluids]. *Monatsberichte der Königlichen Preussische Akademie der Wissenschaften zu Berlin*, *23*, 215–228.

Helwani, Z., Othman, M. R., Aziz, N., Kim, J., & Fernando, W. J. N. (2009). Fernando,W.J.N. Solid heterogeneous catalysts for transesterification of triglycerides with methanol: A review. *Applied Catalysis A, General*, *363*(1-2), 1–10. doi:10.1016/j.apcata.2009.05.021

Hoang, D., Bensaid, S., & Saracco, G. (2013). Supercritical fluid technology in biodiesel production. *Green Process Synth.*, *2*, 407–425.

Hosseinizadeh, S. F., Darzi, A. A., & Tan, F. L. (2012). Numerical investigations of unconstrained melting of nano-enhanced phase change material (NEPCM) inside a spherical container. *International Journal of Thermal Sciences*, *51*, 77–83. doi:10.1016/j.ijthermalsci.2011.08.006

Huesemann, M. H., & Benemann, J. R. (2009). Biofuels from microalgae: Review of products, processes, and potential, with special focus on Dunaliella sp. In A. Ben-Amotz, J. E. W. Polle, & D. V. Subba Rao (Eds.), *The Alga Dunaliella: Biodiversity, Physiology, Genomics, and Biotechnology*. Science Publishers. doi:10.1201/b10300-20

Jiang, P. X., Zhang, Y., & Shi, R. F. (2008). Experimental and numerical investigation of convection heat transfer of CO_2 at supercritical pressures in a vertical tube at low Reynolds numbers. *International Journal of Thermal Sciences*, *47*(8), 998–1011. doi:10.1016/j.ijthermalsci.2007.08.003

Justin, J. Z. (2002). *Quantum field theory and critical phenomena* (4th ed.). Oxford, UK: Oxford University Press. doi:10.1093/acprof:oso/9780198509233.001.0001

Kays, W. M., Crawford, M. E., & Weigand, B. (2005). *Convective heat and mass transfer New York*. McGraw-Hill.

Khodadadi, J. M., & Hosseinzadeh, S. F. (2007). Nanoparticle-enhanced phase change materials (NEPCM) with great potential for improved thermal energy storage. *International Communications in Heat and Mass Transfer*, *34*(5), 534–543. doi:10.1016/j.icheatmasstransfer.2007.02.005

Köhler, W., & Hein, D. (1986). *Influence of the wetting state of a heated surface on heat transfer and pressure loss in an evaporator tube, NUREG/IA-0003*. US-NRC.

Kritzer, P., & Dinjus, E. (2001). An assessment of supercritical water oxidation (SCWO): Existing problems, possible solutions and new reactor concepts. *Chemical Engineering Journal*, *83*(3), 207–214. doi:10.1016/S1385-8947(00)00255-2

Li, H., Kruizenga, A., Anderson, M., Corradini, M., Luo, Y., Wang, H., & Li, H. (2011). Development of a new forced convection heat transfer correlation for CO2 in both heating and cooling modes at supercritical pressures. *International Journal of Thermal Sciences*, *50*(12), 2430–2442. doi:10.1016/j.ijthermalsci.2011.07.004

Li, H., Wang, H., Luo, Y., Gu, H., Shi, X., Chen, T., & Zhu, Y. et al. (2009). Experimental investigation on heat transfer from a heated rod with a helically wrapped wire inside a square vertical channel to water at supercritical pressures. *Nuclear Engineering and Design*, *239*(10), 2004–2012. doi:10.1016/j.nucengdes.2009.05.011

Lovett, A., Sunnenberg, G., & Dockerty, T. (2014). The availability of land for perennial energy crops in Great Britain. *Global Change Biology Bioenergy*, *6*(2), 99–107. doi:10.1111/gcbb.12147

Luong, T. D., Phan, V. N., & Nguyen, N. T. (2011). High-throughput micromixers based on acoustic streaming induced by surface acoustic wave. *Microfluidics and Nanofluidics*, *10*(3), 619–625. doi:10.1007/s10404-010-0694-0

Mokry, S., Pioro, I. L., Farah, A., King, K., Gupta, S., Peiman, W., & Kirillov, P. (2011). Development of Supercritical Water Heat-Transfer Correlation for Vertical Bare Tubes. *Nuclear Engineering and Design*, *241*(4), 1126–1136. doi:10.1016/j.nucengdes.2010.06.012

Mosteiro-Romero, M., Vogel, F., & Wokaun, A. (2014). Liquefaction of wood in hot compressed water. Part1 - Experimental results. *Chemical Engineering Science*, *109*, 111–122. doi:10.1016/j.ces.2013.12.038

Oka, Y., Koshizuka, S., Ishiwatari, Y., & Yamaji, A. (Eds.). (2010). *Super Light Water Reactors and Super Fast Reactors*. New York, NY: Springer. doi:10.1007/978-1-4419-6035-1

Oka, Y., & Koshizuka, S. (2000). Design concept of once-through cycle supercritical-pressure light water cooled reactors.*Proceedings of SCR-2000*. The University of Tokyo.

Pioro, I., & Kirillov, P. (2013). *Generation IV Nuclear Reactors as a Basis for Future Electricity Production in the World. In Materials and Processes for Energy: Communicating Current Research and Technological Developments* (pp. 818–830). Formatex Research Center.

Pioro, I. (2011). The Potential Use of Supercritical Water-Cooling in Nuclear Reactors. Chapter. In S. B. Krivit, J. H. Lehr, & Th. B. J. Kingery (Eds.), *Nuclear Energy Encyclopedia: Science, Technology, and Applications*. Hoboken, NJ: Wiley & Sons. doi:10.1002/9781118043493.ch28

Rostami, A. A., Saniei, N., & Mujumdar, A. S. (2000). Liquid flow and heat transfer in microchannels: A review. *Health Technology*, *18*, 59–68.

Saka, S., Campanelli, P., Banchero, M., & Manna, L. (2010). Synthesis of biodiesel from edible, non-edible and waste cooking oils via supercritical methyl acetate transesterification. *Fuel*, *89*(12), 3675–3682. doi:10.1016/j.fuel.2010.07.033

Sandler, S. I. (1999). *Chemical and Engineering Thermodynamics*. John Wiley & Sons, Inc.

Sari, A., & Karaipekli, A. (2009). Preparation, thermal properties and thermal reliability of palmitic acid/expanded graphite composite as form-stable PCM for thermal energy storage. *Solar Energy Materials and Solar Cells*, *93*(5), 571–576. doi:10.1016/j.solmat.2008.11.057

Saxena, S. C. (1995). Bubble column reactors and Fischer-Tropsch synthesis. *Catalysis Reviews. Science and Engineering*, *37*(2), 227–309. doi:10.1080/01614949508007096

Schenk, P. M., Thomas-Hall, S. R., Stephens, E., Marx, U., Mussgnug, J. H., Posten, C., & Hankamer, B. et al. (2008). Second generation biofuels: High-efficiency microalgae for biodiesel production. *Bioenergy Res.*, *1*(1), 20–43. doi:10.1007/s12155-008-9008-8

Schulenberg, T., & Starflinger, J. (2012). *High performance light water reactor, design and analyses*. KIT Scientific Publishing.

Sheehan, J., Camobreco, V., Duffield, J., Graboski, M., & Shapouri, H. (1998). *Life cycle inventory of biodiesel and petroleum diesel for use in an urban bus, NREL/SR-580-24089*. National Renewable Energy Laboratory.

Shen, B., & Zhang, P. (2011). Thermoacoustic waves along the critical isochore. *Physical Review E: Statistical, Nonlinear, and Soft Matter Physics*, *83*(1), 011115. doi:10.1103/PhysRevE.83.011115 PMID:21405669

Sierra-Pallares, J., Marchisio, D. L., Alonso, E., Parra-Santos, M. T., Castro, F., & Cocero, M. J. (2011). Quantification of mixing efficiency in turbulent supercritical water hydrothermal reactors. *Chemical Engineering Science*, *66*(8), 1576–1589. doi:10.1016/j.ces.2010.12.039

Tokanai, H., Ohtomo, Y., Joriguchi, H., Harada, E., & Kuriyama, M. (2010). Heat transfer of supercritical CO_2 flow in natural convection circulation system. *Heat Transfer Engineering*, *31*(9), 750–756. doi:10.1080/01457630903500924

White, F. M. (1974). *Viscous Fluid Flow New York*. McGraw-Hill.

Williams, P. T., & Onwudili, J. (2005). Composition of products from the supercritical water gasification of glucose: A model biomass compound. *Industrial & Engineering Chemistry Research*, *44*(23), 8739–8749. doi:10.1021/ie050733y

Xi, X., Xiao, Z. J., Yan, X., Li, Y., & Huang, Y. (2013). The axial power distribution validation of the SCWR fuel assembly with coupled neutronics-thermal hydraulics method. *Nuclear Engineering and Design*, *258*, 157–163. doi:10.1016/j.nucengdes.2013.01.031

Xi, X., Xiao, Z. J., Yan, X., Li, Y., & Huang, Y. (2014). An experimental investigation of flow instability between two heated parallel channels with supercritical water. *Nuclear Engineering and Design*, *278*, 171–181. doi:10.1016/j.nucengdes.2014.06.034

Xiong, T., Yan, X., Huang, S., Yu, J., & Huang, Y. (2013). Modeling and analysis of supercritical flow instability in parallel channels. *International Journal of Heat and Mass Transfer*, *57*(2), 549–557. doi:10.1016/j.ijheatmasstransfer.2012.08.046

Yermakova, A., Anikeev, V., & Froment, G. (2000). Influence of non-ideality of reaction mixture on the Fischer-Tropsch reaction rate in supercritical solvent.[in Russian]. *Theoretical Foundation of Chemical Technology*, *34*(2), 203–211.

Yermakova, A., & Anikeev, V. (2000). Thermodynamic calculations in modeling of multiphase processes and reactors. *Industrial & Engineering Chemistry Research*, *39*(5), 1453–1472. doi:10.1021/ie9905761

Zang, J. G., Yan, X., Huang, S. F., Zeng, X., Li, Y., Huang, Y., & Yu, J. (2014). A method of extending subcritical heat transfer formulas to supercritical conditions. *Nuclear Engineering and Design*, *266*, 186–193. doi:10.1016/j.nucengdes.2013.11.043

Zhang, L., Wang, H., Gu, H., Qiao, S., Shi, X., & Luo, Y. (2012, October). Experimental investigations on flow instabilities in a forced circulation loop at near-critical and supercritical pressures.*AIP Conference Proceedings*, 1547, 312.

Zhao, J. (2005). *Stability analysis of supercritical water cooled reactors. Doctor of Philosophy in Nuclear Science and Engineer*. Massachusetts Institute of Technology.

Zhong, F., & Meyer, H. (1995). Density equilibration near the liquid-vapor critical point of a pure fluid: Single phase T>Tc. *Physical Review E: Statistical Physics, Plasmas, Fluids, and Related Interdisciplinary Topics*, *51*(4), 3223–3241. doi:10.1103/PhysRevE.51.3223 PMID:9963001

Lord Kelvin (William Thomson). (1871). Hydrokinetic solutions and observations. *Philosophical Magazine*, *42*, 362–377.

T'Joen, C. T., & Rohde, M. (2011, August). Experimental study on the stability of a natural circulation driven supercritical water cooled reactor. *22nd International Symposium on Transport Phenomena, Delft*, The Netherlands.

Anikeev, V. I., Parmon, V. N., Kirillov, V. A., & Zamaraev, K. I. (1990). Theoretical and fundamental studies of solar catalytic power plants based on reversible reactions with participation of methane and syn gas. *International Journal of Hydrogen Energy*, *15*(4), 275–285. doi:10.1016/0360-3199(90)90048-4

Antal, M. J. (1978). *Energy from Biomass and Wastes* (D. L. Klass, Ed.). Chicago, IL: IGT.

Antal, M. J. (1982). *Solar Energy* (K. W. Boer & J. A. Duffie, Eds.). New York: Plenum Press.

Chatoorgoon, V. (2008). Static instability in supercritical parallel-channel systems.*Proceedings of the 16th International Conference on Nuclear Engineering (ICONE 16)*.

Chen, L., Zhang, X. R., Okajima, J., & Maruyama, S. (2013). Numerical simulation of near-critical fluid convective flow mixing in microchannels. *Chemical Engineering Science*, *97*, 67–80. doi:10.1016/j.ces.2013.04.010

Dittus, F. W., & Boelter, L. M. K. (1985). Heat Transfer in Automobile Radiators of the Tubular Type. *International Communications in Heat and Mass Transfer*, *12*(1), 3–22. doi:10.1016/0735-1933(85)90003-X

Fu, D. X., Ma, Y. W., & Li, X. L. (2010). *Direct numerical simulation to compressible turbulence (Chinese)*. Beijing: Science Publisher.

Galan, M. I., Bonet, J., Sire, R., Reneaume, J. M., & Plesu, A. E. (2009). From residual to useful oil:revalorization of glycerine from the biodiesel synthesis. *Bioresource Technology*, *101*(15), 3775–3778. doi:10.1016/j.biortech.2009.01.066 PMID:19269813

Hartnett, J. P., Koh, J. C. Y., & McComas, S. T. (1962). A comparison of predicted and measured friction factors for turbulent flow through rectangular ducts. *Trans.ASME J. Heat Transfer*, *84*(1), 82–88. doi:10.1115/1.3684299

Jounet, A., Mojtabi, A., Ouazzani, J., & Zappoli, B. (2000). Low-frequency vibrations in a near critical fluid. *Physics of Fluids*, *12*(1), 197–205. doi:10.1063/1.870295

Jounet, A., Zappoli, B., & Mojtabi, A. (2000). Rapid thermal relaxation in near-critical fluids and critical speeding up: Discrepancies caused by boundary effects. *Physical Review Letters*, *84*(15), 3224–3228. doi:10.1103/PhysRevLett.84.3224 PMID:11019056

Kruizenga, A., Li, H., & Anderson, M. (2012). Supercritical carbon dioxide heat transfer in horizontal semicircular channels. *Jouranl of Heat Transfer-Transactions of the ASME*, *134*(8), 081802-1–10.

Manarungson, S., Mok, W. S. L., & Antal, M. J. (1993). *Advances in Thermochemical Biomass Conversion* (A. V. Bridgwater, Ed.). London: Blackie Academic and Professional.

Marcel, C. P., Rohde, M., Masson, V. P., & Van der Hagen, T. H. J. J. (2009). Fluid-to-fluid modeling of supercritical water loops for stability analysis. *International Journal of Heat and Mass Transfer*, *52*(21-22), 5046–5054. doi:10.1016/j.ijheatmasstransfer.2009.03.022

Meleterova, Y., Kostanek, F., Rouskova, M., Matejkova, M., Kostanek, P., & Solcova, O. (2015). Microalgae for bioenergy: Key technology nodes. *TheScientificWorldJournal*, *2015*, 1–6. doi:10.1155/2015/597618

Ni, M., & Dyc, L. et al.. (2006). An overview of hydrogen production from biomass. *Fuel Processing Technology*, *87*(5), 461–472. doi:10.1016/j.fuproc.2005.11.003

Pastircakova, K. (2004). Determination of trace metal concentrations in ashes from various biomass materials. *Energy Educ. Sci. Technol.*, *13*, 97–104.

Pioro, I. (2014). Application of supercritical pressure in power engineering. Specifics of thermophysical properties and forced - convective heat transfer. In V. Anikeev & M. Fan (Eds.), *Supercritical Fluid Technology for Energy and Environmental Applications* (pp. 201–234). Elsevier B.V. doi:10.1016/B978-0-444-62696-7.00011-3

Savage, P. E. (1999). Organic chemical reactions in supercritical water. *Chemical Reviews*, *99*(2), 603–621. doi:10.1021/cr9700989 PMID:11848994

Sharabi, M. B., Ambrosini, W., & He, S. (2008). Prediction of unstable behaviour in a heated channel with water at supercritical pressure by CFD models. *Annals of Nuclear Energy*, *35*(5), 767–782. doi:10.1016/j.anucene.2007.09.019

Sharma, Y., Singh, B., & Korstad, J. (2011). Latest developments on application of heterogenous basic catalysts for an efficient and eco friendly synthesis of biodiesel: A review. *Fuel*, *90*(4), 1309–1324. doi:10.1016/j.fuel.2010.10.015

Su, K., & Freire, A. P. S. (2002). Analytical prediction of friction factors and Nusselt numbers of turbulent forced convection in rod bundles with smooth and rough surfaces. *Nuclear Engineering and Design*, *217*(1-2), 111–127. doi:10.1016/S0029-5493(02)00155-3

Susanti, R. F., Kim, J., & Yoo, K. (2014). *Supercritical water gasification for hydrogen production: current status and prospective of high temperature operation.*In V. Anikeev & M. Fan (Eds.), *Supercritical Fluid Technology for Energy and Environmental Applications* (pp. 111–138). Elsevier B.V.

Zhang, Y., & Wang, T. H. (2012). Micro magnetic gyromixer for speeding up reactions in droplets. *Microfluidics and Nanofluidics*, *12*(5), 787–794. doi:10.1007/s10404-011-0922-2 PMID:22707926

Anikeev, V. I., Belobrov, N. S., & Piterkin, R. N. (2006). Results of testing the plant for supercritical water oxidation of nitroglycerin and diethylene glycol dinitrate. *I&ECR*, *45*, 7977–7981.

Arai, K., Smith, R. L. Jr, & Aida, T. M. (2009). Decentralized chemical processes with supercritical fluid technology for sustainable society. *The Journal of Supercritical Fluids*, *47*(3), 628–636. doi:10.1016/j.supflu.2008.11.008

Byrd, A. J., Pant, K. K., & Gupt, R. B. (2008). Hydrogen production from glycerol by reforming in supercritical water over Ru/Al_2O_3 catalyst. *Fuel*, *87*(13-14), 2956–2965. doi:10.1016/j.fuel.2008.04.024

Carles, P., & Dadzie, K. (2005). Two typical time scales of the piston effect. *Physical Review E: Statistical, Nonlinear, and Soft Matter Physics*, *71*(6), 066310. doi:10.1103/PhysRevE.71.066310 PMID:16089870

Chen, L., Deng, B. L., & Zhang, X. R. (2013). Experimental Study of Trans-critical and Supercritical CO_2 Natural Circulation Flow in a Closed Loop. *Applied Thermal Engineering*, *59*(1-2), 1–13. doi:10.1016/j.applthermaleng.2013.05.017

Frohlich, T., Beysens, D., & Garrabos, Y. (2006). Piston effect induced thermal jets in near-critical fluids. *Physical Review E: Statistical, Nonlinear, and Soft Matter Physics*, *74*(4), 046307. doi:10.1103/PhysRevE.74.046307 PMID:17155173

Frohlich, T., Guenoun, P., Bonetti, M., Perrot, F., Beysens, D., Garrabos, Y., & Bravais, P. et al. (1996). Adiabatic versus conductive heat transfer in off-critical SF_6 in the absence of convection. *Physical Review E: Statistical Physics, Plasmas, Fluids, and Related Interdisciplinary Topics*, *54*(2), 1544–1549. doi:10.1103/PhysRevE.54.1544 PMID:9965226

Garrabos, Y., Beysens, D., Lecountre, C., Dejoan, A., Polezhaev, V., & Emelianov, V. (2007). Thermoconvectional phenomena induced by vibrations in supercritical SF_6 under weightlessness. *Physical Review E: Statistical, Nonlinear, and Soft Matter Physics*, *75*(5), 056317. doi:10.1103/PhysRevE.75.056317 PMID:17677174

Harradine, et al.. (1993). Oxidation chemistry of energetic material in SCW. *Haz. Waste Mat.*, *10*, 223.

Podowski, M. Z., & Gallaway, T. (2010). Stability analysis of Generation IV supercritical water reactors. In *Technical Meeting on Heat Transfer.* Pisa, Italy: Thermal-Hydraulics and System Design for Supercritical Water.

Yokota, K., & Fujimoto, K. (1993). Supercritical-phase Fischer-Tropsch synthesis reaction. 3. Extraction capability of supercritical fluids. *Fuel*, *70*(8), 989–994. doi:10.1016/0016-2361(91)90056-G

Zang, J. G., Yan, X., Huang, S. F., Zeng, X., Li, Y., Huang, Y., & Yu, J. (2014). A general method of developing frictional formulas under supercritical conditions and in different geometries. *Annals of Nuclear Energy*, *65*, 262–271. doi:10.1016/j.anucene.2013.10.026

Chen, L., Zhang, X., & Jiang, B. (2013b). Effects of heater orientations on the natural circulation and heat transfer in a supercritical CO_2 rectangular loop. *Journal of Heat Transfer*, *136*(5), 052501. doi:10.1115/1.4025543

Rohde, M., Marcel, C. P., TJoen, C., Class, A. G., & van der Hagen, T. H. J. J. (2011). Downscaling a supercritical water loop for experimental studies on system stability. *International Journal of Heat and Mass Transfer*, *54*(1-3), 65–74. doi:10.1016/j.ijheatmasstransfer.2010.09.063

Dragunov, A. Saltanov, E., Pioro, I., Kirillov, P., & Duffey, R. (2015). Power Cycles of Generation III and III+ Nuclear Power Plants. ASME Journal of Nuclear Engineering and Radiation Science, 1(2).

Dry, M. E. (1981). The Fischer-Tropsch synthesis. In Catalysis-Science and Technology. Springer-Verlag.

IAEA TECDOC Series. (2014). IAEA-TECDOC-1746. Author.

Jackson, J. D., & Hall, W. B. (1979b). Influences of buoyancy on heat transfer to fluids flowing in vertical tubes under turbulent conditions.Turbulent forced convection in channels and bundles, (pp. 613-640). Hemisphere Publishing Corporation.

Kiran, E. (2000). Supercritical Fluids. Kluwer Academic Publishers. doi:10.1007/978-94-011-3929-8

Levelt Sengers, J. M. H. L. (2000). Supercritical fluids: Their properties and applications. In Supercritical Fluids. Kluwer Academic Publishers.

Mokry, S., Naidin, M., & Baig, F. (2008). Conceptual thermal-design options for pressure-tube SCWRs with thermochemical co-generation of hydrogen. *Proceedings of the 16th International Conference on Nuclear Engineering (ICONE-16)*. doi:10.1115/ICONE16-48313

Nuclear News. (2016). Publication of the American Nuclear Society (ANS).

Oka, Y., Koshizuka, S., Ishiwatari, Y., & Yamaji, A. (2010). Super Light Water Reactors and Super Fast Reactors. New York: Springer.

Pioro, I. (Ed.). (2016). Handbook of Generation IV Nuclear Reactors. Duxford, UK: Elsevier - Woodhead Publishing (WP).

Razumovskiy, V. G. Pis'mennyi, E. N., Sidawi, K., Pioro, I. L., & Koloskov, A. E. (2016). Experimental Heat Transfer in an Annular Channel and 3-Rod Bundle Cooled with Upward Flow of Supercritical Water. ASME Journal of Nuclear Engineering and Radiation Science, 2(1).

Saltanov, E., Pioro, I., Mann, D., Gupta, S., Mokry, S., & Harvel, G. (2015). Study on Specifics of Forced-Convective Heat Transfer in Supercritical Carbon Dioxide. ASME Journal of Nuclear Engineering and Radiation Science, 1(1).

Vadillo, V., Sánchez-Oneto, Portela, J.R., & Martínez de la Ossa, E.J. (2014). Supercritical water oxidation for wastewater destruction with energy recovery. In Supercritical Fluid Technology for Energy and Environmental Applications. Elsevier B.V.

Bald, A., & Heusinger, K. (1996, April). Energieerzeugung mit fortschrittlichen Dampfkraftwerken entlastet die Umwelt. *SIEMENS Power Journal*, 5-11.

Cao, H., Zhang, Z., Wu, X., & Miao, X. (2013). Direct biodiesel production from wet microalgae biomass of chlorella pyrenoidosa through in situ transesterification. BioMed Research International.

Chen, L., & Zhang, X. R. (2011). Simulation of heat transfer and system behavior in a supercritical CO2 based thermosyphon: effect of pipe diameter. ASME Journal of Heat Transfer, 133(12).

Chen, L., Zhang, X. R., & Jiang, B. (2014). Effects of Heater Orientations on the Natural Circulation and Heat Transfer in a Supercritical CO2 Rectangular Loop. ASME Journal of Heat Transfer, 136(5).

EN14214 European Committee for Standardization (CEN), EN 14214. (2003). *Automotive fuels - fatty acid methyl esters (FAME) for diesel engines - requirements and test methods*. Author.

EPCOR Genesee Generating Station Phase 3. Project Background. (2007). *CIV E 240. Technical Communications*. W. Kindzierski, Dept. Civil and Environmental Engineering.

Generation IV International Forum. (2015). *2015 GIF Annual Report*. Printed by the OECD Nuclear Energy Agency for the Generation IV International Forum.

Generation IV International Forum. (2016). *Updated Project Plan (2016-2020) On Thermal hydraulics And Safety For The International Research And Development of The Supercritical Water-Cooled Reactor Nuclear Energy System. Version 01*. Author.

IAEA Nuclear power technology development section. (2014). *Heat Transfer Behaviour and Thermohydraulics Code Testing for Supercritical Water Cooled Reactors (SCWRs)*. Printed by the IAEA in Austria. IAEA-TECDOC-1746.

Jackson, J.D., & Hall W.B. (1979a). *Forced convection heat transfer to fluids at supercritical pressure. Turbulent forced convection in channels and bundles*. New York: Hemisphere Publishing Corporation.

Kefer, V., & Seiter, C. (1996, April). Ein 2x660-MW-Kohlekraftwerk mit niedrigen Life-Cycle-Kosten. *SIEMENS Power Journal*, 12-15.

Khabensky, V. B., & Gerliga, V. A. (1995). Hydrodynamic flow instabilities in power equipment components. CBS Publisher & Distributors.

Kirillov, P. L., Lozhkin, V. V., & Smirnov, A. M. (2003). Investigation of borders of deteriorated regimes of a channel at supercritical pressures (In Russian). State Scientific Center of Russian Federation Institute of Physics and Power Engineering by the name of A.I. Leypunskiy.

Lakeh, R. B., Lavine, A. S., Kavehpour, Ganapathi, G.B., & Wirz, R. (2013, July). Effect of natural conversion on thermal energy storage in supercritical fluids. *Proceedings of the ASME 2013 7th International Conference on Energy Sustainability ES2013*. doi:10.1115/ES2013-18079

Meister, J. J. (1996). Chemical modification of lignin. In Chemical modification of lignocellulosic materials. Marcel Dekker Inc.

Mentes, A., Kakac, S., Veziroglu, T.N. et al. (1989). Effect of inlet subcooling on two-phase flow oscillations in a vertical boiling channel. *Waerme- und Stoffuebertragung, 24*, 25-36.

NIST Reference Fluid Thermodynamic and Transport Properties–REFPROP. (2007). *Ver. 8.0 NIST Standard Reference Database 23*. U.S. Dept. of Commerce.

NIST Standard Reference Database-REFPROP, Version 8.0, 2006.

Pioro, I., & Duffey, R. (2015). Nuclear Power as a Basis for Future Electricity Generation. ASME Journal of Nuclear Engineering and Radiation Science, 1(1).

Poette, C., Guedeney, P., & Stainsby, R. (2013). *Gas Cooled Fast Reactors: Recent Advances and Prospects, French Alternative Energies and Atomic Energy Commission (CEA)*. FR13 Conference, Paris, France.

Saltanov, E., & Pioro, I. (2011). World Experience in Nuclear Steam Reheat. In *Nuclear Power: Operation, Safety and Environment*. INTECH. Retrieved from http://www.intechopen.com/books/nuclear-power-operation-safety-and-environment/world-experience-in-nuclear-steam-reheat

Sengers, J. M. H., & Levelt, L. (2000). Supercritical fluids: Their properties and applications. Supercritical Fluids. Springer.

Swenson, H. S., Carver, J. R., & Kakarala, C. R. (1965). Heat transfer to supercritical water in smooth-bore tubes. *Journal of heat transfer,Transactions of the ASME, Series C, 87*, 1289-1292. doi:10.1115/1.3689139

Yu, J., Che, S., Li, R., & Qi, B. (2010). *Instability analysis of natural circulation at supercritical pressure.* The 8th International Topical Meeting on Nuclear Thermal-Hydraulics, Operation and Safety (NUTHO-8), Shanghai, China.

Zhao, J. Y. (2005). *Stability analysis of supercritical water cooled reactors* (PhD thesis). Massachusetts Institute of Technology.

Bae, J.H., & Park, J. H. (2011). Analytical prediction of turbulent friction factor for a rod bundle. *Annals of Nuclear Energy, 38*, 348-357.

Bauer, T., Tamme, R., Christ, M., & Ottinger, O. (2006). PCM-graphite composites for high temperature thermal energy storage. *Proceedings of ECOSTOCK the 10th International Conference on Thermal Energy Storage.*

Bishop, A.A., Sandberg, R.O., & Tong, L.S. (1964). *Forced Convection heat transfer to water at near-critical temperature and supercritical pressures.* Westinghouse Electric Corporation WCAP-2056.

Chatoorgoon, V., Shah, M., & Duffey, R. (2007). *Linear predictions of supercritical flow instability in parallel channels.* 3rd Int. Symposium on SCWR – Design and Technology, Shanghai, China.

Chen, L., Zhang, X.R., Okajima, J., & Maruyama, S. (2013d). Abnormal microchannel convective fluid flow near the gas–liquid critical point. *Physica A, Statistical Mechanics and its Applications, 398*, 10-24.

Ganapathi, G. B., & Wirz, R. (2012, July). High density thermal energy storage with supercritical fluids. *Proceedings of the ASME 6th International Conference on Energy Sustainability ES2012*, 1-9. doi:10.1115/ES2012-91008

Gauthier, J., Brinkmann, G., Copsey, B., & Lecomte, M. (2004). ANTARES: The HTR/VHTR Project at Framatome ANP. *Proceedings of the 2nd International Topical Meeting on High Temperature Reactor Technology (HTR 2004).*

Gupta, S., Saltanov, E., & Pioro, I. (2013b). Heat-Transfer Correlation for Supercritical Carbon Dioxide Flowing in Vertical Bare Tubes. *Proceedings of the 21st International Conference on Nuclear Engineering (ICONE-21).* doi:10.1115/ICONE21-16453

Jackson, J. D. (2002). *Consideration of the heat transfer properties of supercritical pressure water in connectin with the cooling of advanced nuclear reactors.* Paper presented at the 13th Pacific Basin Nuclear Conference, Shenzhen City, China.

Ji, S., Shirahama, H., Koshizuka, S., & Oka, Y. (2001). Stability analysis of supercritical pressure light water-cooled reactor in constant pressure operation. *Proc. of the Ninth International Conference on Nuclear Engineering (ICONE9).*

Kruglikov, P. A., Smolkin, Y. V., & Sokolov, K. V. (2009). Development of engineering solutions for thermal scheme of power unit of thermal power plant with supercritical parameters of steam. *Proceedings of the International Workshop "Supercritical Water and Steam in Nuclear Power Engineering: Problems and Solutions".* (in Russian)

Kuang, G., Ohadi, M. M., & Zhao, Y. (2004). Experimental study on gas cooling heat transfer for supercritical CO2 in microchannels. *Proc.: The 2nd Int Conf on Microchannels and Minichannels*, 325-332.

Misawa, T., Nakatsuka, T., & Yoshida, H. (2009). *Heat Transfer Experiments and Numerical Analysis of Supercritical Pressure Water in Seven-rod Test Bundle.* Paper presented at the 13th International Topical Meeting on Nuclear Reactor Thermal Hydraulics (NURETH-13) Kanazawa City, Ishikawa Prefecture, Japan.

Modell, M. (1982). *Processing methods for the oxidation of organics in supercritical water.* US Patent 4,338,199.

Modell, M., & Amin, S. I. (1978). *Gasification process.* US Patent 4113446.

Moisseytsev, A., & Sienicki, J. (2009). Investigation of Alternative Layouts for the Supercritical Carbon Dioxide Brayton Cycle for a Sodium-Cooled Fast Reactor. *Proceedings of the Supercritical CO_2 Power Cycle Symposium.* doi:10.1016/j.nucengdes.2009.03.017

NIST (National Institute of Standards and Technology). (2013). *NIST Standard Reference Database 23.* NIST Reference Fluid Thermodynamic and Transport Properties - REFPROP. Ver. 9.1. Technology Administration, U.S. Department of Commerce: http://www.nist.gov/srd/nist23.cfm

Nuclear Power Institute of China. (2016). *Status report – Chinese Supercritical Water-Cooled Reactor (CSR1000).* Retrieved from https://aris.iaea.org

Ortega Gómez, T., Class, A., Lahey, R. T., Jr., & Schulenberg, T. (2007). *Stability analysis of heated flow channels with supercritical water.* 3rd Int. Symposium on SCWR-Design and Technology, Shanghai, China.

Schlagenhaufer, M., Starflinger, J., & Schulenberg, T. (2009). Plant control of the High Performance Light Water Reactor. *Proc. Global 2009.*

Schmidt, E., Eckert, E., & Grigull, V. (1946). *Heat transfer by liquids near the critical state.* AFF Translation, No. 527. Air Materials Command.

US Environmental Protection Agency. (2002). *A comprehensive analysis of biodiesel impacts on exhaust emissions.* Draft Technical Report, EPA420- P-02-001.

Zahlan, H., Groeneveld, D. C., Tavoularis, S., Mokry, S., & Pioro, I. (2011). Assessment of supercritical heat transfer prediction methods. *Proceedings of the 5th International Symposium on SCWR (ISSCWR-5).*

Zahlan, H., Groeneveld, D., & Tavoularis, S. (2010). Look-Up Table for Trans-Critical Heat Transfer. *Proceedings of the 2nd Canada-China Joint Workshop on Supercritical Water-Cooled Reactors (CCSC-2010).*

Zang, J. G., Yan, X., & Huang, S. F. (2012). *Analytical Prediction of Turbulent Friction Factor In Circular Pipe Under Supercritical Conditions.* Paper presented at the ICONE20POWER2012, Anaheim, CA. doi:10.1115/ICONE20-POWER2012-55159

Zuber, N. (1966, May). *An analysis of thermally induced flow oscillations in the near-critical and super-critical thermodynamic region.* Final Report Nas8-11422. NASA.

Zuber, N., (1966). *An analysis of thermally induced flow oscillations in the near-critical and super-critical thermodynamic region.* Final Report Nas8-11422. NASA.

Accary, G., Bontoux, P., & Zappoli, B. (2007). Convection in a supercritical fluid: A reduced model for geophysical flows. *Physics of Fluids, 19*(1), 014104. doi:10.1063/1.2432159

Accary, G., Raspo, I., Bontoux, P., & Zappoli, B. (2005). Reverse transition to hydrodynamic stability through the Schwarzschild line in a supercritical fluid layer. *Physical Review E: Statistical, Nonlinear, and Soft Matter Physics, 72*(3), 035301. doi:10.1103/PhysRevE.72.035301 PMID:16241504

Adschiri, T., Takami, S., Suzuki, A., Hojo, D., Aoki, N., Aida, T., & Gimyeong, S. (2015). Recent advances in science and technology of supercritical hydrothermal synthesis—towards super hybrid materials and green nanotechnology (Review). *Review of High Pressure Science and Technology/Koatsuryoku No Kagaku To Gijutsu, 25*, 225-233.

Ahmed, T. (2006). *Reservoir engineering handbook.* Gulf Professional Publishing.

Aionicesei, E., Škerget, M., & Knez, Ž. (2008). Measurement of CO_2 solubility and diffusivity in poly (l-lactide) and poly (d, l-lactide-co-glycolide) by magnetic suspension balance. *The Journal of Supercritical Fluids*, *47*(2), 296–301. doi:10.1016/j.supflu.2008.07.011

Akalın, M. K., Tekin, K., & Karagöz, S. (2012). Hydrothermal liquefaction of cornelian cherry stones for bio-oil production. *Bioresource Technology*, *110*, 682–687. doi:10.1016/j.biortech.2012.01.136 PMID:22342032

Akizuki, M., Fujii, T., Hayashi, R., & Oshima, Y. (2014). Effects of water on reactions for waste treatment, organic synthesis, and bio-refinery in sub- and supercritical water. *Journal of Bioscience and Bioengineering*, *117*(1), 10–18. doi:10.1016/j.jbiosc.2013.06.011 PMID:23867097

Aljbour, S. H., & Kawamoto, K. (2013). Bench-scale gasification of cedar wood – Part I: Effect of operational conditions on product gas characteristics. *Chemosphere*, *90*(4), 1495–1500. doi:10.1016/j.chemosphere.2012.08.029 PMID:22980961

Ambrosini, W. (2007). On the analogies in the dynamic behavior of heated channels with boiling and supercritical fluids. *Nuclear Engineering and Design*, *237*(11), 1164–1174. doi:10.1016/j.nucengdes.2007.01.006

Amiroudine, S., & Beysens, D. (2008). Thermovibrational instability in supercritical fluids under weightlessness. *Physical Review E: Statistical, Nonlinear, and Soft Matter Physics*, *78*(3), 036323. doi:10.1103/PhysRevE.78.036325 PMID:18851161

Amiroudine, S., Bontoux, P., Larroudé, P., Gilly, B., & Zappoli, B. (2001). Direct numerical simulation of unsteady instabilities inside a near-critical fluid layer heated from below. *Journal of Fluid Mechanics*, *442*, 119. doi:10.1017/S0022112001004967

Amiroudine, S., Boutrouft, K., & Zappoli, B. (2005). The stability analysis of two layers in a supercritical pure fluid: Application to Rayleigh-Taylor like instabilities. *Physics of Fluids*, *17*(5), 054102. doi:10.1063/1.1868033

Amiroudine, S., Caltagirone, J.-P., & Erriguible, A. (2014). A Lagrangian–Eulerian compressible model for the trans-critical path of near-critical fluids. *International Journal of Multiphase Flow*, *59*, 15–23. doi:10.1016/j.ijmultiphaseflow.2013.10.008

Amiroudine, S., Ouazzani, J., Carles, P., & Zappoli, B. (1997). Numerical solutions of 1-D unsteady near-critical fluid flows using finite volume methods. *European Journal of Mechanics - B/Fluids*, *16*(5), 665–680.

Anastasakis, K., & Ross, A. B. (2015). Hydrothermal liquefaction of four brown macro-algae commonly found on the UK coasts: An energetic analysis of the process and comparison with bio-chemical conversion methods. *Fuel*, *139*, 546–553. doi:10.1016/j.fuel.2014.09.006

Andersen, W. C., & Bruno, T. J. (2005). Rapid Screening of Fluids for Chemical Stability in Organic Rankine Cycle Applications. *Industrial & Engineering Chemistry Research, 44*(15), 5560–5566. https://doi.org/10.1021/ie050351s

Anderson, T. N., & Morrison, G. L. (2007). Effect of Load Pattern on Solar-Boosted Heat Pump Water Heater Performance. *Solar Energy*, *81*(11), 1386–1395. doi:10.1016/j.solener.2007.01.018

Anisimov, M. A. (2011). Letter to the editor: Fifty years of breakthrough discoveries in fluid criticality. International Journal of Thermophysics, 32, 2001–2009.

Anisimov, M. A. (1991). *Critical Phenomena in Liquids and Liquid Crystals*. Philadelphia, PA: Gordon & Breach Science Publishers.

Archana, V., Vaidya, A. M., & Vijayan, P. K. (2015). Numerical modeling of supercritical CO_2 natural circulation loop. *Nuclear Engineering and Design*, *293*, 330–345. doi:10.1016/j.nucengdes.2015.07.030

Aryasetiawan, F., & Karlsson, K. (1994). Energy Loss Spectra and Plasmon Dispersions in Alkali Metals: Negative Plasmon Dispersion in Cs. *Physical Review Letters*, *73*(12), 1679–1682. doi:10.1103/PhysRevLett.73.1679 PMID:10056856

Assael, M. J., Antoniadis, K. D., & Wakeham, W. A. (2010). Historical evolution of the transient hot-wire technique. *International Journal of Thermophysics*, *31*(6), 1051–1072. doi:10.1007/s10765-010-0814-9

Azadi, P., & Farnood, R. (2011). Review of heterogeneous catalysts for sub-and supercritical water gasification of biomass and wastes. *International Journal of Hydrogen Energy*, *36*(16), 9529–9954. doi:10.1016/j.ijhydene.2011.05.081

Bae, S. J., Lee, J., Ahn, Y., & Lee, J. I. (2015). Preliminary studies of compact Brayton cycle performance for Small Modular High Temperature Gas-cooled Reactor system. *Annals of Nuclear Energy*, *75*, 11–19. doi:10.1016/j.anucene.2014.07.041

Bailly, D., & Zappoli, B. (2000). Hydrodynamic theory of density relaxation in supercritical fluids. *Physical Review E: Statistical Physics, Plasmas, Fluids, and Related Interdisciplinary Topics*, *62*(2), 2353–2368. doi:10.1103/PhysRevE.62.2353 PMID:11088715

Bakhtar, F., Henson, R. J. K., & Mashmoushy, H. (2006). On the Performance of a Cascade of Turbine Rotor Tip Section Blading in Wet Steam. Part 5: Theoretical Treatment. *Proceedings of the Institution of Mechanical Engineers. Part C, Journal of Mechanical Engineering Science*, *220*(4), 457–472. doi:10.1243/09544062JMES151

Bakhtar, F., & Mahpeykar, M. R. (1997). On the performance of a cascade of turbine rotor tip section blading in nucleating steam Part 3: Theoretical treatment. *Proceedings of the Institution of Mechanical Engineers. Part C, Journal of Mechanical Engineering Science*, *211*(3), 195–210. doi:10.1243/0954406971521773

Bakhtar, F., Mashmoushy, H., & Buckley, J. R. (1997). On the performance of a cascade of turbine rotor tip section blading in wet steam Part 1: Generation of wet steam of prescribed droplet sizes. *Proceedings of the Institution of Mechanical Engineers. Part C, Journal of Mechanical Engineering Science*, *211*(7), 519–529. doi:10.1243/0954406971521908

Bakhtar, F., Mashmoushy, H., & Jadayel, O. C. (1997). On the performance of a cascade of turbine rotor tip section blading in wet steam Part 2: Surface pressure distributions. *Proceedings of the Institution of Mechanical Engineers. Part C, Journal of Mechanical Engineering Science*, *211*(7), 531–540. doi:10.1243/0954406971521917

Bakhtar, F., Rassam, S. Y., & Zhang, G. (1999). On the performance of a cascade of turbin rotor tip section blading in wet steam Part 4: Droplet measurements. *Proceedings of the Institution of Mechanical Engineers. Part C, Journal of Mechanical Engineering Science*, *213*(4), 343–353. doi:10.1243/0954406991522301

Barnes, D. I. (2015). Understanding pulverised coal, biomass and waste combustion–a brief overview. *Applied Thermal Engineering*, *74*, 89–95. doi:10.1016/j.applthermaleng.2014.01.057

Baron, A. Q. R., Tanaka, Y., Goto, S., Takeshita, K., Matsushita, T., & Ishikawa, T. (2000). An X-ray scattering beamline for studying dynamics. *Journal of Physics and Chemistry of Solids*, *61*(3), 461–465. doi:10.1016/S0022-3697(99)00337-6

Basu, D. N., Bhattacharyys, S., & Das, P. K. (2008). Effect of geometric parameters on steady-state performance of single-phase NCL with heat loss to ambient. *International Journal of Thermal Sciences*, *47*(10), 1359–1373. doi:10.1016/j.ijthermalsci.2007.10.010

Behringer, R. P., Onuki, A., & Meyer, H. (1990). Thermal equilibration of fluids near the liquidvapor critical point: ^{3}He and ^{3}He-^{4}He mixtures. *Journal of Low Temperature Physics*, *81*(1-2), 71–102. doi:10.1007/BF00683153

Bellmore, C. P., & Reid, R. L. (1983). Numerical Prediction of wall temperature for near critical para-hydrogen in turbulent upflow inside vertical tubes. *Journal of Heat Transfer*, *105*(3), 22–29. doi:10.1115/1.3245618

Benavente, V., Calabuig, E., & Fullana, A. (2015). Upgrading of moist agro-industrial wastes by hydrothermal carbonization. *Journal of Analytical and Applied Pyrolysis*, *113*, 89–98. doi:10.1016/j.jaap.2014.11.004

Berche, B., Henkel, M., & Kenna, R. (2009). Critical phenomena: 150 years since Cagniard de la Tour. Journal of Physical Studies, 13(3).

Besarati, S. M., & Goswami, D. Y. (2014). Analysis of Advanced Supercritical Carbon Dioxide Power Cycles With a Bottoming Cycle for Concentrating Solar Power Applications. *Journal of Solar Energy Engineering*, *136*(1), 010904–010904.

Beysens, D., Chatain, D., Nikolayev, V. S., Ouazzani, J., & Garrabos, Y. (2010). Possibility of long-distance heat transport in weightlessness using supercritical fluids. *Phys. Rev. E, 82*.

Beysens, D., Fröhlich, T., & Garrabos, Y. (2011). Heat can cool near-critical fluids. *Physical Review E: Statistical, Nonlinear, and Soft Matter Physics*, 84, 051201 (8p).

Beysens, D., Garrabos, Y., Chatain, D., & Evesque, P. (2009). Phase transition under forced vibrations in critical CO_2. *Europhysics Letters*, *86*(1), 16003. doi:10.1209/0295-5075/86/16003

Beysens, D., Garrabos, Y., & Fröhlich, T. (2008). Heat can cool near critical fluids. *Proceedings of 59th IAC Congress.*

Beysens, D., Garrabos, Y., Wunenburger, R., & Lecoutre-Chabot, C. (2002). Can heat flow backward? Unusual thermal phenomena observed in near-critical fluids. *Physica A*, *314*, 427.

Biggs, F., Mendelsohn, L. B., & Mann, J. B. (1975). Hartree-Fock Compton profiles for the elements. *Atomic Data and Nuclear Data Tables*, *16*(3), 201–309. doi:10.1016/0092-640X(75)90030-3

Biller, P., & Ross, A. B. (2011). Potential yields and properties of oil from the hydrothermal liquefaction of microalgae with different biochemical content. *Bioresource Technology*, *102*(1), 215–225. doi:10.1016/j.biortech.2010.06.028 PMID:20599375

Biller, P., Ross, A. B., Skill, S. C., Lea-Langton, A., Balasundaram, B., Hall, C., & Llewellyn, C. A. et al. (2012). Nutrient recycling of aqueous phase for microalgae cultivation from the hydrothermal liquefaction process. *Algal Research*, *1*(1), 70–76. doi:10.1016/j.algal.2012.02.002

Bioenergy, I. (2015). *Thermal Gasification of Biomass*. Task 33 Database. Retrieved May 05, 2016 from http://www.ieabioenergy.com/task/thermal-gasification-of-biomass/

Boukari, H., Shaumeyer, J. N., Briggs, M. E., & Gammon, R. W. (1990). Critical speeding up in pure fluids. *Physical Review A.*, *41*(4), 2260–2263. doi:10.1103/PhysRevA.41.2260 PMID:9903350

Boutrouft, K., Amiroudine, S., & Ambari, A. (2006). Stability diagram and effect of initial density stratification for a two-layer system in a supercritical fluid. *Physics of Fluids*, *18*(12), 124106. doi:10.1063/1.2424497

Brazhkin, V. V., Fomin, Yu. D., Lyapin, A. G., Ryzhov, V. N., & Trachenko, K. (2012). Two liquid states of matter: A dynamic line on a phase diagram. *Physical Review E: Statistical, Nonlinear, and Soft Matter Physics*, *85*(3), 031203. doi:10.1103/PhysRevE.85.031203 PMID:22587085

Brazhkin, V. V., & Trachenko, K. (2012). What separates a liquid from a gas. *Physics Today*, *65*(11), 68–69. doi:10.1063/PT.3.1796

Brennecke, J. F., & Eckert, C. A. (1989). Phase equilibria for supercritical fluid process design. *AIChE Journal. American Institute of Chemical Engineers*, *35*(9), 1409–1427. doi:10.1002/aic.690350902

Brondel, D., Edwards, R., Hayman, A., Hill, D., Mehta, S., & Semerad, T. (1994). Corrosion in the oil industry. *Oilfield Review*, *6*(2), 4-18.

Brown, A. (1980). Engineering thermodynamics work and heat transfer. *Rogers, GFC and Mayhew, YR, Longman, 1980*, 667.

Brunner, G. (2005). Supercritical fluids: Technology and application to food processing. *Journal of Food Engineering*, *67*(1), 21–33. doi:10.1016/j.jfoodeng.2004.05.060

Brunner, G. (2015). Supercritical process technology related to energy and future directions–An introduction. *The Journal of Supercritical Fluids*, *96*, 11–20. doi:10.1016/j.supflu.2014.09.008

Budihardjo, I., & Morrison, G. L. (2009). Performance of Water-in-Glass Evacuated Tube Solar Water Heaters. *Solar Energy*, *83*(1), 49–56. doi:10.1016/j.solener.2008.06.010

Byrappa, K. (1991). Hydrothermal growth of crystals. *Progress in Crystal Growth and Characterization of Materials*, *21*, R7–R10.

Byrappa, K., & Yoshimura, M. (2013). *Handbook of hydrothermal technology* (2nd ed.). Oxford, UK: William Andrew.

Cabeza, P., Queiroz, J. P. S., Criado, M., Jiménez, C., Bermejo, M. D., Mato, F., & Cocero, M. J. (2015). Supercritical water oxidation for energy production by hydrothermal flame as internal heat source. Experimental results and energetic study. *Energy*, *90*, 1584–1594. doi:10.1016/j.energy.2015.06.118

Calderazzi, L., & di Paliano, P. C. (1997). Thermal stability of R-134a, R-141b, R-13I1, R-7146, R-125 associated with stainless steel as a containing material. *International Journal of Refrigeration*, *20*(6), 381–389. doi:10.1016/S0140-7007(97)00043-1

Calvignac, B., Rodier, E., Letourneau, J.-J., Almeida dos Santos, P. M., & Fages, J. (2010). Cocoa butter saturated with supercritical carbon dioxide: Measurements and modelling of solubility, volumetric expansion, density and viscosity. *International Journal of Chemical Reactor Engineering*, *8*(1). doi:10.2202/1542-6580.2191

Calvignac, B., Rodier, E., Letourneau, J.-J., Vitoux, P., Aymonier, C., & Fages, J. (2010). Development of an improved falling ball viscometer for high-pressure measurements with supercritical CO_2. *The Journal of Supercritical Fluids*, *55*(1), 96–106. doi:10.1016/j.supflu.2010.07.012

Calvo, L., Gil, M., Otero, M., Morán, A., & García, A. (2012). Gasification of rice straw in a fluidized-bed gasifier for syngas application in close-coupled boiler-gasifier systems. *Bioresource Technology*, *109*, 206–214. doi:10.1016/j.biortech.2012.01.027 PMID:22297044

Cammarata, L., Fichera, A., & Pagano, A. (2003). Stability maps for rectangular circulation loops. *Applied Thermal Engineering*, *23*(8), 965–977. doi:10.1016/S1359-4311(03)00027-9

Cantero, D. A., Álvarez, A., Bermejo, M. D., & Cocero, M. J. (2015). Transformation of glucose into added value compounds in a hydrothermal reaction media. *The Journal of Supercritical Fluids*, *98*, 204–210. doi:10.1016/j.supflu.2014.12.015

Cardenas-Toro, F. P., Forster-Carneiro, T., Rostagno, M. A., Petenate, A. J., Maugeri Filho, F., & Meireles, M. A. A. (2014). Integrated supercritical fluid extraction and subcritical water hydrolysis for the recovery of bioactive compounds from pressed palm fiber. *The Journal of Supercritical Fluids*, *93*, 42–48. doi:10.1016/j.supflu.2014.02.009

Carstens, N. A., Vilim, R. B., Hejzlar, P., & Driscoll, M. J. (2008). Dynamic modeling of the S-CO2 recompression cycle. In *International Conference on Advances in Nuclear Power Plants, ICAPP 2008* (Vol. 2, pp. 784–798).

Cartmell, E., Gostelow, P., Riddell-Black, D., Simms, N., Oakey, J., Morris, J., & Pollard, S. J. et al. (2006). Biosolids a fuel or a waste? An integrated appraisal of five co-combustion scenarios with policy analysis. *Environmental Science & Technology*, *40*(3), 649–658. doi:10.1021/es052181g PMID:16509299

Cassetti, G., & Colombo, E. (2013). Minimization of local impact of energy systems through exergy analysis. *Energy Conversion and Management*, *76*, 874–882. doi:10.1016/j.enconman.2013.08.044

Castello, D., & Fiori, L. (2015). Supercritical water gasification of biomass: A stoichiometric thermodynamic model. *International Journal of Hydrogen Energy*, *40*(21), 6771–6781. doi:10.1016/j.ijhydene.2015.03.120

Cayer, E., Galanis, N., & Nesreddine, H. (2010). Parametric study and optimization of a transcritical power cycle using a low temperature source. *Applied Energy*, *87*(4), 1349–1357. doi:10.1016/j.apenergy.2009.08.031

Cengiz, N. Ü., Yıldız, G., Sert, M., Selvi Gökkaya, D., Sağlam, M., Yüksel, M., & Ballice, L. (2015). Hydrothermal gasification of a biodiesel by-product crude glycerol in the presence of phosphate based catalysts. *International Journal of Hydrogen Energy*, *40*(43), 14806–14815. doi:10.1016/j.ijhydene.2015.08.097

Chacartegui, R., Sánchez, D., Muñoz, J. M., & Sánchez, T. (2009). Alternative ORC bottoming cycles FOR combined cycle power plants. *Applied Energy*, *86*(10), 2162–2170. doi:10.1016/j.apenergy.2009.02.016

Chaikina, Yu. A. (2011). Distinctive features of random local motions in critical and supercritical fluids. *Russian Journal of Physical Chemistry B*, *5*(7), 1116–1129. doi:10.1134/S1990793111070128

Chaikina, Yu. A. (2012). Development of finite fluctuations in local density, velocity, and temperature for real fluids. *Russian Journal of Physical Chemistry B*, *6*(7), 834–843. doi:10.1134/S1990793112070020

Chandrasekar, S. (1961). *Hydrodynamic and Hydromagnetic Stability*. Oxford, UK: Clarendon Press.

Chatoorgoon, V. (2001). Stability of supercritical fluid flow in a single-channel natural-convection loop. *International Journal of Heat and Mass Transfer*, *44*(10), 1963–1972. doi:10.1016/S0017-9310(00)00218-0

Chatoorgoon, V., Voodi, A., & Upadhye, P. (2005). The stability boundary for supercritical flow in natural-convection loops Part II: CO_2 and H_2. *Nuclear Engineering and Design*, *235*(24), 2581–2593. doi:10.1016/j.nucengdes.2005.06.004

Chen, L., Zhang, X. R., & Jiang, B. (2014). Effects of heater orientations on the natural circulation and heat transfer in a supercritical CO2 rectangular loop. Journal of Heat Transfer-Transactions of ASME, 132.

Chen, Y. (2006). *Novel cycles using carbon dioxide as working fluid: new ways to utilize energy from low-grade heat sources*. KTH, Stockholm, Sweden. Retrieved from https://www.diva-portal.org/smash/get/diva2:10578/FULLTEXT01.pdf

Chen, G., Dresselhaus, M. S., Fleurial, J.-P., & Caillat, T. (2003). Recent Developments in Thermoelectric Materials. *International Materials Reviews*, *48*(1), 45–66. doi:10.1179/095066003225010182

Chen, G.-Y., Li, W.-Q., Chen, H., & Yan, B.-B. (2015). Progress in the aqueous-phase reforming of different biomass-derived alcohols for hydrogen production. *Journal of Zhejiang University*, *16*(6), 491–506. doi:10.1631/jzus.A1500023 PMID:26055913

Cheng, L., Ribatski, G., & Thome, J. R. (2008). Analysis of supercritical CO_2 cooling in macro- and micro-channels. *International Journal of Refrigeration*, *31*(8), 1301–1316. doi:10.1016/j.ijrefrig.2008.01.010

Chen, H., Goswami, D. Y., Rahman, M. M., & Stefanakos, E. K. (2011). A supercritical Rankine cycle using zeotropic mixture working fluids for the conversion of low-grade heat into power. *Energy*, *36*(1), 549–555. doi:10.1016/j.energy.2010.10.006

Chen, H., Goswami, D. Y., & Stefanakos, E. K. (2010). A review of thermodynamic cycles and working fluids for the conversion of low-grade heat. *Renewable & Sustainable Energy Reviews*, *14*(9), 3059–3067. doi:10.1016/j.rser.2010.07.006

Chen, H., Yogi Goswami, D., Rahman, M. M., & Stefanakos, E. K. (2011). Energetic and exergetic analysis of CO_2- and R32-based transcritical Rankine cycles for low-grade heat conversion. *Applied Energy*, *88*(8), 2802–2808. doi:10.1016/j.apenergy.2011.01.029

Chen, L., Deng, B. L., Jiang, B., & Zhang, X. R. (2013d). Thermal and hydrodynamic characteristics of supercritical CO_2 natural circulation in closed loops. *Nuclear Engineering and Design*, *257*, 21–30. doi:10.1016/j.nucengdes.2013.01.016

Chen, L., & Zhang, X. R. (2011). Simulation of heat transfer and system behavior in a supercritical CO_2 based thermosyphon: Effect of pipe diameter. *Journal of Heat Transfer-Transactions of ASME*, *133*(12), 2505–2513. doi:10.1115/1.4004434

Chen, L., Zhang, X. R., Deng, B. L., & Jiang, B. (2013g). Effects of inclination angle and operation parameters on supercritical CO_2 natural circulation loop. *Nuclear Engineering and Design*, *265*, 895–908. doi:10.1016/j.nucengdes.2013.06.037

Chen, L., Zhang, X. R., Okajima, J., & Maruyama, S. (2013b). Abnormal microchannel convective fluid flow near the gas-liquid critical point. *Physica A*, *398*, 10–24. doi:10.1016/j.physa.2013.11.002

Chen, M. F., Yamaguchi, H., Zhang, X. W., & Niu, X. D. (2015). Performance analyses of a particularly designed turbine for a supercritical CO2-based solar Rankine cycle system. *International Journal of Energy Research*, *39*(13), 1819–1827. doi:10.1002/er.3299

Chen, W.-T., Zhang, Y., Zhang, J., Yu, G., Schideman, L. C., Zhang, P., & Minarick, M. (2014). Hydrothermal liquefaction of mixed-culture algal biomass from wastewater treatment system into bio-crude oil. *Bioresource Technology*, *152*, 130–139. doi:10.1016/j.biortech.2013.10.111 PMID:24287452

Chen, Y., Lundqvist, P., Johansson, A., & Platell, P. (2006). A comparative study of the carbon dioxide transcritical power cycle compared with an organic rankine cycle with R123 as working fluid in waste heat recovery. *Applied Thermal Engineering*, *26*(17–18), 2142–2147. doi:10.1016/j.applthermaleng.2006.04.009

Chen, Y., Lundqvist, P., & Platell, P. (2005). Theoretical research of carbon dioxide power cycle application in automobile industry to reduce vehicles fuel consumption. *Applied Thermal Engineering*, *25*(14–15), 2041–2053. doi:10.1016/j.applthermaleng.2005.02.001

Chiwata, Y., & Onuki, A. (2001). Thermal Plumes and Convection in Highly Compressible Fluids. *Physical Review Letters*, *87*(14), 144301. doi:10.1103/PhysRevLett.87.144301 PMID:11580652

Clementoni, E. M., Cox, T. L., & Sprague, C. P. (2014). Startup and Operation of a Supercritical Carbon Dioxide Brayton Cycle. *Journal of Engineering for Gas Turbines and Power*, *136*(7), 71701. doi:10.1115/1.4026539

Cohen, M. H., & Jortner, J. (1973). Inhomogeneous Transport Regime in Disordered Materials. *Physical Review Letters*, *30*(15), 699–702. doi:10.1103/PhysRevLett.30.699

Conboy, T., Pasch, J., & Fleming, D. (2013). Control of a Supercritical CO 2 Recompression Brayton Cycle Demonstration Loop. *Journal of Engineering for Gas Turbines and Power*, *135*(11), 111701. doi:10.1115/1.4025127

Conboy, T., Wright, S., Pasch, J., Fleming, D., Rochau, G., & Fuller, R. (2012). Performance characteristics of an operating supercritical CO_2 Brayton cycle. *Journal of Engineering for Gas Turbines and Power*, *134*(11), 111703. doi:10.1115/1.4007199

Corma, A., García, H., & Leyva, A. (2005). Comparison between polyethylenglycol and imidazolium ionic liquids as solvents for developing a homogeneous and reusable palladium catalytic system for the Suzuki and Sonogashira coupling. *Tetrahedron*, *61*(41), 9848–9854. doi:10.1016/j.tet.2005.06.119

Crook, A. W., & Engineers, I. of E. (1994). Profiting from Low-grade Heat: Thermodynamic Cycles for Low-temperature Heat Sources. IET.

Cusack, N., & Enderby, J. E. (1960). A Note on the Resistivity of Liquid Alkali and Noble Metals. *Proceedings of the Physical Society*, *75*(3), 395–401. doi:10.1088/0370-1328/75/3/310

Dahl, S. D., & Davidson, J. H. (1997). Performance and Modeling of Thermosyphon Heat Exchangers for Solar Water Heaters, *ASME. Journal of Solar Energy Engineering*, *119*(3), 193–200. doi:10.1115/1.2888018

Dai, Y., Wang, J., & Gao, L. (2009). Parametric optimization and comparative study of Organic Rankine Cycle (ORC) for low grade waste heat recovery. *ResearchGate*, *50*(3), 576–582. https://doi.org/10.1016/j.enconman.2008.10.018

Dai, B., Li, M., Dang, C., & Ma, Y. (2015a). Effects of lubricating oil on thermal performance of water-cooled carbon dioxide gas cooler. *Applied Thermal Engineering*, *80*, 288–300. doi:10.1016/j.applthermaleng.2015.01.071

Dai, B., Li, M., Dang, C., Tian, H., & Ma, Y. (2015b). Thermodynamic performance assessment of carbon dioxide blends with low GWP working fluids used for heat pump water heater. *International Journal of Refrigeration*, *56*, 1–14. doi:10.1016/j.ijrefrig.2014.11.009

Dang, C., & Hihara, E. (2004 a). In-tube cooling heat transfer of supercritical carbon dioxide, part 1: Experimental measurement. *International Journal of Refrigeration*, *27*(7), 736–747. doi:10.1016/j.ijrefrig.2004.04.018

Dang, C., & Hihara, E. (2004 b). In-tube cooling heat transfer of supercritical carbon dioxide, part 2: Comparison of turbulence models in numerical calculation. *International Journal of Refrigeration*, *27*, 748–760. doi:10.1016/j.ijrefrig.2004.04.017

Dang, C., & Hihara, E. (2010). Numerical study on in-tube laminar heat transfer of supercritical fluids. *Applied Thermal Engineering*, *13*(13), 1567–1573. doi:10.1016/j.applthermaleng.2010.03.010

Dang, C., & Hihara, E. (2012). Predicting the cooling heat transfer coefficient of supercritical CO_2 with a small amount of entrained lubricating oil by using the neural network method. *International Journal of Refrigeration*, *35*(4), 1130–1138. doi:10.1016/j.ijrefrig.2012.01.022

Dang, C., Hoshika, K., & Hihara, E. (2012). Effect of lubricating oil on the flow and heat transfer characteristics of supercritical carbon dioxide. *International Journal of Refrigeration*, *35*(5), 1410–1417. doi:10.1016/j.ijrefrig.2012.03.015

Dang, C., Iino, K., Fukuoka, K., & Hihara, E. (2007). Effect of lubricating oil on cooling heat transfer of supercritical carbon dioxide. *International Journal of Refrigeration*, *30*(4), 724–731. doi:10.1016/j.ijrefrig.2006.09.006

Dang, C., Iino, K., & Hihara, E. (2008). Study on two-phase flow pattern of supercritical carbon dioxide with entrained PAG-type lubricating oil in a gas cooler. *International Journal of Refrigeration*, *31*(7), 1265–1272. doi:10.1016/j.ijrefrig.2008.01.014

Dang, C., Iino, K., & Hihara, E. (2010). Effect of PAG-type lubricating oil on heat transfer characteristics of supercritical carbon dioxide cooled inside a small internally grooved tube. *International Journal of Refrigeration*, *33*(3), 558–565. doi:10.1016/j.ijrefrig.2009.12.010

Davidson, J. H., Mantell, S. C., & Jorgensen, G. (2002). *Status of the Development of Polymeric Solar Water Heating Systems.* In D. Y. Goswami (Ed.), *Advances in Solar Energy* (Vol. 15, pp. 149–186). Boulder, CO: American Solar Energy Society.

de Bruijn, R., van Diest, R. J. J., Karapantsios, T. D., Michels, A. C., Wakeham, W. A., & Trusler, J. P. M. (1997). Heat transfer in pure critical fluids surrounded by finitely conducting boundaries in microgravity. *Physica A*, *242*(1-2), 119–140. doi:10.1016/S0378-4371(97)00197-0

De Swaan Arons, J., & Diepen, G. (1963). Thermodynamic study of melting equilibria under pressure of a supercritical gas. *Recueil des Travaux Chimiques des Pays-Bas*, *82*(3), 249–256. doi:10.1002/recl.19630820308

Debrah, S. K., Ambrosini, W., & Chen, Y. (2013). Discussion on the stability of natural circulation loops with supercritical pressure fluids. *Annals of Nuclear Energy*, *54*, 47–57. doi:10.1016/j.anucene.2012.10.015

Delgado-Torres, A. M., & García-Rodríguez, L. (2007). Preliminary assessment of solar organic Rankine cycles for driving a desalination system. *Desalination*, *216*(1), 252–275. doi:10.1016/j.desal.2006.12.011

Demuth, O. J. (1981). Analyses of Mixed Hydrocarbons Binary Thermodynamic Cycles for Moderate Temperature Geothermal Resources. *ResearchGate, 2*. Retrieved from https://www.researchgate.net/publication/253167221_Analyses_of_Mixed_Hydrocarbons_Binary_Thermodynamic_Cycles_for_Moderate_Temperature_Geothermal_Resources

Demuth, O. J. (1983). *Preliminary assessment of condensation behavior for hydrocarbon-vapor expansions which cross the saturation line near the critical point* (No. EGG-M-00283; CONF-830812-54). EG and G Idaho, Inc. Retrieved from http://www.osti.gov/geothermal/biblio/5673705-preliminary-assessment-condensation-behavior-hydrocarbon-vapor-expansions-which-cross-saturation-line-near-critical-point

Deniz, I., Vardar-Sukan, F., Yüksel, M., Saglam, M., Ballice, L., & Yesil-Celiktas, O. (2015). Hydrogen production from marine biomass by hydrothermal gasification. *Energy Conversion and Management*, *96*, 124–130. doi:10.1016/j.enconman.2015.02.048

Devi, L., Ptasinski, K. J., & Janssen, F. J. J. G. (2003). A review of the primary measures for tar elimination in biomass gasification processes. *Biomass and Bioenergy*, *24*(2), 125–140. doi:10.1016/S0961-9534(02)00102-2

Diebold, J. P. (2000). *A review of the chemical and physical mechanisms of the storage stability of fast pyrolysis bio-oils*. National Renewable Energy Laboratory.

Dindar, C., & Kiran, E. (2002). High-pressure viscosity and density of polymer solutions at the critical polymer concentration in near-critical and supercritical fluids. *Industrial & Engineering Chemistry Research*, *41*(25), 6354–6362. doi:10.1021/ie0108999

Ding, N., Azargohar, R., Dalai, A. K., & Kozinski, J. A. (2014). Catalytic gasification of glucose to H_2 in supercritical water. *Fuel Processing Technology*, *127*, 33–40. doi:10.1016/j.fuproc.2014.05.014

DiPippo, R. (2004). Second Law assessment of binary plants generating power from low-temperature geothermal fluids. *Geothermics*, *33*(5), 565–586. doi:10.1016/j.geothermics.2003.10.003

Dobashi, K., Kimura, A., Oka, Y., & Koshizuka, S. (1998). Conceptual design of a high temperature power reactor cooled and moderated by supercritical light water. *Annals of Nuclear Energy*, *25*(8), 487–505. doi:10.1016/S0306-4549(97)00079-0

Dong, L., Asadullah, M., Zhang, S., Wang, X.-S., Wu, H., & Li, C. Z. (2013). An advanced biomass gasification technology with integrated catalytic hot gas cleaning: Part I. Technology and initial experimental results in a lab-scale facility. *Fuel*, *108*, 409–416. doi:10.1016/j.fuel.2012.11.043

Dostál, V. (2004). *A supercritical carbon dioxide cycle for next generation nuclear reactors*. Massachusetts Institute of Technology.

Dostal, V., Driscoll, M. J., Hejzlar, P., & Todreas, N. E. (2002, January). A supercritical CO2 gas turbine power cycle for next-generation nuclear reactors. In *10th International Conference on Nuclear Engineering* (pp. 567-574). American Society of Mechanical Engineers. doi:10.1115/ICONE10-22192

Dostal, V., Hejzlar, P., & Driscoll, M. J. (2006). The supercritical carbon dioxide power cycle: Comparison to other advanced power cycles. *Nuclear Technology*, *154*, 283–301.

Dubey, S., & Tiwari, G. N. (2008). Thermal Modeling of a Combined System of Photovoltaic Thermal PV/T Solar Water Heater. *Solar Energy*, *82*(7), 602–612. doi:10.1016/j.solener.2008.02.005

Duffey, R. B., & Pioro, I. L. (2005). Experimental heat transfer for supercritical carbon dioxide flowing inside channels (survey). *Nuclear Engineering and Design*, *235*(8), 913–924. doi:10.1016/j.nucengdes.2004.11.011

Dumas, A., Claverie, M., Slostowski, C., Aubert, G., Carême, C., Le Roux, C., & Aymonier, C. et al. (2016). *Fast-geomimicking using chemistry in supercritical water* (International Edition). Angewandte Chemie; doi:10.1002/anie.201604096

Dunham, M. T., & Iverson, B. D. (2014). High-efficiency thermodynamic power cycles for concentrated solar power systems. *Renewable & Sustainable Energy Reviews*, *30*(0), 758–770. doi:10.1016/j.rser.2013.11.010

Dyreby, J. J., Klein, S. A., Nellis, G. F., & Reindl, D. T. (2013). Modeling Off-Design and Part-Load Performance of Supercritical Carbon Dioxide Power Cycles. In *ASME Turbo Expo 2013: Turbine Technical Conference and Exposition. Volume 8: Supercritical CO_2 Power Cycles; Wind Energy; Honors and Awards* (Vol. 8, p. V008T34A014). ASME. https://doi.org/10.1115/GT2013-95824

El-Hanany, U., & Warren, W. W. Jr. (1975). Knight Shift in Expanded Liquid Mercury. *Physical Review Letters*, *34*(20), 1276–1279. doi:10.1103/PhysRevLett.34.1276

Elliott, D. C., Hart, T. R., Schmidt, A. J., Neuenschwander, G. G., Rotness, L. J., Olarte, M. V., & Holladay, J. E. et al. (2013). Process development for hydrothermal liquefaction of algae feedstocks in a continuous-flow reactor. *Algal Research*, *2*(4), 445–454. doi:10.1016/j.algal.2013.08.005

Elvassore, N., Vezzù, K., & Bertucco, A. (2005). Measurement and modeling of CO_2 absorption in poly (lactic-co-glycolic acid*). The Journal of Supercritical Fluids*, *33*(1), 1–5. doi:10.1016/j.supflu.2004.04.002

Esen, M., & Esen, H. (2005). Experimental Investigation of a Two-Phase Closed Thermosyphon Solar Water Heater. *Solar Energy*, *79*(5), 459–468. doi:10.1016/j.solener.2005.01.001

Etemoglu, A. B. (2008). Thermodynamic Evaluation of Geothermal Power Generation Systems in Turkey. *Energy Sources, Part A: Recovery, Utilization, and Environmental Effects*, *30*(10), 905–916. https://doi.org/10.1080/15567030601082589

Faber, T. E. (1972). *Introduction to the theory of liquid metals*. Cambridge, UK: Cambridge University Press.

Faeth, J. L., Valdez, P. J., & Savage, P. E. (2013). Fast hydrothermal liquefaction of Nannochloropsis sp. to produce biocrude. *Energy & Fuels*, *27*(3), 1391–1398. doi:10.1021/ef301925d

Feher, E. G. (1968). The supercritical thermodynamic power cycle. *Energy Conversion*, *8*(2), 85–90. doi:10.1016/0013-7480(68)90105-8

Feng, S., Cheng, S., Yuan, Z., Leitch, M., & Xu, C. C. (2013). Valorization of bark for chemicals and materials: A review. *Renewable & Sustainable Energy Reviews*, *26*, 560–578. doi:10.1016/j.rser.2013.06.024

Fernandez, N., Hwang, Y., & Radermacher, R. (2010). Comparison of CO_2 heat pump water heater performance with baseline cycle and two high COP cycles. *International Journal of Refrigeration*, *33*(3), 635–644. doi:10.1016/j.ijrefrig.2009.12.008

Finšgar, M., & Jackson, J. (2014). Application of corrosion inhibitors for steels in acidic media for the oil and gas industry: A review. *Corrosion Science*, *86*, 17–41. doi:10.1016/j.corsci.2014.04.044

Finšgar, M., & Merl, D. K. (2014). 2-Mercaptobenzoxazole as a copper corrosion inhibitor in chloride solution: Electrochemistry, 3D-profilometry, and XPS surface analysis. *Corrosion Science*, *80*, 82–95. doi:10.1016/j.corsci.2013.11.022

Fiori, L., Bassoa, D., Castelloa, D., & Baratierib, M. (2014). Hydrothermal carbonization of biomass: Design of a batch reactor and preliminary experimental results. *Chemical Engineering (Albany, N.Y.)*, 37.

Fonts, I., Gea, G., Azuara, M., Ábrego, J., & Arauzo, J. (2012). Sewage sludge pyrolysis for liquid production: A review. *Renewable & Sustainable Energy Reviews*, *16*(5), 2781–2805. doi:10.1016/j.rser.2012.02.070

Franz, G., Freyland, W., Gläser, W., Hensel, F., & Schneider, E. (1980). Structure of expanded liquid rubidium by neutron diffraction. *Journal de Physique. Colloque*, *C8*, 194–198.

Franz, G., Freyland, W., & Hensel, F. (1980). Thermodynamic and electric transport properties of fluid cesium and rubidium in the M-NM transition region. *Journal de Physique. Colloque*, *C8*, 70–73.

Freyland, W. (1979). Magnetic susceptibility of metallic and nonmetallic expanded fluid cesium. *Physical Review B: Condensed Matter and Materials Physics*, *20*(12), 5104–5110. doi:10.1103/PhysRevB.20.5104

Freyland, W. (1980). Electronic correlation and the M-NM transition in fluid alkali metals. *Journal de Physique. Colloque*, *C8*, 74–76.

Freyland, W. (1981). Metal-nonmetal transition in expanded fluid alkali metals. *Comments Solid State Physics*, *10*, 1–10.

Gandikota, G., Amiroudine, S., Chatain, D., Lyubimova, T., & Beysens, D. (2013). Rayleigh and parametric thermo-vibrational instabilities in supercritical fluids under weightlessness. *Physics of Fluids*, *25*(6), 064103. doi:10.1063/1.4811400

Gandikota, G., Chatain, D., Amiroudine, S., Lyubimova, T., & Beysens, D. (2014a). Frozen-wave instability in near-critical hydrogen subjected to horizontal vibration under various gravity fields. *Physical Review E: Statistical, Nonlinear, and Soft Matter Physics*, *89*(1), 012309. doi:10.1103/PhysRevE.89.012309 PMID:24580229

Gandikota, G., Chatain, D., Amiroudine, S., Lyubimova, T., & Beysens, D. (2014b). Faraday instability in a near-critical fluid under weightlessness. *Physical Review E: Statistical, Nonlinear, and Soft Matter Physics*, *89*(1), 013022. doi:10.1103/PhysRevE.89.013022 PMID:24580335

Gao, L., & Honda, T. (2002). Experiments on heat transfer characteristics of heat exchanger for CO2 heat pump system. In *Proc. of the Asian Conference on Refrigeration and Air Conditioning*. Kobe, Japan: Japan Society of Refrigeration and Air Conditioning Engineers.

Gao, Y., Yu, B., Wu, K., Yuan, Q., Wang, X., & Chen, H. (2016). Physicochemical, Pyrolytic, and Combustion Characteristics of Hydrochar Obtained by Hydrothermal Carbonization of Biomass. *BioResources*, *11*(2), 4113–4133. doi:10.15376/biores.11.2.4113-4133

Garg, P., Kumar, P., Dutta, P., Conboy, T., & Ho, C. (2014). Design of an Experimental Test Facility for Supercritical CO2 Brayton Cycle. In *ASME 2014 8th International Conference on Energy Sustainability collocated with the ASME 2014 12th International Conference on Fuel Cell Science, Engineering and Technology* (p. V001T05A005). ASME.

Garg, P., Kumar, P., & Srinivasan, K. (2013). Supercritical carbon dioxide Brayton cycle for concentrated solar power. *The Journal of Supercritical Fluids*, *76*, 54–60. doi:10.1016/j.supflu.2013.01.010

Garrabos, Y., Bonetti, M., Beysens, D., Perrot, F., Fröhlich, T., Carlès, P., & Zappoli, B. (1998). Relaxation of a supercritical fluid after a heat pulse in the absence of gravity effects: Theory and experiments. *Physical Review E: Statistical Physics, Plasmas, Fluids, and Related Interdisciplinary Topics*, *57*(5), 5665–5681. doi:10.1103/PhysRevE.57.5665

Ginosar, D. M., Petkovic, L. M., & Guillen, D. P. (2011). Thermal Stability of Cyclopentane as an Organic Rankine Cycle Working Fluid. *Energy & Fuels*, *25*(9), 4138–4144. doi:10.1021/ef200639r

Gitterman, M., & Steinberg, V. A. (1970). Criteria for the commencement of convection in a liquid close to its critical point. *High Temperature (USSR)*, *8*(4), 754.

Gitterman, M., & Steinberg, V. A. (1970). Criteria for the commencement of convection in a liquid close to the critical point. *High Temperature, USSR*, *8*(4), 754.

Giuliani, G., & Vignale, G. (2005). *Quantum Theory of the Electron Liquid.* Cambridge, UK: Cambridge University Press. doi:10.1017/CBO9780511619915

Glisic, S., & Skala, D. (2009). The problems in design and detailed analyses of energy consumption for biodiesel synthesis at supercritical conditions. *The Journal of Supercritical Fluids*, *49*(2), 293–301. doi:10.1016/j.supflu.2008.12.011

Gnielinski, V. (1976). New equations for heat and mass transfer in turbulent pipe and channel flow. *Chemical Engineering Journal.*, *16*(2), 359–368.

Gökkaya, D. S., Saglam, M., Yuksel, M., & Ballice, L. (2016). Hydrothermal gasification of xylose: Effects of reaction temperature, pressure, and K_2CO_3 as a catalyst on product distribution. *Biomass and Bioenergy*, *91*, 26–36. doi:10.1016/j.biombioe.2016.04.013

Goldmann, K. (1954). Heat transfer to supercritical water and other fluids with temperature-dependent properties. *Chemical Engineering Progress Symposium Series, 50*(11), 105-110.

Gonzalez, D. J., Ng, D. A., & Silbert, M. (1990). Variational MHNC calculations of the structure of expanded liquid rubidium. *Journal of Non-Crystalline Solids*, *117/118*, 469–472. doi:10.1016/0022-3093(90)90976-S

Gorbaty, Yu. E., & Bondarenko, G. V. (2007). Water in the supercritical state. *SCF-TP Online, 2*(2) 5–19. Retrieved from http://scf-tp.ru/articles/2007_02/index.html

Goswami, D. Y., Hingorani, S., & Mines, G. (1991). A Laser-Based Technique for Particle Sizing to Study Two-Phase Expansion in Turbines. *Journal of Solar Energy Engineering*, *113*(3), 211–218. doi:10.1115/1.2930495

Gourgouillon, D., Avelino, H., Fareleira, J., & Da Ponte, M. N. (1998). Simultaneous viscosity and density measurement of supercritical CO 2-saturated PEG 400. *The Journal of Supercritical Fluids*, *13*(1), 177–185. doi:10.1016/S0896-8446(98)00050-3

Greenfield, A. J. (1964). Hall coefficients of liquid metals. *Physical Review*, *135*(6A), A1589–A1595. doi:10.1103/PhysRev.135.A1589

Greffet, J. J., Carminati, R., Joulain, K., Mulet, J. P., Mainguy, S., & Chen, Y. (2002). Coherent Emission of Light by Thermal Sources. *Nature*, *416*(6876), 61–64. doi:10.1038/416061a PMID:11882890

Grief, R. (1998). Natural circulation loops. *Journal of Heat Transfer*, *110*(4b), 1243–1257. doi:10.1115/1.3250624

Guan, Q., Mao, T., Zhang, Q., Miao, R., Ning, P., Gu, J., & Chai, X.-S. et al. (2014). Catalytic gasification of lignin with $Ni/Al_2O_3–SiO_2$ in sub/supercritical water. *The Journal of Supercritical Fluids*, *95*, 413–421. doi:10.1016/j.supflu.2014.10.015

Guenoun, P., Beysens, D., Khalil, B., Garrabos, Y., Kammoun, B., Le Neindre, B., & Zappoli, B. (1993). A thermo cycle around the critical point of CO_2 under reduced gravity. *Physical Review E: Statistical Physics, Plasmas, Fluids, and Related Interdisciplinary Topics*, *47*(3), 1531–1540. doi:10.1103/PhysRevE.47.1531

Guildner, L. A. (1958). The thermal conductivity of carbon dioxide in the region of the critical point. *Proceedings of the National Academy of Sciences of the United States of America*, *44*(11), 1149–1153. doi:10.1073/pnas.44.11.1149 PMID:16590326

Güngören Madenoğlu, T., Sağlam, M., Yüksel, M., & Ballice, L. (2016). Hydrothermal gasification of biomass model compounds (cellulose and lignin alkali) and model mixtures. *The Journal of Supercritical Fluids*, *115*, 79–85. doi:10.1016/j.supflu.2016.04.017

Guo, L., Jin, H., & Lu, Y. (2015). Supercritical water gasification research and development in China. *The Journal of Supercritical Fluids*, *96*, 144–150. doi:10.1016/j.supflu.2014.09.023

Guo, S., Guo, L., Cao, C., Yin, J., Lu, Y., & Zhang, X. (2012). Hydrogen production from glycerol by supercritical water gasification in a continuous flow tubular reactor. *International Journal of Hydrogen Energy*, *37*(7), 5559–5568. doi:10.1016/j.ijhydene.2011.12.135

Guo, Y., Wang, S., Xu, D., Gong, Y., Ma, H., & Tang, X. (2010). Review of catalytic supercritical water gasification for hydrogen production from biomass. *Renewable & Sustainable Energy Reviews*, *14*(1), 334–343. doi:10.1016/j.rser.2009.08.012

Gupta, S., Saltanov, E., Mokry, S. J., Pioro, I., Trevani, L., & McGillivray, D. (2013). Developing empirical heat-transfer correlations for supercritical CO_2 flowing in vertical bare tubes. *Nuclear Engineering and Design*, *251*, 116–131. doi:10.1016/j.nucengdes.2013.02.048

Gutiérrez, C., Rodríguez, J., Gracia, I., De Lucas, A., & García, M. (2013). Development of a strategy for the foaming of polystyrene dissolutions in $scCO_2$. *The Journal of Supercritical Fluids*, *76*, 126–134. doi:10.1016/j.supflu.2013.01.020

Hämäläinen, K., Huotari, S., Laukkanen, J., Soininen, A., Manninen, S., Kao, C.-C., & Mezouar, M. et al. (2000). Free electron gas under high pressure. *Physical Review B: Condensed Matter and Materials Physics*, *62*(2), R735–R738. doi:10.1103/PhysRevB.62.R735

Harrison, W. A. (1966). *Pseudopotential in the Theory of Metals*. New York: Benjamin.

Hasan, M. M., Afroz, A. M., & Tsubaki, K. (2008). Heat Transfer Coefficients and Pressure Drops during In-tube Condensation of CO_2/DME Mixture Refrigerant. *International Journal of Refrigeration*, *29*(6), 76–80.

Hasegawa, M. (1971). Theory of Plasmon Damping in Metals. II. Effects of Electron-Ion Interaction. *Journal of the Physical Society of Japan*, *31*(3), 649–667. doi:10.1143/JPSJ.31.649

Hashimoto, K., & Saikawa, M. (2004). Trend Report of Development of Residential CO2 Heat Pump Hot Water Heater in Japan. *Proceedings for the JSME Ninth Power Energy Technical Symposium*, 425–430.

Hashimoto, K., Tanahashi, M., & Tomita, M. (2002). Direct numerical simulation on the near wall heat transfer of carbon dioxide. *Proceeding of 5th IIR-Gustav Lorentzen Conference on Natural Working Fluids.* Guangzhou, Guangzhou, China: International Institute of Refrigeration.

Heberle, F., Preißinger, M., & Brüggemann, D. (2012). Zeotropic mixtures as working fluids in Organic Rankine Cycles for low-enthalpy geothermal resources. *Renewable Energy*, *37*(1), 364–370. doi:10.1016/j.renene.2011.06.044

He, C., Giannis, A., & Wang, J.-Y. (2013). Conversion of sewage sludge to clean solid fuel using hydrothermal carbonization: Hydrochar fuel characteristics and combustion behavior. *Applied Energy*, *111*, 257–266. doi:10.1016/j.apenergy.2013.04.084

Hejzlar, P., Dostal, V., & Driscoll, M. J. (2006). A supercritical CO2 cycle- a promising power conversion system for Generation IV reactors. In *Proceedings of the 2006 International Congress on Advances in Nuclear Power Plants, ICAPP'06* (Vol. 2006, pp. 722–731).

Hensel, F., & Warren, W. W. Jr. (1999). *Fluid Metals: Liquid–Vapor Transitions of Metals*. Princeton, NJ: Princeton University Press. doi:10.1515/9781400865000

He, S., Jiang, P. X., Xu, Y. J., Shi, R. F., Kim, W. S., & Jackson, J. D. (2005). A computational study of convection heat transfer to CO_2 at supercritical pressures in a vertical mini tube. *International Journal of Thermal Sciences*, *44*(6), 521–530. doi:10.1016/j.ijthermalsci.2004.11.003

Hill, J. P., Kao, C.-C., Caliebe, W. A. C., Gibbs, D., & Hastings, J. B. (1996). Inelastic X-Ray Scattering Study of Solid and Liquid Li and Na. *Physical Review Letters*, *77*(17), 3665–3668. doi:10.1103/PhysRevLett.77.3665 PMID:10062277

Ho, C. K., Conboy, T., Ortega, J., Afrin, S., Gray, A., Christian, J. M., … Wani, P. (2014). High-Temperature Receiver Designs for Supercritical CO_2 Closed-Loop Brayton Cycles. In *ASME 2014 12th International Conference on Fuel Cell Science, Engineering and Technology* (Vol. 1, p. V001T02A003). ASME. https://doi.org/10.1115/ES2014-6328

Ho, C. K. (2015). *Cost and Performance Tradeoffs of Alternative Solar-Driven Supercritical CO_2 Power Cycles (presentation). (No. SAND2015-4055PE). Sandia National Laboratories*. Albuquerque, NM: SNL-NM.

Hoekman, S. K., Broch, A., Robbins, C., Purcell, R., Zielinska, B., Felix, L., & Irvin, J. (2014). Process development unit (PDU) for hydrothermal carbonization (HTC) of lignocellulosic biomass. *Waste and Biomass Valorization*, *5*(4), 669–678. doi:10.1007/s12649-013-9277-0

Hoekman, S. K., Broch, A., Robbins, C., Zielinska, B., & Felix, L. (2013). Hydrothermal carbonization (HTC) of selected woody and herbaceous biomass feedstocks. *Biomass Conversion and Biorefinery*, *3*(2), 113–126. doi:10.1007/s13399-012-0066-y

Holman, J. P., & Boggs, J. H. (1960). Heat transfer to Freon 12 near the critical state in a natural circulation loop. *Journal of Heat Transfer*, *82*(3), 221–226. doi:10.1115/1.3679913

Hoogwijk, M., Faaij, A., van den Broek, R., Berndes, G., Gielen, D., & Turkenburg, W. (2003). Exploration of the ranges of the global potential of biomass for energy. *Biomass and Bioenergy*, *25*(2), 119–133. doi:10.1016/S0961-9534(02)00191-5

Hosokawa, S., Sinn, H., Hensel, F., Alatas, A., Alp, E. E., & Pilgrim, W.-C. (2002). Collective dynamics of liquid Hg investigated by inelastic X-ray scattering. *Journal of Non-Crystalline Solids*, *312–314*, 163–167. doi:10.1016/S0022-3093(02)01677-0

Hough, S. (2009). *Supercritical Rankine Cycle.* University of Idaho. Technical Paper. ME-517. Retrieved from http://www. if. uidaho. edu/~ gunner/ME443-543/HW/rankine.pdf

Howell, J. R., & Lee, S. H. (1999). Convective heat transfer in the entrance region of a vertical tube for water near the thermodynamic critical point. *International Journal of Heat and Mass Transfer*, *42*(7), 1177–1187. doi:10.1016/S0017-9310(98)00263-4

Huang, B. J. (1980). Similarity theory of solar water heater with natural circulation. *Solar Energy*, *25*(2), 105–116. doi:10.1016/0038-092X(80)90465-X

Huber, G. W., Iborra, S., & Corma, A. (2006). Synthesis of Transportation Fuels from Biomass: Chemistry, Catalysts, and Engineering. *Chemical Reviews*, *106*(9), 4044–4098. doi:10.1021/cr068360d PMID:16967928

Hung, T.-C. (2001). Waste heat recovery of organic Rankine cycle using dry fluids. *ResearchGate, 42*(5), 539–553. https://doi.org/10.1016/S0196-8904(00)00081-9

Hung, T. C., Shai, T. Y., & Wang, S. K. (1997). A review of organic rankine cycles (ORCs) for the recovery of low-grade waste heat. *Energy*, *22*(7), 661–667. doi:10.1016/S0360-5442(96)00165-X

Inui, M., Hong, X., & Tamura, K. (2003). Local structure of expanded fluid mercury using synchrotron radiation: From liquid to dense vapor. *Physical Review B: Condensed Matter and Materials Physics*, *68*. 094108. doi:10.1103/PhysRevB.68.094108.

Inui, M., Kajihara, Y., Ohara, K., Matsuda, K., Ishikawa, D., & Hosokawa, S. (2016). Critical Indices ν and γ Experimentally Obtained at the Liquid-Vapor Critical Point in Fluid Hg. *Journal of Physical Society of Japan*, 85, 035001. doi:10.7566/JPSJ.85.035001.

Inui, M., Matsuda, K., Ishikawa, D., Tamura, K. & Ohishi, Y. (2007). Medium-Range Fluctuations Accompanying the Metal-Nonmetal Transition in Expanded Fluid Hg. *Physical Review Letters*, *98*, 185504. doi:10.1103/PhysRevLett.98.185504.

Inui, M., Ishikawa, D., Matsuda, K., Tamura, K., Tsutsui, S., & Baron, A. Q. R. (2005). Observation of fast sound in metal–nonmetal transition in liquid Hg. *Journal of Physics and Chemistry of Solids*, *66*(12), 2223–2229. doi:10.1016/j.jpcs.2005.09.021

Ishikawa, D., Inui, M., Matsuda, K., Tamura, K., Tsutsui, S., Baron, A. Q. R. (2004). Fast sound in expanded fluid Hg accompanying the metal-nonmetal transition. *Physical Review Letters*, *93*, 097801. doi:10.1103/PhysRevLett.93.097801.

Isshiki, M., Ohishi, Y., Goto, S., Takeshita, K., & Ishikawa, T. (2001). High-energy X-ray diffraction beamline: BL04B2 at SPring-8. *Nuclear Instruments & Methods in Physics Research. Section A, Accelerators, Spectrometers, Detectors and Associated Equipment*, *467–468*, 663–666. doi:10.1016/S0168-9002(01)00441-7

Itou, M., Hiraoka, N., Ohata, T., Mizumaki, M., Deb, A., Sakurai, Y., & Sakai, N. (2001). Present status of the Cauchois-type Compton scattering spectrometer at SPring-8. *Nuclear Instruments & Methods in Physics Research. Section A, Accelerators, Spectrometers, Detectors and Associated Equipment*, *467-468*, 1109–1112. doi:10.1016/S0168-9002(01)00592-7

Iverson, B. D., Conboy, T. M., Pasch, J. J., & Kruizenga, A. M. (2013). Supercritical CO_2 Brayton cycles for solar-thermal energy. *Applied Energy*, *111*(0), 957–970. doi:10.1016/j.apenergy.2013.06.020

Jackson, J. D. (2001). Some striking features of heat transfer with fluids at pressures and temperatures near the critical point. *Keynote Paper for International Conference on Energy Conversion and Application (ICECA)*, Wuhan, China.

Jackson, J. D., & Hall, W. B. (1979). Forced convection heat transfer to fluids at supercritical pressure. In S. Kakac & D. B. Spalding (Eds.), *Turbulent Forced Convection in Channels and Bundles* (pp. 563–611). New York: McGraw-Hill.

Jacobson, M. Z. (2009). Review of solutions to global warming, air pollution, and energy security. *Energy & Environmental Science*, *2*(2), 148–173. doi:10.1039/B809990C

Jain, A., Balasubramanian, R., & Srinivasan, M. P. (2016). Hydrothermal conversion of biomass waste to activated carbon with high porosity: A review. *Chemical Engineering Journal*, *283*, 789–805. doi:10.1016/j.cej.2015.08.014

Jain, P. K., & Rizwan-uddin, . (2008). Numerical analysis of supercritical flow instabilities in a natural circulation loop. *Nuclear Engineering and Design*, *238*(8), 1947–1957. doi:10.1016/j.nucengdes.2007.10.034

Jany, P., & Straub, J. (1987). Thermal diffusivity of fluids in a broad region around the critical point. *International Journal of Thermophysics*, *8*(2), 165–180. doi:10.1007/BF00515200

Jena, U., McCurdy, A. T., Warren, A., Summers, H., Ledbetter, R. N., Hoekman, S. K., & Quinn, J. C. et al. (2015). Oleaginous yeast platform for producing biofuels via co-solvent hydrothermal liquefaction. *Biotechnology for Biofuels*, *8*(1), 1. doi:10.1186/s13068-015-0345-5 PMID:26468320

Jones, W. P., & Launder, B. E. (1972). The prediction of laminarization with a two-equation model of turbulence. *International Journal of Heat and Mass Transfer*, *15*(2), 301–314. doi:10.1016/0017-9310(72)90076-2

Jordan, C. A., & Akay, G. (2012). Occurrence, composition and dew point of tars produced during gasification of fuel cane bagasse in a downdraft gasifier. *Biomass and Bioenergy*, *42*, 51–58. doi:10.1016/j.biombioe.2012.03.014

JRAIA. (2016). *About 'Eco Cute' Exceeding 5 million Units*. Retrieved from https://www.jraia.or.jp/product/heatpump/i_broke.html

Judex, J. W., Gaiffi, M., & Burgbacher, H. C. (2012). Gasification of dried sewage sludge: Status of the demonstration and the pilot plant. *Waste Management (New York, N.Y.)*, *32*(4), 719–723. doi:10.1016/j.wasman.2011.12.023 PMID:22284442

Jue, Y., Yoshiaki, O., Yuiki, I., Jie, L., & Jaewoon, Y. (2007). Numerical investigation of heat transfer in upward flow of supercritical water in circular channels and tight fuel rod bundles. *Nuclear Engineering and Design*, *237*(4), 420–430. doi:10.1016/j.nucengdes.2006.08.003

Juhaszl, A. J., Rarick, R. A., & Rajmohan, R. (2009). High Efficiency Nuclear Power Plants Using Liquid Fluoride Thorium Reactor Technology. In *Proceedings of the 7th International Energy Conversion Engineering Conference*, 1-12. doi:10.2514/6.2009-4565

Jüngst, S., Knuth, B., & Hensel, F. (1985). Observation of Singular Diameters in the Coexistence Curves of Metals. *Physical Review Letters*, *55*(20), 2160–2163. doi:10.1103/PhysRevLett.55.2160 PMID:10032064

Kakac, S., & Ishil, M. (1983). *Advances in two-phase flow and heat transfer fundamentals and applications volumes I and II*. Academic Press.

Kalogirou, S. A. (2004). Solar thermal collectors and applications. *Progress in Energy and Combustion Science*, *30*(3), 231–295. doi:10.1016/j.pecs.2004.02.001

Kambo, H. S., & Dutta, A. (2014). Strength, storage, and combustion characteristics of densified lignocellulosic biomass produced via torrefaction and hydrothermal carbonization. *Applied Energy*, *135*, 182–191. doi:10.1016/j.apenergy.2014.08.094

Kambo, H. S., & Dutta, A. (2015). Comparative evaluation of torrefaction and hydrothermal carbonization of lignocellulosic biomass for the production of solid biofuel. *Energy Conversion and Management*, *105*, 746–755. doi:10.1016/j.enconman.2015.08.031

Kamm, B., Gruber, P. R., & Kamm, M. (2007). Biorefineries–Industrial Processes and Products Ullmann's Encyclopedia of Industrial Chemistry. Wiley-VCH Verlag GmbH & Co. KGaA.

Kaneko, M., Ikeda, H., Tokiai, T., Nagao, S., Suto, H., & Tamano, M. (2008). The development of PAG-PVE copolymer (ECP). *Proc. Int. Symp. on New Refrigerants and Environmental Technology*, 376-388.

Kaneko, M., Ikeda, H., Tokiai, T., Yoshii, A., & Suto, H. (2006). The development of PAG refrigeration lubricants for automotive A/C with CO2. JSAE Automotive Air-Conditioning Conference, 142-159.

Karellas, S., & Schuster, A. (2008). Supercritical Fluid Parameters in Organic Rankine Cycle Applications. *International Journal of Thermodynamics*, *11*(3), 101–108. https://doi.org/10.5541/ijot.1034000217

Kassoy, D. R. (1979). The response of a confined gas to a thermal disturbance). I. Slow transient. *SIAM Journal on Applied Mathematics*, *36*(3), 3. doi:10.1137/0136044

Kawasaki, K. (1970). Kinetic equations and time correlation functions of critical fluctuations. *Annals of Physics*, *61*(1), 1–56. doi:10.1016/0003-4916(70)90375-1

Kays, W. M., & Crawford, M. E. (1980). *Convective heat and mass transfer*. New York: McGraw-Hill.

Kesić, Ž., Lukić, I., Brkić, D., Rogan, J., Zdujić, M., Liu, H., & Skala, D. (2012). Mechanochemical preparation and characterization of CaO· ZnO used as catalyst for biodiesel synthesis. *Applied Catalysis A, General*, *427*, 58–65. doi:10.1016/j.apcata.2012.03.032

Kimball, K. J., & Clementoni, E. M. (2012). Supercritical Carbon Dioxide Brayton Power Cycle Development Overview. In *ASME Turbo Expo 2012: Turbine Technical Conference and Exposition. Volume 5: Manufacturing Materials and Metallurgy; Marine; Microturbines and Small Turbomachinery; Supercritical CO_2 Power Cycles* (Vol. 5, p. 931). ASME. https://doi.org/10.1115/GT2012-68204

Kim, M. H., Pettersen, J., & Bullard, C. W. (2004). Fundamental Process and System Design Issues in CO_2 Vapor Compression Systems. *Progress in Energy and Combustion Science*, *30*(2), 119–174. doi:10.1016/j.pecs.2003.09.002

Kim, M. S., Ahn, Y., Kim, B., & Lee, J. I. (2016). Study on the supercritical CO_2 power cycles for landfill gas firing gas turbine bottoming cycle. *Energy*, *111*, 893–909. doi:10.1016/j.energy.2016.06.014

Kim, S. G., Lee, J., Ahn, Y., Lee, J. I., Addad, Y., & Ko, B. (2014). CFD investigation of a centrifugal compressor derived from pump technology for supercritical carbon dioxide as a working fluid. *The Journal of Supercritical Fluids*, *86*, 160–171. doi:10.1016/j.supflu.2013.12.017

Kimura, K., Matsuda, K. & Yao, M. (2013). Plasmon line width in liquid metals. *Journal of the Physical Society of Japan*, *82*, 115001. doi:10.7566/JPSJ.82.115001

Kimura, K., Matsuda, K., Hiraoka, N., Fukumaru, T., Kajihara, Y., Inui, M. & Yao, M. (2014). Inelastic x-ray scattering study of plasmon dispersions in solid and liquid Rb. *Physical Review B: Condensed Matter and Materials Physics*, *89*, 014206. doi:10.1103/PhysRevB.89.014206

Kiran, E. (1994). Polymer formation, modifications and processing in or with supercritical fluids. In Supercritical Fluids (pp. 541-588). Orono, ME: Springer Netherlands. doi:10.1007/978-94-015-8295-7_22

Kirkels, A. F., & Verbong, G. P. J. (2011). Biomass gasification: Still promising? A 30-year global overview. *Renewable & Sustainable Energy Reviews*, *15*(1), 471–481. doi:10.1016/j.rser.2010.09.046

Kittel, C. (1986). *Introduction to Solid State Physics* (6th ed.). New York: John Wiley and Sons.

Knez Hrnčič, M., Markočič, E., Kramberger, B., & Knez, Ž. (2011). *Supercritical fluids as product and process solvents in emerging applications*. Paper presented at 13th European Meeting on Supercritical Fluids, Hague, The Netherlands.

Knez Hrnčič, M., Markočič, E., Trupej, N., Škerget, M., & Knez, Ž. (2014). Investigation of thermodynamic properties of the binary system polyethylene glycol/CO_2 using new methods. *The Journal of Supercritical Fluids*, *87*, 50–58. doi:10.1016/j.supflu.2013.12.021

Knez Hrnčič, M., Venderbosch, R. H., Škerget, M., Ilić, L., & Knez, Ž. (2013b). Phase equilibrium data of hydrogen in pyrolysis oil and hydrogenated pyrolysis oil at elevated pressures. *The Journal of Supercritical Fluids*, *80*, 86–89. doi:10.1016/j.supflu.2013.04.015

Knez Hrnčič, M., Venderbosch, R. H., Škerget, M., & Knez, Ž. (2013a). Observation of phase behavior for bio-oil+ diesel+ carbon dioxide and bio-oil+ tail water+ carbon dioxide system. *Journal of Chemical & Engineering Data*, *58*(3), 648–652. doi:10.1021/je301146g

Knez, Ž., Škerget, M., & Knez Hrňič, M. (2010). Principles of supercritical fluid extraction and applications in the food, beverage and nutraceutical industries. In S. Rizvi (Ed.), Separation, extraction and concentration processes in the food, beverage and nutraceutical industries (pp. 3-38). Elsevier

Knez, Ž., Markočič, E., Leitgeb, M., Primožič, M., Knez Hrnčič, M., & Škerget, M. (2014). Industrial applications of supercritical fluids: A review. *Energy*, *77*, 235–243. doi:10.1016/j.energy.2014.07.044

Kobayashi, N., Tanaka, M., Piao, G., Kobayashi, J., Hatano, S., Itaya, Y., & Mori, S. (2009). High temperature air-blown woody biomass gasification model for the estimation of an entrained down-flow gasifier. *Waste Management (New York, N.Y.)*, *29*(1), 245–251. doi:10.1016/j.wasman.2008.04.014 PMID:18653324

Kohno, H., & Yao, M. (2002). Dynamic critical phenomena of mercury. *Journal of Physics Condensed Matter*, *14*(6), L171–L176. doi:10.1088/0953-8984/14/6/106

Koshizuka, S., Takano, N., & Oka, Y. (1995). Numerical analysis of deterioration phenomena in heat transfer to supercritical water. *International Journal of Heat and Mass Transfer*, *38*(16), 3077–3084. doi:10.1016/0017-9310(95)00008-W

Koyama, S., Takato, N., Kuwahara, K., Jin, D., Xue, J., & Miyara, A. (2006), Experimental study on the performance of a refrigerant mixture CO2/DME system. *Proceedings of JSRAE Annual Conference,* 133-136. (In Japanese)

Kramberger, B., Markočič, E., & Knez, Ž. (2013). Phase equilibria of binary mixture of carbon monoxide and water at elevated temperatures and pressures. *Chemical Engineering Science*, *99*, 77–80. doi:10.1016/j.ces.2013.05.048

Kreitlow, D. B., Reistad, G. M., Miles, C. R., & Culver, G. G. (1978). Thermosyphon models for d(ownhole heat exchanger application in shallow geothermal systems. *Journal of Heat Transfer*, *100*(4), 713–719. doi:10.1115/1.3450883

Kresse, G., & Hafner, J. (1997). Ab initio simulation of the metal/nonmetal transition in expanded fluid mercury. *Physical Review B: Condensed Matter and Materials Physics*, *55*(12), 7539–7548. doi:10.1103/PhysRevB.55.7539

Kruse, A. (2008). Supercritical water gasification. *Biofuels. Bioproducts and Biorefining*, *2*(5), 415–437. doi:10.1002/bbb.93

Kruse, A. (2011). Behandlung von Biomasse mit überkritischem Wasser. *Chemie Ingenieur Technik*, *83*(9), 1381–1389. doi:10.1002/cite.201100040

Kruse, A., Bernolle, P., Dahmen, N., Dinjus, E., & Maniam, P. (2010). Hydrothermal gasification of biomass: Consecutive reactions to long-living intermediates. *Energy & Environmental Science*, *3*(1), 136–143. doi:10.1039/B915034J

Kruse, A., & Dahmen, N. (2015). Water – A magic solvent for biomass conversion. *The Journal of Supercritical Fluids*, *96*, 36–45. doi:10.1016/j.supflu.2014.09.038

Kruse, A., Funke, A., & Titirici, M.-M. (2013). Hydrothermal conversion of biomass to fuels and energetic materials. *Current Opinion in Chemical Biology*, *17*(3), 515–521. doi:10.1016/j.cbpa.2013.05.004 PMID:23707262

Kukkonen, C. A., & Wilkins, J. W. (1979). Electron-electron scattering in simple metals. *Physical Review B: Condensed Matter and Materials Physics*, *19*(12), 6075–6093. doi:10.1103/PhysRevB.19.6075

Kulhánek, M., & Dostál, V. (2011). Thermodynamic Analysis and Comparison of Supercritical Carbon Dioxide Cycles. *Proceedings of Supercritical CO_2 Power Cycle Symposium.*

Kumar, K. K., & Gopal, M. R. (2009). Effect of system pressure on the steady state performance of a CO_2 based natural circulation loop. *Applied Thermal Engineering*, *29*(16), 3346–3352. doi:10.1016/j.applthermaleng.2009.05.010

Kumar, K. K., & Gopal, M. R. (2011). Experimental studies on CO_2 based single and two-phase natural circulation loops. *Applied Thermal Engineering*, *31*(16), 3437–3443. doi:10.1016/j.applthermaleng.2011.06.029

Kumar, K., & Tuckerman, L. S. (1994). Parametric instability of the interface between two fluids. *Journal of Fluid Mechanics*, *279*(-1), 49. doi:10.1017/S0022112094003812

Kurganov, V. A., Zeigarnik, Yu. A., & Maslakova, I. V. (2014). Heat transfer and hydraulic resistance of supercritical pressure coolants. Part IV: Problems of generalized heat transfer description, methods of predicting deteriorated heat transfer; empirical correlations; deteriorated heat transfer enhancement; dissolved gas effects. *International Journal of Heat and Mass Transfer*, *77*, 1197–1212. doi:10.1016/j.ijheatmasstransfer.2014.06.014

Kuwahara, T., Niu, X. D., Yamaguchi, H., Iwamoto, Y., & Zhang, X. R. (2011). Performance study of supercritical CO_2-based solar Rankine cycle system with a novel-concept thermally driven pump. *Proceedings of the Institution of Mechanical Engineers. Part A, Journal of Power and Energy*, *225*(4), 413–419. doi:10.1177/0957650910396420

Landau, L. D., & Lifschitz, E. M. (1959). *Fluid Mechanics* (Vol. 6). Pergamon Press.

Landau, L. D., & Zeldovich, J. (1943). On the relation between the liquid and the gaseous states of metals. *Acta Physico-Chimica URSS*, *18*, 194–196.

Larjola, J. (1995). Electricity from industrial waste heat using high-speed organic Rankine cycle (ORC). *International Journal of Production Economics*, *41*(1), 227–235. doi:10.1016/0925-5273(94)00098-0

Launder, B. E., & Sharma, B. I. (1974). Application of the energy dissipation model of turbulence to calculation of flow near a spinning disc. *Letters of Heat Mass Transfer.*, *1*(2), 131–138. doi:10.1016/0094-4548(74)90150-7

Lee, J., Lee, J. I., Ahn, Y., & Yoon, H. (2012). Design Methodology of Supercritical CO_2 Brayton Cycle Turbomachineries. In *ASME Turbo Expo 2012: Turbine Technical Conference and Exposition. Volume 5: Manufacturing Materials and Metallurgy; Marine; Microturbines and Small Turbomachinery; Supercritical CO_2 Power Cycles* (Vol. 5, p. 975). ASME. https://doi.org/10.1115/GT2012-68933

Lee, S. H., & Howell, J. R. (1998). Turbulent developing convective heat transfer in a tube for fluids near the critical point. *International Journal of Heat and Mass Transfer*, *41*(10), 1205–1218. doi:10.1016/S0017-9310(97)00217-2

Lemmon, E. W., Huber, M. L., & McLinden, M. O. (2013). *NIST Standard Reference Database 23: Reference Fluid Thermodynamic and Transport Properties-REFPROP, Version 9.1*. NIST.

Liao, B., Guo, L., Lu, Y., & Zhang, X. (2013). Solar receiver/reactor for hydrogen production with biomass gasification in supercritical water. *International Journal of Hydrogen Energy*, *38*(29), 13038–13044. doi:10.1016/j.ijhydene.2013.03.113

Liao, S. M., & Zhao, T. S. (2002a). Measurements of heat transfer coefficients from supercritical carbon dioxide flowing in horizontal mini/micro channels. *Journal of Heat Transfer-Transactions of ASME*, *124*(3), 413–420. doi:10.1115/1.1423906

Liao, S. M., & Zhao, T. S. (2002b). An experimental investigation of convection heat transfer to supercritical carbon dioxide in miniature tubes. *International Journal of Heat and Mass Transfer*, *45*(25), 5025–5034. doi:10.1016/S0017-9310(02)00206-5

Linzer, W., & Walter, H. (2003). Flow reversal in natural circulation systems. *Applied Thermal Engineering*, *23*(18), 2363–2372. doi:10.1016/S1359-4311(03)00214-X

Liu, G. X., Huang, Y. P., Wang, J. F., Lv, F., & Leung, L. K. H. (2016). Experiments on the basic behaviour of supercritical CO_2 natural circulation. *Nuclear Engineering and Design*, *300*, 376–383. doi:10.1016/j.nucengdes.2016.01.021

Liu, Q., Duan, Y., & Yang, Z. (2014). Effect of condensation temperature glide on the performance of organic Rankine cycles with zeotropic mixture working fluids. *Applied Energy*, *115*, 394–404. doi:10.1016/j.apenergy.2013.11.036

Liu, Z., & Balasubramanian, R. (2012). Hydrothermal Carbonization of Waste Biomass for Energy Generation. *Procedia Environmental Sciences*, *16*, 159–166. doi:10.1016/j.proenv.2012.10.022

Liu, Z., & Balasubramanian, R. (2014). Upgrading of waste biomass by hydrothermal carbonization (HTC) and low temperature pyrolysis (LTP): A comparative evaluation. *Applied Energy*, *114*, 857–864. doi:10.1016/j.apenergy.2013.06.027

Liu, Z., Quek, A., Kent Hoekman, S., & Balasubramanian, R. (2013). Production of solid biochar fuel from waste biomass by hydrothermal carbonization. *Fuel*, *103*, 943–949. doi:10.1016/j.fuel.2012.07.069

Lomperski, S., & Cho, D. (2004). Stability of a natural circulation loop with a fluid heated through the thermo dynamic pseudo critical point. *Proceedings of ICAPP'04*.

Lorentzen, G. (1990). *Trans-Critical Vapour Compression Cycle Device*. International Patent Publication, Patent No. WO 90/07683.

Lorentzen, G. (1995). The use of natural refrigerants: A complete solution to the CFC/HCFC predicament. *International Journal of Refrigeration*, *18*(3), 190–197. doi:10.1016/0140-7007(94)00001-E

Lorentzen, G., & Pettersen, J. (1992). New Possibilities for Non-CFC Refrigeration. *Proceedings of the IIR International Symposium on Refrigeration, Energy and Environment*, 147–163.

Lv, P., Yuan, Z., Ma, L., Wu, C., Chen, Y., & Zhu, J. (2007). Hydrogen-rich gas production from biomass air and oxygen/steam gasification in a downdraft gasifier. *Renewable Energy*, *32*(13), 2173–2185. doi:10.1016/j.renene.2006.11.010

Lynam, J. G., Reza, M. T., Yan, W., Vásquez, V. R., & Coronella, C. J. (2015). Hydrothermal carbonization of various lignocellulosic biomass. *Biomass Conversion and Biorefinery*, *5*(2), 173–181. doi:10.1007/s13399-014-0137-3

Ma, Z., & Turchi, C. (2011). Advanced supercritical carbon dioxide power cycle configurations for use in concentrating solar power systems. *Proceedings of Supercritical CO_2 Power Cycle Symposium.*

Madhawa Hettiarachchi, H. D., Golubovic, M., Worek, W. M., & Ikegami, Y. (2007). Optimum design criteria for an Organic Rankine cycle using low-temperature geothermal heat sources. *Energy*, *32*(9), 1698–1706. doi:10.1016/j.energy.2007.01.005

Maebashi, H., & Takada, Y. (2009). First-Principles Understanding of the Anomalous Structural Change in an Expanded Liquid Alkali Metal. *Journal of the Physical Society of Japan*, *78*, 053706. doi:10.1143/JPSJ.78.053706

Mago, P. J., Chamra, L. M., Srinivasan, K., & Somayaji, C. (2008). An examination of regenerative organic Rankine cycles using dry fluids. *Applied Thermal Engineering*, *28*(8–9), 998–1007. doi:10.1016/j.applthermaleng.2007.06.025

Mahan, G. D. (2000). *Many-Particle Physics*. New York: Kluwer Academic/Plenum. doi:10.1007/978-1-4757-5714-9

Maizza, V., & Maizza, A. (1996). Working fluids in non-steady flows for waste energy recovery systems. *Applied Thermal Engineering*, *16*(7), 579–590. doi:10.1016/1359-4311(95)00044-5

Major, J. (2014). *Astronauts Work to Make Water That Burns*. Retrieved November 11, from http://news.discovery.com/space/astronauts-work-to-make-water-that-burns-140103.htm

Mäkelä, M., Benavente, V., & Fullana, A. (2015). Hydrothermal carbonization of lignocellulosic biomass: Effect of process conditions on hydrochar properties. *Applied Energy*, *155*, 576–584. doi:10.1016/j.apenergy.2015.06.022

Mäkelä, M., Benavente, V., & Fullana, A. (2016). Hydrothermal carbonization of industrial mixed sludge from a pulp and paper mill. *Bioresource Technology*, *200*, 444–450. doi:10.1016/j.biortech.2015.10.062 PMID:26519695

Mancini, T., Heller, P., Butler, B., Osborn, B., Wolfgang, S., Vernon, G., & Moreno, J. et al. (2003). Dish-Stirling Systems: An Overview of Development and Status, *ASME. Journal of Solar Energy Engineering*, *125*(2), 135–151. doi:10.1115/1.1562634

Marcout, R., Zwilling, J. F., Laherrere, J. M., Garrabos, Y., & Beysens, D. (1995). ALICE 2, an advanced facility for the analysis of fluids close to their critical point in microgravity. *Microgravity Quarterly*, *5*, 162–170.

Markočič, E., Kramberger, B., van Bennekom, J. G., Heeres, H. J., Vos, J., & Knez, Ž. (2013). Glycerol reforming in supercritical water; a short review. *Renewable & Sustainable Energy Reviews*, *23*, 40–48. doi:10.1016/j.rser.2013.02.046

Markočič, E., Škerget, M., & Knez, Ž. (2011). Solubility and diffusivity of CO_2 in poly (l-lactide)–hydroxyapatite and poly (d, l-lactide-co-glycolide)–hydroxyapatite composite biomaterials. *The Journal of Supercritical Fluids*, *55*(3), 1046–1051. doi:10.1016/j.supflu.2010.10.001

Mason, A. A., & Davidson, J. H. (1995). Measured Performance and Modeling of an Evacuated-Tube, Integral-Collector-Storage Solar Water Heater, *ASME. Journal of Solar Energy Engineering*, *117*(3), 221–228. doi:10.1115/1.2847803

Materazzi, M., Lettieri, P., Mazzei, L., Taylor, R., & Chapman, C. (2015). Fate and behavior of inorganic constituents of RDF in a two stage fluid bed-plasma gasification plant. *Fuel*, *150*, 473–485. doi:10.1016/j.fuel.2015.02.059

Matsuda, K., Fukumaru, T., Kimura, K., Tamura, K., Katoh, M., Kajihara, Y., . . . Sakurai, Y. (2015). X-ray Compton scattering experiments for fluid alkali metals at high temperatures and pressures. AIP Conference Proceedings, 1673. doi:10.1063/1.4928258

Matsuda, K., Nagao, T., Kajihara, Y., Inui, M., Tamura, K., Nakamura, J., Kimura, K., Yao, M., Itou, M., Sakurai, Y. & Hiraoka, N. (2013). Electron momentum density in liquid silicon. *Physical Review B: Condensed Matter and Materials Physics*, 88, 115125. doi: 10.1103/PhysRevB.88.115125

Matsuda, K., Tamura, K. & Inui, M. (2007). Instability of the Electron Gas in an Expanding Metal. *Physical Review Letters*, *98*, 096401. doi:10.1103/PhysRevLett.98.096401

Matsuda, K., Inui, M., & Tamura, K. (2006). Structural studies of expanded fluid rubidium up to the supercritical regions. *Science and Technology of Advanced Materials*, *7*(6), 483–489. doi:10.1016/j.stam.2006.07.003

Matsuda, K., Tamura, K., Inui, M., Kajihara, Y., Nagao, T., Yao, M., & Sakurai, Y. et al. (2011). Synchrotron radiation studies of an expanded fluid alkali metal: From the structure towards the electronic state. *The European Physical Journal. Special Topics*, *196*(1), 95–107. doi:10.1140/epjst/e2011-01421-4

Matsuda, K., Tamura, K., Katoh, M., & Inui, M. (2004). Molybdenum cell for x-ray diffraction measurements of fluid alkali metals at high temperatures and high pressures. *The Review of Scientific Instruments*, *75*(3), 709–712. doi:10.1063/1.1646769

Mazaheri, H., Lee, K. T., Bhatia, S., & Mohamed, A. R. (2010). Sub/supercritical liquefaction of oil palm fruit press fiber for the production of bio-oil: Effect of solvents. *Bioresource Technology*, *101*(19), 7641–7647. doi:10.1016/j.biortech.2010.04.072 PMID:20510608

McMahan, A., Klein, S. A., & Reindl, D. T. (2007). A Finite-Time Thermodynamic Framework for Optimizing Solar-Thermal Power Plants. *Journal of Solar Energy Engineering*, *129*(4), 355–362. doi:10.1115/1.2769689

Mehrani, R., Barati, M., Tavasoli, A., & Karimi, A. (2015). Hydrogen production via supercritical water gasification of bagasse using Ni–Cu/γ-Al2O3 nano-catalysts. *Environmental Technology*, *36*(10), 1265–1272. doi:10.1080/09593330.2014.984771 PMID:25387488

Meng, X., de Jong, W., Fu, N., & Verkooijen, A. H. M. (2011). Biomass gasification in a 100 kWth steam-oxygen blown circulating fluidized bed gasifier: Effects of operational conditions on product gas distribution and tar formation. *Biomass and Bioenergy*, *35*(7), 2910–2924. doi:10.1016/j.biombioe.2011.03.028

Meyer, H., & Kogan, A. B. (2002). Onset of convection in a very compressible fluid: The transient toward steady state. *Physical Review E: Statistical, Nonlinear, and Soft Matter Physics*, *63*(5), 056310. doi:10.1103/PhysRevE.66.056310 PMID:12513599

Mikielewicz, D. P., Shehata, A. M., Jackson, J. D., & McEligot, D. M. (2002). Temperature, velocity and mean turbulence structure in strongly heated internal gas flows - Comparison of numerical predictions with data. *International Journal of Heat and Mass Transfer*, *45*(21), 4333–4352. doi:10.1016/S0017-9310(02)00119-9

Miller, R. W. (1983). *Flow measurement engineering handbook*. Academic Press.

Mirzoev, A. A., Mirzoev, Jr., A. A., Sobolev, A. N., & Gelchinski, B. R. (2008). Analysis of the electronic structure of liquid rubidium by the methods of ab initio moleculardynamics, linear muffin-tin orbitals and recursion. *Journal of Physics: Condensed Matter*, *20*, 114104. doi:10.1088/0953-8984/20/11/114104

Misale, M., Frogheri, M., DAuria, F., Fontani, E., & Garcia, A. (1999). Analysis of single-phase natural circulation experiments by system codes. *International Journal of Thermal Sciences*, *38*(11), 977–983. doi:10.1016/S1290-0729(99)00106-4

Misale, M., Garibaldi, P., Passos, J. C., & Bitencourt, G. G. (2007). Experiments in a single-phase natural circulation mini-loop. *Experimental Thermal and Fluid Science*, *31*(8), 1111–1120. doi:10.1016/j.expthermflusci.2006.11.004

Miura, Y., Yoshihara, S., Ohnishi, M., Honda, K., Matsumoto, M., Kawai, J., & Onuki, A. et al. (2006). High-speed observation of the piston effect near the gasliquid critical point. [R]. *Physical Review E: Statistical, Nonlinear, and Soft Matter Physics*, *74*(1), 010101. doi:10.1103/PhysRevE.74.010101 PMID:16907038

Möbius, A., Boukis, N., Hindersin, S., & Kerner, M. (2015). Cultivation and Hydrothermal Gasification of Microalgae Scenedesmus Obliquus–First Experimental Results. *Chemie Ingenieur Technik*, *87*(12), 1720–1725. doi:10.1002/cite.201500002

Moisseytsev, A., Sienicki, J., & others. (2008b). *Performance improvement options for the supercritical carbon dioxide Brayton cycle*. Academic Press.

Moisseytsev, A., & Sienicki, J. et al. (2008a). *Performance improvement options for the supercritical carbon dioxide Brayton cycle. Argonne National Laboratory (ANL). Funding organisation*. USDOE Office of Science. doi:10.2172/935094

Moisseytsev, A., & Sienicki, J. (2011). *Investigation of plant control strategies for the supercritical CO_2 Brayton cycle for a sodium-cooled fast reactor using the plant dynamics code. Argonne National Laboratory (United States). Funding organisation*. USDOE Office of Nuclear Energy, Science and Technology.

Moisseytsev, A., & Sienicki, J. J. (2008a). Controllability of the supercritical carbon dioxide brayton cycle near the critical point. In *International Conference on Advances in Nuclear Power Plants, ICAPP 2008* (Vol. 2, pp. 799–809).

Moisseytsev, A., & Sienicki, J. J. (2008b). Transient accident analysis of a supercritical carbon dioxide Brayton cycle energy converter coupled to an autonomous lead-cooled fast reactor. *Nuclear Engineering and Design*, *238*(8), 2094–2105. doi:10.1016/j.nucengdes.2007.11.012

Moisseytsev, A., & Sienicki, J. J. (2012). Dynamic simulation and control of the S-CO_2 cycle: From full power to decay heat removal. In *Advances in Thermal Hydraulics 2012* (pp. 52–60). ATH.

Mori, H., Kaida, T., Ohno, M., Yoshida, S., & Hamamoto, Y. (2012). Heat transfer to a supercritical pressure fluid flowing in sub-bundle channels. *Journal of Nuclear Science and Technology Journal of Nuclear Science and Technology*, *49*(4), 373–383. doi:10.1080/00223131.2012.669240

Mori, K., Onishi, J., Shimaoka, H., Nakanishi, S., & Kimoto, H. (2002). Cooling heat transfer characteristics of CO2 and CO2-oil mixture at supercritical pressure conditions. In *Proc. of the Asian Conference on Refrigeration and Air Conditioning*. Kobe, Japan: Japan Society of Refrigeration and Air Conditioning Engineers.

Morley, J. (2001). *The Future of HFCs as Refrigeration Working fluids*. Hemel Hempstead, UK: Dupont Fluoroproducts.

Morrison, G. L., Budihardjo, I., & Behnia, M. (2004). Water-in-Glass Evacuated Tube Solar Water Heaters. *Solar Energy*, *76*(1–3), 135–140. doi:10.1016/j.solener.2003.07.024

Morrison, G. L., Budihardjo, I., & Behnia, M. (2005). Measurement and Simulation of Flow Rate in a Water-in-Glass Evacuated Tube Solar Water Heater. *Solar Energy*, *78*(2), 257–267. doi:10.1016/j.solener.2004.09.005

Mostaghim Besarati, S. (2014). *Analysis of Advanced Supercritical Carbon Dioxide Power Cycles for Concentrated Solar Power Applications*. Graduate Theses and Dissertations.

Mott, N. F. (1966). The electrical properties of liquid mercury. *Philosophical Magazine, 13*(125), 989–1014. doi:10.1080/14786436608213149

Mukherjee, A., Bhattacharyya, S., & Bagchi, B. (2002). Pressure and temperature dependence of viscosity and diffusion coefficients of a glassy binary mixture. *The Journal of Chemical Physics, 116*(11), 4577–4586. doi:10.1063/1.1445747

Musgrove, G., Conboy, T. M., & Wilkes, J. (2014). *Fundamentals of Supercritical CO_2*. Presented at the ASME Turbo Expo 2014, Düsseldorf, Germany.

Myong, H. K., & Kasagi, N. (1990). A new approach to the improvement of k-ε turbulence model for wall-bounded shear flows. *JSME International Journal, Series II., 33*(1), 63–72.

Myong, H. K., Kasagi, N., & Hirata, M. (1989). Numerical prediction of turbulent pipe flow heat transfer for various Prandtl number fluids with the improved k-ε turbulence model. *JSME International Journal, Series II., 32*(4), 613–622.

Nakano, A., & Shiraishi, M. (2004). Numerical simulation for the Piston effect and thermal diffusion observed in supercritical nitrogen. *Cryogenics, 44*(12), 867–873. doi:10.1016/j.cryogenics.2004.06.002

Nekså, P. (2002). CO_2 Heat Pump Systems. *International Journal of Refrigeration, 25*(4), 421–427. doi:10.1016/S0140-7007(01)00033-0

Nekså, P., Rekstad, H., Zakeri, G. R., & Schiefloe, P. A. (1998). CO_2-Heat Pump Water Heater: Characteristics, System Design and Experimental Results. *International Journal of Refrigeration, 21*(3), 172–179. doi:10.1016/S0140-7007(98)00017-6

Ngo, S. I., Nguyen, T. D. B., Lim, Y.-I., Song, B.-H., Lee, U.-D., Choi, Y.-T., & Song, J.-H. (2011). Performance evaluation for dual circulating fluidized-bed steam gasifier of biomass using quasi-equilibrium three-stage gasification model. *Applied Energy, 88*(12), 5208–5220. doi:10.1016/j.apenergy.2011.07.046

Nikitin, K., Kato, Y., & Ngo, L. (2006). Printed circuit heat exchanger thermal-hydraulic performance in supercritical CO_2 experimental loop. *International Journal of Refrigeration, 29*(5), 807–814. doi:10.1016/j.ijrefrig.2005.11.005

Nikolayev, V., Garrabos, Y., Lecoutre, C., Charignon, T., Hitz, D., Chatain, D., & Beysens, D. et al. (2015). *Boiling crisis dynamics: Low gravity experiments at high pressure*. Micogravity Sci. Technol; doi:10.1007/s12217-015-9447-8

NIST. (2000). *Thermophysical properties of pure fluids database*. NIST.

NIST. (2006). *NIST Standard Reference Database-REFPROP*. Version 8.0. NIST.

Niu, X. D., Yamaguchi, H., Iwamoto, Y., & Zhang, X. R. (2013). Optimal arrangement of the solar collectors of a supercritical CO_2-based solar Rankine cycle system. *Applied Thermal Engineering, 50*(1), 505–510. doi:10.1016/j.applthermaleng.2012.08.004

Niu, X. D., Yamaguchi, H., Zhang, X. R., Iwamoto, Y., & Hashitani, N. (2011). Experimental study of heat transfer characteristics of supercritical CO_2 fluid in collectors of solar Rankine cycle system. *Applied Thermal Engineering, 31*(6), 1279–1285. doi:10.1016/j.applthermaleng.2010.12.034

Nowak, W., Borsukiewicz-Gozdur, A., & Stachel, A. A. (2008). Using the low-temperature Clausius–Rankine cycle to cool technical equipment. *Applied Energy, 85*(7), 582–588. doi:10.1016/j.apenergy.2007.09.001

Noyori, R. (1999). Supercritical Fluids. *Chemical Reviews, 99,* 353-634.

Ochs, T. L., & O'Connor, W. K. (2006, March 7). *Energy recovery during expansion of compressed gas using power plant low-quality heat sources*. Retrieved from http://www.google.com/patents/US7007474

Ochsner, K. (2008). Carbon dioxide heat pipe in conjunction with a ground source heat pump (GSHP). *Applied Thermal Engineering*, *28*(16), 2077–2082. doi:10.1016/j.applthermaleng.2008.04.023

Olgun, H., Ozdogan, S., & Yinesor, G. (2011). Results with a bench scale downdraft biomass gasifier for agricultural and forestry residues. *Biomass and Bioenergy*, *35*(1), 572–580. doi:10.1016/j.biombioe.2010.10.028

Oliveira, I., Blöhse, D., & Ramke, H.-G. (2013). Hydrothermal carbonization of agricultural residues. *Bioresource Technology*, *142*, 138–146. doi:10.1016/j.biortech.2013.04.125 PMID:23735795

Onaka, Y., Miyara, A., Tsubaki, K., & Koyama, S. (2007), Performance analysis on heat pump cycle of CO2/DME mixture refrigerant. *Proceedings of JSRAE Annual Conference,* 333-336. (in Japanese)

Onuki, A. (1991). Sound propagation in phase-separating fluids. *Physical Review A.*, *43*(12), 6740–6755. doi:10.1103/PhysRevA.43.6740 PMID:9905025

Onuki, A., & Ferrell, R. A. (1990). Adiabatic heating effect near the gasliquid critical point. *Physica A*, *164*(2), 245–264. doi:10.1016/0378-4371(90)90198-2

Onuki, A., Hao, H., & Ferrel, R. A. (1990a). Fast adiabatic equilibration near the liquid-vapor critical point. *Physical Review A.*, *41*(4), 2256–2259. doi:10.1103/PhysRevA.41.2256 PMID:9903349

Padilla, R. V., Soo Too, Y. C., Benito, R., & Stein, W. (2015). Exergetic analysis of supercritical CO_2 Brayton cycles integrated with solar central receivers. *Applied Energy*, *148*, 348–365. doi:10.1016/j.apenergy.2015.03.090

Padilla, R. V., Too, Y. C. S., Benito, R., McNaughton, R., & Stein, W. (2016). Thermodynamic feasibility of alternative supercritical CO_2 Brayton cycles integrated with an ejector. *Applied Energy*, *169*, 49–62. doi:10.1016/j.apenergy.2016.02.029

Pantoula, M., & Panayiotou, C. (2006). Sorption and swelling in glassy polymer/carbon dioxide systems: Part I. Sorption. *The Journal of Supercritical Fluids*, *37*(2), 254–262. doi:10.1016/j.supflu.2005.11.001

Paolucci S. (1982). *On the filtering of sound from the Navier-Stokes equations.* SAND 82-8257.

Paris COP21. (2015). *Climate Change Conference.* Retrieved October 14, 2016 from http://www.cop21.gouv.fr/en/

Parshetti, G. K., Kent Hoekman, S., & Balasubramanian, R. (2013). Chemical, structural and combustion characteristics of carbonaceous products obtained by hydrothermal carbonization of palm empty fruit bunches. *Bioresource Technology*, *135*, 683–689. doi:10.1016/j.biortech.2012.09.042 PMID:23127830

Pasch, J., Conboy, T., Fleming, D., Carlson, M., & Rochau, G. (2014). Steady State Supercritical Carbon Dioxide Recompression Closed Brayton Cycle Operating Point Comparison With Predictions. In *ASME Turbo Expo 2014: Turbine Technical Conference and Exposition* (p. V03BT36A008). ASME. doi:10.1115/GT2014-25777

Patankar, S. V. (1980). *Numerical Heat Transfer and Fluid flow*. New York: Hemisphere.

Patankar, S. V. (1980). *Numerical heat transfer and fluid flow*. New York: McGraw Hill.

Patel, V. C., Rodi, W., & Shceuerer, G. (1984). Turbulence models for near-wall and low Reynolds number flows. A review. *AIAA Journal*, *23*(9), 1308–1319. doi:10.2514/3.9086

Pattison, P., & Williams, B. (1976). Fermi surface parameters from the Fourier analysis of Compton profiles. *Solid State Communications*, *20*(6), 585–588. doi:10.1016/0038-1098(76)91066-8

Patzelt, D. J., Hindersin, S., Elsayed, S., Boukis, N., Kerner, M., & Hanelt, D. (2015). Hydrothermal gasification of Acutodesmus obliquus for renewable energy production and nutrient recycling of microalgal mass cultures. *Journal of Applied Phycology*, *27*(6), 2239–2250. doi:10.1007/s10811-014-0496-y

Pavlovič, I., Knez, Ž., & Škerget, M. (2013). Hydrothermal Reactions of Agricultural and Food Processing Wastes in Sub- and Supercritical Water: A Review of Fundamentals, Mechanisms, and State of Research. *Journal of Agricultural and Food Chemistry*, *61*(34), 8003–8025. doi:10.1021/jf401008a PMID:23848589

Pecnik, R., Rinaldi, E., & Colonna, P. (2012). Computational Fluid Dynamics of a Radial Compressor Operating With Supercritical CO_2. In *ASME Turbo Expo 2012: Turbine Technical Conference and Exposition. Volume 5: Manufacturing Materials and Metallurgy; Marine; Microturbines and Small Turbomachinery; Supercritical CO_2 Power Cycles* (Vol. 5, p. 985). ASME. https://doi.org/10.1115/GT2012-69640

Persichilli, M., Kacludis, A., Zdankiewicz, E., & Held, T. (2012). *Supercritical CO_2 Power Cycle Developments and Commercialization: Why SCO_2 can Displace Steam.* Presented at the Power-Gen India & Central Asia.

Peter, L., David, W., Michael, F., Fincke, E., Leroy, C., & William, M. (2007). Microgravity Phase Separation Near the Critical Point in Attractive Colloids. *45th AIAA Aerospace Sciences Meeting and Exhibit.*

Peterson, A. A., Vogel, F., Lachance, R. P., Fröling, M., Antal, M. J. Jr, & Tester, J. W. (2008). Thermochemical biofuel production in hydrothermal media: A review of sub-and supercritical water technologies. *Energy & Environmental Science*, *1*(1), 32–65. doi:10.1039/b810100k

Petrov, N. E., & Popov, V. N. (1988). Heat transfer and hydraulic resistance with turbulent flow in a tube of water under supercritical parameters of state. *Thermal Engineering*, *35*(6), 577–580.

Pfaltzgraff, L. A., Cooper, E. C., Budarin, V., & Clark, J. H. (2013). Food waste biomass: A resource for high-value chemicals. *Green Chemistry*, *15*(2), 307–314. doi:10.1039/c2gc36978h

Pfeifer, H. P., Freyland, W., & Hensel, F. (1979). Equation of state and transport data on expanded liquid rubidium up to 1700 °C and 400 bar. *Berichte der Bunsengesellschaft fur Physikalische Chemie*, *83*(3), 204–211. doi:10.1002/bbpc.19790830305

Phillips, T. (2014). *Starting Fire With Water.* Retrieved November 11, from https://science.nasa.gov/science-news/science-at-nasa/2014/10jan_firewater

Pilgrim, W. C., Ross, M., Yang, L. H., & Hensel, F. (1997). Monatomic-molecular transition in expanded rubidium. *Physical Review Letters*, *78*(19), 3685–3688. doi:10.1103/PhysRevLett.78.3685

Pioro, I. L., Khartabil, H. F., & Duffey, R. B. (2004). Heat transfer to supercritical fluids flowing in channels - empirical correlations (survey). *Nuclear Engineering and Design*, *230*(1-3), 69–91. doi:10.1016/j.nucengdes.2003.10.010

Pitla, S. S., Ramadhyani, S., & Groll, E. A. (2001). Convective heat transfer from in-tube flow of turbulent supercritical carbon dioxide: Part 1-numerical analysis. *International Journal HVAC&R Research.*, *7*(4), 345–366. doi:10.1080/10789669.2001.10391280

PROPATH Group. (1997). *PROPATH: a program package for thermophysical properties.* Author.

Pumaneratkul, C., Iwamoto, Y, Yamaguchi, H. (2015). Nozzle test to Supercritical CO_2 turbine in Rankine cycle system. *International Journal of Advances in Science and Technology*, 167-174.

Pumaneratkul, C., Iwamoto, Y., Niu, X. D., Kuwahara, T., & Yamaguchi, H. (2015). Characteristics of thermally driven pump in supercritical CO_2 solar Rankine cycle system. *International Journal of Mechanical and Production Engineering*, *3*(4), 1–6.

Qiu, G., Liu, H., & Riffat, S. (2011). Expanders for micro-CHP systems with organic Rankine cycle. *Applied Thermal Engineering*, *31*(16), 3301–3307. doi:10.1016/j.applthermaleng.2011.06.008

Quoilin, S. (2007, June). *Experimental study and modeling of a low temperature Rankine Cycle for small scale cogeneration*. Université de Liège. Retrieved from http://orbi.ulg.ac.be/handle/2268/1273

Redmer, R., & Warren, W. W. (1993). Magnetic-Susceptibility of Cs and Rb from the Vapor to the Liquid-Phase. *Physical Review B: Condensed Matter and Materials Physics*, *48*(20), 14892–14906. doi:10.1103/PhysRevB.48.14892 PMID:10008020

REFPROP. (n.d.). *NIST Standard Reference Database 23, Version 7.0*. NIST.

Reiter, G. F., Deb, A., Sakurai, Y., Itou, M., Krishnan, V.G., & Paddison, S. J. (2013). Anomalous Ground State of the Electrons in Nanoconfined Water. *Physical Review Letters*, *111*, 036803. doi: 10.1103/PhysRevLett.111.036803.

Reza, M. T., Lynam, J. G., Uddin, M. H., & Coronella, C. J. (2013). Hydrothermal carbonization: Fate of inorganics. *Biomass and Bioenergy*, *49*, 86–94. doi:10.1016/j.biombioe.2012.12.004

Román, S., Nabais, J. M. V., Laginhas, C., Ledesma, B., & González, J. F. (2012). Hydrothermal carbonization as an effective way of densifying the energy content of biomass. *Fuel Processing Technology*, *103*, 78–83. doi:10.1016/j.fuproc.2011.11.009

Rosengarten, G., Morrison, G., & Behnia, M. (1999). A Second Law Approach to Characterising Thermally Stratified Hot Water Storage With Application to Solar Water Heaters, *ASME. Journal of Solar Energy Engineering*, *121*(4), 194–200. doi:10.1115/1.2888166

Rutin, S. B., & Skripov, P. V. (2012). Apparatus for studying heat transfer in nanofluids under high-power heating. *Journal of Engineering Thermophysics*, *21*(2), 144–153. doi:10.1134/S1810232812020063

Rutin, S. B., & Skripov, P. V. (2013a). Investigation of not fully stable fluids by the method of controlled pulse heating. 1. Experimental approach. *Thermochimica Acta*, *562*, 70–74. doi:10.1016/j.tca.2013.03.030

Rutin, S. B., & Skripov, P. V. (2013b). Heat transfer under supercritical parameters of pulse-heated liquid. *Russian Journal of Physical Chemistry B*, *7*(8), 943–949. doi:10.1134/S1990793113080101

Rutin, S. B., & Skripov, P. V. (2013c). Heat transfer in supercritical fluids under pulse heating regime. *International Journal of Heat and Mass Transfer*, *57*(1), 126–130. doi:10.1016/j.ijheatmasstransfer.2012.10.027

Rutin, S. B., & Skripov, P. V. (2014). Heat transfer under high-power heating of liquids. 2. Transition from compressed to supercritical water. *International Journal of Heat and Mass Transfer*, *79*, 526–531. doi:10.1016/j.ijheatmasstransfer.2014.08.043

Rutin, S. B., & Skripov, P. V. (2016). Controlled high-power heat release as a tool to selecting working pressure for supercritical water. *Journal of Engineering Thermophysics*, *25*(2), 166–173. doi:10.1134/S1810232816020028

Rutin, S. B., Smotritskiy, A. A., Starostin, A. A., Okulovsky, Yu. S., & Skripov, P. V. (2013). Heat transfer under high-power heating of liquids. 1. Experiment and inverse algorithm. *International Journal of Heat and Mass Transfer*, *62*, 135–141. doi:10.1016/j.ijheatmasstransfer.2013.02.060

Rutin, S. B., Volosnikov, D. V., & Skripov, P. V. (2015). Heat transfer under high-power heating of liquids. 3. Threshold decrease of heat conduction in supercritical region. *International Journal of Heat and Mass Transfer*, *91*, 1–6. doi:10.1016/j.ijheatmasstransfer.2015.07.001

Sakai, N. (1987). Simulation of Compton Double Scatterings of Circularly Polarized γ-Rays by Magnetic Electrons. *Journal of the Physical Society of Japan*, *56*(7), 2477–2485. doi:10.1143/JPSJ.56.2477

Sargent and Lundy LLC. (2003). *Assessment of Parabolic Trough and Power Tower Solar Technology Cost and Performance Forecasts*. DIANE Publishing.

Sarkar, M. K. S., & Basu, D. N. (2017). Numerical appraisal on the suitability of supercritical condition in natural circulation loop with isothermal boundary conditions. *International Journal of Thermal Sciences*, *111*, 30–40. doi:10.1016/j.ijthermalsci.2016.08.002

Sarkar, M. K. S., Tilak, A. K., & Basu, D. N. (2014). A state-of-the-art review of recent advances in supercritical natural circulation loops for nuclear applications. *Annals of Nuclear Energy*, *73*, 250–263. doi:10.1016/j.anucene.2014.06.035

Schmidt, E. (1960). Warmetransport durch naturliche konvektion in stiffen bei kritischem zustand. *International Journal of Heat and Mass Transfer*, *1*(1), 92–101. doi:10.1016/0017-9310(60)90012-0

Schmieder, H., Abeln, J., Boukis, N., Dinjus, E., Kruse, A., Kluth, M., & Schacht, M. et al. (2000). Hydrothermal gasification of biomass and organic wastes. *The Journal of Supercritical Fluids*, *17*(2), 145–153. doi:10.1016/S0896-8446(99)00051-0

Selvi Gökkaya, D., Saglam, M., Yüksel, M., & Ballice, L. (2015). Supercritical water gasification of phenol as a model for plant biomass. *International Journal of Hydrogen Energy*, *40*(34), 11133–11139. doi:10.1016/j.ijhydene.2015.04.078

Sengers, J. V. (1962). *Thermal conductivity measurements at elevated gas densities including the critical region* (Unpublished thesis). Universiteit van Amsterdam, Amsterdam, The Netherlands.

Sengers, J. V. (1965). Thermal conductivity and viscosity of simple fluids. *International Journal of Heat and Mass Transfer*, *8*(8), 1103–1116. doi:10.1016/0017-9310(65)90138-9

Sermyagina, E., Saari, J., Kaikko, J., & Vakkilainen, E. (2015). Hydrothermal carbonization of coniferous biomass: Effect of process parameters on mass and energy yields. *Journal of Analytical and Applied Pyrolysis*, *113*, 551–556. doi:10.1016/j.jaap.2015.03.012

Shabani, N., Akhtari, S., & Sowlati, T. (2013). Value chain optimization of forest biomass for bioenergy production: A review. *Renewable & Sustainable Energy Reviews*, *23*, 299–311. doi:10.1016/j.rser.2013.03.005

Shan, X. Y., Schmidt, D. P., & Watkins, J. J. (2007). Study of natural convection in supercritical CO_2 cold wall reactors: Simulations and experiments. *The Journal of Supercritical Fluids*, *40*(1), 84–92. doi:10.1016/j.supflu.2006.03.023

Shimoji, M. (1977). *Liquid Metals An Introduction to the Physics and Chemistry of Metals in the Liquid State*. London: Academic Press.

Shiralkar, B. S., & Griffith, P. (1969). Deterioration in heat transfer to fluids at supercritical pressure and high heat fluxes. *Journal of Heat Transfer*, *91*(1), 27–36. doi:10.1115/1.3580115

Shülke, W. (2007). *Electron Dynamics by Inelastic X-Ray Scattering*. Oxford University Press.

Singh, R., Kearney, M. P., & Manzie, C. (2013). Extremum-seeking control of a supercritical carbon-dioxide closed Brayton cycle in a direct-heated solar thermal power plant. *Energy*, *60*, 380–387. doi:10.1016/j.energy.2013.08.001

Singh, R., Miller, S. A., Rowlands, A. S., & Jacobs, P. A. (2013). Dynamic characteristics of a direct-heated supercritical carbon-dioxide Brayton cycle in a solar thermal power plant. *Energy*, *50*, 194–204. doi:10.1016/j.energy.2012.11.029

Skripov, P. V., Smotritskiy, A. A., Starostin, A. A., & Shishkin, A. V. (2007). A method of controlled pulse heating: Applications. *Journal of Engineering Thermophysics*, *16*(3), 155–163. doi:10.1134/S1810232807030071

Skripov, V. P., & Potashev, P. P. (1962). Heat exchange with carbon dioxide along the supercritical isobars under free convection. [in Russian]. *Journal of Engineering Physics*, *5*(2), 30–34.

Söderlind, P., & Ross, M. (2000). Rubidium at high pressure and temperature. *Journal of Physics Condensed Matter*, *12*(6), 921–931. doi:10.1088/0953-8984/12/6/314

Spiegel, E. A. (1965). Convective instability in a compressible atmosphere. *I. The Astrophysical Journal*, *141*, 1068. doi:10.1086/148197

Stanley, H. E. (1971). *Introduction to Phase Transitions and Critical Phenomena*. New York: Oxford University Press.

Stanley, H. E. (1971). *Introduction to Phase Transitions and Critical Point Phenomena*. Oxford, UK: Oxford University Press.

Stauffer, D., Ferer, M., & Wortis, M. (1972). Universality of second-order phase transitions: The scale factor for the correlation length. *Physical Review Letters*, *29*(6), 345–349. doi:10.1103/PhysRevLett.29.345

Stene, J. (2005). Residential CO_2 Heat Pump System for Combined Space Heating and Hot Water Heating. *International Journal of Refrigeration*, *28*(8), 1259–1265. doi:10.1016/j.ijrefrig.2005.07.006

Straub, J., Eicher, L., & Haupt, A. (1995). Dynamic temperature propagation in a pure fluid near its critical point observed under microgravity during the German Spacelab mission D-2. *Physical Review E: Statistical Physics, Plasmas, Fluids, and Related Interdisciplinary Topics*, *51*(6), 5556–5563. doi:10.1103/PhysRevE.51.5556 PMID:9963290

Sturm, K., & Oliveira, L. E. (1981). Wave-vector-dependent plasmon linewidth in the alkali metals. *Physical Review B: Condensed Matter and Materials Physics*, *24*(6), 3054–3062. doi:10.1103/PhysRevB.24.3054

Sugathapala, A. G. T. (2013). *Technologies for Converting Waste Agricultural Biomass to Energy*. Osaka: Academic Press.

Supercritical fluids - YouTube. (n.d.). Retrieved October 27, 2016, from https://www.youtube.com/watch?v=yBRdBrnIlTQ

Su, S., Fang, X., Li, L., Wu, J., Zhang, J., Xu, W., & Hu, J. (2015). HFC-134a emissions from mobile air conditioning in China from 1995 to 2030. *Atmospheric Environment*, *102*, 122–129. doi:10.1016/j.atmosenv.2014.11.057

Swapnalee, B. T., Vijayan, P. K., Sharma, M., & Pilkhwal, D. S. (2012). Steady state flow and static instability of supercritical natural circulation loops. *Nuclear Engineering and Design*, *245*, 99–112. doi:10.1016/j.nucengdes.2012.01.002

Tamura, K., & Hosokawa, S. (1991). X-ray diffraction measurements for expanded liquid mercury around the metal-nonmetal transition region. *Journal de Physique. IV*, *C5*, 39–44.

Tamura, K., & Hosokawa, S. (1998). Structural studies of expanded fluid mercury up to the liquid-vapor critical region. *Physical Review B: Condensed Matter and Materials Physics*, *58*(14), 9030–9038. doi:10.1103/PhysRevB.58.9030

Tamura, K., Inui, M., & Hosokawa, S. (1999). Energy-dispersive x-ray diffraction equipment for fluids at extreme conditions of high temperatures and high pressures. *The Review of Scientific Instruments*, *70*(1), 144–152. doi:10.1063/1.1149556

Tanaka, H., Nishiwaki, N., & Hirata, M. (1967). Turbulent heat transfer in vertical tubes at supercritical pressures. *Proc. J.S.M.E. Semi-International Symposium*, 127-134.

Tchanche, B. F., Papadakis, G., Lambrinos, G., & Frangoudakis, A. (2009). Fluid selection for a low-temperature solar organic Rankine cycle. *Applied Thermal Engineering*, *29*(11–12), 2468–2476. doi:10.1016/j.applthermaleng.2008.12.025

Thermodynamic Tables Project. (1976). International Thermodynamic Tables of the Fluid State: Carbon Dioxide. Pergamon Press.

Tian, Y., & Zhao, C.-Y. (2013). A review of solar collectors and thermal energy storage in solar thermal applications. *Applied Energy*, *104*, 538–553. doi:10.1016/j.apenergy.2012.11.051

Titirici, M.-M., Thomas, A., & Antonietti, M. (2007). Back in the black: Hydrothermal carbonization of plant material as an efficient chemical process to treat the CO_2 problem? *New Journal of Chemistry*, *31*(6), 787–789. doi:10.1039/b616045j

TJoen, C., & Rohde, M.TJoen. (2012). Experimental study of the coupled thermo-hydraulic-nertronic stability of a natural circulation HPLWR. *Nuclear Engineering and Design*, *242*, 221–232. doi:10.1016/j.nucengdes.2011.10.055

Toor, S. S., Rosendahl, L., & Rudolf, A. (2011). Hydrothermal liquefaction of biomass: A review of subcritical water technologies. *Energy*, *36*(5), 2328–2342. doi:10.1016/j.energy.2011.03.013

Turchi, C. S., Ma, Z., & Dyreby, J. (2012a). Supercritical Carbon Dioxide Power Cycle Configurations for Use in Concentrating Solar Power Systems. In *ASME Turbo Expo 2012: Turbine Technical Conference and Exposition. Volume 5: Manufacturing Materials and Metallurgy; Marine; Microturbines and Small Turbomachinery; Supercritical CO_2 Power Cycles* (Vol. 5, pp. 967–973). ASME. https://doi.org/10.1115/GT2012-68932

Turchi, C. S., Ma, Z., & Dyreby, J. (2012b). Supercritical carbon dioxide power cycle configurations for use in concentrating solar power systems. In *ASME Turbo Expo 2012: Turbine Technical Conference and Exposition* (pp. 967–973). San Diego, CA: American Society of Mechanical Engineers. doi:10.1115/GT2012-68932

Turchi, C. S., Ma, Z., Neises, T., & Wagner, M. (2012). Thermodynamic Study of Advanced Supercritical Carbon Dioxide Power Cycles for High Performance Concentrating Solar Power Systems. In *ASME Turbo Expo 2012: Turbine Technical Conference and Exposition* (pp. 375–383). San Diego, CA: American Society of Mechanical Engineers. doi:10.1115/ES2012-91179

Turchi, C., Ma, Z., & Dyreby, J. (2009). Supercritical CO2 for application in concentrating solar power systems. *Proceedings of Supercritical CO_2 Power Cycle Symposium.*

Turchi, C. S., Ma, Z., Neises, T. W., & Wagner, M. J. (2013a). Thermodynamic Study of Advanced Supercritical Carbon Dioxide Power Cycles for Concentrating Solar Power Systems. *Journal of Solar Energy Engineering*, *135*(4), 41007. doi:10.1115/1.4024030

Turchi, C., Mehos, M., Ho, C. K., & Kolb, G. J. (2010). *Current and Future Costs for Parabolic Trough and Power Tower Systems in the Us Market: Preprint (No. NREL/CP-5500-49303). National Renewable Energy Laboratory*. Golden, CO: NREL.

Umeki, K., Yamamoto, K., Namioka, T., & Yoshikawa, K. (2010). High temperature steam-only gasification of woody biomass. *Applied Energy*, *87*(3), 791–798. doi:10.1016/j.apenergy.2009.09.035

Urban, M., & Sadlej, A. J. (1995). Electronic structure and electric properties of the alkali metal dimers. *The Journal of Chemical Physics*, *103*(22), 9692–9704. doi:10.1063/1.469984

Üremek Cengiz, N., Eren, S., Sağlam, M., Yüksel, M., & Ballice, L. (2016). Influence of temperature and pressure on hydrogen and methane production in the hydrothermal gasification of wood residues. *The Journal of Supercritical Fluids*, *107*, 243–249. doi:10.1016/j.supflu.2015.09.018

Vadillo, V., Sánchez-Oneto, J., Portela, J. R., & Martínez de la Ossa, E. J. (2013). Problems in supercritical water oxidation process and proposed solutions. *Industrial & Engineering Chemistry Research*, *52*(23), 7617–7629. doi:10.1021/ie400156c

Van Bennekom, J., Vos, J., Venderbosch, R., Paris Torres, M., Kirilov, V., Herres, H.J., … Penninger, J. (2009). *Supermethanol: reforming of crude glycerine in supercritical water to produce methanol for re-use in biodiesel plants*. Paper presented at 17th European Biomass Conference and Exhibition, Hamburg, Germany.

van de Beld, B., Elliott, D. C., Neuenschwander, G. G., & Kruse, A. (2005). Biomass gasification in near-and supercritical water: Status and prospects. *Biomass and Bioenergy*, *29*(4), 269–292. doi:10.1016/j.biombioe.2005.04.006

Van Driest, E. R. (1956). On turbulent flow near a wall. *Journal of the Aeronautical Sciences*, *23*(11), 1007–1011. doi:10.2514/8.3713

Vardon, D. R., Sharma, B. K., Blazina, G. V., Rajagopalan, K., & Strathmann, T. J. (2012). Thermochemical conversion of raw and defatted algal biomass via hydrothermal liquefaction and slow pyrolysis. *Bioresource Technology*, *109*, 178–187. doi:10.1016/j.biortech.2012.01.008 PMID:22285293

Vardon, D. R., Sharma, B., Scott, J., Yu, G., Wang, Z., Schideman, L., & Strathmann, T. J. et al. (2011). Chemical properties of biocrude oil from the hydrothermal liquefaction of Spirulina algae, swine manure, and digested anaerobic sludge. *Bioresource Technology*, *102*(17), 8295–8303. doi:10.1016/j.biortech.2011.06.041 PMID:21741234

Venderbosch, R. H., Ardiyanti, A. R., Wildschut, J., Oasmaa, A., & Heeres, H. J. (2010). Stabilization of biomass-derived pyrolysis oils. *Journal of Chemical Technology and Biotechnology (Oxford, Oxfordshire)*, *85*(5), 674–686. doi:10.1002/jctb.2354

Vera, D., Jurado, F., Margaritis, N. K., & Grammelis, P. (2014). Experimental and economic study of a gasification plant fuelled with olive industry wastes. *Energy for Sustainable Development*, *23*, 247–257. doi:10.1016/j.esd.2014.09.011

Victor, D. G., & Kennel, C. F. (2014). Climate policy: Ditch the 2 C warming goal. *Nature*, *514*(7520), 30–31. doi:10.1038/514030a PMID:25279903

Vijayan, P. K., Bade, M. H., Saha, D., Sinha, R. K., & Venkat, R. V. (2004). A generalized flow correlation for single-phase natural circulation loops. *Proceedings of the 17th National and 6th ISHMT/ASME Heat and Mass Transfer Conference*.

Vijayan, P. K., Austregesilo, H., & Teschendorff, V. (1995). Simulation of the unstable oscillatory behavior of single-phase natural circulation with repetitive flow reversals in a rectangular loop using the computer code ATHLET. *Nuclear Engineering and Design*, *155*(3), 623–641. doi:10.1016/0029-5493(94)00972-2

vom Felde, A., Sprösser-Prou, J., & Fink, J. (1989). Valence-electron excitations in the alkali metals. *Physical Review B: Condensed Matter and Materials Physics*, *40*(15), 10181–10193. doi:10.1103/PhysRevB.40.10181 PMID:9991563

Wagner, W., & Prußb, A. (2002). The IAPWS formulation 1995 for the thermodynamic properties of ordinary water substance for general and scientific use. *Journal of Physical and Chemical Reference Data*, *31*(2), 387–535. doi:10.1063/1.1461829

Wakefield, C. B. (1996). *Non ideal gas behavior (No. CONF-960220--)*. Edmond, OK: University of Central Oklahoma.

Waldner, M. H., & Vogel, F. (2005). Renewable production of methane from woody biomass by catalytic hydrothermal gasification. *Industrial & Engineering Chemistry Research*, *44*(13), 4543–4551. doi:10.1021/ie050161h

Wallace, J. F., & Linning, A. W. (1970). *Basic engineering thermodynamics*. Pitman Publishing Ltd.

Wang, F., Chang, Z., Duan, P., Yan, W., Xu, Y., Zhang, L., & Fan, Y. et al. (2013). Hydrothermal liquefaction of Litsea cubeba seed to produce bio-oils. *Bioresource Technology*, *149*, 509–515. doi:10.1016/j.biortech.2013.09.108 PMID:24140857

Wang, H., Bi, Q., & Leung, L. K. H. (2016). Heat Transfer from a 2 × 2 Wire-Wrapped Rod Bundle to Supercritical Pressure Water. *International Journal of Heat and Mass Transfer*, *97*, 486–501. doi:10.1016/j.ijheatmasstransfer.2016.02.036

Wang, H., Wang, W., Bi, Q., & Wang, L. (2015). Experimental study of heat transfer and flow resistance of supercritical pressure water in a SCWR sub-channel. *The Journal of Supercritical Fluids*, *100*, 15–25. doi:10.1016/j.supflu.2015.02.011

Wang, J., Sun, Z., Dai, Y., & Ma, S. (2010). Parametric optimization design for supercritical CO_2 power cycle using genetic algorithm and artificial neural network. *Applied Energy*, *87*(4), 1317–1324. doi:10.1016/j.apenergy.2009.07.017

Wang, K., Xu, X., Wu, Y., Liu, C., & Dang, C. (2015). Numerical investigation on heat transfer of supercritical CO_2 in heated helically coiled tubes. *The Journal of Supercritical Fluids*, *99*, 112–120. doi:10.1016/j.supflu.2015.02.001

Wang, L., Shahbazi, A., & Hanna, M. A. (2011). Characterization of corn stover, distiller grains and cattle manure for thermochemical conversion. *Biomass and Bioenergy*, *35*(1), 171–178. doi:10.1016/j.biombioe.2010.08.018

Wang, Y., Wang, S., Zhao, G., Guo, Y., & Guo, Y. (2016). Hydrogen production by partial oxidation gasification of a phenol, naphthalene, and acetic acid mixture in supercritical water. *International Journal of Hydrogen Energy*, *41*(4), 2238–2246. doi:10.1016/j.ijhydene.2015.12.115

Warren, W. W. Jr, Brennert, G. F., & El-Hanany, U. (1989). NMR investigation of the electronic structure of expanded liquid cesium. *Physical Review B: Condensed Matter and Materials Physics*, *39*(7), 4038–4050. doi:10.1103/PhysRevB.39.4038 PMID:9948738

Wasteshipment. (2016). Retrieved May, 10, 2016, from wasteshipment.com/waste-management

Welander, P. (1967). On the oscillatory instability of a differentially heated fluid loop. *Journal of Fluid Mechanics*, *29*(01), 17–30. doi:10.1017/S0022112067000606

White, C. M. (1932). Friction Factor and Its Relation to Heat Transfer. *Transactions of the Institution of Chemical Engineers*, *18*, 66–86.

Wichtmann, A., Deckers, M., & Ulm, W. (2005). *Ultra-supercritical steam turbine turbosets – best efficiency solution for conventional steam power plants*. Paper presented at International Conference on Electrical Engineering, Kunming, China.

Wigner, E. P. (1934). On the Interaction of Electrons in Metals. *Physical Review*, *46*(11), 1002–1011. doi:10.1103/PhysRev.46.1002

Wilkinson, R. A. (1998). Density relaxation of liquid-vapor critical fluids in Earths gravity. *International Journal of Thermophysics*, *19*(4), 1175–1183. doi:10.1023/A:1022650011721

Wilkinson, R. A., Zimmerli, G. A., Hao, H., Moldover, M. R., Berg, R. F., Johnson, W. L., & Gammon, R. W. et al. (1998). Equilibration near the liquidvapor critical point in microgravity. *Physical Review E: Statistical Physics, Plasmas, Fluids, and Related Interdisciplinary Topics*, *57*(1), 436–448. doi:10.1103/PhysRevE.57.436

Wilson, A. H. (1931). The theory of electronic Semi-Conductors. *Proceedings of the Royal Society of London. Series A, Containing Papers of a Mathematical and Physical Character*, *133*(822), 458–491. doi:10.1098/rspa.1931.0162

Winterbone, D. E. (1997). *Advanced thermodynamics for engineers*. London: Arnold.

Winter, R., Hensel, F., Bodensteiner, T., & Gläser, W. (1987). The static structure factor of Cesium over the whole liquid range up to the critical point. *Berichte der Bunsengesellschaft fur Physikalische Chemie*, *91*(12), 1327–1330. doi:10.1002/bbpc.19870911205

Woolf, D., Amonette, J. E., Street-Perrott, F. A., Lehmann, J., & Joseph, S. (2010). Sustainable biochar to mitigate global climate change. *Nature Communications*, *1*(5), 56. doi:10.1038/ncomms1053 PMID:20975722

Wright, S. A., Conboy, T. M., & Rochau, G. E. (2011). Supercritical CO2 Power Cycle Development Summary at Sandia National Laboratories. In *International Conference on ORC Power Systems Delft.*

Wright, S. A., Fuller, R., Noall, J., Radel, R., Vernon, M. E., & Pickard, P. S. (2008). Supercritical CO2 brayton cycle compression and control near the critical point. In *International Conference on Advances in Nuclear Power Plants, ICAPP 2008* (*Vol. 2*, pp. 810–819).

Wright, S. A., & Pickard, P. S. (2009). Supercritical CO2 test loop operation and first test results. In *International Congress on Advances in Nuclear Power Plants 2009, ICAPP 2009* (Vol. 1, pp. 351–360). Atomic Energy Society of Japan.

Wright, S. A., Pickard, P. S., Vernon, M. E., Radel, R. F., & Fuller, R. (2009). Description and test results from a supercritical CO2 Brayton cycle development program. In *7th International Energy Conversion Engineering Conference*. doi:10.2514/6.2009-4607

Wunenburger, R., Garrabos, Y., Lecoutre, C., Beysens, D., Hegseth, J., Zhong, F., & Barmatz, M. (2002). Thermal response of a two-phase near-critical fluid in low gravity: Strong gas overheating as due to a particular phase distribution. *International Journal of Thermophysics*, *23*(1), 103–115. doi:10.1023/A:1013948925906

Wunenburger, R., Garrabos, Y., Lecoutre-Chabot, C., Beysens, D., & Hegseth, J. (2000). Thermalization of a two-phase fluid in low gravity: Heat transferred from cold to hot. *Phys. Rev. Lett. American Physical Society*, *84*, 4100–4103.

Xiao, L.-P., Shi, Z.-J., Xu, F., & Sun, R.-C. (2012). Hydrothermal carbonization of lignocellulosic biomass. *Bioresource Technology*, *118*, 619–623. doi:10.1016/j.biortech.2012.05.060 PMID:22698445

Xi, X., Xiao, Z., Yan, X., Xiong, T., & Huang, Y. (2014). Numerical simulation of the flow instability between two heated parallel channels with supercritical water. *Annals of Nuclear Energy*, *64*, 57–66. doi:10.1016/j.anucene.2013.09.017

Xu, X., Wu, Y., Liu, C., Wang, K., Dang, C., & Bai, W. (2016). Experimental investigation on heat transfer characteristics of supercritical CO_2 cooled in horizontal helically coiled tube. *International Journal of Refrigeration*, *67*, 190–201. doi:10.1016/j.ijrefrig.2016.03.010

Yadav, A. K., Gopal, M. R., & Bhattacharyya, S. (2012). CO_2 based natural circulation loops: New correlations for friction and heat transfer. *International Journal of Heat and Mass Transfer*, *55*(17-18), 4621–4630. doi:10.1016/j.ijheatmasstransfer.2012.04.019

Yakaboylu, O., Harinck, J., Smit, K., & de Jong, W. (2015). Supercritical water gasification of biomass: A literature and technology overview. *Energies*, *8*(2), 859–894. doi:10.3390/en8020859

Yamaguchi, H., Zhang, X. R., & Fujima, K. (2008). Basic study on new cryogenic refrigeration using CO 2 solid–gas two phase flow. *International Journal of Refrigeration, 31*(3), 404-410.

Yamaguchi, H. (2008). *Engineering fluid mechanics* (Vol. 85). Springer Science & Business Media.

Yamaguchi, H., Sawada, N., Suzuki, H., Ueda, H., & Zhang, X. R. (2010). Preliminary study on a solar water heater using supercritical carbon dioxide as working fluid. *Journal of Solar Energy Engineering-Transactions of ASME, 132*(1), 101–106. doi:10.1115/1.4000350

Yamaguchi, H., Zhang, X. R., Niu, X. D., & Hashitani, N. (2013). A novel thermally driven pump and its test in a supercritical CO_2 loop system. *International Journal of Energy Research*, *37*(11), 1331–1338. doi:10.1002/er.2934

Yamamoto, T., Furuhata, T., Arai, N., & Mori, K. (2001). Design and testing of the Organic Rankine Cycle. *Energy*, *26*(3), 239–251. doi:10.1016/S0360-5442(00)00063-3

Yin, S., & Tan, Z. (2012). Hydrothermal liquefaction of cellulose to bio-oil under acidic, neutral and alkaline conditions. *Applied Energy*, *92*, 234–239. doi:10.1016/j.apenergy.2011.10.041

Yoshida, S., & Mori, H. (2000). Heat transfer to supercritical pressure fluids flowing in tubes. *Proceedings of SCR-2000*. Tokyo: The University of Tokyo.

Yoshida, T., Oshima, Y., & Matsumura, Y. (2004). Gasification of biomass model compounds and real biomass in supercritical water. *Biomass and Bioenergy*, *26*(1), 71–78. doi:10.1016/S0961-9534(03)00063-1

Yoshikawa, S., Smith, R. L. Jr, Inomata, H., Matsumura, Y., & Arai, K. (2005). Performance of a natural convection circulation system for supercritical fluids. *The Journal of Supercritical Fluids*, *36*(1), 70–80. doi:10.1016/j.supflu.2005.02.007

Yun, R., Hwang, Y., & Radermacher, R. (2006). Gas cooler heat transfer and pressure drop characteristics of CO2/oil mixture in a microchannel. *Proc. of the 7th IIR-Gustav Lorentzen Conference on Natural Working Fluids*, 503-505.

Yunus, C. A., & Afshin, J. G. (2011). *Heat and Mass Transfer: Fundamentals and Applications*. Academic Press.

Yu, S.-C., Chen, L., Zhao, Y., Li, H.-X., & Zhang, X.-R. (2015). A brief review study of various thermodynamic cycles for high temperature power generation systems. *Energy Conversion and Management*, *94*, 68–83. doi:10.1016/j.enconman.2015.01.034

Zappoli B., Amiroudine S., Gauthier S. (1997). Instabilité gravitationnelle dans un fluide supercritique pur. *C.R. Acad. Sci. Paris*, *324*, Série II b.

Zappoli, B. (1992). The response of a nearly supercritical pure fluid to a thermal disturbance. *Physics of Fluids. A, Fluid Dynamics*, *4*(5), 1040–1048. doi:10.1063/1.858255

Zappoli, B., Amiroudine, S., Carles, P., & Ouazani, J. (1996). Thermoacoustic and buoyancy-driven transport in a square side-heated cavityfilled with a near-critical pure fluid. *Journal of Fluid Mechanics*, *316*(-1), 53–72. doi:10.1017/S0022112096000444

Zappoli, B., Bailly, D., Garrabos, Y., Le Neindre, B., Guenoun, P., & Beysens, D. (1990). Anomalous heat transport by the piston effect in supercritical fluids under zero gravity. *Physical Review A.*, *41*(4), 2264–2267. doi:10.1103/PhysRevA.41.2264 PMID:9903351

Zappoli, B., Beysens, D., & Garrabos, Y. (2015). *Heat Transfer and related effects in supercritical fluids*. Dordrecht: Springer. doi:10.1007/978-94-017-9187-8

Zappoli, B., Beysens, D., & Garrabos, Y. (2015). *Heat Transfer and Related Phenomena in Supercritical Fluids*. Berlin: Springer-Verlag.

Zappoli, B., & Carles, P. (1995). The thermoacoustic nature of the critical speeding up. *Europ. J. Mech. B*, *14*, 41–65.

Zeton, P. V., Tews, I., Onarheim, K., Beckman, D., Zhu, Y., Solantausta, Y., & Snowden-Swan, L. et al. (2014). *Biomass Direct Liquefaction Options: TechnoEconomic and Life Cycle Assessment*. Richland: Pacific Northwest National Laboratory.

Zhang, F., Chen, S., Xu, C., Chen, G., Zhang, J., & Ma, C. (2012). Experimental study on the effects of operating parameters on the performance of a transpiring-wall supercritical water oxidation reactor. *Desalination*, *294*, 60–66. doi:10.1016/j.desal.2012.03.013

Zhang, J., Chen, W.-T., Zhang, P., Luo, Z., & Zhang, Y. (2013). Hydrothermal liquefaction of Chlorella pyrenoidosa in sub- and supercritical ethanol with heterogeneous catalysts. *Bioresource Technology*, *133*, 389–397. doi:10.1016/j.biortech.2013.01.076 PMID:23454385

Zhang, X. R., Yamaguchi, H., Fujima, K., Enomoto, M., & Sawada, N. (2005). A Feasibility Study of CO_2-Based Rankine Cycle Powered by Solar Energy, *JSME Int. J. Ser. B*, *48*, 540–547.

Zhang, X. R., Yamaguchi, H., Fujima, K., Enomoto, M., & Sawada, N. (2005). A feasibility study of CO_2-based Rankine cycle powered by solar energy. *JSME International Journal. Series B, Fluids and Thermal Engineering*, *48*(3), 540–547. doi:10.1299/jsmeb.48.540

Zhang, X. R., Yamaguchi, H., Fujima, K., Enomoto, M., & Sawada, N. (2006). AIP Conference Proceedings: Vol. 832. *Experimental Performance Analysis of Supercritical CO2 Thermodynamic Cycle Powered by Solar Energy* (pp. 419–424). AIP Publishing. doi:10.1063/1.2204534

Zhang, X. R., Yamaguchi, H., Fujima, K., Enomoto, M., & Sawada, N. (2007). Theoretical analysis of a thermodynamic cycle for power and heat production using supercritical carbon dioxide. *Energy*, *32*(4), 591–599. doi:10.1016/j.energy.2006.07.016

Zhang, X. R., Yamaguchi, H., Uneno, D., Fujima, K., Enomoto, M., & Sawada, N. (2006). Analysis of a novel solar energy powered Rankine cycle for combined power and heat generation using supercritical carbon dioxide. *Renewable Energy*, *31*(12), 1839–1854. doi:10.1016/j.renene.2005.09.024

Zhang, X. R., Zhang, Y. L., & Chen, L. (2014). Experimental study on solar thermal conversion based on supercritical natural convection. *Renewable Energy*, *62*, 610–618. doi:10.1016/j.renene.2013.08.025

Zhao, H., Penninckx-Sans, A., Lee, L.-T., Beysens, D., & Jannink, G. (1995). Probing the universal critical-adsorption profile by neutron reflectometry. *Physical Review Letters*, *75*(10), 1977–1980. doi:10.1103/PhysRevLett.75.1977 PMID:10059177

Zhao, M., Gu, H. Y., & Cheng, X. (2014). Experimental study on heat transfer of supercritical water flowing downward in circular tubes. *Annals of Nuclear Energy*, *63*, 339–349. doi:10.1016/j.anucene.2013.07.003

Zhao, P., Shen, Y., Ge, S., & Yoshikawa, K. (2014). Energy recycling from sewage sludge by producing solid biofuel with hydrothermal carbonization. *Energy Conversion and Management*, *78*, 815–821. doi:10.1016/j.enconman.2013.11.026

Zhong, F., & Meyer, H. (1996). Density equilibration near the liquid-vapor critical point of a pure fluid. II. Coexisting phases for $T<T_c$. *Physical Review E: Statistical Physics, Plasmas, Fluids, and Related Interdisciplinary Topics*, *53*(6), 5935–5948. doi:10.1103/PhysRevE.53.5935 PMID:9964952

Zhong, F., & Meyer, H. (1999). Heat transport in a pure fluid near the critical point: Steady state and relaxation dynamics. *Journal of Low Temperature Physics*, *114*(1/2), 231–255. doi:10.1023/A:1021862122617

Zhu, Z., Toor, S. S., Rosendahl, L., & Chen, G. (2014). Analysis of product distribution and characteristics in hydrothermal liquefaction of barley straw in subcritical and supercritical water. *Environmental Progress & Sustainable Energy*, *33*(3), 737–743. doi:10.1002/ep.11977

Zia, J., Sevincer, E., Chen, H., Hardy, A., Wickersham, P., Kalra, C., … Vandeputte, T. (2013). *High-potential Working Fluids for Next Generation Binary Cycle Geothermal Power Plants* (Report). Retrieved from http://digital.library.unt.edu/ark:/67531/metadc846464/m1/14/

Ziman, J. M. (1961). A theory of the electrical properties of liquid metals. I: The monovalent metals. *Philosophical Magazine*, *6*(68), 1013–1034. doi:10.1080/14786436108243361

Ziman, J. M. (1979). *Principles of the Theory of Solids*. Cambridge, UK: Cambridge University Press.

Znini, M., Majidi, L., Bouyanzer, A., Paolini, J., Desjobert, J.-M., Costa, J., & Hammouti, B. (2012). Essential oil of Salvia aucheri mesatlantica as a green inhibitor for the corrosion of steel in 0.5 MH_2 SO_4. *Arabian Journal of Chemistry*, *5*(4), 467–474. doi:10.1016/j.arabjc.2010.09.017

Zoueshtiagh, F., Amiroudine, S., & Narayanan, R. (2009). Experimental and numerical study of miscible Faraday instability. *Journal of Fluid Mechanics*, *628*, 43. doi:10.1017/S0022112009006156

Zvirin, Y. (1981). A review of natural circulation loops in pressurized water reactors and other systems. *Nuclear Engineering and Design*, *67*(2), 203–225. doi:10.1016/0029-5493(82)90142-X

About the Contributors

Lin Chen is now a granted JSPS Research Fellow in Institute of Fluid Science, Tohoku University, Japan. He obtained his B.Eng and Ph.D. in Mechanics (Energy and Resources Engineering) from Peking University. His current research topics include energy resources, oceanic methane hydrate utilization, supercritical fluids, microscale compressible fluid flows, and multiscale heat/mass transfer. He has authored over 80 international journal papers and/or conference presentations, and books/chapters. His doctoral thesis won the outstanding award and was published as an English book by Springer (Title: "Microchannel Flow Dynamics and Heat Transfer of Near-Critical Fluid"). He received the Outstanding Young Scholar of the MOE (China) in 2012 and 2013, the Innovation Award in 2011, 2012, and 2014, the Best Paper of the Chinese Association of Refrigeration in 2015, and the Elsevier Excellent Reviewer of the year in 2013, 2014, and 2015. He is also a Reviewer, an Editorial/Advisory Member, and the Guest Editor of many renowned international journals. He is currently an Executive Editor of the Journal of Natural Gas Science and Engineering (Elsevier).

Yuhiro Iwamoto is currently an Assistant professor of Department of Electrical and Mechanical Engineering, Nagoya Institute of Technology, Japan since June 2016. He is also a Collaborative Researcher of Institute of Fluid Science (IFS) in Tohoku University, Japan since April 2014. He received B.Eng. (2008), M.Sc. (2010) and Ph.D. (2013) in Doshisha University. He has received several awards, e.g. Research Fellow of The Japan Society for the Promotion of Science from 2011 to 2013 and so on. He has engaged in researches on experimental and computational fluid mechanics and engineering, involving basic study and application for supercritical CO_2 and magnetic functional fluids. He is currently interested in and engages in research on effective utilization of CO_2, e.g. a Solar Rankine System and Solar Water Heater using Supercritical CO and Ultra-Low-Temperature Refrigeration System using Dry Ice. He also numerically approaches these topics using Lattice Boltzmann Method as numerical scheme and General-Purpose of Computing on Graphics Processing Units (GPGPU) as a technique for high performance computing. Many reviewed journal papers have been published in these fields.

* * *

Sakir Amiroudine is a Professor in Fluid Dynamics in the Institute of Mechanics and Engineering at the University of Bordeaux (France). He received his Master of Science, at the Center for Microgravity and Materials Research, University of Alabama in Huntsville (USA) and his PhD from CNES (French Space Agency). His main research topics include: thermal and thermo-vibrational instabilities in supercritical fluids miscible fluids, spectral analysis of vortex shedding around bluff bodies, porous medium saturated with supercritical fluid and electro-osmotic flows in microfluidic systems. He has been the chief scientist for many projects and principal investigator of the national French research group CNES/CNRS. He has authored more than 50 refereed international publications (journal papers, conferences, books) and has a wide international collaboration with researchers in the world.

Dipankar N. Basu is an Assistant Professor in the department of Mechanical Engineering at Indian Institute of Technology Guwahati since June 2012. He received his master's degree in Mechanical Engineering from the Jadavpur University, India in 2004. He completed his Ph.D. from Indian Institute of Technology Kharagpur. He served as an Assistant Professor at IIEST Shibpur for nearly four years before joining IIT Guwahati. His principal research interest is in the field of nuclear thermalhydraulics, two-phase flow, supercritical heat transfer, optimization of thermal systems and microfluidics. He is currently working on computational tool development for simulation of flows with free-surfaces and phase-change. He has co-authored more than 50 referred journal and conference publications. He is a regular reviewer of number of reputed international journals and also the principal investigation of several sponsored projects.

Saeb Besarati graduated with PhD degree in Chemical Engineering from the University of South Florida. His dissertation was entitled as "Analysis of Advanced Supercritical Carbon Dioxide Power Cycles for Concentrating Solar Power plants". He has published more than 20 research papers and book chapters in the fields of solar energy, thermodynamics, heat transfer, and optimization. He has frequently served as the reviewer for prestigious journals and international conferences. After graduating from the university, he joined eSolar, Inc. as an application engineer. He works at SolarReserve, LLC now, where his main responsibility is to model the operation of molten salt receivers.

Daniel Beysens obtained an engineering degree from Institut d'Optique, Paris-Orsay in 1969 and a PhD in 1973 from University Pierre & Marie Curie, Paris. Beysens worked as a researcher at CEA-Saclay since 1973, becoming Director of research to lead the Laboratory of Condensed State Physics (1991-1995), then the Institute of Fundamental Research on Condensed Matter at CEA-Grenoble (1995-1999) and the ESEME joint CEA-CNRS-ESPCI laboratory at Bordeaux and Paris (2000-2015). Beysens operated as PI or co-PI experiments on supercritical fluids in sounding-rockets, space shuttle, MIR and International Space stations and became President of the European Low Gravity Association (2003-2007). He now works as Emeritus Director of research at ESPCI; he also serves as President of OPUR (since 1999). Beysens is the recipient of various awards; he has published near 450 research papers and coauthored or coedited 11 books, including a recent book on heat transfer in supercritical fluids (Springer-Verlag, 2015).

Huijuan Chen is a research engineer at the Global Research Center of General Electric Company, with focus on thermal energy systems for aviation and power applications. She holds 5 US patents and have published more than 20 research papers, among which some have been cited for more than 600 times. She received her doctoral degree (2010) at University of South Florida in the Department of Chemical and Biomedical Engineering under the supervision of Dr Yogi Goswami and Dr Elias Stefanakos.She has served as the reviewer for International J. of Heat and Mass Transfer, International J. of Solar Energy, Energy, etc.

Maohong Fan is from University of Wyoming, USA. He has extensive experience and scientific groundwork on supercritical fluids and chemical topics. He is a professional in the study of hydrothermal synthesis of simple and complex inorganic crystals and materials. Prof. Maohong Fan has developed the unique devices that allow analyzing the phase composition, morphology, and other properties of the resulting nanomaterials. He has edited and authored 125 books, book chapters, journal special issues, and journal papers in recent years.

Yves Garrabos obtained an Arts et Métiers engineering degree from the Ecole Nationale Supérieure des Arts et Métiers, Paris, in 1969 and received his PhD in 1982 from the Université Pierre & Marie Curie, Paris-6. Garrabos joined the Centre National de la Recherche Scientifique (CNRS) in 1971 as researcher at the Laboratoire des Hautes Pressions in Meudon-Bellevue (1971-1986) and as a Director of research at the Laboratoire d'Ingéniérie des Matériaux et des Hautes Pressions at the University Paris XIII-Villetaneuse (1987-1992). In 1993, Garrabos joined the Institut de Chimie de la Matière Condensée de Bordeaux (ICMCB) in Pessac to create and lead the Supercritical Fluid Group (1993-2012). Since 2012, Garrabos works as Emeritus Director of Research at ICMCB. Garrabos participated as a Principal or coPrincipal Investigator for experiments dedicated to near critical fluids and granular matter in sounding-rockets, space shuttle, MIR and International Space stations. Garrabos has published more than 150 research papers. With Zappoli and Beysens, Garrabos received the Gaz de France award in 2000 from the French Academy of Sciences and recently coauthored a book on Heat Transfer in Supercritical Fluids (Springer-Verlag, 2015).

Khaled Gasem has been actively involved in the energy sector for over 30 years. His activities include: a standing consultant for a number of energy companies, co-founded a number of industrial research consortia, co-founded a number of civic organizations, strategic planner for a number of universities and private firms, recipient of a number patents and invention disclosures. Dr. Gasem is author/co-author of about 400 publications and presentations in his areas of interest. He is a member of the American Institute of Chemical Engineers, the American Chemical Society, Phi Kappa Phi Honor Society, Omega Chi Epsilon Honor Society, and Tau Beta Pi Engineering Honor Society. He has received a number of honors and awards throughout his career. His academic honors and awards include the Regents Professor (highest academic rank and honor at Oklahoma State University), and the Bartlett Chair. His was also honored and recognized by the oil and gas industry through the R. N. Maddox Gas Processors Professorship, the AMOCO Chair, and the Halliburton Award of Excellence. Dr. Gasem has mentored / graduated over 60 scientists, engineers, and graduate students, including 11 visiting scholars and postdoctoral fellows, 24 PhD candidates, and 30 MS graduates.

Eiji Hihara received his doctoral degree from University of Tokyo in 1982. He became a Lecture (1982) and an Associate Professor (1983) in Mechanical Engineering Department, and a Full Professor (1999) in Institute of Environmental Studies at the University of Tokyo. His research interests include promotion of energy efficiency in residential sector, cooling systems using solar thermal, heat pump-desiccant hybrid air-conditioning systems, estimation method of energy consumption of domestic heat pump water heaters and package air conditioners, supercritical carbon dioxide cooling and boiling heat transfer, Low GWP refrigerant heat pumps (Vapor-liquid equilibrium for mixed-refrigerant), etc.

Yanping Huang, research fellow and Ph. D supervisor in Nuclear Power Institute of China (NPIC), obtained his Ph. D degree on nuclear science and technology in Xian Jiaotong University in 2003. He is the chief engineer of the reactor engineering sub-institute of NPIC, and served as the director of the CNNC Key Laboratory on Nuclear Reactor Thermal Hydraulics Technology from 2009 till now. He is the winner of National Science Fund for Distinguished Young Scholars in China at 2013. Dr. Huang was the technical program committee chair of the 21st International Conference On Nuclear Engineering (ICONE21). He was also the initiator and organizer of China-France, China-Korea bilateral international academic conference, and the technical program committee member and session chair of NURETH and other international conferences in nuclear engineering field. Dr. Huang published more than 200 papers (more than 30 were indexed by SCI, more than 150 were indexed by EI) in nuclear reactor R&D up to now. Dr. Huang is the vice chief-editor of the transaction of the ASME journal of Nuclear Engineering and Radiation Science.

Masanori Inui is a Professor in Graduate School of Integrated Arts and Sciences at Hiroshima University since April 2009. He obtained Dr. of Science in Department of Physics at Kyoto University in January 1989. He served as a Research Assistant in College of General Education at Kyushu University from June 1989, and became an Associate Professor in Faculty of Integrated Arts and Sciences at Hiroshima University in April 1994. His principal research interest is in the field of disordered materials such as liquid metals and semiconductors that valence electrons play an important role. He started his research using a synchrotron radiation from his master course student at Kyoto University, and currently conducts x-ray scattering experiments at SPring-8 in Japan. He has co-authored more than 100 referred journal and conference publications. He also conducts liberal arts education for first-year students at Hiroshima University.

Željko Knez (male), born in 1954, PhD of chemical engineering is a leading expert in the field of supercritical fluids. He was vice rector, dean of Faculty, currently he is in a position as a Vice-dean for research on the Faculty of Chemistry and Chemical Engineering. In 1984, he introduced the supercritical fluid technologies in the Slovenian space and he is an expert in separation processes and product design using sub- and supercritical fluids. Prof. Knez is a member of several eminent associations: The Council of Science RS (from 1997 to 2004), Council for Higher Education (from 2002 to 2003), ARRS Governing Board (from 2006 to 2009), EFCE – member of WP High pressure technologies, EFCE – president of WP High pressure technologies, EFCE – member of executive board, Vice-President of the Slovenian Chemical Society, member of AOCS (American Oil Chemists Society), ISASF (International Society for Advancement of Supercritical Fluids), ProcessNet - the German platform for process engineering Chemical Engineering and Technical Chemistry, Member of European Academy of Sciences and Arts (EASA). His bibliography covers 226 scientific articles and 15 review articles, 28 book chapters, 22 patents (including the EU, the USA, Canada and Japan). His "h" index is 34. He cooperate with leading industrial partners from all over the world.

Maša Knez Hrnčič was born in 1984 in Maribor of Slovenia. She got his B.E (2008) and PhD (2014) from University of Maribor, Slovenia. Since October 2014, she joined the University of Maribor, Faculty of Chemistry and Chemical Engineering as a Teaching/Research Associate and in March 2016 he became an Assistant Professor there. Her research topics include chemical engineering, green chemistry, energy engineering, etc.

Hongzhi Li is a deputy director of the coal-based clean energy division in Xi'an Thermal Power Research Institute. He received a joint PhD in Power Engineering and Engineering Thermal Physics from Xi'an Jiao Tong University and University of Wisconsin-Madison at 2010. He was also a visiting scholar of AREVA and Siemens technical center at 2009. He works on thermal hydraulics of supercritical fluids associated with several advanced energy systems for almost ten years. He currently leads a program on design, construction and operation of a 5MWe supercritical CO_2 fossil-fired power generation integral test facility. Dr. Li was recently sponsored by the Young Talent Supporting Program from the Chinese Society for Electrical Engineering. He has been awarded more than 10 patents and has published over 30 papers.

Mohammed Mahdi is a student at the Faculty of Energy Systems and Nuclear Science (FESNS), University of Ontario Institute of Technology (UOIT) (Oshawa, Ontario, Canada).

Kazuhiro Matsuda is an Associate Professor in the division of Physics and Astronomy, Graduate school of Science, Kyoto University since April 2009. He received his master's degree in Engineering from Kyoto University, Kyoto in 1997. He obtained Dr. of Engineering from the department of Materials Science and Engineering, Graduate School of Engineering, Kyoto University in November 2006. He served as an Assistant Professor at the department of Materials Science and Engineering, Graduate School of Engineering, Kyoto University from 2003 to 2008. His main research interest is on the physical properties of fluid metals, especially on those in the expanded state. He is currently carrying out experimental studies of fluid metals using a synchrotron radiation mainly focusing on the electronic state, such as, investigation for the electron momentum density in fluid metals by X-ray Compton scattering technique, and investigation for the excitation and damping of plasma oscillations in fluid metals by X-ray Raman scattering.

Roman Popov is a student at the Faculty of Energy Systems and Nuclear Science (FESNS), University of Ontario Institute of Technology (UOIT) (Oshawa, Ontario, Canada).

Sergey Rutin graduated Ural Federal Institute with Specialist degree in Technical Physics in 1980. PhD thesis was devoted to heat transfer in supercritical fluids in a scale of small characteristic times and sizes. Currently working in the Institute of Thermal Physics, Ural Branch of Russian academy of sciences as a scientific researcher.

Thomas Schulenberg studied physics and mechanical engineering at University of Karlsruhe. Dissertation in mechanical engineering at University of Karlsruhe. Gas turbine design for combined cycle power plants with SIEMENS. Since 2000 head of the Institute for Nuclear and Energy Technologies at KIT.

Pavel Skripov graduated Ural Federal Institute with Specialist degree in Technical Physics in 1978. PhD thesis and Doctoral thesis were devoted to spontaneous boiling-up of fluids under pulse heating regimes. Currently working in the Institute of Thermal Physics, Ural Branch of Russian academy of sciences as a leading scientific researcher.

Xi Sui got his PhD in nuclear engineering from from the Nuclear Power Institute of China in 2014. His doctor thesis is about the coupled neutron-thermal hydraulic instability of supercritical water in two parallel channels. Now his main work is about the hydraulic performance of nuclear reactor, such as flow distribution at the core inlet.

Ricardo Vasquez Padilla, PhD, MSc(ME), BS(ME), is a senior lecturer and course coordinator of the Bachelor in Mechanical Engineering in the School of Environment, Science and Engineering at Southern Cross University. Prior to that position, He worked as Research Fellow at CSIRO Energy Centre in Newcastle (2014-2016) and as Assistant Professor at Universidad Del Norte (2011-2014). He received his doctoral degree (2006-2011) at University of South Florida in the Department of Chemical and Biomedical Engineering under the supervision of Dr Yogi Goswami and Dr Elias Stefanakos. His research interests include modelling and experimentation of advanced power blocks operating with different heat sources, Concentrated Solar Thermal (CST), hybrid renewable energy systems and applications of renewable energy for heating, cooling and desalination.

Zejun Xiao, male, born in July 1967, Ph. D in nuclear science and technology, research fellow and Ph. D supervisor, worked on nuclear reactor R&D for 25 years. Dr. Zejun Xiao got his Bachelor's degree and Master's degree of nuclear power engineering in 1989 and 1992, respectively. He then worked in Nuclear Power Institute of China (NPIC) in 1992. During 2001~2004, he got his Ph. D degree in Xian Jiaotong University. During his research career, Dr. Zejun Xiao has solved numerous scientific and engineering issues in nuclear reactor R&D, published more than 100 papers up to now, and attained different kinds of Chinese science and technology awards for 20 times.

Ting Xiong got a bachelor's degree at the Tsinghua University in 2008 and his doctor's degree of Nuclear Science and Technology at the Tsinghua University in 2013.

Hiroshi Yamaguchi obtained his Ph.D. in Mechanical Engineering from University of Manchester Institute of Science and Technology (UMIST), U.K. in 1982. He is currently Professor in Department of Mechanical Engineering, Doshisha University, Japan. His research expertise is on supercritical CO_2 flow, energy conversion system and functional fluid flows.

Aleksandr Yampol'skiy graduated Ural Federal Institute with Specialist degree in Automation and Control of Technical Processes in 2010. PhD student in Institute of Thermal Physics in 2010-2013. Currently working in a software company. Specializes in automation of experiments, digital signal processing, numerical data analysis.

Xiao Yan has long expertise in nuclear reactor thermal hydraulics and safety areas. He got his doctor degree in Xi'an Jiaotong University and has worked in Nuclear Power Institute of China since 2005. Dr. Yan has published over 90 SCI/EI indexed papers and obtained about 20 patents in relative field. Dr. Yan has won about 10 national defense rewards for his contribution and excellent work.

Jinguang Zang got his doctor degree in Tsinghua University in 2014. His main research topic includes flow and heat transfer of supercritical water, advanced numerical techniques, experimental studies about safety relative issues. Dr. Zang has published several papers in this field and serves as the GIF SCWR TH/S PMB member of China.

Bernard Zappoli, born in 1951, obtained his PhD in 1979. He worked on the asymptotic solutions of the Boltzmann equation. He was hired in 1975 in The Sociéte Européenne de Propulsion where he worked on the rocket engines development of the European launcher Ariane. In 1981 Zappoli moved to the French Space Agency (Centre National d'Etudes Spatiales, CNES) in Toulouse where he led the French Microgravity Research Program in physical sciences. From 1992 to 2005 he was Associate Professor of Mechanical Engineering at the Paul Sabatier University in Toulouse. From 2005 to 2009 Zappoli served as Associated Research Director at the Laboratoire de Physique et Mécanique des Milieux Hétérogènes (PMMH) of the French Centre National de la Recherche Scientifique in Paris. He published more than 60 peer-reviewed articles and co-authored a book on heat transfer in supercritical fluids (Springer-Verlag, 2015). Bernard Zappoli is now Emeritus Engineer at the French Space Agency.

Index

A

Acoustic Filtering 34, 38, 45, 51
Alkali Metals 102-106, 108, 110-111, 114-115, 118, 128

B

Bio Oils 556, 587
Biobased Chemicals 554-555, 587
Biofuels 554-555, 558, 560-561, 575, 587-592, 595-596, 598-600, 606
Boundary Heat Flux 70, 83, 262, 268
Boundary Layer 2, 6-8, 10, 15, 19, 21-22, 24-25, 28, 33-34, 36, 38, 44, 46, 51, 54, 62, 66, 68, 80, 83-84, 86, 238-239, 246-247, 258, 271, 286, 288, 302, 306, 311, 482, 488-489, 491-492, 495, 501, 505-506, 543
Bounding Power 194, 198
Brayton Cycle 383-384, 387, 389-394, 402, 440-441, 444, 452
Buoyancy Convection 5, 25, 28, 31-32, 58, 90

C

Carbon Dioxide 176, 215, 235, 237, 292-293, 295-297, 309, 313-316, 318-319, 321-323, 335-336, 342, 371, 373-374, 376, 381-383, 387, 404-406, 408-410, 412-413, 418, 420-421, 432-433, 435-438, 441, 443, 449, 452-453, 488, 561, 593, 596, 606
CFD 206, 257, 482, 538, 544
Choked Flow 356
Cnes 52
CO2 Turbine 346, 352, 357
Coal Fired Power Plant 458, 463-464, 466, 470-475, 477, 479
Collected Heat Quantity 224-225, 234
Compressed Fluid 334, 404, 453
Compressed Liquid 272, 275, 288, 291
Compressibility Factor 340, 354, 366, 408
Compressible 1, 5, 35-36, 39, 42, 45, 62-63, 65, 83, 87, 89, 99, 340, 343, 352-353, 571
Concentrated Solar Power 372, 382, 402
Cooling 6-8, 10, 15, 20, 24-27, 34, 54, 56-57, 79, 83, 136, 140, 142-146, 149-152, 156, 159, 163, 166-168, 175-178, 185, 188-191, 195-196, 204-205, 211-212, 216, 238-239, 243, 256-258, 282, 292, 294, 296, 300, 306, 309, 313-314, 316-319, 321-322, 342, 358, 370, 375, 382-384, 389, 391-393, 405, 423, 429, 464, 469, 473, 475, 511
Core Design 475, 477, 483, 485, 487
Correlation 3-4, 64, 66-67, 73, 113-114, 121, 123, 133, 140, 165-166, 195, 198, 211, 239-240, 243, 246-247, 249, 257-258, 266, 296, 304-307, 354, 417-418, 420-421, 491-492, 494-498, 500, 502, 505, 509, 516, 543
corrosion 470, 562, 573
Critical Distance 10, 65, 89, 99
Critical Diverges 99, 137, 184
Critical Point (Also Called a Critical State) 1-2, 7, 10, 28, 31, 33-36, 38-39, 42, 44, 47-48, 51-52, 56-58, 61, 64-65, 87, 99, 103, 114-115, 121, 123, 128, 137, 143, 146, 152, 155, 161, 204, 215-217, 220, 222, 226, 237, 250-251, 253-254, 263, 274-275, 284, 291, 339, 342, 350, 366, 373, 376, 378, 384, 386-387, 389, 402, 405, 409, 412-413, 452, 481, 540, 549, 554-555, 559, 562, 571, 589, 591, 602, 604
Critical Pressure 40, 54, 106, 120, 204, 219-220, 236-237, 271-272, 282, 288, 334, 336, 344, 352, 355, 370, 378, 404, 413, 422, 453, 465, 469, 549, 559, 600
Critical Slowing Down 2, 4-5, 7, 28, 31, 34-35, 58, 70
Critical Speeding Up 7, 28, 31, 70
Critical Temperature 7, 9-10, 19, 21, 35-36, 54, 65, 90, 105, 124, 133, 160, 185, 204, 216-217, 227, 269, 279, 282, 284, 293, 334, 336, 338, 340, 351, 376, 378, 381, 388, 453, 469, 549, 561, 571, 600

D

Deteriorated Heat Transfer (DHT) 155, 168, 275, 283, 287, 418, 420-421, 452, 482, 549
Dryer 191, 480, 482
Dynamic Structure Factor 114, 124-125, 133

E

Economizer 460-461, 464-466, 474, 480
Electron gas 105, 114, 116, 118, 120, 128
Electronic Properties 102, 109, 128
Energy Conversion 56, 93, 99, 136-137, 171, 179, 216-217, 219, 234-235, 337, 341, 363, 458, 558, 588, 606
Environment Friendly 366
Equation of State (EOS) 5, 33-35, 45, 48, 63-64, 104, 108, 195-199, 203-204, 211, 214, 354-355, 366, 528
Evacuated Solar Collector 215-216, 219-222, 224-228, 230-232, 234, 344-348, 350, 352-353, 358-359
Evaporator 376, 382, 460-461, 464-466, 468-470, 474, 477, 479
Exergy 371-372, 393, 402, 410
Experiment 5, 26, 51, 58, 111, 115, 118, 143, 149, 152, 155, 211, 221-222, 225-226, 230-231, 257, 263, 277-281, 286, 304, 306, 312, 353, 469, 510, 535, 538-541

F

Faraday Instability 54
Feedwater 423, 425, 453, 460-461, 463-464, 466, 468, 470-471, 475, 477, 479-480, 485
Finite Volume 4, 38, 45, 51, 64, 203, 257, 279
Finite-Difference Method 198, 214
Flow Instability 140, 161, 166, 249-250, 481-483, 510-511, 514, 518-522, 524, 526, 535, 538-539, 543-544, 549
Flow Pattern 74, 292, 294, 321-330, 485
Fluid Metals 102-103, 106, 109-110, 128, 133
Fluoro Carbon 216, 234
Friction Factor 65-67, 144, 193, 199, 211, 214, 248-249, 266, 296, 500, 516, 528
Fuel Assembly 480, 485-487

G

Global Warming 215-217, 234, 236, 293, 330, 335-336, 363, 377, 557, 591

H

Heat Flux 5-6, 24, 62, 68, 70, 77, 79, 81-83, 140, 142, 169, 207-210, 236-239, 241, 247, 252, 256-257, 259, 262, 268, 272, 275-277, 282, 287-288, 292, 298, 301-302, 304, 308-309, 311, 313, 316, 318-319, 321-322, 334, 350, 405, 418-422, 466, 468-469, 475, 479-480, 482, 488-489, 496-497, 506, 510, 512, 514, 517-518, 520-521, 532, 535, 543, 605
Heat Pipe 10
Heat Pump 56, 171, 292-294, 297, 319, 321, 323, 327, 330, 334, 337, 342
Heat Recovery Efficiency 225-228, 230, 232, 234
Heat Transfer Coefficient 79, 145-146, 152-153, 178, 191, 195, 199, 202, 205, 207-208, 210-211, 214, 237, 239, 256-257, 271, 275, 277, 282, 292, 295, 298-306, 308-311, 315-316, 318-323, 325, 328-330, 334, 344, 350-352, 405, 452, 481-482, 488, 495, 543, 549
Heat Transfer Correlation 140, 246, 266, 491-492, 494-498, 543
Heat Transfer Enhancement 75, 155, 184
High Pressure Phase Equilibrium 587
High Pressure View Cell 587
Hot Spot 72, 83-84, 87, 99
Hydrodynamic Instability 54
Hydrothermal Carbonization 560, 563, 587, 595
Hydrothermal Gasification 563, 587, 595
Hydrothermal Liquefaction 560-561, 563, 587
hydrothermal treatment 559, 595

I

Improved Heat Transfer (IHT) 452
Inelastic X-Ray Scattering 109-111, 118-119, 124, 133
Irreversibility 402

K

Kassoy 6-7
Kelvin Helmholtz Instability 83, 85, 87, 99

L

Lagrange-Euler 48, 50-51
Live Steam 459-460, 465-466, 473-475, 477, 479-480
Low-grade Heat 217, 219, 369-370, 392, 394, 604

M

Mass Flux 192, 196-197, 237-239, 241, 248, 250, 257, 260-261, 275, 292, 298-300, 304, 306, 313, 322-323, 326-327, 330, 420-421, 466, 468-469, 480, 496, 506, 512, 514
Mercury 102, 105-106, 120
Metal-Nonmetal Transition 128, 133
Meyer 14-15, 34, 62, 68-70
Microchannel 56-62, 65-87, 89-90, 322
Mixing 55-57, 67-68, 70-75, 78-79, 84-86, 88, 92, 309-310, 458, 470-471, 476-477, 479, 511
multicomponent systems 554

N

Natural Circulation Loop 136, 140, 167, 178, 185, 188-189, 214
Natural Convection 54, 87, 90-91, 136, 140, 169-170, 174, 178, 184, 188, 466-467, 474
Natural Refrigerant 234
Navier-Stokes 5, 33-34, 36, 51, 62-63, 157, 298
Near Supercritical Fluids 2, 28, 31
Near-critical Flow 62, 65, 69, 71, 79
Near-Critical Fluid 55-62, 64, 66, 68-70, 75, 77, 83-85, 87, 89-90, 92, 99, 137, 168
Near-Critical Point 405, 549
Near-Critical Region 59-60, 62, 64-65, 137, 144, 174, 177-178, 272, 280, 283, 288, 291
Normal Heat Transfer (NHT) 453
Nuclear Power 56, 137, 235, 249, 272, 275, 378, 383, 405, 427, 430, 432, 452-453, 460, 475, 479, 481, 483, 496, 505, 511, 605-606
Nuclear Power Plant 56, 137
Numerical Analysis 60, 353, 366, 481, 510, 515, 525, 532, 543-544
Numerical Simulation 44, 50, 198, 292, 306, 309, 478, 538-540

O

Oil 215, 292, 294, 297, 319-330, 335, 377, 427, 473, 556-557, 561-562, 573, 592-593, 596-600
Onuki 7, 13-14, 34, 59-60, 62, 65-66, 68-70, 83-84, 86-90

P

Parametric 33-34, 44-48, 52, 205, 372, 482, 511, 519-520, 535, 543-544
Perturbation 5, 9, 42-43, 60-61, 68, 70, 72, 79-80, 83, 86-88, 90, 103-104, 166, 194, 199, 214, 527, 540
phase equilibria 556, 561, 571
Piston Effect 1-2, 5, 7, 9-11, 17, 28, 31, 34, 36, 39, 45, 59-60, 70, 79, 87, 89-90, 99
Pressure Drop 27, 65, 73, 78, 173, 192-193, 236, 248, 292, 298, 304, 306, 312, 321, 384-385, 418, 421, 466, 468, 470, 488-489, 500, 511, 514-516, 518, 524-526, 530, 533-536, 540, 542, 544
Pressure Loss 57, 73, 77, 79, 236-237, 248-249, 293, 481-482, 500, 505, 539, 543
Probability Density Function 241, 243, 268
Properties of CO2 159, 217-218, 244, 257, 326, 335, 337, 342, 346, 387, 390
Pseudo Critical Steam 252, 269
Pseudo Critical Temperature 269
Pseudo Critical Water 252, 269
Pseudocritical Point 191-192, 202, 205-207, 210, 295, 298, 306, 308, 334, 342-344, 352, 411-413, 418-419, 453, 482, 488, 495, 505, 549
Pseudocritical Temperature 152, 191, 204, 206, 214, 295, 299-304, 306, 308, 313, 315, 322, 326, 328, 413
Pulse Heating 271, 277, 282-283, 286, 288

R

Rankine Cycle 335-339, 341-346, 350, 358-359, 361, 363, 369-372, 374, 376-378, 381-382, 384, 393, 402, 404, 437, 453, 460-465, 479, 604-605
Rayleigh-Taylor 34, 39, 41, 43-44, 51, 54
Rayleigh-Taylor Instability 54
Rayleigh-Vibrational Instability 46, 52, 54
Reactor Core 191, 282, 474-477, 479-480, 485, 511
Recovered Heat Quantity 224-225, 234
Recuperator 384, 387, 389-391, 402, 443, 452
Refrigerant 171-172, 204, 216-217, 231, 234, 292-293, 320, 323, 327, 334, 358, 377
Renewable Energy 56, 215-216, 231, 234, 336, 370, 558, 604
Reynolds Number 21, 61, 65-66, 83, 136, 145-146, 151-152, 158, 161-162, 164-165, 168, 171, 178, 203, 211, 222, 230, 239, 304, 309-311, 354, 366, 417, 482, 500, 533, 535

S

Saturated Steam 251, 269, 460, 466, 474-475, 480
Saturation Line 273, 291, 376, 409, 411, 466
Schwarzchild Criterion 36, 39, 51, 54

Separator 323, 480, 482
Simulation Method 538
Small-Angle X-Ray Scattering 109-110, 112, 134
Solar Energy 176, 185, 215-216, 337, 344-345, 363, 427, 588, 595, 598-599, 604-606
Solar Radiation 173-174, 221-222, 224-226, 228, 230-232, 347-350, 394, 402, 604-605
Solar Receiver 382, 384, 390-391, 402
Solar Thermal 171-172, 174, 185, 216-217, 220-222, 224-227, 231-232, 234, 369-370, 382, 394, 402
Solar Thermal Energy 216-217, 220-222, 224-227, 231-232, 234, 402
Solar Water Heater 169-171, 176, 215-217, 219-222, 226-228, 231-232, 234
Stability 1, 19-22, 33-35, 38, 42-43, 46-48, 51-52, 54, 56-57, 60-61, 64-66, 68, 70, 83, 87, 99, 114, 128, 136-137, 140-141, 144, 146, 149, 160, 166-168, 175, 177-178, 184, 188, 190-193, 195, 199-200, 209-212, 214, 236, 250-251, 262, 264, 313, 320, 374, 376-377, 381, 384, 393, 481-483, 510-511, 514, 519-527, 531-533, 535, 543-544, 561, 572, 602
Stability Analysis 34, 42, 51, 61, 83, 199, 214, 511, 524-526
Stability Map 149, 184, 195, 199-200, 214, 521
Stability Threshold 137, 192-193, 195, 199-200, 210, 214
Stability Transition 146, 184
Static Structure Factor 134
Steam Generator 422, 453, 464, 466, 474-475, 477, 479-480, 482
Structural properties 128
Supercritical Carbon Dioxide 176, 215, 235, 237, 292, 295-296, 309, 313-316, 318-319, 321-322, 335, 373, 382, 412, 606
Supercritical Fluids 1-2, 17, 28, 31, 33, 35, 44, 48, 51-52, 56, 61, 137, 140, 165, 167-168, 177-178, 184-185, 203-204, 222, 235-237, 239-240, 246, 249-251, 263, 266, 269, 271-272, 274-275, 282, 292, 309, 369-370, 372, 392, 404, 410, 510-511, 522, 554-557, 561, 573, 575, 588-590, 595, 600, 603-605
Supercritical Heat Exchanger 185
Supercritical Nuclear Cooling 185
Supercritical Pressure 62, 214, 250-251, 273, 278, 286-288, 291, 293-296, 301, 314, 323, 370, 465-470, 475, 479, 510-511, 517, 527
Supercritical Rankine cycle 337-339, 341, 343-344, 359, 369-372, 376-378, 381-382, 393, 402, 465, 479, 604-605
Supercritical Refrigeration 185
Supercritical Water 2, 137, 140, 149, 164, 191, 204, 235, 237, 249-251, 259, 262, 266, 269, 272-273, 275, 282-283, 288, 309-311, 316-317, 404-405, 411, 453, 469, 474-475, 479, 481-483, 486, 488, 491, 495-496, 498, 500-501, 510-511, 518-522, 524, 528, 538, 543-544, 555, 557, 559, 561-563, 574-575, 589, 591-595, 605
Supercritical Water Cooled Reactor 475, 479
Superheated Steam 420, 443, 453, 463, 475
Superheater 423, 460-461, 464-466, 468, 470-471, 474, 477, 479
Synchrotron radiation 109, 111, 115, 120-121, 128
Synthesis 2, 55-57, 61, 74, 83, 559, 588-593, 596-598, 600-602
System Design 56, 61, 137, 141, 155, 172, 185

T

Thermal Efficiency 191, 271, 283, 338, 363, 372, 382-384, 389, 391, 393-394, 404-405, 422, 426-427, 429, 432, 434, 444, 463, 475, 479, 481, 485, 604-606
Thermal Effusivity 287, 291
Thermal Hydraulics 235-236, 263, 266, 483, 544
Thermal Physical Properties 84, 155, 157, 159-160, 174, 217-218, 236-238, 482-484, 500
Thermal Power Plant 283, 337, 423, 427
Thermalhydraulics 195, 206, 211
Thermalization 1-2, 4-5, 7, 9, 24, 28, 32
Thermal-Mechanical Effect 60, 68, 73, 84, 87, 89-90, 99
thermodynamic and transport properties 1-2, 5, 171, 222, 554
Thermodynamic Cycle 334, 336-338, 366, 384, 402
Thermophysical Properties 14, 50-51, 191-192, 197, 205, 216, 222, 272, 277-279, 283, 288, 295-296, 304, 309, 311, 313, 322, 326, 343, 390, 404-406, 408, 411-412, 417-420, 496, 510
Thermosyphon 136-137, 140-141, 156-157, 160, 177-178, 185
Thin Hot Boundary 66, 68, 84, 89, 99
Transcritical Cycle 334, 337, 345
Transformation 27, 103, 262, 370, 500, 543, 588, 591, 595, 597, 600, 606
Turbine 189, 337, 339, 344-346, 352-354, 356-358, 363, 370, 372-376, 378, 383-385, 389-390, 393, 402, 405, 422-423, 425, 427, 429, 435-436, 452-453, 459-461, 463-466, 468, 472-475, 479-480, 485, 604, 606

U

Ultra-Supercritical Parameters 291, 427

V

Van der Waals 5, 33-34, 36-38, 51, 86, 102, 124, 127

W

Wall Function 481-482, 500-501, 503-505, 543, 549
Weightlessness 2, 5, 10-11, 28, 32, 44
Wire Probe 271, 276, 287-288

X

X-Ray Compton Scattering 111, 115-116, 133-134
X-Ray Diffraction 106-107, 109-111, 114, 118, 120, 134
X-Ray Raman Scattering 118, 128, 133-134
X-ray Scattering 102, 106, 108-112, 118-119, 121, 124, 128, 133-134

Z

Zeotropic Mixture 382, 393-394, 402, 604

www.ingramcontent.com/pod-product-compliance
Lightning Source LLC
LaVergne TN
LVHW082052011026
849162LV00019B/1572